AF393372

M. J. Kupersmith

In Collaboration with
A. Berenstein

Neuro-vascular Neuro-ophthalmology

With 284 Figures in 594 Separate Illustrations,
Some in Color, and 61 Tables

Springer-Verlag
Berlin Heidelberg New York
London Paris Tokyo
Hong Kong Barcelona
Budapest

Mark J. Kupersmith, M.D.
Professor of Neurology and Ophthalmology
New York University Medical Center
Director of Neuro-ophthalmology New York Eye and Ear Infirmary
530 First Avenue, New York, NY 10016, USA

In Collaboration with
Alejandro Berenstein, M.D.
Professor of Radiology and Neurosurgery
Director of Surgical Neuroangiography Service
New York University and Bellevue Medical Center
560 First Avenue, New York, NY 10016, USA

Contributions by InSup Choi, M.D., Associate Professor of Radiology,
Harvard University

ISBN-13: 978-3-642-77622-9   e-ISBN-13: 978-3-642-77620-5
DOI: 10.1007/978-3-642-77620-5

Library of Congress Cataloging-in-Publication Data. Kupersmith, M.J. (Mark J.), 1951–   Neurovascular neuro-ophthalmology/M.J. Kupersmith, A. Berenstein.   p.   cm. Includes bibliographical references and index. ISBN-13: 978-3-642-77622-9   1. Eye – Blood-vessels – Diseases. 2. Neuroophthalmology. 3. Ocular manifestations of general diseases. I. Berenstein, Alex, 1947–   . II. Title. [DNLM: 1. Eye Diseases – complications. 2. Eye Diseases – diagnosis. 3. Eye Diseases – therapy. 4. Eye Manifestations. 5. Neurologic Manifestations. 6. Vascular Diseases – complications. WW 460 K96n] RE720.K87 1993 617.7–dc20

Photography: Martha Helmers
Line drawings: Craig Luce
Production editor: Meike Seeker, Heidelberg
Reproduction of the figures: Gustav Dreher GmbH, Stuttgart
Typesetting: Data conversion by K+V Fotosatz GmbH, Beerfelden

To my wife, Geri, and my children, Dana and Matthew,
whose time I stole to write this book.

Preface

Neuro-ophthalmologic complications frequently result from disorders which alter the intracranial and intraorbital circulation. Due to widespread systems involved in sensory visual and cognitive visual functions and eye movements, visual disturbances are typically found in the common vascular disorders such as atherosclerosis, migraine, and aneurysms. Neuro-ophthalmologic dysfunction can occur as part of a larger syndrome or as the predominant clinical abnormality resulting from common or rare local or systemic vascular disorders. Many of the concepts concerning the diagnosis and treatment of neurovascular disorders have been facilitated by evolving techniques of neuroimaging, such as magnetic resonance imaging; methods for measuring systemic coagulation and inflammation; and improved capability to perform superselective catheterization of normal and abnormal blood vessels and vascular lesions. The neuro-ophthalmologic evaluation of clinical signs and symptoms often provides accurate localization and diagnosis of these lesions. Many of these clinical abnormalities, such as visual field defects and ocular misalignments, can be quantified and followed in order to assess either the natural history of the disorder or the effects of therapies.

No one medical speciality can effectively manage these neurovascular neuro-ophthalmologic disorders alone. The complexity of diagnosis and treatment planning requires a multidisciplinary team. This approach, which brings ophthalmologists, neurosurgeons, neurologists, and neuroradiologists together to confer in order to diagnose and treat these patients, has been pioneered at New York University Medical Center by Joseph Ransohoff, M.D. Dr. Ransohoff's guidance and inspiration has led to the formation of a NYU Neurovascular Service, which has evolved and developed new treatments for these often complex problems.

We would also like to thank A. Rhoton, N. Charles, K. Noble, J. Olver, A. Litt, J. Bogousslavsky, B. Grimson, J. Selhorst, S. Marinkovic, M. Milisavljevic, S. Alecksic, L. Frohman, M. Slavin, and R. Koplin, who provided original photographs and drawings of cases and anatomy.

M. J. KUPERSMITH
A. BERENSTEIN

Contents

Chapter 3

Neuro-ophthalmic Manifestations
of Intracranial Dural Venous Disorders

Chapter 4

Orbital Vascular Lesions

Chapter 5

Vascular Optic Neuropathies

Chapter 6

Aneurysms Involving the Motor and Sensory Visual Pathways

Chapter 7

Vascular Malformations of the Brain

Chapter 8

Embolic and Noninflammatory Occlusive Vascular Disease
of the Visual and Ocular Motor Systems

Chapter 9

Inflammatory Vascular Disorders of the Brain and Eye

Chapter 10

Migraine

CHAPTER 1

Circulation of the Eye, Orbit, Cranial Nerves, and Brain

1.1 Introduction

The diagnosis and management of the numerous vascular disorders that affect the sensory and motor visual systems cannot be considered without knowledge of the circulation to the eye, optic nerve, orbit, chiasm and optic tract, lateral geniculate nucleus, visual radiation, occipital lobe, cranial nerves, and brainstem. Diseases that cause neuro-ophthalmologic dysfunction can alter the arterial/capillary blood supply, venous drainage, and the hemodynamic balance of almost all intraorbital and intracranial structures. A brief overview of the cranial circulation provides a foundation to which details can be added in order to understand the circulation of each segment of the visual and oculomotor pathways. Variations of the normal circulation between individuals and in a single subject are numerous. Only anomalies which can have neuro-ophthalmologic implications are considered here. The figures illustrate cases with these important or common variants. The diagrams of the normal arterial and venous systems should be considered only as starting points in the evaluation of normal and abnormal studies.

1.2 Carotid Arterial System

1.2.1 Common Carotid Artery

The left common carotid artery (CCA) originates directly from the aorta while the right CCA arises from the bifurcation of the innominate artery. Each CCA is surrounded by the carotid sheath fascia that also envelops the internal jugular vein and the vagus nerve. At approximately the fourth or fifth cervical vertebra, the CCA bifurcates into a smaller, anteriorly located external carotid artery (ECA) and a larger internal carotid artery (ICA).

Table 1.1. The external carotid artery system

Branch arteries	Anastomoses
Facial	
Angular	Palpebral
Occipital	Vertebral (at C1, C2), superficial temporal
Stylomastoid (inconsistently to seventh nerve)	MMA
Superficial temporal	
Transosseous twigs to dura	
Internal maxillary	(see Sect. 1.5.1)
Middle meningeal (MMA)	
Anterior deep temporal	Lacrimal
Infraorbital	

1.2.2 External Carotid Arterial System

The ECA has significant import in neuro-ophthalmologic disorders because of its role in providing the blood supply to part of the orbit, the dura and cranial nerves, and skull (Table 1.1). The ECA also supplies the thyroid gland, larynx, and face (Fig. 1.1). ECA anastomose with the arteries of the ICA, many through the internal maxillary arterial system (Fig. 1.2), are an important source of collateral blood supply to the latter system if developmental defects or acquired occlusive disease disrupts the normal orbital or intracranial circulation. The orbital segment of the ophthalmic artery and the branch arteries of the cavernous sinus are rich in these anastomoses.

1.2.2.1 Middle Meningeal Artery (Fig. 1.3)

The middle meningeal artery (MMA) is the most important source of ECA anastomoses with both the intracranial and orbital circulations (see Sects. 1.5.1.1.1, 1.5.1.1.2, 1.5.3.1.2). From its internal

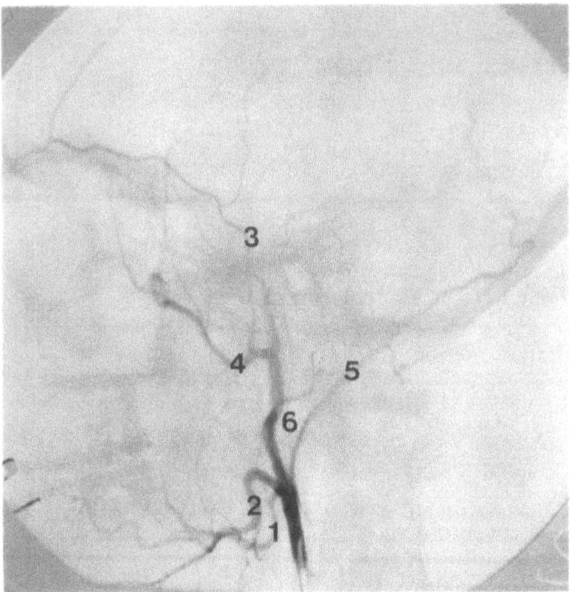

Fig. 1.1. Lateral external carotid artery, subtraction angiogram. *1,* lingual artery; *2,* facial artery; *3,* superficial temporal artery; *4,* internal maxillary artery; *5,* occipital artery; *6,* ascending pharyngeal artery

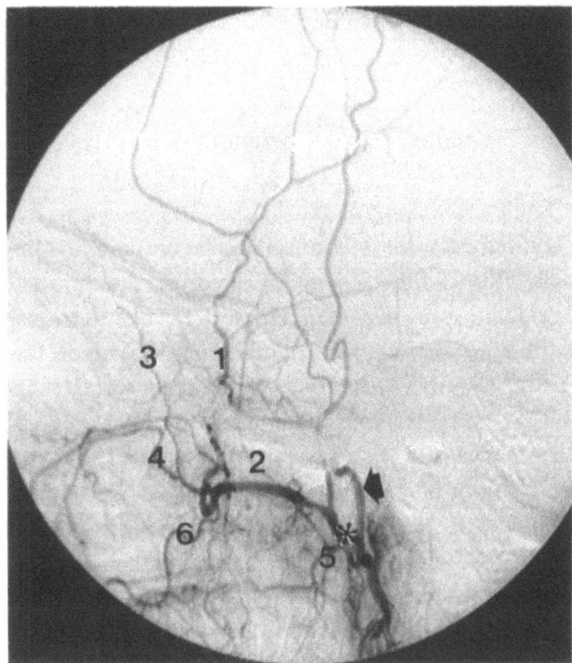

Fig. 1.2. Lateral internal maxillary artery, subtraction angiogram. *1,* middle meningeal artery; *2,* middle deep temporal artery; *3,* anterior deep temporal artery; *4,* infraorbital artery; *5,* sphenopalatine arteries; *6,* descending palatine artery. The superficial temporal artery (*arrow*) originates from the continuation of the external carotid artery after the internal maxillary artery (*asterisk*)

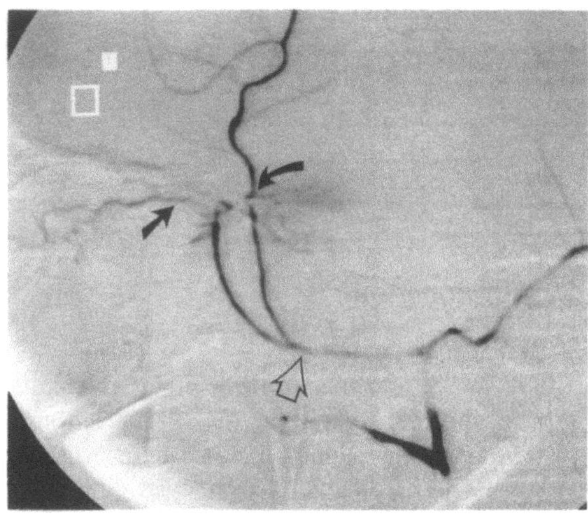

Fig. 1.3. A lateral projection subtraction angiogram of a middle meningeal artery (*open arrow*) shows one variation with the origin of the meningeal lacrimal artery (*straight black arrow*). The middle meningeal artery enters the skull through the foramen spinosum (*curved arrow*)

maxillary artery origin, the MMA courses rostrally and enters the intracranial space through the foramen spinosum, where it makes a characteristic sharp bend posteriorly and divides (Fig. 1.3). The MMA supplies most of the dura overlying the cerebrum all the way to the midline, including the superior sagittal sinus. The MMA supply to the dura and orbit is in balance with branches from the ethmoidal arteries so that if one system is dominant to a region, the other is smaller or hypoplastic. For example, in some individuals the anterior falx is supplied by an anterior falcine artery via an anastomosis with the anterior ethmoidal artery, rather than the MMA.

The posterior branches of the MMA supply the dura covering the parietal and occipital areas. The posterior fossa dura and meninges receive some blood supply from the MMA, but the occipital artery, the ascending pharyngeal artery, and less commonly branches from the vertebral arteries provide the majority of the arterial contribution to the dura in this region.

1.2.3 Internal Carotid Artery System
(Tables 1.2 – 1.5)

Table 1.2. The internal carotid artery system

ICA segments and branch arteries	Relationship to important landmarks
Cervical	
Petrous	Anterior to jugular foramen through foramen lacerum
Cavernous	(see Sect. 1.5.3.1)
Intradural	Pierces dura, beneath anterior clinoid into suprasellar cistern
Ophthalmic	Exits intracranial cavity via optic canal
Hypophyseal	Passes medially to hypophysis
Posterior communicating	Lateral to third nerve
Anterior choroidal	Posterior path to choroid plexus
Bifurcation to middle/anterior cerebral	Lateral and superior to chiasm

Table 1.3. Posterior communicating artery (modified from [1, 2])

Structural relationships	Lateral to hypophyseal stalk
	Medial to uncus
	Adjacent to third nerve
Neural structures supplied (variable) from	Anterolateral thalamus
tuberothalamic artery (polar)	Hypothalamus, subthalamus
	Cerebral penduncle
	Posterior perforating substance
	Tuber cinereum
	Posterior limb internal capsule
	III nerve
Collaterals	Anterior choroidal artery
	Hypophyseal artery

Table 1.4. Anterior cerebral artery

Branch	Structures supplied
Anterior communicating, via	Anterior basal ganglia,
Heubner (rarely completely nonfunctional)	genu and posterior internal capsule, anterior thalamus, hypothalamus
Perforating arteries	Anterior perforating substance/anterior basal ganglia
Medial lenticular striate	
Recurrent artery of Heubner	
Orbitofrontal	Frontal lobe
Frontal	Frontal lobe
Pericallosal (above corpus callosum) ⎫ Callosomarginal (above pericallosal) ⎭	Midline frontal/parietal lobes, corpus callosum, fornix, septum pellucidum

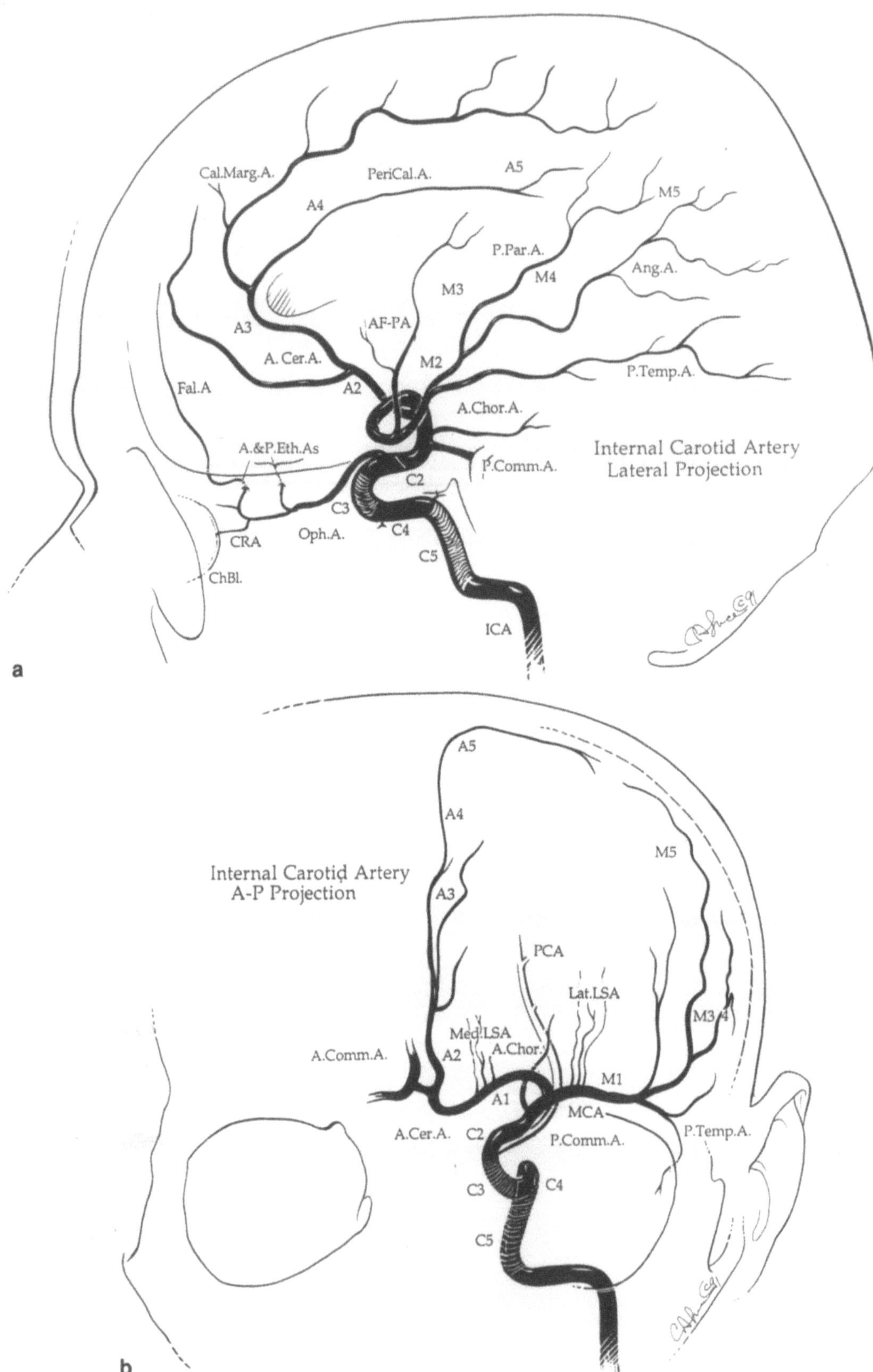
Cal.Marg.A.
PeriCal.A.
A5
A4
M5
P.Par.A.
Ang.A.
M3
M4
A3
AF-PA
A. Cer.A.
M2
P.Temp.A.
Fal.A
A2
A.Chor.A.
Internal Carotid Artery
Lateral Projection
A.&P.Eth.As
C2
P.Comm.A.
C3
CRA
Oph.A.
C4
C5
ChBl.
ICA
a
A5
A4
M5
Internal Carotid Artery
A-P Projection
A3
PCA
Lat.LSA
Med.LSA
M3/4
A.Comm.A.
A2
A.Chor.
A.Cer.A.
C2
A1
M1
MCA
P.Comm.A.
P.Temp.A.
C3
C4
C5
b

Table 1.5. Middle cerebral artery

Branch[a]	Structures supplied
Lateral lenticulostriate	Basal ganglia, anterior portion, internal capsule
Anterior temporal	Temporal lobe
Posterior temporal	Temporal lobe
Ascending frontal	Frontal lobe
Orbitofrontal	Frontal lobe
Anterior parietal	Parietal lobe
Posterior parietal	Parietal lobe
Angular	Parietal/temporal lobes

[a] Note that there is a variable number and pattern of cortical branches (Fig. 1.4).

◀ **Fig. 1.4a, b.** Lateral (**a**) and anteroposterior (**b**) projections of a typical internal carotid artery (*ICA*) and its major cerebral branches. Internal carotid artery segments: C5, C4, C3, C2. Proximal internal carotid artery branches: ophthalmic (*Oph. A.*), posterior communicating (*P. Comm. A.*), anterior choroidal (*A. Chor. A.*), anterior communicating (*A. Comm. A.*), anterior cerebral (*A. Cer. A.*) [and its branches: A1, medial lenticulate striate (*Med. LSA*), A2, A3, callosomarginal (*Cal. Marg. A.*), A4, pericallosal (*Peri Cal. A.*), A5], and middle cerebral (*MCA*) [and its branches: M1, lateral lenticulostriate (*Lat. LSA*), M2, M3, M4, posterior temporal (*P. Temp. A.*), posterior parietal (*P. Par. A*), angular (*Ang. A.*), M5]. *AF-PA,* anterior frontal precentral artery; *A. & P. Eth. As.,* anterior and posterior ethmoidal arteries; *Ch. Bl.,* choroidal blush; *CRA,* central retinal artery; *Fal. A.,* falcine artery

1.2.4 Cerebral Arteries (Fig. 1.4; Tables 1.4, 1.5)

1.3 Vertebrobasilar Arterial System

1.3.1 Vertebral and Basilar Arteries
(Fig. 1.5; Table 1.6)

The posterior fossa arterial circulation arises from the vertebrobasilar arterial system. Each vertebral artery arises from the subclavian artery and courses rostrally, behind the jugular vein, to enter the transverse foramen of the sixth cervical vertebra. The vertebral artery, accompanied by a venous plexus and sympathetic nerves, ascends through the transverse foramen of each vertebra. At the second cervical vertebra, the vertebral artery turns laterally and then superiorly, passing through the transverse process of the first cervical vertebra to enter the intracranial cavity at the level of the foramen magnum.

1.3.2 Posterior Cerebral Artery
(Fig. 1.5, Table 1.7; see also Fig. 1.30)

Though the posterior cerebral artery (PCA) appears to arise from the basilar artery by the time of birth in most subjects, the PCA actually develops from the embryological ICA. An embryological PCA that originates from the ICA, with an associated hypoplastic posterior communicating artery, is a common normal variant. The PCA arises from the ICA in approximately 15% of angiograms [3, 4]. In most subjects, the proximal PCA remains connected to the ICA via the posterior communicating artery. Various classifications have been proposed to identify the different segments of the proximal PCA (P1, P2, P3, P4, from proximal to distal portions), which is particularly useful when disease has affected or surgery is contemplated in this region [5].

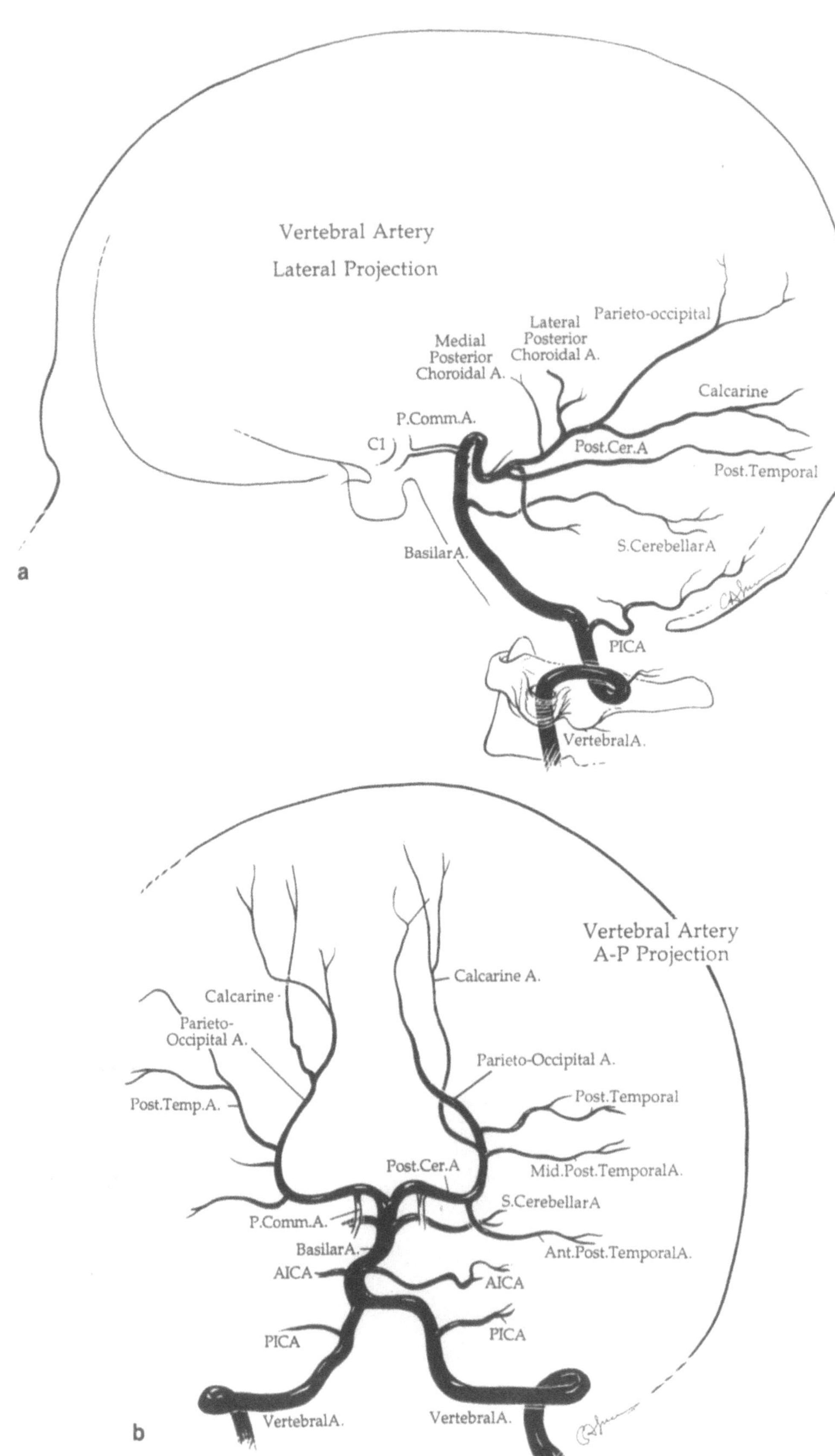
Vertebral Artery
Lateral Projection
Medial Posterior Choroidal A.
Lateral Posterior Choroidal A.
Parieto-occipital
Calcarine
P.Comm.A.
C1
Post.Cer.A
Post.Temporal
Basilar A.
S.Cerebellar A
PICA
Vertebral A.
a

Vertebral Artery
A-P Projection
Calcarine A.
Calcarine
Parieto-Occipital A.
Parieto-Occipital A.
Post.Temp.A.
Post.Temporal
Post.Cer.A
Mid.Post.Temporal A.
S.Cerebellar A
P.Comm.A.
Ant.Post.Temporal A.
Basilar A.
AICA
AICA
PICA
PICA
Vertebral A.
Vertebral A.
b

Table 1.6. The vertebral and basilar arteries

Artery	Structures supplied
Branches from medullary segment	Cervical spinal cord, meninges, medulla
Posterior inferior cerebellar (PICA)	Medulla
Medial division	Cerebellar vermis
Lateral division	Cerebellar hemisphere
Lateral spinal	Spinal cord
(or directly from PICA)	
Anterior spinal	Spinal cord
Basilar	
Midline small perforators	Pons, midbrain
Small circumferential	Pons, midbrain
Anterior inferior cerebellar	Seventh, eighth nerves, cerebellar penduncle and hemisphere
Superior cerebellar	Cerebellar vermis, cerebellar hemisphere, dorsal midbrain
Posterior cerebral	[a] Midbrain, thalamus, posterior corpus callosum, temporal, parietal, and occipital lobes

[a] See Sects. 1.3.2, 1.5.3.3.

Table 1.7. Posterior cerebral artery (modified from [6, 7])

Segment	Branch	Structures supplied
P1[a]	Interpeduncular perforators	Third nerve, midbrain
	Thalamoperforators	Midbrain, posterior diencephalon
	Short circumflex	Midbrain
	Long circumflex	Superior colliculus, pretectum
P2	Medial posterior choroidal	Thalamus, medial geniculate nucleus, atrium, fornix
	Lateral posterior choroidal	Thalamus, lateral geniculate nucleus, atrium, fornix
	Thalamogeniculate	Posterior thalamus, medial geniculate nucleus, pulvinar
	Interpeduncular perforators	Third nerve, midbrain
	Quadrigeminal	Quadrigeminal plate
P3, P4	Hippocampal	(see Sect. 1.5.2.10.1)
	Posterior pericallosal	
	Posterior temporal	
	Parieto-occipital	
	Calcarine	

[a] See Sect. 1.5.3.3.3.

◄ **Fig. 1.5 a, b.** Lateral (**a**) and anteroposterior (**b**) projections of the vertebrobasilar arterial system and major arteries with common variations of the cortical branches of the posterior cerebral artery (*Post. Cer. A.*). The branches from the vertebrobasilar artery include: posterior inferior cerebellar (*PICA*), anterior inferior cerebellar (*AICA*), superior cerebellar (*S. Cerebellar A*), and posterior cerebral (*Post. Cer. A.*). The branches of the posterior cerebral artery include: anterior posterior temporal (*Ant. Post. Temporal A.*), middle posterior temporal (*Mid. Post. Temporal A.*), posterior temporal (*Post. Temporal*), parieto-occipital, and calcarine

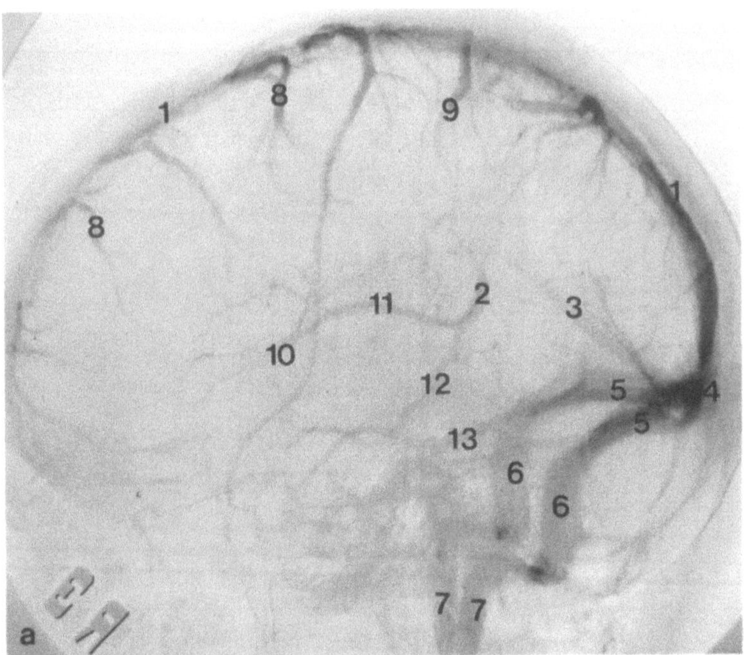

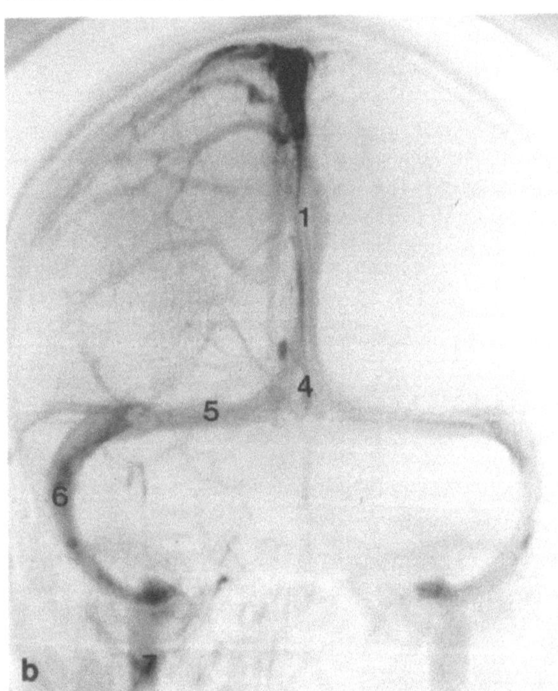

Fig. 1.6. a The late phase of the lateral projection subtraction angiogram demonstrates the superficial and deep venous systems. *1*, superior sagittal sinus (inferior sagittal sinus not seen); *2*, great vein of Galen; *3*, straight sinus; *4*, confluence of sinuses; *5*, transverse sinus (both sides seen); *6*, sigmoid sinus (both sides seen), cavernous sinus (not seen); *7*, internal jugular veins (both seen); *8*, frontal veins; *9*, vein of Troland; *10*, septal vein; *11*, internal cerebral vein; *12*, basal vein of Rosenthal; *13*, vein of Labbe.

b The venous phase of the frontal projection angiogram shows: *1*, superior sagittal sinus; *4*, confluence of sinuses; *5*, transverse sinus; *6*, sigmoid sinus; *7*, internal jugular vein; and cerebral veins (not numbered)

1.4 Intracranial Venous Drainage
(Figs. 1.6, 1.7)

The venous system of the brain generally parallels the arterial system. The cerebral veins are grouped into either the superficial or deep venous systems. Since there are numerous anastomoses between venous channels, occlusion of a single large vein infrequently causes a neurological deficit. However, one such example of neuro-ophthalmologic significance is the hemianopia that develops with occlusion of the internal occipital vein (see Sect. 1.5.2.10.2) [8, 9].

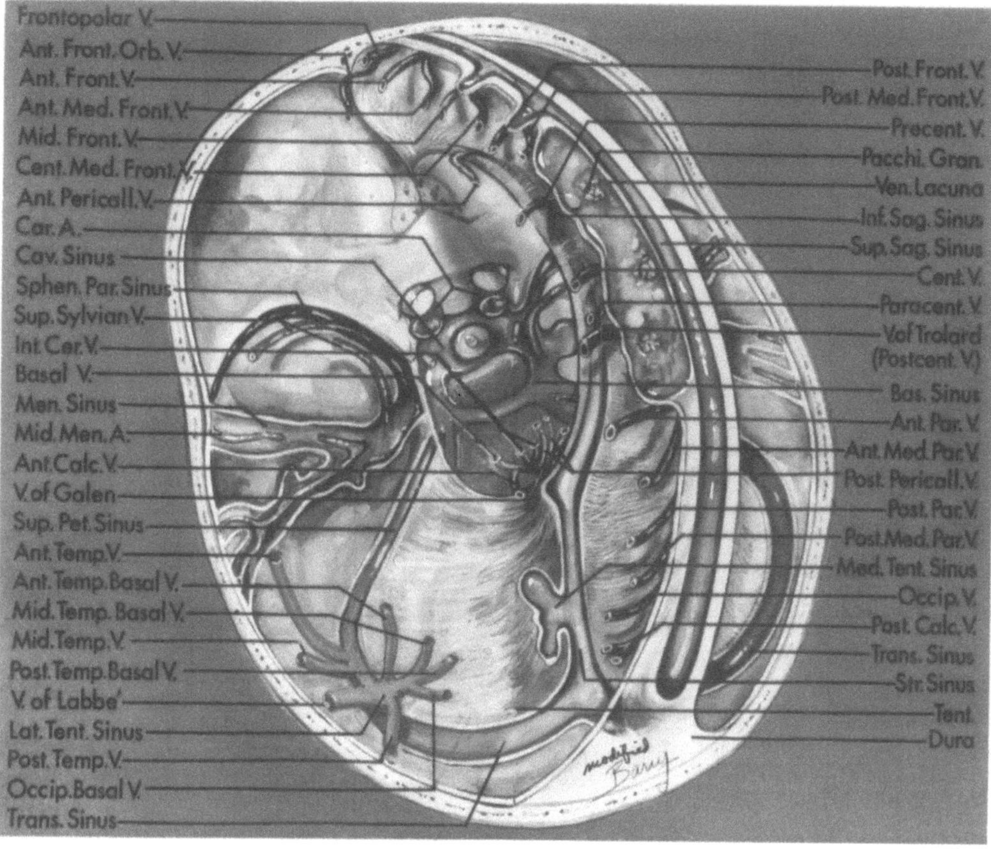

Fig. 1.7. Dural sinuses in the falx, base of the skull, and bridging veins. Terminating venous channels include: *Sup. Sag. Sinus,* superior sagittal sinus; *Trans. Sinus,* transverse sinus; *Lat. Tent. Sinus,* lateral tentorial sinus; *Cav. Sinus,* cavernous sinus; *V. of Galen,* vein of Galen; *Int. Cer. V,* internal cerebral; *Inf. Sag. Sinus,* inferior sagittal sinus; and *Basal V.,* basal vein of Rosenthal. The veins from the lateral surfaces of the frontal, parietal, and occipital lobes are: frontopolar; *Ant. Front. V.,* anterior frontal; *Mid. Front. V.,* middle frontal; *Post. Front. V.,* posterior frontal; *Precent. V.,* precentral; *Cent. V.,* central; *Ant. Par. V.,* anterior parietal; *Post. Par. V.,* posterior parietal; *Occip. V.,* occipital; and vein of Trolard. The veins from the medial surface that drain into the superior sagittal sinus are: *Ant. Med. Front. V.,* anterior medial frontal; *Cent. Med. Front. V.,* centromedial frontal; *Post. Med. Front. V.,* posteromedial frontal; *Paracent. V.,* paracentral; *Ant. Med. Par. V.,* anteromedial parietal; *Post. Med. Par. V.,* posteromedial parietal; and *Post. Calc. V.,* posterior calcarine veins. The veins from the orbital surface that drain into the superior sagittal sinus are the anterior fronto-orbital veins (*Ant. Front. Orb. V.*). The veins that drain into venous channels in the tentorium (*Tent.*) drain the lateral and basal surfaces of the temporal lobe: *Ant. Temp. V.,* anterior temporal; *Mid. Temp. V,* middle temporal; *Post. Temp. V.,* posterior temporal; vein of Labbe; *Ant. Temp. Basal V.,* anterior temporobasal; *Mid. Temp. Basal V.,* middle temporobasal; *Post. Temp. Basal V.,* posterior temporobasal vein and the basal surface of the occipital lobe; *Occip. Basal V.,* occipitobasal. Parts of the frontal, parietal, and temporal lobes drain into the sphenoparietal and cavernous sinuses via the superior sylvian vein (*Sup. Sylvian V.*) and its tributaries, the frontal, parietal, and temporal sylvian veins. The veins emptying into the straight sinus (*Str. Sinus*) or its tributaries drain the limbic region: part of the frontal and parietal lobes surrounding the corpus callosum and the medial temporal lobe. These veins include: paraterminal; posterior fronto-orbital; olfactory; *Ant. Pericall. V.,* anterior pericallosal; *Post. Pericall. V.,* posterior pericallosal; uncal; anterior hippocampal; medial temporal; and *Ant. Calc. V.,* anterior calcarine. The anterior end of the basal vein also empties the deep sylvian vein and the anterior cerebral veins. The internal carotid artery (*Car. A.*) passes through the cavernous sinus. The meningeal sinuses (*Men. Sinus*) in the floor of the middle cranial fossa course with the middle meningeal arteries (*Mid. Men. A.*). The medial tentorial sinuses (*Med. Tent. Sinus*) receive tributaries from the cerebellum and join the straight sinus. A small basilar sinus (*Bas. Sinus*) is located behind the dorsum on the clivus. Pacchionian granulations (*Pacchi. Gran.*) drain the cerebrospinal fluid into venous lacunae (*Ven Lacuna*). *Sphen. Par. Sin.,* sphenoparietal sinus; *Sup. Pet. Sin.,* superior petrosal sinus. (Modified from [12])

The cerebral veins drain into the dural venous channels, through which most of the orbital and intracranial venous blood normally passes (Figs. 1.6, 1.7). These venous sinuses are located along the inner table of the cranium in the folds of the dura. The two internal jugular veins provide the predominant outflow from this system. In most normal subjects, the other venous routes have a minor function for exit of the blood from the intracranial cavity. These include the emissary veins from the dural venous channels through the skull to the extracranial veins, the venous plexus at the base of the brain and clivus into the spinal canal venous plexus, and the orbital to facial veins.

1.4.1 Supratentorial Deep and Superficial Venous Systems

1.4.1.1 Superficial Cerebral Veins
(Fig. 1.7; see also occipital lobe veins, Fig. 1.33)

The superficial cerebral veins lie on the pial surface of the cortex (Table 1.8). The veins from the dorsal surface of the brain drain into the superior sagittal sinus. The veins from the ventral surface of the brain drain into the cavernous and the transverse sinuses. The superficial veins are connected to the basal vein, via the deep middle cerebral vein from the insula, and to the subependymal veins through minute medullary veins in the white matter. The pattern of the cortical veins varies among brains and even between the cerebral hemisphere of an individual's brain. However, three large veins, which anastomose at the posterior aspect of the sylvian fissure, are consistently present. The principal dural venous sinuses (Figs. 1.6, 1.7; Table 1.9) are also constant.

1.4.1.2 Deep Cerebral Veins (Table 1.10)

The deep venous system is divided into veins that drain the ventricular regions, veins that drain the areas of the cisterns at the base of the brain, and veins that receive drainage from both groups [13]. An important angiographic venous landmark, the

Table 1.8. Superficial cerebral veins (modified from [11, 12])

Frontal lobe veins
   Lateral frontal surface
   Medial frontal surface
   Inferior frontal surface
Parietal lobe veins
   Lateral parietal surface
   Medial parietal surface
Temporal lobe veins
   Lateral temporal surface
   Inferior temporal surface
Occipital veins (see Sect. 1.5.2.10.2)
   Lateral occipital surface
   Medial occipital surface
   Inferior occipital surface
Anastomotic veins (consistently present)
   Trolard: Superior sagittal sinus[a]
   Labbé: Transverse sinus[a]
   Superficial middle cerebral: Sphenoparietal sinus[a]
   Transcerebral venous system

[a] Termination.

thalamostriate vein (paired), passes anteriorly towards the foramen of Monro, where it makes a sharp posterior turn to merge with the anterior septal vein and form the internal cerebral vein. The thalamostriate and internal cerebral veins form a well-recognized venous angle just posterior to the foramen of Monro that is closed when there is hydrocephalus. The internal cerebral veins join to become the single great vein of Galen. Other subependymal veins, which are variably present and of inconsistent size, include the thalamocaudate, posterior caudate, and posterior septal veins [14]. There are numerous normally occurring variations in the pattern of deep venous drainage.

1.4.2 Venous Drainage of the Posterior Fossa

The veins of the posterior fossa (Table 1.11) drain into the great vein of Galen, the petrosal sinuses, or the transverse and straight sinuses.

Table 1.9. Dural venous sinuses (modified from [10])

Sinus	Venous channel drained
Superior sagittal	Superficial cortical veins, sphenoparietal sinus channel, meningeal veins
Confluence of Herophili	Superior sagittal sinus, straight sinus, inferior sagittal sinus, vein of Galen, cortical veins (particularly inferior occipital)
Transverse (right/left)	Confluence, cortical veins (occipital, posterior temporal, brainstem, cerebellar), superior petrosal sinus
Sigmoid (right/left)	Transverse sinus, cortical veins (posterior temporal, brainstem, cerebellar), inferior petrosal sinus
Internal jugular (right/left)	Sigmoid sinus
Superior petrosal	Cavernous sinus
Inferior petrosal	Cavernous sinus, small veins from pons, medulla, internal auditory vein
Cavernous sinus	(see Sect. 1.5.3.1.3)

Table 1.10. Deep venous system (modified from [13])

Ventricular veins	Cisternal veins
Frontal horn	Anterior group
Lateral ventricle	Middle group
Atrium/occipital horn	Posterior group: includes internal
Temporal horn	cerebral, great vein,
Choroidal	basal vein (posterior),
Thalamic	occipital cortical

Table 1.11. Venous drainage of the posterior fossa

Vein	Termination
Superior vermian	Great cerebral or superior cerebellar vein
Precentral cerebellar (single)	Great cerebral or superior cerebellar vein
Mesencephalic	Great cerebral or superior cerebellar vein
Anterior pontomesencephalic	Petrosal veins at anterolateral cerebellar margin and transverse sinus
Transverse pontine	Petrosal veins at anterolateral cerebellar margin and transverse sinus
Cerebellar hemisphere	
Superolateral	Petrosal veins at anterolateral cerebellar margin and transverse sinus
Inferolateral	Petrosal veins at anterolateral cerebellar margin and transverse sinus
Midline superior	Straight sinus
Midline inferior	Straight sinus
Inferior vermian	Straight, transverse, or confluence of sinuses

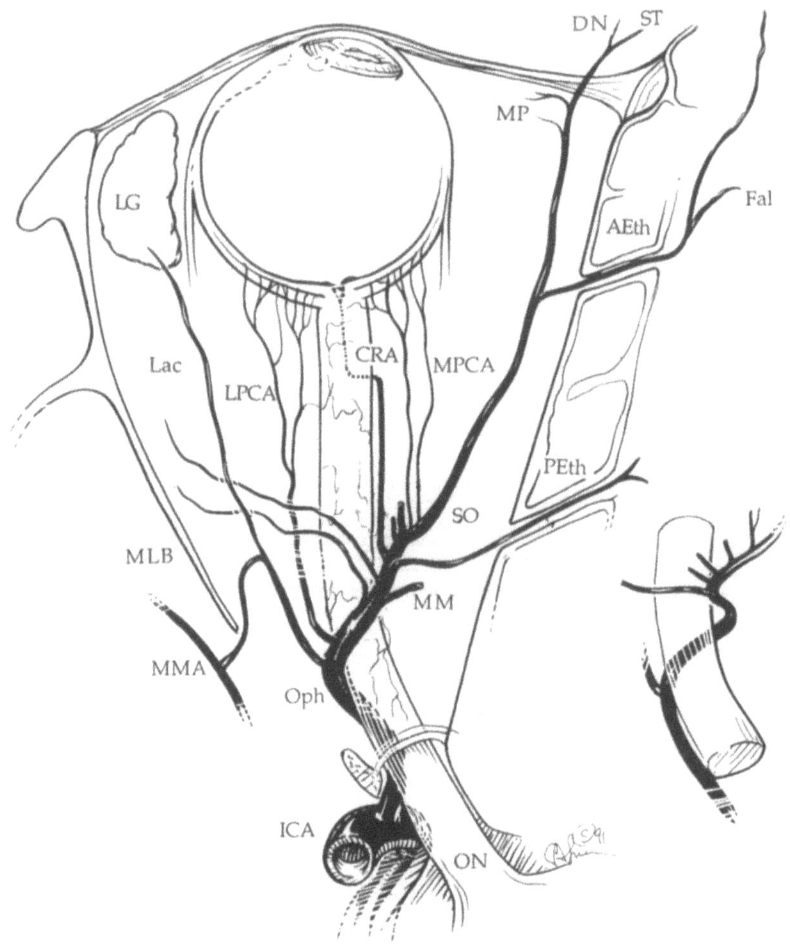

Fig. 1.8. Superior view of the intraorbital branches of the ophthalmic artery. The ophthalmic artery passes over the optic nerve in more than 80% of individuals. The vascular pattern differs if the ophthalmic artery passes beneath the optic nerve (*inset* on *right*). The ophthalmic (*Oph*) artery branches include: medial muscular (*MM*), superior oblique (*SO*), posterior ethmoidal (*PEth*), anterior ethmoidal (*AEth*), falcine (*Fal*), medial palpebral (*MP*), dorsal nasal (*DN*), supratrochlear (*ST*), medial (*MPCA*) and lateral ciliary (*LPCA*), central retinal (*CRA*), and lacrimal (*Lac*). In this orbit, the middle meningeal artery (*MMA*) provides a collateral branch, the meningolacrimal branch (*MLB*), through the superior orbital fissure to the lacrimal artery

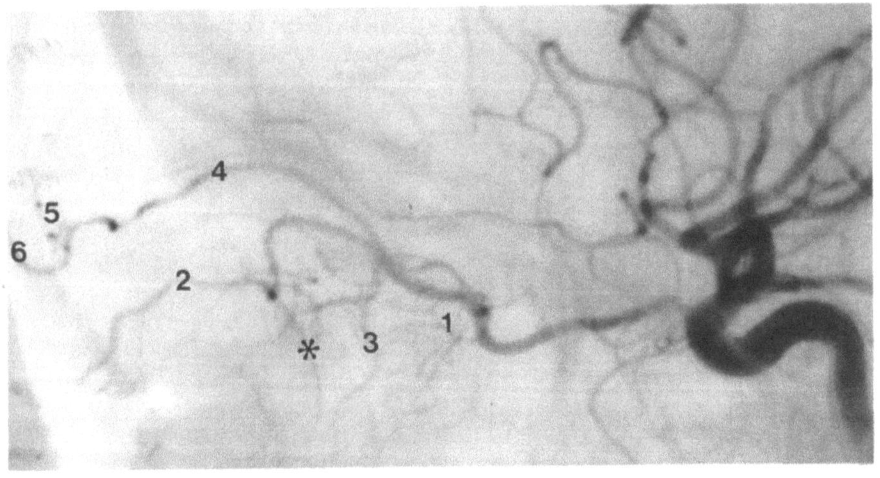

1.5 Circulation in Regions of Neuro-ophthalmologic Importance

1.5.1 Orbital Circulation

1.5.1.1 Arterial System

The orbit is supplied by several arterial systems including the internal maxillary artery (MMA, anterior deep temporal artery), the ICA (ophthalmic artery), and the facial artery. In order to properly analyze the vascular anatomy of the orbit, the variations in the embryological development of the vascular supply to this region must be understood. The differences in development can account for the anatomical disposition and functional role of the collateral circulation between the ophthalmic and external carotid arteries.

1.5.1.1.1 Ophthalmic Artery

Among individuals and even between the orbits of a single subject, the intraorbital ophthalmic artery, crossing either over or under the optic nerve, has a variable branching pattern (Figs. 1.8, 1.9) [15]. As the ophthalmic artery passes anteriorly, the branch arteries supply either the medial or lateral orbital tissues (Table 1.12).

Multiple arteries can send collaterals to the ophthalmic arterial system (Figs. 1.10, 1.13). The anastomoses in the orbit provide both latent and manifest connections between the ICA (ophthalmic artery) and the ECA (see below and Table 1.13) systems. A collateral may develop when the two arteries have a prominent anastomosis or a common embryological origin is shared between the vessels. In the orbit, if the development of the origin of the ophthalmic artery is delayed or if a portion of the developing artery atrophies, the ophthalmic artery territory will receive the majority of its blood supply via one or more branches of the ECA (Table 1.13) (Fig. 1.11).

◄ **Fig. 1.9.** A lateral subtraction internal carotid artery injection demonstrates the origin of the ophthalmic artery from the supracavernous carotid artery. While in the orbit, the ophthalmic artery branches include: *1,* posterior ciliary arteries; *2,* lacrimal artery; *3,* muscular branches; *4,* medial nasal ophthalmic artery; *5,* supratrochlear artery; *6,* palpebral artery. The choroidal blush (*asterisk*) is seen as well

Table 1.12. Ophthalmic artery branch arteries

Medial orbital branches	Lateral orbital branches
Anterior ethmoidal	Lacrimal
Posterior ethmoidal	Muscular
Supraorbital	
Dorsal nasal	
Supratrochlear	
Medial palpebral	
Posterior ciliary arteries	
Central retinal artery	

The most important collaterals from the MMA to the orbit are its lacrimal and ethmoidal anastomoses. Rarely, the ophthalmic artery actually originates from the MMA and not from the ICA. In this situation, the ophthalmic artery enters the orbit through the superior orbital fissure (Fig. 1.11d) instead of the optic canal.

1.5.1.1.2 Orbital Arterial Embryology

Several key embryological events are significant in deciding the dominant vascular pattern in the orbit [16a, 17] (Table 1.14). The primitive ophthalmic artery is formed by the fusion of the dorsal ophthalmic artery, arising from the siphon of the ICA, and the ventral ophthalmic artery, arising from the anterior cerebral artery (ACA). The ventral ophthalmic artery enters the orbit through the optic canal while the dorsal ophthalmic artery enters the orbit via the superior orbital fissure. As the ventral ophthalmic artery is recruited by the primitive ophthalmic artery, the proximal portion of the dorsal ophthalmic normally regresses. This leaves a single origin of the ophthalmic artery arising from the supraclinoid ICA that enters the orbit with the optic nerve through the optic canal. In approximately 10% of individuals, the ophthalmic artery originates from the cavernous carotid artery and enters the orbit via the superior orbital fissure, because the dorsal ophthalmic artery persists (Fig. 1.12a). Rarely, the ophthalmic artery origin remains near the ACA because of failure of migration of the ventral ophthalmic artery (Fig. 1.12b). There are other rare variations, such as persistence of both origins of the ophthalmic artery, resulting in dual origins of the ophthalmic artery. Failure or regression of the intracranial origin of the ophthalmic artery re-

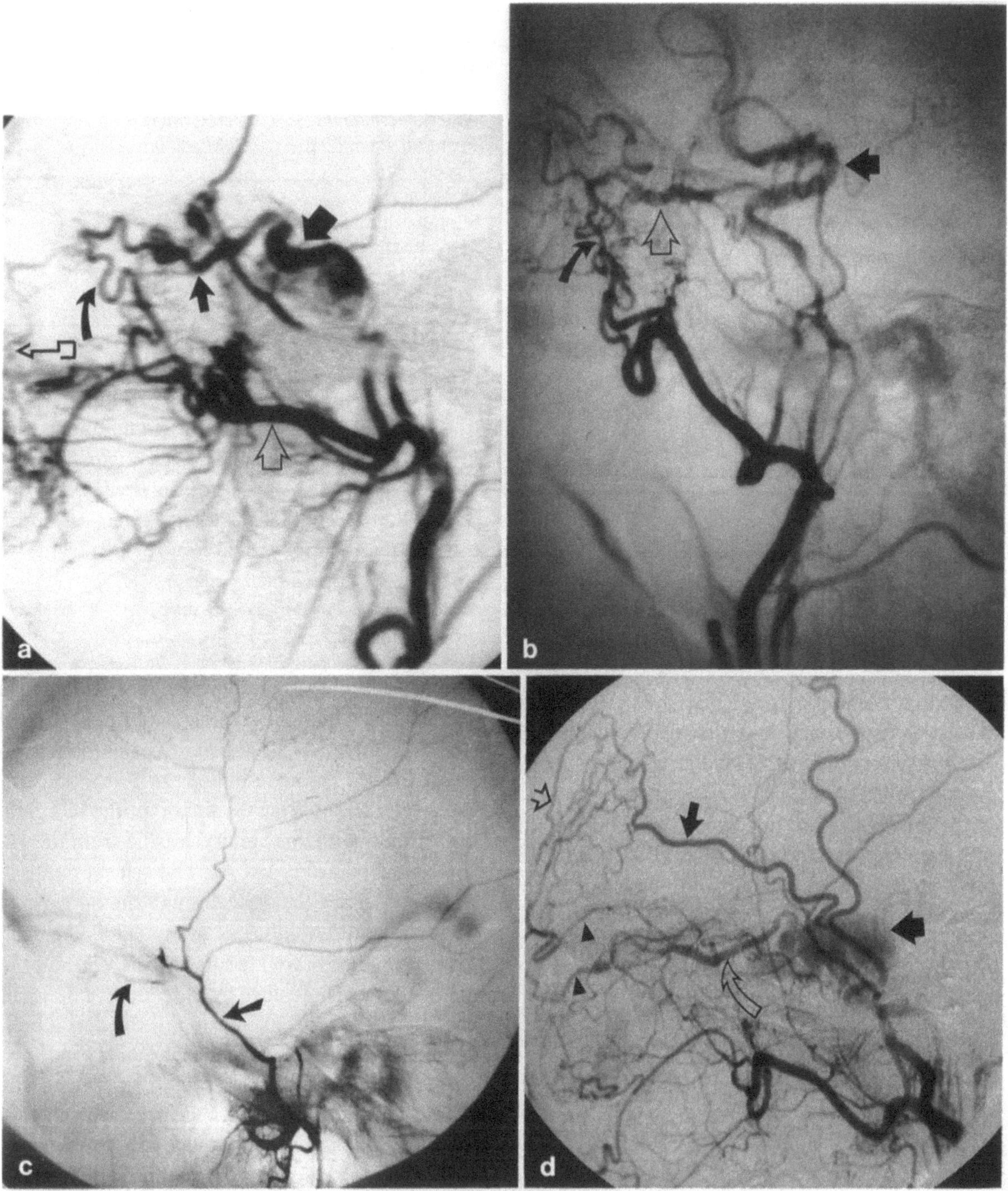

Table 1.13. Anastomoses of the internal carotid (ophthalmic) and external carotid artery systems (modified from [16])

Ophthalmic artery collateral	Anastomotic channels from ECA
Frontal or palpebral artery	Superficial temporal
Nasal artery	Angular artery from facial
Supraorbital artery	Superficial temporal
Inferior medial muscular artery	Infraorbital artery (IMA)
Lateral muscular artery	Anterior deep temporal (IMA)
Anterior ethmoidal artery (anterior nasal branch)	Sphenopalatine (IMA)
Anterior ethmoidal artery (artery of the falx)	MMA (IMA)[a]
Posterior ethmoidal artery meningeal branch	MMA (IMA)[a]
Posterior ethmoidal artery septal branch	Sphenopalatine (IMA) Greater palatine (IMA)
Deep recurrent ophthalmic artery	Inferior lateral trunk and its anastomoses[a]
Lacrimal artery	MMA (MMA to lacrimal table)[a]
Palpebral arteries	Facial artery

ECA, External carotid artery; IMA, internal maxillary artery; MMA, middle meningeal artery.
[a] More common anastomoses.

◀ **Fig. 1.10a–d.** Ophthalmic artery collaterals from the external carotid artery system.
a Following a trapping procedure which closed the internal carotid artery above a carotid cavernous fistula and in the cervical region, a lateral view external carotid artery subtraction angiogram was performed. It demonstrates reconstitution of the ophthalmic artery (*arrow*) to the carotid artery (*broad arrow*) via the anterior deep temporal artery (*curved arrow*) to lacrimal anastomosis which arises from the internal maxillary (*open arrowhead*). A branch of the facial artery (*open arrow*) also anastomoses with the ophthalmic artery.
b A lateral view common carotid artery subtraction angiogram, performed after closure of the cavernous carotid artery in the treatment of a cavernous carotid aneurysm, demonstrates the anterior deep temporal artery (*curved arrow*) anastomosis to the ophthalmic artery (*open arrowhead*). Retrograde filling in the ophthalmic artery reconstitutes the internal carotid artery (*large arrow*).
c A lateral internal maxillary artery injection demonstrates a middle meningeal artery (*arrow*) and a meningeal lacrimal artery to the orbit (*curved arrow*).
d A lateral subtraction external carotid artery angiogram demonstrates the superficial temporal artery (*narrower solid arrow*) providing anastomosis (*open arrowhead*) with the palpebral branches of the ophthalmic artery. These anastomoses fill the distal branches of the ophthalmic artery (*solid arrowheads*) which then opacify the ophthalmic artery (*curved arrow*). The ophthalmic artery then fills the internal carotid artery, which supplies a spontaneous carotid cavernous fistula, causing dilatation of the cavernous sinus (*broad solid arrow*)

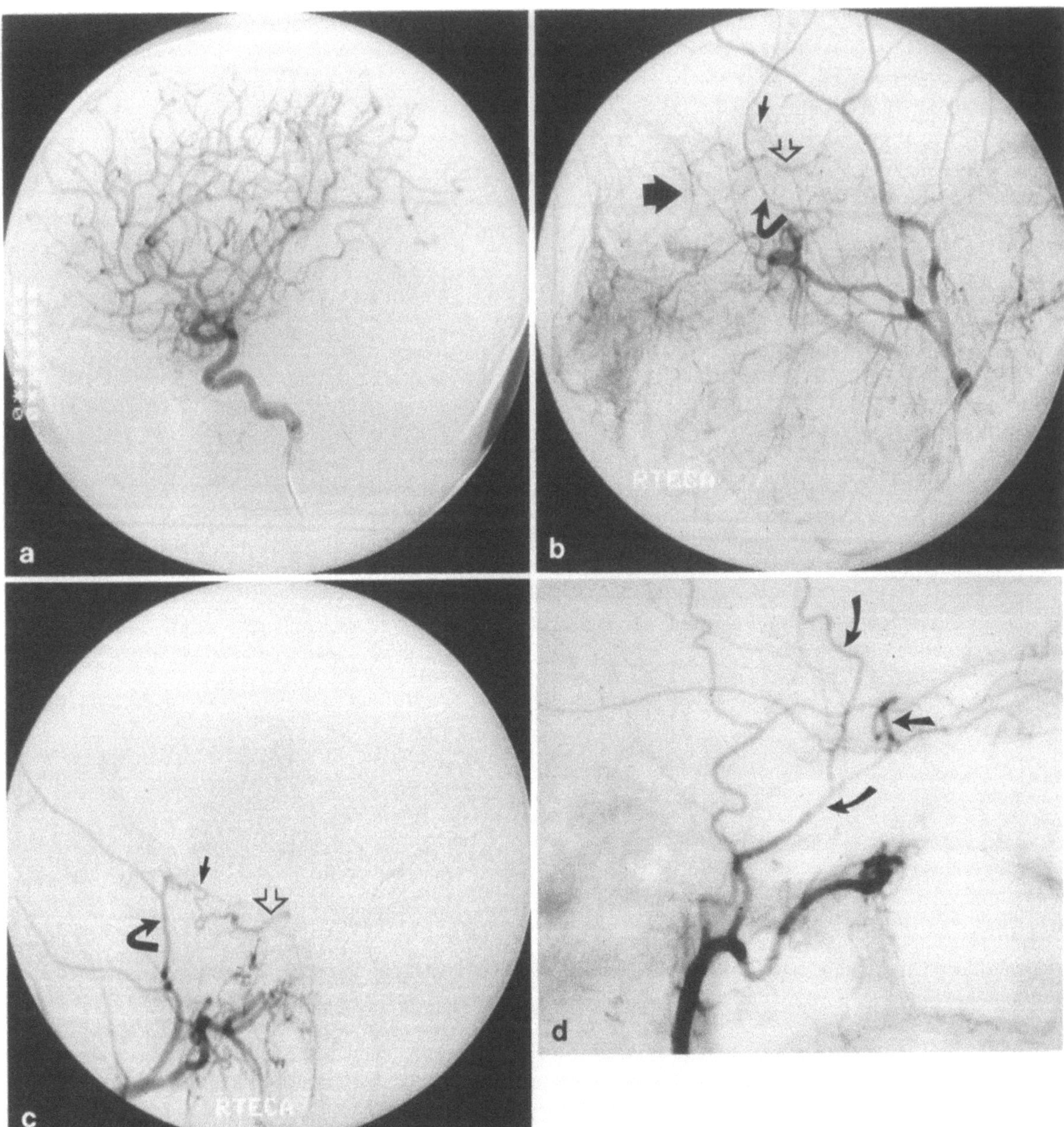

Fig. 1.11 a–d. Ophthalmic artery origin from external carotid artery. **a** The lateral subtraction internal carotid angiogram fails to visualize an origin of the ophthalmic artery. **b** In the same case, the lateral view external carotid artery injection demonstrates the middle meningeal artery (*curved arrow*) providing a small branch artery in Hyrtl's canal (*small arrow*) to supply the ophthalmic artery (*open arrowhead*). The choroidal blush (*large solid arrow*) is seen in this injection. **c** The frontal projection external carotid artery angiogram of the same indi-vidual demonstrates the middle meningeal artery (*curved arrow*), the Hyrtl's canal branch artery (*small arrow*), and the ophthalmic artery (*open arrowhead*). **d** In another individual, the lateral view (front to right) external carotid artery subtraction angiogram demonstrates the ophthalmic artery (*arrow*) originating from the middle meningeal artery (*curved arrows*). In this situation, there is no origin from the internal carotid artery for the ophthalmic artery, which enters the orbit via the supraorbital fissure

Table 1.14. Embryological mechanisms of ophthalmic arterial system variants (modified from [16b])

Anatomic variant: site of origin of ophthalmic artery	Migration of ventral ophthalmic artery	Anastomosis of ventral/dorsal ophthalmic arteries	Regression of dorsal ophthalmic artery	Orbital anastomosis	Transsphenoidal regression
C3 carotid siphon	+	+	+	+	+
ACA	−	+	+	+	+
Double: C3-C4 carotid siphon, anastomosed	+	+	−	+	+
Double: C3-C4 carotid siphon, not anastomosed	+	−	−	with dorsal ophthalmic	+
C4 carotid siphon	+	+	regression of ventral ophthalmic	+	+
Double: C3-C4 carotid siphon, not anastomosed	+	−	−	ventral ophthalmic +	+
Double: C3 portion siphon and MMA	+	+	+	−	−
Triple: C3-C4 portion siphon and MMA	+	+	−	double	−
MMA	+	+	+	+	regression of proximal primitive ophthalmic

+, present; −, absent. MMA, middle meningeal artery; ACA, anterior cerebral artery; C3, C4, segments of the internal carotid artery.

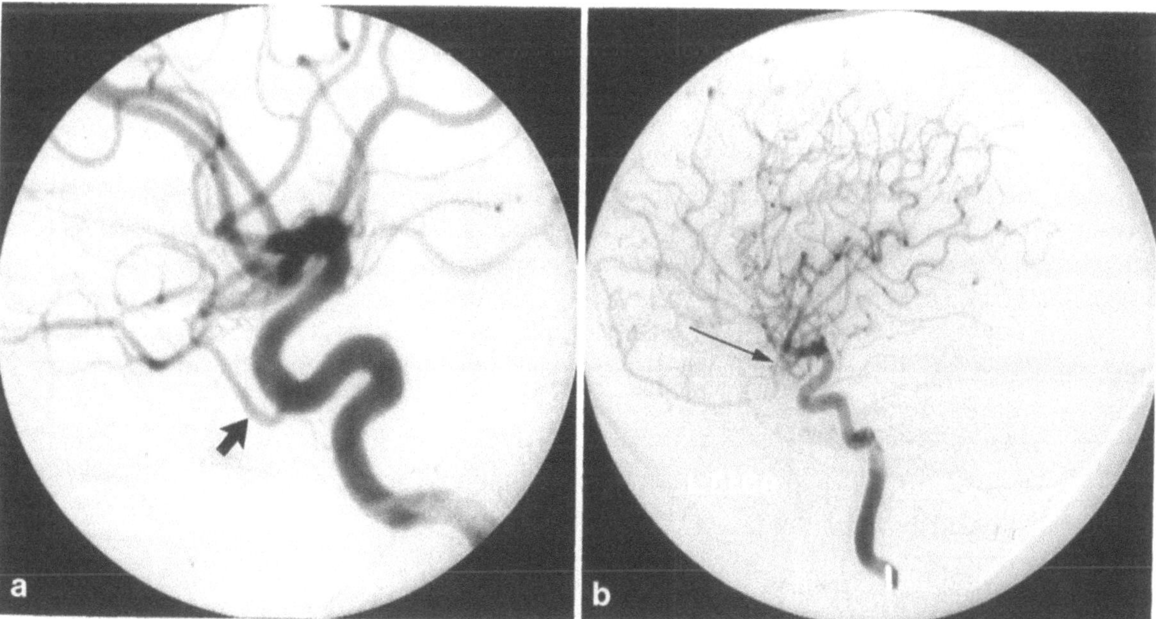

Fig. 1.12 a – b. Variants of the ophthalmic artery origin from the internal carotid artery. **a** The lateral view internal carotid artery subtraction angiogram reveals an intracavernous origin of the ophthalmic artery (*arrow*). **b** In another individual, the lateral projection internal carotid artery subtraction angiogram demonstrates a persistent ventral origin of the ophthalmic artery from the anterior cerebral artery (*arrow*)

Table 1.15. Middle meningeal supply to the lacrimal area (from [16])

Lacrimal territory	Lacrimal variant	Meningolacrimal variant
Site of origin of main artery for glandular, glandulopalpebral, musculoanastomotic branches	Ophthalmic artery	MMA
Anastomoses		
Palpebral	Ophthalmic-ophthalmic (IC/IC)	Meningo-ophthalmic (EC/IC)
Muscular	Temporo-ophthalmic (EC/IC)	Meningo-temporal (EC/EC)
	Infraorbital-ophthalmic (EC/IC)	Meningo-infraorbital (EC/EC)
Transsphenoidal anastomosis	Recurrent meningeal	Meningo-ophthalmic
Arteries anastomosed to transsphenoidal branch	Lacrimal-MMA	Ophthalmic by artery-MMA
Site of origin of recurrent tentorial artery	intraorbital lacrimal	Second portion of intraorbital ophthalmic
Bone changes	No Hyrtl's canal	Hyrtl's canal

MMA, middle meningeal artery; EC, external carotid; IC, internal carotid.

sults in the dominance of the ECA system in the orbit.

The orbit is the site of various potential anastomoses between branches that develop from the embryonic ECA system and the ophthalmic artery. A branch of the embryonic stapedial artery provides the most prevalent anastomosis (an anterior branch of the MMA) between the ophthalmic artery and the internal maxillary artery system. The stapedial artery initially develops from the ICA system, but it becomes annexed to become part of the ECA system and serves as a potential collateral between the two systems. The stapedial artery also frequently provides arterial supply to the orbit (through the superior orbital fissure) via a single vessel, the embryonic orbital artery. This vessel arises from the embryological supraorbital artery, which later becomes the MMA (Table 1.15).

Several other clinically important anastomoses with the orbital segment of the ophthalmic artery arise from arterial contributions from the embryonic orbital artery and branches that arise from the ICA. One branch of the embryonic orbital artery, the anterior deep temporal artery, can anastomose with the lacrimal artery or the muscular artery to the lateral rectus (Fig. 1.10a, b). As it enters the orbit through the infraorbital canal, the infraorbital artery (another branch of the embryonic orbital artery) can also supply the orbit through branch vessels to the lacrimal sac, the inferior rectus, and the lower eyelid. The infraorbital artery has branches that anastomose with the inferior muscular and dorsal nasal arteries. The distal internal maxillary artery branch, the sphenopalatine artery, prior to its entry into the sphenopalatine foramen, provides a small branch vessel that supplies a portion of the inferior extraconal orbit (Fig. 1.13).

The ethmoidal arteries are another route of potential ophthalmic artery collaterals. The posterior ethmoidal artery sends an ascending branch through the cribiform plate to anastomose with the MMA and small branches from the supracavernous ICA. The anterior ethmoidal artery ascends intracranially to supply the anterior falx and anastomose with the MMA.

Branches from the cavernous ICA are also potential collaterals with the orbital circulation (see Sect. 1.5.3.1).

If embryological branches fail to regress, they may supercede the development of the common patterns of blood supply to the orbit. For example, if the embryonic orbital artery divides intracranially rather than in the orbit: (1) the lacrimal artery will arise from the MMA (Fig. 1.10c); and (2) the MMA will anastomose with other branches from the intraorbital ophthalmic arterial system, rather than the lacrimal artery. In another rare variation, the ophthalmic artery originates from the MMA as a result of regression of the proximal primitive ophthalmic artery (Fig. 1.11d) [18].

The facial artery or superficial temporal artery generally has a minor role in providing collaterals

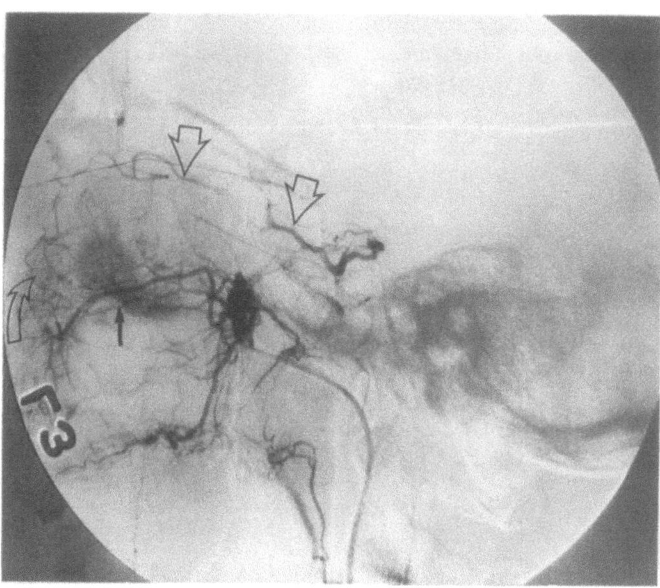

Fig. 1.13. In this patient with a dural arteriovenous malformation, the contrast injection of the internal maxillary artery demonstrates anastomoses through the infraorbital foramen (*arrow*) to the palpebral arteries (*curved arrow*), which then opacify the ophthalmic artery (*open arrowheads*)

to the orbit. However, the anastomoses with the ophthalmic artery branches via the nasal and palpebral arteries can be functional (Fig. 1.10d).

1.5.2 Sensory Visual Pathway Circulation

The vascular supply to the intraorbital optic nerve, retina, and choroid arises predominantly from the ophthalmic arterial circulation [19] via the posterior ciliary arteries, the central retinal artery, and the pial vascular network along the optic nerve (Fig. 1.8). Two, less commonly three, posterior ciliary arteries course anteriorly from their ophthalmic artery origin. Each posterior ciliary artery divides into 10–20 branches (short posterior ciliary arteries) which pierce the sclera. Branches from the medial long posterior ciliary artery enter the globe medially and inferiorly to the optic nerve. The short posterior ciliary arteries arising from the lateral long posterior ciliary artery enter the globe laterally and superiorly to the optic nerve (Fig. 1.14a). When present, branches from a superior long posterior ciliary artery enter the sclera above the optic nerve. An arterial and arteriolar network, the circle of Haller–Zinn, typically surrounding either the superior or inferior aspect of the optic nerve, is in-

consistently present. This incomplete arterial ring is usually intrascleral, but on occasion this network may be extrascleral. The circle is supplied by pial arteries and short posterior ciliary arteries and sends recurrent choroidal arteries to the choroidal circulation (Fig. 1.14b,c). Occasionally, eyes have macroscopic extrascleral anastomoses between short posterior ciliary arteries (Fig. 1.14a).

The blood supply to the intraocular portion of the optic nerve is divided into four regions: the optic disc surface or nerve fiber layer, the prelaminar area, the laminar area, and the retrolaminar segment. Controversy surrounds several aspects of the vascular anatomy to the optic nerve and choroid (see Sect. 1.5.2.4.5 for a discussion of these issues).

1.5.2.1 Retinal Arteries and Capillaries
(Fig. 1.15)

After originating from the central retinal artery, the retinal arteries branch on the surface of the disc and in the adjacent peripapillary retina. There are four main trunks, one to each quadrant of the retina (superonasal, superotemporal, inferonasal, and inferotemporal). On occasion, the central retinal artery divides into superior and inferior trunks prior to reaching the disc surface. In addition to the

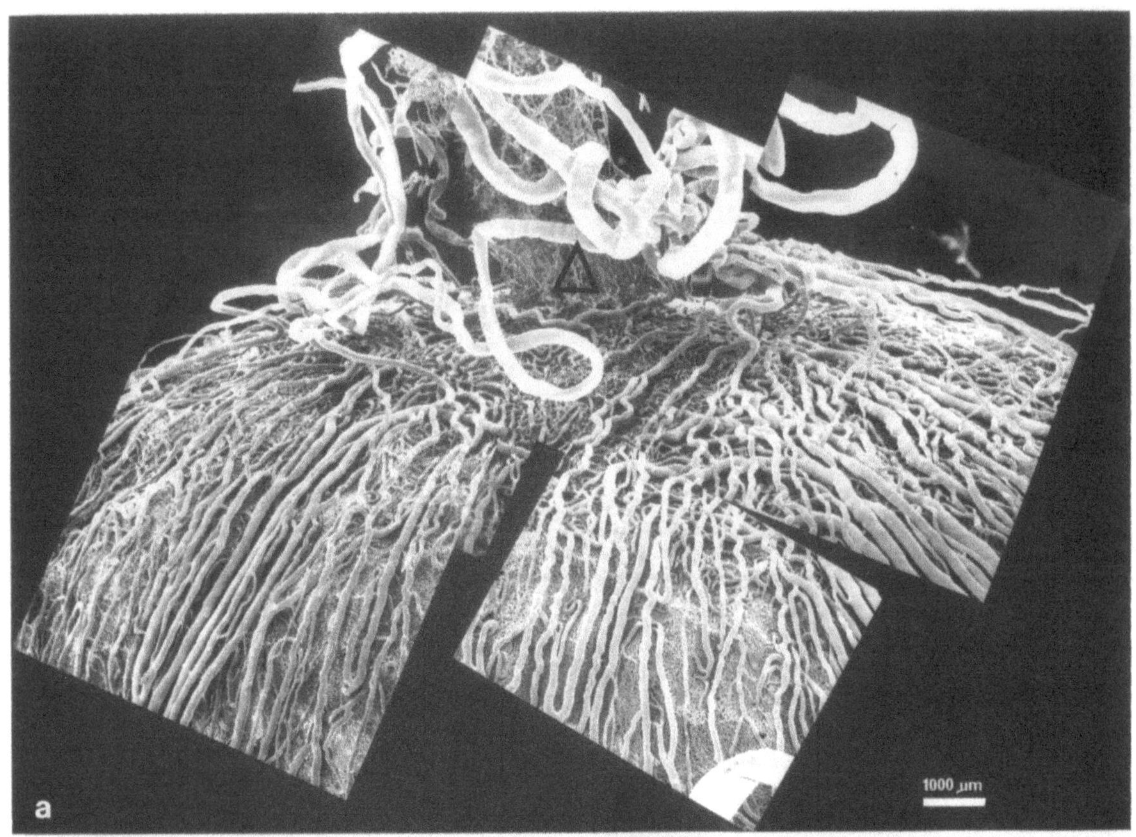

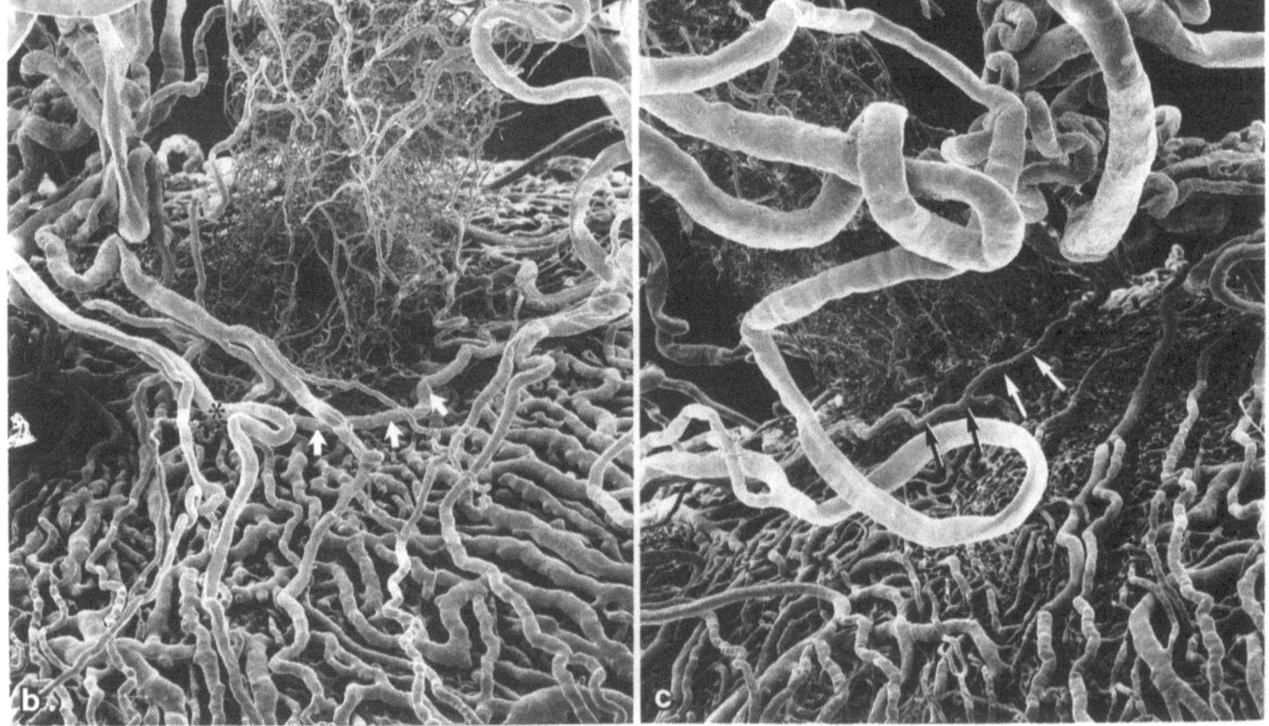

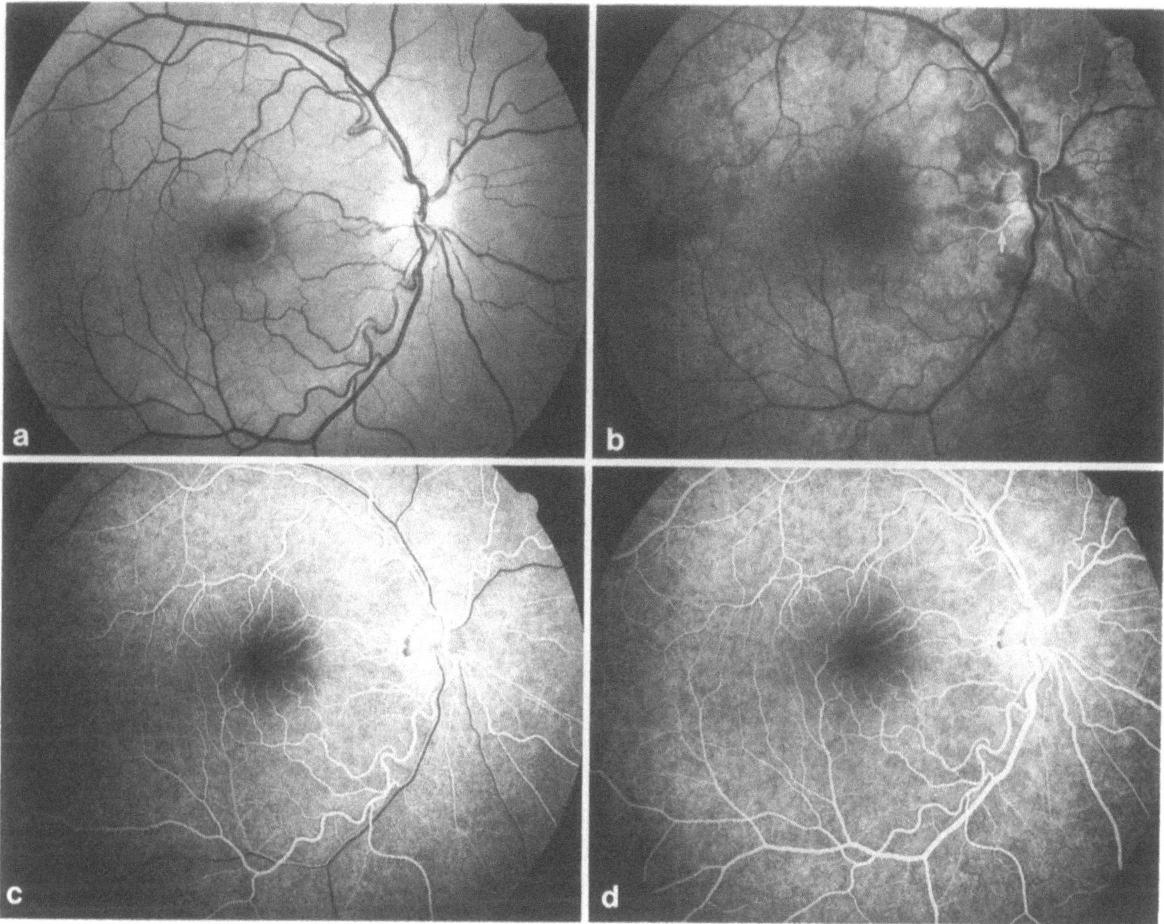

Fig. 1.15 a–d. In the normal fluorescein angiogram, first the choroidal circulation fills, and then opacification of the arteries, then capillaries, and veins of the retina occurs.

a The red-free photograph demonstrates the light gray arteries and darker veins in the retina. The nerve fiber layer is readily seen along superior and inferior poles of the optic disc.

b Twelve seconds after the intravenous injection of fluorescein dye, the choroidal circulation shows patchy filling. There is early opacification of the retinal arteries. A cilioretinal artery (*arrow*) fills intensely and it was the

only artery to the retina that was opacified when the choroid began to fill (not seen). The deep portion of the temporal aspect of the optic disc is also opacified.

c Early filling of the retinal veins shows laminar flow at 17 s. The choroidal circulation and retinal arteries are completely opacified. The deep and superficial vessels of the temporal optic disc are also well filled. The deep portion of the nasal segment of optic disc has filled.

d Within 20 s, all of the vessels of the choroid, retina, and optic nerve are opacified. (Photographs provided by Kenneth Noble)

◀ **Fig. 1.14 a–c.** Scanning electronmicrographs of the arteries, arterioles, and capillaries in the region of the laminar and retrolaminar optic nerve. **a** A montage of photomicrographs of a vascular cast of the posterior globe choroid and optic nerve supplied by lateral and medial short posterior ciliary arteries. Note the extrascleral anastomosis (*open arrowhead*) between the short posterior

ciliary arteries. **b** Higher magnification shows a partial "circle" of Haller–Zinn (*arrows*), which is just below the capillary network of the optic nerve in the photomicrograph. A choroidal branch (*asterisk*) originates from the lateral short posterior ciliary artery. **c** Higher magnification illustrates a superior arteriolar anastomosis of the "circle" of Haller–Zinn (*arrows*). (From [46]) large retinal arteries, a variable number of smaller arteries cross the disc, predominantly over the temporal border (some are cilioretinal arteries, see Sect. 1.5.2.2). The retinal arteries are located in the outer nerve fiber layer or the adjacent ganglion cell layer. Each main trunk supplies an arterial tree that continues to branch. The vessels gradually narrow as they pass to the peripheral retina (Fig. 1.15). The arteries, distal to the first retinal branching, lose their muscular layer to become arterioles.

Each distal arteriole supplies a segmental network of radial capillaries which enter the deeper layers of the retina and eventually join to venules. The capillary network of the inner retina has few communications with the deeper capillaries that supply the outer retinal layers [20]. The outer layers of retina are nourished by diffusion and active transport of electrolytes, nutrients, and fluids across the retinal pigment epithelium (a blood – ocular barrier) from the choroidal circulation. The inner two thirds of the retina receives nutrients from the branch arteriole/capillary network of the retinal arteries (a blood – ocular barrier).

Except for several exceptions, the inner layers of the retina, from the bipolar cells to the nerve fibers, are supplied exclusively by central retinal artery-fed retinal arterioles. One exception is the far peripheral nasal retina that can receive blood via the choroidal circulation. Additionally, there are recurrent branches to the choriocapillaris from arterial circles of the iris or ciliary muscle that provide blood to the far peripheral retina. These anterior arterial circles are supplied by both the anterior ciliary and posterior ciliary arteries [21]. On occasion, the posterior ciliary arteries may also provide some blood to the immediate peripapillary retina. However, the peripapillary retinal capillaries do not receive blood via the same arterioles that supply the surface of the optic disc [22]. Of major clinical significance, in some subjects one or more cilioretinal arteries supply the papillomacula and macula regions of the retina (see below).

1.5.2.2 Choroidal Arteries and Capillaries
(Fig. 1.16)

Each posterior ciliary artery or branch from the circle of Haller – Zinn provides blood supply to multiple choroidal arterioles [23]. One or more arterioles supply a segmental area of choriocapillaris. The choriocapillaris vessels provide no blood – ocular barrier: these vessels have large fenestrations that allow the blood and nutrients to freely leak into the extravascular spaces of the choroid. The medial posterior ciliary artery supplies the inferior and medial choroid. The lateral posterior ciliary artery supplies the superior and lateral choroid except when a superior posterior ciliary artery is present, in which case the latter vessel supplies the choroid superior to the optic disc [24]. In addition to providing blood supply to the choroid and the optic nerve, the posterior ciliary arteries, along with the muscular arteries (in the body of the extraocular muscles), contribute to the anterior arterial circle in the iris [25]. The rate of blood flow per gram of tissue to the choroid is higher than to any other organ in the body, and it is almost 10 times the rate of the cerebral blood flow [26].

Each choroidal artery supplies an independent segment of choriocapillaris [24, 27]. Although the post mortem injection of a single posterior ciliary artery with a liquid agent fills the entire choroid, indicating there is an extensive anastomotic network between the segments of choriocapillaris [28], an in vivo functional role for these anatomical connections has not been demonstrated.

Fluorescein angiography has provided most of the current evidence for understanding the dynamics of the human choroidal circulation. Fluorescein angiographic studies confirm the segmental nature of the choroidal blood supply. Each segment appears to have little or no overlap with adjacent regions, causing watershed territories between the areas supplied by each posterior ciliary branch artery (see Sect. 5.1). The size and shape of the area of the choroid and optic nerve supplied by each posterior ciliary artery is variable between subjects and even between the eyes of a single individual [24]. Additionally, there are no direct anastomoses between posterior ciliary arteries except at the level of the intrascleral circle of Haller – Zinn. Intra-arterial

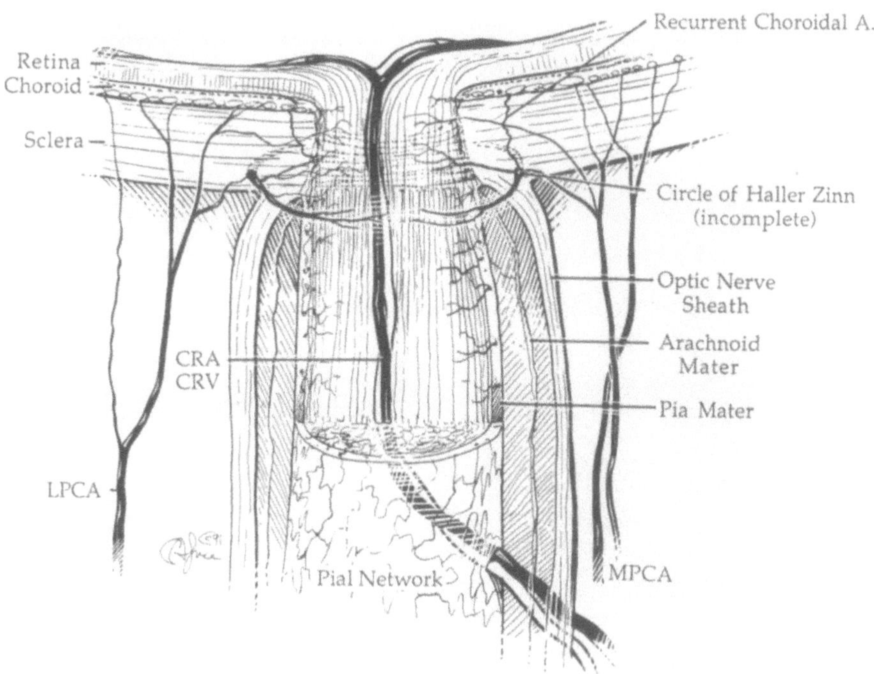

Fig. 1.16. The blood supply to the proximal optic nerve and choroid arises from the branches of the medial (*MPCA*) and lateral (*LPCA*) posterior ciliary arteries, the incomplete circle of Haller – Zinn, the pial network, and recurrent choroidal arteries. The posterior ciliary arteries send branches to the circle of Haller – Zinn and the pial network. The central retinal artery (*CRA*) provides minute branches to the optic nerve capillaries as it passes anteriorly to supply the retina. The central retinal vein (*CRV*) parallels the course of the CRA

anastomoses in the choroid are infrequent and choroidal arteriovenous anastomoses do not occur in normal eyes [23]. Fluorescein angiography also confirms the wide variation of the extent and area of choroid supplied by each posterior ciliary artery or by branches from the circle of Haller – Zinn [29]. A functional watershed zone, usually between the fovea and the nasal border of the optic disc, is typically seen on angiography. These watershed areas have different shapes and sizes and often include the optic disc.

Arteries that provide a choroidal arterial blood contribution to the retinal circulation, specifically true cilioretinal arteries, are inconsistently present on fluorescein angiography. Ophthalmoscopy cannot adequately distinguish a retinal branch that originates from the central retinal artery deep within the disc from an artery that truly arises from the choroidal circulation. Though the existence of this uveal – retinal arterial anastomosis was described by many early investigators [30, 31], others have failed to find it or have rarely demonstrated such an artery [19, 32]. In addition, vessels that appear to be cilioretinal arteries on funduscopy frequently do not have fluorescein confirmation of a choroidal origin [33]. No matter what their origin, one or more "cilioretinal" arteries course anteriorly in the optic nerve and cross the temporal (rarely nasal) disc border into the retina and provide partial or all of the blood supply to the papillomacula area. Each "cilioretinal" artery is an end artery because the area of retina supplied receives no additional arterial input [34].

Some authors have also suggested that there is a specific submacular artery from a branch of a posterior ciliary artery that passes through the choroid to the supply the area beneath the macula [23]. However, the existence of a specific choroidal artery that selectively provides blood to the macula has not been confirmed in other investigations [35].

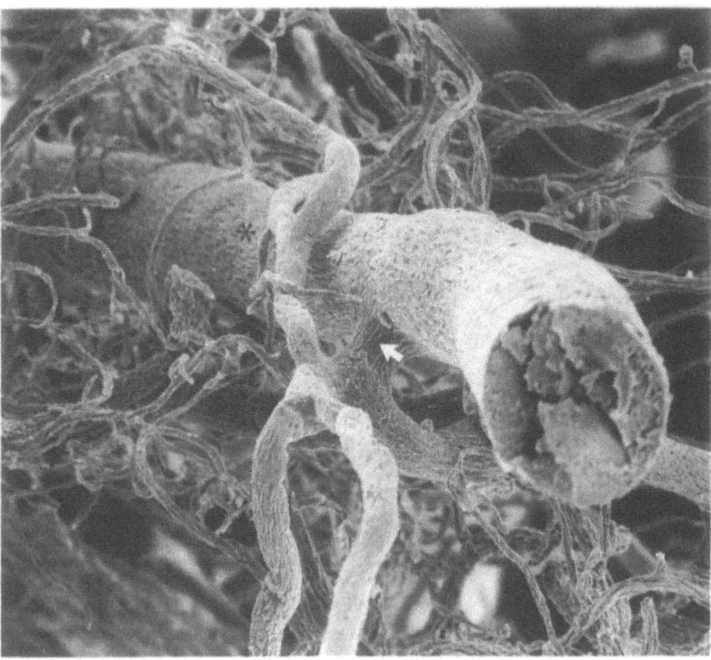

Fig. 1.17. Scanning electronmicrograph demonstrates a branch (*arrow*) from the intraneural central retinal artery (*asterisk*), which arises approximately 1.5 mm posterior to the lamina and anastomoses with optic nerve capillaries. (From [46])

1.5.2.3 Central Retinal Artery

The principal arterial supply to the retina originates from the central retinal artery and not the posterior ciliary arteries. The central retinal artery arises directly from the ophthalmic artery or rarely from the medial posterior ciliary artery (Fig. 1.8). The central retinal artery enters the dural sheath that surrounds the intraorbital optic nerve, approximately 5–15 mm behind the globe, usually at a site inferior and medial to the opening in the sclera for the optic nerve. The central retinal artery has a course of 1–2.5 mm in the subdural or subarachnoid space before penetrating into the substance of the optic nerve. The central retinal artery contributes up to eight small twigs within the nerve that anastomose with the capillaries of the retrolaminar optic nerve supplied by the pial blood network (see below and Fig. 1.17) [36]. The central retinal artery also inconsistently sends a branch that courses, for a variable distance, posteriorly in the orbital optic nerve [19, 36]. There are no direct anastomoses between the central retinal artery and any posterior ciliary artery or the arterial circle of Haller–Zinn, even though the retrobulbar pial vascular network has connections to the arterial circle of Haller–Zinn. Contrary to the results of early studies, there is no separate artery of the optic nerve that arises from the ophthalmic artery and provides significant blood supply to the retrobulbar portion of the optic nerve [37]. The optic nerve, immediately behind the globe, is not dependent on the central retinal artery for a significant blood supply contribution [37].

1.5.2.4 Optic Nerve

1.5.2.4.1 Optic Disc Surface

The surface of the optic disc is composed of the axons from the retinal ganglion cells and a network of small vessels. Most of the vessels are capillaries

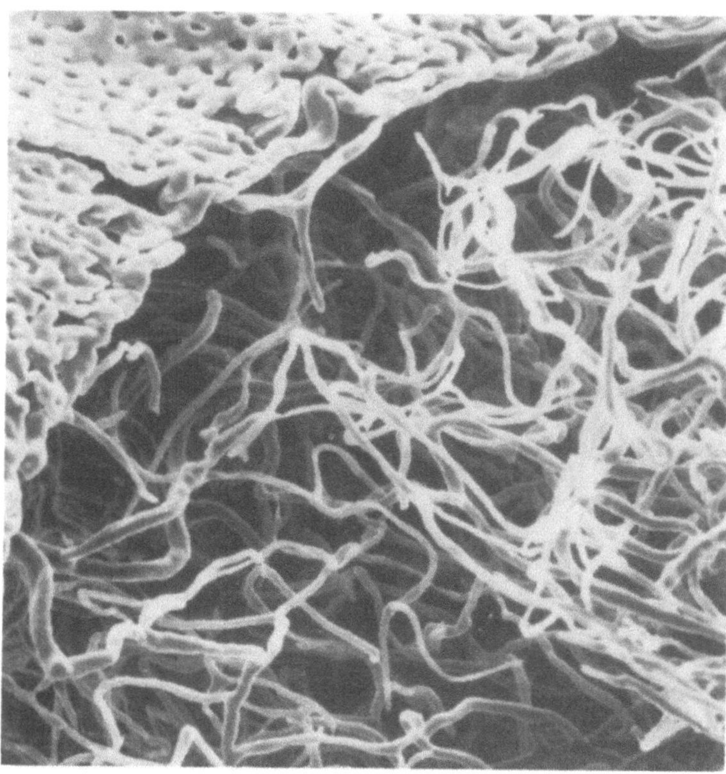

Fig. 1.18. A scanning electronmicrograph of retinal view at the edge of the juxtapapillary choroid. The choriocapillaris (*above*) does not send branches to the prelaminar portion of the optic nerve. (From [50])

with diameters from 7 to 10 µm, whose walls consist of endothelial cells and a basement membrane. The tight junctions between the endothelial cells constitute a blood–ocular barrier in the superficial layer of the optic disc (and in the retina) that is the physiological equivalent to the blood–brain barrier (see [38] for a discussion of the blood–ocular barriers). Anatomical studies show that the major arterial input to these capillaries is from the arterioles in the adjacent retina. These peripapillary arterioles may not be apparent on ophthalmoscopy in subjects where the highly reflective Bruch's membrane appears white and opaque, similar to the bare sclera [39]. The retinal arteriole contribution to the disc surface is suggested by angiographic demonstration of filling of the superficial vessels during the retinal arterial phase [40].

In some individuals, the temporal portion of the disc receives a significant blood supply from a posterior ciliary arterial contribution via the prelaminar region of the optic nerve. However,

since the capillaries of the choroid do not anastomose with the capillaries of the disc surface, the choriocapillaris has no role in the circulation of this area.

1.5.2.4.2 Prelaminar Optic Nerve

The prelaminar region of the optic nerve, located immediately beneath the disc surface, has fewer collagen septae and more glia than the laminar portion of the optic nerve. The prelaminar optic nerve receives arterial supply via centripetal end vessels originating from the choroidal arterial network and the incomplete circle of Haller–Zinn (posterior ciliary arteries provide the principal input to both). Most authors agree that the choriocapillaris does not provide significant blood to this region of the optic nerve (Fig. 1.18). This segment of the optic nerve is susceptible to ischemia because there are few if any arterial collaterals that overlap the area supplied by each posterior ciliary artery.

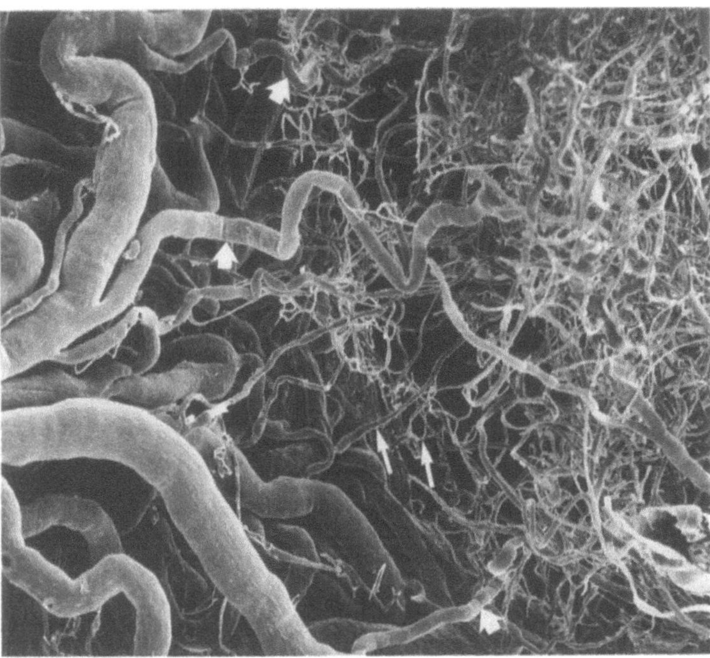

Fig. 1.19. A high magnification electronmicrograph demonstrates the medial short posterior ciliary artery branches where they enter the sclera to form the superior and inferior parts of the circle of Haller–Zinn. Pial branches *(large arrows)* and small centripetal branches from both the choroid and recurrent choroidal branches *(small arrows)* enter the lamina and retrolaminar optic nerve. (From [46])

1.5.2.4.3 Laminar Optic Nerve

The lamina cribrosa portion of the optic nerve contain numerous vascular septae between bundles of axons [41]. The capillaries in the connective tissue of the lamina cribosa are part of a continuous network of capillaries across the optic nerve at this level. The short posterior ciliary arteries, either directly or indirectly via recurrent choroidal arteries and/or arterial input from the circle of Haller–Zinn, provide the principal arterial input to this portion of the optic nerve [15, 40] (Figs. 1.19, 1.20). The arterial input ends predominantly in a transverse vascular network surrounding the laminar portion of the optic nerve [42].

Though minute branches of the central retinal artery anastomose with the pial network, the central retinal artery has no significant role in the circulation in this area. The anastomoses between the central retinal artery and the circle of Haller–Zinn, originally described by Zinn [43, 44] and Haller [44, 45], do not exist [19, 44].

1.5.2.4.4 Retrolaminar Optic Nerve

The optic nerve behind the lamina cribrosa (retrolaminar) is surrounded by meninges and contains myelinated axons and delicate septae between nerve fascicles [42]. Intraseptal arteries and arterioles provide the blood supply to the capillaries in this region, principally via the meningeal– pial network. Arterioles to the optic nerve can originate from the pial network before this vascular network reaches the sclera. Small (typically less than 2 mm) intraneural branches of the central retinal artery provide a variable and minor contribution to this portion of the optic nerve [46] (Fig. 1.17). Anterior-

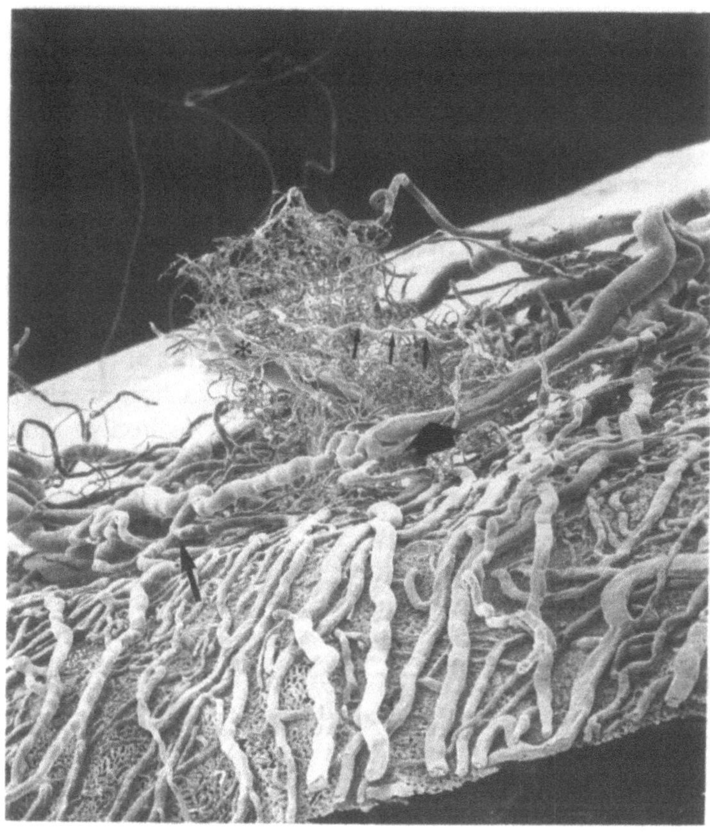

Fig. 1.20. Scanning electronmicrograph showing the axial position of the anterior "circle" (intrascleral) with anastomosis (*broad arrow*), recurrent choroidal branches (*fine arrows*) and pial branches (*long arrow*), and the central retinal artery (*asterisk*) surrounded by residual optic nerve capillaries. (From [46])

ly, adjacent to the lamina cribosa, no single vascular pattern is found [47].

The pial vascular network, which is continuous to the sclera, is supplied by several sources (Fig. 1.20). Before entering the sclera, the posterior ciliary arteries supply the pial network. Recurrent arteries from the peripapillary choroid and the circle of Haller–Zinn provide additional anastomoses with the pial vessels [46, 48] (Fig. 1.19).

1.5.2.4.5 Optic Nerve Controversy

Debate exists concerning several aspects of the optic nerve circulation. Not all investigators agree on the role of a choroidal arterial blood supply to the prelaminar and laminar optic nerve, the existence of recurrent choroidal arteries to the retrolaminar optic nerve, or whether there is a functional longitudinal, rather than a transverse, arterial vascular

network along the length of the optic nerve. Some of the discrepancies among the investigations are attributable to the differences in the histological methods used in the various studies, whether monkeys or other animals versus human specimens were examined, whether experimental intra-arterial injections of latex, particles, or dye were given to live or post mortem animals, and whether serial sections or scanning electron microscopy of post mortem anatomical preparations was performed.

Some investigations using serial section techniques failed to demonstrate the peripapillary choroidal arterial supply to the laminar portion of the optic nerve [42, 49]. Though these studies concur that the posterior ciliary arteries provide arterial blood to this area, they suggested that these vessels do not pass through the choroid. In some individuals the posterior ciliary artery contribution may principally be indirect through the circle of

Haller–Zinn. In one study, Lieberman et al. could not identify recurrent choroidal arteries to the laminar or retrolaminar regions of the optic nerve [42]. It is easy to see how the serial section method could fail to demonstrate the posterior ciliary artery branches to the choroid and both laminar and retrolaminar portions of the optic nerve. In contrast, Hayreh's extensive evaluations have showed that the retrolaminar portion of the optic nerve receives a major blood supply from the peripapillary choroidal blood network or from the posterior ciliary arteries or both.

A choroidal contribution to the retrolaminar optic nerve has been confirmed by more recent investigations in both humans and primates using scanning electron microscopy that demonstrated recurrent choroidal arteries anastomosing with the retrolaminar pial network [46, 50]. This technique allows a three-dimensional visualization of the angioarchitecture, which can only be inferred from serial section techniques.

Although anatomical collaterals exist along the length of the optic nerve, the functional pial circulation to the optic nerve is segmental. The pial network in each segment does not provide a significant contribution to adjacent regions [51]. However, Lieberman et al. [42] suggested that a continuous longitudinal, rather than a transverse, network of capillaries and precapillaries from the pia and central retinal artery provides the blood supply to the optic nerve between the lamina cribosa and the superficial disc surface. They also hypothesized that the pial-derived longitudinal vessels and branches of the central retinal artery in the prelaminar and laminar regions might be analogous to the so-called "artery of the optic nerve" of Francois [42] and provide significant blood supply to this region. Though capillary anastomoses between the vessels in the superficial nerve fiber layer of the disc and the prelaminar network of the optic nerve (similar to those reported by Henkind and Levitsky [40]) exist, their functional significance has not been demonstrated. Also, the central retinal artery provides a very minor contribution to the blood supply of the prelaminar and retrolaminar segments of the nerve. It is unlikely that a longitudinal capillary system provides significant collateral blood flow to prelaminar, laminar, or retro-laminar portions of the optic nerve. Though the differences between the two concepts concerning the blood supply to the prelaminar and laminar optic nerve seem minor, the exact source and route of the blood supply to this region may be relevant to a vascular theory of glaucoma, low tension glaucoma (see Sect. 5.6), and anterior ischemic optic neuropathy.

The predominant blood supply to the optic nerve, beneath the optic disc surface, originates from the arteries of the choroidal circulation, from the posterior ciliary arteries that pass through the sclera, or indirectly through the circle of Haller–Zinn. Clearly, the role of the retinal circulation is minor. Although the capillaries on the disc surface anastomose with those in both the prelaminar and laminar portions of optic nerve [40, 47], the retinal arteries do not provide a significant (longitudinal network) collateral blood flow to the region of the optic nerve posterior to the nerve fiber layer.

Fluorescein angiography (Fig. 1.15) demonstration of the appearance of fluorescence in the deeper region of the disc concomitantly with the filling of the choroidal vessel suggests that this region of the nerve receives the same arterial input as the choroid. This disc fluorescence is still seen following occlusion of the central retinal artery, further supporting the concept that the central retinal artery or retinal arterial contribution to the blood supply of this area is not significant [52]. The superficial radial vascular network on the disc normally appears later, during the early filling of the retinal arteries, than the choroid [53]. Debate exists on the interpretation of fluorescein studies. Lieberman et al. argued that synchronous filling of the prelaminar disc and the peripapillary choroidal vessels on fluorescein angiography does not conclusively prove that the disc fluorescence arises via the choroidal circulation. They concluded that the fluorescein results were indicative of a common posterior ciliary artery supply to the adjacent peripapillary choroid and the prelaminar optic nerve via scleral, but not choroidal, branches [54].

1.5.2.4.6 Blood Flow in the Retina, Choroid, and Anterior Optic Nerve

Most of the studies of blood flow regulation in response to alterations of various physiological parameters are performed in experimental animal preparations. As previously discussed, the rate and volume of blood flow through the choroid is much greater than through the retina. Autoregulation of blood flow exists in the retina but it is not present in the choroidal circulation. As a result, elevation of the intraocular pressure (IOP) reduces the choroidal blood flow, while the blood flow in the retina remains constant over a large range of IOPs [55].

Choroid. Systemic blood pressure and IOP fluctuations affect the blood flow in both the posterior ciliary arteries as well as the choroidal circulation. In animals, an experimentally induced fall in the systemic blood pressure reduces the fluorescein angiographic filling in the choroidal watershed zones, between the posterior ciliary arteries, prior to other alterations in the circulation. Systemic hypertension or atherosclerosis increase the vascular resistance and decrease the choroidal blood flow. Hyperviscosity also reduces blood flow in the choroid.

The influence of autonomic neurotransmission on the ocular circulation is complex. For example, in monkeys stimulation of either the sympathetics from the cervical sympathetic chain or parasympathetics from the oculomotor nerve reduces choroidal blood flow [56].

Retina. In contrast to the choroidal circulation, the retinal arterial system has a remarkable capacity for autoregulation. Human retinal arteries vasoconstrict in response to acute elevation in systemic blood pressure in order to prevent supranormal blood flow to the retinal tissue [57]. Human retinal blood flow remains constant over a wide range of systemic blood pressure elevation, up to a diastolic pressure of 115 mmHg. Raising the blood pressure beyond this level, results in an abnormal increase of the blood flow [58]. In humans, the blood flow in the optic nerve posterior to the globe is generally unaltered by moderate elevation of the systemic blood pressure [59].

The autoregulation in the retina can adjust the blood flow in order to maintain the appropriate metabolic environment in the retina. For example, human retinal blood flow increases by 38% and both retinal arteries and veins dilate in response to isocapnic hypoxia (80% oxygen saturation) in order to maintain the normal O_2/CO_2 ratio in the retina. Hyperoxia (100% oxygen saturation) has the opposite affect, decreasing blood flow by 36% and causing narrowing of retinal arteries and veins [60]. Hypercapnia, caused by carbon dioxide inhalation, results in a dilatation of human retinal vessels that is not reversed by concomitant hyperoxia [61].

Perivascular innervation seems to have no role in altering the function of the central retinal artery or other retinal arteries.

Optic Nerve. As mentioned above, elevation of the IOP reduces blood perfusion in the choroid. In contrast, the autoregulation in the optic disc head circulation maintains blood flow when the IOP is elevated as long as it does not exceed the critical perfusion pressure [62, 63]. Beyond this level, elevation of the IOP also reduces the perfusion of the peripapillary retinal capillaries [20]. In animal studies, when the IOP is lowered below normal, the blood flow to the retina and the prelaminar optic nerve is also reduced. When the IOP is normal, the blood flow to the prelaminar and laminar regions of the optic nerve is greater than to the retrolaminar optic nerve. One possible explanation to account for this difference is that the retrolaminar region, which is myelinated, may have a lower metabolic activity, and demands less oxygen and nutrients [64].

When the IOP is markedly elevated, the blood flow to the retrolaminar area increases, in part because of dilatation of the vessels. When the IOP exceeds the arterial pressure, the blood flow to the anterior, but not the posterior, portion of the laminar segment of the optic nerve is arrested [65]. The inverse relationship between a marked elevation in IOP and a fall in the arterial perfusion in choroid and laminar optic nerve is significant in the pathophysiology of tissue injury in several clinical entities (see below). In contrast, autoregulation maintenance of a constant blood flow in the retina and optic disc surface for IOPs of more than 32 mmHg protects these regions from damage by the same entities [66].

Laser doppler determination of blood velocity and fluorescein angiography of the human optic disc have confirmed the experimentally determined relationships between retinal and choroidal blood flow and IOP elevation. For example, these studies show that increasing the IOP only transiently reduces optic nerve blood flow speed before it normalizes within 3 min, indicating the effectiveness of autoregulation [67]. Fluorescein angiography studies show that elevation of the IOP reduces the choroidal blood supply to the optic disc and the peripapillary choroidal blood supply in both monkeys [68, 69] and humans [70].

For the clinician, an understanding of the functional blood circulation is more important than absolute agreement on the details of the discrepancies concerning anatomical dissections when considering the vascular disorders affecting the optic nerve. In sum, the distal (anterior) optic nerve seems to demonstrate reasonable autoregulation for maintaining constant blood flow in the face of a wide range of alterations in the systemic blood pressure or IOP (the two most important determinants for tissue perfusion pressure). The susceptibility to ischemia in this portion of the optic nerve may be related in part to an increased metabolic demand by the unmyelinated optic nerve, so that even relatively minor drops in local perfusion may result in tissue injury and cause visual dysfunction. Experimental data in cats lends support to this hypothesis, since the normal blood supply is greatest in the prelaminar and laminar regions, and it gradually diminishes along the optic nerve as the nerve passes posteriorly to the chiasm [59]. The importance of understanding the role of acute ischemia of the optic nerve is obvious for conditions such as anterior ischemic optic neuropathy, shock-induced optic neuropathy, and the vasculitides. In addition, the identification of an incomplete circle of Haller–Zinn located predominantly either superiorly or inferiorly to the optic nerve may create an anatomical susceptibility to transient drops in posterior ciliary artery perfusion pressure. The resulting segmental ischemia in the optic nerve could ac-

count for the occurrence of altitudinal defects with anterior ischemic optic neuropathy [46]. Chronic hypoperfusion and ischemia may be significant in other disorders of the optic nerve such as low tension glaucoma and possibly chronic open angle glaucoma [47].

1.5.2.4.7 Intraorbital Optic Nerve
(Figs. 1.8, 1.14a)

A pial vascular network, via collaterals from small branch arteries arising from the ophthalmic artery, supplies the retrobulbar optic nerve. This pial network sends microscopic centripetal vessels into the optic nerve along its course through the orbit. These vessels supply segmental units of an intraneuronal capillary network that incompletely encircle nerve fiber bundles across the width of the nerve [71].

When the ophthalmic artery crosses above (75% of orbits) the optic nerve, 75% of these orbits have a direct ophthalmic arterial supply (arising predominantly superior to the nerve) to the pial vascular network of the optic nerve. In the other 25%, the pial network is supplied by an arterial branch arising from the central retinal artery or the medial posterior ciliary or muscular arteries. In cases where the ophthalmic artery crosses under (25% of orbits) the optic nerve, 50% of these orbits have some direct intraorbital ophthalmic arterial branch (arising predominantly inferior to the nerve) to the optic nerve pial network. In 25% of the latter orbits, one of the previously mentioned branch contributions to the optic nerve is present. Other branch arteries, such as the posterior ethmoidal artery, uncommonly supply this arterial capillary network [72].

It is doubtful whether a true longitudinal axial arterial system, such as an artery to the optic nerve, that supplies the intraorbital optic nerve posterior to the entrance of the central retinal artery exists [19]. The posterior intraneuronal branch of the central retinal artery is only a small insignificant vessel that travels posteriorly for a short distance. However, as mentioned previously, the central retinal artery, prior to its entry into the subarachnoid space, does send some arterial twigs to the pial network. Ethmoidal, anterior deep temporal, and middle meningeal arterial collaterals to the ophthalmic artery can indirectly provide blood supply to the orbital optic nerve (see Sect. 1.5.1.1.1).

1.5.2.5 Venous Drainage of the Orbit

1.5.2.5.1 Retinal Veins

Each retinal venule drains a segment of the capillary network. The venules then drain into progressively larger vessels that merge into the central retinal vein which exits from the eye. Latent collaterals between the central retinal vein and the choroidal venous system are located at the border of the optic nerve and the retina (when manifest, they are called a "optociliary shunt").

1.5.2.5.2 Choroidal Veins

Each sectoral area of choriocapillaris is drained by a venule that empties into a vortex vein [27]. After exiting the eye, the vortex veins collect into the posterior ciliary veins. Though the anterior uvea can also drain via the vortex veins, the predominant venous drainage is via the scleral and episcleral venous plexuses into the veins in the ocular muscles.

1.5.2.5.3 Venous Drainage of the Optic Nerve

The veins of all the intraocular segments of the optic nerve and disc surface predominantly empty into the central retinal vein. A plexus of veins located along the pia receives a minor amount of blood from the nerve.

Though the prelaminar portion of the optic nerve principally drains through the central retinal vein, the venous outflow can also empty into the choroidal venular circulation. The choroidal venous channels provide a potential route for draining the optic nerve and the retina when the central retinal vein is occluded. The venules in the optic nerve sheath are an additional route of venous drainage for the prelaminar optic nerve [22].

The venous drainage of the laminar portion of the optic nerve empties into the central retinal vein [42] or the choroidal venous circulation. Collecting veins in the optic nerve course in the intraneuronal septae. Rarely, septal veins from the laminar region drain into the pial veins [42].

Most of the septal veins in the retrolaminar region drain into the central retinal vein. The pial venous plexus generally has a minor role in the venous drainage of this area [36, 42].

Table 1.16. Orbital veins

Supraorbital
Supratrochlear
Superior and inferior palpebral
Angular
Anterior facial
Ophthalmic
Pterygoid plexus
Ethmoidal

1.5.2.5.4 Orbital Venous System (Table 1.16)

The central retinal vein exits the optic nerve at the entrance of the central retinal artery or in close proximity to it [19, 36]. After entering the orbit, the central retinal vein connects to a plexus of veins in the orbital fat.

The orbital venous plexus empties into the superior and inferior divisions of the ophthalmic vein, but there are variable connections with veins that drain the anterior orbit (Table 1.16; also see examples in Figs. 2.1, 2.8 and 2.22 b in Chap. 2). The superior division of the ophthalmic vein (SOV), the dominant vein in the orbit, passes posteriorly from the medial superior orbital region. The superior palpebral, supratrochlear, and supraorbital veins are directly or indirectly connected to the SOV. The SOV is located superior to the optic nerve and beneath the superior rectus, outside of the muscle cone. The SOV exits the orbit via the superior orbital fissure to empty into the cavernous sinus. The inferior division of the ophthalmic vein (IOV), which is smaller than the SOV, is connected anteriorly to the facial vein and inferiorly through the inferior orbital fissure to the pterygoid plexus. The IOV courses posteriorly over the inferior rectus, exits the orbit via the superior orbital fissure, and usually empties into the SOV prior to exiting the orbit. Less frequently, the IOV directly enters into the anterior cavernous sinus. The pattern of venous connections within the orbit is variable between individuals and orbits.

1.5.2.6 Intracanalicular and Intracranial Optic Nerve and Chiasm

1.5.2.6.1 Arteries (Figs. 1.21–1.24)

The intracanalicular portion of the optic nerve receives arterial blood principally from the intracanalicular segment of the ophthalmic artery. Additional contributions to the intracanalicular optic nerve blood supply are variably present. They include branches from the anterior cerebral, anterior communicating, and internal carotid arteries when these arteries provide blood supply to the intracranial portion of the optic nerve [71, 73]. The various arterial branches enter the intracanalicular optic nerve pial vascular network, which sends centripetally oriented terminal vessels into the optic nerve. The shearing effect of trauma, with or without a fracture of the optic canal, can tear this delicate vascular network. Depending on whether the vascular damage is segmental or diffuse, the injury may cause partial or total visual loss.

The intracranial optic nerve receives a variable blood supply from several potential sources. The internal carotid, anterior cerebral, and anterior communicating arteries provide the predominant arterial supply to this portion of the optic nerve and chiasm [73–76]. The intracranial ophthalmic artery can contribute some blood supply to the inferior portion of the optic nerve [77]. Dawson claimed that the ophthalmic artery sends a major vessel to the prechiasmal optic nerve [78], but most individuals have only minute branches from the ophthalmic artery to this portion of the nerve [19, 36]. A more recent investigation demonstrated a significant ophthalmic artery contribution to the chiasm in only 4% of individuals [79].

In most subjects the predominant arterial supply to the posterior intracranial optic nerve and anterior chiasm is via an anterior superior hypophyseal artery arising from the C4 portion of the ICA. The superior surface of the chiasm receives arterial input via branches from the anterior cerebral and anterior communicating arteries. Most of the anterior cerebral branches originate from the proximal, A1, segment of the artery [80]. The inferolateral chiasm often receives a small branch artery from the medial ICA [81]. The central chiasm is supplied by branches of arteries that penetrate the inferior chiasm [82]. Posteroinferiorly, the chiasm is supplied by the superior hypophyseal or posterior communicating arteries. All of these arteries end in pial arteries. These pial arteries then terminate in a pial vascular network of continuous transverse centripetal pial arterioles and capillaries that surrounds each optic nerve and chiasm and anastomoses in the midline. The pial capillary network is part of the same vascular system that surrounds the entire anterior visual pathway, starting at the exit of the optic nerve from the globe and extending to the chiasm and along the optic tracts to the lateral geniculate nucleus [71].

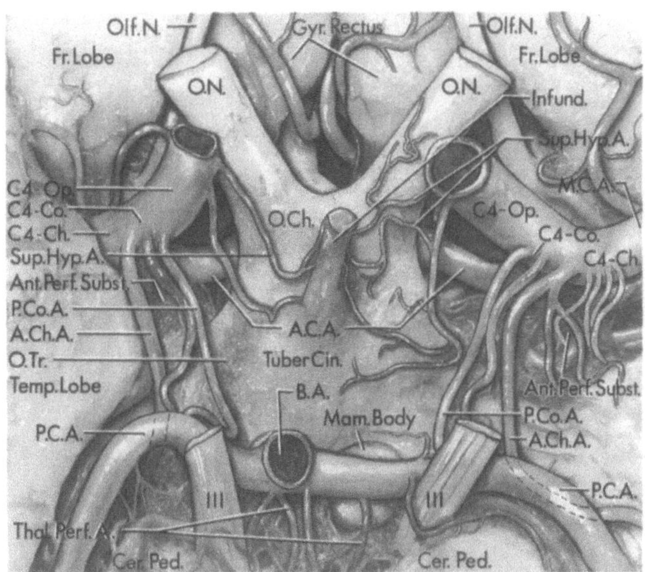

Fig. 1.21. Arterial supply to the region of the chiasm and intracranial optic nerves. Inferior view of the perforating branches of the supraclinoid portion of the internal carotid artery. The supraclinoid portion of the artery gives rise to the posterior communicating (*P. Co. A.*), anterior choroidal (*A. Ch. A.*), middle cerebral (*M.C.A.*), and anterior cerebral arteries (*A.C.A*). The supraclinoid portion of the artery is divided into three segments based on the site of origin of these branches: an ophthalmic segment (*C4 – Op.*) that extends from the origin of the ophthalmic artery (not shown because the internal carotid artery was divided above the level of origin of the ophthalmic artery to the origin of the posterior communicating artery; a communicating segment (*C4 – Co.*) that extends from the origin of the posterior communicating artery to the origin of the anterior choroidal artery; and a choroidal segment (*C4 – Ch.*) that extends from the origin of the anterior choroidal artery to the level of the bifurcation of the internal carotid artery into the anterior cerebral and middle cerebral arteries (the label *C4* from the original is not the same as the C4 segment in the carotid cavernous artery in Fig. 1.35). The ophthalmic segment sends perforating branches to the optic nerves (*O.N.*) and optic chiasm (*O. Ch.*) and the tuber cinereum (*Tuber Cin.*). The superior hypophyseal arteries (*Sup. Hyp. A.*) pass to the infundibulum of the hypophysis (*Infund.*). The communicating segment sends one perforating branch on each side to the optic tracts (*O. Tr.*) and the region around the mammillary bodies (*Mam. Body*). The choroidal segment sends its perforating branches into the anterior perforated substance (*Ant. Perf. Subst.*). The posterior cerebral arteries (*P.C.A.*) arise from the basilar artery (*B.A.*) and pass laterally around the cerebral peduncles (*Cer. Ped.*). The temporal lobe (*Temp. Lobe*) is lateral to the carotid artery. The middle cerebral arteries pass laterally into the sylvian fissure below the anterior perforated substance. The frontal lobes (*Fr. Lobe*), gyrus rectus (*Gyr. Rectus*), and olfactory nerves (*Olf. N.*) are above the optic nerves. The thalamoperforating arteries (*Thal. Perf. A.*) pass posteriorly between the oculomotor nerves (*III*). (From [79])

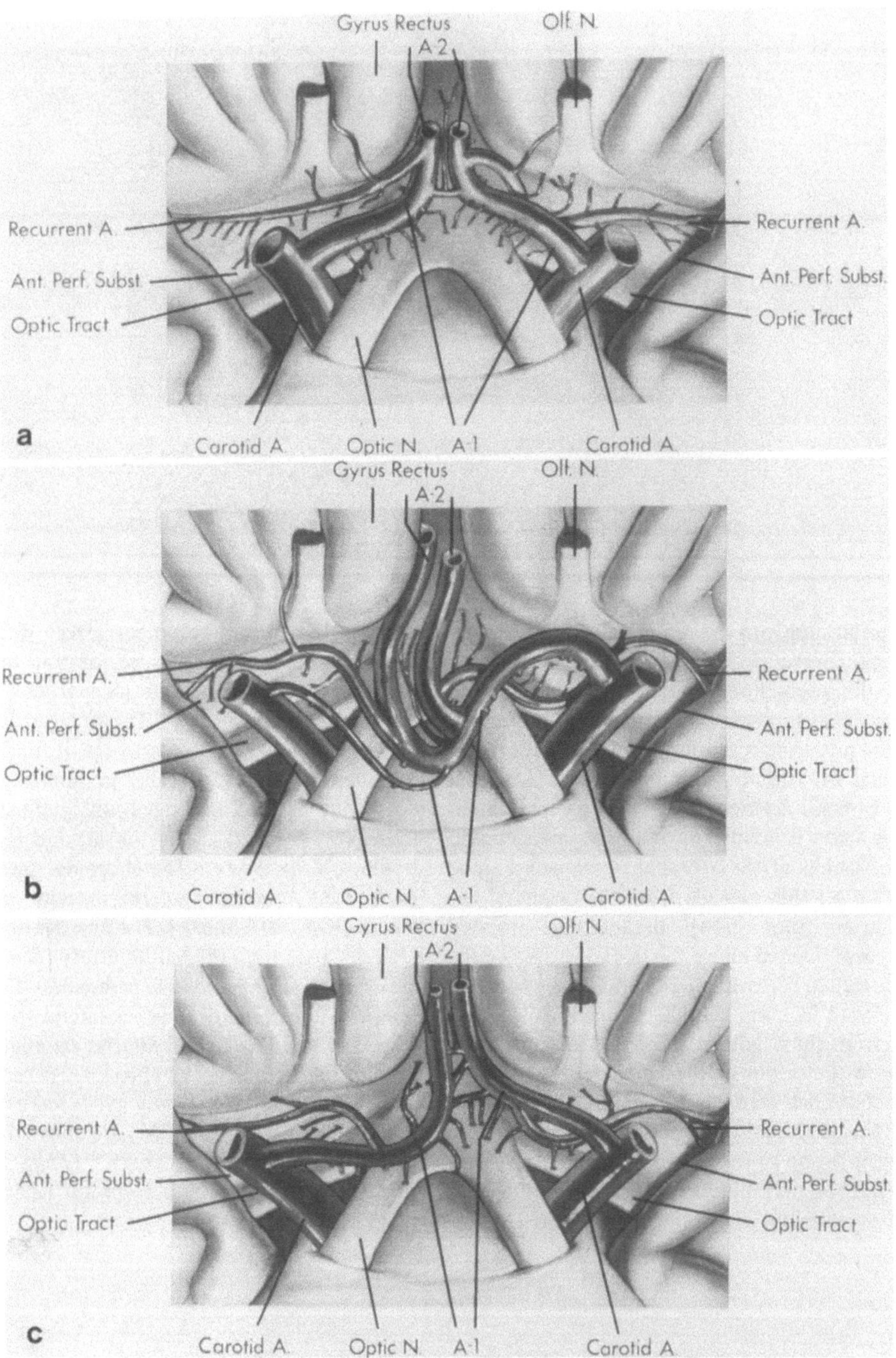

Fig. 1.22 a–c. Anterior views of A1 and proximal A2 segments of the anterior cerebral (*ACA*), anterior communicating (*A. Co. A.*), and recurrent arteries showing three of the many variations in the blood supply from this system to the chiasm. Gyrus rectus, olfactory tract, and frontal lobe above; optic nerves and chiasm below. Arterioles to optic nerves, chiasm and tracts, and lamina terminalis arise from the anterior cerebral and anterior communicating arteries. **a** A1 segments of equal size and small communicating artery pass above the optic chiasm. Recurrent arteries arise from lateral side of A2. Recurrent arteries pass anterosuperior to A1. **b** Both A2 segments arise from a large left A1. The right A1 is small. A1 segments pass above the optic nerves. Recurrent arteries arise from A2 segments. The right recurrent artery is longer than the A1 segment. Left recurrent artery passes superior to A1. Branches of the anterior communicating artery supply the lamina terminalis above the optic chiasm. **c** A1 segments are connected by three small communicating arteries. Both recurrent arteries arise from A1 segments; the left passes superior and posterior to A1, and the right courses anterior and superior to A1. (From [80])

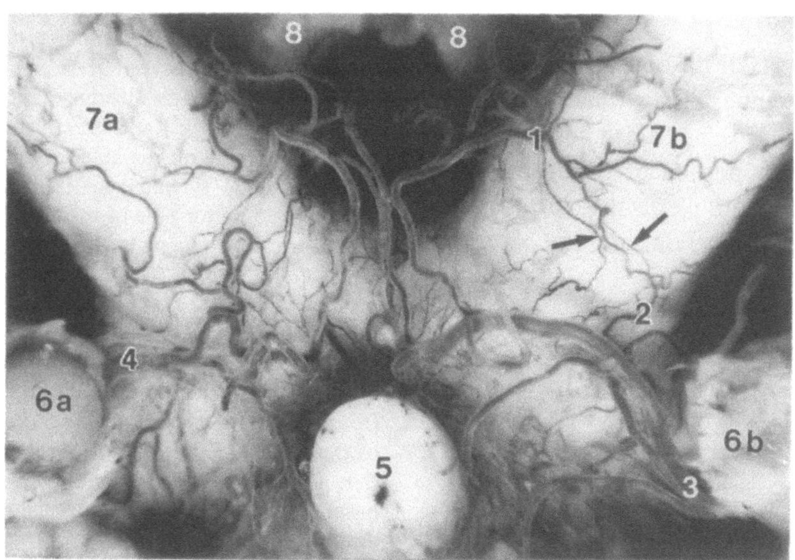

Fig. 1.23. Basal view of vascular network on the posterior optic nerves and chiasm showing an anastomosis (*arrows*) between two optic branches (*1, 2*) of left superior hypophyseal artery (*3*). The right superior hypophyseal artery (*4*), pituitary stalk (cut) (*5*), right and left internal carotid arteries (*6a, 6b*), right and left optic nerves (*7a, 7b*), and anterior cerebral arteries (*8*) are seen. *x 5.1 (Modified from [76])

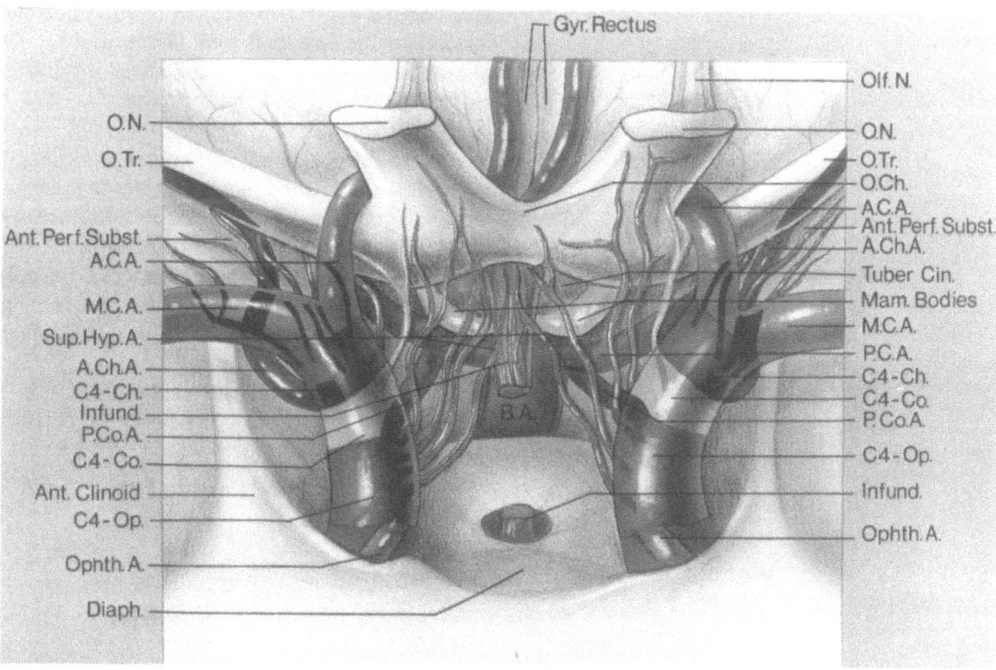

Fig. 1.24. Anterior view of the optic nerves (*O.N.*) and chiasm (*O. Ch.*) with both optic nerves divided and elevated to show the lower surface of the floor of the third ventricle and the perforating branches passing to it. The infundibulum (*Infund.*) has been divided above the diaphragma sella (*Diaph.*). *O. Tr.,* optic tract; *Ant. Perf. Subst.,* anterior perforating substance; *A.C.A.,* anterior cerebral artery; *M.C.A.,* middle cerebral artery; *Sup. Hyp. A.,* superior hypophyseal artery; *A. Ch. A.,* anterior choroidal artery; *C4–Ch.,* choroidal segment of the internal carotid artery; *Infund.,* infundibulum; *P. Co. A.,* posterior communicating artery; *C4–Co.,* communicating segment of the carotid artery; *C4–Op.,* ophthalmic segment of the internal carotid artery; *Ophth. A.,* ophthalmic artery; *Olf. N.,* olfactory nerve; *Tuber. Cin.,* tuber cinereum; *Mam. Bodies,* mammillary bodies; *P.C.A.,* posterior cerebral artery. (From [80])

1.5.2.6.2 *Venous Drainage* (Fig. 1.25)

The superior aspects of the optic nerve and chiasm are drained by a venous plexus which empties into the anterior cerebral veins (which course with the ACA). The inferior portion of these optic structures drains via a venous plexus that passes posteriorly (via the peduncular vein) into the basal veins [12, 19].

Fig. 1.25. a An anterior view with the frontal lobes retracted demonstrates the venous system in the region of the exposed optic nerves (*Optic N.*), optic tracts (*Optic Tr.*), and chiasm. The olfactory (*Olf. V.*), deep sylvian, peduncular (*Ped. V.*) and anterior cerebral veins (*Ant. Cer. V.*) converge on the anterior end of the basal vein. The superficial sylvian vein (*Sup. Sylvian V.*) passes around the anterior pole of the temporal lobe to join the venous sinuses coursing in the floor of the middle fossa. *Orb. Gyr.,* orbital gyrus; *Post. Front. Orb. V.,* posterofrontal orbital vein; *Car. A.,* carotid artery; *Temp. Sylvian V.,* temporal sylvian vein.
b An anterior view of the posterior and superior aspects of the chiasm and optic tracts with the frontal lobes (*Front. Lobe*) retracted demonstrates the venous system of the area. The deep middle cerebral veins (*Deep Mid. Cer. V.*) converge with the anterior cerebral vein at the basal vein (*Basal V.*). The deep middle cerebral veins drain from the sylvian fissure (*Sylvian Fiss.*). The olfactory vein (*Olf. V.*) drains posteriorly along the olfactory tract (*Olf. Tr.*) near the gyrus rectus (*Gyr. Rectus*) and the anterior cerebral vein and its tributary, the anterior pericallosal (*Ant. Pericall. V.*), and the paraterminal veins (*Paraterm. V.*). The anterior communicating vein (*Ant. Comm. V*) connects the anterior cerebral veins across the lamina terminalis (*Lam. Term.*). The peduncular vein (*Ped. V.*) crosses the cerebral peduncle (*Ped.*) medial to the temporal lobe (*Temp. Lobe*). The infundibulum (*Infund.*) passes inferiorly around the optic nerves (*Optic N.*) and optic chiasm.
c An inferior posterior view of the veins and the optic tract with the optic and oculomotor nerves (*III*) and the infundibulum divided shows the veins in the region of the anterior and posterior perforated substances (*Ant.* and *Post. Perf. Subst.*). The anterior cerebral veins (*Ant. Cer. V.*) drain the posterior chiasm and anterior parts of the optic tracts. The median anterior pontomesencephalic vein (*Med. Ant. Pon. Mes. V.*) joins the peduncular veins in the midline. Inferior striate veins (*Inf. Str. V.*) exit the anterior perforated substance, and inferior thalamic veins (*Inf. Thal. V.*) exit the posterior perforated substance. The premammillary veins (*Premam. V.*) arise anterior to the mammillary bodies (*Mam. Body*). (**a** from [12], **b, c** from [13])

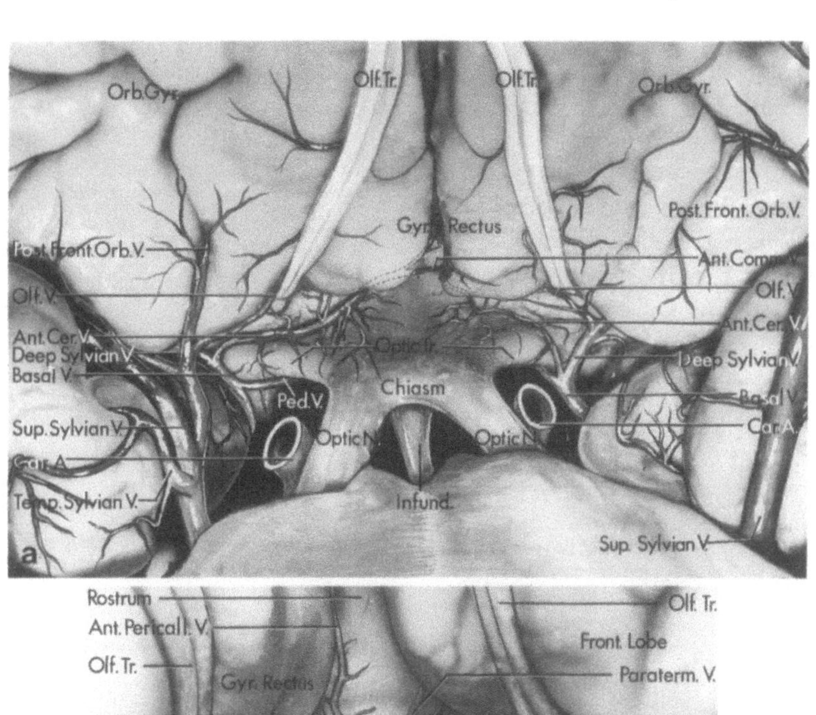
Orb.Gyr.
Olf.Tr.
Olf.Tr.
Orb.Gyr.
Gyr. Rectus
Post. Front. Orb.V.
Post. Front. Orb.V.
Ant. Comm.
Olf. V.
Olf. V.
Ant. Cer. V.
Ant. Cer. V.
Deep Sylvian V.
Optic Tr.
Deep Sylvian V.
Basal V.
Chiasm
Basal V.
Ped. V.
Car. A.
Sup. Sylvian V.
Optic N.
Optic N.
Car. A.
Temp. Sylvian V.
Infund.
Sup. Sylvian V.
a

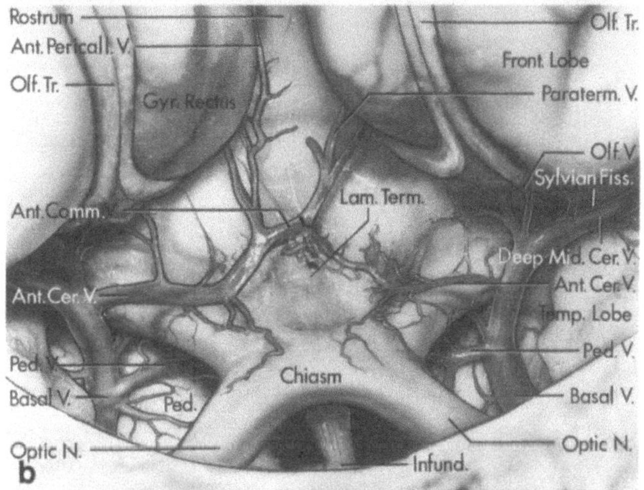
Rostrum
Olf. Tr.
Ant. Pericall. V.
Front. Lobe
Olf. Tr.
Paraterm. V.
Gyr. Rectus
Olf. V.
Sylvian Fiss.
Lam. Term.
Ant. Comm.
Deep Mid. Cer. V.
Ant. Cer. V.
Ant. Cer. V.
Temp. Lobe
Ped. V.
Ped. V.
Basal V.
Chiasm
Basal V.
Ped.
Optic N.
Optic N.
Infund.
b

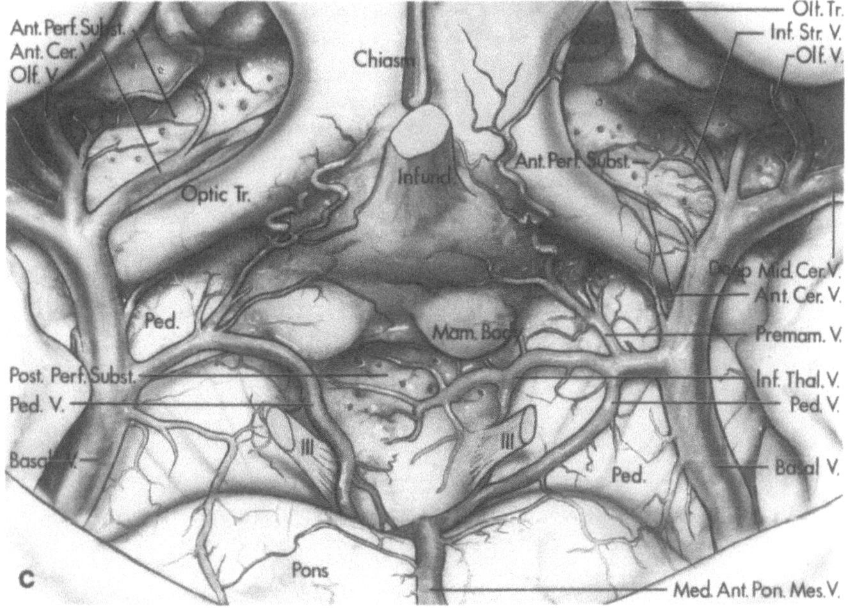
Olf. Tr.
Ant. Perf. Subst.
Inf. Str. V.
Ant. Cer. V.
Chiasm
Olf. V.
Olf. V.
Infund.
Optic Tr.
Ant. Perf. Subst.
Deep Mid. Cer. V.
Ant. Cer. V.
Ped.
Mam. Body
Premam. V.
Post. Perf. Subst.
Inf. Thal. V.
Ped. V.
Ped. V.
III
III
Basal V.
Basal V.
Ped.
Pons
c
Med. Ant. Pon. Mes. V.

1.5.2.7 Optic Tract

The blood supply to the optic tract is diverse owing to the variability of branch arteries from the circle of Willis. No single artery consistently provides the major arterial input to the optic tract. Branches of arteries that course near the optic tract tend to supply the pial vascular network of this structure (Fig. 1.26). Intra-arterial Indian ink injection studies demonstrate consistent contributions to the optic tract from the posterior communicating and anterior choroidal arteries, one of which is usually dominant.

Arterioles from the pial network penetrate the optic tract perpendicularly to its location on the tract surface. As in other areas of the anterior visual pathway, collateral arterioles join together to supply a transverse capillary network that courses in both the anterior and posterior directions along the tract. The transverse capillary network which incompletely encircles the nerve fascicles becomes more abundant near the lateral geniculate body [71]. Contributions from both the anterior choroidal and posterior communicating arteries provide overlapping blood supply to these capillaries [83]. However, there seems to be no direct anastomoses between the posterior communicating and the anterior choroidal arteries.

The optic tract drains via the deep middle cerebral, inferior striate, olfactory, and peduncular veins [13].

1.5.2.7.1 Anterior Choroidal Artery

The anterior choroidal artery is an essential vessel that supplies both the optic tract and lateral geniculate nucleus. The anterior choroidal artery arises from the ICA in approximately 77% of subjects, from the middle cerebral artery (MCA) in 12%, from the posterior communicating artery in 7%, and from the bifurcation of the ICA to middle and anterior cerebral arteries in 3% [84, 86, 87].

In most subjects, the anterior choroidal artery supplies the posterior optic tract to its termination in the lateral geniculate nucleus, the posterior limb of the internal capsule, the globus pallidus and part of the putamen, the anterior segment of the optic radiation, the anterior commissure, the uncus, the cerebral peduncle, the paramedian territory of the midbrain, and the choroid plexus [79, 84, 85, 87]

Fig. 1.26a, b. The arterial supply to the optic tract and ▶ lateral geniculate body (nucleus).

a A lateral view of the right middle fossa with temporal lobe removed demonstrates the arteries to the optic tract (*O. Tr.*) and lateral geniculate body in addition to the relationship of the posterior cerebral artery (*P.C.A.*) to the brainstem, cranial nerves, and adjacent vascular structures. The optic nerve (*O.N.*) can be traced posteriorly to the optic tract to its termination in the lateral geniculate body (*L. Gen. Bo.*). The anterior choroidal (*A. Ch. A.*) and lateral posterior choroidal arteries (*L. P. Ch. A.*) enter the temporal horn of the ventricle to supply the choroid plexus (*Ch. Pl.*). The oculomotor (*III*) and trochlear nerves (*IV*) course between the posterior cerebral and superior cerebellar arteries (*S.C.A.*). A large premammillary artery (*Premam. A.*) passes upward from the posterior communicating artery (*P. Co. A.*) to enter the posterior perforated substance and tuber cinereum. The trigeminal nerve and its ophthalmic (*V₁*), maxillary (*VI₂*), and mandibular (*V₃*) divisions are seen in the lateral wall of the cavernous sinus. The cavernous and supraclinoid segments of the carotid artery (*C.A.*), the proximal (A1) and adjacent distal (A2) segments of the anterior cerebral artery, and the anterior communicating artery (*A. Co. A.*) and the middle cerebral arteries (*M.C.A.*) are only a short distance from the proximal posterior cerebral artery. The medial posterior choroidal artery (*M. P. Ch. A.*) gives perforating branches to the cerebral peduncle and encircles the brainstem medial to the posterior cerebral artery. The peduncular perforating (*Ped. Pe. A.*) and thalamogeniculate (*Th. Gen. A.*) arteries send branches to the cerebral peduncle and geniculate bodies, respectively. The anterior and middle temporal arteries (*A.T.A., M.T.A.*) course above the edge of the tentorium (*Tent.*) to the inferior surface of the temporal lobe. *Rec. A.*, recurrent artery of Huebner. (From [90])

b A basal and slightly caudal view shows the network of small arteries on the pia of the optic tract supplied by branches from the superior hypophyseal, tuberoinfundibular and posterior communicating arteries. Anastomoses (*small arrows* to the *right*) are seen between the right superior hypophyseal artery (*large arrow*) and a branch (*curved arrow*) of the tuberoinfundibular artery (*white arrow*) which arises from the internal carotid artery. Anastomoses (*small asterisks*) connect two branches from this same tuberoinfundibular artery. The optic tract (*large asterisk*) is medial to the posterior communicating (*1*) and posterior cerebral (*2*) arteries and lateral to the pituitary stalk (*3*) (cut). *x13.8. (Modified from [76])

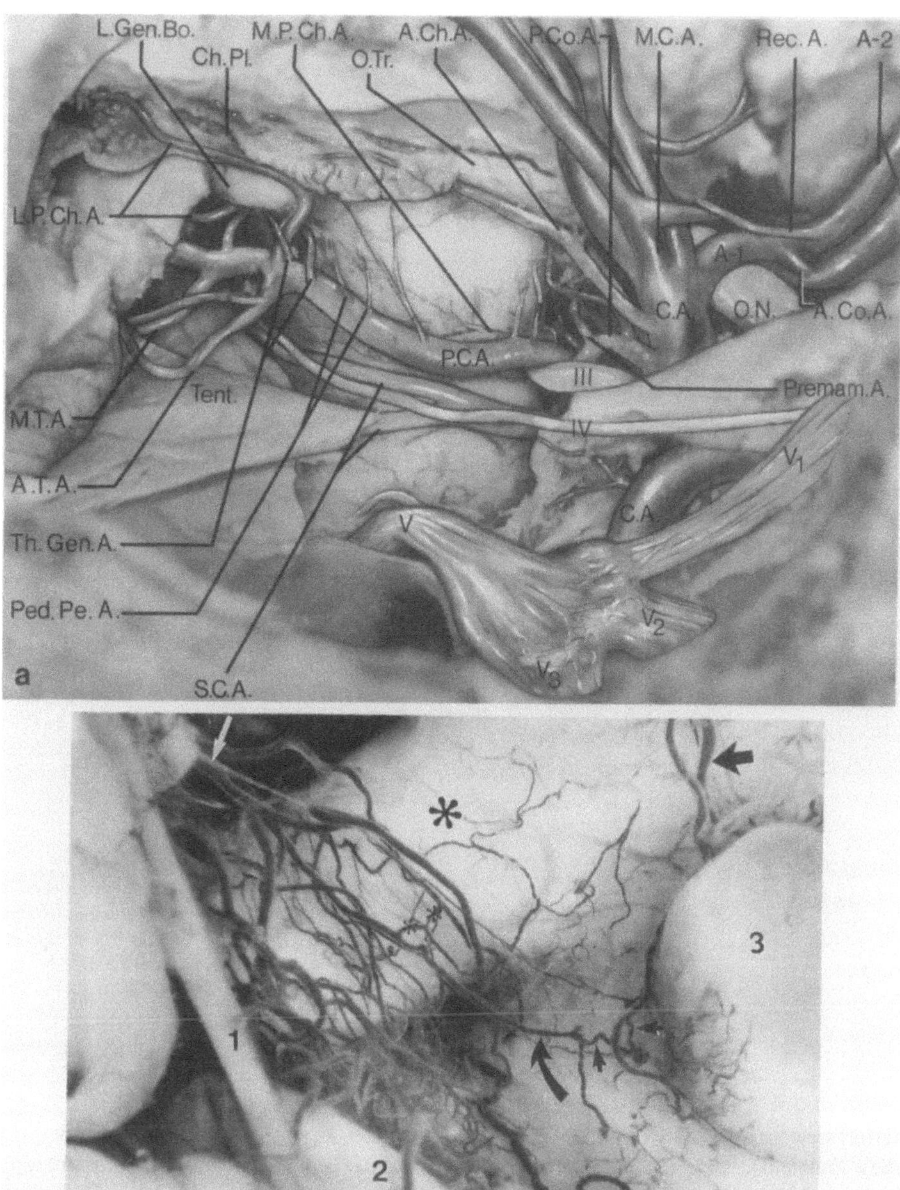
L.Gen.Bo.
Ch.Pl.
M.P.Ch.A.
O.Tr.
A.Ch.A.
P.Co.A.
M.C.A.
Rec.A.
A-2
L.P.Ch.A.
A-1
C.A.
O.N.
A.Co.A.
P.C.A.
III
Premam.A.
Tent.
IV
M.T.A.
V₁
A.T.A.
C.A.
Th.Gen.A
V
Ped.Pe.A.
V₂
S.C.A.
V₃
a
1
3
2
b

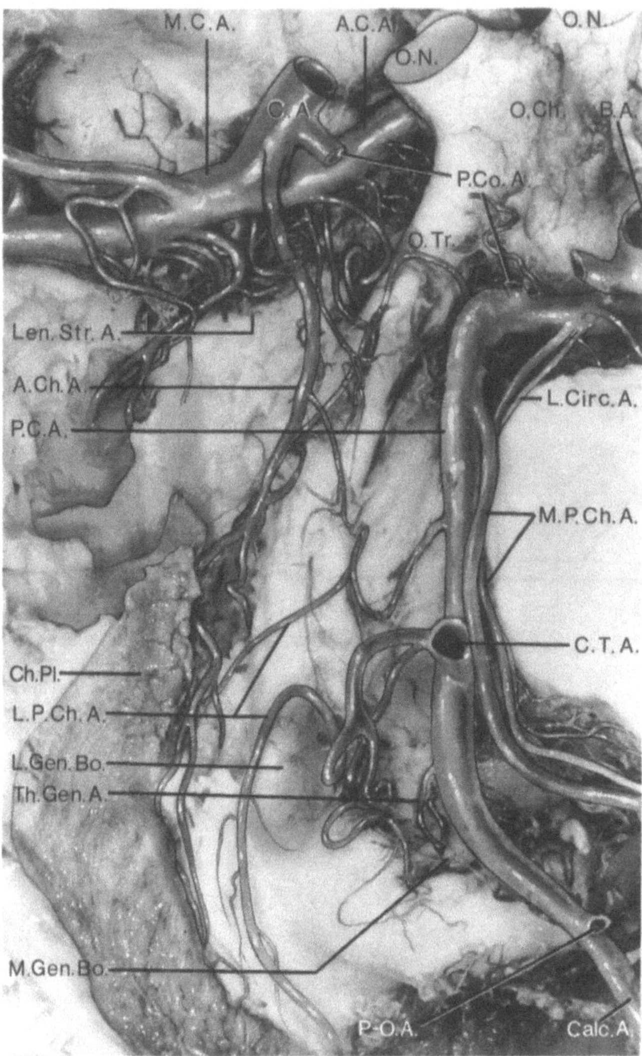

Fig. 1.27. Inferior view of the right optic tract, lateral geniculate body (nucleus), cerebral hemisphere, and brainstem with part of the temporal lobe removed to expose the temporal horn, choroid plexus, and posterior cerebral artery branches. The course of the anterior choroidal artery and the lateral posterior choroidal arteries to the optic pathway is shown. The optic nerve (*O.N.*) can be followed to the optic chiasm (*O. Ch.*), optic tract (*O. Tr.*), and lateral geniculate body (*L. Gen. Bo.*). The lateral posterior choroidal artery (*L. P. Ch. A.*) supplies the posterior optic tract and the lateral geniculate body before passing into the lateral ventricle to reach the choroid plexus (*Ch. Pl.*) The anterior choroidal artery (*A. Ch. A.*) arises from the internal carotid artery (*C.A.*) just distal to the posterior communicating artery (*P. Co. A.*). The anterior choroidal artery passes through the choroidal fissure lateral to the optic tract and supplies it and passes into the temporal horn to anastomose with the lateral posterior choroidal arteries. Lenticulostriate arteries (*Len. Str. A.*) proceed posterosuperior from the middle cerebral artery (*M.C.A.*). The anterior cerebral artery (*A.C.A.*) passes superior to the optic nerves and chiasm. Thalamogeniculate arteries (*Th. Gen. A.*) pass superiorly from the posterior cerebral artery to the medial geniculate body (*M. Gen. Bo.*). The long circumflex artery (*L. Circ. A.*) arises from P1 and encircles the midbrain medial to the posterior cerebral artery. One medial posterior choroidal artery (*M. P. Ch. A.*) arises from the proximal segment of the posterior cerebral artery (P1) and the other from a more distal portion (P2). One lateral posterior choroidal artery arises directly from the posterior cerbral artery and another from the common temporal artery (*C.T.A.*). The basilar (*B.A.*), parieto- occipital (*P.O.A.*), and calcarine (*Calc. A.*) arteries are also seen. (From [90])

(Fig. 1.27). The lateral posterior choroidal artery provides collaterals to the region supplied by the posterior branches of the anterior choroidal artery. Anastomoses between two arteries are found in the area near the lateral geniculate nucleus (see Sect. 1.5.2.8).

1.5.2.8 Lateral Geniculate Nucleus

The lateral geniculate nucleus (LGN) receives blood supply from the anterior choroidal artery and two or three posterior choroidal arteries [84]. The lateral half of the LGN is almost always supplied by the anterior choroidal artery. The medial aspect of the LGN is supplied by lateral posterior choroidal arterial branches [87]. The intervening region of the LGN, which includes the hilum, receives blood from the anastomotic network supplied by both arterial systems (Fig. 1.28). The lateral posterior choroidal arteries originate directly from the PCA in the majority of cases (Fig. 1.27). Less commonly, a lateral posterior choroidal artery arises from the parieto- occipital artery. Either of the anterior or the posterior choroidal arterial systems can be dominant in this region if the other is less well developed [88, 89].

In as many as 48% of cases, small portions of the LGN are also supplied by other PCA branches, including the hippocampal, anterior temporal, posterior temporal, middle temporal parieto-occipital, medial posterior choroidal, and calcarine arteries [90]. The geniculothalamic artery, which arises from the distal (P2 or P3) portion of the PCA, provides a minor contribution to the LGN and a major portion of the blood supply to the other thalamic nuclei [6, 91].

Arterioles from the various arteries anastomose in the pial–arachnoid area over the LGN forming a vascular network [90, 92]. Arterioles from this network perforate the LGN perpendicularly to its surface. The projections from the nasal retina of the contralateral eye and the temporal retina of the ipsilateral eye terminate in the six vertically arranged layers of the LGN, which contain columns that correspond to segments of the homonymous visual field. Each of the six cell layers contains an individual capillary network, but anastomoses exist between the adjacent capillary networks [71]. The hilum, which receives the macula field projection, is supplied by both the anterior choroidal and lateral posterior choroidal arteries. Rare patients with infarcts of the LGN have contributed to the understanding of the varied blood supply to this structure (see also Chap. 8).

1.5.2.9 Optic Radiation

The blood supply to the optic radiation varies as the postgeniculate pathway courses from the LGN, loops into the anterior temporal lobe, sweeps posteriorly in the depths of the parietal lobe around the occipital horn, and ends in the calcarine fissure. Small arteries and arterioles from the LGN accompany and supply the axons that originate from ganglion cells in the LGN. The anterior choroidal artery, lenticulostriate arteries, and other MCA branches provide the blood supply to the carrefour and anterior temporal lobe portions of the visual pathway, including Meyer's loop [83].

One branch of the MCA, called the deep optic branch by Abbie, also provides blood to the visual radiation. This vessel passes through the anterior perforated substance and the white matter to reach the ventricle deep to the optic radiation [82, 92]. This deep optic branch actually appears to be a long lateral lenticulostriate artery [93]. More distal MCA branches, the angular and posterior temporal arteries, penetrate the cortex to supply the upper visual radiations.

The proximal segment of the PCA gives rise to the thalamogeniculate arteries (variable number from 1 to 7, average 2.4) [90] that also contribute blood supply to the region of the carrefour. The calcarine artery branches provide the predominant blood supply to the subcortical optic radiation in the occipital lobe. Leptomeningeal anastomoses between the middle and posterior cerebral arteries contribute to an overlapping blood supply on the cortical surface that supplies the optic radiation via perforating arteries. However, few arteries or arterioles actually enter the optic radiation. These vessels terminate in precapillary vessels which follow the visual pathway fibers to supply the visual radiation [94] (Fig. 1.29).

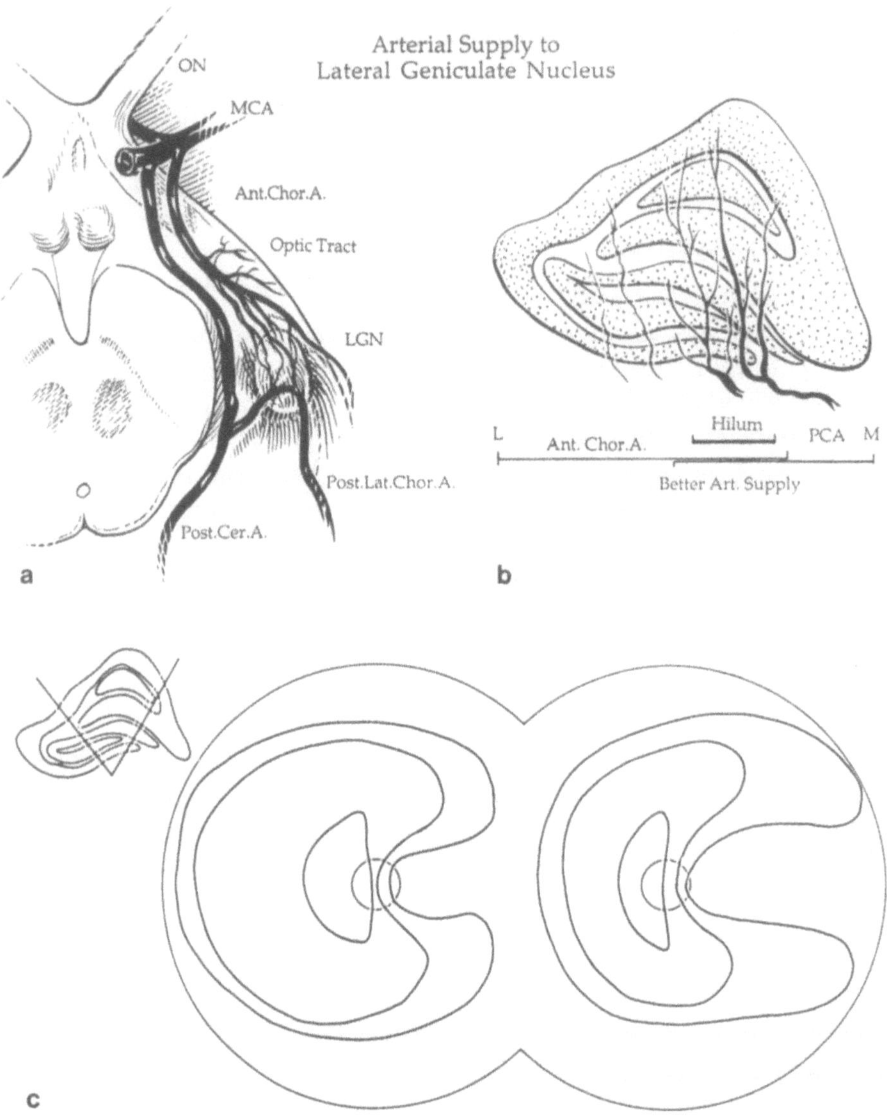

Fig. 1.28. a The anterior choroidal (*Ant. Chor. A.*) and lateral posterior choroidal arteries (*Post. Lat. Chor. A.*) provide the principal blood supply to the vascular network on the surface of the lateral geniculate nucleus. Either vessel may be dominant. *ON*, optic nerve; *MCA*, middle cerebral artery; *LGN*, lateral geniculate nucleus; *Post. Cer. A.*, posterior cerebral artery. **b** Section of the lateral geniculate body shows the six layers with a rich blood supply to the hilar region. Note the sparse blood supply to the lateral and medial regions of the lateral geniculate nucleus. *PCA*, posterior carotid artery; *L*, lateral; *M*, medial. **c** When the posterior lateral choroidal artery is dominant, occlusion of this vessel results in a hilar lesion with a homonymous horizontal sector defect. (See Fig. 8.18 for a comparison infarct when the posterior lateral choroidal artery is not dominant)

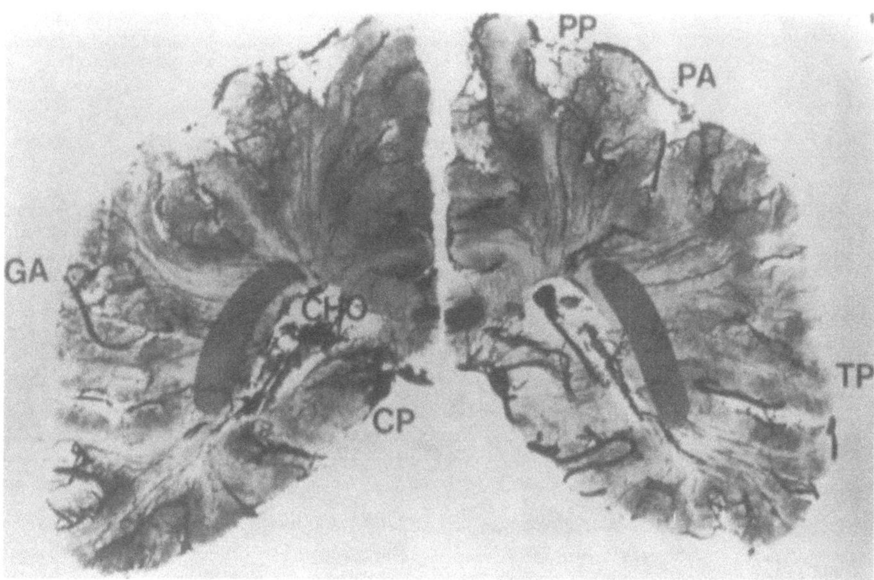

Fig. 1.29. Coronal section through the visual radiation after injection of black ink into the cerebral vessels. The optic radiations are seen in the shaded area. The arteries to this region include the posterior parietal artery (*PP*), the anterior parietal artery (*PA*), the posterior temporal artery (*TP*), the artery to the angle of gyrus (*GA*), the anterior choroidal artery (*CHO*) and the posterior cerebral artery (*CP*). (From [83])

1.5.2.10 Occipital Lobe

1.5.2.10.1 Arteries (Figs. 1.30–1.32)

The occipital cortex is predominantly supplied by arteries that originate from the PCA. The distal segment (P3) commonly divides into a parieto-occipital artery (more medial) and a calcarine artery. As these two arteries course posteriorly along the medial border of the hemisphere, the parieto-occipital artery passes superiorly between the occipital and parietal lobes (Figs. 1.5, 1.30). On the lateral projection of the vertebral angiogram, the posterior temporal artery is inferior, the calcarine artery is in the middle, and the parieto-occipital artery is superior as the three vessels course peripherally [95].

The calcarine artery originates directly from the PCA in 78% of subjects, from the parieto-occipital artery in 16%, and from a posterior temporal artery in 6% [96, 90] (Fig. 1.31). Though classically described as following the calcarine fissure, the calcarine artery more often than not does not follow this sulcus (75% of cases). The calcarine artery can be located on the floor of the calcarine fissure, on the medial surface of the hemisphere paralleling the fissure, or on the medial surface coursing on an oblique course upward and posterior from the PCA, or it can divide into superior and inferior branches that pass on the superior and inferior aspects of the calcarine fissure, respectively [97]. In approximately 50% of subjects, the calcarine artery passes along the medial occipital surface or on the floor of the calcarine sulcus as a main trunk, providing multiple branch arteries to the superior and inferior striate cortex. In approximately one third of subjects, the calcarine artery divides prior to reaching the striate area, sending one branch artery superiorly and another artery inferiorly to the superior and inferior regions of the visual cortex, respectively. In less than 10% of subjects, the calcarine artery principally supplies either the occipital lobe superior or inferior to the calcarine sulcus. In these cases either a posterior temporal branch or a parieto-occipital branch artery supplies the region of the sulcus not supplied by the calcarine artery.

As discussed above, the calcarine artery is supplemented by contributions from the posterior temporal or parieto-occipital arteries, two additional branches from the PCA. There are many possible

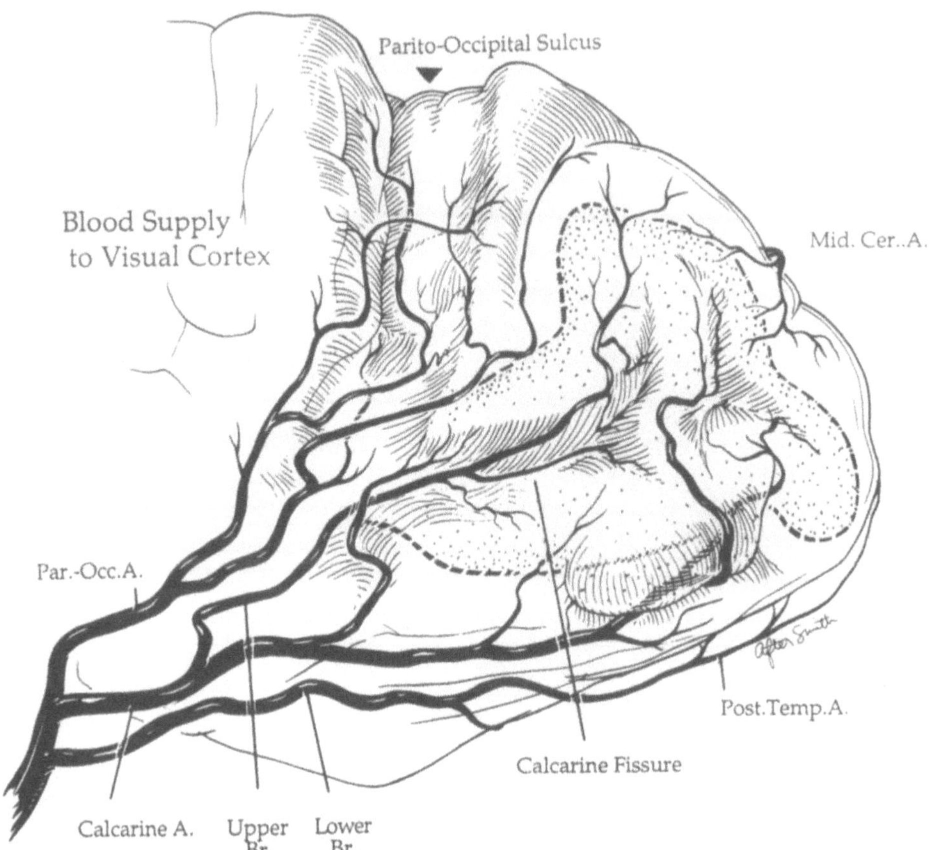

Fig. 1.30. The medial surface of the occipital portion of a left hemisphere illustrates the features of the visual area and one variation of the arterial pattern. The walls of the calcarine and parieto-occipital fissures are separated. The posterior cerebral artery provides the predominant blood supply to this region via parietal occipital (*Par. Occ. A.*), posterior temporal (*Post. Temp. A.*), and upper and lower branches (*Br.*) of the calcarine artery. Note the potential collateral from the middle cerebral artery (*Mid. Cer. A.*)

collaterals between the branches of the PCA and between the MCA and PCA (Fig. 1.32). For example, occipital branches from the MCA, most commonly via leptomeningeal anastomoses, connect the temporal branches of the MCA to the occipital circulation. However, MCA branches that provide terminal branch arteries to the striate area are found in less than 10% of hemispheres [96, 98].

In up to 53% of hemispheres, the posterior temporal artery provides blood supply to the occipital lobe. In these individuals terminal arteries from the posterior temporal artery pass to the inferior posterior calcarine fissure or to both the superior and inferior aspects of the posterior calcarine fissure [97]. In other cases, angiography demonstrates an anas-

tomosis between the posterior temporal artery and branches from the calcarine artery in the region of the posterior calcarine fissure [95].

In approximately 27% of cerebral hemispheres, the parieto- occipital artery provides significant blood supply to the occipital lobe, mainly the anterior superior portion. In 25% of subjects, the parieto-occipital artery provides a significant accessory calcarine artery to the striate area [96, 98].

There are numerous other arterial anastomoses with the PCA system, but they are generally ineffective. Less frequent potential collaterals to the PCA blood supply of the occipital lobe arise from anterior temporal branches of the MCA, from the distal ACA through posterior pericallosal branches of the proximal PCA, or from the anterior

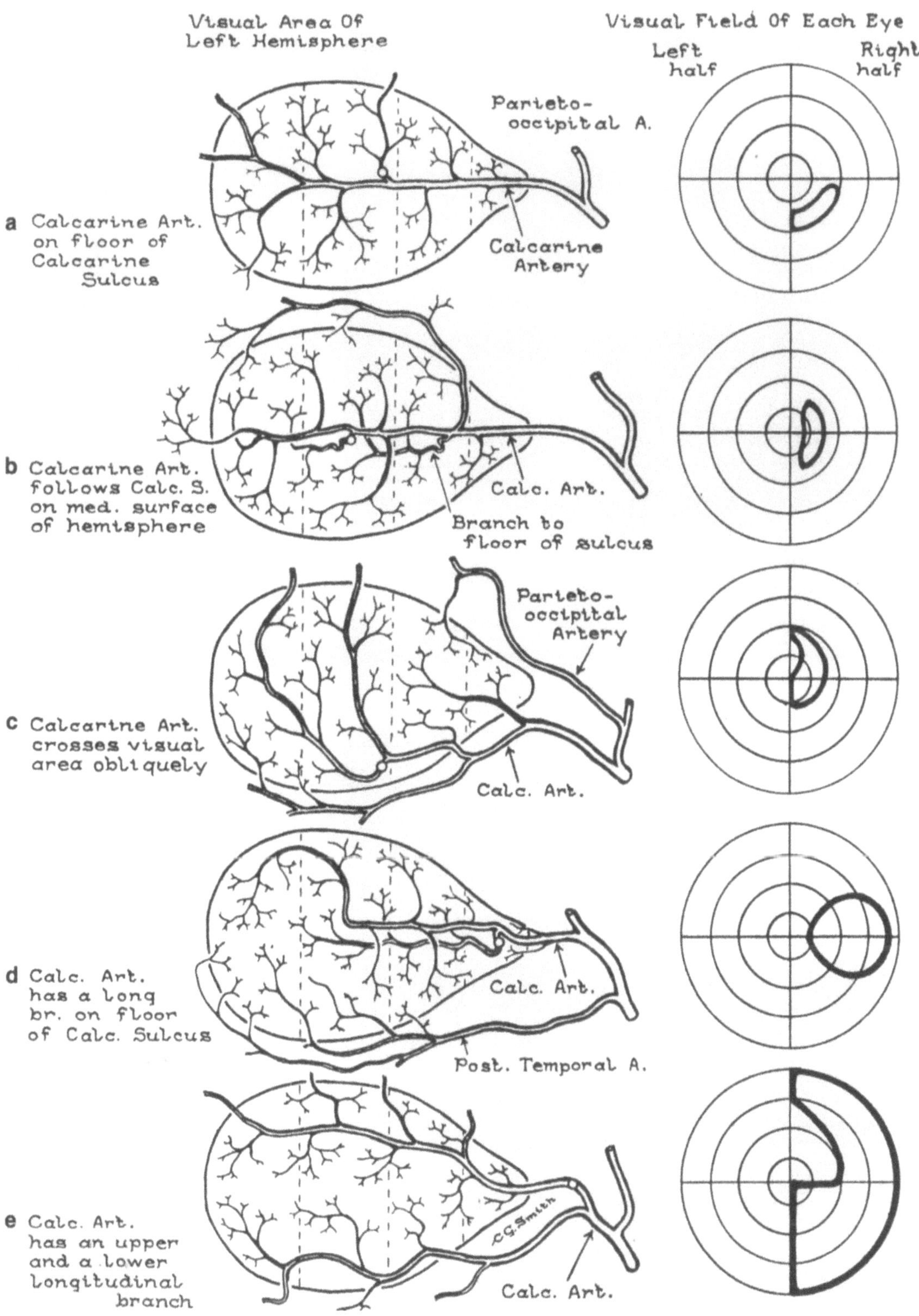

Fig. 1.31a–e. Diagrams illustrating variations in the course and branching of the calcarine artery. The *heavy line* in the diagram of the visual field to the right of each example outlines the portion of the field that projects to the patch of the cortex supplied by the vessel marked by a *small circle* close to its origin. (From [97])

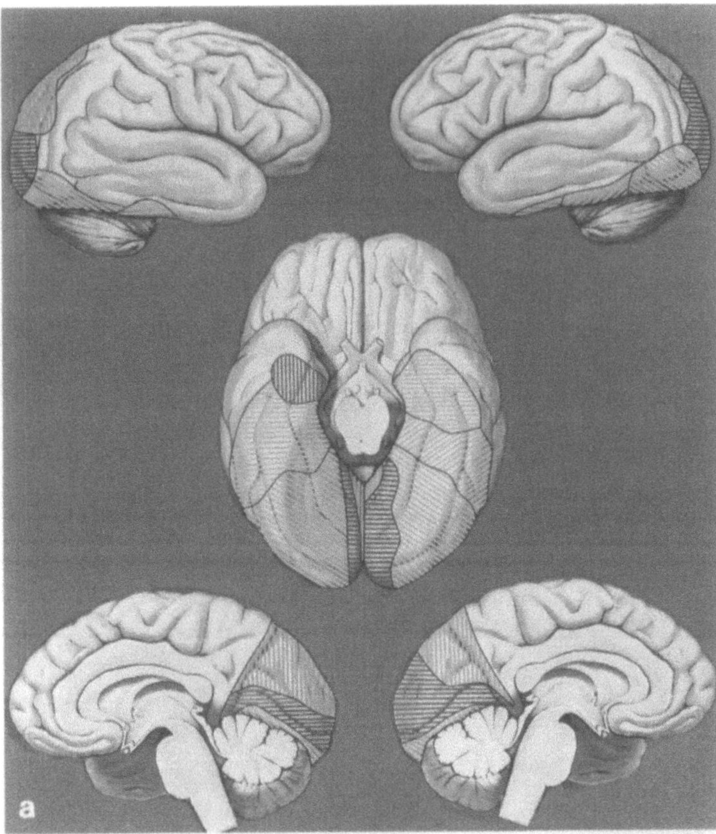

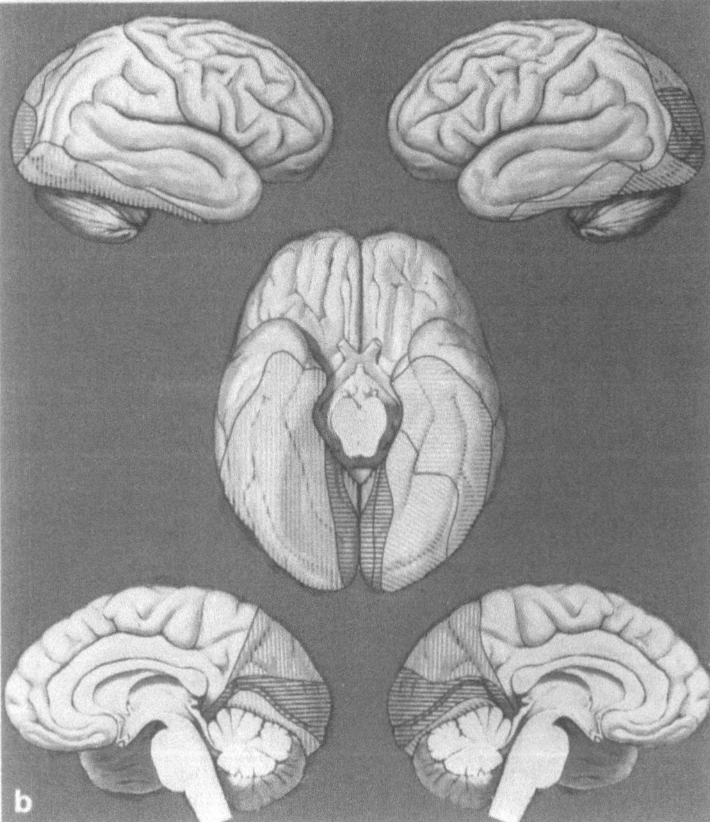

Fig. 1.32a–c. Lateral, medial, and basal views of the brain with coded sectors representing specific cortical branch distributions of the posterior cerebral artery. Note the variations among brains and between the hemispheres of individual cases. The code corresponding to each branch is as follows: *vertical dark bars*, hippocampal artery; *diagonal and horizontal light bars*, temporal arteries; *horizontal dark bars*, calcarine arteries; *vertical light bars*, parieto-occipital artery. The temporal arteries are further subdivided: *light horizontal stripes*, anterior temporal artery; *light vertical stripes* (only in right temporal lobe figure B.), common temporal artery; *light diagonal stripes, angled upward to right*, anterior temporal artery; *light diagonal stripes, angled down to right*, posterior temporal artery.

a The most common pattern (44% of hemispheres) is represented on the right cerebral hemisphere (*upper and lower left* and *left half* of basal view, *center*). This pattern includes hippocampal, anterior temporal, and posterior temporal arteries. The cortical distribution of the parieto-occipital artery is larger than that of the calcarine artery.

The second most common pattern (20% of hemispheres) is represented on the left cerebral hemisphere (*upper and lower right* and *right half* of basal view, *center*). This pattern includes anterior, middle, and posterior temporal calcarine and parieto-occipital arteries. In this pattern the anterior temporal artery supplies the region usually supplied by the hippocampal artery.

b The third most common pattern (16% of hemispheres) is shown on the right hemisphere (*upper and lower left* and *left half* of basal view, *center*). In this pattern there is a common temporal artery that supplies the entire inferior surface of the temporal lobe. The calcarine and parieto-occipital arteries are also present.

The fourth most common pattern (10% of hemispheres) is depicted on the left hemisphere (*upper and lower right* and *right half* of basal view, *center*). This arrangement includes anterior and posterior temporal, calcarine, and parieto-occipital arteries but no hippocampal or middle temporal branches of the posterior cerebral artery. The area of the calcarine artery is split into two sectors to illustrate that there were two calcarine arteries arising from the posterior cerebral artery, as occurred in 10% of hemispheres.

c The fifth most common pattern (10% of hemispheres) is illustrated on the right cerebral hemisphere (*upper and lower left* and *left half* of basal view, *center*) and includes hippocampal, anterior, middle and posterior temporal, calcarine, and parieto-occipital arteries. The area supplied by the posterior temporal artery is split into two parts to show that two posterior temporal arteries arose from the posterior cerebral artery and was noted in 6% of cerebral hemispheres. The parieto-occipital artery supplied the greater portion of the medial surface.

The last pattern illustrates some notable variants (*upper and lower right* and *right half* of basal view, *center*). Two hippocampal arteries arose from the posterior cerebral artery, a finding present in 12% of cerebral hemispheres. The anterior temporal artery supplied a smaller than usual amount of the anterior lateral temporal surfaces, the remainder being supplied by the middle cerebral artery. The calcarine artery supplies an unusually large area on the medial surface. (From [90])

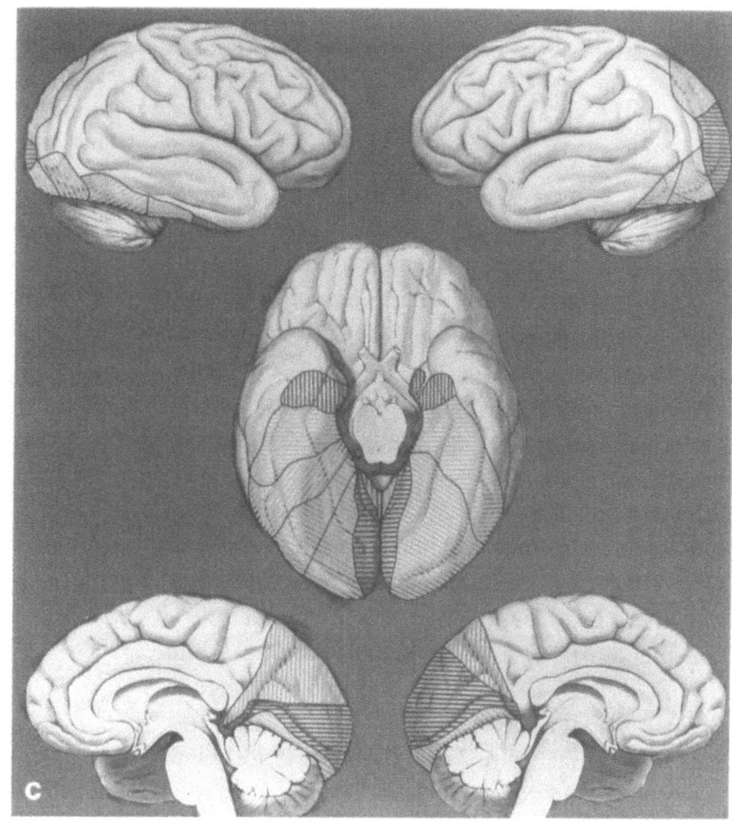

choroidal artery to the lateral posterior choroidal artery [95].

The clinical phenomenon of macula sparing in the setting of a homonymous hemianopia, following occipital trauma or occlusion of the PCA, may occur because patients have adequate collateral circulation from the MCA or an unaffected branch of the PCA (Fig. 1.30). Macula sparing does not occur if the occluded calcarine artery is the sole source of blood supply to the striate cortex (25% of cerebral hemispheres).

Large arteries supply the pia–arachnoid arterial network on the surface of the striate cortex. Multiple arteriolar anastomoses are found on the cortical surface in the pia–arachnoid layer. Penetrating arteries arise from this vascular complex and pass perpendicularly to the cortical surface to supply the gray matter cortex and white matter subcortical regions [94, 99]. Prior to its termination, each penetrating artery gives off numerous branches which vascularize the capillaries of a specific cortical segment [100]. Numerous anastomoses exist at the precapillary and capillary levels between the capillary network supplied by each penetrating artery [101]. However, these anastomoses cannot provide the quantity of blood required to prevent an infarct of the cortex in the territory of an occluded artery. The capillary network is thickest in the cellular layers of the cortex and thinnest in the white matter, possibly reflecting the difference in metabolic demands for these two tissues. The capillaries course in the same direction as the nerve fibers.

Clinicopathological studies, initially performed on trauma victims [102, 103], later on patients with vaso-occlusive disorders (with MR correlation), provide the basis for understanding the topographic organization of the visual field in the striate cortex (see Chap. 8).

1.5.2.10.2 Veins (Fig. 1.33)

In the striate cortex, small surface veins drain the layers of the superficial cortex as deep as layer III. The deeper veins receive blood from the venules in cortical layer IV or deeper regions. The veins in layers I–IV empty into intracortical veins. The intracortical veins pass in a direction perpendicular to the cortical surface, adjacent to parallel coursing arteries, and give a bushlike appearance to the surface of the striate cortex [104]. Arteriovenous shunts between intracortical arteries and veins, which are probably vestiges of the primitive anastomotic vascular system, are rarely found in normal brains.

The veins on each occipital cortical surface terminate in main cortical veins that are variably present. The inferior occipital cortical veins drain deep into the occipitobasal vein, which empties into a confluence of veins, the lateral tentorial sinus (Fig. 1.7) [105]. The lateral occipital surface is drained via superficial cortical veins which empty into the occipital vein. As it courses anteriorly into the superior sagittal sinus, the occipital vein may be joined by the posterior calcarine or posterior parietal veins. Less commonly, the occipital vein passes downward into the lateral tentorial sinus. The lateral tentorial sinus drains into the transverse sinus. The medial surface of the occipital lobe drains via the anterior and posterior calcarine veins. The anterior calcarine and parieto-occipital sulci veins drain into an internal occipital vein, which lies in the calcarine fissure and terminates either in the internal cerebral vein or the basal vein [12, 105]. The deep regions of the occipital lobe drain via subependymal veins.

Fig. 1.33 a–c. Venous drainage in the region of the occipital lobe.
a Following removal of the dura mater over the lateral tentorial sinus (*Tent. Sinus*), the occipital (*Occip. V.*) and bridging veins are demonstrated. The temporal (*Mid./Ant./Post. Temp. V.*), temporal basal (*Mid./Ant./Post. Temp. Basal V.*), occipital (*Occip. V.*), and occipital basal (*Occip. Basal V.*) veins empty into the tentorial (*Tent. Sinus*) and transverse (*Trans. Sinus*) sinuses. *Occip. Lobe,* occipital lobe; *Front. Lobe.,* frontal lobe; *Temp. lobe.,* temporal lobe; *Sylvian Fiss.,* sylvian fissure; *Sphen.*

Ridge, sphenoid ridge; *Mid. Fossa,* middle fossa; *Tent.,* tentorium.

b Following removal of the falx and division of the splenium callosum, the occipital poles (*Occip. Pole*) was separated to expose the straight sinus (*Str. Sinus*) and tentorium (*Tent.*). The internal cerebral veins (*Int. Cer. V.*) course in the roof of the third ventricle and are joined by the posterior pericallosal (*Post. Pericall. V.*) and anterior calcarine veins (*Ant. Calc. V.*). The medial atrial (*Med. Atr. V.*) and lateral atrial (*Lat. Atr. V.*) veins arise in the lateral ventricle and join the internal cerebral vein and basal veins (*Basal V.*). The occipital, posterior temporal (*Post. Temp. V.*), and occipital basal veins converge on the lateral tentorial sinus (*Tent. Sinus*). The cuneus is bordered anteriorly across the parieto-occipital sulcus (*Par. Occip. Sulc.*) by the precuneus and posteriorly by the calcarine fissure (*Calc. Fiss.*). The following are also seen: *Post. Med. Par. V.,* posterior medial parietal vein; *Cing. Sulcus,* cingulate sulcus; *Cing. Gyr.,* cingulate gyrus; *Post. Calc. V.,* posterior calcarine vein.

c The veins of the medial cerebral hemisphere are seen following removal of the falx and division of the corpus callosum and the brainstem. The straight sinus (*Str. Sinus*) courses in the junction of the falx and tentorium (*Tent.*) and the superior sagittal sinus (*Sup. Sag. Sinus*) courses in the upper margin of the falx. The medial surface of the occipital lobe is drained by the posterior medial parietal (*Post. Med. Par. V.*) and the anterior (*Ant. Calc. V.*) and posterior calcarine (*Post. Calc. V.*) veins. The veins crossing the medial surface of the hemisphere include the anterior pericallosal (*Ant. Pericall. V.*), posterior pericallosal (*Post. Pericall. V.*), posteromedial frontal (*Post. Med. Front. V.*), posteromedial parietal (*Post. Med. Par. V.*), anterior calcarine, and posterior calcarine veins. The posterior region veins drain into the vein of Galen (*V. of Galen*). The lateral tentorial sinus (*Tent. Sinus*) drains the anterior temporobasal (*Ant. Temp. Basal V.*), middle tempoorbasal (*Mid. Temp. Basal V.*), posterior temporobasal (*Post. Temp. Basal V.*), posterior temporal (*Post. Temp.*), and occipitobasal (*Occip. Basal V.*) veins and the vein of Labbé (*V. of Labbé*). The internal cerebral vein (*Int. Cer. V.*) and basal vein of Rosenthal (*Basal V.*) empty into the vein of Galen. The paracentral vein (*Paracent. V.*) drains the paracentral lobule and the adjacent part of the cingulate gyrus. The anteromedial parietal vein (*Ant. Med. Par. V.*) courses along the ascending ramus of the cingulate sulcus. The paraterminal vein (*Paraterm. V.*) empties into the veins that course in front of the lamina terminalis (*Lam. Term.*) and third ventricle (*3rd Vent.*). The lateral mesencephalic vein (*Lat. Mes. V.*) courses along the dorsal edge of the cerebral peduncle (*Ped.*). The following are also seen: *Cent. Med. Front. V.,* centromedial frontal vein; *Ant. Med. Front. V.,* anteromedial frontal vein; *Massa Inter.,* massa interna; *Optic N.,* optic nerve; *Car. A.,* carotid artery. (From [12])

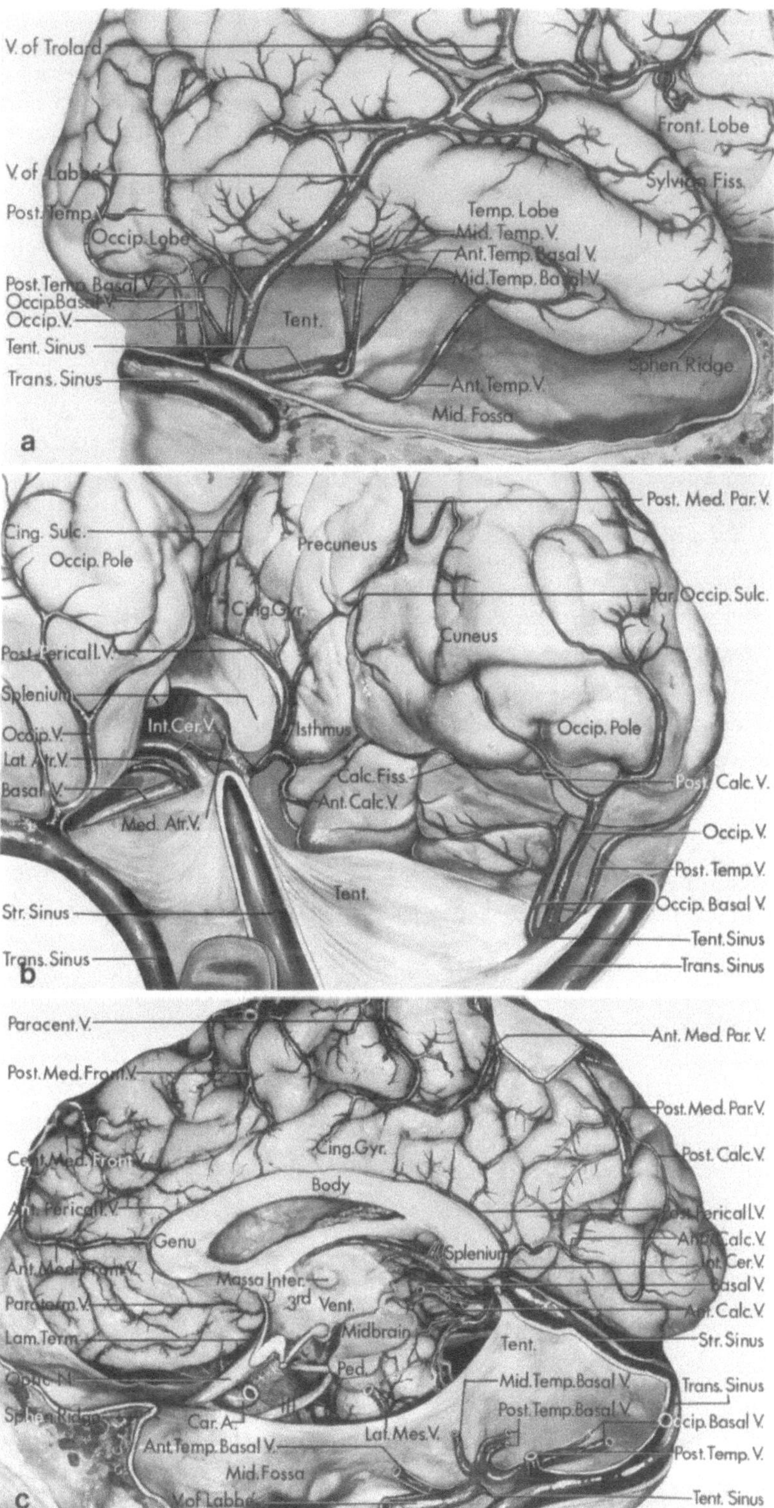
V. of Trolard
Front. Lobe
V. of Labbé
Sylvian Fiss.
Post. Temp. V.
Temp. Lobe
Occip. Lobe
Mid. Temp. V.
Ant. Temp. Basal V.
Mid. Temp. Basal V.
Post. Temp. Basal V.
Occip. Basal V.
Occip. V.
Tent.
Tent. Sinus
Trans. Sinus
Ant. Temp. V.
Sphen. Ridge
Mid. Fossa
a

Cing. Sulc.
Post. Med. Par. V.
Occip. Pole
Precuneus
Cing. Gyr.
Par. Occip. Sulc.
Cuneus
Post. Pericall. V.
Splenium
Occip. V.
Int. Cer. V.
Isthmus
Occip. Pole
Lat. Atr. V.
Basal V.
Calc. Fiss.
Post. Calc. V.
Ant. Calc. V.
Med. Atr. V.
Occip. V.
Post. Temp. V.
Occip. Basal V.
Str. Sinus
Tent.
Tent. Sinus
Trans. Sinus
Trans. Sinus
b

Paracent. V.
Ant. Med. Par. V.
Post. Med. Front. V.
Post. Med. Par. V.
Cent. Med. Front. V.
Cing. Gyr.
Post. Calc. V.
Ant. Pericall. V.
Body
Genu
Splenium
Pericall. V.
Ant. Med. Front. V.
Ant. Calc. V.
Massa Inter.
Int. Cer. V.
Paraterm. V.
3rd Vent.
Basal V.
Lam. Term.
Midbrain
Ant. Calc. V.
Tent.
Str. Sinus
Optic N.
Ped.
Mid. Temp. Basal V.
Trans. Sinus
Sphen. Ridge
Car. A.
Post. Temp. Basal V.
Occip. Basal V.
Ant. Temp. Basal V.
Lat. Mes. V.
Post. Temp. V.
Mid. Fossa
V. of Labbé
Tent. Sinus
c

1.5.3 Efferent Pathway Circulation

1.5.3.1 Cavernous Sinus Region

1.5.3.1.1 Internal Carotid Artery

A thorough understanding of the variations of the normal vascular anatomy at the base of the skull and the cavernous sinus region is essential in order to proceed with angiographic studies that will identify all the potentially involved arteries and venous drainage in high flow vascular lesions, arteriovenous shunts, or occlusive vascular diseases. This area is a critical region for many neuro-ophthalmic disorders because of the complex relationship among the cranial nerves, arteries, venous channels, and the base of the skull (Table 1.17, perisellar region and neurovascular contents; Fig. 1.34). The arrangement of the ICA passing within a venous plexus is unique in normal human vascular anatomy. The MMA and other branches from the internal maxillary artery provide many potential anastomoses with the branches of the ICA in this region.

Fig. 1.34. Dissection showing relationship of the cranial ▶ nerves to the cavernous sinus. Superolateral view of the pituitary gland and right cavernous sinus and intracavernous structures including the carotid artery and third, fourth, fifth and sixth cranial nerves (*CN III, IV, V, VI*). Lateral dural wall of cavernous sinus removed. Tortuous carotid artery bulges superiorly, pushing the cavernous sinus roof upward, and indents the lateral margin of the pituitary gland. The inferior hypophyseal artery passes to the pituitary gland. CN III and IV enter the roof of the cavernous sinus by passing through the interclinoid ligament. CN VI on the left side enters the dura posterior to the dorsum and passes laterally around the left carotid artery. CN VI is above the superior margin of the trigeminal sensory (*CN V_S*) and motor roots (*CN V_M*) and only the first division (*CN V_1*) is exposed distal to the ganglion. (From [114])

The ICA within the cavernous sinus (S- or C-shaped on the lateral projection angiogram) can be considered in three parts, the ascending C5, horizontal C4, and clinoid area C3 segments. Specific intracavernous arterial branches arising from each segment (Fig. 1.35) are constant, while others are variably present. The C5 segment contains a constant meningohypophyseal trunk (also called dorsal hypophyseal artery) which supplies the clivus, meninges, and the posterior hypophysis. A lateral artery of the clivus also commonly arises from this portion of the ICA and provides branches to the sixth nerve below the petroclinoid ligament and the petrous ridge.

An anterior capsular artery and a posterior capsular artery pass extradurally from the medial sur-

Table 1.17. Perisellar region and neurovascular contents (modified from [16e])

Foramen of perisellar region	Neurovascular contents
Optic canal	Optic nerve, ophthalmic artery
Superior orbital fissure	III, IV, V1, VI, sympathetic nerve, ophthalmic vein, recurrent ophthalmic artery, meningeal collaterals
Foramen rotundum	Artery of foramen rotundum, V2, emissary vein
Foramen ovale	Accessory MMA to the cavernous ICA, emissary vein, V2, V3
Foramen lacerum	ICA, sympathetic network, carotid branch of ascending pharyngeal artery, vidian nerve
Foramen spinosum	MMA
Vesalius foramen	Accessory MMA branch to ICA, emissary vein

ICA, internal carotid artery; MMA, middle meningeal artery.

Fig. 1.35. A superior view of the left intracavernous inter- ▶ nal carotid artery shows the important branch arteries which anastomose with external carotid artery branches: middle meningeal artery, artery to the foramen rotundum, and accessory meningeal artery from the internal maxillary (*IMA*) or the ascending pharyngeal artery (*Asc. Phar. A.*). The inferolateral trunk arises from the C4 segment and sends an anterior branch (*Ant. Br.*), posterior branch (*Post. Br.*), and middle branch to the middle meningeal artery. A C5 trunk (meningo-hypophyseal trunk) can provide inferior hypophyseal (*Inf. Hypoph. A.*), clival, basal tentorial (*Bas. Tent. A.*), and marginal tentorial (*Marg. Tent. A.*) arteries. Minute capsular arteries pass along the floor of the sella after originating from the medial surface of C4. *Oph. A.,* ophthalmic artery

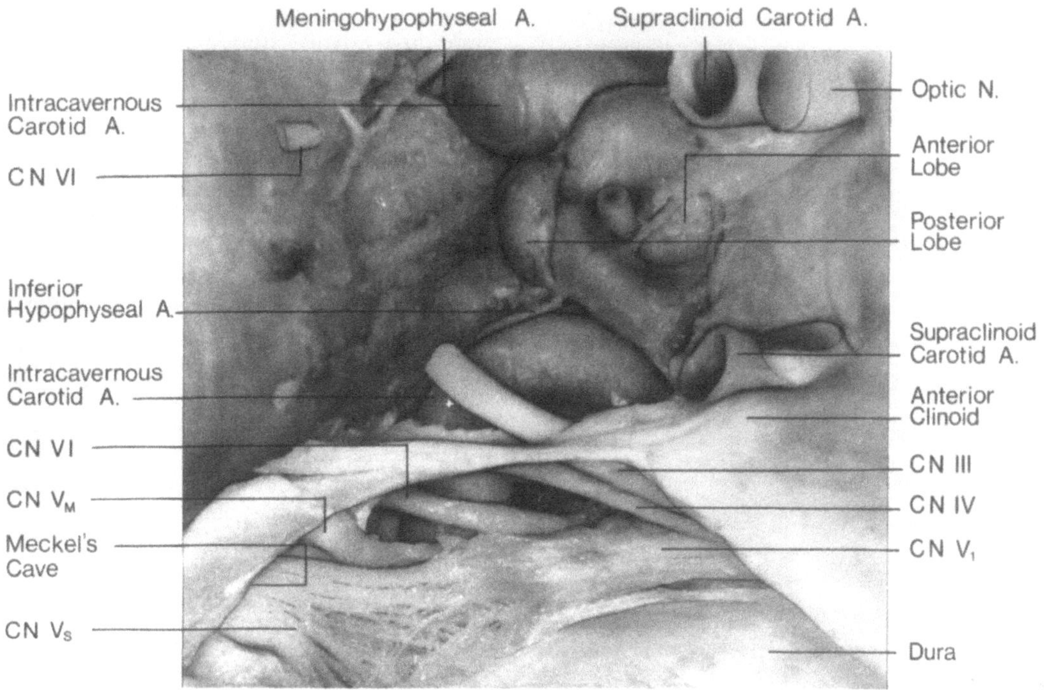

ICA Cavernous Sinus Branches
Left Side, From Above

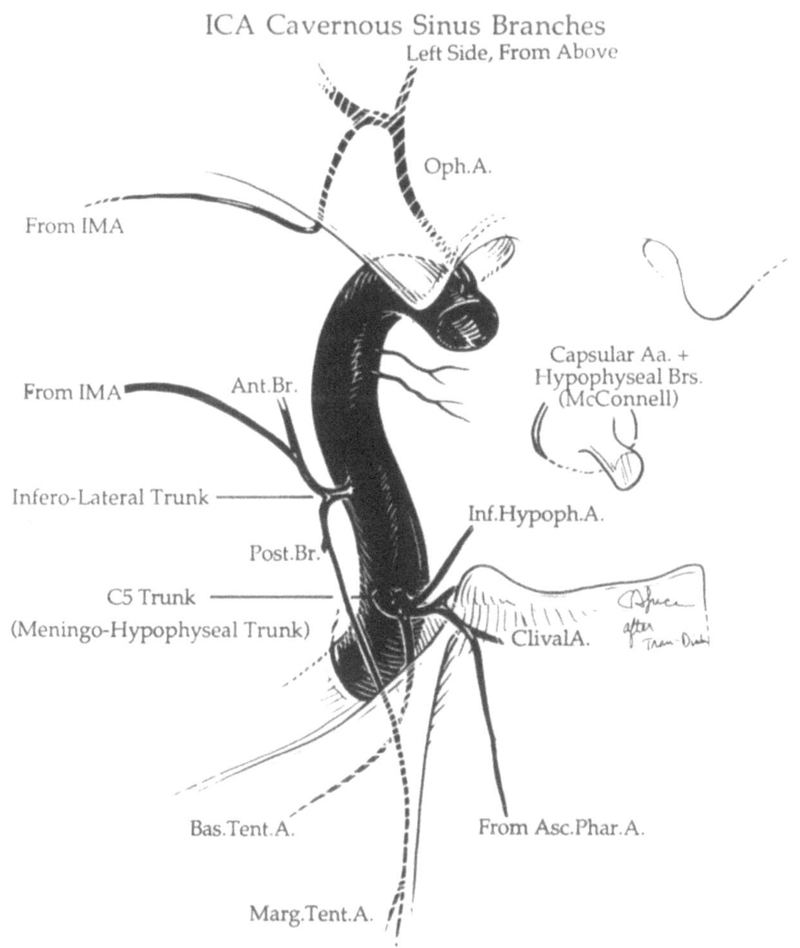

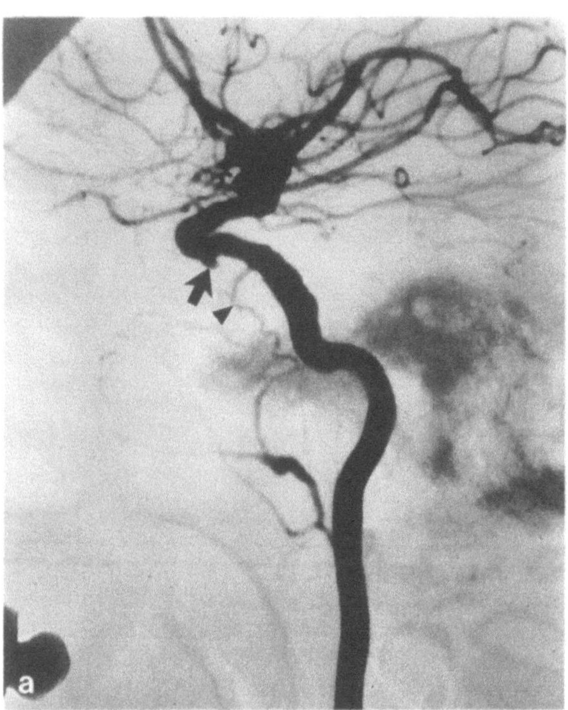

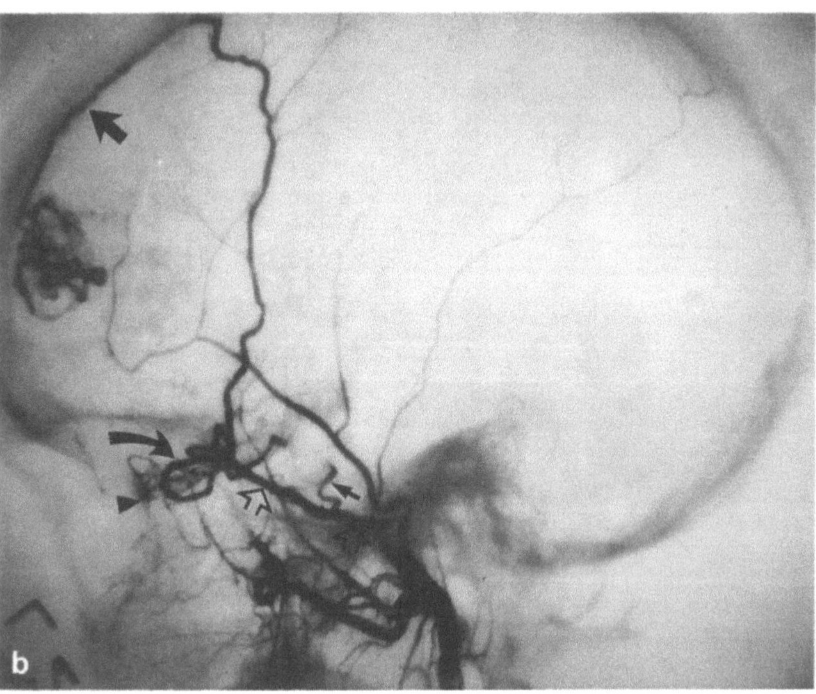

Fig. 1.36a, b. Angiography of pathological cases demonstrates the external carotid artery system collaterals through the inferolateral trunk of the cavernous internal carotid artery.

a The lateral projection internal carotid artery subtraction angiogram demonstrates a dilated inferior lateral trunk (*large arrow*) retrogradely filling the accessory meningeal artery (*arrowhead*), which passes through the foramen ovale to eventually supply a small arteriovenous malformation involving the internal maxillary artery. This case demonstrates the functional collateral from the internal carotid artery to the external carotid artery system via the inferior lateral trunk and the accessory meningeal artery.

b In another case, a lateral view subtraction angiogram of the external carotid artery shows vascular alterations because of the increased blood flow to the frontal region dural vascular malformation. A prominent falcine artery (*large arrow*) supplies the lesion and an enlarged inferior lateral trunk (*small arrow*) carries blood from the internal carotid artery into the dilated meningeal arterial system (*open arrows*). In addition, there is a meningolacrimal variant (*curved arrow*) to the ophthalmic circulation which causes a choroidal blush (*arrowhead*)

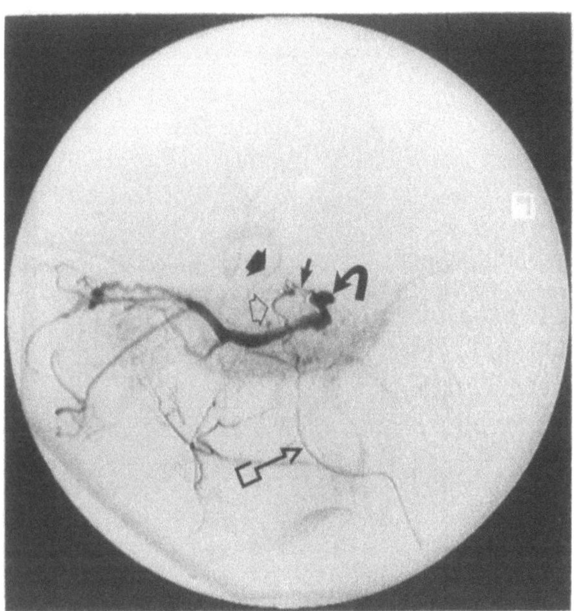

Fig. 1.37. In a patient with a cavernous sinus area dural arteriovenous malformation, a contrast injection via a catheter placed in the accessory meningeal artery (*open arrow*) opacifies the inferior lateral trunk (*open arrowhead*) as well as the meningeal branch artery from it (*small arrow*) to fill the cavernous sinus shunt (*curved arrow*). There is faint filling of the internal carotid artery (*solid arrowhead*) via the inferior lateral trunk

face of the C4 segment. Each anterior capsular artery and posterior capsular artery anastomoses to its contralateral counterpart in the midline and each provides blood to the structures of the hypophyseal fossa.

The inferior lateral trunk, also called the inferior artery of the cavernous sinus by Parkinson [106], is a constant artery arising from the C4 segment. The inferior lateral trunk most frequently divides into four branches (Figs. 1.35, 1.36). A superior branch supplies the roof of the cavernous sinus. An anteromedial branch enters the superior orbital fissure. An anterolateral branch passes into the foramen rotundum. A posterior branch courses medially and beneath the trigeminal ganglion. The dominance of the inferior lateral trunk supply to each area depends on whether potential collateral arteries from the ECA are significant. For example, the accessory meningeal artery can supply the same territories as the artery of the foramen rotundum (Fig. 1.37) (see Sect. 1.5.3.1.2)

The meningohypophyseal trunk, arising from the C5 segment, often has several branches. These include the posterior inferior hypophyseal, medial clival, and marginal tentorial (medial and lateral) arteries (Fig. 1.35). A lateral clival artery, which supplies the dura in this region, can also originate from the C5 segment.

As mentioned previously, the ophthalmic artery generally arises from the subarachnoid C3 segment of the ICA, but in up to 10% of subjects it originates from the C4 segment (see Sect. 1.5.1.1.2, Fig. 1.12a).

1.5.3.1.2 Collaterals in the Cavernous Sinus Region (Fig. 1.38)

The potential collaterals between the external and internal carotid artery systems in the region of the cavernous sinus depend on the persistence or absence of embryonic vessels in this region [16f] (Table 1.18). The principal collaterals include:

1. The internal maxillary artery anastomoses with the inferior and lateral branches (Fig. 1.2) of the cavernous carotid artery through the inferior lateral trunk. These collaterals are found anteriorly, through the artery of the foramen rotundum (arising from the termination of the internal maxillary artery in the pterygopalatine fossa); inferiorly, via the MMA (Figs. 1.2, 1.3) through the foramen spinosum; and anterolaterally, through the accessory meningeal artery (arising from the internal maxillary artery, either with the MMA or just distal) via the foramen ovale. The MMA may entirely supply the structures in the cavernous sinus if the inferior lateral trunk is poorly developed. In other individuals, the inferior lateral trunk vascularizes the meninges and cranial nerves (III, IV, VI) cavernous sinus as well as adjacent areas.

2. The ascending pharyngeal artery, which originates from the ECA (Fig. 1.1) (less commonly from the occipital artery or ICA), gives a superior pharyngeal artery that infrequently anastomoses with C5 branches. Each ascending pharyngeal artery sends a vessel from a hypoglossal arterial branch medially to anastomose with the medial clival arterial arcade. This vascular plexus connects the right and left C5 segments of both cavernous carotid arteries (Fig. 1.35) [16d]. On occasion, the MMA is the dominant blood supply to the clival and posterior cavernous sinus region.

3. Another source for a potential collateral is through an anastomosis between the two ICAs provided by the capsular artery. This vessel courses along the floor of the sella turcica, connecting the medial aspects of the two C4 ICAs [107] (Fig. 1.35; see also Fig. 2.14b in Chap. 2).

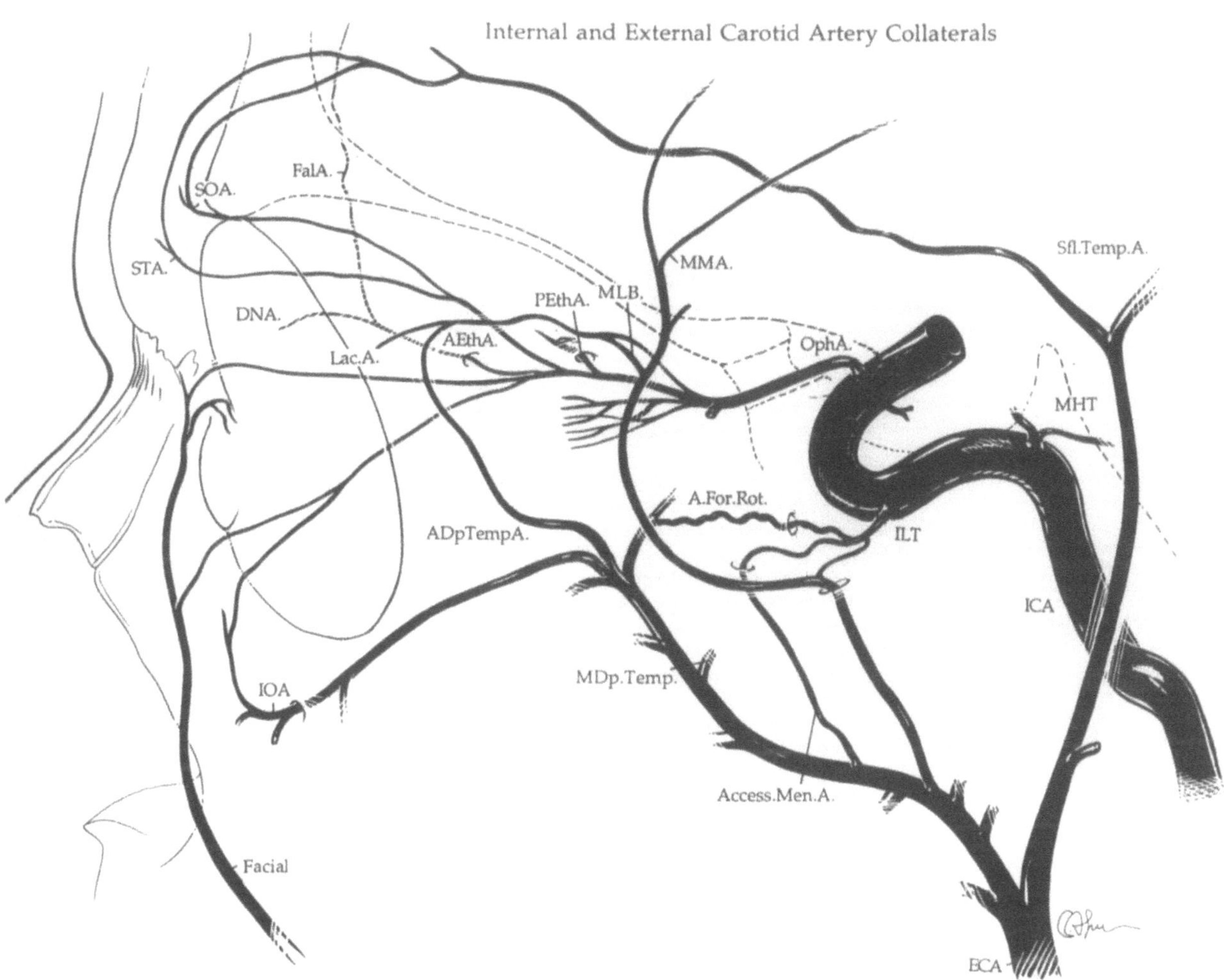

Fig. 1.38. The external carotid artery (*ECA*) anastomoses with the ophthalmic artery (*Oph A.*) in the orbit and with cavernous carotid branches in the region of the cavernous sinus are sites for potential collaterals to the internal carotid artery (*ICA*) circulation. Diagrammed anastomoses include accessory meningeal artery (*Access. Men. A.*) to middle branch of inferior lateral trunk (*ILT*), middle meningeal artery (*MMA.*) to posterior branch of ILT, artery to foramen rotundum (*A. For. Rot.*) to anterior branch of inferior lateral trunk, middle meningeal artery to ophthalmic (*Oph. A.*) or lacrimal artery (*Lac. A.*) via the meningolacrimal branch (*MLB.*), anterior deep temporal artery (*A Dp Temp A.*) to lacrimal artery, anterior ethmoidal artery (*A. Eth. A.*) to falcine artery (*Fal A.*) to middle meningeal artery, superficial temporal artery (*STA., Sfl. Temp. A.*) to ophthalmic artery, infraorbital artery (*IOA*) to muscular branch of ophthalmic artery, facial artery to palpebral artery. *SOA.,* supraorbital artery; *M Dp Temp A.,* middle deep temporal artery; *P Eth A.,* posterior ethmoidal artery; *DNA.,* dorsal nasal artery; *MHT,* meningohypophyseal trunk

Table 1.18. Intracavernous internal carotid artery branches (modified from [16])

Embryonic vessel and site of origin	Arterial remnant	Anastomoses	Anatomical variants	Meningeal and peripheral nervous territories concerned
	Anterolateral branch (ILT)	Artery of the foramen rotundum		V2
	Posterior branch (ILT)	Accessory meningeal artery (AMA)	Unique feeding pedicle to the area, or main supply in association with an hypotrophic ILT AMA origin of the ophthalmic artery AMA origin of the anterior cerebral artery	V3 + Vm ± VII
Dorsal ophthalmic artery (C4 portion of the ICA)		Petrous branch of meningeal artery	ILT origin of the middle meningeal artery	Gasserian ganglion ± VII
	Anteromedial branch (ILT)	Deep recurrent ophthalmic artery	ILT origin of ophthalmic artery Ophthalmic origin of the middle meningeal artery	III, V1, IV, VI, pericarotid plexus ± II
	Superior branch (ILT)	Superficial recurrent ophthalmic artery	Opthalmic and lacrimal origin of the marginal tentorial artery	Marginal third of the tentorium cerebelli
		Marginal tentorial artery of the C5 portion of the ICA	Middle meningeal origin of the marginal tentorial artery Accessory meningeal origin of the marginal tentorial artery Posterior origin of the marginal tentorial artery	Cavernous sinus roof, III, IV
C4 portion of the ICA (medial)	Capsular arteries			Sella turcica floor, sphenoidal sinus roof
Primitive maxillary artery (C5 portion of the ICA)	Posteroinferior hypophyseal artery	Contralateral counterpart	Preclival interinternal carotid anastomosis	Posterior lobe of the pituitary gland and sellar dura
		Contralateral counterpart	Retroclival interinternal carotid anastomosis	
		Clival branch of the hypoglossal division of the neuromeningeal trunk (ascending pharyngeal artery)		Dura over the clivus

Table 1.18 (cont.)

Embryonic vessel and site of origin	Arterial remnant	Anastomoses	Anatomical variants	Meningeal and peripheral nervous territories concerned
Trigeminal artery (C5 portion of the ICA)	Lateral clival artery	Clival branch of the jugular division of the neuromeningeal trunk (ascending pharyngeal artery)	Trigeminal artery persistence Trigeminal reconstitution of an agenetic cervical ICA C5 portion origin of cerebellar arteries	Inferior petrosal sinus, cerebellopontine angle, VI
		Middle meningeal artery: petrous branch (basal tentorial branches)	Basilar origin of the middle meningeal artery	Basal two-thirds of the tentorium cerebelli Superior petrosal sinus V ± IV and III
C5 portion of the ICA	Recurrent artery of the foramen lacerum	Carotid branch of the ascending pharyngeal artery	Ascending pharyngeal origin of the middle meningeal artery	Upper portion of the foramen lacerum, posterior face of the gasserian ganglion, pericarotid nervous plexus
		Cavernous branch of the middle meningeal artery		
C5 portion of the ICA	Lateral artery of the Gasserian ganglion	Dural branches of the middle meningeal artery to the lateral cavernous sinus	Dural covers	Lateral Gasserian ganglion

ICA, internal carotid artery; ILT, inferior lateral trunk; Vm, motor division trigeminal.

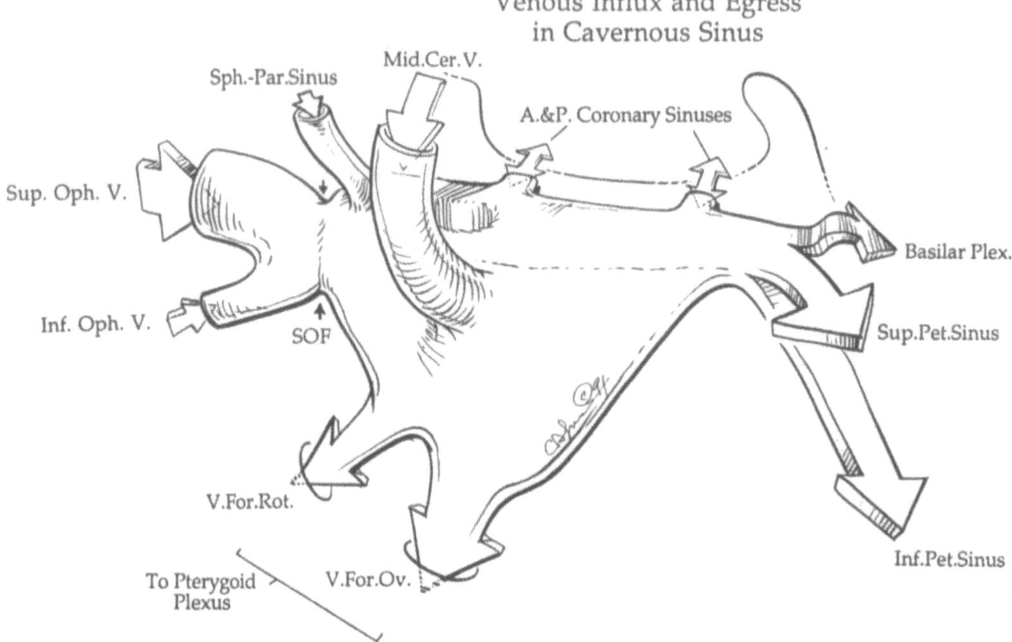

Fig. 1.39. Lateral projection of the left cavernous sinus shows the normal influx and efflux of venous blood (indicated by the direction of the *arrows*). The superior (*Sup. Oph. V.*) and inferior (*Inf. Oph. V.*) divisions of the ophthalmic vein join to form a single vein which passes through the superior orbital fissure (*SOF*) into the anterior compartment. Cerebral venous blood flows into the sphenoparietal sinus (*Sph. Par. Sinus*) and the middle cerebral vein (*Mid. Cer. V.*) (also sylvian vein), which drain into the anterior compartment. The anterior and posterior coronary sinuses connect the two cavernous sinuses. Blood flow drains from the posterior compartment via the superior (*Sup. Pet. Sinus*) and inferior (*Inf. Pet. Sinus*) petrosal sinuses and to a minor extent via the basilar venous plexus (*Basilar Plex.*), the venous outflow through the foramen rotundum (*V. For. Rot.*) and the foramen ovale (*V. For. Ov.*) to the pterygoid plexus

1.5.3.1.3 Venous Structures

The cavernous sinus (Fig. 1.39) is made up of a series of extradural connecting veins, or fenestrated septae-formed channels. Cranial nerves III–VI and pericarotid sympathetics on the ICA course in the cavernous sinus or the leaves of the dura (see below). Fat is also found within the cavernous sinus, particularly in the region of the roof and lateral wall [108]. The cavernous sinus normally has multiple routes of drainage (Table 1.19)

The anterior portion of the cavernous sinus drains the common ophthalmic vein. Rarely, both the superior ophthalmic (SOV) and inferior ophthalmic (IOV) divisions empty individually into the cavernous sinus. The sphenoparietal sinus and the middle cerebral veins empty into the anterior compartment of the cavernous sinus. The sphenoparietal sinus is located lateral to the cavernous sinus and passes along the inferior border of the sphenoid ridge. Rostrally, the sphenoparietal sinus becomes a diploic channel that also empties into the superior sagittal sinus. The pterygoid venous plexus drains the anterior cavernous sinus through the emissary veins that pass through the foramen in the floor of the middle cranial fossa.

The posterior compartment of the cavernous sinus principally receives blood from the anterior cavernous sinus. The blood in the posterior compartment normally drains via the superior and inferior petrosal sinuses. The inferior petrosal sinus connects to the jugular vein/sigmoid sinus junction. The superior petrosal vein courses in the tentorium to empty into the transverse/sigmoid sinus junction.

The right and left cavernous sinuses are connected via an anterior and a posterior coronary venous sinus, which are located on the diaphragma sellae and anterior to the dorsum sella, respectively. The posterior portions of each cavernous sinus are

Table 1.19. Cavernous sinus outflow

Direction	Venous channel
Normal	
Posterior	Superior and inferior petrosal sinuses
Inferior	Emissary veins to the foramen rotundum, foramen ovale, foramen of Vesalius
Opposite cavernous sinus	Anterior and posterior coronary sinuses, basal venous plexus
Abnormal	
Anterior	Opthalmic vein
Superior	Sphenoparietal sinus, middle cerebral vein

also connected via a small basal venous plexus lying on the clivus posterior to the dorsum sellae (Fig. 1.7).

1.5.3.2 Cranial Nerves

1.5.3.2.1 Third Nerve

The third nerve fascicles and nuclei in the midbrain receive blood supply from the posterior thalamoperforating arteries that originate from each proximal (P1) segment of the PCA, or arise as a common trunk from one P1.

As the third nerve exits between the fibers of the crus cerebri into the subarachnoid space of the interpeduncular cistern, it receives blood via small caliber (500–600 µm) branches. These minute arteries originate from the basilar, posterior cerebral, or posterior communicating arteries (Fig. 1.40) [90]. In general, the nutrient arteries arise from the small arteries that pass adjacent to the nerve. Cases with infarcts of the fascicular portion of the nerve that selectively spare or damage specific third nerve function suggest that in some individuals there is specific blood supply to certain nerve fascicles (see Sect. 8.5.2.1.4). There are no such cases of infarcts with an isolated loss of function once the fibers of the third nerve have merged within the midbrain.

In contrast, a functional anatomical arrangement of the blood supply to the nerve as it passes in the subarachnoid space [109] and the cavernous sinus [110, 111] (see Chap. 8) has been revealed by the study of diabetic third nerve palsies. The nutrient vessels to the intracranial portion of the nerve

anastomose within the epineurium and send branches to the perineurium. These vessels also terminate in arterioles and capillaries located between and within the third nerve fascicles [110].

Once the third nerve is in the subarachnoid space, it may actually be pierced by a branch of the long or short circumflex mesencephalic arteries (also called collicular and accessory collicular arteries). The arterioles from these vessels can also supply the nerve [112]. While in the interpeduncular cistern, the oculomotor nerve passes between the superior cerebellar artery and the PCA. In most cases, as the third nerve continues in the subarachnoid space, the third nerve passes slightly lateral and inferior to the posterior communicating artery (Fig. 1.41). On occasion, the nerve courses medially or beneath this vessel.

Just prior to its entry into the dural roof of the cavernous sinus, lateral to the fourth nerve, the third nerve is supplied by a branch of the artery of the free margin of the tentorium cerebelli (also called the artery of Bernasconi) [113]. The artery of the free margin commonly originates from the meningohypophyseal trunk at the C5 segment of the ICA. While passing in the roof of the cavernous sinus, in close apposition to the fourth nerve, the third nerve is supplied by branches from the inferior lateral trunk or equivalent vessels from the ECA (Figs. 1.34–1.36). Small branch arteries, with a caliber of approximately 350–400 µm, arise from the lateral aspect of the ICA in the inferior cavernous sinus to supply three or four microscopic branches to the third nerve in the cavernous sinus [106, 111]. At the mid-cavernous sinus level, the superior branch from the inferior lateral trunk is an important nutrient vessel to the nerve. In approximately 10% of subjects where the MMA provides significant blood supply to the area ordinarily supplied by the inferior lateral trunk, the MMA will also supply the cranial nerves in this territory. In the superior orbital fissure, the third nerve is supplied by the anteromedial branch of the inferior lateral trunk or a recurrent branch from the ophthalmic artery. The recurrent branch, which is often microscopic, with a caliber of 50–75 µm, may also anastomose with the MMA in this region. The recurrent branch courses along with the intraorbital portion of the third nerve to also supply the extraocular muscles near the annulus of Zinn.

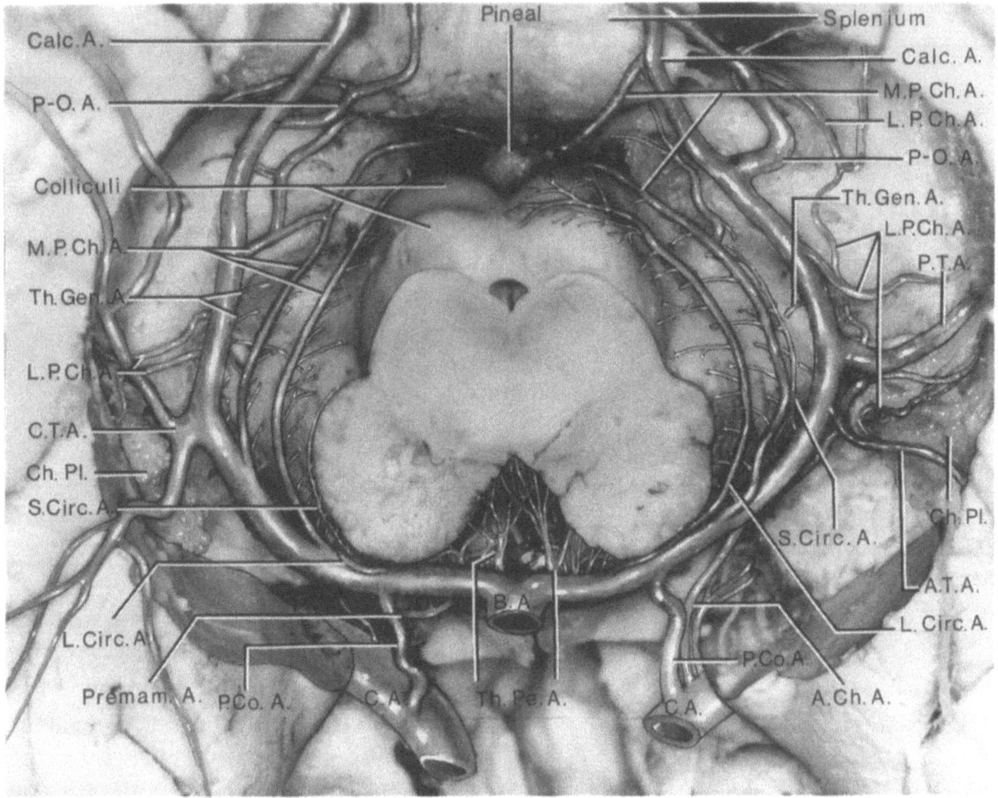

Fig. 1.40. Inferior view of the origin of both posterior cerebral arteries from the basilar artery (*B.A.*) and their course around the midbrain. The medial part of the temporal lobe has been removed to expose the pulvinar (above midbrain) and choroid plexus (*Ch. Pl.*) of the temporal horn. The central branches of the posterior cerebral artery include the thalamoperforating (*Th. Pe. A.*), long circumflex (*L. Circ. A.*) and short circumflex (*S. Circ. A.*) arteries. The right anterior choroidal artery (*A. Ch. A.*) passes above the posterior cerebral artery to the choroid plexus of the temporal horn. A premammillary branch (*Premam. A.*) arises from the left posterior communicating artery (*P. Co. A.*). The thalamogeniculate arteries (*Th. Gen. A.*) penetrate the brain in the region of the geniculate bodies. The lateral posterior choroidal arteries (*L. P. Ch. A.*) can be traced into the temporal horn of the ventricle. Two medial posterior choroidal arteries (*M. P. Ch. A.*) on the *right* arise from the posterior cerebral artery and the calcarine artery (*Calc. A.*); on the *left* there are three medial posterior choroidal arteries, two of which anastomose to run superior to the colliculi and lateral to the pineal gland to enter the third ventricle. A common temporal artery (*C.T.A.*) supplies the left temporal lobe and the right side is supplied by anterior (*A.T.A.*) and posterior temporal branches (*P.T.A.*). The terminal posterior cerebral artery bifurcates into the parieto-occipital (*P-O.A.*) and calcarine arteries. *C.A.*, carotid artery. (From [90])

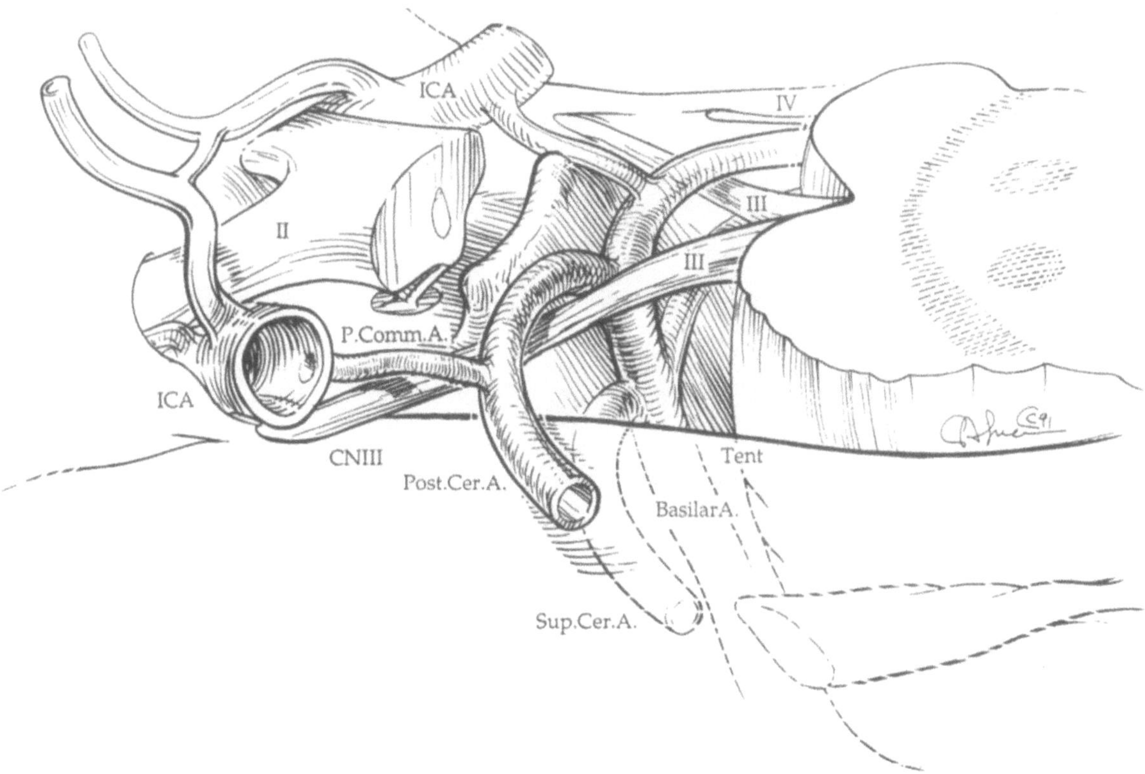

Fig. 1.41. After exiting the midbrain through the crus cerebri, the subarachnoid third nerve passes under the proximal posterior cerebral artery (*Post. Cer. A.*) before coursing anteriorly beneath and the posterior communicating artery (*P. Comm. A.*). In most individuals the third nerve is lateral to the posterior communicating artery along the anterior two-thirds of the length of this vessel. *ICA,* internal carotid artery; *Sup. Cer. A.,* superior cerebellar artery; *Tent,* tentorium; *CN II, III, IV,* cranial nerves

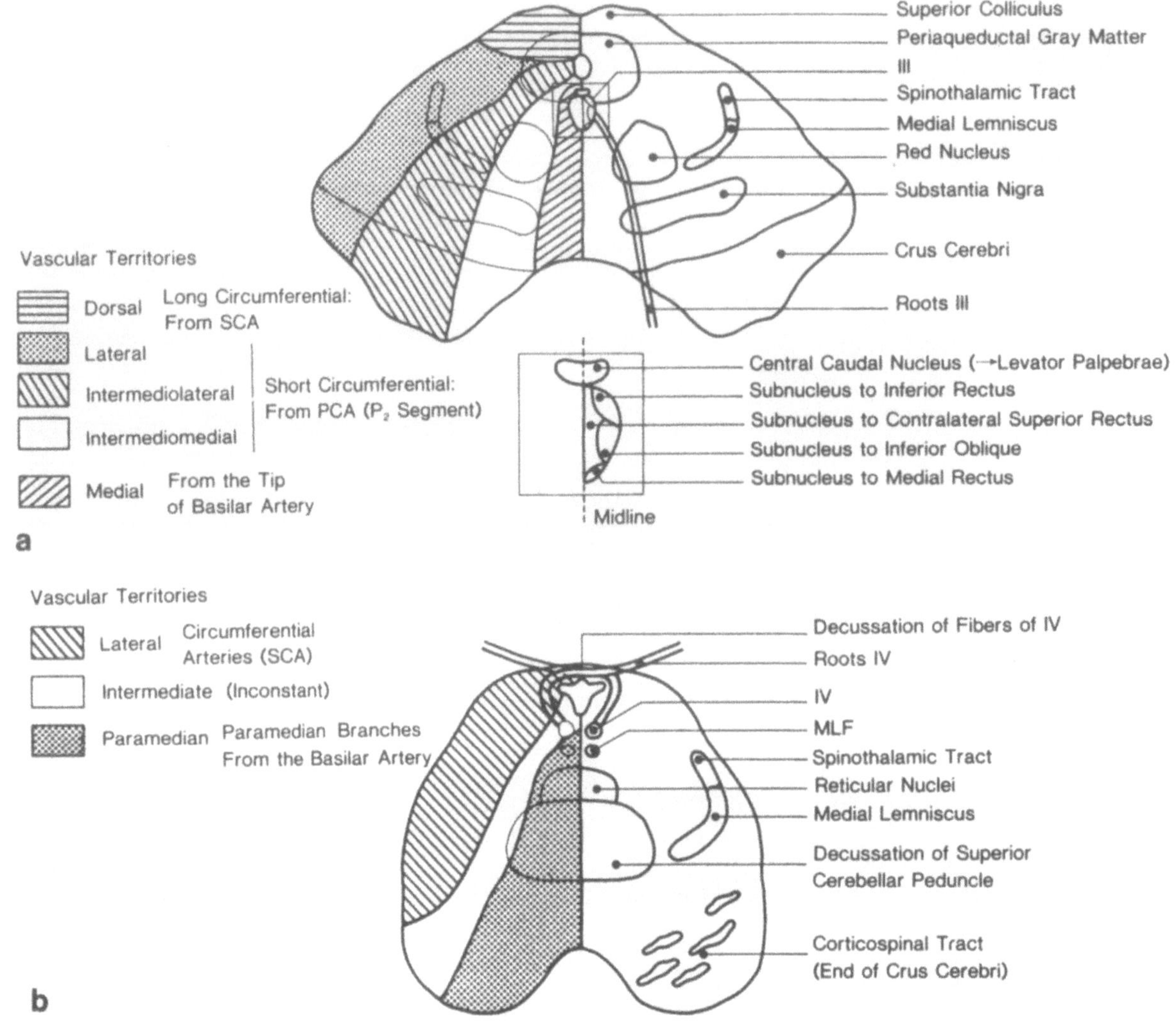

Fig. 1.42 a–d. Schematic transverse cuts from midbrain to medulla with arterial territories: posterior cerebral artery (*PCA*); superior cerebellar artery (*SCA*); anterior inferior cerebellar artery (*AICA*); and posterior inferior cerebellar artery (*PICA*). **a** Level of the third nerve nucleus. **b** Level of a fourth nerve nucleus. **c** Level of the sixth nerve nucleus. **d** Level of eighth (vestibular), tenth, and twelfth cranial nerve nuclei, medial longitudinal fasciculus (*MLF*), and pontine paramedian reticular formation (*PPRF*). (From [119])

1.5.3.2.2 Fourth Nerve

The fourth nerve nucleus receives blood from the thalamoperforating arteries [90]. Within the midbrain, the fourth nerve is supplied by long circumferential arteries. The fourth nerves decussate and exit from the dorsal surface of the lower midbrain. The nerve courses parallel to the superior cerebellar artery, but in the opposite direction, around the lateral aspect of the midbrain (Fig. 1.42b).

The artery of Bernasconi also supplies the fourth nerve as it enters the posterior cavernous sinus. While the fourth nerve is in the cavernous sinus, the superior branch of the inferior lateral trunk is its principal nutrient vessel. In the superior orbital fissure, the fourth receives blood from the anteromedial branch of the inferior lateral trunk. If the MMA is dominant, it also provides the blood supply to the fourth nerve within the cavernous sinus and superior orbital fissure.

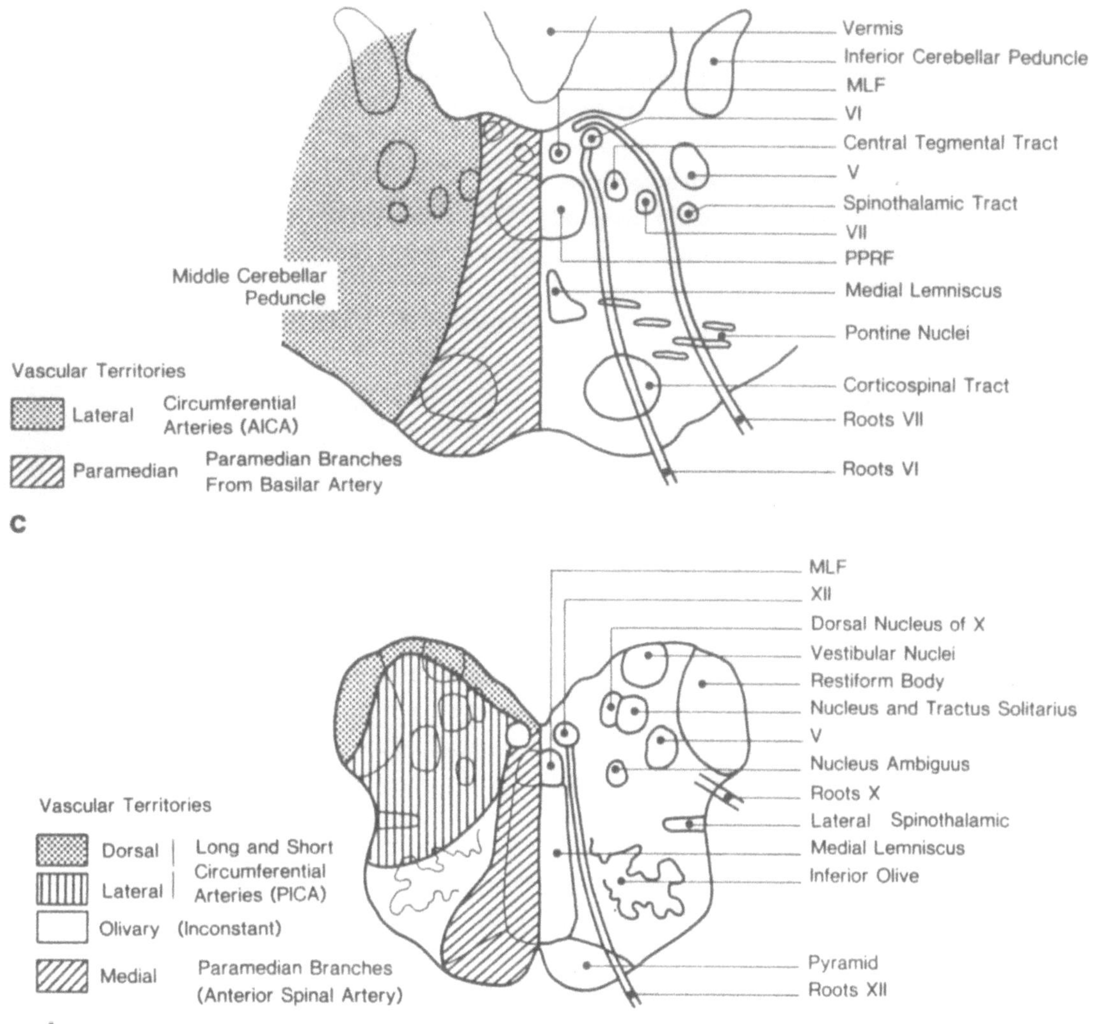

1.5.3.2.3 Sixth Nerve

As the sixth nerve (abducens) exits the basis pontis, it is supplied by arterial twigs arising from the basilar artery (Fig. 1.42c). The abducens nerve is also infrequently penetrated by the anterolateral pontine vein or a tributary of the anteromedian pontine vein [112]. As the nerve passes extradurally adjacent to the dorsum sella, it is supplied by arteries from the jugular branch of the ascending pharyngeal artery and a lateral artery of the clivus, a branch of the dorsal meningeal artery [16f, 114]. The sixth nerve actually passes within the cavernous sinus between the carotid artery and the lateral dural wall [106]. A branch from the inferior lateral trunk supplies the sixth nerve, which is located in the inferior cavernous sinus. In the superior orbital fissure, the sixth nerve is supplied by the anteromedial branch of the inferior lateral trunk. If the MMA is the dominant vessel, it also supplies the sixth nerve in the superior orbital fissure.

1.5.3.2.4 Fifth Nerve (Trigeminal)

As the trigeminal nerve exits from the pons, it receives blood from a branch vessel that arises from a segment of the basilar artery between the anterior superior and anterior inferior cerebellar arteries. This small artery is the vestige of the embryological trigeminal artery and it courses with the fifth nerve towards the cavernous sinus. Both the ICA and ECA contribute to the lateral artery of the trigeminal ganglion in Meckel's cave, which is posterolateral to the cavernous sinus (Fig. 1.34). Branches from the carotid branch of the ascending pharyngeal artery and the cavernous sinus region supply from the MMA contribute to the rich vascular input to the trigeminal ganglion [16g]. The first, or ophthalmic, division (V1) is supplied by the inferior lateral trunk or the MMA as it enters the inferior lateral dural wall of the cavernous sinus. The artery of the foramen rotundum from the anterolateral branch of the inferior lateral trunk supplies the V2 division as it passes to this foramen. The V3 sensory division and the motor nerve receive blood from the cavernous branches of the MMA, the accessory meningeal artery, as well as the posterior branch of the inferior lateral trunk which anastomoses with the accessory meningeal artery and the MMA [16h]. The V1 division in the superior orbital fissure receives blood from the anteromedial branch of the inferior lateral trunk.

1.5.3.2.5 Seventh Nerve

As the seventh cranial nerve emerges from the lateral aspect of the pons, it is supplied by minute branches from the basilar artery. The internal auditory artery, a branch of the anteroinferior cerebellar artery, provides small arteries that supply both the seventh and eighth nerves while they are in the internal auditory meatus. There are several variations of the arterial distribution to the intrapetrous portion of the seventh nerve. All include some combination of arterial contribution from the petrous branch of the MMA and the stylomastoid artery. The stylomastoid artery arises from the occipital artery, the posterior auricular artery, or directly from the ECA [16i]. The second and third divisions of the trigeminal nerve, the Gasserian ganglion, and the seventh nerve can share a common blood supply from the MMA [115, 116]. Thus, in some individuals closure of the proximal MMA can cause facial paralysis and numbness.

1.5.3.2.6 Ninth, Tenth, Eleventh, and Twelfth Nerves

The blood supply to the extra-axial intracranial ninth, tenth, eleventh, and twelfth cranial nerves is briefly mentioned because of the clinical importance of the ascending pharyngeal arterial system in providing collaterals to some of the previously discussed cranial nerves. These cranial nerves receive blood from the neuromeningeal branch of the ascending pharyngeal artery [117, 118, 16j]. The neuromeningeal artery has three branches, the musculospinal, hypoglossal, and jugular arteries (caudally to rostrally). The vertebral artery also provides blood to the ninth, tenth, and twelfth nerves. The spinal portion of the eleventh nerve is supplied by a cervical artery and the musculospinal artery. Two other branches of the ascending pharyngeal artery, the pharyngeal and tympanic arteries, do not supply the third, fourth, fifth, sixth, or seventh cranial nerves but these vessels may provide important collaterals to branches of the MMA or the cavernous ICA (see Sect. 1.5.3.1.2) [16k].

1.5.3.3 Brainstem Oculomotor, Lid, and Pupillary Control

Each level of the brainstem can be divided into midline or medial, lateral, and dorsal regions with respect to the various arterial input and vascular occlusive syndromes [119].

1.5.3.3.1 Medulla (Fig. 1.42d)

The anterior spinal artery supplies the medial region of the medulla. Medullary branches from the posterior inferior cerebellar artery, principally via its posterior lateral spinal artery, provide blood to the lateral and dorsal aspects of the medulla [120, 121].

1.5.3.3.2 Pons (Fig. 1.42c)

The medial territory of the pons receives blood supply from paramedian twigs arising from the

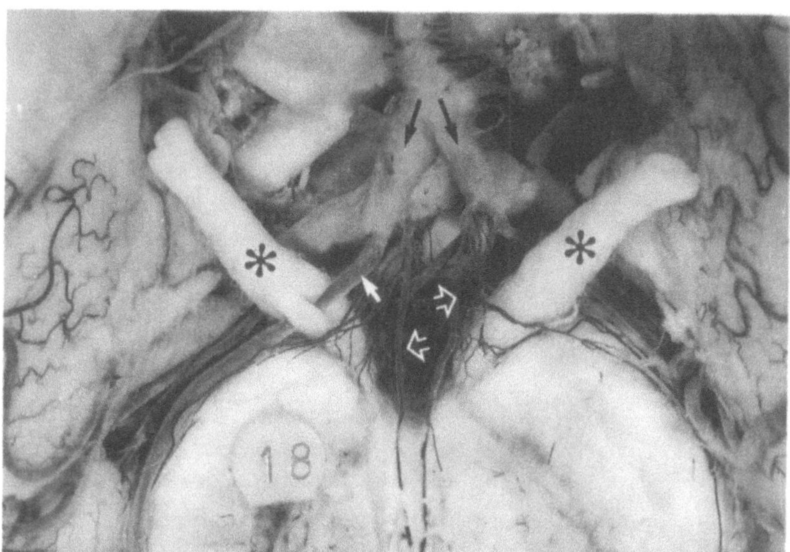

Fig. 1.43. Ventral view of the oculomotor nerves (*asterisks*) exiting the brainstem into the subarachnoid space. The right third nerve is penetrated by a long circumflex mesencephalic artery (*white arrow*). The basilar artery at the *top* divides into the proximal posterior cerebral arteries (*black arrows*). Numerous perforating branches (*open white arrows*) arise from the proximal segments of each posterior cerebral artery to supply the third nerve as it exits the midbrain. (From [112])

basilar artery and from branches of the short circumferential arteries. The lateral region of the pons is supplied by the anterior inferior cerebellar artery and the superior cerebellar artery, which are both long circumferential arteries.

1.5.3.3.3 Midbrain

The caudal midbrain median territory, including the medial longitudinal fasciculus and third nerve nucleus (Fig. 1.42b), receives blood supply from minute branches originating directly from the rostral middle third of the basilar artery. Short circumferential branches from the P1 segment of the PCA supply the intermediolateral territory at this level, including a portion of the medial longitudinal fasciculus. Long circumferential branches, arising from the superior cerebellar artery, provide blood to the lateral surface and the superior colliculus [122, 90].

The rostral paramedian midbrain, subthalamus, and thalamus (Figs. 1.42a, 1.43) receive blood supply from paramedian thalamic arteries (interpeduncular and thalamoperforating arteries) and from the posterior thalamoperforators (also called thalamosubthalamic artery). The medial portion of the rostral red nucleus, periventricular nuclei adjacent to the rostral cerebral aqueduct and the third ventricle, and the subthalamic area containing the rostral interstitial nucleus of the medial longitudinal fasciculus, all receive blood from paramedian and short circumferential arterial branches [88, 89, 123].

As previously discussed, the paramedian vessels originate directly from the P1 segment of the PCA [124a, 124b]. The understanding of the blood supply to this area is somewhat confused because these vessels have been given different names by various authors [125]. Also, the configuration of the paramedian thalamic arteries is variable (Fig. 1.44): (1) A symmetrical arrangement with one vessel from each proximal PCA. (2) An asymmetrical arrangement with a common trunk or paramedian arteries to both sides of the midbrain arising from a single PCA. (3) The paramedian arteries arise from an arterial arcade supplied by both PCAs.

In addition to the structures discussed under the blood supply to the sensory visual system (Sect. 1.5.2.7.1), the anterior choroidal artery can also provide blood supply to the paramedian territory

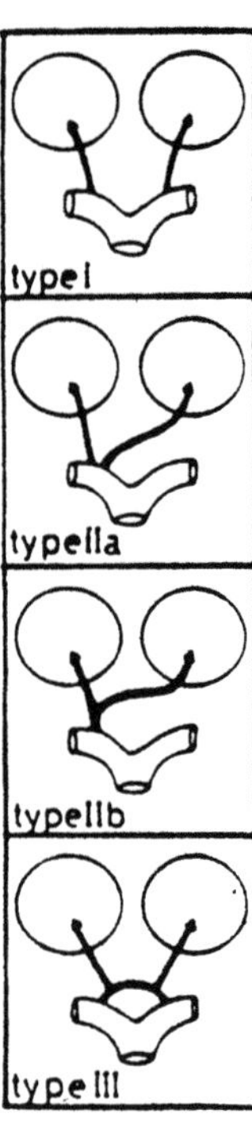

Fig. 1.44. The distribution of branches of the basilar communicating artery showing the variations of the posterior thalamosubthalamic paramedian artery as described by Percheron. (From [128])

of the midbrain [84, 85, 126]. However, oculomotor dysfunction is generally not considered a part of the syndrome of anterior choroidal artery occlusion, which typically includes hemisensory disturbances, a variety of homonymous field defects, hemiparesis or hemifacial weakness alone, and dysarthria.

The lateral regions of the dorsal midbrain and the superior colliculus are supplied by long circumferential branches originating from the superior cerebellar artery or the P1 or P2 segments of the PCA [90, 127]. These branch arteries, also called collicular arteries, anastomose in a tectal vascular network over the dorsal midbrain, including the superior and inferior colliculi [124b].

1.5.3.3.4 Brainstem Veins

The brainstem drains via an intrinsic network of veins, which provides an axial system for segmental blood drainage, similar to the spinal cord. The venous network empties either into an anteromedial medullary vein or a dorsal medullary vein, which pass longitudinally on the anterior and posterior surfaces of the brainstem, respectively [128].

1.6 Cerebral Blood Flow Physiology

For a discussion of neurotransmitter and neuropeptide influences of vasomotor control, see Chap. 10.

Knowledge of the intracranial vascular anatomy must be correlated with an understanding of the factors that maintain and alter cerebral blood flow (CBF) for the study and treatment of the neurologic and ophthalmologic vascular disorders. The factors which alter blood flow vary with the location and the level of the affected circulation (larger cerebral arteries, penetrating arteries, arterioles, and capillaries). Because the brain has a high energy requirement, the various processes that control intraorbital and intracranial circulations (Table 1.20) must provide for homeostasis of local tissue nutrients, oxygen, and electrolytes in order to maintain normal neuronal metabolism over a wide range of hypotension and hypoxemia. The blood flow to a specific region varies with the metabolic requirement of that area (ocular blood flow discussed in Sects. 1.5.2.1 – 1.5.2.4).

Normally, the brain receives from 50 – 100 ml blood per 100 grams of tissue. Glucose and oxygen consumption vary with the region of the brain. The CBF and perfusion of the brain tissue depend on the volume and velocity of the blood flow, the resistance in the arteries, arterioles, capillaries, venules, and veins, and the pressure in the perfused tissue. Autoregulation in the normal circulation maintains a relatively constant CBF over a range of range of blood pressure (50 – 130 mmHg), arterial/capillary blood oxygen pressure (50 – 80 mmHg pO_2), arterial/capillary blood carbon dioxide pressure (30 – 40 mmHg pCO_2) [129, 130], and hematocrit (in terms of blood viscosity) [131]. The compensatory vasomotor response to an alteration in one of these factors may be independent of alterations induced by changes in the other factors. Tissue perfusion is maintained by local autoregulation, independent of perivacular innervation, by either constricting or dilating the small arteries and arterioles (Table 1.21). Effective autoregulation has particular importance in the watershed regions where the cerebral perfusion is normally reduced. The watershed borders are located in the regions between the MCA, ACA, and PCA territories, and between the inferior and superior cerebellar arteries. Similar

Table 1.20. Factors affecting cerebral circulation

Cardiac output
Arterial lumen patency
Blood viscosity: hematocrit, macromolecules, erythrocyte, rigidity
Arterial oxygen pressure
Arterial carbon dioxide pressure
Neurotransmitters: perivascular innervation, intraluminal
Vasoactive peptides
Local autoregulation factors
Venous pressure
Intracranial pressure
Arachidonic acid metabolites
Neuronal activity level

Table 1.21. Normal arterial and arteriolar responses

Altered factor	Response
Systemic hypertension	Constriction
Systemic hypotension	Dilation
Hypercapnia	Dilation
Hypoxia	Dilation
Hyperoxia	Constriction
Elevated hematocrit	Constriction
Hypoglycemia	Dilation

watershed zones can also be found in the choroidal circulation between posterior ciliary arteries.

Even when the autoregulation is present, it fails to maintain normal cerebral perfusion pressure in many pathological conditions. In other diseases, autoregulation is lost. If the cerebral perfusion pressure falls below 40 mmHg, the CBF is reduced. In contrast, at perfusion pressures above 140 mmHg arterioles and capillaries paradoxically dilate, causing an increase in blood flow, and the endothelial cells are damaged, permitting extravasation of fluids and exudates. Chronic systemic hypertension shifts the range of autoregulation so

that the CBF is not maintained when perfusion pressure is reduced below 80 mmHg. However, if the patient becomes normotensive prior to development of vascular damage, the response range can be normalized in many, but not all, cases [132].

The acute systemic hypotension associated with severe blood loss also reduces the CBF. The resulting overactivity of the sympathetic system and release of angiotensin can result in vasoconstriction, which worsens the existing cerebral ischemia.

The vascular response to changes in the pCO_2 or pO_2 can be quite complex when additional factors are also altered. For example, if the arterial pCO_2 exceeds 70 mmHg, further vasodilatation does not occur. Lactic acidosis and severe hypoglycemia impair the vascular response to hypoxia [133]. Hypoxia also causes an increase in erythrocyte rigidity, increasing the blood viscosity. This is particularly important in the sickle cell hemoglobinopathies, where this change can induce thrombosis. Platelet aggregation also slows the microcirculation [134]. Autoregulation is frequently absent in the region of an infarct or a transient ischemic attack. This impairment in autoregulation can persist long after the ischemic event has resolved [135]. The combination of focal cerebral ischemia and elevation of the pCO_2 can be synergistically detrimental because the pCO_2-induced dilatation in the adjacent normal vessels diverts blood away from the ischemic region, resulting in more profound ischemia [136].

The effects of extravascular pressure exerted by both elevated intracranial pressure and brain edema increase the resistance and reduce the blood flow in the capillaries. The resistance to flow in the capillaries is also increased and the tissue perfusion is reduced if the pressure in the venous system is above normal, such as occurs with increased intracranial pressure or obstruction in the major dural venous sinuses.

CHAPTER 2

Carotid Cavernous Fistulas

Carotid cavernous fistulas (CCF) are acquired pathological direct shunts from the cavernous portion of the internal carotid artery (ICA) into the enveloping cavernous sinus (Fig. 2.1). These arteriovenous (AV) shunts can develop spontaneously, but the overwhelming majority (80%) are the result of a traumatic injury to the ICA or a branch artery [1]. In most cases, clinical diagnosis is not difficult, but the variable nature of the signs and symptoms can result in a confusing presentation.

Though the clinical signs and symptoms can fluctuate, most of the abnormalities progressively worsen with the evolution of various hemodynamic processes. Except for the additional direct traumatic injuries, the findings are similar for cases with either traumatic or spontaneous CCFs or cavernous sinus region dural arteriovenous malformations (Tables 2.1, 2.2; see Sects. 2.2.1, 3.2.7.1). The patients complain of visual blur, diplopia, headache, ocular or orbital pain, and subjective bruit. Precise diagnosis depends on the correct interpretation of the symptoms and findings as they relate to the anatomical abnormalities. The clinical examination and laboratory evaluation, including ultrasonography, computed tomography (CT), magnetic resonance (MR) imaging, and cerebral angiography, demonstrate the various hemodynamic processes that account for the neuro-ophthalmologic abnormalities.

2.1 Epidemiology and Causes

2.1.1 Trauma

Motor vehicle accidents are the inciting event in most cases. Trauma from falls or penetrating inju-

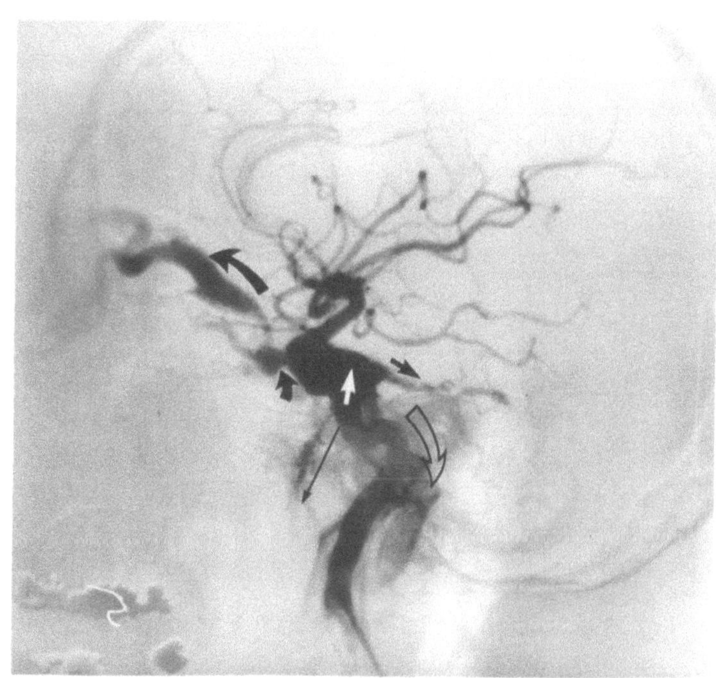

Fig. 2.1. A lateral subtraction angiogram of the left internal carotid artery in a patient with a traumatic carotid cavernous fistula (*white arrow*) draining towards the common ophthalmic vein (*small curved arrow*), and then draining into a markedly dilated superior division of the ophthalmic vein (*long curved arrow*), superior petrosal sinus (*small black arrow*), inferior petrosal sinus to jugular vein (*open curved arrow*) and to the vein of the foramen ovale towards the pterygomaxillary plexus (*thin black arrow*). Note that, despite the arteriovenous shunt, there is normal filling of the left middle, posterior, and both anterior cerebral arteries

Table 2.1. Neuro-ophthalmologic findings: 30 cases of spontaneous cavernous carotid fistula

Elevated intraocular pressure	27
Third nerve paresis	10[a]
Fourth nerve paresis	5[a]
Sixth nerve paresis	8[a]
Ptosis	8[a]
Arterialized conjunctival vessels	17[a]
Chemosis	5[a]
Lid engorgement	7[a]
Proptosis	17[a]
Bruit	18[a]
Venous retinopathy	4
Visual loss	4
Anterior chamber reaction	2
Pupillary dilation	0

[a] Bilateral findings as well; in most cases multiple cranial nerves involved.

Table 2.2. Neuro-ophthalmologic abnormalities with traumatic carotid cavernous fistulas: 95 cases

Elevated intraocular pressure	85
Third nerve paresis	65[a]
Fourth nerve paresis	50[a]
Sixth nerve paresis	60[a]
Ptosis	70[a]
Arterialized conjunctival vessels	95[a]
Chemosis	70[a]
Lid engorgement	50[a]
Proptosis	95[a]
Bruit	93[a]
Venous retinopathy	33
Disc edema	12
Visual loss	20
Anterior chamber reaction	0
Pupillary dilation	9
Thrill/pulsation of globe	15
Dilated iris vessels	3

[a] Bilateral findings as well; in most cases multiple cranial nerves involved.

ries is less frequent. The incidence in the population of CCFs is unknown but it is a relatively uncommon disorder. Since there is a male predominance for most traumatic injuries, the incidence of CCFs is higher in men than in women. CCFs in children are predominantly caused by penetrating injuries or falls.

Iatrogenic complications are an unusual cause of a CCF. Damage to the ICA, sustained during en-darterectomy, particularly when a Fogarty catheter is used above the operative site [2], endovascular procedures for thromboendarterectomy, or angioplasty of the ICA (currently in rare use), can be the inciting injury [3–5]. A CCF can also develop from injury to the ICA during transphenoidal surgery to remove a pituitary tumor. Laceration of the cavernous carotid artery or one of the cavernous branch arteries is likely to occur if the ICA is located medially on the floor of the sella [6]; or if the surgeon mistakenly courses eccentric to the midline. A preoperative MR examination can determine whether the location of the cavernous ICA contraindicates transphenoidal surgery.

2.1.2 Spontaneous Carotid Cavernous Fistulas

Spontaneous CCFs arise from a variety of etiologies that predispose towards a weakness in the wall of a cavernous branch artery or the ICA [7,8]. The arterial wall dilates or balloons outward at this defective area. Trivial trauma encountered in daily living (e.g., bumping against a cabinet door), a valsalva maneuver, vomiting, or sneezing might be a triggering, but not the causal event for rupture of the artery.

Specific disorders that cause a defect in the arterial wall media and a CCF can be identified in approximately 60% of spontaneous cases [7]. These include Ehlers–Danlos syndrome (Fig. 2.2) [9, 10], pseudoxanthoma elasticum [11], fibromuscular dysplasia (see Sect. 8.9) [12], aneurysm of the cavernous ICA (Fig. 2.3; see Chap. 6), a persistent embryological trigeminal artery, and other nonspecific angiodysplasias. Recognizing that a CCF has resulted from one of these etiologies is essential in order to properly manage the AV shunt and the systemic abnormalities seen with some of these disorders.

Since few physicians have extensive experience in the management of the rare connective tissue disorders, a brief review of the clinical manifestations is useful. Ehlers–Danlos syndrome is an inherited disorder with autosomal dominant and recessive varieties that causes abnormalities in collagen metabolism and results in defects in the media of arteries [13]. Musculoskeletal deformities and peripheral venous varicosities are frequent. Bleed-

Fig. 2.2. A lateral subtraction angiogram of the left internal carotid artery in a patient with Ehlers–Danlos syndrome and a spontaneous carotid cavernous fistula. Note the dysplastic appearance of the cervical internal carotid artery (*curved arrow*) in addition to the arteriovenous fistula draining anteriorly to the ophthalmic venous system and posteroinferiorly to the inferior petrosal sinus

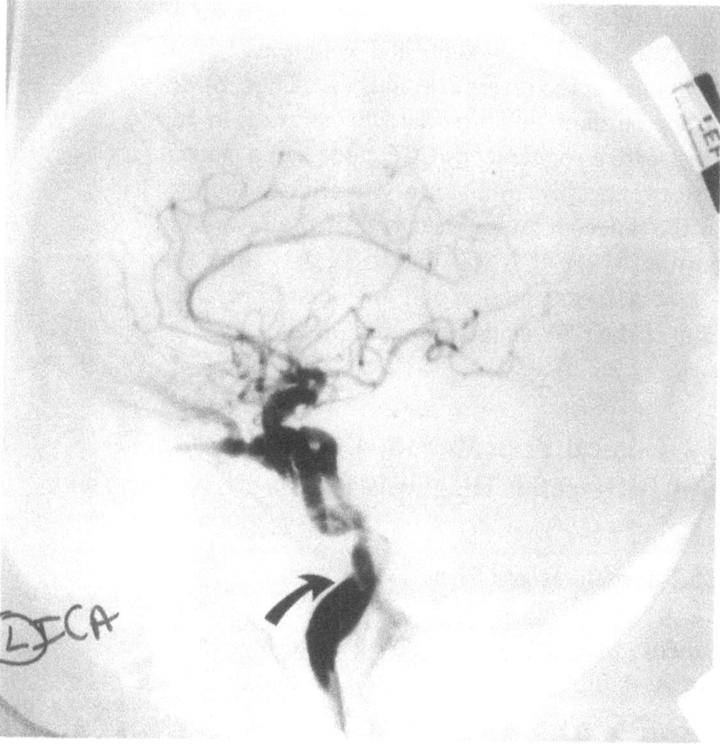

ing is common due to the fragility of the blood vessel walls. Dissection of the major arteries such as the aorta infrequently occurs. Rarely, dissection of the ICA may be the event that causes a CCF. Associated ocular defects are common and include angioid streaks in the retina, macular degeneration, vitreous hemorrhage, ectopic lens, posterior staphyloma, blue sclera, excessive skin folds in the lids, and glaucoma.

Pseudoxanthoma elasticum is another rare connective tissue disorder with both autosomal recessive and dominant forms that disrupts the elastic fibers in the skin, eye, and arteries. Thickening and laxity of the skin is characteristic. Breaks in Bruch's membrane cause flat angioid streaks in the retina emanating from the optic disc that may lead to secondary subretinal neovascularization. Stenosis of major arteries causes claudication in the legs and angina. Cerebral ischemic events are less common [14]. Rupture of an intestinal artery can result in severe gastrointestinal hemorrhage. Rupture of a cerebral artery is rare but it can cause an intracranial hemorrhage or a CCF.

A cavernous carotid aneurysm associated with a CCF, unlike the typical presentation of an unruptured cavernous aneurysm, is typically asymp-

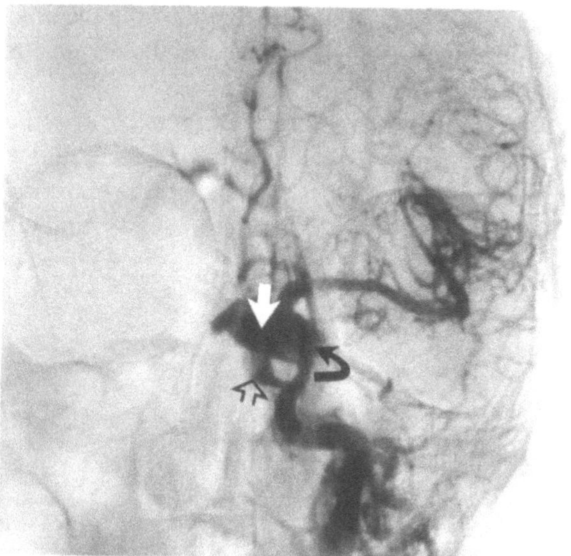

Fig. 2.3. Rupture of an aneurysm of the internal carotid artery producing a carotid cavernous fistula. The anteroposterior view subtraction angiogram of the left internal carotid artery (*curved arrow*) demonstrates the aneurysm lumen (*white arrow*), but the thrombus which is not opacified has been subtracted. The posterior cavernous sinus (*open arrow*) is opacified because a hole in the aneurysm has created a low flow arteriovenous fistula. Because the patient had only a mild sixth nerve paresis, a slow flow shunt, and only posterior drainage, treatment with carotid compression (see Sect. 2.4) resulted in a cure

tomatic prior to rupture into the cavernous sinus. In our series of 78 cavernous aneurysms, none of the patients who had signs and symptoms typical of a mass in the cavernous sinus (see Chap. 6) developed a shunt at a later date. By contrast, in our cases with a spontaneous CCF, none had a clinical history suggestive of an aneurysm prior to rupture. In the latter group an aneurysm of the cavernous portion of the ICA was diagnosed only after the CT or MR examination and angiographic evaluation of their AV shunts were performed.

2.2 Clinical Presentation, Evaluation, and Differential Diagnosis [15]

2.2.1 Symptoms and Signs

Patients with CCFs complain of visual blur (59% of cases), diplopia (53% of cases), headache (53% of cases), ocular or orbital pain (35% of cases), and subjective bruit (80% of cases). Following trauma, the clinical signs that result from the original injury can complicate the analysis of some of these cases (see Tables 2.1, 2.2).

Signs of orbital congestion (Fig. 2.4) may become immediately obvious following head trauma, but the presence of local orbital damage may mask these signs of a CCF. Trauma may also initially

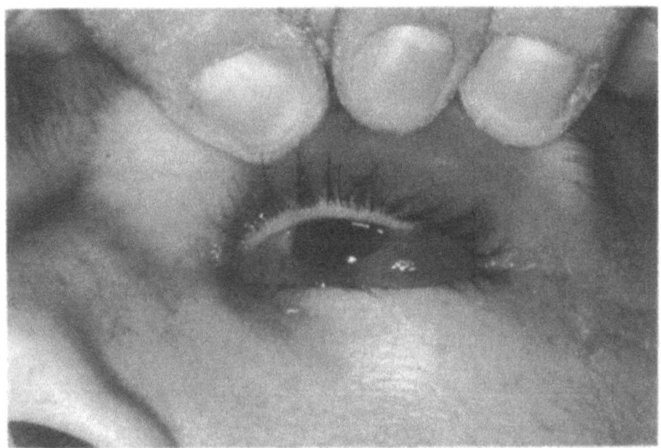

Fig. 2.4. An example of a traumatic carotid cavernous fistulas with significant drainage into the ipsilateral ophthalmic venous system: a teenage boy fell off a building, sustaining a brief loss of consciousness and an immediate left carotid cavernous fistula (corresponds to Fig. 2.19). Chemosis and arterialization of the conjunctiva are evident

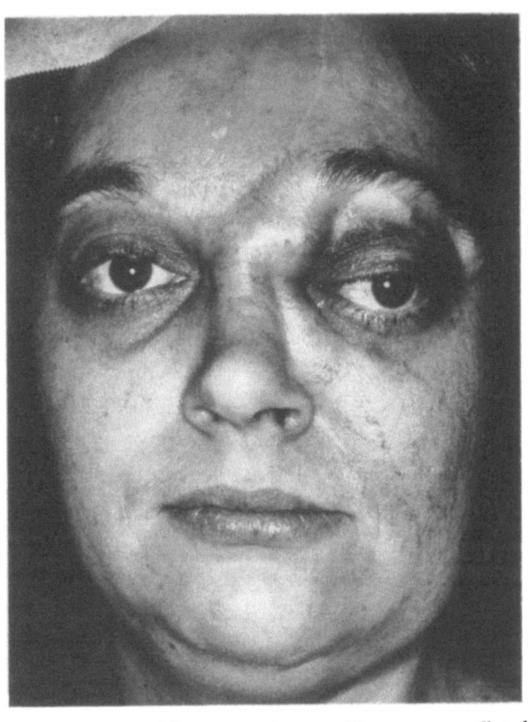

Fig. 2.5. Patient with traumatic carotid cavernous fistula who had previously undergone ligations of both the carotid artery and internal jugular vein in the left cervical region and developed massive venous collaterals through the angular vein and lateral scalp veins (corresponds to Figs. 2.14, 2.27)

cause a small pseudoaneurysm in a branch artery of the cavernous ICA that ruptures only after a latency of days to weeks. If this pseudoaneurysm ruptures into a venous channel in the cavernous sinus, a CCF with orbital signs and a bruit typically develops. Early descriptions in the literature of the clinical examination reported that more than 90% [16] of cases had features of severe or chronic orbital congestion (Fig. 2.5). More recent reports described a lower incidence of patients with severe orbital signs because the diagnosis was made earlier, and cases with CCFs that drained predominantly posteriorly have been recognized (Figs. 2.3, 2.6) [5].

Congestion of the orbital contents from arterialization of the ophthalmic venous system causes elevation of the intraocular pressure (IOP) or secondary glaucoma, venous retinopathy (Fig. 2.7), weakness and mechanical limitation of the eye muscles, arterialization and dilatation of the conjunctival vessels, chemosis, and dilation of the veins and edema in the lids. Proptosis, with the

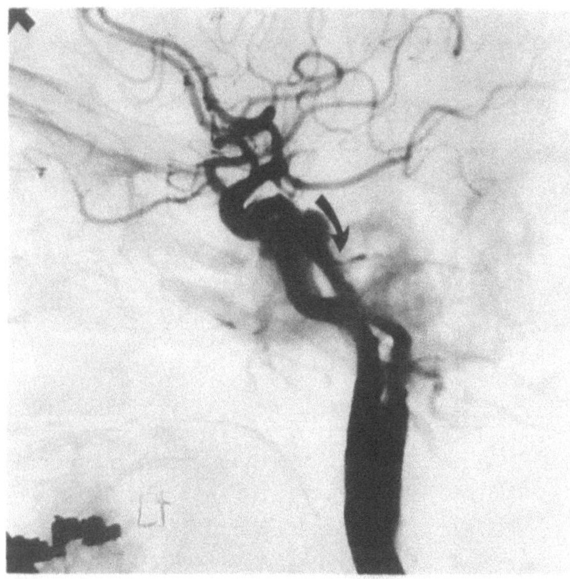

Fig. 2.6. Traumatic carotid cavernous fistula draining almost entirely posteriorly. The left internal carotid artery shows an arteriovenous fistula with almost exclusive drainage to the posterior portion of the cavernous sinus into the inferior petrosal sinus (*curved arrow*). This patient had a complete sixth nerve paresis and a post-auricular region bruit as the only clinical signs

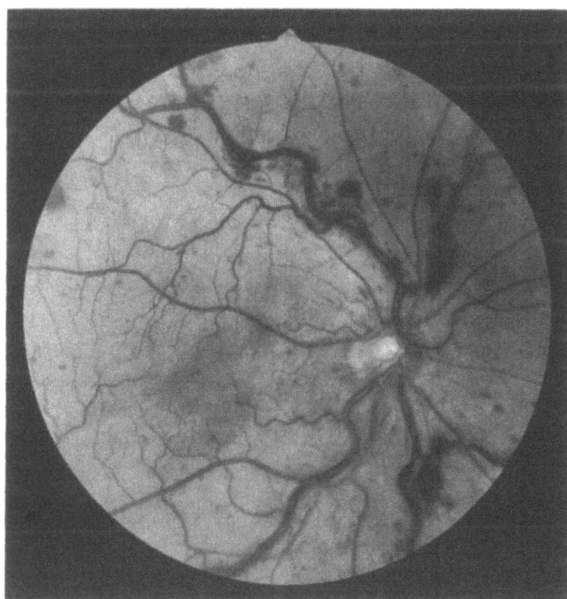

Fig. 2.7. Ophthalmoscopy in this case of traumatic carotid cavernous fistula reveals a nonspecific appearance of a venous retinopathy with moderate dilatation of the retinal veins and multiple areas of retinal hemorrhage. Most cases have fewer findings or require ultrasonography to demonstrate choroidal involvement (see Chap. 3, Fig. 3.13)

globe usually displaced downward and laterally, results because of massive dilatation of the ophthalmic vein. Pulsations that can actually be seen or palpated occur in less than 30% of cases [15, 5].

Cranial and sympathetic neuropathies can also be secondary to the effects of the CCF on the nerves within the cavernous sinus or the direct effects of trauma. Isolated third or sixth nerve or combinations of third, fourth, and sixth nerve dysfunction are common. Trigeminal nerve sensory deficits in the first and second divisions are seen in less than 20% of cases. Fifth nerve defects of motor function or of the third division sensation are almost never a complication of a CCF and are generally a result of a traumatic neuropathy. A complete ptosis, with an absence of levator function, is indicative of a third nerve palsy unless there is direct damage to the levator muscle or aponeurosis. Partial ptosis may arise from a partial third nerve or sympathetic paresis, local lid edema, or a combination of all three factors. The ocular motor dysfunction can affect all the eye movements or it can be limited, such as a weakness in upgaze or abduction. However, the superior oblique muscle is never affected alone.

In trauma patients, it is often difficult to assess to what extent the neuro-ophthalmologic findings are the direct results of the trauma or whether they are secondary to the effects of the CCF on the cranial nerves or orbital tissues. A forced duction test (see Sect. 4.2) can determine if local extraocular muscle congestion or hemorrhage is mechanically limiting the eye movements. However, since arterialized vessels and congestion of the conjunctiva increase the risk of subconjunctival hemorrhage, performing this test may be unjustified.

Following trauma, direct injury of the third, fourth, sensory fifth, and sixth nerves is common, with or without an appreciated fracture through the sphenoid sinus. There are often associated cranial neuropathies of the seventh and eighth nerves resulting from basilar skull fractures through the petrous bone. A seventh nerve palsy complicates the management of the proptosis and related corneal exposure because paresis of the orbicularis oculi prevents the effective movement of tears across the conjunctiva and cornea as well as proper eyelid closure over the cornea. Anesthesia of the cornea further decreases the patient's ability to maintain a clear cornea as trophic epithelial changes from the trigeminal nerve dysfunction develop. The patient, unable to feel corneal drying or a foreign body, does not blink to clear the corneal surface, causing further compromise of the cornea, and possibly leading to an abrasion or ulcer.

The pupil can have normal function and appearance (even with severe ocular motor dysfunction), or it may be larger or smaller than the unaffected eye. The combination of poor light reactivity and poor dilation in the dark of the pupil suggests the presence of both parasympathetic and sympathetic dysfunction. The pupil may be difficult to evaluate if the trauma has damaged the iris sphincter or if a traumatic uveitis is present. An afferent pupillary defect develops in those eyes with a unilateral optic neuropathy.

The IOP is raised in the majority of patients with CCFs that have significant drainage anteriorly from the cavernous sinus. At least 20% of patients will develop a glaucomatous type of visual field loss if the AV shunt is not closed. The IOP is high enough to cause a central retinal artery occlusion in less than 2% of cases. The glaucoma is typically, but not exclusively, unilateral. The biomicroscopic examination and gonioscopy should exclude other causes of unilateral glaucoma such as angle recession, pseudoexfoliation, congenital angle abnormalities, pigmentary abnormalities, uveitis and iris synechiae, narrow angles, essential iris atrophy, and angle neovascularization. If the opposite cavernous sinus is affected by the shunting of blood across the coronary sinus (Fig. 2.8), both eyes have arterialized conjunctival vessels, signs of orbital congestion, and elevated IOP.

In patients with a unilateral glaucoma, the CCF elevation of the episcleral venous pressure contributes to the elevation of the IOP. The elevation of episcleral venous pressure (for discussion see Sect. 2.3.4.2.5) forces blood into Schlemm's canal that can be seen with gonioscopy of the angle. This can be found in cases with a CCF, dural arteriovenous malformation (DAVM), congenital vascular anomalies, or occlusion of the superior vena cava. Some patients with congenital glaucoma, such as develops in cases with Sturge–Weber syndrome, have elevated episcleral venous pressure.

An optic neuropathy, typically with a mild reduction of Snellen acuity, dyschromatopsia, an af-

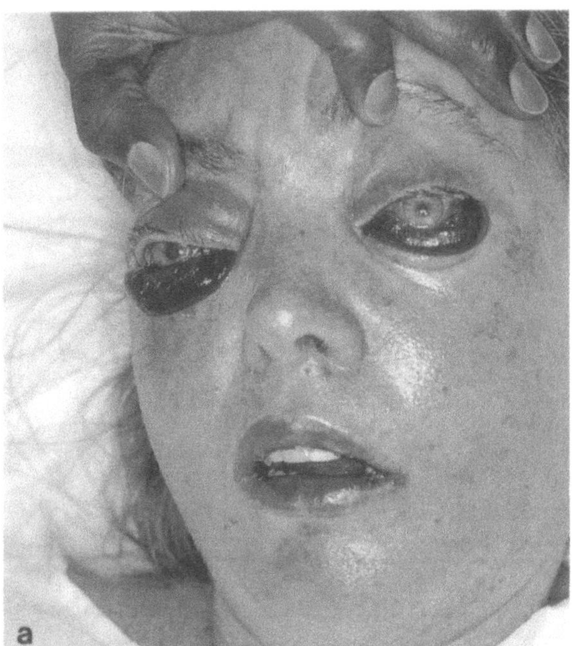

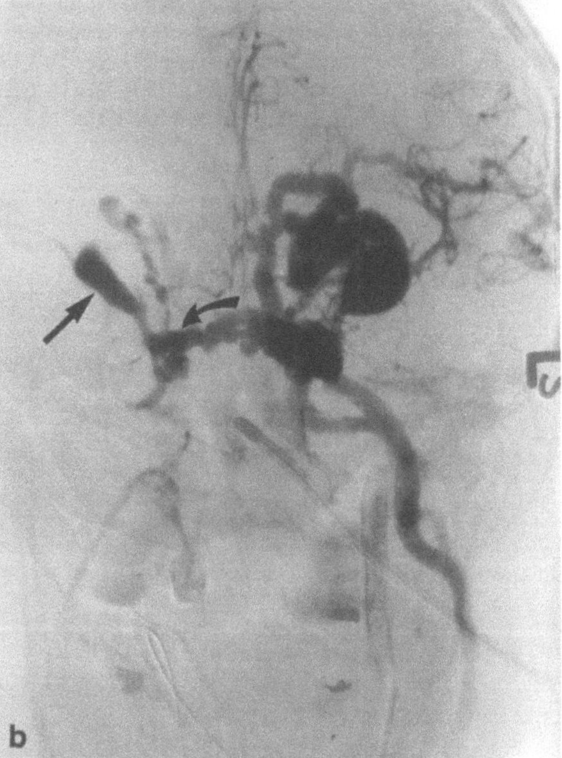

Fig. 2.8. a Clinical picture of a patient with signs and symptoms of bilateral eye involvement with a single traumatic carotid cavernous fistula. b Frontal view of the left internal carotid artery angiogram demonstrating bilateral orbital drainage in a purely left sided fistula. Note filling of the coronary sinus (*curved arrow*) and draining to the opposite cavernous sinus and ophthalmic venous system (*arrow*)

Table 2.3. Pre-embolization causes of visual loss in the 95 cases of traumatic carotid cavernous fistula from Table 2.2

Traumatic optic neuropathy	5
Postsurgical optic neuropathy	2
Reversible optic neuropathy (steal?)	4
Venous hypoxic retinopathy	3
Ruptured globe	2
Surgical anophthalmos	1
Central retinal artery occlusion secondary to elevated intraocular pressure	1
Corneal ulcer/perforation	1

ferent pupillary defect, and a generally constricted visual field (less commonly a central or paracentral scotoma can be documented) can occur. The fundus usually appears normal suggesting a retrobulbar process. The optic nerve dysfunction seems to result from superior ophthalmic vein (SOV) compression of the optic nerve or a steal phenomenon arising from the decreased AV blood flow gradient. As mentioned previously, traumatic injury to the globe, the retina and choroid, and the optic nerve are common with head injuries and are the predominant causes of acute visual loss in these patients (Table 2.3). The retinopathy caused by the CCF must be distinguished, if possible, from local retina trauma (see below).

The traumatic injury of the brain, orbit, eye, and cranial nerves is often more devastating than the consequences of the CCF. Depending on the severity of these injuries, they usually require treatment prior to treating the CCF. Obviously, any life-threatening intracranial hematoma or rupture of the globe is given the first priority. Major injuries to the chest, abdomen, spine, and extremities also take precedent.

2.2.2 Red Eye Differential Diagnosis

The differential diagnosis for a patient with a congested orbit and a proptosed red eye must be considered in those cases who develop the CCF, spontaneously or weeks or months following trauma. (1) A dural AVM that involves the cavernous sinus, (2) dysthyroid orbitopathy, (3) orbital cellulitis or inflammatory disease, (4) infiltrating or metastatic neoplasm such as a lymphoma, (5) or an orbital tumor that has bled such as a lymphangioma, can all cause a congested orbit but can be differentiated on the clinical examination and imaging studies.

An enlarged SOV and swelling of the extraocular muscles can be seen on CT or MR imaging in most of these diseases. Diffuse swelling of the extraocular muscles occurs with a dural arteriovenous malformation, a CCF, or myositis. Focal swelling in the muscle that spares the tendon, with a predilection for the medial and inferior recti, is seen with dysthyroid disease. Metastatic or infiltrative neoplasia can cause focal lumps in one or more muscles.

A soft tissue mass or diffuse changes in the tissues of the orbit on CT or MR imaging is suggestive of neoplasia or pseudotumor. Soft tissue abnormalities in the sinuses, adjacent to the orbit, are suggestive of infectious or infiltrating processes. The clinical examination and history almost always distinguishes an AV shunt in the region of the cavernous sinus from all the other disorders. The presence of an orbital bruit in a congested orbit is highly suggestive of a high-flow shunt (see Sect. 2.2.4).

If the proptosis is subtle and there is mild or no oculomotor dysfunction, then various other entities that cause a red eye are often considered. Not infrequently, these particular case are misdiagnosed as having chronic conjunctivitis. The red eye caused by allergic, viral, or bacterial conjunctivitis is typically associated with a discharge. The dilated conjunctival vessels appear darker and do not have loops near the limbus, unlike the bright red limbal loop vessels seen with an AV shunt to the cavernous sinus (see Fig. 3.11). However, the conjunctival congestion caused by a CCF may predispose the patients to developing a secondary bacterial infection with the associated discharge.

The differential diagnosis for the patient with both a red eye and glaucoma includes chronic uveitis with secondary glaucoma or one of the primary glaucomas (as discussed previously). Both entities can have a limbal flush, but neither causes limbal loop vessels or the arterialized bright red vessels (Fig. 2.9). Though some patients with glaucoma, particularly those with the congenital types, may have enlarged conjunctival vessels, these vessels generally have a dark red appearance. When

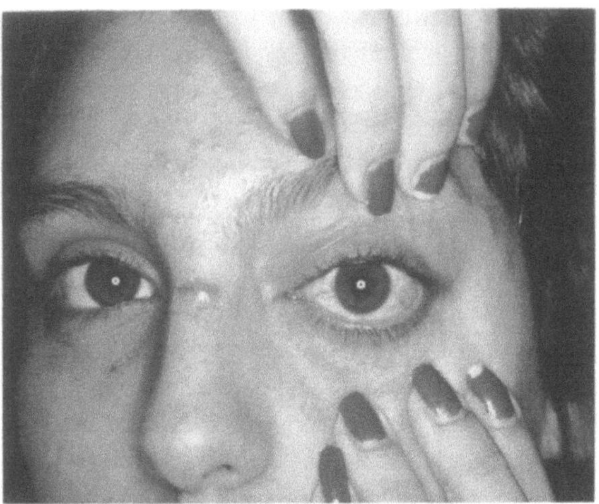

Fig. 2.9. Cases of traumatic carotid cavernous fistulas can demonstrate subtle orbital and red eye signs. In this case, proptosis was only evident with formal measurement and there was no chemosis. Wedge-shaped or single vessel arterialization of the conjunctival vessels is seen. An audible orbital bruit, ocular motor dysfunction, and elevated intraocular pressure were also present

chemosis (Figs. 2.4, 2.5, 2.8) is also present, orbital venous hypertension or an aseptic or infectious orbital inflammatory process must be suspected.

Chronic AV shunts in the region of the cavernous sinus can cause chronic hypoxia and ischemia of the anterior segment. In such patients a mild white blood cellular and flare (protein transudate) reaction in the anterior chamber can be viewed by biomicroscopy. Iris vessel dilation is seen in a minority of AV shunts, but is generally not found with idiopathic uveitis. Though iris neovascularization can develop with profound ischemia, we have never seen this complication in a patient with a CCF. Corneal edema rarely develops.

2.2.3 Venous Ischemic-Hypoxic Retinopathy: Differential Diagnosis

Another important finding in patients with CCFs is venous ischemic-hypoxic retinopathy. It is rarely found without concomitant elevation of the IOP and congestion in the ipsilateral orbit. In fact, the opposite is often found: severe orbital congestion occurs with little or no retinopathy or only mild retinal vein engorgement or disc edema.

All the retinal veins can be slightly or severely dilated and engorged (Fig. 2.7). The retinal hemorrhages of venous retinopathy, from any etiology, are located predominantly in the inner layer of retina, less frequently in the middle, and infrequently in the deeper retinal layers. The hemorrhagic picture may appear similar to a central retinal vein occlusion [17]. Swelling of the optic disc is commonly present when there is a venous retinopathy. Retinal dysfunction results from reduced arterial blood flow into the retina, congestion and dilatation of the retinal veins and the chronic hypoxia and ischemia of the retina. This results in retinal edema which seems to commonly affect the macula. This high predilection may reflect the fact that the macula is the location of the retina where edema can be best seen on ophthalmoscopy. Alternatively, the abnormality of central vision brings the patient to medical attention, while edema in the peripheral retina might not cause a noticeable disturbance in vision.

CCF-related ischemia of the retina also causes cotton wool spots in the nerve fiber layer and rarely neovascularization [18]. A similar picture of venous stasis retinopathy and retinal ischemia may occur with any cause of severe ophthalmic venous outflow compromise or poor ophthalmic artery perfusion pressure. Examples include severe atherosclerotic stenosis or occlusion of the ICA (see Sect. 8.2.5) and partial central vein occlusion (usually associated with more retinal hemorrhages).

The associated direct eye trauma can contribute to the retinal hemorrhages in the periphery or posterior pole, Berlin's edema in the macula, exudates, retinal tears and detachments, and choroidal rupture. The traumatic retinal injury can be seen immediately following the trauma and does not evolve over weeks as occurs with a CCF.

2.2.4 Orbital Bruit Differential Diagnosis

Orbital region bruits that are heard by the patient and the examiner are prevalent with high-flow AV shunts in the cavernous sinus. Bruits are less frequent with the low-flow shunts such as occurs with a dural arteriovenous malformation. Though the patient with a dural low-flow shunt may report hearing a bruit while in a quiet environment or reclining, the examiner usually auscultates nothing. Orbital bruits are uncommonly caused by atherosclerosis in remote sites of cerebrovascular disease, such as from stenosis of the carotid siphon. Less common etiologies for bruits transmitted to the orbit include vertebral fistula or occlusive disease of the subclavian artery, orbital vascular tumors (extremely rare), and hypoplasia or aplasia of the sphenoid wing that permits the intracranial brain pulsations to be heard over the orbit. In the latter cases, the plain skull X-ray and CT reveal the defect in the sphenoid bone responsible for the orbital signs and the patient typically has neurofibromatosis [19–22].

2.3 Laboratory Evaluation

2.3.1 Ultrasonography

Ultrasonography performed in real-time reveals signs of an abnormal hemodynamic process in the orbit. The standard B-scan demonstrates thickening of the choroid, elevation and widening of the optic nerve shadow if papilledema is present, widening of the dark shadows created by the extraocular muscles in the white-looking orbital fat, and a dilated division of the SOV (see Fig. 3.14 in Chap. 3). The abnormal SOV appears as a curved dark sonolucent tubular mass shadow, medial and superior to the optic nerve, located in the region of the superior fat pad. The ophthalmic vein is echo-free because of the lack of acoustic interfaces within the vessel. The A-scan of the SOV shows low internal reflectivity, or small-amplitude echospikes, because of the absence of the interfaces within the vessel. The arterialized blood flow rate in the SOV causes rapid echospikes within the vessel that are seen during the examination when performed in real time [23, 24]. Doppler ultrasonography demonstrates this rapid blood flow more easily than on B-scan [25].

In addition to a CCF, enlargement of the SOV can result from any entity that compromises the venous outflow from the posterior orbit. These include: (1) Orbital venous compromise that develops from a neoplasm-, infection-, or inflammation- related thrombosis in the cavernous sinus. (2) An AV

Table 2.4. Differential diagnosis of enlargement of the superior ophthalmic vein

---

Cavernous sinus obstruction or thrombosis
Arteriovenous shunt of the cavernous sinus
Arteriovenous shunt of the orbit
Orbital varix
Inflammation or neoplasm of posterior orbit
Dysthyroid orbitopathy

---

shunt that directs high-pressure arterial blood flow into the normally low pressure cavernous sinus, effectively reversing the venous blood flow back towards the orbit. 3. An orbital AV malformation, which can mechanically compress the SOV or shunt a high volume of blood into the ophthalmic venous system causing venous hypertension. (4) A varix of the ophthalmic vein associated with thrombosis in the SOV, which can obstruct the blood flow or mechanically compresses the ophthalmic venous outflow. This compression occurs because of either direct pressure by the varix with thrombosis or via a retrobulbar hemorrhage. (5) An inflammatory or neoplastic process in the orbit which, particularly when it is located in the apex or in the superior orbital fissure, can also mechanically compress the SOV. (6) Dysthyroid orbitopathy, which enlarges the extraocular muscles, causing them to compress the SOV in the orbital apex. CT will differentiate most of the above causes of an enlarged SOV from a CCF (Table 2.4).

2.3.2 Computed Tomography

Enlargement of the ophthalmic venous system, most often the superior division of the ophthalmic vein, is seen on both axial and coronal views of the contrast-enhanced CT. The orbital segment of the ophthalmic vein is imaged better than the intracranial precavernous portion of this vessel (Fig. 2.10a). Orbital congestion with soft tissue swelling, particularly of the extraocular muscles and lacrimal gland, is apparent even on noncontrast CT. A high-flow AV shunt will often aneurysmally dilate the involved cavernous sinus (Fig. 2.10b). Convex bulging of the cavernous sinus develops as a result of significant congestion or a

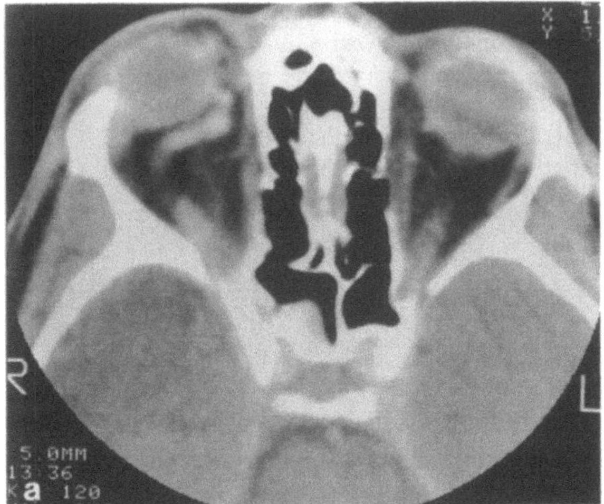

Fig. 2.10a, b. Although computed tomography (CT) is rarely diagnostic of a carotid cavernous fistula, it can demonstrate abnormalities in the venous system (see also Fig. 2.21). **a** Axial contrast enhanced CT demonstrates the marked enlargement of the superior division of the right ophthalmic vein in a characteristic, but nonspecific manner. **b** Axial CT scan after intravenous injection of contrast material demonstrating left cortical venous drainage from a traumatic carotid cavernous fistula

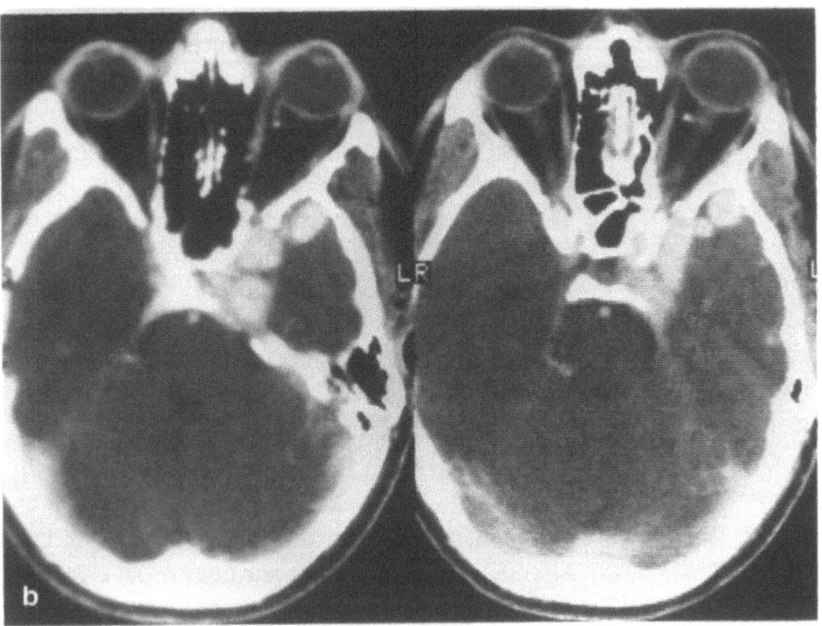

mass in the cavernous sinus. If this change is subtle, it may be apparent only with coronal views of the contrast-enhanced CT. Though exophthalmos can be seen on CT, clinical measurement of proptosis is superior to CT evaluation. CT may be less accurate because the eyes are often not in the primary position when the scan is performed or the CT image slices may visualize the two eyes at different levels, so the comparison is not exact. If cortical venous drainage from the cavernous sinus is prominent, dilatation of one or more cerebral veins may be demonstrated.

In patients with traumatic CCFs, CT often reveals other sequelae of the head injury. Fractures through the base of the skull along the sphenoid sinus in this location or the petrous bone are best demonstrated with bone windows that eliminate the adjacent soft tissues. Paranasal opacification and/or sinus fluid levels are seen particularly in the sphenoid sinus. The CT will also reveal if there are additional intracranial lesions such as epidural or subdural hematomas, an intracerebral hematoma, or a focal contusion of the brain (Fig. 2.11). Because these are potentially life-threatening, the latter problems should always be attended to before initiating a definitive treatment of the CCF. One exception is the CCF associated with a pseudoaneurysm of the ICA that extends into the sphenoid sinus, which is found in approximately 2% of patients with a traumatic CCF. A life-

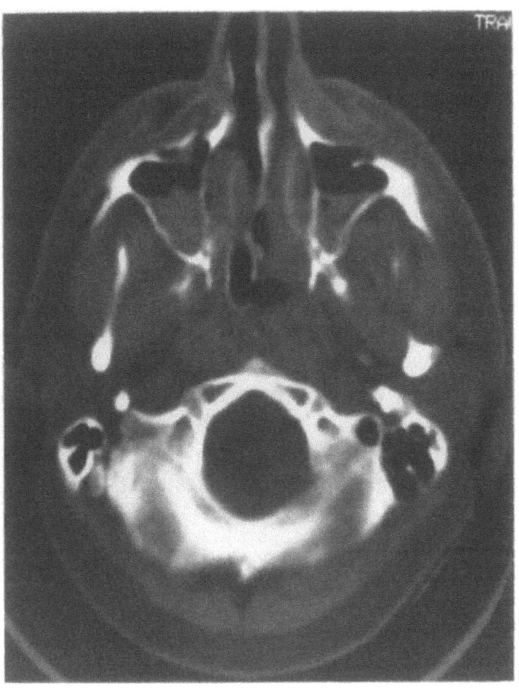

Fig. 2.11. Axial computed tomography with bone windows demonstrates extension of a soft tissue mass from the cavernous sinus to the sphenoid sinus (see also Fig. 2.24)

threatening catastrophic epistaxis can arise if this pseudoaneurysm ruptures. In contrast, facial and orbital fractures are probably best treated after closure of the CCF, so the surgeon avoids an arterialized venous system and the associated soft tissue congestion.

2.3.3 Magnetic Resonance Imaging

MR imaging is not an essential investigation if CT has been performed in a patient with a traumatic CCF. The MR findings are often similar to those seen with CT, but there are important differences. MR imaging may be helpful in patients with a spontaneous CCF caused by an aneurysm (Fig. 2.3). MR imaging can show the high-speed blood flow in the arterialized veins, but the technique to reveal this alteration of flow is not routinely performed. Trauma-induced thrombosis in the cavernous sinus is better seen on MR imaging than on CT (see Sect. 3.2). An MR image with a high-field magnet (1 – 1.5 tesla) is best for revealing these

findings. Unless a program that suppresses the orbital fat is used, the orbital abnormalities, including the soft tissue swelling and congestion of the muscles and the lacrimal gland, may not be apparent. CT is superior to MR imaging for demonstrating bone fractures. MR imaging and MR angiography are the modalities for following any post-treatment complications of the orbital or intracranial blood vessels (see Sect. 2.4.9.2).

2.3.4 Angiography

Selective cerebral angiography remains the diagnostic modality of choice. The previously discussed imaging procedures do not provide the complete analysis of the anatomy and hemodynamics of the CCF necessary for planning therapeutic intervention. A thorough understanding of the variations of the normal vascular anatomy at the base of the skull and the cavernous sinus region is essential in order to proceed with angiographic studies that will identify all involved arteries and the venous drainage of the AV shunt. This must include an analysis of potential collaterals between the external and internal carotid artery systems in the orbit and the cavernous sinus area (Table 2.5; Figs. 2.12 – 2.15; see also Chap. 1).

Table 2.5. Important collaterals of the external and internal carotid arteries in cavernous carotid fistula analysis

---

IMA anastomoses to the inferior and lateral branches of
   ILT (Figs. 2.12 – 2.14)
   Artery of the foramen rotundum
   MMA
   Accessory meningeal artery
Ascending pharyngeal artery
   Superior pharyngeal artery anastomoses to C5 segment of ICA
   Hypoglossal artery anastomoses to medial clival arterial
Capsular artery (Fig. 2.14)
Anterior deep temporal artery (Fig. 2.15)
Recurrent ophthalmic artery to MMA

ICA, internal carotid artery; IMA, internal maxillary artery; ILT, inferior lateral trunk.

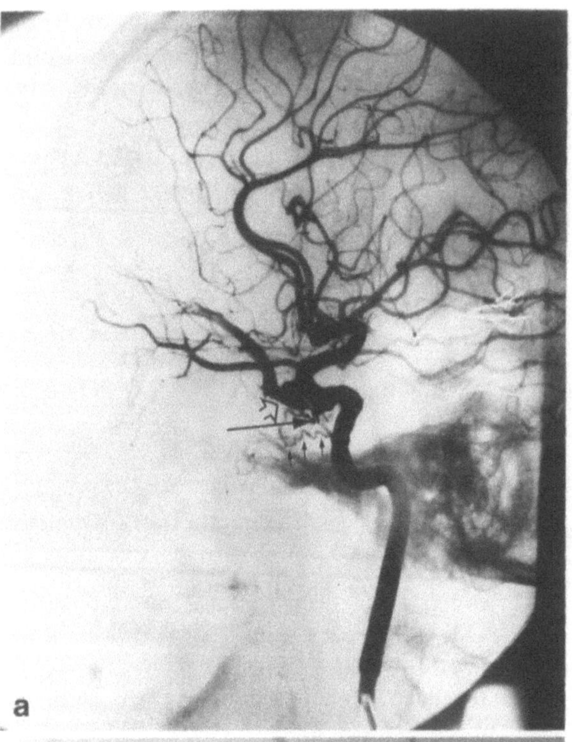

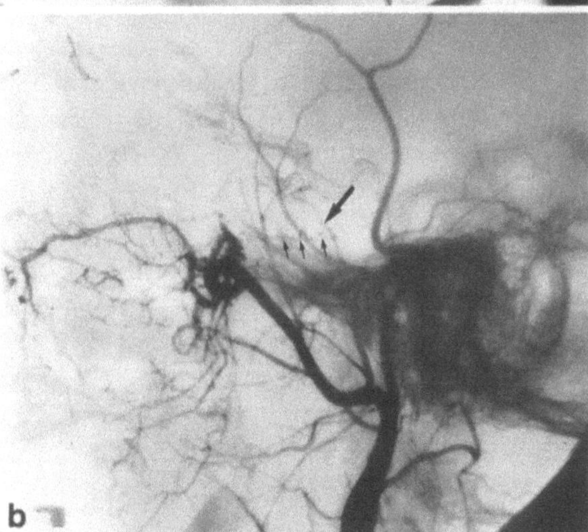

Fig. 2.12a, b. Careful analysis of the angiogram in a case referred with a traumatic carotid cavernous fistula, originally thought to have a pseudoaneurysm of the cavernous internal carotid artery, revealed that there was actually inferior lateral trunk involvement in the fistula (see also Fig. 2.29). a The lateral left internal carotid artery subtraction angiogram demonstrates a traumatic carotid cavernous fistula caused by rupture of the inferior lateral trunk (*long arrow*) that drains into the anterior cavernous sinus (*open arrow*) and ophthalmic venous system. A meningeal branch (*small arrows*) to the inferior lateral trunk is also seen. b The lateral left external carotid artery subtraction angiogram shows the same curved meningeal branch (*small arrows*) involved with an arteriovenous shunt into the cavernous sinus channel (*arrow*). Selective catheterization of the middle meningeal artery (not shown) was used for embolization treatment

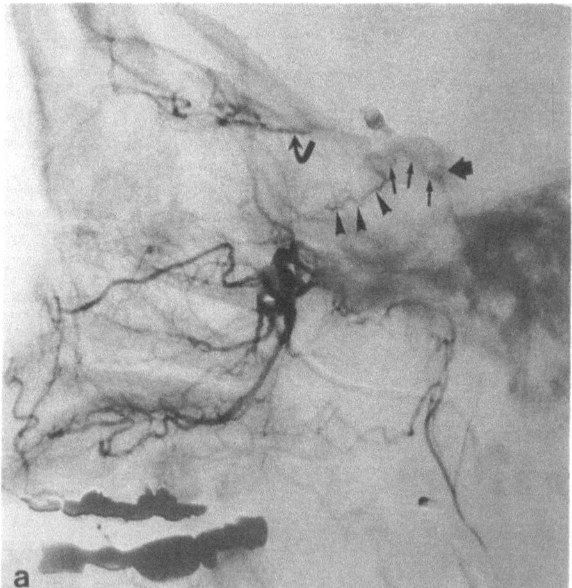

Fig. 2.14a, b. Persistent carotid cavernous fistula after a trapping procedure (corresponds to Fig. 2.5). a Lateral subtraction angiogram of the distal internal maxillary artery demonstrates filling of the internal carotid artery via the artery of the foramen rotundum (*arrowheads*), which reconstitutes the inferior lateral trunk, and the cavernous internal carotid artery (*arrows*), as well as the fistula site (*broad arrow*). Note also filling of the ophthalmic artery

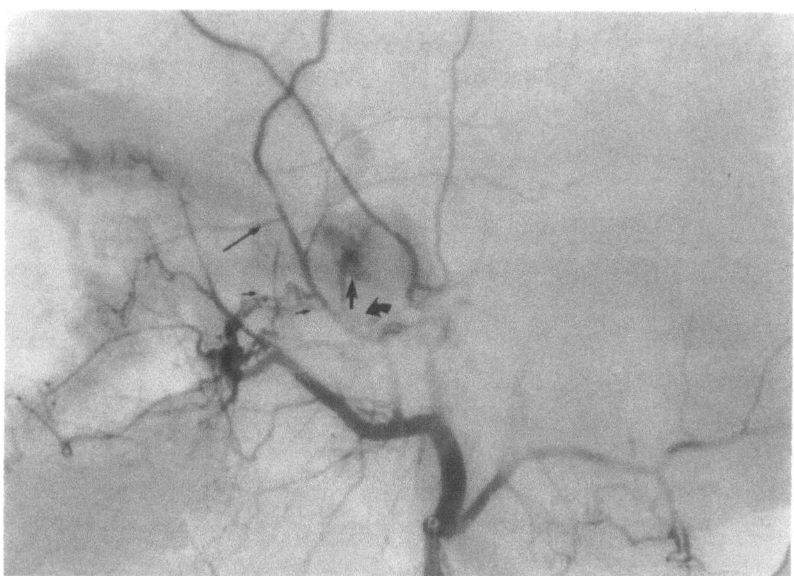

Fig. 2.13. Hypertrophy of collateral circulation in a case of a traumatic carotid cavernous fistula. Lateral subtraction angiogram of the external carotid artery ipsilateral to the fistula. Note the prominence of the artery of the foramen rotundum (*two smallest arrows*) and artery of the foramen spinosum (*curved arrow*) branch of the middle meningeal artery in the inferior lateral trunk (*arrow*) which fills the internal carotid artery and traumatic fistula. Note the important filling of the ophthalmic artery (*long narrow arrow*) from the external carotid artery which, in this case, fills preferentially from the external carotid artery system, as can be seen from the excellent opacification of the ophthalmic artery with only poor opacification of the internal carotid artery

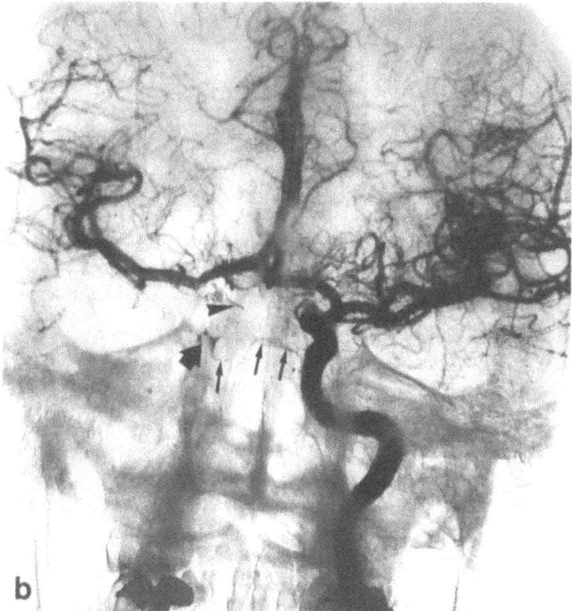

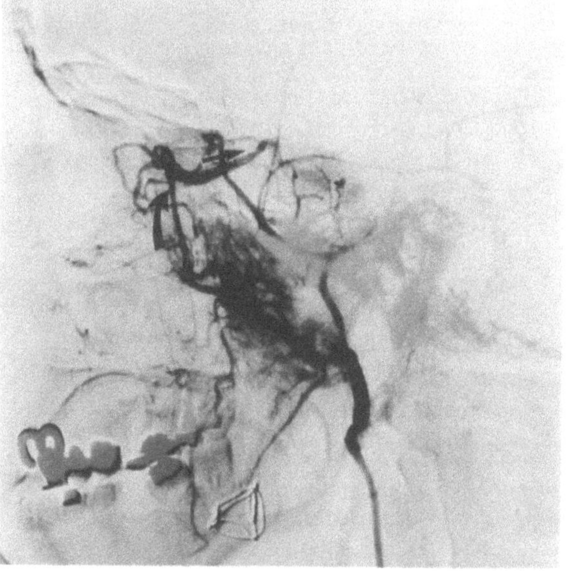

(*curved arrow*). The surgical clip on the supraclinoid internal carotid artery is seen superior to the cavernous internal carotid artery. **b** Frontal subtraction angiogram after injection of the contralateral (left) internal carotid artery demonstrates a transsellar anastomosis (*arrows*) reconstituting the internal carotid artery on the right (*broad arrow*). Note the surgical clip on the supraclinoid carotid artery (*arrowhead*)

Fig. 2.15. An anastomosis from the anterior deep temporal artery to the ophthalmic artery is seen on the lateral subtraction angiogram of the left external carotid artery in a case of traumatic carotid cavernous fistula. Note marked hypertrophy of the anterior deep temporal artery (*curved arrow*) reconstituting the ophthalmic artery (*long arrow*), which is also hypertrophied due to the increased blood flow through this collateral pathway to the carotid cavernous fistula

2.3.4.1 Localization of the Fistula and Involved Arteries

The angiographic localization of the fistula site is accomplished with rapid sequential angiography. If this approach fails, injection of the vertebral artery with manual compression of the affected artery can pinpoint the exact site of the fistula [25a] (Fig. 2.16). A prerequisite for this technique is the presence of a competent posterior communicating artery. As an alternative for evaluation of high-flow AV shunts, a double lumen occlusion balloon catheter is placed in the cervical ICA. Under systemic heparinization, the balloon is inflated and contrast material is injected. Retrograde nonopacified blood flow dilutes the contrast column at the site of the fistula (Fig. 2.17). Usually, a traumatic rupture develops in the C4 or C5 segments of the carotid siphon (the ICA is relatively fixed in this location and susceptible to injury). In patients with spontaneous CCFs, fistulas are located at the cavernous portion of the ICA in 60% of cases. In 40% of patients, fistulas are found at branches of the cavernous ICA, including the inferior lateral trunk or the C5 branches (meningohypophseal trunk) [8].

The anatomical competency of the circle of Willis will also be revealed by angiography. Potential collateral arterial supply to the ICA ipsilateral to the fistula must be identified if the shunt cannot be closed without sacrificing the involved ICA. The anterior communicating artery usually provides the predominant route between both ICAs. Less commonly, the ICAs are connected via a transellar artery that courses anterior to the hypophysis within the sella. This vessel is too small to provide significant collateral blood flow to the brain, but it can reconstitute the ICA and fill the fistula after proximal ligation or a trapping procedure (Fig. 2.14). The posterior communicating artery can provide blood supply to the supraclinoid ICA from the vertebral basilar arterial system.

Selective arteriography of the external carotid artery (ECA) and ICA must be performed in order to determine whether the shunt is principally supplied by ECA or ICA branches, or from both systems. Angiography will demonstrate if the ECA system provides a collateral contribution, via the anterior or middle meningeal arteries (Fig. 2.12b), the artery to the foramen rotundum, or the

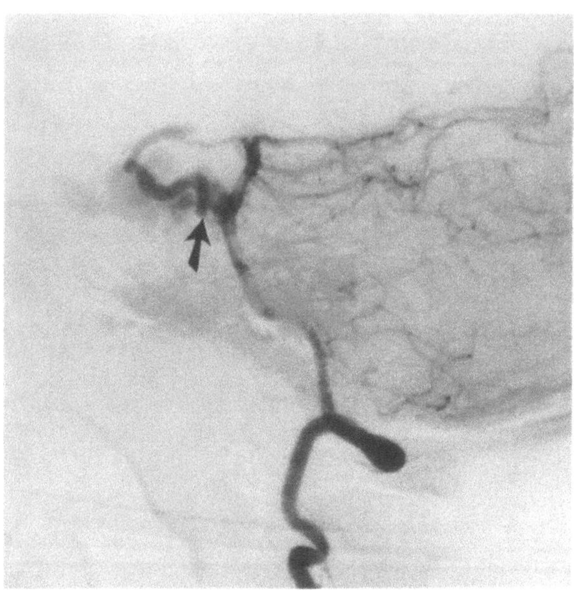

Fig. 2.16. The Huber maneuver localizes the site of the arteriovenous fistula. A left vertebral artery injection while the left internal carotid artery is manually compressed demonstrates filling of the posterior communicating to the left internal carotid artery, thus pinpointing the site of fistulization (*arrow*). For this maneuver to be successful a patent posterior circle of Willis is required

Fig. 2.17. The Berenstein maneuver can pinpoint the site ▶ of an arteriovenous fistula by using a double lumen balloon technique to reduce the flow in the affected internal carotid artery and taking advantage of the preferential flow through the shunt. **a** A lateral subtraction angiogram of a right internal carotid artery demonstrates a carotid cavernous fistula, but the exact site of fistulization cannot be appreciated. Note the symmetric filling of both the superior and inferior divisions of the ophthalmic vein (*curved arrow*) as well as a significant drainage to the sphenoparietal sinus and pial veins (*arrows*). **b** A double lumen balloon catheter with an inflated balloon (*arrow*) has temporarily occluded the internal carotid artery, while contrast is injected pinpointing the exact site of fistulization (*curved arrow*).

neuromeningeal branch of the ascending pharyngeal artery, to the AV shunt. A common carotid arterial injection may only diagnose the presence of an AV shunt, contributes to the misdiagnosis of a pseudoaneurysm (Fig. 2.12a), and fails to identify all the arteries supplying the shunt and the exact site of the AV shunt or hole in the artery.

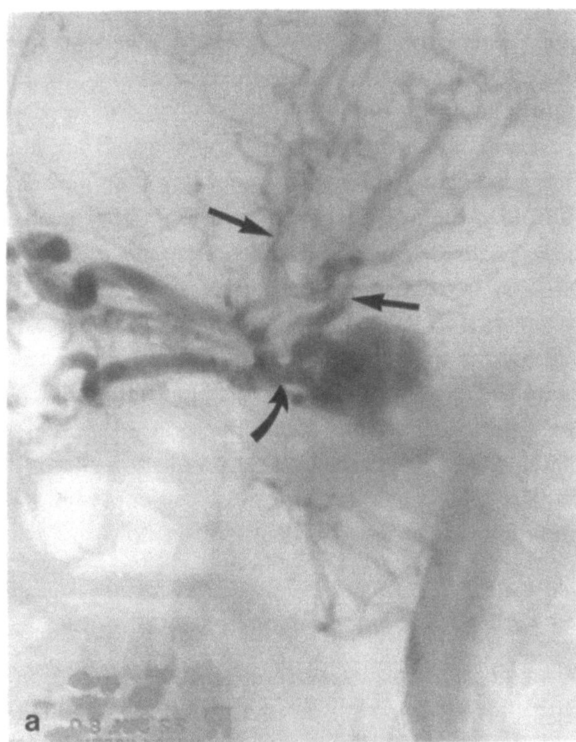

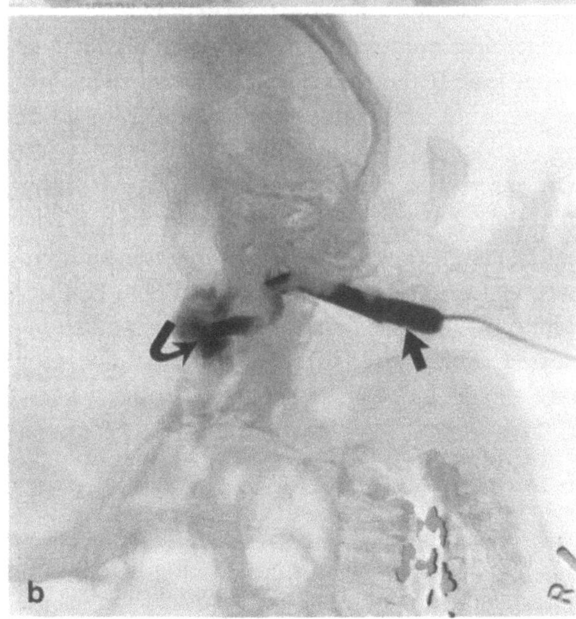

Selective arteriography will also uncover whether a traumatic pseudoaneurysm is present in the cavernous or cervical segment of the ICA ipsilateral to the fistula. Once a pseudoaneurysm is diagnosed, MR imaging may be used to follow the progress of the lesion noninvasively if treatment is not immediately instituted.

The late phase of the angiogram will also reveal the direction of the shunt drainage into and away from the cavernous sinus. The presence of partial thrombosis or aneurysmal dilatation of the cavernous sinus or veins can also be visualized.

Other traumatic AV fistulas involving the ophthalmic (Fig. 2.18), internal carotid, or vertebral arteries may continue to cause symptoms following closure of the CCF. These shunts may only become apparent on angiography after occluding the CCF, because the high flow and marked venous dilatation caused by the CCF may mask the presence of these other lesions. These fistulas are often of minor clinical importance and can close spontaneously, particularly when located in the orbit. Angio-

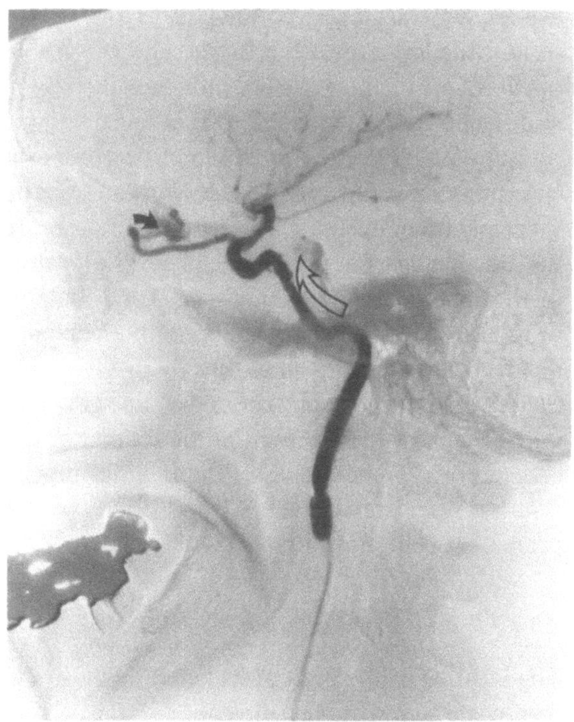

Fig. 2.18. An additional arteriovenous fistula may be present following trauma. After repair of a carotid cavernous fistula with a detachable balloon (*open curved arrow*), the lateral subtraction angiogram of the left internal carotid artery demonstrates significant enlargement of the ophthalmic artery supplying an additional fistula from a posterior ethmoidal artery (*curved arrow*). This fistula was not apparent on the pretreatment angiogram because of the presence of a high flow carotid cavernous fistula. The signs of orbital congestion resolved within weeks of occluding the carotid cavernous fistula and the orbital shunt closed spontaneously

graphy will also uncover the uncommon (< 10% of subjects) anatomical variant of an ophthalmic artery origin from the C4 cavernous segment of the ICA. Because this variation may be associated with a deficiency of collateral circulation to the distal ophthalmic artery, the ICA cannot be occluded at this level without risk of visual loss. (see Fig. 1.12a in Chap. 1). The presence of a cavernous origin of the ophthalmic artery may also account for the visual loss in some patients with thrombosis of the cavernous sinus (see Sect. 3.3.2).

2.3.4.2 Relationship of Clinical Findings to Angiography

The clinical neuro-ophthalmologic signs, bruit, and in some cases, the lack of physical findings, can be explained by angiographic analysis of the arterial inflow, the location of the fistula, and the venous outflow. All of these factors influence the abnormal hemodynamic condition. The predominant route by which the abnormal blood flow exits the cavernous sinus can vary, so the symptoms are diverse and often fluctuate. The nature and degree of dysfunction depends on the extent and the location of thrombosis in the cavernous sinus and draining veins, as well as on the location of the compartment of the cavernous sinus directly involved with the AV shunt. The alteration in the venous system may be iatrogenic or a result of the trauma or disease process. For example, if the internal jugular vein has been ligated, the CCF will drain through prominent collateral facial veins (Fig. 2.5).

2.3.4.2.1 Bruit and Pulsation

The location of the bruit auscultated by the examiner correlates with the direction of the venous drainage of the fistula. Less frequently, the bruit results from transmission through the skull from turbulent blood flow at a distant site. A bruit that is audible over the orbit ipsilateral to the CCF is most common because of the high frequency of anterior drainage towards the ophthalmic vein. When the bruit is audible over the contralateral orbit, it suggests drainage of arterialized blood across the intracavernous sinuses to the contralateral cavernous sinus. When heard over the posterior auricular area or near the jugular vein on either side, the bruit

suggests the cavernous sinus has significant drainage posteriorly. No bruit is heard over the ipsilateral orbit when the anterior cavernous sinus or ophthalmic vein is thrombosed.

Thrills and pulsations of the globe that are either seen or palpated were frequently described in the older literature [16]. More recent reports observed a thrill or pulsation in less than 5% of cases, possibly because of earlier diagnosis and referral for treatment. When present, the pulsations suggest arterialization of the ophthalmic venous system because the cavernous sinus drains predominantly in the anterior direction. More commonly (approximately 20% of patients), pulsation of the upper lid veins or pulsating fluctuations of the applanation tonometry determined IOP are observed.

2.3.4.2.2 Orbital Congestion

A small CCF can cause severe orbital congestion with marked injection, edema, proptosis, swelling and dysfunction of the extraocular muscles, and dangerous elevation of the IOP with a high risk of visual loss. If the shunted blood flows exclusively or primarily in the anterior direction, with all of the arterialized flow into ophthalmic venous system, the venous hypertension in the orbit is maximal (Figs. 2.4, 2.19). This occurs when there is partial or complete thrombosis of the posterior compartment of the cavernous sinus and poor connection to the opposite anterior compartment. The increased orbital arterialized venous flow does not have an adequate outflow pathway for decompres-

Fig. 2.19a–d. Even a small arteriovenous shunt will ▶ cause marked orbital signs if there is exclusive drainage of a carotid cavernous fistula into the ophthalmic venous system (corresponds to Fig. 2.4). a Lateral subtraction angiogram of the left internal carotid artery demonstrates a traumatic carotid cavernous fistula with exclusive anterior drainage (*arrow*). b Plain radiograph of the skull in lateral projection after a radiopaque silicone-filled balloon with a metal clip has been placed on the venous side of the arteriovenous shunt. c Immediate postembolization angiogram of the common carotid artery demonstrates preservation of the internal carotid artery flow and obliteration of the fistula. d Clinical picture 3 weeks after embolization with complete resolution of all signs and symptoms (compare with Fig. 2.4)

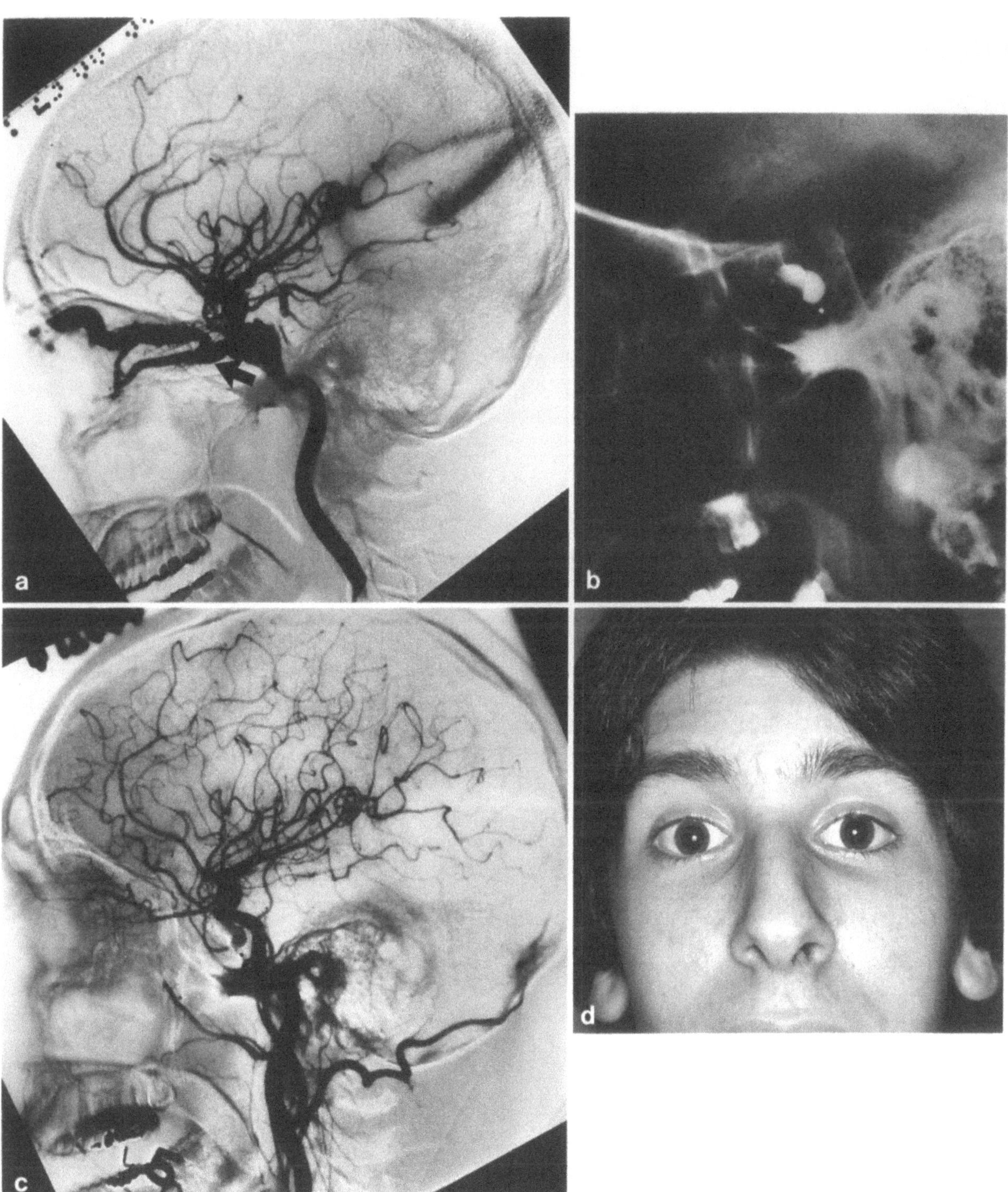

sion. Transmission of the hypertension in the ophthalmic venous system to the vortex veins that drain the choroid causes congestion and edema of the choroid. This contributes to the infrequent development of choroidal effusion and serous retinal and choroidal detachments [26, 27] (see Fig. 3.13a in Chap. 3).

The position of the head in relation to the heart will affect the orbital congestion because of dependent venous stasis. Patients almost always note that their orbits are more swollen with an increased pressure sensation after sleeping with the head down. Sleeping with the head raised to more than 30 degrees allows the effect of gravity to increase the venous outflow from the orbit and reduces orbital congestion.

2.3.4.2.3 Bilateral Signs

The contralateral cavernous sinus frequently receives the arterialized blood via the anterior and posterior aspects of the coronary venous plexus (Figs. 2.8, 2.14). Orbital congestion and cranial neuropathy will develop on the side opposite to the AV shunt if the arterialized blood engorges and causes significant venous hypertension in the contralateral cavernous sinus and ophthalmic venous system. Bilateral signs of varying severity are found with a unilateral fistula in approximately 20% of cases. Bilateral traumatic fistulas occur in less than 1% of cases. The orbital findings are worse on the side ipsilateral to the fistula unless thrombosis in the ipsilateral anterior cavernous sinus results in the shunting of most of the arterialized blood into the opposite cavernous sinus and ophthalmic venous system.

2.3.4.2.4 Cranial Nerve Dysfunction

The patient may have few or no findings if the AV shunt involves the posterior portion of the cavernous sinus, if thrombosis of the anterior cavernous sinus has occurred, or if incompetent communication with the other areas of the cavernous sinus prevents pathological flow into the ophthalmic venous system. When the cavernous sinus principally drains via the petrosal circulation only a sixth nerve paresis and a retroauricular bruit may be present (Fig. 2.6). The sixth nerve paresis may result from

compromise of its blood supply from the cavernous branches of the ICA. A "steal" of the arterial blood through the shunt or compression of the branch artery or the vasovasorum of the nerve may cause ischemia of the cranial nerve. Cranial nerve dysfunction can also arise from venous congestion or thrombosis which directly compresses the nerve in the dural wall of the cavernous sinus (see Sect. 1.5.3.2). Similar mechanisms may account for the third, fourth, and sixth nerve pareses and the rare sensory disturbance of the first division of the fifth nerve found in patients with a CCF in either the anterior or posterior compartment of the cavernous sinus.

Traumatic cranial neuropathies occur immediately after the injury. In contrast, a delayed onset and progressive cranial nerve deficits indicate that the CCF is the cause, rather than a traumatic injury. Distinguishing a traumatic neuropathy is important as these injuries have a poor prognosis for recovery, particularly when the palsy is complete. A hemorrhage into or a shearing of the cranial nerves may develop because the nerves are fixed as they enter the cavernous sinus and are susceptible to trauma.

2.3.4.2.5 Secondary Glaucoma

Elevation of the episcleral venous pressure contributes to the raised IOP and glaucoma in patients with a CCF [28] (Fig. 2.20). The episcleral venous pressure can be determined by methods that deliver external pressure to the globe in an amount required to just collapse the episcleral veins [29–31]. Assuming that the vessel wall has no inherent tension, the intraluminal pressure of the episcleral vessels is equal to the applied pressure when the veins change in caliber or indent [32].

The relationship between the intraocular pressure p and the episcleral venous pressure p_v is revealed by the Goldmann formula for aqueous humor dynamics. Where f is the aqueous flow and r is the resistance for aqueous outflow, the equation is: $p = f \times r + p_v$. Unfortunately, the measurement (range 6.3–10.6 mmHg) fluctuates widely with the method used and the position of the head in relation to the heart. In an extreme example, inverting the body significantly elevates the episcleral venous pressure as well as raising the IOP [33]. If the aque-

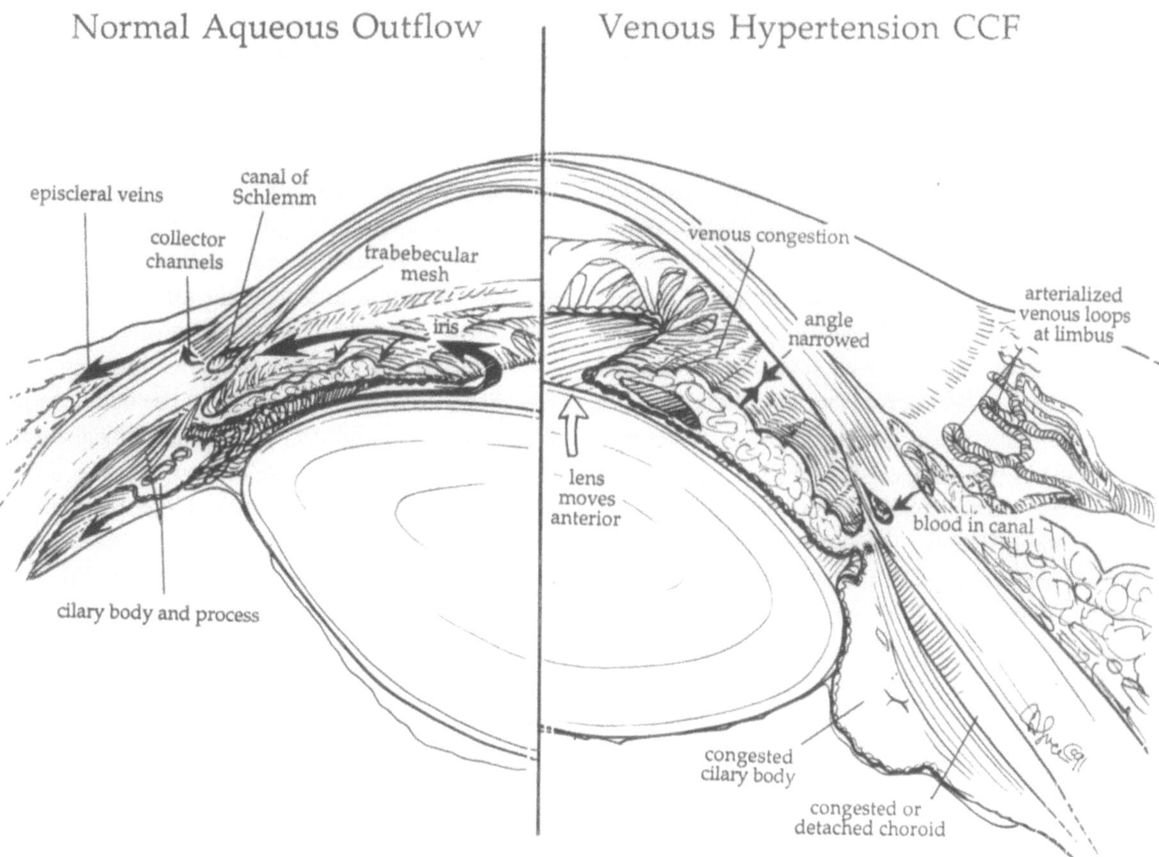

Fig. 2.20. Several mechanisms account for ophthalmic venous hypertension-induced elevation of the intraocular pressure. The normal rate of aqueous drainage through Schlemm's canal Is slowed when the episcleral venous pressure is raised. This reduces aqueous outflow from the eye, raises intraocular pressure, and forces blood into Schlemm's canal. The aqueous outflow is compromised further in cases with choroidal congestion or effusion that bows the lens–iris diaphragm forward and narrows the inlet to the trabecular meshwork

ous production remains constant, elevation of the episcleral venous pressure decreases the aqueous outflow through the trabecular meshwork and Schlemm's canal. This causes a secondary elevation of the IOP.

In 95% of our patients, all of whom had some degree of anterior drainage from the CCF, the IOP was elevated from 4 mmHg to 40 mmHg above normal in the eye ipsilateral to the shunt. Additionally, the venous congestion of the iris, ciliary muscle, and ciliary body and, when present, choroidal effusion (clinically apparent in approximately 1% – 2% of cases) all contribute to shallowing of the anterior chamber. This causes narrowing of the inlet to the trabecular meshwork, which increases the resistance to aqueous outflow and further elevates the IOP [34]. Another mechanism that may contribute to the development of glaucomatous damage re-

sults from the collapse of the intraocular veins secondary to the increased ophthalmic venous pressure, which slows the intraocular blood flow [35]. This further narrows the abnormal arterial–venous oxygen gradient to the eye, predisposing to ischemic-hypoxic damage of the ganglion cell layer of the retina and increasing the probability and severity of visual field loss. Rarely, the IOP rises above the intraluminal pressure in the central retinal artery and an acute occlusion of the central retinal artery develops [5].

2.3.4.2.6 Ischemic-Hypoxic Disease

The CCF increases the ophthalmic venous pressure and shunts blood away from the normal arteries directly into the veins, bypassing the capillaries in the eye and orbit. The decreased arterial–venous blood and oxygen gradient to the ocular tissues [36], including the retina and choroid, anterior segment, eye muscles, pupillary sphincter, and possibly the intracranial optic nerve, is a major mechanism for the dysfunction in these tissues. In patients with CCFs, decreased ophthalmic artery perfusion pressure and increased ophthalmic venous pressure account for the venous stasis retinopathy, which varies in severity. In approximately 70% of cases with retinopathy, funduscopy demonstrates only dilatation of the retinal veins. Other patients will also have several nerve fiber layer hemorrhages. Still fewer patients (in less than 10% of cases with retinopathy) will develop a fulminant retinopathy with severe venous engorgement and multiple large retinal hemorrhages. The slow arterial flow and hypoxia and ischemia of the retinal tissue and capillaries cause edema and increased retinal dysfunction. Rarely, the retinal ischemia caused by a chronic CCF is severe enough to result in proliferative neovascularization of the retina [18]. One such case with a CCF was successfully treated with panretinal laser photocoagulation, the treatment used for neovascular diabetic retinopathy. However, laser therapy should be considered only in patients where direct treatment of the closure of the shunt is unsuccessful (see Sect. 2.4).

Fluorescein angiography demonstrates a delay in the filling of both the retinal and choroidal arteries, confirming the presence of slow arterial inflow into the eye and stasis in the retinal arterioles. Areas of capillary nonperfusion and other areas with capillary dilatation and microaneurysms (see Chaps. 3, 8) develop in cases with a chronic CCF [17]. Ophthalmodynamometry-determined systolic pressure is lower in the ipsilateral eye than in the eye opposite the CCF, providing indirect evidence of the reduced ophthalmic artery perfusion pressure.

Less than 10% of cases have been reported to have signs of anterior segment hypoxia and ischemia [37]. Increased permeability of both the iris and ciliary vessels results from the hypoxia. Biomicroscopy of the anterior chamber in these patients reveals leukocytes and proteinaceous flare, both of which are usually mild. Corneal edema or striae are rare, and hypoxia-induced anesthesia of the cornea has not been observed in any our 125 cases. Dilatation of the iris radial and circumferen-

tial arteries also causes extravasation of dye from these vessels into the anterior chamber following the intravenous injection of fluorescein. The ischemia of the iris muscles can account for pupillary mid-dilation and sphincter paralysis [17]. True neovascularization of the iris that leads to absolute glaucoma is rare [38,39], occurring in none of our cases. A cataract can develop if the anterior segment ischemia and hypoxia are severe, but we have never observed this complication.

Ischemia and hypoxia of the extraocular muscles may also account for some of the ocular motor dysfunction, in addition to the mechanical limitation from congestion of the eye muscles (Fig. 2.21a) [40] or the cranial neuropathy. Once the shunt is closed, restoring the normal orbital arterial flow, the eye movements often recover more rapidly than would be expected from resolution of the orbital congestion or from recovery of a compressive or ischemic cranial neuropathy. Also, there are cases with total ophthalmoplegia, normal pupillary function, a normal corneal reflex, and normal sensation of the cornea, face, and forehead. This suggests that in these patients the cranial nerves are intact and the cause of the eye movement limitation is a local ocular muscle dysfunction.

2.3.4.2.7 Optic Neuropathy

The optic neuropathy associated with a CCF is retrobulbar and often subtle, with relative sparing of the Snellen acuity (20/30 or better), mild to moderate dyschromatopsia, a mild afferent pupillary defect, and generally only a mild constriction of the visual field. A venous infarct or compression of the optic nerve in the orbit or intracranially (Fig. 2.21b) or an arterial "steal" of the intracanalicular or intracranial blood supply to the optic nerve are possible causes. A severe loss of vision rarely occurs. The anterior optic nerve seems to be spared by the CCF, but we have had one patient with an anterior ischemic optic neuropathy (see Sect. 5.1) associated with a dural arteriovenous malformation involving the cavernous sinus (see Sect. 3.2.7).

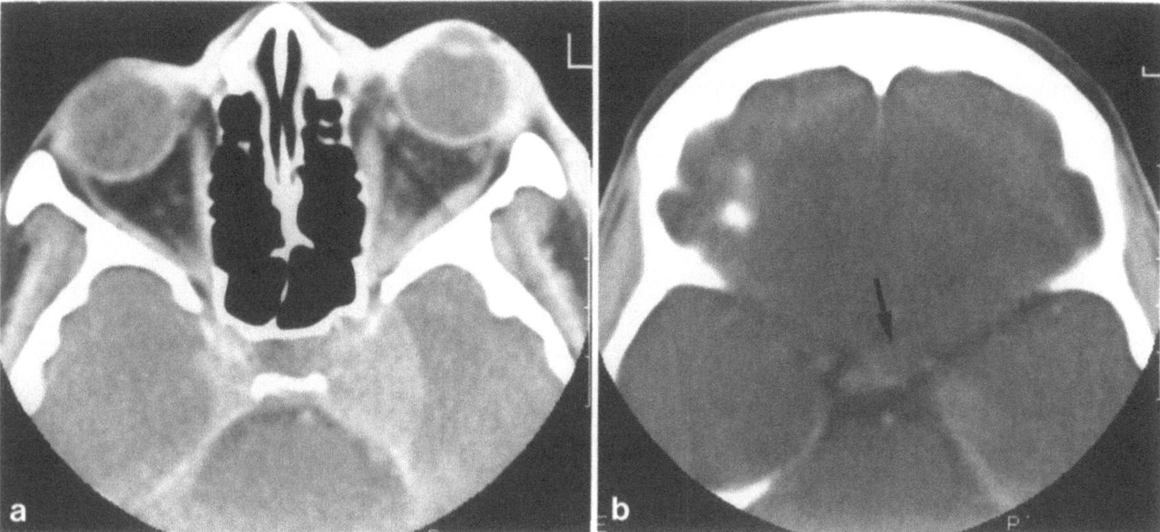

Fig. 2.21a, b. Axial computed tomogram in a traumatic carotid cavernous fistula with a compression of the left intracranial optic nerve from the venous mass in the suprasellar region. The patient had 20/200 acuity, a central scotoma, a relative afferent pupillary defect in the left eye, and a slight superior temporal field depression in the right eye with both eyes normal on ophthalmoscopy.

a Note the engorgement of the extraocular muscles and orbital fat in the left orbit, bulging of the left cavernous sinus to the superior orbital fissure. **b** A higher slice demonstrates a venous mass (*arrow*) predominantly on the left side of the suprasellar cistern. Within hours of balloon closure of the arteriovenous fistula, the acuity had improved to 20/80 in the left eye

2.3.4.2.8 Cerebral Dysfunction

A CCF can drain from the cavernous sinus superiorly (Figs. 2.17a, 2.22a,b) to the sphenoparietal sinus or the deep sylvian veins into the cortical veins. Though these cerebral veins may appear tortuous and dilated at angiography, the vessels infrequently rupture and cause an intracerebral or a subarachnoid bleed [41, 42]. A risk of hemorrhage is probably present only in those cases where all the other routes of outflow from the cavernous sinus are occluded [43] (Fig. 2.23). We have seen this complication in 2 of 150 cases. Following closure of the shunt, arteriography demonstrates restoration of the normal cerebral circulation. The arterialized veins fill in the normal sequence, and within several weeks to months the vessel caliber normalizes (Fig. 2.22c).

2.3.4.2.9 Epistaxis

When the CCF drains inferiorly (rarely the sole drainage), a life-threatening situation may arise. The shunted blood can empty anteroinferiorly from the cavernous sinus via the emissary veins through the base of the skull or through the veins in the sphenoid sinus, ending in the dilated pterygomaxillary venous plexus. If a pseudoaneurysm (found in 1% – 2% of traumatic CCFs) is present in the ICA or a branch artery, rupture of the pseudoaneurysm can cause uncontrollable epistaxis (Fig. 2.11).

2.3.4.2.10 Red Eye

The conjunctiva and sclera develop arterialized and dilated blood vessels and chemosis because of the ophthalmic venous hypertension induced by the arterialized venous system or venous thrombosis, or both. Bright red vessels with limbal loops are

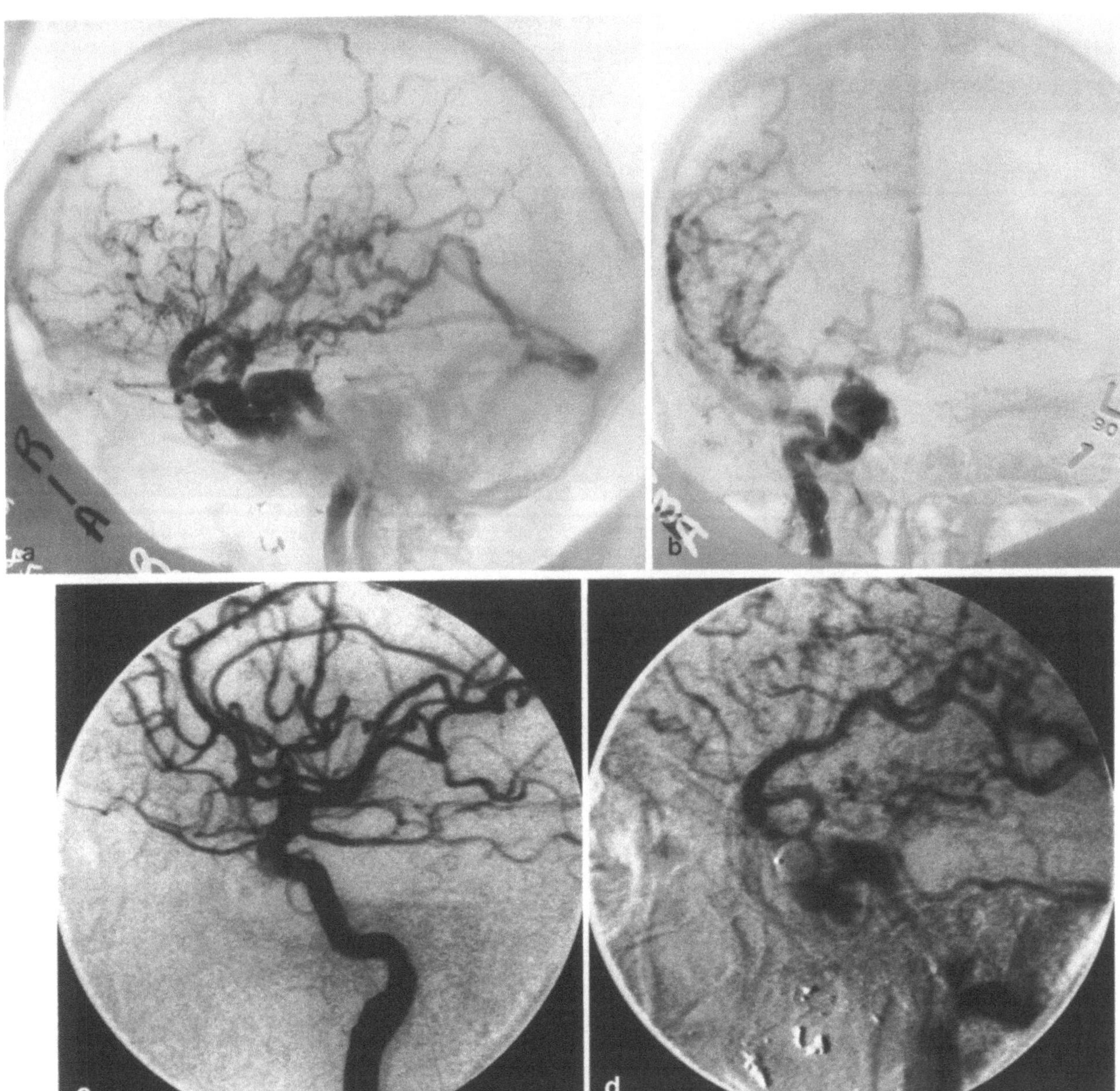

Fig. 2.22 a–d. A complete angiographic steal of the blood flow through the affected internal carotid artery and cortical venous hypertension can occur without cerebral dysfunction, as in this patient with a traumatic carotid cavernous fistula. Subtraction angiograms in the lateral (**a**) and frontal (**b**) views following a right internal carotid artery injection demonstrates that the abnormal venous drainage is almost exclusively towards the sphenoparietal sinus and cortical veins as well as the inferior petrosal sinus. There is little, if any filling of the cerebral arteries distal to the fistula. **c** After balloon occlusion of the fistula note the preservation of the internal carotid artery and the presence of filling of cerebral arteries. **d** A later phase of the same injection shows that the prominent venous structures now drain in the normal direction and with the expected time sequence. The venous hypertensive changes resolved in a follow-up angiogram which was performed several weeks later (not shown)

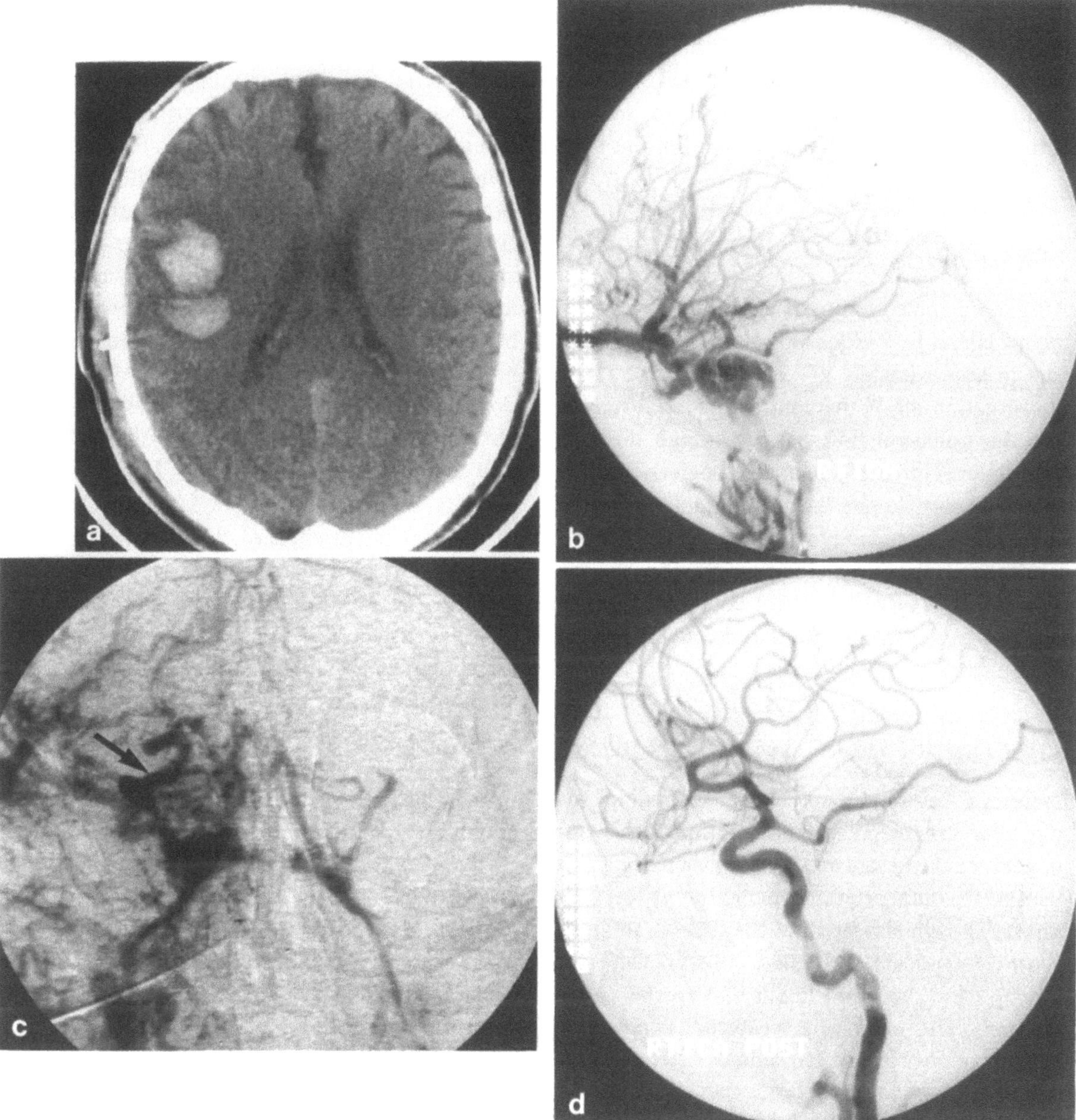

Fig. 2.23 a – d. The cortical venous drainage associated with a carotid cavernous fistula will result in a neurologic deficit only when the abnormal venous drainage causes an intracerebral hemorrhage (occurred in only two of our cases). **a** An axial view computed tomogram demonstrates a spontaneous intracerebral hematoma in a patient complaining of a pulsatile bruit and a red eye following a motor vehicle accident. **b** The lateral digital subtraction angiogram of the right internal carotid artery demonstrates the typical appearance of a traumatic carotid cavernous fistula with abnormal venous drainage into the ophthalmic, pterygomaxillary, and cortical veins. **c** The frontal view of the same angiogram shows the significant cortical venous drainage (*arrow*) with less filling of the ophthalmic and pterygomaxillary veins and the contralateral cavernous sinus. **d** Postembolization angiogram of the internal carotid artery in a lateral projection demonstrates preservation of the internal carotid artery and repair of the fistula

seen when there is arterialized flow in the conjunctiva (Fig. 2.9). If thrombosis occurs in the ophthalmic veins, the vessels are dilated with a darker, less bright red color. In long-standing cases, the same hemodynamic changes cause dilation of the veins in the skin of the face and lids and pulsations in the lid veins (Fig. 2.5).

2.3.4.2.11 Angiographic Cerebral "Steal"

An angiographic picture of a complete "steal" of the ipsilateral ICA blood flow above the fistula occurs in approximately 2% of cases (Fig. 2.22). In this situation all of the contrast material injected into the ipsilateral ICA passes through the high-flow shunt. However, cerebral or cognitive deficits do not develop despite the angiographic suggestion of an abnormal hemodynamic state in the involved hemisphere. Obviously, blood supply from the contralateral ICA or the posterior communicating artery prevents ischemia of the ipsilateral cerebral hemisphere.

2.3.4.3 Special Considerations for Angiography

Each CCF has certain vascular abnormalities that may alter the treatment considerations. The exact location of the fistula and whether there is involvement of the dural arterial collaterals must be determined. Though almost all of the fistulas are in the C4 or C5 segments of the ICA, a fistula can rarely develop from the supraclinoid ICA to the cavernous sinus. The presence of a cavernous carotid aneurysm, a persistent trigeminal artery, or an angiodysplasia can complicate therapy (Figs. 2.2, 2.3). Additional vascular injuries (see below) will raise specific management questions and each has its own unique problems when planning therapy. Collateral arterial circulation to the ipsilateral cerebral hemisphere must also be visualized.

Traumatic fistulas are frequently associated with other vascular injuries, in addition to the soft tissue and bone injuries of orbital, facial, and intracranial trauma. A pseudoaneurysm of the ICA at the site of the fistula usually necessitates a planned closure of the ICA as well as the level AV shunt (see Chap. 6). These false aneurysms are composed of clot and adventitia and do not have a true arterial wall. The pseudoaneurysm often increases in size and ruptures, particularly if the local hemodynamic pressure increases. This can occur following closure of the AV shunt with elimination of the site of "decompression" of the arterial blood flow. Pseudoaneurysms that bulge into the sphenoid sinus (Fig. 2.24) carry a significant risk of arterial epistaxis which can be lethal [44]. Thrombus in the pseudoaneurysm lumen can cause emboli to the intracranial circulation. When a pseudoaneurysm is found in the cervical ICA, it will also frequently, but not always, require ICA occlusion in order to prevent rupture. In each of these patients functional angiography must determine the patency of the circle of Willis and each patient must be tested for his or her capacity to tolerate a planned ICA occlusion (see Sect. 6.2.4.3).

A persistent trigeminal artery, an embryonic vessel that connects the internal carotid and basilar arteries, can be the site of the hole in the ICA. In addition to demonstrating the posterior communicating artery, angiographic evaluation of the vertebrobasilar system is performed to exclude this anomaly. Due to an inherent weakness in the wall of the ICA, a fistula may develop in individuals with a trigeminal artery following trivial trauma [7]. Failure to identify this vascular variation might lead to unsuccessful closure of the AV shunt because treatment will be directed only at the ICA contribution to the fistula.

When a CCF is spontaneous or results from trivial trauma, the patient must be investigated for additional vascular lesions or anatomical variants. For example, a CCF can develop from rupture of an aneurysm of the cavernous carotid artery. Caution should be exercised since these aneurysms often contain thrombus (Figs. 2.4, 2.25a) that can be dislodged into the cerebral circulation during therapy. The site of the hole in the aneurysm wall responsible for the shunt must be identified, as for CCFs of any etiology.

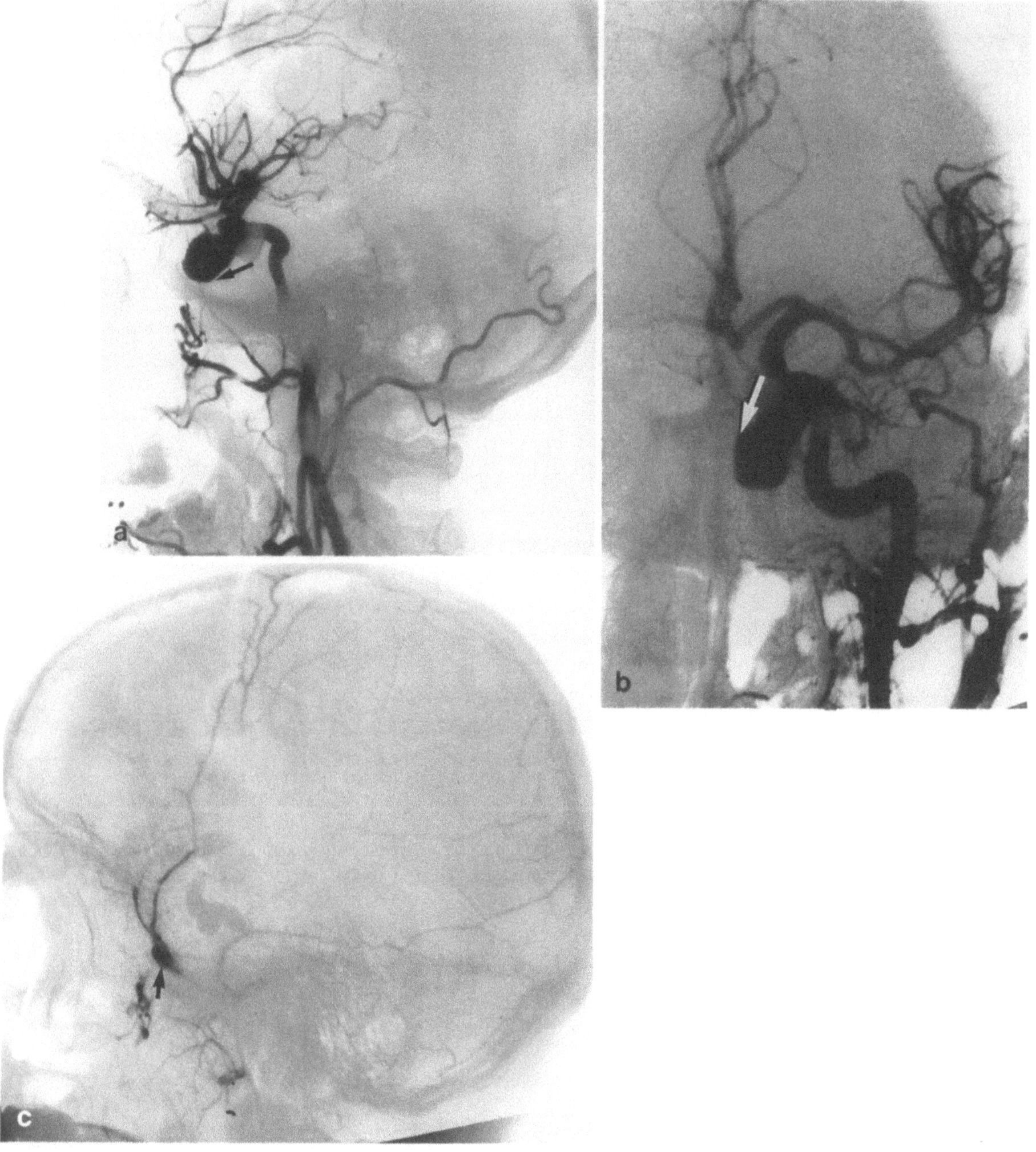

Fig. 2.24a – c. Life-threatening epistaxis can result from rupture of a traumatic pseudoaneurysm of the internal carotid artery (corresponds to Fig. 2.11). This patient had undergone balloon repair of a traumatic carotid cavernous fistula 1 year before. **a,b** Lateral (**a**) and frontal (**b**) subtraction angiograms of the left common carotid artery demonstrate a large post- traumatic aneurysm (*arrows*) of the internal carotid artery projecting anteriorly and medially into the sphenoid sinus. This represents an urgent situation for treatment. **c** In the same patient a lateral subtraction angiogram of the middle meningeal artery on the opposite side demonstrates a post- traumatic pseudoaneurysm of the middle meningeal artery (*arrow*). This lesion also required closure prior to rupture and the development of a subdural hematoma

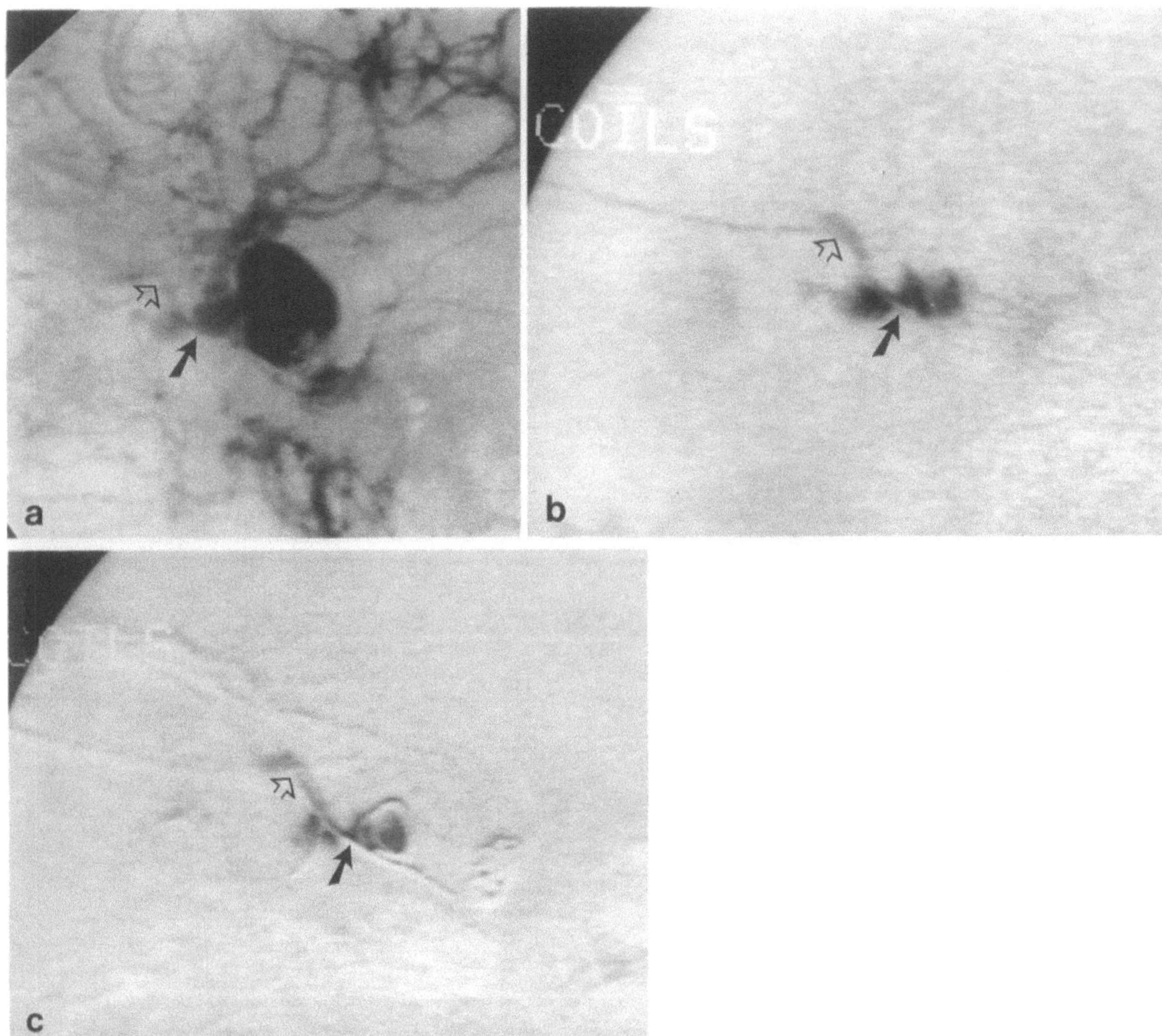

Fig. 2.25a–c. A transorbital approach may be necessary to close a carotid cavernous fistula.

a An 80-year-old woman awoke with a congested painful right eye and orbit. Within 3 days she had no light perception vision and a hypopyon. There was marked clouding and thinning of the cornea, total ophthalmoplegia, and ptosis. A lateral right internal carotid artery angiogram demonstrates a carotid cavernous fistula from a carotid cavernous aneurysm which drained anteriorly (*dark arrow*) into the ophthalmic venous system and into the ophthalmic vein. This superior ophthalmic vein (*open arrow*) appeared to be partially thrombosed. The inferior ophthalmic vein (*below the arrow*) also appeared to be thrombosed.

b After incomplete closure of the carotid cavernous fistula with coils, a transvenous approach through the superi-

or ophthalmic vein was used for embolization. The lateral view shows the long radiopaque catheter in the superior ophthalmic vein (*open arrow*). Following injection of contrast material, there is opacification of the cavernous sinus (*dark arrow*).

c The thin flexible coils were passed through the ophthalmic vein (*open arrow*) and into the cavernous sinus (*dark arrow*). The coils can be seen posteriorly in the cavernous sinus and the controlled angiogram shows a marked reduction in the filling of the cavernous sinus. The control angiogram showed complete closure of the fistula. Unfortunately, the corneal ischemia resulted in perforation of the globe which required enucleation. This case represents a severe ischemic complication of a carotid cavernous fistula

2.4 Therapy

2.4.1 Visual Loss, Cranial Neuropathy

Most traumatic or high-flow CCFs require intervention. In contrast to the low- or slow-flow shunts (see Sect. 3.2.7), which are often associated with partially thrombosed ruptured cavernous aneurysms or dural arteriovenous malformations, spontaneous closure of a high-flow CCF is unlikely. Thrombosis will not occur in a shunt when the hole in the ICA is large and the blood flow through the shunt is rapid. Prior reports that described spontaneous thrombosis of CCFs in 5% – 10% of cases [45] mistakenly included patients with the low-flow type of dural arteriovenous malformation in the parasellar region that drained into the cavernous sinus.

Visual disability persists as long as the CCF remains patent. Visual loss, including complete unilateral blindness, is reported in from 26% [46] to 89% of untreated patients [16]. The causes of visual loss include glaucoma, retinopathy, optic neuropathy, and corneal ulcer from proptosis-related exposure. Patients without severe visual loss frequently have persistent diplopia.

The primary goal of therapy is to occlude the AV shunt in order to restore the normal arterial – venous blood flow and oxygen gradient, predominantly to the orbit, by resolving the venous hypertension and increasing the inflow in the ophthalmic artery. All other treatments are necessarily compromises, and should only be considered if the primary goal is not obtainable. Once the ischemia and hypoxia of the orbit and the compression of the cranial nerves are eliminated, the neuro-ophthalmologic findings will resolve, provided extensive tissue infarction has not occurred. Resolution of the bruit is a secondary, and usually minor goal.

2.4.2 Glaucoma

The management of the elevated IOP or resulting glaucoma is usually conservative and temporary, until the CCF is closed. Though filtration surgery may be beneficial in patients with other causes of glaucoma with elevated episcleral venous pressure (e.g., Sturge – Weber syndrome), intraocular surgical procedures are to be avoided in patients with CCFs. There is an increased risk of choroidal effusion and hemorrhage that results from a surgical procedure that suddenly lowers the IOP in the setting of an arterialized venous system and orbital venous hypertension.

It is uncertain whether topical pharmacological agents alter the episcleral venous pressure or the IOP in these patients. β-Adrenergic blockers do not lower the episcleral venous pressure [47] but the IOP is generally lowered by between 5 and 10 mmHg by topical agents in normal subjects. These agents reduce the aqueous production but do not alter the ocular hypertension induced by altering body position [47a]. Topical epinephrine may [48] or may not [49] lower the episcleral venous pressure in normal eyes. Topical pilocarpine may [50] or may not [49] lower the episcleral venous pressure in addition to affecting other mechanisms that cause a reduction of the IOP. However, in the congested eye pilocarpine can paradoxically raise the IOP by moving the lens – iris diaphragm forward, which causes a narrowing of the inlet to the trabecular meshwork. In an eye with choroidal effusion (see case described in Sect. 3.2.7.1.5), pilocarpine is relatively contraindicated. Oral or intramuscular carbonic anhydrase inhibitors will not lower the episcleral venous pressure in normals and in patients with CCFs, but the IOP is lowered because of decreased aqueous production.

Management of the IOP is individualized for each patient with a CCF. The following are some guidelines for therapy. Patients with a positive family history of glaucoma are usually treated more intensively. In patients with an IOP of more than 24 mmHg but without visual field loss, β-adrenergic topical drops are administered twice daily. Epinephrine-type drops such as dipivefrin are added in those cases where the pressure is greater than 30 mmHg or if the β-adrenergic drops do not significantly lower the IOP. If the IOP remains elevated, topical pilocarpine 4%, every 6 h, is added to the regimen. Unfortunately, the pilocarpine-induced blurring associated with the small pupil, the alteration of accommodation, and the problem in examining the fundus make it difficult to evaluate the patient with this dynamic pathology. Also, pilocarpine is not suggested in patients with choroidal detachment or effusion or where the lens – iris diaphragm is forward. If the IOP remains above 30 mmHg, an oral carbonic anhydrase inhibitor (acetazolamide or methazolamide) is prescribed in divided doses.

Progressive elevation of the IOP or the development of visual field loss necessitate earlier intervention to close the shunt. If the fistula cannot be occluded (unlikely with the various transarterial or transvenous approaches), laser trabeculoplasty or iridotomy (particularly in eyes with a narrow angle) can stabilize or lower the IOP, but topical and systemic pharmacological agents will continue to be required.

For example, one patient with a spontaneous CCF who had bilateral proptosis, uncontrolled glaucoma with a shallow anterior chamber, narrow iris–trabecular meshwork angles, and forward bowing of the lens–iris diaphragm could not be embolized. The patient could not tolerate the occlusion of the ICA which would have been required in order to close her specific fistula. She required timoptic 0.5%, pilocarpine 4%, acetazolamide, and laser peripheral iridotomies in order to stabilize her glaucoma. However, her acuity remained at 20/200 in each eye because of cataracts. Intraocular surgery to remove the cataracts has been avoided because of the risk of a choroidal hemorrhage [51].

Since none of the glaucoma therapies actually relieve the pathological mechanisms that cause visual loss, the benefits are marginal. This is especially true for those cases with severe venous hypertension and a markedly reduced orbital arterial–venous gradient.

When the shunt is closed the IOP is significantly lowered within 48 h. Normalization of the IOP will occur even if there is an increase in the proptosis and chemosis as a result of pooling of the contrast material in the orbital tissues after angiography and embolization.

2.4.3 Corneal Abnormalities

Management of the exposure of the conjunctiva and cornea is complicated when traumatic motor seventh nerve and sensory fifth nerve pareses are present. Orbicularis oculi weakness prevents eyelid closure and adequate distribution of the tears across the cornea. Corneal anesthesia allows a foreign body to remain on the cornea, damaging the corneal epithelium. Deficient innervation to the cornea also causes a loss of trophic influence on the corneal epithelium, with recurrent erosions and poor reepthelialization. A procedure such as lateral or total tarsorrhaphy can protect the cornea from a foreign body, lubricate the cornea, and promote normal growth and adhesiveness of the corneal epithelium. These patients are not candidates for a bandage contact lens. Though prolapse of the conjunctiva also interferes with the movement of the tears, the conjunctiva usually normalizes within weeks after fistula closure. Frequent topical lubrication with drops and ointment is necessary. Resection of prolapsed, necrotic conjunctiva is rarely required.

2.4.4 Manual Compression of the Internal Carotid Artery

Manual compression of the involved ICA is an alternative therapy which can be used in cases with low-flow shunts who have few minor symptoms and signs [7,52]. After the patient feels the carotid pulse of the affected ICA in the neck, the *contralateral* hand is used to compress the cervicle ICA. The arterial perfusion pressure distal to the compression site is reduced and the blood flow through the fistula is slowed. If the perfusion pressure in the cerebral arteries becomes too low, a paresis will develop in the contralateral hand, releasing the carotid compression site. In order to prevent a cerebral infarction, carotid compression is never performed with the ipsilateral hand or by a person other than the patient (see Sect. 3.2.7.3.3.1).

2.4.5 Emergency Treatment

Emergency embolization to close the CCF or treat the associated vascular injury is necessary in less than 20% of cases. Visual loss from severe exposure and proptosis, optic neuropathy, glaucoma, or retinopathy generally deteriorates over weeks, but once any degree of visual loss develops the shunt should be closed within 1 week. If the IOP increases rapidly despite antiglaucoma medications, the fistula should be occluded prior to severe elevation of the IOP.

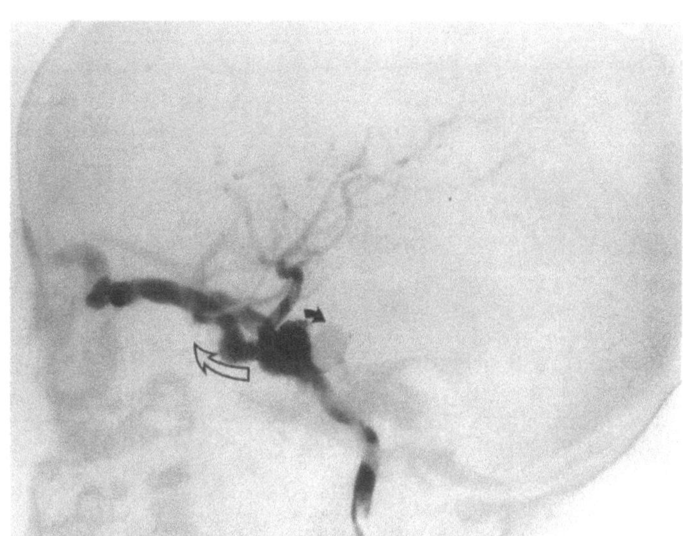

Fig. 2.26. Emergency need for embolization: lateral view of the left internal carotid artery 24 h after successful repair of a carotid cavernous fistula. Acute aggravation of the patient's symptoms was due to posterior migration of the original balloon (*curved arrow*), which diverted all the fistula flow anteriorly towards the ophthalmic venous system. This situation requires immediate occlusion of the fistula, even if this is at the expense of sacrificing the internal carotid artery. If not treated, acute visual loss is likely

As previously discussed, epistaxis from an arterial hemorrhage, often from an associated injury to the internal maxillary artery or a traumatic pseudoaneurysm of the ICA extending into the sphenoid sinus, requires closure of the damaged artery (Fig. 2.24). Occlusion of the fistula can often be accomplished during the same procedure. Ideally, closure of the pseudoaneurysm and the involved artery should be performed before rupture and epistaxis.

If the posterior compartment of the cavernous sinus spontaneously thromboses or if the posterior outflow from the cavernous sinus is iatrogenically blocked by a balloon, the orbital signs can dramatically worsen. All the pathological arterial flow suddenly passes into the ophthalmic venous system, requiring immediate closure of the shunt (Fig. 2.26). However, in this situation the flow-guided catheter may not pass into the anterior compartment of the cavernous sinus because of the hemodynamic change (Sect. 2.4.7). In this situation ICA closure at the level of shunt may be necessary.

2.4.6 Surgery

Surgery is not considered a primary modality in the treatment of a CCF despite that the earliest successful therapy was accomplished by ligation of the common carotid artery [53]. Ligation of the cervical carotid artery is designed to reduce the blood flow through the shunt and precipitate thrombosis of the fistula. However, cervical ligation of either the common carotid artery or the ICA only has a cure rate of approximately 40% [54].

Carotid ligation combined with intracranial trapping of the supraclinoid ICA, above the fistula [1], has also had limited success. The extensive arterial collaterals frequently continue to provide arterial supply to the cavernous carotid artery and the shunt following this procedure (Figs. 2.12, 2.14, 2.27). Dandy first recognized this possibility during a surgical observation. He described backflow into the carotid artery that kept the CCF patent following a trapping procedure [55]. Later, Adson realized that if the ophthalmic artery were not surgically occluded, this vessel could provide the route for this backflow into the trapped portion of the carotid artery. Though the patent ophthalmic artery perpetuated the shunt, it also prevented an iatrogenic cause of acute visual loss [56]. Some authors have improperly concluded that carotid ligation when coupled with shunt closure will cause ischemic neovascular glaucoma or neovascularization of the optic disc and retina [57], but our long-term follow-up of patients with cavernous carotid occlusion for aneurysms in this location did not show any cases of this complication. However, if the shunt remains patent, proximal ligation causes worsening of the ocular hypoxia and ischemia [17], so that proliferative retinopathy, retinal hemorrhages, and visual loss develop following the procedure [58].

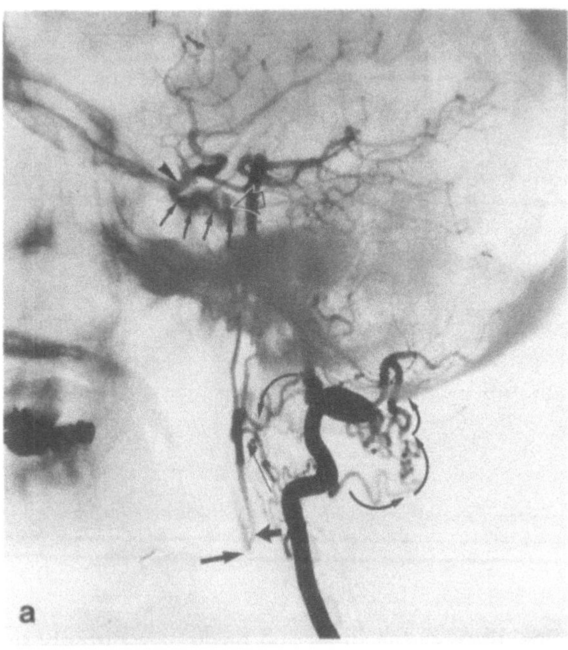

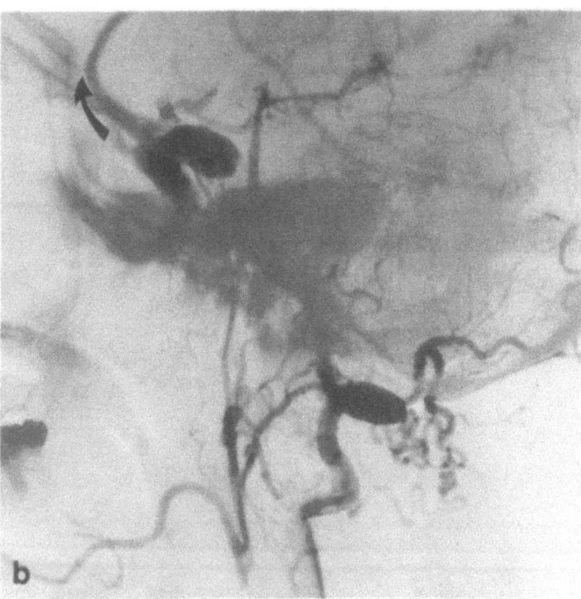

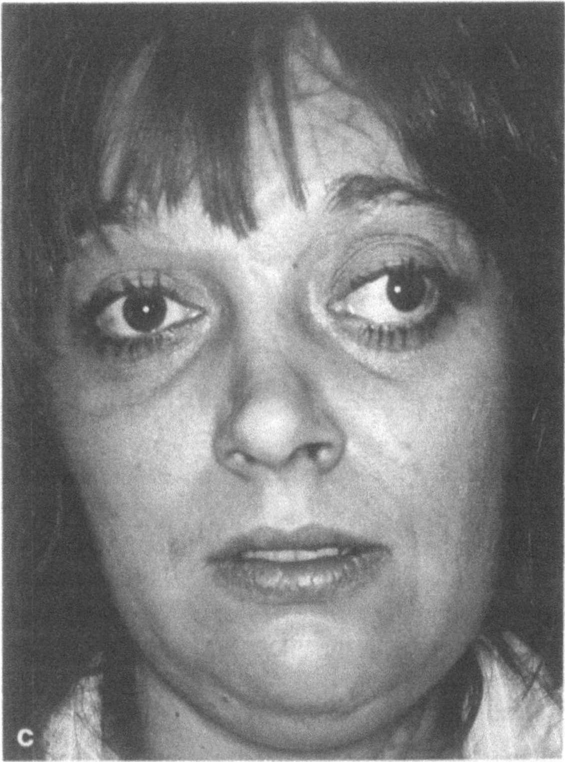

a Lateral subtraction angiogram of the left vertebral artery demonstrates collateral circulation from the C2 cervicle artery to the occipital artery (*curved arrows*). This reconstitutes the ascending pharyngeal artery via the spinomuscular collateral (*very thin straight arrow*). Note the stump of the ascending pharyngeal artery (*broad arrow*) reconstituting the external carotid artery (*arrow*). Also note the filling of the supraclinoid carotid (*arrowhead*) via the posterior communicating artery. The cavernous internal carotid artery (*multiple arrows*) remains patent with filling of the fistula (*open arrow*). Recall that in Fig. 2.14, the left artery of the foramen of rotundum and the contralateral right cavernous internal carotid artery via a transellar artery also reconstituted the involved left cavernous internal carotid artery.

b A later phase of the same injection demonstrates significant thrombosis within the cavernous sinus and drainage superiorly to the sphenoparietal sinus (*curved arrow*).

c The control angiogram (not shown) demonstrated complete occlusion of the fistula and the face and orbital signs resolved (compare to Fig. 2.5). However, the patient was left with poor vision in the left eye and a secondary exotropia

Fig. 2.27a–c. This patient, who was referred with a persistent carotid cavernous fistula and a surgically induced posterior ischemic optic neuropathy after multiple ligations including intracranial trapping, demonstrates why surgical procedures that trap a carotid cavernous fistula or ligate supplying arteries typically fail (corresponds to Figs. 2.5, 2.14).

Prior to the development of percutaneous embolization techniques, a combined approach of ligating the cervical carotid artery, clipping the ophthalmic and supraclinoid carotid arteries, and introducing muscle emboli via the ECA was advocated by Hamby [59]. Currently, surgical interventions involving ligation or trapping procedures have no place in the treatment of CCFs.

In contrast to trapping of the ICA, surgical procedures that directly approach the cavernous sinus can accomplish the goal of shunt closure [60]. One technique is really a surgical embolization of the cavernous sinus. Thrombogenic copper-clad steel wires are passed through the lateral wall of the cavernous sinus, towards the shunt, or in the venous sac that contains the increased blood flow, in order to induce thrombosis [61,62].

In patients who are unable to tolerate ICA occlusion and the artery cannot be preserved, a bypass surgical procedure is required before embolization (see Sect. 6.2.4.3 for example).

2.4.7 Embolization

2.4.7.1 Arterial Route

The first cure of a CCF with embolization was reported by Brooks in 1931. He floated a muscle embolus into the ICA and the fistula to occlude the shunt [63].

Modern treatment of CCFs, with preservation of the ICA, using a detachable balloon was first introduced by Serbenenko in 1974 [64]. Subsequently, numerous series have demonstrated the benefit and safety of this technique [5, 65]. Most endovascular procedures are performed via a percutaneous route through the femoral artery. A route via a direct puncture of the cervical carotid artery is used in a few cases. These include patients in whom catheter placement distal to the occluded or stenotic segment of the ICA cannot be accomplished because of tortuosity or stenosis of the aortic arch or the origin of the common carotid artery, or prior ligation of the common carotid artery.

Once the fistula site is reached, a flow-guided catheter with a detachable balloon is introduced into the shunt from the arterial side. The high flow creates the hemodynamic conditions which carry the partially inflated balloon attached to the catheter into and through the shunt. Once the balloon is on the venous side of the fistula, the balloon is inflated in the cavernous sinus, occluding the shunt without compromising the ICA. The balloon is then detached and the catheter is removed (Fig. 2.19c).

Different types of detachable balloons can be attached to a catheter of variable stiffness or to a flow-guided catheter that uses the hemodynamic advantage of the rapid blood flow into the shunt to overcome the turbulence or tortuous arterial anatomy. The balloon is used to occlude the fistula at the exact site of the shunt. Even a shunt with a large hole can be closed since there are balloons that can be inflated to a size of up to 2 cm or more than one balloon can be used (Fig. 2.28). In the deflated state, the balloon can be introduced through a relatively small percutaneous arterial puncture site via a coaxial catheter system. Compared to other embolic agents, balloons are the most versatile for treatment of a CCF. Balloons can be inflated, deflated, and reinflated in order to obtain the proper positioning in the cavernous sinus. Prior to detachment, a detachable balloon can be removed if necessary.

Balloons are constructed from latex [66] or silicone [67]. In most cases, latex balloons are used because their distensibility and hysteresis permits them to be made into different sizes and shapes. They retain their original size and shape even after multiple inflations and deflations. Currently available silicone balloons have similar characteristics. The balloons can be filled with contrast material that is radiopaque on a plain skull radiograph (Fig. 2.19b). Though the contrast material eventually disappears and the balloon deflates, usually within several weeks, this is an adequate time interval to seal the fistula.

A venous aneurysm or pouch, which may develop at the site of balloon deflation, is often asymptomatic, but a cranial neuropathy with pain can develop (see Sect. 2.4.9.2). Closing the fistula with a balloon filled with silicone [68] or 2-hydroxyethyl-methacrylate (HEMA) [69], another polymerizing agent mixed with contrast, prevents deflation, avoiding a venous pseudoaneurysm. However, a cranial nerve palsy which is present prior to embolization may worsen or fail to recover, because the mass of the silicone-filled balloon compresses

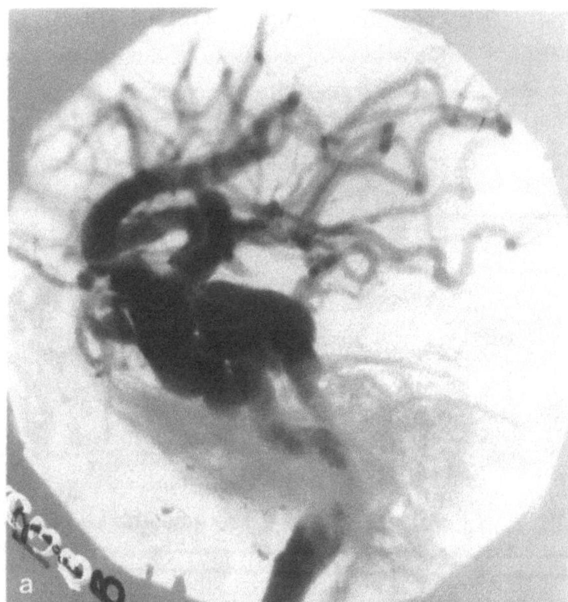

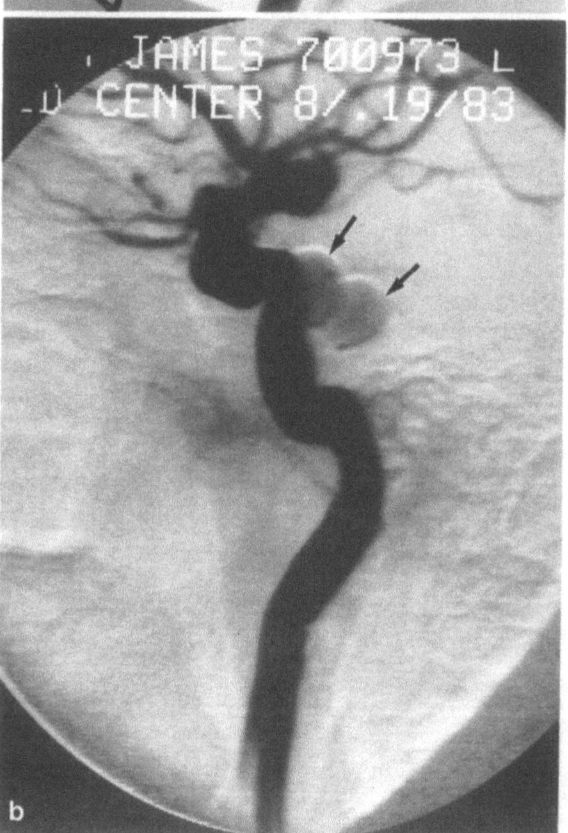

Fig. 2.28a, b. In some cases multiple balloons may be required to prevent movement of the detached balloon into a large size venous side of a carotid cavernous fistula. **a** Lateral subtraction angiogram of a traumatic carotid cavernous fistula in a case of a gunshot wound producing a large laceration of the internal carotid artery. The fistula drains primarily to the sphenoparietal sinus and into posterior superior compartment of the cavernous sinus to the inferior petrosal vein. **b** Two balloons were needed to close this fistula while still maintaining normal flow in the affected internal carotid artery

the cranial nerves in the cavernous sinus [70, 71]. The silicone mass in the inflated balloon may also project into the ICA lumen, compromising the blood flow. Technical advances such as the introduction of a silastic balloon by Hieshima in 1981 [70] have reduced the risk of premature balloon deflation. In addition, detachable metallic coils have been recently used to close these AV shunts (see Chap. 6).

The ECA rarely has an arterial contribution to either a spontaneous or traumatic CCF. However, if a branch of the ECA is involved, it represents a collateral to a branch of the cavernous carotid artery so that a balloon cannot be passed through the shunt. The ECA supply is selectively embolized with particles (polyvinyl alcohol) or a liquid fast polymerizing agent, cyanoacrylate, to induce thrombosis in the arterial supply to the shunt (Fig. 2.29, postembolization of Fig. 2.12). In order to prevent passage of the agent into the ICA circulation through collaterals, the embolizing material is not injected under pressure [72] and a balloon catheter may be used to temporarily close the ICA in cavernous sinus.

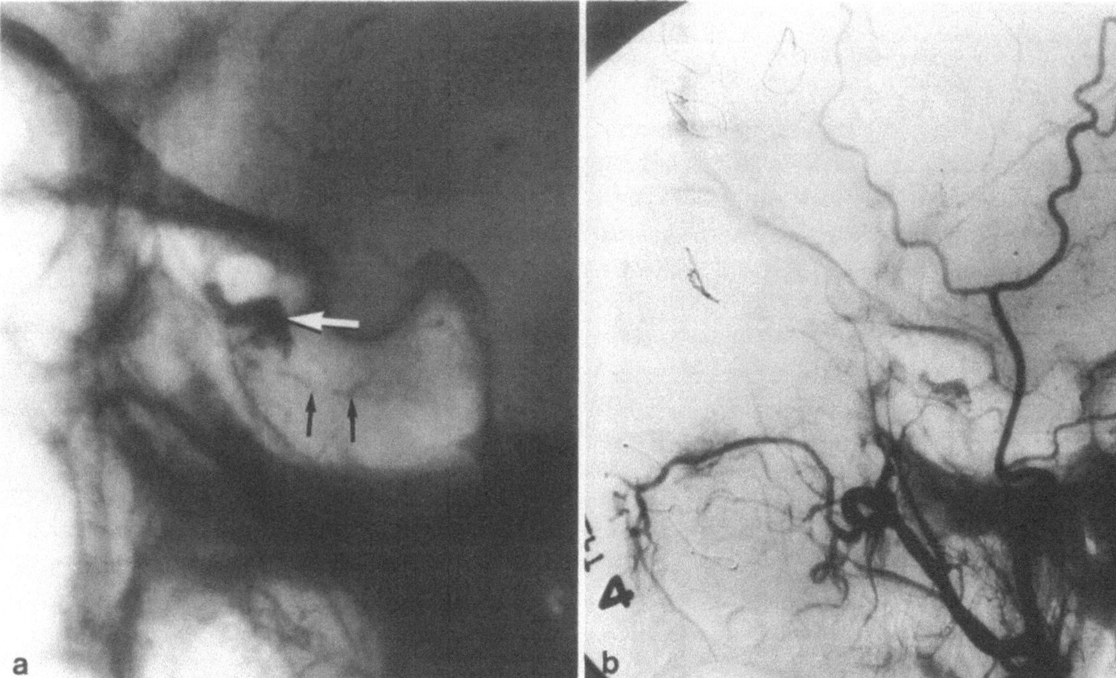

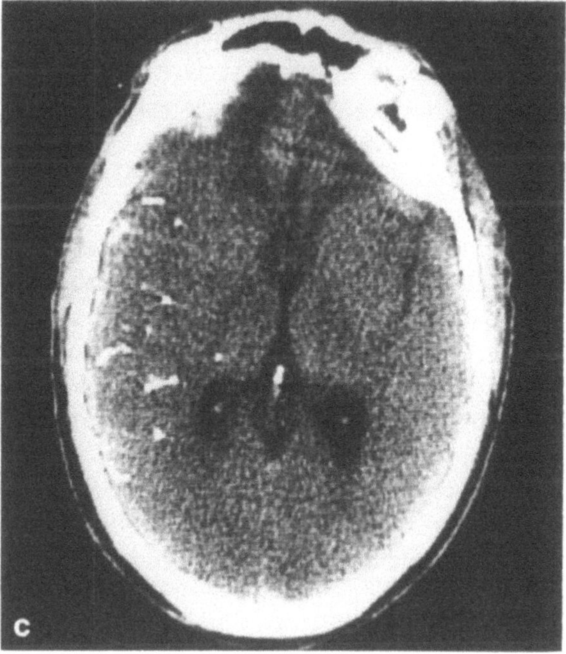

Fig. 2.29 a – c. The external carotid artery may be used to embolize a traumatic carotid cavernous fistula (corresponds to pre-embolization Fig. 2.12). In this case cyanoacrylate was superselectively injected into the middle meningeal artery anastomosis to the inferior lateral trunk, which supplied the fistula.

a The skull radiograph shows the radiopaque acrylic cast in the anterior cavernous sinus (*large arrow*) and the middle meningeal artery (*small arrows*). Small flecks of acrylic are also seen in the middle meningeal and middle cerebral arteries in the area above the sella.

b The postembolization lateral subtraction external carotid angiogram and internal carotid artery (not shown) angiograms showed no arteriovenous shunting. A radiopaque cast in the anterior cavernous sinus is also seen.

c Because the pressure used to embolize, acrylic passed into the internal carotid artery via the inferior lateral trunk. The patient experienced a transient hemiparesis and the axial computed tomogram showed acrylic emboli in the middle cerebral artery territory

2.4.7.2 Venous Route

The venous route for embolization of the cavernous sinus side of a CCF was first used surgically, as discussed earlier. Various agents that induce thrombosis, including balloons, metal coils, cyanoacrylate, or particles such as polyvinyl alcohol, have been injected directly into the cavernous sinus with successful closure of the shunt. If a percutaneous flow- guided catheter cannot be passed via the ICA into the shunt and there is no ECA collateral supply to embolize, a venous route embolization should be considered. The catheter can be placed through the internal jugular vein and the inferior petrosal sinus. The cavernous sinus posterior compartment, and on occasion the anterior compartment, can be reached. Most importantly, if the shunt is located in the anterior compartment and the posterior cavernous sinus drainage is iatrogenically occluded, the orbital symptoms will worsen because of the blocked posterior drainage. Therefore, the exact site of the fistula must be known when using the posterior venous approach. Metal coils may be superior to detachable balloons for this lesion since the coils can be packed to close all of the cavernous sinus if necessary.

In patients with Ehlers – Danlos syndrome, the abnormal vessel wall increases the risk of the arterial approach. The possible complications included hemorrhage, iatrogenic cerebral emboli, and a venous pouch. Thus, a percutaneous venous approach may be safer in these cases. A detachable balloon or coils are passed percutaneously on a stiff catheter via the jugular vein into the cavernous sinus [73] (Fig. 2.30).

The anterior compartment of the cavernous sinus can also be approached by the venous route. The catheter can be passed percutaneously or via a surgically exposed branch of the supraorbital vein through the ophthalmic vein (Fig. 2.25b, c) [74, 75]. This route should be reserved for patients with a chronic CCF. The orbital veins are normally thin walled. In an acute CCF, the massively dilated hypertensive ophthalmic veins may rupture easily when catheterized. In contrast, a chronically dilated vein with arterialized flow will develop a thickened, arterialized wall that decreases the risk of rupture by the catheter.

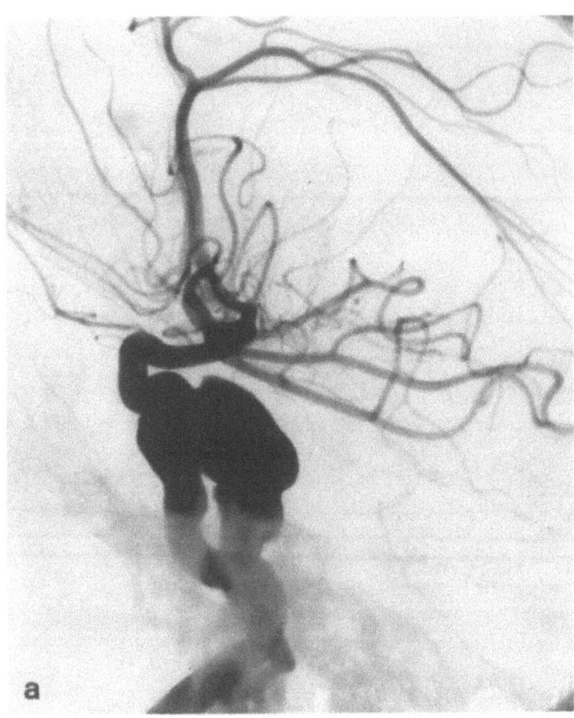

Fig. 2.30a – c. A percutaneous transvenous approach can be used to treat a carotid cavernous fistula when the site of the arteriovenous shunt is in the posterior compartment of the cavernous sinus. **a** Lateral subtraction angiogram of the internal carotid artery showing a fistula draining exclusively into the posterior compartment and downwards filling of the petrosal sinus. **b** Coaxial introduction of a catheter into the jugular vein (*arrow*) and inferior petrosal sinus junction allows the retrograde advancement of a zephyr balloon catheter to the posterior compartment of the cavernous sinus to seal the fistula (*long arrow*). **c** The postembolization internal carotid artery angiogram shows occlusion of the fistula with preservation of the involved artery

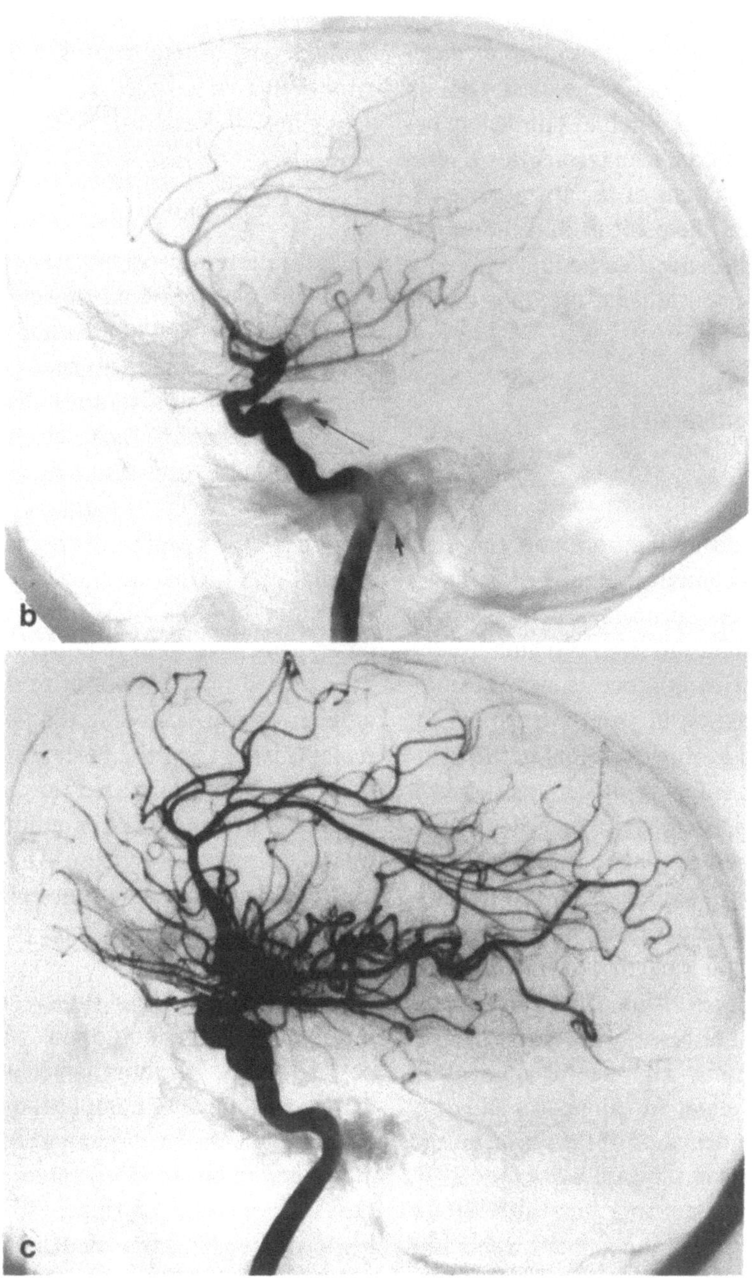

2.4.8 Management of Associated Injuries

As mentioned previously, neurosurgical management of associated epidural, subdural, or intracerebral hematomas or cerebral contusion has the highest priority, before management of a traumatic CCF. However, repair of facial fractures adjacent to or involving the orbit should follow the closure of the CCF, because the arterialized venous system can cause uncontrollable bleeding during surgery.

2.4.9 Results and Complications

2.4.9.1 Surgery

Neurosurgical procedures that occlude the ICA have led to cerebral infarcts and hemorrhages [76–78]. The high rate of neurologic complications may relate to an inability to properly monitor the patient's tolerance of carotid occlusion, as can be done during embolization in an awake patient.

Following carotid ligation proximal to the fistula or an intracranial trapping procedure, a fulminant hypoxic-ischemic retinopathy can develop as early as within 1 month of the procedure [17]. The ocular hypoperfusion is also associated with postural amaurosis. The reduced ophthalmodynamometry on the side of the surgery confirms that decreased ocular arterial perfusion and increased retinal ischemia cause the increase in retinal hemorrhages, microaneurysms, and neovascularization following surgery. Retinal function also fails because local metabolic demands of the retina exceed the amount of oxygen in the capillary blood [17]. Absolute neovascular glaucoma may also follow unsuccessful surgery [15]. Intracranial trapping of the supraclinoid carotid can also cause an optic neuropathy because of damage to the arterial supply (see Sect. 1.5.2.6) to the intracranial optic nerve. As mentioned previously, following trapping surgery, approximately 50% of patients have a recurrence or unsuccessful closure of the fistula because of persistent arterial supply to the shunt via the collaterals to the ICA in the cavernous area (see Sects. 2.3.4.1, 1.5.3.1.2) (Fig. 2.27). Following a carotid occlusive procedure, the reopening of the fistula causes elevation of the venous pressure and a narrowing of the arterial–venous oxygen gradient in

the orbit and ocular tissues that is worse than prior to surgery. Since a CCF is rarely life-threatening and the goal of treatment is to preserve vision, any procedure that has significant risk of causing visual loss should be avoided.

2.4.9.2 Embolization

Embolization procedures have their own set of problems. The retrograde injection of embolic material through the ophthalmic vein or superior petrosal sinus to the cavernous sinus may induce thrombosis in the cavernous sinus with worsening orbital congestion [79]. A detachable balloon placed into the cavernous sinus via a direct puncture or percutaneously via the jugular vein can alter the direction of the fistula outflow to compromise the normal ophthalmic or cortical venous outflow. If the shunt involves the posterior cavernous sinus compartment or drains posteriorly, caution must be exercised that this posterior drainage is not occluded without occluding the fistula. If the shunt remains patent and the posterior drainage is blocked, then all the arterialized flow will be to the anterior compartment of the cavernous sinus and to the ophthalmic venous system. This results in severe and progressive orbital congestion and proptosis, which may require emergency intervention (Fig. 2.26).

Following a subtotal transvenous embolization, if the cavernous sinus drains superiorly into the cortical veins, an intraparenchymal hemorrhage can occur [80]. This complication is rare and none of our patients have demonstrated this problem. Another complication is rupture by the catheter of the inferior petrosal sinus with a subarachnoid hemorrhage [80]. Immediate intervention with surgical exposure and placement of minicoils at the rupture site is one approach to treat this complication. As mentioned earlier, the supraorbital vein and ophthalmic venous system may be used to approach the anterior compartment, but rupture of these vessels with arterialized bleeding that is difficult to control can also occur. Arterial balloon embolization has also caused an intracerebral hemorrhage in a patient with Ehler– Danlos syndrome, possibly because of the increased fragility of these vessels [10].

When the shunt is partially closed by embolization, complete closure of the fistula is likely to occur if the blood flow appears stagnant or slow in the control angiogram. In these cases, the IOP should normalize within 72 h. If the IOP does not fall, it is unlikely that the shunt is closed. For example, one of our cases with a persistent trigeminal artery was not cured by embolization despite markedly slowed flow through the shunt. Her elevated IOP and proptosis remained, prompting an airplane trip for additional consultation. Following the flight, the fistula closed spontaneously. Her examination normalized, and there was no recurrence during the 8-year follow-up period.

With current techniques, premature deflation of the balloon and recurrence of the fistula is seen in less than 5% of cases. In these patients, deflation occurs within the first 48 h after embolization and the CCF requires repeat embolization. The deflated balloon may remain in the cavernous sinus or may pass through the jugular vein to lodge in the lung without side effects.

To date te overall preservation rate of the ICA is approximately 60% – 70% for all embolized CCFs, but with more recent techniques and experience the rate is approaching 90%. A 100% preservation will never be feasible since there are patients with an associated pseudoaneurysm or other vascular lesions that necessitate a planned closure of ICA at the level of thc shunt. Thc embolization protocol should dictate that if the ICA cannot be preserved during the first endovascular procedure, the procedure is aborted. A second attempt at embolization should be performed on another day. This will often lead to preservation of the ICA and closure of the fistula. If the ophthalmic artery arises from the cavernous ICA, occlusion of the carotid artery at this level is contraindicated as it can cause visual loss.

If closure of the ICA is required to cure the fistula or other vascular abnormalities, the patient's ability to tolerate this occlusion must be determined before. Postembolization, the patient is kept flat until the collateral blood supply through the circle of Willis prevents symptoms such as dizziness or cerebral deficits. The head is gradually elevated over several days while the patients neurologic status is monitored. Even when occlusion of the ICA is necessary, embolization is superior to surgical procedures that trap the shunt by closing the ICA

above and below the fistula, because collaterals may still reconstitute the trapped cavernous carotid artery and the fistula following surgery. In contrast, a balloon placed at level of the shunt closes the fistula.

Another potential complication is the premature detachment of the balloon while it is still in the arterial circulation [71, 81, 82, 84]. The balloon migrates into the cerebral circulation, blocking the distal blood flow into the supraclinoid ICA or middle or anterior cerebral arteries to cause a transient ischemic episode or a cerebral infarct [84]. This complication has occurred in less than 1% of balloon-embolized cases. Immediate iatrogenic elevation of the systemic blood pressure, systemic heparinization, and possibly isovolemic hemodilution will diminish the symptoms temporarily by permitting blood to pass the embolus to the ischemic hemisphere. An emergency arteriotomy is then performed by the neurosurgeon for removal of the balloon. One patient with this complication died in Kendal's series [82]. We have had one patient who developed dysphasia and a dense right hemiparesis when the balloon lodged at the bifurcation of the ICA, but it improved on raising the blood pressure. Following a craniotomy for balloon embolectomy, the patient recovered without a deficit (Fig. 2.31).

Migration of cyanoacrylate or particles from the ECA through the inferior lateral trunk into the ICA can also cause cerebral ischemia. We had one such patient who experienced dysphasia and a hemiparesis for 24 h following liquid acrylic injection into a fistula of the inferior lateral trunk. The cyanoacrylate embolized into the branches of the middle cerebral artery (Fig. 2.29c) [85]. This complication can be avoided with a low-pressure injection and using a balloon to temporarily occlude the cavernous ICA. The embolic material should also be injected distal to ECA collaterals to the ophthalmic arterial circulation, otherwise it can occlude the ophthalmic artery and its branches. Liquid acrylic agents should not be injected into fistula via the ICA because of the risk of cerebral embolization [86].

The incidence of a symptomatic venous pouch following delayed deflation of a detachable balloon ranges from 2.4% (in our 125 cases) to 21% [65]. Following partial balloon deflation, a pouch, which causes first division trigeminal pain or a cranial neuropathy or both, may develop on the venous side of the shunt. In symptomatic cases, the venous pouch is closed with a detachable balloon placed into the cavernous sinus around the deflated balloon. Two of our cases with a symptomatic venous pouch spontaneously resolved their sixth or third nerve paresis, respectively, over several weeks. One or our patients who had Ehler–Danlos syndrome and was embolized via a transarterial approach had a third nerve paresis with mild pain that has persisted for at least 4 years. In Kendal's series, one case with a symptomatic pouch was associated with persistence of the fistula and a painful sixth nerve paresis [82]. MR imaging is the best noninvasive method of following the venous pseudoaneurysm (Fig. 2.32). The flow in the defect is slow because the fistula is closed. The T1 and T2 studies reflect the extent of thrombosis in the pouch. Expansion or shrinkage of the mass in the cavernous sinus is easily demonstrated.

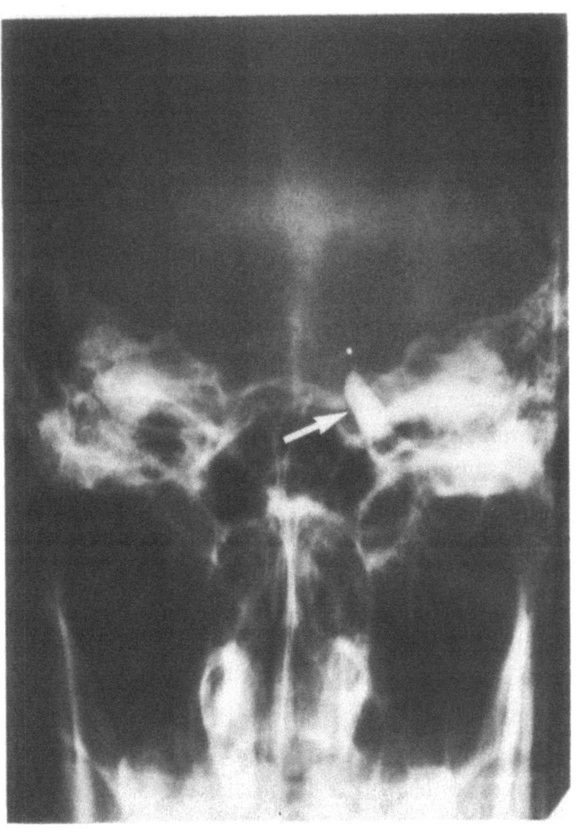

Fig. 2.31. Premature detachment of a balloon resulted in migration of the balloon to the bifurcation of the internal carotid artery. The patient experienced a hemiparesis and dysphasia. The frontal view skull X-ray shows the contrast- filled balloon (*arrow*) with its metal clip above the region of the cavernous sinus. Following a surgical arteriotomy and removal of the balloon, the patient recovered without a neurologic deficit

Fig. 2.32a, b. Magnetic resonance imaging of a pseudoaneurysm or venous pouch following percutaneous embolization of a carotid cavernous fistula via an arterial route using a balloon inflated with contrast material. The coronal (**a**) and sagittal views (**b**) show the thrombosed fistula (*large white arrow*) as well as the pseudoaneurysm (*curved arrow*) and the internal carotid artery flow (*small arrows*). This patient with Ehler–Danlos syndrome had a symptomatic third nerve paresis from this venous sac which remained untreated because of excessive fragility of her vessels. In order to avoid this problem a hydroxymethylacrylate (HEMA)-filled balloon can be used. If the patient does not develop a cranial neuropathy within several weeks of embolization and if a large defect in the bone of the sphenoid sinus is not present, it is unlikely that a venous pouch will become symptomatic or require embolization to close the sac

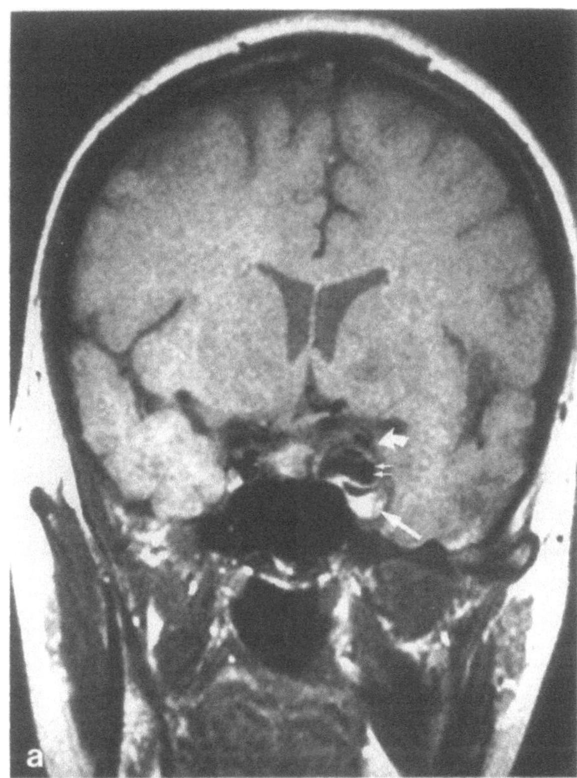

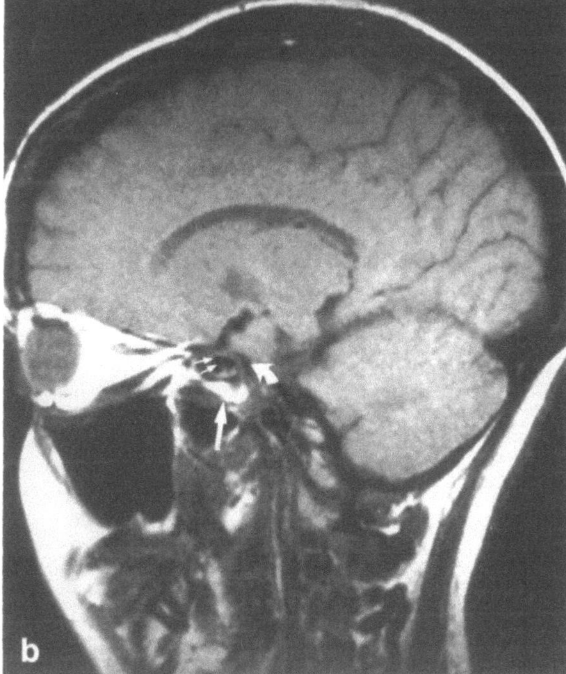

2.4.9.3 Rate of Resolution of Clinical Signs Following Shunt Closure

Following closure of the shunt there is immediate resolution of the subjective and objective bruit, and this can be used to clinically confirm the angiographic evidence of CCF occlusion. If a bruit is present or recurs, it suggests balloon movement or deflation or that an additional hole exists with persistence of an AV shunt.

The elevated IOP becomes markedly lowered within 48 h of the CCF closure in more than 90% of cases with shunt closure. In all cases normal pressure returns within 1 week. If the IOP does not normalize or if it rises after falling, reopening of the fistula should be suspected even if the immediate postembolization angiogram did not reveal a patent shunt. The visual field defects related to long-standing secondary glaucoma do not resolve.

The signs of orbital congestion resolve within several weeks. Oculomotor and lid dysfunction that is not secondary to a traumatic cranial nerve paresis will improve as the orbital and cavernous sinus congestion diminishes. Normal extraocular function returns within weeks of shunt closure. Except for one case with a venous pouch, our patients with spontaneous CCFs occluded by embolization completely resolved their deficits within 1 month.

Following partial shunt closure, even if slow flow is still present through the shunt on the immediate postembolization angiogram, the orbital signs will improve. However, if the orbital venous outflow has been compromised, the findings will be temporarily unaltered or acutely worse. With thrombosis of the remaining arterial flow through the shunt, usually within days to weeks, the symptoms will also resolve. In some cases the shunt will completely reopen. If the remaining fistula shunts blood into the posterior cavernous sinus only the bruit may return. If the remaining abnormal drainage is predominantly anteriorly then the neuro-ophthalmologic findings will worsen.

In some cases with severe orbital signs, MR imaging will reveal thrombosis within the cavernous sinus. Systemic heparinization may prevent further thrombosis and clinical deterioration, but most cases are treated symptomatically and gradually improve. If there is any doubt as to the mechanism of clinical deterioration, then angiography should be performed in order to identify whether a possible significant AV shunt persists.

2.4.10 Therapy Is Not Always Warranted

There are few situations where therapy of the AV shunt may not be warranted. When the fistula is small and drains posteriorly through the petrosal sinus, the patient may have no symptoms and signs except for a bruit. Expectant follow-up is indicated unless the bruit is debilitating. Low-flow shunts that cause only mild proptosis and redness of the eye may also be observed. However, even if the symptoms are mild an angiogram or MR angiogram is suggested to completely evaluate each case prior to deciding whether therapy of the shunt is to be avoided or undertaken. For example, one of our patients had only a mild sixth nerve paresis, and angiography demonstrated stasis of contrast material in a ruptured cavernous carotid aneurysm. Embolization was not performed, but manual compression therapy (see Sect. 2.4.4) was associated with resolution of the cranial neuropathy. If manual compression is contemplated, the carotid bifurcation should be investigated by angiography, high-definition ultrasonography, or MR angiography to determine whether this treatment is contraindicated because of significant atherosclerosis. Angiography is also essential in all cases of a traumatic CCF in order to exclude the possibility of a pseudoaneurysm.

CHAPTER 3

Neuro-ophthalmic Manifestations of Intracranial Dural Venous Disorders

3.1 Introduction

Diseases of the venous channels cause symptoms by producing venous hypertension and congestion in the tissues due to obstruction or arterialized blood flow in the affected venous channel. Any dural venous sinus can be affected. The disorders that affect the dural venous channels include primary thromboses, arteriovenous (AV) shunts, inflammation and infections, and compression and invasion by neoplasms. Neuro- ophthalmologic signs and symptoms of increased intracranial pressure (ICP) are seen when the blood flow through a major venous channel is acutely disrupted or when the remaining venous outflow is inadequate to drain the brain. Focal neurologic dysfunction arises when the venous outflow from a limited region is disturbed. When the cavernous sinus is prominently involved, the clinical features are of major interest to the neuro-ophthalmologist.

3.2 Dural Arteriovenous Malformations and Shunts

3.2.1 Epidemiology

The true incidence of pathological AV shunts that drain into the dural venous sinuses is unknown. Despite the description, as early as 1931, of cases with intracranial arteriovenous malformations (AVMs) that were supplied primarily by the external carotid artery (ECA) system [1], prior to 1971 there were less than 100 cases with a dural arteriovenous malformation (DAVM) reported in the literature. The increasing frequency of this diagnosis reflects the heightened awareness of these lesions. This is due in part to a better understanding of the vascular anatomy of the collaterals between the ECA and the internal carotid artery (ICA), the dural arterial

system, and the intracranial venous network. Technical advancements such as selective angiography, particularly of the ECA [2], techniques that subtract the background skull and magnify the images, and improved catheters for superselective injections have improved the ability to diagnose and delineate the extent of these DAVMs. Clearly effective endovascular and surgical therapies in the region of the dural venous sinuses have encouraged more extensive evaluation of these treatable lesions.

Approximately 10% – 15% of all clinically apparent intracranial AVMs are of the dural type [3]. The incidence of symptomatic lesions varies with the location of the involved sinus [4]. DAVMs are probably more common than diagnosed, because asymptomatic lesions are not routinely investigated. Additionally, patients who are symptomatic will only be diagnosed if the clinician suspects an AV shunt that involves the dural venous system. For example, the presence of a minor symptom such as a bruit, suggestive of abnormal dural venous drainage near the ear, would be cause for further investigation in a patient with an intracerebral hemorrhage even if angiography of the ICA and vertebral angiogram was unrevealing [5]. The real incidence of DAVMs may not even be revealed by autopsy series among the general population because the dura is not always excised in its entirety when the brain is removed for study.

Except for a posterior fossa location, where the incidence is reported to be greater in men, approximately two thirds of the patients with intracranial DAVMS are women [6]. Patients usually present between the ages of 20 and 79 years. Although rare, symptomatic lesions can occur in infants or toddlers [7- -9]. DAVMs that drain into the cavernous sinus are more frequently found in women 50 years of age or older or in young women during or immediately following pregnancy. The onset of symptoms in association with pregnancy may be related to alterations in blood volume or coagulation that

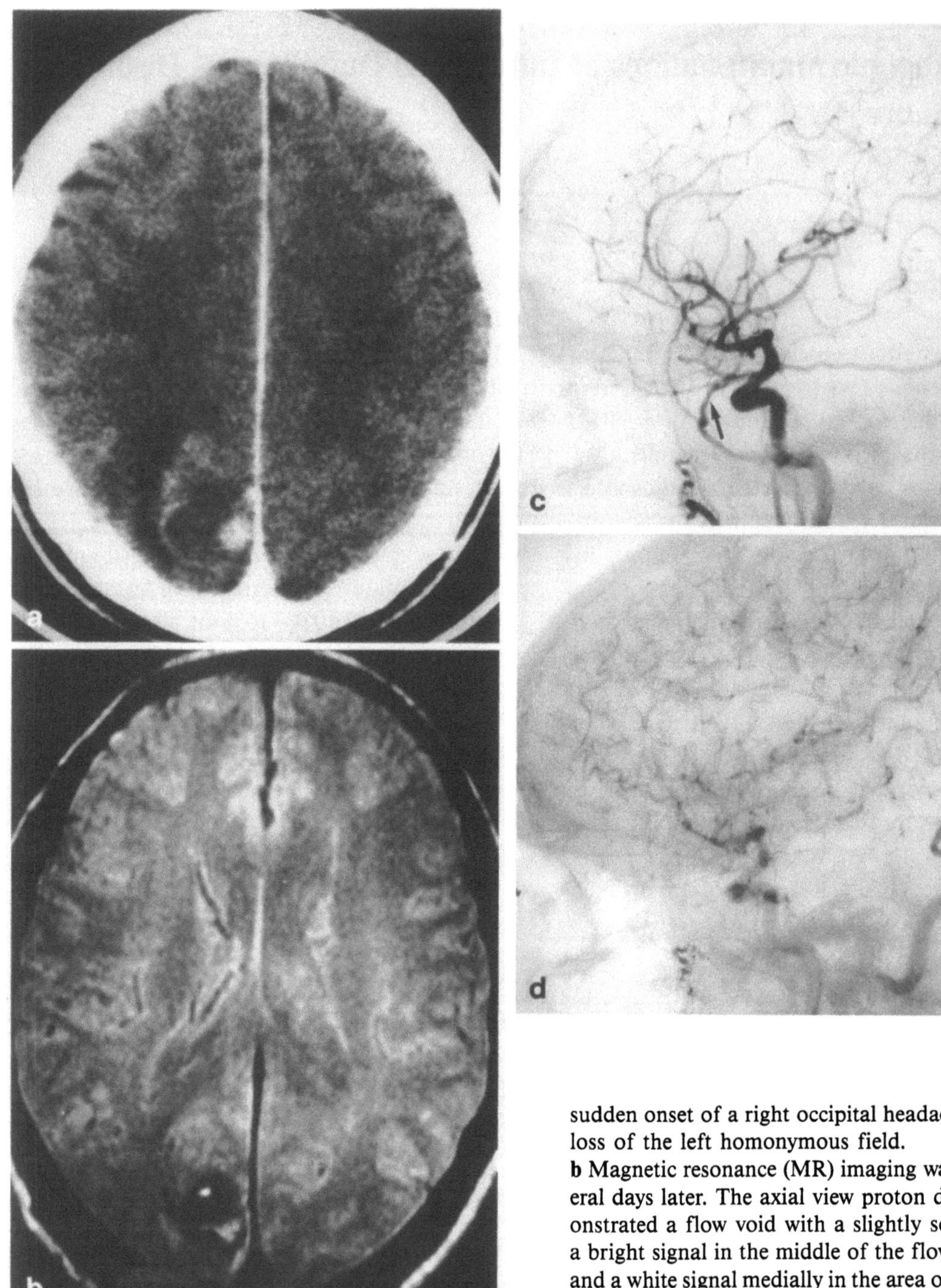

Fig. 3.1a–d. An intracerebral hemorrhage caused by rupture of the venous ectasia associated with a DAVM was originally considered to be a neoplasm based on the computed tomographic (CT) appearance of the lesion.
a The axial view noncontrast CT scan demonstrates a mass in the right occipital lobe with an area of increased signal medially suggestive of a hemorrhage. The lateral extent of the lesion has the appearance of a ring with increased signal. This 45-year-old man complained of the sudden onset of a right occipital headache and complete loss of the left homonymous field.
b Magnetic resonance (MR) imaging was performed several days later. The axial view proton density scan demonstrated a flow void with a slightly serpiginous shape, a bright signal in the middle of the flow void anteriorly, and a white signal medially in the area of the hemorrhage demonstrated on the CT.
c A lateral common carotid artery subtraction angiogram demonstrated a dural arteriovenous shunt which was supplied by a posterior branch of the middle meningeal artery (*small arrows*). Note the marked venous dilatation (*arrow*) with partial thrombosis.
d In the later phase, the arteriovenous shunt is seen to drain predominantly into the transverse sinus (*arrow*). Though cyanoacrylate embolization completely closed the dural arteriovenous shunt and posterior middle meningeal artery, the patient only made a partial recovery of the field defect.

Table 3.1. Major manifestations of spontaneous dural arteriovenous malformations based on numbers at each location (% occurrence) (modified from [21])

Anterior fossa		Cavernous sinus		Tentorial		Lateral, sigmoid sinuses, torcular	
Intradural bleed	84%	Intradural bleed	0%	Intradural bleed	70%	Intradural bleed	15%
SAH	63%	Bruit	42%	SAH	80%	Bruit	70%
ICH	50%	Vision loss	33%	ICH	60%	Vision loss	12%
SDH	25%	Proptosis	83%	SDH	10%	Papilledema	22%
		CNS	3%	CNS	42%	Headache	46%
		PNS	50%	PNS	14%	CNS	13%
						PNS	7%

SAH, subarachnoid hemorrhage; ICH, intracerebral hemorrhage; SDH, subdural hematoma; CNS, central nervous system; PNS, peripheral nervous system.

predispose to dural venous sinus thrombosis and opening of an AV fistula [2].

3.2.2 Clinical Features

The clinical features depend on the location of the DAVM and the abnormalities of venous drainage. The natural history of these lesions is variable and many patients have few or no symptoms for years. Patients with DAVMs can present with symptoms indistinguishable from a brain AVM such as headaches, diplopia, blurred vision, or focal or diffuse neurologic dysfunction [10, 11]. The headache can be mild and focal or severe and generalized depending on whether there is dilatation of a single venous channel or diffuse venous hypertension, respectively. Focal neurologic deficits will also develop in relation to the location of the disturbance in the cortical venous drainage or an intra-axial hemorrhage or both. However, in some patients a subjective pulsatile bruit may be the only complaint.

In approximately 20% of patients with a DAVM in any intracranial location, rupture occurs because of cortical venous drainage into the subarachnoid or subdural spaces or into the brain (Fig. 3.1) [12]. Intradural hemorrhages have been reported in from 6.7% [13] to 37% of DAVMs with cortical venous drainage (38/104 cases in our series) [14]. Hemorrhages are most frequent with DAVMs located in the anterior fossa or in the region of the tentorium [4, 13–15].

Pediatric DAVMs are often the high-flow congenital fistula type which can cause high-output cardiac failure. Focal or diffuse cortical atrophy and hydrocephalus are additional findings that are seen at any age [9, 16, 17].

Spontaneous resolution of the clinical signs of the DAVM occurs in up to 50% of patients with cavernous sinus lesions, but in far fewer patients with DAVMs in other locations [18– 20].

The clinical features of DAVM in various locations are listed in Table 3.1.

3.2.2.1 Posterior Fossa Dural Arteriovenous Malformation

Approximately 35% of intracranial DAVMs are located in the posterior fossa [22]. DAVMs in this location are more frequent in men than women.

Patients may initially present with few symptoms. Chronic disturbances of balance and hearing and tinnitus send the patient for medical consultation, which frequently fails to identify the problem because of the paucity of clinical findings. When other symptoms such as a progressive visual disturbance from papilledema, brainstem or cerebellar dysfunction, dementia, or a posterior fossa hemorrhage develop, a more complete investigation should establish the diagnosis (see Sect. 3.2.7.2; Fig. 3.2). The symptoms and findings are clearly dependent on the direction of the abnormal venous drainage and the impairment of the venous drainage from the normal structures. Though the examination may elicit signs of neurologic dysfunction, the symptom that strongly suggests the presence of the AV shunt is the subjective pulsatile bruit.

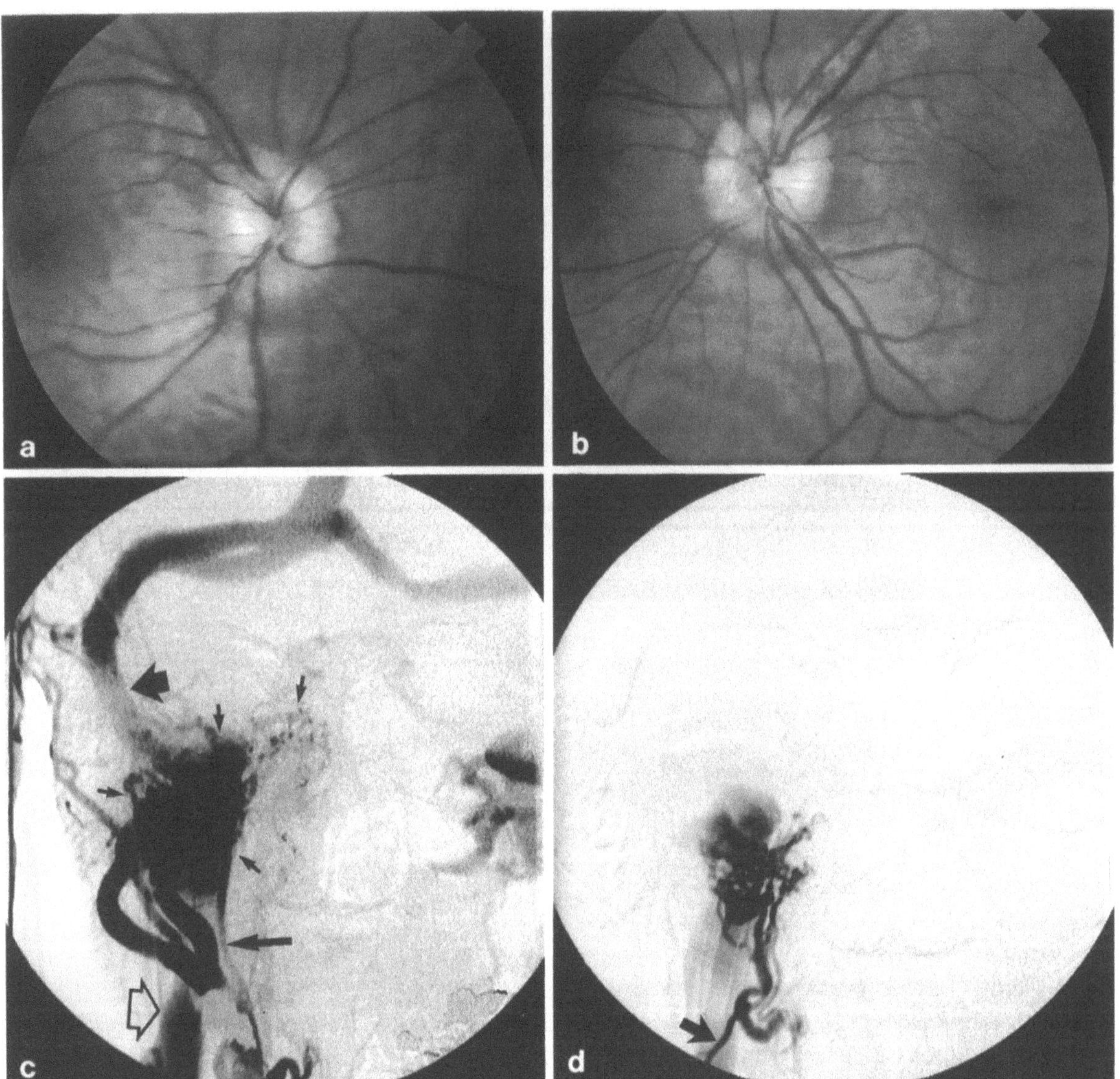

Fig. 3.2 a–e. This 55-year-old man complained of a pulsatile bruit and nonspecific dizziness, gait imbalance, headaches that were diffuse, and some mild blurring of vision for approximately 6 months. A bruit was heard behind the right ear.

a,b The patient was neurologically intact except for an inability to tandem walk and swelling of the right (**a**) and left (**b**) optic discs.

c The anteroposterior view of the right external carotid artery angiogram demonstrates an extensive network of arteriovenous shunting (*small arrows*) which is principally supplied by the ascending pharyngeal artery (*medium size solid arrow*). The AVM drains upward (*large solid arrow*) into the sigmoid sinus, which then fills the torcular, the posterior aspect of the sagittal sinus, and the oppo-

site transverse sinus, in addition to draining downward into the internal jugular vein (*open arrow*).

d Following two embolization sessions where cyanoacrylate was placed into the DAVM nidus and a detachable balloon was placed into the sigmoid sinus via a transvenous route from the opposite internal jugular vein, the lower portion of the DAVM still persists. It is principally supplied by the ascending pharyngeal artery (*arrow*).

e Plain skull radiograph in the anteroposterior view following cyanoacrylate (*small arrows*) and balloon (*large arrow*) embolization clearly demonstrates the material. The patient's papilledema resolved, the tandem gait normalized, and there was no bruit or headache. The patient has remained without symptoms over the follow-up of 2 years

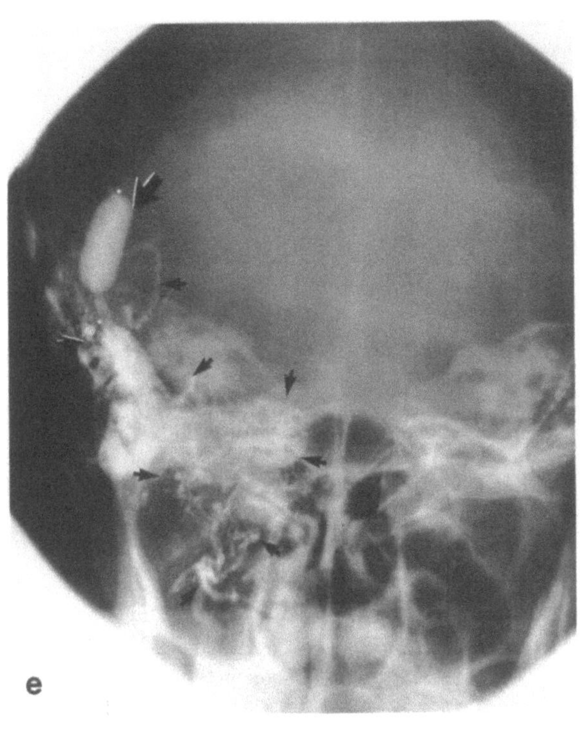

3.2.2.1.1 Bruit

A subjective bruit is often the first manifestation of a DAVM. The patient hears the pulsating noise when the environment is quiet or when the ear ipsilateral to the DAVM is directed downward, as for sleep. Frequently, patients do not mention this symptom because they feel it is a sign of stress or nonorganic disease. Though the bruit may be quite bothersome, patients commonly remain undiagnosed and may even be subjected to surgical procedures designed to relieve tinnitus [23]. The bruit results from the shunting of blood into the sigmoid or lateral sinus and is unilateral unless the DAVM drains bilaterally.

Auscultation in the retroauricular area or over the superior aspects of the cervical internal jugular vein will reveal the pulsatile bruit. The bruit is often diminished with compression of the ipsilateral common carotid artery, because of decreased blood flow through the ECA to the DAVM. Less commonly, a thrill is felt over contributing ECA branches, such as the occipital artery. If the venous drainage is predominantly anteriorly into the ophthalmic vein, a bruit can be auscultated over the orbit, but the patient may not hear the noise himself.

In some cases, the symptomatic bruit may be ignored because a self-audible bruit which cannot be auscultated can also occur in subjects with a normal high jugular bulb [24, 25]. Other rare causes of a subjective bruit include an aneurysm of the petrous ICA, a temporal lobe AV malformation, and a destructive neoplasm in the area adjacent to the middle ear [26].

3.2.2.1.2 Increased Intracranial Pressure

Increased ICP, with or without hydrocephalus, can develop in adults with a DAVM. The resulting chronic papilledema can lead to optic atrophy and visual loss [27–29] (Fig. 3.2).

3.2.2.1.3 Cranial Nerve Dysfunction

Diplopia can result from several mechanisms. A sixth nerve paresis arises from the increased ICP. One or more cranial nerves can be impaired if the AV shunt drains into the cavernous sinus (Fig. 3.3) [30]. A paresis of the third nerve is rare with a posterior fossa DAVM but it has been described in cases following a subarachnoid hemorrhage [31].

3.2.2.1.4 Intracranial Hemorrhage

Posterior fossa DAVMs can rupture, particularly when they are located in the region of the tentorium. A brainstem or cerebellar hemorrhage results if the cortical venous drainage of the DAVM to one of these areas ruptures (Fig. 3.4). Rupture of a mesencephalic vein with a tentorial DAVM nidus can cause a subarachnoid hemorrhage. The natural history of patients with DAVMs who, at presentation, have no neurologic symptoms except for a bruit and have no abnormal cortical venous drainage is unknown. However, we have observed one patient who had a tentorial DAVM without parenchymal venous involvement that was discovered in the evaluation for a meningioma in another location. Seven years later the patient had a spontaneous subarachnoid hemorrhage and the new angiogram revealed cortical venous drainage from the DAVM. The implication of this case is that DAVMs are dynamic lesions. Thrombosis in a draining vein may alter the venous drainage to shunt the abnormal flow into previously uninvolved venous channels.

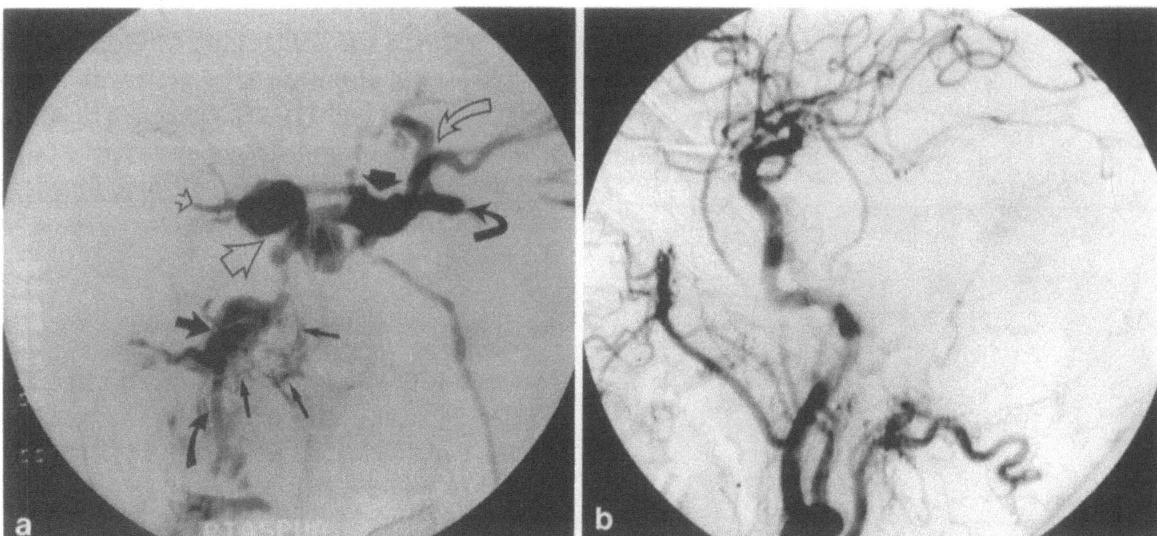

Fig. 3.3a, b. A 70-year-old right-handed woman developed some mild signs of orbital congestion in the left eye. On examination she had signs of mild congestion and several arterialized vessels in both conjunctiva and sclera that were suggestive of a cavernous sinus area DAVM. Because of her history of metastatic breast cancer and arteriosclerotic heart disease and the mild nature of the signs and symptoms, it was elected to follow the patient expectantly. Several months later, she had sudden-onset visual loss in the left eye with worsening orbital congestion and an acute episode of expressive dysphasia. She was found to have an increase in the congestion in the left eye with acuity reduced to 20/400 and a choroidal detachment (see ultrasound, Fig. 3.13a). The angiogram of the left internal and external carotid arteries failed to demonstrate an arteriovenous shunt.

a The frontal view of a right external carotid artery subtraction angiogram demonstrates a right-sided DAVM (*small arrows*) which drains into the right inferior petrosal sinus (*large arrow*) and is supplied by the ascending pharyngeal artery (*slightly curved solid arrow*). The abnormal venous drainage passes upward into the right cavernous sinus (*open arrow*). Because of thrombosis in the right anterior cavernous sinus, including the right ophthalmic artery (*small open arrowhead*), the abnormal blood flow drains across the intercavernous channels to the left cavernous sinus (*solid arrow head*). There is significant drainage into the left sphenoparietal sinus (*curved open arrow*) and to the left ophthalmic vein (*strongly curved solid arrow*). Thus, this case shows how a posterior fossa DAVM can present with signs of a cavernous sinus area DAVM as well as cortical dysfunction because of significant occlusive disease in the dural sinuses.

b Following several sessions of embolization where cyanoacrylate was placed into the DAVM nidus and the arterial supply, the lateral view right common carotid artery subtracted angiogram does not opacify the DAVM. The vision improved to 20/40 as the choroidal effusion resolved (see Fig. 3.13b). The patient had persistent slight injection of the left conjunctiva and sclera and mild elevation of the intraocular pressure in this eye, all of which resolved in 6 months

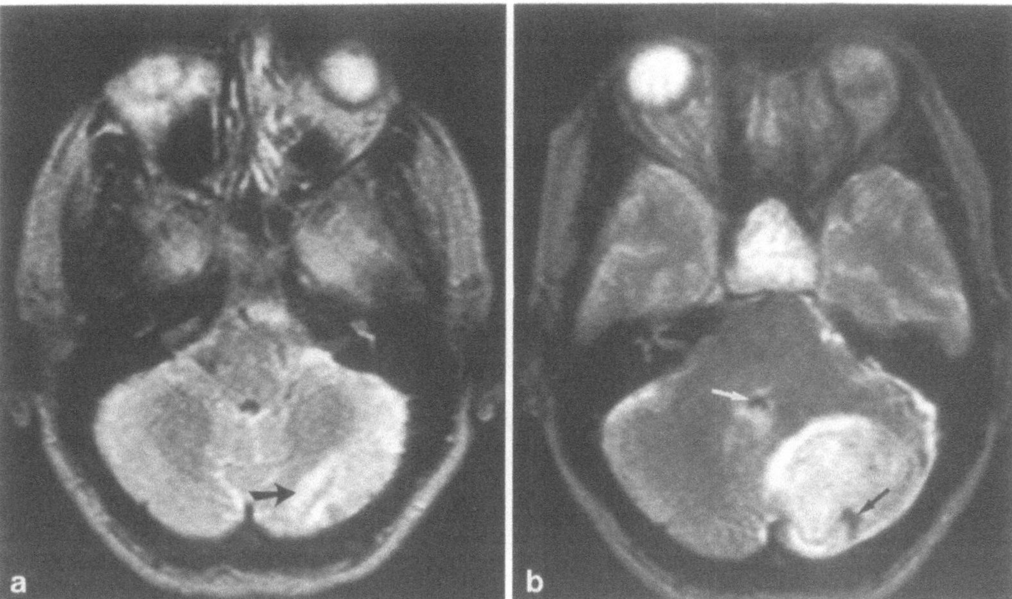

Fig. 3.4a, b. Acute thrombosis of cortical draining vein may cause a focal neurologic deficit, seizure, or nonspecific symptoms. A 65-year-old woman who was followed for an asymptomatic nonsecretory pituitary adenoma complained of the onset of acute dizziness and profuse sweating. Her neurologic examination was normal.
a The T2-weighted magnetic resonance imaging demonstrated a linear gray lesion surrounded by a white density compatible with a venous infarct (*arrow*).

b Several days later, she experienced a worsening of her dizziness and a severe suboccipital headache on the left side. The magnetic resonance image at this time demonstrated a cerebellar hemorrhage, an obvious flow void (*black arrow*), and a shift of the fourth ventricle (*white arrow*). The arteriogram (not shown) demonstrated a DAVM with an arterial supply from the occipital artery as well as branches of the vertebral artery and venous drainage into the cerebellum. Following embolization and surgical excision, the lesion was eliminated

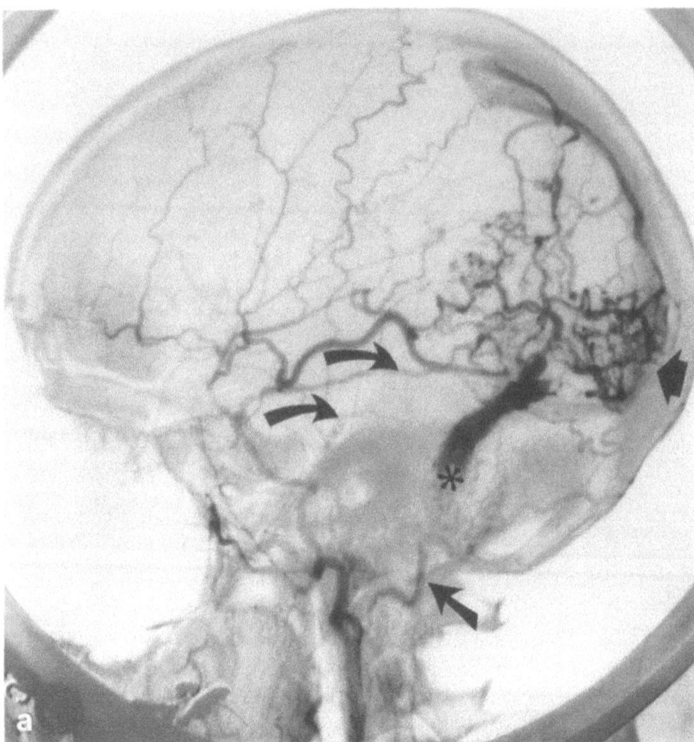

Fig. 3.5 a-c. Cortical venous drainage of a DAVM caused episodes of left-sided headaches accompanied by right visual field fortification-type imagery that would last for minutes in a 40-year-old woman.

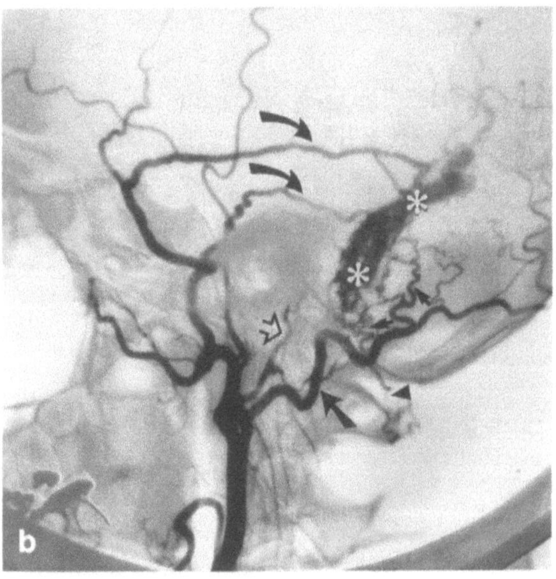

a A lateral view left common carotid subtracted arteriogram demonstrates an extensive DAVM (*asterisk*) that is supplied by posterior branches of the middle meningeal artery (*curved arrows*) and the occipital artery (*straight arrow*). There is no direct supply from the internal carotid artery. However, the venous drainage from the DAVM involves the cortical veins of the occipital lobe (*solid arrowhead*).

b The lateral view external carotid artery angiogram demonstrates the posterior branches of the middle meningeal artery (*curved arrows*) supplying the DAVM (*asterisk*). The occipital artery (*straight arrow*) supplies the shunt via stylomastoid and mastoid branches (*small arrows*). A potentially dangerous anastomosis with the cervical arterial system (*arrowhead*) is also seen. An enlarged posterior auricular artery (*open arrowhead*) also supplies the shunt.

c Following embolization of the middle meningeal, ascending pharyngeal, and occipital arterial blood supply to the DAVM, the lateral view common carotid artery angiogram does not opacify the DAVM. The episodes of headache and visual imagery abated

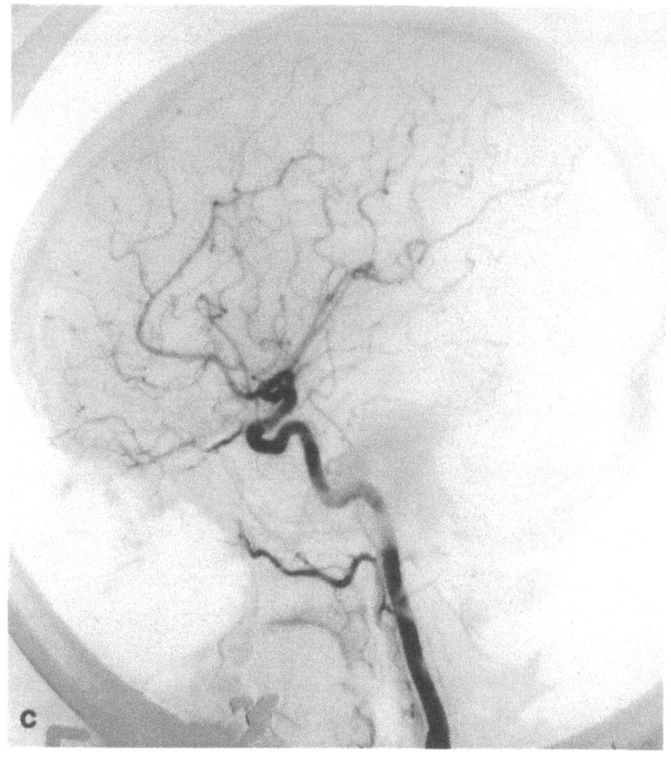

3.2.2.2 Supratentorial Dural Arteriovenous Malformation

Supratentorial DAVMs, excluding those lesions in the region of the cavernous sinus, are found less frequently than DAVMs located in the posterior fossa or near the cavernous sinus. The mean age of symptomatic patients is 55 years [32]. When a DAVM is located rostral to the tentorium, the type of focal cerebral dysfunction depends on the location of the cortical venous dysfunction (local tissue venous hypertension, thrombosis or rupture; Fig. 3.1). For example, recurrent transient epileptic or migrainous-type positive visual phenomenon develop when there is abnormal drainage into the veins of the occipital lobe (Fig. 3.5). Focal symptoms can also include dysphasia, hemiparesis, hemisensory defects, and seizures. Unilateral visual loss can occur from a "steal" when the AV shunt receives significant supply from the ophthalmic artery ethmoidal branches [32]. Severe or permanent neurologic deficits are most frequent following an intracerebral hemorrhage [33, 34]. Approximately 91% of the reported cases with DAVM in this location have presented with an intracerebral or subarachnoid hemorrhage [32].

3.2.3 Mechanisms of Dysfunction of Dural Arteriovenous Malformations
(Excluding the Cavernous Sinus Region)

Symptoms can be related to both the DAVM nidus and the abnormal venous drainage [34, 35]. Local and generalized venous hypertension probably accounts for most of the neurologic and visual dysfunction (Figs. 3.3, 3.5).

When the normal brain venous outflow or drainage is slowed or obstructed, venous hypertension can cause dysfunction in areas that are adjacent to or remote from the AV shunt. Chronic passive congestion from a retrograde increase in venous pressure slows the blood flow, disrupts the local metabolism, and causes ischemia and necrosis in focal areas of any region of the brain [36]. A posterior fossa DAVM can increase the ICP by elevating the superior sagittal sinus pressure [37] and decreasing the absorption of cerebrospinal fluid (CSF) through the arachnoid villi. Congestion in the cavernous sinus raises the pressure in the ophthalmic veins, with resulting orbital signs (see below). Venous hypertension also results from the shunting of arterialized blood into areas remote

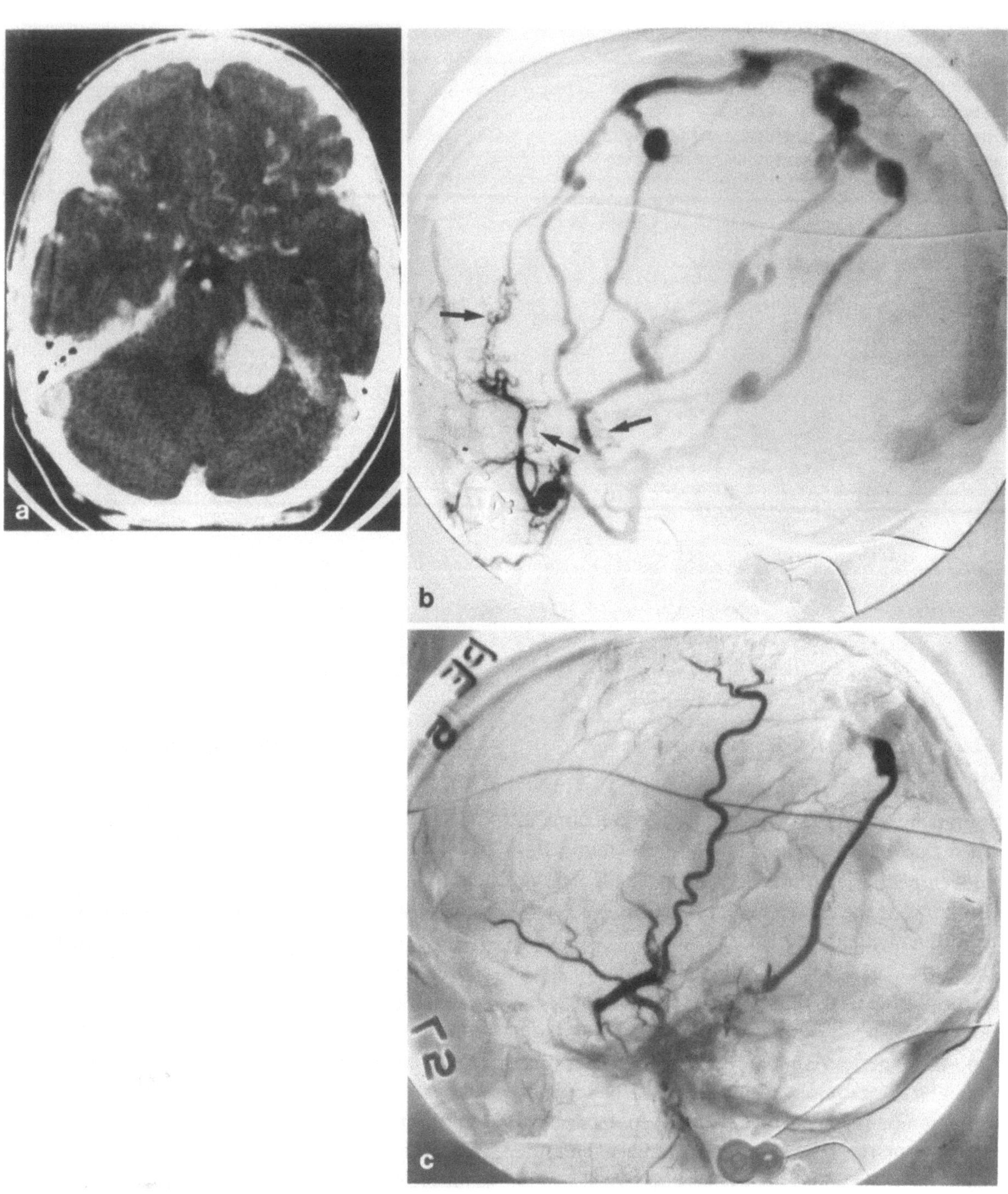

◄ **Fig. 3.6a–c.** Dementia can be a prominent symptom when an extensive DAVM is associated with compromise of the venous system because of occlusion or venous hypertension.

a A 30-year-old woman had progressive dementia and gait imbalance over a 6-month period. The axial view contrast enhanced computed tomogram demonstrates a marked increase in the appearance of the blood vessels in the brain as well as an area of aneurysmal dilatation of a vein near the petrous ridge. Angiography demonstrated multiple dural arteriovenous shunts on both sides of the intracranial cavity supplied by multiple vessels. There was marked abnormal venous drainage over both hemispheres with poor drainage of the normal vessels because of diffuse venous hypertension.

b An example of the arteriovenous shunts (*arrows*) and abnormal cortical venous drainage is seen on this lateral left internal maxillary artery angiogram.

c Following embolization, the left distal external carotid artery angiogram shows a marked reduction in the abnormal venous drainage. The patient underwent embolization on the opposite side as well. The lesion was not cured but the patient had a marked improvement in her memory and cognitive function

from the DAVM. For example, exophthalmos can develop in patients with a transverse sinus area DAVM. In these cases, stenosis or occlusion in the sigmoid sinus or jugular vein results in redirection of the abnormal drainage anteriorly into the cavernous sinus and ophthalmic vein [22, 38, 39] (Fig. 3.3).

Dementia can result when there is severe compromise of the deep venous drainage of the brain or diffuse intracranial venous hypertension. In the cases with venous occlusion, the shunted blood compounds the venous hypertension in the cerebral hemisphere. Closure of the AV shunt(s) successfully reverses the dementia only when there is adequate venous channels to carry the venous outflow from the entire brain (Fig. 3.6).

If thrombosis occurs in veins that drain normal structures, further compromise of the normal brain drainage results. This mechanism is clinically most apparent with DAVMs that drain into the cavernous sinus and may account for cases that have sudden worsening of orbital signs [40, 35] or others that eventually have a spontaneous cure [41, 42, 19]. In order to effect a cure, the thrombosis must include the DAVM nidus.

Spontaneous thrombosis of one draining vein of a DAVM may cause the abnormal flow to shift to a previously uninvolved vein or venous sinus with the onset of new or worse of symptoms and findings. For example, thrombosis of the cavernous sinus that is involved with a DAVM results in the shunting of arterialized blood to the contralateral cavernous sinus, posteriorly through the inferior petrosal sinus, inferiorly to pterygomaxillary plexus, or superiorly through the sphenoparietal sinus (Fig. 3.3). Thrombosis can also be induced by minor alterations in the circulation such as occur during angiography or following exposure to a significant change in the atmospheric pressure (see Sect. 3.2.7.3.3.3). Because of the many available venous channels, isolated thrombosis of the cavernous sinus is not typically associated with cerebral symptoms or increased ICP.

DAVMs with cortical venous drainage are subject to marked dilatation and to rupture of these involved veins (Fig. 3.1). In contrast, patients who do not hemorrhage or develop central nervous system deficits are less likely to have drainage into the cortical veins [43]. On occasion, the disappearance of the pulsatile bruit that develops with dural sinus thrombosis precedes a hemorrhage or a focal neurologic deficit because of the sudden alteration of the venous drainage. An intraparenchymal hemorrhage results in focal tissue damage or a more diffuse mass effect. Recurrent microscopic hemorrhages may be clinically silent. If the blood is subarachnoid, diffuse cerebral dysfunction similar to that occurring in patients with ruptured brain AVMs develops. Secondary vasospasm does not occur. Even though many DAVMs have numerous venous pathways for potential decompression, if drainage to cortical veins exists [44], hemorrhage is common, occurring in 25% [13] to 37% of cases [33].

Intradural hemorrhages are most frequent with DAVMs located in the anterior fossa or along the tentorium. Hemorrhage is less common with DAVMs that drain only into the posterior fossa venous sinuses and practically unheard of with parasellar DAVMs that drain solely into the cavernous sinus (see Table 3.1). Though rebleeding occurs, the frequency of this complication is unknown. Spontaneous regression of DAVMs is uncommon following an intracerebral hemorrhage [19].

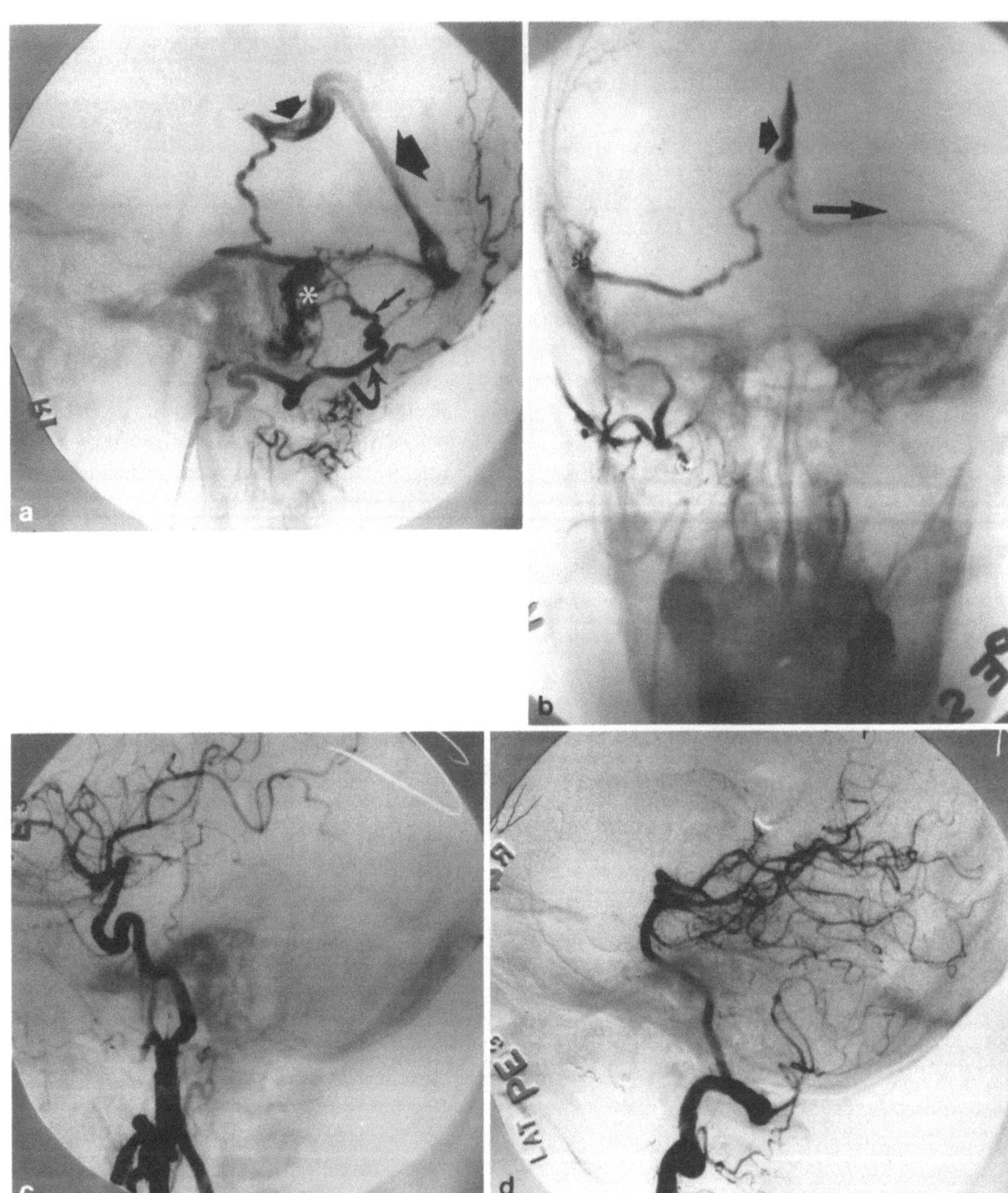

◄ **Fig. 3.7a–d.** Even if therapy eliminates the AV shunting caused by a DAVM, if there is significant occlusive disease in the venous sinuses, the increased intracranial pressure will persist.
a The lateral view subtraction angiogram with a selective injection into the right occipital artery (*curved arrow*) demonstrates a mastoid branch (*straight arrow*) supplying a DAVM (*asterisk*) in a patient with chronic bilateral papilledema. The prominent cortical venous drainage passes upward into the great vein of Galen (*small solid arrowhead*) which drains into the straight sinus (*large solid arrowhead*). The ipsilateral transverse and sigmoid sinuses were not opacified by any arterial injection, suggesting that these sinuses were thrombosed.
b The anteroposterior view of the right occipital artery injection in this case shows the DAVM draining to the great vein and inferior sagittal sinus (*arrowhead*) across the midline in the direction of the opposite transverse sinus (*arrow*).
c,d Following cyanoacrylate embolization of the nidus and the arterial feeders, neither right common carotid arteriography (**c**) nor vertebral angiography (**d**) demonstrates a residual DAVM nidus. However, the right transverse and sigmoid sinuses still fail to opacify because of the prior thrombosis. The patient continued to have the headaches and papilledema with constricted visual fields, so a lumbar peritoneal shunt was required to relieve the symptoms

Table 3.2. Mechanisms of increased intracranial pressure associated with dural arteriovenous malformation

---

Thrombosis, narrowing, obstruction of dural venous sinus preventing normal venous drainage
Reduced cerebrospinal fluid absorption via arachnoid villi
Arterialized blood flow in dural venous sinuses[a]
Obstructive hydrocephalus[a]

---

[a] Infrequent.

Aneurysmal venous dilatation unusually causes symptomatic mechanical compression of adjacent structures [9, 16]. Even when massive dilatation of draining veins is present, focal disturbances of the brain are rare. However, dilatation and thrombosis in the cavernous sinus might cause the cranial neuropathy that is associated with DAVMs that drain into these venous compartments (see Sect. 3.2.7.1.7). Symptomatic compression of the intracranial optic nerves or chiasm by dilated veins in the suprasellar cistern is also rare.

The signs and symptoms of increased ICP are examples of dysfunction that can develop from more than one mechanism of venous compromise. When the outflow of the blood in the dural venous sinuses is delayed or blocked, the dural sinus pressure and the cerebral venous blood volume is increased. The CSF absorption through the arachnoid villi into the distended venous system is reduced, with a resulting increase in the CSF pressure. Most patients who experience diffuse cerebral dysfunction or develop papilledema have a combination of sinus thrombosis in addition to AV shunting (Fig. 3.7) [23]. The CSF absorption is also impaired when there is blood in the subarachnoid space, which contributes to the development of hydrocephalus [10]. Obstruction of the cerebral aqueduct or compression of the brain by aneurysmal dilatation of draining veins is an additional, but unusual, mechanical mechanism for the increased ICP in this disorder (Table 3.2).

DAVMs are not typically high-volume shunts unless a direct fistula is present. However, when a large fistula is present in pediatric or elderly patients, the rapid blood flow through the shunt may induce high-output cardiac failure [45]. Since most DAVMs are the low-flow type, a symptomatic arterial "steal" in either the ICA or the vertebral arterial circulations rarely occurs. However, an example of this complication was reported in two cases of amaurosis fugax caused by a steal from the ophthalmic artery [35, 45a]. Angiography suggested that the visual disturbance resulted from shunting of blood away from the ophthalmic artery into a meningeal arterial feeder to the DAVM. In the second report a recurrent branch from the ophthalmic artery supplied the DAVM [45a]. The first report also included a case of diplopia and cerebellar dysfunction, without large parenchymal veins or compression of the brainstem or cerebellum, suggesting a mechanism of "steal" from the vertebral basilar arterial system [35]. In contrast to the central nervous system, a steal mechanism associated with a DAVM supplied by the middle meningeal artery that causes a cranial neuropathy, such as a facial nerve palsy, is not rare [34].

3.2.4 Pathology and Etiology

Few DAVMs have been studied pathologically, and most of our understanding is based on angiographic investigations. Patients with a nontraumatic single artery dural fistula provide support for a congenital origin of the abnormality. The presence of DAVMs in infancy also supports a congenital etiology, but neither is conclusive proof. AV shunts, approximately 50–90 μm in diameter, are found in the dura of normal subjects [46, 47]. In some cases, the opening of these AV shunts seems to be acquired following thrombosis in the adjacent dural sinus [5, 6]. Pathological studies suggest that a latent true congenital AV shunt, with a deficiency or absence of the capillaries between the arteries and veins of the dura, is present prior to the thrombosis [48, 49]. In some cases, the dura between the abnormal vessels may fail to develop [50]. Additionally, the presence of cavernous spaces containing flowing blood that bulge into segments of the lumen of normal venous sinuses may contribute to the development of AV shunting in these areas [51].

Once the shunted blood causes arterialized venous flow and local venous hypertension, the thin-walled emissary venules and veins, which normally drain into the dural sinuses, can massively dilate [52]. The shunting of blood also further increases the arterial blood flow, causing the involved dural arteries to dilate.

Support for an acquired etiology of clinically apparent disease is also found in patients where angiography was performed for other reasons, but did not demonstrate a DAVM, and who later developed this lesion [5]. The hypothesized inciting venous thrombosis is typically found in, but not limited to, the dural sinus adjacent to the AVM on angiography [53]. Areas of thrombosis may be more apparent in patients with a history of prior trauma [54]. In summary, it is likely that in most cases a congenital microscopic dural AV shunt or malformation [55] is present, but nonfunctional and initially beyond the resolution of angiography. Following an acquired thrombosis in the adjacent dural venous sinus, the shunts increase in size, resulting in a clinically and angiographically apparent syndrome.

3.2.5 Neuroimaging

3.2.5.1 Computed Tomography and Magnetic Resonance Imaging

Though computed tomography (CT) or magnetic resonance (MR) imaging is needed in all patients suspected of harboring a DAVM, the scans commonly appear normal unless aneurysmal venous dilatation, thrombosis of cortical veins, or an intracerebral hemorrhage is present (Fig. 3.1). Focal or generalized atrophy of the brain or hydrocephalus are nonspecific secondary alterations. Thrombosis of a draining vein or a dural venous sinus is more easily demonstrated by MR imaging (see Sect. 3.3.2.3.2; Figs. 3.4, 3.15). A chronic enlargement of the ECA branches can sometimes widen the middle meningeal artery vascular markings on the inner table of the skull and enlarge the foramen for penetrating arteries (which can also be seen on CT bone windows). Though the clinical signs of a cavernous sinus area DAVM are sanguine (see below), DAVMs in other locations require a high index of clinical suspicion to consider performing angiography in the face of a normal CT or MR image.

3.2.5.2 Angiography

The angiographic evaluation must include superselective arterial catheterization. The study is tailored to the clinically suspected location of the DAVM in order to accurately pinpoint the nidus and all potential arterial feeders [21]. An unsubtracted common carotid artery angiogram will rarely localize the nidus and may even fail to demonstrate the DAVM. When the DAVM is in the anterior fossa the angiogram should evaluate the ICA, internal maxillary artery, middle meningeal artery, and if required the contralateral ICA and internal maxillary artery branches. When a tentorial lesion is suspected, the ICA (tentorial branch from the cavernous ICA), the internal maxillary artery, the ascending pharyngeal artery, the occipital artery, the vertebral artery, and the contralateral ICA, internal maxillary artery, and ascending pharyngeal artery should be investigated. DAVMs that involve the infratentorial dural sinuses (posterior fossa) require visualization of the ICA (tentorial branches), internal maxillary artery, ascending pharyngeal artery, occipital artery, vertebral artery, and contralateral arteries specific for each lesion. A single vessel fistula or multiple arterial feeders may be present. The late phase of the angiogram demonstrates the direction of the venous drainage from the DAVM as well as the venous outflow from the normal intracranial structures. Venous dilatation and thrombosis in the dural venous sinuses can also be seen.

When planning therapy, the entire arterial supply to the nidus must be identified, and any potential dangerous anastomoses between these vessels and arteries that supply normal structures must be considered. This is particularly pertinent for preventing emboli to the eye or the ICA circulation or cranial nerve dysfunction. Potential routes for transvenous embolization must also be evaluated. One side benefit of performing angiography is the occasional induction of thrombosis of the lesion, though the incidence of this benefit is unknown.

3.2.6 Therapy and Course

Except for the cavernous sinus region lesions, the natural history for DAVMs in other intracranial locations is unknown. Since the signs of cavernous sinus involvement are easily documented, more is known of the natural history of these lesions and 30% – 50% spontaneously normalize (see Sect. 3.2.7.3). The signs and symptoms of DAVMs in any location can remain static, deteriorate gradually or suddenly, or spontaneously resolve [56, 57].

Immediately following a hemorrhage, ICA angiography may fail to visualize a DAVM. Following resolution of the hematoma, a selective angiogram is required to confirm that the nidus is permanently thrombosed. As mentioned above, lesions with cortical drainage are more prone to rupture. Patients with a subarachnoid or intraparenchymal hemorrhage can have permanent deficits similar to those experienced by patients with parenchymal AVMs. Cases without cortical drainage and few symptoms do not necessarily require treatment. Untreated patients must be periodically examined, both clinically and by neuroimaging, because a change in the venous drainage can produce cortical venous involvement and subsequent hemorrhage.

The goals of therapy are to occlude the nidus of the DAVM, to eliminate any compromise of venous outflow from normal intracranial structures, and to increase the arterial flow to normal tissue in order to prevent a clinical "steal." When dural sinus thrombosis occurs, eliminating the DAVM may not immediately result in clinical improvement because of the persistent defect in venous drainage from normal structures. For example, when thrombosis of the sigmoid sinus is present, the papilledema will not resolve despite closure of a posterior fossa DAVM. Eventually, the brain will find another route of venous drainage, but visual loss and optic atrophy commonly develop [23, 29].

3.2.6.1 Embolization

3.2.6.1.1 Arterial Route

Improvements in arterial catheter systems and embolization agents, such as liquid cyanoacrylate (isobutyl-2-cyanoacrylate, IBCA, or *N*-butylcyano-acrylate, NBCA) and polyvinyl alcohol (PVA) particles, (140 – 1000 μm) [54, 55, 35], have allowed superselective embolization into distal, rather than proximal, arterial feeders and the nidus of the DAVM. Occlusion of the embolized vessels and DAVM can be more effectively accomplished with these methods. The liquid acrylic agents produce a cast of the nidus and are more likely to result in permanent occlusion of the lesion (see embolization of cavernous DAVM). Partial embolization that significantly reduces the abnormal arterial flow into the DAVM can promote thrombosis of the remaining shunt and adjacent venous sinus. Though there are reports that embolization with gelfoam, another particulate agent, can successfully treat DAVMs [60], the rapid resorption of this agent often results in recurrence of the shunt, so this material is not recommended.

3.2.6.1.2 Venous Route

Transvenous embolization with metallic coils or detachable balloons can be used to close DAVMs that involve the transverse, sigmoid, and cavernous sinuses (see Sect. 3.2.7) [61]. The embolization materials can be introduced percutaneously via the jugular vein to induce thrombosis in the affected sinus (see Fig. 3.2d, e).

Other techniques can be used to thrombose the DAVM nidus. Electrothrombosis [62] has been used with copper needle insertion directly into the cavernous sinus or through the ophthalmic vein to treat cavernous sinus area DAVMs, but this procedure requires a surgical exposure.

3.2.6.2 Treatment of Papilledema

If sinus thrombosis leads to continued or progressive signs despite embolization or DAVM resection, treatment should be directed at relieving the symptoms. Chronic papilledema that causes progressive visual loss may treated with carbonic anhydrase inhibitors to decrease CSF production and corticosteroids, but a lumboperitoneal shunt or an optic nerve sheath window or fenestration is a more definitive therapy. If untreated, the papilledema will eventually resolve, but permanent visual loss and optic atrophy often remain. Ventricular shunting should be performed for persistent symptomatic hydrocephalus.

3.2.6.3 Surgery

Partial embolization which is limited to arteries that can be safely closed may be effective when combined with the surgical resection of the remaining nidus. This approach is useful in patients with a tentorial DAVM that has a significant contribution from the C5 segment of the ICA, a vessel that cannot be embolized without high risk. Despite closure of numerous arterial feeders, if the tentorial supply is prominent the tentorial DAVM remains patent. If cortical venous drainage persists, excision of the remaining lesion and the affected wall of the adjacent dural sinus is curative [28, 13, 63]. In these cases, closure of the sinus at the level of the DAVM via a percutaneous transvenous embolization is also an alternative to posterior fossa surgery. Once the DAVM is occluded, surgical resection of a venous aneurysm that has caused focal neurologic deficits or uncontrollable seizures is infrequently necessary [28, 64].

Surgical excision of a DAVM in the anterior fossa is required in most symptomatic patients [59]. Prior embolization is not always effective or possible. For example, in some cases branches of the

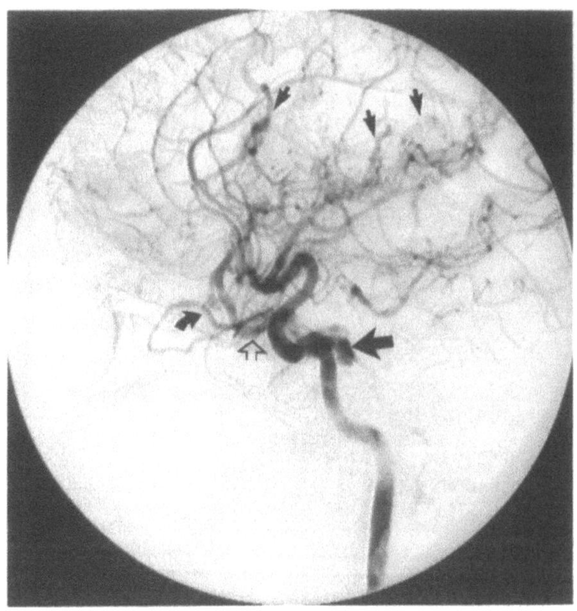

Fig. 3.8. A 47-year-old man had a right intracerebral hemorrhage and seizures 6 months after a 3-month period of a right sixth nerve paresis. Angiography demonstrated a DAVM to the right cavernous sinus with thrombosis of the anterior and posterior aspects of the cavernous sinus causing the abnormal drainage upward through the spheno-parietal sinus. The lateral internal carotid artery subtraction angiogram, following embolization of the middle meningeal artery supply to the DAVM, demonstrates a branch of the ophthalmic artery (*curved arrow*) passing intracranially to supply the DAVM. The DAVM opacifies the cavernous sinus posteriorly (*solid arrow*), and the abnormal drainage passes via the sphenoparietal sinus (*open arrow*) into the cortical veins (*small arrows*). Embolization of the ophthalmic artery was not feasible for fear of blinding the patient. The sphenoparietal sinus was clipped at surgery and a small segment of DAVM removed. The postoperative angiogram demonstrated an absence of the DAVM

ophthalmic artery supply the DAVM, and embolization carries a significant risk of causing visual loss in the ipsilateral eye [65] (Fig. 3.8).

3.2.6.4 Radiation

Radiotherapy may be an alternative treatment in cases of extensive lesions that cause progressive deficits and are not amenable to embolization or surgical resection [66–68]. A total dose of radiation of as little as 4000 cGy may be beneficial.

3.2.6.5 Treatment Failure and Complications

Closure of the proximal feeding arteries by any method, without occluding the nidus, may not be effective because collateral vessels may be recruited to supply the remaining lesion [50]. This explains why surgical ligation of the branches of the ECA is generally not successful. After arterial embolization, recanalization of the nidus is also more likely to occur in cases treated by particulate rather than liquid adhesive embolization.

If a significant amount of the thrombogenic embolic material passes through the nidus into the venous drainage (which also drains normal structures), thrombosis in the vein can cause a deterioration in the clinical signs. The worsening is temporary and warrants symptomatic treatment. Heparinization or thrombolytic agents are rarely required. However, if the nidus remains partially patent with continued shunting of arterial blood, the signs will not resolve and immediate reembolization is necessary.

Another complication arises if thrombosis is induced in nutrient arteries that supply the cranial nerves, en passant (see Sect. 3.2.7.3.2.2). Provocative testing prior to embolization should prevent this complication. A superselective infusion of 50 mg cardiac xylocaine into the artery to be tested will determine whether a cranial neuropathy will develop from occlusion of the vessel [67]. PVA particles rather than liquid adhesives can be used to avoid a potential neuropathy when the test is positive. Alternatively, further advancement of the embolization catheter, distal to the nutrient vessel, may allow the use of the liquid adhesive in these cases. A cranial neuropathy induced by a liquid acrylic is more likely to be permanent because of penetrance into nutrient vessels. In contrast, if a neuropathy develops following PVA embolization, recanalization of the nutrient vessel is likely and so is recovery of function.

More devastating complications develop when errant embolic material occludes arteries that supply normal central nervous system or ocular structures. In these cases, the emboli pass into the normal artery through an anastomosis with the embolized abnormal artery (Fig. 3.8) [21]. Transient and permanent ischemic episodes or infarcts of the cerebral hemisphere, brainstem, or eye can occur. For example, passage of the acrylic agent or PVA through a lacrimal anastomosis into the ophthalmic artery circulation can cause retinal, choroidal, or diffuse ocular ischemia (Fig. 3.9). Prior knowledge of dangerous anastomoses can often be gained from selective angiography. This allows the implementation of specific measures to avoid this complication (Table 3.3).

Transvenous embolization approaches can also cause serious complications. A hemorrhagic infarction or an intradural hemorrhage may result if embolization occludes the dural venous sinus that is the dominant channel for venous outflow from the brain. This complication should only occur when there is significant thrombosis in other venous channels or if the dural venous sinus on the contralateral side is hypoplastic. Hemorrhage could also occur if the embolizing catheter ruptures the venous sinus. If the DAVM remains patent, occlusion of an existing venous drainage can redirect the shunted blood to another area, causing new and even worse dysfunction (Fig. 3.10).

3.2.7 Dural Arteriovenous Malformation of the Cavernous Sinus Area
(see also Chap. 2)

3.2.7.1 Clinical Signs and Symptoms

DAVM involvement of the cavernous sinus may not occur more frequently than DAVMs in other locations, but the signs and symptoms of cavernous sinus and orbital disease commonly bring these patients to early neuro-ophthalmologic evaluation. However, as mentioned below, many of the patients have only subtle signs so the correct clinical diagnosis may not be made until significant orbital congestion is present. In our series, 68% of all spontaneous AV shunts that involved the cavernous sinus were caused by a DAVM. These lesions are more frequent in postmenopausal women and in women during or immediately following pregnancy. The symptoms and signs fluctuate with alterations of the orbital venous outflow that develop from thrombosis and changes in head position. The clinical features are listed in Table 3.4.

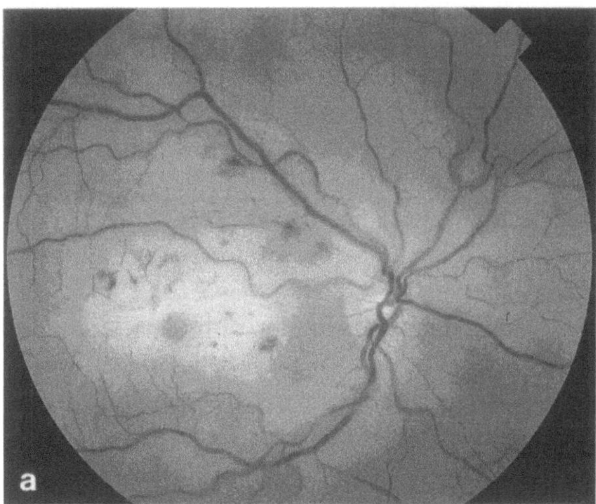
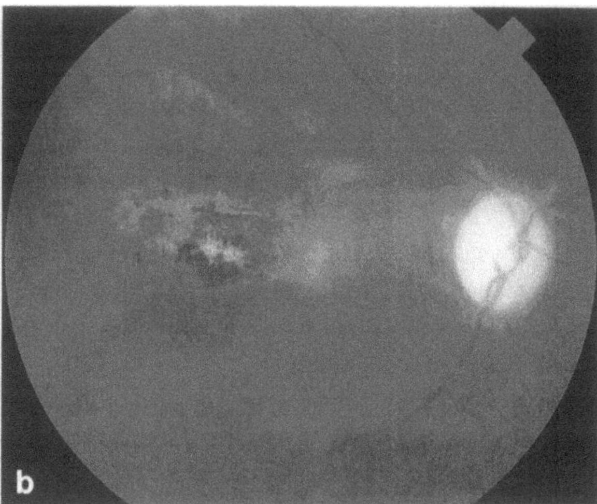

Fig. 3.9a,b. A 35-year-old woman with an anterior fossa DAVM that drained into the frontal lobe had cyanoacrylate embolization of the lesion. One hour after the deposition of the acrylic agent, the patient experienced visual loss in the right eye. Initially, the fundus was normal despite finger counting vision and a dense relative afferent pupillary defect. In approximately an hour, an embolus appeared on the disc surface in the central retinal artery and the retinal vessels narrowed. Ischemic changes were seen in the retina but the macula appeared to be spared. Despite ocular massage, measures to elevate the blood CO_2, and an anterior chamber paracentesis, the ischemic changes persisted. The patient was anticoagulated with heparin sulfate immediately. Over the next several days, only bare hand motion vision was noted and the macula demonstrated ischemic changes as well.
a Ophthalmoscopy, 1 week after the iatrogenic emboli, demonstrates a small embolus on the disc, narrowing of the arteries, and diffuse swelling of the inner layer of the retina including the macula with numerous faint hemorrhages.
b The visual loss persisted and ophthalmoscopy 6 months later demonstrated diffuse pallor of the optic disc, markedly narrowed arteries and veins, a chorioretinal scar in the macula, and a diffuse retinal pigment epithelial disturbance. The embolus is still seen on the optic disc. The cyanoacrylate had occluded the posterior ciliary arterial circulation as well as the central retinal artery

Table 3.3. Methods of avoiding errant emboli via anastomoses

1. Mechanically block the potential dangerous anastomosis
2. Advance the catheter distal to the anastomosis
3. Reverse the blood flow away from the normal artery towards the DAVM with an intra-arterial balloon catheter
4. Use PVA particles larger than the lumen of the anastomosis

Table 3.4. Frequency of clinical features of cavernous sinus dural arteriovenous malformations (89 of our patients)

Elevated intraocular pressure	+ + + +
Third nerve paresis	+ +
Fourth nerve paresis	+ [a]
Sixth nerve paresis	+ +
Ocular motor dysfunction	+ + +
Ptosis	+ +
Arterialized conjunctival vessels	+ + + +
Chemosis	+ +
Lid engorgement	+ + +
Proptosis	+ + + +
Bruit	+ ±
Venous retinopathy	+ +
Visual field defect	+ +
Pain	+ +
Glaucomatous cupped optic disc	+

[a] Never occurs without other ocular motor dysfunction.

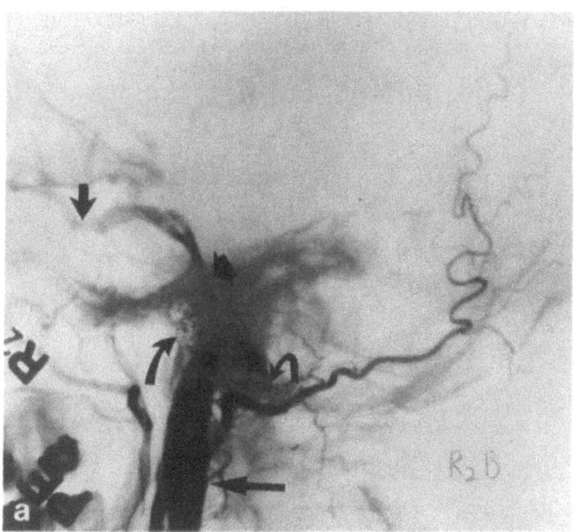 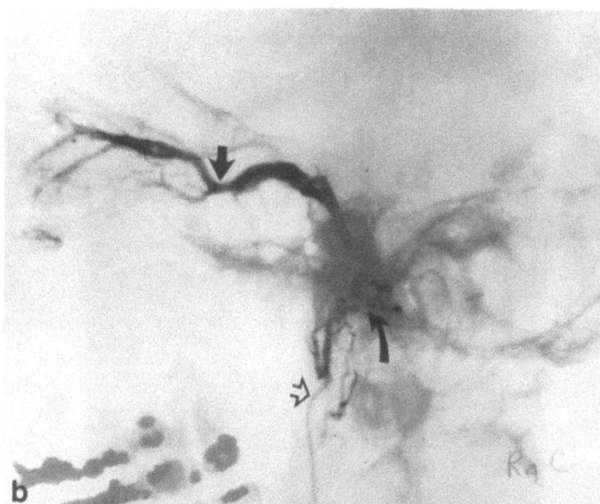

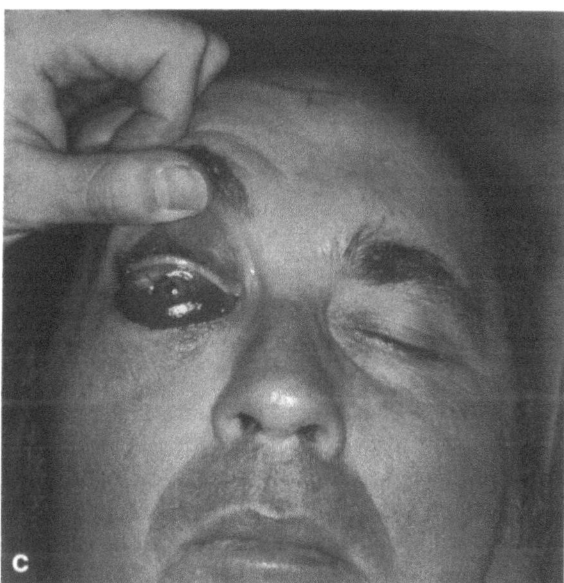

Fig. 3.10a – c. Partial embolization may actually worsen the existing condition or create new problems. A 55-year-old man complained of several months of vertigo, tinnitus, and gait imbalance. He was found to have an extensive DAVM that involved the foramen magnum.

a The lateral external carotid artery subtracted angiogram demonstrates the network of DAVM (*curved arrows*) whose principal drainage is downward into the internal jugular vein (*large arrow*). There is also some drainage via the inferior petrosal sinus (*solid arrowhead*) into the cavernous sinus. The ophthalmic vein (*medium sized arrow*) is faintly opacified.

b Following partial intra-arterial embolization and transvenous embolization of the drainage into the jugular vein, the arteriogram with the catheter (*open arrow*) in the ascending pharyngeal artery demonstrates a small amount of remaining DAVM nidus (*curved arrow*) with all of its abnormal venous drainage into the cavernous sinus and the ophthalmic venous system which is now markedly opacified (*dark arrow*).

c The sudden onset of venous hypertension in the ophthalmic venous system caused marked orbital congestion and limitation of eye movements, as well as a severe elevation of the intraocular pressure in the right eye, none of which had been present prior to embolization. Immediate closure of the remaining AVM nidus using cyanoacrylate embolization through the ascending pharyngeal artery was curative, but a twelfth nerve paresis resulted because of occlusion of the nutrient vessel to this peripheral nerve

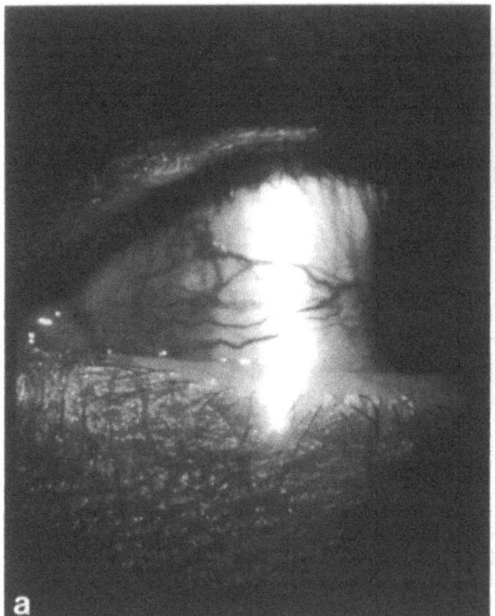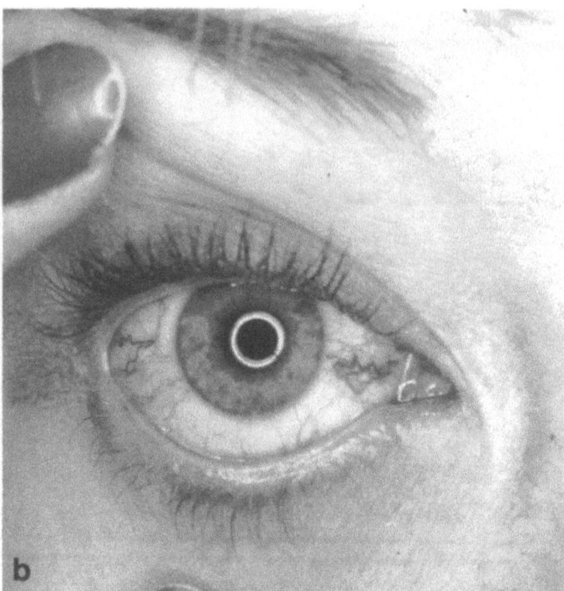

Fig. 3.11a, b. The orbital signs may be subtle in cases of a DAVM that involves the cavernous sinus. **a** This slit lamp photograph demonstrates an arterialized conjunctiva and vessels causing a limbal loop in a patient with a cavernous area DAVM. **b** This 30-year-old woman had the onset of a mild red eye during pregnancy and was treated with local ophthalmic antibiotics and steroids without success. She was found to have a meningeal arteriovenous fistula. The arterialized conjunctival vessels with limbal loops are characteristic of an arteriovenous shunt which drains into the cavernous sinus. The intraocular pressure in the affected eye was 26 mmHg, but there was no evidence of orbital congestion, proptosis, or limitation of eye movement

3.2.7.1.1 Red Eye

Unlike the explosive or sudden presentation of a traumatic or high-flow spontaneous carotid cavernous fistula, patients typically present with a usually gradual onset "red-eyed shunt syndrome" [69]. Many are treated with topical antibiotics or corticosteroids for the chronic eye redness, which can be a focal or diffuse injection of the conjunctiva and sclera. The red eye appears fundamentally different from the injection caused by external inflammation or the ciliary flush of uveitis. When there is shunted arterialized blood in the ophthalmic vein, close inspection reveals dilated, tortuous, bright red conjunctival and epibulbar vessels. These arterialized vessels make an acute angle near the limbus, called "limbal loops" [70] (Fig. 3.11). When the abnormal arterial inflow is reduced, the vessels appear darker. However, persistent dilatation suggests continued compromise of the orbital venous outflow. Dilated dark conjunctival vessels are also seen in eyes with congenital glaucoma and Sturge–Weber syndrome. In these last two disorders, the vessels tend to be less tortuous and lack "limbal loops."

3.2.7.1.2 Secondary Glaucoma

Following the onset of a red eye, the elevation of the episcleral venous pressure causes a rise of the intraocular pressure (IOP) in the affected eye. If both cavernous sinuses become involved the ocular hypertension becomes bilateral. The chronic elevation of the IOP combined with the reduced ocular blood perfusion pressure contribute to the development of glaucomatous visual deficits in approximately 20% of patients and blindness in fewer cases [69, 40, 71]. Normalization of the IOP is often difficult to achieve, despite antiglaucoma therapy, including topical β-blocking agents, epinephrine agonists, and cholinergic agonists (see Chap. 2 for a cautionary note), systemic carbonic anhydrase inhibitors, and laser iridotomy and trabeculoplasty [70, 72, 72a]. Clearly, the principal therapy of the secondary glaucoma must be directed at correcting the abnormal hemodynamic condition. Even when arterial embolization worsens the IOP, this complication is usually temporary, and endovascular retreatment via other arterial feeders or the venous route is the proper approach. Failure to realize this often leads to more aggressive antiglaucoma therapy [72a].

3.2.7.1.3 Orbital Congestion

Signs of orbital congestion similar to, but not usually as severe as, those found in cases with traumatic carotid cavernous fistulas develop (Fig. 3.12). The proptosis is frequently less than 5 mm. The chemosis can be minimal or absent. In contrast to patients with traumatic fistulas, the proptosis caused by a DAVM is often mild so an exposure keratitis is infrequent. When congestion of the orbital tissues develops, the patient may be misdiagnosed as having dysthyroid orbitopathy or orbital inflammatory disease. When a red eye and elevated IOP are present without proptosis, the patient may be thought to have chronic anterior uveitis, open or closed angle glaucoma, or ischemic oculopathy from carotid occlusive disease (for discussion see Sect. 2.3.4.2.6).

The orbital signs can suddenly worsen following thrombosis in the ophthalmic veins or cavernous sinus. The severe orbital signs persist if the DAVM remains patent. However, the orbital signs may resolve within weeks [41] because of presumed propagation of thrombosis into the nidus of the DAVM and closure of the shunt.

3.2.7.1.4 Bruit

A pulsatile bruit is heard less frequently by these patients than in patients with a carotid cavernous fistula. A bruit can be auscultated by the examiner in approximately 25% of cases [67]. The auscultated bruit is quieter than with a carotid cavernous fistula and is often barely audible. The bruit may be heard over the orbit, in the retroauricular region, or both.

3.2.7.1.5 Retinal and Choroidal Dysfunction

Hypoxic-ischemic retinopathy, similar in appearance to the retinal alterations caused by a carotid cavernous fistula, develops in approximately 15% of cases. We have rarely encountered patients with

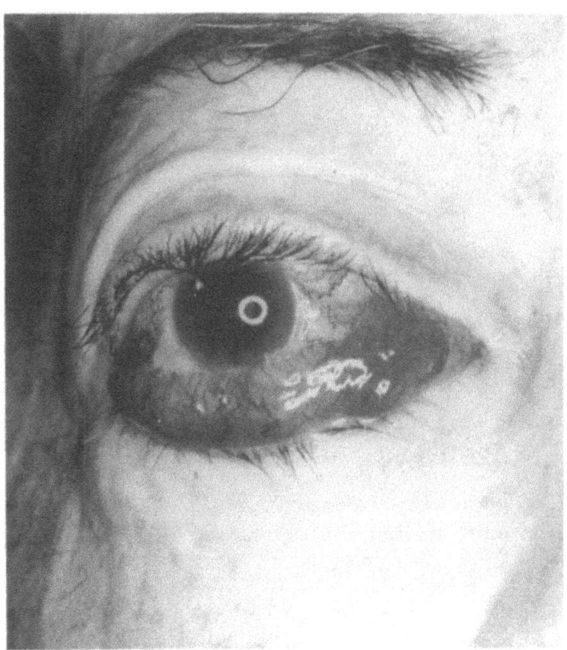

Fig. 3.12. In contrast to the cases in Fig. 3.11, some DAVMs cause severe ocular and orbital abnormalities. This 72-year- old woman had progressive orbital congestion and visual loss over a 4-month period. The clinical evaluation of the affected right eye revealed a visual acuity of 20/200, an inferior altitudinal field defect, a relative afferent pupillary defect, swelling of the optic nerve and venous hypoxic retinopathy, and an intraocular pressure of 35 mmHg despite maximal medical antiglaucoma therapy. There was a right third nerve paresis and significant orbital congestion and proptosis. The orbital congestion and arterialized vessels and prolapse of the chemotic conjunctiva are evident. This patient was seen prior to the development of microcatheter systems and due to the presence of severe arteriosclerosis selective catheterization could not be performed. Approximately 18 months later, the orbital signs, third nerve paresis, and elevated intraocular pressure normalized, but the vision never recovered

other signs of chronic ocular ischemia. One of our patients had an anterior ischemic optic neuropathy as well as retinopathy. None of our cases developed ischemic corneal alterations, iritis, iris atrophy or rubeosis, or retinal neovascularization, typical of an ischemic oculopathy.

Spontaneous symptomatic exudative retinal detachments or serous retinal detachments have been rarely reported [73, 74, 74a]. However, a choroidal detachment or effusion seems to develop in as many as 4% of cases with cavernous sinus AV

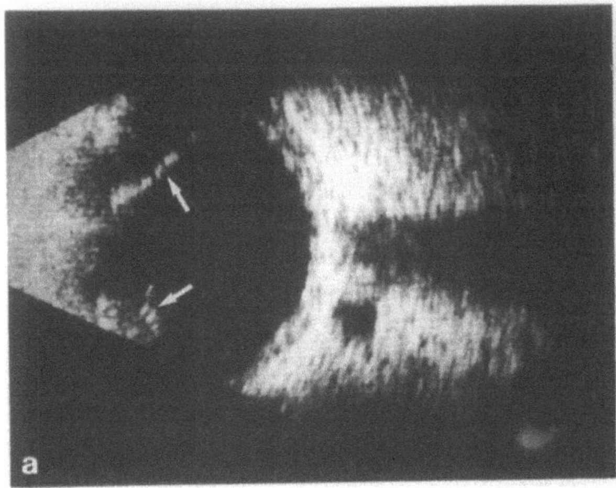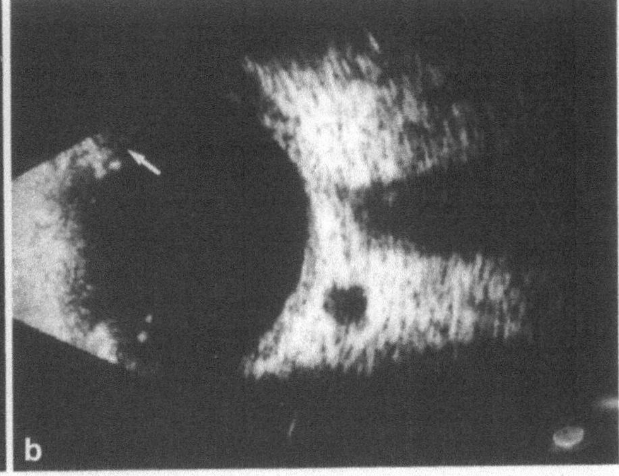

Fig. 3.13 a, b. Ultrasonography may reveal an unsuspected choroidal effusion or detachment in a patient with a cavernous sinus area DAVM. **a** Ultrasonography of the patient shown in Fig. 3.3 demonstrates a choroidal detachment (*arrows*). **b** Following embolization and improvement in the clinical signs and symptoms, ultrasonography shows almost complete resolution of the choroidal detachment (*arrow*)

shunts. A small area of peripheral choroidal effusion or partial detachment that does not cause symptoms may be overlooked. Indirect ophthalmoscopy following wide pupillary dilation will reveal the choroidal involvement in these cases. Ultrasonography will also demonstrate the choroidal effusion or detachment (Fig. 3.13a). Patients with large, near total or complete detachments are diagnosed more frequently because the diminished vision stimulates a more thorough examination. These patients always have additional signs, such as abnormal episcleral vessels or venous retinopathy, which suggest the presence of compromised ophthalmic vein outflow. An associated reduction of ciliary body production of aqueous humor will lower the IOP. Rarely, a ciliochoroidal effusion and detachment can cause acute angle closure that does not respond to medical therapy or laser iridotomy [75]. The same mechanism also caused the displacement of a posterior chamber synthetic intraocular lens into the anterior chamber in one of our patients.

Though the choroidal detachment resolves and the vision improves with occlusion of the DAVM, a visual deficit frequently persists. One of our patients with a DAVM had the visual acuity reduced to finger counting secondary to a choroidal effusion as well as a third nerve paresis. Following spontaneous thrombosis of the DAVM after an air-

plane excursion, his third nerve paresis and the effusion resolved with improvement in the vision, but the acuity was never better than 20/70 [67]. A second patient had her acuity improve to 20/40 from 20/100 following the resolution of the choroidal effusion after embolization (Fig. 3.13b).

3.2.7.1.6 Pain

Discomfort or frank pain in the eye or orbit or frontal area is often transiently present with thrombosis in the draining venous system, but the pain is infrequently severe or persistent. We have seen only one patient who presented because of constant severe focal pain without any signs of an AV shunt. Months after his initial presentation, he developed a mild seventh nerve paresis. This patient's DAVM was located posterior to the dorsum sellae, along the clivus. Despite the paucity of clinical signs, angiography demonstrated that the DAVM drained into both cavernous sinuses as well as both petrosal sinuses.

3.2.7.1.7 Cranial Nerve Dysfunction

Many patients have oculomotor disturbances similar to those found in patients with a carotid cavernous fistula. Venous congestion, hypoxia, and ischemia contribute to the extraocular muscle dys-

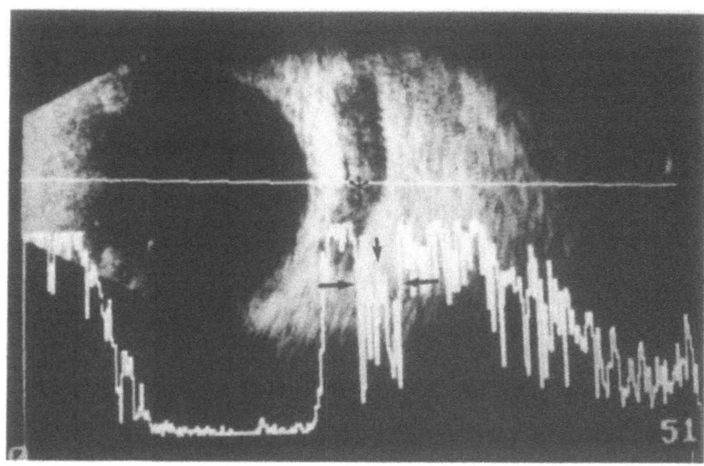

Fig. 3.14. Ultrasonography was performed in this patient with a cavernous sinus DAVM and orbital congestion. The B-scan demonstrates an enlarged superior ophthalmic vein. The A-scan contains a segment with small spikes (*small arrow*), representing flowing blood in the ophthalmic vein, which is sharply delineated (*large arrows*) from the orbital fat

function. However, cranial neuropathies definitely occur. This is evidenced by patients with a DAVM that drains entirely into the posterior cavernous sinus who have a sixth nerve paresis without signs of orbital congestion. A misdirection syndrome (see Sect. 6.3) has also been described in a patient following resolution of the shunt [76], indicating that the dysfunction is not solely a manifestation of local eye muscle involvement. The sixth nerve is the most frequently affected cranial nerve. Rarely, the third nerve, with pupillary dilation, is the only abnormal cranial nerve affected [77]. An isolated fourth nerve palsy has not been described. Cranial nerves seven to twelve are unusually involved even when the DAVM is extensive [78, 34].

3.2.7.1.8 Cerebral Dysfunction

Rarely, if thrombosis in the cavernous sinuses is extensive, abnormal pial drainage into the cerebral hemisphere or the posterior fossa veins causes a hemorrhage, a cortical infarct, or a local metabolic disturbance that results in cerebral deficit or seizure (Figs. 3.3, 3.5, 3.8). Because of the high incidence of significant thromboses in the normal outflow from the cavernous sinus, cortical venous involvement is more likely to develop in patients with bilateral ocular or orbital findings.

3.2.7.2 Neuroimaging

3.2.7.2.1 Ultrasonography, Computed Tomography, and Magnetic Resonance Imaging

The neuroimaging studies carried out are similar to those performed on patients with a carotid cavernous fistula. The orbital ultrasonographic findings are similar to the abnormalities in a patient with a carotid cavernous fistula (see Sect. 2.3.1) except for an increased incidence of thrombosis and a lower rate of blood flow in the ophthalmic veins of patients with a DAVM. The B-scan often shows a large superior ophthalmic vein sharply delineated from the orbital tissues. On A-scan, there are steeply rising and ending spikes on either surface of the vessel wall. Low blurred spikes are found between the large spikes representing flowing blood in the ophthalmic vein (Fig. 3.14). However, these findings are seen in many patients with orbital congestion of other causes. As mentioned previously, B-scan ultrasonography will also reveal choroidal effusion or detachments which might be difficult to see with ophthalmoscopy (Fig. 3.13a).

The appropriate CT or MR study should be performed in all patients suspected of having an AV shunt in the region of the cavernous sinus. Frequently, the contrast CT is normal or reveals only the nonspecific enlargement of the superior division of the ophthalmic vein (bilateral if clinical

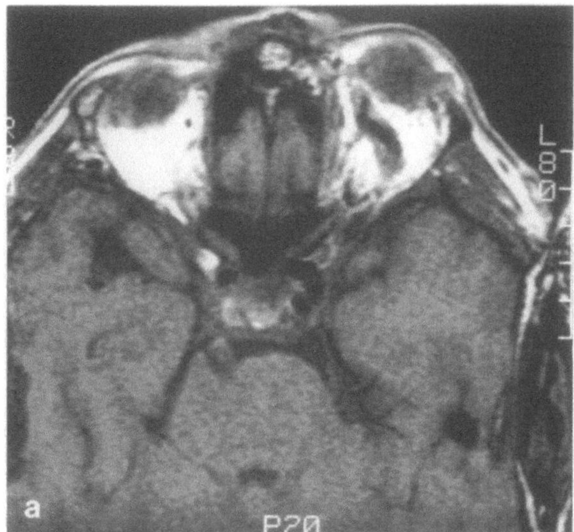

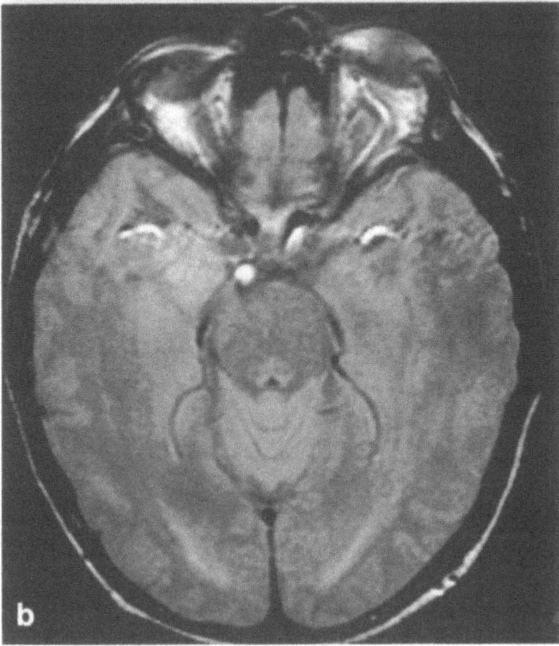

Fig. 3.15. a T1-weighted axial magnetic resonance imaging demonstrates partial thrombosis in the left superior ophthalmic vein in a patient with acute worsening of the orbital signs associated with a cavernous sinus DAVM. **b** The proton-weighted magnetic resonance image also shows a mixed signal response compatible with partial thrombosis in the affected superior ophthalmic vein

signs are bilateral). Other signs of orbital congestion such as swelling of the extraocular muscles can also be demonstrated. Dilatation and thrombosis of the involved cavernous sinus are not easily visualized.

The MR findings are similar for all AV shunts that drain into the cavernous sinus, except that patients with a DAVM may show thrombosis in the cavernous sinus or ophthalmic venous system. Arterialized blood flow in the superior ophthalmic vein typically causes a darker or hypointense appearance (signal void) on both T1- and T2-weighted pulse sequences because of rapidly moving protons in flowing blood. When thrombosis of the ophthalmic vein is present, the blood, and therefore the protons, have little movement so the signal appears as a white hyperintensity on the T1-weighted sequence [41] (Fig. 3.15). Rarely, MR imaging demonstrates dilatation or thrombosis of cortical veins (see Fig. 3.4).

3.2.7.2.2 Angiography

An angiogram is not essential for diagnosis in a patient suspected of having a symptomatic spontaneous low-flow shunt. However, if the orbital or cranial nerve dysfunction cannot be managed conservatively or the presence of cortical venous drainage is suspected, then angiography is warranted in order to determine the best route for treating the abnormal AV communication. Additionally, in our series, patients with bilateral orbital disease more often had multiple arteries supplying the lesion and tended to have extensive venous sinus occlusions. The latter group of patients also seemed more prone to develop intradural venous drainage, so that an early angiographic evaluation is required in these cases. Since the risk of a neurologic complication during angiographic examination is less than 0.1% when it is performed by a neuroradiologist who is trained to perform superselective catheterization, angiography should be performed in every patient who warrants a definitive diagnosis and treatment.

Superselective angiography is performed in all cases considered for treatment to confirm the diagnosis and determine the entire arterial supply to the DAVM, potential anastomoses to important normal arteries (see Table 3.3, Figs. 3.16–3.18), the ab-

Fig. 3.16a–c. A 55-year-old woman with Osler–Weber–Rendu disease had progressive signs of orbital congestion typical of a cavernous sinus DAVM. On angiography, a DAVM that was supplied by a small C5 meningeal branch, the middle meningeal artery, and the artery of foramen rotundum to the inferior lateral trunk was found. However, a potentially dangerous anastomosis was also recognized, so additional treatment planning was required.

a The lateral view distal internal maxillary artery injection subtraction arteriogram demonstrates that the anterior deep temporal artery (*small arrows*) supplies the ophthalmic artery (*dark slightly curved arrow*). The artery of the foramen rotundum (*small open arrow*) to the inferior lateral trunk (*large open arrow*) supplies the DAVM, which drains into the cavernous sinus (*dark strongly curved arrow*).

b In order to prevent errant embolization into the ophthalmic artery via the anterior deep temporal artery while the catheter tip (*long narrow arrow*) was in the internal maxillary artery, large particles of gel foam were used to block access to the distal internal maxillary artery branches. This maneuver redirected the blood flow so that there was more prominent opacification of the artery to the foramen rotundum (*small open arrow*), the inferior lateral trunk (*larger open arrow*), and the cavernous sinus. The altered flow allowed embolization with cyanoacrylate to close the DAVM except for the small portion supplied by the C5 branch from the internal carotid artery.

c The lateral view internal carotid artery subtraction angiogram demonstrates the small C5 contribution (*large arrow*) to the cavernous sinus. The opaque glue in the artery of the foramen rotundum, inferior lateral trunk, and cavernous sinus (*small arrows*) obscures the contrast column in the internal carotid artery

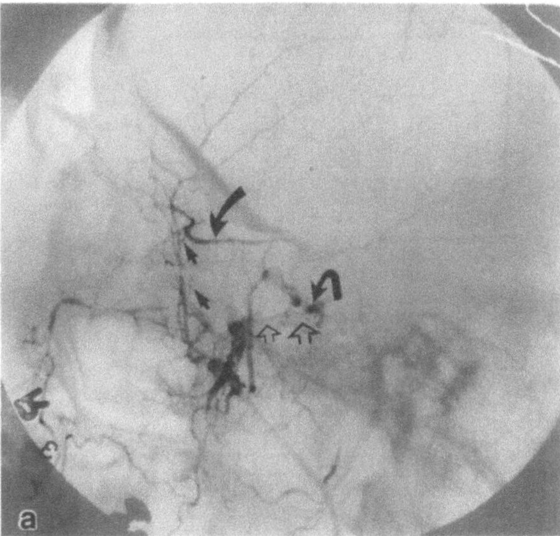

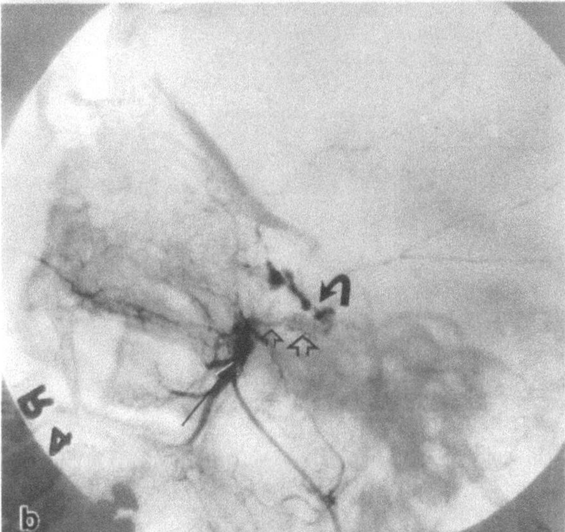

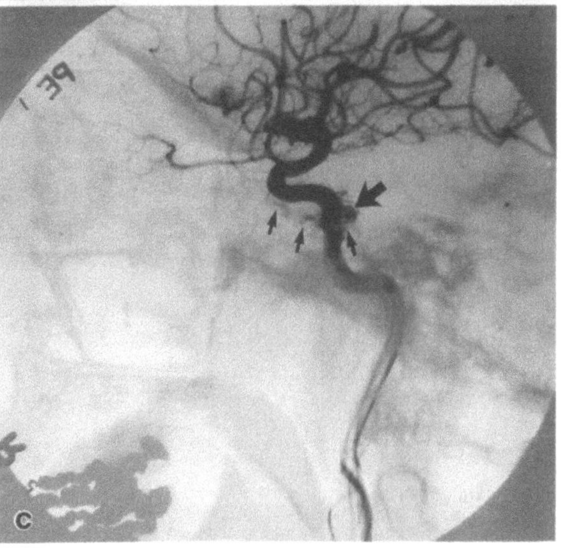

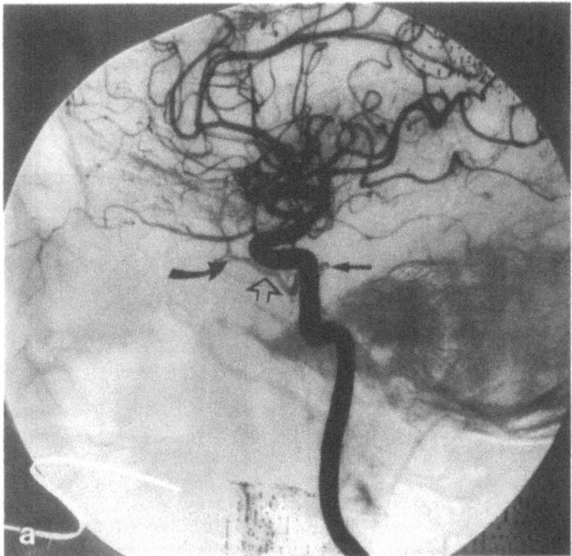

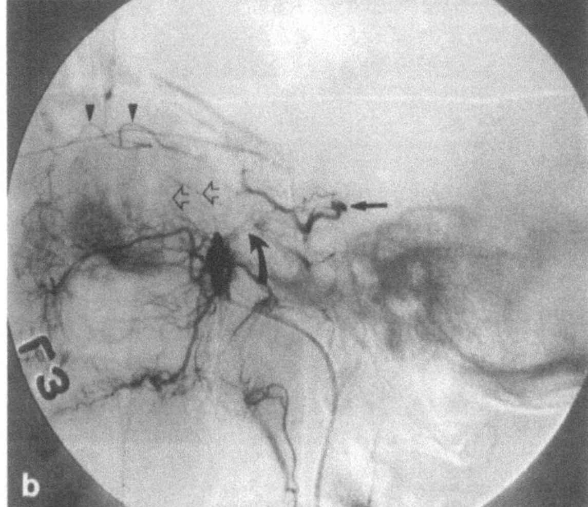

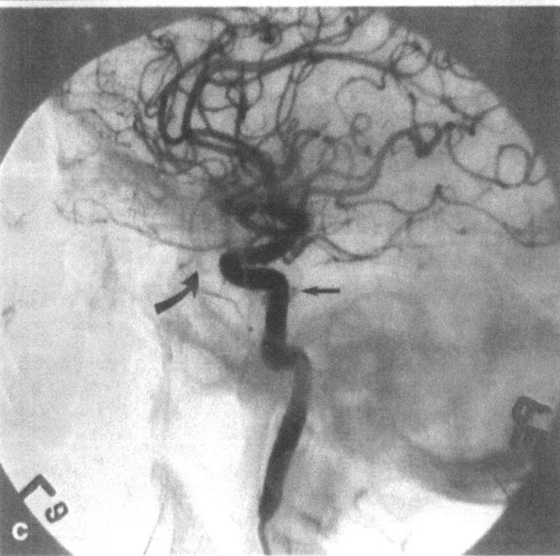

Fig. 3.17a–c. Pre-embolization selective angiography will identify all of the arteries supplying the DAVM as well as potential dangerous anastomoses during treatment.

a A lateral view subtraction angiogram of the internal carotid artery demonstrates a C5 branch (*arrow*) of the cavernous carotid artery to the DAVM. Note the filling and partial thrombosis of the cavernous sinus (*open arrow*) and the ophthalmic vein (*curved arrow*).

b A lateral view subtraction angiogram of the distal internal maxillary artery shows the same meningeal artery (from C5). An AV shunt (*arrow*) to the cavernous sinus channel is opacified through the artery of the foramen rotundum (*curved arrow*). Note the opacification of the ophthalmic artery (*arrowheads*) through ethmoidal arteries (*open arrowheads*).

c Embolization was performed with PVA particles that were larger than the lumen of the ethmoidal to ophthalmic artery anastomosis in order to prevent visual loss. A postembolization lateral view internal carotid artery angiogram demonstrates persistence of a faint C5 branch (*arrow*) without filling of the cavernous sinus. The radiopaque PVA particles are seen in the region of the anterior cavernous sinus and ophthalmic vein (*curved arrow*) normal venous channels, and possible routes for transvenous embolization. For the purpose of an angiographic evaluation, the parasellar region can be divided into three areas: (1) the orbital apex and lateral cavernous sinus; (2) the posterior cavernous sinus and dorsum sellae; and (3) the anterior tentorium cerebelli.

The arterial supply to a DAVM in the first of these regions typically includes a contribution from the inferior lateral trunk from the C4 segment of the cavernous portion of the ICA. The dura in this region, the structures of the cavernous sinus, and the DAVM often share blood supply from the middle meningeal artery, accessory meningeal artery, or artery of foramen rotundum (Figs. 3.16, 3.17). In the second region, additional arterial contribution to the DAVM originates from the C5 segment of the cavernous carotid artery, the ascending pharyngeal artery, or both, from one or both sides (Fig. 3.3). If the DAVM extends posteriorly to the medial anterior tentorium (third region), branches from the occipital artery, the petrous portion of the middle meningeal artery (constant supply to the seventh nerve), or the artery to the free margin can provide additional blood supply the lesion (Fig. 3.18). Arterial contributions to DAVMs arise from the contralateral internal carotid, middle meningeal, or ascending pharyngeal arteries in approximately 25% of cases. Bilateral AV shunts are found infrequently, in contrast to the 33% of cases with bilateral arterial blood supply to a DAVM.

Collateral circulation in the parasellar area and the patency of the circle of Willis must also be evaluated (see Sects. 1.5.3.1.2 and 2.3.4.3 for discussions of the anatomy and angiography of shunts in this region). The cervical ICA is investigated for the presence of atherosclerotic stenosis or an ulcerated plaque in the event that manual compression therapy is warranted (see Sect. 3.2.7.3.3.1). Angiography will reveal whether the patient has a single AV fistula or whether a network of minute AV connections with the appearance a "net" is present. Some DAVMs are extensive and involve almost the entire base of the skull (Fig. 3.18). The large majority of DAVMs receive direct arterial supply from two or more arteries. In most cases, the venous phase of the angiogram reveals some degree of thrombosis in the cavernous sinus ipsilateral to the DAVM [40].

The area of thrombosis and obstruction of outflow in one or both cavernous sinuses determines the direction of the abnormal venous drainage and dictates the clinical manifestations (see Figs. 3.3, 3.8, 3.10c).

Unlike patients with a carotid cavernous fistula, the angiographic analysis of the arterial pattern and venous compromise generally cannot be correlated with the degree of the severity of the orbital signs, glaucoma, and other causes of visual loss. The only exception is when there is spontaneous sudden worsening of the orbital congestion because of an acute thrombosis of the ophthalmic venous system [41].

3.2.7.3 Course, Treatment, and Complications

Since Newton's and Hoyt's classic article on cavernous sinus area DAVMs, numerous reports have improperly emphasized a benign course for these lesions [71]. Some authors suggested that angiographic evaluation [69] and treatment [40] of these DAVMs may be unwarranted. Though the prognosis in approximately 40% – 50% cases includes spontaneous recovery, usually within 6 months, or a stable nonprogressive course [69, 40], visual disability may result from ocular motility dysfunction or visual loss, or both, even when the shunt spontaneously thromboses.

Treatment of low-flow spontaneous AV shunts must be considered when visual disability is progressive or does not resolve despite conservative or medical therapy.

3.2.7.3.1 Surgery

Surgical ligation or proximal embolization of feeding arteries from the ECA is not effective in this disorder. The nidus of the DAVM remains patent because most lesions are supplied by a branch of the cavernous ICA as well as multiple branches in the ECA system which are not surgically accessible [79]. Only a surgical procedure that occludes the shunt can give comparable results to percutaneous embolization. Surgery of the cavernous sinus requires a craniotomy, is technically difficult, and may cause damage to the cranial nerves.

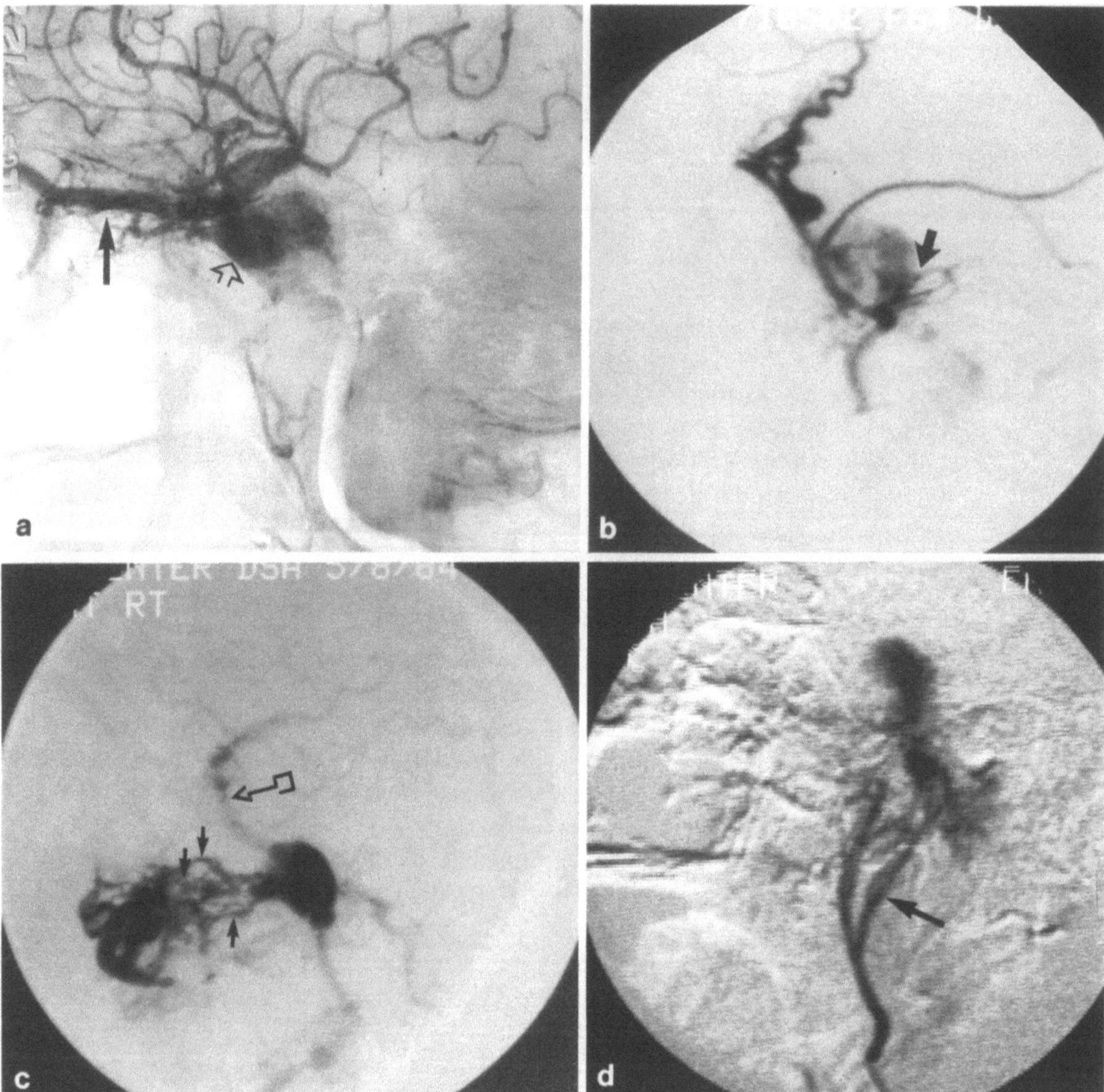

Fig. 3.18a–g. In some patients, such as the one discussed in in the section on radiation treatment who had progressive bilateral optic neuropathy (Sect. 3.2.7.3.3.2), angiography will reveal a cavernous sinus DAVM with bilateral supply from numerous arteries with meningeal branches.

a A lateral view left internal carotid artery subtraction angiogram shows immediate filling of the cavernous sinus (*open arrow*) and the ophthalmic vein (*solid arrow*) as a result of an extensive DAVM which involved the entire base of the skull.

b A lateral view left middle meningeal artery subtracted angiogram demonstrates a middle meningeal artery branch artery (*solid arrow*) supplying the DAVM.

c A lateral view distal left internal maxillary artery angiogram demonstrates small branches of the artery of the foramen rotundum (*small arrows*) supplying the DAVM. There is abnormal venous drainage superiorly into the pial veins (*open arrow*) via the sphenoparietal sinus.

d A lateral left ascending pharyngeal artery subtraction angiogram showed the neuromeningeal trunk (*arrow*) supplying a portion of the DAVM in the region of the clivus and the anterior margin of the foramen magnum.

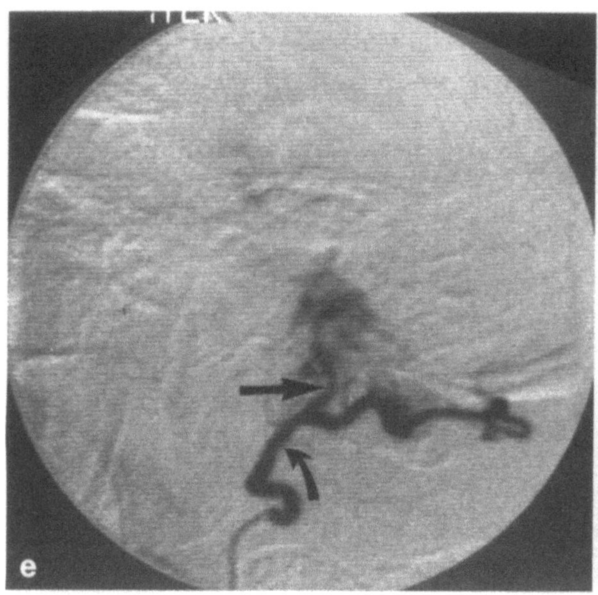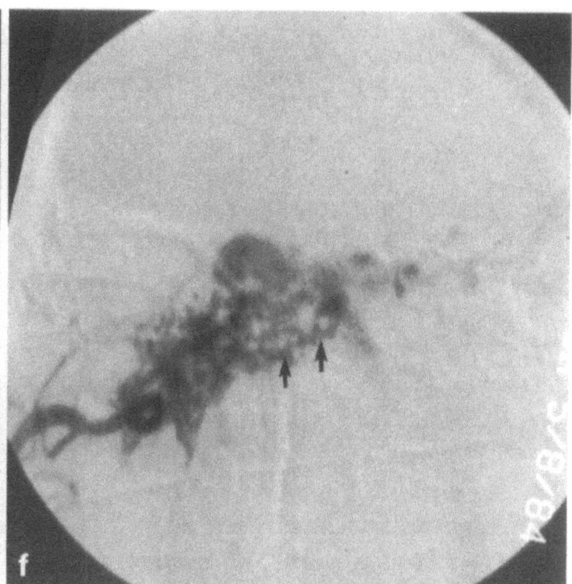

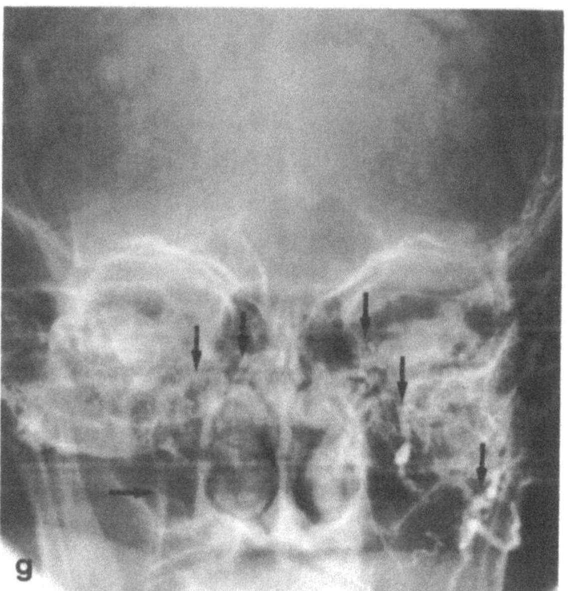

e The left lateral occipital artery angiogram (*curved arrow*) shows an enlarged stylomastoid artery (*arrow*) feeding the DAVM in the region posterior to the petrous bone.

f A frontal view right internal maxillary artery angiogram shows the extension of the DAVM into the right cavernous sinus. The artery of the foramen rotundum supplies the right side of the AVM and, via a transcavernous anastomosis (*small arrows*), the left side of the lesion as well.

g A postembolization frontal view skull radiograph shows the extensive radiopaque cyanoacrylate cast in the DAVM nidus and feeding artery (*arrows*). Despite multiple attempts at embolization, the patient had progressive visual loss and was referred for radiotherapy. (From [67])

3.2.7.3.2 Embolization

The development and improvement of superselective angiographic catheter systems, including coaxial catheters and microguide wires [80–82], embolic agents, and calibrated balloon systems, have made percutaneous endovascular embolization the treatment of choice for DAVMs that shunt into the cavernous sinus [83, 84, 67]. Embolization can be combined with manual compression therapy (see below) in cases where the postembolization angiogram reveals persistent AV shunting via branches from the cavernous ICA.

The embolization materials used for these DAVMs is a liquid cyanoacrylate or PVA or a combination of both. A liquid acrylic agent is the first choice for therapy, when possible, since it induces permanent occlusion of the nidus and feeding pedicles more often.

The branch arteries arising directly from the cavernous ICA to the DAVM need not be embolized for this therapy to be successful [85]. Embolization of these vessels is technically difficult because of the sharp turn at their origin and their small caliber. Also, any errant emboli that remain in the ICA can pass distally to cause cerebral ischemia. In contrast, the majority, if not all, of the ECA supply to the DAVM must be embolized to effect a clinical cure, particularly in cases where the nidus is not completely embedded with the embolizing agent. When PVA is used, recanalization of the nidus through incompletely embolized or remaining feeding arteries is likely to occur. Thus, embolization of all ECA feeders from both sides, preferably with a liquid adhesive agent, should be performed if appropriate.

As discussed previously, the penetrating property of liquid acrylic agents can also cause complications. These can be anticipated from the provocative xylocaine test (see Sect. 3.2.6.5). If the catheter cannot be advanced beyond the origin of the arterial pedicle to the cranial nerve, PVA is used [79]. The flow pattern will carry the particles from a proximal location into the nidus. The particle size can be large enough so that these emboli remain in the larger abnormal vessels. Also, since the large PVA particles do not penetrate into the arteriole–capillary network, there is less chance of occluding the normal nutrient vessels. As discussed

previously, the drawback of using PVA is that embolization with PVA has a lower cure rate than acrylic because of failure to penetrate the nidus and recanalization of the DAVM [21].

In some cases, more effective embolization can be accomplished via a transjugular or transophthalmic venous approach. However, care must be taken not to occlude the posterior cavernous sinus without thrombosing the nidus (Fig. 3.10), because the orbital congestion will markedly worsen as the shunted blood flows entirely into the ophthalmic venous system.

3.2.7.3.2.1 Results of Embolization

In reviewing our cases, 69 patients with DAVMs were considered for embolization because of progressive proptosis, increasing IOP despite topical antiglaucoma agents, debilitating diplopia in primary gaze, and, in one case, severe headache.

Selective catheterization of the middle meningeal artery (the primary blood supply to the DAVMs in these patients) could not be performed in three cases because severe atherosclerosis- related stenosis and tortuosity of the proximal arteries (all prior to 1988 when significant advances in catheter technology were introduced). One of these patients (Fig. 3.12) could not receive embolization treatment despite uncontrolled glaucoma, hypoxic venous retinopathy, optic neuropathy with an altitudinal field defect, severe proptosis, and a third nerve paresis

Embolizations were attempted in 66 cases (two had previously undergone unsuccessful gelfoam embolization at other institutions). This was combined with postembolization manual compression (see below) where necessary. Angiographically, 60 DAVMs were considered to have successful arterial or transvenous embolization, 50 after a single hospitalization. The angiogram performed after one, two, or rarely three embolizations (Figs. 3.16–3.18) demonstrated occlusion that was complete in 36 and partial in 24 of the DAVMs (10 cases required further subsequent embolizations; see below). Stasis of flow in the remaining nidus or arterial feeders (mostly from C4 or C5 branches of the cavernous ICA) was noted in these 24 cases. The radiopaque cast of the nidus and embolized vessels was readily apparent on skull X-ray (Fig. 3.18). A transvenous approach, alone or in combination with arterial embolization, was performed in three cases, with a cure in two and improvement in one. The clinical neuro-ophthalmologic abnormalities, except for a glaucomatous cup, resolved in 60 embolized cases. The IOP normalized within 2 days and the diplopia resolved within 1 month following successful occlusion of a DAVM. Partially occluded lesions clinically improved within 2 months except for the six unsuccessfully embolized cases and the ten cases mentioned below.

Additional embolization was necessary in ten cases with DAVMs initially considered successfully but incompletely embolized after the first embolization. Four cases, all of whom had a C5 carotid contribution to their DAVMs, experienced a clinical recurrence 3 weeks after the first embolization despite PVA occlusion of two of three and three of four ECA feeders in each case. A second embolization in these cases occluded the DAVMs without recurrence. One of these two patients had Osler- -Weber–Rendu disease and had three AVMs in the lungs that were occluded with detachable balloons and IBCA (see Sect. 4.13.5). A fifth case required a second embolization because of an iatrogenic thrombosis of the posterior cavernous sinus (discussed below). Five other patients with extensive DAVMs supplied by bilateral internal carotid, bilateral middle meningeal, bilateral ascending pharyngeal, and left artery of foramen rotundum had a planned staged approach to embolization. The orbital findings persisted in two of these cases, but the signs completely resolved following airplane travel.

In six patients with severe atherosclerosis, only proximal embolizations with PVA could be performed. Significant improvement in the clinical findings resulted in only two cases, so four were failures.

Two additional patients, both with extensive DAVMs (bilaterally supplied by more than six arteries) that extended along the whole base of the skull (Fig. 3.18), had unsuccessful extensive embolization in three procedures. Neither was clinically cured. One patient had resolution of his headache, mild reduction in the IOP, and no visual disability that remained in the 6-year follow-up. The other patient had a progressive bilateral optic neuropathy (diminished acuity and color vision,

central and paracentral scotomas), without significant elevation of the IOP, despite embolization. She was then referred for radiotherapy (see Sect. 3.2.7.3.3.2).

3.2.7.3.2.2 Complications of Embolization

Potentially serious complications from cyanoacrylate treatment of DAVMs are infrequent because liquid acrylic agents are only injected superselectively into arteries that should not supply normal structures. As discussed previously, cyanoacrylate embolization of an artery that supplies a cranial nerve can cause a permanent palsy. Prior to initiating the xylocaine tolerance test, we had one patient who developed a permanent unilateral paralysis of the tongue from occlusion of an ascending pharyngeal arterial branch. Another developed a facial nerve palsy from occlusion of the petrous branch of the middle meningeal artery. Middle meningeal artery embolization with a large volume or a high-pressure rapid injection of the liquid adhesive agent can result in errant emboli passing through anastomoses with the ICA into the cerebral or ophthalmic circulations. We had one patient who had this complication and developed a hemiparesis and hemianopia which lasted 72 h. Another patient had a permanent nasal visual field defect with preserved visual acuity, despite embolization with a small volume and low-pressure injection of acrylic, and the route of abnormal embolization was never apparent.

Complications from PVA embolization are less frequent and are usually transient. Transient worsening in ocular motor or lid function occurred in several patients, but this may have resulted from thrombosis in the cavernous sinus (see below). We have had no patients who experienced a cerebral or ocular ischemic episode following PVA treatment.

If either embolizing agent passes through the nidus into the posterior cavernous sinus, thrombosis will block the venous outflow posteriorly. If the DAVM nidus and any arterial feeders remain patent, sudden progressive orbital congestion that threatens vision develops as all of the arterialized blood is shunted into the anterior cavernous sinus, causing severe hypertension in the ophthalmic veins (Fig. 3.10). We have encountered this problem twice. Immediate embolization of the remaining DAVM and arterial supply is required.

If embolization induces thrombosis in the anterior cavernous sinus and the ophthalmic veins, the orbital congestion worsens, the IOP becomes markedly elevated, and the vision can be severely compromised. The deterioration can worsen over several days. We have seen this complication in only two patients. Maximal medical treatment to lower the IOP is recommended. It is unclear whether heparin or aspirin therapy is beneficial in this circumstance. However, as long as the AV shunt is closed, there is no arterialized blood to cause venous hypertension, and the orbital and ocular tissues receive oxygenated arterial blood. After several days to weeks, the ophthalmic veins recanalize or the orbital venous blood exits the orbit via alternative routes (e.g., facial veins, pterygoid venous plexus). Improvement in all signs is noted in approximately 1 month. Within 3 months of the thrombosis, the neuro-ophthalmologic abnormalities resolve.

Additional minor complications include a headache which is ipsilateral to the embolized dural vessels. The pain resolves within days in all cases. Because of pooling of the contrast material or partial venous thrombosis in the orbital tissues, transient slight worsening of the orbital congestion commonly occurs following angiography.

3.2.7.3.3 Alternative Therapies

Patients were not considered for embolization if they had only minimally elevated IOP, a normal retina and optic nerve evaluation, and fusion in primary gaze. Twenty two additional patients (includes three cases mentioned above in whom selective catheterization could not be performed) were initially managed without embolization (two were embolized later, see below).

3.2.7.3.3.1 Manual Compression

Other therapeutic modalities can be used alone or in combination with embolization. Intermittent manual compression of the cervical carotid artery [87, 67] ipsilateral to the shunt supplied by branches of the cavernous ICA is prescribed as the sole therapy in patients with mild findings. Manual compression is also appropriate when selective embolization is not possible or when there are dangerous anastomoses.

The presence of atherosclerotic disease, such as an ulcerated plaque, in the cervical carotid artery is a contraindication to this therapy. Intermittent compression of a diseased carotid artery can dislodge an embolus into the cerebral circulation [88]. Thus, manual compression can only be implemented in patients who have had angiographic or detailed ultrasonographic demonstration of a normal cervical carotid artery. Poor cross-filling to the ipsilateral hemisphere from the contralateral ICA system is only a relative contraindication to this treatment.

Initially, the ipsilateral carotid artery is compressed using the contralateral hand for 5 min every 2 h for 1–3 days. If this is tolerated, the compression time is then increased by 5 min every 3 days until 20 min of compression per 2-h interval is reached. If cerebral ischemia develops, the motor weakness in the contralateral arm causes release of the hand performing the carotid compression as a safeguard. We have not seen this complication as long as the patient followed these directions. Carotid compression is never performed with the ipsilateral hand or by any person but the patient.

Six of our patients with DAVMs were given self-administered carotid compression as the sole therapy. Two patients had unilateral mild proptosis, sixth nerve paresis, and arterialized episcleral vessels with otherwise normal neuro- ophthalmic evaluations. The third was the patient who refused embolization and had a partial third nerve paresis, glaucoma, and proptosis. Two patients had mild proptosis and slight elevation of the IOP in the affected eye. Despite the presence of secondary glaucoma, venous retinopathy, limitation of eye movements, and proptosis, selective catheterization for embolization could not be accomplished in this case. This last patient had the only complication with this therapy. The patient compressed her cervical carotid artery continuously for 2 h, resulting in a paresis of the contralateral leg which lasted 12 h.

One patient did not comply and one patient was lost to follow- up. All the clinical signs resolved within 2 months except for mild injection of the conjunctiva in three cases.

3.2.7.3.3.2 Radiation

Radiation is an adjunct therapy that is recommended only for patients who have a progressive visual loss despite medical therapy and multiple, but incomplete, embolizations [66, 67]. We irradiated one such patient who had a bilateral optic neuropathy and normal IOP because of an extensive DAVM (Fig. 3.18). The patient had embolization of six arterial feeders, but continued to have failing vision. Several months following therapy with oral dexamethasone and 4000 cGy to the DAVM at the base of the skull, the acuity improved from finger counting at 3 feet OD and 20/60 OS to 20/100 and 20/40, respectively [67]. Her vision remained stable over a 2-year follow-up period.

3.2.7.3.3.3 Atmospheric Pressure Change

As mentioned previously, following cerebral angiography, resolution of the clinical abnormalities may develop. Other factors that may temporarily alter the local circulation may result in thrombosis of the DAVM. Changes in the atmospheric pressure associated with high-altitude travel or an airplane excursion may be therapeutic, particularly in patients who have had partial closure of the shunt by prior embolization. However, uncontrolled venous system thrombosis may result in a permanent visual defect (from choroidal detachment or effusion) despite complete resolution of the DAVM.

One case with a third nerve palsy developed visual loss with finger counting vision from a choroidal detachment and complete unilateral ophthalmoplegia following the plane flight to New York for therapy. On the following day, angiography revealed spontaneous thrombosis of the DAVM, the ipsilateral cavernous sinus, and the ophthalmic vein. The patient was treated with intravenous heparin (partial thromboplastin time elevated to 1.5 times control levels) and the cavernous sinus syndrome resolved over 1 week. The eye movements normalized but the acuity never improved to better than 20/70. Another patient had complete resolution of the bruit and bilateral sixth nerve pareses following a plane flight. Angiography on the next day showed spontaneous occlusion of the DAVM which had been demonstrated on a pri-

or angiogram. A third patient with bilateral orbital congestion and sixth nerve palsies despite two embolizations became completely normal following two airplane flights.

3.2.7.3.4 Therapy of Arteriovenous Shunt Not Warranted

Sixteen cases with DAVMs initially received no therapy directed at their vascular lesion. One patient had no change in his mild proptosis and red eye, which were the only clinical signs, over 3 years. Two patients deteriorated over 6 months and were embolized successfully. In one patient, the DAVM resolved on the airplane flight to our institution. Ten other DAVMs that caused mild proptosis, episcleral injection, and mild elevation in the IOP resolved in several months without complication. Two additional patients had severe visual loss. The first was the patient in whom treatment was considered but could not be performed because of severe arteriosclerosis. She remained with acuity of 20/100 OD and 20/30 OS, moderate color vision loss OD, an inferior altitudinal field loss OD, and an afferent pupillary defect OD. The second was the patient previously discussed with spontaneous occlusion of the DAVM and the ophthalmic venous system following airplane travel.

3.3 Dural Venous Sinus Thrombosis

3.3.1 Introduction

Thrombosis of a dural venous sinus (for venous anatomy see Sects. 1.4, 1.5.3.1.3) is an unusual clinical problem that is frequently overlooked, except when the cavernous sinus is involved. The symptoms and findings are related to the location of the involved sinus and secondary thrombosis of adjacent cortical veins. Thrombosis can develop in any of the dural venous sinuses. Septic causes arise in conjunction with infections of adjacent structures such as the paranasal sinuses or meninges. Aseptic etiologies are rare and develop in relation to chronic debilitating illness such as metastatic cancer, congestive heart failure, and dehydration or disorders that predispose to an increase in systemic venous thrombosis (Table 3.5) [89–97].

In healthy individuals, dural sinus thrombosis develops in association with pregnancy in only approximately one in every 2500 pregnancies. This complication occurs during the pregnancy or in the immediate postpartum period [98]. Numerous cases have been described with temporary occlusion of the posterior superior sagittal sinus or the adjacent cerebral veins and bilateral occipital hemorrhages in the immediate postpartum period (Fig. 3.19). The resulting cortical blindness recovered in most cases without therapy (see Sect. 8.11.3).

3.3.2 Cavernous Sinus Thrombosis

3.3.2.1 Etiology

When associated with a traumatic carotid cavernous fistula or a DAVM, thrombosis in the cavernous sinus is rarely a life- threatening illness. In contrast, when the thrombosis is related to a septic process the consequences can be catastrophic. Previously, this disorder resulted from suppurative facial skin and sinus bacterial infections that propagated intracranially along the venous channels or directly from the adjacent sinus, respectively [99]. The use of antibiotics has essentially eliminated these etiologies [100–103, 103a]. Currently, a hematogenous spread of bacterial infection or local invasion from a paranasal sinus fungal infection are the predominant causes of septic thrombosis, particularly in immune-compromised patients. Rarely, orbital cellulitis-caused thrombophlebitis in the ophthalmic veins leads to direct spread of bacteria into the cavernous sinus [104, 105].

Aseptic cavernous sinus thrombosis is even rarer, but it has been described in association with the many chronic illnesses and systemic thrombotic disorders that can cause thrombosis in any dural venous sinus (Table 3.5)

3.3.2.2 Clinical Signs and Symptoms

The signs and symptoms arise from the congestion in the ophthalmic venous system and the damage to the cranial nerves within the cavernous sinus and

Table 3.5. Conditions associated with aseptic thrombosis in the dural venous sinuses

Chronic debilitating illnesses
Dehydration
Right heart failure
Metastatic cancer
Polycythemia vera
Pregnancy
Oral contraceptives
Sickle cell trait
Systemic coagulation disorders
Compression or invasion by a neoplasm

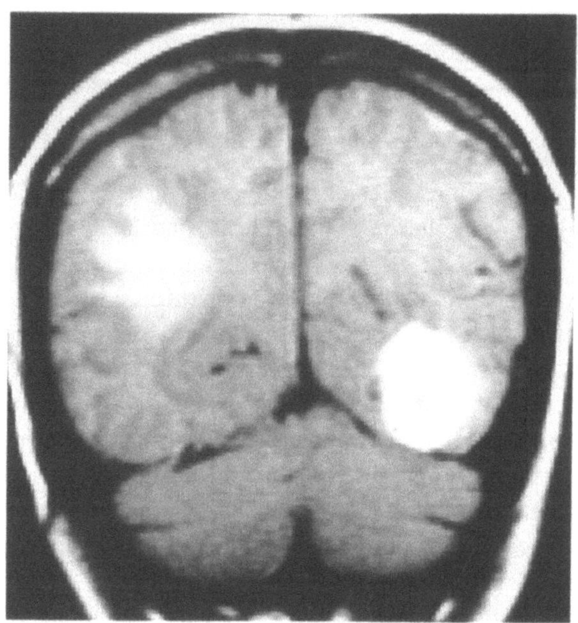

Fig. 3.19. This patient had a sudden onset of headache and bilateral visual loss 2 days post partum. The T1-weighted coronal magnetic resonance scan shows bilateral occipital area hemorrhages presumably secondary to a transient posterior superior sagittal sinus thrombosis. She had a complete recovery of the visual loss without any therapy

superior orbital fissure. The patient complains of a deep retro-orbital pain, proptosis, a red eye, and diplopia. Blindness develops in the ipsilateral eye in approximately 10% of cases [102]. Hypesthesia, but not anesthesia, of the first and second divisions of the trigeminal nerve is common. Partial or complete ophthalmoplegia and ptosis out of proportion to the amount of orbital congestion is also frequent. The venous thrombosis, ischemia from secondary arteritis of the nutrient vessels, and infections neuritis can all cause dysfunction of the cranial nerves. The lid, periorbital skin, and conjunctiva become erythematous and edematous [106]. Though the findings may begin in one orbit, in some patients, even if treated with antibiotics and heparin, the thrombosis can extend through the intercavernous connections within the dura [107] to the contralateral cavernous sinus [108, 109]. This results in bilateral proptosis, lid edema, and ophthalmoplegia.

If the red eye is subtle or overlooked, it may be difficult to clinically distinguish cavernous sinus thrombosis from pituitary apoplexy, sudden expansion or thrombosis within a cavernous aneurysm, or the Tolosa–Hunt syndrome. If the injection of the eye is prominent, orbital cellulitis or inflammation may be suspected. The presence of arterialized limbal loop conjunctival vessels and a bruit are useful in differentiating an AV shunt. Similar findings of ophthalmoplegia and central retinal artery occlusion with subtle injection of the conjunctiva and sclera are also seen with the "Saturday night retinopathy" (see Chap. 8). This disorder can be distinguished by the history of the onset after substance abuse, the lack of fever or progression of findings, and an MR demonstration of a normal cavernous sinus.

Most of the cases of visual loss in the eye ipsilateral to the thrombosis develop in patients with locally invasive fungal infections from adjacent sinuses, while cases of ophthalmoplegia and no visual loss are more likely to develop from a bacterial infection or an aseptic cavernous sinus thrombosis. Occlusion of the central retinal artery (Fig. 3.20) [110, 111], emboli to the retinal arteries [112], ischemic optic neuropathy [108], and an optic neuropathy from extension of the infection into the optic canal are all causes of visual loss. In one case,

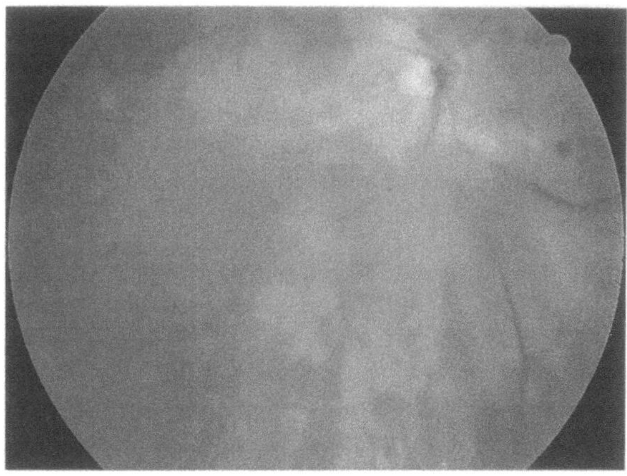

Fig. 3.20. Cavernous sinus thrombosis can be diagnosed without neuroimaging confirmation. The patients are frequently too ill to undergo angiography, and computed tomography and low-field magnetic resonance imaging are often nondiagnostic. A 53-year- old diabetic man had the onset of mild pain and congestion in the right orbit. He sought medical attention because he experienced sudden visual loss several days later in the right eye. At the time of the evaluation, there was mild limitation of eye movement in all directions, several millimeters of proptosis, and slight injection of the conjunctiva and sclera. There was bare hand motion vision in this eye. His temperature was 101 °F (38.3 °C) and he appeared septic. The view of the fundus is obscured by a cataract but diffuse swelling of the inner layers of the retina with narrowing of the retinal arteries is apparent

both posterior ciliary and retinal artery occlusions occurred in the same eye [112a]. Ophthalmic artery thrombosis (recall that the ophthalmic artery can originate from the cavernous portion of the ICA) develops in patients with secondary arteritis of the ICA [113, 114] and causes ischemia to the ocular tissues. Corneal ulceration from exposure can also occur [115]. If there is severe orbital congestion, the marked elevation of the IOP may contribute to the ischemia of the retina or optic nerve [108].

The secondary arteritis and thrombosis within the ICA can cause a single or multiple cerebral infarcts. The lethargy or somnolence commonly observed in patients with cavernous sinus thrombosis, even if aseptic, may also develop from pituitary insufficiency because the venous thrombosis extends to the hypophysis [113, 116, 117].

3.3.2.3 Neuroimaging

Until high-field MR imaging became available, CT, angiography, and venography inconsistently demonstrated thrombosis in the cavernous sinus and ophthalmic venous system.

3.3.2.3.1 Computed Tomography

The contrast-enhanced CT may show nonenhancing enlargement of the cavernous sinus. The superior ophthalmic vein does not enhance, but it may be enlarged. Soft tissue changes, including opacification of the air cells, can be seen if sinusitis in the adjacent paranasal sinuses is the cause. Swelling of the paranasal mucous membranes, the lack of air – fluid level, and bone destruction are characteristic of mucormycosis [118]. If the orbit has been invaded, both extraconal and intraconal soft tissue changes, usually next to an infected sinus, with enlargement of extraocular muscles occur. Unusually, the sinusitis appears minor and the dysfunction is related to extension into the orbital apex, an area which is extremely difficult to visualize [119]. Multiple lucencies in the frontal lobe may be present when cerebral infarction has occurred (Fig. 3.21). Air may also be seen in the cavernous sinus when gas-producing bacteria are the cause of the thrombosis [120].

3.3.2.3.2 Magnetic Resonance Imaging

By virtue of the exquisite sensitivity of high-field MR imaging to the different stages of venous thrombosis, it is the imaging study of choice to investigate patients suspected of having this clinical entity [121, 122]. In the cavernous sinus and the ophthalmic veins, the normal flow void, a sign of flowing blood that causes the vessels to appear black on a T1-weighted study, is absent, causing a gray appearance of these structures. The T2-weighted image may be dark because of the deoxyhemoglobin in the acutely thrombosed veins and venous sinus. After several days the T1 and T2 studies show the thrombosis as white or hyperintense because of conversion of deoxyhemoglobin to methemoglobin (Fig. 3.23). In cases where thrombus is only in short segments of the veins or is only partial, special spin-echo techniques may be required in order to demonstrate the pathology [123,

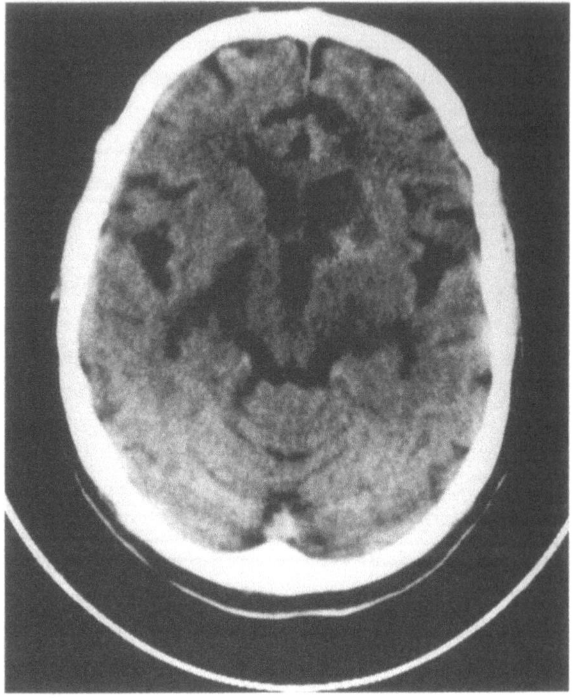

Fig. 3.21. Culture from the nasal cavity and sinuses from the patient in Fig. 3.20 demonstrated *Curvularia dermitiaceous* fungi. The patient was presumed to have a cavernous sinus thrombosis and he was begun on antifungal and anticoagulation therapy. No significant abnormalities were seen on the original computed tomogram scan (not shown). Despite treatment, the patient developed multiple cerebral infarcts (shown) from fungal involvement of the cerebral vessels and died within 3 weeks

124]. If a routine investigation of the brain is performed or low-field MR imaging is used, the images may fail to demonstrate thrombosis within the cavernous sinus.

3.3.2.3.3 Angiography and Venography

The late phase of the cerebral angiogram may demonstrate a partial filling or a complete absence of contrast material in the cavernous sinus. The cortical veins reverse flow and drain blood away from the cavernous sinus towards the superior sagittal sinus (for examples see Figs. 3.3, 3.8). Early in the arterial phase, the ICA may appear totally occluded or a spiral-shaped filling defect, suggestive of a secondary arteritis, may be present [125]. Even if the ICA remains patent, a distal thrombosis may prevent filling of the ophthalmic artery or cerebral arteries.

A venogram performed via an injection of contrast material into the supraorbital vein to the ophthalmic veins, while compressing the facial veins, will not fill the occluded portion of the ipsilateral cavernous sinus [126, 127].

3.3.2.4 Treatment

Immediate treatment with the appropriate antibiotics, based on the gram stain and culture from the inciting skin furuncle, blood, or paranasal sinus, is imperative. Corticosteroids are required if phlebitis affects the pituitary gland, but if fungi are the infecting agent, pharmacological doses of corticosteroids can spread the infection.

The role of anticoagulation in bacterial or aseptic cavernous sinus thrombosis remains controversial. Early complete heparinization decreases the residual morbidity in survivors, but does not alter the mortality rate. The use of low-dose heparin or a delayed institution of anticoagulation seems to be of little value [128]. Anticoagulation does not significantly increase the risk of intracranial hemorrhage or the dissemination of infection [129]. However, if warfarin is used and the prothrombin time rises as high as three times the control, a cerebral hemorrhage is likely to result [130]. Successful treatment with intravenous fibrinolytic agents has only been reported in anecdotal cases.

3.3.2.5 Special Circumstance Phycomycosis

Mucormycosis is the most commonly reported infection that causes both catastrophic orbital and cavernous sinus invasion. The Mucorales fungi are ubiquitous and only become pathogens in patients who are immunosuppressed because of chronic debilitation, diabetes mellitus, or immunosuppressive drugs [131, 122, 123]. Diabetic ketoacidosis is a particularly significant risk factor [124, 125]. Other fungi such as *Aspergillus* spp. and *Curvularia dermitiaceous* can cause a similar but usually less severe syndrome. Fungal spores are inhaled with dust into the nasal cavity where hyphae form and proliferate within the muscular walls of the arteries, veins, and lymphatic channels. The soft tissues and blood vessels become invaded by nonseptating

hyphae [136]. The affected mucous membranes in the nasal cavity appear dark, almost black, from coagulation necrosis. Biopsy of this tissue reveals the hyphae within the vessels and tissues.

The infection rapidly spreads from the nasal cavity via the arteries, directly into the adjacent sinuses and orbit, and via the veins into the cavernous sinus [131]. At this point the patient may have a low-grade fever and will usually complain of head or orbital pain. Invasion into the wall of the ICA causes partial or complete thrombosis of this vessel [137]. In this situation, angiography will often fail to opacify the ipsilateral ophthalmic artery [138, 139]. Once the ICA is involved, cerebral infarcts in either frontal lobe are likely to occur [140].

Thrombosis develops as a consequence of the invasion of the hyphae into the cavernous sinus. Spread of the hyphae to the contralateral cavernous sinus will result in bilateral thromboses [108, 141].

An immediate diagnosis, facilitated by early biopsy of the nasal cavity, leads to early intervention and a higher rate of successful treatment. Even when immediate intensive treatment is delivered, the cavernous sinus thrombosis caused by mucormycosis is often fatal [142]. As mentioned previously, direct invasion into the cerebral arteries results in multiple cerebral infarcts and blindness. Intravenous amphotericin B and radical excision of the involved sinuses and orbit may be life-saving. Although some patients have been spared orbital exenteration and still managed to survive, recovery of visual loss does not occur [143–145]. Despite this aggressive approach, the outcome may not be satisfactory because the underlying failure in immunity prevents control of the infection.

3.3.3 Superior Sagittal Sinus Thrombosis

3.3.3.1 Clinical Findings and Symptoms

Acute occlusion of the superior sagittal sinus can induce headaches, seizures, confusion, and coma as a result of cerebral edema and hemorrhages caused by venous infarction [146]. The overall mortality rate for this disorder is approximately 10% [147], but it is significantly higher in patients with a clinically apparent underlying systemic illness [148]. The thrombus is rarely limited to the superior sagittal sinus, and other dural venous sinuses and cortical veins are also commonly thrombosed [149]. The nature of the focal neurologic deficits depends on the location of the venous cerebral infarcts. Papilledema is common because impairment of the normal brain venous drainage causes increased ICP. Aseptic etiologies (Table 3.5) are much more common than infections spread from the blood or frontal sinusitis. The thrombosis from frontal sinusitis typically involves the anterior portion of the superior sagittal sinus.

3.3.3.2 Neuroimaging

Unenhanced CT can demonstrate a white thrombus within the acutely affected sinus. Depending on the stage of the thrombus, following intravenous contrast material, no abnormality may be seen [150] or a dark triangle in the superior sagittal sinus ("delta" sign) may occur from nonfilling of the sinus [151–153]. Thrombosis of adjacent cortical veins may appear acutely as white cords on the noncontrast CT. In acute cases, multiple small subcortical hemorrhages can be visualized in one or both cerebral hemispheres [154, 155].

Even when the CT does not demonstrate the sinus occlusion, MR imaging and scintigraphy with indium-111-labeled platelets can show the area of thrombosis [156–160]. MR imaging can show acute thrombosis and a lack of blood flow in an occluded sinus.

Cerebral angiography following injection of contrast material into the ICA fails to opacify the thrombosed portion of the superior sagittal sinus and other affected dural sinuses (Fig. 3.22). The venous phase shows dilatations of the medullary veins and reversal of blood flow, with drainage predominantly through the internal cerebral and basilar veins and cavernous sinus [161].

Venography is rarely required to establish the diagnosis. However, the sinus can be catheterized via a small burr hole that permits the introduction of a catheter directly into the superior sagittal sinus which can be used for the direct injection of a fibrinolytic agent (see below).

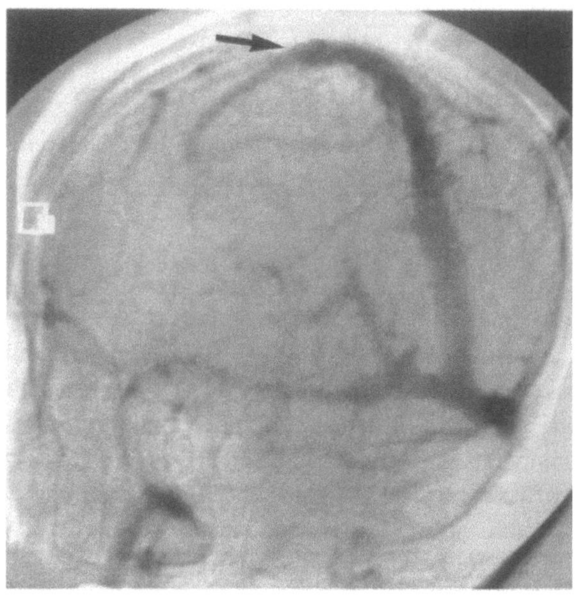

Fig. 3.22. An obliquely oriented digital intravenous angiogram demonstrates thrombosis of the anterior sagittal sinus (*arrow*) in a young woman with a hypercoaguable state and papilledema

3.3.3.3 Treatment

Treatment with early complete heparinization may prevent extension of the thrombosis [149]. Though cases have been reported where anticoagulation is associated with an intracranial hemorrhage as a complication [162], controlled studies have found no increase in this complication when the dose of heparin was monitored to maintain a partial thromboplastin time of twice the control. When aseptic sinus thromboses in all locations were evaluated, heparin treatment resulted in fewer deaths and permanent neurologic defects than no treatment [162a]. Despite these findings, the variable natural history of superior sagittal sinus thrombosis, which can include a benign course (more benign when in an anterior location) in some untreated cases, makes it difficult to recommend full anticoagulation in all cases with occlusion on MR imaging.

The use of systemic intravenous fibrinolytic agents, tissue plasminogen activator, urokinase, and streptokinase, has been reported to be successful in anecdotal cases but an intracerebral hemorrhage may complicate this therapy [163, 164]. Clinical recovery has also been described with a combination of intravenous heparin and urokinase [165], though systemic hemorrhages can occur as a complication [166]. Direct infusion of urokinase into the superior sagittal sinus can lyse the acute thrombus and open the occluded sinus [167, 168].

Any therapy that normalizes the venous circulation will result in a decrease in the raised ICP [167]. Corticosteroids may be helpful in reducing the neurologic deficits caused by secondary cerebral edema. Anticonvulsants are administered if seizures develop.

3.3.4 Lateral Sinus Thrombosis

3.3.4.1 Etiology

In addition to the typical causes of venous sinus thrombosis, the lateral sinuses can be inflamed and thrombosed because of chronic mastoiditis and otitis media. The latter two entities can cause petrositis, which was more prevalent in the preantibiotic era [169].

3.3.4.2 Clinical Signs and Symptoms

Occlusion of the dominant lateral sinus raises the ICP without causing hydrocephalus (called otitic hydrocephalus). This results in bilateral papilledema and headaches (Fig. 3.23) [170–172]. Previously, otitic hydrocephalus was an overdiagnosed condition and most of the patients with this diagnosis probably had idiopathic increased ICP (also called pseudotumor cerebri of childhood) [173, 174].

Infectious thrombosis of the lateral sinus may be associated with chronic mastoiditis and inflammation of the mastoid emissary vein, which causes edema and tenderness behind the ear [175]. Headache, drowsiness, papilledema, and pareses of one or both sixth nerves are frequent. The presence of focal neurologic deficits suggest a spread of the infection with thrombosis of the cerebral veins or the formation of an abscess. Dysfunction of the lower cranial nerves (ninth, tenth and eleventh) occurs if the thrombophlebitis also affects the jugular vein in the jugular foramen. An aseptic thrombosis of the lateral sinus generally has a less virulent and often a variable course, with many of these cases improving spontaneously.

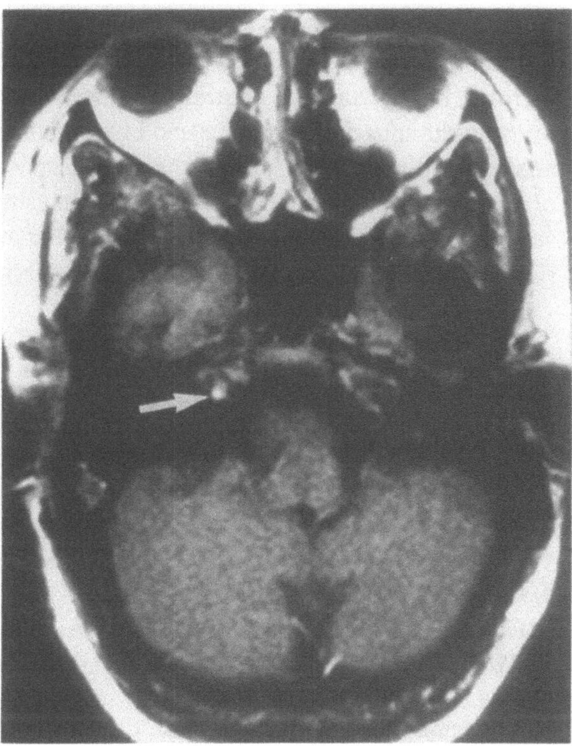

Fig. 3.23. A 23-year-old overweight woman who presented with visual obscurations and headaches was found to have bilateral papilledema. A lumbar puncture was normal except for an opening pressure of 450 mmH$_2$O. She was presumed to have idiopathic increased intracranial hypertension but the magnetic resonance scan demonstrated an area of acute thrombus (*white arrow*) where the sigmoid sinus and internal jugular vein meet. An extensive coagulation evaluation revealed no abnormalities and the patient was successfully treated with oral acetazolamide

The CT, MR, and angiographic findings show changes typical of thrombosis in the specific dural sinus.

3.3.4.3 Treatment

The treatment should be directed towards relieving the elevated ICP, preventing visual loss, and, in the suppurative cases, controlling the infection. In the patient with progressive visual loss, a lumboperitoneal shunt or optic nerve sheath decompression is performed (see Sect. 5.3 for management of this complication). Systemic anticoagulation or antifibrinolytic therapy or both have also been associated with good clinical outcomes.

3.3.5 Thrombosis in Other Dural Sinuses

Thrombosis in the inferior petrosal, superior petrosal, inferior sagittal, or straight sinuses develop from the spread of thrombus and inflammation arising in adjacent venous sinuses. Acute occlusion of the straight sinus is often catastrophic, causing seizures and coma (see Fig. 7.11 in Chap. 7). Extension of the thrombosis into the internal cerebral vein is frequently lethal.

3.3.6 Mechanical Occlusion of the Dural Venous Sinuses

When a neoplasm occludes the dominant venous drainage of the brain, increased ICP and papilledema result. Tumor invasion of a single sigmoid sinus, even when the occlusion is acute, or at the jugular foramen only rarely results in papilledema [176]. Obstruction of the posterior segment of the superior sagittal sinus by a meningioma can also result in ICP elevation and chronic papilledema (Fig. 3.24). If the sinus remains obstructed following tumor removal, the redirection of the venous drainage from the brain will normalize the ICP in 2 – 4 months, but visual loss and optic atrophy develop in the interim [177, 178]. A similar course can develop in cases when a neoplasm or histiocytosis occludes the dominant lateral sinus (Fig. 3.25) [179] or compresses the torcula. In rare instances, removal of the tumor decompresses the sinus, permitting restoration of the normal venous drainage from the brain [180]. In most cases, the affected sinus remains occluded so that a lumboperitoneal shunt or optic nerve sheath decompression is required to prevent the visual loss.

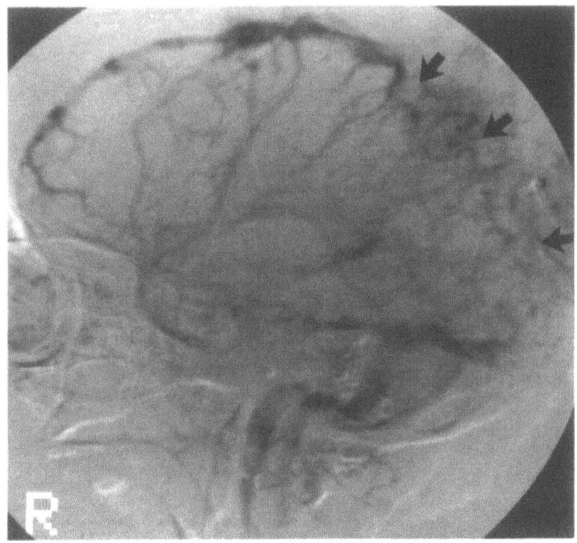

Fig. 3.24. A 35-year-old man had a bilateral midline meningioma that involved both superior occipital lobes and the posterior superior sagittal sinus. Despite several surgical resections and radiation therapy, he developed chronic low- grade papilledema. The late phase of the digital intravenous angiogram demonstrates occlusion of the posterior superior sagittal sinus (*arrows*). There was diffuse prominence of the cortical veins. Acetazolamide therapy successfully controlled the papilledema and stopped his episodes of visual obscuration

Fig. 3.25 a, b. A 29-year-old woman had a history of 6 ▶ months of headaches and visual obscurations that were worse with change of head position. She also complained of hearing loss. **a** A T2- weighted axial view magnetic resonance scan demonstrates bilateral petrous bone lesions which were found to be eosinophilic granulomas (*white arrow*). Thrombosis was noted (*black arrow*) in the region of the right sigmoid sinus to jugular vein connection. A lumbar puncture was normal except for the elevated opening pressure. **b** In the same patient, the late phase of a digital intravenous angiogram demonstrates thrombosis in the right sigmoid sinus (*arrow*). The eosinophilic granulomas were treated with low-dose radiotherapy and the papilledema was controlled with oral acetazolamide

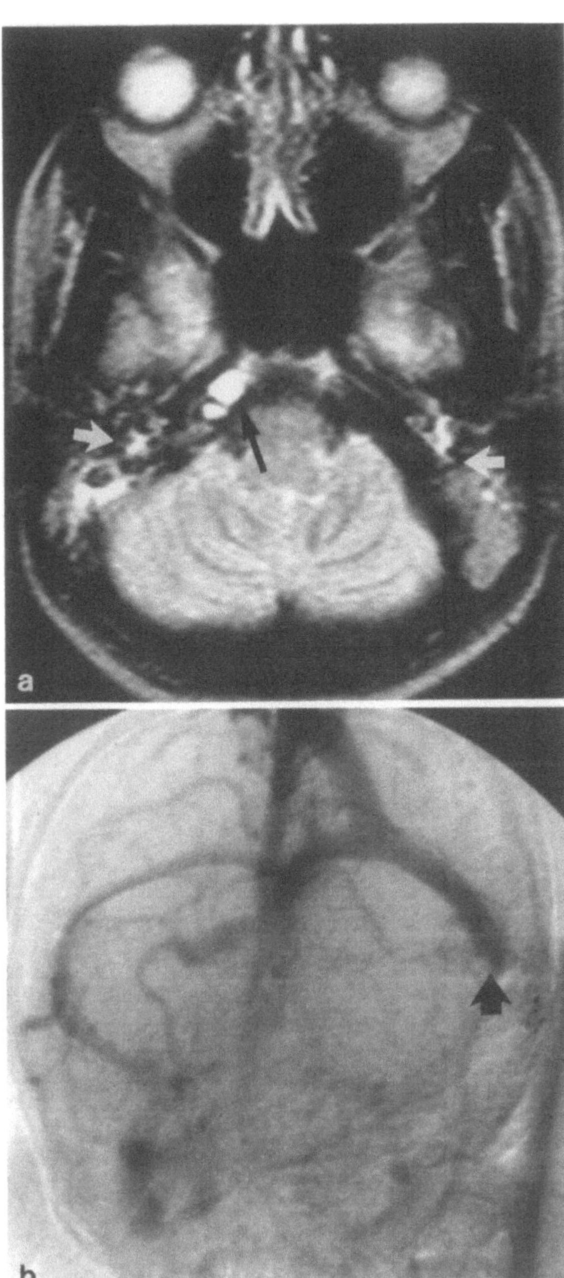

CHAPTER 4

Orbital Vascular Lesions

4.1 Introduction

Vascular lesions of the orbit are infrequent. Though the entities discussed in this chapter are to be considered in the differential diagnosis when a patient presents with the appropriate signs and symptoms, nonvascular lesions such as dysthyroid orbitopathy, orbital pseudotumor, metastatic neoplasm, lymphoma, and mucocele are much more commonly found. Many of the clinical presentations of the vascular lesions develop because of an alteration in the orbital hemodynamics (see Chap. 1) as well as the local mass effect.

4.2 Clinical Evaluation of Patients with Orbital Disease

The signs and symptoms are often similar among patients with a variety of orbital diseases. These include proptosis, ocular motor restriction, ptosis or lid edema, corneal exposure, conjunctival injection, globe indentation, optic neuropathy, lacrimal gland swelling or enlargement, and uncommonly glaucoma. The direction of the displacement of the globe may suggest the location of the lesion. For example, intraconal lesions push the eye straight out. As might be expected, the globe is displaced downward by a mass in the superior orbit and a mass in the inferior orbit causes an upward displacement. The proptosis caused by orbital vascular lesions can be mild to severe. This is contrasted with the slight proptosis that results from severe paralysis of all the eye muscles. In the latter case, relaxation of the orbital tissues allows the globe to bulge forward approximately 3 mm or less without any resistance to manual retropulsion.

Enophthalmos is an unusual sign of orbital disease which is subtle and often overlooked. Non-pathological orbital asymmetry and traumatic orbital floor fractures are the causes in the majority of cases. Less frequent etiologies include the absence of the sphenoid wing, maxillary mucocele, orbital metastasis, scleroderma, and orbital varix [1]. The enophthalmos reported in Horner's syndrome is only apparent.

Orbital congestion is caused by a pathological increase in arterial blood flow into the venous system or a decrease in venous outflow in the orbit, or both. The result is nonspecific diffuse lacrimal injection and swelling (the examiner must elevate the lid and have the patient look downward to see the gland), a generalized mild restriction of all the extraocular excursions, and injection and edema of the conjunctiva. Resistance to manual retropulsion of the globe can be caused by orbital congestion. This resistance is usually greater with orbital mass lesions than with a carotid cavernous fistula or dural arteriovenous malformation (DAVM).

Ocular misalignment from edema and ischemia of the eye muscles causes diplopia in one or all directions. Forced duction testing can be performed to differentiate a neuropathic mechanism from a restrictive myopathy. Following the application of topical anesthetic solution to the conjunctiva, the conjunctiva over the muscle to be tested or the limbus (where the conjunctiva and Tenon's capsule fuse) opposite the direction to be tested is anesthetized with a swab of proparacaine or lidocaine. The patient looks in the direction to be tested for mechanical restriction. Using a tooth forceps, the anesthetized limbus or the muscle and its insertion are grasped and the globe is rotated. For example, if the right lateral rectus is to be investigated, the medial limbus or the right lateral rectus is grasped with the forceps as the subject abducts the right eye. Normally, there should be no resistance to this maneuver. If the result is equivocal, the yoke muscle (for example, the left medial rectus for the right lateral rectus) and the same muscle in the other orbit should be tested for comparison. A patient with an arteriovenous shunt to the cavernous sinus can have a paralytic cranial nerve dysfunction component, an orbital congestion component, and ischemia and hypoxia of the muscles as causes of the ocular motor dysfunction. However, it is not essential to distinguish the mechanism prior to treatment of an arteriovenous shunt.

Ptosis develops if an orbital lesion involves the levator palpebrae or its aponeurosis. Levator weakness is usually associated with underaction of the superior rectus because the two muscles share a common sheath. Orbital lesions also commonly cause lid edema or a pseudoptosis when proptosis is present. A neuropathic ptosis from a lesion of the third nerve in the orbit is unusual. A lesion that involves the ophthalmic artery can cause sympathetic dysfunction with a slight (2–3 mm) downward ptosis of the upper lid and upward ptosis of the lower lid.

Marked proptosis leads to an exposure keratopathy. The resulting corneal abrasion or ulcer is likely to become secondarily infected. A protein flare and cellular anterior chamber reaction are present when the keratopathy is severe. An anterior chamber reaction and corneal thickening with folds in Descemet's membrane can also result if there is significant anterior segment ischemia. A ciliary flush is seen in patients with either an iritis or a keratopathy. The straight dilated vessels of the ciliary flush and the dark dilated vessels of conjunctivitis are easily distinguished from the arterialized bright red limbal loops in the conjunctiva in patients with an arteriovenous shunt in the cavernous sinus or the orbit.

A mass or aneurysmally dilated veins in the anterior orbit indent and compress the globe to cause choroidal folds. Edema of the sclera and choroid also causes folds in the choroid and retina. If this compression and edema are chronic, the overlying area of the retinal pigment epithelium atrophies. Obstruction or compression of the vortex veins that drain the choroidal veins can result in choroidal congestion and hemorrhage. This blood can rupture through the retinal pigment layer into the retina.

An optociliary shunt vessel on the optic disc is a sign of chronic compromise of the venous outflow through the central retinal vein. In this situation, the retinal venous blood passes out of the eye via collaterals (which are normally latent and not apparent on ophthalmoscopy) that appear to "dive" deeply at the disc border to connect to the choroidal venous system. Optociliary shunt vessels have been described most frequently in patients with an optic nerve sheath meningioma or following a central retinal vein occlusion. However, these venous shunts can be associated with any lesion that causes chronic obstruction or slowing of blood flow in the central retinal vein, including increased intracranial pressure, most orbital tumors, and optic nerve gliomas. Ophthalmic vein hypertension or thrombosis by itself is not associated with the development of these shunt vessels.

Dilatation of the ophthalmic vein, usually the superior division of the ophthalmic vein (SOV), can be caused by any lesion that disturbs the venous outflow from the orbit, alters the direction of the venous blood flow, or causes venous hypertension. The different possible causes of this venous dilatation include a carotid cavernous fistula, DAVM or mass that involves the cavernous sinus, dysthyroid orbitopathy, orbital varix, and pseudotumor in the posterior orbit. These entities can be really distinguished by the clinical presentation and examination, computed tomography (CT), magnetic resonance (MR) imaging, and if necessary angiography (see Sects. 2.3.4, 3.2.7.2).

Optic nerve dysfunction is one of several reasons for visual loss in orbital disease. A mass that chronically compresses the optic nerve can cause local ischemia, demyelination, and axonal transmission blockade. Acute orbital hemorrhage, hemorrhage into a mass, and acute ophthalmic venous thrombosis can acutely compress and infarct the optic nerve and severely elevate the intraocular pressure. The rise in intraocular pressure and optic nerve compression may be rapidly relieved by a lateral canthotomy and intravenous mannitol (Fig. 4.1). If the optic neuropathy or the extreme rise in the intraocular pressure are not alleviated by the canthotomy, a more extensive orbital decompression, usually via a lateral orbitotomy with removal of a piece of the zygomatic bone, is emergently performed.

Mass lesions which are confined to the orbit infrequently cause ocular hypertension. This is contrasted with the significant incidence of glaucoma

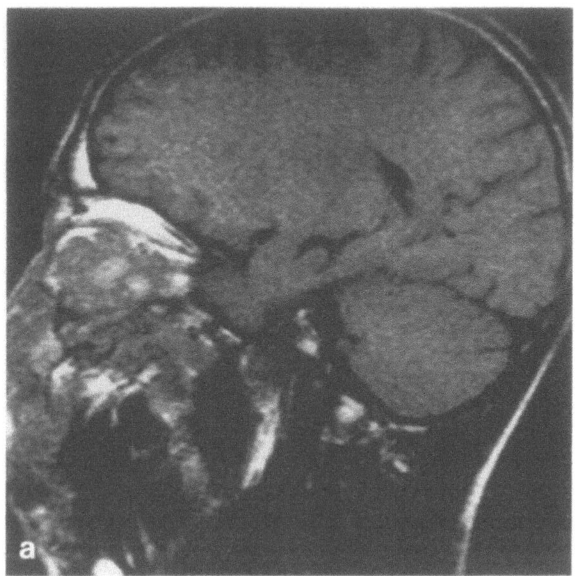

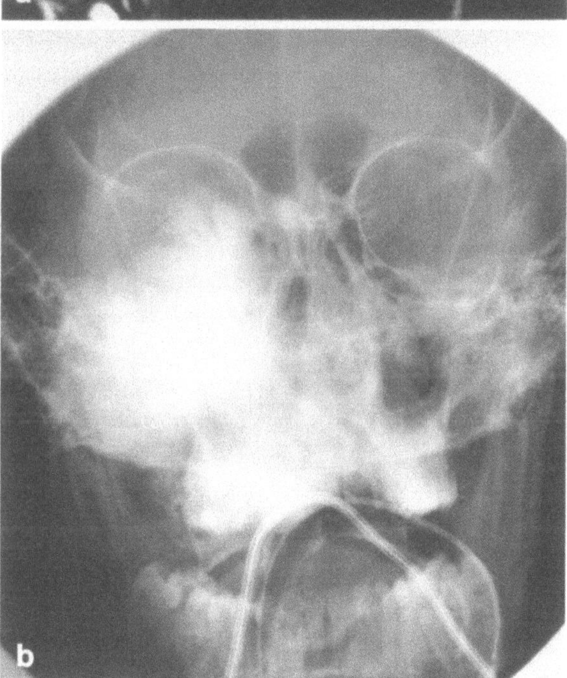

Fig. 4.1a, b. This 13-year-old boy had an venous vascular malformation that involved the palate and orbit with recurrent bleeding from the mouth and nose and mild proptosis. **a** The sagittal T1-weighted magnetic resonance (MR) image shows the mass involving the midface and extending within the muscle cone of the orbit. **b** A frontal view skull radiograph following contrast material injection via direct puncture of the venous malformation shows filling of the lesion from the palate to the orbit. After treatment by direct ethanol injection into the lesion, acute thrombosis caused expansion of the mass, resulting in orbital congestion, marked elevation in intraocular pressure, and an acute reduction in visual acuity to 20/100. A relative afferent pupillary defect was present and the fundus was normal. Immediate therapy with intravenous mannitol and a lateral canthotomy was effective in normalizing intraocular pressure and restoring the vision to 20/20. There was no residual change in the optic nerve nor an afferent pupillary defect and the mass eventually regressed

that develops with arteriovenous shunts of the cavernous sinus (see Sects. 2.3.4.2.5, 3.2.7.1.2). When there is swelling, infiltration, or restriction of the inferior rectus, the intraocular pressure is elevated on upgaze because of compression against the globe.

The presence of an orbital bruit or a palpable or visible pulsation almost always indicates the presence of a pathological increase of the arterial blood flow into the orbit or of the rapid flow in the ophthalmic vein from the shunting of arterial blood (see Sect. 2.2.4). If the bruit is auscultated only over the orbit, it most often arises from an arteriovenous shunt in the cavernous sinus with venous drainage predominantly to the orbit or a high-flow shunt within the orbit. However, an orbital bruit does not always arise from rapid blood flow (see Sect. 8.3.1). Stenosis and turbulence in the carotid siphon from atherosclerosis is an infrequent cause of a bruit over the globe or orbit. Transmitted pulsations from the brain can also cause an orbital bruit in patients with the nonvascular lesion of congenital agenesis of the sphenoid wing [2–4]. The inexperienced examiner can improperly conclude that the presence of an orbital bruit and proptosis in the latter case is caused by a carotid cavernous fistula, particularly if the patient is examined for the first time following trauma [5]. The lack of arterialized conjunctival vessels will distinguish sphenoid wing agenesis from a carotid cavernous fistula without a radiograph. The direction of the proptosis is probably not useful in differentiating these two entities. However, a skull X-ray (anteroposterior view), CT, or MR imaging will definitively demonstrate the developmental bone anomaly. Occasionally, a neurofibroma or an optic nerve glioma is found in an orbit with sphenoid wing agenesis.

Specific orbital signs and symptoms are dependent on the hemodynamics of each lesion. The shunting of arterial flow into the ophthalmic venous system causes an audible bruit and creates venous hypertension and a "steal," with relative hypoxia and ischemia of the orbital structures. Pulsations are rarely visible, except in the veins of the lid, but palpable pulsations are more frequently present. There are numerous signs of pathological elevation of the pressure in the ophthalmic venous system (see Sect. 2.2). These include proptosis, dilated veins in the lid, conjunctiva, and episclera (brighter red if the blood flow is arterialized), elevated intraocular pressure, chemosis, swelling of the lacrimal gland and muscles and ophthalmic veins, and dilatation of retinal and choroidal veins.

Except for the absence of a bruit, patients with lesions that slow or occlude the normal venous outflow from the orbit generally have signs of orbital congestion which are similar to those caused by an arteriovenous shunt. The clinical dysfunction is also usually less severe, except when acute thrombosis of the ophthalmic vein or a retrobulbar hemorrhage obstructs the venous outflow and causes a sudden dramatic worsening. Proptosis also develops when the abnormal large orbital veins are distended by transiently increasing the pressure within the systemic venous circulation as occurs with a Valsalva maneuver. Increasing the local pressure in the ophthalmic venous system by the force of gravity when the head is flexed forward below the heart has a similar effect. Not all orbital vascular lesions show this response. For example, the proptosis caused be a lymphangioma does not worsen with elevation of the systemic venous pressure despite the presence of numerous pathologic venous channels. Another hemodynamic factor, episodic spontaneous hemorrhages in the lymphangioma, accounts for some cases with acute worsening of the proptosis, as well as hemorrhage in the conjunctiva, ocular motor dysfunction, and visual loss.

4.3 Imaging of Orbital Lesions

CT, MR imaging, angiography, venography, and ultrasonography are all used for imaging patients with orbital disease. Each study has its benefits and weaknesses, so that two or more of the studies are often necessary to diagnose and plan management of an orbital vascular lesion. There are few reasons to order a routine X-ray series or tomography of the orbit, since the soft tissues are not visualized and the orbital bones are better shown with other modalities such as CT. An exception to this rule occurs when the lesion originates in the orbital bones and secondarily affects the tissues of the orbit. For example, an interosseous capillary hemangioma of the frontal bone has a diagnostic pattern of a lytic lesion with a fine radiating reticular pattern seen on the radiographs [6].

4.3.1 Computed Tomography

Following the administration of intravenous contrast material, the CT of the normal orbit demonstrates slight enhancement of the sclera and choroid, lacrimal gland, and extraocular muscles with moderate enhancement of the ophthalmic artery and veins [7, 8]. Both axial and coronal views are necessary to demonstrate the relationship of a lesion to the normal orbital structures. Examination of the orbital bones may aid in determining the nature of an orbital lesion. For example, acute lesions rarely alter the orbital bones. In contrast, a chronic mass lesion can widen the optic canal and the superior and inferior orbital fissures. Chronic pressure on the bone from within the orbit can thin and focally displace the walls of the orbit outward, but the lamina of the bone usually remains intact. In contrast, a malignant lesion can invade the orbital wall, eroding and destroying the bone. Mass lesions that are congenital or begin during infancy frequently enlarge the entire orbit. One drawback in the use of CT is the diagnosis of subtle degrees of proptosis. Minor rotations of the head or eccentric gaze cause one eye to appear to protrude compared to the other on the axial images, so that misdiagnosis by CT is possible.

4.3.2 Magnetic Resonance Imaging

High-field MR imaging provides resolution equal to or superior to CT for anatomical detail of the soft tissues in the orbit [9]. A fat suppression program will reduce the chemical-shift artifact of the orbital fat [10]. Improving the signal to noise ratio by the use of a surface coil enhances the MR demonstration of orbital structure fine detail. Sagittal, coronal, and axial planes are imaged without having to reposition the patient as is necessary for CT. The entire length of the optic nerve to the chiasm can be seen in one axial plane image. The orbital fat has a short T1 value, giving it a high signal (appears white). However, with fat suppression, the intraconal fat has a low intensity (dark appearance). The extraocular muscles, optic nerve, lacrimal gland, and globe have a longer T1 value (appear darker). The lens appears slightly brighter on a T1-weighted study. The T2-weighted spin echo technique produces a high signal from the vitreous (appears white) similar to cerebrospinal fluid, while the fat becomes darker relative to water. The ophthalmic artery in the posterior orbit and the ophthalmic veins are easily seen on T1-weighted images because they are dark from the flowing blood. The choroid may also appear dark on a T1 study because of the rapid blood flow. One drawback of MR imaging is that the information provided by viewing the orbital bones is lost.

Following intravenous gadolinium, structures in the orbit that lack a significant blood–tissue barrier will enhance. For example, the T1-weighted images demonstrate lighter eye muscles and marked whitening of the choroid [11].

The resolution of high-field MR imaging is particularly useful for demonstrating the relationship of a lesion to the normal orbital tissues. High-flow vascular orbital lesions often appear heterogeneous with areas of rapid blood flow, slower blood flow, thrombosis, and hemorrhage (see individual entities). Similar to a contrast CT, intravenous gadolinium will cause many lesions to enhance (appear more white on T1) [11].

4.3.3 Ultrasonography

Orbital ultrasonography in patients suspected of having vascular lesions is performed through the closed eyelids using the A-mode, B-mode, and Doppler techniques. The A-mode is helpful in characterizing the nature of the scanned tissue [12]. The B-mode reveals the anatomical relationships of a lesion to the normal orbital structures. Ultrasono-

graphy can be helpful in differentiating certain neoplasms that appear solid on CT and MR imaging. However, both CT and MR imaging are superior for localizing and determining the relationship of the lesion to the normal orbital tissues. Ultrasonography is not helpful when lesions are located in the posterior half of the orbit. Alterations in the direction or rate of blood flow can be demonstrated with Doppler sonography [13].

4.3.4 Angiography

Angiography of orbital vascular lesions must be performed with selective catheterization and high-resolution techniques in order to visualize the ophthalmic artery and the important branch arteries to the eye and optic nerve and the abnormal blood supply to the lesion (see Sects. 1.5.1, 1.5.2). Bone subtraction techniques, optimal patient positioning, and magnification of the visualized vessels enhance the diagnostic yield. When there is chronic increased blood flow to the orbit, the ophthalmic artery can enlarge. Displacement of the ophthalmic artery is difficult to detect unless the mass effect is significant. Straightening of the ophthalmic artery may also develop from a large lesion. Arterial branches to the lesion must be identified, particularly when the abnormal vessels arise en passage from arteries that supply the eye and optic nerve. The angiogram can also indicate whether other intracranial arteries are involved with the pathological process. Arteriography of the external carotid artery system may also visualize direct or indirect blood supply to the lesion and important orbital anastomoses to be avoided in order to prevent potential ocular complications of therapy.

4.3.5 Venography [14]

Since the development of noninvasive orbital imaging by CT and MR, orbital venography is performed only in specific situations. The contrast material is injected into the ophthalmic venous system via a collateral midline forehead vein or the angular vein. Less commonly, the ophthalmic venous system is investigated via a catheter that is introduced into the internal jugular vein and passed

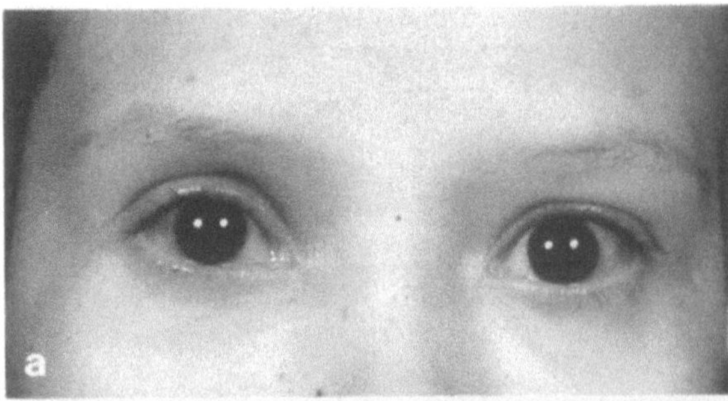

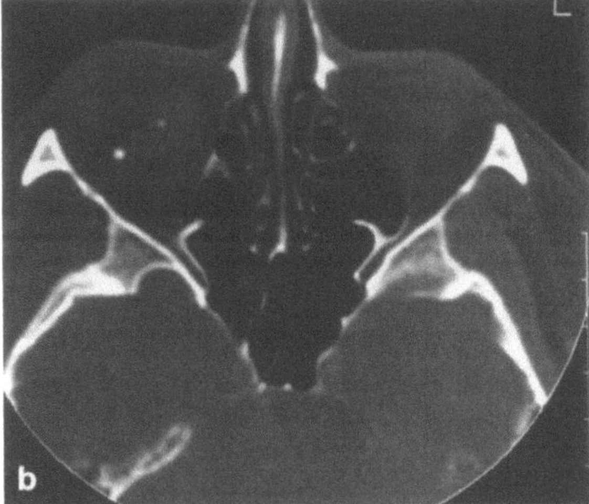

Fig. 4.2. a Frontal view of a patient with a right orbital varix reveals enophthalmos of the right eye. Following a Valsalva maneuver, the globes were in the same position in the orbit (not shown). **b** In this patient, the axial plane unenhanced computed tomography scan with bone windows demonstrates a sausage-shaped mass in the superior orbit involving the superior ophthalmic vein with two punctate areas of calcification representing calcified thrombus within a varix

through the cavernous sinus. Venography of any orbital mass lesion can reveal nonspecific compression and displacement of the ophthalmic vein. The direction of displacement depends on the location of the lesion. The veins proximal to the stenosis, compression, or thrombosis of the ophthalmic vein dilate. Massive venous dilatation occurs with a varix or a high-volume arteriovenous shunt in the orbit.

4.4 Orbital Varix

4.4.1 Clinical Manifestations

A varix of the ophthalmic vein is probably the most common orbital vascular anomaly [15], but it is an infrequent orbital malformation. Varices are not associated with other intraorbital or intracranial vascular malformations or neoplasms.

The clinical history and features salient to this illness include: unilateral (extremely rare bilateral lesions) intermittent exophthalmos (see Table 4.1 for differential diagnosis), proptosis brought out or worsened by maneuvers that elevate the ophthalmic venous system pressure such as change in head position, coughing or straining; and intermittent swelling of the lids or venous engorgement of the lids conjunctiva or both. On occasion, the affected eye may actually be enophthalmic (Fig. 4.2a). The lesion may be clinically apparent within the first or second decade of life but subtle degrees of exophthalmos may not be noted early. Careful review of old photographs usually demonstrates that the proptosis and a widened palpebral fissure are longstanding.

An acute orbital hemorrhage occurs in less than 5% of patients. Thrombosis of the varix or an orbital hemorrhage accounts for episodes of sudden worsening of the proptosis, pain, and visual loss [16]. The acute deterioration in vision occurs as a result of a compressive or thrombotic central retinal vein occlusion or a retrobulbar optic neuropathy. Direct compression of the globe results in choroidal folds (Fig. 4.3a). Mechanical restriction of the extraocular eye muscles can also develop with an acute hemodynamic alteration. Many varices are asymptomatic and are only diagnosed when CT is performed for an unrelated problem. Patients can have recurrent thromboses or bleeds that cause little, if any, permanent visual loss. Less commonly, there may be a step-wise deterioration in the vision or a catastrophic, severe, permanent visual loss.

Few of the diseases listed in Table 4.1 cause episodic proptosis without some degree of antecedent orbital signs, and the CT and MR scans are frequently diagnostic. Intermittent obstruction of the orbital venous drainage as it passes through the su-

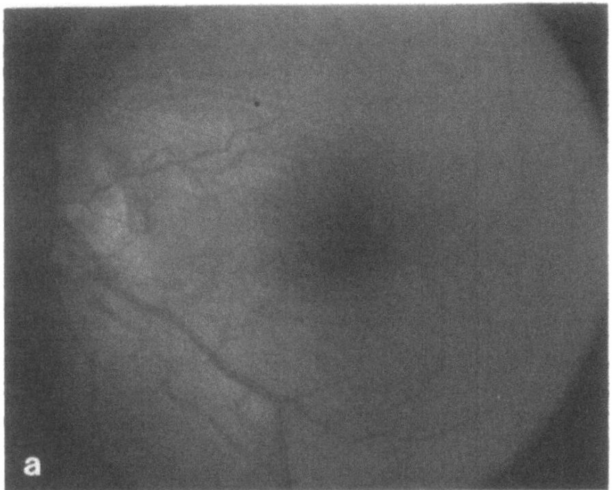

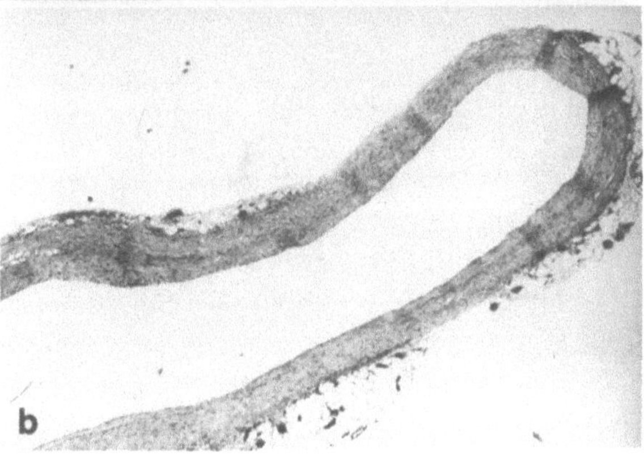

Fig. 4.3. a Ophthalmoscopy of the left eye demonstrated choroidal folds involving the macula which were caused by external compression of the globe by a varix. **b** The histopathology of the orbital varix demonstrates a single-lumen vascular channel with smooth muscle in the walls. (Case provided by Richard Kieler)

Table 4.1. Differential diagnosis of intermittent exophthalmos

---

Varix
Recurrent sinus infections (ethmoiditis)
Mucocele
Lymphangioma
Hypoplasia or agenesis of the sphenoid wing and
    encephalocele
Neurofibromatosis
Ruptured dermoid
Dysthyroid ophthalmopathy
Orbital inflammatory pseudotumor
Carotid cavernous fistula
Dural arteriovenous malformation to
    the cavernous sinus
Chronic hydrocephalus with herniation of the sub-
    arachnoid space through the superior orbital fissure

---

perior orbital fissure is the cause of exophthalmos in many of these lesions.

4.4.2 Histopathology

The varix is either a predominantly single-channel vascular structure or composed of several dilated channels (Fig. 4.3b). The vessel walls are thin and contain smooth muscle and areas of fibrosis. Thrombus with varying degrees of organization and areas of calcification are found within the vessel lumen.

4.4.3 Imaging

If skull radiography is performed (not necessarily recommended), increased vascular markings in the orbital region of the frontal bone and small round calcifications within the orbit from organized thrombi may be seen (Fig. 4.2b). However, cavernous hemangiomas and other structures that contain areas of venous thrombosis can have similar appearing phleboliths.

On occasion, a CT scan, even when performed with contrast, may fail to reveal a varix. Due to the compression of the normal soft tissue structures in the orbit and the frequent location in the lateral orbit adjacent to the lateral rectus, a varix is sometimes difficult to differentiate from an enlarged or infiltrated extraocular muscle. If the patient performs a Valsalva maneuver or the ipsilateral jugular vein is compressed during the performance of the scan, a characteristic dilatation of the orbital varix and the ophthalmic veins will reveal a sausage-shaped mass [17]. On occasion, the varix has a CT appearance similar to that of many different orbital mass lesions.

An MR image obtained without a surface coil, using the standard T1- and T2-weighted protocols, can appear normal because the hyperintensity of the orbital fat may prevent the varix from being distinguished. The varix does not have a signal void because of the lack of rapid blood flow in the lumen. A mixed signal with both hypointense and hyperintense areas on MR images is suggestive of a partially thrombosed varix [18].

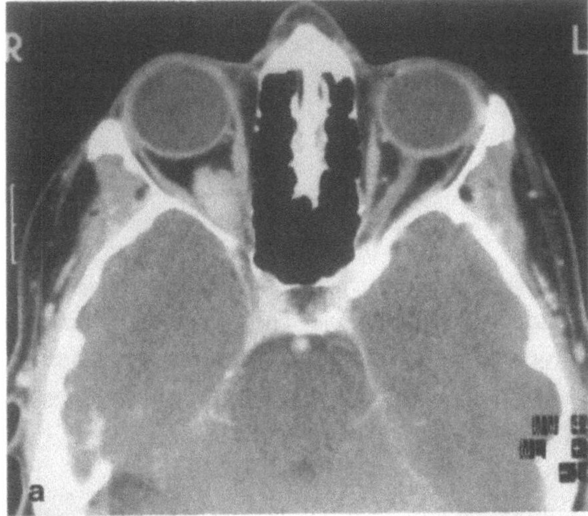

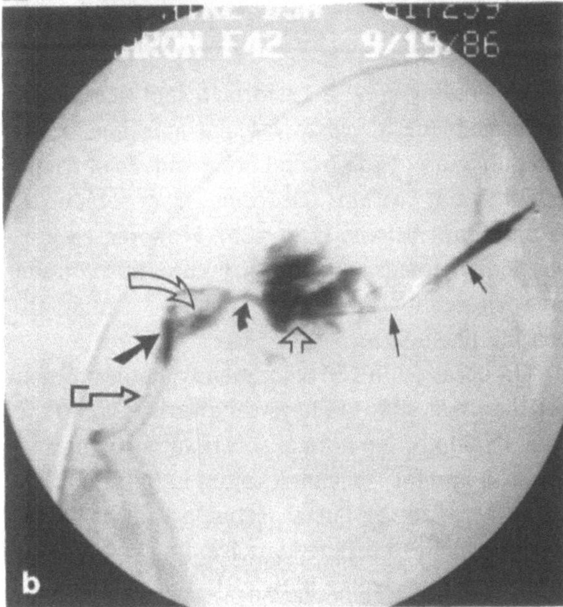

Fig. 4.4a, b. This 45-year-old lady had experienced intermittent orbital signs and episodes of visual loss over the course of 15 years until she was completely blind in this right eye with a pale atrophic optic nerve. She complained of chronic pain that was worse with change of head position, particularly when she was face down.
a The contrast-enhanced axial view CT scan demonstrates an enhancing smooth wall mass in the superior orbit extending to the orbital apex. In the same patient, orbital MR imaging was entirely normal.
b A subtracted lateral view (anterior to the right) orbital venogram from this patient demonstrates the introduction of a microcatheter passed through a larger angiocatheter (*small arrows*) into the opacified varix (*open arrow*). The drainage via a small ophthalmic vein through the superior orbital fissure (*curved arrow*) into the cavernous sinus and the internal carotid artery within the cavernous sinus (*open curved arrow*) is seen. Drainage from the cavernous sinus passed posteriorly into the basilar venous plexus (*angled arrow*) and then inferiorly into both inferior petrosal sinuses (*open arrow with tail*)

If treatment of the lesion is considered, carotid angiography is required to rule out the presence of an arteriovenous fistula, an orbital AVM or an intracranial AVM with orbital venous drainage (see Sect. 4.5). Except in cases where the lesion contains a capillary component, a varix may not be visualized by the injection of contrast material into the ipsilateral internal carotid artery [19].

Venography remains the diagnostic procedure of choice in most cases. Venography is accomplished either by the injection of contrast material directly into the varix (if the varix is located far anteriorly in the orbit and reachable with a small exci-

sion under the supraorbital rim) or indirectly through catheterization of a superior orbital or angular vein. A massively dilated ophthalmic vein with one or more areas of constriction is typical (Fig. 4.4b) [14, 20–22]. An alternative catheterization route, via the internal jugular vein through the petrosal sinus, then through the cavernous sinus to the ophthalmic vein, will demonstrate the relationship of the varix to the cavernous sinus and the intracranial portion of the ophthalmic venous system.

The B-scan ultrasound often shows dilated venous structures, but a dilated ophthalmic vein is di-

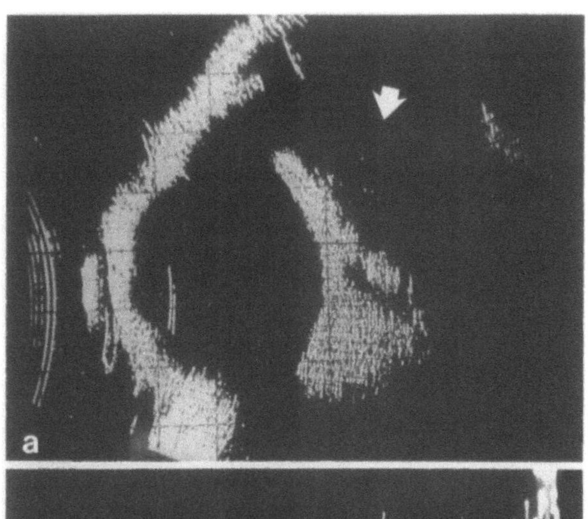
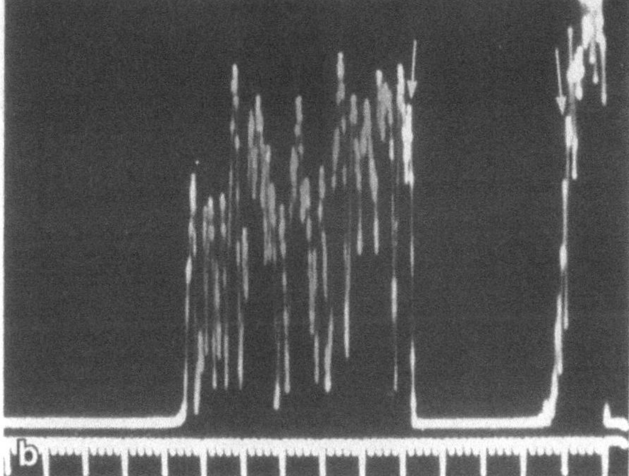

Fig. 4.5. a Following a Valsalva maneuver, a large echo-free area is seen superior and posterior to the globe (*white arrow*) on the B-scan in a case of orbital varix. **b** The A-scan shows a sharply delineated (*white arrows*) echo-free area of low flow within the same orbital varix. (Case provided by Richard Koplin)

agnostic (see Sect. 4.2) only if the enlargement of the echo-free area is only seen with a Valsalva maneuver or compression of the internal jugular vein (Fig. 4.5a). An A-scan shows a smooth-walled structure which is clearly delineated from surrounding structures (Fig. 4.5b). The internal reflectivity is low because of the slow-moving congested blood flow. Blood flow activity is difficult to measure on a Doppler study because of this lack of significant blood flow.

4.4.4 Course and Treatment

In general these lesions have a benign course. The patients typically have intermittent mild proptosis that is prominent when bending forward, lifting, straining, and performing other maneuvers associated with elevation of the venous pressure. Episodes of moderate or severe exophthalmos are due to recurrent thrombosis or hemorrhage or both within the lumen of the abnormal vessels. Diplopia and visual loss can occur during these episodes and the vision may remain impaired despite a spontaneous reduction of the proptosis. Severe visual loss is unusual. A surgical decompression of the orbit may be required if an acute hemorrhage or thrombosis causes severe orbital compression that threatens the vision because of marked elevation in the intraocular pressure, corneal exposure or compression of the optic nerve. In some cases, the optic nerve dysfunction and the elevated intraocular pressure can be relieved by a large lateral canthotomy (Fig. 4.1). However, if this minor procedure is not immediately successful, then a lateral orbitotomy must be performed. Continuous pain or disfigurement can also be treated surgically, but these are less frequent indications for a surgical procedure.

The surgical evacuation of blood deep within the orbit and excision of the varix can be curative, but sometimes the varix cannot be found without extensive dissection and damage to the normal orbital structures. Surgery itself may also cause a significant hemorrhage into the orbit if the blood flow into the varix cannot be controlled. Embolization treatment using electrocautery [20], polyvinyl alcohol particles or metal coils advanced through a tributary vein, or direct injection into the varix can induce thrombosis and involution of the mass and prevent hemorrhage. However, the acute thrombosis of the lesion induced by embolization may cause severe, though temporary, orbital compression that could require an emergency canthotomy.

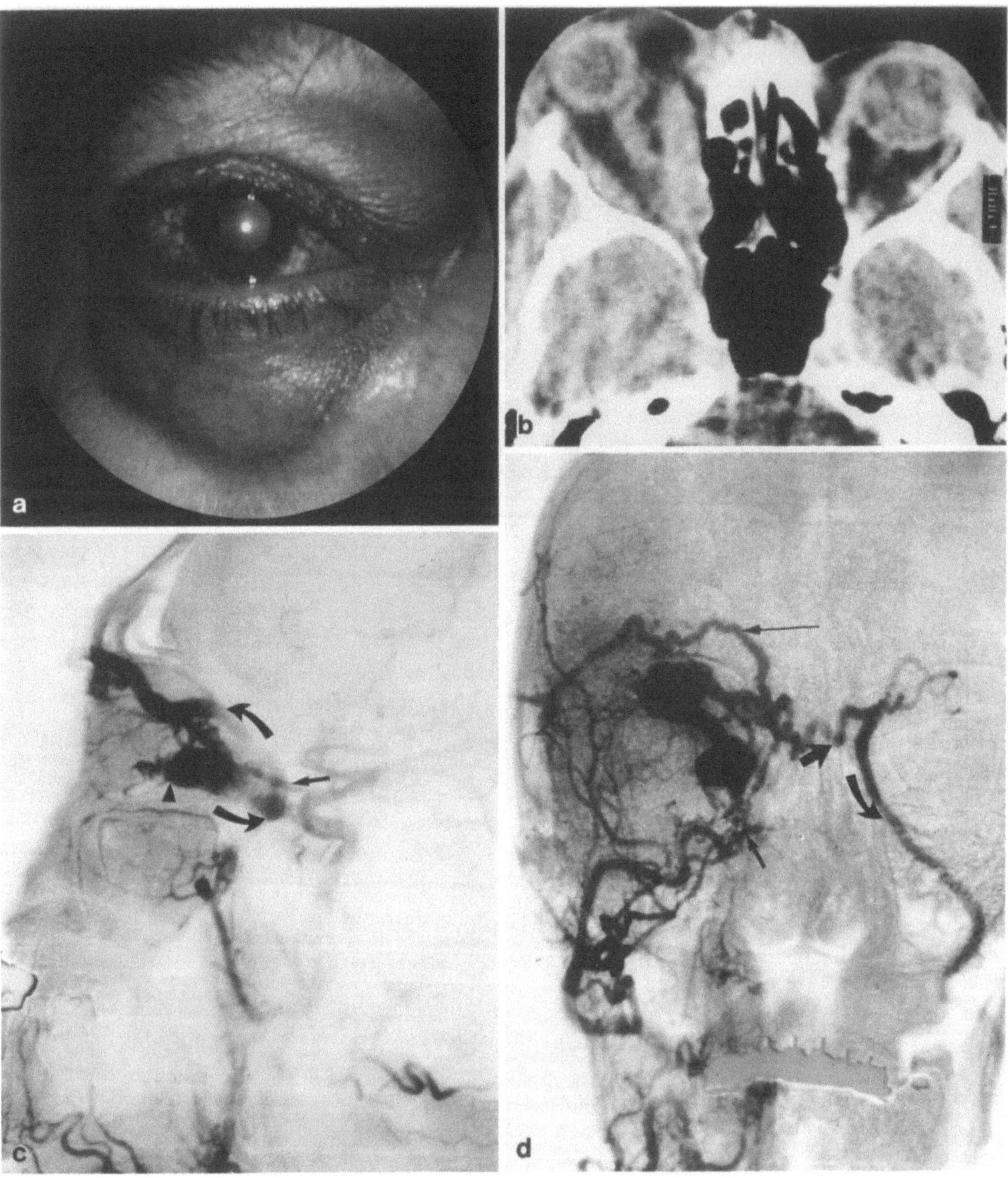

4.5 Arteriovenous Malformation

4.5.1 Clinical Manifestations

True congenital AVMs are rarely isolated within the orbit. Most arteriovenous shunts in the orbit are part of a more extensive intracranial or facial AVM. The abnormal vessels are often part of the dilated ophthalmic venous system, carrying shunted arterial blood from a brain or DAVM or a carotid cavernous fistula. Additionally, some intracranial AVMs receive arterial blood supply via hypertrophied collaterals from the orbital or the ethmoidal branches of the ophthalmic artery or from branches of the external carotid artery which pass through the orbit. In these cases, a "steal" from the ophthalmic artery rarely causes amaurosis fugax [23]. Venous hypertension in the ophthalmic vein accounts for most of the signs of orbital and intraocular congestion (Fig. 4.6). Some orbital AVMs are associated with the vascular neurocutaneous disorders such as Wyburn–Mason (Sect. 4.13.3) or Osler–Weber–Rendu syndromes (Sect. 4.13.5).

Patients with a true congenital orbital AVM (Fig. 4.7) can have a variable course. While the proptosis in some cases is progressive, in others it remains stable. Signs of orbital congestion, arterialized conjunctival vessels with limbal loops and conjunctival edema, are also variably present. Pulsation in the upper lid can be seen and palpated, and a bruit is frequently auscultated over the globe. The course may be punctuated by episodes of spontaneous orbital hemorrhage which suddenly cause a catastrophic exacerbation of the proptosis and visual compromise. The retrobulbar hemorrhage compresses the optic nerve or decreases the intraorbital blood supply to the optic nerve. Visual compromise also results from marked elevation of the intraocular pressure, central vein occlusion, or an exposure keratitis.

Traumatic arteriovenous shunts of the orbit are not real AVMs because they develop as a result of focal damage to an artery in the orbit. One such shunt can occur following an injury to an ethmoidal artery due to fracture of the ethmoid bone with subsequent rupture of the artery into the ophthalmic venous system [24]. The presence of an intraorbital shunt may be masked by the finding of an associated injury to the intracranial vessels, such

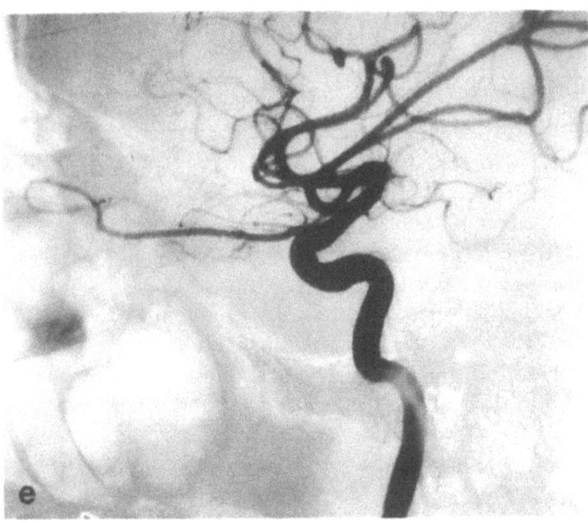

Fig. 4.6a–e. A patient with an orbital arteriovenous malformation experienced episodes of sudden proptosis and reduced vision in the affected eye.
a Signs of orbital congestion including lid swelling and arterialized vessels on the conjunctiva and sclera are in evidence. A venous stasis retinopathy was also present (not shown).
b An axial view CT scan demonstrates the obvious proptosis and signs of orbital congestion including swelling of both the lateral and medial rectus muscles.
c The lateral common carotid artery subtraction angiogram shows that the hypertrophied ophthalmic artery (*arrow*) supplies a large arteriovenous shunt to the inferior division of the ophthalmic vein (*arrowhead*). Because of flow restriction anteriorly, the contrast from the inferior division flows retrograde into the superior division of the ophthalmic vein (*curved arrows*).
d The frontal right external carotid artery injection demonstrates the infraorbital artery supply (*small arrow*) to the arteriovenous malformation. Note the drainage from the malformation to the nasal venous arcade (*arrow*) which fills the contralateral angular vein (*curved arrow*) and an ipsilateral superficial vein (*long arrow*) above the orbit
e Following embolization and surgical therapy, the lateral internal carotid artery angiogram demonstrates a normal ophthalmic artery without any signs of an arteriovenous malformation remaining in the orbit

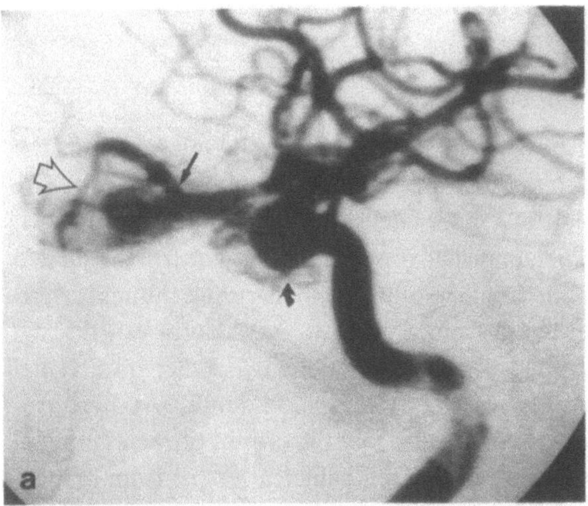

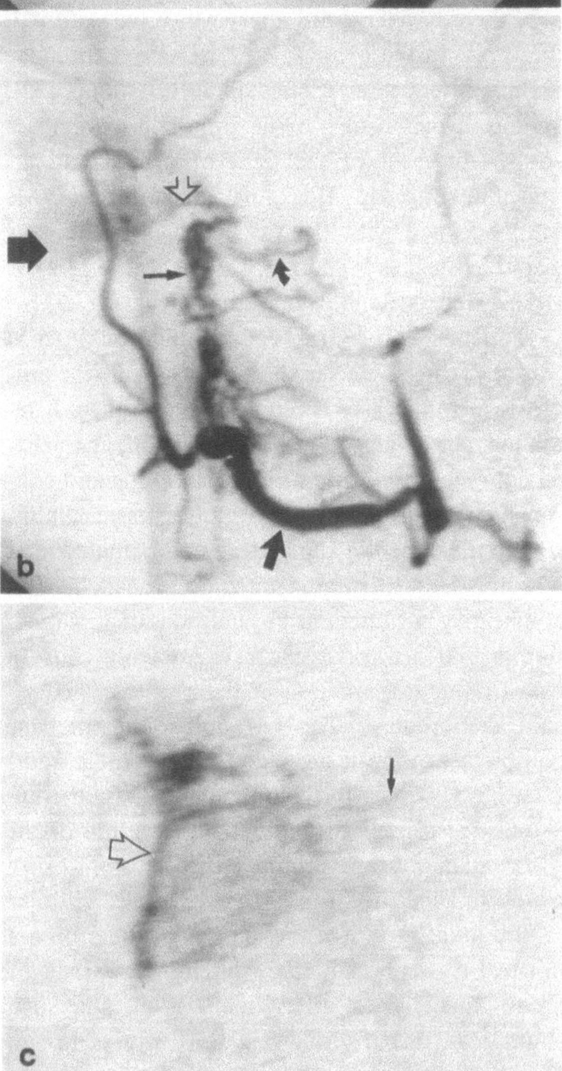

Fig. 4.7a–c. A 54-year-old man presented with slowly progressive orbital congestion punctuated by episodes of acute worsening.

a The lateral right internal carotid artery angiogram demonstrates an orbital arteriovenous malformation (*open arrow,* distal portion) that is fed from the ophthalmic artery (*arrow*) as well as via branches arising from the cavernous internal carotid artery (*curved arrow*).

b A lateral view of the right internal maxillary artery (*arrow*) demonstrates a network (*thin arrow*) of abnormal vessels that anastomose with the branch of the inferior lateral trunk of the internal carotid artery (*curved arrow*) and supply to the arteriovenous malformation. A branch from the middle meningeal artery (*open arrow*) also supplies the arteriovenous malformation (*broad arrow*).

c A magnified lateral distal superselective catheterization of the ophthalmic artery system with a catheter placed at the location of the *small arrow* (corresponding to the *small arrow* in **a**). This injection did not fill the posterior ciliary arteries nor the central retinal artery and the lesion was embolized with polyvinyl alcohol particles with the catheter in this position. The distal portion of the arteriovenous malformation (*open arrow*) corresponds to the *open arrow* in **a**

as a carotid cavernous fistula (see Chap. 2). Following treatment that closes the carotid cavernous fistula, the persistence of a second lesion, the orbital arteriovenous fistula, may spuriously suggest a recurrence of the carotid cavernous fistula. Fortunately, most of these traumatic orbital fistulas close spontaneously. However, surgical ligation or embolization of the supplying artery and occlusion of the shunt may be required in those cases where hemorrhage or progressive enlargement of the lesion occurs.

4.5.2 Histopathology

The histopathology of a congenital AVM shows an abnormally thickened muscularis in the walls of arteries and veins without intervening capillaries (Fig. 4.8). Fibrosis and areas of endothelial proliferation are seen in the areas between the abnormal arteries and veins [25].

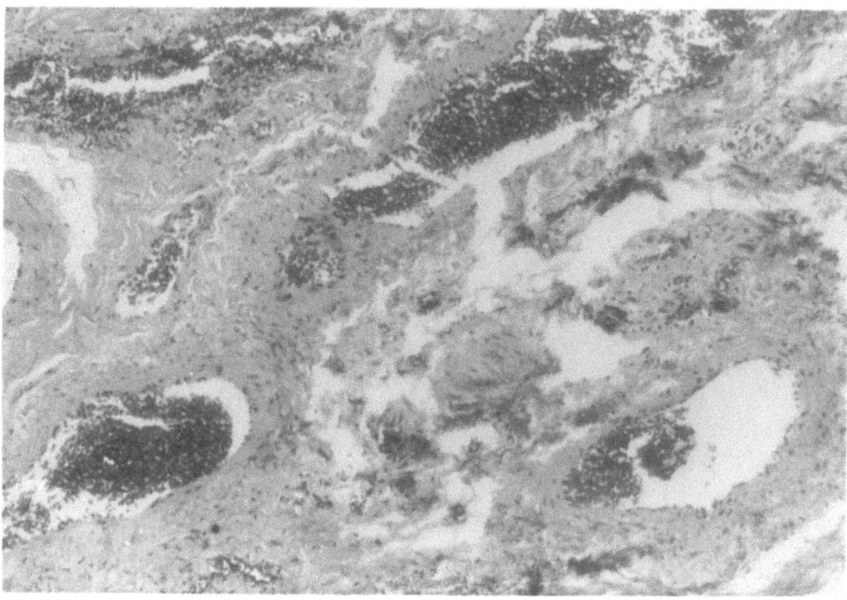

Fig. 4.8. Both thin- and thick-walled vascular channels, representing arteries and veins without any intervening capillaries, are seen in this orbital vascular arteriovenous malformation. (Case provided by Norman Charles)

4.5.3 Imaging

B-scan ultrasonography shows the dilated vascular channels and swelling of the orbital tissues, particularly the extraocular muscles. The A-scan shows that some areas of the AVM are sharply demarcated from the orbital structures. There are echo-free regions because of the blood-induced congestion of the lesion. The rapid blood flow in the lesion causes a measurable sound attenuation.

The contrast-enhanced CT visualizes an irregularly shaped, high-attenuation lesion with vascular channels, which can be small and focal or as large as the entire orbit with extension into the lids (Fig. 4.6b). The unenhanced CT will reveal areas of intraluminal high-attenuation signal caused by the intravascular blood. Extravascular blood is seen when a recent clinical hemorrhage has occurred. The MR image provides additional hemodynamic information such as a flow void resulting from the arterialized blood flow.

Angiography demonstrates that most AVMs receive arterial supply from branches originating from both the external carotid artery to the orbit (see Chap. 1) and the ophthalmic artery (Fig. 4.6c,

d). When the AVM is supplied by branches from the ophthalmic artery "en passage" to supplying the posterior ciliary arteries or the central retinal artery, treatment is likely to cause visual loss (Fig. 4.7).

4.5.4 Treatment

Patients with recurrent hemorrhage, progressive orbital dysfunction, or visual loss should have treatment directed at the AVM. The therapy of a small localized AVM is surgical, but most symptomatic lesions are more extensive and require staged embolization, or embolization followed by surgical excision (Fig. 4.6e). If embolization through the ophthalmic artery is to be performed, in order to preserve vision, the catheter must be placed distal to the origin of the posterior ciliary arteries or directly into the arterial feeder of the AVM (Fig. 4.7c). In some patients embolization can be accomplished through internal maxillary branch arteries to the orbit, such as the middle meningeal artery and the anterior deep temporal artery (Figs. 4.6, 4.9). When percutaneous embolization is not possible, surgical

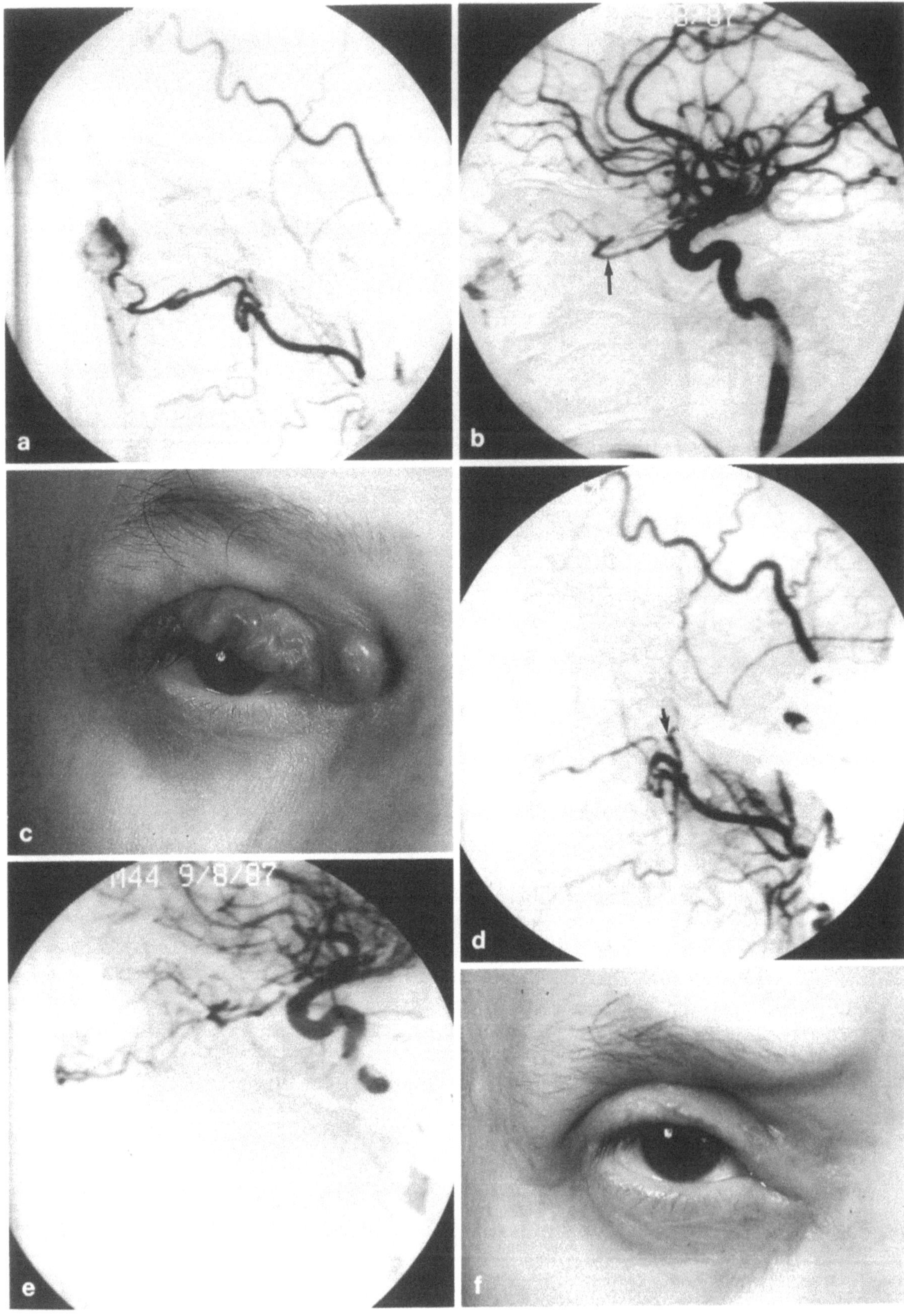

exposure of the AVM with direct puncture of the lesion and injection of an embolic agent or electrocautery can thrombose the nidus.

4.6 Capillary Hemangioma

4.6.1 Clinical Manifestations

Capillary hemangioma, sometimes called a juvenile or strawberry hemangioma, occurs as a single tumor lesion located in the skin of the lid or face. Superficial capillary hemangiomas can be located in any site of the dermis. These lesions commonly present before 3 months of age or even at birth. They are more common in girls than boys and are usually not associated with other skin blemishes or systemic lesions. Capillary hemangiomas are rarely limited to the orbit but they can invade the orbit from an origin in the eyelid (Fig. 4.10).

Capillary hemangiomas must be distinguished from other elevated reddish-colored skin blemishes, such as the port-wine stain (nevus flammeus) that is associated with Sturge–Weber syndrome (Sect. 4.13.1). A capillary hemangioma is a spongy blue or reddish lesion which often feels rubbery. In contrast, a port-wine stain is flatter and more purple in color and does not blanch with local compression. Also, the nevus flammeus does not enlarge except in proportion to the child's growth, while a capillary hemangioma typically grows more rapidly than the infant (see below).

The clinical natural history of these lesions most often includes a period of rapid growth during the first 6 months of life. The mass involutes in approximately 30% of cases by 3 years of age, and in 60% of patients by 4 years of age. The particular clinical presentation depends on the location of the lesion. When located in the lid, the lesion can grow in any quadrant of the lid and can extend into the conjunctiva or posteriorly through the orbital septum into the retrobulbar space. When the mass invades the orbit, the globe can be displaced or distorted. Larger extensive lesions distort and deform the face and alter the development of the orbit [26]. When the mass is in the lid, the ptosis can distort the cornea, inducing astigmatism and myopia. Ob-

◄ **Fig. 4.9. a** A lateral view external carotid artery injection demonstrates filling of a lid arteriovenous malformation via anterior branches to the lid from the infraorbital artery, a branch artery from the internal maxillary artery. **b** In the same patient, the lateral internal carotid artery angiogram demonstrates ophthalmic artery (*arrow*) filling of the arteriovenous malformation. **c** Obvious right upper lid vascular malformation caused recurrent hemorrhages, necessitating therapy. **d** Following superselective embolization of the infraorbital artery with cyanoacrylate, the lateral subtraction external carotid artery angiogram demonstrated an absence of filling of the branch artery (*arrow*) that had supplied the arteriovenous malformation and the arteriovenous malformation was not seen. **e** The ophthalmic artery was not embolized. The lateral internal carotid artery angiogram demonstrates the ophthalmic artery no longer supplied an arteriovenous malformation (compare to **b**). **f** Following embolization, regression of the lid lesion is obvious

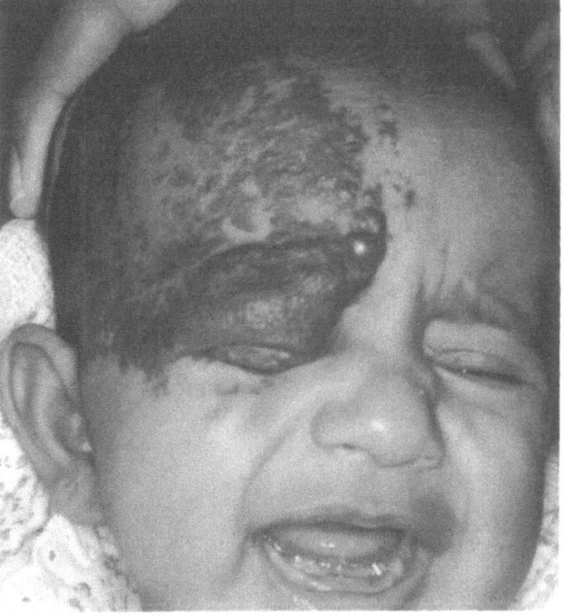

Fig. 4.10. An extensive capillary hemangioma involves the right upper lid, causing an occlusion amblyopia in this infant. The lesion is elevated, has a reddish color, and feels spongy on palpation

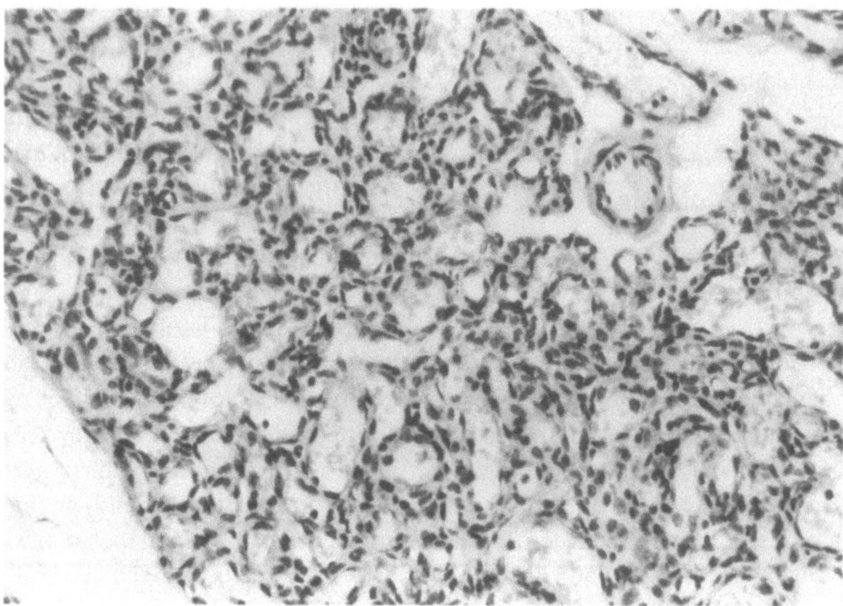

Fig. 4.11. Histopathological examination of this capillary hemangioma demonstrates numerous endothelial cells which have the appearance of forming multiple vascular lumens, with pericytes surrounding many of these vascular channels. (Case provided by Norman Charles)

struction of the visual axis leads to an occlusion amblyopia and secondary strabismus [27]. Spontaneous thrombosis in the hemangioma can disrupt the normal arterial and venous circulation of the involved area with swelling, necrosis, or hemorrhage of the lid. In infants and toddlers, the mass often increases in size while crying or breath holding [28].

Systemic problems can develop as a result of secondary thrombocytopenia if there is significant entrapment of platelets and erythrocytes within the lesion. The coagulopathy can lead to hemorrhage from the hemangioma or at remote sites (Kasbach–Merrit syndrome) [29–31]. Infants rarely develop high-output congestive heart failure because of the rapid blood flow into the hemangioma.

4.6.2 Histopathology

The histopathology of this hamartoma demonstrates lobules that contain engorged endothelial cells (Fig. 4.11). Mitoses are present only during the early rapid proliferative phase. Rapid tumor growth can also be correlated with congestion of the mass because of numerous endothelial cells. These endo-

thelial cells are surrounded by a basement membrane, best seen with a reticulin stain. The cells are organized with the appearance of forming lumens, which become more obvious when the cell growth has slowed. Pericytes also surround the endothelial cells [32]. An older lesion is often surrounded by a capsule and the lobules are separated by fibrous septa or dilated vascular channels. As lesions involute over months and years, fewer endothelial cells are seen and larger vascular channels and an increase in the areas of fibrosis, collagen deposition, and fat deposition are noted. The lesion can infiltrate the adjacent orbital fat, lacrimal gland, or extraocular muscles.

4.6.3 Imaging

When there is a question of tumor extension into the orbit, the imaging evaluation should include CT or MR imaging. Ultrasonography may be helpful in distinguishing this lesion from other types of masses, because a capillary hemangioma frequently has multiple channels or lumens (see below).

Ultrasound B-scans show the extent of the mass in the subcutaneous and anterior orbital tissues.

The A-scan provides the acoustic criteria for diagnosis. Highly reflective echospikes arise from the interface between the mass of tumor cells, the septa between lobules, and the vascular channels. This is contrasted with areas of low internal reflectivity when there is a solid mass of endothelial cells. The rapid blood flow to the lesion can be demonstrated with a doppler study.

The CT and MR images usually show the lesion to have well- defined borders only after involution has begun. During the growth phase, the contrast-enhanced CT often demonstrates the tumor borders to be indistinct. The mass may become more white following contrast administration. The lesion may appear uniformly white or heterogeneous. Even in cases where there are no clinical orbital signs, CT frequently shows that a lid mass extends across the preseptal area and perforates through orbital septum into the intra- and extraconal spaces. Distortion of the orbital tissues, including the globe, and enlargement of the orbital cavity are seen when the hemangioma invades into the orbit.

Angiography is not essential in the evaluation unless surgical or embolization treatment is planned. Selective catheterization will demonstrate the specific large arterial contributions to lesions in the lid and orbit that arise from the ophthalmic artery branches. The external carotid artery, particularly the angular artery, frequently supplies the hemangioma [24]. Angiography will often reveal additional arterial feeders from branches of the internal maxillary artery. The veins appear early in the angiogram and drain through both the facial and orbital venous systems. The hemangioma opacifies intensely.

4.6.4 Treatment

The treatment depends on the severity of the problems discussed previously. Superficial lesions that do not impair the vision or bleed are not treated, since the majority will spontaneously regress. Intervention is warranted if progressive enlargement of the lesion causes orbital signs, when lid involvement causes progressive astigmatism or occlusion amblyopia, or if lid necrosis develops. Treatment is essential in patients with high-output cardiac failure or significant platelet consumption. Platelet transfusion and antifibrinolytic therapy (aminocaproic acid) are administered as required.

Corticosteroids frequently control or reduce the size of a capillary hemangioma [33]. Systemic steroids are administered with an initial daily dose of prednisone of up to 2.5 mg/kg. The dose is gradually reduced after several weeks to the smallest amount that prevents regrowth of the lesion. Since the corticosteroids can have significant systemic side effects and complications in infants, only brief periods of treatment are used. Rebound growth of the lesion often follows dose reduction or complete withdrawal of the corticosteroids.

In infants, the intralesional injection of 40 mg triamcinalone often profoundly reduces the size of the hemangioma with fewer systemic complications [34, 35]. Repeated injections are usually, but not always, effective in aborting tumor regrowth. A future increase in the size of the mass is not necessarily prevented by the injections. The injection into the lesion produces swelling and tension in the local tissue that dissipates within a week. A smaller mass then remains. Necrosis of the skin or eyelid is a rare complication of intralesional injection in the eyelid [36]. Corticosteroid emboli that end in retinal arteries and cause visual loss is another rarer complication.

Surgery is performed if the lesion that compromises the vision fails to respond to corticosteroids or recurs following withdrawal, or if significant systemic side effects limit the use of corticosteroid treatment [37, 38]. If the lesion is not completely excised, surgery can actually exacerbate the tendency for growth. The surgery of an orbital or invasive lid hemangioma is often difficult to perform. The entire arterial blood supply to the mass cannot be easily eliminated by preoperative embolization or by surgical ligation without visual loss. This is particularly relevant in cases where the feeding vessels arise from important branches of the ophthalmic artery. If an associated Kasbach – Merrit syndrome is present, antifibrinolytic drugs such as aminocaproic acid may prevent excessive intra- and postoperative bleeding [39].

Local irradiation of the tumor with external cobalt (500 – 900 cGy in divided doses) or proton beam or implanted radon seeds has been reported to reduce the size of the lesion [26]. However, radiotherapy is not the treatment of choice because of potential complications. The long-term risk of developing a malignancy at the irradiated site and the risk of a visual disturbance from radiation damage to the eye (conjunctiva, cornea, retina, optic nerve) and surrounding structures (see Sect. 5.9) make this therapy unwarranted in most cases.

Superficial lesions that bleed frequently or compromise vision can also be treated with a CO_2 laser or cryotherapy as alternative treatments [40]. Local injection of sclerosing solutions such as alcohol can also shrink a capillary hemangioma. Care must be taken not to induce blindness by introducing the sclerosing agent into the ophthalmic artery circulation. Following alcohol injection into a lid lesion, there is temporary swelling of the lid and painful tension in the overlying skin.

Arterial embolization with polyvinyl alcohol particles has been rarely used [41].

4.7 Cavernous Hemangioma (Cavernoma)

4.7.1 Clinical Manifestations

Cavernous hemangioma is the most common intraconal orbital tumor. Women are affected twice as often as men. Patients usually present between the second and fourth decades of life with the complaint of an increasing painless unilateral proptosis. Since the size of the tumor generally increases slowly or is stable, the symptoms slowly worsen or remain static. Asymptomatic lesions are typically diagnosed when a scan that includes views of the orbits (obvious on CT, may not be obvious on routine MR imaging) is performed for some other reason. The onset of the proptosis may be so insidious that it is only first noticed by an individual who has not seen the patient in years. Additional cavernous hemangiomas are rarely found either systemically or in the brain, except in the equally unusual familial type of this disorder (see Chap. 7).

The clinical features [42, 43] develop in relation to the local mass effect of the hemangioma on the orbital structures. The blood flow into the tumor is slow and there are no large arterial feeding vessels or draining veins. Extralesional hemodynamic al-

terations, except for infrequent hemorrhage, cause few clinical signs. Since the clinical defects are often static or asymptomatic, a review of old photographs will often reveal the length of time that the proptosis has been present. When proptosis develops, the globe is often deformed medially because the mass is most commonly located in the temporal portion of the retrobulbar space.

If the mass indents the globe, retinal and choroidal striae can be seen on funduscopy. An optically correctable reduction in the visual acuity results from the hyperopia induced by the choroidal folds. Displacement of the optic nerve on CT is not always associated with an optic neuropathy. However, chronic compression of the optic nerve, particularly if the mass is located in the posterior orbit or in the orbital apex, can reduce the vision and cause secondary optic atrophy. Mechanical obstruction of the venous outflow from the eye will cause unilateral disc swelling and edema or dilated retinal veins and hemorrhages reminiscent of a venous stasis retinopathy. A frank central retinal vein occlusion picture rarely occurs. The ocular movements may also be limited by local restriction if the mass is large or if the cranial nerves or the extraocular muscle insertions are distorted near the annulus of Zinn by a lesion located in the orbital apex. Infrequently, thrombosis or hemorrhage within the tumor will acutely block the ophthalmic venous outflow from the orbit, causing a sudden severe exacerbation of the proptosis.

Rarely, amaurosis fugax may develop with extreme positions of gaze because the intraconal tumor mechanically induces a transient ischemia or axonal blockade in the optic nerve [44, 45]. This phenomenon is more related to tumor location than to its size. In fact, proptosis may not be present in some cases. Gaze-evoked amaurosis has also been described in patients with other mass lesions that involve the orbital optic nerve, such as an optic nerve sheath meningioma [46–48].

4.7.2 Histopathology

The histopathology of this hamartoma clearly points to a vascular cell origin (Fig. 4.12). The mass is typically composed of flattened endothelial cells with large, dilated, irregularly shaped vascular

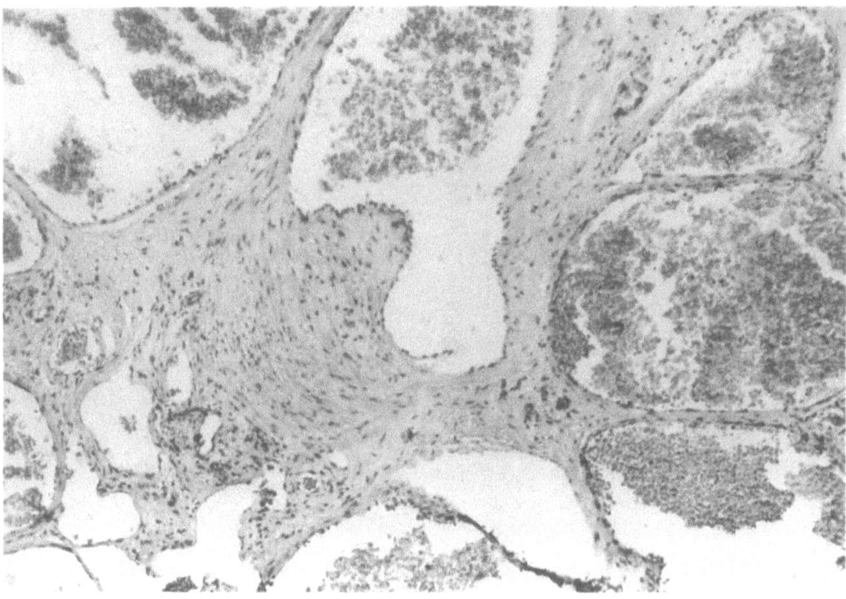

Fig. 4.12. An orbital cavernoma contains areas of thin-walled vascular channels with intervening mucoid stroma. (Case provided by Norman Charles)

channels approximately 0.5 – 1.0 mm in diameter. Smooth muscle is found in the walls of these channels [32]. Fibrous interstitial tissue develops between the lumens. No mitoses are seen. A fibrous capsule surrounds the tumor. Calcification, hemosiderin deposition, and thrombosis within the lesion are often present. These alterations suggest that this tumor enlarges because of expansion of the vascular channels and intralesional thrombosis and hemorrhage.

4.7.3 Imaging

Imaging studies should include an unenhanced and a contrast-enhanced CT or MR imaging. Ultrasonography can be diagnostic. Venography is not helpful. Angiography is generally not indicated because of the lack of a major arterial supply to cavernomas. However, if the tumor is massive, angiography may be needed to locate the ophthalmic artery. Following the injection of contrast material into the internal carotid artery ipsilateral to the tumor, late pooling (in the venous phase) of the contrast within a lumen of the tumor may be seen. The mass effect from a tumor will often straighten and stretch the ophthalmic artery. Subtle degrees of displacement of the ophthalmic artery may be difficult to demonstrate because of the normally variable course of this vessel within the orbit.

B-scan ultrasonography reveals regions with cystic or lumen changes within the lesion (Fig. 4.13). The cavernoma usually appears as a smooth-bordered intraconal mass. Indentation of the globe is seen no better than on CT or MR scans. The extraocular muscles may be shadowed by the mass. A-scan ultrasonography reveals more about the nature of the tumor by demonstrating internal reflectivity between the large vascular channels and areas with mild wave attenuation. The attenuation is fairly typical of this lesion because of the presence of blood in the lumens. The area of the tumor capsule has high reflectivity compared to the surrounding orbital structures. Blood flow cannot be appreciated.

A well-encapsulated lesion with smooth margins that is usually, but not always, intraconal, is seen on CT (Fig. 4.14a) [44, 49, 50]. A contrast-enhanced CT demonstrates only mild if any discernible enhancement of the tumor. However, if the scan is delayed, approximately 30 min after contrast injection, the mass enhances. The mass can appear homogeneous or variegated. Rarely, the lesion contains minute areas of calcification which

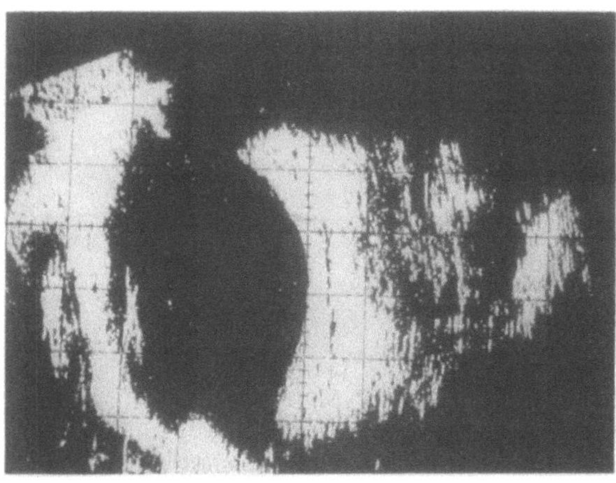

Fig. 4.13. B-scan ultrasonography demonstrates a retrobulbar cavernoma inside the muscle cone with areas that have channel or lumen characteristics. (Case provided by Richard Koplin)

Fig. 4.14a, b. Axial contrast CT scans from two cases of orbital cavernoma. **a** A small, oval, intraconal, slightly enhancing lesion is seen to displace the optic nerve temporally. Despite the displacement, the patient's visual performance was entirely normal and there was no relative afferent pupillary defect. **b** A mass filled the entire orbit, bowing the ethmoid bone slightly medially and expanding the intraconal space. This eye became blind over several years and developed white optic atrophy

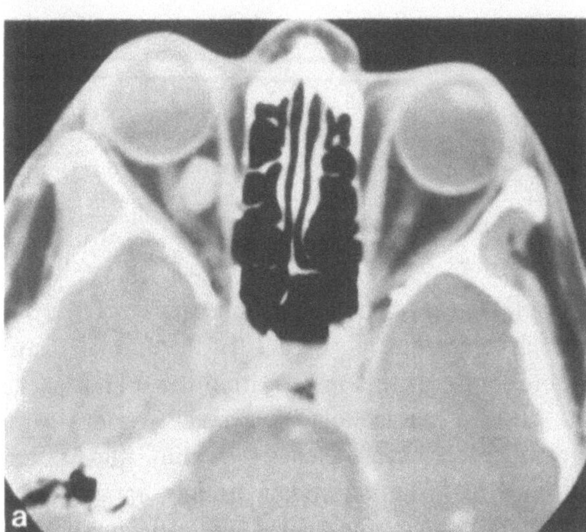

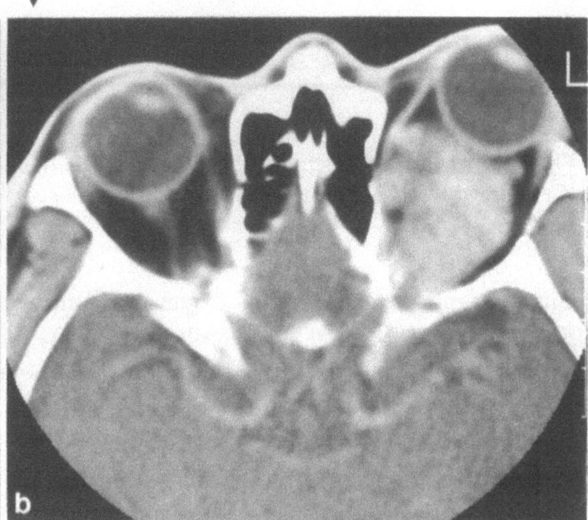

are best demonstrated using the bone windows. The optic nerve, even in patients without optic nerve dysfunction, is typically displaced (but not surrounded or infiltrated) by the mass. When the cavernoma encroaches upon the anterior orbit, an indentation of the posterior margin of the globe can be visualized. These lesions often cause gradual continuous pressure against the orbital bones, which results in outward bowing, but not destruction, of the lateral orbital wall (Fig. 4.14b).

MR imaging has not added to the diagnostic evaluation of these lesions except to reveal areas of thrombosis within the cavernoma. Subtle changes in the orbital bones may not be obvious on MR images. The lesion has a relatively hypointense signal on a T1-weighted study. A heterogeneous and relatively hyperintense signal is seen on T2-weighted images [51].

4.7.4 Treatment

The patient can be followed without treatment as long as the course is static and there is no visual loss, optic disc edema, venous retinopathy, or exposure keratopathy. However, a minority of authors feel these tumors should be removed prophylactically, unless the patient is a poor surgical candidate [52].

The treatment is surgical excision and the approach depends on the location of the cavernoma. A lateral orbitopathy is usually the best approach since most tumors are found in the lateral aspect of the muscle cone, with the optic nerve located medial to the mass. A traction suture placed through the tumor may facilitate manipulation and allow reduction in the size of the mass as blood escapes from the lesion. If the mass extends into the orbital

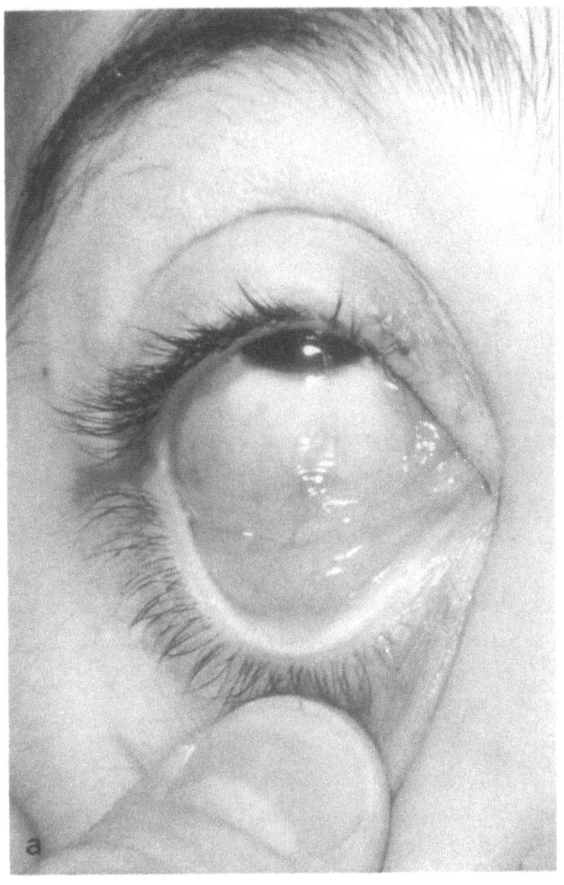

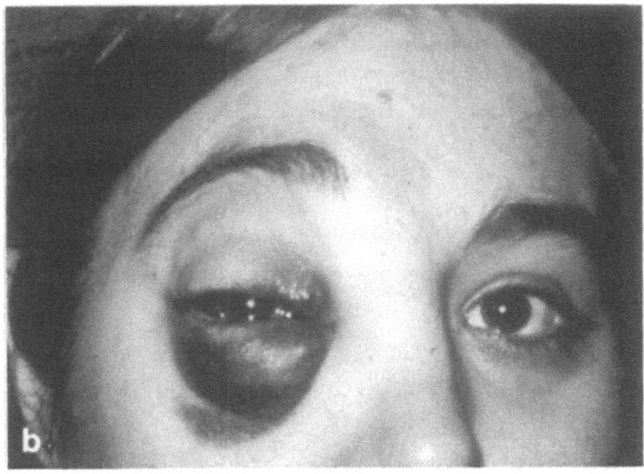

Fig. 4.15. a An orbital lymphangioma extended through the orbital septum into the subtenon's space. **b** The same patient experienced acute ecchymosis and swelling of the lower lid, proptosis, and subconjunctival hemorrhage when an intraorbital "chocolate cyst" ruptured. (Case provided by Norman Charles)

apex, the ocular motor nerves may adhere to the posterior surface of the capsule, requiring a fine dissection for their preservation. Iatrogenic disruption of a posterior ciliary artery can cause infarcts of the choroid and optic nerve. Rarely, surgical manipulation damages the orbital pial blood supply to the optic nerve, resulting in an optic neuropathy. If total excision of the cavernoma is accomplished, there is no recurrence of the tumor.

4.8 Lymphangioma

4.8.1 Clinical Manifestations

Lymphangioma is another infrequently occurring vascular tumor which is often clinically apparent in the first decade of life. The course is typified by slowly progressive symptoms related to lid or orbital involvement and punctuated by acute exacerbating episodes [53]. One fourth of the lesions are diagnosed during infancy; one fourth are discovered between the ages of 2 and 10 years; one fourth are found during the second decade; and the remaining one fourth are diagnosed in patients older than 20 years. Eyelid and conjunctival lesions occur alone or with extension into the orbit (Fig. 4.15a). Lymphangiomas in this location are seen at birth as well as in those individuals who develop signs between the age of 1 and 5 years. When there is a rapid onset during infancy, a biopsy of the lesion is needed to exclude a hemorrhagic rhabdomyosarcoma. Orbital involvement is less commonly found when the presentation is during adulthood. These tumors can also be found in other locations of the head and neck, including the pharynx and the sinus cavities.

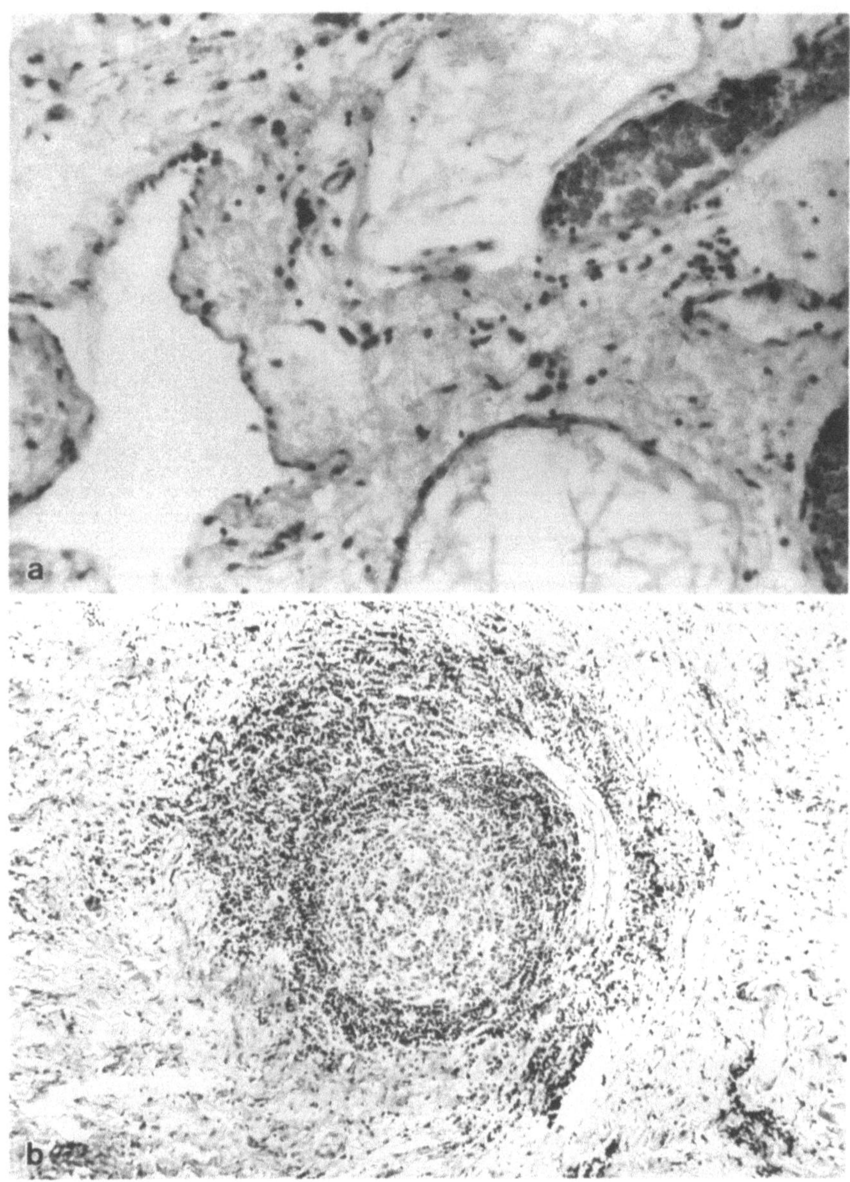

Fig. 4.16. a The histopathology of a lymphangioma revealed endothelial cell-lined lumen with loose connective tissue stroma between the areas of lumen as well as small blood vessels. **b** In this portion of the lymphangioma, a lymphoid germinal center was found surrounded by stroma. (Case provided by Norman Charles)

Diffusely infiltrating lesions that involve the orbit and lid are common. Lesions that are limited to the conjunctiva or lids are more unusual. Proptosis is usually slowly progressive. Intermittent proptosis, without obvious superficial signs, occurs in approximately 25% of patients. A visible lesion in the lids or beneath the conjunctiva, without proptosis, is found in 15% of cases. At least 50% of patients have proptosis and visible lid or subconjunctival lesions, or both. More than 50% of cases have lesions that involve some portion of the orbit posterior to the orbital septum. There is also frequent involvement of the structures surrounding the orbit, such as the lids, conjunctiva, and face.

Lymphangiomas cause other typical orbital signs which include extraocular eye muscle imbalance and limitation, occlusion amblyopia when lid involvement blocks the visual axis during infancy, and progressive optic atrophy from acute or chronic compression of the optic nerve.

Several mechanisms account for the variable clinical features and course of this tumor. Acute exacerbation of the orbital signs develop from one of several causes. Hyperplasia of the lymphoid tissue or venous congestion in the tumor occurs in association with an upper respiratory tract infection, a local sinusitis, or an orbital or facial cellulitis. A patient with a lymphangioma that involves the pharynx or an adjacent sinus is more prone to have recurrent cellulitis. Another common mechanism that accounts for sudden worsening of the proptosis is the rupture of a cyst and hemorrhage within the mass and orbit. Leakage of blood into the anterior orbit also leads to a subconjunctival hemorrhage (Fig. 4.15b). Since the lymphangioma is not supplied by large high-flow arteries and it is not drained by a few large venous channels, these patients do not demonstrate clinical findings typical of a highly vascular lesion, such as a bruit or worsening proptosis with maneuvers that alter the ophthalmic venous pressure.

4.8.2 Histopathology

The histopathology of a lymphangioma demonstrates extensive areas of large lumens and cysts, which are often bloodless and lined by endothelium (Fig. 4.16a). Lymphoid tissue is present between these lumens or channels, and sometimes lymphoid germinal center follicle formation is seen (Fig. 4.16b). Hemosiderin-laden macrophages, cholesterol clefts, and phleboliths are found in the cysts. There are no pericytes or smooth muscle cells in the wall of these cavernous channels. Multiple small blood vessels are scattered throughout the lesion. Fresh and old hemorrhage are identified within "chocolate cysts" that are surrounded by fibrotic walls. A lymphangioma is a choristoma because lymphatic tissue is not found within the normal orbit and it is uncertain how these lesions can arise within the orbit. One theory suggests the tumor originates from a mesenchymal cell line which, instead of becoming endothelial vascular channels, develops into lymph channels [53, 54].

4.8.3 Imaging

The imaging evaluation includes CT or MR imaging and possibly ultrasonography. All of these techniques are capable of demonstrating the cystic and diffuse infiltrative nature of the tumor. The B-scan clearly shows the soft tissue mass with cystic spaces infiltrating into the lids and orbital structures. The A-scan has highly reflective echospikes with areas of low attenuation because of the lack of blood in some cyst spaces.

The CT shows the lesion to have a heterogeneous appearance with low-density cystic areas (Fig. 4.17a). When the lymphangioma extends into the orbit, it infiltrates into both the intraconal and extraconal spaces. The lesion frequently infiltrates through the orbital septum into the preseptal space. The extraocular muscles can be indistinguishable from the mass. Focal areas within the mass or along the rim enhance after intravenous contrast material. Following an acute hemorrhage, high-attenuation areas representing collections of blood can be identified. Optic nerve displacement is frequent, and the optic nerve is often enveloped by the tumor. Because the mass is present from infancy, enlargement of the entire orbit is frequently seen even with a plain orbital radiograph.

MR imaging can add to the CT findings by demonstrating cystic cavities and areas of nonflowing intraluminal blood (Fig. 4.18). Feeding arteries in the mass may also be identified by their flow void appearance [55].

Angiography is not diagnostic and is only performed in planning surgery. Typically, multiple small arterial feeders supplying the tumor are visualized. Deep orbital lesions receive blood from branches of the ophthalmic artery, while lesions that extend into the lid or the face have an extensive blood supply via the arteries from the external carotid artery (Fig. 4.17b). There are no major venous channels directly involved in the process. Venography can show dilated channels which communicate with the ophthalmic venous system, but a true orbital varix is not seen [56]. On occasion, venography may fail to demonstrate any abnormal vascularity [57].

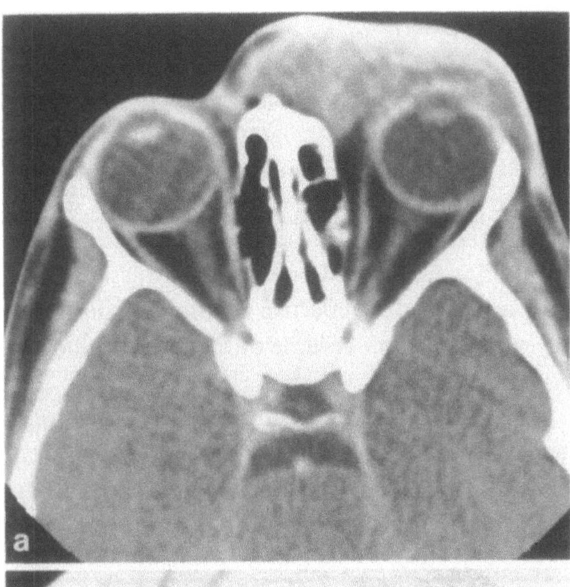

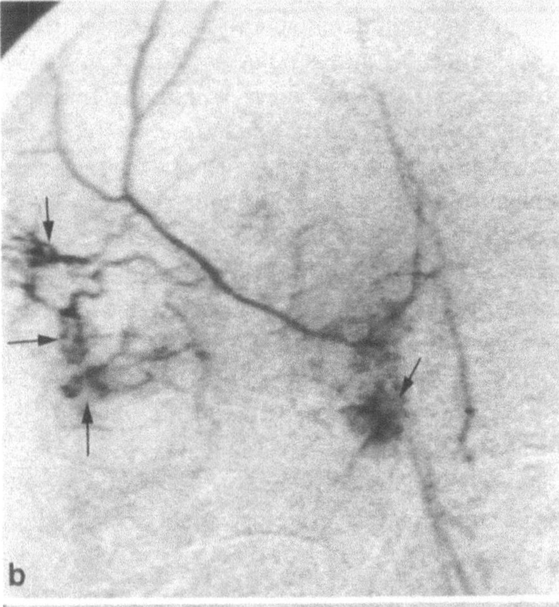

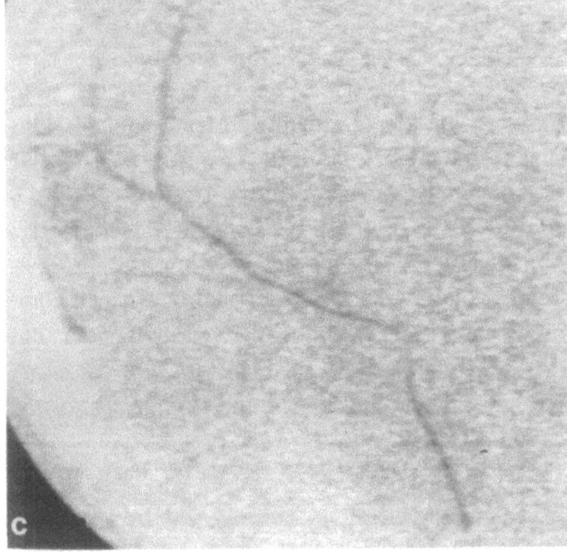

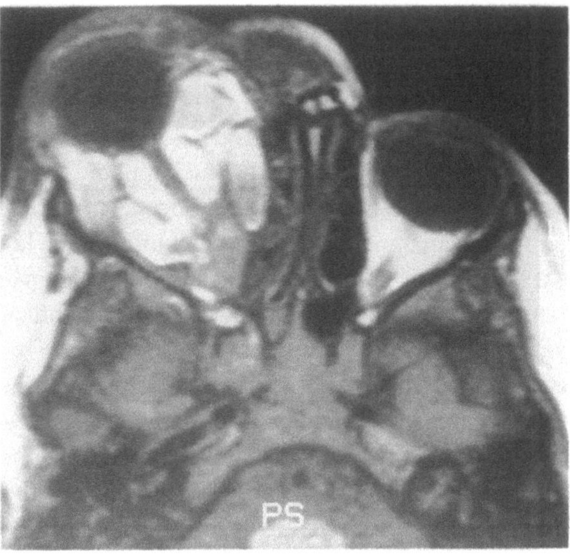

Fig. 4.18. An axial view T1-weighted image orbital MR scan demonstrates a multilobulated lymphangioma that has caused proptosis, enlarged the orbital cavity, and bowed the lateral wall of the orbit outward

Fig. 4.17a–c. A 2-year-old boy with an extensive lymphangioma involving the orbit developed occlusion amblyopia from this mass, necessitating treatment. **a** An axial view contrast- enhanced CT scan visualized a mildly enhancing, diffusely infiltrating mass extending along the whole preseptal area to the nose. Lower views (not shown) demonstrated extension of the lymphangioma into the soft tissues of the maxillary region. **b** A lateral view of a superselective catheterization of the superficial temporal artery showed multiple tufts of vascular supply to the lymphangioma (*arrows*). Multiple other branch arteries including the facial artery and internal maxillary artery branches also supplied the lymphangioma. **c** Following polyvinyl alcohol particle embolization in the affected external carotid artery vessels, a repeat lateral view superficial temporal artery catheterization failed to demonstrate the previously visualized tumor vascularity. The lesion became reduced in size and softer. Follow-up treatment was suggested but the family refused further intervention

4.8.4 Treatment

The clinical course of the lymphangioma is variable and often marked by progressive orbital signs and severe visual compromise. However, if the lesion is superficial, patients not infrequently remain relatively stable, with only minor degrees of proptosis and cosmetic difficulties. When the vision is threatened, intervention is warranted. Surgery has been useful for removing the chocolate cyst associated with acute hemorrhage. Not all children require decompression of acute hemorrhage since visual recovery often occurs spontaneously as the acute hemorrhagic proptosis resolves [58]. Complete excision of an infiltrative lesion without causing significant damage to the orbital tissues is uncommon. The vessels of the tumor are friable and bleed easily during manipulation. The tumor is removed by piecemeal dissection, and the excision is usually limited to the anterior orbit in order to relieve severe proptosis. Removal of the orbital veins can cause visual loss. Recurrent visual loss and proptosis are common even following extensive surgical debulking of a lymphangioma.

Despite the lack of arteriovenous shunting or major arterial blood supply to a lymphangioma, we have recently used embolization therapy. Devascularization of superselective arteries to the tumor can reduce the size of the lid lesion in order to prevent occlusion amblyopia or facilitate postembolization surgical excision (Fig. 4.17c).

Radiotherapy has been helpful only in temporarily reducing the size of the lesion, but long-term control of these lesions with this modality has not been demonstrated. On occasion, a brief course of systemic corticosteroids can reduce the clinical signs secondary to inflammation and swelling in the orbit.

4.9 Hemangiopericytoma

4.9.1 Clinical Manifestations

Orbital hemangiopericytomas are rare retrobulbar vascular tumors [59, 60] that are found in individuals of any age. The peak incidence seems to be in the fourth to fifth decades of life. The majority of hemangiopericytomas are located outside of the orbit, predominantly in the musculoskeletal tissue of the pelvis and lower extremities [61]. The diagnosis of this specific neoplasm is rarely established preoperatively. The clinical course is very variable and typically includes features of a progressive orbital mass. Proptosis is usually present at the time of the initial presentation. Diplopia and decreased visual acuity follow as the tumor grows [62]. Lid swelling and ecchymosis develops in approximately 10% of patients. Local and intracranial invasion are common. Distant metastatic lesions occur in up to 15% of cases.

4.9.2 Histopathology

Hemangiopericytomas are considered in this chapter on orbital vascular lesions because the neoplastic cells are derived from the pericytes of blood vessels and the tumors are highly vascularized. This neoplasm is very cellular and contains a network of reticulin-staining basement membrane surrounding the cells (Fig. 4.19). The tumor cells may form branching sinusoidal channels that appear similar to blood vessels. The mass can contain areas of myxomatous and cystic degeneration as well as frank necrosis. Pleomorphic cells and mitoses are present in approximately one third of lesions. Hemorrhage and infiltration of the tumor into normal tissue at margins of the resection are usually signs of a more malignant lesion and are associated with tumor recurrence [63–65].

4.9.3 Imaging

The orbital imaging evaluation includes CT or MR imaging and possibly ultrasonography, but the lesions may be deep in the orbit so ultrasonography may not be helpful. When the lesion is in the anterior or mid-orbit, the B-scan shows a solid mass. On A-scan, the lesion appears echo free because of the lack of interfaces within the mass. The tumor has low internal reflectivity and minimal attenuation on A-scan. In some cases, blood flow can be demonstrated within or to the tumor, but it is not as significant as with an AVM.

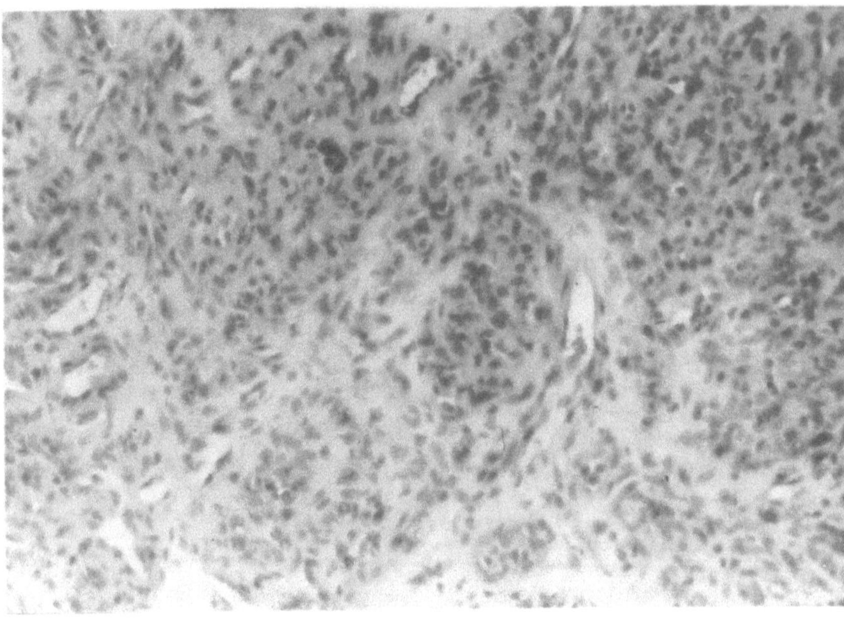

Fig. 4.19. Histopathology of an orbital hemangiopericytoma reveals a highly cellular tumor with multiple areas of vascular-type channels. (Case provided by Norman Charles)

The contrast-enhanced CT shows a homogeneously enhancing mass which is usually located in the superior aspect of the orbit, posterior to the globe. The tumor can appear discrete or it can have borders that appear to infiltrate the orbital tissues. MR imaging cannot differentiate this mass from other orbital neoplasms. Preoperatively, angiography may be useful in identifying the large arterial feeding vessels with rapid blood flow to the tumor. An early tumor blush is frequently seen. The contrast material is rapidly shunted through the mass into the orbital venous system.

4.9.4 Treatment

Surgical excision remains the principal treatment, but local recurrences are frequent, particularly when the tumor is incompletely removed [62, 66, 67]. Partial resections tend to be performed when the surgeon attempts to preserve the normal orbital tissues and vision. Locally invasive lesions in the orbit require exenteration even if usable vision is present preoperatively. When there is intracranial extension, the tumor is essentially incurable. A local recurrence should be treated by wide surgical excision. Preoperative embolization of the large vessel arterial supply to the tumor facilitates the surgery and prevent major blood loss.

Alternative therapies have not effected cures. Radiotherapy has been used for intracranial hemangiopericytomas with limited success, and a clear benefit in the treatment of orbital tumors has not been established. However, tumors smaller than 5 cm in diameter have been reported to respond to radiotherapy [68]. Embolization with sclerosing agents such as alcohol may be temporarily palliative.

4.10 Malignant Hemangioendothelioma

4.10.1 Clinical Manifestations

Malignant hemangioendothelioma is an extremely rare malignant neoplasm. There are fewer than 20 cases reported in the literature. This tumor occurs predominantly in young adults, without a predilection for one sex. Symptoms arise from progressive local growth of the tumor in the orbit as well as from metastatic lesions, of which there is a high incidence. The orbital signs of proptosis, diplopia, and visual loss develop rapidly, usually over several months.

4.10.2 Histopathology

Histopathologically, the tumor has a typical appearance of a sarcoma. The malignant cells are pleomorphic and hyperchromatic. The neoplasm can contain irregular blood- filled lumens as well [69]. The neoplastic cells can appear similar to endothelial cells, with tight junctions, basal lamina, Weibel – Palade bodies, and factor VIII antigen. These findings suggest that malignant hemangioendothelioma may arise from vascular cellular elements. This would be very unusual since less than 1% of all sarcomas are felt to originate from vascular tissues.

4.10.3 Imaging

The imaging studies, including CT or MR imaging, and ultrasonography demonstrate features consistent with a solid, locally invasive neoplasm. CT and MR imaging will also visualize erosion of the tumor through the bones of the orbit into the temporal fossa and intracranial cavity.

4.10.4 Treatment

Treatment results have been poor because of the high recurrence rate and the infiltrative and aggressive nature of the tumor. Not infrequently, the tumor extends intracranially, requiring a wide surgical excision for palliation. Long-term survival has been described in cases with intraorbital tumors following orbital exenteration and radiotherapy [70, 71].

4.11 Leiomyoma

4.11.1 Clinical Manifestations

Another rare vascular neoplasm, leiomyoma, originates from the smooth muscle cells of blood vessels [72]. There is no clear predilection for either sex. Slow tumor growth causes proptosis that gradually worsens only after many years. As a result, the lesion is usually not noted until adulthood [73, 74].

4.11.2 Histopathology

The tumor is composed of spindle-shaped cells that have a tendency to organize into vascular channels. The lesion can be extremely vascular. The mass is surrounded by a capsule and does not infiltrate normal tissue [75].

4.11.3 Imaging

The imaging studies, including CT or MR imaging and ultrasonography, demonstrate a solid mass with smooth sharply delineated borders that is limited to the orbit. The contrast CT demonstrates marked uniform tumor enhancement. Remodeling, but not destruction, is seen in the adjacent orbital bone.

4.11.4 Treatment

Total surgical excision is possible and local recurrences develop only if a complete removal is not accomplished.

4.12 Vascular Tumors of the Sphenoid Wing and the Cavernous Sinus

Though many neoplasms of the sphenoid wing and cavernous sinus areas are derived from nonvascular elements, these tumors can encase the internal carotid artery and ophthalmic artery and diminish the orbital arterial inflow. Compression of the intracranial ophthalmic vein and cavernous sinus slows or occludes the orbital venous outflow. Lesions in this area are common and most neuro-ophthalmologic findings result from direct compression and infiltration of the cranial nerves in the cavernous sinus and superior orbital fissure and the optic nerve as it passes intracranially. However, many signs develop in relation to the compromise of the vascular system. Most of these neoplasms are meningiomas with arterial supply principally arising from branches of the internal carotid artery. However, some lesions also receive blood from the ophthalmic artery (particularly from the ethmoidal arteries) and the middle meningeal artery. Analysis of the blood supply to these lesions and the involvement of the internal carotid artery is necessary in planning therapy.

4.12.1 Meningioma

Meningiomas in all locations occur in women more often than in men and are usually first noted in individuals older than 30 years of age. If the mass is located entirely within the cavernous sinus, varying degrees of ptosis, pupillary abnormalities, ophthalmoplegia, and first and second division trigeminal dysfunction are present. Compromise of the ophthalmic vein at the entrance into cavernous sinus does not usually cause orbital congestion and proptosis. In contrast, orbital signs are more frequent when the ophthalmic vein compromise occurs in the superior orbital fissure (Fig. 4.20). Secondary hyperostosis of the sphenoid wing can also cause cranial nerve dysfunction and proptosis by shallowing of the orbit. If the tumor involves the

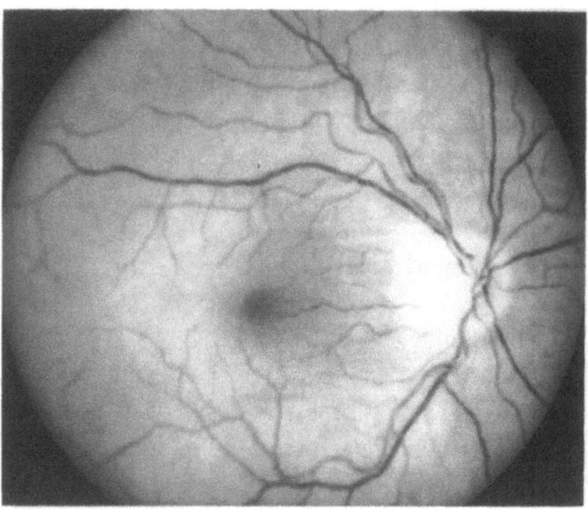

Fig. 4.20. Ophthalmoscopy of the right eye of this 37-year-old woman with blurred vision and a sphenoid wing meningioma that involved the supraorbital fissure revealed horizontal choroidal folds. There was no extension of the tumor into the orbit. The visual acuity was mildly reduced to 20/40 and there was a trace relative afferent pupillary defect. Following partial removal of the tumor and radiation therapy, the choroidal folds resolved over the course of a year

Fig. 4.21a–c. A 31-year-old woman had progressive unilateral ophthalmoplegia and ptosis. At the time of the original examination, the patient was unable to move her lid or her eye in any direction and the pupil was in a middle position and fixed. There was an absent corneal reflex as well.

a The lateral right internal carotid artery angiogram injection shows a massive meningioma blush involving the entire intracranial portion of the carotid artery. The supraclinoid internal carotid artery is pushed upward and backward.

b A lateral view of the middle meningeal artery (*arrows*) injection demonstrates the blood supply to the tumor from the middle meningeal artery and a significant tumor blush.

c Following embolization of the middle meningeal artery with polyvinyl alcohol particles, the lateral view of the external carotid artery does not demonstrate any evidence of tumor blush. Note that the normal middle meningeal artery is still patent (*arrows*)

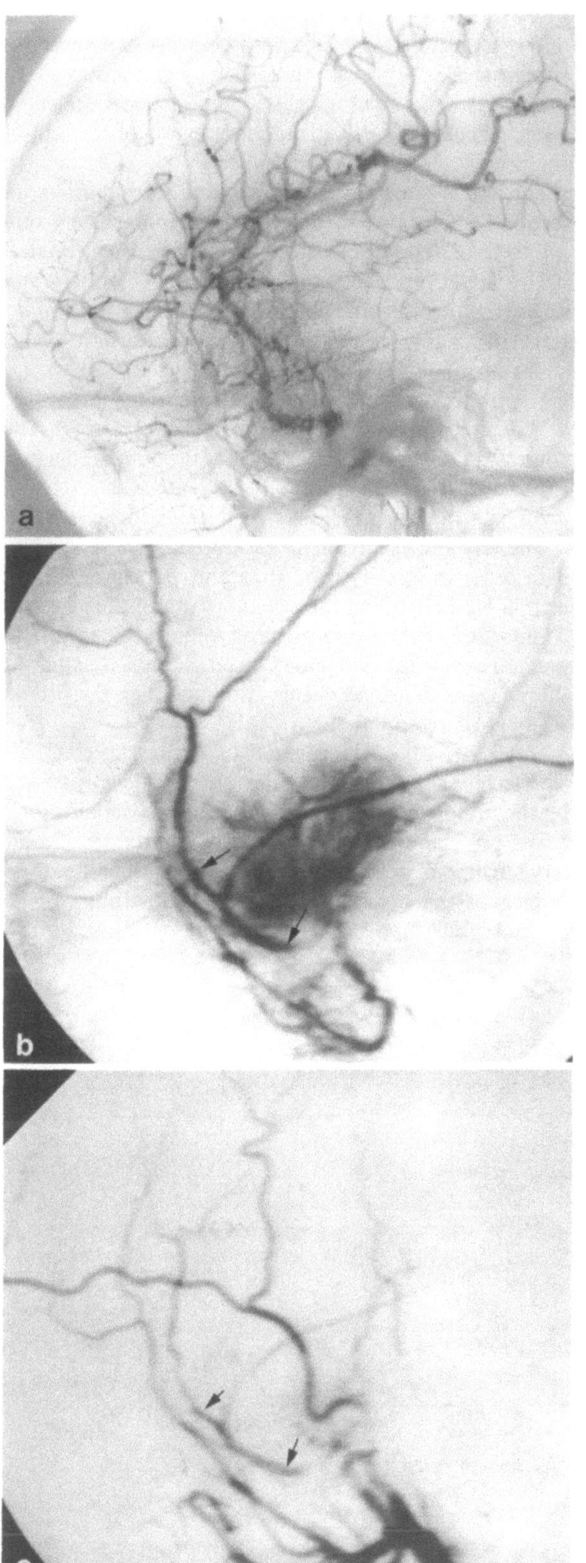

medial portion of the sphenoid wing, a slowly progressive optic neuropathy occurs. Complete surgical excision of meningiomas that involve the cavernous sinus often worsens the cranial neuropathy. Conventional linear accelerator radiation therapy or radiosurgery with high-dose focused radiation seems to yield better functional results. In cases where surgery is warranted, preoperative embolization can reduce the vascularity of the tumor so that the excision is facilitated (Fig. 4.21).

4.12.2 Fibrous Dysplasia

The sphenoid wing is a unusual site for a monostotic fibrous dysplasia [76]. Most cases of fibrous dysplasia have multiple bones involved, the long bones being the most frequent site in the monostotic form. Fibrous dysplasia has a predilection for women, particularly the presentation of a sphenoid wing area mass, and is typically found in young adults. This lesion is sometimes confused with the hyperostosis that develops adjacent to a sphenoid wing meningioma, particularly if the meningioma is small or en plaque (Fig. 4.22). Rapid blood flow and arteriovenous shunting within the fibrous dysplasia can cause an audible bruit (Fig. 4.22a, b). Slowly progressive proptosis develops from shallowing of the orbit as the thickened bone mass increases. Orbital congestion or an optic neuropathy occur from venous outflow obstruction or narrowing of the optic canal, respectively [77, 78]. Rarely, the bone proliferation extends to involve the chiasm [79] or both optic foramina [80]. However, many patients have lesions that appear significant on CT but show few clinical signs or symptoms. Progressive shallowing of the orbit which leads to corneal exposure or compromises the optic nerve requires widely extensive and reconstructive surgery. If there is an optic neuropathy, opening the optic canal will decompress the optic nerve [81]. When significant arteriovenous shunting is present, preoperative embolization will reduce the blood loss at surgery (Fig. 4.22e, f).

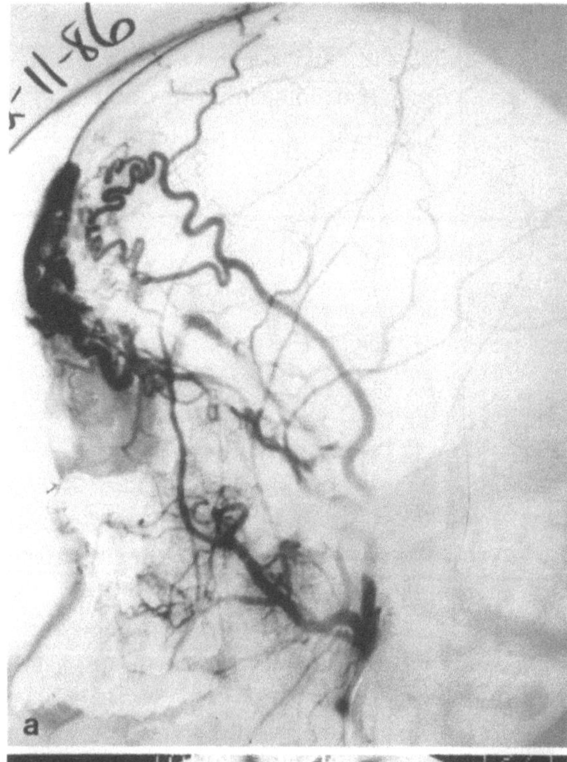

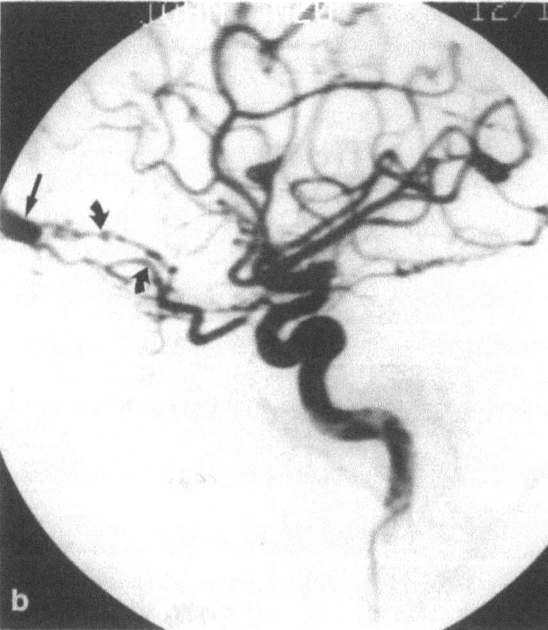

Fig. 4.22 a–f. Fibrous dysplasia of the sphenoid bone and orbit can be associated with extensive arteriovenous shunting. The CT scan (not shown) demonstrated fibrous dysplasia thickening the frontal bone, sphenoid wing, orbital fissures, and anterior portion of the roof of the orbit.

a The lateral projection of the external carotid artery injection demonstrates the superficial temporal artery supply to the arteriovenous shunting in the frontal bone.

b The lateral view subtracted right internal carotid artery angiogram demonstrates ethmoidal branches (*curved arrows*) from the ophthalmic artery to an arteriovenous shunt (*arrow*) in the diploic space.

c The frontal view of a direct puncture of a supraorbital vein demonstrates the catheter tip in fistula (*arrow*). The contrast injected at this site opacifies the ophthalmic vein (*open arrow*) and drains into the cavernous sinus (*curved arrow*) and across the midline via the anterior (*double arrowheads*) and posterior (*arrowhead*) coronary sinuses.

d In order to opacify the fistula and associated venous ectasia without filling the cavernous sinus, the vein was compressed at the supratrochlear foramen by an instrument. The frontal view shows increased opacification of the venous side of the fistula without filling the orbital segment of the ophthalmic vein.

e Acrylic was injected into the fistula using this compression technique. The venous side of the fistula is opacified by the radiopaque glue cast without any complication.

f The postembolization lateral view internal carotid artery angiogram demonstrates the continuation of a distal branch of the ophthalmic artery and no filling of an arteriovenous shunt

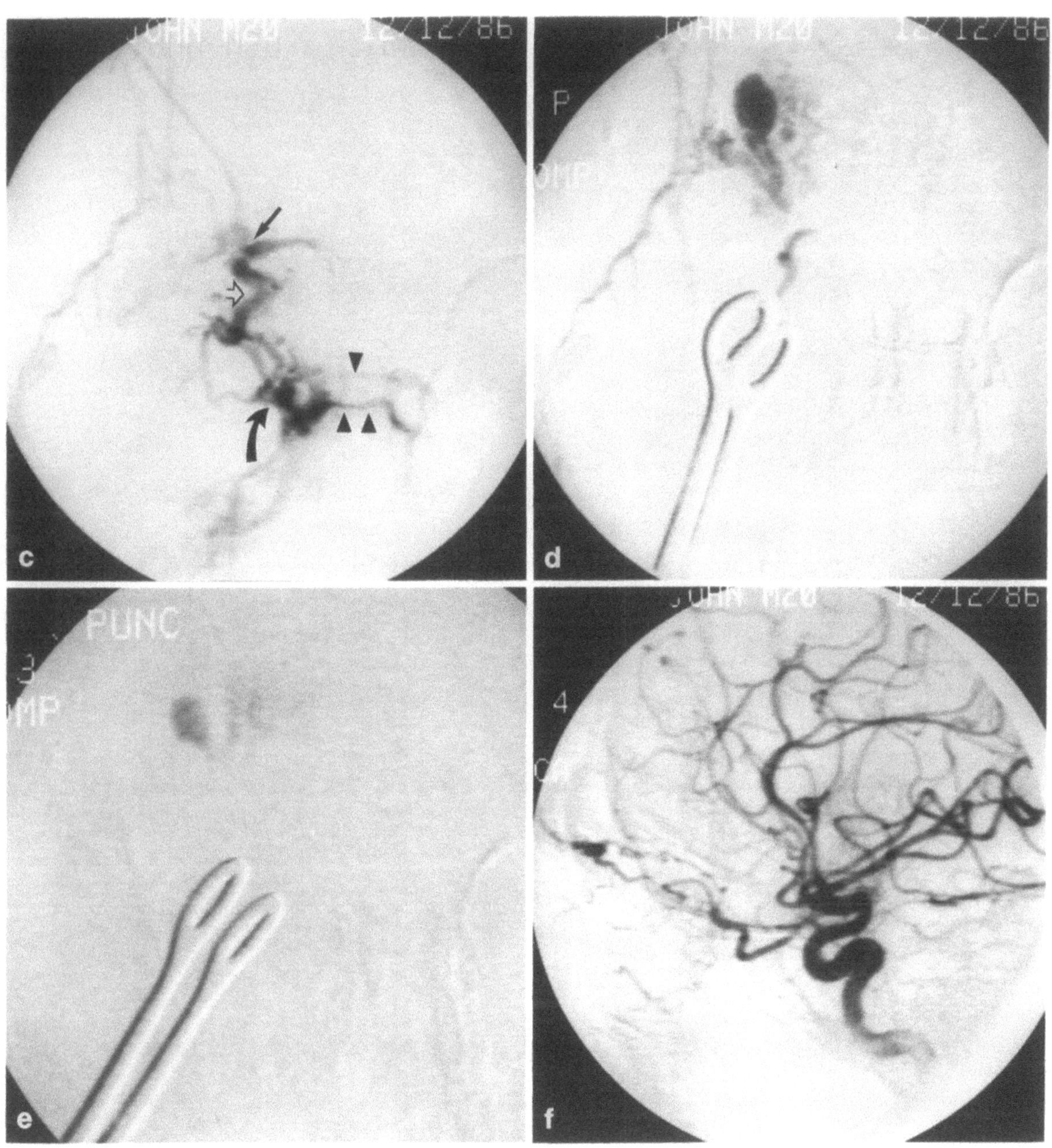

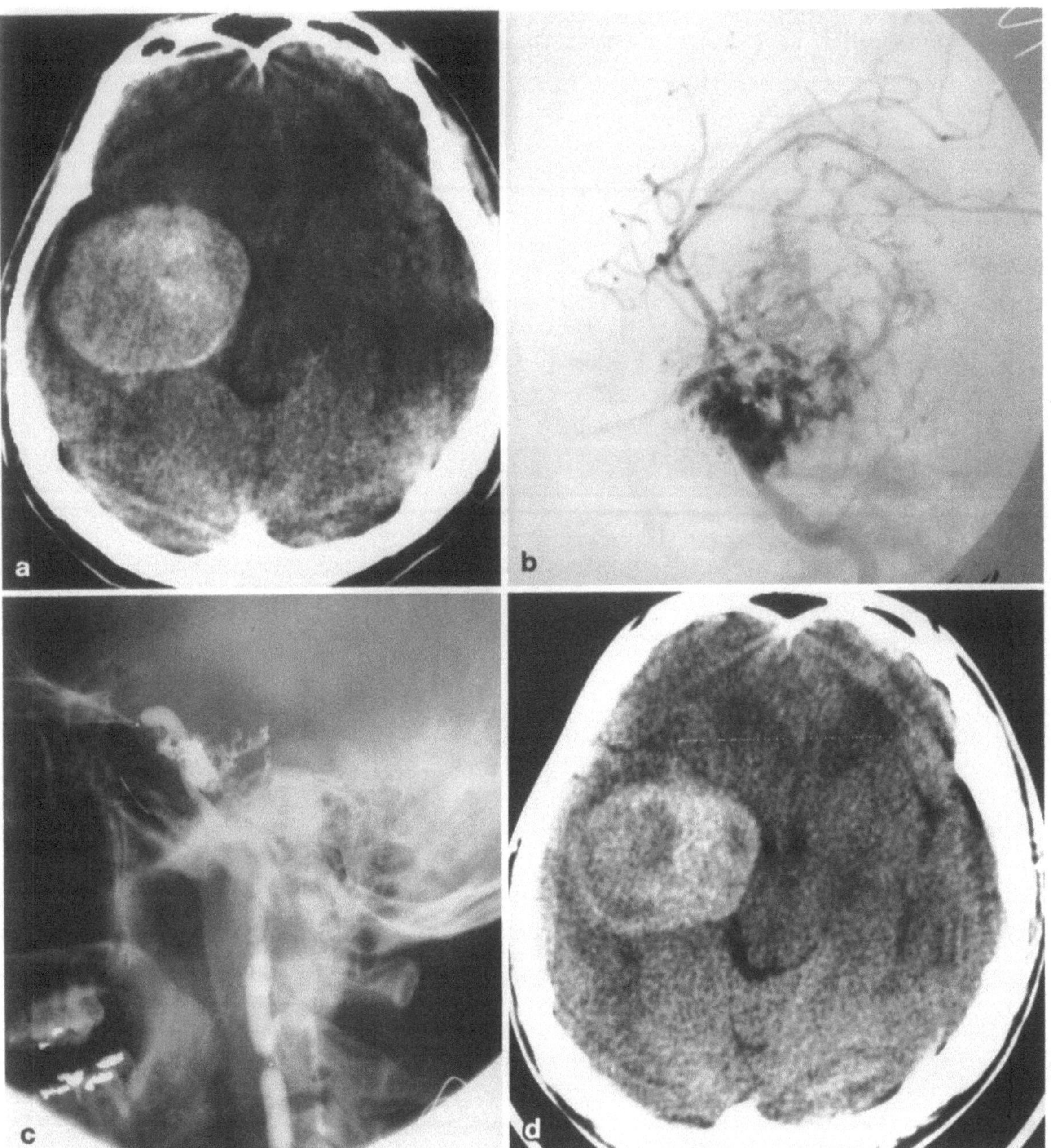

4.12.3 Hemangiopericytoma

When a tumor in the region of the cavernous sinus appears extremely vascular, hemangiopericytoma should be considered in the differential diagnosis (Fig. 4.23). Hemangiopericytomas are unusual, but not rare, and are locally invasive. Recurrences are frequent since total excision is rarely accomplished. Preoperative embolization reduces the extensive blood supply, facilitating a more complete tumor removal. Unfortunately, radiotherapy has not been useful in the treatment of this neoplasm.

◄ **Fig. 4.23a–d.** A 25-year-old man presented with a vague history of blurring of vision. He was found to have findings of a very mild right optic neuropathy and a right optic tract lesion with an incongruous field defect and reduced acuity to 20/30 with a very mild right relative afferent pupillary defect in the right eye. In addition, he had a mild ptosis and slight underaction of upgaze in both abduction and adduction with the right eye
a The unenhanced axial view CT scan demonstrated a huge solid mass extending from the base of the skull and cavernous sinus into the hemisphere and displacing his lateral ventricular system.
b The lateral view subtracted right internal carotid artery angiogram demonstrates extensive tumor neovascularization which is supplied by branches of the internal carotid artery. The internal carotid artery is elevated, displaced posteriorly, and narrowed.
c A lateral skull radiograph demonstrates the cast of cyanoacrylate in tumor blood vessels as well as multiple detachable balloons which were used to occlude the carotid artery from its cervical origin to below the origin of the ophthalmic artery.
d A repeat axial plane noncontrast CT shows a large area of low-attenuation signal within the lateral aspect of the tumor representing ischemic necrosis. The presurgical embolization therapy facilitated the removal of this tumor and only a small portion remained with the cavernous sinus. The patient remained entirely intact except for reduced right corneal reflex, absent right fifth nerve function, and a sixth nerve paresis. However, 1 year later the tumor regrew, necessitating repeat surgery

4.13 Vascular Phakomatoses (Neurocutaneous Syndromes)

Many of the neurocutaneous syndromes, also termed phakomatoses, contain vascular lesions (Table 4.2). Two of the better recognized entities, neurofibromatosis and tubersclerosis, are not considered here since they are not generally associated with vascular lesions (infrequently arterial ectasias). The discussion concentrates on those syndromes which develop primarily because of, or in relation to, a congenital vascular anomaly or from lesions that originate from vascular cellular elements.

Table 4.2. Neurocutaneous syndromes

Sturge-Weber
Von Hippel-Lindau
Wyburn-Mason
Osler-Weber-Rendu
Klippel-Trenaunay-Weber
Neurofibromatosis [a]
Tuberous sclerosis [a]

[a] Not typically associated with vascular lesions.

4.13.1 Sturge–Weber Syndrome (Encephalotrigeminal Angiomatosis)

4.13.1.1 Clinical Manifestations

Sturge–Weber syndrome is the most frequent vascular neurocutaneous disorder. It affects men and women equally without a recognizable hereditary pattern. A port-wine stain (called nevus flammeus) is noted at birth and is associated with a venous malformation in the intracranial meninges and dura [82]. Though port-wine stains occur in approximately 3 children per 1000 births, only 5% of patients have the constellation of symptoms that constitutes the Sturge–Weber syndrome [83, 84]. There is a high incidence of epilepsy, which commonly begins in childhood. The port-wine stain is a flat violet-colored skin lesion, which may blanch slightly or not at all with locally applied pressure. Typically, the lesion is located in any area of the skin on the face or scalp innervated by the trigeminal nerve, but it can occasionally be found underneath the hairline in the suboccipital area

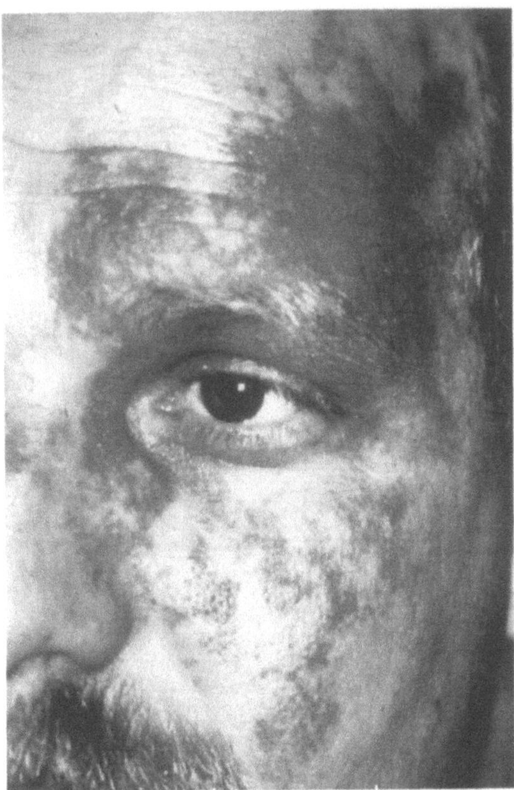

Fig. 4.24. Facial skin lesion in Sturge–Weber syndrome: an extensive port-wine stain involving both the V1 and V2 areas. (Case provided by Roy Geronomus)

(where it may be missed). The lesions are commonly unilateral, infrequently bilateral. The skin lesion often has slight extension across the midline (Fig. 4.24).

Focal and generalized seizures are the most common problems and seem to be related to injury of the cerebral hemisphere beneath the meningeal venular malformation. Though grand mal seizures are common, episodes with visual disturbances such as positive visual phenomena or transient blindness have been reported. The cerebral lesions rarely occur without the cutaneous facial or neck lesions [85, 86].

Static cerebral lesions account for cases with a nonprogressive hemiparesis, hemisensory defect, or homonymous hemianopia. The severity and combination of the deficits are variable. In the extreme case, the extremities contralateral to the damaged hemisphere are poorly developed and small. Patients with a homonymous hemianopia may have

optic disc pallor (with preserved acuity and color vision), despite the fact that the in utero injury is of the postgeniculate visual radiation. In these cases, transynaptic degeneration across the lateral geniculate nucleus to the axons of the optic tract with retrograde degeneration into the optic nerve results (Fig. 4.25) [87]. The optic disc in the eye with temporal field loss, which is contralateral to the brain atrophy, develops the atrophy with a distribution of a horizontal bow- tie pattern. This develops because there is a loss of the nasal nerve fiber layer. This includes atrophy of the portion of the papillomacula bundle that directly passes into the temporal border of the disc. The midnasal and midtemporal disc appears pale. The temporal retinal nerve fiber layer (carrying nasal field information) entering the optic disc along its vertical borders is preserved. Thus, there is a contrasting pink color of the superior and inferior poles of the disc.

Glaucoma of varying degrees of severity is common in the eye ipsilateral to the skin lesion. The risk of glaucoma is greatest in patients with port-wine stain involvement of the upper lid. Open angle glaucoma results from developmental angle anomalies that block the normal aqueous outflow from the anterior chamber through the trabecular meshwork, or from conjunctival or episcleral vascular anomalies. In the latter cases, the episcleral pressure is elevated and the aqueous outflow at the sclera into the veins of the orbit is secondarily decreased. The epibulbar blood vessels are frequently tortuous. Progressive visual loss with buphthalmos in infants, and visual loss without an enlarged globe in young adults, occurs in greater than 25% of cases. The glaucoma is often difficult to control, even with maximal medical therapy and laser trabeculoplasty. In these cases, trabeculectomy and filtering surgery are often required.

Occasionally, a choroidal hemangioma is seen as an elevated gray–yellow or brown–red lesion, located most commonly in the posterior pole (Fig. 4.26). Though the hemangioma can be seen with a direct ophthalmoscope, the larger view and depth perception afforded by binocular viewing may reveal a lesion or the extensive nature of the lesion that is not seen with the monocular direct ophthalmoscope. A visual disturbance can result if there is pressure degeneration of the overlying retinal pigment epithelium or elevation of the retina

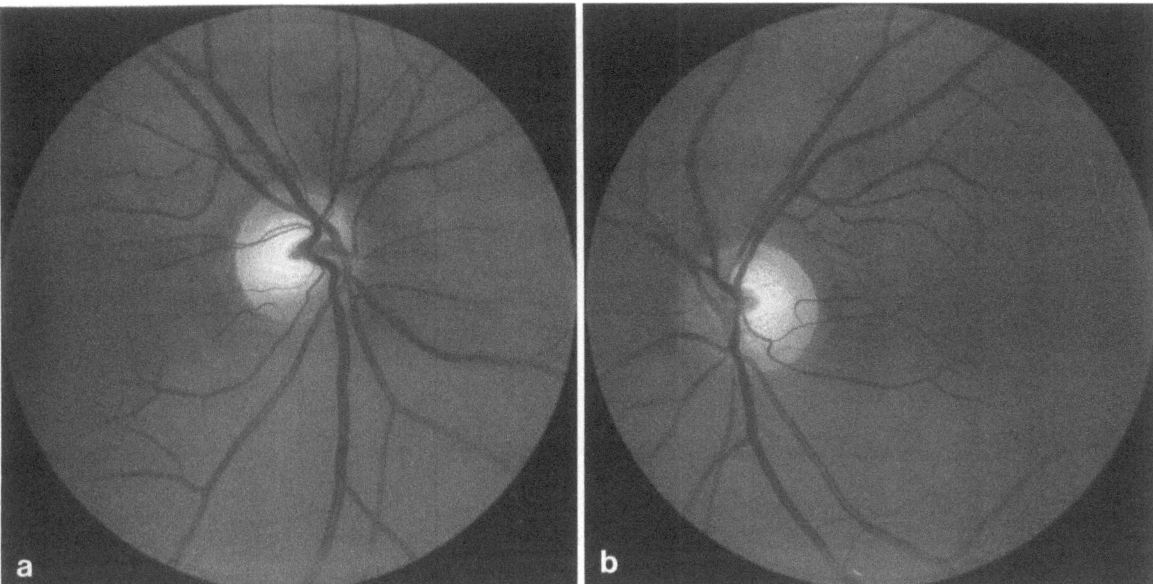

Fig. 4.25 a, b. A 42-year-old woman had a history of a left cerebral Sturge–Weber lesion with a seizure disorder and a right hemiparesis that had been present her entire life. She was also known to have a complete right homonymous hemianopia. Her evaluation showed that her acuity was 20/20 in both eyes with normal examination except for the ophthalmoscopy. **a** On ophthalmoscopy of the right eye she was found to have bow-tie atrophy with a loss of the nerve fiber layer directly temporal along the papillomacular bundle and an absence of the nasal nerve fiber layer. **b** The left eye had mild temporal pallor with some thinning of the nerve fiber layer temporally. This case represents the effects of transynaptic degeneration across the lateral geniculate nucleus because the in utero cerebral lesion has disrupted the retrogeniculate visual pathway (see text)

from subretinal serous fluid around this hamartoma (for evaluation see Sect. 4.13.1.3) [88]. Approximately 50% of choroidal hemangiomas are isolated lesions [89]. Some patients have asymptomatic arteriovenous shunts in the peripheral retina even when there are no choroidal lesions.

4.13.1.2 Histopathology

The histopathology of the facial port-wine lesion is composed of markedly dilated capillary and venular vascular channels, which are lined by endothelium. This malformation is located in the deeper layers of the dermis [90] (Fig. 4.27).

The choroidal hemangioma is a benign hamartoma. It has a similar appearance to the port-wine lesion with dilated endothelium-lined vascular channels that are separated by fibrous connective tissue.

The leptomeningeal malformation is composed of dilated venules with thickened, often hyalinized, walls which frequently contain calcifications (Fig. 4.28) [91]. It is incorrect to call this lesion an angioma since it contains no tumorous elements. There are no true arteriovenous shunts. The malformation lies in thickened meninges overlying an atrophic cerebral cortex which contains many areas of degeneration and loss of neurons and subcortical gliosis and calcification. The cerebral hemisphere frequently also has regions with heterotopia, microgyria, polygyria, agyria, or some combination of all of the above congenital defects. This suggests that the brain anomalies originate because of abnormal early development.

4.13.1.3 Imaging

A choroidal hemangioma can be demonstrated by ultrasonography of the eye even if it has not been visualized on ophthalmoscopy. The B-scan shows a flat subretinal mass that may be associated with elevation of the overlying retina (Fig. 4.26b). The A-

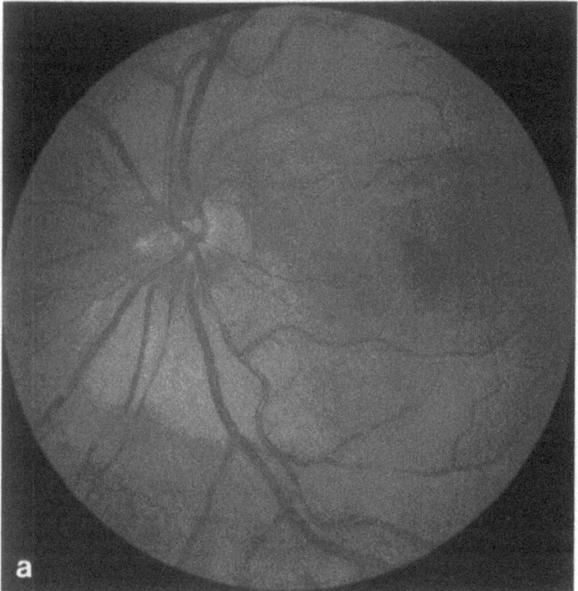

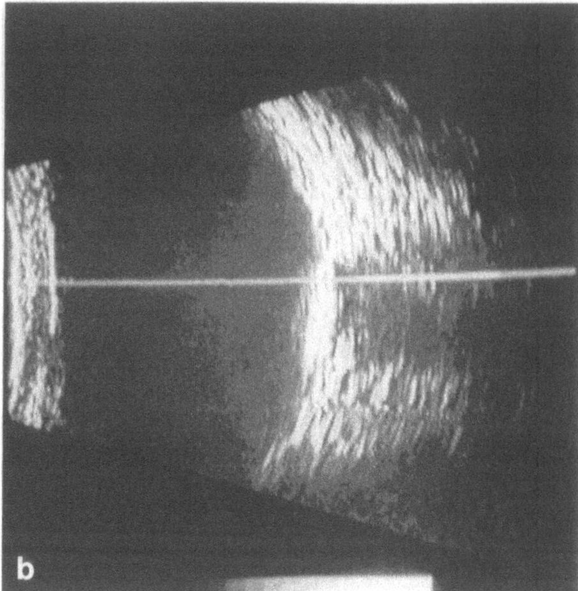

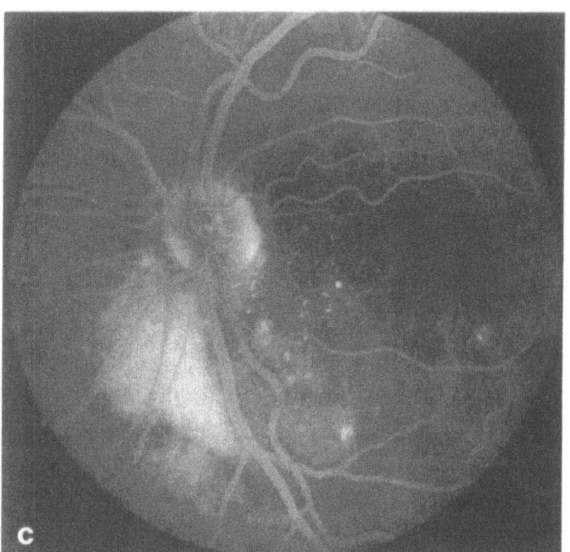

Fig. 4.26. a Ophthalmoscopy reveals an extensive choroidal hemangioma from the inferior and temporal disc border to the macula. The entire retina over the lesion is elevated. The patient has a scotoma that corresponded to this lesion. **b** B- scan ultrasonography of the same case demonstrates the elevated, flat, solid-appearing lesion adjacent to the optic nerve shadow. **c** On fluorescein angiography, an area of leakage of fluorescein dye is seen inferior to the disc, and there are multiple areas of retinal pigment epithelial defects in the retina overlying the lesion. (Case provided by Kenneth Noble)

scan shows the lesion to have high internal reflectivity because of the large interfaces between the surfaces of the many vascular channels. The mass has areas of a solid consistency. A choroidal hemangioma is distinguished from a solid mass such as a melanoma because the hemangioma has acoustical high reflectivity, minor sound attenuation, a flat shape, and of course a lack of growth. CT performed with bone windows will also show a choroidal hemangioma if it is calcified.

On fluorescein angiography, the choroidal hemangioma fills early, before the retinal arteries (Fig. 4.26c). The lesion demonstrates a vascular network pattern. Window defects are seen in areas with retinal pigment epithelium loss. Late leakage is seen in the area with subretinal fluid [92].

After 2 years of age, the skull X-ray will show an area of unilateral intracranial linear calcification, just inside the calvarium. The pattern suggests that the lesion follows the cortical sulci. When hemiatrophy of the cerebrum is present, the petrous pyramid can be elevated and the intracranial vault contralateral to the hemiparesis is smaller.

CT is performed to image the intracranial ab-

Fig. 4.27. The histopathology of the nevus flammeus demonstrates a vascular lesion composed of capillaries and venules in the deeper layers of the dermis. (Case provided by Roy Geronemus)

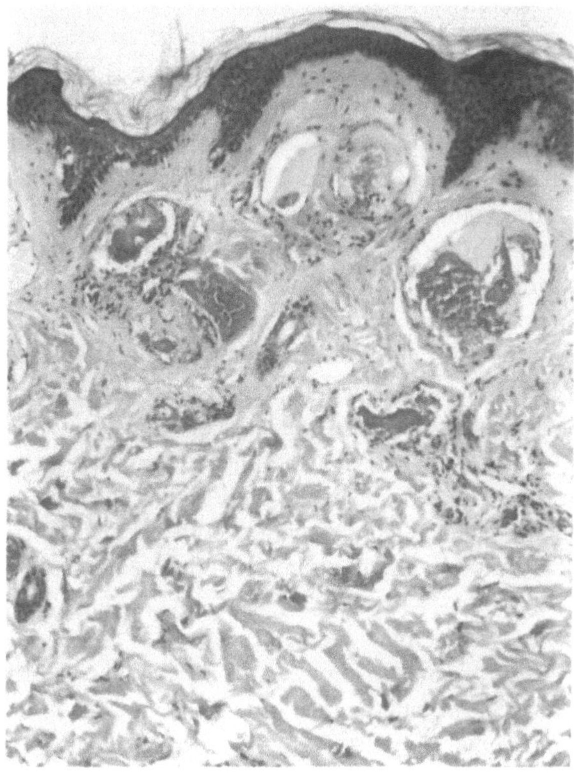

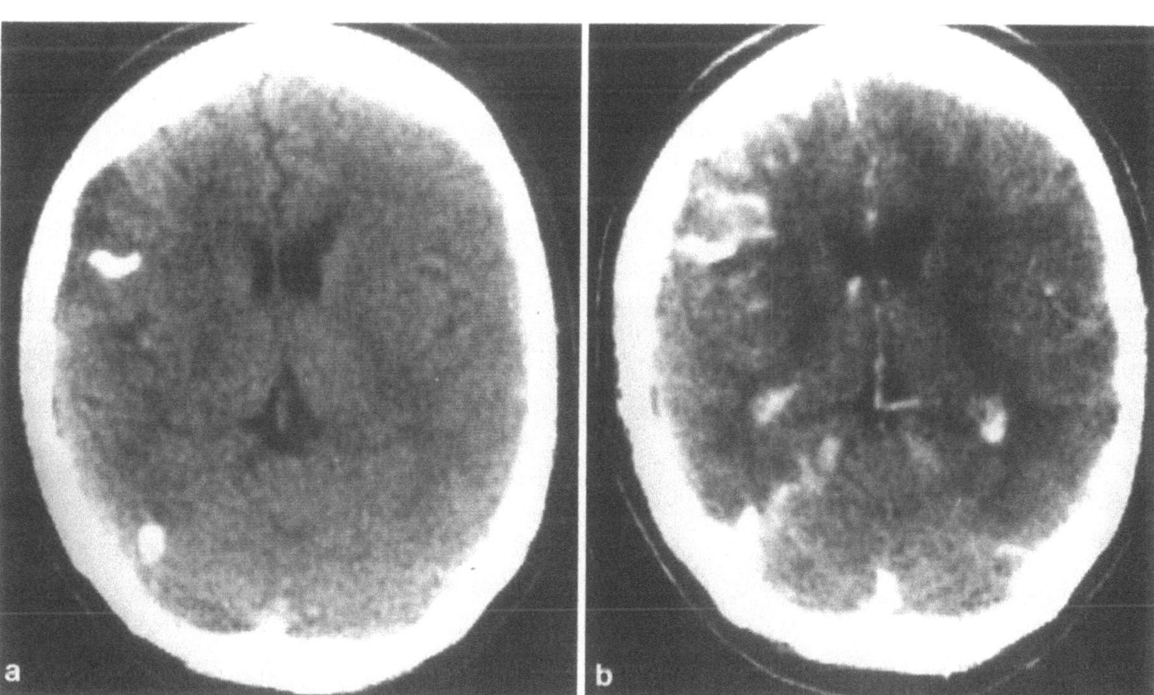

Fig. 4.28a,b. A patient with Sturge–Weber syndrome. **a** The axial view unenhanced CT scan demonstrates areas of calcification along the right cerebral gyri. **b** Following the administration of contrast in the same case, the axial view CT scan demonstrates enhancement along the gyri, particularly in the frontal region of the right hemisphere

normality. The venous malformation contains areas of calcification and follows the cortex along the gyri. The affected cerebral hemisphere is typically, but not exclusively, ipsilateral to the port-wine stain (Fig. 4.28). Bilateral cacifications are unusually seen in patients with a unilateral facial lesion [93]. The cerebral atrophy is often found in parietal or occipital areas, but the entire hemisphere is rarely atrophic. Contrast enhancement of the pial venous anomaly may not always be demonstrated.

The MR image shows cerebral atrophy, enlargement of deep cerebral veins (dark on T1- and T2-weighted images), and areas of calcification [94, 95]. In some patients the unenhanced MR image may appear normal. Following intravenous gadopentetate dimeglumine, both the leptomeningeal cerebral and choroidal lesions enhance [96, 97].

Cerebral angiography often fails to opacify the entire cerebral lesion because of the slow rate of blood flow and the lack of arteriovenous shunting in the malformation. Few normal cortical veins are found in the area of the lesion, further supporting the hypothesis that the vascular malformation may be a developmental anomaly of the cortical venous system (see Chap. 7).

4.13.1.4 Treatment

Seizure control is obtained with anticonvulsant medications. Rare patients with uncontrolled seizures require epilepsy surgery. The therapy of the glaucoma depends on the mechanism and extent of progression of the visual disturbance. When symptomatic, laser photocoagulation of the choroidal hemangioma can prevent serous detachment of the retina [92]. The port- wine stain can be cleared by flashlamp-pulsed tunable dye laser therapy [98, 99].

4.13.2 Von Hippel–Lindau Disease (Retinocerebellar Angiomatosis)

4.13.2.1 Clinical Manifestations

Von Hippel–Lindau disease is an autosomal dominant disorder with incomplete penetrance that causes hemangioblastomas in the central nervous system. One or more hemangioblastomas, typically located in the posterior fossa, less commonly in the spinal canal, and single (most often) or multiple retinal angiomas are present [100, 101]. The full spectrum of this disorder is seen in approximately 25% of patients with retinal hemangiomablastomas [102]. Intracranial or spinal hemangioblastomas (termed Lindau's disease) or retinal hemangioblastoma alone (termed von Hippel's disease) [103] can be the only manifestations of the disorder. When a patient has the full spectrum of disease, members of his immediate family should have a retinal and neurologic examination [104]. A systemic evaluation (for polycythemia, cysts, and neoplasms) and MR imaging or CT of the head might also be considered [105].

4.13.2.1.1 Visual System Disorder

Early on, the retinal hemangioblastoma, which is located in the midperipheral retina (rarely in the posterior pole), is asymptomatic. The tumor has a gray–red color [104]. Because of its location, this retinal lesion is not easily seen with a direct ophthalmoscope. On occasion, the hemangioblastoma is not seen with indirect ophthalmoscopy. However, the presence of a retinal artery that remains large instead of, as normal, gradually tapering as it passes peripherally, suggests there is a high-flow vascular lesion. The hemangioblastoma eventually increases in size and becomes apparent (Fig. 4.29). Retinal hemangioblastomas are found in both eyes in up to 50% of cases. Sudden or progressive visual loss can occur as a consequence of retinal edema, exudates, and hemorrhage in the retina, particularly when the macula is involved, or when the hemorrhage breaks into the vitreous (Fig. 4.29a). In most cases, both the exudates and the size of the tumor increase in size over time. Retinal detachment associated with vitreous traction and secondary glaucoma eventually develops in many eyes with a prior bleed. In these cases, a phthisical globe ultimately results.

Other vascular lesions of the anterior visual pathway are unusual in this disorder. An asymptomatic presumed hemangioblastoma of the optic disc has been described [106]. A hemangioblastoma of the retrobulbar optic nerve is also extremely rare. In the latter case, signs of a progressive orbital mass with monocular blindness may result [107, 108].

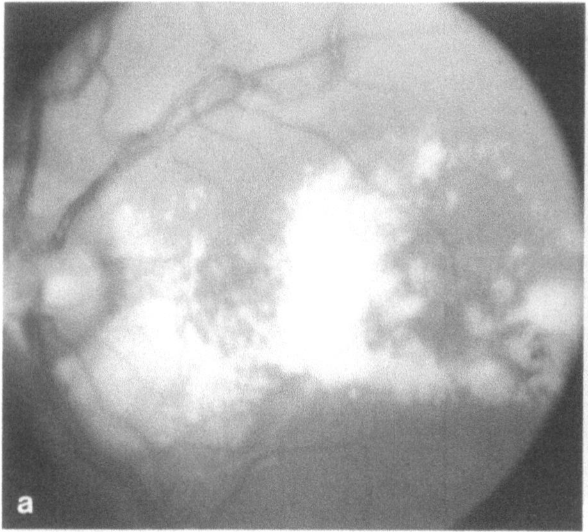
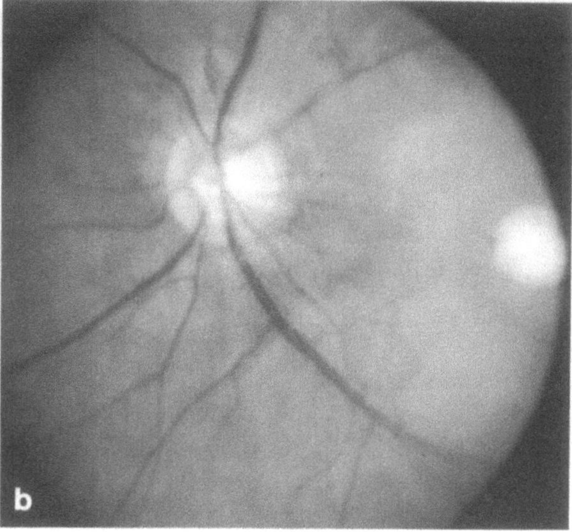

Fig. 4.29a, b. A patient with von Hippel disease and hand motion vision. **a** Ophthalmoscopy demonstrates an extensive exudative reaction in the entire posterior pole, including the macula, secondary to a hemangioblastoma which was temporal to the macula. Both the superior and inferior temporal arteries provide blood supply to the lesion and so are of increased caliber. **b** Following laser therapy of the distal arterial feeders from the superotemporal artery as well as the hemangioblastoma, the exudative reaction has resorbed and the temporal arteries are of smaller caliber. (Case provided by Kenneth Noble)

4.13.2.1.2 Central Nervous System Hemangioblastomas

One or more posterior fossa hemangioblastomas commonly cause suboccipital headaches which are often present for years prior to establishing the diagnosis. The pain becomes more diffuse or frontal and the vision blurs if increased intracranial pressure results from ventricular obstruction caused by growth of the posterior fossa hemangioblastomas. As the brainstem and cerebellum become compressed, neurologic dysfunction, such as various gaze palsies, ataxia, hemiparesis, dysphagia, dysarthria, or all of the above, develops [109–113]. In addition to the visual loss that occurs with chronic papilledema (see Sect. 5.3), the obstructive hydrocephalus- induced ballooning of the third ventricle can compress the chiasm and the intracranial optic nerves [114].

A clinical presentation of epilepsy occurs in patients with the extremely rare supratentorial hemangioblastoma. We have seen only one such case over the last 10 years. When the hemangioblastoma is in the spinal canal, the symptoms depend on the level at which the spinal cord is compromised by the mass [115].

4.13.2.1.3 Systemic Manifestations

Systemic lesions are common in patients with both ocular and intracranial lesions [116]. However, the patients are usually asymptomatic from the associated polycythemia or the cysts and angiomas located in the pancreas, kidney, epididymis, or liver [117]. These patients also have an increased incidence of renal cell carcinoma and pheochromocytoma [118].

4.13.2.2 Histopathology

The histopathology of the angiomatosis retinae hamartoma appears similar to that of the posterior fossa hemangioblastoma (Fig. 4.30a). The retinal hemangioblastoma is composed of endothelial cells and pericytes which form abnormal vessels. Stroma that contains vacuolar lipid inclusions within astrocytes is found between the vascular channels [119, 120]. The retinal lesion replaces the immediately adjacent retina. An exudative retinal detachment and vitreous rhegmatogenous detachment are seen in eyes removed for painful phthisis.

The central nervous system hemangioblastoma

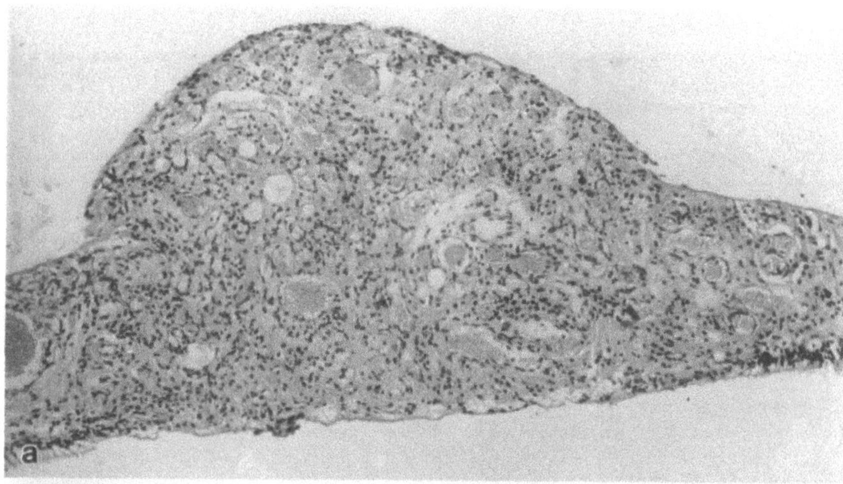

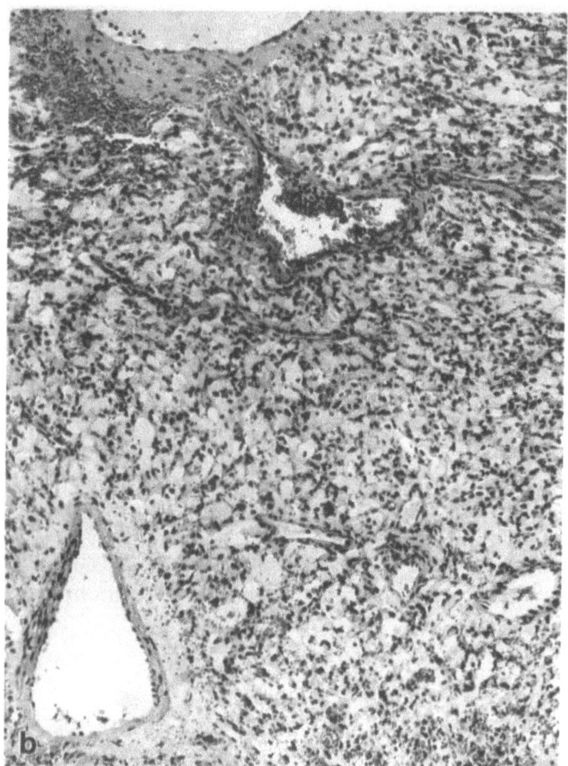

Fig. 4.30 a, b. Histopathology of hemangioblastoma. **a** The histopathology of a retinal hemangioblastoma demonstrates a neoplasm composed of endothelial cells and pericytes with abnormal appearing lumen. The retina in the area has been replaced by the mass. On the lower right and left margins of the tumor, retinal elements are seen (case provided by Norman Charles).
b A cerebellar hemangioblastoma has similar histological characteristics to the retinal angioma in that it is composed of large areas of endothelial cells with vascular lumen

is a well-encapsulated cystic mass with a neoplastic nodule (Fig. 4.30b). Sheets of clear neoplastic cells are separated by numerous capillaries in the wall of the tumor. The capsule is extremely vascular.

4.13.2.3 Imaging

The imaging studies include CT or MR imaging and cerebral angiography of the intracranial lesions, fluorescein angiography of the ocular lesions, and CT or MR imaging and spinal angiography for the spinal cord hemangioblastomas. Ultrasonography, CT, and MR imaging are used to evaluate the abdominal cavity lesions (in addition to urine evaluation for catecholamines and metabolites).

Fluorescein angiography demonstrates the retinal angioma even if it is small. The tumor is supplied by one or more efferent arteries that arise from a retinal artery (Fig. 4.31) [121, 122]. The angioma fills early and demonstrates progressive hyperfluorescence. The angioma leaks profusely into the retina and vitreous during the middle and late phases of the study. The draining veins fill early because of the rapid arteriovenous shunting through the angioma.

The contrast CT typically shows multiple tumors located in the posterior fossa [123]. A small area of enhancement is visualized in the nodule of the wall of each low-density cystic hemangioblastoma. Each mass can compress and distort the brainstem, fourth ventricle, or cerebellum. Dilation

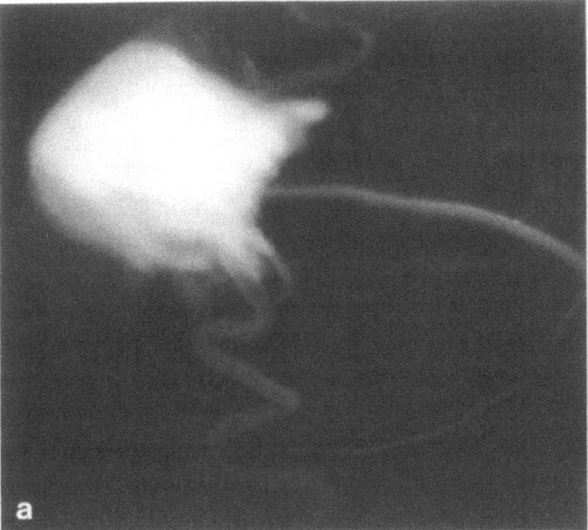 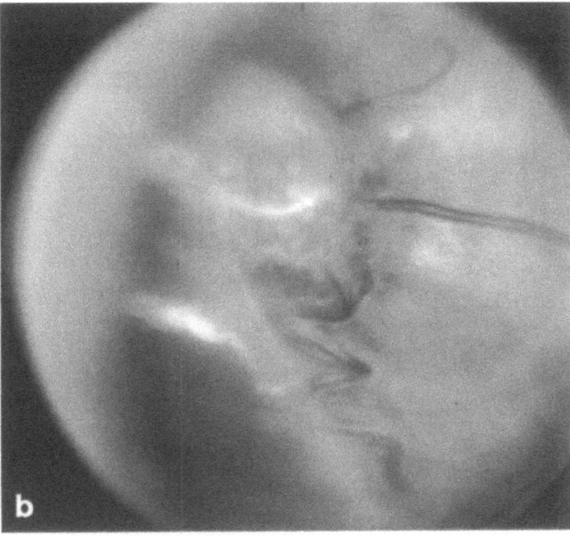

Fig. 4.31a, b. Retinal hemangioblastoma. **a** On fluorescein angiography, the retinal hemangioblastoma is seen to enhance profusely and the details of the supplying arteries and draining veins are visualized. **b** The corresponding ophthalmoscopic picture of this peripheral retinal hemangioblastoma with gliosis and involved vessels. (Case provided by Kenneth Noble)

of the third and lateral ventricles and periventricular edema are seen when there is obstruction of the ventricular system at the cerebral aqueduct or the fourth ventricle.

In addition to abnormalities similar to those demonstrated by CT, MR imaging reveals the high blood flow characteristic of a hemangioblastoma (Fig. 4.32) [124].

Cerebral angiography demonstrates that each tumor is supplied by one or more large arterial feeders. Opacification of the tumor nodule and an adjacent relatively avascular area are pathognomonic of a hemangioblastoma (Fig. 4.33). In some cases, the entire hemangioblastoma opacifies. When the hemangioblastoma has several large high-flow arterial feeders and large vein rapid outflow, the lesion can appear similar to an AVM.

4.13.2.4 Treatment

The ocular manifestations of von Hippel's disease are best treated by preventive therapy of the retinal hemangioblastoma [125, 126]. Photocoagulation is the preferred treatment for the retinal tumor when

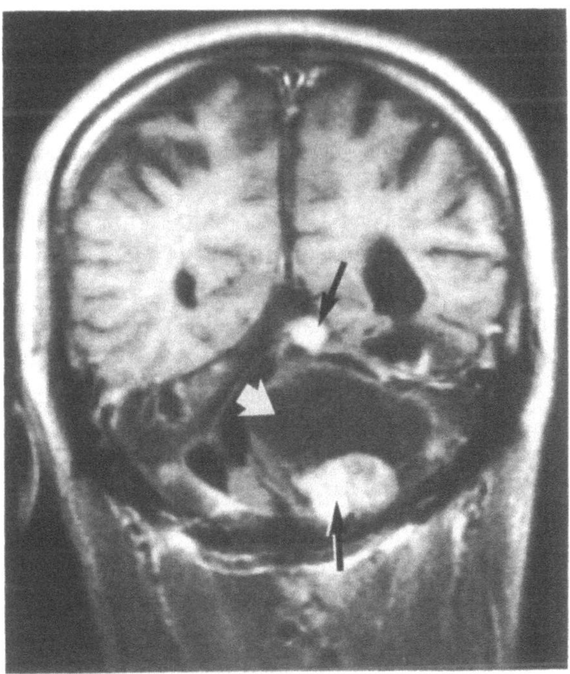

Fig. 4.32. The gadolinium enhanced coronal T1-weighted MR scan shows a large cystic area (*white arrow*) as well as two enhancing nodules (*arrows*) typical of hemangioblastomas in the posterior fossa

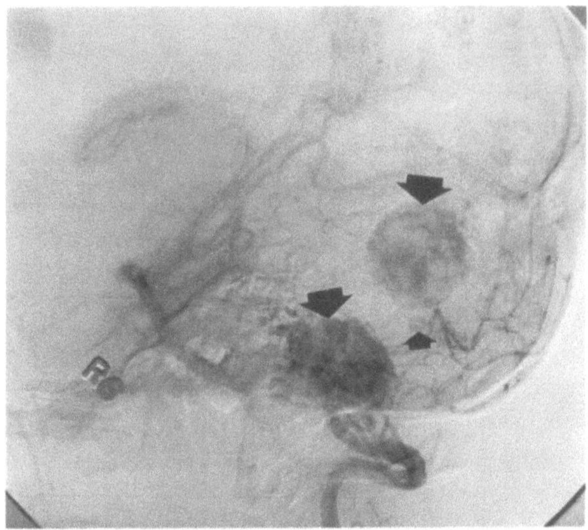

Fig. 4.33. The lateral subtraction right vertebral angiogram shows three posterior fossa hemangioblastomas (*arrows*). The smaller lesion was not clearly seen on a CT scan. In this case, the two larger hemangioblastomas were almost entirely opacified by the contrast material

there is no associated retinal detachment or fibrosis and no neovascularization that extends into the vitreous. Prior to directly treating the lesion, light laser burns are initially made to the distal aspect of the efferent arteries in order to reduce the probability of an iatrogenic hemorrhage or a transudative retinal detachment. Recently, it has been suggested that the dye yellow laser may be the most effective of the various lasers available for therapy [127]. Therapy is repeated until obliteration of the angioma and its feeding arteries is accomplished. Exudates which surround the angioma or the macula edema resorb in most cases following successful photocoagulation (Fig. 4.29a, b). If the lesion is obliterated it will not fill in the subsequent fluorescein angiography.

Cryotherapy is applied to larger angiomas, when the lesions are in an anterior location or following a significant hemorrhage or an exudative reaction. Initially, a relatively high temperature (-60 to $-70\,^{\circ}$C) probe is applied and a colder temperature ($-80\,^{\circ}$C) is used for subsequent treatments until the lesion is obliterated [125].

Retinal hemangioblastoma has also been successfully treated with other modalities. Diathermy can be applied to an angioma with a retinal detachment [128]. A scleral excision has been performed over the area of the angioma. The tumor is identified by indirect ophthalmoscopy and a diathermy needle with the current switched on is introduced to effect hemostasis. Large hemangioblastomas, which cannot successfully be treated with laser or cryotherapy, can be managed by surgical excision [129].

If possible, hemangioblastomas of the posterior fossa should be totally excised. These neoplasms are frequently difficult to remove surgically because of invasion into the brainstem, involvement of the cranial nerve roots, and the intimate relationship with the blood vessels that supply and drain the brainstem and cerebellum. The rapid blood flow to the lesions is frequently associated with increased risk of significant blood loss during surgery. Partial resection is often followed by regrowth of the tumor with a recurrence or worsening of the neurologic signs and symptoms. Surgical excision of a hemangioblastoma in any intracranial location can be facilitated by preoperative embolization using particles, such as polyvinyl alcohol, to decrease the arterial blood flow into the lesion.

A ventricular shunt is frequently placed to relieve the obstructive hydrocephalus as the first procedure. Providing a drainage pathway for the cerebrospinal fluid can reverse the lethargy, diffuse cerebral dysfunction, or papilledema and visual disturbance secondary to the increased intracranial pressure. Upward herniation of posterior fossa structures through the tentorium following shunting is not a major consideration even though the mass lesions are almost always located in the posterior fossa.

When a hemangioblastoma compresses the spinal cord surgical excision is also indicated. Preoperative embolization is also effective for the lesions in this location.

4.13.3 Wyburn–Mason Syndrome

4.13.3.1 Clinical Manifestations

Wyburn–Mason syndrome is a rare entity. It is perhaps the most dramatic of the neurocutaneous disorders, with multiple large tortuous arteries and veins forming one or more arteriovenous communications (racemose) in the retina and through the optic nerve (Fig. 4.34a) and an associated intracranial AVM [130, 131]. The retinal vascular anomaly is often connected to a meningeal and parenchymal AVM that involves the ipsilateral anterior visual pathway [132]. Both sexes are equally often affected. The vascular malformation may arise from a disturbance which prevents the normal division of the anterior plexus of the embryonic primitive vascular mesoderm. Ordinarily, the vascular mesoderm separates into the hyaloid vascular system to the eye and the vascular system to the midbrain [133].

Associated cutaneous abnormalities are infrequent, but occasionally flat faint red–blue colored lesions with a pinpoint shape or a small port-wine defect (see Sect. 4.13.1) can be present.

The clinical features include poor vision in the involved eye from birth in patients with extensive retinal lesions (Fig. 4.34a). In some cases there is no measurable visual loss (Fig. 4.34b). The visual loss can be acquired when the abnormally high arterial blood inflow or compromise of the retinal venous outflow causes retinal hemorrhages, exudates, and edema, especially if the macula is affected. An intracranial or intraorbital AVM may cause optic nerve dysfunction, with or without a bruit or proptosis [134]. If the lesion is extensive and involves the entire cerebral hemisphere, a hemiparesis and hemisensory defect can be present. If the AVM extends to the optic radiation or the visual cortex, a hemianopia may result. When the AVM is also located deeper in the brain parenchyma, involving the thalamus, midbrain, pons, or cerebellum, clinical signs develop in relation to the specific areas affected and the extent of the AVM. Typical of AVMs in the brain, some patients with an intracranial AVM are relatively asymptomatic, while others have one or more catastrophic hemorrhages (see Sect. 7.1). In approximately 20% of cases, additional AVMs involve the face or pharynx with local tissue swelling or hemorrhage.

4.13.3.2 Imaging

Fluorescein angiography of the retina reveals the rapid blood flow through the central retinal artery. A direct shunt from the large retinal arteries into the large retinal veins is demonstrated by the early appearance of laminar flow in the veins of the reti-

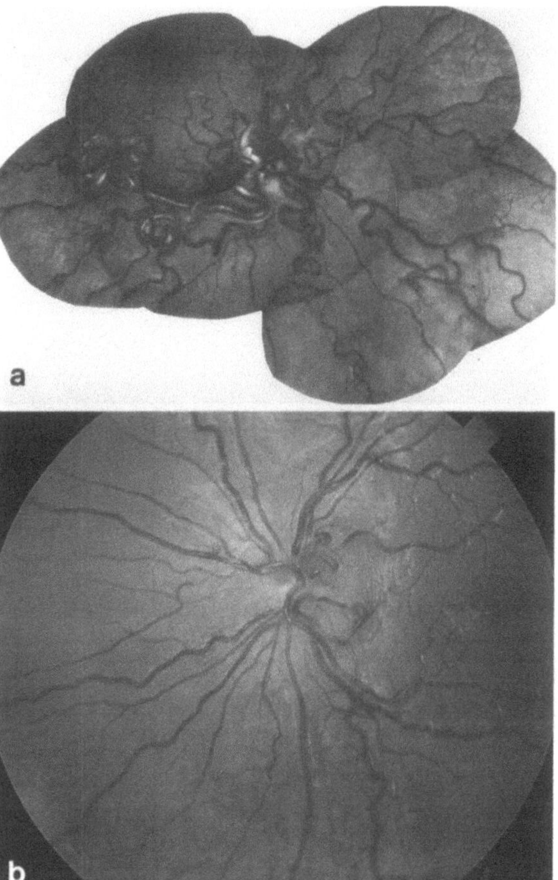

Fig. 4.34. a Ophthalmoscopy demonstrates the marked dilatation of the arteries and veins from this arteriovenous shunt in a patient with Wyburn–Mason syndrome and 20/800 vision in this affected eye. **b** In contrast to the patient in **a**, ophthalmoscopy of this second patient shows a small area of arteriovenous shunt on the optic disc in a patient with normal visual function. This 9-year-old girl, who had complained of recurrent spontaneous bleeding from the palate and slight swelling and increased warmth over the left cheek, was found to have a Wyburn–Mason syndrome as well; her MR scan and angiogram are seen in Figs. 4.35, 4.36.

na. In most cases, there is no extravascular leakage of fluorescein.

Features of the orbital arteriovenous fistula can be demonstrated by the various imaging techniques. Ultrasonography shows the high-flow nature of the shunt with dilated vascular channels in the orbit. CT of the head and orbit shows serpiginous AVM vessels that enhance following intravenous contrast. The orbital lesion may or may not

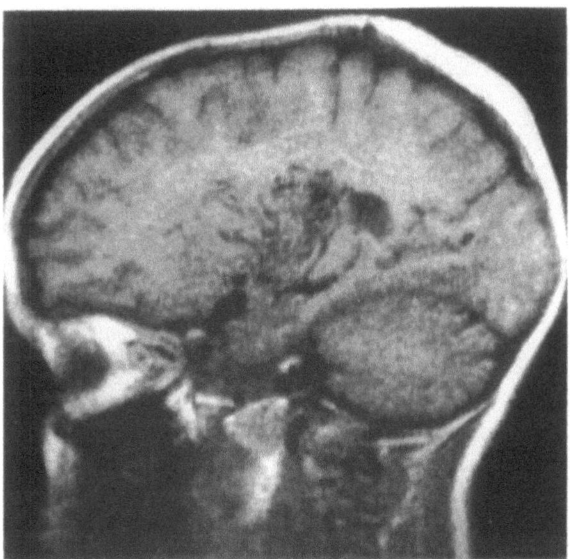

Fig. 4.35. The sagittal T1-weighted MR scan of the patient in Fig. 4.34b demonstrates abnormal flow voids, suggestive of an arteriovenous malformation, extending from the posterior orbit and involving the thalamus and the cerebral hemisphere

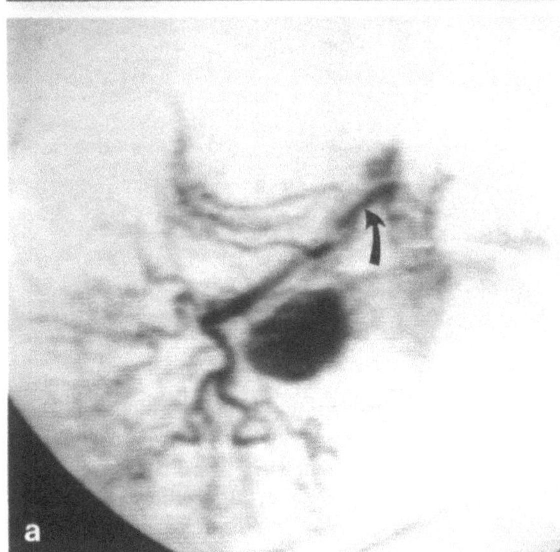

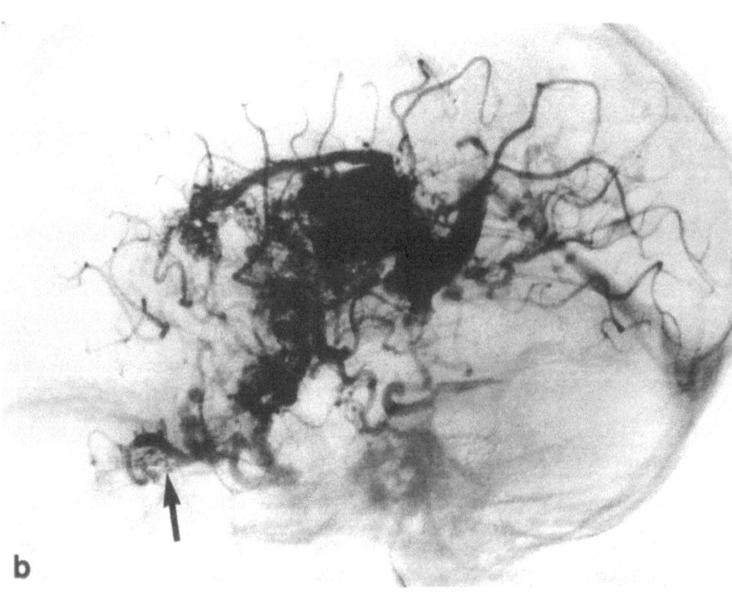

Fig. 4.36. a The cause of the recurrent gum bleeding in the case of Fig. 4.34b was an arteriovenous malformation with a large venous sac that was supplied by the distal palatine artery (*arrow*), a branch from the external carotid artery. **b** The lateral internal carotid artery angiogram demonstrates the extensive arteriovenous malformation intracranially as well as in the orbit apex (*arrow*) in the same case

appear continuous with the intracranial AVM. MR imaging shows a picture similar to the CT but the hemodynamic alterations characteristics of the lesion are more obvious (see Sect. 7.1) (Fig. 4.35). Because of chronic pressure on the bone induced by dilatation of the vessels accompanying the optic nerve, a plain skull X-ray may show the ipsilateral optic canal to be enlarged compared to the contralateral canal. Cerebral angiography visualizes the large arterial feeders as well as a network of small arterial feeders supplying the AVM in the orbit, face, and intracranially (Fig. 4.36) (see Sect. 7.1 for examples of angiography).

4.13.3.3 Treatment

There is no treatment of the retinal and optic nerve dysfunction. Treatment of the brain AVM is discussed in Chap. 7.

4.13.4 Klippel–Trenaunay–Weber Syndrome

Klippel–Trenaunay–Weber syndrome is another rare disorder that is associated with vascular abnormalities in the skin, but ocular or orbital lesions are infrequent. Most cases are inherited but the pattern of clinical expression is variable. Hemangiomas, varicosities, and soft tissue hypertrophy are found in the dermal and subcutaneous layers of the skin, most often in an extremity. Telangiectasia and angiomas of the conjunctiva, varicosities of the veins in the retina or orbit, or angiomas of the sclera and choroid have all been rarely described [135, 136]. Infrequently, vascular malformations are found in the spinal cord or brain. Treatment is geared to the specific symptomatic lesions.

4.13.5 Osler–Weber–Rendu Syndrome (Hereditary Hemorrhagic Telangiectasia)

4.13.5.1 Clinical Manifestations

Osler–Weber–Rendu syndrome is a rare autosomal dominant disorder with incomplete penetrance that causes telangiectasia of the skin and mucus membranes, including the conjunctiva. The signs are often first noticed at puberty. There is an equal sex predilection. Characteristic dermal

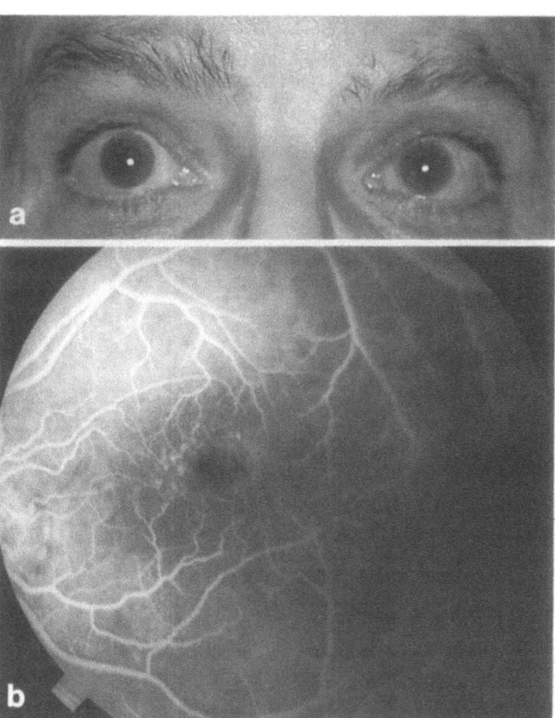

Fig. 4.37. a This patient with Osler–Weber–Rendu disease has bilateral conjunctival telangiectases which caused diffuse chronic injection of both conjunctiva. b In the same patient, telangiectases of the retina around the macula are best seen with fluorescein angiography. No leakage was seen from these vessels. The patient had a complaint of slight metamorphopsia but there were no deficits in acuity or the visual field

and mucosal telangiectases that are either punctate or irregularly shaped and larger than 1 cm cause spontaneous hemorrhages. Patients are especially prone to episodes of epistaxis, without any underlying disorder of coagulation [137–140]. Episodes of bleeding from telangiectases and arteriovenous fistulas that are found in the viscera can be life-threatening. AVMs and fistulas in the lungs occur in approximately 16% of patients and predispose the patient towards paradoxical emboli to the brain, brain abscesses, hemoptysis, and air emboli. Less frequently, emboli involve the liver, spleen, pancreas, and kidney.

An associated retinal or orbital AVM or an orbital varix is rare. However, conjunctival telangiectases which cause focal or diffuse chronic injection of the conjunctiva are common (Fig. 4.37a). Hemorrhage from the conjunctiva rarely cause bloody tears. Retinal telangiectases are also rare (Fig. 4.37b) [141].

Intracranial vascular abnormalities include arteriovenous fistulas, which are true AVMs (8.1% of patients) that frequently involve the dura. Parenchymal telangiectases or cavernomas (16.7% of patients) are also common. Rupture of a meningeal telangiectasia causing a hemorrhage or a spontaneous carotid cavernous fistula is rare. We have only seen one patient with Osler–Weber–Rendu syndrome and a cavernous sinus area DAVM out of 75 cases with DAVMs in this location (see Sect. 3.2). Spinal cord AVMs are also found in approximately 7.1% of patients [142]. Despite the significant incidence of intracranial vascular anomalies, most neurologic deficits result from complications of the vascular lesion in the lungs [143, 144].

Hereditary hemorrhagic telangiectasia must be differentiated from ataxia telangiectasia (Louis–Bar syndrome). Though the latter disorder can also have conjunctival telangiectasia, it is characterized by a progressive gait ataxia [145]. Ataxia telangiectasia is also associated with chromosomal abnormalities and immunodeficiency. T-cell function and numbers are diminished and selective deficiencies of immunoglobulin G (IgG), IgA, and IgE are found [146–148]. These alterations cause the patient to be more prone to develop lymphoid leukemia and upper respiratory tract infections [145].

4.13.5.2 Histopathology

The histopathology of telangiectasia in any location reveals that the lesions are composed of a collection of small venules, without an arterial elastic or muscular layer. This accounts for the fragile nature of the vessel walls and the tendency to hemorrhage. These vascular malformations cause pressure necrosis of the overlying mucus membrane epithelium as well. The pathology of the cerebral AVM, DAVM, and venous varicosities does not differ from that found in patients without Osler–Weber–Rendu syndrome.

4.13.5.3 Treatment

The treatment of the orbital and intracranial lesions is similar to that used for each specific abnormality. Percutaneous embolization with detachable balloons can also be used to close the pulmonary arteriovenous shunts.

4.14 Cavernous Hemangioma of the Brain and Retina

The rare association of cavernous hemangiomas of the retina and the brain was first noted in 1940 but it was not considered a distinct entity until 1971 [149, 150] (see Sect. 7.2.2). Though this disorder may have a familial tendency, no definite hereditary pattern has been established. A few patients also have skin hemangiomas, most commonly on the neck. Some patients will have retinal and cutaneous angiomas, without vascular lesions in the central nervous system [151]. Both sexes are equally often affected. The onset of neurologic symptoms, usually seizures, is in the second to sixth decade.

The ocular hemangioma is unilateral and typically involves the retina or the surface of the optic disc [152]. Visual loss is unusual except in cases of vitreous hemorrhage or if the hemangioma is located in the macula [153]. The retinal lesion is composed of a cluster of thin-walled dilated veins which incompletely fill during the venous phase of the fluorescein angiogram. Arteriovenous shunting is not seen. Late extravasation of the dye is infrequently seen. Twin vessels, a paired retinal arteriole and venule separated by less than a single venule diameter, can be seen in familial cases of retinal hemangioma as well as in cases of von Hippel–Lindau disease [154].

Another ocular vascular lesion, a hemangioma of the optic disc, is composed capillary endothelial cells and pericytes. This capillary hemangioma is not associated with other vascular lesions in the central nervous system. Exudates in the macula can cause a visual disturbance in these cases. The lesion may have a fine capillary appearance on fluorescein angiography [155]. A rare juxtapapillary capillary hemangioma can be associated with intraretinal and subretinal exudates and a serous retinal detachment overlying the lesion [156].

CHAPTER 5

Vascular Optic Neuropathies

Knowledge of the vascular supply to the optic nerve, retina, and choroid (see Chap. 1) is essential for the understanding of visual defects that are caused by segmental or focal ischemia of the optic nerve or retina. Recall that the optic nerve and the retina have *different* blood supplies: the retina is supplied by the branches from the central retinal artery and the optic nerve is supplied principally by the posterior ciliary/circle of Haller–Zinn circulation.

5.1 Anterior Ischemic Optic Neuropathy

Anterior ischemic optic neuropathy (AION) is a relatively common disturbance of the optic nerve principally affecting the elderly. The average age of patients is approximately 66 years [1, 2]. Except for diabetics, AION only unusually occurs in patients less than 50 years of age [3]. There is a slight female preponderance.

Loss of vision from optic nerve ischemia was first recognized as a complication of acute severe blood loss [4–6]. Spontaneously occurring AION, which is related to arteriosclerosis, is a relatively recently recognized entity first described in 1948 by Kurz [7]. In fact, the first series of cases of AION was not reported until 1966 [8].

5.1.1 Clinical Presentation

AION causes a painless unilateral visual loss that may be sudden or preceded, by days to weeks, by minor blurring of vision. Bilateral simultaneous visual loss is extremely rare. However, some patients experience a subclinical ischemic episode in the presumed normal eye prior to presentating with symptomatic acute visual loss in the contralateral eye. These patients may have a superior altitudinal defect or a small arcuate field defect, preserved or relatively spared acuity, and a flat pale disc in one eye and disc swelling with acute visual loss in the newly involved eye. The combination of findings may lead the examiner to use the poor term "pseudo-Foster–Kennedy" syndrome or misdiagnose a "Foster–Kennedy" syndrome (see below).

On ophthalmoscopy, the acute AION disc is diffusely or segmentally swollen and often pale. In some cases, secondary capillary dilatation may cause the disc to appear injected (Fig. 5.1). Flame-shaped hemorrhages are frequently observed within the peripapillary nerve fiber layer. The number of hemorrhages is not indicative of the extent of optic nerve damage, nor does it reflect the severity of visual loss unless a hemorrhage fortuitously involves the macula. The altitudinal or arcuate field loss corresponds to the area of segmental disc swelling when the disc changes are focal [9] (Fig. 5.2). Optic disc swelling may actually precede the visual loss or cause only minimal visual dysfunction, but disc swelling without any visual symptoms is rare [10]. In those cases where a unilateral disc is swollen without visual loss, other clinical entities should be considered (see Sect. 5.1.6).

AION is not an inflammatory papillitis. The vessels are not significantly sheathed, nor are cells seen in the vitreous overlying the swollen disc. Confusion over this issue probably arises from terms such as vascular "pseudopapillitis" used by Francois [11], and ischemic optic "neuritis" used by Hollenhorst [12].

The central visual function is affected in most patients. Initially, the Snellen visual acuity is 20/200 or worse in 42% of cases, while only 18% of cases have 20/40 acuity or better (Table 5.1). The visual field defect (Fig. 5.3) is predominantly altitudinal, occurring in approximately 60% of cases. The inferior field is affected three times more commonly than the superior field [9]. The preponderance of defects in the lower field may be partial-

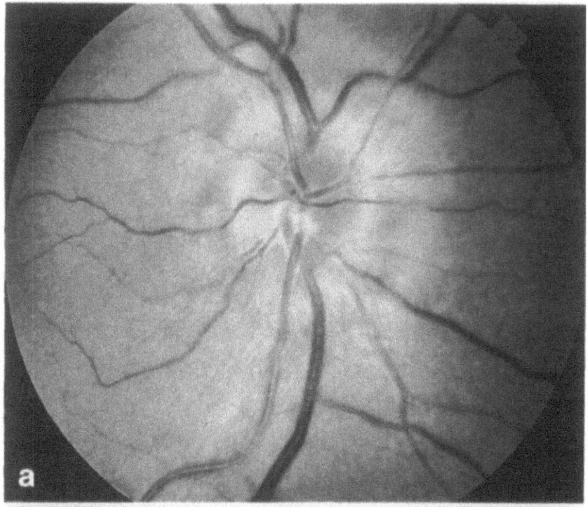

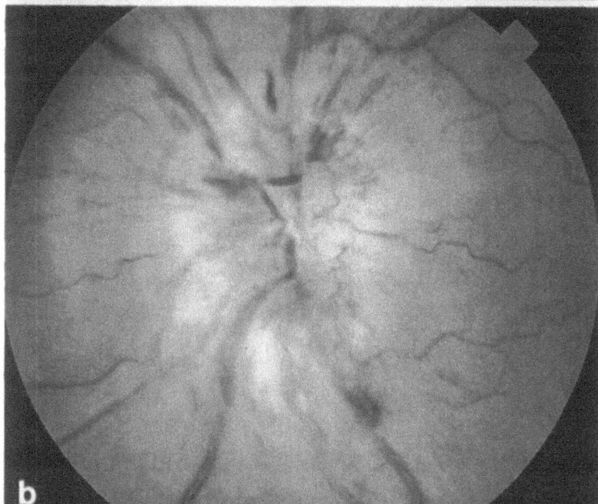

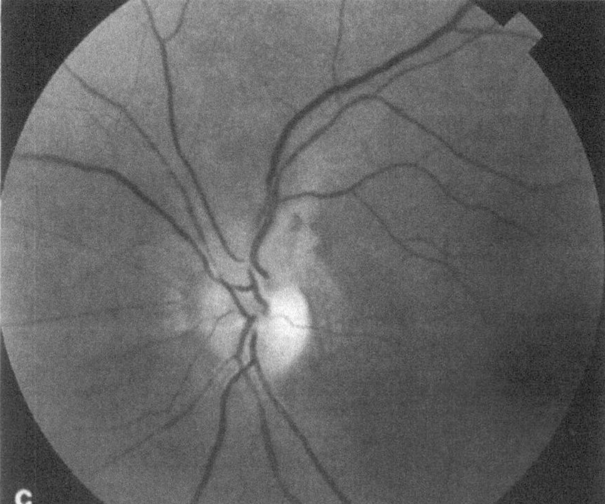

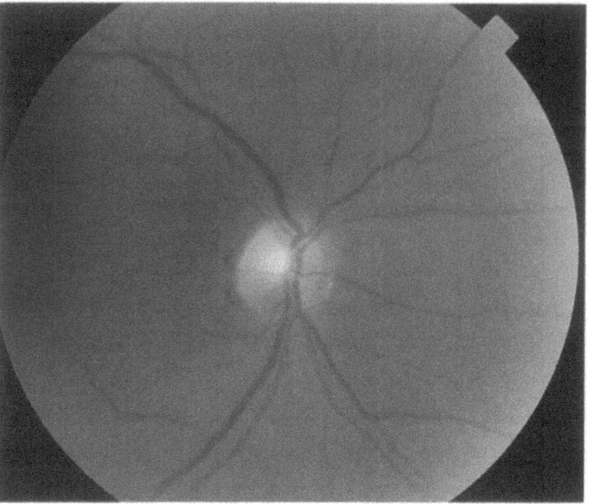

Fig. 5.2. This 65-year-old man complained of a vague visual disturbance in the right eye. He was not clear for how long a period of time this had been present. His visual acuity was 20/20 in each eye and color vision was intact. There was a relative afferent pupillary defect in the right eye. The visual field revealed an inferior altitudinal field defect from the right eye. Ophthalmoscopy revealed yellow pallor of the upper portion of the disc which spared the papillomacular bundle. The artery on the superotemporal aspect of the disc was focally narrowed. This patient was felt to have had anterior ischemic optic neuropathy. His visual performance remained unchanged during the 3-year follow-up

Fig. 5.1a–c. This 55-year-old man had noted the onset of blurred vision in the left eye on awakening. He had 20/20 acuity in the right eye and could only count fingers at 2 feet in the left eye in the superior field. **a,b** Ophthalmoscopy of the right eye (**a**) was normal with a full disc without an obvious physiological cup. The left optic nerve (**b**) is diffusely swollen with nerve fiber layer hemorrhages on and just off the disc. The inferonasal nerve fiber layer has cotton-wool spots. **c** Ten months later, the left optic disc is flat and pale and there is some gliosis of the vessels on the disc. The peripapillary choroidal crescent is now obvious because of the atrophy of the overlying nerve fiber layer. There was no return of vision. This is clearly a case of an anterior ischemic optic neuropathy

Table 5.1. Visual loss in eyes with acute anterior ischemic optic neuropathy (modified from [2, 8, 9, 14, 16])

	Miller and Smith [8]	Cullen [1]	Kurz [2]	Ellenberger et al. [14]	Boghen and Glaser [9]
Snellen acuity					
>20/80	9	12	26	35	22
20/80-20/400	–		19	10	17
<20/400	5	9	29	15	10
Visual field defects					
Altitudinal	7	6	28	51	16
Central scotoma or centrocecal	2	5		8	6
Arcuate	2	–	22	4	18
Other	3	10		1	

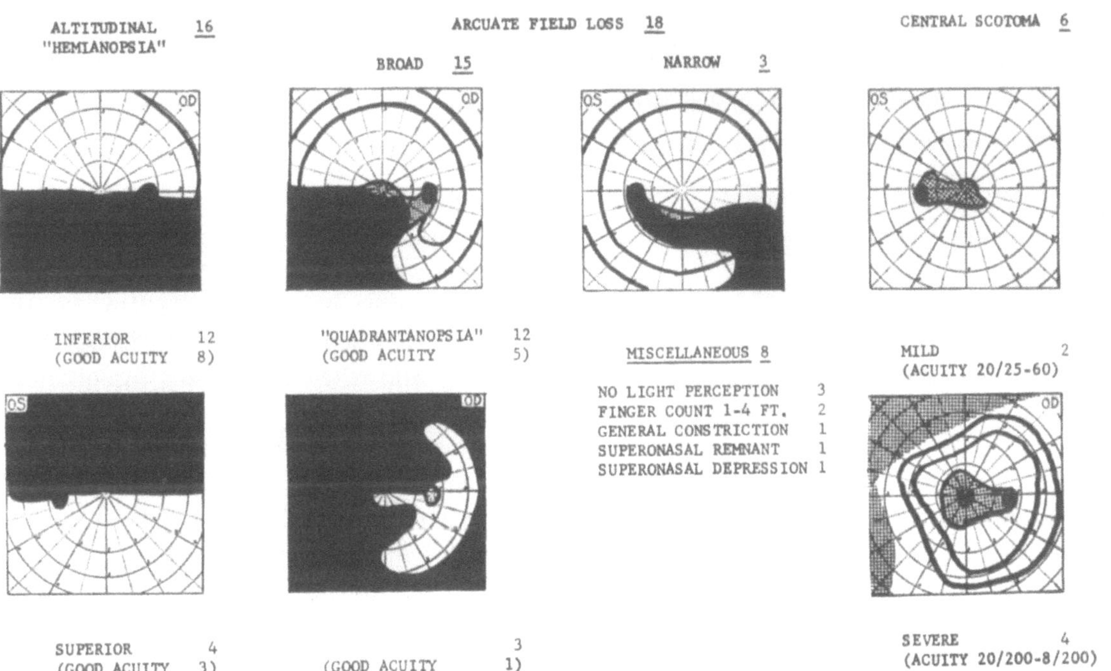

Fig. 5.3. Visual field defects in nonarteritic anterior ischemic optic neuropathy in a study of 48 affected eyes in 34 patients. (From [9])

ly explained because the loss of this particular field seems to be more bothersome to the patient, even when the central acuity is preserved. These patients would be more likely to seek medical attention. An isolated central or centrocecal scotoma occurs in approximately 20% of cases; less commonly, a quadrantic or an arcuate field defect is present [13–15]. Dyschromatopsia frequently develops when the ischemia affects the papillomacular bundle. However, the color vision is normal when the ischemia is segmental and affects only the superior or inferior aspects of the optic nerve.

A relative afferent pupillary defect is found in unilateral cases. However, if the field loss is minimal or absent, an afferent pupillary defect may not be readily apparent. A relative afferent pupillary defect may also be absent in cases with bilateral optic neuropathy, even if the episode in the previously affected eye was subclinical.

5.1.2 Clinical Course

Following the onset of the visual disturbance, the acuity may worsen over a period of, usually, from several to 24 h. Less often, in approximately 30% of patients, the loss of visual acuity or field, or both, worsens over an entire week or in a stepwise fashion over several weeks [9]. The cause of the progressive deterioration is unknown. It is possibly related to continuous or recurrent hypoperfusion of the optic nerve because of an anatomical variation of blood supply. A reduction in the blood perfusion pressure to the optic nerve, such as occurs during sleep, may also precipitate or worsen an episode of ischemia. Not infrequently, patients are first aware of their visual loss when they wake from sleep. Another mechanism for progressive optic nerve injury may be that the lack of distensibility of the collagen scleral canal compresses the ischemic swollen optic nerve, causing further compromise of the nerve with stasis and blockade of axoplasmic flow. This results in extracellular edema and further damage to the optic nerve. The shift in extracellular electrolytes from the injured axons also adds to the edema. Recent experimental work has also shown that extracellular calcium can worsen the effects of anoxia on the optic nerve [17] and contribute to the death of injured axons.

Though long-term follow-up evaluation may show minimal improvement or fluctuation in the visual fields, most field defects remain unchanged over years [9]. Snellen acuity improvement rarely exceeds 1 or 2 lines, except in those patients who have a relative central scotoma at presentation. In these latter patients, visual recovery of equal to or greater than 2 Snellen acuity lines is not unusual [14]. Rare patients who initially have severe visual dysfunction will demonstrate dramatic recovery (Fig. 5.4).

One to three months following the acute event, diffuse or segmental flat disc pallor (segmental atrophy is almost pathognomonic for AION), often with one or more narrowed arteries on the disc, is seen on funduscopy (Figs. 5.1c, 5.4b). A mild increase in the cup/disc ratio eventually develops in approximately 15% of affected optic discs. The cup is shallow and surrounded by pallor of the remaining neural rim. This differs from the alterations of the optic disc caused by glaucoma where there is a deep cup, which is typically surrounded by a healthy appearing pink neural rim [14, 18, 19]. Contrary to most reports, in one small study, three of ten eyes with AION developed a cup/disc ratio greater than 0.6, similar to the cupping in glaucomatous eyes [20]. Even in the normal population, there is an increase in the cup/disc ratio with aging. It is easy to see how, in elderly patients with a mild elevation of the intraocular pressure (IOP), it is sometimes difficult to ascertain when the visual field loss or optic disc damage has developed because of ischemia or glaucoma.

The low incidence of significant excavation of the disc generally reported in AION (Fig. 5.5) can be contrasted with the significant incidence of cupping of the disc in cases of ischemic optic neuropathy resulting from giant-cell arteritis (GCA). This probably reflects the severity of the ischemic necrosis caused by multiple arterial occlusions seen in the latter disease (see Sect. 5.2.6). In one study of patients with arteritic damage of the optic nerve, five of ten affected eyes had marked disc excavation. However, in the same study the unaffected eye often had a cup/disc ratio of 0.5 or larger [18], suggesting that the increased cup/disc ratio in the affected eyes was only partially acquired.

A recurrence of AION in a previously affected

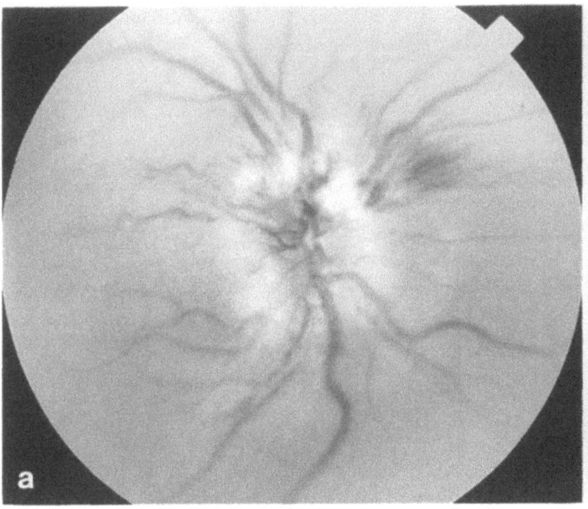
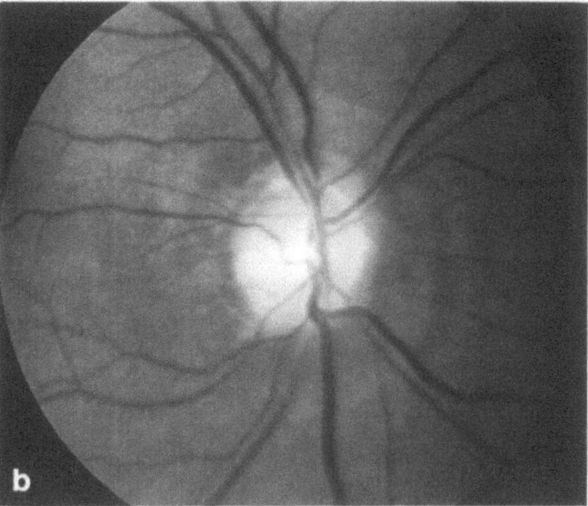

Fig. 5.4a, b. Major recovery of visual performance is unusual with severe anterior ischemic optic neuropathy. **a** A 57-year-old man with no significant previous medical problems complained of 3 days of progressive visual loss in the right eye without any associated pain. The visual acuity was 10/100 OD and 20/20 OS. There was a marked dyschromatopsia in the right eye and a relative afferent pupillary defect in the right eye. There was marked swelling of the optic nerve, hemorrhages on the disc and superonasally, cotton-wool spots, and superior enlargement of the nerve fiber layer. The left eye (not shown) had a full disc with a very small physiological cup but was otherwise normal. **b** One month following the initial evaluation, the acuity returned to 20/30 in the right eye. A mild color vision defect, a mild relative afferent pupillary defect, and a small inferonasal field defect remained. Funduscopy shows the diffuse pallor of the optic nerve with narrowing of the arteries on the inferotemporal portion of the disc

eye, classically thought never to occur, is now known to occur with a low incidence [9, 21–24]. Patients with a recurrence have two discrete episodes of ischemia, separated in time by weeks or months [25, 26]. Thus, a recurrence differs from the situation where, following the onset of AION, the vision deteriorates in a stepwise or stuttering course over 1–3 weeks. All cases with either progressive or recurrent optic neuropathy should have a computed tomography (CT) or magnetic resonance (MR) scan that visualizes the entire length of the optic nerve in order to exclude a compressive or infiltrative etiology. An examination of the cerebrospinal fluid may be indicated in order to determine that an inflammatory or carcinomatous disease is not the cause. An erythrocyte sedimentation rate test and, possibly, a temporal artery biopsy should be performed to exclude GCA in the patients of the appropriate age.

5.1.2.1 Second Eye Involvement

Second eye involvement is not unusual but it is rarely immediate. According to one study, in approximately 24% of patients the second eye will develop AION within 2 weeks to 20 years (mean 2.9 years) following the attack in the first eye [13]. In another study with a mean follow-up period of 3.5 years, the second eye became involved in 33% of cases [15]. Severe visual loss in both eyes is unusual, so generally there is a good prognosis for maintaining usable vision for daily living activities.

As mentioned previously, if a history suggestive of a prior AION is not elicitable, when the second eye becomes involved the unsuspecting physician may diagnose a case of Foster–Kennedy syndrome. The examination should readily differentiate the patient with both an old and an acute AION from the patient harboring a meningioma that causes unilateral visual loss with optic atrophy in the eye ipsilateral to the tumor and papilledema in the second eye due to increased intracranial pressure

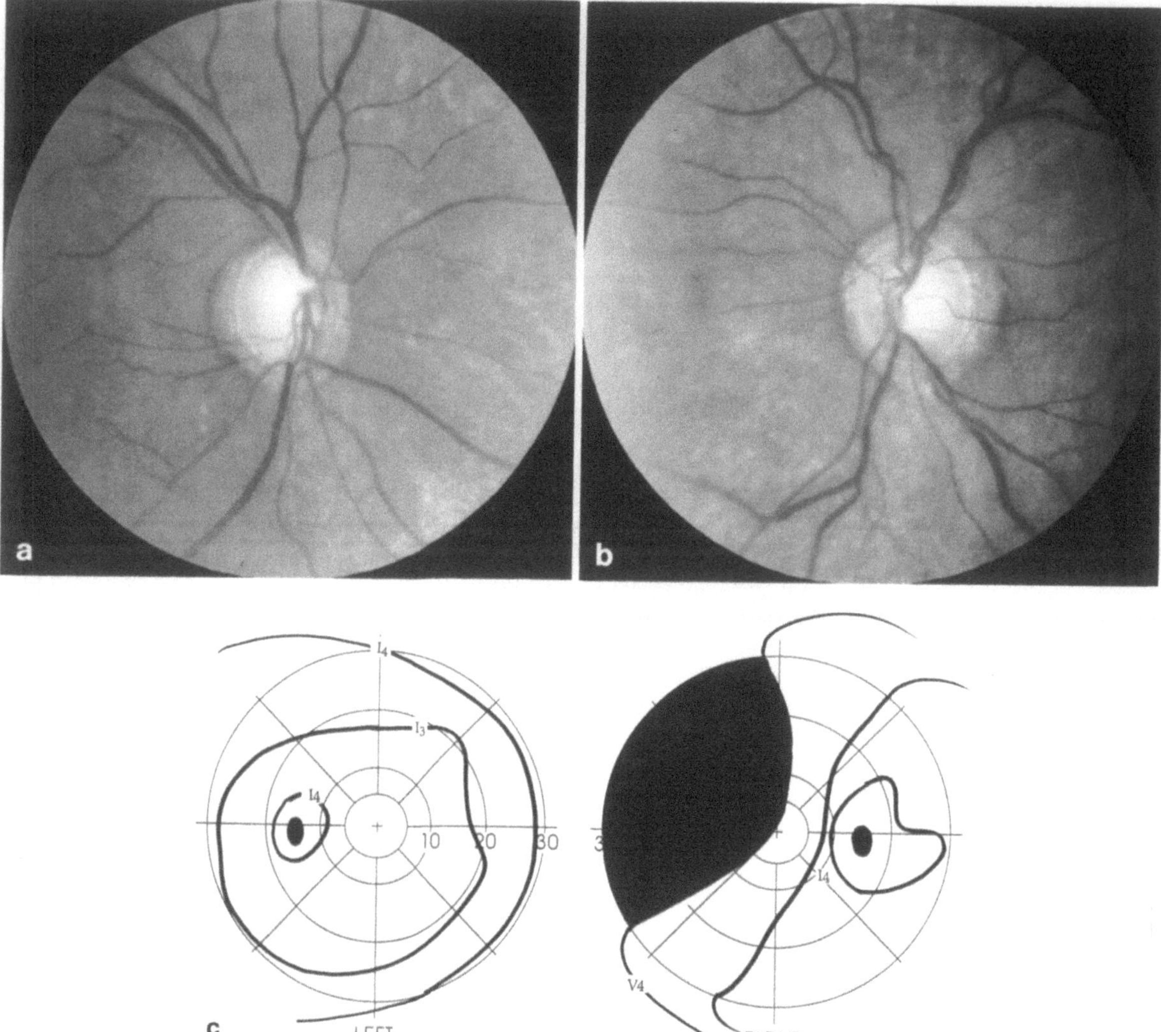

Fig. 5.5 a – c. Determining the role of ischemia in cavern-ous degeneration of the optic nerve is often difficult. A 72-year- old man with 5 years of diabetes controlled by oral hypoglycemic agents and background retinopathy had been followed as a glaucoma suspect without ever having elevated intraocular pressure. He had no defini-tive complaints of a visual disturbance. The visual acuity was 20/20 and the color vision was normal when mea-sured from each eye. There was a mild but definite affer-ent pupillary defect in the right eye.
a Ophthalmoscopy of the right eye revealed a 0.7 cup ex-cavated temporally and narrowing of the arteries on the temporal aspect of the disc. There was some very mild background retinopathy.

b The left eye had similar retinopathy with a 0.5 shallow cup.
c Goldmann perimetry revealed a dense loss of the nasal field, predominantly in the superonasal quadrant in the right eye. In the left eye, the visual field was entirely in-tact. Examinations remained stable over the next 9 years and it was presumed that the focal excavation of the disc and the nasal field loss was secondary to a focal anterior ischemic optic neuropathy. This case demonstrates the difficulty in distinguishing the injury of glaucoma from ischemic disease

(ICP). In cases of AION, the eye with the swollen disc has visual acuity loss, the appropriate visual field defect, or both. Dyschromatopsia is also often present. The optic disc with the previous episode will have diffuse or segmental, yellow or white, sharp-bordered pallor, often with one or more focally narrowed arteries on the surface. In contrast, early in the course, the eye with papilledema will have normal color vision and acuity and only an enlarged blind spot on perimetry.

5.1.3 Risk Factors

The risk factors associated with developing AION include advancing age (over 50 years) (Table 5.2). In one series, only 21 of 212 patients with AION were less than 50 years of age [3]. Patients with AION also have an increased prevalence of hypertension (36% of cases) and diabetes mellitus (20% of cases), both factors that increase the risk of developing a small vessel vasculopathy [13, 9]. Approximately 50% of AION patients younger than 40 years of age at the onset have diabetes mellitus [3]. An optic nerve disturbance similar to AION can also develop in patients with significant anemia (typically the hemoglobulin is less than 7 g/100 ml) [27]. Though 11% of AION patients less than 45 years of age also have a history of migraine [3], this incidence of migraine may not be higher than is found in the general population (see Sect. 10.4.3). However, there are clearly cases of AION associated with migraine episodes (see Chap. 10). In young women, preeclampsia is another rare cause of ischemia of one or both optic nerves [28].

Table 5.2. Systemic factors associated with ischemia of the anterior optic nerve in patients less than 45 years of age

---

Diabetes mellitus
Migraine
Anemia
Oral contraceptives
Vasculitis
Pre-eclampsia and eclampsia
Malignant hypertension
Platelet and coagulation disorders
Chronic papilledema

Crowding of the vessels and tissue in the optic nerve of patients with AION seems to cause a structural predisposition to the development of an infarct of the nerve. Following the original observation by William Hoyt, M.D., most examiners have noted that ophthalmoscopy of the second uninvolved eye (approximately 60% of cases) or of the affected eye prior to an attack frequently revealed congenital fullness of the optic disc with a cup/disc ratio ranging from less than 0.2 to a complete absence of a physiological cup [25, 29, 30] (Fig. 5.1b).

5.1.4 Cataract Surgery

There is an association between developing AION and cataract extraction [31, 32]. The episode can begin within hours of the cataract surgery or there may be a delay, with the visual loss first noted from days to a month into the postoperative period. This disorder was originally misnamed a postcataract optic "neuritis" or "papillitis" but there is no evidence suggesting an inflammatory etiology. The risk of developing an AION does not seem to be related to whether an intracapsular or extracapsular procedure is performed, whether local or general anesthesia is administered, whether a retrobulbar injection is given, or whether ocular massage or other mild measures to lower the IOP are performed preoperatively. It is theoretically possible that vigorous measures, such as preoperative administration of intravenous mannitol, designed to acutely lower the IOP can adversely alter the hemodynamic state in an eye. Following cataract extraction, the development of optic nerve ischemia may be related to a postoperative rise in the IOP which compromises an existing marginal blood perfusion pressure to the optic nerve [33]. However, not all of these AION cases have had this perioperative elevation of IOP. In addition, since cataract patients are typically elderly, many of them have atherosclerosis and some would undoubtedly develop typical AION even without an intraocular procedure. It is certainly possible that the optic nerve circulation may be adversely altered by merely opening the eye in an elderly subject who has systemic arteriosclerosis.

Following cataract extraction, there is on occasion a mild swelling of the optic disc, associated with cystoid macula edema, that may appear similar to a mild case of AION. However, careful ophthalmoscopy will often reveal the macula edema. On fluorescein angiography, the optic disc will leak in both disorders but the macula will be stained by the dye only in the eye with cystoid macula edema [34]. An afferent pupillary defect, present in most eyes with AION, is rarely found with cystoid macula edema. Also, dyschromatopsia on pseudo-isochromatic plate testing is typically not present with cystoid macula edema unless the patient cannot distinguish the control pattern. Though the central field can be depressed because of the edema, a nerve bundle or altitudinal defect does not occur with cystoid macula edema. The patient with this edema, but not the one with AION, will frequently complain of metamorphopsia.

Up to 50% of patients who experience AION in relation to cataract surgery will develop AION in the second eye following cataract extraction. In these patients, cataract surgery should be considered only if the visual dysfunction from the remaining cataract warrants the risk. Measures may be taken to prevent a fall in the optic nerve blood perfusion pressure in the perioperative period. Certainly, maintaining normal IOP and preventing systemic hypotension are essential. Also, recall that patients with systemic hypertension may require a higher blood pressure than normotensive subjects in order to maintain normal tissue perfusion.

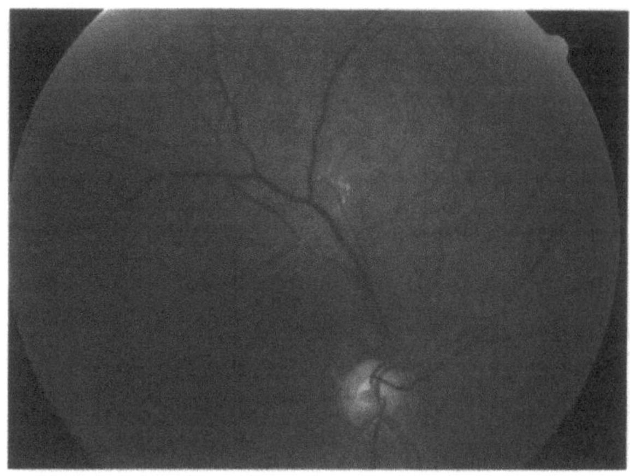

Fig. 5.6. Embolic ischemic optic neuropathy is rare but has been reported in association of cardiac catheterization. A 57-year-old man underwent cardiac catheterization for evaluation of arteriosclerotic heart disease. Immediately upon awakening, he noted reduced vision from the right eye. The evaluation performed several months later revealed a visual acuity of 20/80 OD and 20/20 OS. There was dyschromatopsia and a relative afferent pupillary defect in the right eye. Threshold perimetry revealed generalized constriction in the field with depression of the central field. The left eye visual field was normal.
Ophthalmoscopy of the right eye demonstrates a diffuse pallor of the optic disc, worse temporally, with an embolus in the superotemporal retinal artery. Ophthalmoscopy of the left eye (not shown) was entirely normal. Both eyes had physiological cups of approximately 0.4. This case was presumed to be embolic anterior ischemic optic neuropathy as a complication of cardiac catheterization

5.1.5 Emboli

It is unclear why emboli from the internal carotid artery (ICA) or the heart are an extremely rare cause of AION. Single case reports provide virtually all that is known about this problem. Most of these reports concerned patients with evidence of multiple or massive embolization, as exemplified in the following:

A patient with sepsis, a right cerebral infarct, and visual loss in the right eye was found at autopsy to have thrombosis, presumably caused by emboli, in two pial arterioles that supplied the retrolaminar portion of the right optic nerve and in one arteriole in the vascular septa separating the nerve bundles. An associated infarct of a segment of the retrolaminar optic nerve was demonstrated pathologically, but had not been noted clinically. However, the optic nerve damage was not the sole cause of the visual disturbance because fibrin thrombi were also present in the retina and choroid [35].

Visual loss developed in a patient with metastatic chondrosarcoma and tumor emboli. Multiple emboli were seen in the retinal vessels at the time that the AION was noted. At autopsy, an infarct of the optic nerve as well as infarcts of the choroid and retina were noted. Emboli were seen in the central retinal artery, posterior ciliary arteries, and choroidal vessels [36].

Other cases with an embolic cause of AION have been described following coronary artery bypass surgery or cardiac catheterization [37] (Fig. 5.6). These eyes typically have retinal emboli as well because of massive embolization in the ophthalmic circulation.

Cholesterol emboli from carotid atherosclerosis have been seen in the retinal arteries in two cases of AION [38]. We have also seen one such case.

A temporal association between total occlusion of the ICA and the onset of AION is another rare occurrence and has been documented in only three cases [39, 40]. Symptoms and findings of ischemia of the ipsilateral cerebral hemisphere indicated the occlusion of the ICA or a middle cerebral artery. Emboli were seen in the retina of only one of these three cases [37]. Patients with an AION picture associated with an ischemic oculopathy syndrome caused by severe carotid stenosis [43] or a dural arteriovenous malformation are not rare but these cases are readily differentiated from cases of AION alone.

Before considering the rare diagnosis of embolic AION, concomitant cerebral ischemia or retinal emboli on ophthalmoscopy should be obvious.

5.1.6 Differential Diagnosis of Unilateral Visual Loss and Optic Disc Swelling

The differential diagnosis of unilateral visual loss with unilateral or extremely asymmetric bilateral disc swelling is extensive (see entities in this chapter, Table 5.3). Visual loss at presentation usually distinguishes AION from the papilledema of increased ICP, though visual loss can occur with chronic papilledema (see Sect. 5.3). Other causes of an ischemic optic neuropathy include GCA and collagen vascular diseases (see Sect. 5.2 and Chap. 9). Orbital tumors, such as an optic nerve sheath meningioma or an optic nerve glioma, which block axoplasmic flow in the optic nerve and the venous outflow from the eye cause swelling of the disc with fluctuating or progressive visual dysfunction. An acute luetic optic neuropathy usually occurs during the secondary phase syphilis. Cytomegalovirus involvement of the optic nerve accompanies the typical cytomegalovirus retinitis in immune-compromised individuals. Following a prior central vein occlusion or with diabetes mellitus, optic disc elevation or swelling may also result from neovascularization of the optic disc.

Table 5.3. Unilateral disc swelling

Condition	Visual acuity loss
Papillophlebitis or partial central retinal vein occlusion	−
Optic nerve sheath meningioma or glioma	±
Orbital tumors	±
Hypercoaguable state	±
Early unilateral papilledema from increased intracranial pressure	−
Sarcoid of the optic disc	±
Buried drusen of the optic disc	±
Partial posterior vitreous detachment	−
Syphilitic perineuritis	−
Congenital full disc	−
Hypotony	±
AION	±
Leber's optic neuropathy	+
Neuroretinitis	+
Cystoid macula edema	+
Vitritis	±
Diabetic neovascularization	±
Infections or demyelinating papillitis	+

±, With or without visual acuity loss; −, without visual acuity loss; + with visual acuity loss.

In patients less than 45 years of age, a true papillitis with visual loss may also be caused by demyelinating disease, sarcoidosis, or secondarily with idiopathic uveitis. A demyelinating optic neuritis can acutely swell the disc and cause varying degrees of acuity and visual field loss (or even an altitudinal field defect). However, more than 90% of the patients are younger than 50 years, pain on eye movement is present in approximately 90% of cases, and pale swelling of the disc is not seen. Neuroretinitis causes disc swelling, an exudative macula star, and visual loss that resolve spontaneously. A papillophlebitis with mild or no visual loss is frequently associated with cells in the vitreous over an injected, rather than a pale disc (see Sect. 5.8 for differential diagnosis of unilateral disc swelling without visual loss).

5.1.7 Pathology and Localization of the Ischemic Process

AION seems to develop when the perfusion pressure in the optic nerve falls below a critical level as a result of hypoperfusion or temporary occlusion of the arteries or arterioles in the arterial system supplied by the posterior ciliary arteries. If the posterior ciliary arteries are narrowed or incapable of vasodilating because of atherosclerosis, a transient fall in the systemic blood pressure that normally occurs during sleep or sitting or standing may become pathological. A permanent occlusion of a major vessel may not be a necessary event. Similarly, permanent thrombosis of a cerebral artery is usually not seen following a cerebral infarct. AION does not result from thrombosis of a "central artery of the optic nerve" as originally proposed [42], since it is unlikely that this vessel exists (see Sect. 1.5.2.4). There is little evidence that a complete occlusion of a particular artery occurs in this disorder.

Few cases of AION have had pathological verification of the location of infarct in the optic nerve. Cogan reported a case of severe focal axonal and myelin loss in an optic nerve, but the defect could not be related to an occlusion or hypoperfusion in the territory of a specific artery [43]. A later study revealed a preferential loss of axons in the peripheral region of the nerve, but again there was no correlation with occlusive disease in a specific artery or arteriole [44]. In a third report, one patient with AION of approximately 2 weeks' duration was found to have swollen axons, loss of myelin, and foam-laden macrophages without inflammatory cells in the laminar and retrolaminar portions of the optic nerve. Again, there was no demonstration of occlusion of a specific vessel in the periocular area. Additionally, this patient had an otherwise clinically silent occlusion of both the common and external carotid arteries ipsilateral to the visual loss [39]. The vast majority of reported pathological studies of optic nerve infarction have been performed in patients who have had GCA (see Sect. 5.2).

The lack of pathological verification of occlusion of a specific artery in AION contributes to the difficulty in conclusively localizing the arterial involvement that can cause the observed wide spectrum of visual dysfunction. The variability of the blood supply to the optic nerve among individuals and even between eyes in a given subject also makes it difficult to accurately pinpoint the arterial supply to the optic nerve responsible for the ischemia.

Fluorescein angiography results suggest that AION may be similar to the ischemic optic neuropathy that occurs following experimental occlusion of the posterior ciliary artery in monkeys [45a]. Fluorescein angiography performed within the first few days after the onset of AION shows delayed perfusion or an absence of filling of the peripapillary choroid and the deeper portion of the optic disc in a sectoral pattern [19, 45b] (Fig. 5.7). The extent of the area of nonperfusion in the choroid depends on the number of posterior ciliary arteries and the variation of the area supplied by each vessel (see Sect. 1.5.2.4). When there is no superior posterior ciliary artery, a filling defect in the nasal choroid and disc is suggestive of hypoperfusion of the medial posterior ciliary artery. Poor fluorescein filling of the temporal portion of the disc and choroid suggests reduced flow through the lateral posterior ciliary artery. Nonfilling of a watershed area in the choroid between the lateral and medial posterior ciliary artery blood supplies can also be seen in acute cases [45c, 46, 47]. The common occurrence of an altitudinal field defect could reflect hypoperfusion of watershed areas that include either the superior or inferior aspects of the optic nerve. On occasion, the entire disc lies within a watershed zone, and this may account for those cases where there is profound reduction of central acuity and extensive field loss.

5.1.8 Laboratory and Systemic Disease Evaluation

Guyer et al. reported that AION patients more than 45 years of age had an increased prevalence of cerebrovascular disease, which was defined as the presence of infarcts of brain or eye and transient ischemic attacks, including amaurosis fugax [3]. However, no infarcts of the brain caused death in these patients. These additional ischemic episodes of the eye or brain occurred before or after the attack of AION, and were equally distributed between the cerebral circulations, both ipsilateral and contralateral, to the affected optic nerve. Prior

Fig. 5.7. a Fluorescein angiography, performed on the same day as the onset of an anterior ischemic optic neuropathy, demonstrates hypoperfusion of a large segment of the choroidal circulation temporal to the disc. In this case, there was mild swelling of the optic disc, 5/400 acuity, and a large relative afferent pupillary defect in the affected right eye. The left eye had normal vision and had no discernible physiological cup. **b** Perimetry revealed a dense central scotoma of 10°, worse superonasally, to all isopters. The patient could only detect hand motion in the depressed field. The field of the left eye was normal. The visual defect was permanent

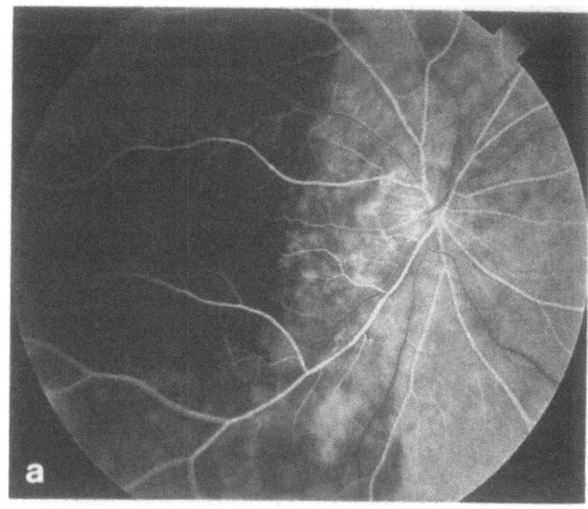

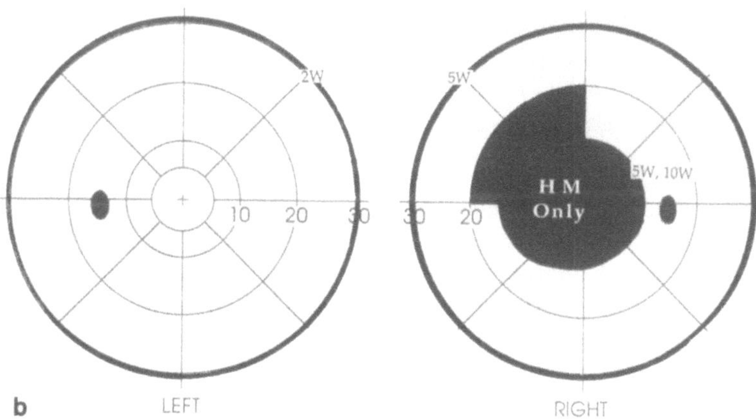

studies did not find similar significant incidence of retinal or cerebral ischemic events in patients with AION. More recently, a case control study failed to demonstrate significant carotid stenosis in patients with AION. Thirteen patients with AION had a mean stenosis of 18% in the ICA ipsilateral to the involved eye and 12% in the contralateral ICA as determined by ultrasonography. This was compared to matched subjects with amaurosis fugax who had a mean stenosis of 69% in the ipsilateral ICA [48].

The increased prevalence of arteriosclerotic cardiac disease is of greater consequence to the patient with AION than the risk of developing a cerebral vascular episode [49]. This is particularly significant for those individuals older than 65 years. One study observed 43 myocardial infarcts in 200 patients of all ages with AION, which compares with 25 expected for the control population [3]. However, though myocardial infarction caused death in nine of these 43 patients, when all patients with AION were considered the mortality rate was not higher than for the overall population [3].

Procedures that can be performed in the office such as ophthalmodynamometry may be helpful in determining whether there is significant stenosis or occlusion of the ICA ipsilateral to the AION (see Sect. 8.3.2). In patients with typical AION, a CT or MR scan is unnecessary and cerebral angiography is rarely, if ever, warranted.

The systemic evaluation of patients over 50 years of age should include determining the erythrocyte sedimentation rate, even if they have none of the systemic complaints referable to GCA (see Sect. 5.2). Fasting and a 2-h postprandial blood glucose level determination, a glucose tolerance test or a glycosylated hemoglobin, a hemogram (to exclude severe anemia as a contributing cause), blood pressure measurement, and a clinical and electrocardiographic evaluation for arteriosclerotic heart disease should be included. In patients less than 50 years old, additional blood tests should exclude syphilis, hypercoaguable syndromes, or collagen vascular diseases as potential causes of vaso-occlusive disease in the optic nerve.

5.1.9 Treatment

There is no effective treatment for preventing or reversing the visual defects of AION. Anticoagulants and diphenylhydantoin [50] have been tried unsuccessfully. It has been suggested that corticosteroids might prevent secondary damage to the optic nerve that could develop because of strangulation of the optic nerve caused by edema and swelling of the axons in the tough collagen scleral canal. Corticosteroids might be effective in patients without a significant visual defect or prior to significant infarction of the optic nerve [45c, 51]. Few patients are evaluated prior to infarction with visual loss, so it has not been possible to collect a large controlled series to test this hypothesis. Once visual loss occurs an antiedema or anti-inflammatory agent is unlikely to be effective because AION is not an inflammatory disease and tissue infarction cannot be reversed by corticosteroids. Megadoses of corticosteroids might theoretically be beneficial in preventing secondary injury to the injured axons. However, since not every patient with ischemia-induced disc swelling will develop a clinical visual deficit, it would be difficult to determine whether corticosteroids could truly modify the natural history in these cases.

Nevertheless, low (20 mg) to moderate (40 mg) daily doses of prednisone may be given to a patient with ischemic disc swelling, normal acuity, and full suprathreshold fields. The prednisone is gradually withdrawn as the disc swelling and hemorrhages resolve, usually over a 1-month period. Monitoring of the patients, who are typically elderly, for any corticosteroid complications is essential. Despite the unproven efficacy, corticosteroids are also frequently prescribed as a desperation therapy to patients with a prior AION that caused severe visual loss if AION then affects the second eye.

Measures that lower the IOP may improve the blood perfusion pressure/IOP ratio to the optic nerve in patients with acute AION or prior to cataract surgery [45c]. However, because of the induced diuresis, drugs like carbonic anhydrase inhibitors and mannitol may also lower the systemic blood pressure, with a net effect of lowering the blood perfusion pressure to the eye.

Recently, surgical decompression of the optic nerve sheath has been reported to halt the progression and improve the vision in patients experiencing progressive visual loss [52].

Aspirin may be effective in reducing the incidence of second eye involvement by AION.

5.2 Giant-Cell Arteritis

Giant-cell arteritis (GCA) is a systemic illness causing severe visual loss in patients over 60 years of age, most cases being older than 70 years [53]. The incidence is 17.4/100,000 people; the prevalence is 133/100,000 for those older than 50 years with an occurrence rate of 24/100,000 in the overall population [54]. The prevalence increases to 843/100,000 for the population aged 80 years and over [55]. GCA infrequently attacks persons younger than 60 and it is rare below the age of 50 years. The youngest reported case of GCA with visual loss and a temporal artery biopsy demonstrating active GCA was probably a 48-year-old man with occlusion of the central retinal artery, an erythrocyte sedimentation rate (ESR) of 7 mm/h, and a 15-year history of Raynaud's phenomenon prior to the visual loss [56]. The youngest documented case of giant cells in a temporal artery biopsy specimen was 35 years old. This patient had a normal ESR and complained only of headache and tenderness over the temporal artery [57]. GCA is found predominantly in caucasians and less than 10% of patients are black [58, 59]. Men and women are at equal risk of developing the disease.

Various aspects of GCA were described centuries ago. The association of visual loss and biopsy-proven giant-cell inflammation of the temporal arteries was first noted by Jennings [60]. However, the relationship between temporal artery pain and the development of visual loss was noted as early as 1000 a.d. [61]. The first description of the clinical systemic aspects of GCA was probably made by Hutchinson in 1890 [62]. Because of the predilection to involve the superficial temporal artery and the associated pains around this vessel described in these early reports, GCA is often called temporal arteritis.

5.2.1 Clinical Presentation

5.2.1.1 Visual Disturbance

Ocular motor paresis or unilateral visual loss or both will develop in from 17% [54] to 55% of untreated patients [63]. In approximately 20% of patients, recurrent monocular amaurosis fugax precedes the permanent visual loss in the same eye by several days [12, 54, 64]. Amaurosis can occur with a change in body position from supine to upright. The normal orthostasis in combination with the narrowing of posterior ciliary, central retinal, and ophthalmic arteries, as well as concomitant anemia, result in a significant reduction of the ocular blood perfusion pressure [65]. One extreme example of the effect of postural hypotension on vision occurred in a patient who had multiple episodes of a central retinal artery occlusion in the same eye, seen on funduscopy, that were directly related to the change in his posture. When he lowered his head the vision would return and the retina would appear normal [66]. Transient visual loss has also been rarely described in association with external hyperthermia, Uhthoff's sign, which is ordinarily associated with demyelinating disease [67, 68].

When permanent visual loss occurs it is sudden, evolving over the course of hours to a day, but rarely over weeks [69]. Severe reduction in vision, with frequent complete blindness or only bare light perception, is common. Following the onset of visual loss, the second eye will lose vision within 1 day to 3 weeks in 10% – 33% of untreated patients. Approximately 72% of the cases with second eye involvement experience the attack within 7 days of the visual loss in the first eye. After 2 months, second eye involvement is infrequent, though not rare [12, 64].

Though the visual loss is almost always a result of retinal or optic nerve ischemia, on rare occasions a visual disturbance, such as homonymous field dysfunction, can result from infarction of the visu-

Table 5.4. Criteria for polymyalgia rheumatica (modified from [80])

1. Pain and stiffness, shoulder or pectoral and pelvic and hip muscles
2. Lassitude, fever, weight loss, anorexia, no infection
3. No objective evidence of synovitis
4. Elevated ESR over 50 mm/h using Westergreen method
5. Normal serum levels of enzymes that reflect muscle damage
6. No severe headache, temporal artery swelling or tenderness, joint claudication, or visual disturbance
7. Complete rapid clinical response to a daily dose of less than or equal to 15 mg prednisone
8. Age greater than 50 years

al cortex. Cortical blindness from bilateral occipital infarction is even more rare [70]. Infarction of the brainstem is also unusual [71, 72].

5.2.1.2 Systemic Involvement

Systemic symptoms are frequent but variable. Headache, particularly in the area over the superficial temporal artery, is common. The pain, which is either constant or intermittent, is described as deep-boring or lancinating. The temporal artery can be swollen and is often tender to touch. The patient frequently states that he cannot rest his head on the pillow or brush his hair over this area. Claudication causes fatigue of the tongue while talking or eating and pain of the muscles of mastication after chewing for several minutes. Sternocleidomastoid pain and tenderness are less common.

Polymyalgia rheumatica (PMR) (Table 5.4) often begins prior to the onset of GCA or the visual loss in patients with GCA [71]. PMR is commonly called "arthritis" by the patient. It is typified by proximal muscle and periarticular pain, particularly with use of the extremities, morning stiffness, fever, anorexia, and weight loss. On occasion, the patient only has the nonspecific complaint of "just not feeling well." It may be difficult to determine when a patient with PMR develops GCA since up to 50% of PMR patients, even without headache, have been reported to have a positive temporal artery biopsy [72 – 75]. However, one must question these findings because temporal artery biopsies are usually performed on individuals who are clinically suspected of having GCA. In fact, when patients with symptoms such as jaw claudication, swollen temporal artery, aortic arch syndrome, and occipital artery tenderness are eliminated, only 8% of patients with PMR have positive biopsies [74].

Obscurations or blurring of the vision, lasting seconds, or true amaurosis fugax, lasting seconds to minutes, have been described in patients with PMR, but these symptoms suggest the progression of the illness to GCA. Except in the patient who has been treated with low doses of corticosteroids for PMR, transient amaurosis, visual obscurations, or permanent visual loss are uncommon in a patient with GCA who does not have concomitant headaches or other systemic complaints. GCA and loss of vision may develop months or years after the onset of PMR [78]. However, not all cases with PMR are at risk to develop visual loss and PMR clearly exists without progression to GCA. In one major series of 500 patients with PMR, diagnosed with the criteria in Table 5.4 and followed up for 1–10 years, only three developed GCA and none had visual loss [79]. In a second study on 56 patients with PMR, the same authors randomly performed 14 temporal artery biopsies, all of which were negative [80].

Severe visual loss occurs in 33%–50% of patients with GCA who have systemic symptoms [81]. Unquestionably, all elderly patients with AION or a central retinal artery occlusion, or even amaurosis fugax, should be questioned closely for symptoms of GCA and their ESR should be measured as part of their laboratory evaluation. The rarest situation occurs when the patient has a normal ESR and no systemic complaints, but has a positive temporal artery biopsy and visual loss. Fortunately, the overwhelming majority of patients who experience a loss of vision have systemic complaints or headache, that typify GCA, for weeks to months (mean 3.5 months) prior to the visual episode [12]. Patients can also present with a syndrome of nonspecific symptoms all suggestive of a neoplastic process, including fever, weight loss, and an elevated ESR, but without a headache [82]. Infrequently, cases of GCA will have an acute ischemic optic neuropathy or a central retinal artery occlusion, minimal or no systemic complaints, a normal

histological appearance to the temporal artery, and an elevated ESR [83]. Another unusual, but not rare presentation is one where the visual loss and systemic complaints develop but the ESR remains normal or only minimally elevated [84].

Other rare presentations of GCA include ICA or middle cerebral artery distribution transient ischemic attacks or a cerebral infarct or an encephalopathy of abrupt onset [85]. Basilar and posterior cerebral artery territory infarcts have been described but are even less frequent [86]. In these cases the cerebral angiographic demonstration of arterial beading and occlusions can support the diagnosis of arteritis, but the findings are often nonspecific [87].

Angina and renal dysfunction can also develop from arteritis of the medium or large coronary and renal arteries, respectively [53, 88]. Peripheral vascular disease is rarely caused by GCA, but when present gangrene of the legs can result [89]. However, major clinical signs from occlusion of these vessels are uncommon compared to the arteritis involvement of the temporal artery and the ophthalmic circulation.

5.2.2 Clinical Examination

5.2.2.1 Systemic Clinical Examination

The clinical examination will often reveal a tender, nodular, swollen temporal artery with a decreased or absent pulse. Erythema or edema of the skin overlying the artery is also common. However, approximately one third of patients with an abnormal temporal artery biopsy will appear to have a normal temporal artery on clinical examination [90, 91]. Rarely, ischemic necrosis of the skin and muscle over the inflamed temporal artery develops [62]. The proximal muscles of the extremities can be tender but true weakness is generally not present. Loss of the deep tendon reflexes is also not associated with GCA.

Ischemia of the cranial nerves VII–XII or the muscles of the head and neck is infrequent. Though also rare, when lingual infarction is present it is almost pathognomonic for GCA [92] (Fig. 5.8a).

Fig. 5.8a, b. The ischemic complications of giant cell arteritis can affect almost any locale. A 70-year-old woman complained of acute visual loss in the right eye. She had numerous additional systemic symptoms suggestive of giant cell arteritis. Her sedimentation rate was 100 mm/h and she had a pale swollen optic nerve with no light perception vision from the right eye. She was treated with prednisone 100 mg/day. Despite this therapy, she continued to have systemic complaints and began to develop visual obscurations in the left eye over the next week. She was hospitalized and treated with intravenous methylprednisolone 1 g/day in divided doses. While hospitalized, the patient developed ischemia of the tongue as well as a return of visual obscurations in her left eye when the methylprednisolone dose was reduced to 500 mg/day. Raising the dose back to 1 g/day, keeping the patient reclining with the head flat, and instituting a liquid diet was associated with no further ischemic episodes.
a A lateral view left common carotid artery subtraction angiogram reveals severe occlusive disease in the external carotid artery circulation (*arrow*). The remaining vessels, including the lingual artery (*curved arrow*), are extremely narrowed. The right common carotid angiogram showed a similar picture. The internal carotid artery circulation was relatively normal in comparison. Giant cell arteritis diffusely affected the external carotid artery systems on both sides which resulted in ischemia of the tongue. After an additional week of high dose methylprednisolone, the dose was tapered over 1 week. She was discharged on an oral dose of prednisone without a recurrence of visual or head and neck ischemic difficulties.
b Three months later, the affected eye had a diffusely pale, flat optic nerve with an excavated cup of 0.8 and no temporal rim. The left eye (not shown) remained normal and the physiological cup was less than 0.3

5.2.2.2 Ophthalmologic Findings

Though Wagener and Hollenhorst [93] claimed that most patients with acute visual loss have normal fundi at presentation 1−2 days after visual loss, this is probably more the exception than the rule. In patients with unilateral visual loss, the severe ischemia of the optic nerve causes a swollen pale optic disc, sometimes with splinter hemorrhages in the peripapillary nerve fiber layer [94]. After several weeks to months, the hemorrhages resolve and the disc flattens and becomes more pale. Because of the severe loss of optic disc tissue, an increased cup/disc ratio can be seen on funduscopy, but the cup is relatively flat (Fig. 5.8b; not deep as with glaucoma; see Sect. 5.6) [18, 95]. The cupping or cavernous degeneration of the optic nerve has been confirmed pathologically [96].

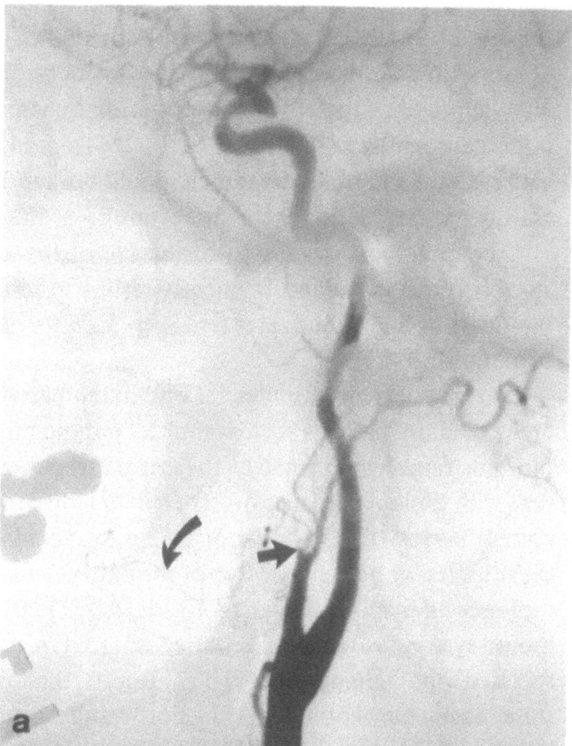

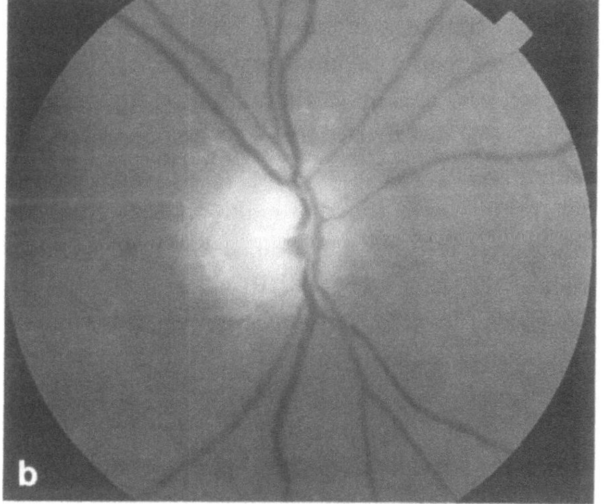

The retina is also commonly affected and retinal ischemia causes cotton wool spots, either focally or diffusely, in approximately 8% of patients with optic nerve infarction [93]. A central retinal artery occlusion, with typical swelling of the inner layer of the retina and the "cherry" red spot in the macula (see Sect. 8.2.3.4), is the cause of visual loss in less than 10% of cases. In contrast, venous stasis with venous engorgement and white-centered hemorrhages without frank retinal ischemia rarely occurs [97].

Rarely, the retrobulbar optic nerve is the site of the infarct. In this circumstance, the unilateral visual loss with a normal fundus and an afferent pupillary defect are suggestive of a posterior ischemic optic neuropathy (PION) [93]. A PION from any cause is rare and other diagnoses should always be considered (see Sect. 5.9 for discussion).

The visual loss, regardless of the site of the lesion, is often severe and frequently results in total blindness of one or both eyes (see Sect. 5.2.6 for explanation).

Varying degrees of unilateral ophthalmoparesis are found in patients but complete ophthalmoplegia is rare. Approximately 10% of patients present with diplopia because of ischemia to either the cranial nerves III, IV, or VI or to the extraocular eye muscles, or both [98]. Most of the patients with diplopia have subtle ocular motor defects. Oculomotor synkinesis has been reported after GCA associated ophthalmoparesis. This suggests that in some cases the motility dysfunction results from ischemia of the cranial nerve or brainstem, rather than the extraocular muscles [99]. Two cases have been described with brainstem lesions that caused an internuclear ophthalmoplegia [100]. Almost 50% of untreated patients who complain of diplopia will develop an optic neuropathy, usually in the eye with the ophthalmoparesis. However, once there is a loss of vision, a subsequent cranial neuropathy rarely occurs [101].

Except for the relative afferent pupillary defect, the pupil is usually spared even if the third nerve is paretic [81]. A tonic pupil has been reported in patients with [102] or without visual loss [103]. Pupillary dysfunction may also develop from ischemia of the anterior segment. In the latter case, the examination reveals white blood cells and elevated protein in the aqueous humor as well as dysfunction of both the iris dilator and sphincter muscles that results in pupillary paralysis and a medium- or small-sized pupil [55]. A Horner's pupil is rare, but when ocular ischemia exists, sympathetic dysfunction can be difficult to diagnose without pharmacological testing (see Sect. 6.1.7.1).

5.2.3 Laboratory Evaluation

5.2.3.1 Erythrocyte Sedimentation Rate

The blood erythrocyte sedimentation rate (ESR) is the standard laboratory screening test for monitoring GCA activity used in most patients. The ESR is elevated in approximately 90% of patients with active GCA [104]. The ESR depends on the concentration in the blood of erythrocytes and asymmetric macromolecules, predominantly γ-globulin and fibrinogen. GCA causes an increase in the concentration of these macromolecules, which elevates the ESR. When measuring a rapidly rising ESR — rates greater than 100 mm/h are characteristically found in GCA — the Westergreen method is superior to the other methods [105]. Though the Westergreen method needs to be adjusted for anemia, it is the most sensitive technique because it has a nonlinear response to a minimal increase of these macromolecules. The Westergreen method also uses a 200-mm column that expands the scale and delays early packing so that small rises in the ESR will be obvious. However, the citrate used in this method dilutes the concentration of macromolecules and artificially lowers the ESR. The Wintrobe method (which also requires adjustment for anemia) has a rapid packing phase that makes it insensitive to a marked elevation in the macromolecule concentration, but it is more sensitive in the normal to minimally elevated ESR range [106]. When using the first two methods, it is important to recall that approximately 63% of patients with GCA have an associated anemia [107]. A third technique, the Zeta sedimentation ratio, may also be useful in GCA because it is insensitive to alterations in the hematocrit in the range from 20% to 40% and is therefore not artificially elevated by anemia. The Zeta sedimentation ratio is also linear over a wide range of elevated concentrations of the asymmetric macromolecules.

The normal ESR varies with the age of the individual. The normal ESR is less than or equal to the age of the patient divided by 2 for men or the sum of the age plus 10 divided by 2 for women [108]. In general, the ESR is higher in patients with PMR who have a positive temporal artery biopsy (mean 79 mm/h) than in those patients whose biopsies are negative (mean 53 mm/h) [109]. Thus, the higher the ESR, the more secure the clinical diagnosis is likely to be. However, the presence of an ESR above 100 mm/h is not diagnostic by itself since it can be associated with occult neoplasms, collagen vascular disease, multiple myeloma or other macroglobulinemias, leukemia, and tuberculosis. In these last diseases an optic neuropathy can develop because of direct infiltration by neoplasia, a paraneoplastic syndrome (rare), inflammation, or ischemia. Thus, every patient with an elevated ESR and systemic symptoms will not have GCA [110].

Once a diagnosis of GCA is established and corticosteroid treatment has begun, the trend of the ESR is more important than the absolute rate. Clearly, a fall in the ESR or stabilization at a lower level is superior to a rise. The ESR is never the sole indicator of disease activity (see Sect. 5.2.7).

5.2.3.2 Additional Blood Abnormalities

The C-reactive protein also reflects the presence of systemic inflammation, but it rises and falls earlier than the ESR. Thus, the C-reactive protein may suggest an early exacerbation of the GCA or that a dose of corticosteroids is inadequate prior to a rise of ESR [111].

A normochromic normocytic anemia and an increase in the plasma concentration of fibrinogen and α_2 globulin are all commonly found in patients with active GCA (these will normalize with corticosteroids). Though there is certainly a systemic inflammatory process, elevation of the immunoglobulin G (IgG) level or complement factors C3 and C4 occurs less frequently [112]. A positive speckled or homogeneous antinuclear antibody titer of 1:64 or greater has been reported in as many as 50% of patients with positive temporal artery biopsies [113].

5.2.4 Temporal Artery Biopsy

A histopathological description of the effects of the arteritis in the temporal artery was first given in 1932 [114]. The demonstration of arteritis in this vessel has been the gold standard for establishing a diagnosis of GCA, but difficulties exist with using this as the single criteria. The artery biopsy should be taken from the superficial temporal artery ip-silateral to the visual loss and should include the segment that is tender and swollen, if present. A specimen longer than 1 cm, and preferably 2 cm in length, is essential in order to increase the probability of identifying a site of inflammation (see below). Though angiography of the ECA circulation may demonstrate localized beading in the affected segments of the temporal artery, this method is rarely necessary to identify an abnormal arterial segment [115]. Doppler studies can detect diminished blood flow in the affected temporal artery but ultrasonography is also rarely used to localize the specimen for biopsy.

Histopathological examination of most of the biopsy specimens reveals segments where the entire vessel wall is involved by a chronic inflammatory process. The inflammation is most intense in the media, with multinucleated giant cells, monocytes, and eosinophils, as well as polymorphonuclear leukocytes (Fig. 5.9). Necrosis of the media or muscular layer is seen best on ultrastructural examination [116]. Arterial occlusion and recanalization of thrombosed areas are seen at sites of active inflammation [53]. Intimal damage, such as internal elastica disruption or proliferation or both, is seen best by using special stains for elastic tissue [117]. Isolated noninflammatory intimal abnormalities are not diagnostic of GCA since they can occur with other entities such as atherosclerosis. Surprisingly, there are no uniform pathological criteria established for diagnosing GCA in a biopsy with minimal abnormalities.

5.2.4.1 Skip Areas

Because the inflammatory involvement of the vessel is multifocal and not diffuse, there may be skip areas where there is minimal or no inflammation. These skip areas are found in approximately 27% of the serial sections in most specimens. Therefore, numerous serial sections from a large arterial specimen are examined in order to increase the diagnostic yield from each temporal artery biopsy [118, 119]. Few segments are entirely normal and the presence of a low- grade nonspecific inflammation should alert the pathologist to look at more sections. In untreated cases, the false-negative biopsy rate ranges from 9% to 61% [50, 120–124].

If the first biopsy is negative in a patient with

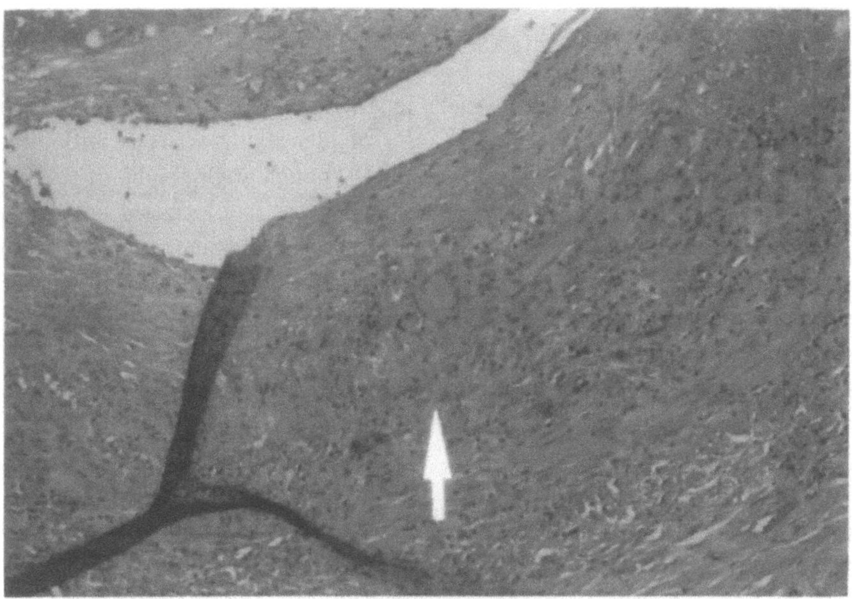

Fig. 5.9. Biopsy of the temporal artery in a patient with giant cell arteritis shows severe narrowing of the vascular lumen with marked inflammatory necrosis in the media and giant cells (*arrow*). (Case provided by Larry Frohman)

active GCA, a biopsy of the temporal artery on the opposite side is frequently [125], but not always positive [126]. Before committing the patient to long-term corticosteroids, an additional biopsy should be considered in cases where GCA is a real consideration but the clinician has doubts. However, in the case where the patient gives a history typical of GCA, has an elevated ESR, and has no other demonstrable systemic illness that accounts for the alteration in the ESR, a second biopsy is not absolutely mandatory in order to diagnose GCA.

5.2.4.2 Corticosteroid Effects

Corticosteroids will alter the inflammatory changes in the arterial biopsy. One very alarming report reviewed the temporal artery biopsies from patients begun on corticosteroid therapy prior to the biopsy. When the biopsy was negative, the patient was considered to have GCA if the appropriate clinical symptoms were present, if the plasma viscosity or the ESR was significantly elevated (approximately 70 mm/h or higher), or if there was a rapid clinical improvement to corticosteroid medication. A positive biopsy was defined as containing some combination of a chronic lymphocytic and giant-cell in-

flammation, vascularization of the vessel wall, atrophy and fibrosis of the media, or significant loss of the elastica. A biopsy was considered negative if there was no inflammation and only fibroelastic intimal thickening was present. The biopsies were positive in 82% of specimens within 2 days, in 60% of specimens within 1 week, and in only 10% of specimens after 10 days of corticosteroids. However, the temporal artery segments were frequently shorter (mean 6.2 mm) than the recommended biopsy length so it is possible that longer specimens could have increased the diagnostic yield [127]. However, another series established that signs of active arteritis will be eradicated within 6–8 weeks of corticosteroid therapy, even when a lengthy biopsy is investigated [126]. The inflammatory process in each artery of an individual may not be in phase and equally suppressed by corticosteroids. Despite these reports, any patient suspected of temporal arteritis should have corticosteroid therapy initiated immediately (see Sect. 5.2.7), even if a delay in performing the biopsy is anticipated.

5.2.4.3 Risk of Temporal Artery Biopsy Versus Diagnostic Uncertainty

A biopsy of a temporal artery is a low-risk procedure. There is only a remote chance of a local infection developing at the biopsy site. A significant hemorrhage is also rare, but bleeding can develop when the vessel has no arteritis and the lumen is patent or if the biopsy is performed without the use of a local vasoconstrictive agent [128]. A biopsy is contraindicated in the rare case where the temporal artery supplies a surgical functioning superficial temporal artery to middle cerebral artery bypass (see Sect. 6.2.4.3 for use and Sect. 8.6.5 for failure). The superficial temporal artery is only rarely a naturally occurring provider of a significant collateral to the intracranial circulation (see Fig. 8.13 in Chap. 8).

Though, on occasion, a patient with a classic history of systemic GCA can be treated without a temporal artery biopsy, before committing an elderly patient to the required long-term corticosteroid therapy with all of the risks, especially if osteoporosis, diabetes mellitus, or peptic ulcer are present, the diagnosis should be certain. A biopsy is mandatory in any case with an atypical clinical presentation.

5.2.4.4 Temporal Artery Inflammation Without Giant-Cell Arteritis

On occasion, a temporal artery biopsy may reveal inflammatory changes in a patient who does not have PMR or GCA, nor the clinical profile of GCA. For example, inflammatory changes in the temporal artery have been described in four young patients, ages 7–22 years, who had soft painless nodules palpable along the length of the temporal artery. The temporal artery biopsies showed granulomatous inflammation and intimal proliferation, but there were no giant cells. In addition, none of the patients developed systemic signs or symptoms typical of GCA during the 2-year follow-up period [129]. Biopsy of the temporal artery in patients with a systemic vasculitis such as polyarteritis nodosa (see Sect. 9.2) may also reveal inflammation in this vessel.

5.2.5 Pathophysiology

The pathophysiology of GCA has remained elusive. An elevation of serum immunoglobulin and complement in a few patients suggest circulating immune complexes may have some role in this disorder [130, 131]. Though immune complex deposition is not found in the walls of affected arteries, immunoglobulins are identified in some temporal artery biopsy specimens [132, 133]. In one study, three of 16 positive temporal artery biopsies contained deposits of IgG or IgM [113].

5.2.6 Ocular Pathology

Reports on individual autopsied cases of GCA have demonstrated histopathological evidence of a granulomatous arteritis in one or more of the following arteries: ophthalmic, posterior ciliary, and central retinal [134, 135]. Active inflammation is frequently found in the superficial temporal, ophthalmic, and posterior ciliary arteries in the patient who has symptomatic disease at the time the specimens are obtained [71] (Fig. 5.10)

Giant cells, arterial necrosis, and thrombosis, which are diagnostic of GCA, are not found in all autopsied specimens, particularly if the patient had been treated with corticosteroids. This is exemplified by the report of the histopathology of a patient treated with corticosteroids for 2 months. Though lymphocytic aggregates were found in the adventitia surrounding the posterior ciliary arteries and many arteries had areas of internal elastic lamina fragmentation, giant cells, thrombosis, or necrosis was not found in these vessels [96].

An infarct of the optic nerve, accounting for the visual loss in most cases, has been demonstrated in the area of the nerve immediately posterior to the lamina cribrosa [96]. It is difficult to relate the visual loss to involvement of a particular artery with any degree of certainty. Cases with mild or limited visual loss such as an altitudinal or arcuate visual field defect have not been studied pathologically. Though cases with only a focal infarction of the optic nerve have been demonstrated, most of the reported specimens have shown signs of inflammation and thrombosis in multiple arteries that supply the eye. The tendency for GCA to involve multiple

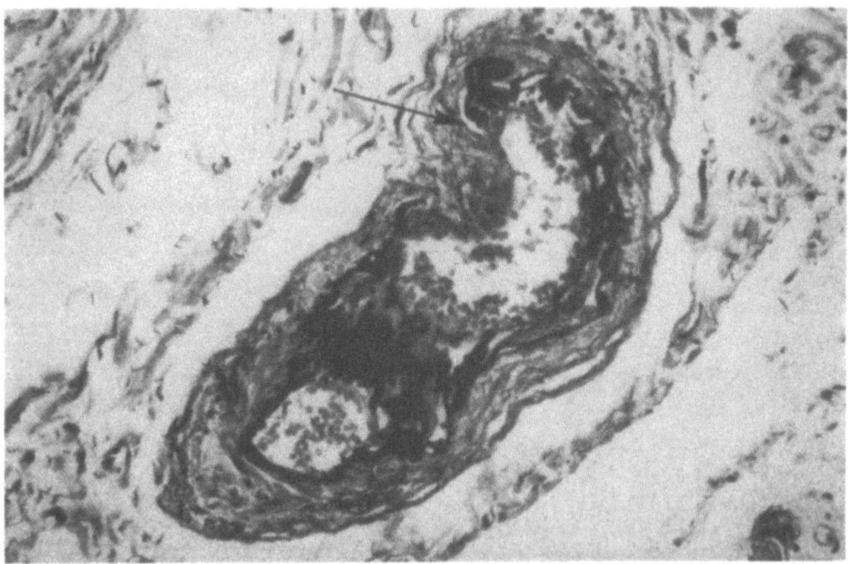

Fig. 5.10. The ciliary artery behind the left eye is from a 67-year-old woman who developed blindness in the left eye with systemic complaints of giant cell arteritis 4 months prior to her death from a pulmonary embolus. She had been treated with prednisone during the interval prior to her death. This vessel contains fragmented internal elastica lamina (*arrow*). Some chronic inflammatory cells and histiocytes are present in the adventitia and periadventitia. Verhoeff elastic stain, ×320. (From [8])

arteries and cause ischemic damage to areas of the retina, choroid, and optic nerve in a single patient probably accounts for the high incidence of profound visual loss in affected eyes.

The inflammatory process seems to affect the posterior ciliary arteries more intensely and more frequently than the other branches of the ophthalmic artery. A perivascular lymphocytic infiltration typically involves the posterior ciliary arteries and may extend to other vessels of the circle of Haller–Zinn [136]. It is unclear why inflammation of the choroidal arteries has not been a feature of this disease despite the high frequency of arteritis found in the short posterior ciliary arteries which are continuous with the choroidal vessels in the sclera.

Another patient who illustrates the extensive nature of arteritis and the problems of clinical localization was described as having a clinical central retinal artery occlusion. Yet microscopic examination revealed no involvement of the central retinal artery, but severe inflammation and intimal proliferation which caused compromise of the lumens of the posterior ciliary arteries were present. Additionally, in this patient the inner layer of the retina was relatively intact and without signs of retinal infarction [137]. However, as mentioned previously, other cases with inflammation and thrombosis of the central retinal artery have been documented pathologically [134].

The arteritis can be seen in the arterial supply to the extraocular eye muscles, but ischemic muscle damage has been rarely documented pathologically, even in cases who have had clinically apparent ophthalmoparesis [98].

Less frequently, arteritis occurs in other branches of the ophthalmic artery or branches of the external carotid artery (ECA) or ICA that do not supply the orbit (Fig. 5.8a). There seems to be a predilection for GCA involvement of the branches of the ECA. This increased susceptibility may be related to the relative abundance of elastic fibers in the ECA system [71].

5.2.7 Treatment and Course

GCA is a self-limiting disease that typically persists for several months to up to 3 years. The disease usually becomes quiescent within 1 year, and remission is infrequent before 6 months. Rarely, GCA remains active for longer than 3 years. Relapses are rare following a complete remission, but recurrences have been described as long as 9 years after the initial diagnosis [138]. Delayed visual loss, occurring 3–7 years after the diagnosis of GCA, which is not clearly related to a recurrence of GCA, has been rarely reported [139, 140].

If GCA is suspected, because of the risk of catastrophic visual loss, corticosteroids are administered immediately after the blood for the hemogram, ESR, and C-reactive protein is withdrawn. Though a corticosteroid drug is the treatment of choice, the dose should vary with the clinical presentation. Rheumatologists typically recommend an initial daily dose of 60–80 mg prednisone or an equivalent dose of another corticosteroid for the systemic disease or after unilateral visual loss in order to prevent second eye involvement [141]. However, until the inflammation is under control, even daily doses of 100 mg prednisone may fail to prevent visual loss in the first or second eye [142–145]. Therefore, we recommend an initial daily dose of 60–80 mg prednisone in patients with GCA who have no visual disturbance. Higher doses, 100–120 mg daily to megadose therapy (see below), are suggested for patients with transient or permanent visual loss depending on whether the defect is acute or older than several days.

Diplopia and ophthalmoparesis often fluctuate early in the course and respond dramatically to corticosteroids [146].

Though low-dose prednisone (5–20 mg) usually controls the symptoms of PMR, it is inadequate in preventing progression to GCA and visual loss. In most cases of GCA without prior treatment, 80–100 mg prednisone daily will result in improvement of the systemic complaints within 4–72 h of therapy [144], but recovery of vision is rare [147, 148]. The dose of corticosteroid is maintained until the C-reactive protein becomes negative and the ESR normalizes or demonstrates a significant trend towards normalization and the systemic complaints remit.

5.2.7.1 Megadose Corticosteroids

If administered acutely, intravenous megadose corticosteroids (500–1000 mg methylprednisolone every 12 h) may reverse acute visual loss. Recent application of this therapy restored the vision in an eye with 20/50 Snellen acuity and retinal ischemia. The visual loss had developed despite 60 mg prednisone following optic nerve infarction and blindness in the first eye [149]. We observed a patient who complained of monocular visual loss for 3 h with 20/200 Snellen acuity and a pale swollen optic disc in the affected eye, whose vision recovered to 20/30 following intravenous methylprednisolone (250 mg every 6 h for 4 days). A similar case with only light perception at the initiation of therapy recovered following intravenous methylprednisolone [150]. In another of our patients with a blind eye from GCA optic nerve infarction, methylprednisolone (250 mg every 6 h) stopped the visual obscurations in the second eye as well as progression of lingual infarction. When the dose was lowered to 125 mg every 6 h, the symptoms recurred, but were controlled following reinstitution of the higher dose (Fig. 5.8).

Despite these successes, the probability of visual recovery is low, particularly if infarction of the retina or optic nerve is complete. Additionally, megadose corticosteroids may not prevent blindness in all cases [151]. However, since the clinical evaluation cannot conclusively distinguish relative from absolute ischemic damage to the optic nerve or retina (see Sect. 5.2.6), we suggest that intravenous megadose corticosteroids should be administered to patients who are evaluated within 24 h of visual loss. If visual recovery is to occur with therapy, it will be seen within 48 h. The corticosteroid dose is not reduced as long (usually 3–7 days) as the patient experiences visual obscurations in the second eye or signs of recurrent or progressive ischemia in other organs.

Megadose therapy is followed by standard high doses (80–100 mg) of oral prednisone or an equivalent corticosteroid. The medication dose is then adjusted in relation to the ESR, C-reactive protein, and the control of systemic and ocular symptoms.

5.2.7.2 Daily Treatment

There is no single formula for adjusting the corticosteroid dose. A daily dose (usually 20–60 mg) of oral prednisone is required for the first month. After 1 month the dose is gradually reduced as long as the ESR is stable (hopefully near or in the normal range) and the symptoms are quiescent. Though there are no real data, in patients with negative temporal artery biopsies who do not have absolutely diagnostic criteria for GCA, the corticosteroids may be gradually withdrawn within 1–3 months as long as the appropriate reevaluation has been carried out. It is unusual for patients with biopsy-proven GCA to remain asymptomatic if corticosteroids are completely withdrawn prior to 6 months. In one major series, patients required corticosteroids for a mean of 7 months [54]. Though alternate day therapy has fewer complications, it is not as effective as daily corticosteroid for GCA and it is not recommended. Divided doses or a single daily dose of corticosteroid are equally effective maintenance therapies [152]. After 3 months, if there is a mild rise in the ESR but no recrudescence of systemic complaints, it is probably not necessary to increase the corticosteroid dose (there are other benign causes of an increase in the ESR) [79]. If systemic symptoms recur, the prednisone is increased in increments of 10 mg/day until control is obtained. If neurologic or visual symptoms develop, 100 mg to megadose therapy should be instituted.

5.2.7.3 Adjunct Therapy

There are no alternative treatments to corticosteroids. Dapsone (100 mg daily), an antimicrobial and antiinflammatory agent commonly used to treat leprosy, has been used as an adjunct therapy. Once the acute inflammatory phase is controlled, in some patients, a lower dose of corticosteroid may be more effective in combination with the dapsone [153, 154]. Dapsone-induced methemoglobinemia or hepatic abnormalities may limit the use of this agent. Patients with glucose-6-phosphate dehydrogenase or methemoglobin reductase deficiency or an abnormal hemoglobin are more prone to hemolysis by dapsone.

Cytotoxic drugs have also been reported to have a steroid sparing effect in GCA [155], but few studies have been performed.

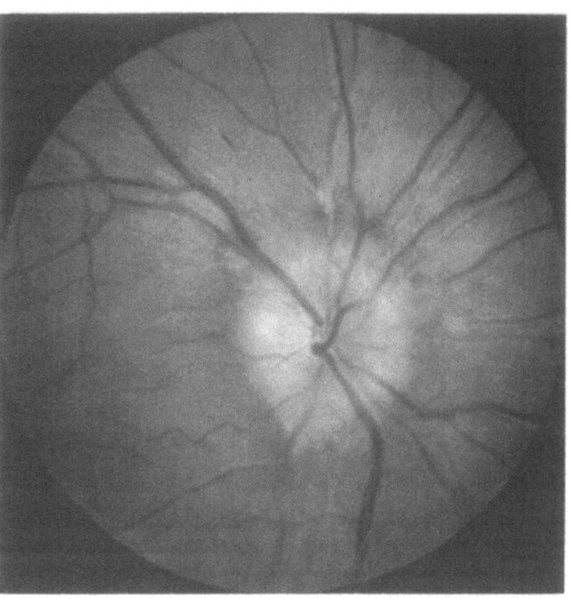

Fig. 5.11. This 26-year-old female student noted the onset of blurred vision 3 weeks after a rapid withdrawal of chronic prednisone therapy for Crohn's disease. She had a normal neurologic examination and computed tomogram of the head. Her lumbar puncture had an opening pressure of 450 mmH$_2$O with normal analysis of the cerebrospinal fluid. The Snellen acuity was 20/20, color vision was intact, and the visual fields were normal except for mild enlargement of the blind spots in both eyes. Ophthalmoscopy demonstrates diffuse mild swelling of the right and left (not shown) optic nerves typical of early low- grade papilledema. The vessels in the immediate peripapillary nerve fiber layer were mildly obscured by the edema. Reinstitution and gradual withdrawal from the prednisone was associated with resolution of the papilledema and visual complaints

5.3 Papilledema

Papilledema caused by increased ICP is generally not considered to be a primary vascular disorder but it is included here because the visual loss associated with papilledema is frequently secondary to infarction of the anterior optic nerve [156]. The treatment is always designed to lower the ICP or reduce the effect of the transmitted pressure in the subarachnoid space on the optic nerve. Chronic untreated papilledema associated with hydrocephalus, an intracranial mass lesion, or venous dural sinus thrombosis causes transient visual obscurations and leads to visual loss. The factors that increase the risk of visual loss in patients with idiopathic increased ICP are vascular related and include anemia and systemic hypertension [157].

Fig. 5.12a–c. A 68-year-old man with longstanding severe but controlled high blood pressure complained of visual loss over the course of a week in his remaining seeing right eye. He had a long-term history of intermittent swelling of both optic nerves associated with visual loss over a 30-year period. He had never been treated for this condition. At the time of evaluation, the acuity was a poor 20/200 OD and finger counting at 2 feet OS. There were no additional neurologic abnormalities. A magnetic resonance scan demonstrated aqueductal stenosis with marked dilatation of the third and lateral ventricles.
a Ophthalmoscopy of the right eye demonstrated pale swelling of the optic nerve with a large macular and papillomacular area subretinal and retinal scar with atrophy of the surrounding retinal pigment epithelium. Nerve fiber layer hemorrhages were noted inferiorly along the retinal veins.
b Ophthalmoscopy of the left eye revealed questionable swelling of the optic disc with diffuse yellow pallor and areas of peripapillary retinal pigment epithelium atrophy.
c Fluorescein angiography of the right eye demonstrated dilatation of the capillaries on the superior aspect of the disc without any significant leakage from the scar. A lumbar puncture revealed an opening pressure of 320 mmH$_2$O with a normal cerebrospinal fluid analysis. The day following the lumbar puncture, the vision returned to 20/80 in the right eye and was unchanged in the left eye. The patient underwent a ventricular pleural shunt for treatment of chronic hydrocephalus that caused atrophic papilledema. The chorioretinal scar was presumed to be secondary to this same process, which had presumably impaired the venous outflow from the eye intermittently over the years

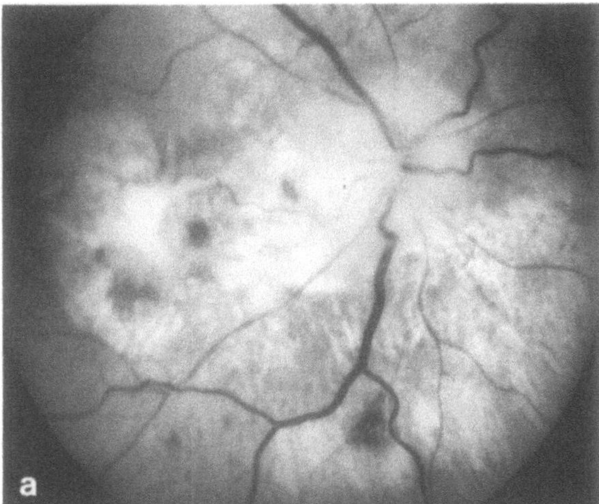
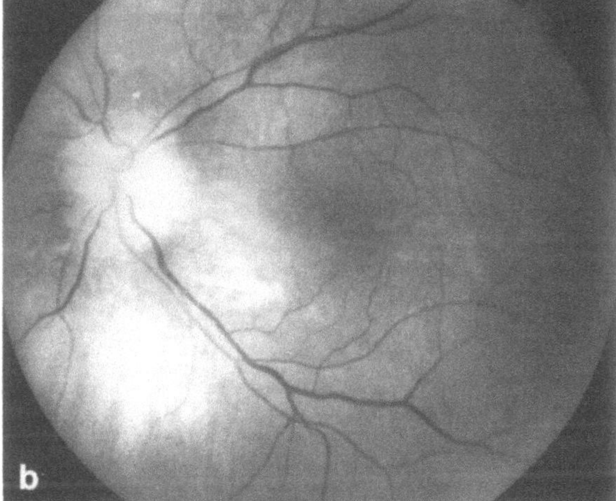
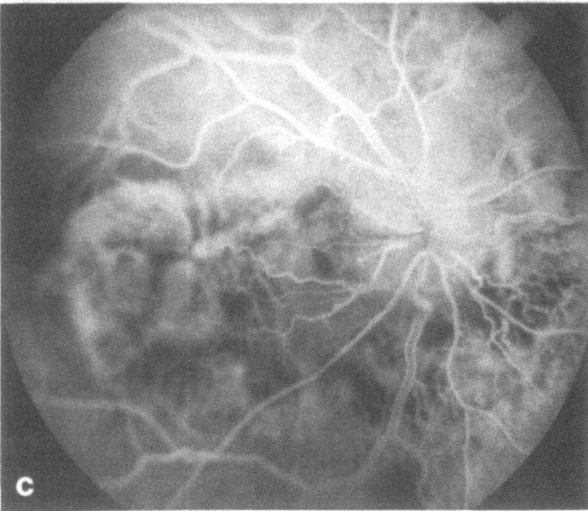

5.3.1 Ophthalmologic Findings

Bilateral, occasionally asymmetrical, swelling of the optic discs, dilated disc surface capillaries, peripapillary edema, and preservation of the physiological cup are seen on funduscopy with early papilledema (Fig. 5.11). The retinal veins dilate and hemorrhages develop in the retina around the disc and posterior pole. Edema in the peripapillary retina causes retinal folds, predominantly in the retina adjacent to the temporal border of the disc. Choroidal folds result from congestion in the choroidal circulation. Edema and hemorrhage in the macula accounts for visual loss in some cases. Capillary occlusions cause cotton wool spots in the peripapillary nerve fiber layer and on the optic disc. Chronic papilledema leads to grayish "gliotic" ap-

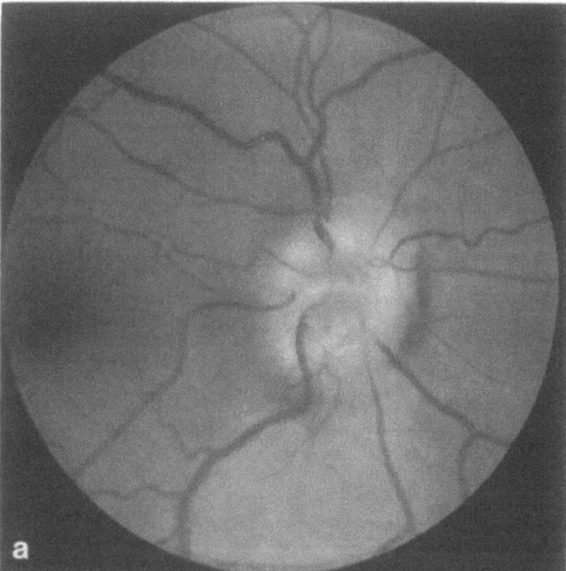

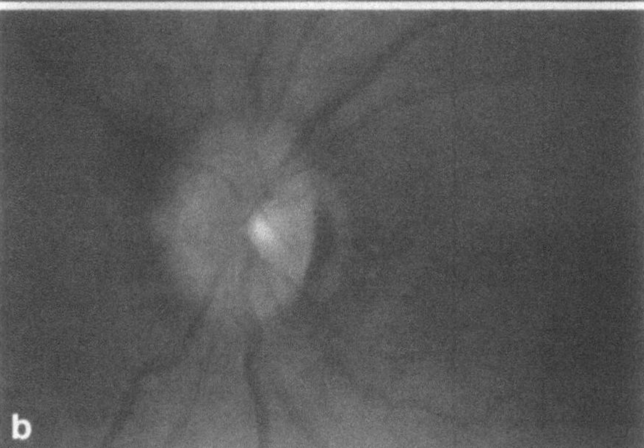

Fig. 5.13a, b. A 28-year-old moderately overweight nurse had the onset of very mild headaches with significant blurring of vision, particularly in the left eye, over the course of 4 months. Her obscurations in the left eye became markedly worse in the immediate premenstrual period. The visual acuity was 20/20 in the right eye and 20/30 in the left, and there was a relative afferent pupillary defect and dyschromatopsia in the left eye. The visual field in the right eye was intact except for a mild enlargement of the blind spot and the left eye had a dense inferonasal field defect in addition to the mild enlargement of the blind spot.
a Ophthalmoscopy of the right eye demonstrated mild to moderate swelling of the optic disc and elevation of the vessels over the disc with a nasal peripapillary retinal hemorrhage. The capillaries on the disc surface were dilated.
b Ophthalmoscopy of the left eye demonstrated swelling and elevation of the disc, particularly nasally, in addition to a large area of grey yellow temporal pallor. The computed tomogram of the head was normal. The lumbar puncture revealed an opening pressure of 550 mmH$_2$O with normal analysis of the cerebrospinal fluid. On a weight-reducing low-sodium diet and acetazolamide 1 g/day, the patient's headaches and visual symptoms were reduced. There was no further visual field loss. However, over the next 10 years, the patient noted that in the premenstrual period there were episodes of visual obscurations in the left eye. These obscurations were prevented if the patient limited her sodium intake and was treated with hydrochlorothiazide during these premenstrual intervals

pearing disc swelling. Rarely, chronic papilledema causes subretinal neovascularization which involves the macula and disrupts the central vision because of traction and hemorrhage into the macula [158] (Fig. 5.12).

Early in the course, enlargement of both blind spots reflects the dysfunction caused by edema and elevation of the peripapillary retina. Nasal depression or diffuse constriction of the visual fields characteristically develop with the optic neuropathy of chronic papilledema (Fig. 5.13). Dyschromatopsia and a decrease in visual acuity also occur. A paracentral scotoma can develop in the temporal field as a result of massive swelling of the optic discs with edema and folds in the peripapillary retina or the choroid. If the macula or the papillomacular bundle is affected, a central scotoma can be mapped (Fig. 5.14).

Fig. 5.14a–c. Three cases with typical visual field defects associated with papilledema. **a** Constriction of the central isopters and enlargement of the blind spots in an arcuate pattern are typically reversible with reduction of the intracranial pressure. **b** Chronic papilledema caused an altitudinal field defect in the right eye and the dense arcuate scotoma in the left eye. **c** Blind spot enlargement, central isopter constriction, and nasal field defects (step, arcuate scotoma) are seen. (From [157a])

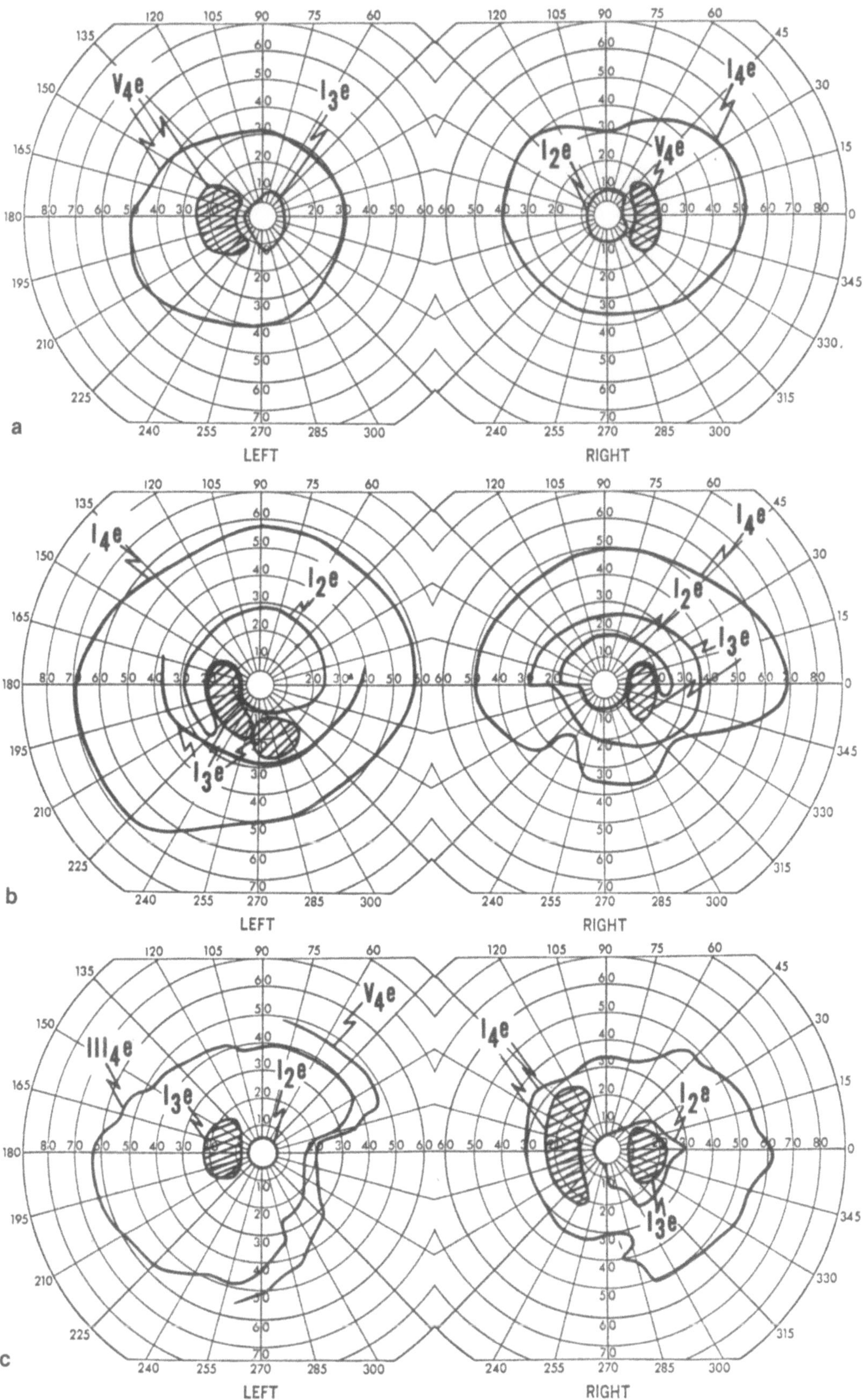
V4e
I3e
I4e
I2e
V4e
LEFT
RIGHT
a
I4e
I2e
I4e
I2e
I3e
I3e
LEFT
RIGHT
b
V4e
III4e
I2e
I3e
I4e
I2e
I3e
LEFT
RIGHT
c

5.3.2 Mechanisms of Visual Disturbance

Transient visual obscurations that occur with pathological elevation of the optic discs develop because of a disturbance in the relation between the arterial circulation and optic nerve tissue perfusion pressure that results in ischemia of this region of the optic nerve. For example, the elevated tissue pressure that results from axonal swelling and edema occludes the small compressible arterioles and capillaries to the optic disc. Central retinal venous outflow is compromised from the transmission of the ICP down the subarachnoid space to the sclera. This causes an elevation of the optic disc venous pressure, distends the veins, and blocks axoplasmic flow, further raising the local tissue pressure [159–165]. Relieving the increased ICP reduces the elevation of local tissue pressure, improves optic nerve blood flow, and prevents obscurations and progression of the visual loss. Transient visual obscurations, which can result from any pathological elevation of the optic disc, such as optic nerve drusen, may all result from a similar mechanism, an alteration of local blood perfusion to the optic nerve tissue [166].

There are numerous other causes (see Sects. 8.2.3.3, 10.2.4.1) of transient visual obscurations or temporary amaurosis. Bilateral transient scotomas, similar to transient visual obscurations, lasting seconds develop with systemic hypotension or vertebral insufficiency. If significant stenosis is present in both extracranial vertebral arteries, bilateral amaurosis can even occur with a change in head position [167].

5.4 Leber's Optic Neuropathy

Leber's optic neuropathy is a rare painless progressive optic neuropathy, without an identifiable cause. Patients are typically younger than 30 years old but rare cases in the sixth decade have been described. Though both sexes are affected, most symptomatic patients are men in the second or third decade of life. Leber's neuropathy is predominantly found in caucasians. Though a vasculopathic etiology has not been proven, this disorder is associated with abnormalities of the microvasculature of the anterior optic nerve and peripapillary retina.

5.4.1 Molecular Aspects

Studies show exclusive maternal transmission to the offspring, possibly via abnormal cytoplasmic mitochondria. Approximately 50% of the sons of female carriers develop optic atrophy. Though half of the daughters have ophthalmoscopic abnormalities as a marker of the disease, only 21% become symptomatic [168]. Mitochondrial abnormalities are found on electron microscopy of peripheral muscle biopsy [169]. A mitochondrial (cytoplasmic) defect in the genes that encode for the mitochondrial electron transport chain NADH-Q1 oxioreductase activity has been identified in these patients [170]. A point mutation at position 11,778 or other positions of mitochondrial DNA is found in many families with this disorder [171].

5.4.2 Cardiac and Neurologic Abnormalities

Leber's optic neuropathy may be associated with cardiac and various other neurologic disorders including spinal cord, peripheral nerve, and basal ganglia degeneration [169, 172]. Electrocardiogram alterations such as prolongation of the QRS complex occur but most are asymptomatic. Palpitations do occur, and Wolf–Parkinson–White (WPW) or Lown–Ganong–Levine (LGL) syndromes are present in approximately 18% of women but in only 5% of men [168].

5.4.3 Ophthalmologic Findings and Clinical Course

Intraretinal telangiectatic capillaries and microangiopathy in the peripapillary area are typically seen in the fundus of patients and asymptomatic carriers. These abnormal vessels are competent and do not leak on fluorescein angiography [173]. A thickened, white–gray, opaque nerve fiber layer is often present for months before the onset of visual loss in the patient and in asymptomatic family members (Fig. 5.15a). The fundus abnormalities

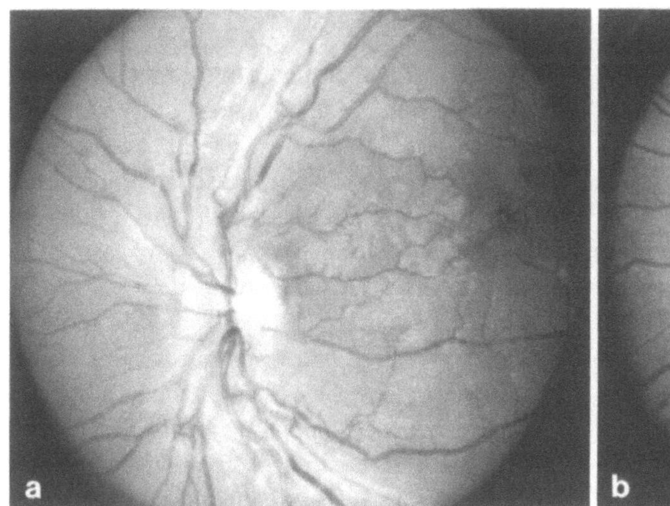
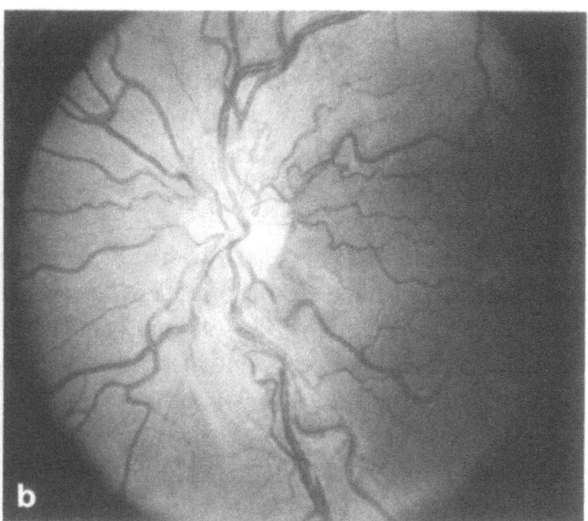

Fig. 5.15. a Prior to visual loss, ophthalmoscopy demonstrates thickening of the nerve fiber layer, tortuosity of the vessels, and areas of peripapillary intraretinal vascular abnormalities in this patient who had lost vision in the other eye from Leber's optic neuropathy. **b** This 20-year-old man with Leber's optic neuropathy had the onset of visual loss for several months and was found to have 20/200 vision in this eye. Ophthalmoscopy demonstrated temporal pallor and loss of the corresponding papillomacular nerve fiber layer. The remaining nerve fiber layer along the superior and inferior margins of the disc appears thickened. (Cases provided by Michael Slavin)

are frequently milder in females than males [174]. Immediately prior to the onset of visual symptoms, the retinal vessels around the optic disc dilate, the nerve fiber layer swells, and occasionally splinter hemorrhages in the peripapillary retina develop. The optic disc in one eye swells first so the two fundi typically appear asymmetrical. The second eye develops disc swelling and visual loss within weeks to months following the visual loss in the first eye.

The microangiopathy and peripapillary nerve fiber layer disappear as the disease progresses. First, the papillomacular nerve bundle degenerates. Eventually, pallor develops, predominantly on the temporal aspect of the disc from the loss of the papillomacula bundle (Fig. 5.15b). When the history and findings of the acute period are unknown, narrowed or tortuous retinal arteries may remain as a clue to the etiology of the bilateral optic atrophy. In some cases, the optic atrophy may appear nonspecific and diffuse.

The visual acuity is decreased to a range from 20/80 to 20/400 and it often remains at 20/200. Acuity loss worse than 10/400 is rare. Dyschromatopsia occurs in all cases and testing with the Farnsworth–Munsell 100 hue uncovers color vision defects prior to significant acuity loss. Prior to the acuity loss, visual field testing often reveals enlargement of the blind spot because of peripapillary retinal dysfunction. As the acuity fails, a central or centrocecal scotoma develops. Prior to visual loss, the pattern of visual evoked potentials may be normal in some cases [175] and abnormal in others [176].

The visual loss may appear insidiously and painlessly, and it progresses over weeks or months. The vision in the second eye is often slightly abnormal at the time of presentation of the visual disturbance in the first eye. A significant decrease in acuity develops in the second eye within months of the visual loss in the first eye. In the vast majority of cases, the patients are left with bilateral acuity of 20/200 or worse, central scotomas, and optic atrophy. Patients can often see better if they are shown how to view eccentrically in order to avoid the central scotoma. These young patients can adjust to the loss of central vision and with rehabilitation learn to use their remaining normal peripheral field.

When performed before the atrophic stage, fluorescein angiography clearly demonstrates the vascular abnormalities. Arteriovenous shunting is seen in the telangiectatic peripapillary vessels. As nerve fiber layer atrophy develops, less shunting is seen, the arterioles narrow, and the telangiectasia disappear [177].

Initially, one study in the United States noted a marked spontaneous recovery in one case [178]. However, studies outside of the United States have reported that as many as 11% of patients recovered central vision [179]. It is possible that some cases of Leber's optic neuropathy have been misdiagnosed as optic neuritis on the basis of their recovery. Clearly, some patients with the disease experience a degree of visual recovery. A recent report correlated a poor prognosis for return of vision with the presence of the 11778 DNA mutation and a better prognosis for significant visual recovery with the absence of this genetic defect [180].

5.4.4 Treatment

No treatment is effective. Corticosteroids administered in conventional or megadoses have not improved the vision. Vitamin B_{12} administered prophylactically or after the onset of visual symptoms has not prevented the visual loss or reduced the deficit.

5.5 Postshock Optic Neuropathy

5.5.1 Causes

An acute bilateral optic neuropathy which develops as a sequela of acute blood loss such as a gastrointestinal hemorrhage or an episode of severe hypotension is uncommon [5, 6, 181–187]. Acute bilateral optic neuropathy has also been described following cardiac surgery, which is obviously associated with a prolonged period of sustained hypotension and significant time spent on the cardiopulmonary bypass machine [188, 189]. The presence of systemic arteriosclerosis, typically found in coronary artery bypass patients, may affect the segmental vascular supply to the prelaminar or retrolaminar portions of each optic nerve. This predisposes the patient to the effects of systemic hypotension with significantly lowered lo-

cal tissue blood perfusion pressure and ischemic infarction.

Visual loss may be noted immediately following the hypotension, or there may be a delay of up to several days before a visual disturbance is discovered. It has been hypothesized that circulating angiotensin that is released in response to the hypotension or blood loss causes vasospasm of the arteries to the optic nerve and a delayed ischemia in the latter situation [190]. However, in some patients, particularly following coronary artery bypass surgery where the patient often has a transient postoperative encephalopathy, neither the patient nor the medical team might be aware of the exact onset of the visual disturbance.

Hypotension without an underlying anatomical predisposition is unlikely to cause damage to the anterior visual pathway. In a study of a small group of patients who had been in shock (defined as systemic hypotension with a systolic blood pressure no greater than 80 mmHg in hypertensives or lower in otherwise normal patients, or a fall in systolic pressure greater than 50 mmHg in hypertensives), the optic discs were normal in the 35 eyes of the 18 patients investigated (one eye with a prior central retinal artery occlusion was eliminated from the study). In addition, none of the patients had visual field defects typical of glaucoma or AION [191].

5.5.2 Ophthalmologic Findings

The visual loss is bilateral in more than 90% of the cases [6]. Acutely, ophthalmoscopy of the optic discs is either bilaterally normal or shows bilateral mild pallid disc swelling with splinter peripapillary nerve fiber layer hemorrhages typical of AION [192]. Optic atrophy, with or without an increase in the cup/disc ratio (cupping or excavation), develops after months. When significant excavation is seen, this must be differentiated from glaucoma or low-tension glaucoma (LTG). The decreased acuity, dyschromatopsia, pallor of the remaining neural rim, nonglaucomatous visual field defects, lack of progression of the field loss, and normal IOP differentiate the postshock optic neuropathy from glaucoma (see Sect. 5.6 for discussion) [193].

In our series of 10 patients with hypotension and blood loss or severe atherosclerosis, the long-term evaluation showed bilateral diffuse optic disc pallor, without significant cupping of the disc, and bilateral severe loss of acuity and central scotomas as the most common syndrome. Our results are biased towards those cases with profound visual loss who were referred to a neuro-ophthalmology service because of major visual complaints. All of our patients were elderly or had prior significant systemic hypertension or diffuse atherosclerosis or acute blood loss, so that systemic hypotension was not the sole risk factor for the optic nerve ischemia. None of our acute cases, all of whom had normal IOP, demonstrated significant recovery or worsening of vision after the blood pressure and the hematocrit were stabilized.

When the retrolaminar optic nerve or the chiasm is the site of the ischemia, acutely the fundi appear normal. Pupils which are poorly reactive to light or amaurotic can differentiate anterior visual pathway dysfunction from bilateral occipital or parietal occipital ischemia. The cortical lesions carry a better prognosis for visual recovery. The differential diagnosis in cases where ophthalmoscopy does not demonstrate swelling of the optic nerves but the pupils are not reactive to light must also include a previously undiagnosed pituitary tumor that developed apoplexy following hypotension or cardiac surgery [194, 195].

5.6 Low-Tension Glaucoma

Low-tension glaucoma (LTG) is an uncommon disorder that can cause progressive visual loss. LTG is almost always seen in elderly patients, most of whom have a history of systemic atherosclerosis or hypertension [196].

Confusion has been prevalent since von Graefe first reported the presence of optic disc cupping without elevation of the IOP [197]. The excavation or cupping of the optic disc seen on ophthalmoscopy is typically a sign of glaucoma when it is accompanied by elevation of the IOP and the appropriate visual field loss. Cupping of the optic disc, associated with glaucoma and LTG, develops because there is severe loss of the axons, myelin, and glial support tissue of the optic disc and the vitreous pushes into the atrophic disc head to bow the

lamina cribrosa outwards [198]. Alternatively, a weakness in the connective tissue of the lamina might result in poor structural support predisposing to the pathological increase in the physiological cup and injury to the axons in the optic nerve.

5.6.1 Ophthalmologic Findings

5.6.1.1 Comparison with Glaucoma

LTG is a bilateral disease that is diagnosed in eyes with glaucomatous cupping of the optic disc that meet the criteria listed in the Table 5.5 (Fig. 5.16). As in high-pressure glaucoma the nasal rim of disc is spared early on and appears pink on ophthalmoscopy.

Table 5.5. Criteria for low-tension glaucoma

Glaucomatous excavation of the optic disc
Visual field loss similar to the defects typical of
  glaucoma[a]
Early preservation of visual acuity
Normal clinically determined color vision
  (by pseudoisochromatic plate testing)
Normal intraocular pressure, no prior elevation
Normal diurnal intraocular pressure testing
Normal aqueous outflow facility

[a] Visual field defects may not be exactly the same in low-tension glaucoma and glaucoma.

5.6.1.2 Visual Fields

The visual field dysfunction typically includes arcuate nerve bundle and nasal step defects. Subtle differences in the field loss pattern such as a paracentral field defect may distinguish LTG from glaucoma. Patients with LTG have a higher frequency of dense paracentral scotomas or nerve bundle defects located within the central 30° of the field. In particular, field loss in the central 5° is more prevalent earlier in the course of LTG [199]. The scotomas caused by LTG are often denser at the time of presentation [200]. In cases of LTG with severe hemifield loss, the remaining portion of the field, which appears normal on suprathreshold testing, has higher sensitivity (lower threshold) on threshold perimetry when compared to similar areas of field in patients with high- pressure glaucoma [201].

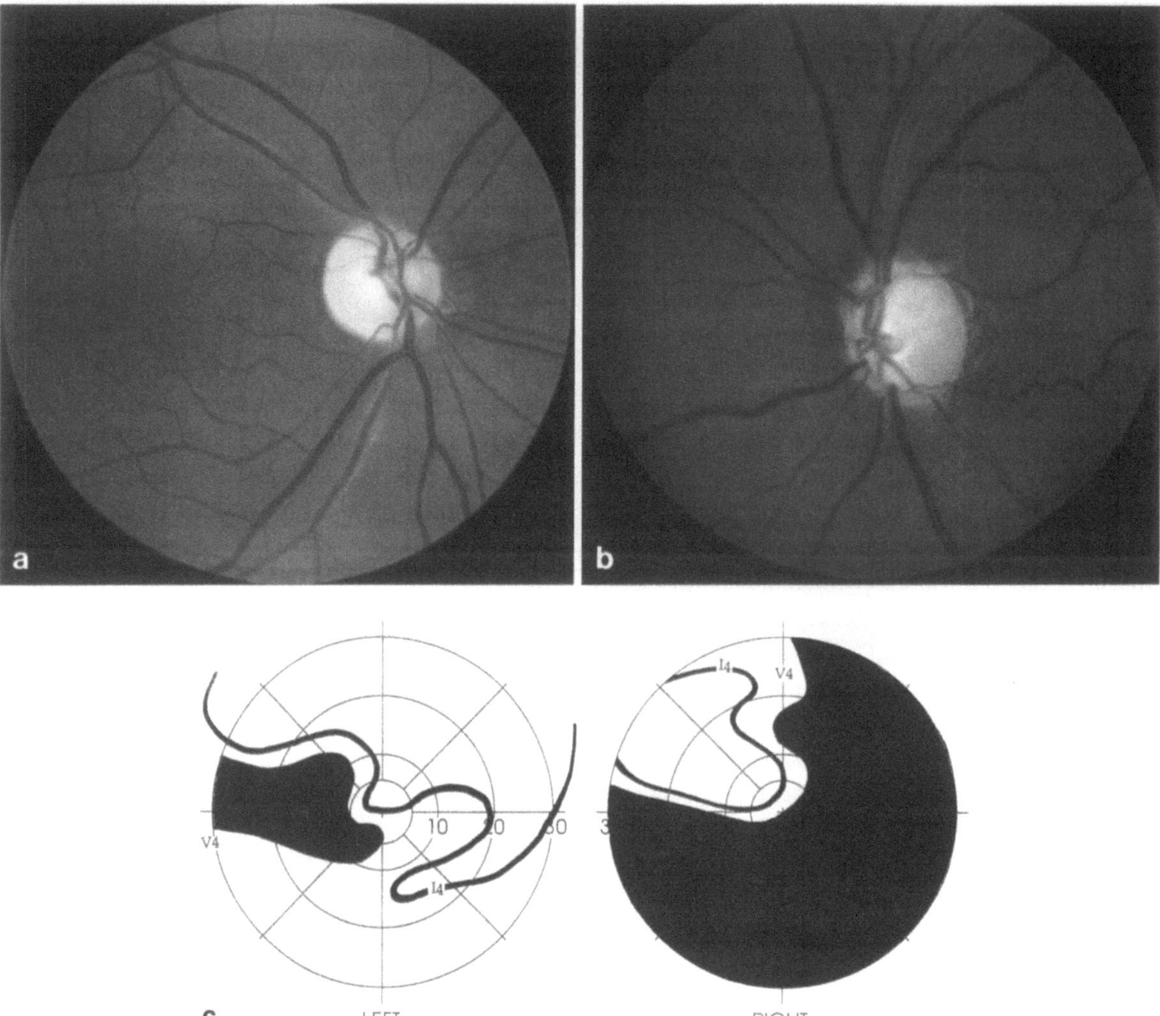

Fig. 5.16a–c. This 70-year-old man with a history of arteriosclerotic heart disease complained of blurred vision for several years. The patient had never had a history of elevated intraocular pressure nor were there any signs of inflammation in the eye. He had never experienced any episodes of shock nor required a blood transfusion. At the time of evaluation, he had 20/30 acuity in each eye with normal color vision.

a,b Ophthalmoscopy demonstrated marked excavation of both optic discs with no real temporal rims remaining which is seen better in the left eye (**b**) than the right eye (**a**). There was a questionable thin rim of nasal neural rim remaining in both eyes.

c Only the superonasal field remained in the right eye. In the left eye, the perimetry demonstrates a paracentral inferior arcuate defect and a wedge-shaped field loss to all isopters extending into the superior temporal field. Over the next 6 years there was gradual worsening of his visual field despite therapies which maintained the intraocular pressures at 12 mmHg or below. A computed tomogram of the optic nerves and chiasm was also normal. This case is typical of low-tension glaucoma

5.6.1.3 Optic Nerve Appearance

Though subtle alterations in the optic nerve may distinguish a group of eyes with LTG from eyes with glaucoma, the differentiation is often difficult for individual cases. Compared to the glaucomatous optic nerve, the optic disc with LTG seems to have worse cupping, which is greatest inferiorly and inferotemporally [202]. The optic disc may also have more diffuse mild pallor and diminished arterial blood flow as determined by fluorescein angiography [204]. However, in some cases with high-pressure glaucoma, the fluorescein angiography may also show areas of focal disc hypofluorescence or decreased perfusion that correspond to areas of significant field loss [204, 205]. In fact, not all glaucoma specialists agree that the extent of optic nerve cupping or the type of visual field loss at a given exam can differentiate LTG from primary open-angle glaucoma [206].

5.6.2 Pathogenesis

A definitive cause of LTG is unknown. Ischemia may be an important mechanism for the development of the cavernous degeneration of the optic nerve. Diminished blood perfusion of the retrolaminar optic nerve may arise from a disturbance in the relationship between choroidal arterial blood flow and the IOP [159, 160, 207]. In addition, factors that lower the blood pressure in the ophthalmic artery seem to worsen LTG [208]. Clinical support for a vascular etiology for LTG is found in the occasional patient with acute hypotension- or blood loss- associated cavernous degeneration of the optic disc who has field defects similar to the field loss found in LTG. As previously mentioned, LTG is predominantly found in the elderly with atherosclerosis, who are prone to ischemic disease. Also, there is higher incidence of migraine (another disorder with vasculopathic effects) in patients with LTG than in the normal population [209]. In one study, a similar incidence of nasal step field defects was described in patients with AION, open angle glaucoma, and LTG [210]. Experimental and clinical evidence suggests that optic nerve ischemia develops when autoregulation fails to maintain normal blood flow to the optic nerve from the posteri-

or ciliary and choroidal circulation. When the perfusion pressure falls below minimal levels, a relative ischemia of watershed areas results. When the optic nerve falls within the affected watershed, the ischemia causes infarction and a loss of optic nerve tissue [211].

Evidence against a vascular mechanism include post mortem studies that demonstrate the persistence of capillaries on the optic disc in cases of early and moderate glaucoma [212]. Additionally, chronic experimental glaucoma in monkeys does not alter capillary blood flow in the optic disc [213]. In contrast, in a study of 40 human eyes with high-tension glaucoma a significant reduction of capillaries in the region of the lamina was found, and some eyes contained areas with fewer capillaries in the peripapillary choroid as well [214]. Most importantly, the presence of capillaries does not necessarily indicate that the amount of arterial blood flow into these capillaries is normal (see Sect. 5.1). Despite the suggestion that LTG has a vascular etiology, there is no evidence that proves that a more generalized vasculopathy or coagulopathy exists in these patients [215]. In reference to a possible vascular etiology for glaucoma, the distribution of the loss of optic nerve axons is different for glaucoma and AION [216].

There are many clinical similarities between LTG and high- pressure glaucoma, and LTG may represent a bridge syndrome between AION and true glaucoma. We have seen numerous patients without a history of a sudden visual disturbance who have excavation of the optic cup, diffuse pallor, and narrowed arteries on the optic disc, progressive visual field defects typical of glaucoma, and elevated IOP. These patients seem to have a combination of chronic or episodic ischemia as well as ocular hypertension as the cause of their visual loss and optic disc disturbance.

5.6.3 Differential Diagnosis

Clinically apparent cupping or cavernous degeneration of the optic nerve can also occur in patients with severe ischemia of the optic nerve which causes extensive loss of optic nerve axons and the glial support tissue. This is seen following occlusion of one or more posterior ciliary arteries in patients with GCA [45b]. Long after the acute episode, patients with GCA can be differentiated from those with LTG by the severity of the visual failure, the history of a sudden onset of visual loss, and diffuse optic disc pallor beyond the cupping in GCA (see Sect. 5.2). Patients with GCA typically have less extensive or shallower optic cups than those with LTG or glaucoma.

Cavernous degeneration of the optic disc nerve can also be seen in the elderly with atherosclerosis, even if there is no definitive field loss [217]. At autopsy, these cases have been found to have atherosclerotic changes diffusely throughout the body, and the posterior ciliary arteries and the circle of Haller–Zinn are specifically affected [218].

Intradural fusiform [219] or saccular aneurysms [220] rarely cause cavernous degeneration of the optic nerve (see Sects. 6.4.2.2, 6.9.1.2). Though chronic vascular insufficiency to the distal optic nerve has been proposed as the mechanism for acquiring an abnormally large optic cup, similar cupping has also been rarely found with neoplasms that compress the intracranial optic nerves [221]. In the case reported by Portney and Roth, a 51-year-old with a unilateral progressive optic neuropathy caused by a carotid–ophthalmic artery aneurysm, excavation of the superotemporal portion of the optic disc was found on the clinical examination. The pathology revealed a loss of optic nerve axons with preservation of the glia. The glial cells had collapsed into the area of axonal degeneration in the optic cup but ischemic changes were not seen [220].

Most patients with compressive lesions can be distinguished from patients with LTG or glaucoma solely on the basis of the clinical examination. Intracranial compressive lesions cause visual field loss which affects the temporal field or the central 5°, early impairment of acuity and color vision, and diffuse optic disc pallor of the neural rim outside the area of excavation. Some of the patients with compressive lesions undoubtedly have an existing congenital increase in the size of the optic cup, so that any loss of axons results in a marked enlargement of the cup. In contrast, until late in the course, patients with glaucoma often have a pink remaining neural rim of the disc, normal clinically measured (tested with pseudoisochromatic plates) color vision, normal visual acuity, and relative

preservation of the central field. In all types of glaucoma, the temporal field is relatively spared until the late stages of the disease when a total cup replaces the optic disc. Though cases of unilateral LTG have been reported, a thorough search for other causes of optic neuropathy is essential before the exceedingly rare diagnosis of a unilateral LTG can be established. Differentiation between the clinical entities cannot be made on clinical findings alone in patients with end-stage disease who have total cupping accompanied by profound compromise of the acuity and visual field. In these cases neuroimaging of the intracranial optic nerves and chiasm, and possibly a cerebrospinal fluid evaluation, is warranted.

5.6.4 Treatment and Course

Medical and surgical therapies are designed to lower the IOP to the low teens but the benefit of reducing the IOP has yet to be established [222].

Many, but not all, patients with LTG experience progressive deterioration in the field defects and worsening of the extent of optic nerve cupping despite topical antihypertensive drops, oral carbonic anhydrase inhibitors, surgical filtering procedures, or laser trabeculoplasty [223–225]. In some patients with systemic hypertension, medically induced hypotension may produce chronic or episodic ischemia to the optic nerve, particularly while the patient is asleep. Reducing the medical antihypertensives may prevent worsening of vision in these cases.

Some patients who do not demonstrate the expected progressive deterioration in the optic disc or visual fields may not have LTG. These patients with a static course often have a well- defined episode of systemic hypotension, blood loss, or cardiovascular collapse that caused infarction of the axons and support tissues of the optic disc (see Sect. 5.5).

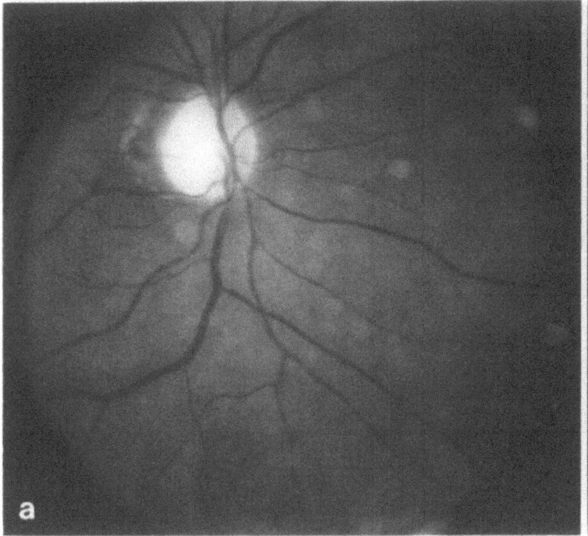

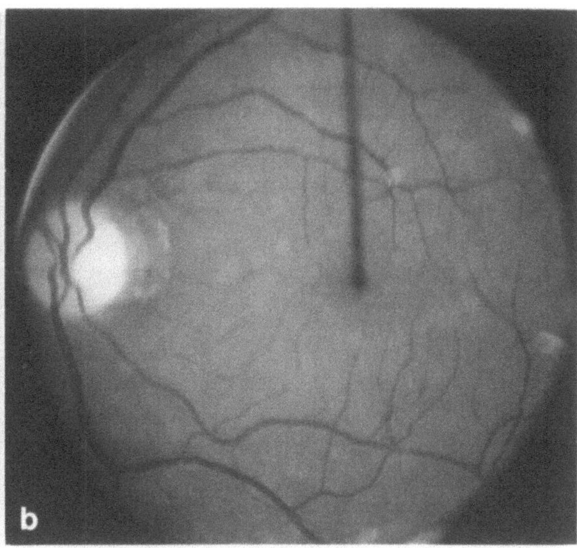

Fig. 5.17a, b. A 45-year-old woman complained of blurred vision in both eyes for at least 1 year. Her visual acuity was 20/50 and there was mild dyschromatopsia in both eyes. The visual fields revealed small arcuate shaped field defects in both eyes. Ophthalmoscopy demonstrated diffuse pallor, worse temporally, of the right optic disc (a) and more focal temporal pallor of the left optic disc

(b). A computed tomogram of the optic nerves and chiasm and serological evaluation for inflammatory and infectious diseases were entirely normal. This patient was thought to have bilateral anterior ischemic optic neuropathy associated with juvenile diabetes. The patient's vision remained stable over the next 6 years

5.7 Juvenile Diabetes Mellitus

Adults with juvenile onset diabetes mellitus can develop a self-limiting syndrome of ischemic swelling of one or both optic nerveheads. The capillaries on the disc dilate and there is visual loss of varying severity. The incidence of this disorder is unknown since only those patients with profound visual loss are routinely seen by an ophthalmologist. Diabetic patients also commonly complain of blurred vision that is often attributed to fluctuation of the blood glucose. There are many patients who have subtle visual dysfunction, and they are infrequently examined during the acute period. Their optic disc pallor is only discovered at a later date.

This type of ischemic optic neuropathy causes dyschromatopsia, Snellen acuity that ranges from 20/20 to 20/100, and visual field abnormalities. Nasal or altitudinal visual field defects and large blind spots or large blind spots alone may be found. In acute cases, symmetrical or asymmetrical disc swelling is present. The visual acuity and the visual fields defects often improve spontaneously.

After several months, there is flat pallor of the involved optic disc, but the dilated capillaries and edema can persist for as long as 12 months [226] (Fig. 5.17). Microvascular arterial occlusion or hypoperfusion, possibly affecting a posterior ciliary artery, may be the cause of this type of ischemic optic neuropathy [227]. Some of these patients also have an associated sensorineural deafness and diabetes insipidus [228, 229].

5.8 Optic Disc Vasculitis or Papillophlebitis

Papillophlebitis, often called the "big blind spot" syndrome, develops in young healthy patients, most often women, without evidence of systemic disease. The patients present with the complaint of blurred vision or transient obscurations in one eye. On funduscopy a unilateral optic disc swelling, swollen tortuous veins, and hemorrhages limited to the peripapillary retina are seen (Fig. 5.18). A mild cellular reaction may be detected in the vitreous over-

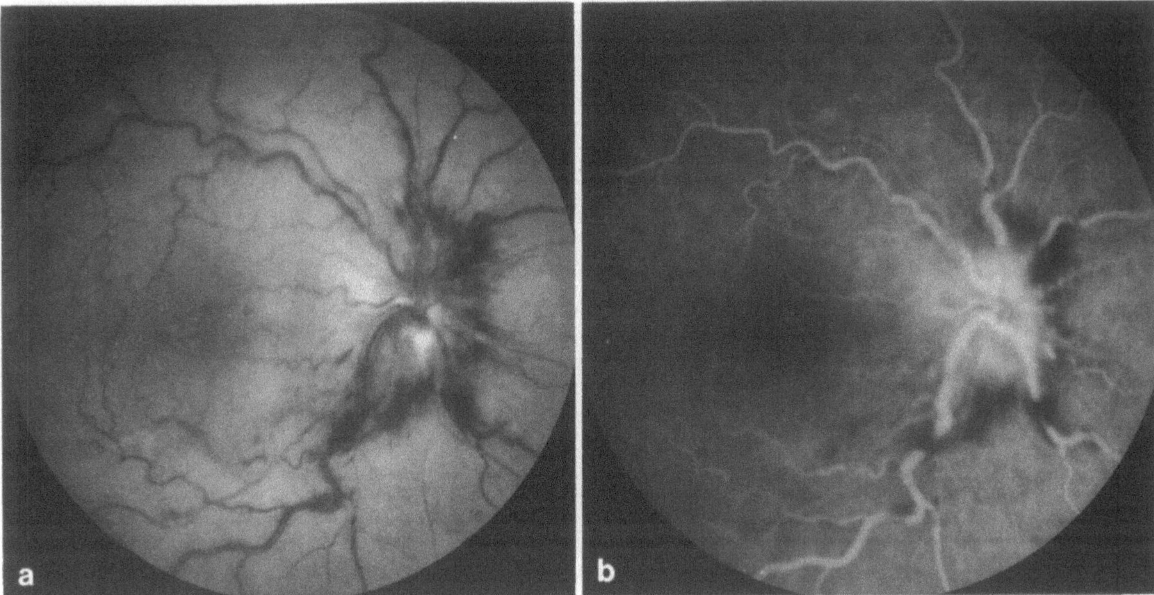

Fig. 5.18 a, b. A 28-year-old woman complained of blurred vision in the right eye for several weeks. At the time of evaluation, she was noted to have 20/20 acuity, normal color vision, and no afferent pupillary defect in either eye.

a There was moderate enlargement of the right blind spot but otherwise the visual fields were normal. Ophthalmoscopy demonstrates a swollen optic nerve with peripapillary hemorrhages. The veins were mildly to moderately engorged. The hemorrhages were noted to the mid-peripheral areas of the retina. Magnetic resonance scan and the blood evaluation for inflammatory, infectious, and coagulation disorders were entirely normal.

b The fluorescein angiogram demonstrated some mild leakage from the swollen portion of the disc but otherwise there were no abnormalities seen. After several months the visual blurring began to resolve, and within 1 year the fundus had normalized and there were no visual complaints. This case was diagnosed as papillophlebitis

lying the affected optic disc. The visual field of the affected eye reveals an enlarged blind spot. Signs of optic nerve dysfunction such as dyschromatopsia, an afferent pupillary defect, or a central scotoma are absent [230–234]. The visual impairment is not significant unless the peripapillary edema or hemorrhages extends into the macula. The disc swelling and a fluctuating disturbance of vision generally lasts 3–18 months. The course is benign with eventual normalization of the vision. Perivenous sheathing and slight dilatation of the small vessels on the disc frequently remain. This disorder rarely recurs.

5.8.1 Differential Diagnosis

The differential diagnosis of unilateral disc swelling without visual loss or signs of an optic neuropathy is extensive (see Sect. 5.1.6). An impending occlusion of the central retinal vein can be differentiated from papillophlebitis by fluorescein angiography, the lack of a vitreal cellular response over the optic disc, and the extensive hemorrhages in the peripheral retina in the former disorder [235]. An orbital mass which causes swelling of the optic disc by compromising the venous outflow is easily seen on the appropriate contrast CT or MR images. Lesions that involve the orbital optic nerve, such as a sheath meningioma, glioma or nerve sheath hydrops, or block axoplasmic flow, as well as ocular venous outflow via the central retinal vein, result in optic disc swelling. These patients have signs of optic neuropathy, visual loss which frequently progress-

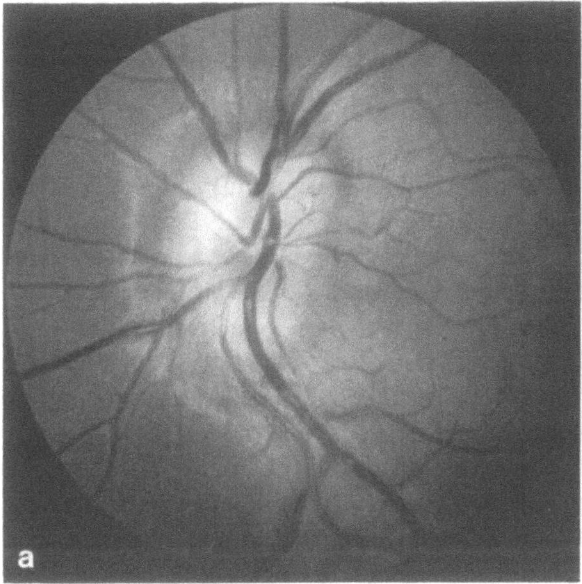
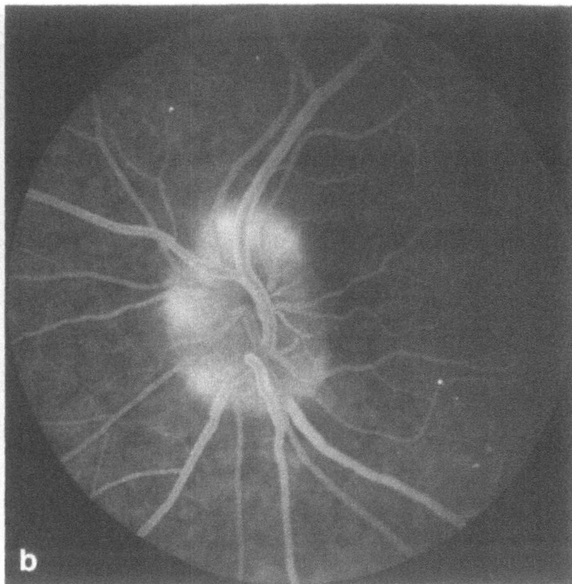

Fig. 5.19a, b. A 17-year-old woman complained of intermittent blurring of vision in the left eye over the course of a month without any headaches. Her visual acuity of 20/20, color vision, and visual fields were normal from both eyes except for some mild enlargement of the blind spot from the left eye. There was no relative afferent pupillary defect.
a Ophthalmoscopy of the right eye was normal. In the left eye, there was diffuse elevation of the optic nerve without any edema, exudates, hemorrhages, or dilation of the veins.
b Fluorescein angiography of the left eye demonstrated focal oval-shaped areas of fluorescein staining of the disc. Ultrasonography demonstrated calcified drusen buried in the optic nerve

es, proptosis, and a demonstrable lesion on CT or MR. Rare patients with increased ICP will have unilateral or markedly asymmetrical papilledema [236]. The Foster–Kennedy syndrome and AION (see Sect. 5.1) also cause a unilateral swollen disc, but the contralateral eye has optic disc pallor and visual loss. In elderly patients, AION can cause unilateral segmental swelling of the optic disc, with little or no visual loss. Vitritis, particularly when the inflammation is just over the optic disc and there is ocular hypotony, causes disc swelling. Hypertensive retinopathy causes disc swelling which is infrequently unilateral, but often asymmetric. The severe elevation of blood pressure and the presence of retinal arterial vasospasm, cotton wool spots, and hemorrhages in the retina away from the optic disc suggest hypertension as the etiology. The perineuritis of tertiary syphilis will also enlarge the blind spot without signs of an optic neuropathy.

The optic disc will temporarily appear swollen following a partial vitreous detachment.

Buried drusen elevate one or both discs focally or diffusely, without edema so the vessels in the peripapillary nerve fiber layer are not obscured (Fig. 5.19). The physiological cup is often eradicated and excessive vascular branching is seen on the disc. No hemorrhages are seen unless there is associated deep juxtapapillary neovascularization. The drusen usually remain buried in patients below 30 years of age. Eventually the drusen appear at the disc margins as yellow irregular "hard" exudates (Fig. 5.20). Patients with optic nerve drusen complain of visual blurring or obscurations that can occur intermittently over months. Defects in the visual field include an enlargement of the blind spot or fluctuating arcuate scotomas. Because the drusen are often calcified, ultrasonography shows increased reflectivity in the optic disc and CT with

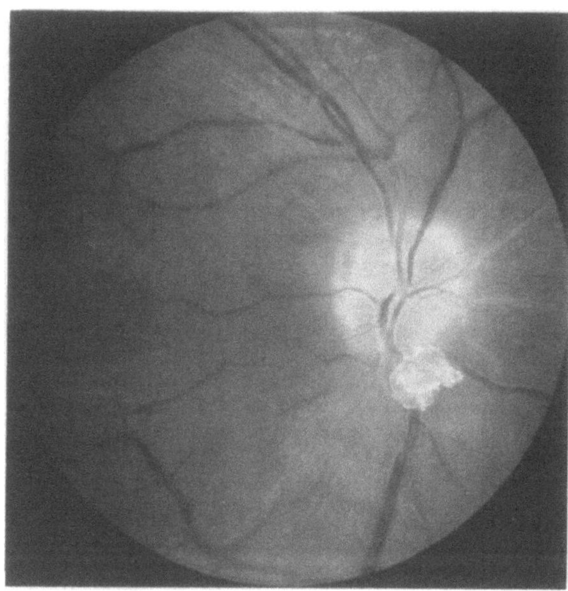

Fig. 5.20. A large drusen is seen on the inferior optic disc border in a patient who had visual obscurations some years previously

bone windows shows a high-density lesion in the optic nerve near the globe. The drusen often autofluoresce and cause late staining following intravenous fluorescein.

An enlarged blind spot can be caused by entities that disturb the function of the peripapillary retina but not that of the optic disc. These disorders include optic nerve coloboma and the retinal – choroidal changes of myopia such as peripapillary atrophy, myopic conus, or staphyloma. A big blind spot that persists for several months has also been described in patients with an idiopathic peripapillary retinal dysfunction [237]. Enlargement of the blind spot has also been reported in multiple envanescent white-dot syndrome without disc edema [238].

5.8.2 Pathology

Few cases of papillophlebitis have been verified pathologically and some clinical series have included young patients with partial central vein occlusion of other etiologies [239]. As described by Coats [240], the histopathology of papillophlebitis demonstrates an inflammatory infiltrate, predominantly lymphocytic, in the wall of the central retinal vein and in the surrounding tissues [241, 242].

5.8.3 Diagnostic Evaluation

The purpose of the diagnostic evaluation is to eliminate the other causes of optic disc swelling. Fluorescein angiography will determine if venous obstruction is present. A venous outflow delay may be seen because who have been diagnosed as having the congestion from disc edema. Some cases of papillophlebitis actually have partial thrombosis of the central retinal venous system [243]. Extensive laboratory evaluation fails to reveal a systemic vasculitis or collagen vascular disease or coagulopathy. CT or MR imaging are performed in patients where there are orbital signs, optic nerve dysfunction, or an atypical course.

5.8.4 Treatment

No treatment has been shown to shorten the course of visual disturbance or disc swelling or prevent macula dysfunction. Corticosteroids [231] and anticoagulation or antiplatelet therapy have not been beneficial.

5.9 Posterior Ischemic Optic Neuropathy

Posterior ischemic optic neuropathy (PION) is a rare disorder caused by several entities. PION is a diagnosis of exclusion which is considered in a patient who presents with a painless sudden unilateral visual loss, an afferent pupillary defect, and a normal optic disc in the affected eye. The visual disturbance may range from a mild impairment to severely diminished visual acuity. Central scotomas are the most prevalent type of visual field defect, but altitudinal or nerve fiber bundle visual field defects also occur [244]. CT or MR imaging is performed to exclude a mass (neoplasm, mucocele, or aneurysm) that has acutely expanded to compress the optic nerve (see Sects. 6.1.7.2, 6.4.2.2).

Infarction of the retrobulbar optic nerve can occur as a result of occlusion of the centripetal pial vascular plexus, which provides the blood supply to the intraorbital optic nerve [245]. The central optic nerve may be more susceptible to the ischemia induced by hypoperfusion in the retrobulbar optic nerve [246]. PION has been reported in patients with GCA, systemic lupus erythematosus [247, 248], atherosclerosis [244], and polyarteritis nodosa. Acute occlusion of the ICA is rarely associated with PION [249]. PION has also been described in one or both eyes after general anesthesia in patients with systemic atherosclerosis and peripheral vascular disease (similar to postshock optic neuropathy) [192].

An ischemic etiology may be clinically differentiated from an inflammatory optic neuritis. Unlike optic neuritis, PION is not associated with pain on eye movement, occurs in the elderly when GCA or atherosclerosis is the cause, and does not spontaneously recover. In patients with systemic lupus erythematosus or polyarteritis nodosa, a retrobulbar optic neuropathy can develop from a vasculopathy or a demyelinating process. These mechanisms can only be clinically differentiated on the basis of recovery following treatment (see Sects. 9.3, 9.4).

5.10 Rare Causes of Ischemic Optic Neuropathy

Rare causes of ischemic optic neuropathy include collagen vascular diseases (see Chap. 9), migraine (see Chap. 10), emboli (see Chap. 8), and radiation microvasculopathy (see below).

5.11 Radiation Damage to the Eye and Brain

5.11.1 Introduction

Therapeutic irradiation of neoplasms in the head and neck areas can compromise the blood vessels that supply the central nervous system and no portion of the visual system is immune to this complication. The beneficial effect of radiotherapy is obtained by the destruction of neoplastic cells which are in a proliferative stage as well as by damaging

Table 5.6. Factors which increase the risk of radiation complications (modified from [251–254])

Age less than 10 years
Underlying vascular disease
  Diabetes mellitus
  Hypertension
  Collagen vascular disease
  Concomitant systemic or intrathecal chemotherapy

the endothelium of the vasculature that provides the tumor blood supply. The latter effect is probably the mechanism by which radiation damages the visual pathway since the neurons do not normally proliferate after birth and are relatively radiation insensitive.

Fractionated focused radiotherapy delivered by a high-voltage linear accelerator through multiple ports with a low daily dose causes fewer complications than if the radiation is administered via a low-voltage machine with a large daily dose. Radiation damage of the normal brain seldom occurs in patients given less than 200 cGy/day or less than 6000 cGy total dose [250]. However, certain factors increase the risk of a radiation complication even when delivered in daily doses not usually associated with these complications (Table 5.6).

The central nervous system damage, loosely termed radionecrosis, typically develops 6–24 months following conventional radiotherapy. Radiation necrosis may develop sooner or acutely when more massive doses of radiation are administered over a shorter time period [255, 256]. In particular, radiation seeds implanted for 12 days in the sella to treat a pituitary adenoma can deliver excessive doses. There are reports that radiation doses of 12,500 cGy [257], 15,000 cGy delivered over 28 days [258], and 10,000 cGy delivered in four courses over a 3.5-year period [259] resulted in damage to the arteries in the arachnoid surrounding the intracranial optic nerves, chiasm, and the adjacent brain.

The severe complications mentioned above do not include the temporary visual dysfunction caused by swelling of the irradiated neoplastic tissue or transient demyelination or edema in the adjacent regions of the brain. In the former case, the neurologic deficits transiently worsen because of compression of the brain or the chiasm in patients with cerebral hemisphere tumors or pituitary adenomas with significant suprasellar extension, respectively [260]. Several days of withholding the radiation and administering corticosteroids rapidly reverses the deficit or the defect may remit spontaneously.

5.11.2 Radiation Retinopathy

Radiation retinopathy, described first in the 1930s [261, 262], is a common sequela of external beam irradiation of neoplasms in the orbit or the paranasal sinuses [263, 264] and external beam and implanted plaques for neoplasms of the retina (retinoblastoma) [265, 266] and choroid (melanoma) [267, 268].

Radiation retinopathy is rare following irradiation of intracranial tumors [269] unless the eyes are not shielded and are included in the therapeutic portals. Some authors have suggested that patients with growth hormone-secreting adenomas are more prone to develop retinopathy, but we have not found this to be true [270]. We have observed no cases of retinopathy at our institution in 150 patients with pituitary adenomas treated by high-voltage focal irradiation (4500–5500 cGy delivered via 3 portals, 2 temporal, 1 superofrontal), in 100 patients treated for various other neoplasms of the intracranial anterior visual pathway, or in 300 patients given 5500–6000 cGy to the whole brain for gliomas.

The retinopathy has all the signs of chronic retinal hypoxia and ischemia from a vaso-occlusive disorder including retinal hemorrhages, cotton-wool patches, microaneurysms, macula edema, and occasionally neovascularization [263–265, 271]. Fluorescein angiography demonstrates leakage from the injured capillaries of the optic disc and the retina and from the areas of retinal neovascularization. Microaneurysms and areas of nonperfusion or occlusion of retinal capillaries are also visualized (Fig. 5.21). Histopathologically, the affected vessels have thickened hyalinized walls and segments occluded by a fibrillary material and thrombus [266, 272].

Visual loss may result from capillary loss, hemorrhage or edema in the macula, occlusion of a branch retinal or central retinal artery, or hemor-

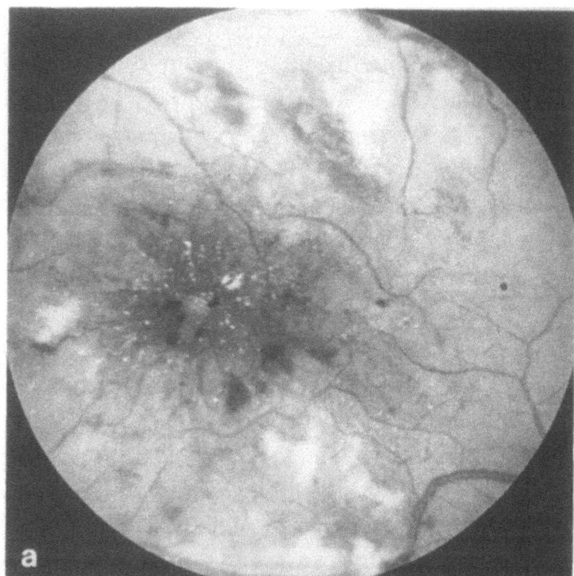

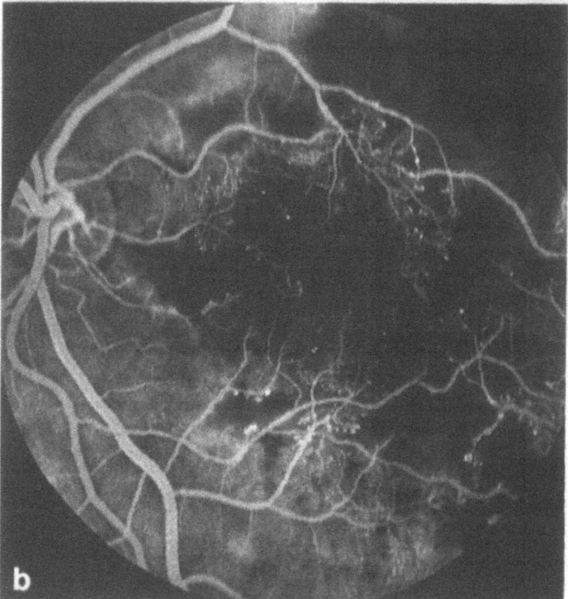

rhage into the vitreous from retinal neovascularization. The ocular hypoxia and ischemia may also lead to neovascularization of the iris and absolute glaucoma. Spontaneous recovery of the retinal circulation and vision is rare [273].

◄ **Fig. 5.21a, b.** A 45-year-old woman with acromegaly and a pituitary adenoma was treated with radiotherapy (5000 rad). Approximately 18 months later, she began to complain of visual difficulty, the left eye being worse than the right. The vision deteriorated over the course of months and at the time of her evaluation the visual acuity was 20/30 OD and 20/400 OS. Ophthalmoscopy of the right eye demonstrated areas of hemorrhages, edema and microaneurysms throughout the posterior pole (not shown).
a In the left eye, the macula region is particularly affected by the changes of radiation retinopathy with marked hemorrhages, edema and exudates, and cotton-wool spots.
b Fluorescein angiography of the macula region demonstrates numerous areas of either blockage or loss of capillaries with microaneurysms. There are multiple areas of fluorescein leakage from the retinal vessels in the area. The patient had an abnormal glucose tolerance test associated with acromegaly, and following transphenoidal adenomectomy the growth hormone normalized, as did the glucose tolerance test. This was followed by marked improvement in the radiation retinopathy in both eyes. (From [273])

5.11.3 Radiation Optic Nerve and Chiasm Neuropathy

Radiation damage of the anterior optic nerve will usually complicate external beam or local ocular plaque irradiation if the dose exceeds 9000 cGy to this area. However, a cumulative radiation dose as low as 3500 cGy may be associated with a neuropathy if the patient is diabetic [272]. Myointimal proliferation causes the narrowing of the posterior ciliary arteries that leads to ischemia of the anterior optic nerve [266]. When the anterior optic nerve is affected, hyperemic disc swelling and peripapillary hemorrhages, exudates, and subretinal fluid are seen on funduscopy in addition to the signs of radiation retinopathy. The optic disc and retina have areas of capillary nonperfusion, as revealed by fluorescein angiography [272]. The hyperemia resolves, the optic disc atrophies, and the macula edema may resorb after several months.

The visual loss ranges from as poor as no light perception to a subtle disturbance with 20/25 acuity, with progressive visual deficits frequently occurring. The vision rarely improves but there are reports of patients whose acuity returned to 20/25 over months even when the vision at presentation was as poor as 20/200. If the eye is blind, the prognosis for visual recovery is practically nonexistent.

Radionecrosis of the orbital portion of the optic nerve is a rare complication following irradiation of malignant paranasal sinus or intracranial neoplasms [264, 274, 275]. In these case radiation doses of 6,500 cGy and greater are usually required to damage this portion of the optic nerve.

Radiation injury to the intracranial optic nerves and chiasm predominantly develops following irradiation of tumors which are in an intrasellar or parasellar location. If the visual loss occurs, it typically develops within 3 years of the irradiation of pituitary adenomas. Rare cases have been described where visual deterioration and cerebral necrosis began 10 years or more after therapy [276]. External beam irradiation to the parasellar region which does not exceed 5000 cGy, delivered over 4.5 − 5.5 weeks, unusually causes this complication [250, 277]. Daily doses of 200 cGy rarely cause visual loss [278]. None of our adult patients with anterior visual pathway tumors who received a total dose of 4500 − 5500 cGy in daily fractions of less than 200 cGy through three portals developed optic nerve or chiasm damage. One patient with hypertension had a central retinal artery occlusion 3 years following irradiation [279]. After daily doses greater than 250 cGy, visual deterioration can develop in patients treated for craniopharyngiomas or pituitary adenomas, regardless of whether the patients have had prior surgery [280].

Optic nerve damage is even rarer when irradiation, in doses up to 6000 cGy, is delivered to the whole brain of patients with gliomas or metastatic lesions. We have seen only one patient among 300 glioma cases who received 5500 − 6000 cGy for treatment of an astrocytoma who had a retrobulbar optic neuropathy. The eye ipsilateral to the neoplasm was affected first; this was followed by a similar problem in the contralateral eye 1 month later that left the patient completely blind. In contrast, radon implants in the parasellar region that delivered massive levels of radiation over a short period of time (now all but abandoned for use in this area) commonly caused visual loss (see Sect. 5.11.1).

The histopathological examination in these cases shows that the arachnoid surrounding the intracranial anterior visual pathway has areas of fibrosis and small vessel occlusions. There are also

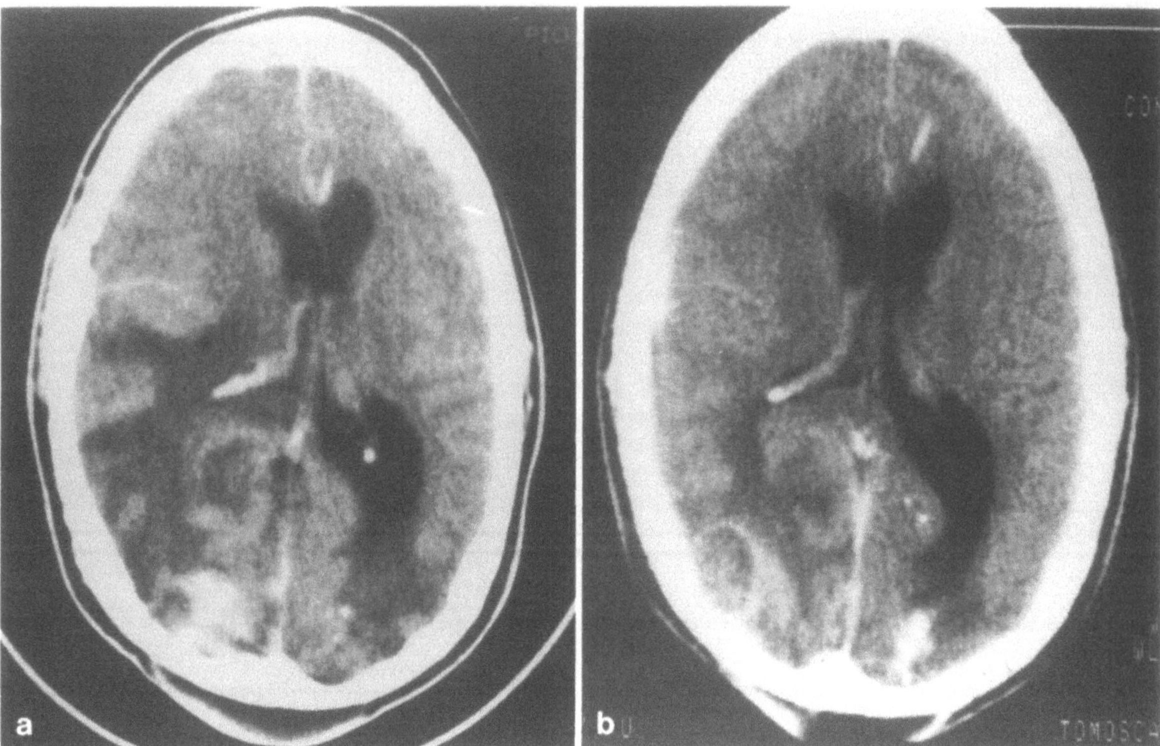

Fig. 5.22a, b. This 21-year-old man underwent radiation therapy followed by chemotherapy for a brainstem glioma. The treatment stabilized his previously progressively deteriorating neurologic course. Beginning approximately 2 years after the completion of therapy, the patient started to have recurrent episodes of occipital ischemia and hemorrhages. The patient was left with a visual acuity of 20/200 in each eye with limited field after multiple episodes over the course of 1 year.

a The noncontrast axial computed tomogram shows an area of edema involving the entire posterior aspect of the right cerebral white matter as well as an area of lucency in the occipital white matter extending to the occipital horn of the ventricular system. On the same side, there is also an area of increased signal as a result of a hemorrhage in the right occipital lobe. There is a mass effect in the right hemisphere. The left occipital lobe also shows an area of low attenuation typical of an infarct.

b Following intravenous contrast, there is enhancement of the left occipital lobe suggestive of a breakdown of the blood–brain barrier. A biopsy of the right occipital mass performed at this time was compatible with a diagnosis of radiation encephalopathy.

areas of demyelination and gliosis in the optic nerves and chiasm [280].

5.11.4 Retrochiasmal Radiation Damage

Radiation necrosis and vasculopathy which affect the optic tracts, lateral geniculate nucleus, optic radiations, or occipital cortex are rarer than the retinopathy or optic neuropathy. In most affected cases the total dose of irradiation has exceeded 6000 cGy. Bilateral occipital damage is even rarer. One of our patients developed cortical blindness from recurrent occipital infarcts and hemorrhages

following 5500 cGy irradiation and chemotherapy treatment for a brainstem glioma (Fig. 5.22). Ischemic damage arises from obliteration of the capillaries and arterioles in the affected regions of the brain [281]. Histological examination of the injured tissue reveals fibrinoid necrosis, endothelial damage, fibrosis of the adventitia, and perivascular lymphocytes [282]. Areas of demyelination and gliosis are seen in addition to the small vessel vasculopathy.

In the rare cases that recover, it is likely that edema (possibly from a local blood–brain barrier breakdown) rather than radionecrosis is the cause. An example of this problem is demonstrated by one of our patients with a midbrain diencephalon glioma who had radiation seeds placed into the rostral brainstem that delivered 22,000 cGy locally. She developed bilateral blindness from possible demyelination or edema of both optic tracts and lateral geniculate bodies (on MR imaging) which improved over months following corticosteroid therapy.

5.11.5 Arterial Occlusive Disease

Radiation also rarely induces occlusive disease of the large intracranial arteries in adults, but children younger than 10 years of age are more prone to this complication. Severe stenosis or occlusion of one or both ICAs leads to a "moya moya" type vascular pattern with enlargement of the lenticulostriate arteries and other small arteries which serve as collaterals to supply the middle and anterior cerebral arteries (see Sect. 8.10). These children have symptoms and findings of ischemia in the affected cerebral hemisphere. This complication can develop with a radiation dose as low as 1000 cGy or as long as 6 years after radiotherapy of an orbital hemangioma in an infant [283]. Thrombosis in the intracranial ICA and occlusion of other cerebral arteries occurs more frequently following irradiation of a chiasmal glioma with 3000–5500 cGy [284–286].

Uncommonly, Premature atherosclerosis can also develop in irradiated major extracranial and intracranial arteries [287]. In some adults, therapeutic irradiation of the cervical area can induce or accelerate atherosclerosis in the large extracerebral arteries, while sparing the intracranial circulation, after a latency of up to 20 years [288]. However, most cases develop atherosclerosis and fibrosis in the irradiated vessels within 5 years of treatment [289, 290]. Severe stenosis and thrombosis in these large arteries are rare [291–294]. Cranial irradiation-induced damage or severe atherosclerosis of the intracranial intermediate-sized arteries is even rarer and probably develops only in adults with a predisposing risk factor such as hyperlipidemia [295]. However, in the future this cerebral vascular complication may be reported more frequently with more widespread application of radiosurgery techniques (Fig. 5.23).

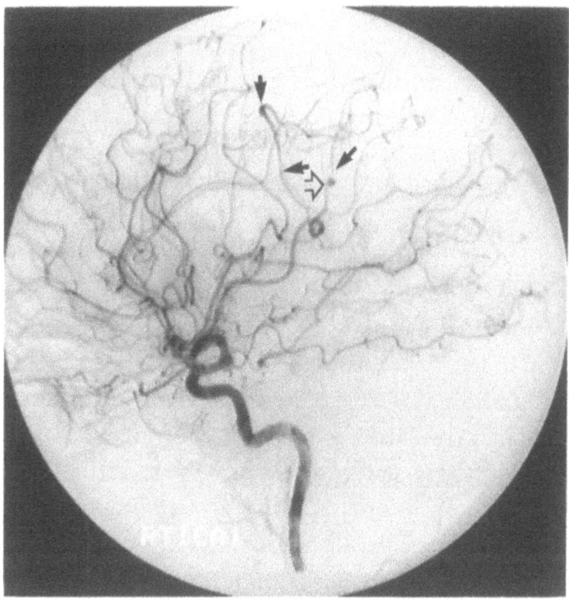

Fig. 5.23. Six months following radiotherapy of a large posterior frontal arteriovenous malformation of the brain with a linear accelerator, progressive symptoms of headache and focal sensory seizures developed. The magnetic resonance scan several months later revealed holohemispheric edema. One year after treatment, the lateral view internal carotid artery subtraction angiogram demonstrated areas of focal stenosis and dilatation of the middle cerebral arteries (*arrows*). An area of ectasia or an aneurysm (*open arrow*) is seen. In this case, this application of radiotherapy thrombosed the arteriovenous malformation but a diffuse vasculopathy resulted

5.11.6 Nonvascular Complications

Not all the complications of therapeutic radiation are secondary to vascular damage. Malignant neoplasms can arise at the site of treatment many years after irradiation. Sarcomas [296, 297], meningiomas [298], and gliomas of the brain [299–301] and chiasm [302] have all occurred within the portals of the radiation.

Nonvascular radiation injuries to the eye include cataract, keratitis, mild loss of eyelashes and eyebrows, and a necrotizing adenitis of the lacrimal gland. Conjunctival injection is typically mild unless significant keratopathy is present [303, 304].

5.11.7 Treatment

Radiation retinopathy typically deteriorates over months, but it may remain stable. Focal laser photocoagulation of the retina can reduce the macular edema [263, 305]. Panretinal laser photocoagulation can induce regression of neovascularization affecting the optic nerve and the retina [306]. Control of factors such as systemic hypertension and diabetes mellitus that accelerate small vessel occlusive disease may be helpful in controlling the retinopathy. In one case with acromegaly, lowering the growth hormone by surgically excising the growth hormone-secreting pituitary tumor normalized the glucose tolerance curve and the radiation retinopathy resolved [273].

In contrast to the retinopathy, no therapy will stabilize or restore the visual loss from damage to the optic nerve or chiasm. Systemic corticosteroids have been unsuccessful except in cases where edema or compression of the normal structures by necrotic tumor or brain caused the visual loss [307]. Successful treatment with hyperbaric oxygen, administered immediately following the onset of visual dysfunction, has been reported in anecdotal cases [308], but others have found no such benefit [309]. Similarly, no therapy exists for capillary occlusion radionecrosis in any area of the brain.

Various therapies have been attempted to reduce or prevent the consequences of large artery damage. Anecdotal reports describe successful treatment of patients who have large artery stenotic lesions and strokes or transient ischemic attacks with warfarin [310] or aspirin [283]. However, the variable course of this complication, which often includes long periods of stabilization, and the small number of cases in each report make analysis of treatment inaccurate. Our case with recurrent occipital infarcts (see Fig. 5.22) was not helped by aspirin, but he seemed to benefit from minidose heparin. Reconstructive surgery of the damaged cervical carotid artery may prevent transient ischemic attacks or cerebral infarction, though surgery of these vessels is a complicated procedure due to the arterial wall fragility, fibrosis of the adjacent tissue, and healing difficulties [311, 312].

CHAPTER 6

Aneurysms Involving the Motor and Sensory Visual Pathways

6.1 General Discussion

6.1.1 Introduction

Intracranial aneurysms can cause patients to present with blurring, decreased vision, diplopia, or ptosis so that ophthalmologic consultation is sought early. However, most patients with saccular aneurysms of the intradural arteries present with focal or diffuse headaches, a stiff neck, and an altered state of consciousness, because of a subarachnoid hemorrhage (SAH). Depending upon the location of the aneurysm and the direction of the rupture, a hemorrhage will develop in the brain parenchyma, the ventricular system, the subarachnoid space, or in some combination of these. The high frequency of severe morbidity and mortality following a hemorrhage and secondary vasospasm makes diagnosis prior to a bleed essential [1]. This is often possible because many aneurysms cause signs and symptoms of sensory visual loss or lid, pupil, or oculomotor dysfunction before rupturing.

Thus, early diagnosis and avoiding misdiagnoses such as tension headache, viral syndromes, migraine, and syncope are critical. Few will fail to recognize the constellation of symptoms of a SAH. In contrast, the nonspecific symptoms of a moderate headache, neck or back pain, and dizziness that develop days to weeks prior to a clinical hemorrhage may not be diagnostic. When present, specific neuro-ophthalmologic signs of visual loss or an ophthalmoparesis can be highly suggestive.

The intraluminal environment of the aneurysm is in a dynamic state because of alterations of local blood flow and thrombus formation and lysis. If untreated most intradural aneurysms will remain unchanged or increase in size. Repeat angiography, performed over a period of 1–9 years, demonstrates a slowly progressive increase in the size in approximately 48% of all unruptured aneurysms [2]. Following rupture, most untreated aneurysms do not become larger in the first 6 months. However, these aneurysms can expand later, and when all ruptured aneurysms are considered anterior communicating artery aneurysms (29%) are the most prone to increase in size following an initial hemorrhage.

Not all intracranial aneurysms (cavernous sinus, petrous regions) are located in the subarachnoid or intradural space, nor are they all associated with a SAH or life-threatening complications. Rarely, patients with saccular aneurysms present with transient focal neurologic symptoms, which are most likely caused by cerebral emboli originating from thrombus within the aneurysm sac [3]. There are also nonsaccular aneurysms which have different clinical presentations and natural history (Sect. 6.9).

6.1.2 Incidence

In general autopsy series, the incidence of developmental saccular or berry aneurysms is approximately 8% and the prevalence is approximately 5% [4]. Aneurysms are slightly more frequent in women than in men. The aneurysms are located in the arteries of the anterior circle of Willis in 90% of cases. In autopsy studies, the most frequent location of aneurysms is in the middle cerebral arteries [4, 5]. This is contrasted with clinical investigations that report the anterior cerebral artery is the most commonly affected vessel [6]. Approximately 25%–30% of patients with a known aneurysm have one or more additional intracranial aneurysms.

Rupture of a saccular aneurysm infrequently occurs in patients younger than the fourth decade. It is even rarer in patients younger than 20 years. This suggests that factors other than a congenital defect in the wall at a branching of the artery are important in the development of a symptomatic outpouching of a cerebral artery. Though systemic

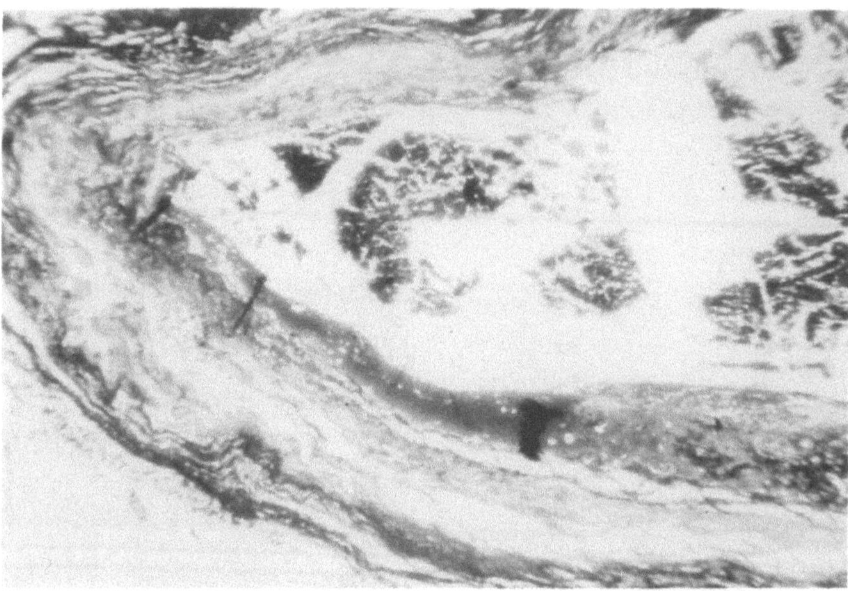

Fig. 6.1. Histological section through an unruptured aneurysm with defective media and fibrinoid changes in the wall

hypertension is commonly considered to be a significant risk factor in the genesis of saccular aneurysms or to contribute to the tendency to rupture, neither hypothesis has been proven [7].

6.1.3 Etiology

There is no known definite single etiology that accounts for the development of all saccular aneurysms. The large majority of arterial aneurysms are of the developmental saccular variety and occur in otherwise healthy patients. Saccular aneurysms typically have absent elastica at the site of rupture and the media is barely discernible in the aneurysm wall (Fig. 6.1) [8, 9]. Recent studies have suggested that patients with ruptured aneurysms may have a deficiency of collagen type III in the cerebral arteries [10, 11] that might predispose to a defect in the arterial media.

There is an increased incidence of aneurysms in patients with the various connective tissue disorders. A defect in the arterial wall media found in patients with fibromuscular dysplasia (see Sect. 8.9) [12], Marfan's syndrome [13, 14], and Ehlers–Danlos syndrome [15] must be a significant predisposing factor. It is difficult to explain why progeria, an extremely rare connective tissue

disorder, is rarely associated with cerebral aneurysms [16]. Cerebral aneurysms are also more frequent in patients with polycystic kidney disease [17, 18] and coarctation of the aorta [19]. Except for the various connective disorders, a familial predilection to develop cerebral aneurysms is rare [20–23, 14].

6.1.4 Rupture

The probability that an intracranial aneurysm will rupture and cause a SAH seems to vary with the location and size of the aneurysm. Generally, moderate to large aneurysms (larger than 1 cm and less than 2.5 cm) have a higher incidence of bleeding than small (2–5 mm) or giant aneurysms (greater than 2.5 cm) [24, 25]. However, both giant and small aneurysms do rupture.

Giant aneurysms are relatively uncommon, making up approximately 5% of all intracranial aneurysms [26]. Because of their large size, giant aneurysms often have a presentation suggestive of a progressive intracranial mass lesion. However, a hemorrhage is the presenting problem in approximately 33% of giant aneurysms (data from two series with 202 cases) [26, 27]. Other studies report an even higher incidence of rupture of these aneurysms [28]. As with all intradural aneurysm bleeds,

Fig. 6.2a–c. Vasospasm that follows a subarachnoid hemorrhage can cause focal or diffuse neurologic defects. **a** An axial noncontrast computed tomogram shows an infarct in the medial frontal region, a territory supplied by the left anterior cerebral artery, in a patient with a ruptured anterior communicating artery aneurysm. **b** Following surgery the same patient developed aphasia, progressive lethargy, and right arm weakness. The frontal view subtraction angiogram shows vasospasm in the left supraclinoid internal carotid (*open arrow*), the A1 segment of the anterior cerebral (*small arrow*), and the middle cerebral artery (*large arrows*). **c** The frontal subtraction angiogram after balloon catheter mechanical dilatation (angioplasty) of the distal internal carotid artery and middle cerebral artery shows a normal size of these vessels. The patient's neurologic status stabilized after this procedure. Three months later the patient was clinically normal

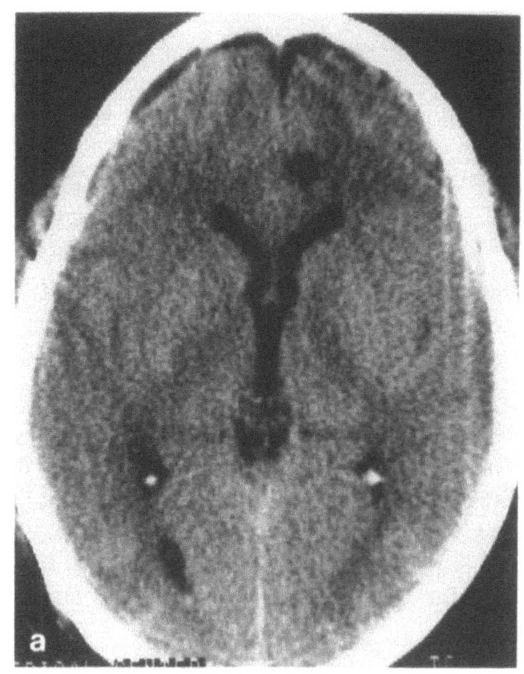

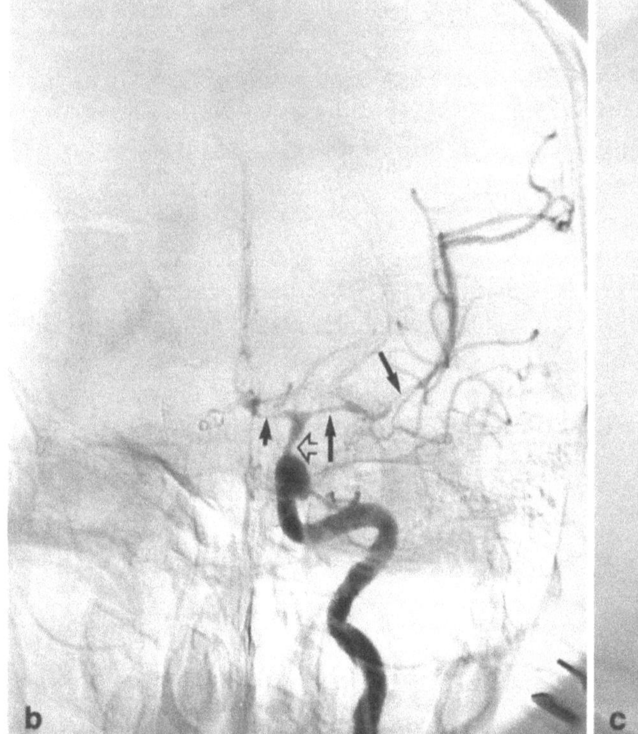

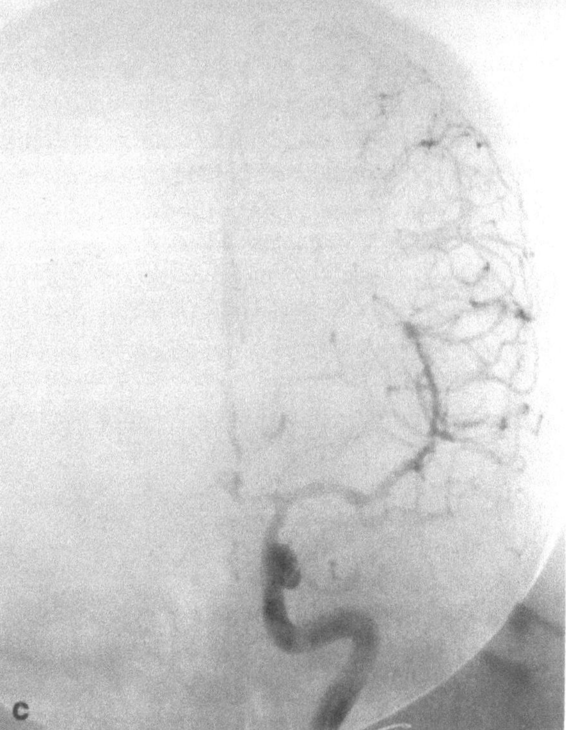

rupture of a giant aneurysm carries a significant risk of causing severe morbidity or mortality. Thus, these aneurysms require therapy prior to rupture if the treatment does not carry a higher risk than the risk of hemorrhage. The treatment risks will vary with the location of the aneurysm (aneurysms of the basilar artery are worse than aneurysms in the anterior circulation), whether the aneurysm wall is actually part of the wall of a large cerebral artery, whether perforators to the brain arise from the aneurysm, and whether an arterial bypass is required.

The overall annual incidence of SAH from all ruptured aneurysms is approximately 9.6–12 per 100,000 [29, 30, 24, 25]. The rate of hemorrhage for all unruptured aneurysms is 1% per year [31].

Vasospasm typically occurs when there is a large amount of subarachnoid blood, and it usually develops within 3–9 days following a SAH. Vasospasm is most often limited to the arteries in areas adjacent to the aneurysm (Fig. 6.2). Angiographic vasospasm can be asymptomatic or it can be associated with neurologic deficits. If multiple arteries develop segments of severe narrowing, the patient can have diffuse or multifocal cerebral dysfunction [32, 33]. Prolonged or severe vasospasm results in cerebral infarcts and permanent dysfunction.

6.1.5 Rebleed

Rebleeding occurs in at least 22% of untreated aneurysms, and 51% of patients who rebleed die [34]. The mortality associated with second and third hemorrhages in patients with either an aneurysm of the anterior communicating or posterior communicating arteries is even higher [35, 31]. After a SAH, the rebleed rate is highest between days 5 and 7. If the aneurysm is not treated, the greatest probability of rebleeding is in the first 6 months following the initial SAH [36a, 36b]. Thereafter, the rebleed rate is 3% per year for all intradural aneurysms. The presence of multiple aneurysms does not increase the late rebleed rate, because the previously ruptured aneurysm is usually the site of rerupture.

6.1.6 Morbidity and Mortality

Following a SAH, approximately 25% of patients will die prior to hospitalization. When these deaths are combined with the deaths and neurologic deficits caused by rebleeds and vasospasm, almost 60% of patients will die or have significant morbidity related to their aneurysm. The patients' outcome can be directly related to the neurologic status at presentation. A moribund patient has a poor prognosis and a patient without a deficit has an excellent

prognosis. Clearly, the best outcome is achieved by those patients who have not had a clinical SAH prior to their treatment.

The overall mortality rate from a ruptured aneurysm has varied from 34% early in the cooperative study (1956–1964) to 46.7% later in the study (1963–1970) [6]. In another study of patients with a SAH, it was found that if the patient was moribund there was a 20% death rate, as opposed to a 9% death rate for patients who were awake and alert [37]. In general, studies from a single institution have reported more encouraging results, possibly reflecting the expertise and interest of the individual neurosurgical service. For example, in a review of 249 cases, Sundt reported 80% of his patients with internal carotid artery (ICA) aneurysms had an "acceptable" outcome with only a 4.4% mortality rate [38, 39].

6.1.7 Neuro-ophthalmologic Symptoms and Signs

The clinical signs and symptoms depend on the location and size of the aneurysm and whether a focal or diffuse hemorrhage has occurred. As discussed previously, giant aneurysms often present as a mass. The symptoms and findings can remain static, gradually worsen from slow expansion, or suddenly deteriorate because of rapid expansion or thrombosis within the aneurysm. The nature of the defect obviously depends on which adjacent neural structures are compromised.

6.1.7.1 Cranial Nerve Pareses

Unilateral cranial nerve pareses typically develop because of compression by critically located large and giant aneurysms of the arteries of the circle of Willis and the vertebral arteries (Table 6.1). Paresis of the tongue, vocal cord, and swallowing can be caused by a vertebral aneurysm. An isolated fourth nerve paresis is almost never caused by an aneurysm. However, fourth nerve dysfunction can develop, in association with varying degrees of third, sixth, fifth, and sympathetic innervation disturbances when an aneurysm is located in the cavernous sinus. An isolated paresis of the sixth nerve can be a false localizer that results from increased intracranial pressure (ICP) or an inflammatory men-

Table 6.1. Aneurysms and cranial nerve pareses

Arterial location	Cranial nerves affected	Frequency
Vertebral	IX, X, XI, XII	+
Basilar	III[a]	+ + +
	VI	+
Posterior communicating	III	+ + + +
Cavernous internal carotid	III, IV, VI sympathetic[b]	+ + + +
	V[c]	+

[a] Unilaterally or bilaterally.
[b] In various combinations.
[c] Sensory, V 1 > V 2.

ingitis caused by subarachnoid blood. When caused by an aneurysm of the cavernous or petrous ICA, a sixth nerve paresis often occurs with some element of a Horner's pupil or slight ptosis, and in most cases there is some degree of third, fourth, or fifth nerve dysfunction. A Horner's syndrome with ipsilateral pain and no other cranial dysfunction is a nonspecific syndrome, sometimes called Raeder's syndrome, which can occur with a lesion in the petrous ICA.

The aneurysm most commonly associated with a unilateral paresis of the third nerve is the posterior communicating artery aneurysm (see Sect. 6.3). A third nerve palsy, in isolation or in conjunction with other cranial nerve dysfunction, can also be found in patients with a cavernous carotid aneurysm. Unilateral or bilateral pareses of varying degrees occur in patients with a basilar aneurysm. The third nerve can also be compressed by a temporal lobe hematoma with uncal herniation following a hemorrhage from a middle cerebral artery aneurysm. Though a selective paresis of the superior or inferior divisions of the third nerve are suggestive of a cavernous sinus lesion, midbrain lesions have caused such dysfunction and in one case a diabetic third nerve paresis only affected the superior branch [39a].

A misdirection syndrome can develop in a paretic third nerve following clipping of a posterior communicating artery aneurysm (see Sect. 6.3.4). A similar misdirection syndrome can also occur in patients who have a longstanding third nerve paresis caused by an aneurysm of the ICA or neoplasm in the cavernous sinus, even if the lesion is not treated [40, 41]. Since a misdirection syndrome rarely develops in patients with a microvascular ischemic third nerve paresis [41a] (see Sect. 8.8.2.1), this suggests that compression rather than an infarct of the nerve is the mechanism of this phenomenon.

Ptosis of the upper lid develops when the third nerve is compressed by an aneurysm in any location. The lid dysfunction ranges from a mild ptosis of 1 – 2 mm to a severe ptosis with a complete loss of levator function. The ptosis is unilateral except when a basilar artery aneurysm is present.

Sympathetic dysfunction is suspected when the upper lid ptosis is mild (approximately 2 mm), the levator function is normal, and the lower lid is ptotic upward (because of Mueller's muscle weakness), decreasing the amount of sclera observed between the lid margin and the lower limbus. The enophthalmos of a Horner's syndrome is only apparent, not real, and results because of the position of the lids. An associated miotic pupil in Horner's syndrome that results from a cavernous carotid aneurysm may not be immediately obvious if there is concomitant mydriasis from a third nerve paresis. A failure of the pupil to dilate in the dark and an absence of a dilation response to topical cocaine or hydroxyamphetamine are indicative of a third-order defect in sympathetic pupillary innervation. Topical phenylephrine 0.5% will dilate this Horner's pupil, because there is denervation supersensitivity of the iris dilator muscle [42]. Additionally, the upper lid ptosis associated with sympathetic dysfunction may be masked by levator weakness from a third nerve paresis.

6.1.7.2 Visual Loss

Giant aneurysms of the supraclinoid or infraclinoid ICA, the anterior cerebral artery, and rarely the basilar artery can present because of symptoms caused by compression of the intracranial anterior visual pathway and the pituitary gland and stalk [43]. The displacement of the optic apparatus may be asymptomatic until a critical amount of distortion develops.

The visual field defects are most often asymmetric, affecting one eye more than the other (see Sect. 6.4.2.2). This is unlike the classic bitemporal hemianopia caused by midline compression of the chiasm by suprasellar extension of a pituitary adenoma. However, the field defects caused by aneurysms cannot always be distinguished from the field loss caused by neoplasms. The visual loss caused by tumors or aneurysms may or may not be associated with headache or pain that is referred to one or both eyes or orbits. Aneurysm- related visual loss usually worsens over months to years. However, the vision may fluctuate, either worsening or improving, possibly because of thrombosis and dilatation of the aneurysm. Sudden visual loss, rarely due to rupture of an aneurysm, results if there is an arterial hemorrhage into the optic nerve or chiasm [44, 45]. Even when aneurysm expansion seems to have caused sudden visual loss, the presence of optic atrophy in most patients (when first examined) clearly demonstrates that a chronic disturbance has been present, though unnoticed until the rapid change.

In contrast to most aneurysms, most neoplasm-related visual loss tends not to fluctuate, and progressive worsening is typical except under the following specific conditions. Hemorrhage or infarction in a pituitary adenoma can cause a dramatic sudden drop in vision that can improve as necrosis of the tumor relieves the compression of the anterior visual pathway. Expansion of a cyst in a craniopharyngioma can also suddenly compress the optic pathway. A rapid loss of vision in pregnant women that improves post partum can arise from a meningioma involving the anterior visual pathway. However, despite the spontaneous improvement, gradual visual loss typically recurs months to years later from growth of the meningioma.

6.1.7.3 Terson's Syndrome

Another frequent cause of unilateral or bilateral visual loss results from hemorrhage into the retina, subhyaloid space, and vitreous. This can occur with a SAH from an aneurysm located in almost any intradural location [46–48]. An anterior communicating artery aneurysm is the most frequent aneurysm associated with this complication, called Terson's syndrome. It is rare following a traumatic SAH. Patients with a subhyaloid hemorrhage have a higher mortality rate (rising to approximately

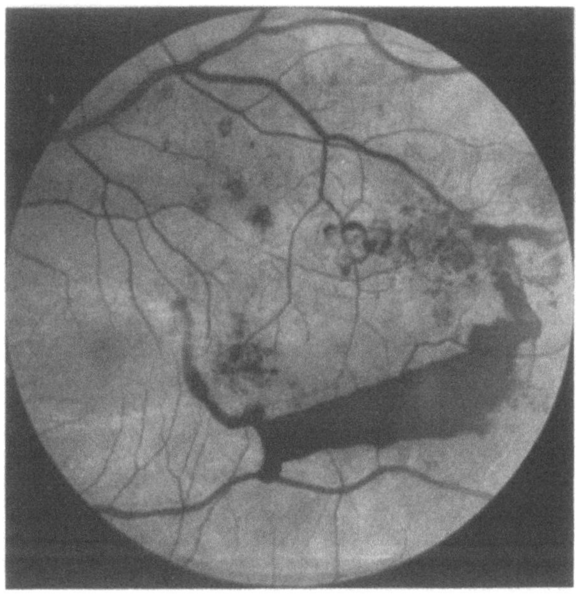

Fig. 6.3. Terson's syndrome can result from the subarachnoid bleed caused by rupture of almost any intradural aneurysm. Both deep and superficial hemorrhages are seen in the retina of this patient who had a subarachnoid hemorrhage. A layered, flat topped subhyaloid hemorrhage is seen inferior and temporal to the macula

50%) than those without a retinal hemorrhage [49–51].

The ophthalmoscopic appearance of the blood depends on the layer of the retina involved (Fig. 6.3). If the blood tracks along the nerve fiber layer, it appears flame-shaped. The blood may dissect between the posterior vitreous face and the inner layer of the retina [52], or between the inner limiting membrane and the nerve fiber layer [53] to form a red mass on the surface of the retina. When the collection of blood is larger, the potential space enlarges, causing a preretinal mass of layered blood. If the posterior vitreous face remains intact and the patient is upright, the top of the bleed may flatten giving a boat-shaped appearance. The blood often covers the underlying retina in an area of the posterior pole and if it blocks the macula the central visual loss is profound. The blood can also rupture into the vitreous and the resulting visual loss is diffuse. Rarely, concomitant hemorrhage into the orbit also results in exophthalmos [54].

Because the gravely ill patient with a SAH may not be examined through a dilated pupil, a large retinal hemorrhage can be missed. The view of the fundus may also be obscured by vitreous blood [55]. Although a delayed onset of Terson's syndrome may suggest another SAH [56], most patients with a latency prior to a vitreous hemorrhage have not rebled. The delayed vitreous hemorrhage most likely occurs in patients who have a mass of subhyaloid blood that ruptures through the posterior vitreous face after several days.

There are several possible mechanisms to explain how a SAH can cause an intraocular bleed. The first explanation is that, following a SAH, the sudden and rapid rise of the intracranial pressure is transmitted in the subarachnoid space of the optic nerve to compress the outflow of the central retinal vein and retinochoroidal venous anastomoses [57]. The central retinal vein and retinal veins dilate and rupture, resulting in the retinal hemorrhage. Another theory postulates that the blood in the subarachnoid space passes down the optic nerve sheath, which is continuous with the posterior portion of Tenon's capsule. The subarachnoid blood enters the eye through the potential space around the optic nerve while the nerve is in the scleral canal. However, the pathology does not confirm this hypothesis since most of the blood is found in the subdural space, rather than in the subarachnoid space of the optic nerve [58]. Also, retinal hemorrhages are frequently observed in patients without SAH who have a sudden elevation of the intracranial pressure [59]. A third hypothesis suggests that the retinal venous hemorrhages develop because the increased intracranial pressure causes intracranial and orbital venous hypertension and impairs the venous outflow from the orbit [60]. This mechanism seems less plausible, since the ophthalmic venous system is usually well decompressed through the collaterals such as the facial and pterygoid veins.

In most cases, the intravitreal blood clots, settles inferiorly, and is eventually resorbed, usually over a 3- to 12-month period. Since there are no associated retinal tears or holes, once the vitreous clears, the vision normalizes, except in the rare case with damage to the macula [51]. Vitrectomy, which is performed to surgically remove the intraocular blood, is rarely necessary if only one eye is affected. If the patient is functionally blind because of bilateral Terson's syndrome or the eye without a hemorrhage has diminished vision, an early pars plana

vitrectomy of at least one eye is indicated [52, 61, 62].

6.1.7.4 Pituitary Dysfunction

Symptoms typical of hypopituitarism, including loss libido, easy fatiguability, amenorrhea, loss of pubic hair, polyuria and polydypsia, gynecomastia, obesity, and impotence, can result if cavernous or intradural ICA aneurysm expands medially into the pituitary fossa and compresses the hypophysis or the infundibulum [63]. Though an aneurysm can cause diabetes insipidus, an untreated pituitary adenoma does not. The intrasella extension of an ICA aneurysm was probably described first in 1889 [64]. Unless angiography or magnetic resonance (MR) imaging is performed, an intrasella mass can be mistaken for a pituitary adenoma. Both masses can appear similar on computed tomography (CT) and both can erode the sella floor and expand the pituitary fossa. Obviously, transphenoidal surgery in the case of an aneurysm can lead to a disastrous outcome. On occasion, an intradural ICA aneurysm can also erode downward into the cavernous sinus to compromise one or more cranial nerves. These cases may be differentiated from those with aneurysms of the cavernous ICA who have pain and cranial nerve dysfunction as an earlier clinical presentation (see Sect. 6.2).

6.1.8 Neuroimaging

6.1.8.1 Computed Tomography

CT or MR imaging or both are the first imaging studies performed in patients with either a chronic or an acute onset of symptoms. An aneurysm must be large or the imaging slice must serendipitously include the plane of the aneurysm in order for the aneurysm to be demonstrated. However, the clinical history and neurologic signs should direct the neuroradiologist to the appropriate area for additional scan images. On the noncontrast CT, partial thrombosis or clot in an aneurysm lumen will appear white to gray, depending on whether the clot is acute or chronic. Calcification in the wall of a giant aneurysm can also be demonstrated. Following intravenous contrast material, with a giant aneurysm the patent portion of the lumen becomes

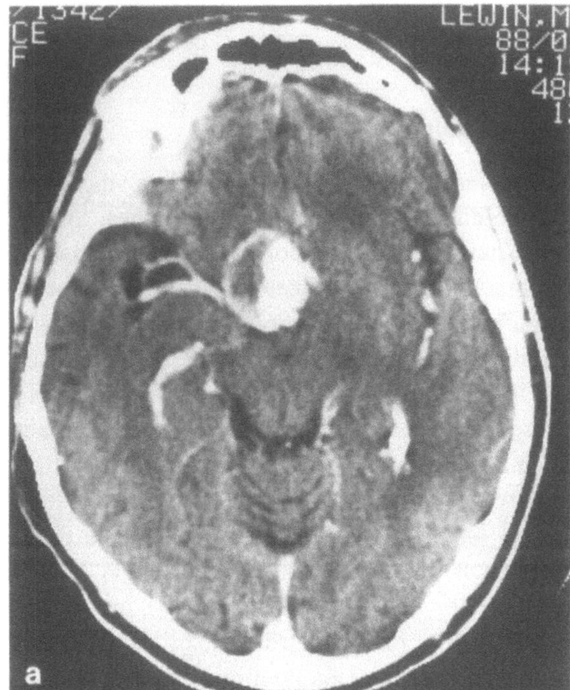

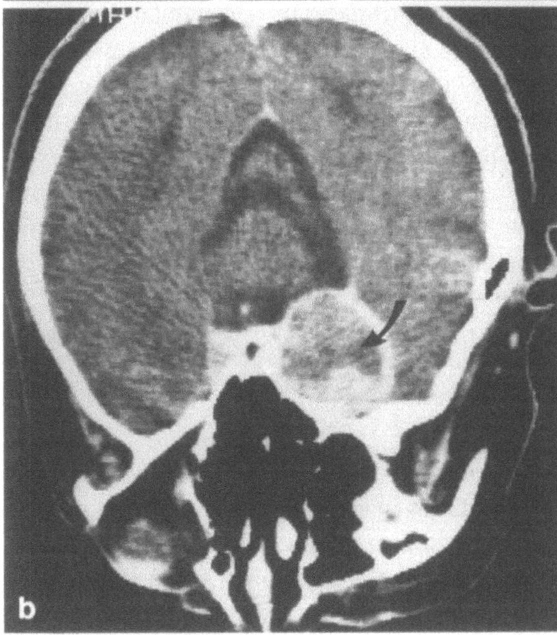

Fig. 6.4a, b. Computed tomography remains helpful in cases of giant aneurysms. a Axial view contrast-enhanced computed tomogram demonstrates areas of thrombus (*gray*) and bright white enhancement representing the patent lumen of a suprasellar bifurcation aneurysm (see Fig. 6.25). b Axial projection of a contrast-enhanced computed tomogram in a case of carotid cavernous aneurysm demonstrates a "target" sign. The *curved arrow* points to the area of thrombus within the aneurysm. Immediately anterior to the darker thrombus is the enhanced lumen. The wall and adventitia of the aneurysm have a bright white appearance due to enhancement

aneurysm is present, compression of the adjacent neural structures can also be visualized on CT.

Following an acute SAH, blood is often demonstrated in the subarachnoid space and in the basilar cisterns for approximately 48 h. When the blood is limited to a focal area, it suggests the ruptured aneurysm is in the same locale. This can be particularly helpful when the aneurysm has caused no focal neurologic signs or in cases where multiple aneurysms are present. For example, blood over a cerebral hemisphere, in the interhemispheric fissure, or in the posterior fossa suggests a ruptured aneurysm in the middle cerebral artery, anterior communicating artery, or a posterior fossa artery, respectively. When present, the demonstration of blood in the subarachnoid space is adequate evidence of a SAH so that an examination of the cerebrospinal fluid (CSF) by lumbar puncture may not be essential.

6.1.8.2 Magnetic Resonance Imaging

As with CT, MR imaging will visualize the aneurysm if the mass is large or the plane of the image passes through the sac (see the figures for individual cases). The multiplane capability, detection of flowing blood, demonstration of the relationship of the aneurysm to adjacent neural structures and the age of thrombus in the aneurysm lumen, and elimination of bone artifact are all reasons why MR imaging is superior to CT in the evaluation of a patient with an aneurysm. MR imaging readily distinguishes an aneurysm from other mass lesions. Flowing blood appears dark on both T1- and T2-weighted studies (Fig. 6.5). When partial thrombosis is present, the clotted area appears white on both T1- and T2-weighted studies (Fig. 6.6). An aerated anterior clinoid may simulate

bright white (Fig. 6.4a). The clot appears darker in comparison and the wall of the aneurysm enhances. The combination may give an appearance of a target (Fig. 6.4b) [65]. If the CT is performed more than half an hour after contrast infusion, or a low dose of contrast medium is administered, or if there is poor cardiac output, enhancement of the aneurysm may be minimally present. When a giant

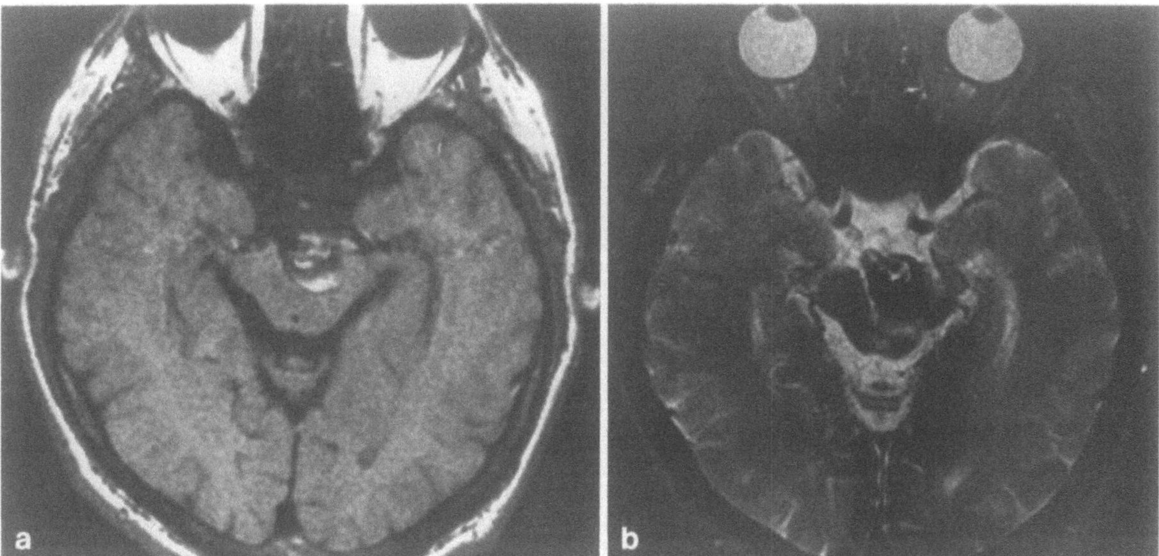

Fig. 6.5 a, b. In most cases magnetic resonance imaging is superior to computed tomography in diagnosing an aneurysm and determining the relationship of the aneurysm to adjacent neural structures. **a** In this case a basilar artery aneurysm invaginates into the midbrain. The T1-weighted image contains areas of low to high signal (black, gray, and white appearance) representing different aspects of blood flow and thrombus. In this situation, the *white* represents an area of acute clot (probably secondary to methemoglobin) and the *black area* represents flowing blood in the lumen. Note the flow artifact on a horizontal plane to the aneurysm. **b** In the same individual, a T2-weighted axial view of the dorsal portion of the aneurysm is black because of thrombus (probably intracellular methemoglobin) within the lumen. The stages of intraluminal thrombus cannot be directly compared to the stages of an intraparenchymal hematoma on magnetic resonance imaging (see Sect. 7.1.9.2)

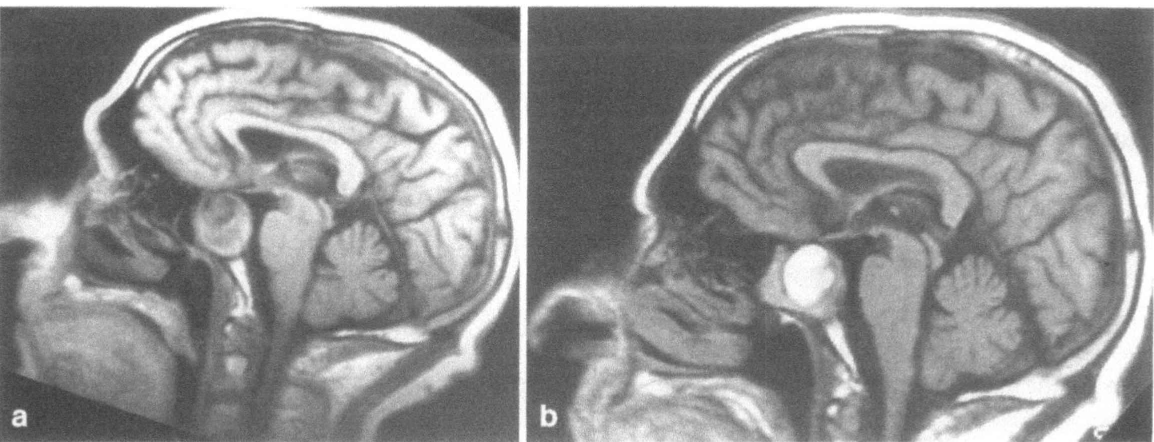

Fig. 6.6. a The T1-weighted sagittal view of this carotid cavernous aneurysm demonstrates an area of peripheral old thrombus posteriorly which has a white–gray appearance representing layering from clot of various stages of thrombosis and lysis. Note the central lumen which has a darker gray appearance because of slow flow. **b** Following closure of the carotid artery at the level of the aneurysm by balloon embolization, the T1-weighted image (similar white appearance on T2-weighted images) shows acute white thrombus (probably secondary to extracellular methemoglobin) in the previously patent lumen

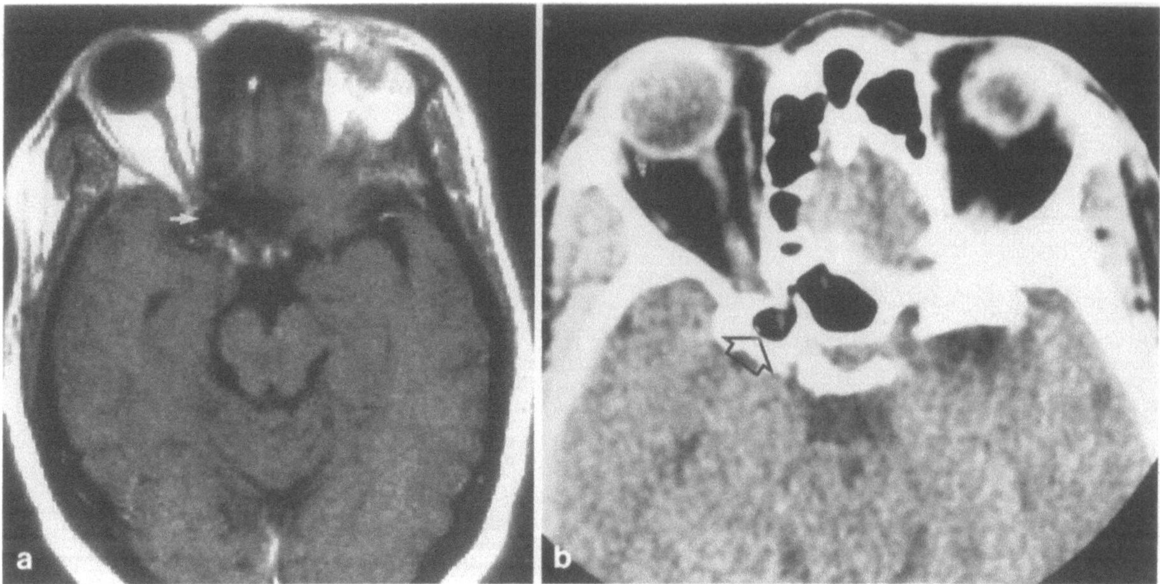

Fig. 6.7a, b. Similar to flowing blood, air appears black on both T1- and T2-weighted magnetic resonance images. **a** The axial T1-weighted image (also on the T2-weighted image) demonstrates what appears to be a possible ophthalmic artery aneurysm (*white arrow*) adjacent to the right optic nerve. **b** A computed tomogram in the same patient demonstrates that the possible flow void was not a vascular lesion, but an aerated anterior clinoid (*open arrow*). In this location a pneumatized area of bone should not be mistaken for signal void of flow

an ophthalmic artery aneurysm because the air appears dark on both sequences (Fig. 6.7). The laminar flow pattern, with slower flow near the aneurysm wall, and the more rapidly flowing blood in the center of the lumen can be seen. In some cases the aneurysm can appear thrombosed, but further investigation with various pulse sequences and rephasing techniques will reveal persistent blood flow in the aneurysm lumen. Although intravenous gadolinium passes through the patent lumen of the aneurysm, the blood flow through the aneurysm lumen often causes a flow void. In some cases, however, the gadolinium does enhance a portion of the aneurysm. Following the infusion of gadolinium, a fast field echo sequence enhances MR imaging's capability to demonstrate blood flow. MR imaging is also the best noninvasive modality to demonstrate enlargement, shrinkage, and thrombosis of the aneurysm mass.

Aneurysm Clips. Metal aneurysm clips cause artifacts in both the CT and the MR image. The beam-hardening effect of the clip causes scattered white line artifacts radiating from the region of the clip on CT. Clips that exhibit considerable ferromagnetism cause the most artifacts and carry the potential hazard of movement or displacement by MR imaging [66, 67]. Slight bending of a phantom image can be seen some distance from the clip with spin-echo or inversion recovery sequences. Artifactual bright signal can be visualized next to the clip. The nature of the artifact also depends upon the pulse sequences, the direction of the frequency-encoding gradient, and the direction and strength of the magnetic field [68]. Though nonferromagnetic metal clips also cause MR artifacts, these distortions are less significant.

6.1.8.3 Cerebral Angiography

Cerebral angiography of both the ICA and vertebral arterial systems is required in patients suspected of having a ruptured aneurysm. Angiography must demonstrate the position of the aneurysm and its neck as well as whether there are additional aneurysms. The angiogram should also provide information concerning the normal cerebral circulation, which is required in order to plan therapy. The goals of the therapy which need to be considered during angiography are: the exclusion of the aneurysm from the circulation, prevention of a rebleed or a primary bleed, decompression of the involved neural structures, reduction of arterial pulsations on neural structures, and preservation of the arterial input to the normal structures. The relationship of the aneurysm and its neck to the parent vessel, other arteries, the brain, and the cranial nerves must be determined. Angiography performed with flow arrest and a balloon catheter allows a controlled leak of the contrast material. This permits a more exact determination of the location of the entrance and exit of the parent vessel from the aneurysm. A complete angiogram must determine whether the parent artery seems to provide the sole or dominant blood supply to an area of the brain, whether there are competent collaterals to the same territory, and whether vasospasm is present in the arteries adjacent to a ruptured aneurysm.

Angiography should also demonstrate whether the contralateral ICA or the ipsilateral vertebral artery will spontaneously fill the anterior and middle cerebral arteries ipsilateral to the aneurysm via the anterior communicating artery or the posterior communicating artery, respectively (Fig. 6.8). Temporary manual compression of the injected cervical ICA decreases the washout effect on the contrast column passing from the injected contralateral ICA. This maneuver will also facilitate the flow of contrast material from the injected ICA into the anterior and middle cerebral arteries ipsilateral to the compression [69]. Injecting the vertebral artery and compressing the ipsilateral cervical ICA will demonstrate whether the posterior communicating artery can provide significant blood supply to the ipsilateral cerebral hemisphere (Heubner maneuver). When the aneurysm is located in the posterior fossa, an ICA injection of contrast will show whether the posterior communicating artery will fill the basilar artery.

In addition to establishing the existence of anatomical competence, angiography can also be used to provide functional information about the status of the cerebral circulation. For example, if closure of a major artery is contemplated, temporary occlusion of the vessel with an intra-arterial balloon will test the patient's ability to function without this artery (see Sect. 6.2.4.3 for discussion of the

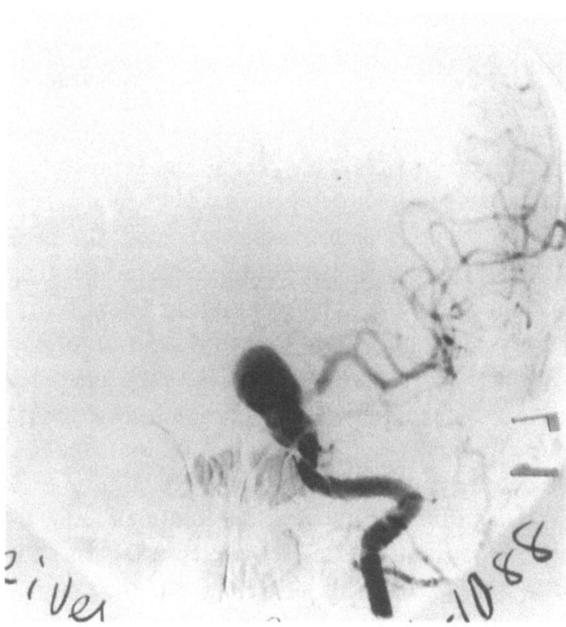

Fig. 6.8. Angiography reveals the functional status of the circle of Willis in addition to the anatomic details of the aneurysm. The frontal view left internal carotid artery angiogram in a case of an ophthalmic artery aneurysm demonstrates a lack of cross-filling to the contralateral middle cerebral artery despite manual compression of the cervical internal carotid artery (ICA) contralateral to the injection. Note the absence of filling of the anterior cerebral arteries with injection of the left ICA. The right ICA angiogram (not shown) filled both anterior cerebral arteries via a right azygous anterior cerebral artery, but no other left cerebral vessels were opacified. Thus, this patient would be unable to tolerate occlusion of the left ICA as a treatment for the aneurysm. (case corresponds to Fig. 6.21; see Fig. 6.23d for example of spontaneous cross-filling)

occlusion test). A second technique, the injection of sodium amytal into the ICA, can be helpful in determining whether the ipsilateral hemisphere is dominant [70]. The superselective injection of amytal may be used to test whether closure of the tested vessel will cause dysfunction [71]. If no neurologic deficit develops following the injection of 50–75 mg, then closing the artery can be safely performed. If a patient has poor collateral blood supply or fails the tolerance test, a surgical procedure that creates a collateral blood supply to the region through an arterial connection distal to the occlusion is sometimes required prior to direct treatment of the aneurysm.

6.1.9 Cerebrospinal Fluid

A lumbar puncture is performed in any case where a SAH is suspected and CT or MR imaging fails to demonstrate blood in the subarachnoid space, ventricles, or brain. The opening pressure of the cerebrospinal fluid (CSF) is recorded while the patient is reclining with the legs extended, the neck straight, and the head at the same level as the spine and breathing normally (not during Valsalva maneuver or breath holding). The presence of bloody CSF that fails to become clear in successive tubes and xanthochromia (yellow coloration) in a centrifuged specimen (in a subject who has not had a prior traumatic lumbar puncture) are diagnostic of a SAH. If a lumbar puncture is performed immediately following a clinically suspected SAH, the CSF may not appear xanthochromic. Xanthochromia develops within 4–6 h of a SAH because the oxyhemoglobin released by lysis of erythrocytes forms methemoglobin. In contrast, crenation of the erythrocytes develops immediately upon contact with the CSF [72, 73]. If the blood remains localized to the adventitia around the aneurysm, or the amount of hemorrhage is minute, blood might not be found in the CSF removed via the lumbar puncture. Since blood is an irritant to the meninges, the CSF protein concentration and white blood cell count are frequently elevated above what is expected from the blood alone [74]. The blood also disrupts the glucose transport across the meninges, causing the CSF glucose concentration to be low.

6.1.10 Therapy and Course

Therapy is directed at preventing a rupture or rebleed and decompressing involved neural structures. Surgical clipping prior to rupture is the ideal treatment. Once the aneurysm ruptures, medical intervention must block or reverse vasospasm and reduce the risk of further hemorrhage in order to keep the patient stable until surgery.

6.1.10.1 Medical Treatment

Antifibrinolytic agents and systemic hypotension have been considered beneficial in decreasing the immediate rebleed rate [75–77] by some, but not all authors. Secondary hydrocephalus may be more frequent without antifibrinolytics because of a delay in the normal breakdown of subarachnoid blood. There is also a higher incidence of thrombophlebitis and pulmonary emboli in patients given antifibrinolytic agents.

In addition to the direct effects of a SAH, the narrowing of the cerebral arteries from vasospasm causes significant morbidity from the resulting cerebral ischemia. Many treatments attempting to prevent and reverse vasospasm have had limited success. Calcium channel antagonists such as nimodipine have been recently shown to reduce the frequency of this complication [78–80]. When possible, early surgical clipping of the aneurysm followed by intravascular volume expansion, hemodilution, and medically induced mild hypertension can decrease symptomatic vasospasm [81, 82]. Technical advances have led to the use of percutaneous mechanical angioplasty of the symptomatic stenotic arteries [82a]. The affected vessels can be dilated, but to be of benefit the procedure is seen only when performed prior to cerebral infarction (Fig. 6.2b).

6.1.10.2 Surgery

Optimally, after careful microsurgical dissection, the neck of the aneurysm is occluded by a metal clip with spring closure. The position and size of the aneurysm and neck, whether the parent vessel and aneurysm can be separated, and whether perforating arteries arise from the aneurysm will influence the probability of completely excluding the aneurysm from the circulation with preservation of the parent artery. Early surgical clipping of the ruptured aneurysm is effective in preventing a rebleed [83, 84], but complications such as vasospasm have limited the successful outcomes. Clipping of the aneurysm will allow more intense therapy in the treatment of vasospasm. In rare cases, the giant aneurysm with hard intraluminal thrombus causes a symptomatic mass effect on neural structures requiring excision of the aneurysm as well. In most cases, the exclusion of the aneurysm from arterial circulation, which ends the pulsation and mass effect, is sufficient.

Surgical clipping that does not completely exclude the aneurysm neck and sac from the arterial

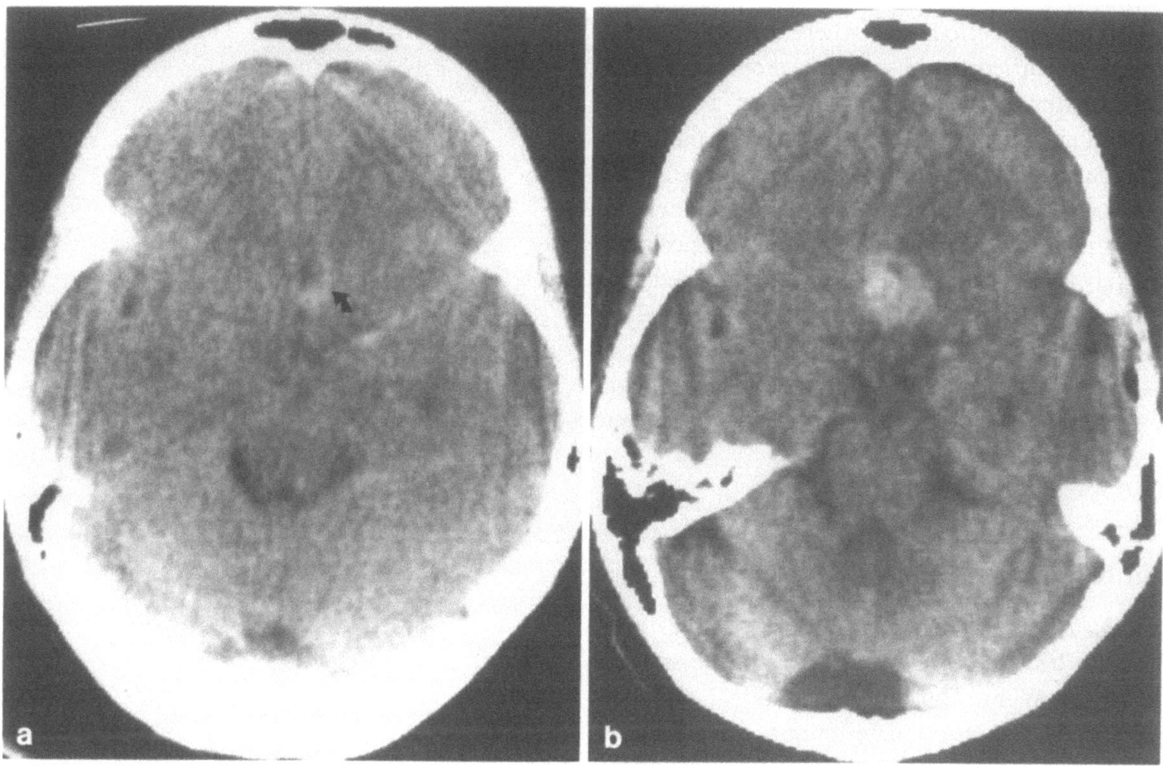

Fig. 6.9. a The axial plane noncontrast computed tomogram shows an area of intravascular increased signal which was shown to be an anterior communicating artery aneurysm (*curved arrow*). The aneurysm could not be clipped safely and it was wrapped and an acrylic coating was placed. **b** A follow-up noncontrast scan 6 months later showed marked enlargement of the aneurysm. At this time, the patient was experiencing progressive visual field loss typical of chiasmal compression and a left optic neuropathy (corresponds to Fig. 6.27)

supply may not prevent a rupture from the remaining portion of the aneurysm. In this situation, repeat angiography must be utilized to demonstrate whether the unclipped segment will thrombose (cure) or increase in size (increased risk of rupture). CT or MR imaging are frequently not useful to visualize the local region of the clip because of metallic artifacts (see Sect. 6.1.8.2).

6.1.10.3 Wrapping and Coating

If an aneurysm cannot be clipped or is not completely obliterated, measures to prevent further expansion of the aneurysm sac or rupture have included procedures that might reinforce the aneurysm wall. Gauze or muslin wrapping or acrylic coating may or may not prevent recurrent bleeding [85, 86] or aneurysm expansion (Fig. 6.9).

The wrapping materials can cause complications by inducing a delayed inflammation. Symptoms develop when the inflammation affects the adjacent neural structures. One striking example is the visual loss that occurs when the intracranial anterior visual pathway is involved. Well-documented cases of progressive optic nerve or chiasm dysfunction that developed approximately 1–18 months after wrapping have been reported [87–90]. The wrapping materials can actually cause an optochiasmatic arachnoiditis [91]. When a clip artifact is present on CT, it is difficult to visualize any area of increased enhancement, suggestive of inflammation. Repeat angiography may demonstrate narrowing of an adjacent artery, possibly from a secondary vasculitis [92]. In some cases systemic corticosteroids can restore the vision. If steroids are not effective, surgical lysis of the adhesions and excision of an inflammatory granuloma (when present) can either stabilize or improve the vision. Cyclophosphamide treatment for 6 months was effective in a patient in whom surgical intervention failed and who became steroid toxic [91]. We have observed one curious case where the patient developed a corticosteroid-sensitive progressive optic neuropathy over 1–6 months after a posterior communicating artery aneurysm was partially clipped and wrapped. Arteriography showed focal stenosis of the ICA and the proximal anterior cerabral artery that had not been present postoperatively. Following 1 month of gradually reducing doses of prednisone, the acuity in the affected eye improved to 20/25 from 20/400. The wrapping of aneurysms near the optic apparatus is probably unwarranted, particularly since there is no concrete evidence that it is effective.

6.1.10.4 Cervical Carotid Occlusion

Occlusion of the cervical carotid artery has been used to treat inoperable aneurysms in the anterior circle of Willis. Current microsurgical and percutaneous embolization techniques have reduced the use of ligation and clamping procedures of the ICA or common carotid artery. In fact, when carotid occlusion is indicated, balloon embolization has essentially replaced the surgical approach to closing the ICA. Nevertheless, a great deal has been learned from surgical occlusion procedures. Ligation of the common carotid artery acutely decreases the distal blood pressure and can induce thrombosis within an aneurysm located in the proximal anterior circulation [93, 94]. However, the blood flow that ordinarily cross-fills, from the contralateral ICA, may maintain the blood flow and pressure within a carotid aneurysm located in the distal segment of the ICA near the bifurcation. Thus, ligation of either the common carotid artery or the ICA may be a temporary therapy and not a definitive cure.

Occlusion of the ICA seems to be superior to cervical common carotid artery occlusion for inducing thrombosis of aneurysms in the ICA. Tindall reported that, following cervical carotid artery ligation, repeat angiography between 1 day and 48 months later showed the aneurysm did not refill in 20%, decreased in size in 45%, was unchanged in 24%, and enlarged in 2% of the patients [95]. In contrast, Mount and Taveras reported that after ICA ligation, 80% of the aneurysms did not refill

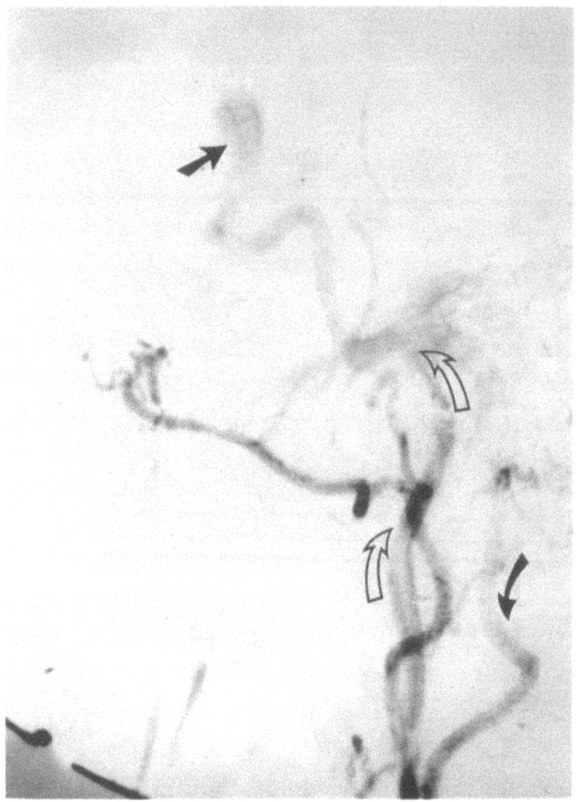

Fig. 6.10. Twenty years after a common carotid artery ligation for a ruptured posterior communicating artery aneurysm, this patient represented with a second subarachnoid hemorrhage. The dorsal cervical artery anastomosis to the external carotid branch, the occipital artery (*filled curved arrow*), reconstituted the common carotid artery on the vertebral artery lateral angiogram. The internal carotid artery (*open curved arrows*) fills the aneurysm (*arrow*)

and 20% were unchanged on the follow-up angiogram [96]. Despite the significant improvement of the angiographic picture of aneurysms reported with carotid ligation, the long- term benefit has been considerably less impressive. For example, though common carotid artery ligation protects patients with a posterior communicating artery aneurysm from rebleeding during the first 1–6 months compared to conservative therapy [97], the late rebleed rate is similar in the two groups [98, 99] (Fig. 6.10).

Additionally, carotid occlusion can cause significant neurologic complications. In the cooperative study, the overall death rate was 22% and the morbidity rate was 36% following ligation of the

ICA after a SAH in patients with an aneurysm in the ICA [100]. The incidence of cerebral infarction is lower in patients who have a common carotid artery ligation than in patients who have an ICA ligation. Gradual rather than acute ligation, closure of the cervical ICA or common carotid artery using a clamp such as a Selverstone clamp, placed around the exposed artery, was initially thought to reduce the risk of a cerebral infarct. Since the blood perfusion pressure distal to the vessel narrowing remains constant until the lumen is approximately 90% occluded, clinical benefit or complications are not seen early in the course of gradual closure. However, the last turn of the clamp suddenly causes a significant decrease in the distal perfusion pressure and is associated with cerebral ischemic complications [101]. The benefits and failures of gradual clamp occlusion of the carotid artery are similar to the results using ligation.

Patients are at risk of developing acute cerebral infarcts if the cerebral blood flow falls below 30 ml/100 g per minute following occlusion of the ICA [102]. However, the competency of the circle of Willis and the result of the occlusion tolerance test (see Sect. 6.2.4.3) are simple predictors of the patient's ability to tolerate an iatrogenic carotid closure [103]. Even if the patient seems capable of tolerating common carotid artery or ICA occlusion, there is an increased long-term risk (26% of patients) of developing cerebral or ocular ischemia if cerebral atherosclerosis develops [104]. An external carotid artery anastomosis to the middle cerebral artery or a petrous to supraclinoid bypass, combined with occlusion of the cervical ICA [105–108], should reduce the incidence of cerebral ischemia. The increased blood flow in the remaining patent ICA may also predispose to de novo aneurysm formation or to an increase in the size of existing aneurysms in this circulation [109]. The long-term risks of carotid artery occlusion in patients with a functioning bypass are unknown.

6.1.10.5 Embolization

The endovascular therapy of intracranial aneurysms is in a state of rapid evolution because of new embolizing agents and delivery systems. Presently, embolization is considered in cases where neurosurgery cannot be performed or carries unreasonable risk. Deconstructive procedures, best performed with detachable balloons, close the aneurysm as well as the parent vessel and have been typically successful in the treatment of aneurysms of the cavernous ICA (see Sect. 6.2). Reconstructive procedures eliminate the aneurysm and preserve the parent artery. Though a detachable balloon can occasionally accomplish this goal, metallic coils seem better suited for reconstructive procedures. No matter which method is used, percutaneous endovascular procedures have the advantage of monitoring the effects on an awake patient.

Detachable balloons can exclude the aneurysm from the parent artery without compromise of this vessel only if a defined neck is present [110–114]. The balloon must be anchored to the aneurysm neck and cannot move inside of the dome. The balloon must maintain its size and induce permanent thrombosis of the aneurysm. Excluding the points of entry and exit of the ICA from a giant aneurysm will lead to thrombosis of the aneurysm. The long-term follow-up of this therapy and the role of balloon embolization in ruptured aneurysms are questions that remain. Rebleeding has occurred in aneurysms that appeared occluded on the postembolization angiogram [115]. A detachable balloon can also be used to exclude the aneurysm and the involved arterial segment when the patient can tolerate ICA occlusion or has a functioning bypass (see Fig. 6.14) (see Sect. 6.2 for discussion).

The embolization of aneurysms with preservation of the parent artery poses its own problem. Embolization with detachable balloons or coils has been well established for treating unruptured aneurysms, but embolizing aneurysms within several days of rupture can actually precipitate a rebleed [116]. Technical difficulties with detachable balloons include anchoring problems within the aneurysm, premature deflation, failure to close the entire neck of the aneurysm, failure of the balloon material to induce thrombosis, bulging of the balloon into the lumen of the parent vessel, and failure of the shape of balloon to mold to the aneurysm. Newer delivery systems and materials for coils have made embolization therapy safer and more effective in the treatment of ruptured and unruptured aneurysms (see Figs. 6.17, 6.19).

Complications of embolization of cavernous and supraclinoid ICA aneurysms include temporary symptoms of reduced vision, headache, or worsening cranial neuropathy from thrombosis of the arterial blood supply to cranial nerves and temporary expansion of the aneurysm. Cerebral or ocular ischemic events may develop from emboli arising from thrombosis within the aneurysm − especially if there is recanalization of the vessel or incomplete closure of the aneurysm. Complete closure of the ICA at the level of the cavernous sinus and at the bifurcation of the common carotid artery will prevent embolic episodes. Even in patients who pass the tolerance test for sacrifice of the ICA, hypoperfusion of the ipsilateral cerebral hemisphere can cause neurologic deficits. This complication can be avoided if, immediately postembolization, the patient is kept flat and the head is gradually elevated, as tolerated, over days. In those patients who fail the tolerance test, a prophylactic bypass procedure is performed prior to embolization (see Sect. 6.2.4.3). ICA closure must be performed soon after the bypass so that a hemodynamic demand will insure the patency of the graft.

6.2 Carotid Cavernous Aneurysm

6.2.1 Incidence and Differential Diagnosis

Aneurysms of the cavernous segment of the ICA are extradural and predominantly cause symptoms, such as pain in the orbit and diplopia, because of compromise of the cranial nerves in the cavernous sinus and superior orbital fissure. Only 3% of all intracranial aneurysms are found in this location [117]. Cavernous carotid aneurysms comprise approximately 15% of all of the symptomatic *unruptured* aneurysms [118]. They are a common cause, along with meningioma, nasopharyngeal cancer, and metastatic cancer, of a cavernous sinus syndrome [119, 120]. Most patients with these aneurysms are more than 50 years old (range 30−75 years). Women are affected 10 times more often than men. Systemic hypertension does not appear to be a causative factor for developing symptoms, nor is there a greater incidence of high blood pressure in these patients than in the normal population.

6.2.2 Clinical Presentation and Neuro-ophthalmologic Dysfunction

As mentioned above, the clinical symptomatology and neuro- ophthalmologic deficits are similar to the dysfunction caused by any mass which is located in the cavernous sinus. Typically, a progressive unilateral disturbance of the local cranial nerves causes diplopia, ptosis, blurred vision from accommodative paresis, orbital, eye, and face pain, and, less frequently, anesthesia of the forehead and cornea [121]. The third, fourth, and sixth nerves are the most frequently involved. Degeneration of these nerves develops from the compression by the aneurysm mass, which is typically located lateral to the nerves [122]. If the onset of ophthalmoplegia is sudden and painful, it results either from compromise of the blood supply to the cranial nerves because of acute compression or thrombosis in a branch artery from the cavernous carotid artery [123, 124].

Though a loss of fifth nerve sensation is less common, trigeminal pain is usually the earliest symptom. Few patients never experience pain. When the pain develops it occurs predominantly in the first trigeminal division, ipsilateral to the aneurysm. Some patients present with a combination of ptosis, diplopia, and chronic pain which develops gradually or abruptly (obviously diplopia is always sudden). The pain can be explosive, possibly because of sudden expansion or hemorrhage into the arterial wall or thrombosis in the lumen of the aneurysm. The pain can be intractable despite the use of narcotics and corticosteroids. Pain along the second trigeminal division, in the maxillary area or upper teeth or jaw, occurs less frequently and never as the sole manifestation of the aneurysm. The pain can remit while other signs of cranial neuropathy remain.

An isolated sixth nerve paresis is common because of the proximity of the nerve to the ICA. An isolated partial third nerve paresis can also develop, but an isolated fourth paresis is extremely rare [125, 126]. In our series of 78 cavernous ICA aneurysms, none had superior oblique weakness without signs of dysfunction in the sixth or third nerves or both. Early on, a partial third nerve paresis, including either an inferior or superior division dysfunction, is possible. Though a total absence of sixth nerve function and a moderate paresis of the third nerve can occur alone, a complete absence of third nerve function without other cranial nerve involvement is rare. Some combination of a partial or complete paresis of the third, fourth, and sixth nerves is common. Longstanding compression of the third nerve may lead to one of the various manifestations of a misdirection syndrome, similar to those found after the surgical clipping of a posterior communicating artery aneurysm (see Sect. 6.3.4.5 for discussion).

On occasion, the pupil is relatively spared in the face of severe paresis of the superior rectus and the levator palpebra, because the pupillary fibers, which course in the inferior division of the third nerve, are not compromised when the superior division is affected. More commonly, because of both parasympathetic and sympathetic dysfunction, the size of the ipsilateral pupil is equal to or slightly larger than the unaffected pupil. Close inspection reveals reduced dilation in the dark and less reactivity to light and near-accommodative stimuli in the affected pupil. A miotic, poorly reactive, pupil with a third nerve paresis usually arises when there is aberrant regeneration (see Sect. 6.3.4.5) [127]. A complete external third nerve ophthalmoplegia with a normal pupil does not occur. Sympathetic dysfunction (see Sect. 6.1.7.1) can occur with a lateral rectus paresis without other cranial nerve dysfunction because the sympathetic innervation passes into the sixth nerve sheath for a short distance while in the cavernous sinus.

A loss of fifth nerve sensation is found only in longstanding cases. The first and second divisions of the trigeminal nerve are located inferiorly to the aneurysm sac, which usually bulges laterally [122]. The aneurysm must extend posteriorly and inferiorly in order to compress the fifth nerve. The corneal reflex is diminished or absent in approximately 20% of patients, and a decrease in sensation in the first division of trigeminal nerve occurs in approximately 25% of patients. Corneal anesthesia and a secondary neurotropic keratopathy occur only when the aneurysm is large and compromises the fifth nerve close to the trigeminal ganglion [128]. Less than 10% of the aneurysms expand far enough posteriorly in the cavernous sinus to cause a loss of sensation in the second division of the trigeminal nerve. Aneurysm expansion towards the petrous bone that causes a loss of third division

sensation and fifth nerve motor dysfunction is extremely rare.

If the aneurysm is very large and extends anteriorly towards the optic foramen or anterosuperiorly through the dura, the patient can develop an optic neuropathy. Visual loss is always preceded by the signs of a cavernous sinus dysfunction and almost all patients have a complete oculomotor and pupillary paresis. In fact, all of our patients with an optic neuropathy had complete ophthalmoplegia in the affected eye. Generalized optic atrophy that is worse in the eye ipsilateral to the aneurysm is typically seen on funduscopy. The visual field loss depends on the direction of optic nerve distortion and on whether the nerve is also compressed by the ICA or anterior cerebral artery, dura of the entrance to the optic canal, or the sphenoid wing (see Sect. 6.4.2.2.1 for types of visual field defects). Once an optic neuropathy begins, the visual loss almost always progresses over months to years, and monocular blindness is inevitable if the aneurysm is untreated. Though chiasm dysfunction can cause a field defect in the eye contralateral to the aneurysm, blindness rarely develops in this eye. In our series, blindness never developed in the second eye of affected patients.

Significant proptosis is also rarely an early finding. The aneurysm must be massive in size and present for years in order to erode through the superior orbital fissure into the orbit. Mild proptosis results from relaxation of the extraocular muscles with complete ophthalmoplegia. The extremely low incidence of proptosis in our patients (one case) is contrasted with the 60% incidence reported in the older literature [129]. This difference is probably reflects earlier diagnosis because of improved neuroimaging with CT and MR imaging.

An orbital or cephalic bruit is ordinarily not found in this disorder because the lesion is not high flow and does not have an arteriovenous shunt like a carotid cavernous fistula. However, if the aneurysm erodes anteriorly an orbital bruit can be auscultated because of turbulent blood flow within the aneurysm.

Other rare presentations result from erosion of the aneurysm into the sella to cause pituitary dysfunction. Pituitary disturbances are always accompanied by symptoms and findings of cavernous sinus dysfunction. Cerebral ischemic episodes are

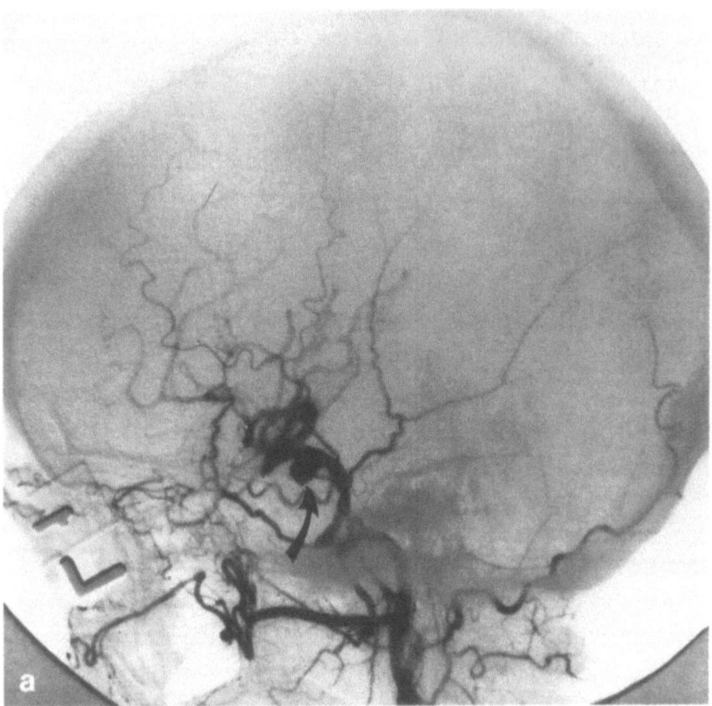

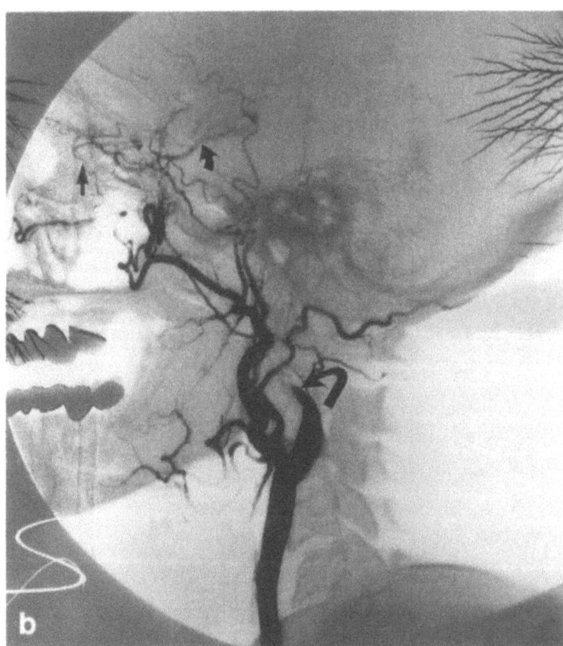

Fig. 6.11a, b. Though often considered, cerebral emboli rarely originate from thrombus within the lumen of a giant aneurysm.

a This 52-year-old woman experienced the onset of left painful ophthalmoplegia and a right hemiparesis during treatment of uterine cancer complicated by secondary disseminated intravascular coagulation. The lateral view left common carotid artery subtraction angiogram demonstrates the lumen of a carotid cavernous aneurysm (*curved arrow*) with a mass effect pushing the cavernous internal carotid artery upward. There is poor opacification of the middle cerebral artery branches. The view is somewhat obscured because of the overlying superficial temporal artery.

b The repeat common carotid angiogram performed 24 h later shows a "beak" sign (*tightly curved arrow*) of acute thrombosis of the left internal carotid artery. The anterior deep temporal artery (*arrow*) anastomosis to the ophthalmic artery (*curved arrow*) fills the supraclinoid internal carotid artery artery. The hemiparesis resolved entirely. Only a minimal partial third nerve paresis remains 5 years later

Table 6.2. Differential diagnosis of painful monocular ophthalmoplegia not restricted to one cranial nerve

Location	Etiology
Cavernous sinus	Aneurysm
	Neoplasm [b]
	Inflammation (idiopathic) [a]
	Infection (from adjacent sinus)
Superior orbital fissure	Neoplasm [b]
	Inflammation (idiopathic) [a]
Orbit	Dysthyroid [b]
	Infection
	Inflammation (idiopathic)
	Neoplasm [b]

[a] Rare.
[b] With or without pain.

also rare and are caused by emboli from thrombus within the aneurysm [130]. Cerebral emboli are more likely to develop when there is also an underlying coagulation disorder (Fig. 6.11).

6.2.3 Differential Diagnosis of Painful Ophthalmoplegia

The differential diagnosis of ophthalmoplegia includes disorders that affect the superior orbital fissure and the orbit and may or may not cause pain (Table 6.2). A painless progressive superior orbital fissure syndrome (which may be clinically indistinguishable from an anterior cavernous sinus syndrome) can be caused by a neoplasm, such as lymphoma or meningioma. Prior to high-resolution CT and MR imaging, patients with aneurysms were often misdiagnosed as having a primary cavernous sinus tumor, cavernous sinus invasion by a neoplasm extending from the adjacent sinuses or nasopharynx, or Tolosa – Hunt syndrome (idiopathic inflammation).

In the case of neoplasms, sixth nerve dysfunction is commonly the first sign. When present, pain usually follows or accompanies the diplopia. Eventually multiple cranial nerves become compromised. CT and MR imaging show a mildly enhancing mass when the lesion is a meningioma, a chordoma, or a sinus or nasopharyngeal cancer. A neurinoma of the third nerve does not enhance on either study. Additionally, bone destruction is seen with nasopharyngeal cancer and chordoma, in contrast to the bone expansion and erosion that occur with an aneurysm. Also, a neoplasm will not have a target sign (see Sects. 6.1.8.1, 6.2.4.1). A small meningioma or lymphoma may be difficult to visualize, even with a thin-section, coronal, high-resolution, contrast- enhanced CT or MR image (even with gadolinium). MR imaging easily distinguishes a neoplasm from an aneurysm. If the ophthalmoplegia is sudden, explosive, painful, and associated with visual loss, an infarct or hemorrhage of a pituitary adenoma with extension into the cavernous sinus must be suspected. MR imaging can easily distinguish an infarct or hemorrhage in a pituitary adenoma from an aneurysm.

An acute painful superior orbital fissure or cavernous sinus syndrome may be caused by the Tolosa – Hunt syndrome, which is rare and frequently overdiagnosed. The CT and MRI may show no obvious changes or the alterations are nonspecific with the demonstration of a soft tissue mass. The Tolosa – Hunt syndrome is pathologically related to pseudotumor of the orbit, an idiopathic orbital inflammatory disease. In most patients with orbital pseudotumor, varying degrees of ophthalmoplegia which does not appear limited to the muscles innervated by a single cranial nerve is present. Painful ophthalmopareses caused by a myositis of the extraocular muscles can be distinguished by the injection over the muscles. Proptosis, chemosis, and conjunctival injection differentiate the orbital process from a lesion located intracranially in the cavernous sinus. A beneficial response to corticosteroids is not diagnostic of an inflammatory disorder.

Orbital cellulitis is another cause of abrupt-onset painful ophthalmoplegia. This frequently occurs secondary to an adjacent sinusitis. The patient usually has fever and malaise, tenderness of the periorbital tissues, lid edema, conjunctival injection, and chemosis, all of which easily distinguish this disorder from an unruptured cavernous sinus aneurysm. CT or MR imaging will also readily distinguish these entities and reveal the presence of orbital inflammation and sinus disease.

When the patient states the diplopia is intermittent, it may only reflect the ocular misalignment not being present in all fields of gaze. The patient experiences diplopia only when looking into the field of the affected extraocular muscle. Another cause of intermittent diplopia which is painless is myasthenia gravis. In the latter disorder, there is no clinical evidence of pupillary dysfunction or sensory disturbances. CT and MR images of the orbital and intracranial structures are normal. A positive edrophonium test is considered diagnostic of myasthenia gravis, though there are rare reports of patients with cavernous sinus lesions who exhibit a partial response to edrophonium [131]. Though myasthenia gravis may initially involve only one eye, with time the dysfunction almost always affects the eyelids or extraocular muscles of both eyes. In contrast, the aneurysm- related muscle dysfunction remains unilateral.

Dysthyroid orbitopathy typically causes a limitation of upgaze or abduction. The presence of other signs, such as proptosis, injection over the horizontal recti, lid retraction, bilateral dysfunction, lagophthalmos, elevation of the intraocular pressure by 4 mmHg or more on upgaze, and a positive forced duction test (see Sect. 4.2) in the direction of the eye movement limitation, will distinguish this entity from cranial nerve dysfunction. Diplopia without significant pain is also caused by a carotid cavernous fistula, a dural arteriovenous malformation which drains into the cavernous sinus, or cavernous sinus thrombosis (see individual entities).

Rarely, microvascular disease of multiple nerves in the cavernous sinus can cause a painful or painless ophthalmoplegia (see Sect. 8.8.5). This is a diagnosis of exclusion following a complete evaluation that includes high- resolution MR imaging of the cavernous sinus, superior orbital fissure, and the orbital apex.

6.2.4 Neuroimaging

6.2.4.1 Computed Tomography and Magnetic Resonance Imaging

On CT any parasellar mass may cause convex bulging of cavernous sinus which is best seen on a coronal view (Fig. 6.12). A target sign (see Sect. 6.1.8.1) (Fig. 6.4b) may be seen or the aneurysm may enhance uniformly if no thrombosis is present (Fig. 6.12). Thinning of the adjacent bones or erosion of

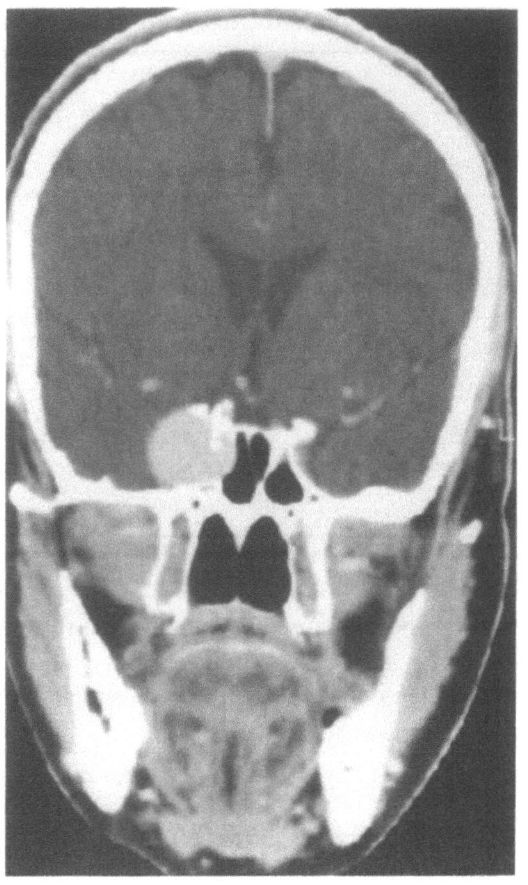

Fig. 6.12. The coronal view contrast-enhanced scan demonstrates erosion of the lateral wall of the sphenoid sinus by the cavernous carotid aneurysm. Because of this finding with the risk of hemorrhage into the sphenoid sinus, this 75-year-old woman who had mildly painful, clinically stable, third and fourth nerve pareses underwent balloon embolization to close the internal carotid artery and the aneurysm at this level

the aneurysm into the sphenoid sinus or superior orbital fissure can be seen in longstanding cases (Fig. 6.12).

MR imaging is the best noninvasive method to visualize a cavernous carotid aneurysm and its relationship to the ICA. As mentioned in Sect. 6.1.8.2, MR imaging readily differentiates an aneurysm from a neoplastic mass in the cavernous sinus. It may also reveal whether the aneurysm has eroded through the dura into the intradural space (Fig. 6.13a).

Fig. 6.13. a Magnetic resonance imaging in this 45-year-old woman with a painless diplopia only on eccentric gaze due to a partial third nerve palsy demonstrates an intradural nipple (*arrow*) projecting through the dura from a left cavernous carotid aneurysm (*broad arrow*).
b The lateral view left internal carotid artery subtracted angiogram was also suggestive of an intradural extension (*arrow*) of the aneurysm.
c The calibrated leak technique lateral view angiogram clearly demonstrates the intradural extension of the aneurysm lumen that appeared to be pinched by the dura (*arrow*). Arteriography of the right internal carotid artery demonstrated a lack of cross-filling to the left cerebral hemisphere (not shown). A bypass surgical procedure to the dominant hemisphere would be necessary for an anticipated closure of the left internal carotid artery at the level of the aneurysm. Because of the paucity of symptoms and signs, the treatment was withheld and 4 months later the patient experienced a subarachnoid hemorrhage. Fortunately, she was not left with any additional neurologic deficit and she was treated with a bypass and a surgical occlusion of the internal carotid artery at the level of the aneurysm

6.2.4.2 Angiography

Cerebral angiography is performed to determine whether treatment can be performed with an acceptable risk. The angiogram may demonstrate a neck of the aneurysm, but more often a definable neck is not seen because the wall of the cavernous ICA is actually part of the aneurysm wall. The mass of the aneurysm exceeds the lumen in most cases and frequently displaces the ICA (Fig. 6.11a). Careful study of the angiogram will reveal whether the aneurysm arises within the cavernous sinus or whether it is actually an intradural lesion that projects into the cavernous sinus [132] to cause the cranial neuropathy. The angiogram may distinguish whether a cavernous ICA aneurysm extends into the subarachnoid space (Fig. 6.13b,c). Angiography will also reveal if the origin of the ophthalmic artery is subarachnoid or from the cavernous ICA. The patency of the anterior circle of Willis is determined as previously discussed in Sect. 6.1.8.3. Angiography demonstrates additional aneurysms, 50% of which are mirror aneurysms in the opposite cavernous ICA, in approximately 20% of patients with aneurysms of the cavernous ICA.

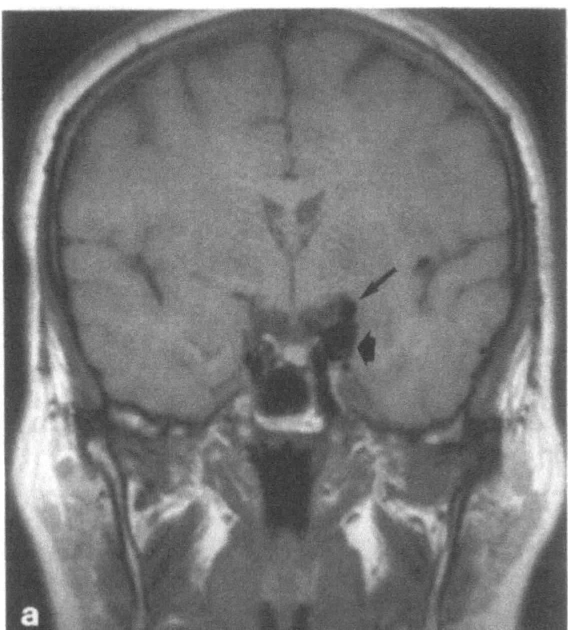

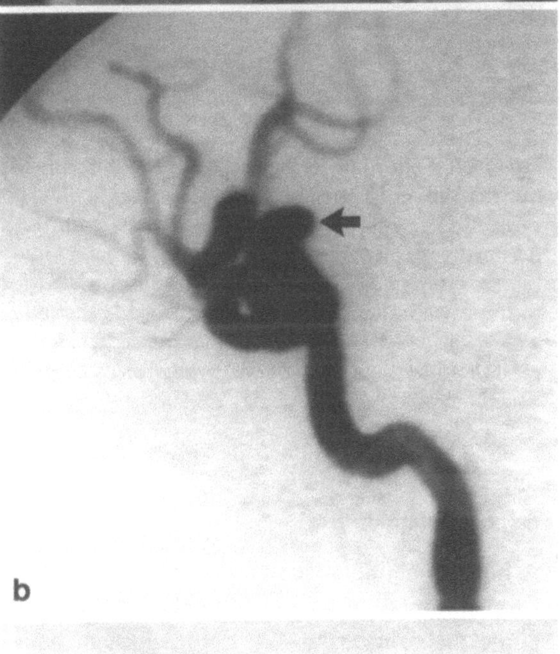

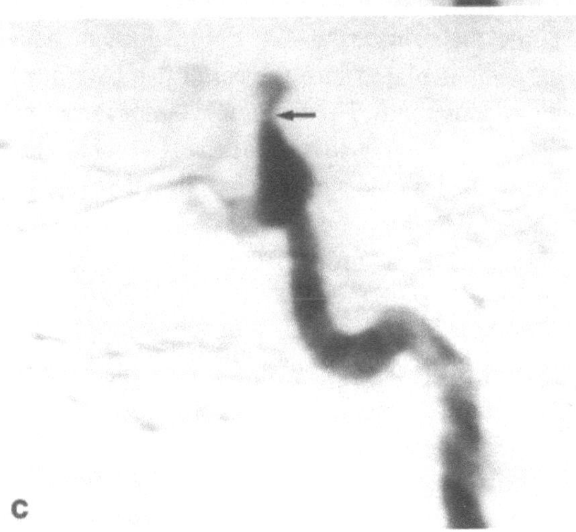

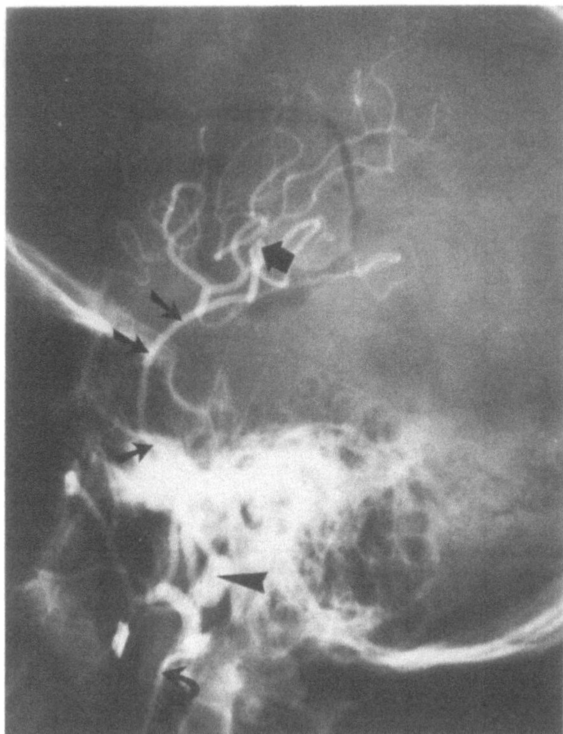

Fig. 6.14. Following treatment with balloon embolization of the left internal carotid artery for a cavernous carotid aneurysm, the functioning bypass from the superficial temporal artery (*small arrows*) to the middle cerebral artery (*large arrow*) is seen. The *arrowhead* demonstrates an inflated balloon still attached to the catheter which is used to test the patient's tolerance to carotid occlusion and the patency of the bypass prior to balloon detachment

6.2.4.3 Tolerance Test

A functional tolerance test of ICA occlusion is performed while the patient is awake and under systemic heparinization. A double-lumen balloon catheter is placed into the cervical ICA via a percutaneous route. The balloon is temporarily inflated for 15–20 min to occlude the ICA while the patient's neurologic status is monitored [133]. If the patient develops a neurologic deficit during the tolerance test, a prophylactic extracranial to intracranial arterial bypass procedure between the superficial temporal and middle cerebral arteries (Fig. 6.14) [134, 135, 103], between the subclavian and middle cerebral arteries via a venous graft, or between the petrous and supraclinoid ICA using an axillary vein graft [136] is performed. Even if the

patient can withstand ICA occlusion, the presence of a mirror aneurysm or lumen compromise from atherosclerosis in the opposite ICA may preclude occluding the ipsilateral ICA without a prophylactic bypass.

6.2.5 Treatment and Course

Most of the patients have progression of a combination of third, fourth, and sixth nerve pareses, but the pain may be episodic or continuous. As discussed previously, patients with an anesthetic cornea have reduced vision because of a trophic keratopathy. Permanent visual loss from corneal scarring is rare and we have seen only one such case.

Approximately 25% of patients will stabilize or improve spontaneously. Spontaneous thrombosis within the aneurysm is marked by episodes of severe pain and worsening cranial neuropathy. These periods are often followed by a reduction or a complete resolution of the pain within several days to weeks and gradual improvement of the cranial nerve pareses over weeks to months. In these cases, the pain typically resolves completely within approximately 6 weeks of the ictus. In other patients, stability or spontaneous improvement of the cranial neuropathy gradually occurs without an acute event or severe pain. Once the symptoms have resolved, the patient can remain asymptomatic.

The small asymptomatic aneurysms that are found at the time of the evaluation for symptomatic aneurysms in other locations infrequently enlarge after years.

Approximately 4% of carotid cavernous aneurysms rupture into a venous compartment of the cavernous sinus, resulting in a carotid cavernous fistula. Prior to rupture, these cases are undiagnosed and typically have small, asymptomatic aneurysms. Similar to other carotid cavernous fistulas (see Chap. 2), the symptoms depend on the venous drainage, possible thrombosis in the cavernous sinus, and the rate of blood flow through the shunt.

Symptomatic aneurysms can progressively enlarge and attenuate the adjacent dura and bone compartment of the cavernous sinus (Fig. 6.12). Rarely, erosion of the lateral wall of the sphenoid sinus results in the aneurysm bulging into the sphenoid sinus. A rupture of this aneurysm could cause severe epistaxis [137]. However, except for rare cases, the reports of this complication have been in patients with traumatic pseudoaneurysms [138, 139, 139a].

Cavernous carotid aneurysms rarely cause SAH, even when the lesions are large enough to cause optic neuropathy [122]. One of these cases has been reported more than once [26, 140–142]. Each of two pathologically studied cases with a subarachnoid space directed rupture had an aneurysm that was definitively shown to have ruptured laterally, in the direction where there is no bone to support the cavernous sinus. We have seen one case with a cavernous aneurysm that had a projection of a small nipple of the aneurysm into the subarachnoid space (Fig. 6.13) that ruptured to cause a SAH. Additional cases of SAH have been described with parasellar aneurysms, but clear documentation that the ruptured aneurysm originated from the cavernous ICA is often lacking [143]. The clinical course of a true carotid cavernous aneurysm differs from that of an aneurysm with an intradural origin and neck that has its sac projecting extradurally into the cavernous sinus [132], because the latter aneurysms frequently cause a SAH. A recent series described SAH in three of 44 patients [138]. The 6.8% incidence of SAH in this report is considerably higher than the 1.3% in our 78 cavernous aneurysms [144]. Clearly, the difference arises because the former report included patients whose aneurysms extended to an intradural position, just proximal to the ophthalmic artery.

Treatment is indicated if the visual loss or cranial neuropathy is debilitating or progressive. Patients who have persistent severe pain that is difficult to control medically also warrant intervention. Patients with a carotid cavernous fistula are treated as outlined in Sect. 2.4.

As with other aneurysms, the ideal therapy excludes the aneurysm from the cavernous ICA and preserves the blood flow through the affected ICA. However, in most cases, the ipsilateral ICA is occluded during treatment [134, 145]. The cerebral circulation is preserved via the anterior communicating artery from the contralateral ICA, through the posterior communicating artery from the posterior circulation, or via a prophylactic extracranial to intracranial arterial bypass [107].

6.2.5.1 Surgery

Until recently, successful direct clipping of an aneurysm in cavernous sinus had been attempted infrequently and accomplished rarely without significant complications, including ischemia in the ipsilateral cerebral hemisphere [146, 147]. Direct surgery of the cavernous aneurysm carries a mortality rate of approximately 20% [146]. Surgical reconstruction of the ICA has been performed with a surgical clip in patients with supraclinoid aneurysms, but it is virtually impossible in the cavernous sinus. Adequate surgical exposure cannot be obtained without severely compromising the venous channels and cranial nerves in the cavernous sinus. We have also seen patients develop an optic neuropathy as a complication of this surgery. A surgical procedure that traps the aneurysm by ligating the ICA at supraclinoid and cervical segments accomplishes no more than percutaneous embolization (see below). Surgery exposes the patient to the risks of a craniotomy and the occlusion of the ICA without the benefit of monitoring an awake patient. No matter the method applied, the same precautions must be taken in a case where occlusion of the ICA is required to treat the aneurysm.

6.2.5.2 Embolization

Percutaneous balloon embolization [134, 145] has had the highest success rate of occluding this aneurysm. Preservation of the ICA is rare (Fig. 6.15) because of the frequent lack of a clearly definable neck, the intimate relation of the aneurysm and the ICA wall, and the large distance between the entrance and exit of the ICA from the aneurysm. Therefore, the tolerance test is performed in every case considered for this therapy. In patients who tolerate temporary ICA occlusion, following the reversal of heparinization, the entrance and exit of the ICA from the aneurysm, and frequently the ICA at the level of the aneurysm, are permanently blocked by one or more detachable balloons (Fig. 6.16). The proximal segment of the cervical ICA is closed with another balloon and a metal coil to prevent emboli from a thrombus that might pass into the external carotid artery (see Sect. 6.2.5.3). Further evolution of endovascular procedures will

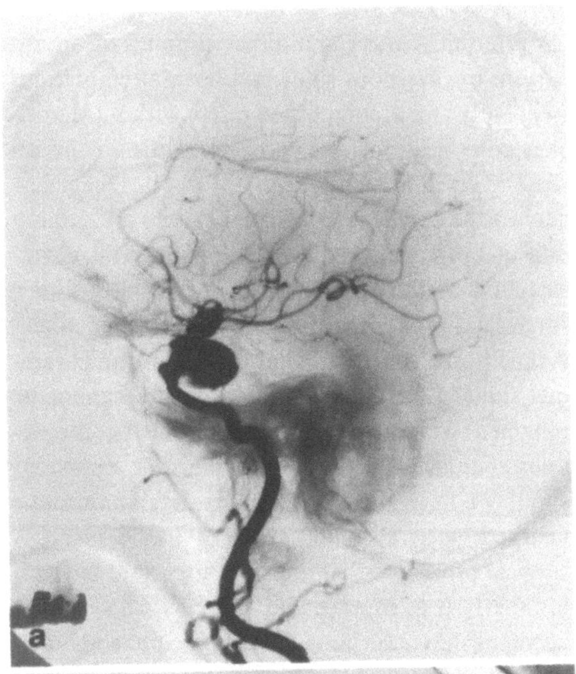

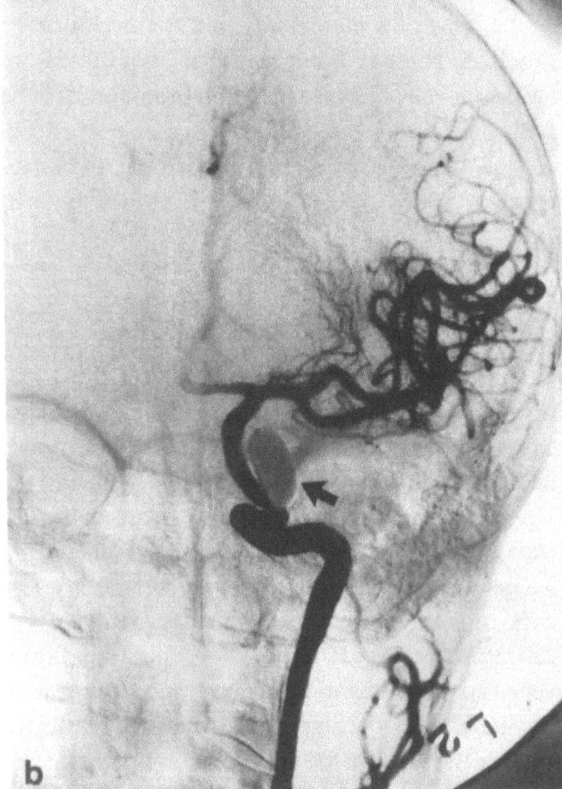

Fig. 6.15 a, b. Embolization with detachable balloons rarely preserves the flow in the internal carotid artery with a cavernous aneurysm. **a** The lateral internal carotid artery subtraction angiogram demonstrates an intracavernous aneurysm in this 75-year-old woman who had experienced intermittently painful ophthalmoplegia over the course of 20 years. She had a severe compromise of the third, fourth, and sixth cranial nerves. **b** The frontal view of the left internal carotid artery demonstrates preservation of the left internal carotid artery and exclusion of the aneurysm from the circulation by a detachable balloon (*arrow*)

Fig. 6.16 a, b. In most cases detachable balloon cure of a ▶ carotid cavernous aneurysm necessitates closure of the parent internal carotid artery. **a** The frontal view left internal carotid artery angiogram demonstrates a huge left cavernous carotid artery aneurysm in this 30-year-old woman who experienced a severe painful third nerve paresis. The symptoms had begun during pregnancy and persisted for 6 months post partum when this study was performed. The entire mass of the aneurysm appeared to be patent lumen. **b** The frontal view right internal carotid artery angiogram demonstrates filling of the left anterior cerebral artery as well as a segment of the left middle cerebral artery (*arrow*) following balloon embolization of the left internal carotid artery with multiple balloons (*curved arrow*). A wide-based neck of the aneurysm with extensive involvement of the internal carotid artery at this level made closure of the cavernous carotid artery essential to exclude the aneurysm from the circulation. The third nerve paresis resolved completely within 6 months

undoubtedly increase the rate of preservation of the ICA (Fig. 6.17). If necessary, an arterial bypass procedure is performed several days prior to embolization (see Fig. 6.14). MR imaging (see Sect. 6.1.8.2) (Fig. 6.6b) is superior to CT in revealing thrombosis in the treated aneurysm. To date, metallic coil embolization that preserves the ICA has not been effective in reversing the cranial nerve pareses.

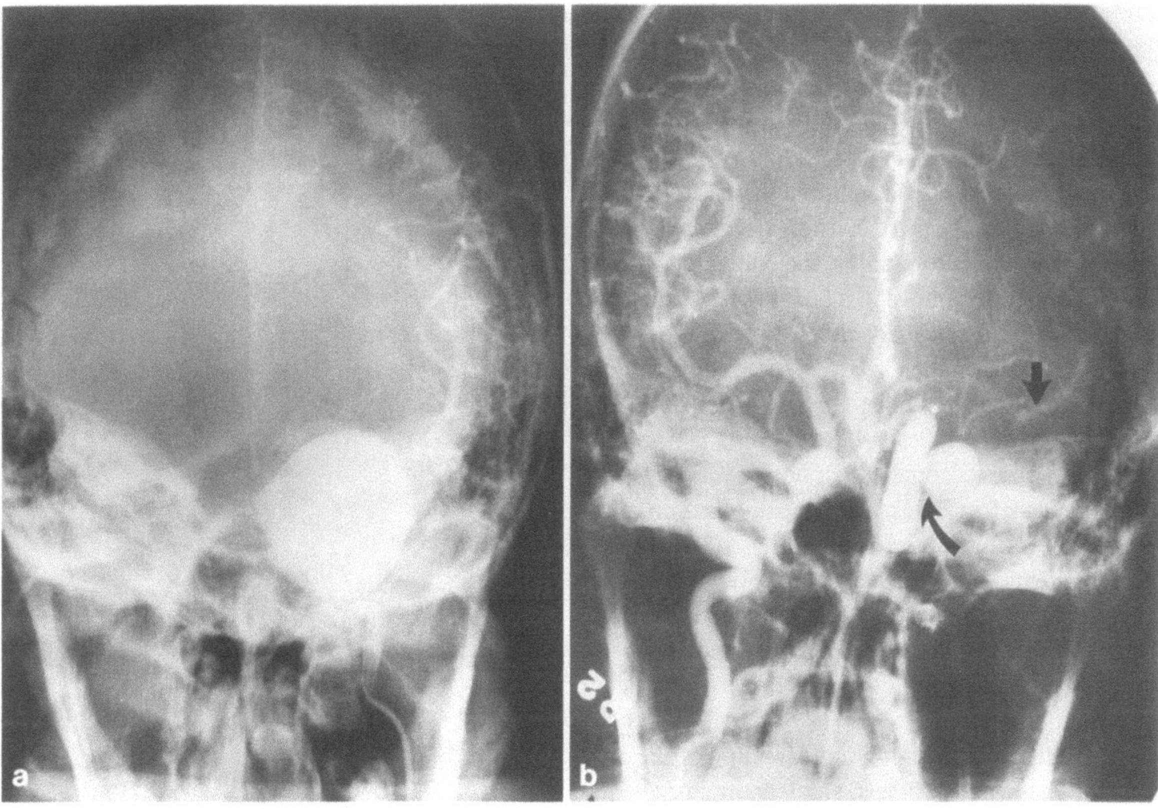

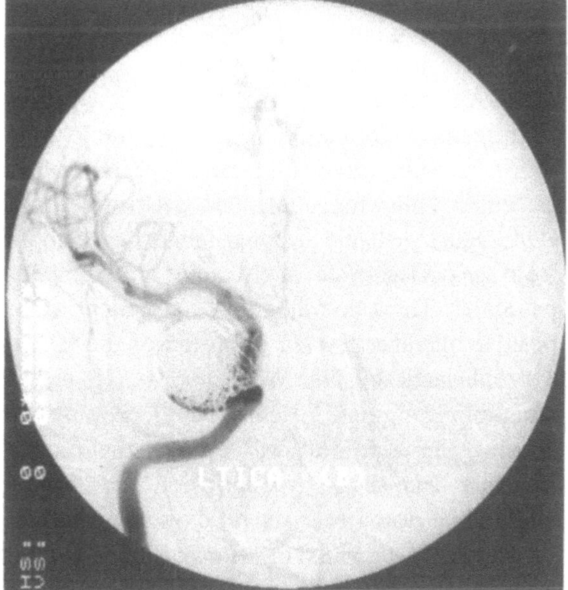

Fig. 6.17. Newer percutaneous embolization techniques, such as electrothrombogenic coils, should permit a higher rate of aneurysm closure without occlusion of the parent artery. This 71-year-old woman with a progressive painful left ophthalmoplegia was treated by placing approximately 4 m of coils in the aneurysm lumen. The immediate postembolization frontal view subtraction angiogram shows preserved blood flow in the left internal carotid artery, the ball of coils in the aneurysm, and a stagnant collection of contrast material in the inferior region of the aneurysm lumen which persisted in the venous phase. Follow-up angiography showed an absence of opacification in any portion of the aneurysm

6.2.5.3 Complications

Following embolization or surgical carotid occlusion, sudden thrombosis with expansion of the aneurysm mass can cause temporary worsening of the cranial neuropathy and pain [27]. In most cases, even without obvious thrombosis, there is a temporary increase in the first division trigeminal pain. Over several weeks to months, the pain and the cranial neuropathy improve in all of these cases. However, if the patient has a complete ophthalmoplegia before treatment, then recovery of ocular motor function is limited. Patients with partial ophthalmoplegia frequently resolve the diplopia in primary gaze, but rarely become orthophoric in all the cardinal fields of gaze. Following embolization, the pre-embolization optic neuropathy stabilizes in most patients, but significant improvement in the vision occurs in few cases. The pupillary function rarely normalizes, and the pupil often remains permanently dilated and poorly reactive to both near and light stimuli.

In addition to the temporary swelling of the aneurysm mass and worse cranial neuropathy and pain mentioned above, there are other complications of balloon embolization. Premature balloon deflation within 48 h of embolization causes the pain and neuropathy to worsen or not to resolve. Premature detachment of the balloon while it is in the ICA lumen can lead to migration of the balloon to the carotid bifurcation or into the middle or anterior cerebral arteries to cause acute cerebral ischemia. Following unplanned occlusion of the ICA, acute systemic hypotension or an embolus from the occlusion site rarely causes a cerebral TIA or infarct. This can even occur in a patient with a positive tolerance test for occlusion of the ICA or a prophylactic bypass. When the ICA is closed, maintaining normal systemic blood pressure and keeping the head flat will prevent the cerebral ischemia from distal hypoperfusion. The head is gradually elevated over several days as tolerated.

If the ICA proximal to the occlusion is not also closed with a stumpectomy, thrombus can propagate to the carotid bifurcation where emboli can dislodge into the external carotid artery. These emboli can pass through collaterals, such as the inferior lateral trunk [148] (see Sects. 1.5.1.1.1, 1.5.3.1.2), to the ophthalmic artery or the cerebral arteries

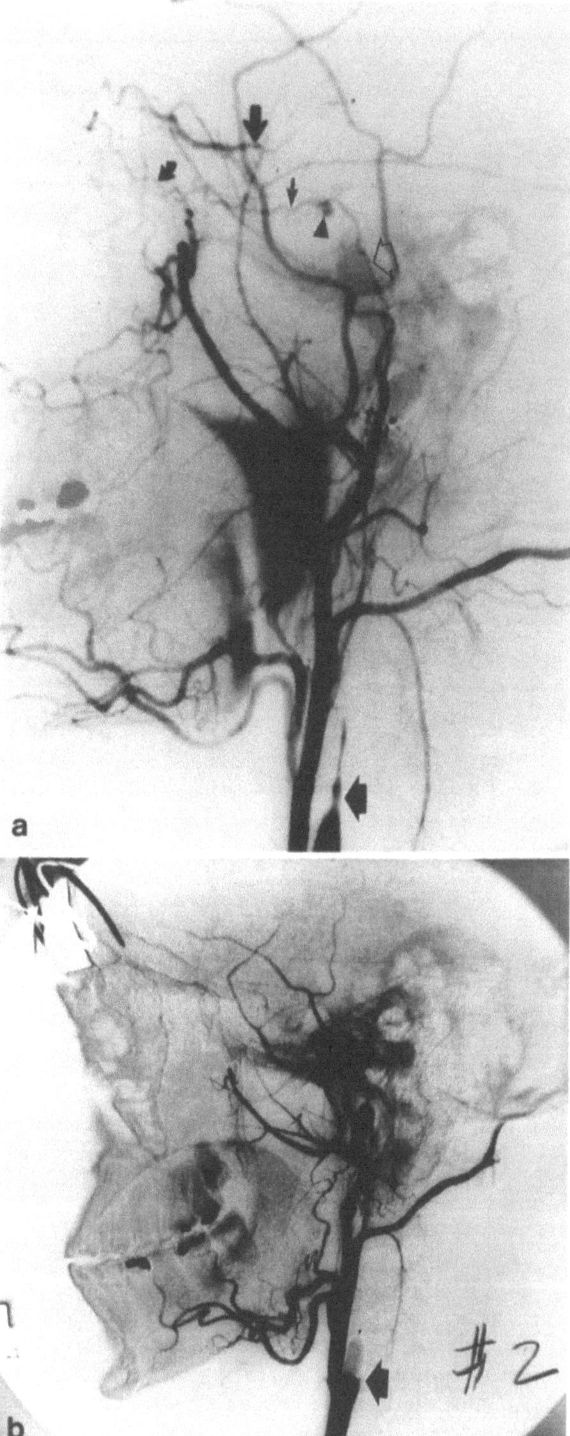

◄ **Fig. 6.18 a, b.** An embolic cerebral transient ischemic attack occurred 24 h after proximal balloon occlusion of the internal carotid artery. **a** Lateral common carotid artery injection on the left following balloon occlusion of the internal carotid artery for a giant cavernous aneurysm, late phase. Note the acute thrombosis of the proximal internal carotid artery (*broad arrow*) and the position of the most distal balloon (*open arrow*), which is proximal to the cavernous carotid artery. Also note filling of the anterior deep temporal artery (*curved arrow*) to the ophthalmic artery (*large arrow*). The cavernous carotid artery remains partially patent via flow from the artery to the foramen rotundum (*small arrow*). Note stagnation of contrast material in the cavernous ICA (*arrowhead*). **b** Following stumpectomy (*arrow*) with a detachable balloon, there were no further embolic events. Preventing thrombus from entering the external carotis artery artery circulation can avoid cerebral or ocular embolic complications

(Fig. 6.18). When balloon embolization does not completely exclude the aneurysm from the circulation, an embolus from thrombus within the aneurysm lumen can dislodge and pass into the distal circulation. An embolus can also develop if the balloon compromises, but does not completely close, the ICA lumen.

Though severe atherosclerotic occlusive disease can cause ischemic oculopathy, patients with therapeutic occlusion of the ICA do not develop this complication (see Sect. 8.2.5). Another theoretical complication is acute monocular visual loss which can result if the ophthalmic artery arises from the occluded cavernous segment of the ICA. However, since none of our patients have had this anatomic variant, we have not observed this complication.

Immediately following successful embolization, a digital intravenous angiogram (infrequently used now), a digital intraarterial angiogram, or MR angiography will show that the aneurysm is excluded from the cerebral circulation and that the cerebral arteries to the ipsilateral cerebral hemisphere fill via the expected collaterals. As discussed previously, MR imaging is the best noninvasive method of following a giant aneurysm to determine whether the mass is reduced or thrombosis has occurred. However, the appropriate MR technique must be used because the aneurysm may appear to be occluded when a patent lumen is still present. Previously, contrast- enhanced CT was used for the same purpose. The presence of ring enhancement without intraluminal enhancement on the contrast CT is strongly suggestive of thrombosis of the aneurysm.

6.2.5.4 Carotid Ligation

An alternative therapy, balloon or gradual common carotid artery occlusion with an externally applied clamp, is reserved for the patient who cannot tolerate ICA occlusion and who is considered a poor surgical risk for an extracranial to intracranial arterial bypass (see Sect. 6.1.10.4).

Common carotid artery ligation is not free of complications, and in one study an acute cerebral infarct with severe hemiparesis developed in up to 40% of cases. Though the aneurysm appears smaller on angiography, the cranial neuropathy may worsen [149], possibly because of intraluminal thrombosis or collapse of the aneurysm sac with further distortion or ischemia to the nerves. Also, in some cases the aneurysm may increase in size despite ligation of the common carotid artery [150].

6.3 Posterior Communicating Artery Aneurysm

6.3.1 Incidence

Aneurysms arising from the ICA at the origin of the posterior communicating artery comprise up to 39% of all ruptured intradural aneurysms [151].

6.3.2 Clinical Presentation

Except for a SAH, the patients present most often with third nerve dysfunction. This results because the posterior communicating artery and third nerve are juxtaposed in the subarachnoid space of the basal cisterns, prior to the nerve entering the dural leaves of the cavernous sinus superiorly to the sixth nerve (see Fig. 1.41 in Chap. 1). Most, but not all, of the patients with third nerve findings have first division trigeminal nerve distribution pain. In some series, an intracranial aneurysm (not all were posterior communicating artery aneurysms) has been reported as the most common cause of a third nerve paresis, occurring in almost 30% of all cases with an oculomotor palsy [152]. The third nerve paresis may precede or develop concomitantly with a posterior communicating artery aneurysm-related SAH. Less than 10% of patients with a posterior communicating artery aneurysm that ruptures lack third nerve findings. Patients may complain of pain in the ipsilateral orbit, eye, or frontal region for months or years prior to any clinical findings. Rarely, a giant aneurysm in this location can also compress the optic nerve or optic tract (see Sects. 6.3.2.2, 6.4.2.2 for discussion of intracranial anterior visual pathway visual loss and visual field defects).

6.3.2.1 The Third Nerve Controversy

The onset of third nerve dysfunction indicates either expansion of the aneurysm to directly compress the third nerve or a localized extravasation of blood into the nerve fascicles [153, 154]. Thus, a third nerve paresis suggests an unstable posterior communicating artery aneurysm that requires immediate evaluation and treatment prior to the occurrence of a major SAH. In one study six of 26 patients with a posterior communicating artery aneurysm had a third nerve paresis an average of 29.6 days prior to a SAH [155]. Some degree of pupillary dilation is almost always associated with the signs of external (lid and eye muscles) oculomotor nerve dysfunction (see below).

Third nerve pareses with pupillary sparing have been reported in 3% − 5% of patients with aneurysms in *all* locations [152, 156, 157]. However, these data are derived from older series that preceded the development of detailed high-magnification cerebral arteriography, which improved the accuracy of diagnosis and localization of the aneurysm. These earlier reports are not useful in determining the incidence of a third nerve paresis presentation in patients with intradural aneurysms, because they suffer from lumping aneurysms in all locations (both intradural and extradural) together. They also included patients who presented with a clinically apparent SAH and not an isolated third nerve lesion [118, 121, 152, 158]. In the only study to document a third nerve paresis prior to a SAH, the presence or absence of pupillary dysfunction was not described [155].

Though the pupil is usually significantly enlarged and poorly reactive to light, if at all, it is important to recognize that minimal pupillary involvement can occur with posterior communicating artery aneurysm compression of the third nerve. While the patient looks in the distance, an anisocoria greater than or equal to 1.0 mm in room light, with slow incomplete reactivity to light in the involved eye when tested in the dark, and no optic nerve dysfunction in the ipsilateral eye can be considered pathological. The pupillary abnormality must accompany additional third nerve findings in order to diagnose a compressive third nerve paresis.

Pupillary dysfunction is not uncommon with all etiologies of a third nerve paresis. An abnormal pupil is even found in between 15% and 28% of patients with microvascular ischemic third nerve palsies [156, 159, 160]. A complete external paresis associated with a minimal pupillary dilation is unlikely to be caused by a posterior communicating artery aneurysm. Nevertheless, any patient, even if diabetic or hypertensive, who does not have an obvious meningitis and develops any degree of pupillary sphincter dysfunction in association with other third nerve signs must be evaluated with CT or MR imaging and cerebral angiography.

Because of the fear of missing the diagnosis of an unruptured intradural aneurysm, controversy exists regarding the extent of the radiographic evaluation and the management of a patient with an isolated *pupillary-sparing* third nerve paresis. Clearly, the overwhelming majority of cases of paresis of the oculomotor nerve with a normal pupil are caused by a microvascular infarct of the peripheral nerve (see Sect. 8.8.2). Approximately 80% of patients with a noncompressive etiology for their third nerve paresis have sparing of the pupil at the time of the initial examination [157, 160].

Until the report by Kissel et al., a pupillary-sparing third nerve paresis was virtually always considered to have an etiology other than an aneurysm [161]. In review of 84 cases with posterior communicating artery aneurysms, four of 12 patients who did not have a SAH had a normal pupil at the time of their presentation of a partial third nerve paresis. Three additional patients with a pupillary-sparing third nerve paresis presented a SAH, not an isolated cranial neuropathy (from a total of 52 patients with a posterior communicating artery aneurysm and a SAH). Progression to pupillary dysfunction occurred in five of seven of these patients (four within 5 days and one within 4 months). Only two patients who did not initially have a SAH had a posterior communicating artery aneurysm and a pupillary-sparing third nerve paresis 1 week after presentation (2.4% of all 84 patients). Interestingly, in this study 33 of 84 patients had no third nerve dysfunction. In contrast, another series reported that none of 34 patients with a posterior communicating artery aneurysm and a third nerve paresis had pupillary sparing [162]. Our own series had one patient (who never bled) with a presentation of a pupillary-sparing partial external third nerve paresis out of 80 cases with a posterior communicating artery aneurysm.

There are a few well-documented cases of patients with a posterior communicating artery aneurysm and a pupillary-sparing third nerve paresis, but these patients have had a partial and never a complete paresis of the levator palpebrae and extraocular muscles [156, 163, 164]. Other reports of pupillary sparing have unconvincing documentation [165]. When there is a complete ophthalmoplegia of all of the oculomotor nerve-innervated muscles (including the levator) except the iris sphincter, then a posterior communicating artery aneurysm cannot be the cause. Although extremely rare, if any degree of pupillary dysfunction is present, even in the face of a total ophthalmoplegia, a posterior communicating artery aneurysm may be present [160].

Practically, only those patients with a pupillary-sparing third nerve paresis without signs and symptoms of a SAH need to be considered in determining the risk of them harboring a posterior communicating artery aneurysm. If the patient is followed expectantly for a week and the pupil becomes abnormal, the patient must be evaluated for a possible posterior communicating artery aneurysm even though a vasculopathic third nerve paresis can follow a similar course [160]. In the most recent investigation, none of the 23 patients that presented with a pupillary-sparing third nerve paresis (19 partial, 4 complete) had aneurysms [160]. When all patients with all the etiologies of a pupillary-sparing third nerve paresis are considered, intradural aneurysms are a rare cause.

Various mechanisms have been postulated to explain how a posterior communicating artery aneurysm can cause a pupillary-sparing partial third nerve paresis. Kasoff postulated that the pupil may be spared if, in these cases, the pupillary fibers are located in the lateral portion of the third nerve [166], which is similar to the third nerve topography in the dog [167]. However, in man the pupillomotor fibers are located in the superior peripheral portion of the nerve [168], where they are ordinarily susceptible to the dorsal medial compression of the third nerve by a posterior communicating artery aneurysm that points posteriorly, inferiorly, and laterally. One case with normal pupillary function, which refutes the Kasoff hypothesis was clearly observed at surgery to have the aneurysm sac enlarged temporally towards the anterior petroclinoid ligament. The aneurysm impinged on the lateral and superolateral aspect of the third nerve, which does not contain pupillomotor fibers [169].

Aneurysms in other locations such as the cavernous ICA or the basilar artery can also cause third nerve dysfunction. Cavernous carotid aneurysms can spare the pupil if the inferior division of the cavernous third nerve is not compressed. In addition, other cranial nerves are typically, but not always, involved [134, 170]. However, as discussed in Sect. 6.2, a delay in this diagnosis is never catastrophic since these aneurysms are not associated with SAH.

An aneurysm of the basilar artery may also cause a pupillary-sparing third nerve paresis [171]. This aneurysm is typically large and should not be missed by high-resolution contrast CT or MR scans that image the appropriate area. There is only one example in the literature of a case of a complete external third nerve paresis and a normal pupil caused by an unruptured basilar aneurysm compression of a third nerve. This patient, who also had systemic hypertension, did not have worsening or resolution of his third nerve paresis. His failure to improve within the expected time period, considered typical of a microinfarct of the peripheral third nerve, prompted a CT to be performed and the aneurysm was diagnosed [172].

Though a paresis limited to the superior division of the third nerve, causing a ptosis and superior rectus dysfunction, is usually suggestive of a lesion in the cavernous sinus, rare cases have been caused by basilar artery aneurysms [170, 173]. Though also rare, a superior division third nerve paresis can be caused by a diabetic microinfarct of the peripheral nerve [174, 175] or a lesion that affects selective fascicles of the nerve in the midbrain. Posterior communicating artery aneurysms are not associated with an isolated superior division paresis.

When a patient who is older than 50 years of age presents with a pupillary-sparing *isolated* third nerve paresis (this means the patient has been assessed for other cranial nerve, motor, sensory, and cognitive deficits and does not have meningismus or fever), he is evaluated for the etiologies of microvascular disease (see Sect. 8.8) and observed closely over the next week. CT or MR imaging can be performed if the patient does not show improvement within 8 weeks, since 96% of the microvascular third nerve pareses demonstrate improvement within this time period [160]. A similar radiological evaluation should be undertaken if other neurologic findings appear during this follow-up period. In adults less than 50 years of age, CT or MR imaging should be performed even when the pupil is normal, because other lesions are more likely to be the cause in this age range. As discussed previously, patients more than 20 years of age who present with pupillary-involving third nerve dysfunction should be considered to have a posterior communicating artery aneurysm until proven otherwise. In patients younger than 18 years, a presentation of a third nerve paresis with pupillary dysfunction caused by a posterior communicating artery aneurysm is exceedingly rare [176]. A SAH is the more common presentation in this age group [177]. Since 14 years of age is the age of the youngest patient to present with a pupillary-involving third nerve paresis from a posterior communicating artery aneurysm, angiography is usually unwarranted in patients younger than 11 years old [178]. However, patients with pupillary-involving third nerve palsies, even if the dysfunction is episodic as in ophthalmoplegic migraine, should undergo CT or MR imaging.

Rare combinations of pupillary dysfunction and external ophthalmoplegia associated with posterior communicating artery aneurysms are scattered in case reports in the literature. There are exceedingly rare cases of aneurysms that cause the pupil to be dilated and poorly reactive to light with a defective accommodative response, similar to the acute ciliary ganglion dysfunction found early in an Adie's pupil, with minimal lid dysfunction and no difficulty with eye movement [158, 179, 180] or pupil and ocular motor dysfunction without an associated ptosis [118, 181]. Another unusual presentation, a dorsal midbrain syndrome, can develop if a giant posterior communicating artery aneurysm points posteriorly and compresses the midbrain [182].

6.3.2.2 Visual Loss

Posterior communicating artery aneurysms are infrequently large enough and extend anterolaterally to compresses an optic nerve or project into the suprasellar cistern to compress the chiasm or optic tract [43, 183]. In these cases, the resulting visual loss is asymmetrical. The visual field loss typically includes a central and temporal scotoma in the eye ipsilateral to the aneurysm, with or without a temporal field defect in the contralateral eye. Massive enlargement of the aneurysm causes lateral chiasm or optic tract compression with an incongruous homonymous hemianopia (see Sect. 6.4.2.2 for discussion of visual field defects). Rupture of the aneurysm into the optic apparatus is rare but an intraocular hemorrhage is not (Terson's syndrome). Another rare cause of unilateral visual loss can occur when acute glaucoma develops because of the mid-dilated pupil. Angle closure with a significant elevation of the intraocular pressure develops in patients who already have existing narrow angles [184].

6.3.3 Neuroimaging

6.3.3.1 Computed Tomography and Magnetic Resonance Imaging

Following a SAH, the CT can reveal blood in the basal cisterns. A hemorrhage is seen in the temporal lobe if the ruptured aneurysm has a dome that projects in the lateral direction. If the image plane does not slice through the aneurysm, a routine head CT or MR scan often fails to visualize a posterior communicating artery aneurysm that is not giant sized.

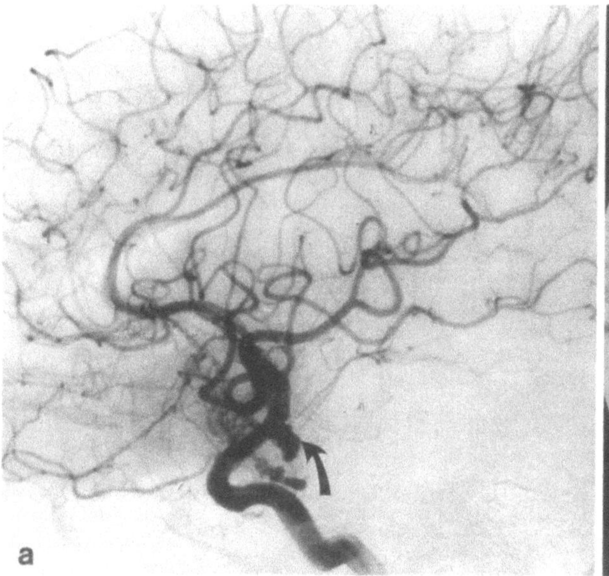 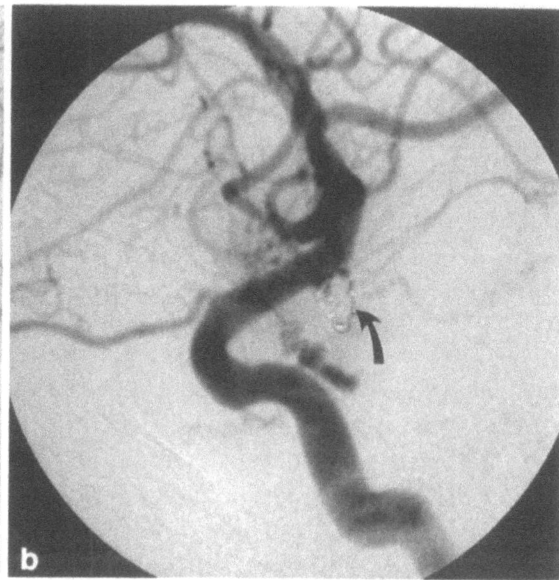

Fig. 6.19a, b. A 55-year-old woman experienced progressive visual loss in both eyes and had a chiasmal type field defect from a subfrontal meningioma. **a** The lateral internal carotid artery angiogram demonstrated a posterior communicating artery aneurysm (*curved arrow*) as well as the tumor vascularity which was supplied by branches off the cavernous internal carotid artery (large vessels seen below the aneurysm). **b** As this aneurysm was not amenable to surgical clipping because of the surrounding tumor, embolization was performed with multiple electrothrombogenic coils. The lateral internal carotid artery angiogram 3 weeks following the embolization demonstrates an absence of the aneurysm and a ring of metal coils (*curved arrow*) where the aneurysm had been. The patient later underwent uneventful surgical removal of the meningioma

6.3.3.2 Angiography and Normal Blood Supply

In most cases, cerebral angiography establishes the diagnosis of a posterior communicating artery aneurysm and visualizes the neck and the direction of the aneurysm projection (Fig. 6.19a). If an ICA contrast injection is negative and a posterior communicating artery aneurysm is clinically suspected, a vertebral arterial injection study must be performed in order to visualize the entire length of the posterior communicating artery. Various projections and stereoscopic angiography may be required to demonstrate the relationship of the neck to the parent vessel and other surrounding arteries. In rare cases with severe vasospasm, the aneurysm may not be visualized. In other rare cases, the aneurysm in this region arises from the origin of the anterior choroidal artery.

The importance of the blood supply to the cerebral hemisphere from the posterior circulation through the posterior communicating artery must be assessed. Arteriography can also demonstrate other potential collaterals to the region (see Table 1.3 in Chap. 1, Sect. 6.3.4.1). Angiography also reveals whether there is an associated developmental anomaly, a posterior cerebral artery origin from the ICA. The posterior cerebral artery originates from the ICA at the level of the posterior communicating artery twice as frequently in patients with a posterior communicating artery aneurysm as in the normal population [151].

An infundibulum, seen on 7% of normal angiograms, is a dilatation of the ICA at the posterior communicating artery origin which can be confused with an aneurysm. An infundibulum is typically no larger than 3 mm, has no neck, and may or may not be a "preaneurysm" state [185a, 185b]. Histology reveals that the junctional dilatation contains normal adventitia, media, internal elastic membrane, and intima, all typical of a normal vascular structure. A junctional dilatation is not necessarily a pathological entity that is destined to become an aneurysm [186].

6.3.4 Therapy and Outcome

6.3.4.1 Surgery

As is the case for all intradural aneurysms, the outcome of surgical clipping is dependant on the clinical status of the patient prior to surgery or after a SAH. If untreated, 60% of ruptured posterior communicating artery aneurysms will rebleed within the first 6 months. Complications can also develop if a surgical clip or dissection to expose the aneurysm neck interrupts the normal blood supply to an area that has no significant collateral circulation. For example, compromise of the posterior communicating artery can be devastating if this vessel is dominant over the anterior choroidal artery in supplying the brain. If the tuberothalamic artery is occluded in the dominant hemisphere, the resulting infarct of the anterolateral thalamus causes visual and constructional apraxias, language and memory difficulties, and disorientation. If the nondominant cerebral hemisphere is affected, visual and constructional apraxia without language or memory problems can develop. An infarct of either hemisphere can also cause a homonymous field defect or a hemiparesis [187]. The surgery may actually worsen the third nerve dysfunction. A delayed-onset third nerve paresis rarely occurs as a result of vasospasm of the posterior communicating artery after clipping of a posterior communicating artery aneurysm [188].

6.3.4.2 Embolization

One clear indication for percutaneous embolization is the case of a giant posterior communicating artery aneurysm that has been previously explored and found to be unclippable by a neurosurgeon experienced in aneurysm surgery. The usual precautions concerning the patient's capacity to tolerate occlusion of the ICA must be taken. The goal, which is similar to that of surgery, is to exclude the aneurysm from the arterial circulation and preserve the normal blood flow in the ICA. Balloon embolization of a ruptured aneurysm can cause an acute rebleed. If the balloon deflates or thrombosis of the aneurysm is not complete, the risk of rebleeding is not reduced. Percutaneous embolization with coils that induce thrombosis within the

aneurysm may be a more definitive therapy (Fig. 6.19b).

6.3.4.3 Carotid Ligation

Prior to the advent of safe techniques for clipping these aneurysms and balloon embolization, surgical closure of the cervical carotid artery was performed. Cervical carotid artery ligation has continued to be used as an alternative therapy in the infrequent case where the aneurysm (usually giant without a definitive neck) is considered unclippable or not embolizable. Morris reported that in 29 cases, 18 aneurysms were definitively smaller, seven were slightly smaller, and two were unchanged after either ICA or common carotid artery ligation [189]. Though the rebleed rate was diminished during the first 6 months [98], the incidence of late rehemorrhaging from posterior communicating artery aneurysms after a common carotid artery occlusion was similar between the patients with ligation and those treated conservatively with only blood pressure management (Fig. 6.10) [190]. Percutaneous balloon embolization is probably the preferred method of occluding the cervical ICA or common carotid artery in these cases.

6.3.4.4 Third Nerve Function

Following surgery, carotid ligation, or embolization, the aneurysmal sac shrinks and decompresses the third nerve. Surgical debulking of the aneurysm mass is rarely required. The diplopia or the ptosis-related monocular vision can be quite annoying if not debilitating. Unless the third nerve is sectioned at surgery or the nerve is severely damaged by hemorrhage directly into the nerve, function gradually recovers over months [191]. In addition to the aneurysm compression of the nerve, surgical dissection, which stretches the nerve and interrupts its blood supply temporarily, worsens the third nerve disturbance. The recovered third nerve has fewer axons, smaller fibers, and less myelination [192] than a normal third nerve.

As long as a complete ptosis persists, the patient does not have spontaneous diplopia. When the lid opens, the patient has diplopia, with the images varying in their relation to each other depending on the direction of gaze. Initially, therapy is limited to alternating a patch between the eyes. A patch is worn over the affected eye most of the day and the normal eye is covered when in familiar surroundings or while sitting or viewing television. Using the affected eye for at least 1 h daily decreases the probability of developing a secondary contracture from shortening of the unaffected antagonist eye muscles. The extraocular eye movements may normalize but residual underaction and secondary overaction of the muscles frequently leave the patient with a tropia and persistent diplopia. If the residual eye movement deficit is severe, the patient often resorts to some type of occlusion therapy over the affected eye. If the deficit is small, prisms, initially temporary fresnels followed by permanently ground prisms, can compensate for the misalignment in primary or downward gaze, or both.

If the deficit remains large, strabismus surgery can restore monocular vision over a limited range of eye movements. Prism therapy and surgery can also be used in combination. Surgery can be contemplated if no recovery is noted by 3–6 months following the aneurysm clipping. Surgery should not be performed in a patient who continues to improve, even after 6 months. A newer approach uses botulinum toxin injection into the antagonist muscle to temporarily weaken it. This results in a reduction of the secondary deviation of the eye and permits limited fusion (often in conjunction with prism therapy). The injections can be initiated as soon as the patient has made some oculomotor recovery. The botulinum toxin injection can be repeated every several months until the recovery is complete or surgery is performed.

Ptosis surgery is performed, after the same time period of waiting for recovery, once the diplopia is controlled. If the ptosis is corrected when the patient still has severe limitation of eye movements, then a patch will be required to avoid the diplopia.

Persistent dilation of the pupil and ciliary muscle dysfunction are common even in patients who recover extraocular movement. Photophobia can be compensated for with tinted lens and the accommodative paresis can be managed with a plus lens for near vision.

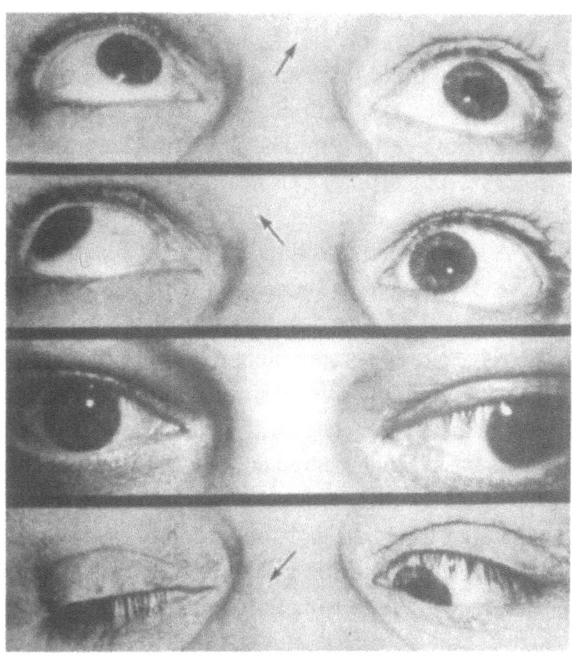

Fig. 6.20. This patient shows signs of an incompletely recovered left third nerve paresis with findings of misdirection. On upgaze to either left or right, there is an inability to completely elevate the left eye. In primary gaze, there is a left exotropia. On downgaze to the right, there is left upper lid retraction. (From [40])

6.3.4.5 Oculomotor Synkinesis

If the recovered third nerve impulses do not arrive at the appropriate muscles, a misdirection syndrome, also called oculomotor synkinesis or aberrant regeneration, develops. This phenomenon was first reported by Gowers [193] (Fig. 6.20). Approximately 3 months following decompression, an atypical pattern of recovery of third nerve functioning appears [194]. This can also occur in patients with longstanding compression from cavernous sinus meningiomas or aneurysms [41, 195]. Typically, the upper eyelid is in the normal position in primary gaze, but on abduction of the eye the lid falls. On adduction, the lid paradoxically elevates. The levator palpebrae overaction occurs when neural fibers from the medial rectus subnucleus innervate the levator, so attempted adduction also contracts the levator. Less frequently, the upper lid elevates on downgaze (pseudo-Graefe phenomenon).

The pupil may constrict on adduction or downgaze because of a similar inappropriate innervation to the pupillary sphincter. Additionally, sector contraction of the iris sphincter is commonly seen with both light stimulation and eye movements [196]. Iris sphincter hypersensitivity to a weak cholinergic stimulus (2.5% methacholine or 0.125% pilocarpine) can also be demonstrated in some, but not all cases [197].

A theoretical mechanism which accounts for these phenomena is based on the observation that following Wallerian degeneration of the axons distal to the site of compression, axons proximal to the disrupted endoneurium sprout and regrow. The nerve fibers regenerate distally [198] and collateral axons pass into preserved endoneurial sheaths (first theorized by Bielchowsky) [199]. However, the regenerated fibers are not topographically organized and can grow into the wrong endoneurial sheaths, reinnervating the wrong muscles. In support of this hypothesis, misdirection is not seen in patients with a severed nerve because there is no apposition of the proximal and distal endoneurial segments. It is also extremely rare following an infarct of the nerve because the endoneurium is not disrupted. This anatomical theory does not account for the rare cases of spontaneous recovery from a misdirection syndrome because the misdirected regrown fibers should remain once they are established [200].

Another theory suggests that ephaptic transmission (electrical cross-talking between adjacent axons because of a lack of adequate myelination) in a recovering peripheral third nerve accounts for the oculomotor synkinesis. A third hypothesis postulates that following retrograde degeneration of damaged third nerve axons, a central reorganization unmasks existing connections that are normally inhibited between third nerve subnuclei nuclei [201].

6.4 Carotid–Ophthalmic Artery Aneurysm

6.4.1 Incidence

Carotid–ophthalmic artery aneurysms frequently cause a disturbance of the intracranial anterior visual pathway and therefore hold a particular interest for the ophthalmologist and neurologist inter-

ested in vision. Aneurysms that develop in the segment of the intracranial ICA adjacent to the origin of the ophthalmic artery have an incidence of between 1.5% to 7.9% of all intracranial aneurysms [202–204]. However, carotid–ophthalmic aneurysms make up a higher percentage, 13%, of all unruptured aneurysms [36a]. Because the ophthalmic artery is the principal branch artery in the area, many authors consider all aneurysms on the ventral surface of the intradural portion of the ICA proximal to the bifurcation to be carotid–ophthalmic or paraophthalmic aneurysms. However, aneurysms can also arise from the origin of large perforating arteries on the ICA.

Up to 20% of cases have bilateral carotid–ophthalmic aneurysms. Between 31% and 64% of the cases have additional aneurysms located at other sites, with the posterior communicating artery and middle cerebral arteries the most frequently involved vessels [205, 206]. Women are clearly more often affected than men, and the ratio of females to males is, at 143/107, higher than for all other aneurysms except those of the cavernous ICA [36a]. The predilection for this aneurysm to affect women is even greater among cases with visual loss, 90% of whom are women [207]. The age of the patients at presentation is similar to the age for other aneurysms, and 80% are diagnosed between the fourth and seventh decades.

6.4.2 Clinical Presentation

6.4.2.1 Subarachnoid Hemorrhage

SAH is the most common clinical presentation, with a hemorrhage reported in 161 of 261 cases combined from three series [36a, 204, 207]. Those patients who experience visual loss have a lower incidence of SAH. Approximately 60% of patients with a complaint of a progressive visual disturbance have unruptured aneurysms (many are giant aneurysms) at the time of diagnosis [207]. If untreated, the incidence of a fatal SAH in these unruptured aneurysms is approximately 22% within 4 years of presentation [36b].

6.4.2.2 Visual Loss

Because of the proximity of the carotid – ophthalmic aneurysm to the optic nerve and chiasm, one would think progressive deterioration of vision should be present in most, if not all, cases, but many patients have no overt visual disturbance [204, 208]. In fact, headache and orbital pain, and not visual loss, are the most common complaints in these patients, despite approximately 28% of these aneurysms being larger than 2.5 cm in diameter [207] and 45% larger than 1.2 cm. However, when the visual disturbance is caused by a slowly expanding mass (of any type), patients are often unaware of the gradual visual loss until the second eye becomes involved. Additionally, many of the reported series on this aneurysm did not include detailed neuro-ophthalmic examinations. Since many patients present with a SAH, their altered state of consciousness or severe neurologic deficit might preclude the discovery of a visual disturbance. For example, Yasargil reported that less than 25% of patients had visual complaints [204] while Drake and others described an even lower rate of visual symptoms [203, 206, 209].

In contrast to the patients who present with a SAH, patients with giant aneurysms have a higher incidence of visual loss. Hosobuchi reported that 17 of 19 patients had visual problems on presentation [210], and Jefferson described 66 cases (it is not clear out of how many total cases) of optic pathway compromise by giant aneurysms [48, 211]. Other reports describe cases with an initial presentation because of various combinations of optic nerve [212, 213], chiasm [44, 214, 215], or optic tract dysfunction from these lesions (Fig. 6.21a – h).

When a supraclinoid aneurysm disturbs the chiasm, the aneurysm is located in the laterochiasmal area in 22% of patients, in the suprachiasmal area in 39%, and in the infrachiasmal area in 34%. When the aneurysm is beneath the chiasm a portion of the aneurysm may be intracavernous and it may be difficult to distinguish whether the origin is intradural or in the cavernous sinus. This differentiation is even more difficult in the 5% of cases where the entire wall of the ICA is included in the wall of the aneurysm.

Fig. 6.21a – i. The visual loss caused by an aneurysm can be insidious. Cases are misdiagnosed as having normal or low- tension glaucoma. However, the field loss is asymmetric and the reduction in visual acuity is out of proportion to the degree of cupping in the patients with aneurysms.

a A 75-year-old woman complained of poor vision in the left eye for an unknown period of time. The visual acuity was 20/40 OD and 20/200 OS. There were cataracts in both eyes, but there was a visually significant cataract in the left eye. Intraocular pressures were both 21 mmHg at that time. Goldmann perimetry shows some mild constriction of the field in the right eye compatible with cataract. In the left eye, there was a left superior paracentral scotoma. Following the cataract surgery, the vision improved to 20/30 in the left eye.

b,c Fundus photography obtained 4 months after cataract surgery reveals myopic discs in both eyes (bright eye) with a slight increase in the cup to disc ratio in the left eye (c).

d A repeat visual field examination 11 months after the fundus photography demonstrates no significant change in the visual field for the right eye except "baring of the blind spot." The left eye has a significant nasal field depression, both inferiorly and superiorly. There was also a superior arcuate defect. At that time, the visual acuity was 20/40 OD and 20/30 OS and the intraocular pressure was 22 mmHg in both eyes. She was suspected of having normal tension glaucoma.

e Ten months later, the patient returned because of progressive visual loss in the left eye. At that time, the visual acuity was 20/40 OD and 20/100 OS. The visual field in the right eye remained unchanged. The left eye visual field had lost almost the entire nasal field. Superior and inferior arcuate and central scotomas were present.

f A coronal view T1-weighted magnetic resonance scan obtained at that time reveals an aneurysm with partial thrombosis (*arrow*). The aneurysm displaces the hypothalamus and third ventricle above it.

g A lateral view internal carotid artery subtraction angiogram shows a left paraophthalmic artery aneurysm. The injection of contrast material into the right internal carotid artery demonstrated no cross-filling and the patient was unable to tolerate carotid compression (corresponds to Fig. 6.8).

h The patient refused surgery or embolization therapy. Over the course of 3 years, the patient became entirely blind in the left eye with an amaurotic pupil and an optic atrophy. She then developed a progressive dementia. The coronal magnetic resonance scan demonstrates a smaller lumen but a much larger aneurysm mass compressing the subfrontal area of the brain.

i A repeat lateral view internal carotid artery angiogram confirms that the aneurysm lumen is indeed smaller (*arrow*). However, the internal carotid artery is displaced posteriorly (*curved arrow*) because of the mass

Fig. 6.21a – i see pp. 274/275

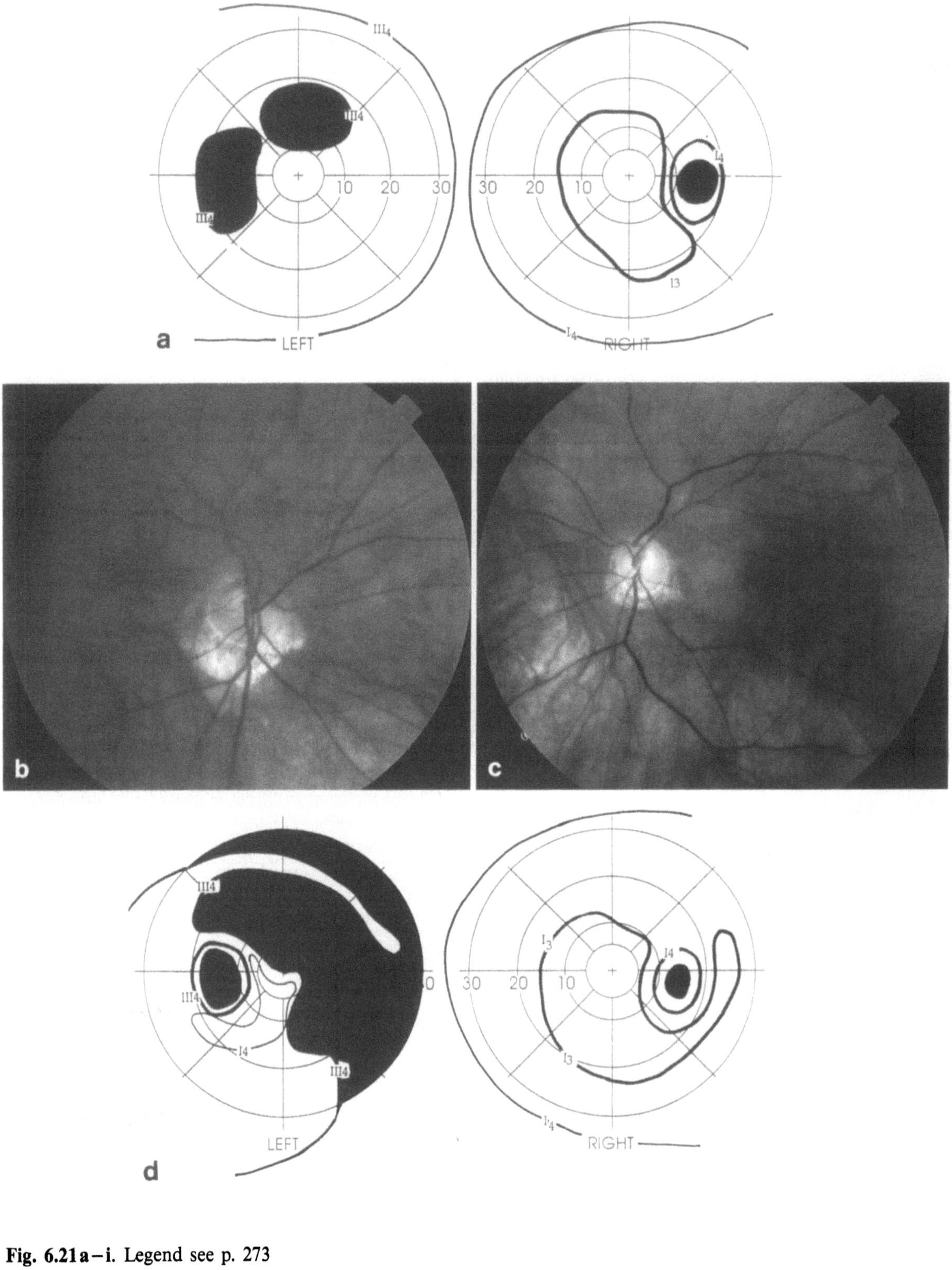

Fig. 6.21 a – i. Legend see p. 273

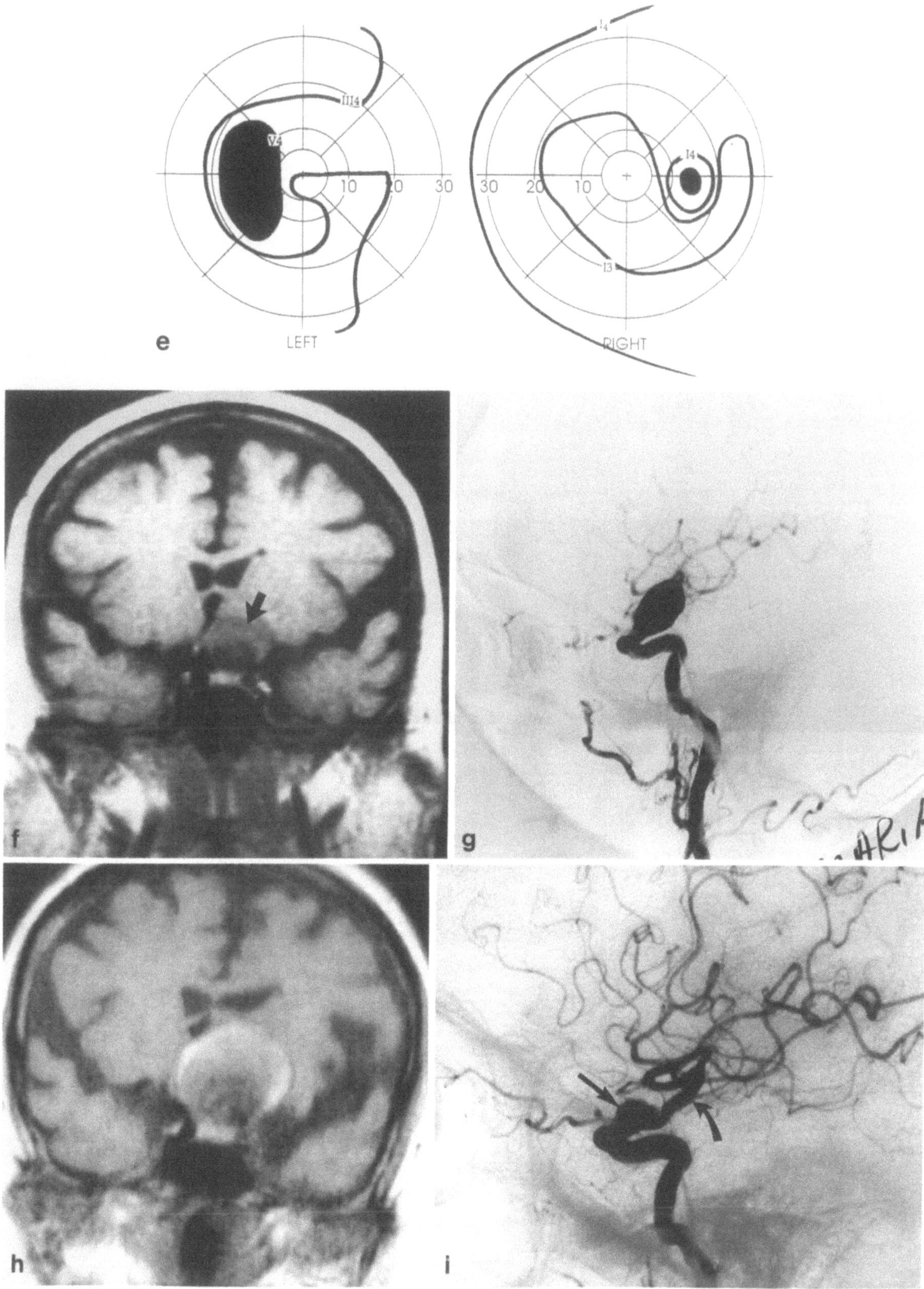
LEFT
RIGHT
e
f
g
h
i

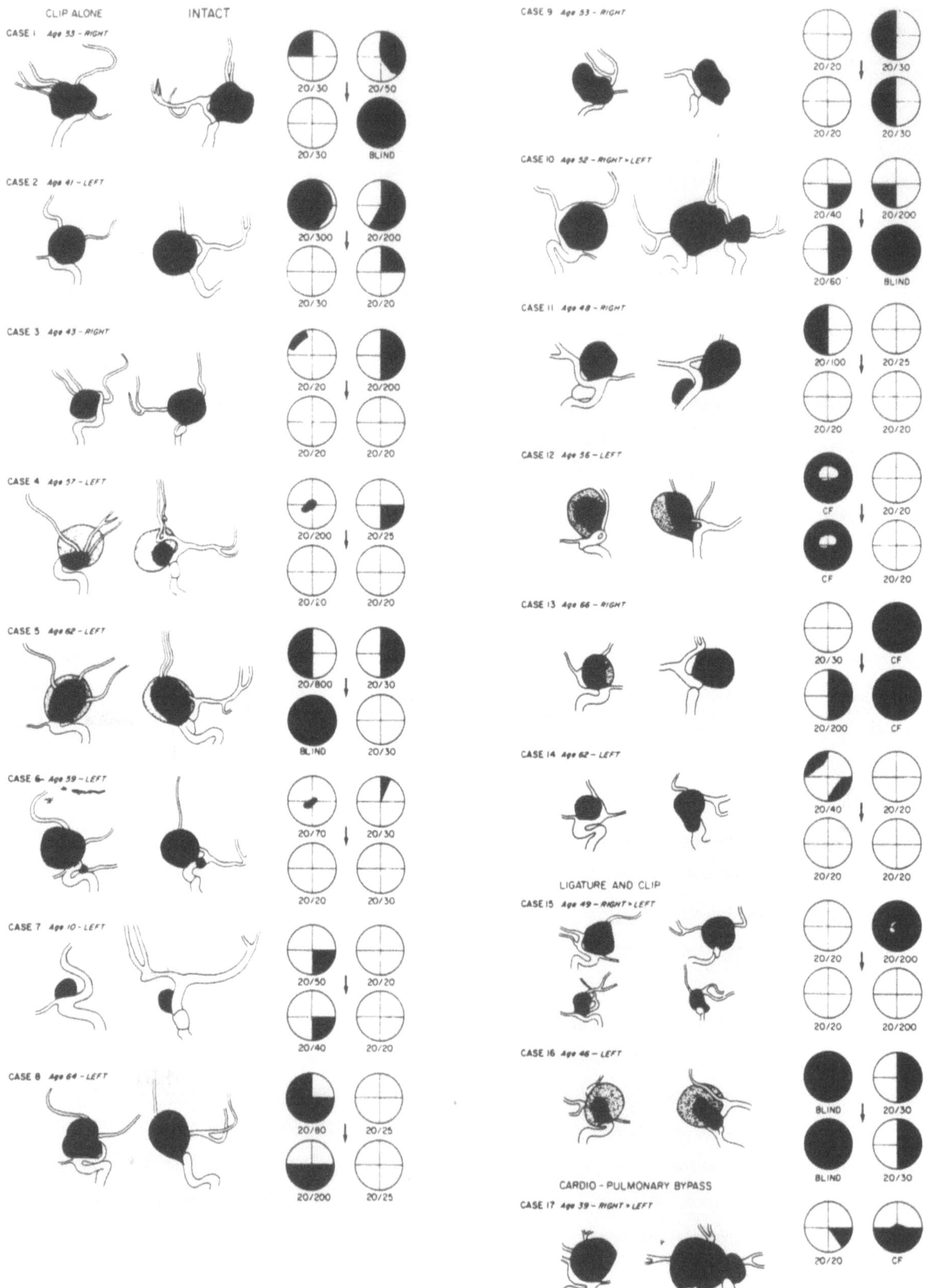
CLIP ALONE
INTACT
CASE 1 Age 53 - RIGHT
20/30   20/50
20/30   BLIND
CASE 2 Age 41 - LEFT
20/300   20/200
20/30   20/20
CASE 3 Age 43 - RIGHT
20/20   20/200
20/20   20/20
CASE 4 Age 57 - LEFT
20/200   20/25
20/20   20/20
CASE 5 Age 62 - LEFT
20/800   20/30
BLIND   20/30
CASE 6 Age 59 - LEFT
20/70   20/30
20/20   20/30
CASE 7 Age 10 - LEFT
20/50   20/20
20/40   20/20
CASE 8 Age 64 - LEFT
20/80   20/25
20/200   20/25
CASE 9 Age 53 - RIGHT
20/20   20/30
20/20   20/30
CASE 10 Age 52 - RIGHT > LEFT
20/40   20/200
20/60   BLIND
CASE 11 Age 48 - RIGHT
20/100   20/25
20/20   20/20
CASE 12 Age 56 - LEFT
CF   20/20
CF   20/20
CASE 13 Age 66 - RIGHT
20/30   CF
20/200   CF
CASE 14 Age 62 - LEFT
20/40   20/20
20/20   20/20
LIGATURE AND CLIP
CASE 15 Age 49 - RIGHT > LEFT
20/20   20/200
20/20   20/200
CASE 16 Age 46 - LEFT
BLIND   20/30
BLIND   20/30
CARDIO - PULMONARY BYPASS
CASE 17 Age 39 - RIGHT > LEFT
20/20   CF

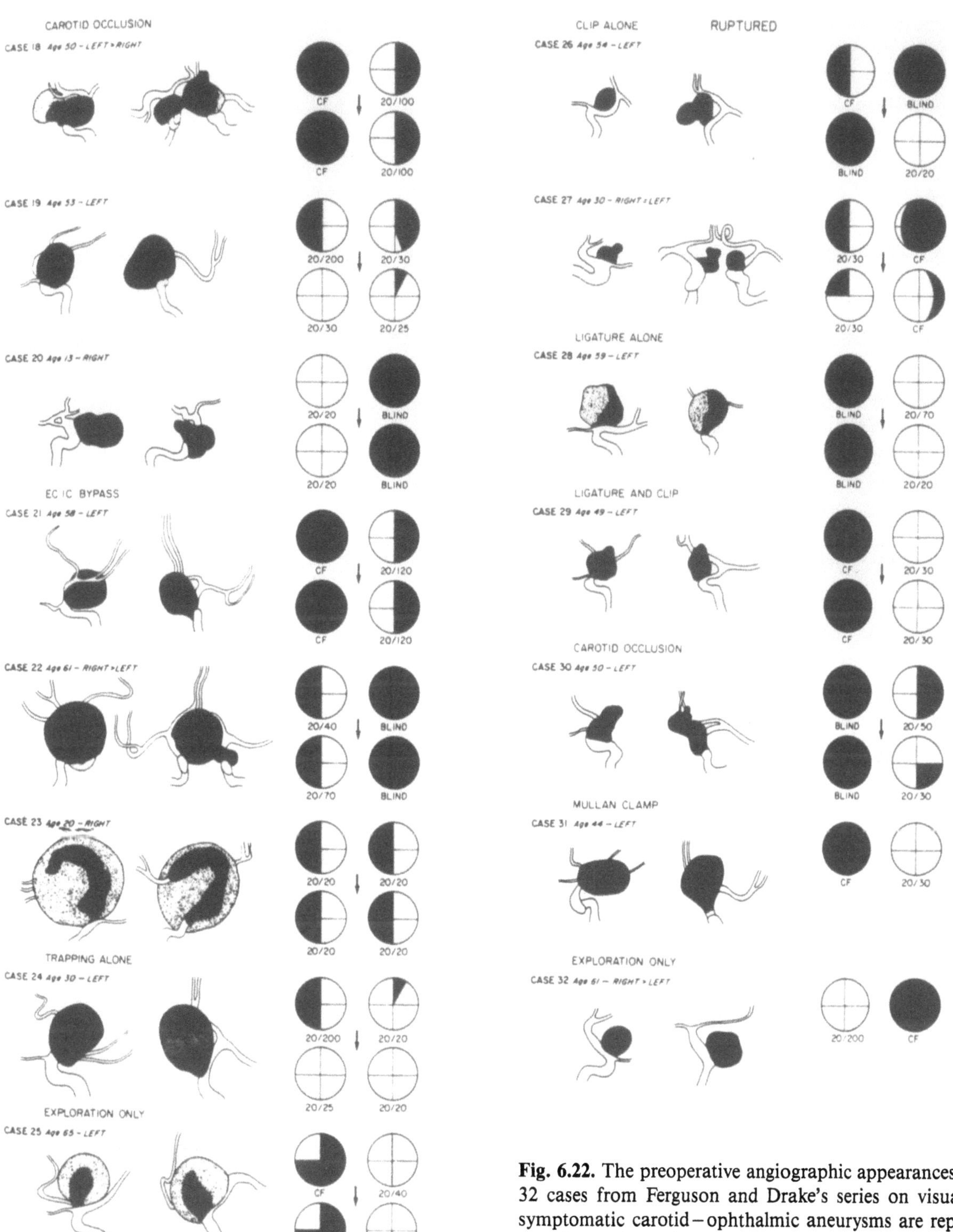

Fig. 6.22. The preoperative angiographic appearances of 32 cases from Ferguson and Drake's series on visually symptomatic carotid–ophthalmic aneurysms are reproduced, together with a schematic representation of the preoperative and postoperative visual fields and acuities. The lateral view of the angiogram is to the *left* and the anteroposterior view is to the *right* in each case. The preoperative visual examinations are *above* and the postoperative results *below* for each case. The right eye is to the *right* and the left eye to the *left*. *CF*, finger counting. (From [207])

6.4.2.2.1 Patterns of Visual Loss

The pattern of visual loss is typical of asymmetric involvement of one or both posterior optic nerves or the chiasm or both (Fig. 6.22). In the eye ipsilateral to the aneurysm, patients often have a central scotoma. Some patients have an altitudinal depression [216, 217], or a temporal or nasal quadranopia or hemianopia. If the chiasm is involved, the contralateral eye will have a temporal quadranopia or hemianopia or a normal field if the crossing fibers of the temporal field are not compromised [207]. The chiasm is more likely to be involved when the aneurysm arises from the superior hypophyseal artery and projects medially [218]. Unlike the field loss caused by a suprasellar extending pituitary adenoma, a symmetrical bitemporal hemianopia is unusual in a patient with a giant aneurysm.

In one study, a central scotoma, alone or in addition to other field defects, occurred in the eye ipsilateral to the aneurysm in 25 of 32 patients [207]. In some cases the aneurysm projected across the midline to compress the contralateral optic nerve, causing a central scotoma in this eye as well. A partial temporal quadranopia to a complete hemianopia in the eye contralateral to the aneurysm was noted in 14 patients. Monocular visual field deficits occurred in eight cases with a central scotoma, in four with nasal field loss, in one with a superior temporal quadranopia, and in one with an altitudinal field defect in the eye ipsilateral to the aneurysm. Binasal field defects occurred in only one patient. A homonymous field defect, typical of optic tract dysfunction, was found in one patient.

In another series, 13 of 14 patients had a central scotoma in at least one eye, often in addition to another type of field defect in the same eye. Bilateral central scotomas were noted in seven of 14 patients. Other types of field defects included partial or complete temporal field depression in seven cases, homonymous hemianopia. In addition, there were bilateral central scotoma in one, and a nasal field defect 2/14 (one with central scotoma), one inferior altitudinal defect with central scotoma, a junctional defect in five patients, but none had a bitemporal field defect [211]. Rarely, an altitudinal field defect without a central scotoma is seen in the ipsilateral eye (Fig. 6.23).

In our series, nine of 14 patients had field loss in both eyes, though the eye contralateral to the aneurysm had an acuity of 20/30 or better in seven of 9 patients. The field defects in the contralateral eye were often subtle and would have been missed without formal perimetry. In contrast, the visual field defects caused by supraclinoid aneurysms that did not arise from the origin of the ophthalmic artery were worse in the eye contralateral to the aneurysm. Significant visual field defects were found in both eyes in nine of 12 of these patients with supraclinoid ICA aneurysms. However, the acuity of the contralateral eye was 20/30 or better in six of 9 of the contralateral eyes. Both types of aneurysms frequently involve the ipsilateral intracranial optic nerve, but the supraclinoid aneurysms that do not arise form the ventral surface of the ICA are more likely to cause additional dysfunction in the chiasm and optic tract.

Fig. 6.23. a Threshold perimetry demonstrates an inferior altitudinal type field defect along with an overall depression over most of the visual field. The field defect had not been previously present in this patient with early open angle glaucoma and normal intraocular pressure who used pilocarpine eye drops in both eyes.
b Because of the asymmetrical visual field loss and adequate control of the intraocular pressure, a computed tomogram was obtained. The axial view noncontrast scan demonstrates a round lesion eroding into the posterior portion of the optic canal (*arrow*).
c The sagittal view T1-weighted magnetic resonance scan reveals an aneurysm (*arrow*) that is displacing the left optic nerve (*arrowhead*) upward.
d The frontal view left internal carotid artery subtraction angiogram demonstrates an ophthalmic artery aneurysm as well as cross-filling across the anterior communicating artery to fill the right anterior cerebral and middle cerebral arteries. An injection of contrast material into the right internal carotid artery visualized both the anterior cerebral and left middle cerebral arteries as well as demonstrating competency of the anterior circle of Willis so that the patient could tolerate carotid occlusion. However, the aneurysm was clipped successfully with the patency of the left internal carotid artery preserved

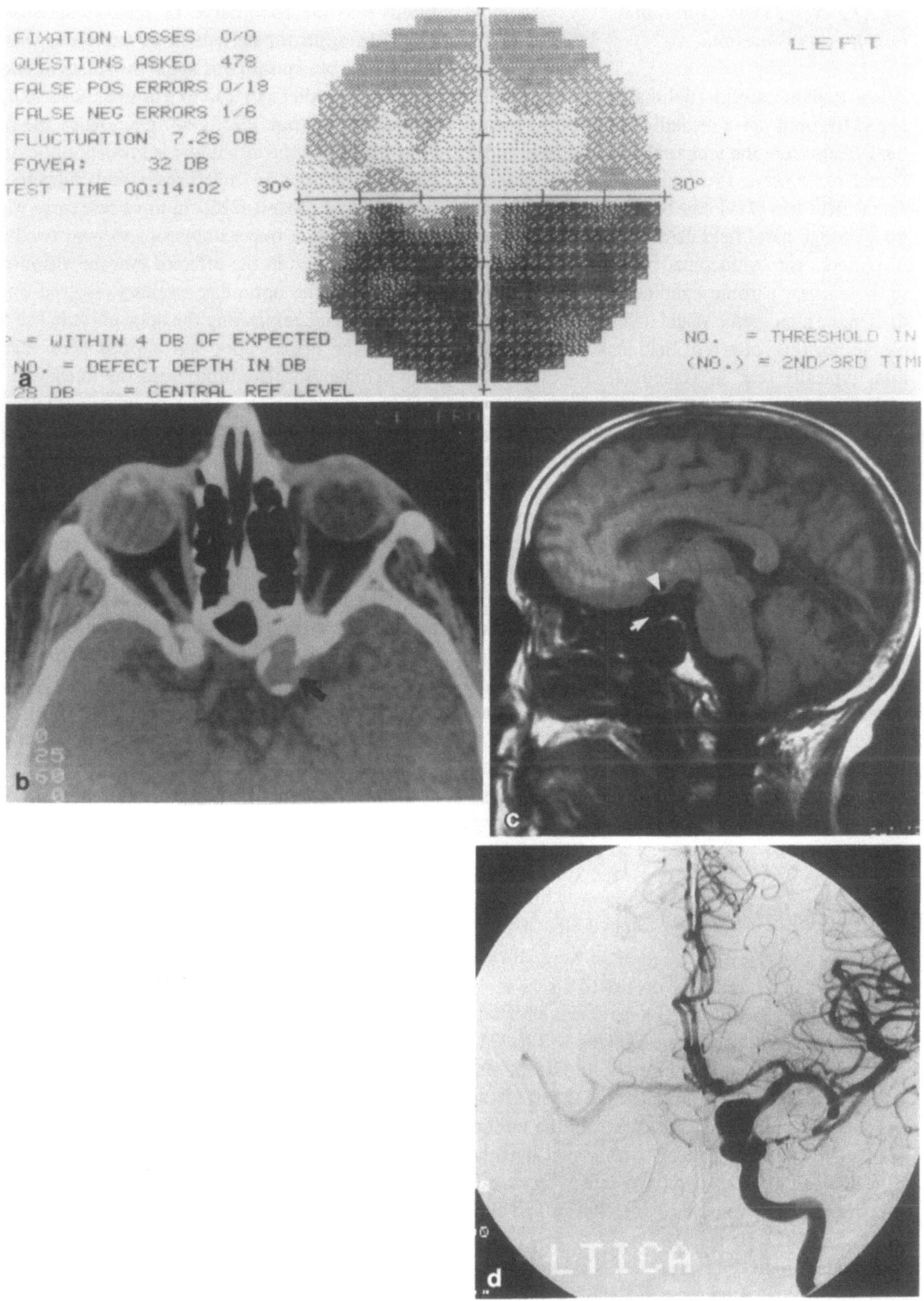
FIXATION LOSSES 0/0
QUESTIONS ASKED 478
FALSE POS ERRORS 0/18
FALSE NEG ERRORS 1/6
FLUCTUATION 7.26 DB
FOVEA: 32 DB
TEST TIME 00:14:02   30°
LEFT
30°
= WITHIN 4 DB OF EXPECTED
NO. = DEFECT DEPTH IN DB
28 DB   = CENTRAL REF LEVEL
NO. = THRESHOLD IN
(NO.) = 2ND/3RD TIM
a
b
c
d
LTICA

6.4.2.2.2 Nasal Field Defect and Low-Tension Glaucoma

When a giant carotid–ophthalmic aneurysm displaces the optic nerve medially, compressing the axonal fibers from the temporal retina located in the lateral optic nerve, the resulting nasal monocular visual field loss [219] can be mistaken for glaucoma. Though nasal field defects are typically found in patients with glaucoma, retinal detachment or retinoschisis, chronic papilledema, optic disc drusen, or nonorganic visual loss are additional etiologies for this pattern of field loss. Bilateral nasal loss, typically seen in glaucoma, can also occur with any of the above conditions.

As discussed under low-tension glaucoma (Sect. 5.6), a carotid–ophthalmic aneurysm can unusually cause cavernous degeneration of the optic nerve [220] that can be mistaken for normal or low-tension glaucoma, particularly if the field loss preferentially affects the nasal field. It is worth repeating that, unlike glaucoma patients, the aneurysm patient suffers from reduced visual acuity and clinically apparent dyschromatopsia, out of proportion to the extent of the cupping of the optic nerve; and the nonexcavated portion of the optic disc has areas of pallor. Fortunately, most of the aneurysm-related visual loss occurs in the temporal or central field, and if a nasal field defect is present, it is usually hemianopic (i.e., respects the vertical meridian) (Fig. 6.21).

6.4.2.2.3 Optic Neuropathy

When the aneurysms are larger than 2 cm, 40% of patients will have unilateral decreased visual acuity and another 40% will have bilateral loss of acuity. Approximately 90% of these patients will have diminished acuity in at least one eye. Rarely, the acuity is worse in the eye contralateral to the aneurysm because the giant mass extends across the midline [214]. Only 40% of patients with visual dysfunction have an afferent pupillary defect because of the frequent involvement of the chiasm and both optic nerves [221].

In addition to the typical slowly progressive visual loss discussed above, a precipitous fall in vision can result from a sudden expansion of an aneurysm against the optic apparatus, a SAH that bleeds into the optic nerve, or Terson's syndrome [222]. In approximately 10% of cases the vision fluctuates, but episodes of spontaneous significant improvement that can lead to a misdiagnosis of optic neuritis are rare [223, 224]. Even when the patient states that the visual loss is sudden, optic disc pallor is present in most cases and suggests a longstanding process [225]. In three rare cases with acute visual loss, ophthalmoscopy showed swelling of the optic disc in the affected eye; the vision recovered and the optic disc swelling resolved once the aneurysm compressing the optic nerve in the region of the orbital apex was clipped [226–228]. Hemorrhage into the orbit from the aneurysm is another extremely rare cause of acute visual loss that is accompanied with proptosis [229].

Most series, including our own, have found optic disc pallor and nerve fiber layer loss in almost all patients with visual loss at the time of clinical presentation. In contrast, there is one report that describes a normal appearance of the optic discs in 60% of patients with visual loss [221]. When optic atrophy is present, the optic nerve pallor can be focal or diffuse and asymmetric between the eyes. The optic disc ipsilateral to the aneurysm is usually more atrophic [230]. As with other compressive lesions of the anterior visual pathway, the degree of pallor does not always correspond to the extent of the visual acuity or field loss.

6.4.2.3 Pituitary Dysfunction

If a bitemporal field defect is present, an aneurysm may not be clinically distinguishable from a nonsecreting pituitary tumor or pituitary apoplexy, and CT may not definitively differentiate an aneurysm from a neoplasm. As mentioned previously, MR demonstration of a flow void within the aneurysm lumen is diagnostic (see Sect. 6.1.8.2). Unless MR imaging or angiography is performed a misdiagnosis of a pituitary adenoma can lead to a catastrophic hemorrhage from transphenoidal surgery [63].

Similar to the discussion concerning the diagnosis of visual dysfunction, a problem of inadequate documentation exists in determining the frequency of pituitary dysfunction from compression or vascular compromise of the hypophyseal stalk or the pituitary gland. Pituitary dysfunction may be transient or subtle. Typically, the patients are often given corticosteroids emergently or preoperatively, so cortisol determinations are frequently inaccurate. Detailed pituitary hormone analysis and stimulation studies may be required in order to uncover subtle abnormalities [231]. Less than 2% of all intradural aneurysms extend into sella. These aneurysms most likely originate from medial ICA branches. In other patients, the aneurysm extends into the sella from a location superior to the cranial nerves. Practically, few patients require long-term replacement hormone therapy, suggesting that chronic symptomatic pituitary insufficiency is unusual.

6.4.3 Neuroimaging

6.4.3.1 Computed Tomography and Magnetic Resonance Imaging

CT performed with axial and coronal views after contrast administration will show the aneurysm if it is large or if thin slices are obtained through the involved area. The relationship of the aneurysm to the anterior clinoid may be demonstrated (Fig. 6.23b, c). In the case of giant aneurysms, the visual examination is crucial to direct the neuroradiologist to the correct area to obtain the appropriate image slices.

If the patient has a SAH alone, blood located in the suprasellar cistern will be apparent on CT for approximately 48 h, but unless the blood is limited to this cistern, the localizing value may be lost. After more than 48 h following a SAH, the CT will also reveal whether there is acute communicating hydrocephalus or a cerebral infarct.

MR imaging gives information similar to the CT, but blood in the cisterns or interhemispheric fissures may not be seen as easily. As mentioned previously, MR imaging easily distinguishes an aneurysm from a neoplasm. Partial thrombosis within the aneurysm, patency, and laminar flow within the aneurysm lumen are readily visualized (Figs. 6.21, 6.23c). Again, if a visual disturbance is present on examination, MR imaging can be better directed to locate the aneurysm.

Skull radiographs are not essential for the management of this aneurysm. Linear calcification in the wall of the parent vessel or aneurysm can be seen above the anterior clinoid, but this is not diagnostic because calcification also occurs in an arteriosclerotic ICA without an aneurysm. When a giant aneurysm is present, depending on the direction of the mass, erosion of the anterior clinoid, enlargement of the sella turcica, or erosion of the intracranial optic canal can be seen.

6.4.3.2 Angiography

The exact location of the aneurysm neck and the parent and the functional vascular anatomy, needed to determine the therapeutic approach, are demonstrated by cerebral angiography. The relationship of the aneurysm to the ophthalmic artery and the ICA, the blood supply to the optic pathway, and the potential collateral supply to these vessels must be known. The ophthalmic artery arises from the intradural ICA above the cavernous sinus and below the anterior clinoid in more than 90% of subjects [232, 233]. The variations of the origin of the ophthalmic artery, which can arise from the intracavernous ICA or via external carotid branches such as the middle meningeal and ethmoidal arteries, are detailed in Chap. 1.

Not all of the aneurysms in this area arise directly from the origin of the ophthalmic artery. Selective angiography with multiple views may reveal an origin from other branches of the ICA [234]. The term "paraophthalmic artery aneurysms" includes all aneurysms in this segment of the ICA (Figs. 6.21, 6.23). In contrast to aneurysms arising from the medial branches of the ICA, ophthalmic artery aneurysms can displace the ICA posteriorly and inferiorly, resulting in closure of the siphon on the lateral angiogram [218].

Partial thrombosis causes the aneurysm (only the lumen is directly visualized on angiography) to appear irregular and smaller on angiography than the aneurysm mass seen with CT or MR imaging. The mass effect of a giant aneurysm can be indirectly determined by the amount of upward or lateral displacement of the anterior cerebral arteries or the posterior or lateral displacement of the ICA. Angiography will also demonstrate the frequent occurrence of multiple aneurysms.

6.4.4 Treatment and Course

The rebleed rate of carotid–ophthalmic aneurysms before surgery is performed may be as low as 10%. Typically, the timing of when surgery is undertaken depends on the grade of the patient following a SAH. The usual aneurysm and SAH considerations include whether the patient has multiple aneurysms, an intracerebral hematoma, an accessible aneurysm neck, or vasospasm. If the aneurysm is unruptured, some authors have advocated following the patient without treatment [235]. However, if the neurosurgeon and the neuroradiologist determine that the aneurysm is readily clippable or embolizable, or the patient can tolerate ICA occlusion, treatment should be implemented.

6.4.4.1 Surgery

The surgical approach depends on the location and size of the neck, the direction in which the aneurysms points, the relationship of the aneurysm to the optic nerves, chiasm, and the ICA, and whether the anterior clinoid obscures the view of the aneurysm or neck. Surgical drilling-off of the anterior clinoid and removing the dura mater in the area may be required to expose the involved structures adequately. If the aneurysm is larger than 1 cm, then decompressing the aneurysm with a catheter and suction may help facilitate the dissection and clip placement [236, 237]. Another factor that complicates surgery occurs when the ICA (parent artery) shares a common wall with the aneurysm so a definitive entrance or exit cannot be demonstrated. Reconstruction of the arterial wall with a clip that preserves the patency of the vessel should be attempted. Temporary occlusion of an exposed carotid artery in the neck or a temporary intracranial clip will decrease the blood flow to the aneurysm and facilitate the dissection. If a giant aneurysm compresses the optic apparatus, then opening the clipped aneurysm and removing the hardened thrombus may help the vision to normalize more rapidly.

If there is no surgical neck or the vessel cannot be reconstructed, an intracranial trapping procedure, which isolates the aneurysm and the involved arterial segment, can be performed. Though surgical intracranial trapping accomplishes this goal, percutaneous embolization can give a similar result without a craniotomy and with the safety of monitoring an awake patient. Cerebral blood flow monitoring, electroencephalography, and spectral electroencephalography are useful modalities to monitor the tolerance of the cerebral hemisphere to occlusion of the ICA, but none is as predictive as the angiographic tolerance test when the patient is awake (see Sect. 6.2.4.3).

Direct surgery on giant aneurysms is often associated with a significant morbidity and mortality rate because of their large size and frequent location in an area where dissection is difficult. Additionally, following a SAH, the patients poor clinical status increases the surgical risk. Though Drake reported that five of 24 patients did poorly (rebleed or cerebral ischemia, but no deaths) with therapeutic proximal occlusion of the carotid artery, direct surgery on the aneurysm was associated with a poor outcome (death) in only two of 15 patients [27]. In another series, Sundt reported that 21 of 23 surgically clipped carotid ophthalmic aneurysms did well neurologically [238]. An optic neuropathy can result from surgical manipulation of the nerve or damage to the arterial supply of the optic nerve. Though less than 5% of patients experience complete loss of vision from intraoperative procedures, there is little data on the incidence of iatrogenic partial visual loss.

6.4.4.2 Carotid Ligation

Indirect therapy with external ligation of the common carotid artery has a lower cure rate than direct surgery, but the reported series generally include only small numbers of cases [27, 239]. Additionally, occlusion of the common carotid artery does not alter the rebleed rate of the giant aneurysms [27]. The results of this treatment on the visual disturbance have been variable. Sengupta described one patient who became blind in the eye ipsilateral to the carotid occlusion, probably from acute thrombosis in the aneurysm [203]. Thurel reported two patients whose vision improved but a third who had worse vision following carotid ligation [240]. White advocated exploring the aneurysms of those patients who fail proximal carotid ligation [215]. In one study, the eyes with severe visual loss failed to improve significantly, but in two of the three patients the less involved eye improved following occlusion of the cervical carotid artery. However, one of the patients suffered a severe cerebral infarct postoperatively [230].

6.4.4.3 Embolization

Embolization of large or giant carotid–ophthalmic artery aneurysms can also be an effective therapy when a clear clippable neck cannot be visualized on angiography. However, when detachable balloons are used occlusion of the parent ICA is often required. Embolization with coils shouls avoid this difficulty. The origin of the ophthalmic artery should be spared to prevent ischemic damage to the intracranial optic nerve.

6.4.4.4 Visual Recovery

The degree of visual field and acuity recovery following exclusion of the aneurysm from the circulation depends on the prior extent of compression-induced ischemia, demyelination, and axonal loss in the optic structures. Following surgery, the visual acuity and field loss improve in less than 50% of patients, and the vision worsens in approximately 15% of patients [207]. Similar results of recovery of visual function can be achieved by successful embolization.

6.5 Aneurysm of the Internal Carotid Artery Bifurcation

6.5.1 Incidence

The incidence, approximately 4% of all aneurysms [36a], of aneurysms in the segment of the ICA that bifurcates into the anterior and middle cerebral arteries is considerably less than that of aneurysms in the middle and anterior cerebral arteries, the posterior communicating artery, or the carotid– ophthalmic artery. However, some of the ICA bifurcation aneurysms are undoubtedly grouped with the carotid–ophthalmic aneurysms in some studies. Most of the patients with this aneurysm are between 40 and 60 years old.

6.5.2 Clinical Presentation

A patient with a bifurcation aneurysm often presents because the mass lesion causes progressive visual loss. SAH occurs in less than 10% of these giant aneurysms, but smaller aneurysms have a risk of rupture and rebleeding similar to the rates for carotid–ophthalmic aneurysms. When the patient has a SAH presentation, visual loss may not be apparent even if the mass appears to displace the anterior visual pathway on CT or MR imaging, or even by surgical observation. A SAH is more common in the small to large sized aneurysms.

As with cases of carotid–ophthalmic artery aneurysms, the visual field loss depends on the direction in which the aneurysm mass points [43, 44, 56]. A given aneurysm can involve the ipsilateral posterior optic nerve, lateral chiasm, or optic tract, or all three structures (Figs. 6.24, 6.25). Many patients have temporal field loss in the eye contralateral to the aneurysm. A central scotoma, in addition to a temporal or nasal quadranopia or hemianopia, is typically found in the ipsilateral eye. When the aneurysm points posteriorly into the optic tract, an incomplete incongruous homonymous field depression develops (Fig. 6.25; see also Sect. 6.4.2.2).

Hypothalamic/pituitary dysfunction develops in a few patients who have the aneurysm mass displacing the hypophyseal stalk or compressing hypothalamus [241]. Patients with an aneurysm in this area should have an endocrinological evaluation, if appropriate symptoms are present or there is unexplained lethargy.

6.5.3 Neuroimaging

6.5.3.1 Computed Tomography and Magnetic Resonance Imaging

The CT and MR appearance is similar to that of carotid–ophthalmic artery aneurysms. The anatomical relationship of the aneurysm to the optic pathway and the brain is best demonstrated by MR imaging (Figs. 6.25, 6.26). Though the bifurcation location can be inferred from MR imaging, angiography is the definitive study.

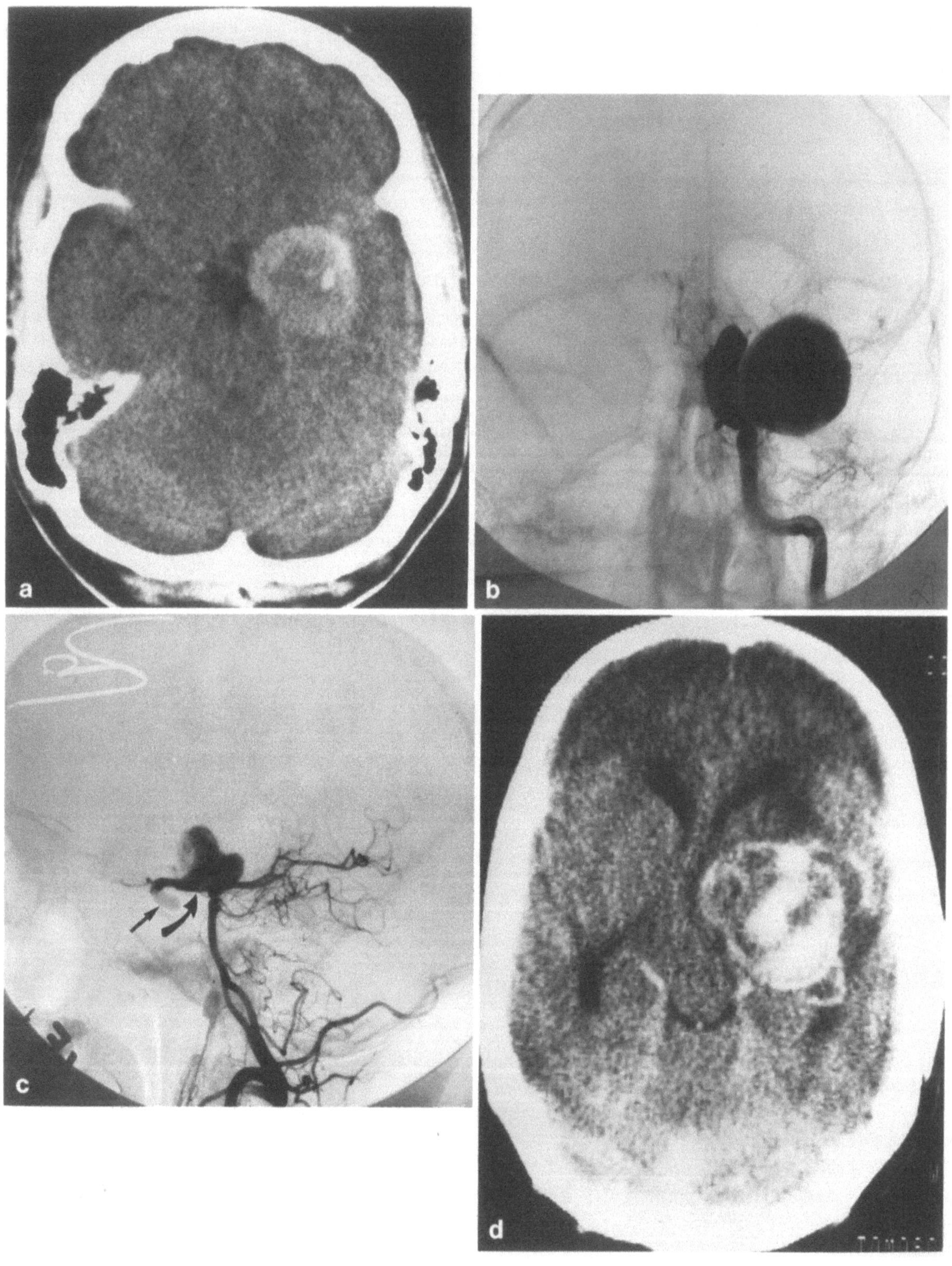

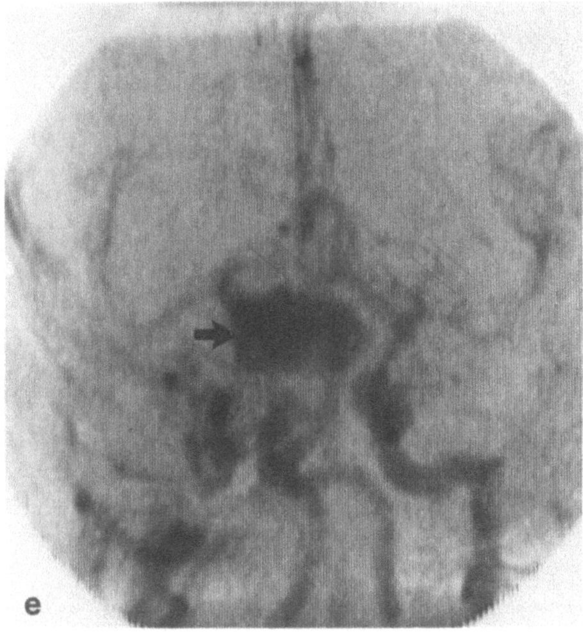
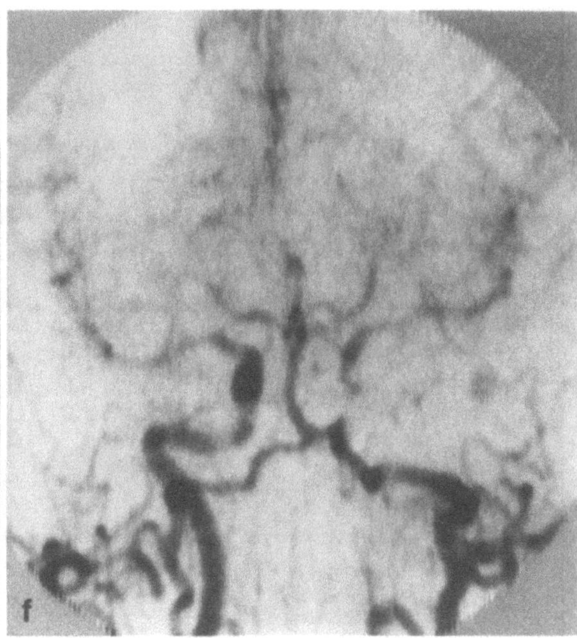

Fig. 6.24a–f. The risk of directly treating some bifurcation aneurysms is extremely high. An 18-year-old man presented with severe headache and acute loss of vision of the entire right homonymous field. The acuity was 20/20 in both eyes and there was a relative afferent pupillary defect in the right eye.

a An axial view unenhanced computed tomogram shows a heterogeneous appearing round mass in the lateral aspect of the left suprasellar cistern.

b A frontal view left internal carotid artery subtraction angiogram reveals a large bifurcation aneurysm extending laterally.

c Following embolization that closed the carotid artery at the level of the aneurysm with a detachable balloon (*arrow*), a lateral view vertebral angiogram demonstrates continued filling of the aneurysm through the posterior communicating artery (*curved arrow*), which is seen as a shadow behind the aneurysm.

d Two years later, the patient presented with a severe headache and a loss of the left field from both eyes. Since he had never recovered the right field, his visual acuity was finger counting at 2 feet in both eyes. The axial contrast enhanced computed tomogram demonstrates enlargement of the mass into the midline region with a heterogeneous signal including clear areas of contrast enhancement that represent a patent lumen of the aneurysm.

e The frontal view of a digital intravenous angiogram reveals the lumen of the aneurysm to be patent (*arrow*). It was presumed that the same posterior communicating artery seen in **c** provided the arterial inflow to the aneurysm.

f The patient was placed on intravenous methylprednisolone and within 5 days began to demonstrate improvement of vision. Within several months, there was a complete normalization of the left homonymous field and a return of the visual acuity to 20/15 in both eyes. A frontal view digital intravenous angiogram performed 1 year later demonstrates an absence of filling of the aneurysm. The patient has subsequently been followed up for 6 years without any additional neurologic or visual difficulties

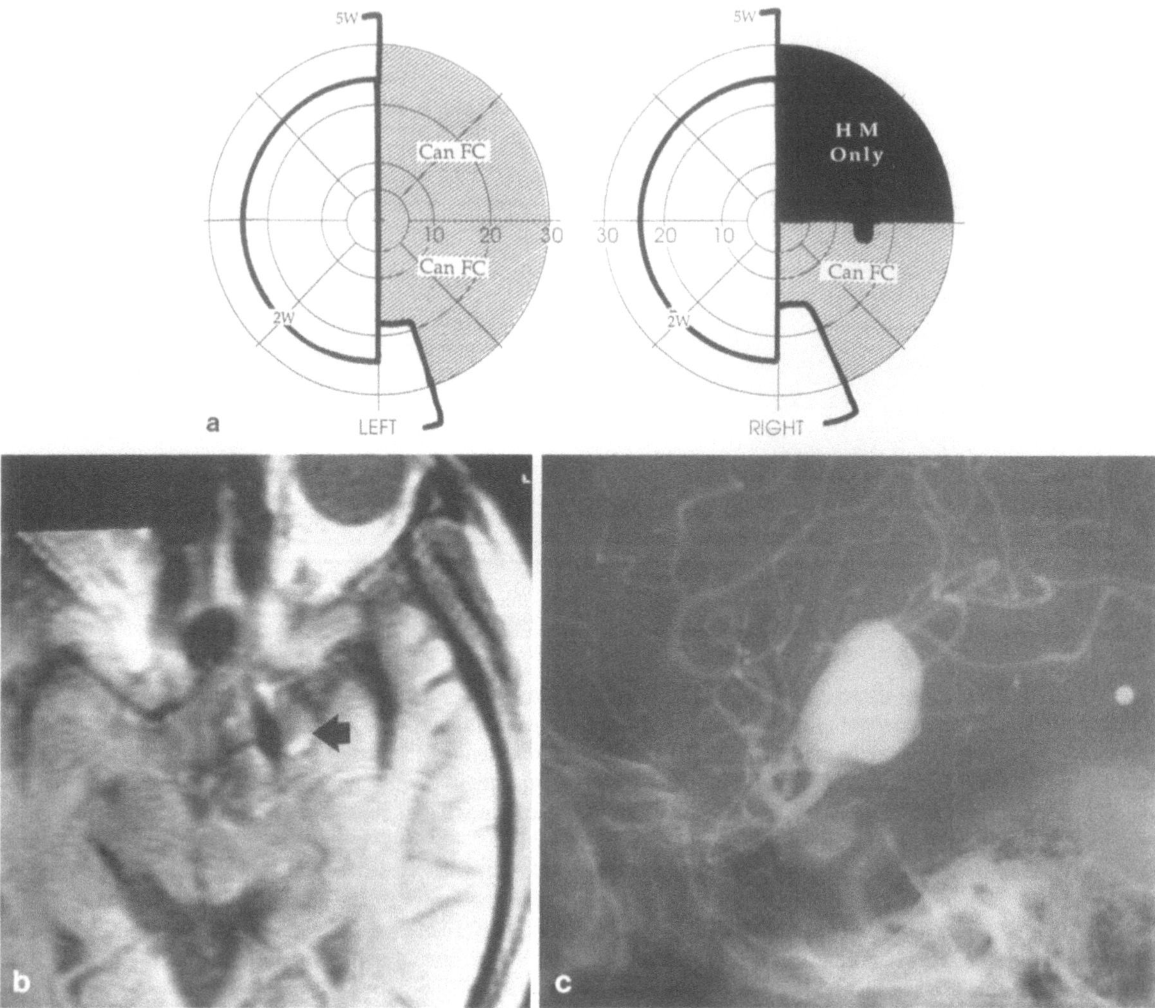

Fig. 6.25 a – c. Prior to the development of the newer embolic coil methods, if treatment was considered precarious, patients with unruptured bifurcation aneurysms were followed up closely.

a A dense incongruous right homonymous field defect, which was worse superiorly, was found in this 50-year-old man who was complaining of blurred vision for several months.

b Magnetic resonance imaging demonstrates an area of heterogeneous signal with dark flow void and areas of thrombus (*arrow*) compressing the left optic tract.

c The lateral view left internal carotid artery angiogram reveals a bifurcation aneurysm. Because the aneurysm involved the dominant hemisphere, neither surgery nor embolization were considered warranted. Over the next 5 years, the patient lost the remaining right homonymous field and developed mild recent memory loss as the aneurysm mass increased in size

6.5.3.2 Angiography

Angiography will demonstrate the location of the aneurysm at the distal segment of the ICA bifurcation into middle and anterior cerebral arteries (Figs. 6.24b, 6.25c). Angiography must determine whether the contralateral ICA can provide collateral arterial blood flow to the cerebral hemisphere ipsilateral to the aneurysm.

6.5.2 Treatment and Course

Typically, unless the giant aneurysm is treated, the visual dysfunction gradually or suddenly worsens over months to years. For example, a unilateral optic neuropathy will deteriorate to blindness and a partial homonymous field defect will become a dense homonymous hemianopia. If the patient has longstanding severe visual loss and optic atrophy

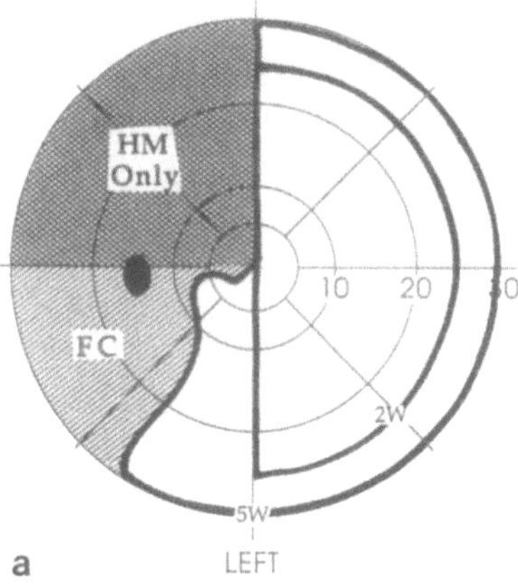

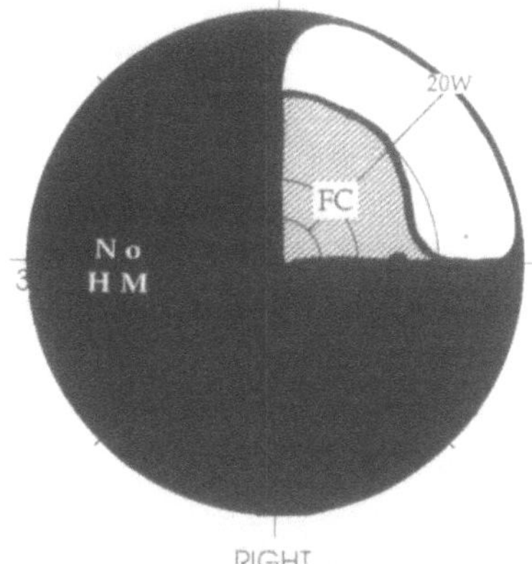

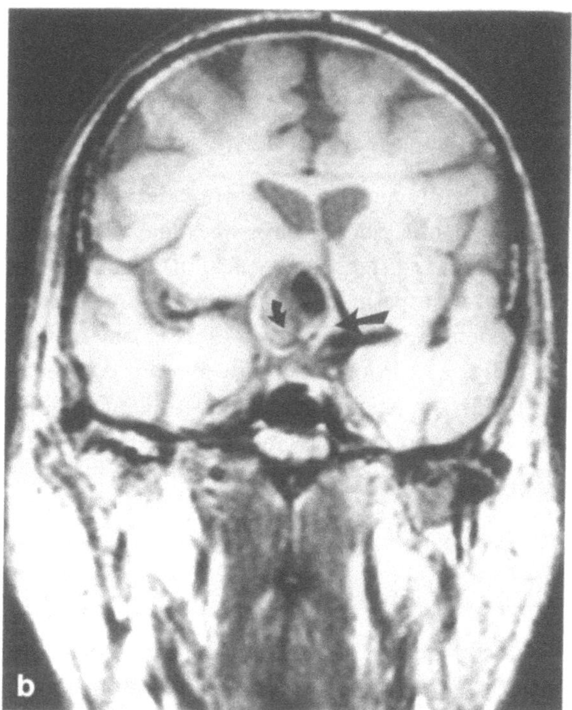

Fig. 6.26a, b. A 75-year-old man complained of progressive visual loss in the right eye over 1 year and the left eye over 3 months. The visual acuity was finger counting at 1 m and 20/25 for the right and left eyes, respectively. Pallor was seen in both optic discs. **a** The visual field reveals minimal preservation only in the superotemporal field of the right eye. The field from the left eye shows a temporal hemianopia, which was worse to a 2-mm white test object than to a 5-mm one. **b** A coronal view T1-weighted magnetic resonance scan demonstrates a giant suprasellar aneurysm (angiography revealed an ICA bifurcation aneurysm) with a flow void and area of thrombus (*curved arrow*). The chiasm is displaced to the left (*arrow*)

that is only monocular at the initial evaluation, and if treatment carries a significant risk of severe morbidity, then treatment may not be warranted.

If the mass is large and compresses the cerebral hemisphere, the patient can exhibit signs of cerebral dysfunction.

The goals of treatment of this aneurysm are typical, and include exclusion of the aneurysm from the arterial circulation while preserving the parent ICA. When the aneurysm is located at the bifurcation, occlusion of the involved arterial segment could prevent the collateral blood flow through the anterior communicating artery from supplying the middle cerebral artery ipsilateral to the aneurysm. In these cases, an extracranial artery bypass to the middle cerebral artery allows the immediate occlusion of the ICA by providing external carotid artery blood to supply the middle cerebral artery. The bypass prevents hypoperfusion of the ipsilateral cerebral hemisphere and decreases the incidence of clinical cerebral ischemic events [103] (see Sect. 6.4.4 for discussion of embolization and surgery and Sect. 6.2.4.3 for discussion of bypass).

6.6 Anterior Communicating Artery Aneurysm

6.6.1 Incidence

Approximately 25% – 35% of all intradural aneurysms (range for six studies) are found in the anterior communicating artery or proximal segments of the anterior cerebral arteries. Additional aneurysms are found in approximately 30% of these cases. Aneurysms in the more distal portions of the anterior cerebral artery are less frequent.

6.6.2 Clinical Presentation

6.6.2.1 Subarachnoid Hemorrhage

A SAH is the most frequent presentation. Confusion, dementia, and lethargy or coma, often without lateralizing signs, are typical of a rupture. Because of vasospasm in the anterior cerebral arteries, the patient can also have weakness in one or both legs (Fig. 6.2). Prior to rupture, anterior communicating artery aneurysms infrequently increase in size to compress a cerebral hemisphere, pituitary gland, or the anterior visual pathway (Fig. 6.27, corresponds to Fig. 6.9) [242, 243].

6.6.2.2 Visual Disturbances

Visual dysfunction is not uncommon in patients, with or without a SAH. As mentioned previously, a ruptured anterior communicating artery aneurysm is the aneurysm most commonly associated with Terson's syndrome (Sect. 6.1.7.3). Visual loss from damage of some portion of the intracranial anterior visual pathway has been reported in as many as 28% of patients with anterior cerebral artery aneurysms [118]. The visual loss can be unilateral or bilateral, in which case it is commonly asymmetric [43, 183, 244]. The eye contralateral to the parent artery of the aneurysm is more frequently affected by optic nerve or chiasm compression [245]. This is particularly prevalent in patients with giant aneurysms [43, 214]. The rare optic tract dysfunction results if a giant aneurysm projects posteriorly (Fig. 6.28) [246]. Compression, ischemia, and direct hemorrhage into the intracranial anteri-

or visual pathway are all possible mechanisms for the visual loss. The differential diagnosis in cases of a SAH and acute visual loss from involvement of the intracranial visual pathway includes pituitary apoplexy and a ruptured craniopharyngioma cyst.

6.6.3 Neuroimaging

6.6.3.1 Computed Tomography and Magnetic Resonance Imaging

Interhemispheric blood is characteristically seen on CT following an acute hemorrhage. Depending on the size of the aneurysm the mass may be visualized in the anterior circle of Willis on contrast CT or MR imaging (Figs. 6.9, 6.28a). Unilateral or bilateral lesion(s) with low attenuation on CT or bright signal on T2-weighted MR images in the midline frontal lobe can suggest a secondary infarct from vasospasm of the anterior cerebral artery.

6.6.3.2 Angiography

A small anterior communicating artery aneurysm is easily missed on angiography when only lateral and anteroposterior views are performed. Obliquely oriented views will show the suspected aneurysm and may reveal the relationship of the aneurysm neck to the parent artery and perforators, which are essential in planning the surgical approach. On occasion, a hypoplastic A1 segment of the anterior cerebral artery (commonly found in cases with anterior communicating artery aneurysms) is mistakenly diagnosed as vasospasm.

6.6.4 Therapy and Outcome

6.6.4.1 Surgery

Microsurgical clipping of the aneurysm remains the mainstay of therapy. The neurosurgeon must determine the individual patient's surgical risk. Surgical exclusion of the aneurysm from the circulation while preserving the normal perforating arteries to the brain is often accomplished with the aid of systemic hypotension. A ventricular drain followed by a shunt may be required to relieve the elevated intracranial pressure and hydrocephalus before ap-

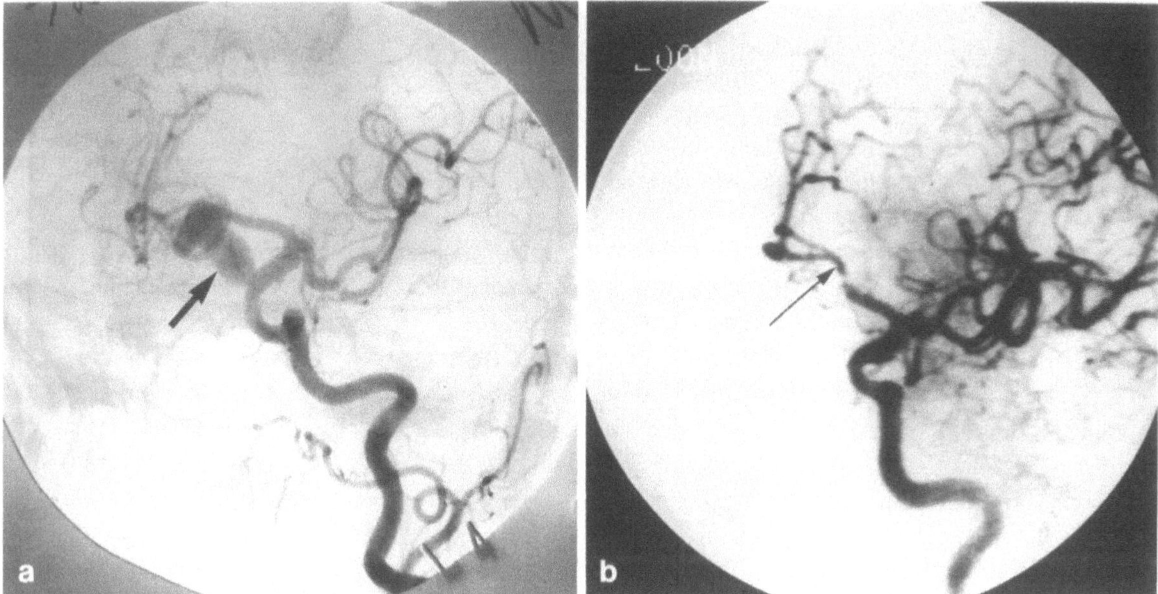

Fig. 6.27a, b. A 53-year-old woman had a subarachnoid hemorrhage and was found to have an anterior communicating artery aneurysm which was unclippable during surgical exploration. Despite wrapping and coating, the aneurysm increased in size. The patient experienced progressive visual loss which began after the surgery. The clinical evaluation revealed that only vague hand motion detection vision remained in the right eye and that the left eye was intact. **a** An oblique view left internal carotid artery subtraction angiogram demonstrates the large anterior communicating artery aneurysm (*arrow*) better than the lateral or frontal projections. **b** Following balloon embolization, the oblique view angiogram reveals the aneurysm is excluded from the circulation and the anterior communicating artery (*arrow*) is preserved (corresponds to Fig. 6.9)

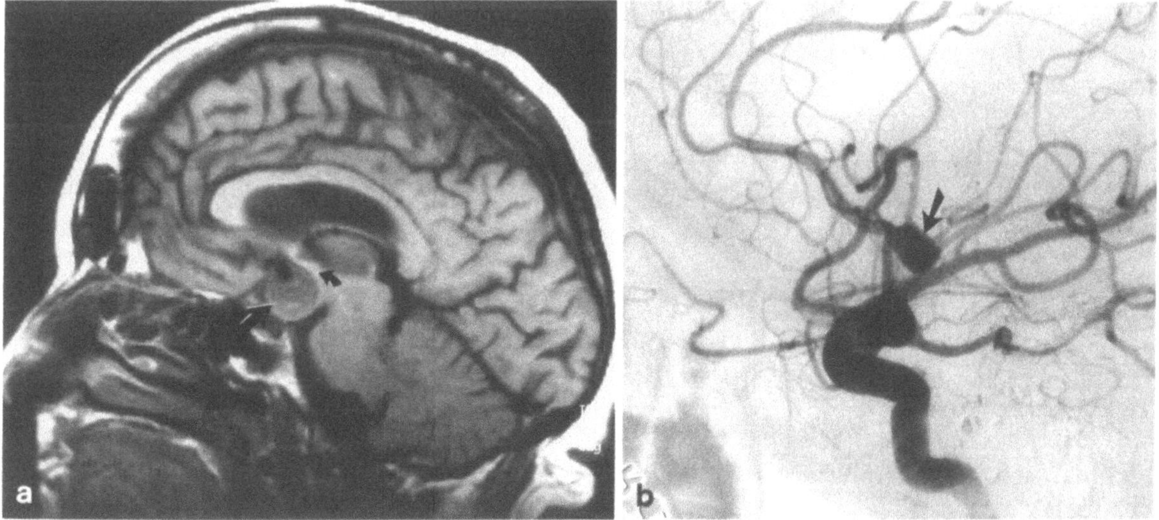

Fig. 6.28a, b. A 42-year-old woman complained of a progressive visual disturbance to the left side. On examination she was found to have a complete left homonymous field defect. **a** The sagittal view T1-weighted magnetic resonance scan shows a large aneurysm (*arrow*) with a small area of flow void and a large area of thrombus displacing the optic tract (*curved arrow*). **b** The lateral view right internal carotid subtraction angiogram demonstrates the relatively small lumen (*arrow*) of the anterior communicating artery aneurysm, which projected posteriorly

proaching the aneurysm. The surgical approach to the aneurysm depends on many factors but the surgery is usually performed on the same side as the parent vessel. However, if there is a second aneurysm which is operable during the same procedure to clip the ruptured anterior communicating artery aneurysm, another approach may be considered. Aneurysms in this locale pose specific problems, such as identifying important perforating arteries (such as Heubner) and the need for temporary clips to facilitate exposure of the aneurysm neck.

The operative morbidity and mortality is about 25% when a SAH has preceded surgery. Following successful clipping of the aneurysm, of those patients who do not die in the perioperative period, 10% are permanently impaired. The dysfunction includes persistent memory loss, a focal deficit, and/or personality changes, all of which arise from secondary infarction of the brain (probably from vasospasm following the SAH). Only 55% of patients return to employment.

Similar to other neurologic outcomes, the visual results depend on the extent and severity of the damage. If the aneurysm points superiorly, when it ruptures the hemorrhage will be directed into the frontal lobe, but if the aneurysm points posteriorly, the hemorrhage can result in irreversible damage to the optic tract or chiasm. Compression from a giant aneurysm that causes severe optic atrophy and longstanding visual loss may not recover following decompression of the optic apparatus. As discussed previously, in the large majority of cases Terson's syndrome is associated with an excellent recovery with conservative therapy alone.

6.6.4.2 Embolization

Embolization can be performed in patients whose medical condition precludes surgery, where the neurosurgeon determines the aneurysm cannot be clipped without an unacceptable risk or where surgery has been attempted and failed. Through a percutaneous route, one or more detachable balloons or coils can be placed into the aneurysm to exclude it from the circulation while preserving the parent artery (Fig. 6.27b). However, if important perforators to the brain are closed, ischemic damage of the deep nuclei can result in a chronic depressed state of consciousness.

6.7 Basilar Artery Aneurysm

6.7.1 Incidence

Basilar artery aneurysms represent less than 5% of all intradural aneurysms. Aneurysms in this location are found equally in men and women, usually between the ages 30 and 80 years. Since most aneurysms of the basilar artery occur near its termination, many patients present with diplopia and ptosis from unilateral or bilateral third nerve dysfunction (see Sect. 6.3.2.1 for discussion) or with more severe neurologic dysfunction or both (see below). When the aneurysm is at a midbasilar level, a less common location, the patient may have unilateral or bilateral sixth nerve pareses and signs of brainstem compression (see below).

6.7.2 Clinical Presentation

The clinical presentation is extremely variable and can include SAH, progressive or intermittent midbrain or pontine dysfunction, unilateral or bilateral third or sixth nerve pareses, vertigo or nonspecific dizziness, and memory loss secondary to thalamic dysfunction or hydrocephalus [247]. SAH and rebleeding are quite common, leading to frequent severe morbidity and mortality. Compression of the pons or midbrain by a giant aneurysm mass may account for the cases with progressive dysfunction (Fig. 6.29). Intermittent or sudden-onset brainstem dysfunction can mimic the attacks of multiple sclerosis. The episodes may develop from partial thrombosis and temporary enlargement of the aneurysm, edema in the wall of the aneurysm, or thrombosis of perforating arteries to the brainstem which arise from the vessel wall included in the aneurysm [248]. Rarely, when this aneurysm is a giant size and points anteriorly into the suprasellar cistern to compress the chiasm, progressive visual field defects may occur.

A "top of the basilar syndrome," which usually results from embolic disease, may result from giant basilar aneurysm compression and ischemia of the rostral brainstem, thalamus, and medial temporal lobes. Emboli from thrombus within the aneurysm occasionally cause posterior cerebral artery area ischemia with homonymous field defects. The components of this "top of the basilar" syndrome include ocular motor and pupillary dysfunction, behavioral disorders, and visual sensory disturbances, with a relative preservation of cortical spinal tract function (see Sect. 8.5.2.1.5).

Disruption of the vestibular system in the midbrain and pons can cause various types of nystagmus (see Sect. 8.5.2.1). Conjugate upbeat nystagmus in primary or upgaze and horizontal or rotatory nystagmus in primary or horizontal gaze are commonly seen.

Midbrain and pontine dysfunction causes supranuclear gaze palsies (see Sects. 8.5.2.1.3, 8.5.2.1.4) such as skew deviation, intranuclear ophthalmoplegia, or horizontal gaze palsy, with or without associated motor or cerebellar deficits [249, 250]. A paresis of upward gaze results from dorsal midbrain compression [251].

A unilateral partial superior division third nerve paresis or bilateral third nerve dysfunction develops from direct compression of the nerves by the aneurysm or disruption of the small arteries to the nerves as they pass in the subarachnoid space or in the midbrain. The lid, pupil, and oculomotor disturbances are usually progressive (see Sect. 6.3.2.1 for discussion of unilateral third nerve dysfunction). When the lesion is in the brainstem, the involvement of the crus cerebri, the cerebellar outflow pathway, or the red nucleus causes either a crossed hemiparesis or an ipsilateral or contralateral dysmetria and tremor, respectively.

6.7.3 Neuroimaging

6.7.3.1 Computed Tomography and Magnetic Resonance Imaging

CT and MR imaging show the location of the aneurysm in relation to the brainstem (Figs. 6.5, 6.29a). When the aneurysm is large the midbrain or pons or both can be compressed, or the mass can be seen to invaginate into the brainstem. Partial thrombosis is typically visualized in the giant aneurysms. Following a SAH, blood can be seen in the cisterns around the upper brainstem and the suprasellar region.

6.7.3.2 Angiography

Cerebral angiography must visualize the relationship of the aneurysm to the important vascular structures such as whether the perforators to the thalamus and upper brainstem (see Sect. 1.5.3.3.3) arise from the aneurysm or from a segment of the proximal posterior cerebral artery involved with the aneurysm (Figs. 6.29b, 6.30a). The position of the aneurysm must also be related to the posterior clinoids. Angiography will also reveal whether the ICA provides significant blood supply to the upper basilar artery via the posterior communicating arteries. A tolerance test of distal basilar artery occlusion, similar to the test for ICA closure, can be performed if sacrificing the basilar artery is anticipated.

6.7.4 Treatment and Course

Since most aneurysms of the basilar artery typically bleed or cause progressive brainstem dysfunction or both, intervention is warranted to prevent the high rate of death.

6.7.4.1 Surgery

The surgical clipping of these aneurysms is frequently partially successful (Fig. 6.30b) and it is associated with significant morbidity and mortality. Many patients have significant brainstem compromise prior to surgery. Occlusion of perforators that exit from the posterior surface of the aneurysm when it involves the proximal posterior cerebral artery causes or worsens the upper midbrain dysfunction. Closure of the posterior cerebral artery will also cause a homonymous hemianopia.

6.7.4.2 Embolization

Alternative treatment with direct balloon embolization of the aneurysm is also limited by the same factors discussed above. Additionally, closure of the basilar artery may be necessary to occlude the aneurysm (Fig. 6.30b, c) or occlusion of the vessel may develop as a complication, so the patient must be capable of tolerating the loss of this vessel. Not infrequently, the parent basilar artery that appears

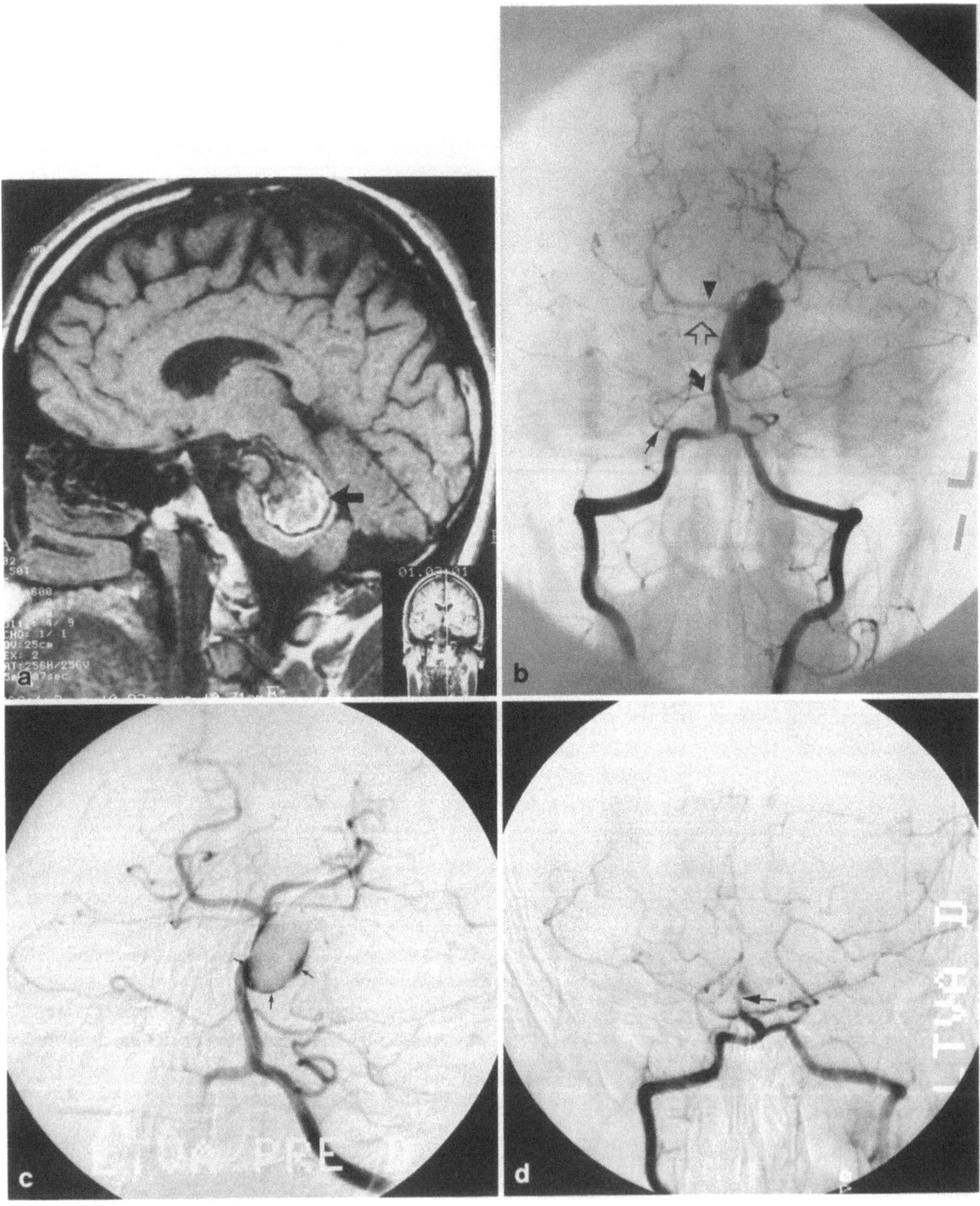

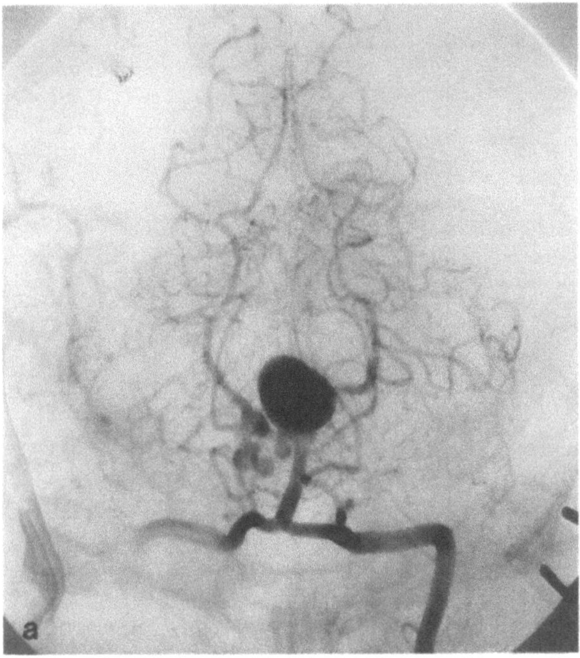

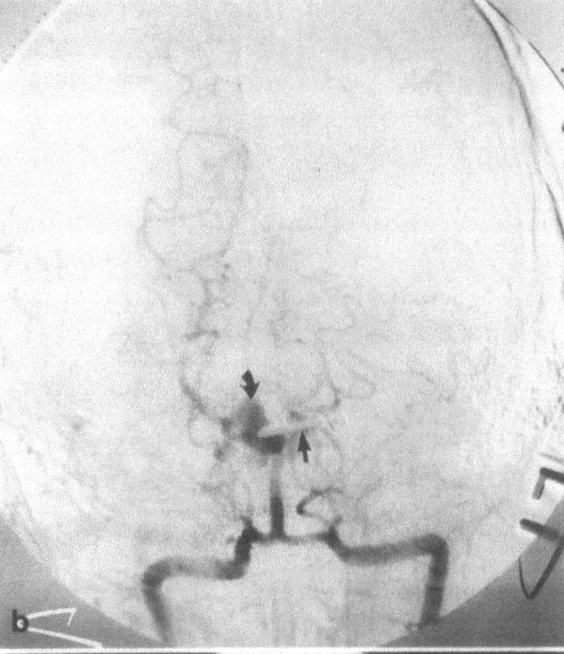

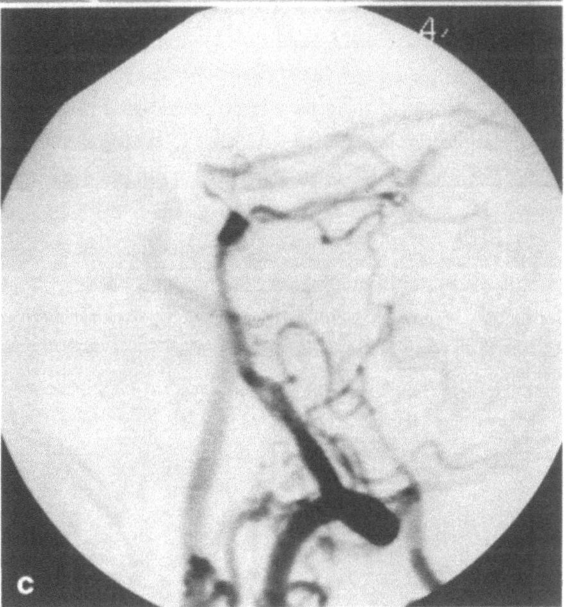

◀ **Fig. 6.29. a** The sagittal view T1-weighted magnetic resonance scan performed in this 39-year-old man with a progressive quadriparesis shows a huge midbasilar artery aneurysm, with thrombus of varying age, invaginating into the brainstem, particularly at the level of the pons (*arrow*).

b The frontal view vertebral artery subtraction angiogram demonstrates a bilobed aneurysm arising from the midbasilar region, between the origin of the superior cerebellar artery (*open arrow*) and the anterior inferior cerebellar artery (*curved arrow*). The posterior inferior cerebellar artery (*small arrow*) and the posterior cerebral artery (*arrowhead*) are seen (marked only on the right side).

c Because of the clinical course, which included progressive four limb weakness, bilateral fourth nerve pareses, and swallowing difficulties, he underwent balloon embolization of the aneurysm. A repeat frontal view vertebral angiogram demonstrates occlusion of most of the aneurysm with a large balloon with slight filling of contrast material around the aneurysm (*arrows*). The basilar artery and all of the previous branch arteries remained patent.

d Forty-eight hours after what was considered successful embolization, the patient experienced a transient lethargy and difficulty in upgaze. He then had progressive lethargy and became unresponsiveness. The anteroposterior view vertebral angiogram shows complete occlusion of the basilar artery above the entrance to the aneurysm (*arrow*). The patient expired 1 week later

Fig. 6.30 a – c. Surgery and embolization in combination can be effective if one mode of therapy has failed to entirely close a basilar artery aneurysm. **a** The frontal view vertebral artery subtraction angiogram revealed an aneurysm at the top of the basilar artery as the cause of a subarachnoid hemorrhage. **b** Surgical exploration lead to partial clipping (*arrow*) with a persistent aneurysmal pouch (*curved arrow*). **c** A detachable balloon was placed into the remaining neck of the aneurysm. The lateral view vertebral angiogram shows that the combination of surgery and embolization accomplished the complete elimination of the aneurysm from the circulation while preserving the parent artery patent after balloon embolization will thrombose (Fig. 6.29c, d). Tolerance testing prior to embolization can demonstrate if the patient has an adequate supply to the distal branches through the posterior communicating arteries via the circulation from each ICA. If the patient cannot tolerate closure of the basilar artery, proximal embolization that closes one or both (performed at different sessions) vertebral arteries (without interfering with the origin of the posterior inferior cerebellar or anterior spinal arteries) is performed. However, no tolerance test or pre-embolization angiogram will predict the patient's ability to tolerate occlusion of perforators to the midbrain and thalamus.

6.8 Other Posterior Circulation Aneurysms

Saccular aneurysms occur at almost any branching of the artery in the posterior circulation. For example, aneurysms are found in the vertebral and posterior inferior cerebellar arteries. These aneurysms are associated with a high mortality rate and approximately 50% die because of a rebleed died within 4–5 years of the first hemorrhage [252]. The majority of posterior fossa aneurysms rebleed during the first 18 months after the first SAH [253]. Generally, these aneurysms cause few neuroophthalmologic signs and symptoms except when brainstem infarction and secondary hydrocephalus occur following a SAH [254, 255]. In rare cases vertebral artery aneurysms cause a paresis of one or both sixth nerves [256]. Aneurysms of the posterior cerebral artery account for less than 1% of aneurysms. These patients present because of a SAH or a bleed into the occipital lobe (Fig. 6.31) [257]. Giant aneurysms can compress or cause ischemia of the posterior visual radiation or calcarine cortex, resulting in a homonymous visual field defect [258–261].

Aneurysms in the posterior fossa require the same medical and surgical care as intradural aneurysms in other locations. However, the morbidity and mortality is greater than for supratentorial aneurysms [262–264]. Surgical clipping depends on obtaining exposure of an aneurysm neck. Embolization with detachable balloons can exclude the

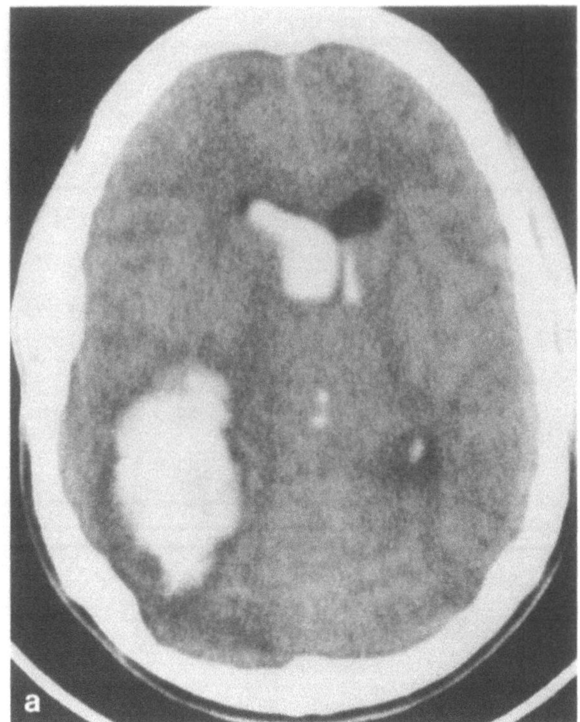

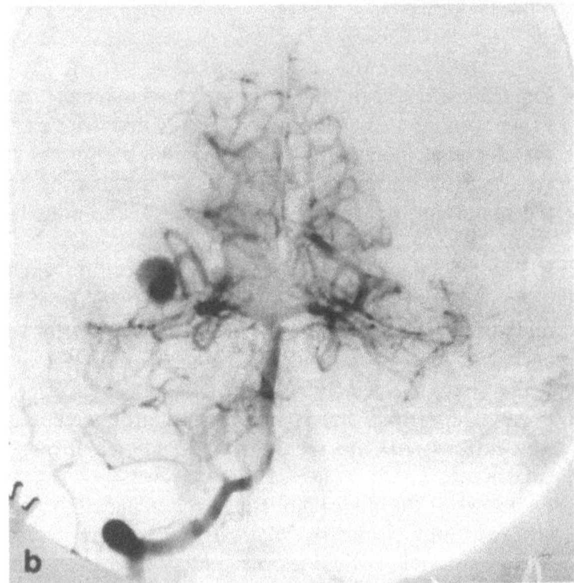

Fig. 6.31a, b. Posterior cerebral aneurysms are rare, but hemorrhages and homonymous visual field defects are frequent with these lesions. **a** The patient had severe headache and acute complete left homonymous hemianopsia. An axial view noncontrast computed tomogram demonstrates a large right occipital and temporal area hemorrhage which extended into the ventricles so that blood is also seen in the anterior horns of the right lateral ventricle. **b** The frontal view right vertebral subtraction angiogram demonstrates an aneurysm originating from a right posterior temporal branch artery, which arose from the right posterior cerebral artery

aneurysm from the circulation or occlude the parent vessel, the vertebral artery. The opposite vertebral artery must be dominant or equal in size to the vertebral artery to be occluded and the origin of the ipsilateral posterior inferior cerebellar artery must be spared.

6.9 Nonsaccular Aneurysms

In comparison to saccular aneurysms, fusiform dilatations (often associated with atherosclerosis), mycotic or vasculitic aneurysmal lesions, and congenital arterial ectasias are infrequent. Congenital fusiform dilatations most commonly affect the intracranial ICA and basilar arteries. Mycotic aneurysms are typically found in the distal branches of the middle cerebral artery. Vasculitis (e.g., collagen vascular disease, drug induced by, for example, amphetamine) can cause inflammatory aneurysms in smaller and medium-sized arteries, including the deeper proximal vessels, such as the lenticulostriate arteries.

6.9.1 Fusiform Aneurysms

6.9.1.1 Incidence

Fusiform aneurysms or dolichoectasias are rare arterial ecatasias. These lesions were found in only 29 cases in a review of 50,000 cerebral arteriograms [265]. Ectasias are predominantly found in patients older than 40 years, but they have been described in children as early as in infancy [266]. They are typically located in the vertebrobasilar circulation [267] and do not originate at the branching of an artery. The ectasia of the vertebral or basilar arteries can span a segment of the artery longer than 1 cm. The involved artery may be minimally or massively dilated.

6.9.1.2 Clinical Syndromes

If the vertebrobasilar aneurysm is large, single or multiple cranial nerve palsies, hydrocephalus, or brainstem compression can occur [268]. Infarction in the brainstem, particularly the pons [269], can develop from occlusion of paramedian penetrating arteries caused by atherosclerosis or local embolization, or a thrombus-related occlusion of the ectatic segment of the basilar artery [270]. Progressive or episodic neuro-ophthalmologic signs and symptoms develop in patients with posterior circulation dolichoectasias because of pareses of the cranial nerves, hydrocephalus [265, 271], or compression or infarction of the brainstem. Hemiparesis, incoordination, facial weakness, nystagmus, and a "one and a half" syndrome signal pontine dysfunction [272–274]. Hemifacial spasm has been described from doliochoectasia compression of the facial nerve in the cerebellopontine angle [268, 275]. Paresis of the oculomotor nerve can also occur [276, 171, 277]. Temporary impairment of the third, fifth, and seventh nerves has been reported in one case [278]. In one unusual case, ectasia of a vertebral artery compressed the lateral medulla to cause a cyclotorsion disturbance of both eyes and an environmental tilt, similar to the complaint of patients with an infarct of the lateral medulla [279] (see Sect. 8.5.2.1.1). Rupture of these arterial ectasias is uncommon [280].

When the fusiform aneurysm is in the anterior circulation, the supraclinoid ICA is the most common locale. Progressive optic nerve dysfunction with optic atrophy or optic disc cupping (see Sect. 5.6) can develop from direct compression (Fig. 6.32) or displacement of the optic structures against a calcified atherosclerotic ICA [281–284]. Dolichoectasia of the middle cerebral artery [285] or the anterior cerebral artery is rare. Depending on the degree of distortion and the location of the involved segment of the ICA, bilateral optic neuropathy or chiasm dysfunction result [286, 287]. Compression of the optic nerve or chiasm by a fusiform aneurysm of the anterior cerebral artery can cause a central scotoma or bitemporal field loss, respectively [288].

6.9.1.3 Etiology

These aneurysms may result from a congenital dysplasia that causes defects in the internal elastic membrane in the vessel wall [289], with subsequent degeneration of the media without systemic disease [290], or from arteriosclerotic degeneration of the arterial wall [280]. However, these lesions have been described in patients with connective tissue disease

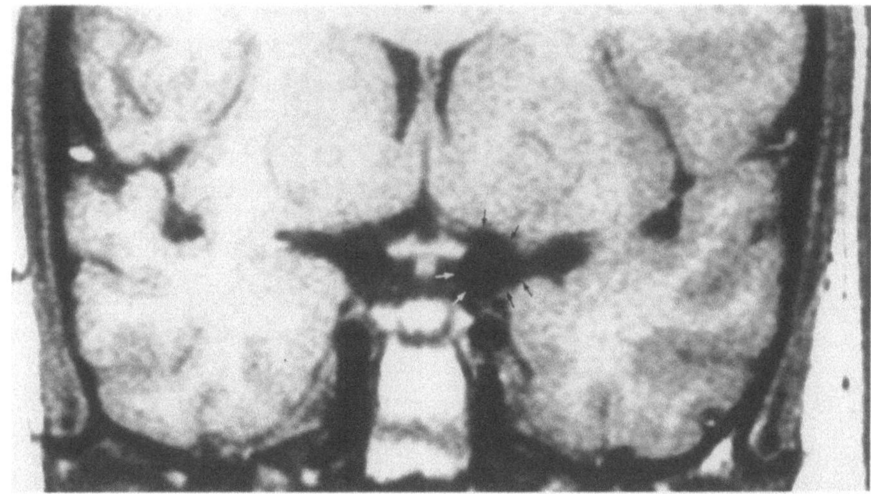

Fig. 6.32. This 10-year-old boy had progressive optic neuropathy and optic atrophy in the left eye without any field defect in the right eye. At the time of the magnetic resonance scan, the Snellen acuity was 20/400 in the left eye and 20/15 in the right eye. The coronal view of a T1-weighted scan demonstrates an area of abnormally large flow void (surrounded by *small black and white arrows*) that was found on angiography to be a large segment of doliochoectasia of the internal carotid artery. Within 2 years only hand motion vision remained in the left eye and the magnetic resonance scan was unchanged. The right eye remained totally unaffected

or a mesodermal dysgenesis such as Marfan's or Ehlers–Danlos syndromes or neurofibromatosis type I, systemic arteritis, or moyamoya disease [291]. Dolichoectasia has also rarely developed following intracranial irradiation [292, 293].

6.9.1.4 Treatment and Course

Fusiform aneurysms can be asymptomatic or cause the above symptoms and findings, but they are infrequently associated with recurrent SAH or death. Patients with thrombotic or embolic episodes are treated with antiplatelet agents or anticoagulation. Direct surgical clipping is not feasible because of the lack of aneurysm neck. Resection, trapping, or balloon occlusion in conjunction with an extracranial bypass procedure may be necessary. Treatment is often difficult because of the large segment of involved basilar artery or ICA.

6.9.2 Traumatic Aneurysms

Traumatic aneurysms (see Sects. 2.1.1, 2.3.2, 2.3.4.3 for discussion of ICA lesions) can affect the cervical, petrous, cavernous, or supraclinoid segments of the ICA. Expansion of the mass causes symptoms appropriate for the location of the lesion. For example, compression of the optic nerve(s) causes visual loss (Fig. 6.33). Rupture of the cavernous aneurysm into the sphenoid sinus causes catastrophic epistaxis, while a SAH results from rupture of a supraclinoid aneurysm [294, 295]. If a hemorrhage occurs, it typically occurs within one to several months of the injury, but epistaxis has been reported as late as 30 years [296].

Traumatic aneurysms in the posterior fossa are rare and associated with a poor outcome if untreated. Motor vehicle accidents and gunshot wounds are the most common traumas. Chiropractic cervical manipulation can also cause damage to the vertebral artery (see Sect. 8.12). Dissection and a pseudoaneurysm that develop in the injured artery cause an ectasia of the arterial wall. The involved segment of arterial wall is composed either of only a thrombus and the arterial adventitia or of the thrombus alone; both of which are predisposed to rupture. Distal embolization from thrombus to the brainstem, thalamus, or occipital lobes can result in ischemic damage of these areas.

Treatment by embolization that excludes the aneurysm and the involved segment of involved artery

Fig. 6.33a–c. In addition to the risk of rupture, traumatic pseudoaneurysms can act as a mass compressing adjacent neural structures. **a** This 24-year-old man was noted to have 20/200 vision in each eye after regaining consciousness from severe head trauma. At that time, a computed tomogram of the head revealed a traumatic subarachnoid hemorrhage but no other lesions. Approximately 1 month later, the acuity in both eyes fell to 20/400. Optic atrophy was noted in both eyes at that time. A T1-weighted coronal magnetic resonance scan revealed a large flow void lesion, compatible with a pseudoaneurysm, extending into the suprasellar cistern to compress the chiasm (*arrows*) upward. **b** The frontal view left internal carotid artery subtraction angiogram demonstrates the same-shaped pseudoaneurysm arising from the left internal carotid artery. **c** Following embolization with a detachable balloon (*arrow*), the aneurysm is eliminated and the internal carotid artery circulation is preserved. However, there was no improvement in the vision

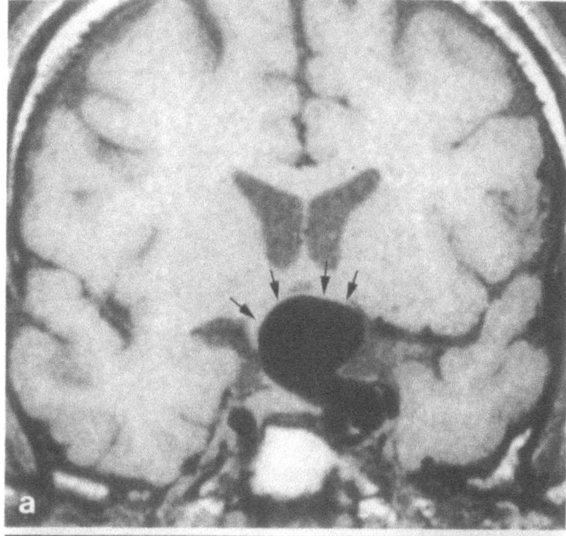

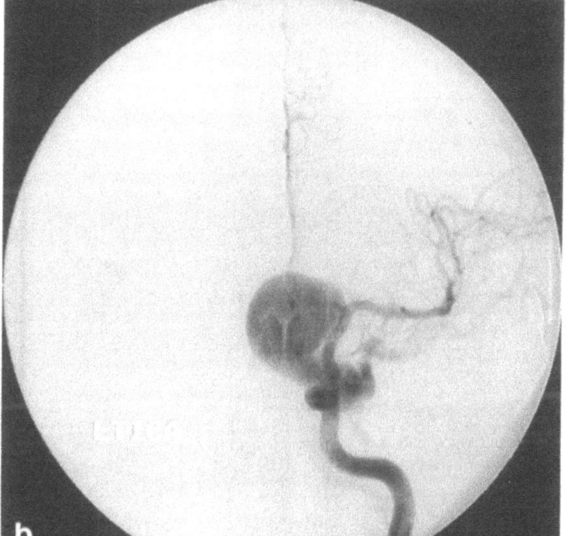

is preferred. A bypass procedure may need to be performed in patients intolerant of occlusion of the affected artery.

6.9.3 Mycotic Aneurysms

Mycotic aneurysms are typically found in the peripheral cerebral arteries as a result of the hematogenous spread of infectious emboli or spread of infection and inflammation from infected meningitis [297]. A chronic inflammatory necrosis and reparative fibrosis develops from a septic thrombus lodged in small arteries or branch points. The resulting thinned wall dilates to form an aneurysmal sac. The majority of mycotic aneurysms are associated with subacute bacterial endocarditis. As many as 33% of patients with this endocarditis will develop transient or permanent neurologic dysfunction, mostly from emboli that arise from the heart, but only a few have a SAH from a ruptured mycotic aneurysm [298]. Although as many as 10% of patients with subacute bacterial endocarditis can develop a mycotic aneurysm [299], most aneurysms go undiagnosed. Fungemia or contiguous spread of fungal infection from adja-

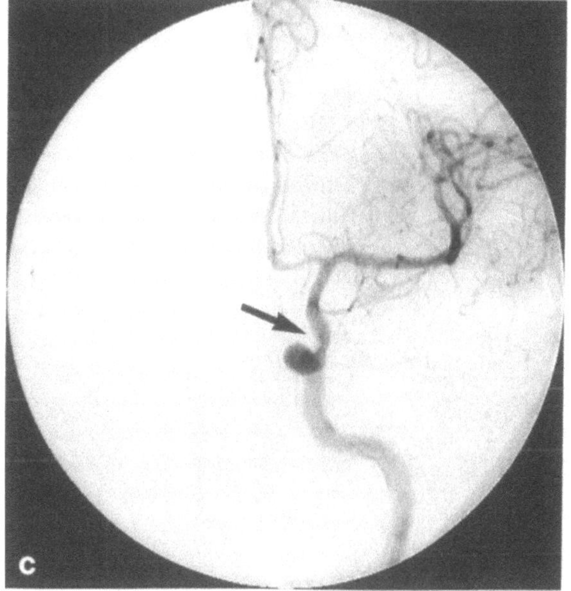

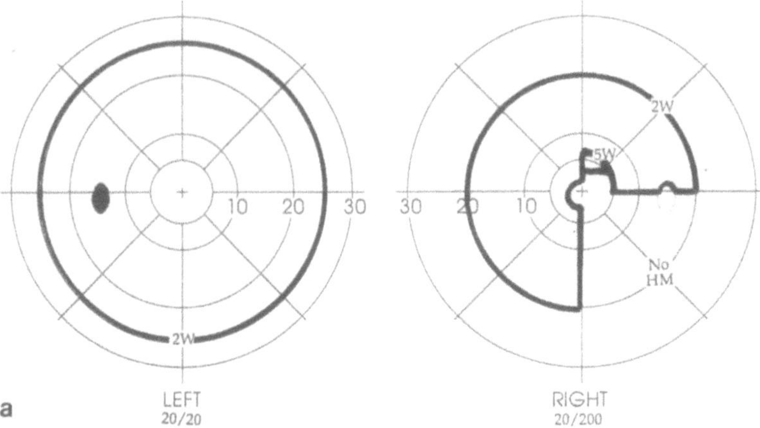

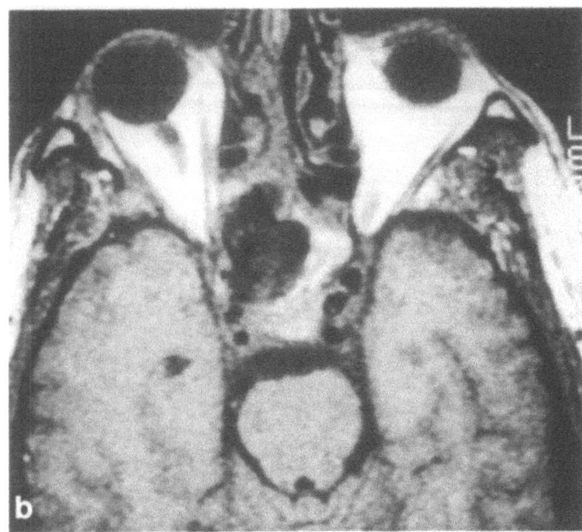

Fig. 6.34a–e. The internal carotid artery is rarely the site of a mycotic aneurysm.

a A 45-year-old man presented with the acute onset of visual loss in the right eye. He had been treated for 6 weeks with an intravenous antibiotic for an acute sphenoid sinusitis prior to the onset of the visual loss. He was found to have a visual acuity of 20/200 in the right eye with a relative afferent defect in the right eye and normal fundi. The visual fields reveal a dense right inferior temporal field defect in which he could not even determine hand motion (*HM*) with the right eye. The field defect extended into the superior temporal central 5° and the nasal 2° to both 2-mm and 5-mm white test objects. The left eye had a normal visual field and 20/20 acuity.
b The axial view T1-weighted magnetic resonance scan demonstrated bright signal in the sphenoid and posterior ethmoid sinuses as well as a large irregular flow void or area of aeration. The low-signal area had not been present on the original scan when the diagnosis of sphenoid sinusitis had been established.
c The lateral view right internal carotid artery subtraction angiogram reveals a large probable mycotic pseudoaneurysm of the intracavernous carotid artery projecting into the sphenoid sinus. Other views demonstrated that a large segment of the internal carotid artery appeared to be involved with this infectious process.
d Following detachable balloon embolization, which excluded the aneurysm from the circulation and occluded the involved segment of the internal carotid artery, the frontal view left internal carotid artery angiogram demonstrates normal filling of both anterior cerebral and middle cerebral arteries.
e Six weeks after embolization and continued antibiotic treatment, the acuity in the right eye had gradually returned to 20/25 and the field defect was greatly reduced in size so that only a partial inferior quadrant field defect to a 2-mm white test object and only a small scotoma (in *black*) to a 5-mm white test object remained

cent sinuses or the orbit usually causes thrombosis in adjacent arteries and veins, but aneurysms rarely occur (see Sect. 3.3.2) [300]. Syphilis is a rare cause of an infectious aneurysm of the cerebral arteries [301, 302].

The diagnosis is established via an examination of the CSF (xanthochromia, blood) and cerebral angiography. A distal middle cerebral artery branch is the most frequently affected vessel. Though less common, the posterior cerebral branches are not spared; an associated intracerebral hemorrhage into the occipital lobe can cause the loss of the contralateral field. MR imaging may reveal edema and inflammation in the brain adjacent to a thrombosed mycotic aneurysm.

A mycotic aneurysm that involves the cavernous segment of the ICA is an extremely rare complica-

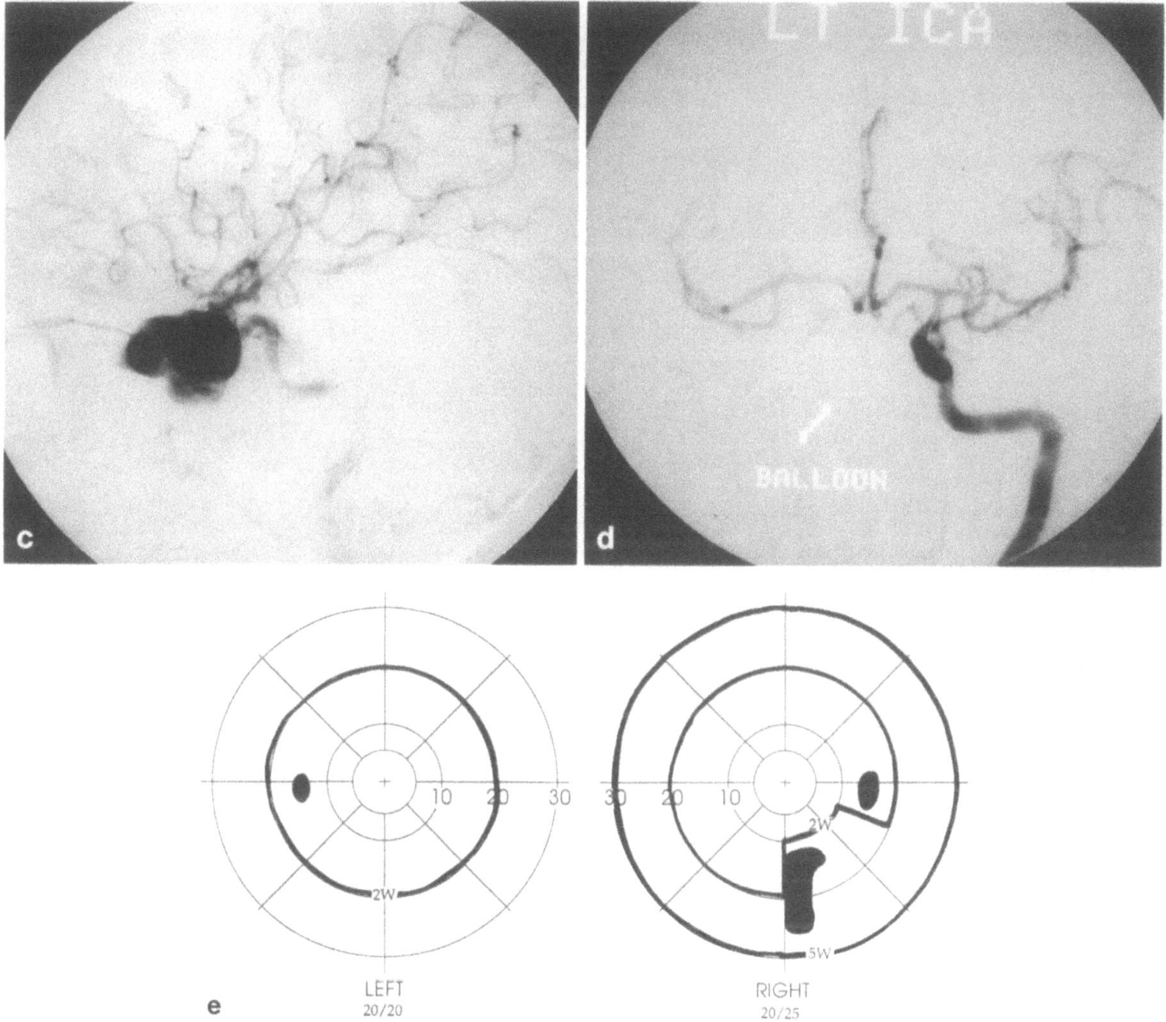

tion of sphenoid sinusitis and secondary cavernous sinus thrombophlebitis [303, 304]. A hematogenous etiology for the cavernous aneurysm is even rarer [305]. These cases are associated with ptosis, ophthalmoplegia, and visual loss ipsilateral to the aneurysm. If the integrity of the medial cavernous sinus is disturbed, rupture of the aneurysm protruding into the sphenoid sinus causes life-threatening epistaxis (as with a traumatic pseudoaneurysm). Embolization of a mycotic aneurysm in this location can be life-saving [306]. We have seen one patient who, despite intravenous antibiotic treatment for sphenoid sinusitis, developed a cavernous ICA mycotic aneurysm. Curiously, despite the lack of concomitant cavernous sinus dysfunction, the patient had visual loss from a retro-orbital optic neuropathy (Fig. 6.34). Following balloon embolization to close the involved segment of the ICA, the aneurysm thrombosed and the vision improved.

The treatment of mycotic aneurysms consists of a full course of antibiotics or antifungal agents appropriate for the infectious agent found in blood cultures. After 10 days of antimicrobial therapy, the aneurysm size is reevaluated with angiography. The aneurysm is surgically excised if the aneurysm progressively enlarges or rebleeds despite medical treatment. The local vasculitis and aneurysm typically improve following the 4- to 6-week course of antibiotics. Repeat angiography should demonstrate a decrease in the size of the aneurysm [307, 308]. For many reasons, including generalized immune compromise, fungal aneurysms have had a uniformly fatal outcome [309].

CHAPTER 7

Vascular Malformations of the Brain

7.1 Arteriovenous Malformations

7.1.1 Introduction

Arteriovenous malformations (AVM) are found throughout the central nervous system and commonly affect the sensory visual and ocular motor systems. Patients can present with symptoms such as headache, migrainelike visual disturbances, visual loss, or diplopia that have brought them to an ophthalmological examination. Once an AVM of the brain (BAVM) is diagnosed, the visual and eye movement disorders must be accurately documented and monitored in order to properly assess the progression of the disease and the effects of therapy. BAVMs are potentially life-threatening lesions that can cause severe neurologic deficits and seizures. Early diagnosis can facilitate the implementation of therapies that can be curative. The literature concerning BAVMs is confused because of the inclusion of many vascular malformations that are not AVMs, inaccurate diagnoses prior to the development of computed tomography (CT) and magnetic resonance (MR) imaging, and incomplete angiography in many reports. Numerous categorizations of intracranial vascular malformations and angiomas exist; Yasargil's classification (Tables 7.1, 7.2) seems to be the most helpful for distinguishing AVMs from other vascular lesions [1].

7.1.2 Incidence

The true incidence of AVMs in the brain is difficult to assess because for years many lesions are relatively asymptomatic or only cause headaches. In the patient with minor symptoms, the appropriate neuroimaging may not seem necessary to the physician. In a 3-year prospective study of cerebrovascular disease in a population of 100,000, 1.8% of the subjects were found to harbor a BAVM [2].

In an autopsy series, 223 of 5276 consecutive autopsies demonstrated brain vascular malformations, a 4% incidence. However, this study included BAVMs, venous malformations, telangiectases, cavernomas, and other vascular malformations [3]. Courville reported an incidence of 0.8% in another autopsy series [4]. In the era prior to CT, 1.8% of 3206 patients initially diagnosed by clinical and angiographic criteria as having a neoplasm were found to have a BAVM or other vascular malformation [5].

There is a 0.14% prevalence of clinically appar-

Table 7.1. Classification of central nervous system vascular lesions (modified from [1])

I. Vascular neoplasms
  1. Hemangioblastoma
  2. Angioblastic meningioma
  3. Hemangiopericytoma
  4. Angiosarcoma

II. Vascular malformations
  1. Telangiectasia
  2. Cavernous malformation (cavernous hemangioma, cavernoma)
  3. Venous anomaly
  4. Arteriovenous malformation

III. Vascular malformation and vascular neoplasm associated with neurocutaneous disorders
  1. Hemangioblastoma (von Hippel-Lindau)
  2. Venous anomaly, encephalofacial angiomatosis (Sturge-Weber-Krabbe-Dimitri)
  3. Hereditary hemorrhagic telangiectasia (Osler-Weber-Rendu), ataxia telangiectasia (Louis-Bar)
  4. Arteriovenous malformation, encephaloretinofacial angiomatosis (Wyburn-Mason)
  5. Congenital venous dysplasia (extremities, spinal) (Klippel-Trenaunay-Weber)
  6. Multiple cavernous hemangioma brain, eye, orbit
  7. Others

Table 7.2. Clinical features of vascular malformations in the brain (modified from [1])

	BAVM	Venous anomaly	Cavernous malformation	Telangiectasia
Pathology	Gliotic brain	Normal brain unless bleed	No intervening parenchyma	Normal brain parenchyma
Pathogenesis	Agenesis of capillaries?	Agenesis of venous connection	Sinusoid change capillary-venule	Capillary dilatation
Locale	All layers and locales	Cerebral, cerebellar	All locales, extrinsic, intrinsic	Brainstem
Heredity	Rare	Doubtful	Seldom	Seldom (Olser-Weber-Rendu)
Associated malformations	Rare	Unknown	Seldom (eye, orbit)	Unknown except Osler-Weber-Rendu
Age	Uncommon <20 years	Uncommon <20 years	35–55 years	>35 years
Multiple	Rare	Solitary	Solitary, rarely multiple	Unknown
Size	Occult-giant	Small-large strip	Oval, 1–5 cm	<1 cm
Course	Variable, silent-death	Most silent, few acute	Most silent, few acute	Silent-acute, recurrent
Contrast CT if no bleed	Often obvious	Umbrella-, mushroom-shape	Honeycomb	Occult
MRI, if no bleed	Flow void, vascular channels, gliosis brain	Low-flow channels	Honeycomb black and white heterogeneity	Irregular gliosis brain
Angiography	High-flow shunts, aneurysms	Seen in venous phase	Rarely visualized	Occult

ent BAVMs in the general population. Prevalence studies of all BAVMs are also hampered by the fact that many lesions are initially asymptomatic, and the diagnosis prior to CT and MR imaging was often inaccurate. On occasion, a malignant glioma with florid neovascularization may appear on angiography to be similar to a BAVM. Although rare, hemorrhage into a glioma or into a metastatic brain tumor such as a hypernephroma may appear similar on CT or MR imaging to a ruptured BAVM that is compressed by the mass of the hematoma.

BAVMs occur with a much lower frequency than aneurysms. In the Cooperative study, the ratio of the number of BAVMs to aneurysms was 1/6.5 [6]. BAVMs are equally distributed with respect to sex and to the right or left cerebral hemisphere. BAVMs are found in a supratentorial location in 90% of cases. The parietal lobe, alone or in combination with other lobes, is the most commonly affected area of the brain (27% of cases). The frontal lobe is the next most frequently involved region, being affected in 22% of all cases [6]. BAVMs are located in the posterior fossa in approximately

Fig. 7.1a, b. Multiple AVMs. **a** Frontal subtraction ▶ angiogram of vertebral artery demonstrates a large malformation in the right parietal region. Note a second small malformation on the left (*curved arrow*). A flow-related P1 aneurysm is also seen (*open arrow*). **b** The magnetic resonance image of the same patient did not demonstrate the arteriovenous malformation on the left. Angiography is still the most sensitive method to analyze and detect these lesions. (From [8])

10% – 18% of reported cases [7], which is in proportion to the volume of tissue in the posterior fossa. The cerebellum is involved 4 times as often as the brainstem, and BAVMs can involve both areas.

Cases with multiple BAVMs (Fig. 7.1) or a BAVM with simultaneous extension into both hemispheres are infrequent. In our series of 250 consecutive cases of BAVM, seven subjects (2.8%) had two malformations. Five patients had the second lesion in the same hemisphere while two others had BAVMs located in the contralateral hemisphere [8]. A familial occurrence of BAVMs is extremely

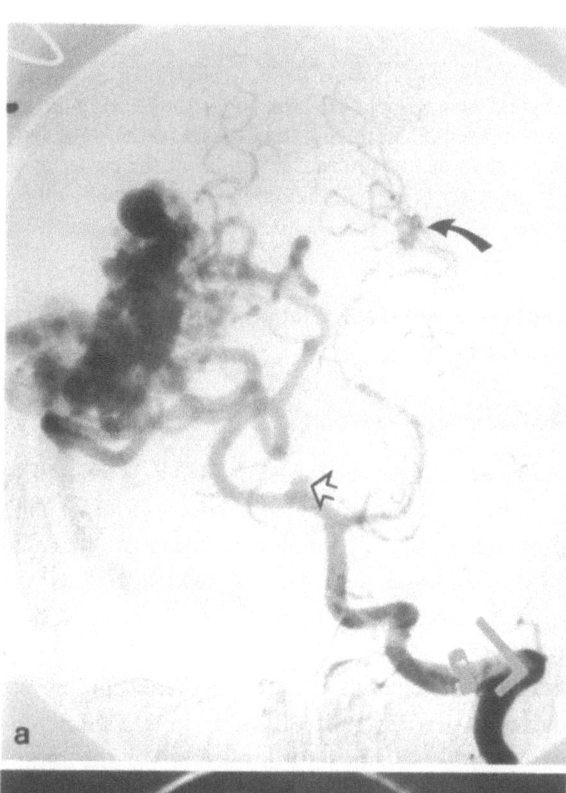

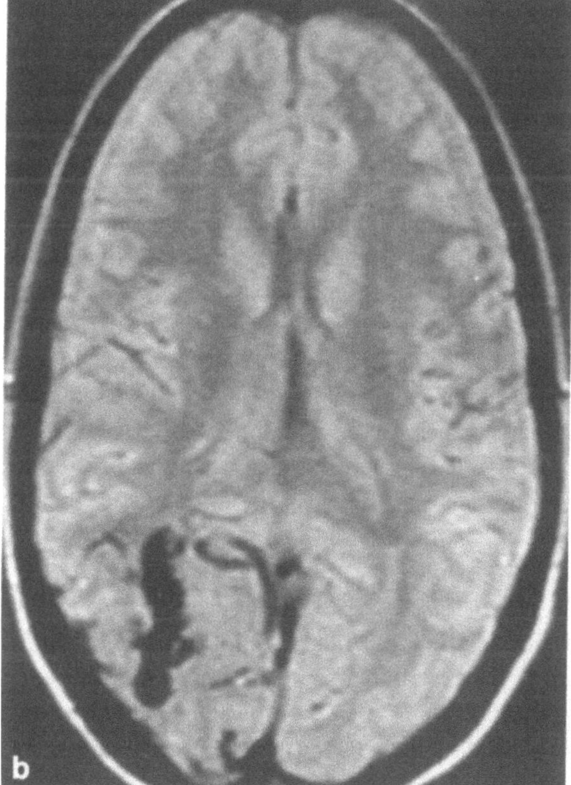

rare, arguing against a genetic defect as the cause of these lesions. This concept is further supported by three of our patients with a BAVM, each of whom had a homozygote twin without a BAVM.

7.1.3 Presentation

Symptoms can begin as early as in the neonatal period and typically continue throughout life. However, less than 10% of BAVMs cause symptoms during the first decade of life. By age 29 years, 55% of these lesions are clinically apparent, and by age 39 years, 71% are manifest [9]. The clinical deficit or complaint often depends on the age of the patient when clinical manifestations begin and the area of the brain involved by the nidus, by the shunting of arterial blood, and by alterations produced by the BAVM on the normal venous system. The presenting clinical symptom is headache in 24% [10] and seizures in 32% [10] to 36% of patients [11]. The percentage of patients who are initially evaluated because their first manifestation is a subarachnoid or intraparenchymal hemorrhage varies with each series, including reports of 30% (15/50 patients) [10], 63% (120/191 patients) [12], and as high as 68% (307/453) in the Cooperative study [6]. However, each series seems biased by referral patterns. Additional less common initial clinical problems included progressive focal or diffuse neurologic dysfunction, hydrocephalus, and cardiac failure.

7.1.3.1 Infants and Preadolescents

Infants are subject to high-output cardiac failure because of the rapid arteriovenous shunting of blood. Cardiac failure rarely occurs in children older than 1 year of age, although cardiomegaly can be seen in older children. Spontaneous hemorrhage, seizures, and headache occur less frequently in this age range, except in cases with high-flow pial arteriovenous fistulas between pial arteries and cerebral veins (Fig. 7.2). Hemorrhage does not occur in vein of Galen malformations, where the shunt is between pial arteries and the vein of Galen or a precursor vessel.

In children who survive the neonatal period, hydrocephalus, with associated cognitive deficits, develops as a result of untreated malformations that

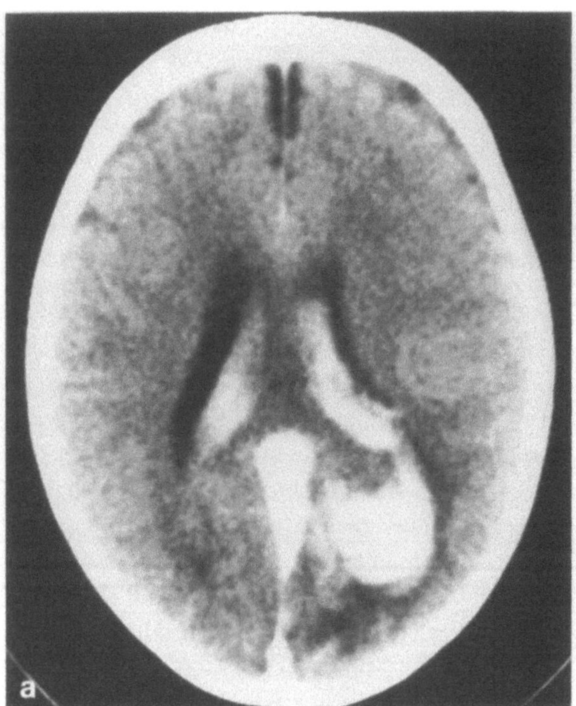

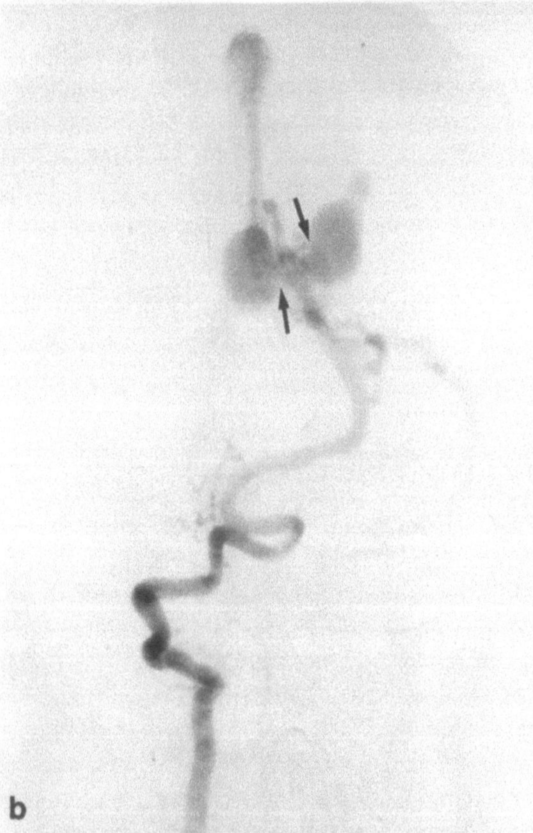

Fig. 7.2a, b. Arteriovenous fistula of the brain presenting in an infant with cerebral hemorrhage. **a** Axial computed tomogram with contrast demonstrates a left intraparenchymal hemorrhage extending into the ventricular system. **b** Frontal subtraction angiogram of the right vertebral artery demonstrates two fistulas of the posterior cerebral artery (*arrows*) draining into aneurysmally dilated veins

cause an aneurysmal dilatation of the vein of Galen (see Sect. 7.1.7).

7.1.3.1.1 Focal Neurologic and Visual Dysfunction

Focal neurologic dysfunction, which is related to the area of brain injury, also occurs in older children but is rare in infants. In some of these patients, progressive subcortical calcification in the affected areas can be seen on CT. Neurologic signs that result from in utero or perinatal brain injury are rare. If ischemia or a developmental failure occurred in utero, one would expect to find numerous cases with extremities contralateral to the motor cortex damage that had failed to develop completely and remained smaller than the normal extremities [13]. We have rarely seen this situation, however. We also have two BAVM patients with visual pathway signs of a possible in utero injury. In patients with congenital, but not acquired posterior parietal or occipital lobe ischemic or developmental lesions [14], a dense complete homonymous hemianopia or quadrantanopia may be associated with transynaptic degeneration across the lateral geniculate nucleus. This causes a retrograde loss of axons in the ipsilateral optic tract to the retinal ganglion cell bodies [15]. The retrograde degeneration also affects the pretectal area-bound axons, resulting in a relative afferent pupillary defect in the eye with the temporal field loss [16]. Ophthalmoscopy in both of our cases (as well as a patient reported by Hoyt [16a]) revealed the typical diffuse optic atrophy in the eye ipsilateral to a parietal/temporal lobe BAVM and bow-tie atrophy in the contralateral disc (Figs. 7.3, 7.4). Since neither of our two patients had obvious ischemic changes in the brain on MR imaging and the abnormal venous drainage in both cases seemed to directly affect the lateral geniculate nucleus and the optic tract, it is possible that the observed optic atrophy and retrograde degeneration might not have resulted from a

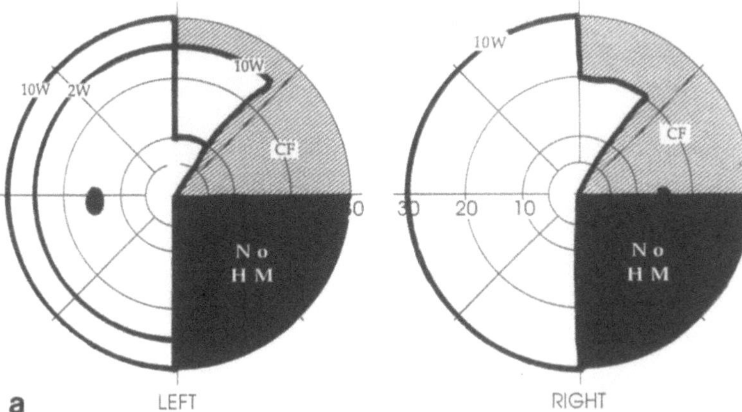

Fig. 7.3a–c. Retrograde degeneration in the anterior visual pathway, possibly transsynaptic across the lateral geniculate, associated with a BAVM. This 63-year-old man with a history of childhood dyslexia complained of progressive speech and memory difficulties over 6 months. He had no visual complaints. **a** In addition to the language and mild global memory disorder, the clinical evaluation revealed a right homonymous, almost complete, hemianopia, which was absolute in the inferior field. **b** Ophthalmoscopy of the right optic disc shows typical bow-tie atrophy. The left eye (not shown) had mild diffuse pallor. These findings were consistent with transsynaptic degeneration across the lateral geniculate body affecting the pregeniculate pathway. **c** The axial view T1-weighted magnetic resonance image shows the left temporal parietal arteriovenous malformation. However, in this one image the deep venous drainage from the arteriovenous malformation seems to extend to the left optic tract (*arrow*), so it is possible that some of the atrophy seen in the optic nerves was not a result of transsynaptic degeneration

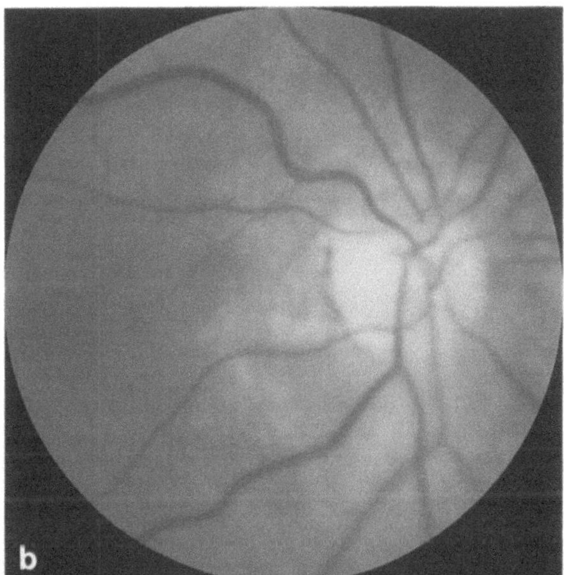

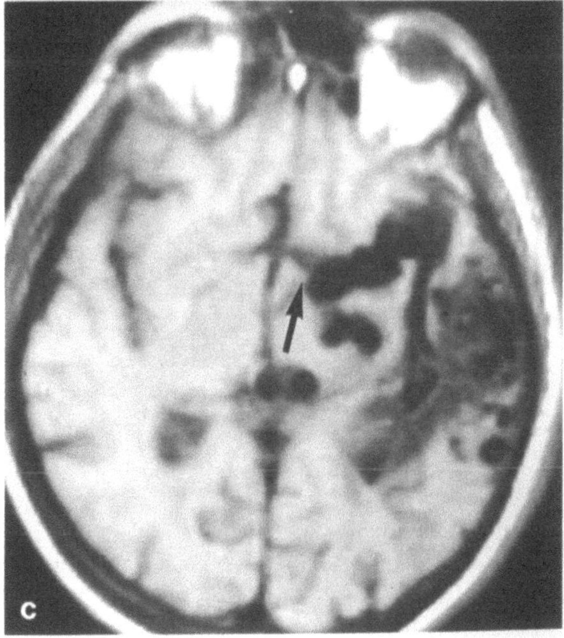

transsynaptic degeneration. No matter the cause, the rarity of this phenomenon in patients with BAVMs would suggest that, except for these cases, the injury to the brain typically occurs after birth.

7.1.3.2 Adolescents

Headaches and seizures increase in frequency in the preteen and teenage years. Patients develop headaches which are commonly misdiagnosed as migraine (see below for discussion). The first hemorrhage can occur in this age group. As in younger children, the risk of hemorrhage is higher when an arteriovenous fistula is present. Review of the clinical history often suggests that some of these patients may have had small bleeds that are over-

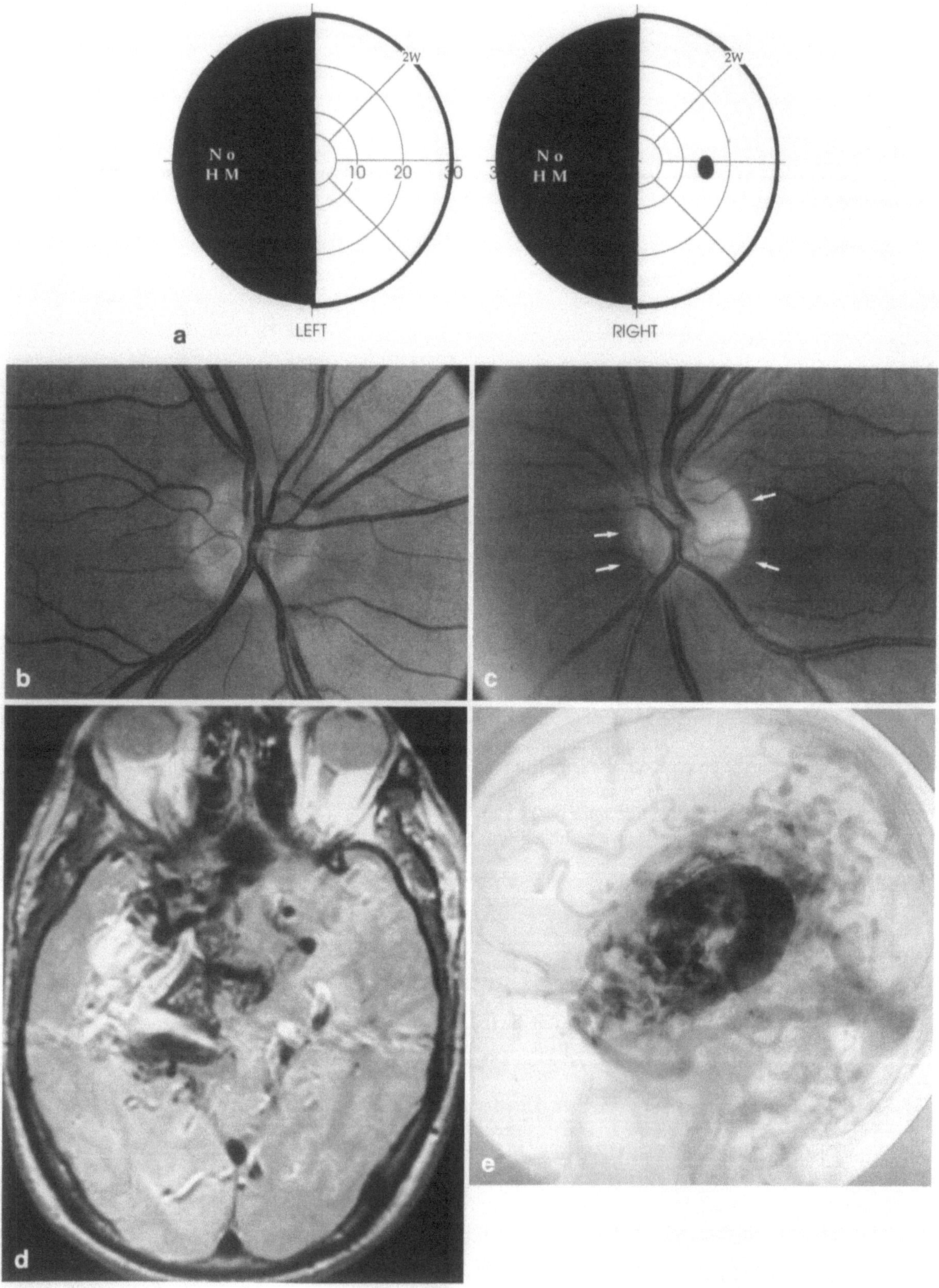

◄ **Fig. 7.4a–e.** Retrograde degeneration in the optic tract caused by a brain arteriovenous malformation.

a A 39-year-old man, known to have a brain arteriovenous malformation, had an intermittently progressive left hemiparesis since 8 years of age. In addition to his motor abnormalities, he had an absolute left homonymous hemianopia.

b, c Ophthalmoscopy reveals mild diffuse atrophy in the right eye (**b**) and bow-tie atrophy with a loss of the nasal and papillomacular nerve fiber layer (*arrows*) in the left eye (**c**), consistent with retrograde degeneration in the right optic tract to the optic nerves.

d The axial view T2-weighted magnetic resonance image shows the heterogeneous signal characteristics of brain arteriovenous malformation deep in the right cerebral hemisphere. The lesion may affect the lateral geniculate body, in which case the atrophy of the optic nerves may not be solely the result of transsynaptic degeneration.

e The lateral view subtraction angiograms of the internal carotid and the vertebral (not shown) arteries demonstrate an extensive right cerebral brain arteriovenous malformation with drainage to the superior sagittal sinus, transverse sinus and straight sinus

looked if the episodes are not associated with neurologic deficits.

7.1.3.3 Adults

Headache which started at a younger age may continue but the incidence of new recurrent headaches is lower in adulthood. However, new headaches can result from a hemorrhage or another hemodynamic event such as thrombosis of a cortical vein (see Sect. 7.1.5). The number of hemorrhages and seizures seems to increase throughout adult life (see below for recurrent episodes of each and a review of large uncontrolled studies).

7.1.4 Hemodynamics: Controversies and Unresolved Questions

7.1.4.1 Overview of Issues

The range of questions and controversies surrounding the understanding of the dynamics and treatment of BAVMs is beyond the scope of this chapter. However, certain key issues will be considered, though some of these have not necessarily been entirely resolved. To this end, in this section the discussion is restricted to cases with true parenchymal BAVMs. Many clinical studies on BAVMs have included a potpourri of vascular lesions such as venous anomalies, venous angiomas, hemangiomas, cavernomas, telangiectasias, and dural AVMs that should not have been classified as BAVMs. Even when only cases of true BAVMs are reported, the bias for invasive treatment or conservative therapy [6] has made accurate study of the natural history of these lesions virtually impossible. The lack of consistent angiographic protocols and quality of these studies further complicates the analysis of reported series. For example, in the Cooperative study, patients with less severe symptoms at entry to the study frequently did not have close scrutiny over the life of the study [6]. Factors that influence the severity and nature of neurologic deficits, the type and frequency of seizures, the bleed rate, and the morbidity and mortality have not been uniformly investigated. Few studies have evaluated such factors as the location of the BAVM, the presence of arteriovenous fistulas within the BAVM, associated aneurysms, and variations and alterations in the venous outflow from the BAVM and brain [17, 18].

The clinical manifestations are related to both the location of the BAVM and the secondary vascular responses to the chronic abnormal arteriovenous shunts which are in a dynamic state of flux with the remaining intracranial circulation. Although these lesions are present from birth, it is the variable changes in the hemodynamic factors that account for the different clinical presentations of BAVMs in similar locales and the frequent delay before symptoms appear. BAVMs can alter both the local and remote intracranial blood circulation.

BAVMs are believed to arise from congenital maldevelopment of the intracranial arteries and veins. The capillaries, which normally develop between the arteries and veins, do not mature leaving the embryological arteriovenous connections or shunts. The majority of BAVMs invaginate into the brain because they follow course of the vessels through the fissures and sulci to appear deep [1]. BAVMs are often wedge-shaped with an apex related to the venous drainage. The veins typically either drain to the deep venous system associated with the ventricles, or the venous drainage of the BAVM follows the pattern of the transcerebral venous system. In the latter pattern, the deep midline veins, the superficial cortical veins, or both are involved.

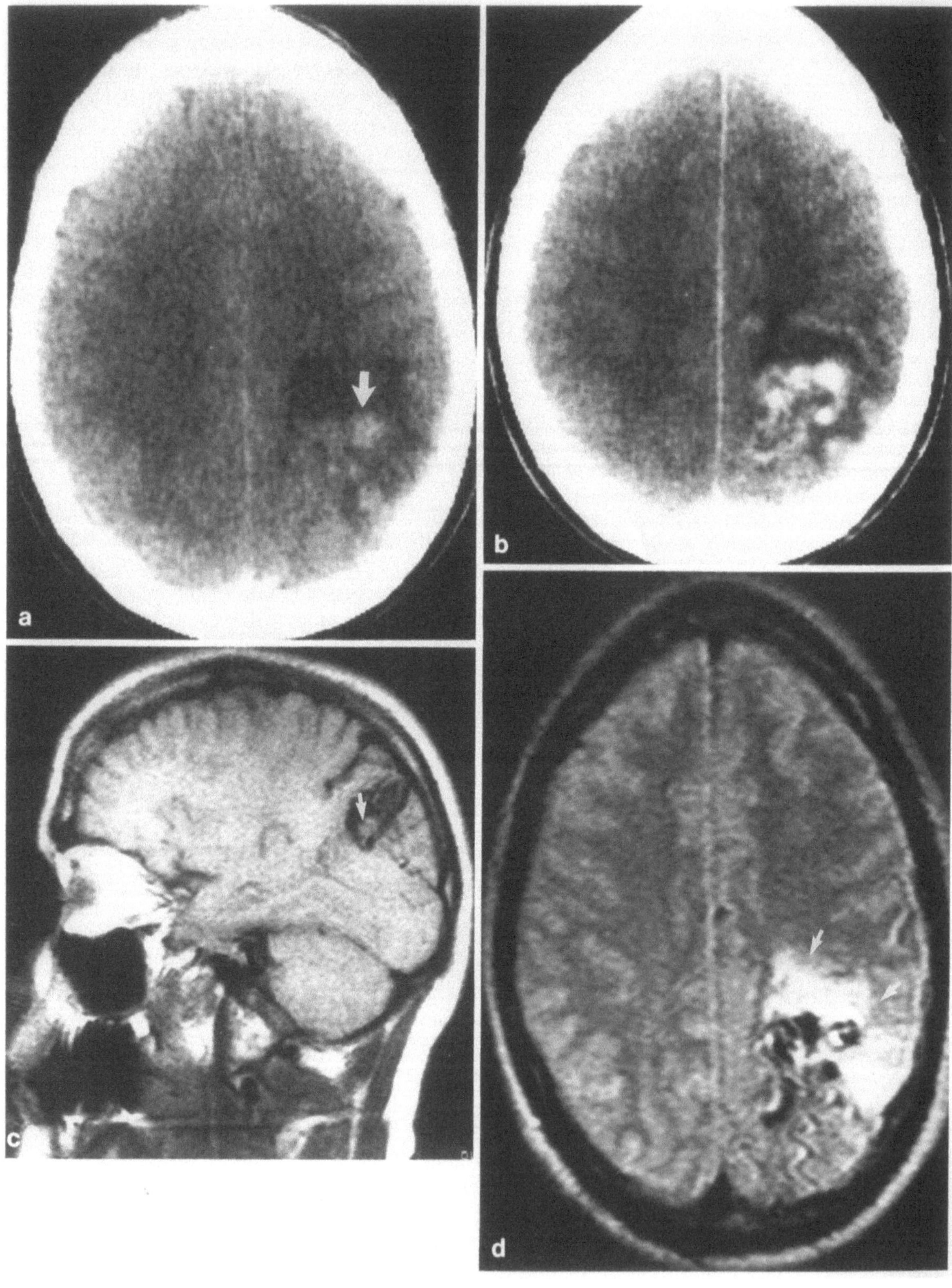

Fig. 7.5 a–e. Acute thrombosis of a cortical vein associated with a parietal lobe brain arteriovenous malformation caused a generalized motor seizure but no visual field defect.

a Axial computed tomogram prior to contrast administration demonstrates an area of increased absorption coefficient (*white arrow*) surrounded by edema anteriorly.
b The contrast-enhanced computed tomogram of the same patient demonstrates the brain arteriovenous malformation and an anterior area of decreased absorption coefficient.
c The sagittal view T1-weighted magnetic resonance image demonstrates that the signal void of the arteriovenous malformation contains an area of increased signal (*white arrow*) suggestive of thrombosis.
d The axial T2-weighted magnetic resonance image demonstrates the arteriovenous malformation as well as an area of increased signal (*white arrows*), suggestive of edema, just anterior to the malformation.
e The late venous phase of the lateral view subtraction angiogram demonstrates a stagnant collection of contrast material in a venous sac (*arrow*) representing a thrombosed venous pouch.

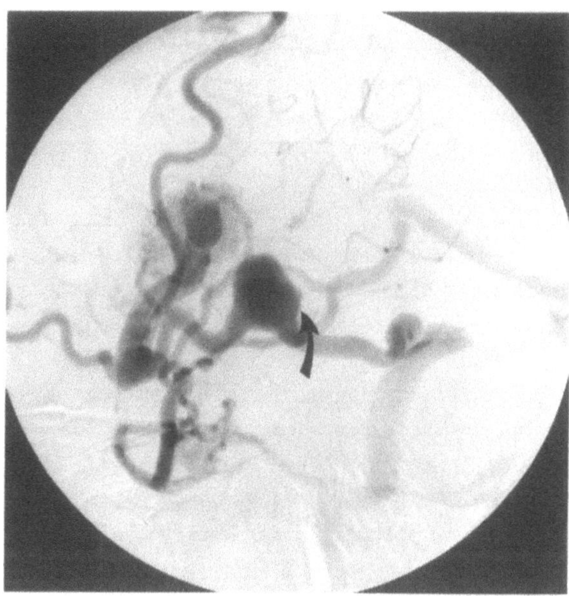

Fig. 7.6. The late phase of a lateral digital subtraction angiogram in a case of temporal lobe brain arteriovenous malformation demonstrates an area of aneurysmal dilatation in the venous drainage due to an outflow restriction (*arrow*). Patients with temporal lobe lesions present more frequently with hemorrhage representing venous restriction

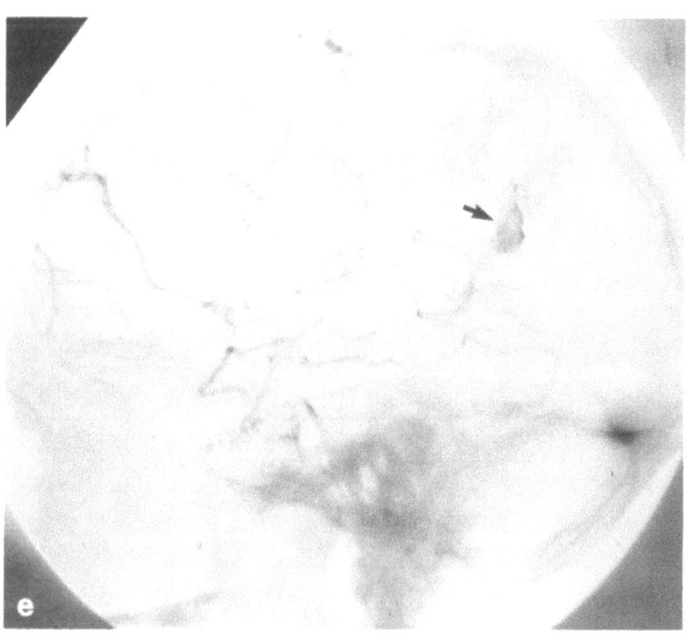

Though there is little disagreement that BAVMs are congenital, the cause or the sequence of events in the maldevelopment of the vascular system and the way the arteriovenous shunt affects the final disposition of the intracranial venous system are unresolved.

The congenital portion of a BAVM contains one or more compartments or niduses, or one or more direct arteriovenous communications (fistula), without an intervening nidus. More frequently, the BAVM contains elements of both. Though angiography may suggest that different areas within a nidus seem to have separate arterial inputs and appear compartmentalized, communication often exists between adjacent regions of the BAVM.

Aneurysmal venous dilatations of the draining veins result from the high flow of shunted blood and restrictive anatomical barriers, such as the dural venous sinuses and the tentorium, and compression by other dilated veins. Venous ectasias can compress and alter the normal venous drainage of the brain. A venous ectasia may also develop following ventricular shunting. While the shunt relieves the increased intracranial pressure, it also decompresses the cerebrospinal fluid external pressure that had limited the dilatation of these veins. (This is particularly pertinent when the vein of Galen malformation compresses the deep venous system.)

Thrombosis in the veins that drain the BAVM, the brain, or both, can also cause clinically apparent episodes (Fig. 7.5). Venous thrombosis that results from mechanical factors such as kinking of outflow veins by the dura at the tentorial edge may account for the increased incidence of bleeding into the insula or medial temporal lobe in patients with temporal lobe BAVMs [18] (Fig. 7.6). Mechanical venous compromise also occurs as the abnormal draining veins enter the dural venous sinuses or as a result of direct compression by the mass of the BAVM or an intraparenchymal hemorrhage. Turbulent blood flow or stenosis in the veins may also contribute to the development of venous thrombosis.

The incidence of arterial aneurysms, which are either high flow related or true congenital berry aneurysms, associated with BAVMs has been variably reported (Fig. 7.7). Though between 2.7% [19] and 16.7% [20] of patients have been described as having arterial aneurysms, 21% of the patients in our series had angiographic evidence of these aneurysms. Prior studies, which described a lower incidence of aneurysms, typically had angiographic studies that were limited. In these evaluations the presence of a BAVM was documented but complete angiography of all the involved (cerebral) vessels was not usually performed.

Rupture of these aneurysms may account for the hemorrhages in some, but not all, patients. Since many arterial ectasias or aneurysms are flow related, they frequently develop in the arteries that supply blood to the BAVM, or in the vessels of the nidus itself. Clearly, the distribution of the sites of these aneurysms differs from the distribution of those berry aneurysms found in patients without BAVMs [21, 21a, 22]. Additionally, flow-related aneurysms are often smaller or resolve when therapy reduces the blood flow through the affected arteries (Fig. 7.7d).

Collateral arteries frequently supply blood to the normal brain parenchyma when the arteries that normally supply this tissue are involved with the BAVM. Superselective angiographic studies in some of these patients may demonstrate that the feeding artery to the BAVM actually terminates in the BAVM. However, in other BAVMs, the feeding vessel gives branches with a "comblike" appearance ("en passage") to supply the BAVM, and the artery continues distally past the BAVM to provide blood to the normal brain (Fig. 7.8a). The arteries that shunt blood into the BAVM may also give branches to the brain parenchyma which concomitantly contribute to to the BAVM. This may result in a "steal" of the blood supply from normal brain [23].

Any interference with the arterial supply (occlusion or steal) to areas of normal brain can stimulate the development of functioning collaterals to supply the affected regions (Fig. 7.8). Subependymal anastomoses which are normally latent remain or become patent. The dural arteries can be recruited to provide collaterals to the leptomeningeal arteries, which supply blood to the parenchyma or the BAVM or both [24] (Fig. 7.9). Leptomeningeal or dural collaterals that supply the BAVM can also develop after a subarachnoid hemorrhage or following partial surgical excision or incomplete embolization where arteriovenous shunting remains (Fig. 7.10) [25]. A failure to develop collateral ar-

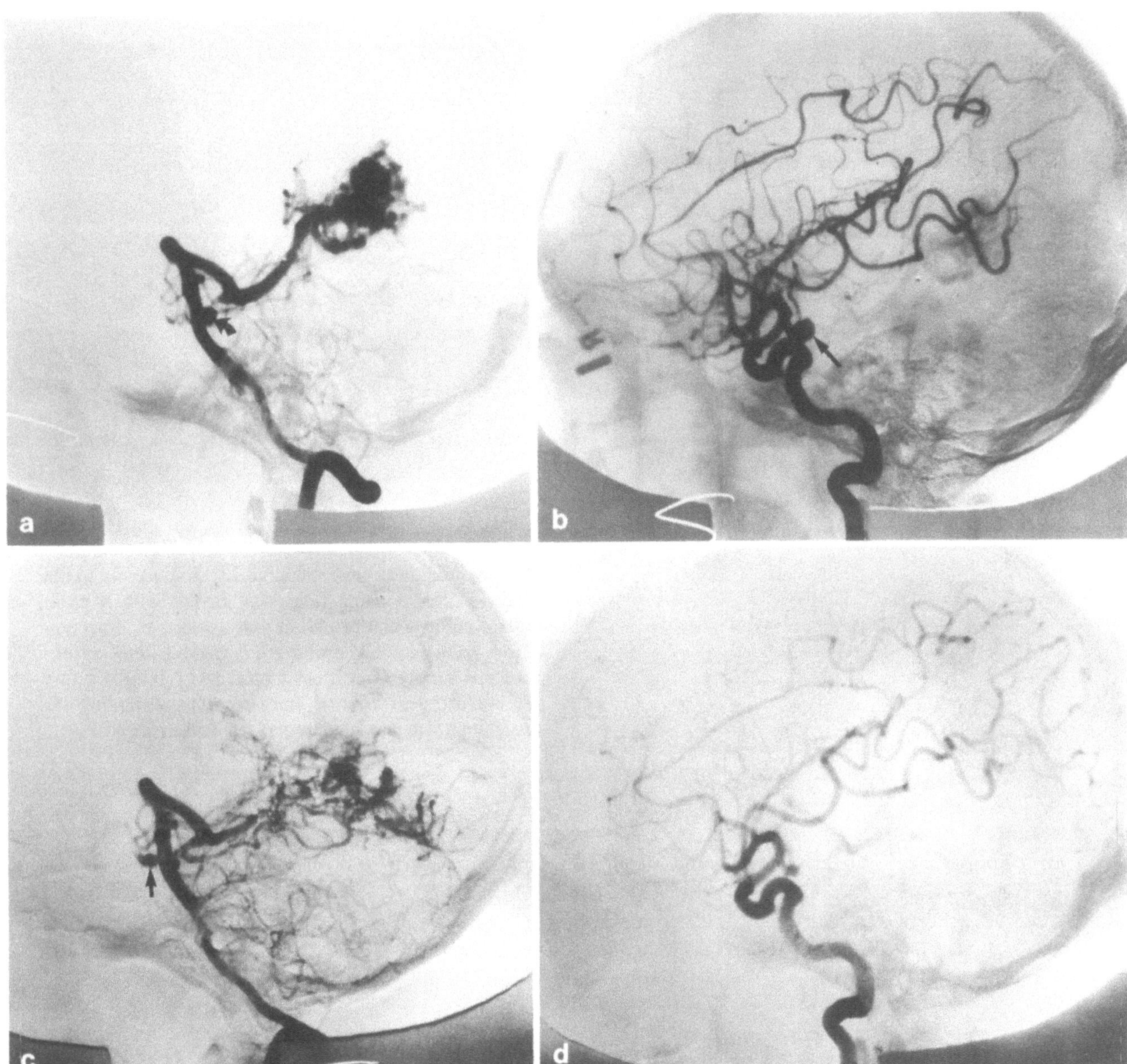

Fig. 7.7a–d. Flow-related aneurysms associated with a brain arteriovenous malformation.

a Lateral subtraction angiography of the left vertebral artery demonstrates a basilar aneurysm (*curved arrow*) at the level of the trigeminal artery associated with an occipital brain arteriovenous malformation supplied by branches of the posterior cerebral artery.

b Lateral subtraction angiogram of the right internal carotid artery in the same patient demonstrates a large posterior communicating artery aneurysm (*arrow*).

c The immediate postembolization angiogram of the left vertebral artery after partial embolization of the brain arteriovenous malformation demonstrates disappearance of the trigeminal aneurysm and filling of the aneurysm from the posterior communicating artery (*curved arrow*).

d The follow-up right internal carotid artery angiogram demonstrates a reduction in the size of the posterior communicating artery aneurysm compared to **b**

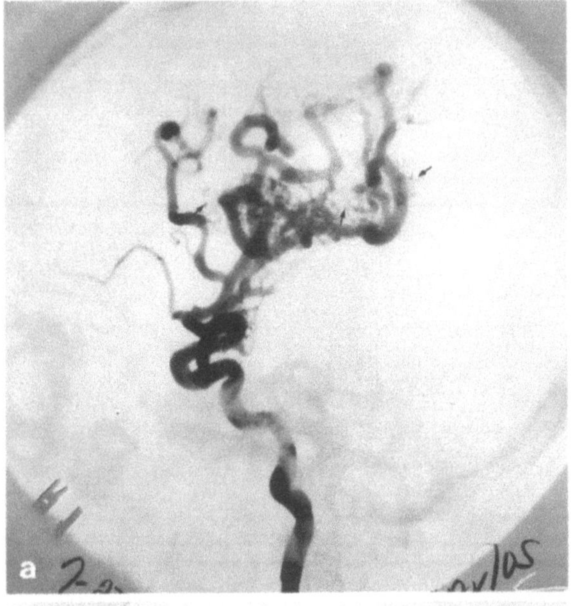

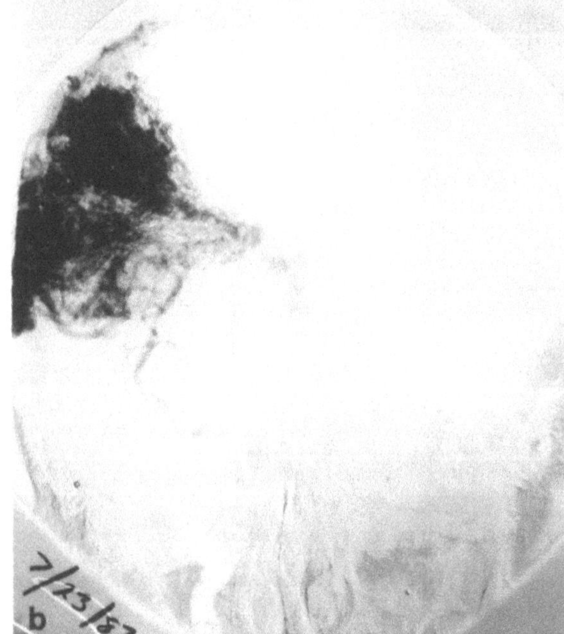

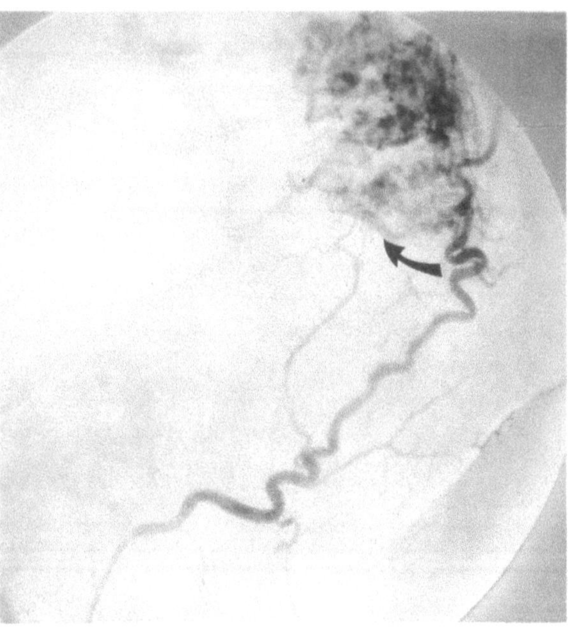

Fig. 7.9. Recruitment of meningeal arterial supply to the cerebral arterial circulation in a patient with a brain arteriovenous malformation. A selective injection through the external carotid artery demonstrates a transosseous supply from the occipital artery reconstituting cerebral arteries (*arrows*). This is an unusual finding unless the patient has had surgery or hemorrhaged

Fig. 7.8a, b. Major cerebral arteries demonstrate "en passage" supply to the brain arteriovenous malformation. The network of dilated small medullary arteries is typically seen in response to local ischemia. **a** The lateral view of the early phase of the right internal carotid artery demonstrates dilatation of multiple middle cerebral artery branches from which multiple branches to the brain arteriovenous malformation originate ("en passage") (*small arrows*). **b** The later phase frontal projection of the angiogram shows an extensive network of small arteries with the apex of the brain arteriovenous malformation pointing deeply

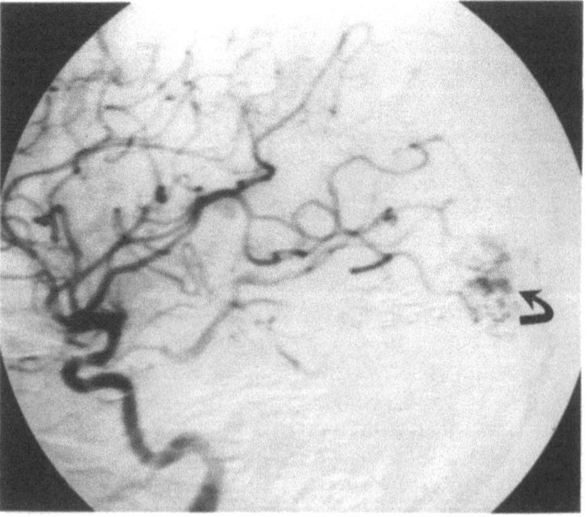

Fig. 7.10. Following embolization of the posterior cerebral artery (corresponds to Fig. 7.12), the lateral subtraction angiogram of the right internal carotid artery reveals reconstitution of the brain arteriovenous malformation via middle cerebral artery leptomeningeal collaterals (*curved arrow*)

teries to the hypoperfused region of the brain in the distribution of arterial flow into the BAVM may be associated with a massive dilated capillary network reminiscent of moyamoya (see Sect. 8.10). Stenotic arterial changes [26] appear similar to the arterial changes noted in experimentally induced high-flow angiopathy [27]. The narrowing of arteries associated with BAVMs affects arteries in locations atypical for atherosclerosis, probably results from a high-flow angiopathy, and may cause symptomatic arterial insufficiency. Arterial narrowing may also develop from vasospasm following a subarachnoid hemorrhage that results from rupture of an associated aneurysm. However, symptomatic or angiographic vasospasm is rarely seen following a BAVM bleed (see Fig. 7.16c).

The reduced tissue perfusion that results from venous hypertension will also cause local tissue ischemia that can stimulate angiogenesis. The subsequent neovascularization is probably similar to the "sprouting" type of vessel growth that arises from the venules in the experimental investigations reported by Folkman [28].

When the shunted blood leaving the BAVM flows into the veins that also drain normal brain (convergence of blood flow), venous hypertension arises in otherwise normal cerebral draining veins. When the draining veins carry blood only from the BAVM, the veins adjacent to the BAVM become collaterals that dilate in order to accommodate the increased blood flow from neighboring areas of the brain. Hemodynamic alterations, including venous thrombosis or congenital hypoplasia or aplasia of the veins, result in further destabilization of the balance between venous pressure and tissue perfusion in the brain. Functional abnormalities such as seizures or neurologic deficits can be caused by these pathological processes in regions near or remote from the BAVM nidus.

If the intraluminal venous pressure increases to a critical level in the BAVM drainage, rupture of the vessel results in an intracranial hemorrhage. Hemorrhages are more likely to develop when the venous outflow has areas of significant restriction or aneurysmal dilatation. The location of the bleed seems dependent on the location of the involved veins. For example, rupture of a subependymal vein results in an intraventricular hemorrhage.

Clearly, the angioarchitecture of the BAVM develops out of a complex, multifactorial, dynamic process. Both congenital and acquired vascular alterations contribute to the evolution of a clinically symptomatic BAVM. Many of the important factors are not entirely understood and further study will be necessary before definitive conclusions are drawn [8].

7.1.4.2 Neurologic Manifestations

The neurologic manifestations arise from damage of nervous system tissue and the neural responses induced by the abnormal hemodynamic state. For example, venous hypertension can disrupt the local brain metabolism and the interstitial electrolyte–water balance in the area adjacent to or remote from the BAVM. The most obvious dysfunction results from the local and diffuse mass effect and tissue destruction caused by a large intraparenchymal hemorrhage. In contrast, the actual vascular insult, such as a subclinical hemorrhage or a venous infarction, that precipitates seizures and causes an epileptogenic focus is often less obvious. Even in patients without a history of clinically apparent hemorrhage, pathological specimens frequently show hemosiderin deposits in the parenchymal tissue surrounding the BAVM, suggestive of prior microscopic bleeds [29, 30]. Mechanical compression of the brain parenchyma or the ventricular system may also cause dysfunction. In unruptured BAVMs, a significant mass effect is unusual but it may result from a massive venous ectasia. If the venous dilatation affects a critical area for cerebrospinal fluid egress, such as the cerebral aqueduct, hydrocephalus may result.

When the BAVM shunts arterial blood away from normal regions in the brain, this steal phenomenon may cause focal neurologic defects. The concept of steal remains controversial because most patients in whom normal arteries cannot be visualized on angiography do not have associated neurologic deficits suggestive of a clinical steal phenomena. An example of this lack of correlation is seen in cases of high-flow carotid cavernous fistula where angiography fails to demonstrate filling of the ipsilateral middle cerebral arterial system, suggestive of a total steal. All of the blood from the internal carotid artery seems to be shunted through the fistula, yet these patients do not develop cere-

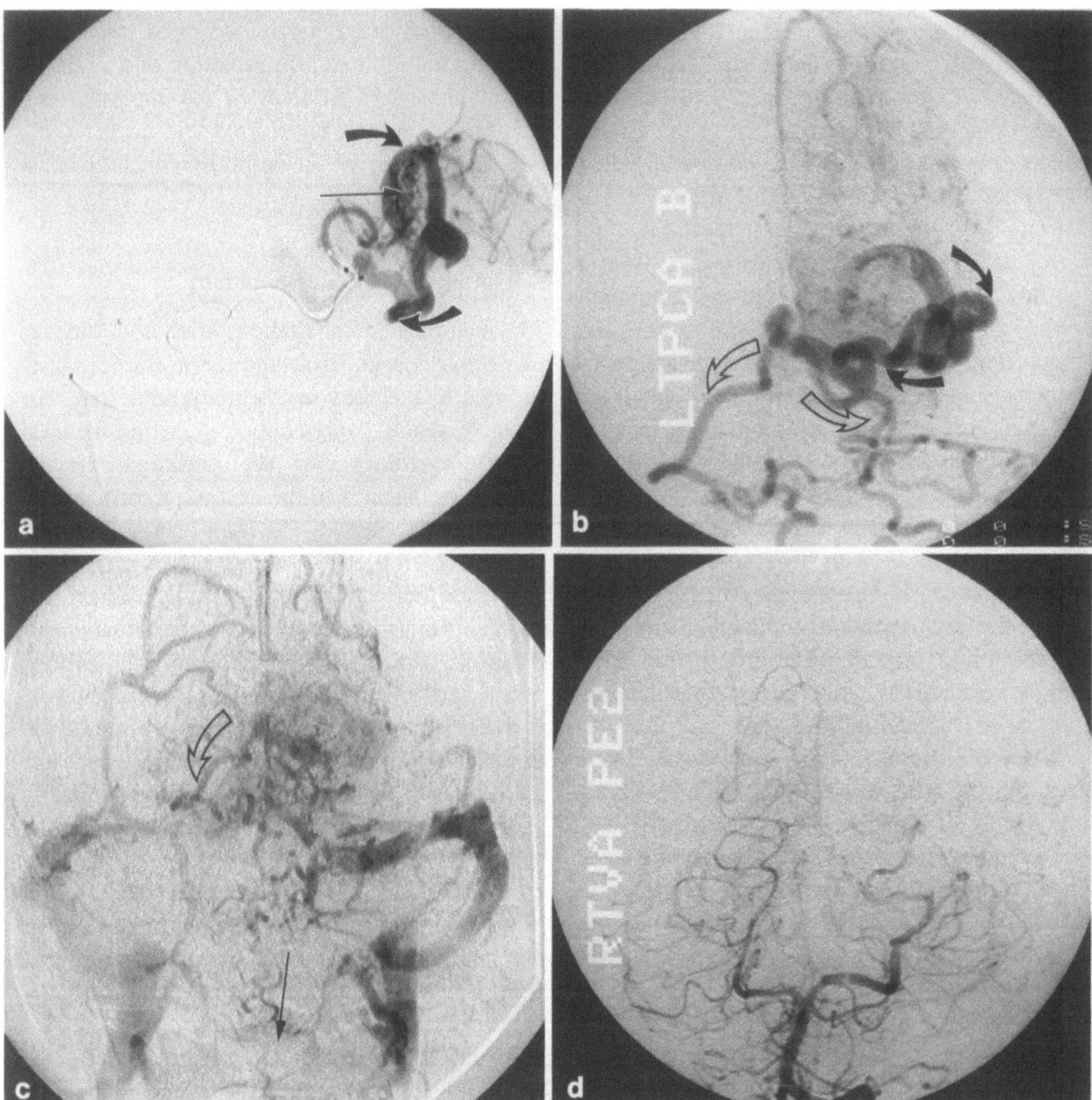

Fig. 7.11 a – e. Venous hypertension can cause diffuse and focal neurologic deficits remote from the brain arteriovenous malformation nidus. A small occipital arteriovenous malformation draining into the posterior fossa presented with multiple lower cranial nerve dysfunction and a depressed level of consciousness.

a The lateral superselective angiogram of the posterior cerebral artery (catheter seen in basilar to distal posterior cerebral artery) shows a relatively small arteriovenous malformation nidus (*thin arrow*) and prominent venous drainage (*curved arrows*).

b Later phase frontal view of the superselective injection shows the dilated venous structures (*filled curved arrows*) and downward drainage to both lateral pontomesencephalic veins (*open curved arrows*).

c The late phase frontal view vertebral artery angiogram in the same patient demonstrates the abnormal venous drainage into the basal veins (*curved open arrow*) and pontomesencephalic venous system in addition to the drainage downwards towards the spinal cord (*long arrow*).

d Following acrylic embolization of the brain arteriovenous malformation nidus, the arterial phase of the vertebral angiogram shows no filling of the malformation.

e The late venous phase demonstrates a marked reduction of the venous hypertension which was associated with significant improvement in the cranial nerve and brainstem dysfunction

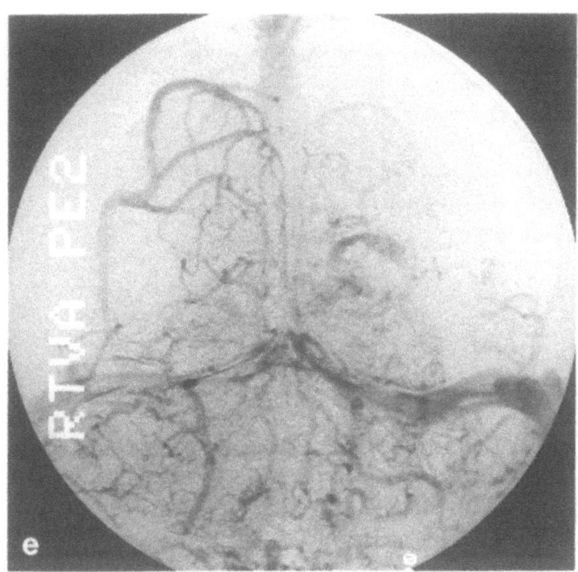

bral hemisphere dysfunction (see Fig. 2.22 in Chap. 2). In some patients a steal mechanism accounts for focal cerebral or monocular visual (see Sect. 7.1.10.1) symptoms, particularly when normal arteries, en passage, indirectly supply the BAVM. The ischemic effect of the BAVM sump becomes more deleterious when it is coupled with other abnormal hemodynamic factors that alter tissue perfusion.

A steal mechanism is also hypothesized to account for gliosis of the brain in areas remote from the BAVM. However, it is more likely that gliosis results from a disturbance in the venous drainage of this region because of a previous thrombosis or hypertension in the venous system. The associated atrophy of the brain usually occurs in a distribution suggestive of a venous, not an arterial abnormality [8]. An example of brain gliosis related to venous dysfunction is the Sturge–Weber syndrome. The vascular lesion associated with the focal brain injury is a venous malformation without arteriovenous shunting. The importance of venous outflow compromise in producing progressive neurologic dysfunction is further demonstrated by patients with a small relatively low flow BAVM who worsen because of deep venous system obstruction (Fig. 7.11). In addition, there are rare patients who, despite an absence of arteriovenous shunting, deteriorate weeks after excision of the BAVM because of venous compromise [31].

7.1.5 Hemodynamics: Clinical Features

The various clinical manifestations of the BAVMs are individually important in terms of the time and risk of later developing a hemorrhage or major neurologic deficit or of dying. Analysis of the BAVM location, angiographic features such as high inflow, poor outflow caused by few or compromised venous channels, associated arterial or venous aneurysms, venous drainage of normal brain, arterial steal, venous infarcts, and venous hypertension must be analyzed in order to understand the pathophysiology of the clinical disorders and to attempt a prognosis for each case.

7.1.5.1 Headache

The accurate assessment of headache pain, a highly subjective symptom, is quite difficult. Constant head pains or definite episodes of throbbing migrainelike headaches or both can occur in the same patient. The headaches tend to remain localized to the side of the BAVM. The incidence (10% – 20%) of headache in patients with BAVMs is comparable to the incidence of migraine in the normal population (see Chap. 10). Headache is the first symptom in approximately 16% of patients. However, the incidence of chronic migrainous type headaches increases following an intracerebral or subarachnoid hemorrhage, particularly if a homonymous hemianopia is present [32]. This suggests that involvement of the occipital circulation predisposes to the development of these headaches. In fact, BAVMs (either ruptured or unruptured) that involve the occipital lobe/posterior cerebral arteries are associated with a higher incidence of headache than BAVMs in other sites. Most series also report that the majority of patients with migrainelike headaches have BAVM involvement of the occipital region [5, 6, 32].

Headache is an outstanding feature, but not the presenting complaint, in approximately 57% of BAVM cases. Headache is a prominent early complaint in approximately 24% of patients, most of whom continue to experience headaches through their lives [10]. In a series of 110 patients with BAVMs, though 15% had episodic headaches, prior to diagnosis, only four patients (3.5%) had migrainelike attacks as the sole clinical manifesta-

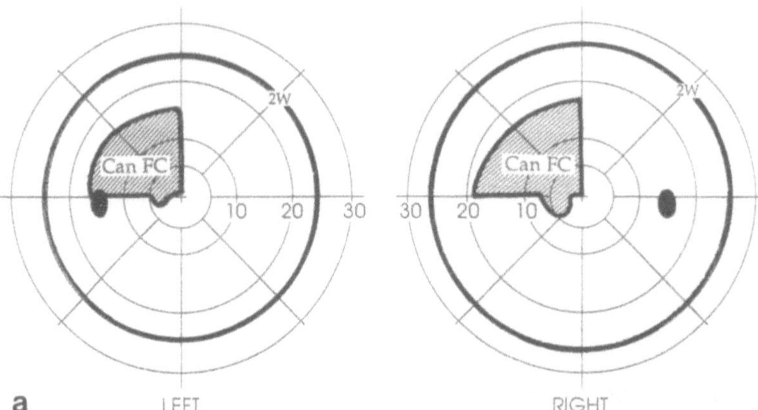

Fig. 7.12 a – e. Migrainelike headache and visual disturbance caused by an occipital brain arteriovenous malformation.

a A 25-year-old woman, with 8 years of episodes of severe throbbing headaches preceded by fortification-type visual imagery or field loss in a left homonymous distribution, had an occipital brain arteriovenous malformation diagnosed on magnetic resonance imaging. During some attacks, there was a complete dense homonymous hemianopia. Between attacks, the visual field showed a left homonymous superior quadrant scotoma to all isopters.

b The sagittal view magnetic resonance image demonstrates signal voids in the occipital lobe from the posterior cerebral artery supply (*arrow*) and a large venous pouch (*curved arrow*).

c The lateral view left vertebral artery subtraction angiogram demonstrates the posterior cerebral artery (*arrow*) supply to the brain arteriovenous malformation near the occipital pole with a large venous sac (*curved arrow*).

d Following acrylic embolization, the lateral skull radiograph demonstrates the radiopaque cast of the malformation and the distal portion of the supplying artery.

e The postembolization control vertebral angiogram demonstrates preservation of the calcarine artery but collateral circulation (within the posterior cerebral artery system and from the middle cerebral artery; see Fig. 7.10) reconstitutes a small portion of the malformation (*curved arrow*). The visual field was unchanged and the remaining area of nidus was treated with radiosurgery. The headaches were reduced in frequency and severity

tion of their BAVM. Cerebral hemorrhage, epilepsy, or homonymous hemianopia occurred before the onset of headaches in all cases, nine of whom had a family history of migraine [19].

A rapid assessment of the headache associated with an undiagnosed BAVM may suggest the patient has migraine. However, careful questioning will reveal significant differences from a typical migraine history [33]. Though the headache is frequently throbbing, unilateral, and episodic, the pain caused by a BAVM is almost always ipsilateral to the lesion or generalized [32]. True migraine episodes typically vary between the right and left sides of the head. As mentioned previously, there seems to be a predilection for headache in patients with a BAVM supplied by the posterior cerebral artery (Fig. 7.12). BAVMs located in the posterior fossa are not associated with migrainelike headaches [34]. The fact that migrainelike episodes more often cease following surgical excision of an occipital

BAVM than after excision of BAVMs in other locations is indirect clinical support of this hypothesized relationship to the posterior cerebral artery [35 – 37]. We have also observed an increase in the headache frequency in patients in whom the middle, but not posterior, cerebral artery contribution to the BAVM has been embolized because the blood flow through the posterior cerebral artery increased, as it was the only remaining blood supply to the BAVM. This suggests that the altered flow in the posterior cerebral artery was responsible for the headache.

Headaches may also occur when there is a meningeal arterial contribution to the BAVM. In these cases, the headache ipsilateral to the BAVM is a prominent symptom. The pain may be made temporarily worse following embolization occlusion of the meningeal blood supply, but eventually the headaches diminish or completely resolve.

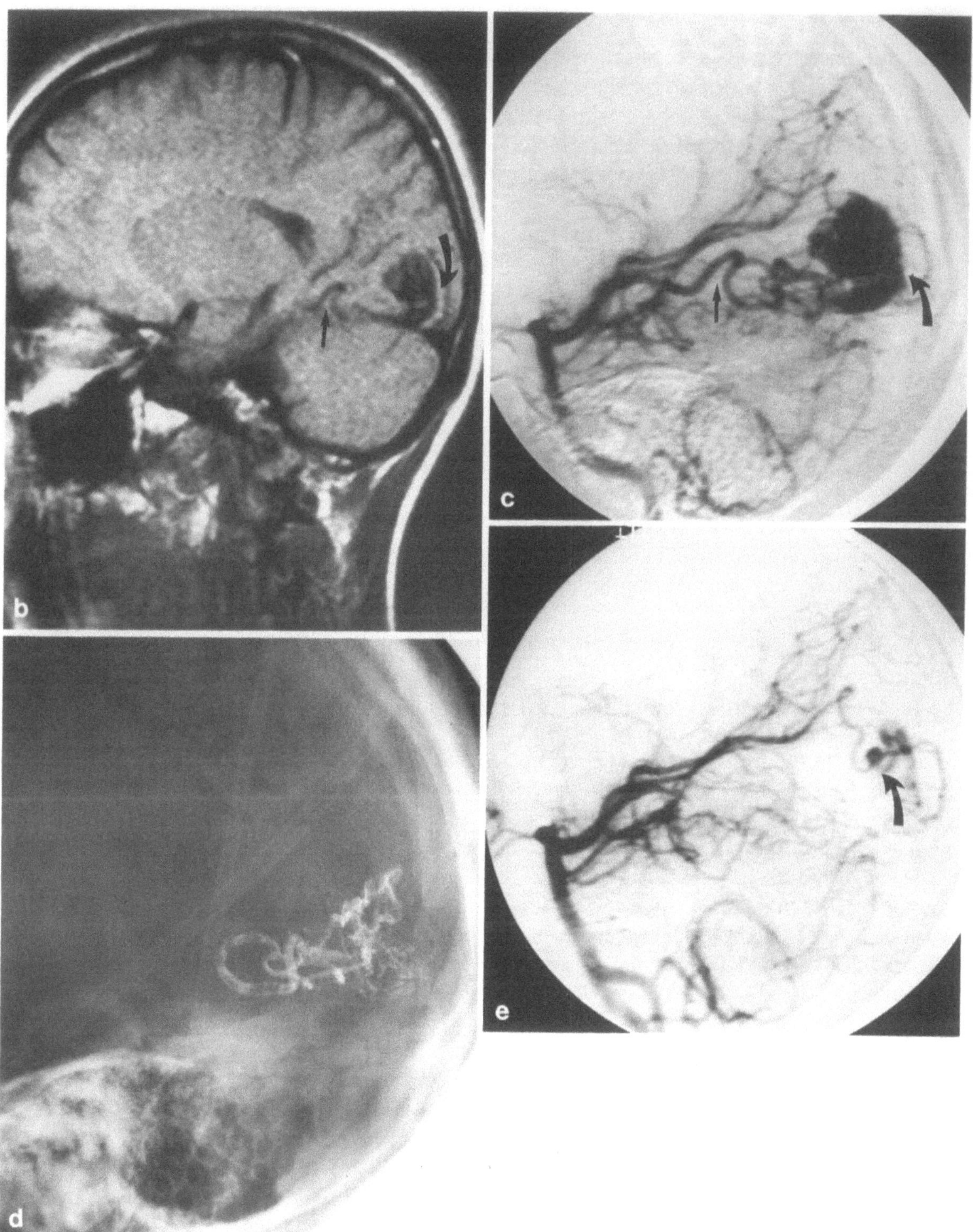

7.1.5.2 Headache: Visual Phenomena

Visual disturbances similar to the visual phenomena of migraine are commonly reported when the BAVM involves the occipital lobe. Positive visual phenomena such as scintillating scotoma and fortification images are almost always seen in the hemifield contralateral to the BAVM [38–40]. Unusually, the positive visual imagery affects both fields simultaneously or only the ipsilateral hemifield. The mechanisms that cause these positive visual phenomena, such as vasomotor alteration (as in migraine) or a seizure discharge, may be difficult to distinguish.

Abnormalities in the cortical occipital veins, rather than an arterial steal, seems to be responsible for these episodes. A similar disturbance also occurs in patients with dural AVMs that receive only dural arterial input and drain through the occipital veins, who experience episodes of fortification imigary followed by homonymous hemianoipia (see Fig. 3.5 in Chap. 3).

Episodes of negative visual phenomena, such as partial or complete homonymous field loss, that are similar to the visual loss of complicated migraine can develop only in the field contralateral to the unruptured BAVM (Fig. 7.12). However, in one series of 46 patients (many with homonymous field loss) evaluated for complicated migraine, none were found to have a BAVM [41]. All patients with complicated migraine, regardless of whether a visual or neurologic deficit is present, require contrast-enhanced CT or MR imaging. Some authors [42] state that a BAVM can be differentiated from migraine because, in contrast to migraine, the visual phenomena associated with a BAVM follow rather than precede the headache. However, the order of the migraine symptoms is an unreliable guide for diagnosis since the visual symptoms may precede, follow, or occur simultaneously with the headache in patients with complicated migraine. The occurrence of other transient neurologic deficits such as hemisensory or hemimotor defects depend on whether the BAVM involves other cerebral areas.

7.1.5.3 Epilepsy

Seizures are the reason for initial presentation or develop soon after other symptoms in 28% [6] to 45% of all cases with BAVMs. In one series, 47% experienced seizures alone and 23% had seizures in addition to one or more hemorrhages [43]. Epilepsy occurs prior to an intracranial hemorrhage in from 18% [6] to 25% of the patients who bleed at a later date [10]. Approximately, one fourth of patients will bleed within 15 years of the onset of seizures [43, 44]. The rate of developing a hemorrhage following the onset of seizures is approximately 4% per year over a 20-year follow-up [45].

The risk of having a seizure seems to be related to the area of brain involved with the BAVM. One study reported that patients with BAVMs in the parietal, temporal, or occipital lobes had an equal risk, while only BAVMs in the frontal lobe caused a higher incidence of developing epilepsy [11]. In contrast, fewer of our patients with occipital BAVMs have had generalized seizures than patients with BAVMs in other lobes. However, the frequency of seizures (mostly focal) was no less in patients with occipital lesions once epilepsy began. In support of our observations, another study reported that seizures were most commonly associated with BAVMs located in the frontal lobe (85%) and least commonly found with occipital lobe BAVMs (55%). In the same investigation, a slightly higher frequency of seizures developed in patients with BAVMs in the parietal (58%) or temporal (56%) lobes [11]. The seizure frequency is unaffected by whether the patients are neurologically intact or have a neurologic deficit during the interictal period.

When BAVMs in all locations are considered together, generalized motor (grand mal) and various types of focal seizures occur with approximately equal frequency [6]. Grand mal episodes occur in patients with supratentorial BAVMs in any location. The character of the focal positive phenomena or neurologic deficits associated with the seizures depends on the location of the BAVM or the abnormal venous drainage (the neurologic signs therefore may occur from dysfunction at a distance from the BAVM nidus) or both. Temporal lobe involvement typically causes psychomotor, uncinate,

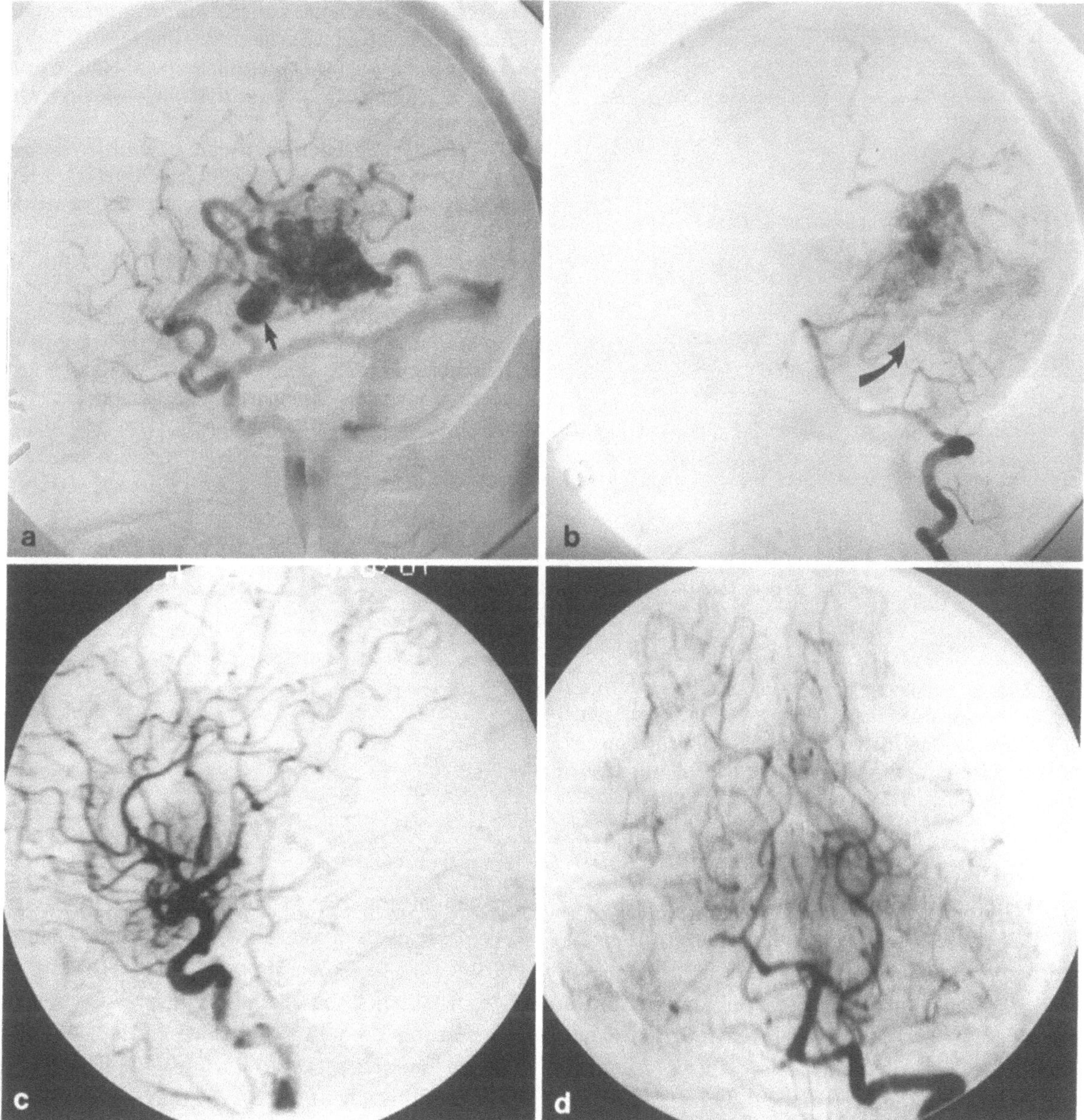

Fig. 7.13 a–e. Seizures caused by a brain arteriovenous malformation.

a The lateral view right internal carotid artery subtraction angiogram reveals a venous ectasia (*arrow*) associated with the brain arteriovenous malformation primarily located in the middle temporal gyrus.

b Lateral subtraction angiogram of the right vertebral artery demonstrates a posterior temporal artery contribution to the malformation from the posterior cerebral artery.

c The lateral view internal carotid artery angiogram after embolization shows complete obliteration of the malformation.

d The frontal view of the vertebral artery injection on the left demonstrates complete obliteration of the malformation.

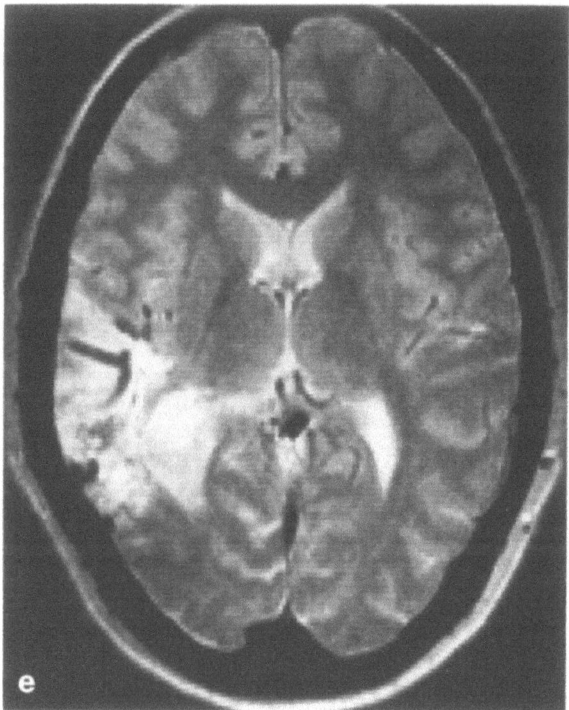

Fig. 7.13. e Following acrylic embolization, the axial magnetic resonance image demonstrates increased signal in the distribution of the embolized malformation. No further seizures occurred while the patient was treated with a conventional dose of phenytoin sodium (Dilantin)

Fig. 7.14a – d. An occipital brain arteriovenous malformation was found in this patient with recurrent episodes of acephalgic left homonymous scintillations.

a The axial T2-weighted magnetic resonance image demonstrates a right occipital vascular malformation with deep venous drainage.

b The lateral subtraction angiogram of the left vertebral artery demonstrates multiple posterior cerebral artery branches supplying a brain arteriovenous malformation in the occipital pole.

c The postembolization control angiogram of the left vertebral artery demonstrates complete obliteration of the malformation.

d Immediately following embolization, a left inferior quadrantanopia that spared the central 5° was noted. The axial magnetic resonance image demonstrates increased signal in the occipital pole. The field defect cleared in 24 h and no further episodes of scintillation occurred

or vertiginous attacks (Fig. 7.13). Frontal lobe BAVMs cause hemisensory, Jacksonian, or hemimotor episodes. Parietal lobe lesions cause hemisensory phenomena (rarely the only manifestation of a seizure). Occipital lobe seizures cause hemifield visual phenomena (see above, migraine-like symptoms) that may appear as formed or unformed visual hallucinations (Fig. 7.14). Similar visual disturbances are rarely found with BAVMs in either the temporal or parietal lobes.

As discussed previously, despite the congenital nature of the BAVM, the mechanisms that lead to epilepsy, beginning in adulthood rather in infancy in many patients, are not always clinically discernible. A parenchymal hemorrhage, thrombosis of a draining vein, a venous infarct, an arterial steal, or a combination of these factors may all contribute to the dysfunction. All of these mechanisms can reduce tissue perfusion and the threshold for abnormal excitatory neuronal discharges.

The seizure frequency can be altered by therapy directed at the BAVM. Surgical excision of the BAVM seems to decrease the frequency of seizures in from 14% to 43.6% of patients [46]. However, in 22% of patients, the seizure frequency increases or new seizures develop postoperatively [43, 46, 47]. The seizure frequency following embolization of the arterial supply of the BAVM may not be altered [48]. However, embolization that effectively penetrates a significant portion of the BAVM nidus seems to reduce the seizure frequency (Figs. 7.13, 7.14). Adjustment of anticonvulsant medication, better patient compliance, and the improved cerebral hemodynamics and tissue perfusion are all factors which contribute to improved seizure control. In our 202 embolized patients, only four (1.9%) developed seizures postembolization during the follow- up period of on average of 6.5 years. Of these four patients, one had an associated hematoma, one had previous surgery, one had additional radiation therapy, and one had no obvious factor to account for the change.

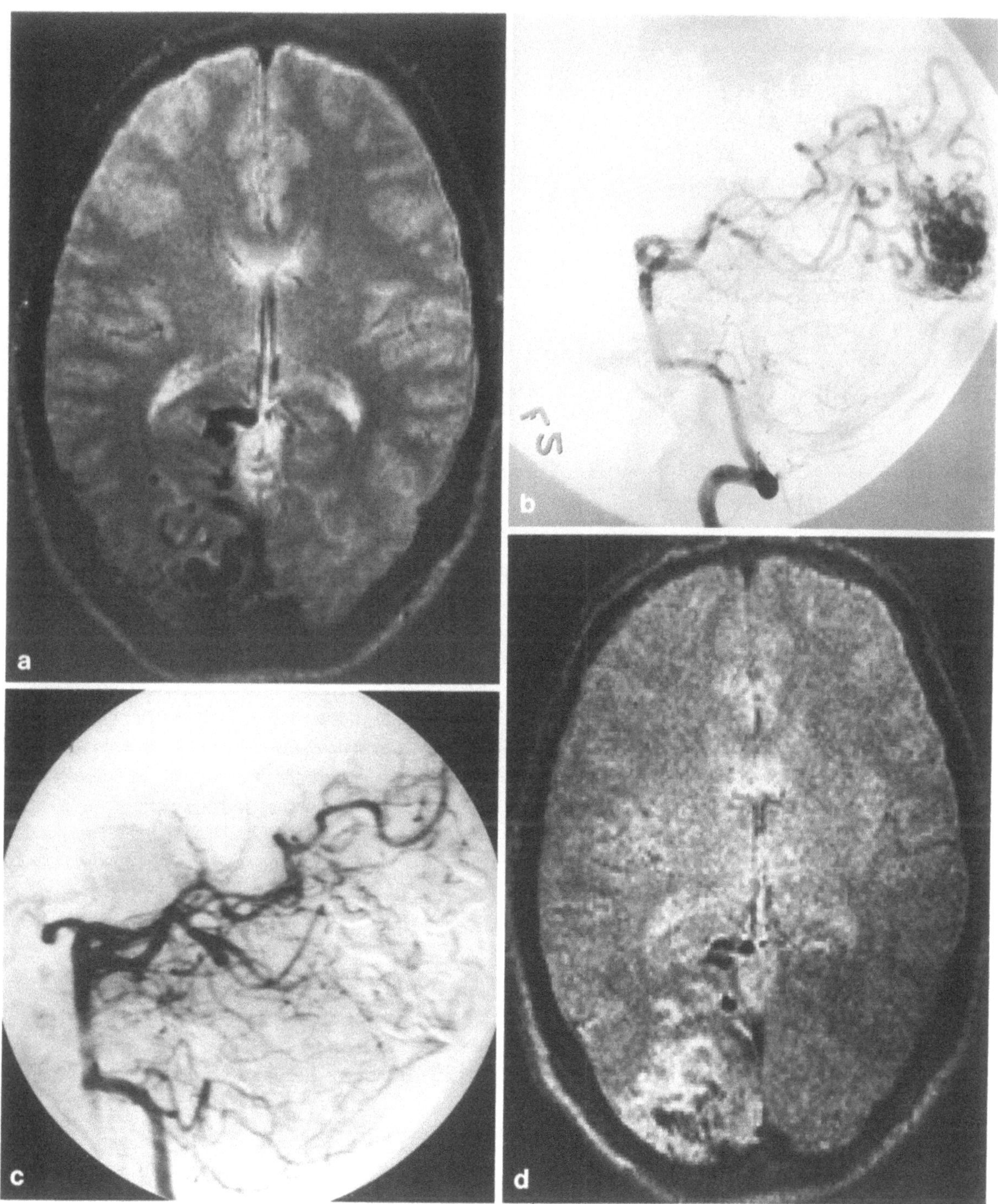

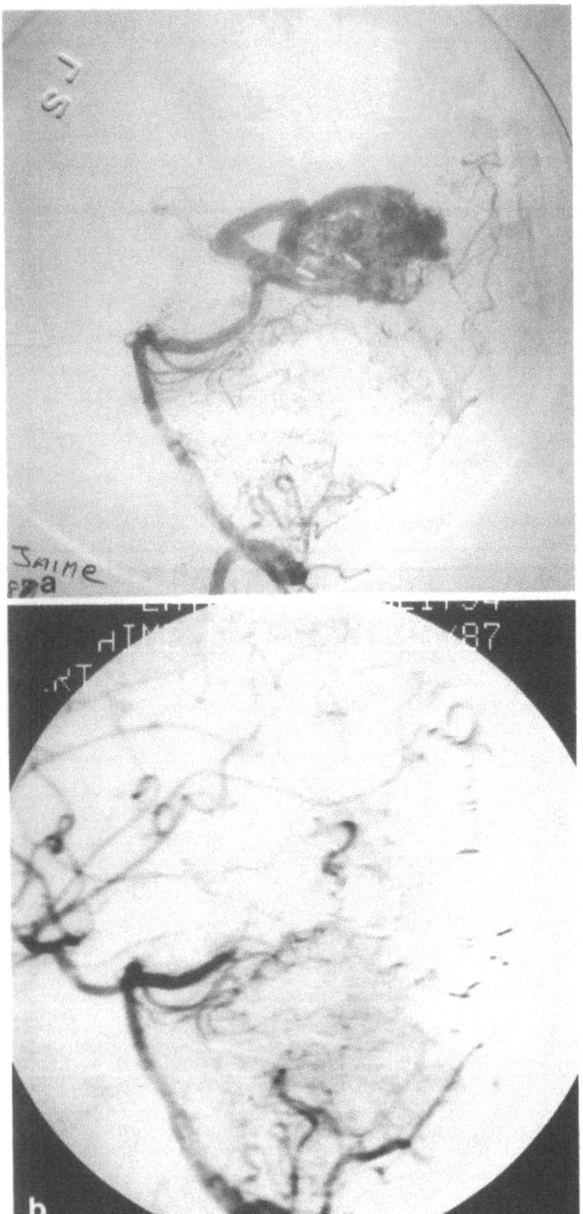

Fig. 7.15 a, b. A right occipital hemorrhage (seen on computed tomography, not shown) associated with a brain arteriovenous malformation caused a complete left homonymous hemianopia. **a** The lateral subtraction angiogram of the left vertebral artery demonstrates a brain arteriovenous malformation supplied by the posterior cerebral artery. Surgical clips, from a previous attempt at excision, are noted overlying the brain arteriovenous malformation. **b** Following cyanoacrylate embolization, the lateral left vertebral artery angiogram demonstrates complete obliteration of the brain arteriovenous malformation. However, the visual field defect never recovered

7.1.5.4 Hemorrhage

The location of the hemorrhage depends on the location of the BAVM (see also Sect. 7.1.6.1.2), its venous drainage, or the presence of aneurysms on feeding arteries. The bleed is subarachnoid in 30%, intraparenchymal in 60%, and intraventricular in 6% of ruptured BAVMs (Fig. 7.15).

7.1.5.4.1 Risk Factors

Aneurysmal dilatation of draining veins seems to increase the risk of hemorrhage. In one report, a bleed occurred in eight of 18 patients (44%) with this finding on angiography, while aneurysmal venous dilatation was found in only 18% of patients who did not hemorrhage [44]. Mechanical constriction or thrombosis of the veins which results in compromise of the venous outflow from the BAVM is seen in 38% [49] to 71% of patients who hemorrhage [43]. The origin of rupture is often at a venous site, even at a distance from the BAVM nidus. The presence of aneurysms on arterial feeders also seems be associated with a greater tendency to bleed since these aneurysms were present in up to 27% of our cases who bled. We can only speculate as to whether the aneurysms are the actual site of hemorrhage or whether their existence merely indicates high flow in these lesions.

In a study of patients with BAVMs who had not had a history of hemorrhage at the time of the diagnosis (surgical cases were excluded prior to considerations), 18% subsequently bled, producing a rate of 2.25% patients bleeding per year over the mean 8.2-year follow-up. Twenty four (77%) of the 31 cases with late bleeds initially presented with epilepsy. In contrast, 22 asymptomatic cases and five others who only had a focal neurologic deficit at the time of diagnosis experienced no hemorrhages during the same time interval [44]. In another study, following presentation, 42% of all cases bled within 20 years, with a 2% patient bleed rate per year after a slightly higher rate during the first 2 years of observation. Thirty three percent of cases bled if the BAVM was unruptured at presentation, while 51% rebled if a hemorrhage had been initially diagnosed. However, only 8% of patients with a neurologic deficit and no patients in whom the BAVM was a coincidental finding had a bleed during the 20 years following the diagnosis [18].

The size of the BAVM may have some role in the tendency to hemorrhage but the reports are inconsistent and it is difficult to analyze size data from studies where complete angiography was not performed [6]. In some studies, smaller BAVMs seem more likely to bleed than giant BAVMs; for example, analysis of the cases that hemorrhaged in one study revealed that 86% were small (less than 2.5 cm) and 46% were large (greater than 2.5 cm) [47]. In contrast, other studies of BAVMs that bled report that up to 64% were large and 36% were small [19, 9]. Still further investigations, including our own series of 202 BAVMs, have found no difference in the frequency of hemorrhage among small, medium, and large BAVMs [10, 18]. It is more plausible that small BAVMs present with a hemorrhagic episode because a small lesion is less likely to cause significant neurologic dysfunction or severe headaches that would bring the patient to earlier medical attention.

Numerous other factors influence the rate of hemorrhage. The risk of hemorrhage may vary with the location of the BAVM. In a 20-year study, when the BAVM was in the parietal lobe there was only a 32% risk of bleeding, while the risk for BAVMs in the temporal or occipital lobes increased to 67% and 52%, respectively [18]. Other investigators report no relationship between the BAVM location and the incidence of bleeding [44]. Age seems to influence the rate of hemorrhage. Patients above 60 years of age have a higher risk of hemorrhage [12, 18]. If the patient is over 60 years old, there is an 89% risk of hemorrhage within 10 years of the diagnosis with a high rate of mortality and morbidity [18].

Most studies have not shown a relationship between the presence of systemic hypertension and the rate of bleeding. However, one study found that only nine (5.4%) of 166 patients with unruptured BAVMS had a history of hypertension at the time of presentation, while five (16.1%) of 31 patients that subsequently bled were hypertensive [44].

Heavy lifting, Valsalva maneuvers, coitus, or emotional stress are associated with hemorrhage in less than 25% of cases, whereas bleeding occurred during sleep in 36% of patients [50]. This suggests there is no definite relationship between hemorrhage and physical exertion [51] or mild head trauma.

7.1.5.4.2 Hemorrhage-Related Morbidity and Mortality

The morbidity associated with rupture of a BAVM is usually not as severe as with a subarachnoid hemorrhage from rupture of an arterial aneurysm. This is probably due to the fact that the BAVM bleed is usually from the venous side of the BAVM and is therefore under less pressure. Vasospasm is rare, possibly because there is a smaller volume of blood in the basal cisterns and the subarachnoid space (Fig. 7.16). It is also conceivable that the physiological alterations in the arterial system induced by chronic high flow to the BAVM prevent severe constriction of these vessels. Despite the lack of vasospasm, 53% of patients with BAVMs have a residual neurologic deficit from the first bleed [6]. Intracerebral hemorrhages cause symptoms in relation to damage of local parenchymal tissue. Infrequently, the hematoma is large enough to cause progressive neurologic dysfunction. The mass effect from hemorrhage can generally be controlled with medical therapy, but an occasional case requires surgical evacuation of the clot. Death results from the first or second hemorrhage in 10% of cases that bleed [6]. Of those patients who bleed, approximately 85% will have a major neurologic deficit or die within a 20-year follow-up period [45].

7.1.5.4.3 Pregnancy

One might empirically surmise that the increased blood volume of pregnancy, the elevation of systemic and intracranial venous pressure associated with the Valsalva of vaginal delivery, the uterine contractions of labor, and hormonally induced vasomotor changes would predispose towards a higher rate of hemorrhage than in nonpregnant patients. There are no definitive prospective studies which compare age-matched women with BAVMs who are pregnant with those who are not. The published studies concern patients selected because they had an intracranial hemorrhage during or immediately following pregnancy [52–54, 55a, b]. Even in these reports, hemorrhage rarely occurred during the delivery. There is no increase in the bleed rate in previously unruptured BAVMs during pregnancy or delivery compared to nonpregnant women

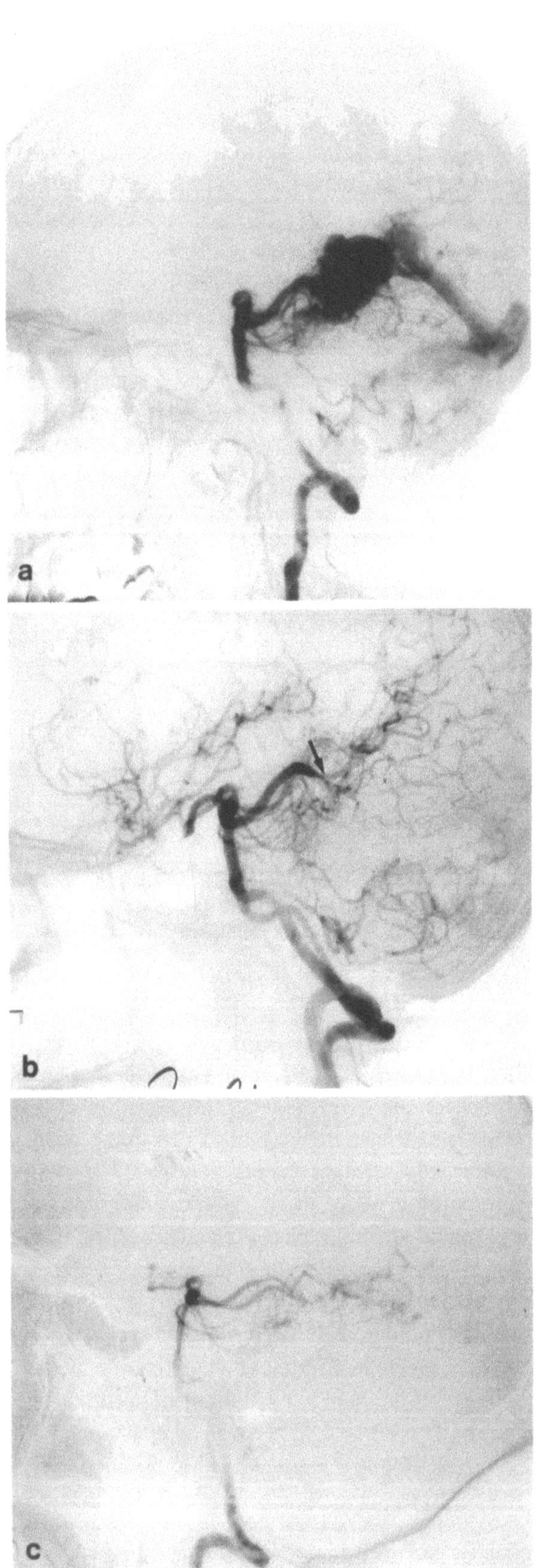

Fig. 7.16a–c. Despite an occipital lobe and intraventricular hemorrhage associated with an occipital brain arteriovenous malformation, this patient had no visual field defect. Vasospasm, which is rarely seen in the case of brain arteriovenous malformations, was noted after a rebleed.
a Following the first bleed, the lateral subtraction angiogram of the left vertebral artery demonstrates a vascular malformation supplied by the posterior cerebral artery.
b The immediate postembolization control angiogram shows the stump of the posterior cerebral artery (*arrow*).
c After repeat hemorrhage, note the spasm in the basilar artery and posterior cerebral arteries

of childbearing age. In addition, the rate of first hemorrhage is similar for pregnant and nonpregnant women with BAVMs [56]. Conservative management, including anticonvulsants in those cases with seizures, is suggested. Early termination of pregnancy is not required. Neither maternal nor fetal mortality rates differ for either vaginal or cesarian deliveries [57].

7.1.5.5 Rebleeding

The reported frequency of rebleeding is approximately 3% – 4% per year, except for a 6% rate during the first year following a hemorrhage [12, 45]. Over 5–8 years, 23% of all BAVMs rebled in the Cooperative study [6]. Another study reported rebleeding in 32 of 134 patients for a similar rebleed rate over the follow-up period, with a mean of 4.8 years for BAVMs that were unruptured at the initial presentation and a mean of 2 years for those which initially presented with a bleed [12]. The location of the BAVM may also influence the rebleed rate. For example, posterior fossa BAVMs generally carry a higher risk of rebleeding early. In one report, 34% of BAVMs involving the cerebellum or brainstem rebled within 6 years [58]. In another study of BAVMs in all locations, 34% of patients surviving the first bleed rebled [59]. The risk of rebleeding seems to increase with age [12].

As in studies on primary hemorrhage, the reports concerning the influence of BAVM size on the rebleed rate are conflicting. One study described the rebleed rate as three to four times higher with medium and large BAVMs compared to the rate for small BAVMS [60], while another report described a similar rebleed rate for BAVMs regardless of the size [18].

7.1.5.6 Increased Intracranial Pressure

Increased intracranial pressure without ventricular dilation can result from arteriovenous shunt-induced intracranial venous hypertension (Fig. 7.17) and thrombosis in the superior sagittal or transverse sinuses. Venous compression of the outflow through the aqueduct of Sylvius or thalamostriate vein obstruction of the foramen of Monro cause obstructive hydrocephalus. Communicating hydrocephalus can also follow a subarachnoid hemorrhage that disturbs cerebrospinal fluid flow out of the basal cisterns and through the arachnoid vili. Acute ventricular obstruction, necessitating emergent surgical intervention is not common.

7.1.5.7 Neurologic Deficits and Dementia

The incidence of progressive focal neurologic deficits is considerably lower than the incidence of hemorrhage or seizures. Visual field depression and motor weakness occur more often than somatic sensory deficits. Progressive neurologic dysfunction is more frequently associated with large BAVMs than small lower flow BAVMs (Fig. 7.3). The smaller AVMs can cause progressive deficits if associated with significant venous thrombosis (Fig. 7.11). In one study, sixteen of 20 patients with gradual neurologic dysfunction had large BAVMs [9]. Patients that present with a focal neurologic defect, without a history of bleed, infrequently bleed at a later date [44, 18]. In many of our patients with progressive neurologic deficits, the angioarchitecture of the BAVM consisted of small vessel disease (Fig. 7.8). These BAVMs were infrequently associated with significant stenosis of the proximal feeding arteries (Fig. 7.18) [27]. In patients with high-flow shunts, both the focal and diffuse cerebral dysfunction can improve following embolization-induced partial occlusion of the major pathological arterial blood supply and the BAVM nidus. In these cases, the reduction of the venous hypertension or an arterial steal from the normal brain or both [61, 48] seems to be correlated with increased filling of the normal brain vessels and improved venous drainage from the brain in

the postembolization angiogram when compared to preembolization studies (Fig. 7.17e, f).

Diffuse cerebral dysfunction and cognitive defects occur most frequently when the BAVM involves the dominant temporal lobe, midline thalamic region, or both cerebral hemispheres [62] (see Chap. 8 for further discussion). Psychomotor slowing and memory loss can also develop when the compromise of the venous outflow from the normal brain is severe enough to cause diffuse intracranial venous hypertension. Following a subarachnoid or an intraventricular hemorrhage, psychiatric disturbances or overt dementia can occur. These signs will generally, but not always, spontaneously improve or resolve completely over weeks to months. Secondary hydrocephalus can cause or aggravate these manifestations.

7.1.6 Posterior Fossa Brain Arteriovenous Malformations: Special Circumstances

Patients with BAVMs located in the posterior fossa typically have signs and symptoms that distinguish them from patients with supratentorial BAVMs. Most posterior fossa BAVMs, approximately 71%, are located in the cerebellum (cerebellar hemisphere, vermis, or both). The brainstem and cerebellum are concomitantly involved in less than 30% of cases. The brainstem is the predominant location in from 5% to 22% of posterior fossa BAVMs [63, 64]. The incidence of vascular lesions in the posterior fossa varies, depending on whether the reports are based on clinical or post mortem studies. For example, in a series of 164 posterior fossa vascular malformations found at autopsy, no symptoms occurred during life in 48% of patients with lesions in the cerebellum and 77% of patients with brainstem lesions. This report also exemplifies the confusion in determining the actual incidence of posterior fossa BAVMs even with pathological investigations, since it included vascular lesions that were not BAVMs such as telangiectases, cavernomas, and venous malformations [65].

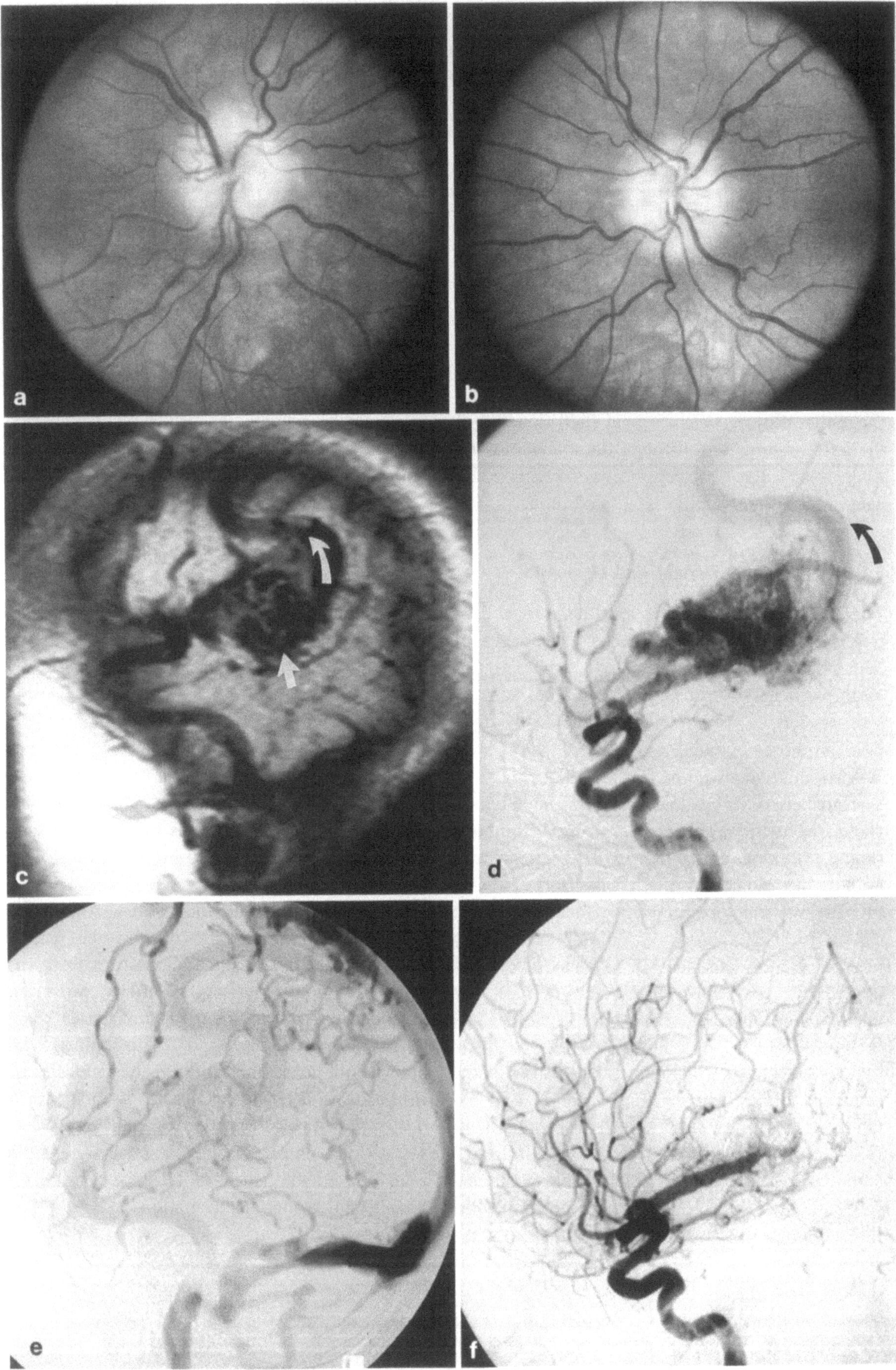

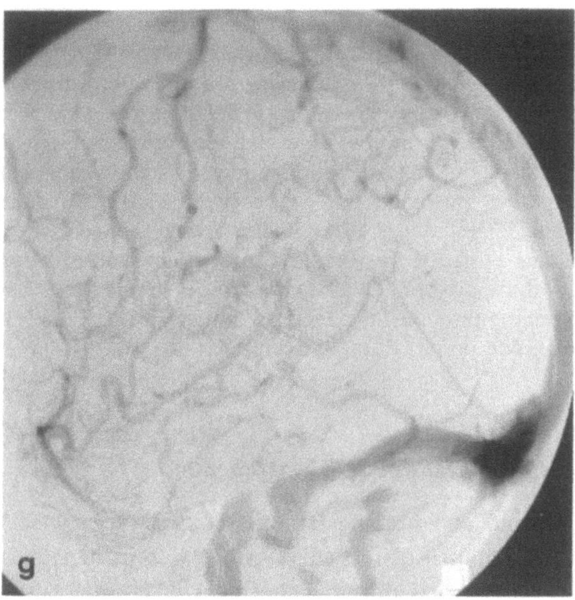

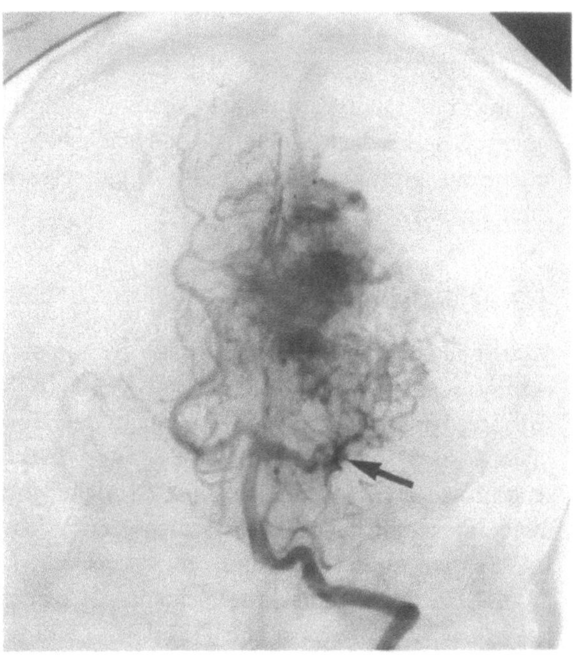

Fig. 7.17a–g. This 30-year-old woman, who presented because of blurred vision, had bilateral diffuse constriction of the visual fields to 15° and papilledema associated with a brain arteriovenous malformation.

a, b Ophthalmoscopy of the right (**a**) and left (**b**) eyes shows chronic papilledema.

c The sagittal magnetic resonance image of the cerebral surface demonstrates the brain arteriovenous malformation (*arrow*) and the large cortical draining vein (*curved arrow*).

d Lateral subtraction angiogram of the left internal carotid artery demonstrates the brain arteriovenous malformation and the draining vein (*curved arrow*).

e A later phase of the contrast injection into the main pedicle (from the middle cerebral artery) supplying the brain arteriovenous malformation demonstrates the significant cortical venous drainage to the superior sagittal sinus as well as the transcerebral venous system. The papilledema was a result of the venous hypertension in the dural venous sinuses.

f Following partial embolization, only a small amount of arteriovenous malformation nidus is seen.

g The later phase of the postembolization angiogram shows normal filling of the cerebral veins and a reduction of the abnormal venous drainage into the superior sagittal sinus. Within 1 week, there was less papilledema and the visual fields expanded to 25°

Fig. 7.18. The frontal view left vertebral angiogram shows marked stenosis (*arrow*) in the left posterior cerebral artery supplying a small vessel arteriovenous malformation in a patient who complained of intermittent dimming of the vision in both eyes

7.1.6.1 Clinical Features

7.1.6.1.1 Headache

Though not usually a prominent symptom, chronic nonspecific headaches may occur. Headache may be the only symptoms in up to 35% of patients who later hemorrhage [63].

7.1.6.1.2 Hemorrhage

Signs and symptoms of hemorrhage are the most common manifestation of posterior BAVMs. In the Cooperative study, sudden hemorrhage developed in 81% of 32 patients with these lesions, five of whom died [6]. Most studies report a higher incidence of hemorrhage for cases of posterior fossa BAVM than for cases of supratentorial BAVM [7, 58, 66]. In three studies, the incidence of hemorrhage was 72% of 32 patients [58], 89% of 18 patients [66], and 92% of 66 patients [7]. In the latter study, the bleed was intraparenchymal in 27 cases (41%) and subarachnoid in 34 cases (52%) [7]. Severe ataxia and brainstem findings often persist following resolution of the hematoma [64] (Fig. 7.19a). In most series, supratentorial BAVMs are more frequently diagnosed prior to hemorrhage because of the frequent presentation with seizures and migrainelike headaches which are not associated with BAVMs limited to the posterior fossa. However, if the abnormal venous drainage passes to the cerebral hemisphere, epilepsy can result.

7.1.6.1.3 Exacerbating and Remitting Course

In patients with an unruptured brainstem BAVM, signs and symptoms of cerebellar or brainstem dysfunction that exacerbate and remit over one or more years are rare. These cases may be misdiagnosed as multiple sclerosis or transient ischemia or occlusive disease of the brainstem [67, 68]. Several mechanisms might explain the fluctuation of the clinical findings in the patients with BAVMs. Brainstem dysfunction may develop as a result of thrombosis of a small penetrating vessel, microscopic hemorrhages that reabsorb, or venous hypertension or thrombosis which slows or prevents the venous drainage of blood from the normal brainstem (Fig. 7.20). An arterial steal is a less likely etiology unless a high-flow shunt is present.

Fig. 7.19a – c. Posterior fossa brain arteriovenous malformation. **a** The axial noncontrast computed tomogram demonstrates an area of hemorrhage (*H*) in the right cerebellar hemisphere. **b** Following contrast administration, the axial computeed tomogram demonstrates a vascular malformation involving the right cerebellar hemisphere with extension to the midline including the vermis but sparing the brainstem. **c** The lateral view subtraction angiogram of the left vertebral artery demonstrates a large posterior fossa brain arteriovenous malformation with supply from all the major arteries in this region, including the superior cerebellar, anterior inferior cerebellar, and posterior inferior cerebellar arteries

7.1.6.1.4 Eighth Nerve Dysfunction

Vertigo and hearing (some secondary to a bruit) and balance disturbances can result from eighth nerve compression by ectatic veins or venous hypertension in the drainage from the nerve.

7.1.6.1.5 Visual Disturbances

Some patients with BAVMs in the posterior fossa describe fluctuating vision or multiple scotomas, visual field distortion, and dyschromatopsia. The visual dysfunction is probably related to the development of ischemia in the occipital lobe. Reduced blood flow through the distal posterior cerebral arteries [69] or venous hypertension in the dural sinuses which prevents the normal blood drainage from the occipital lobe may account for the visual disturbances (see Sect 1.5.2.10).

7.1.6.1.6 Cerebellopontine Angle Features

Facial pain and paresthesias in the distribution of one or more trigeminal nerve roots can occur with posterior fossa BAVMs. Pain that is indistinguishable from trigeminal neuralgia has also been described by patients with BAVMs supplied by the anterior inferior cerebellar artery [70, 71]. However, in a major series concerning tic doloreux, only one of 411 patients had a BAVM, attesting to how infrequent the predominant or only neurologic problem caused by a BAVM is this type of facial pain [72]. Also in contrast to patients with trigeminal neuralgia, patients with a BAVM will often have a fifth nerve sensory or motor deficit that can be elicited by careful examination. In another study, only

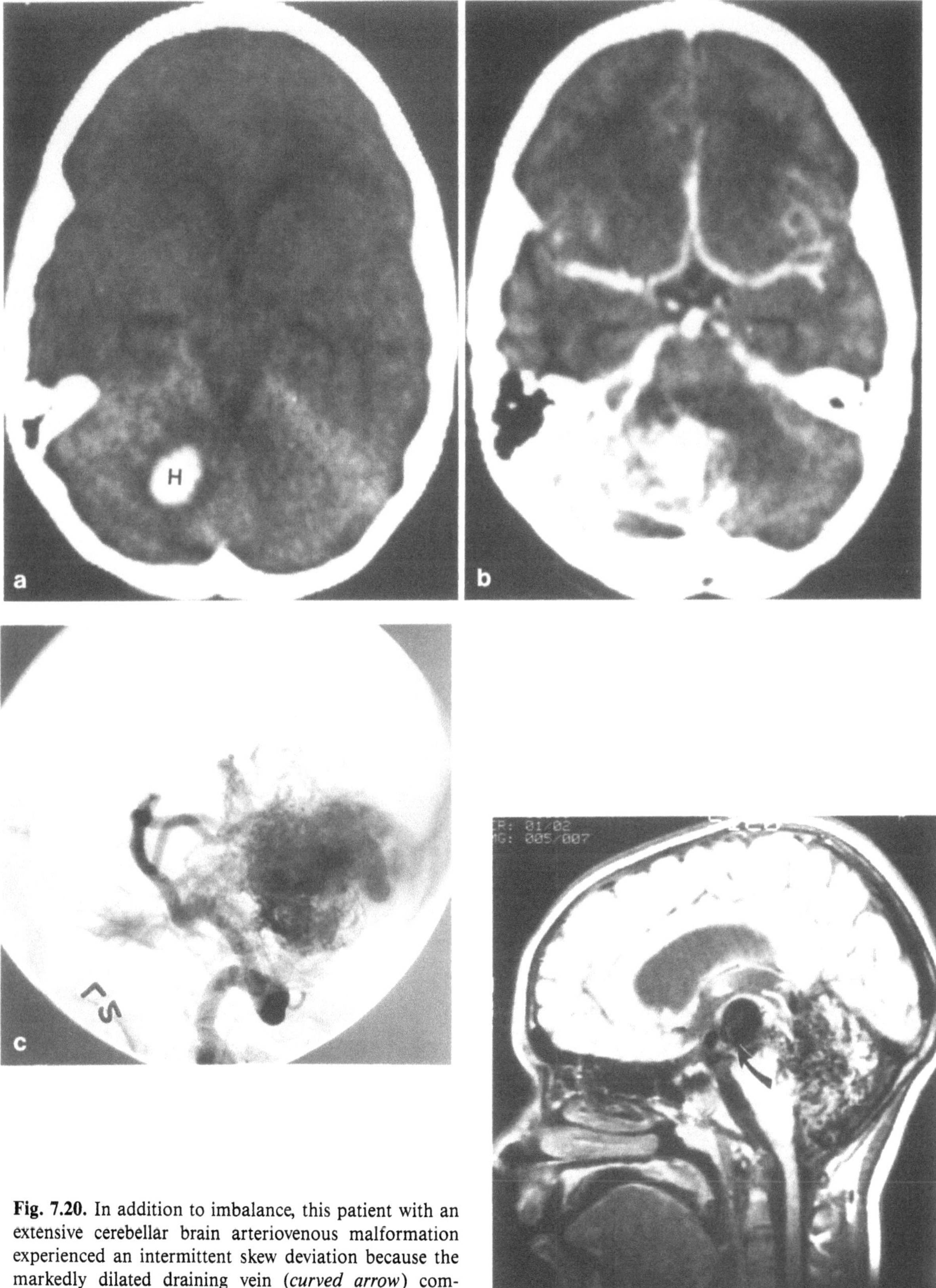

Fig. 7.20. In addition to imbalance, this patient with an extensive cerebellar brain arteriovenous malformation experienced an intermittent skew deviation because the markedly dilated draining vein (*curved arrow*) compressed the neural structures or the venous outflow from the upper brainstem

three of 66 patients with posterior fossa BAVMs located in the cerebellopontine angle that were found at surgery to involve the trigeminal ganglion had facial pain [7]. Hemifacial spasm, which has been associated with an aberrant course of the anterior inferior cerebellar artery, is also rarely a presenting symptom of a posterior fossa BAVM. When hemifacial spasm occurs, other posterior fossa signs are also present.

7.1.6.1.7 Pontine Dysfunction

Pontine vascular malformations frequently present with eye movement and pupillary disorders. Though these lesions are often called cryptic BAVMs because they are not visualized at angiography, more than half are actually telangiectases [73]. For example, McCormick's morphological study of 164 posterior fossa vascular malformations in all locations revealed 70 BAVMs, 38 telangiectases, 31 venous angiomas, 21 cavernous hemangiomas, and four varices. When only pontine malformations were considered, 27 of the 48 lesions were telangiectases and only four were large vessel BAVMs [65]. In another series of 45 autopsied pontine vascular lesions, half were telangiectases, one third were cavernomas, and the remainder were either venous malformations or BAVMs [73a].

The hemorrhage caused by rupture of a low-flow vascular malformation seems to separate or displace the pontine axonal fibers, rather than destroy them, as occurs with a hypertensive hemorrhage. This is probably the reason why surgical evacuation of an acute or chronic clot associated with a malformation (often misdiagnosed as a glioma in cases with chronic or progressive signs) frequently results in clinical improvement of the brainstem dysfunction. As discussed previously, periodic microscopic bleeding and resorption of clot may account for the relapsing and remitting course that can be confused with multiple sclerosis. Horizontal and upbeat nystagmus, Horner's syndrome, unilateral or bilateral medial longitudinal fasciculus syndromes, and sixth nerve paresis commonly occur. Facial weakness, hemimotor deficits, ataxia or other cerebellar disorders, swallowing difficulties, and dysarthria are also frequently encountered. Loss of consciousness or quadriparesis is un-

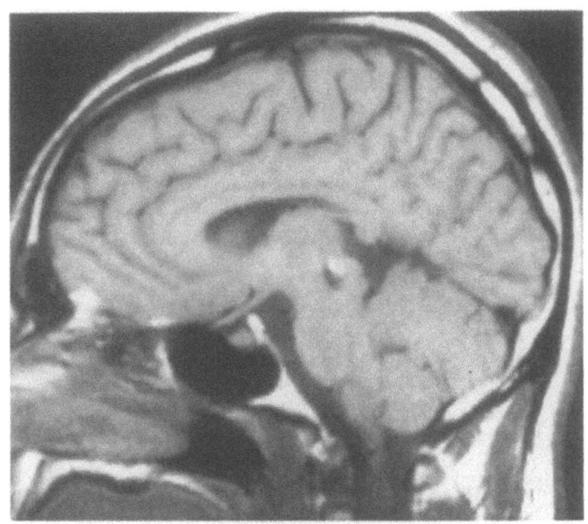

Fig. 7.21. This 45-year-old man presented with an isolated paresis of an inferior rectus muscle. The sagittal T1-weighted magnetic resonance image shows an area of small hemorrhage in the dorsal midbrain, presumably secondary to telangiectasia. Angiography was unrevealing. (Case provided by Larry Frohman)

usual [74]. This course is contrasted with the poor prognosis of cases of pontine hemorrhage associated with systemic hypertension. Coma, quadriparesis, and eventually death occur in up to 60% of patients with the latter lesion [75, 76]. In patients with pontine BAVMs, the spinal fluid is rarely grossly bloody but it may be xanthochromic following an intraparenchymal hemorrhage.

7.1.6.1.8 Midbrain Lesions

The incidence of all vascular malformations within the mesencephalon is clearly lower than the incidence in the pons [65] (Fig. 7.21), hemorrhage into the dorsal midbrain causes eye movement and pupillary disorders typical of lesions in this area. Upgaze or downgaze paresis or both, convergence – retraction nystagmus on attempted upward gaze, unilateral or bilateral fourth nerve pareses, and Horner's pupils or light – near pupillary dissociation can occur with or without a hemorrhage [77].

7.1.7 Vein of Galen Malformations

Vascular malformations of the Galenic venous system are rare [78, 79]. Variable dilatation of the precursor of the vein of Galen, the medial vein of the prosencephalon, has been classically called an "aneurysm." It results from complex fistulas that drain into this vein (Fig. 7.22). This lesion is commonly associated with an anomaly and occlusion of the infratentorial dural venous sinuses. The arteriovenous shunts are supplied by choroidal branches that originate from the basilar or carotid circulations or both. A separate group of older children may have complex BAVM drainage into the true vein of Galen which secondarily dilates because of outflow restriction (not a vein of Galen malformation).

The clinical symptoms vary depending upon the age of presentation. Neonates have life-threatening high-output congestive cardiac failure and failure to thrive. Milder congestive heart failure and macrocephaly develops in some neonates or younger children. Infants frequently have hydrocephalus, epilepsy, developmental delay, and neurologic deterioration. Older children and young adults, who usually have true BAVMs, complain of headaches and can experience subarachnoid hemorrhages, a progressive deterioration in mentation, or both [79, 80].

Either the vein of Galen malformation or AVM of the thalamus often cause extensive neuro-ophthalmologic signs, including a variety of supranuclear gaze palsies, because the lesion nidus or the abnormal venous outflow involves descending cortical pathways to the gaze centers or the upper anterior midbrain or pretectal regions. These findings typically include conjugate gaze palsies, skew deviation, upgaze paresis, and divergence insufficiency. Papilledema results from obstructive hydrocephalus or diffuse venous hypertension from rapid high arterial flow into the venous system, or obstruction of the normal venous outflow. The neuro-ophthalmologic abnormalities may be overlooked because of the severe morbidity and mortality of these lesions.

As for all BAVMs, the goal of therapy for arteriovenous fistulas or BAVMs that involve the vein of Galen is to reduce or eliminate the abnormal arteriovenous shunting of blood. Direct surgical excision of the arteriovenous shunts and fistulas has had a low success rate because of severe complications of the various procedures. Embolization that reduces the arterial feeders to the AVM shunt has reversed the cardiac failure in neonates and the progressive neurologic deterioration in infants. Retrograde transvenous embolization has also been employed when the arterial route was unsuccessful or inaccessible. Embolization can be combined with surgery to occlude or excise the remaining arteriovenous shunts. Some patients require intraventricular shunting to alleviate the symptomatic hydrocephalus that persists despite closure of the shunts [78]. However, most patients who are treated early will not require a ventricular shunt.

7.1.8 Long-Term Follow-up of Brain Arteriovenous Malformations

Few series have reported enough cases of BAVMs without a presenting hemorrhage that were conservatively managed and followed up for an adequate period of long-term evaluation in order to truly determine the natural history of these lesions. For example, only 12 such patients were followed for over 10 years in the entire Cooperative study [6].

Regardless of the clinical presentation, moderate or severe neurologic deficits are found in 33% patients with BAVMs less than 15 years after diagnosis [10]. The rate of severe neurologic abnormalities has been reported to be higher (2%/year) for patients who present with a hemorrhage rather than other symptoms (approximately 0.8%/year) [45].

The risk of death ranges from 6% to 13% from an initial bleed and approximately 6% from a recurrent hemorrhage [59, 12]. Deaths related to a hemorrhage occur in 29% of the patients who bleed [44]. In series that excluded patients who had undergone surgery, conservatively managed cases had a minimum of 20% mortality, most often from recurrent hemorrhage, within 15 years of the diagnosis [60, 43, 81]. In evaluations of all patients with BAVMs, death from all causes will occur in 18% at 10 years, and in 29%–43% at 20 years [18, 45]. The reported incidence of death seems to vary with the type of clinical presentation and the selection of patients in each study. For example, one study

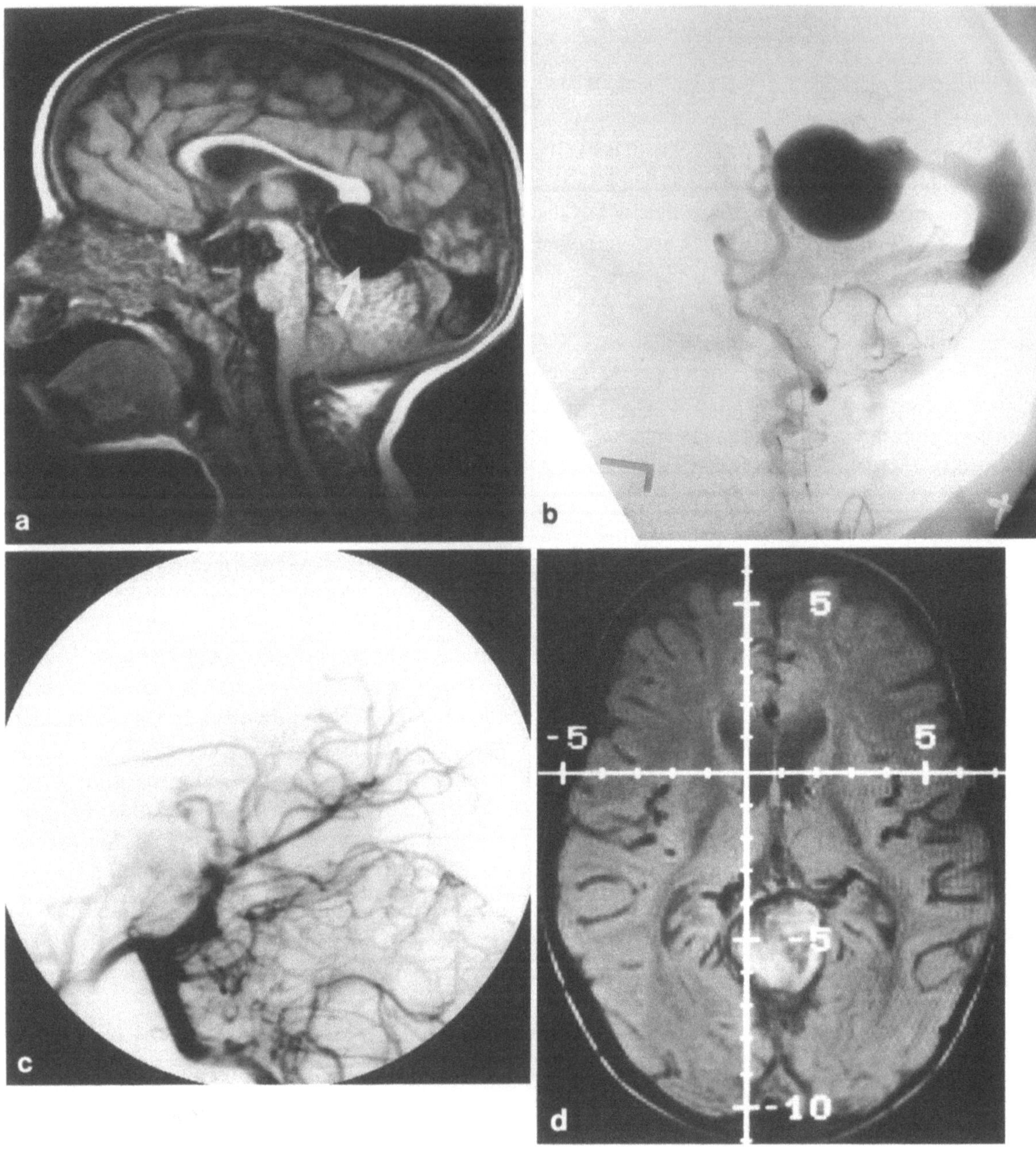

documented that 43% of the patients had died within 20 years regardless of the presentation [45]. In contrast, another investigation reported that the risk of death in the same time interval was 32% if the patient presented with hemorrhage, and 21% if epilepsy was the initial reason for presentation [18]. The risk of death was 22% at 15 years if the presentation was for a neurologic deficit without a hemorrhage. Even harboring an asymptomatic BAVM seems to increase the risk of death since at the end of 20 years 27% of these patients die, though not necessarily from causes related to their central nervous system abnormality [18]. The yearly death rate, 1%, is similar for all symptomatic BAVMs [45].

Persistent neurologic dysfunction can develop relatively early in the clinical course. For example, in one series neurologic deficits developed in 20%

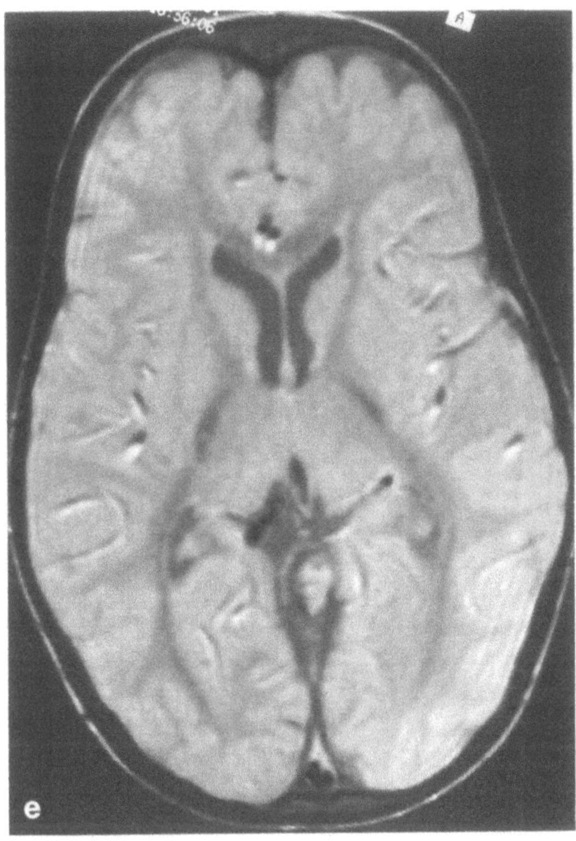

Fig. 7.22 a – e. Vein of Galen malformation. **a** The sagittal T1- weighted magnetic resonance image demonstrates an aneurysmally dilated precursor to the vein of Galen, the mesial vein of the prosencephalon (*arrow*). **b** Similar to the sagittal magnetic resonance image, the lateral subtraction angiogram of the left vertebral artery shows the aneurysmally dilated vein, a narrowed segment in the vein, and drainage into the confluent sinus. **c** After embolization, the lateral subtraction angiogram demonstrates complete obliteration of the malformation. **d** The postembolization axial magnetic resonance image shows the thrombosis of the flow void seen in **a**. **e** The axial magnetic resonance image 1 year after treatment shows involution of the lesion and only a mild dilated flow void immediately posterior to the right thalamus

of 98 patients; the duration of follow-up was not reported, but most patients seemed to have been followed for less than 13 years [9]. In another report, 7% of 135 patients (excluding deaths) who had adequate follow- up and no surgical intervention were incapacitated by their neurologic deficits at the end of the follow-up period, with a minimum of 4 years and a mean of 8.2 years [44]. The risk of developing a neurologic disability is 27% after 20 years and is not dependent on the size, location, or depth of the BAVM in the hemisphere [18]. Cumulative data collected from numerous studies suggest a mortality rate of 10% and a morbidity rate of 30% for the first hemorrhage [82].

7.1.9 Radiographic Evaluation

A complete radiographic evaluation of the BAVM is essential for decision analysis regarding treatment. It is important to determine the feeding vessels, the exact locations of the AVM nidus, and drainage, as well as the vascular disposition of the normal brain and how the normal and abnormal circulations are related. CT and, increasingly, MR imaging are the first modalities usually employed in the diagnosis of these lesions. Significant information is derived from these studies. However, angiographic examination still remains the best available modality to understand the angioarchitecture, flow characteristics, and other parameters when studying AVMs.

7.1.9.1 Computed Tomography

The noncontrast CT is used primarily to demonstrate the size and location of parenchymal or intraventricular hemorrhages, associated hydrocephalus or mass effect, calcifications within the BAVM, and brain edema. However, not all patients who clinically seem to have an intracranial hemorrhage will have blood demonstrable on CT. For example, in one study only eight of 23 patients with clinical signs and symptoms of a hemorrhage had an area of identifiable blood on CT [58]. Subarachnoid blood may be difficult to demonstrate on a routine head scan, particularly if the scan is performed more than 24 h after the ictus.

An acute hemorrhage is obvious as it appears white and the edema in the adjacent brain is dark. As the clot lyses, the lesion has a lower density with a more gray (isodense) appearance. As liquefaction develops, the hemorrhage becomes hypodense (similar to the cerbrospinal fluid). The surrounding tissue may enhance when intravenous contrast is administered, because there is no blood – brain barrier in the granulation tissue and gliotic area in the parenchyma adjacent to a prior clot.

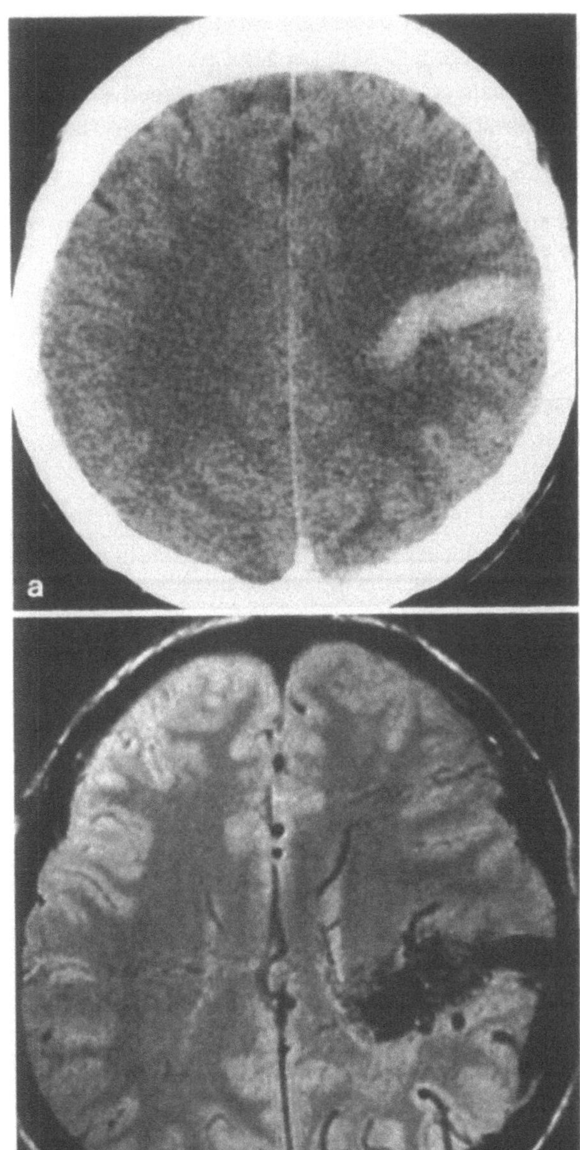

Fig. 7.23 a, b. Intravascular blood has a high-attenuation signal that can appear similar to an intracerebral hemorrhage on computed tomography. **a** The axial noncontrast computed tomogram demonstrates a white sausage-shaped structure without edema in the adjacent brain. Enhancement of the lesion after contrast administration would be suggestive of a vascular malformation (not shown). **b** The axial magnetic resonance image shows the lesion in question is a brain arteriovenous malformation with a large flow void. There was no evidence of a hemorrhage

On occasion, the CT appearance following a bleed may temporarily be similar to that of a malignant glioma with necrosis (see Fig. 3.1 in Chap. 3). Without the presence of a hemorrhage, a BAVM is not always distinguishable from a neoplasm with prominent draining veins. However, the signs of previous venous infarction or encephalomalacia, focal atrophy of the cortex, and focal dilation of the ipsilateral ventricle generally differentiate a BAVM from a glioma. The noncontrast CT can also show areas of calcification, which may be nodular where there has been previous hemorrhage or linear along the abnormal vessels. Areas that contain large vascular structures (usually veins) can have a high absorption coefficient and can be seen on the noncontrast CT. This white–gray lesion is intravascular and should not be confused with an intracerebral hematoma (MR imaging is most useful in this situation; Fig. 7.23).

Following the intravenous injection of contrast material, the CT will reveal individual arteries and veins if they are large enough, but more often a confluence of vessels is seen. The patency of the venous sinuses or signs of sinus thrombosis may also be demonstrated with a contrast CT scan. If only a contrast CT is performed, a small parenchymal hemorrhage may not be distinguishable from areas with slight calcification. Generally, on a noncontrast CT clotted blood does not have an attenuation of more than 60 Hounsfield units (HU), while calcification has more than 80 HU.

The noncontrast CT is abnormal in approximately 84% of posterior fossa BAVMs and the contrast-enhanced CT is abnormal in 96%, with serpiginous vessels seen in 40% of the scans. Many of the noncontrast CT scans have areas that appear as high-density intravascular lesions, while others only show signs of intraventricular blood [63].

7.1.9.2 Magnetic Resonance Imaging

MR imaging is comparable to a contrast CT for diagnosing the presence of a brain BAVM. The sensitivity of MR imaging to paramagnetic properties of blood flow and hemorrhage make it the most useful noninvasive method for diagnosing, locating, and investigating some of the hemodynamic changes in BAVMs. High-field ($>$ 1 tesla, 1 T) techniques are superior to low-field ($<$ 0.5 T) methods for this purpose. Significant supplying arteries and draining veins can be visualized. The BAVM appears as a heterogeneous pattern of vessels because of shifts in blood flow velocity, washout and replacement spins, phase shifts caused by motion along magnetic field gradients, and turbulence of blood flow [83]. MR imaging is also the best modality for determining which neural structures are involved by a BAVM.

High-field MR imaging will demonstrate a change in the appearance of an intracranial hematoma in relation to the evolution of the clot [84, 85]. The acute hemorrhage appears dark on both T1- and T2-weighted images because of the presence of deoxyhemoglobin, unclotted blood, and serum proteins. However, after several days the blood can have a high signal on the T1-weighted images. Subacute hematomas (3 days to 1 month) are hypointense and then develop white, hyperintense areas, predominantly in the periphery of the hematoma. Red blood cell lysis and the formation of extraerythrocyte methemoglobin and hemosiderin, clot matrix formation, and the settling of erythrocytes within the hematoma cause an increase in the intensity seen first on T1-, then on T2-weighted images. An inhomogeneous appearance results because of regions with different magnetic susceptibilities. Over several weeks, the peripheral hyperintense ring thickens until the entire clot is white on all sequences. Edema around the hematoma is also hyperintense on a T2-weighted image. When the clot is older than 1 month, it becomes bright on both T1- and T2-weighted images, while the adjacent brain may appear isodense or hypodense.

Despite the knowledge of the evolution of a hematoma, MR imaging with either low or high fields has been disappointing in determining whether a patient with a BAVM has had a prior hemorrhage. Additionally, MR images can be difficult to interpret following embolization. If the cyanoacrylate has been mixed with tantalum, the metal dephases the protons, causing a decreased signal with a black appearance similar to that of hemosiderin (Figs. 7.13e, 7.14d). In this situation, the gradient echo technique shows flaring of the dark vessel, which is not seen with a flow void. If the cyanoacrylate is mixed with Pantopaque, the oil in the contrast material results in an appearance similar to that of fat [86]. A decrease in the size of the draining veins from an embolized BAVM suggests a reduction in the BAVM nidus or arteriovenous shunting.

Markedly dilated veins can be distinguished from supplying arteries. However, delineating where supplying arteries end in the BAVM nidus and where draining veins begin is best accomplished by angiography. MR angiography permits the assessment of a BAVM and shows the potential blood supply and aneurysmal dilatations of venous structures.

7.1.9.3 Angiography

Cerebral angiography remains the definitive diagnostic procedure for BAVMs. As discussed previously (Sect. 7.1.4), in addition to demonstration of all the arteries supplying the lesion, potential collaterals, angiography will also visualize concomitant arterial aneurysms. Following an interventional procedure such as embolization that successfully reduces the high blood flow through the vessels and nidus of the BAVM, angiography shows regression of the flow-related aneurysm (Fig. 7.7). Angiography will also demonstrate the extent of the venous drainage from the BAVM and how it affects the venous drainage from the normal brain, even at sites remote from the BAVM nidus.

At times, angiography of BAVMs in the posterior fossa may be misleading. It may be difficult with monoplane angiography to be certain whether a BAVM penetrates or lies on the surface of the brainstem [87]. MR imaging has been superior to angiography in demonstrating this distinction. Many posterior fossa vascular malformations detected on CT (better with MR imaging) and verified histologically cannot be visualized by angiography [88]. Though angiographic occult or "cryptic AVMs" (often telangiectases, see Sect. 7.2.1) are rarely described in clinical series of patients with posterior fossa BAVMs, these lesions are more commonly found in autopsy series [89]. Angiographically occult or "cryptic" AVMs can also be found in the supratentorial region [89]. Small arteriovenous shunts may not be visualized if magnification and state of the art angiography techniques are not utilized.

Following intraparenchymal hemorrhage, a

BAVM might not be well visualized or could appear smaller at angiography because of compression by a clot. The BAVM will be better visualized or appear larger following resolution of the hematoma. Therefore, prior to definitive therapy, a repeat angiography may be necessary to assess the true size and hemodynamics of a BAVM following an intra-axial hematoma [90].

7.1.10 Specific Visual Sensory or Ocular Motor Abnormalities Related to Brain Arteriovenous Malformations

Sensory visual disorders result when a BAVM, the abnormalities in venous drainage, or hemorrhage involves the intracranial visual pathway. Additionally, areas remote to the BAVM may also develop ischemia from an arterial steal if associated stenotic changes occur or if the venous drainage interferes with the normal drainage from the optic structures. Visual perception dysfunction can also develop when the temporal, parietal, or occipital lobes are affected. Ocular motor disorders occur with BAVMs that cause dysfunction in the frontal or parietal lobes, diencephalon–thalamus, midbrain, pons, or cerebellum. Eye movement abnormalities can also result from abnormal venous drainage into the cavernous sinus with venous hypertension in the ophthalmic vein.

7.1.10.1 Sensory Visual Disturbances

Homonymous hemianopia is the most common visual field defect associated with BAVMs (Figs. 7.12, 7.15). In one series, this field defect was reported in six of 100 BAVM patients, with all six BAVMs situated near or in the occipital lobe [91]. Rarely, a BAVM that receives supply from the ophthalmic artery causes episodes of amaurosis fugax that are indistinguishable from those caused by embolic disease.

One report described diminished acuity as the presenting symptom in approximately 11% of patients with BAVMs in all locations, after excluding cases with a bleed or seizure presentation. Whether the loss of acuity actually reflected BAVM-related visual dysfunction is not discernible, since no details of the BAVM location or the visual examina-

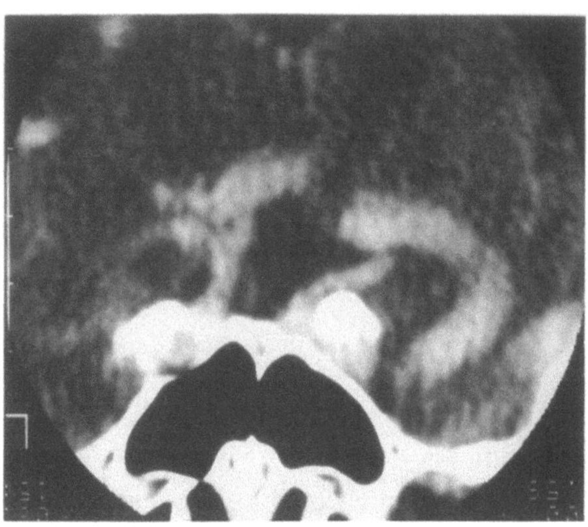

Fig. 7.24. This 35-year-old man complained of sudden visual loss in both eyes over 2 days. When evaluated 3 months later, the patient could only discriminate hand motion at 2 feet and there was marked optic atrophy in each eye. The coronal contrast-enhanced computed tomogram shows marked dilatation of the venous structures in the suprasellar cistern which presumably affected the intracranial optic nerves. Angiography (not shown) demonstrated an extensive right posterior frontal/parietal brain arteriovenous malformation with venous drainage into cortical veins and deep venous system

tion are provided in this report [43]. In our own series of 202 patients, significant reduction of the visual acuity was a rare occurrence, affecting only two patients. One patient had bilateral optic atrophy and decreased acuity to 20/200 because the venous drainage from the BAVM injured the chiasm and the intracranial optic nerves (Fig. 7.24). The second patient had a unilateral optic neuropathy, but there was no clear relationship between the BAVM (Fig. 7.8) and the visual loss.

Though extremely rare, unilateral or bilateral visual loss can have a sudden onset, particularly if a subarachnoid hemorrhage damages the optic nerve or an intraparenchymal bleed causes a mass that compresses the optic nerve. Though any cause of a subarachnoid hemorrhage should be capable of causing a subhyaloid and vitreal hemorrhage, we have seen no cases of Terson's syndrome caused by a BAVM-related bleed (see Sect. 6.1.7.3 for further discussion).

A gradual or stepwise loss of the visual field followed by a visual acuity reduction may result from chronic papilledema related to chronic elevation of intracranial pressure. Bilateral swollen atrophic optic discs with a swollen, edematous, gray nerve fiber layer, caused by the chronic papilledema, may be seen in as many as 5% of BAVM cases, most of whom have significant venous hypertension in the sagittal sinus, hydrocephalus, or an intraparenchymal hemorrhage [92] (Fig. 7.17).

7.1.10.1.1 Positive Visual Phenomena: Scintillations, Palinopsia, Hallucinations

Visual hallucinations, such as scintillating silvery homonymous scotomas or more complicated bright geometric figures that move from the central to the peripheral field, can occur with or without a headache in patients with occipital BAVMs. Episodes of hemifield scintillating scotomas or floaters can be manifestations of focal seizures (see below) in a patient with an occipital lobe BAVM [11] (Figs. 7.12, 7.14). In some, but not all, patients, these episodes are associated with electroencephalographic rhythmic slow waves at 2.5–4.0 Hz. This seizure activity can be suppressed by anticonvulsants [93]. Following surgical excision of the BAVM, the visual phenomena remit but a significant temporary or permanent homonymous hemianopia can develop. When the visual loss is transient, the hemianopia may result from the inhibitory effect of spreading depression in the occipital cortex [36].

Although other positive visual phenomena, such as palinopsia, visual allesthesia, and de novo visual hallucinations, have been described in patients with occipital or parietal– occipital BAVMs, most of the reported cases of these disturbances are associated with a cerebral infarct or trauma. The persistence or recurrence of a visual image after the subject stops viewing an object, usually in the field contralateral to the lesion, is called palinopsia. The image appears in an area of the dysfunctional visual field but the entire field cannot be blind. On occasion, palinopsia develops in a patient with normal visual fields to tangent or Goldmann perimetry determination [94]. However, the presence of hemidyschromatopsia or extinction of a test target on simultaneous bilateral presentation suggests that the field in such a patient is not entirely normal. It has been hypothesized that palinopsia results from distortion of the process that is responsible for creating the normal visual afterimages [95, 96]. Electroencephalographic evidence of paroxysmal occipital activity may occur during an episode of palinopsia. Suppression of this visual disturbance and of the abnormal electroencephalographic activity by anticonvulsants suggests that in some cases palinopsia is a manifestation of focal seizures [97, 98]. Palinopsia may also be a type of "release phenomenon" that possibly results from disruption of the normal cortical inhibition of a visual memory [99, 96]. Even in patients where there is no obvious field defect, palinopsia may develop following surgical excision of an occipital BAVM.

Formed hallucinations (the patient has no immediate prior viewing of the imagined scene) can also occur following surgical excision of an occipital BAVM [100]. Formed hallucinations are generally considered to result from temporal lobe dysfunction, but these phenomena can be caused by dysfunction at most locations in the visual pathway [101]. Visual hallucinations can be caused by BAVMs, neoplasms, and ischemic disease. Following an infarct in the territory supplied by the posterior cerebral artery, patients, all of whom have large homonymous hemianopia, can experience formed visual hallucinations caused by focal seizures [97]. Continuous formed visual hallucinations have also been described following an occipital infarct, without demonstrable seizure activity [100, 102]. The visual hallucinations often consist of elements that reflect prior daily experiences [103].

Another type of episodic visual disturbance that rarely occurs in association with occipital BAVM-related seizures is a flickering feeling or sensation in the contralateral eye without distortion in any portion of the visual field [104]. An explanation for this phenomenon does not exist.

7.1.10.1.2 Parietal Lobe and Occipital Disturbances: Allesthesia

In addition to the field defects from visual radiation dysfunction and palinopsia, allesthesia has been described with BAVMs in the parietal/occipital region. Allesthesia is a visual phenomena where there is transposition of a visual image from the homonymous field where it should be seen to the opposite field. This visual disorder may result from the abnormal transfer of visual imagery from the normal parietoccipital lobe via interhemispheric connections to the abnormal parietoocipital lobes. One such case developed allesthesia following excision of a BAVM located deep within the right anterior occipital and posterior parietal lobes with extension into the thalamus. Postoperatively, the patient had a dense left homonymous hemianopia sparing the central 8°. He experienced palinopsia in the left field and allesthesia in the right field. The allesthesia imagery consisted of seeing people or inanimate objects in episodes lasting from 3 to 15 min. The episodes were associated with seizure activity on the electroencephalogram recorded from the parietooccipital leads ipsilateral to where the BAVM had been. Anticonvulsants stopped the visual episodes, and the epileptic activity on the electroencephalogram resolved [105].

7.1.10.1.3 Homonymous Visual Field Loss

Visual field defects are common in patients with occipital BAVMs that have bled [33], but the field may be normal in subjects with unruptured BAVMs. A permanent homonymous hemianopia following an intracranial hemorrhage was recognized as early as 1930 [106] (Fig. 7.15). In one series of 16 occipital BAVMs, three of seven patients who had no field defect preoperatively had a homonymous scotoma, a wedge sector, or a pie-shaped homonymous field defect following surgical excision. Of the nine patients who had preoperative field abnormalities, worse field defects were found in two and four others had improved fields [37]. Spontaneous recovery of a portion of or the entire field is not infrequent after a bleed (Fig. 7.25).

In the case of a large hemorrhage into the occipital lobe, in addition to the loss of the contralateral field, a paradoxical temporary loss of the field ipsilateral to the clot can develop. This results in complete blindness which usually lasts for several days. The hematoma causes a mass effect that shifts the brain downward through the tentorial incisura and across the midline anteriorly. The resulting compression of the posterior cerebral artery contralateral to the BAVM against the tentorium causes ischemia of the occipital lobe opposite to the bleed [107]. As the hemorrhage resorbs, the ipsilateral field defect improves, provided occlusion

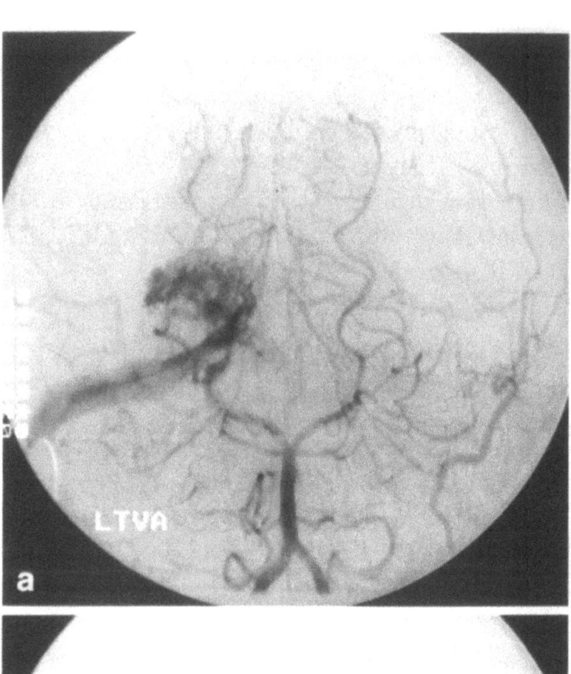

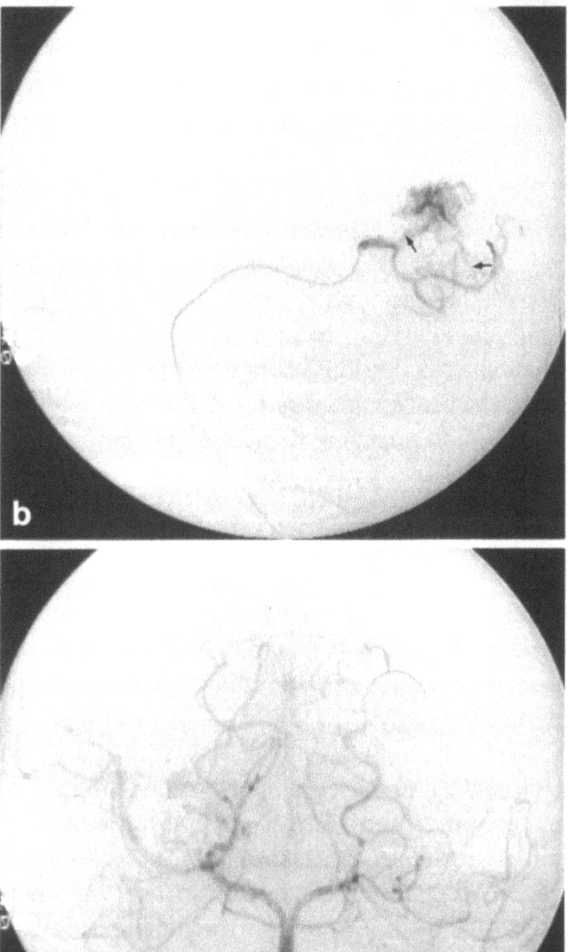

◄ **Fig. 7.25.** This 12-year-old girl had a complete right homonymous hemianopia following a hemorrhage into the left occipital lobe. The visual field normalized without therapy within 3 months. **a** The frontal subtraction angiogram of the left vertebral artery demonstrates a small vessel arteriovenous malformation in the left occipital lobe. **b** The lateral view angiogram, performed with superselective catheterization of the distal posterior cerebral artery, demonstrates a compartment of the arteriovenous malformation supplied by vessels "en passage" (*small arrows*) to the occipital lobe. **c** The frontal view postembolization angiogram demonstrates an absence of arteriovenous shunting and preservation of all the normal arteries. The visual field remained full

of the posterior cerebral artery and significant infarction of the occipital lobe have not occurred.

A BAVM located in the temporal or parietal lobes (Fig. 7.5) can disturb the optic radiations, causing a complete or partial incongruous homonymous defect. The field abnormality is worse in the superior temporal field when the involvement primarily affects Meyer's loop [92]. BAVMs located deep in the hemisphere can cause an optic tract lesion with either a partial incongruous or a complete homonymous hemianopia and typical bow-tie atrophy in the contralateral eye and more diffuse optic atrophy in the ipsilateral eye (Figs. 7.3, 7.4). A relative afferent pupil defect is found in the contralateral eye only if the lesion is limited to the optic tract and the optic nerve ipsilateral to the BAVM is not involved.

7.1.10.1.4 Optic Neuropathy

A decrease in visual acuity has been described with a hypothesized arterial steal to the optic nerves [40], particularly in a rare case where an ethmoidal artery was involved with the shunting of blood [40a]. However, xenon blood flow studies have not demonstrated an arterial steal that can be correlated with the diminished vision. In contrast, transient bilateral blindness lasting from 2 min up to 36 h has been described in patients with infratentorial BAVMs in the cerebellum, probably related to generalized intracranial venous hypertension [63]. We have not seen a case of a posterior fossa BAVM with transient bilateral visual loss unless significant papilledema and increased intracranial pressure were present.

7.1.10.1.5 Orbital Congestion

Orbital congestion and ocular motor dysfunction are unusual developments despite the common occurrence of venous drainage from a BAVM into the cavernous sinus. Unlike with carotid cavernous fistulas, audible bruits are infrequent despite the arterialized venous drainage in the dural venous sinuses. Most of the patients that demonstrate orbital signs have BAVMs with arteriovenous shunts involving the dura in the region of the cavernous sinus as well as venous drainage into the transverse, petrosal or cavernous dural venous sinuses [108]. However, the BAVM associated with the Wyburn–Mason syndrome (see Sect. 4.1.3.3) can drain into the orbit, facial region, and pharynx, resulting in epistaxis or gingivial hemorrhage as well [109]. In our series, all but a few patients with brain BAVMs and clinical signs of cavernous sinus congestion also had a dural arterial supply to the BAVM.

7.1.10.2 Chiasmal Vascular Malformation

A chiasmal syndrome associated with a vascular malformation is rare. The reports of these lesions include a variety of vascular lesions, including AVMs and telangiectases, in the region of the chiasm [110]. Some AVMs may invade the chiasm and cause a progressive hemianopia and optic atrophy, or episodes of sudden transient visual loss following seizures [111]. Recurrent chiasm hemorrhages with episodes of bitemporal field loss can mimic pituitary apoplexy or a hemorrhage into a chiasmal glioma [112]. The visual fields may improve spontaneously or the field loss may remain as a permanent deficit even after the clot has been reabsorbed or surgically evacuated [113]. Recurrent or sudden loss of acuity and altitudinal field defects have been described in patients with hemorrhages into or beneath the intracranial portion of the optic nerve. One report described this injury associated with a subarachnoid hemorrhage, without an obvious AVM, but a vascular malformation was presumed to be the etiology [114].

CT does not easily distinguish an AVM of the chiasm from a neoplasm in the suprasellar cistern. However, in contrast to a neoplasm that does not contain a hemorrhage, the AVM can have a high-attenuation appearance on both the unenhanced and enhanced CT. Carotid angiography can appear normal when the AVM and the hematoma remain completely within the chiasm [115]. The fine resolution capability and the sensitivity of MR imaging to blood flow are useful in visualizing a vascular malformation.

In one patient with an abnormal conglomerate of veins in the suprasellar cistern (surgical observation), a vascular malformation presumed to be a venous "angioma" was diagnosed. The patient had a bitemporal hemianopia and a unilateral central scotoma. He experienced a deterioration in vision over several weeks following a subarachnoid hemorrhage. The arterial phase of both carotid and vertebral injected angiography was normal [116, 117]. It is possible that this patient had a hematoma compressing an AVM nidus and its arterial feeders, or the lesion may have been a cavernoma (see Sect. 7.2.2). Possibly only the draining veins were observed at surgery.

Prior to the development of cerebral arteriography, Cushing described intracranial anterior visual pathway involvement by a vascular malformation, presumed to be a venous "angioma", located at the base of the brain [118]. As more comprehensive angiography was performed, it was recognized that extensive BAVMs or venous malformations that spread along the base of the skull can involve the chiasm (Fig. 7.24). These AVMs are predominantly supplied by the basilar artery circulation, rather than the ICA circulation, and typically cause bilateral optic atrophy with diminished acuity in both eyes and bitemporal hemianopia [92].

7.1.10.3 Ocular Motor, Lid, and Pupillary Dysfunction

7.1.10.3.1 Supratentorial Brain Arteriovenous Malformations

Symptomatic eye movement disorders are unusual in patients with unruptured supratentorial BAVMs. A transient conjugate deviation of the eyes away from the cerebral hemisphere harboring the BAVM can occur with a seizure. A transient conjugate deviation of the eye towards the hemisphere with the BAVM can be seen with an intracerebral hemorrhage. A nonlocalizing sixth nerve paresis may also develop when an intracerebral hematoma or a

subarachnoid hemorrhage causes an elevation in intracranial pressure or the subarachnoid blood causes meningeal irritation. The elevation of the intracranial pressure can also cause a divergence insufficiency. In this supranuclear disorder the esotropia at distance is greater in primary gaze than on viewing to the right or left, and it is unassociated with underaction or slowing of either lateral rectus. A hemorrhage into the temporal lobe that causes uncal herniation results in compression of the ipsilateral third nerve with the expected lid, pupillary, and eye movement disorder. Patients with an unruptured frontal BAVM may have a symptomatic slowing of saccades into the field, away from the involved cerebral hemisphere. Many patients with supratentorial BAVMs who are treated with anticonvulsants have an asymptomatic conjugate jerk nystagmus on right and left horizontal gaze. Upbeat nystagmus on upgaze is usually a sign of toxicity when the anticonvulsant levels rise above the therapeutic range. Rarely, toxic levels of phenytoin cause ophthalmoparesis [119].

7.1.10.3.2 Infratentorial Brain Arteriovenous Malformations

Ocular motor dysfunction is common in patients with BAVMs that are located in the posterior fossa, involve the diencephalon with extension into the upper midbrain, or have abnormal venous drainage into the posterior fossa (Fig. 7.11; see Sect. 8.5.2.1). Nystagmus of all types have been described in patients with posterior fossa BAVMs [69]. Upbeat nystagmus, a medial longitudinal fasciculus syndrome, conjugate or asymmetrical jerk and rotatory nystagmus on eccentric or on primary gaze can all occur depending on the location of the BAVM or whether the abnormal venous drainage passes into the cerebellum or the brainstem. Retraction nystagmus with eyelid retraction occurs with BAVMs in the midbrain [120].

Supranuclear gaze palsies, including a medial longitudianl fasciculus syndrome, skew deviation, alternating skew, and ocular bobbing, have also been described. A temporary, forced conjugate downgaze develops with a diencephalic hemorrhage that compresses the midbrain [121]. Paresis of conjugate upward gaze frequently develops if the BAVM, hemorrhage, or hydrocephalus affects the posterior third ventricular and the dorsal midbrain. The eye movements generally normalize once the dorsal midbrain compression is relieved. Downgaze paresis results if the bleed is more anterior in the midbrain tegmentum (see Fig. 8.20 in Chap. 8). An isolated fourth nerve paresis is not seen but both unilateral and bilateral fourth nerve palsies have occurred with other signs when the BAVM or hemorrhage was located in the anterior midline cerebellum. Nuclear and fascicular involvement of third or sixth nerves are typically associated with other signs of midbrain or pons dysfunction, respectively.

Pupillary dilation results from localized nuclear, fascicular, or subarachnoid third nerve dysfunction. A first-order Horner's syndrome consisting of a miotic pupil, mild ptosis of the upper and lower lids, and sometimes hypohidrosis of the ipsilateral face and body develops when the descending sympathetic pathway, from the diencephalon through the pontine tegmentum, is disrupted.

An ophthalmoplegic migraine-like syndrome from any etiology is extremely rare. A single case has been reported as being caused by a BAVM located in the proximal segment of the posterior cerebral artery [122].

7.1.10.3.3 Papilledema

Neuro-ophthalmologic signs may occasionally have no localizing value. Some of the older literature described papilledema, or what some authors called "blurred disc margins," more frequently in patients with BAVMs located in the posterior fossa and presumed elevation of the intracranial pressure than in patients who had BAVMs in a supratentorial location [123]. However, in one series, a mild degree of "choking" of the optic disc or slight blurring was observed in approximately one third of patients with supratentorial BAVMs who had a subarachnoid hemorrhage [5]. This has not been our experience; considerably fewer patients with a BAVM in any location develop papilledema (see Sect. 7.1.10.4). Most of the patients with intracranial BAVMs and papilledema have significant arterialization of the venous drainage or outflow restriction in the dural venous sinuses (see Sect. 3.2), a distinction which is not made in the older literature (see Sect. 3.2), or hydrocephalus following a subarachnoid hemorrhage. Rarely, young adults with

high-flow BAVMs that drain into the superior sagittal sinus develop papilledema secondary to the intracranial venous hypertension [124, 125]. In these cases, occluding the major arterial input through the arteriovenous shunt will normalize the fundi and reverse the visual loss (Fig. 7.17).

Dilatation and tortuosity of the retinal veins is also described in the older literature [118, 117]. Unfortunately, some of these case descriptions seem similar to what is now considered nonpathological congenital fullness of the optic disc or congenital tortuosity of the retinal veins. Thus, these fundus changes may not have been acquired because of a BAVM-related hemodynamic abnormality or increased intracranial pressure. Nevertheless, some patients do have chronic papilledema that will eventually become pale and atrophic if untreated.

7.1.11 Treatment

7.1.11.1 General Considerations

Surgery, radiation, endovascular embolization, and conservative follow-up with medical therapy for headache and seizure control, or some combination of the above, are options that can be exercised in an individual with a BAVM. The optimal therapy should eradicate the BAVM nidus and preserve the normal circulation without complications. The various modalities of treatment are in continuous evolution, becoming more specific with reduction of the risks. The choice of therapy depends on numerous factors, including the clinical findings and history, the location of the BAVM nidus, arterial feeders, and venous drainage, and the compromise of the normal brain arteries and veins [126–128]. Not all BAVMs require therapy directed at the lesion. For example, therapy may not be warranted in patients with a coincidentally diagnosed asymptomatic BAVM in a difficult location since these lesions have been shown not to hemorrhage or cause neurologic deterioration during a 20-year follow-up [18]. Treatment is recommended more often in patients who have experienced a hemorrhage or have a progressive neurologic deficit. The necessity for immediate direct intervention in patients who only have epilepsy is less obvious. However, since hemorrhage will develop in approximately 25% of these patients within 8 years, interventional therapy of the BAVM is an option if the above factors suggest a low risk of developing a significant treatment complication. As mentioned previously, surgery may relieve or worsen the epilepsy. Embolization does not worsen seizure control. Headache should never be the sole indication for surgery, but embolization of the meningeal arterial supply to a BAVM can safely reduce the frequency and severity of headaches.

The age of the patient can be an important factor when planning treatment. A neonate may require emergency closure of a high-flow shunt that causes congestive heart failure. Since Crawford's study [18] reported a significant incidence of hemorrhage and a high morbidity and mortality rate following a bleed in patients in the fifth and sixth decades of life, we now consider the use of more intensive therapy in these cases.

In women of childbearing age, management of the BAVM must be discussed in relation to a current or prospective pregnancy. Though we have had patients who have had several pregnancies without symptoms from the BAVM prior to discovery of the lesion, once a BAVM is diagnosed, treatment is complicated by pregnancy. Unless an emergency arises, invasive management is withheld until the patient is post partum. A cesarian section delivery is not necessarily recommended and early termination of a pregnancy is rarely indicated.

Therapeutic intervention may be more intensive if the patient seems to have a reversible neurologic deficit that is debilitating. Surgery can be a definitive therapy that is beneficial when complete excision of the BAVM is possible. Embolization can also be used to completely eliminate the BAVM and 16% of our cases have been cured with this modality [129]. However, in cases where the BAVM has single pedicle arterial supply, the embolization cure rate rises to over 90%. Embolization is also used in conjunction with other therapies. It is frequently performed as a preoperative procedure in order to eliminate arterial feeders that could be difficult to manage during surgery [130]. Embolization is also effective in reducing the size of BAVMs so that radiotherapy can be used (see Sect. 7.1.11.3).

In contrast, if potential complications which are debilitating are likely to occur with therapy, treatment of the lesion may be withheld. For example, excision of an occipital BAVM can cause a homonymous hemianopia that can be devastating

to an operator of a motorized vehicle. Obviously, when the BAVM is located in the motor or speech areas, in the thalamus or hypothalamus, or in the brainstem, direct therapy with surgery or embolization may cause a catastrophic deficit. When warranted, embolization may be the best modality for treating the abnormal arteries and the BAVM in these areas, because embolization is performed while the patient is awake. The patient can be clinically monitored to determine whether occlusion of a specific artery can cause a neurologic or visual deficit. For example, prior to embolizing a specific artery, an amytal tolerance test can be performed. If a neurologic deficit results following superselective intraarterial injection of 50–75 mg sodium amytal, the vessel must be preserved [131, 8, 132]. Despite this monitoring, errant cerebral infarcts or hemorrhages can occur during embolization.

7.1.11.2 Embolization

Preoperative embolization of feeding arteries to the BAVM can decrease the rapid blood flow, diminish intraoperative bleeding, and prevent abrupt hemodynamic changes during surgical excision. Polyvinyl alcohol particles and liquid acrylic material are the agents in current use [133–135]. The hardened acrylic, *N*-butylcyanoacrylate, is softer than isobutylcyanoacrylate so the former agent is better suited for surgical cutting of embolized vessels [8, 130].

Percutaneous or intraoperative embolization designed to cure arteriovenous fistulas and BAVMs must obliterate the nidus of the lesion by producing a cast of embolic material and thrombus of the pathological angioarchitecture. Since recanalization of thrombus may lead to a recurrence of the BAVM, a repeat angiogram is performed at 6 months and 2 years after embolization to demonstrate whether a cure has been effected. Although extensive and large BAVMs are often incurable, selective partial embolization of feeding arteries and a portion of the nidus may reverse or prevent progressive neurologic dysfunction [48, 66]. Though embolization with large particles or spheres does not alter the rebleed rate [136], closure of the distal arterial supply and most of the BAVM nidus seems to reduce the rebleed rate. Closure of the associated aneurysms may also prevent a hemorrhage.

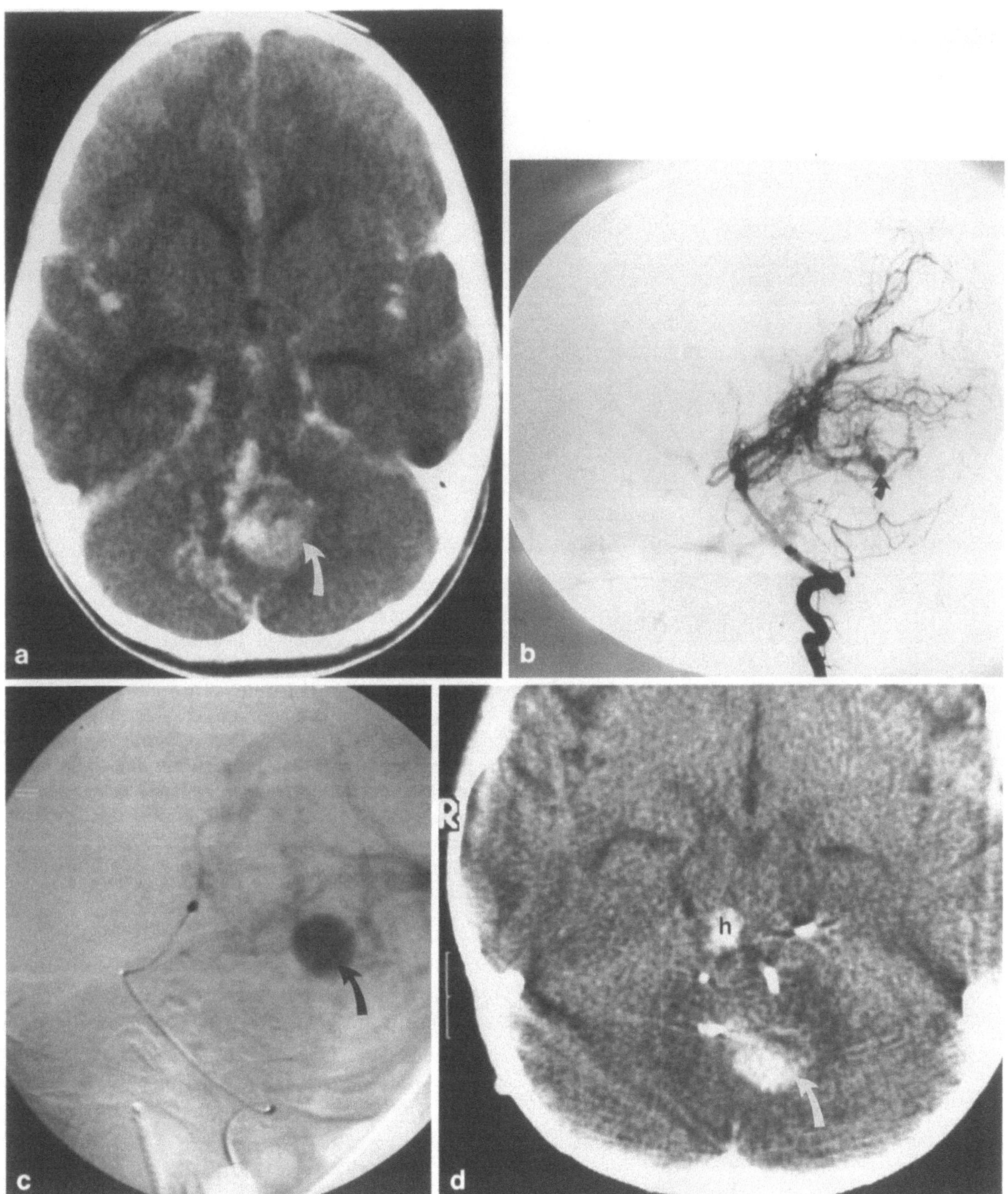

Fig. 7.26 a – d. Rupture of a brain arteriovenous malformation related to embolization.

a The 7-year-old girl had the onset of sudden lethargy, headache, and diplopia due to a skew deviation. The contrast- enhanced axial computed tomogram demonstrates a large area of hematoma associated with a lateral enhancing venous aneurysm (*curved arrow*).

b Lateral subtraction angiogram of the left vertebral artery in the same patient demonstrates drainage of the malformation into a venous developmental anomaly (*curved arrow*).

c Lateral subtraction angiogram of the superior cerebellar artery in late phase. Note the large venous aneurysm (*curved arrow*) which is also seen in **a**.

d Three weeks after embolization the patient had worse diplopia with a new right fourth nerve paresis, ataxia, and right facial weakness. The axial noncontrast computed tomogram performed at that time demonstrates thrombosis in the venous aneurysm (*curved arrow*) and a small hemorrhage in the right posterior lateral portion of the brainstem (*h*)

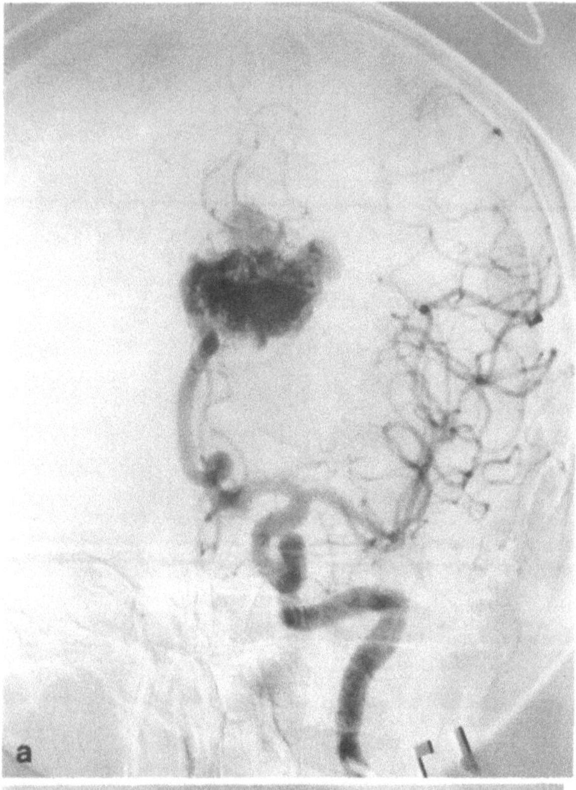

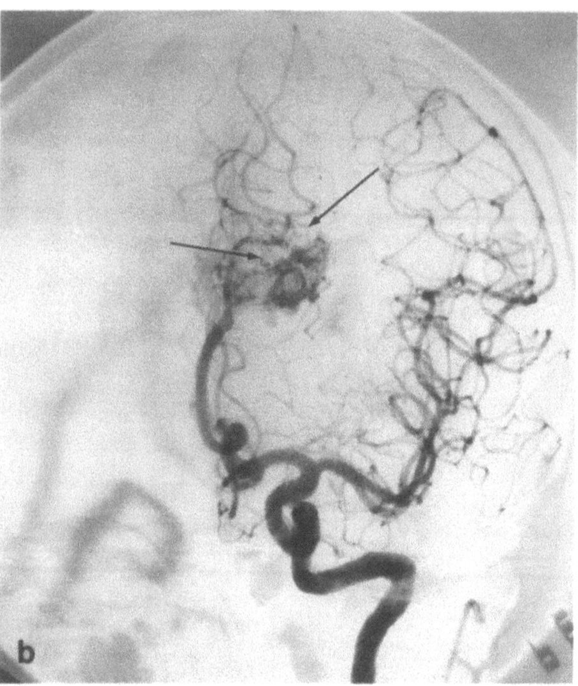

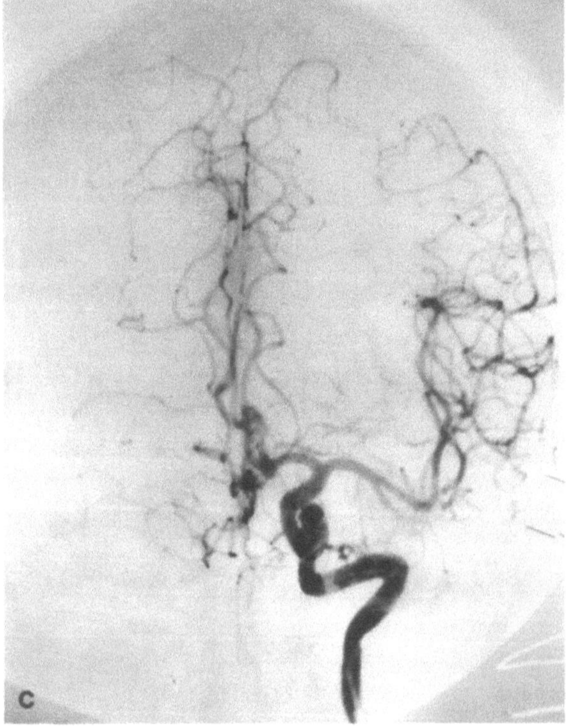

Fig. 7.27a–c. Combined treatment for brain arteriovenous malformation. **a** The frontal subtraction angiogram of the left internal carotid artery shows a brain arteriovenous malformation supplied by the anterior cerebral artery. **b** The immediate postembolization angiogram demonstrates significant reduction in the arteriovenous malformation; however, the presence of some vessels en passage (*small arrows*) prevented complete obliteration of the nidus. **c** Two years after radiosurgery there is complete obliteration of the malformation

The risk of embolization causing an infarct or hemorrhage in the brain is of the order of 5% per patient if all patients with BAVMs in all locations are considered (Figs. 7.16, 7.26). The risk of an embolization-related major permanent neurologic deficit is approximately 2%.

7.1.11.3 Radiation Therapy

Radiotherapy with focused beam irradiation has been employed to successfully treat small BAVMs. Conventional X-ray beams have been of limited utility. Successful treatment using Bragg peak proton beam irradiation has been reported [137]. BAVMs smaller than 4 cm^3 have been reported to have an obliteration rate of 94% within 2 years of proton beam therapy. However, only 39% of

BAVMs larger than 25 cm^3 were obliterated and major neurologic complications occurred in 12% of cases [137, 138]. Steiner has also reported a significant cure rate of small BAVMs ($20\times30\times40$ mm) in approximately 2 years following Leksell gamma knife focused irradiation [139, 140]. However, with either form of irradiation, the risk of hemorrhage is not reduced until the BAVM is obliterated [140a]. Embolization can be used to reduce the size of larger BAVMs to a size small enough to permit the lesions to be treated with irradiation (Fig. 7.27).

7.1.11.4 Surgery

The introduction of microneurosurgery by experienced neurosurgeons, in conjunction with preoperative embolization when needed, has reduced the risk of surgical morbidity and mortality for excision of BAVMs [1, 7, 82, 141 – 143]. Total removal of the BAVM nidus is curative and should be performed when the lesion is not in a critical area, particularly if a hemorrhage has occurred. Because of the poor long-term prognosis if left untreated, surgical excision should also be considered in patients who present with seizures, neurologic deficits, and severe headaches. Complete excision of the BAVM seems to reduce the seizure rate though some patients develop epilepsy postoperatively (see Sect. 7.1.5.3) [46]. Patients with incidental BAVMs should be followed-up expectantly.

The results of partial excision of BAVMs on the frequency of rebleeds and neurologic deficits are unknown.

The complications of surgery have been well described and include cerebral infarct and parenchymal hemorrhage, which can cause a severe neurologic deficit or death. Many factors, including the complexity of the BAVM, its relationship to the arterial feeders and venous drainage, the location of the BAVM, the skill and experience of the neurosurgeon, and significant intraoperative blood loss with resulting cerebral hypoperfusion, influence the surgical outcome. Occlusion of the abnormal arterial supply to the BAVM results in an increase in blood flow to the normal brain. Postocclusive hyperemia, swelling, and hemorrhage in the brain, called normal perfusion breakthrough, is an unusual complication [144 – 146]. Staged surgical

procedures that do not eliminate all major arterial feeders to the BAVM at once may avoid this complication [147].

7.2 Other Vascular Lesions of the Brain

Vascular lesions other than BAVMs which can cause numerous neuro-ophthalmologic signs include cavernoma, venous "angioma" (anomalies), telangiectasia of the brainstem, and the venous malformation of Sturge–Weber syndrome (see Sect. 4.13 for discussion of neurocutaneous vascular lesions).

7.2.1 Telangiectasia

Most telangiectases of the brain are asymptomatic and only cause symptoms when a hemorrhage occurs. A telangiectasia, which is a collection of abnormally dilated venules, is commonly found at autopsy. In the great majority of symptomatic cases, the cerebral angiography is normal. The ratio of occurrences of supratentorial/posterior fossa telangiectasias is approximately 2/3 in autopsy series. Of the two locales, supratentorial telangiectasias are more often clinically silent. These asymptomatic lesions are frequently discovered at autopsy, accounting for the large discrepancy between the frequencies reported for clinical and autopsy studies.

The brainstem is the most frequent site for symptomatic lesions. In fact, telangiectasias are the most common vascular malformations found within the pons, representing 27 of 48 pontine vascular malformations in McCormick's series [60]. The clinical signs can be similar to those found with a pontine glioma, such as vomiting, lethargy, paraparesis, sixth and seventh nerve palsies, and dysphagia (see Sect. 7.1.6.1.7). The patient can deteriorate slowly if a mass is present, or rapidly if a pontine hematoma develops. Following surgical evacuation of a clot, the brainstem deficits often improve [68].

Prior to the introduction of MR imaging, some patients were misdiagnosed as having an intrinsic brainstem neoplasm. High-resolution CT may be normal or may reveal an area of minimal enhancement, or only a swollen brainstem, in a patient with telangiectasia who has not experienced a clinical hemorrhage. Following hemorrhage, a high-attenuation mass that does not enhance with contrast can be seen. High-field MR imaging will show a heterogeneous lesion depending on the number and extent of hemorrhages and the amount of methemoglobin or hemosiderin present. MR imaging imaging in axial, coronal, and sagittal planes is the best method for localization of the clot and vascular malformation. Angiography is typically normal or suggestive of an intaaxial mass following a bleed.

7.2.2 Cavernoma (Cavernous Hemangioma)

7.2.2.1 Epidemiology

Cavernomas, also called cavernous hemangiomas and cavernous angiomas, are benign tumors of congenital origin. These hamartomas are most often asymptomatic and discovered only when a patient has a scan of the head for another reason. These lesions are not true angiomas or hemangiomas. Cavernomas do not have cellular proliferation and increase in size because of intralesional thrombosis and hemorrhage. Men and women have an equal predilection for intracranial cavernomas. These lesions are found in all age groups, from the neonatal period to the eighth decade [148]. The ratio of the number of cavernomas located in the supratentorial to infratentorial regions is 4/1. Cavernomas represent fewer than 15% of all cerebral vascular malformations.

Except in cases with the familial type, a given patient rarely has more than one intracranial cavernoma [149]. Familial cases of cavernomas are extremely rare [150, 151]. However, some of the families with cavernomas have an autosomal dominant condition with a 90% penetrance. These families tend to have multiple cavernous malformations located intracranially, in the spinal cord, and in the retina (posterior pole or peripheral retina) [152–154] (see also Sect. 4.7).

7.2.2.2 Clinical Signs and Symptoms

When symptomatic, cavernomas can cause a variety of neurologic symptoms which may be been present for up to 12 years before the diagnosis is ac-

tually established. These patients present with approximately equal frequencies of seizures, signs of a progressive mass lesion in the brainstem, cerebellum, or cerebral hemisphere, or a sudden or a stepwise worsening because of an acute or a subacute intracerebral hemorrhage [155] (Fig. 7.28). The seizures may be focal or generalized. If the cavernoma is located in the occipital lobe, focal epileptic discharge can cause a recurrent homonymous hemianopia and visual hallucinations [156]. Hemorrhage occurs in 13% – 24% of cases [155, 157]. An intraparenchymal hemorrhage is most common (approximately 40% of cases that bleed) in the fourth decade of life [158]. Papilledema develops when the lesion is deep and obstructs the ventricular system. Hydrocephalus develops in less than 10% of all patients with clinically apparent cavernomas [158]. Although cavernomas can occur in any location in the central nervous system, these lesions are least likely to be found in the thalamus [158].

The natural history of this disorder is not known. In particular, the number of lesions that hemorrhage and the rebleed rate for these lesions are unknown. Even when symptomatic, intracranial cavernomas are rarely associated with death. Most clinical series are predominantly composed of patients with symptomatic lesions, many with hemorrhages. These reports generally include only a few cases with incidentally discovered cavernomas that have been followed prospectively.

7.2.2.2.1 Visual Loss

Except for the cavernomas located in the orbit (see Sect. 4.7), involvement of the anterior visual pathway by these lesions is rare [159]. Compression of the optic nerve by a cavernous hemangioma extending from the anterior cavernous sinus into the superior orbital fissure has been described once [160]. The chiasm can also be compressed if the cavernoma located in the anterior diencephalon becomes exophytic [161].

Few cases with intracranial optic nerve, chiasm, or optic tract invasion by these lesions have been described [162, 163] (Fig. 7.29). One patient was reported to have a third nerve paresis and a quadranopia because of optic tract compression [164]. Recurrent hemorrhage into a cavernoma

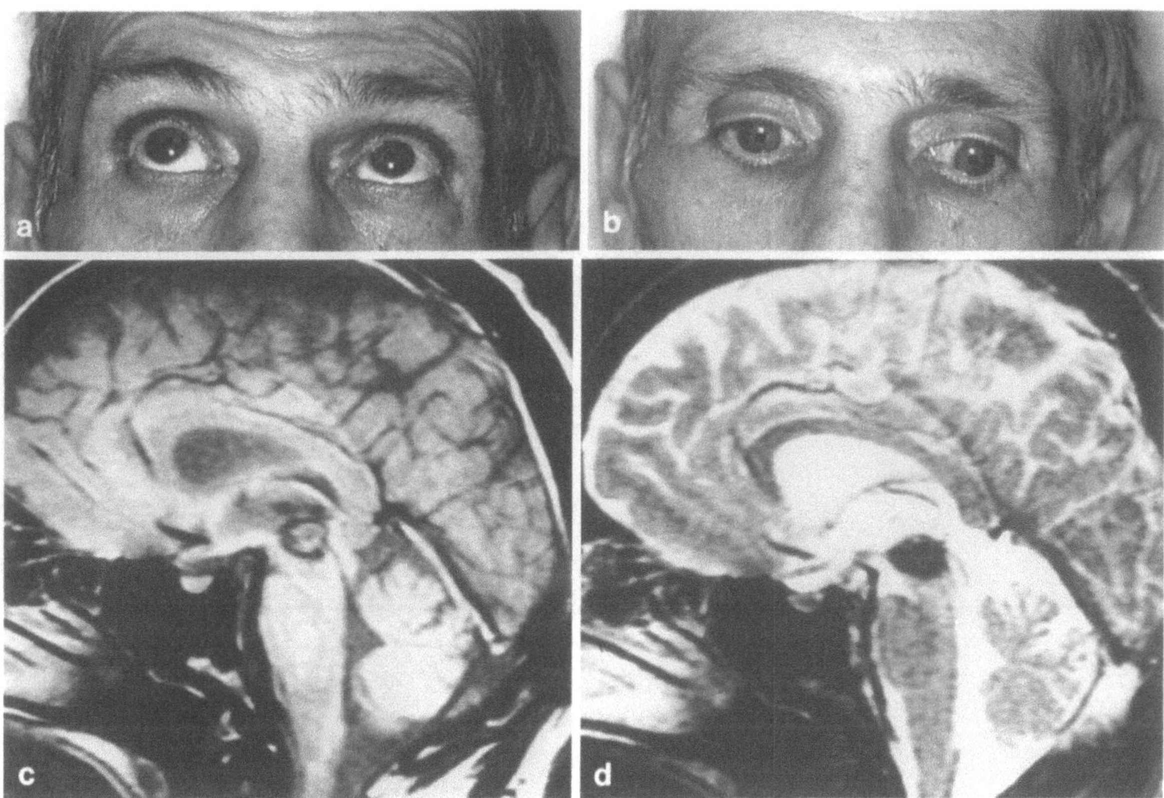

Fig. 7.28 a – d. Cavernoma of the midbrain. This 58-year-old man awoke with vertical diplopia and poor balance when walking. **a,b** In addition to an inability to tandem walk and mild dysmetria in the right upper extremity, the clinical evaluation revealed normal gaze upward (**a**), moderate weakness in downward conjugate gaze (**b**), and a skew deviation with a left hypertropia of 4 diopters. The saccades were slow in up- and downgaze. The left pupil was dilated 2 mm and sluggish. **c** The sagittal view T1-weighted magnetic resonance image demonstrates a lesion with an internal hemorrhage, causing a mixed signal representing different stages of thrombosis, surrounded by a signal void. **d** The T2-weighted magnetic resonance image shows that the lesion had become signal void due to the internal hemorrhage. The patient remained clinically unchanged over the next 6 years

within the intracranial visual pathway can cause episodes of sudden visual loss which may be permanent [113, 165]. If the MR image shows the intrachiasmal hemorrhage without abnormal arteries or draining veins, the diagnosis may be suggested preoperatively [166].

7.2.2.3 Pathology

The histological features of a cavernomas include irregular sinusoids of vascular channels (see Fig. 4.12 in Chap. 4). The walls of these lumens are composed of endothelial cells surrounded by collagen deposition, thick fibrous connective tissue, areas of hyaline degeneration, and small amounts of elastic tissue. Microscopic or gross hemorrhages, both acute and chronic, are seen. Thrombosis within the lumens leads to calcium deposition in the vascular walls. Neuronal tissue is not interposed between areas with the vascular channels. Infiltration into the brain is not seen, and gliotic tissue typically separates the cavernoma from the surrounding neuronal tissue [29]. Major arterial input or large draining veins are rarely found.

7.2.2.4 Imaging

The noncontrast CT demonstrates hypodense or hyperdense areas, or both types of abnormalities in a single cavernoma, depending on whether the le-

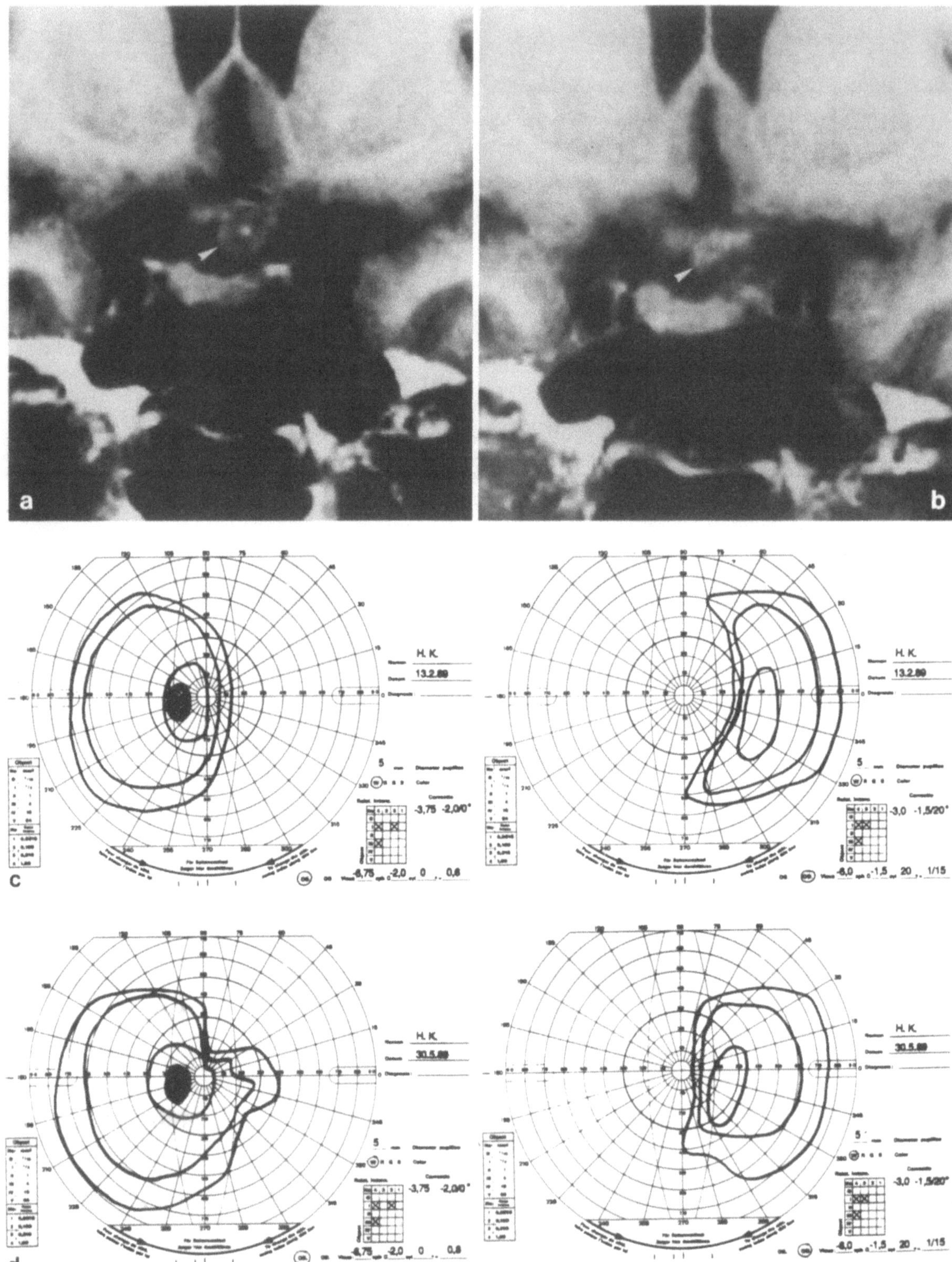

◄ **Fig. 7.29a–d.** Cavernoma of the chiasm. **a** The preoperative coronal view T1-weighted magnetic resonance image shows a suprasellar mass with heterogeneous intensity (*arrowhead*) adjacent to the inferior surface of the chiasm. **b** One week after surgical resection, the magnetic resonance image shows only an area of coagulated cavernoma tissue (*arrowhead*). **c** Preoperatively, the visual fields demonstrate a nasal hemianopia extending to the macular area and into 20° degrees of the temporal field in the right eye. The left eye had a nasal hemianopia. **d** Two months after surgery, the temporal field defect has improved in the right eye and the nasal field defect in the left eye is reduced. (From [166])

sion contains regions with cystic degeneration, thrombosis, hemorrhage, or calcification. The mass effect is slight and extravascular blood is rarely found. Occasionally areas of significant calcification are seen. Following contrast administration, areas of slight enhancement may visualized, giving the cavernoma a heterogeneous appearance. This suggests a minimal local breakdown of the blood–lesion barrier. On occasion the CT shows a ropelike enhancing vein extending to the ventricle or the cortex. Extralesional hemorrhage is rarely seen. A hyperdense lesion with mass effect can be demonstrated in the rare case with an extravascular hemorrhage.

MR imaging of a cavernoma without gross hemorrhage demonstrates a discrete lesion with a centrally located mixed signal [167, 168] (Fig. 7.28). High-field MR imaging demonstrates a signal intensity that varies with the different phase of metabolism of the extracellular methemoglobin. Typically, the cavernoma has a high intensity on T1 weighting that diminishes on T2-weighted images [84]. The paramagnetic properties of hemosiderin in the area around the lesion results in a signal void on both T1- and T2-weighted studies [167, 169].

Cerebral angiography is abnormal in approximately 70% of cases (generally performed only in symptomatic cases). In approximately 75% of these patients, only an avascular mass is seen. An intraparenchymal blush, a large stagnant draining vein, or a suggestion of neovascularization are visualized in less than 15% of angiograms [148, 169]. Direct microscopic observation at surgery has shown slow blood flow within these lesions [170].

7.2.2.5 Treatment

Treatment includes surgical excision of the mass in an attempt to restore neurologic function when a deficit is progressive, especially following a hemorrhage [171, 172]. Surgery has also been recommended by some as a means of improving seizure control [157] or preventing recurrent hemorrhage. There is no conclusive evidence that asymptomatic lesions require excision or that seizures are better controlled by surgery and anticonvulsant medication than with anticonvulsants alone. In one patient with uncontrollable seizures and a normal neurologic examination, fine-beam radiotherapy delivered over 6 weeks was reported to facilitate seizure control without surgery [156]. However, radiation does not effectively eradicate these lesions and surgery is recommended if direct treatment of these lesions is required.

7.2.3 Venous Vascular Anomalies (Venous Angioma)

Venous malformations are usually incidental findings discovered on imaging studies that are performed for unrelated causes because most of them are clinically silent [173, 174]. Venous anomalies can be confused with BAVMs on CT or MR imaging but they do not have arteriovenous shunting. They are developmental anomalies of the venous system which are the most common intracranial vascular malformation found at autopsy anomalies and have been reported in approximately 2.6% of all autopsies [175]. The age of the patients at the time of discovery of the venous angioma ranges from the first to the seventh decade. These anomalies are slightly more frequent in men than in women.

Although venous anomalies are typically asymptomatic, an unusual patient may develop an intraparenchymal hemorrhage. However, in one report up to 15% of cases were described as presenting with a bleed [176]. A rebleed rarely occurs except when the anomaly is located in the cerebellum. Approximately 36% of reported cases of cerebellar venous anomalies that ruptured rebled [177]. Seizures are the most common clinical manifestation

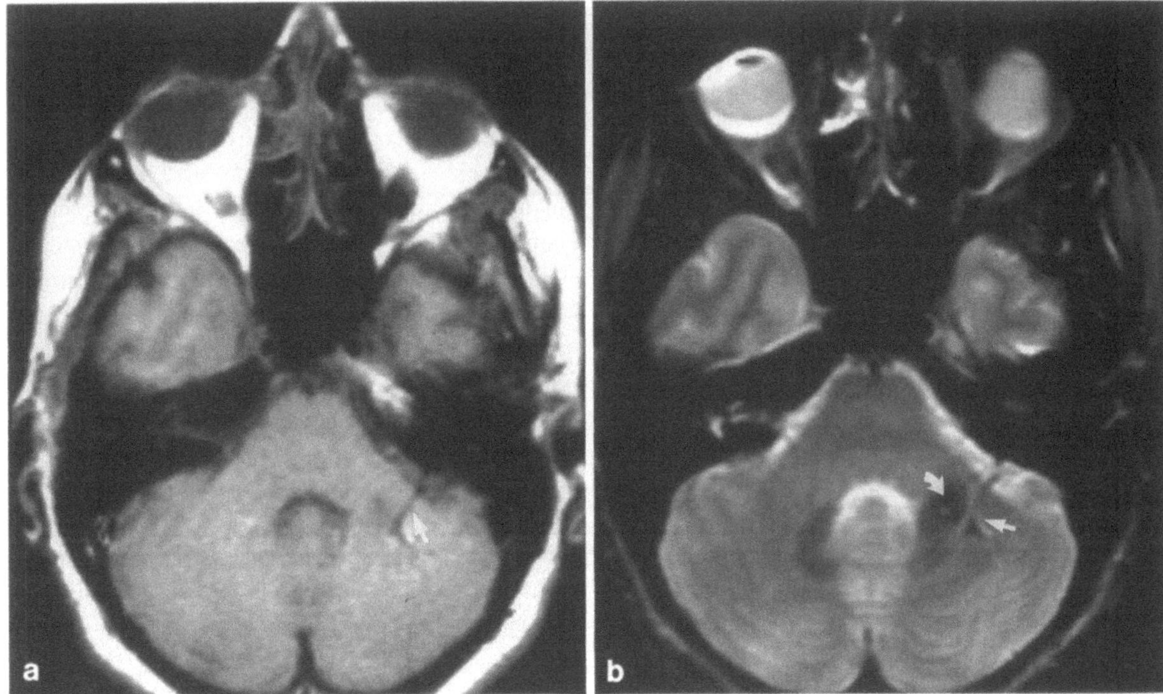

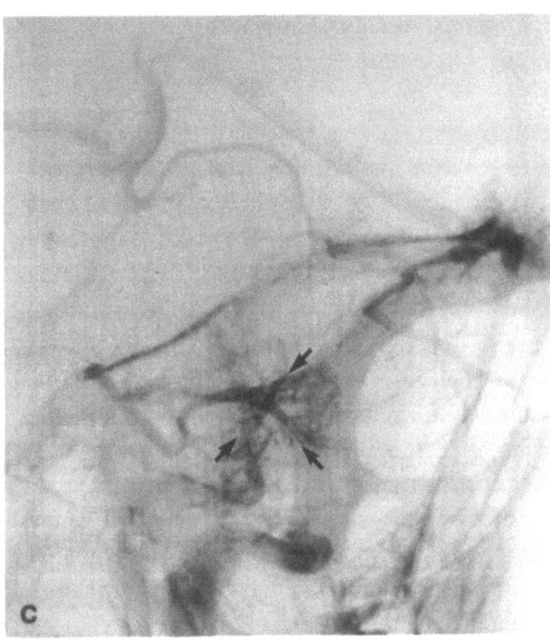

Fig. 7.30a–c. Developmental venous anomalies, sometimes called venous angiomas, can usually diagnosed with magnetic resonance imaging. **a** The axial T1-weighted magnetic resonance image demonstrates a signal void representing a vein (*arrow*) in the cerebellum. **b** The axial T2-weighted magnetic resonance image at the same level demonstrates the venous anomaly has increased signal intensity (*arrow*). Note the presence of a more medial heterogenous signal lesion, suggestive of a cavernoma (*curved arrow*). In patients with a hemorrhage occurs, a developmental venous anomaly is typically associated with a cavernoma. **c** In the same case, the lateral subtraction left vertebral angiogram in venous phase demonstrates the typical spiderlike appearance of a developmental venous anomaly (*arrows*)

when these lesions are supratentorial. As many as 25% of all patients with a venous angioma demonstrated on CT will have epilepsy.

Most patients have a solitary lesion but up to 4% have at least two venous malformations. Any area of the brain can harbor a venous anomaly. In one study, the distribution of lesions included the frontal lobe (26 cases), parietal lobe (5 cases), deep cerebral region (10 cases), and cerebellum (17 cases) [177].

The noncontrast CT is frequently negative, but following contrast administration, a linear enhancement is present in approximately 86% of cases. Similarly, MR imaging often shows a linear signal void area and may reveal the dilated medullary veins (Fig. 7.30a, b). Areas of mixed signal are seen following gadolinium infusion [178]. Angiography is diagnostic and typically reveals the presence of multiple small medullary veins which converge into a large venous trunk (caput medusae; Fig. 7.30c). No arteriovenous shunting is seen.

Direct treatment of a venous anomaly is rarely indicated. Following a hemorrhage, surgical exploration performed to decompress the hematoma often leads to removal of the vascular lesion. When the bleed occurs in the cerebellum, excision may be indicated. Care should be taken not to disrupt the compensatory venous drainage from the normal brain. Embolization or radiation therapy is not indicated.

Embolic and Noninflammatory Occlusive Vascular Disease of the Visual and Ocular Motor Systems

8.1 Introduction

Neuro-ophthalmologic symptoms and signs are common in patients with cerebrovascular disease because the sensory visual and ocular motor systems, spread throughout the cerebrum, deep gray nuclei, brainstem, and cerebellum, are frequently damaged by ischemic disease. The character of the visual field defects and the ocular movement abnormalities, in association with other neurological dysfunction, have classically been used to localize the infarct.

Ischemic disease can cause visual dysfunction at any level of the nervous system. Symptoms and signs of ocular ischemia commonly develop from primary occlusive disease of the branches of the ophthalmic artery and the internal carotid artery (ICA) or from embolic disease from the heart or the ICA. Ocular signs and symptoms are frequent because the ophthalmic arterial system is commonly affected, possibly because it is the first subarachnoid branch artery arising from the ICA and requires a high rate of blood flow in order to support the metabolic demands of the retina and choroid. In some cases the ocular disturbance reflects a more generalized involvement of the arteries in the central nervous system.

8.2 Ocular Ischemia

8.2.1 Intraocular Ischemic Alterations

8.2.1.1 Cotton Wool Spots

Retinal cotton wool spots are the most common, but nonspecific, sign of ocular ischemia. They are also called cytoid bodies. The term "soft exudate" is probably inappropriate since this lesion is not an extravascular exudate. These focal areas of ischemia and infarction of the retinal nerve fiber layer result from stenosis and closure of retinal arterioles [1–3]. Axoplasmic flow, originating from the ganglion cell layer, is prevented from passing the area of infarcted axons. Proximal to the infarct, the axons swell with the dammed axoplasmic contents and become translucent, causing the white, feathery bordered cotton wool spot (see Fig. 9.6 in Chap. 9) [4]. Because of the small size, a single cotton wool spot results in no measurable visual disturbance. Blurring of the vision and a depression in the sensitivity of a threshold measured visual field can occur when multiple retinal infarcts are present. As mentioned above, any disease, including arterial emboli, autoimmune or infectious vasculitis, atherosclerosis of the ophthalmic artery or ICA, that is associated with capillary occlusion and ischemia of the inner retinal layers can cause cotton wool spots. A confluence of cotton wool spots or a focal translucent swelling of the inner layers of the retina and narrowing of the artery are seen on ophthalmoscopy in the affected area with partial or frank occlusion of a retinal branch artery (Fig. 8.1). An extreme example, a central retinal artery (CRA) occlusion, causes more widespread multifocal or diffuse retinal swelling with sparing of the macula area (Figs. 8.2, 8.3). Marked reduction of the retinal blood flow and segmentation of the blood column can be seen in the affected retinal vessels.

8.2.1.2 Retinal Hemorrhage

When the local venous return is compromised because of the slowed and reduced arterial inflow into the affected area, splinter-shaped nerve fiber layer hemorrhages develop in association with cotton wool spots. If the entire branch retinal artery inflow is severely compromised, the retinal veins dilate and the number and size of hemorrhages increase along the involved vascular arcade. On occasion, the marked dilatation of the vein and the den-

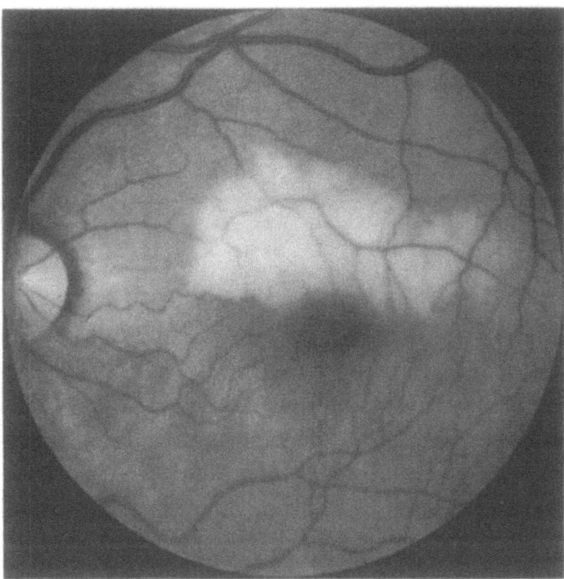

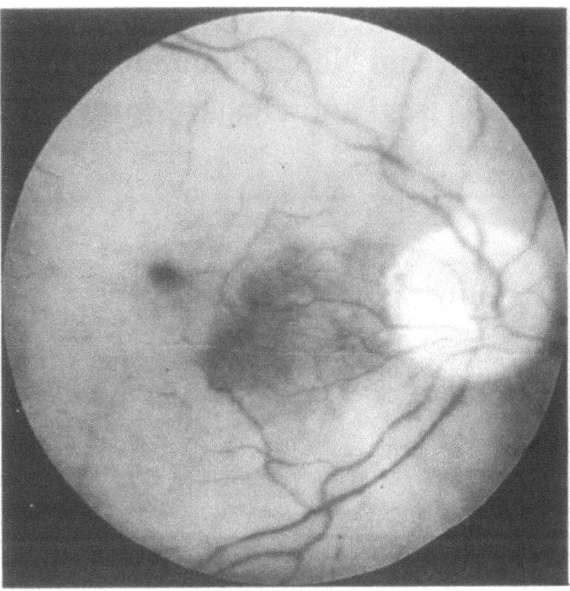

Fig. 8.1. An area of irregularly shaped swelling of the inner layer of the retina is seen just above the macula. The areas of ischemia follow the course of a superotemporal artery suggestive of an acute occlusion of this vessel

Fig. 8.2. The acute occlusion of the central retinal artery causes diffuse swelling of the entire retina except for the papillomacular area just temporal to the disc. However, most of the macula was not spared and only a tiny cherry red spot is seen

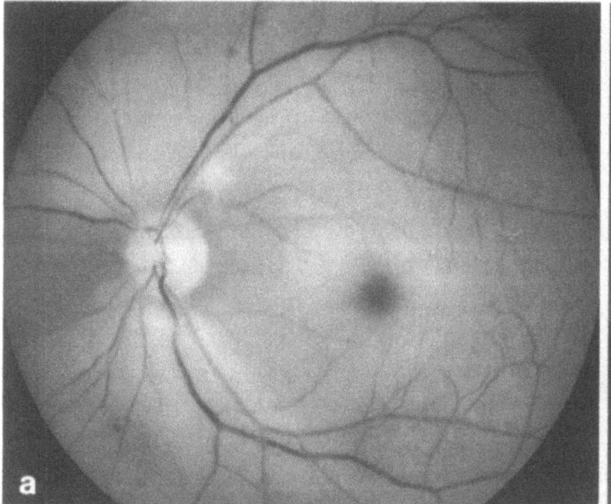

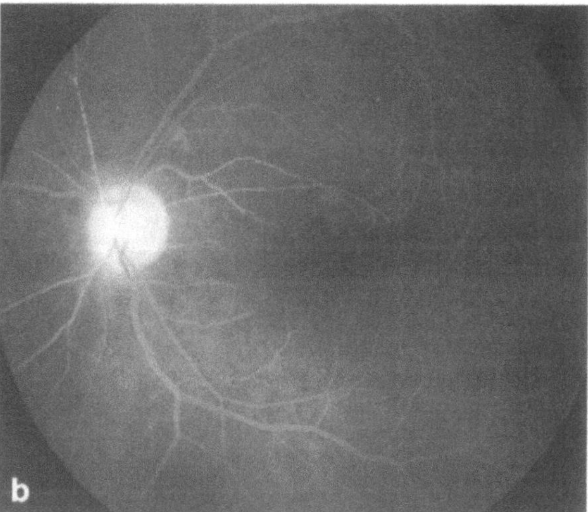

Fig. 8.3. a A second case of a central retinal artery occlusion has more subtle changes in the retina with opacification of focal areas of nerve fiber layer superiorly, temporally, and inferotemporally from the disc. The opacification and swelling of the inner nerve fiber layer is more obvious around the macula. **b** Fluorescein angiography reveals markedly slow flow through the arteries and veins and a breakdown of the dye column in the superonasal artery because of this reduced blood flow

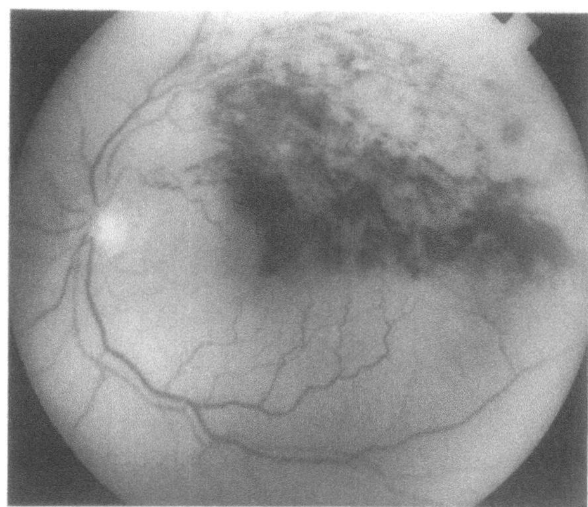

Fig. 8.4. The hemorrhages in the inner layer of the retina follow the course of the axons, giving a feathery-type appearance. The wedge shape of the hemorrhage is typical of a branch vein occlusion. The vision was reduced in this patient temporarily because the hemorrhage involved the macula

Fig. 8.5. a Despite the threatening appearance of the diffusely hemorrhagic disc and peripapillary nerve fiber layer and dilation of the veins, this patient only complained of mild blurred vision. There was no afferent pupillary defect and the visual acuity was reduced to 20/25 because of some mild edema and hemorrhagic involvement of the macula. **b** The hemorrhages extended into the peripheral retina in all quadrants. This patient seemed to have a central retinal vein occlusion without ischemic involvement

▼

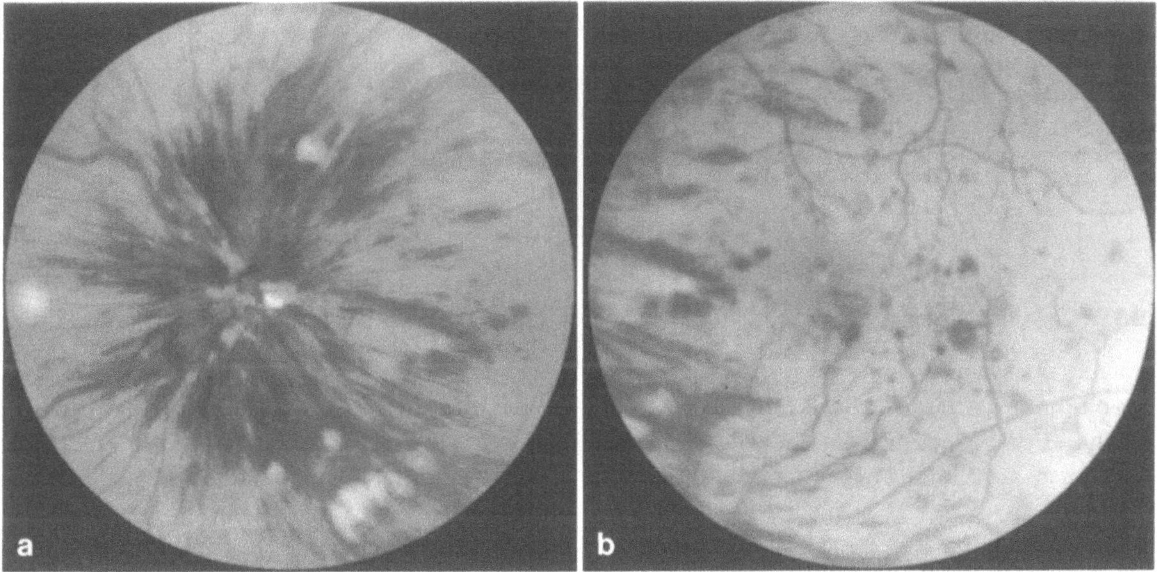

sity of hemorrhages in the inner retinal layers appear similar to a frank occlusion of a branch vein (Fig. 8.4). A partial or complete central retinal vein occlusion is easily distinguished by the presence of multifocal hemorrhages in all four quadrants of the retina, extension into the peripheral retina, and frequent involvement of the macula (Fig. 8.5).

A nerve fiber hemorrhage with a cotton wool spot center, called a Roth spot (Fig. 8.6), is not pathognomonic for a septic embolus, as is classically taught. Similar appearing white-centered retinal hemorrhages can be seen with a microscopic occlusion resulting from any cause in the retina including severe anemia and leukemia.

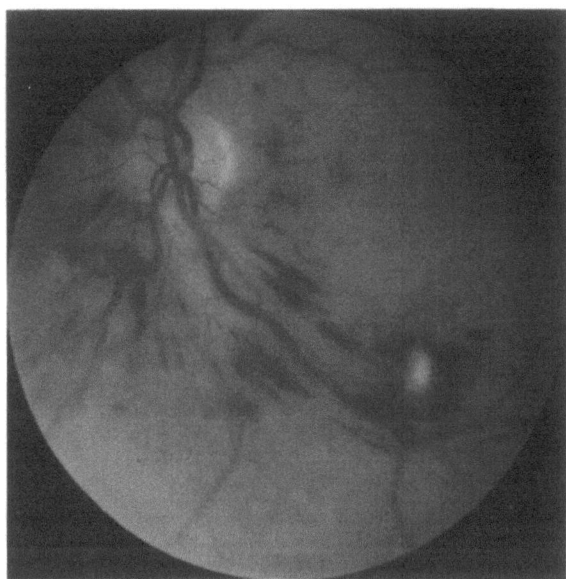

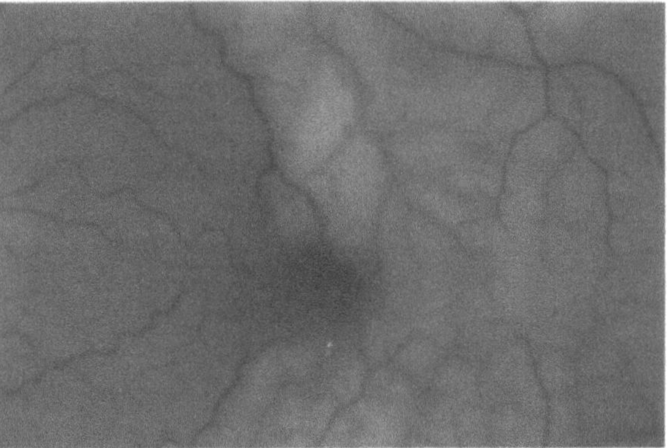

Fig. 8.7. A close-up view of the macula of the left eye demonstrates swelling of the nerve fiber layer superior, inferior, and temporal to the macula because of ischemia. A small highly refractile embolus is seen in one of the inferior paramacular arterioles. An ulcerated plaque found at the origin of the left internal carotid artery was believed to be the source of the embolus

Fig. 8.6. Mild hemorrhages were noted along the retinal veins in all four quadrants. In addition, along the inferotemporal arcade, a white-centered hemorrhage (or Roth spot) was noted in this case of presumed partial central retinal vein occlusion

8.2.1.3 Choroidal Infarct

Ischemic disease of the choroid is difficult to discern on funduscopy. If examined early, subtle focal swelling of the retinal pigment epithelium may be seen with an acute infarct of the choroid. Eventually, focal areas of retinal pigment epithelium loss (called Elschnig pearls or spots) develop in the areas of infarction [5]. Choroidal hypoperfusion, manifested by delayed segmental filling of the choroid, may be demonstrated by fluorescein angiography in acute but not in old cases (see Fig. 5.7 in Chap. 5 and also Sects. 1.5.2.3, 1.5.2.4.6).

8.2.2 Other Signs of Ocular Ischemia

For other signs of ocular ischemia, see Sects. 8.2.3.4.1, 8.2.5, and 8.14.1.

8.2.3 Embolic Ocular Disease

Sudden and recurrent transient or permanent visual loss can occur when emboli from any origin pass into the ophthalmic circulation. Large emboli are seen in the retina at both the proximal and distal branchings of the retinal arteries. Small fragmented emboli lodge in the small distal end arteries, particularly in the paramacular arterioles (Fig. 8.7). These minute emboli may require a stereobiomicroscopic evaluation with a contact lens or a 90 diopter lens to determine their intravascular location. Flowing blood may render an embolus on the optic disc invisible. Raising the intraocular pressure (IOP) by pressing on the globe can alter the blood column, revealing the embolus in this situation [6]. Matter emboli frequently appear larger than the diameter of the involved artery. This is contrasted with air emboli (after chest injury or cardiopulmonary or intracranial surgery), which cause areas of silver-appearing segments between the blood [7]. The true diameter of the artery is underestimated because the width of the vessel is estimated from the apparent size of the intraluminal blood column. If sufficient blood can pass the embolus, an infarction of the retina does not occur. Emboli may cause transient retinal or choroidal ischemia, resulting in amaurosis fugax. Large yellow refractile emboli that lodge at branch points, termed Hollenhorst plaques, are frequently found in patients without a history of a visual disturbance [8, 9]. Sclerosis of the arterial wall around an older embolus is often evident. If the embolus breaks up or migrates, this periarterial sheathing may be the only sign of a prior embolus at that site. Collateral arteries which bypass the site of embolic occlusion rarely develop in the retina [10, 11].

8.2.3.1 Endogenous Emboli

Retinal emboli are generally one of three predominant types of emboli (see Table 8.1). Most emboli lodge in the temporal retinal arteries. Classic teaching suggests that the ophthalmoscopic appearance of the embolus can differentiate between these emboli. Single or multiple, refractile yellow or copper-colored emboli are considered to be cholesterol emboli, which commonly arise from carotid artery atherosclerosis. The blood column in the affected vessel is typically seen both proximal and distal to the cholesterol embolus. The atheromatous nature of these emboli has been confirmed pathologically in only a few autopsied cases [12]. Platelet–fibrin emboli appear white or gray and elongated and tend to take the shape of the artery because they are soft. Platelet emboli fragments pass into several arteries and continually move through the vessels. These emboli are transient; they dissipate, and rarely permanently occlude an artery or cause a retinal infarct. Episodes of transient amaurosis fugax have been documented to be caused by gray-white platelet–fibrin emboli that literally passed through the retinal circulation within minutes [13]. A calcific embolus is hard and irregularly shaped and distorts the involved artery. They are often single, white, and nonrefractile or glistening. These emboli generally arise from the aortic or mitral valves. In practice, the ophthalmoscopic distinction among these emboli is not always clear [10]. For example, white material has been described in the retinal arteries of one patient with an atrial myxoma [14]. Whether an embolus originates from the carotid artery, a cardiac valve or a cardiac chamber thrombus frequently cannot be determined by funduscopy.

Cholesterol retinal emboli can also be seen in the uncommon entity of diffuse disseminated atheroembolism. In this disorder multiple athero-

Table 8.1. Differential diagnosis of ocular and cerebral emboli

Cholesterol[a]
Calcific[a]
Platelet fibrin[a]
Iatrogenic and exogenous materials
Septic
Myxoma, other neoplasias
Marantic
Fat following long bone trauma or pancreatitis
Circulating thrombi with coagulopathy

[a] Most common emboli.

Table 8.2. Causes of cardiac emboli in all patients (modified from [30, 32])

Rheumatic vavular disease
Calcified mitral or aortic valve
Prosthetic valve
Degenerating cloth-covered valve
Left ventricular thrombus following myocardial infarct
Ventricular thrombus from coagulation disorders
Paradoxial embolism (atrial septal defect, patent foramen ovale, ventricular septal defect, pulmonary arteriovenous shunt)
Cardiomyopathy
Infectious endocarditis
Myxomatous degeneration of mitral valve prolapse
Atrial myxoma
Cardiac surgery
Marantic endocarditis
Ventricular aneurysm

matous emboli occlude the small and medium-sized systemic arteries [15, 16]. An elevated sedimentation rate and the ischemic dysfunction of multiple organs may suggest other etiologies such as an infectious endocarditis, an aseptic vasculitis, or a hypercoaguable state. Numerous neuro-ophthalmic symptoms such as homonymous hemianopia and transient monocular visual disturbances occur because of the multifocal central nervous system ischemic disturbance [17].

Cardiac emboli may have any of the previously discussed appearances. They may originate as a result of ventricular, atrial, or valvular diseases (Table 8.2) [18–25]. Transient microembolization in the retinal arteries is frequent in patients undergo-

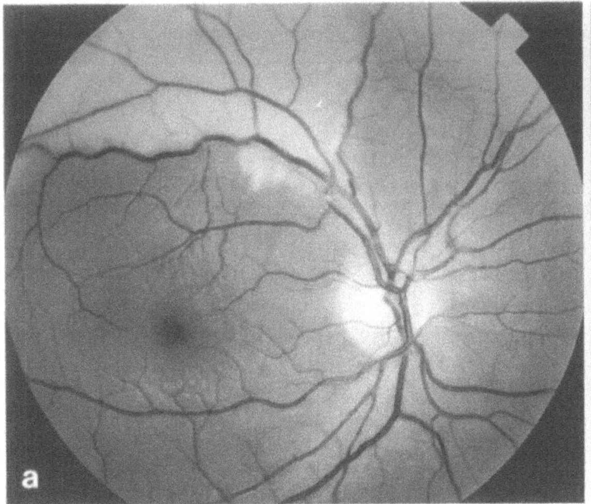
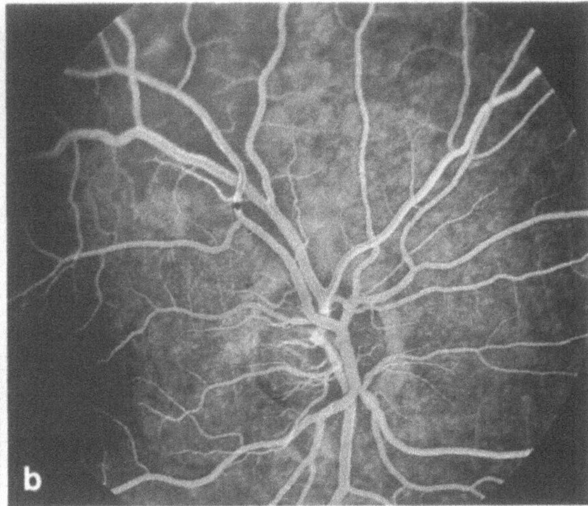

Fig. 8.8. a Following cyanoacrylate intra-arterial embolization of a brain arteriovenous malformation, this patient noted the onset of blurred vision in the right eye. He had a monocular incomplete inferior altitudinal field defect. The cyanoacrylate embolus is noted at the first branching of the superotemporal retinal artery. The inner layer of the retina distal to the embolus shows the obvious ischemic changes. **b** Fluorescein angiography, performed several days after the embolus, reveals that the arterial segment is patent distal to the embolus. The patient made a partial recovery

ing cardiopulmonary bypass or cardiac surgery [26–29].

Emboli such as circulating thrombi, cholesterol, and fat that originate from sources other than the carotid artery or heart are infrequent (see Table 8.1) [33–35].

8.2.3.2 Exogenous or Iatrogenic Emboli

Exogenous emboli can result from either the patient's own actions or as an iatrogenic complication (Fig. 8.8). Intravenous drug abusers often inject, along with illicit drugs, talc and cornstarch particles that can pass through pulmonary shunts into the arterial circulation and lodge in paramacular arterioles, infarcting the retina [36–40]. In rare patients, injection of a corticosteroid suspension into lesions in the orbit, eyelids, nasal mucosa, oral cavity, or the skin of the face or scalp reverses the blood flow in the external carotid artery (ECA) branch collaterals to the ophthalmic artery (see Sect. 1.5.1.1). The suspension particles then pass into the ophthalmic circulation and occlude a retinal branch or central retinal or choroidal artery [41–44]. Local injection of fat into the glabellar area has also been reported to cause visual loss [45]. Iatrogenic emboli rarely occlude the arteries supplying the optic nerve [46]. These types of exogenous emboli can cause either permanent or transient visual loss. In contrast, the visual loss from retinal infarction caused by the intentional intra-arterial injection of an embolization agent, such as cyanoacrylate or polyvinyl alcohol particles, is almost always permanent [47] (see Fig. 3.9 in Chap. 3).

Emboli in the choroid are not seen but the acute choroidal infarction can be detected on ophthalmoscopy or fluorescein angiography [10, 48].

In addition to the iatrogenic injection of materials that cause emboli, transient amaurosis fugax, branch or CRA occlusion, and rarely an ischemic optic neuropathy (see Sect. 5.1.5) can occur as a complication of cardiac catheterization or cerebral arteriography [48–53]. Though vasospasm in the ICA or ophthalmic artery has been postulated as a mechanism in some cases of cerebral angiography complications, catheter- related thromboembolic mechanisms account for most episodes. Transient and permanent amaurosis also result from emboli that enter the circulation during or immediately following carotid endarterectomy (CEA).

8.2.3.3 Transient Visual Disturbances

Patients with transient visual disturbances may note significant blurring or a frank scotoma in a focal area or over the entire visual field. A scotoma can be altitudinal, paracentral or central. It can appear gray, purple, or black. Patients frequently state that a "curtain came down or up" over the top or bottom half of the field. The visual dysfunction does not occur in a quadrantic or hemianopic distribution (as occurs with ischemia of the visual radiation or occipital cortex). However, patients are often unable to determine whether the visual disturbance was in one eye or in the homolateral field because they are anxious and fail to check themselves monocularly. Bilateral visual disturbances can occur in patients with vertebrobasilar insufficiency (see Sect. 8.5.1.1.4). Metamorphopsia, which is usually a sign of macula disease, is not a routine complaint. Momentary mild blurring or symptoms of floaters are not suggestive of embolic events. Minor pain or discomfort in the eye or orbit and positive visual phenomena have been described as accompanying embolic episodes [54], but these symptoms are infrequently volunteered by the patient.

Most, but not all cases of transient monocular visual loss result from emboli in the ophthalmic arterial circulation originating from either an ICA or a cardiac source [55].

When amaurosis fugax or a transient ischemic attack (TIA) develops in patients who have an occluded ipsilateral ICA, the emboli may originate from the common carotid artery or the origin of the ECA. The emboli pass into the ICA circulation via existing collaterals (see Sect. 1.5.3.1.2) such as the middle meningeal artery to the inferior lateral trunk, the anterior deep temporal artery to lacrimal artery, or the middle meningeal artery to ethmoidal or other ophthalmic artery branches [56–58]. Another ECA source for ocular emboli can also occur in the rare variant where the ophthalmic artery originates from the middle meningeal artery [59, 60].

Table 8.3. Causes of transient monocular visual impairment

Emboli
Migraine
Giant cell arteritis
Papilledema
Carotid stenosis or occlusion (postural or light-induced)
Optic nerve drusen
Angle-closure glaucoma
Orbital neoplasms
Mucocele
Papillophlebitis
Retinal detachment (positional)
Congenital optic disc anomalies
Recurrent hyphema
Uthoff's phenomenon optic neuritis
Gaze-evoked with orbital neoplasm

8.2.3.3.1 Differential Diagnosis of Spontaneous Amaurosis Fugax (Transient Monocular Visual Loss)

There are numerous nonembolic causes of transient monocular visual disturbances (see Table 8.3) [61]. For example, amaurosis fugax in patients younger than 45 years is more likely to be secondary to migraine vasospasm (41% of cases of amaurosis in patients younger than 45 years; see Sect. 10.4.2.1), whereas in patients older than 50 years carotid atherosclerotic and cardiogenic emboli are more common [62]. Ocular hypoperfusion associated with ICA stenosis is a less common etiology of transient amaurosis. Patients with giant cell arteritis can have amaurosis fugax or transient visual obscurations from orthostasis or vasospasm [62a] prior to occlusion of the posterior ciliary artery or CRA. The transient visual disturbance of embolic disease or migraine typically have a duration of between minutes and hours. In contrast, the visual obscurations of such entities as papilledema (either monocular or binocular), optic nerve drusen (monocular), or orthostatic hypotension (binocular) are briefer, lasting only seconds (for further discussion of transient visual obscurations see Sect. 5.3.2).

Unilateral transient visual loss from other nonvascular entities is uncommon. A sudden alteration of sphenoid or maxillary sinus pressure can com-

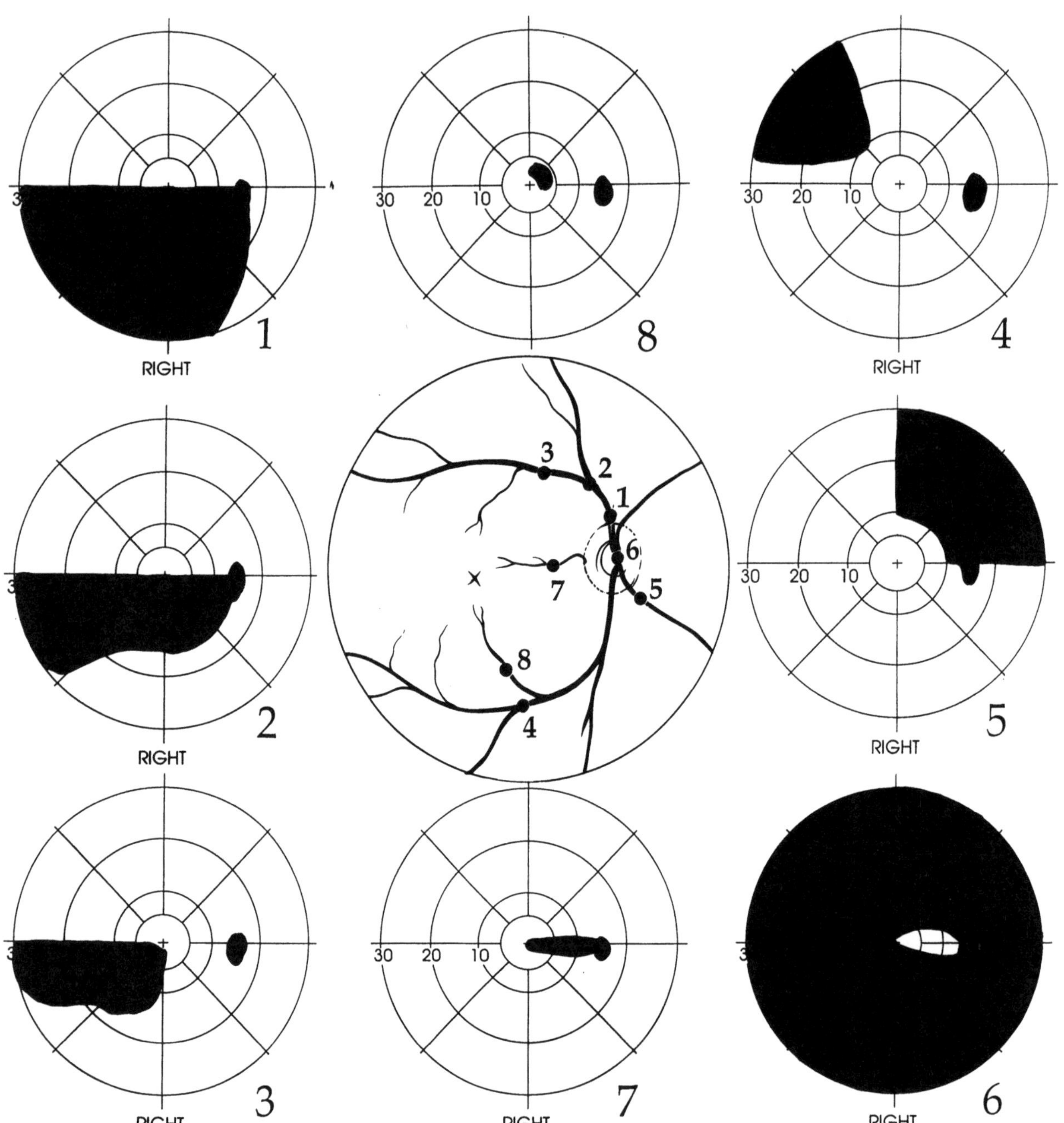

Fig. 8.9. The visual field defect can usually be correlated with the site of retinal arterial occlusion

press the optic nerve in cases where there is no bone separating the sinus mucosa from the optic nerve. Loss of the superior field when the patient is reclining can occur when there is detachment of the inferior retina. The fluid that accumulates beneath the detached portion of the retina shifts downward with gravity as the head is elevated. The affected retina then flattens, resulting in normalization of the visual field. In the presence of an orbital neoplasm, eye movement will occasionally precipitate a transient monocular loss of vision (see Sect. 4.7.1).

A recurrent hyphema related to an intraocular lens positioned in the anterior or posterior chamber after cataract extraction can cause episodic transient fogging of the vision lasting approximately 15 min. This visual disturbance, which may be described as "falling water" over the entire visual field, is accompanied by an ipsilateral headache [63–66]. Unless the affected patient is examined using biomicroscopy within several hours of the onset of the episode, the transient hemorrhage and elevation in the IOP, which are the cause of the visual disturbance and pain, will not be seen.

An extremely rare case of amaurosis fugax has been described with occlusion of a congenital vascular loop that arose from the optic disc [67]. Another extremely rarely reported form of monocular visual loss lasting for hours is a result of transient opacification of the lens in the eyes of diabetic patients [68, 69].

8.2.3.4 Permanent Monocular Visual Loss

Permanent visual loss develops when the ischemia to the segment of the choroid or retina is of sufficient duration and severity to cause an infarct, leaving no overlapping receptor fields in the affected region. Occlusion of the CRA or a branch retinal artery can occur primarily or as a result of an embolus, even if no embolus is seen. Histopathological studies have shown both thrombi and emboli or thrombi alone in the retinal arteries of patients with a CRA occlusion [70–72]. Though retinal and choroidal infarcts are commonly caused by emboli, an optic neuropathy is rarely caused by embolic disease or complete occlusion of the ipsilateral ICA (see Sects 5.1.5).

The nature of the visual field loss can be related to the location of the vascular occlusion (Fig. 8.9).

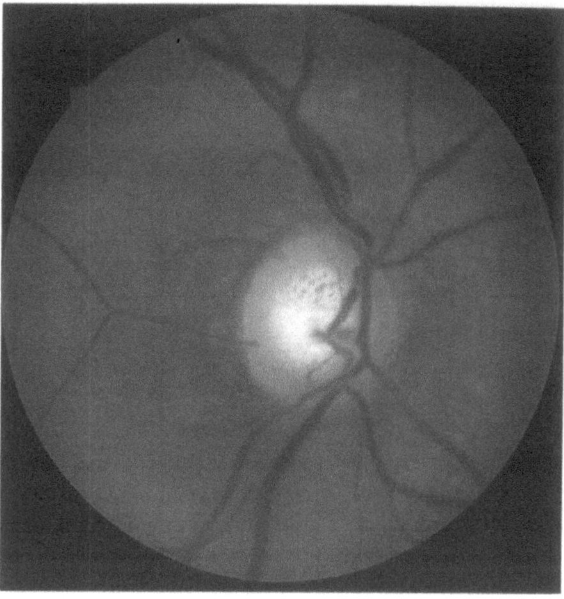

Fig. 8.10. Embolic occlusion of the cilioretinal artery resulted in a bar-shaped horizontal scotoma that involved fixation. The acuity was reduced to 20/50. The temporal border of the disc is pale from retrograde degeneration of the papillomacular bundle axons. There is narrowing of the cilioretinal artery which had been occluded

Occlusion of a nasal or temporal branch retinal artery causes a field defect limited to one quadrant of the field that does not extend beyond the horizontal raphe. If the occlusion of the retinal artery is proximal to the bifurcation into nasal and temporal divisions, the field defect can extend into both the nasal and temporal portions of the field. The acute inner layer swelling disappears several days to weeks after a retinal artery occlusion, but the arterial narrowing frequently remains. The retrograde degeneration of the affected axons causes the corresponding area of the optic disc to appear pale. For example, within several weeks following occlusion of a superotemporal branch artery the superotemporal segment of the disc becomes pale.

Occlusion of a small artery that arises either from the CRA deep within the disc or from the choroidal circulation and that crosses the temporal disc border causes an infarct in the papillomacula area of the retina (Fig. 8.10). The corresponding visual field defect can be a central or centrocecal scotoma or a bar-shaped scotoma that respects the horizontal raphe. On funduscopy, the swelling of this focal area of the retina may be subtle and is often not apparent after several days. Eventually, the retrograde degeneration of the axons to the optic disc causes a focal area of temporal disc pallor.

In some cases, ophthalmoscopy may not distinguish the damage caused by an old retinal artery occlusion from a previous ischemic optic neuropathy [73].

8.2.3.4.1 Central Retinal Artery Occlusion

The peak in the age distribution of patients with a CRA occlusion is in the seventh decade, and only 4% of cases are less than 40 years old [74]. The ophthalmoscopic appearance of an acute CRA occlusion has been well known since von Graefe's description in 1859 [75]. The classic features include pallor of the optic disc, attenuation of the arteries and veins on the disc surface and of the retina, diffuse ischemic clouding of the inner layers of the retina, and a macula that remains reddish brown. The macula appears cherry red—brown because there are few ganglion cells in the fovea to swell and obscure the underlying pigment epithelium coloring of the retina (Fig. 8.2). However, not all patients have this classic appearance to the fundus. For example, some have a relatively normal appearing optic disc, mulifocal ischemic retinal changes, and narrowing of several but not all of the arteries.

Almost half of the patients have unilateral visual loss as severe as only light perception or no light perception [74]. In patients where a visual field can be measured, the temporal field peripheral to 40° is often preserved because the choroidal ciliary circulation continues to provide blood to a portion of the peripheral nasal retina. Approximately 11% of patients have preservation of the central field and visual acuity because of a patent cilioretinal artery. This vessel, which arises either from the choroidal circulation or from the CRA proximal to the occlusion, continues to supply the papillomacula fiber bundle and the macula. The electroretinogram can provide a functional measure of the damage to the inner and middle layers of the retina. The b- wave of the electroretinogram of an eye with an acute CRA occlusion is typically reduced in amplitude and remains abnormal long after the retinal swelling has resolved [76].

Ophthalmodynamometry may be helpful in de-

termining the chances of visual recovery in patients with a CRA occlusion. If the CRA pulsates with exogenous pressure, visual recovery is often better than finger counting. This may signify an incomplete closure of the CRA. In contrast, in cases where pulsations cannot be induced, a poorer visual outcome is found. This may reflect the latter patients are more likely to have had a complete arrest of blood flow in the CRA [76, 77].

Fluorescein angiography demonstrates slowed or absent filling of the artery distal to an CRA acute occlusion (Fig. 8.11) [78]. If the CRA is blocked, the fluorescein dye will not fill the retinal arteries but the choroidal circulation will fill normally [79]. The optic disc vessels will also fill via the posterior ciliary arteries. Extravascular leakage of dye is generally seen only at the focal site of embolus-induced endothelial damage. After days to weeks, the occluded segment usually reopens and the retinal arteries will appear patent. However, the flow through the affected artery and corresponding capillary network often remains slowed as determined by the fluorescein angiogram. Although severe narrowing and sheathing or sclerosis of the arterial walls in longstanding cases cause the intraluminal blood column to appear absent on ophthalmoscopy, the fluorescein angiogram reveals that approximately 91% of these vessels are actually patent [74]. Fluorescein angiography may also reveal arterial occlusion in the choroid only if a large segment of the choroidal circulation is affected.

Eventually in the eye with a CRA occlusion, the nerve fiber layer disappears, the disc atrophies, and the arteries narrow. When there is concomitant occlusive disease in the ophthalmic artery, the retinal pigment epithelium shows signs of atrophy and hyperplasia (Fig. 8.12). Neovascularization of the iris, which can lead to neovascular glaucoma, develops in from 1% to 18.2% of cases. This is considerably less than the incidence of this complication following central retinal vein occlusion. Iris neovascularization can begin as early as 12 days after a CRA occlusion, particularly if fluorescein angiography shows persistent nonperfusion of areas of the retina [80–83]. Late neovascularization of the optic disc, which is usually associated with iris neovascularization, occurs in only 1.8% of patients [84]. The development of secondary ocular neovascularization

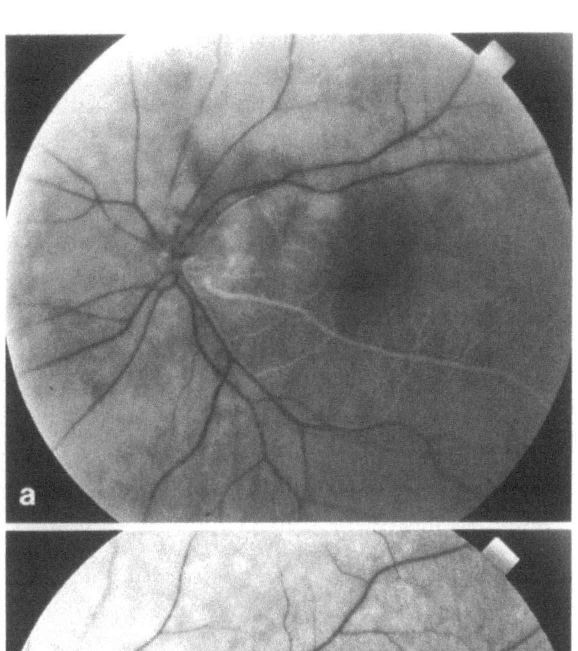

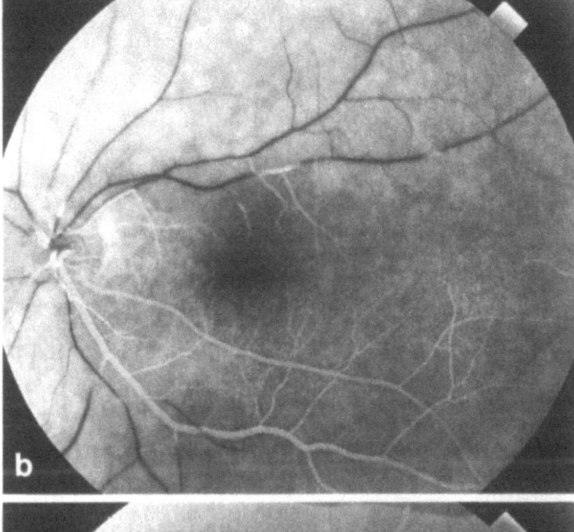

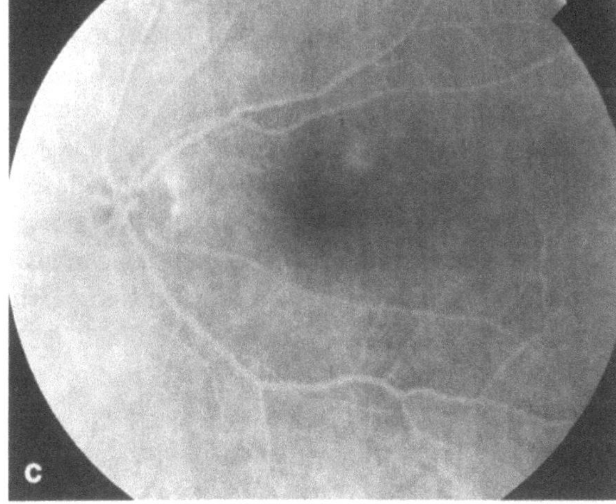

Fig. 8.11. a Fluorescein angiography in this eye with a central retinal artery occlusion demonstrates markedly delayed filling of the temporal arterial tree and absence of filling of the nasal arteries at 31 s. **b** By 48 s, the inferotemporal venous system is well filled and there are focal areas of filling in the superotemporal arteries. The nasal arteries also demonstrate some areas of filling. **c** Restoration of blood flow through the central retinal artery is clearly demonstrated in the fluorescein angiogram performed 1 week later, where the entire retinal arterial and venous circulation is clearly demonstrated within 46 s. Unfortunately, there was no improvement in the visual loss because of the irreversible ischemic damage to the retina. A common carotid artery subtraction angiogram (not shown) demonstrated complete occlusion of the left internal carotid artery

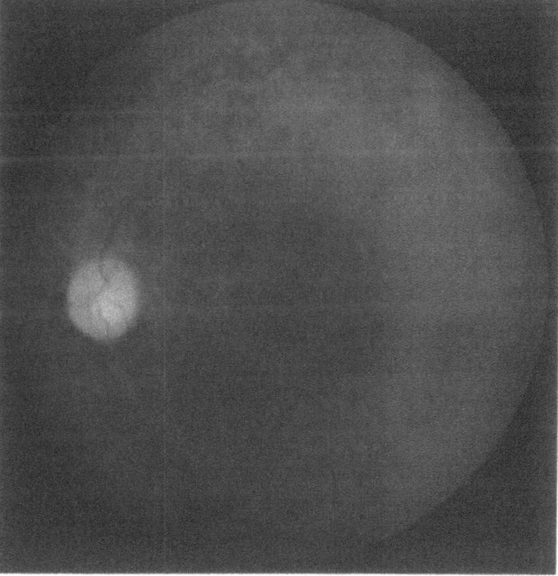

Fig. 8.12. Following occlusion of the central retinal artery and ophthalmic artery, there is diffuse pallor of the optic nerve, narrowing of all the retinal arteries, and a marked disturbance in the retinal pigment epithelium because of the diffuse choroidal ischemia

is not necessarily associated with stenotic disease of the ipsilateral ICA [83].

8.2.4 Relationship of Ocular Ischemic Events to Carotid and Cardiac Disease

8.2.4.1 Emboli

The association of episodes of transient monocular blindness or cerebral hemisphere TIAs with emboli from thrombotic material originating in the ICA was first noted by C. M. Fisher [85, 86]. The realization that transient visual or hemispheric symptoms are harbingers of future cerebral infarcts was reported as early as 1955 [87].

8.2.4.1.1 Carotid Atherosclerosis

Though the overwhelming majority of ocular emboli arise from atheromatous plaques in the cervical carotid arteries, the literature concerning the extent of the carotid atherosclerosis is confusing. Monocular visual loss is often the presenting symptom of patients with atherosclerotic disease of the carotid artery (typified by ulcerated plaques, stenosis, and thrombosis). The incidence and frequency of retinal emboli cannot be absolutely correlated with the presence of ulceration or the severity of stenosis in the ICA. Yet when all patients with a retinal artery occlusion are considered, the incidence of carotid artery atherosclerosis is as high as 75% [88]. In contrast, in most studies significant carotid stenosis occurs in less than 30% of all patients with amaurosis fugax or occlusion of a branch or CRA [89, 90]. Still other reports describe a higher rate of ICA stenosis in patients with amaurosis than in those with transient hemispheric episodes [91, 92]. Despite the presence of carotid atherosclerosis, most patients with transient monocular visual loss do not develop permanent ocular or cerebral infarcts ipsilateral to the amaurosis attacks.

Angiographic confirmation of significant carotid atherosclerosis is not found in the homolateral ICA of all patients with amaurosis fugax or hemispheric TIAs. For example, in a study of 59 patients presenting with amaurosis fugax, the ICA was stenotic in 25% and occluded in 17% of cases, but 33% of cases had no demonstrable carotid le-

sions [93]. In contrast, in another study, only 19% of 95 patients with carotid distribution TIAs or amaurosis or both had a normal cervical ICA on angiography [94]. Studies of patients with a CRA occlusion have described atherosclerosis of the ipsilateral ICA in more than 50% of casess [95].

High-grade carotid stenosis can alter the course of diabetic or hypertensive retinal vaso-occlusive disease. Narrowing of the ICA may actually protect the homolateral retinal arterioles from the effects of uncontrolled severe hypertension. In contrast, severe stenosis may worsen the retinopathy of diabetes (see Sect. 8.14.1.2). Patients who have marked intraocular asymmetry of retinal vascular disease should be investigated for possible stenosis or occlusion of the ICA.

8.2.4.1.2 Cerebral and Ocular Infarct

It is uncertain whether amaurosis fugax related to ICA or common carotid artery atherosclerosis carries the same risk as a cerebral TIA or a previous cerebral infarct of developing a subsequent cerebral infarct or permanent monocular visual loss. The risk factors seem to be similar for patients who stroke, with or without a presentation of ocular ischemic episodes. These factors include systemic hypertension, diabetes mellitus, generalized atherosclerosis, atherosclerotic heart disease, hyperlipidemia, aging, and, for women, being postmenopausal. Cerebral episodes do not occur concomitantly with attacks of monocular visual loss [94] though hemispheric and ocular events can occur in the same patient. Approximately 23% of patients with TIAs will have a cerebral infarct within 3 years of the initial event. Though most of the reported studies are uncontrolled with retrospective analyses, the risk of stroke seems to be less when the transient symptoms are limited to the eye (Table 8.4).

In most studies, the cumulative rate of stroke or permanent visual loss is under 20% in patients with amaurosis fugax, while permanent neurologic or visual complications occur in as many as 40% of patients with cerebral TIAs [96–99]. However, the incidence (approximately 30%) of myocardial infarction and death is similar in the two groups (Table 8.4). Most studies agree, no matter whether the ischemic episode is a transient hemispheric attack or amaurosis fugax, the risk of stroke is greatest

Table 8.4. Development of myocardial infarct, death, or stroke in amaurosis (AmF) and transient ischemic attack (TIA) patients [96, 97, 98]

No. of patients	Symptom	Stroke (%)	Myocardial infarct or death (%)	Duration of study (years)	Reference
93	AmF	14	30	7	96
212	TIA	27	30	7	96
80	AmF	17	24	4	97
34	AmF	12	NS	NS	98
NS	TIA	40	NS	5	99

NS, not stated.

during the first several weeks to 3 months. A reduction in the rate of stroke is seen by the end of the first year, and after 3 years from the last TIA the risk of a cerebral infarct decreases to approximately 3% per year. Cerebral infarcts seem to be more frequent in symptomatic patients who have a greater than 50% stenosis of the ICA [100].

The presence of presumed atheromatous cholesterol retinal emboli (Hollenhorst plaque) seems to suggest that the systemic atherosclerosis is widespread [9]. These emboli are seen in up to 11% patients with carotid atherosclerosis and in approximately 4% of patients with vertebral insufficiency. In one study, 63% of 208 patients had a cerebral TIA or an infarct prior to, concurrent with, or following the discovery of a cholesterol embolus. Yet only 62% of patients had an infarct in the cerebral hemisphere ipsilateral to the retinal embolus [101].

When asymptomatic patients with incidental Hollenhorst plaques are considered, the stroke rate in the homolateral hemisphere is 3.0% or less per year. This is comparable to the 2.4% annual stroke rate ipsilateral to symptomatic plaques in patients who have undergone CEA [102]. Also, patients with asymptomatic Hollenhorst plaques do not have an increased risk for either transient or permanent ocular ischemic events [103].

8.2.4.2 Retinal Arterial Occlusion and Risk of Cerebral Infarct

In contrast to the study of Pfaffenbach and Hollenhorst [101], most reports conclude that a minority of patients with an occlusion of the CRA or

a branch retinal artery will develop cerebral ischemia [104]. Neurologic symptoms occurred contralateral (ipsilateral hemisphere) to the retinal infarct in only six (16%) of 37 patients with a CRA occlusion during a mean follow-up period of 8 years [105]. Following a CRA occlusion, 18% of patients older than 40 years had a cerebral infarct in either the ipsilateral or contralateral cerebral hemisphere after 6.1 years [32].

In two other reports, a cerebral infarct occurred in only 35% of 153 patients with a retinal infarct, and the infarct not infrequently involved the cerebral hemisphere contralateral to the retinal artery occlusion [77, 106]. Additionally, patients without a visible retinal embolus had a survival rate similar to controls matched for age and sex. However, patients in whom an embolus was seen had a decrease in their rate of survival that was highly significant by the fourth year after the retinal vascular occlusion. Most deaths (approximately 38% for patients versus 15% for controls) resulted from cardiac disease [106] (see also below). Clearly, retinal arterial occlusive disease is associated with an increased incidence of cerebral infarcts, but the involved cerebral hemisphere may have no relationship to eye with the infarct. The presence of atherosclerosis in the ICA ipsilateral to the retinal lesion is not predictive of a future infarct in the homolateral hemisphere unless there is a greater than 70% stenosis.

8.2.4.3 Mortality

As alluded to previously, the increased death rate associated with atherosclerosis of the carotid arteries is related to myocardial infarction and the systemic manifestations of atherosclerosis rather than to stroke [107]. Systemic atherosclerosis or coronary artery disease or both are found in 59% of patients with retinal cholesterol emboli [108]. Following the discovery of a Hollenhorst plaque, the survival rate (compared to age-matched normals) is reduced by 13% in the first year, by 27% at the fifth year, and by 40% by the eighth year [101]. The risk of dying of cardiovascular disease is increased if the patient has elevated levels of serum low- density lipids, high-density lipids, or cholesterol [109].

When patients with retinal artery occlusions (not necessarily embolic) in all locations are considered, approximately 36% of them will die because of cardiac, cerebrovascular, or other vascular disease related complications within 11 years of the retinal infarct [74]. The survival rate seems to be poorer for patients with a CRA occlusion. In a study that followed 37 patients for 15 years following the CRA occlusion, 25 (67%) of the patients died within 4–15 years. The mean survival (5.5 years) for these 25 patients was considerably shorter than the mean life expectancy of 15.4 years for the control population [105]. In a comparison study between younger and older subjects, eight (24%) of 34 patients older than 40 years of age died during a 6.1-year follow-up period after CRA occlusion. Ten of the patients older than 40 years with a CRA occlusion had a probable cardiac valvular source for emboli. In contrast to the older age group, the ten patients less than 40 years of age did not have atherosclerosis. In this latter group, collagen vascular diseases, hemoglobinopathies, syphilis, atrial fibrillation, Marfan's syndrome, atrial myxoma, and hypercoagulopathy were the etiologies of the CRA occlusion [32].

8.2.5 Syndrome of Ischemic Oculopathy

In addition to embolic disease to the eye, visual disturbances can result from chronic ocular hypoperfusion that develops with severe stenosis, usually greater than 90%, of the homolateral ICA. An ischemic oculopathy (see Table 8.5 for signs) is rarely the only presenting problem and most of these patients have previous cerebral TIAs or infarcts. Widespread cerebral atherosclerosis is found in approximately 73% of patients with ischemic oculopathy [110]. If the patient has a competent circle of Willis or an ECA system that can supply blood to the ophthalmic artery and the supraclinoid segment of the more proximally occluded ICA, an ischemic oculopathy will not occur. Patients who only have a focal occlusion of the ICA such as following the iatrogenic occlusion of the ICA in the cavernous sinus do not develop ocular ischemia (see Sect. 6.2.5.3). Patients with an ischemic oculopathy syndrome are also more prone to develop subsequent cerebral infarcts than patients with embolic amaurosis fugax from carotid plaques. This increased risk of stroke is a result of diffuse severe cerebral atherosclerosis and a lack of

Table 8.5. Signs of ischemic oculopathy [112, 113]

1. Dilated conjunctival and episcleral vessels
2. Anterior chamber reaction with cells and protein flare
3. Thickened hypoxic cornea
4. Cataract
5. Ocular motor dysfunction from muscle ischemia
6. Retinal microaneurysms, retinal vein dilation, retinal hemorrhages
7. Cotton wool spots
8. Retina edema particularly in the macula
9. Swollen optic disc
10. Neovascularization of the retina
11. Ocular hypotony from ischemia of the ciliary body
12. Iris neovascularization causing absolute glaucoma
13. Mid-dilated, often fixed pupil
14. Permanent monocular severe visual loss
15. Transient amaurosis
16. Bright light-induced amaurosis

adequate ECA collaterals to the ICA system. Rarely, this syndrome is caused by obstructive disease in the ophthalmic artery [111].

Chronic hypoperfusion of the retinal arteries and capillaries causes ischemia and hypoxia of the retinal tissue. The retinal veins dilate, and become congested and tortuous [114]. A venous stasis retinopathy with dilated irregular veins, punctate hemorrhages, microaneurysms, and venous sludging, which is worse in the midperipheral retina, is seen in about 3.7% of all patients with occlusion [115] or a high-grade stenosis [116] of the ipsilateral ICA. The vascular changes are typically best seen in the midperiphery using indirect ophthalmoscopy. Macula edema and cotton wool spots are also commonly seen (Fig. 8.13) [117, 118]. Although an impending central retinal vein occlusion may have a similar fundus appearance, reduced ophthalmodynamometric pressure is diagnostic of the poor arterial inflow and probable ICA stenosis [113]. In addition, fluorescein angiography will often demonstrate areas of capillary nonperfusion in cases with arterial ischemic disease [119]. Retinal neovascularization leads to further retinal and vitreous hemorrhages. In addition to glaucoma, secondary iris neovascularization can rupture, producing a hyphema [120].

Infrequently, a high-grade stenosis or occlusion of the ICA is associated with symptoms of bright

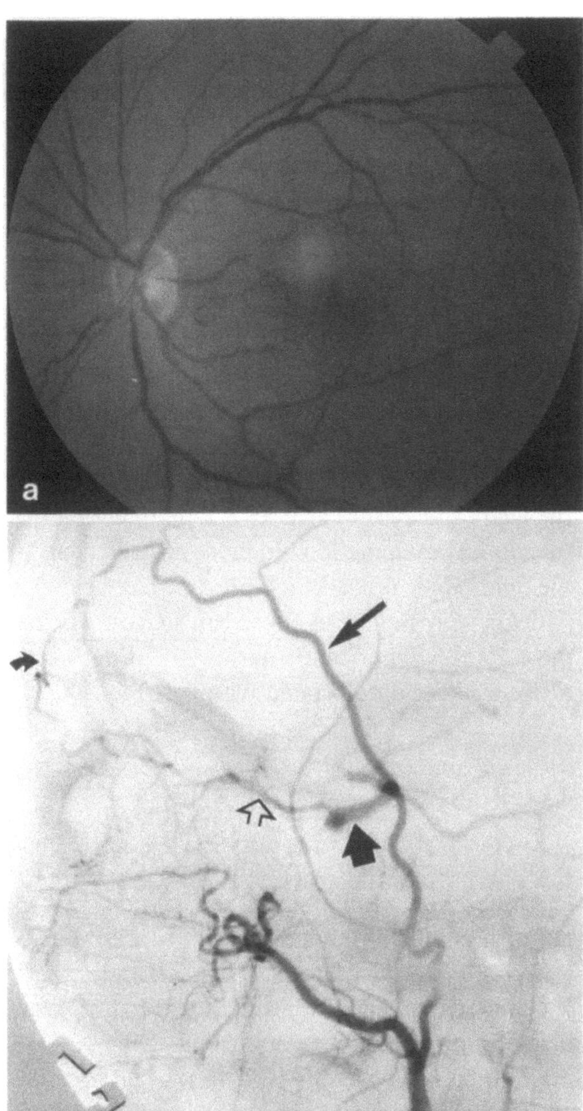

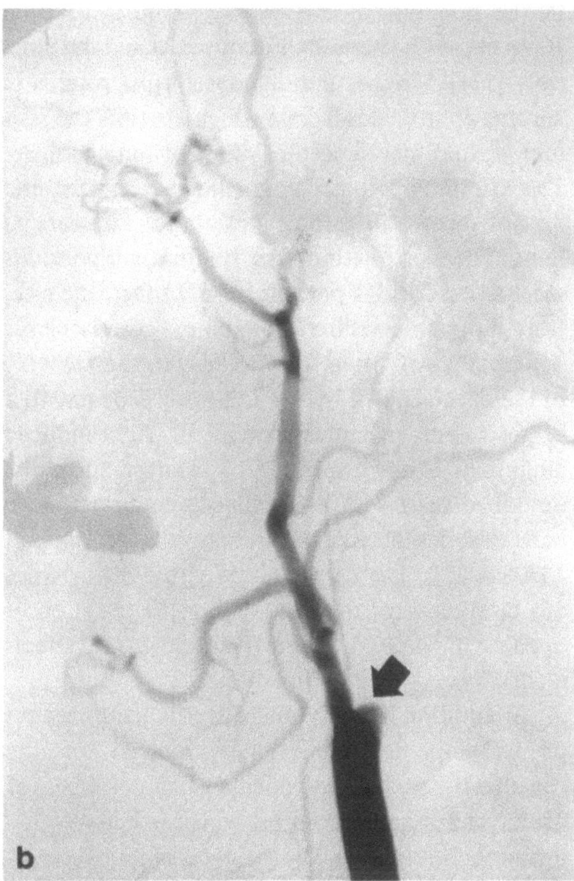

pressure of less than 10 mmHg in the left eye compared to 50 mmHg in the right eye. The clinical picture suggested an ischemic oculopathy syndrome.

b A lateral view left common carotid artery subtraction angiogram demonstrates a complete occlusion of the internal carotid artery (*arrow*). Angiography of the right internal carotid and vertebral arteries (not shown) showed no filling of the left internal carotid artery via either the anterior communicating or posterior communicating arteries.

c The lateral view left lateral external carotid artery injection demonstrates a superficial temporal artery (*broad arrow*) anastomosing (*small curved arrow*) with a palpebral branch of the ophthalmic artery which reconstituted the ophthalmic artery (*open arrowhead*). This ophthalmic artery then supplied the supraclinoid segment of the intracranial portion of the internal carotid artery (*long solid arrow*). Unfortunately, despite this anastomosis, the patient developed a central retinal artery occlusion in the left eye prior to the angiogram

Fig. 8.13. a A 70-year-old man complained of having a vague visual disturbance including flashes for 6 months. Bright light would cause dimming of vision in the left eye. At the time of his evaluation, he had 20/20 acuity in each eye with normal visual fields and color vision and no relative afferent defect. The anterior segment was normal but the intraocular pressure in the left eye was 10 mmHg compared to 15 mmHg in the right eye. Ophthalmoscopy of the left eye demonstrated a suggestion of macula edema, mild venous dilatation, and small hemorrhages throughout the posterior pole and the peripheral retina. Ophthalmodynamometry gave a systolic light-induced transient visual loss or severe blurring in the homolateral eye. Rarely, stenosis of both ICAs causes bilateral light-induced visual dysfunction [121]. The ocular hypoperfusion results in chronic choroidal ischemia which disrupts the normal retinal photoreceptor pigment metabolism. The affected photoreceptors remains bleached and do not recover function rapidly after exposure to light. The electroretinogram findings support this mechanism for this phenomenon. In fact, the electroretinogram measured from an eye with ischemic oculopathy will often show a delay in the recovery of the dark-adapted b-wave following exposure to a bright light stimulus, even if light-induced amaurosis is not present [122]. Almost all of the described cases with light-induced amaurosis have had episodes of recurrent amaurosis or cerebral TIAs as well. The light-induced visual disturbance has been reported to resolve following the surgical opening of the ICA lumen (see Sect. 8.6.8.2 Treatment).

In addition to the visual disturbances, because of ischemia to the orbital branches of the trigeminal nerve, these patients often complain of ocular or orbital discomfort which is better when reclining [112, 123]. The pain becomes severe in those cases that progress to absolute glaucoma because of secondary ocular neovascularization.

Other causes of ICA occlusion such as giant cell arteritis can also be associated with bright light-induced unilateral visual loss [124]. However, not every patient who complains of visual dysfunction in bright sunlight has ICA disease. In fact, an office procedure that employs a bright light to induce unilateral visual dysfunction (photostress test) is used to demonstrate the presence of functional macula disease, such as edema or serous fluid accumulation.

Transient unilateral amaurosis that develops while changing the body posture from a reclining to a sitting or standing position can indicate a hemodynamic rather than an embolic cause for the visual loss. The gravity-related orthostasis can be confirmed by a brachial blood pressure measurement. In these cases, the ophthalmic hypoperfusion is confirmed by a reduced ophthalmodynamometry measurement [125].

Arcus senilis, which is caused by lipid infiltration of the corneoscleral junction, normally occurs with aging, but it may be absent in the eye ipsilateral to severe carotid stenosis [126, 127]. Though pupillary abnormalities may arise from the ocular ischemia, a Horner's pupil and ptosis may also occur in a patient with complete thrombosis of the homolateral ICA [128].

8.3 Carotid and Cardiac Evaluation

In view of the association of ocular ischemic events with coronary artery and cerebrovascular disease, a systemic investigation should be performed in all these patients. Auscultation in the areas overlying the heart and carotid arteries is a pertinent part of the clinical examination. Indirect measurement of the ophthalmic artery blood pressure (see below) in patients with amaurosis fugax can suggest whether the carotid is significantly narrowed but not necessarily whether an ulcerated plaque is present.

8.3.1 Bruit

A carotid bruit results from the presence of turbulent blood flow within the vessel. Carotid bruits are not always present in patients with carotid atherosclerosis, even when they have had ischemic events. When a cervical bruit is heard and later disappears it may suggest complete occlusion of the ICA. A bruit auscultated over the ICA must be differentiated from radiation of a cardiac murmur or a bruit in another extracranial artery.

In patients with arterial occlusive disease (see Sects. 2.3.4.2.1 and 4.2 for discussion of bruits associated with arteriovenous shunting, Sect. 6.2.2 for bruit secondary to aneurysm), an orbital bruit is almost always associated with a bruit of the cervical carotid artery. Orbital bruits often arise from transmission of the cervical carotid bruits [129]. Less commonly, the orbital bruit develops because of increased blood flow through the ECA collaterals to supply the intracranial circulation distal to an occluded ICA [130]. Occasionally, an orbital bruit is a result of transmission from increased blood flow through the posterior communicating artery or the contralateral ICA to the occluded ICA [131]. The altered blood flow through a vertebral artery involved in a subclavian steal syndrome can also cause an orbital bruit [132]. An orbital bruit is often indicative of carotid siphon stenosis and most patients with this sign also have diffuse carotid and vertebral atherosclerosis [133–135].

A bruit can also result from ECA stenosis. The auscultated noise seems to follow the distribution of the ECA branches. In contrast to the bruit of common carotid artery or ICA disease, the intensity of the ECA bruit can be reduced by compressing the ipsilateral superficial temporal artery [136]. An increase in the blood flow through the ECA following complete occlusion of the ICA can also cause a cervical area bruit [137].

An asymptomatic cervical bruit is not a reliable indicator of future cerebrovascular ischemic events. Patients with asymptomatic cervical bruits are more likely to have a myocardial infarction than a TIA or a cerebral infarct [138]. Cerebral infarcts occur ipsilateral to an asymptomatic carotid bruit in approximately 0.25% of patients per year [139]. However, when significant carotid stenosis is found at presentation, the incidence of cerebral infarcts is greater. Additionally, most of these initially asymptomatic patients will have an episode of amaurosis fugax or a cerebral TIA prior to developing a stroke [140]. In general, the bruits associated with atherosclerosis are less useful for clinical localization than the bruits associated with high-flow arteriovenous shunts that drain into the dural venous sinuses.

8.3.2 Ophthalmodynamometry

Ophthalmodynamometry (ODM) is an office procedure that has similar utility to other indirect methods of testing for patency of the ICA [141]. The ODM plunger is pressed into the sclera while the arteries of the disc are viewed via an ophthalmoscope. The diastolic pressure of the ophthalmic artery is reached when the CRA or an artery on the disc pulsates, and this pressure is recorded on the gauge. As the pressure is gradually increased, the systolic pressure is reached when the artery stops pulsating. The ODM-determined systolic pressure should be equal to approximately 50%–67% of the systemic blood pressure and there should be no more than a 20% interocular difference (Table 8.6).

The association of a reduced ODM pressure with carotid atherosclerotic disease was first noted

Table 8.6. Criteria for abnormal ophthalmodynamometry (modified from [142, 143])

Asymmetry of >20% in systolic pressures
Asymmetry of ⩾50% in diastolic pressures
Absolute ophthalmic artery systolic pressure ⩽40 mm Hg
Absolute diastolic pressure ⩽10 mm Hg

in patients with complete ICA occlusion [144]. Abnormally low ODM pressure generally suggests at least a 90% stenosis of the ipsilateral ICA [145, 146]. As with all indirect investigations of the carotid artery, ODM can be normal if there are significant collaterals between the vessels of the ECA and the ICA that provide blood supply distal to the cervical ICA occlusive disease.

8.3.3 Ocular Pneumoplethysmography

Ocular pneumoplethysmography (OPG) is similar to ODM because both tests determine blood flow in the carotid artery–ophthalmic artery–CRA system in relation to an exogenously induced alteration of the intraocular pressure (IOP). The OPG measures pulsation pressure in the ophthalmic artery as the IOP is lowered after first elevating it above the systemic systolic blood pressure. A vacuum deformity of the globe, created by a suction cup placed on the lateral aspect of the sclera, is used to raise the IOP. The IOP is initially elevated above the systolic pressure in the ophthalmic artery, causing the arterial pulsations to cease. As the IOP is gradually lowered, pulsations begin and the pulse pressure and arrival time of the pulse wave are detected by a transducer and recorded. The OPG is abnormal if there is an asymmetry between the two eyes in the systolic pressures of 5 mmHg or more or if the ratio of the ophthalmic artery systolic pressure to the brachial artery systolic pressure falls below the normative data for the laboratory [147, 148].

8.3.4 Carotid Ultrasonography and Thermography

Ultrasonography of the cervical ICA can directly demonstrate segments of stenosis or ulceration

[149, 150]. A Doppler study indirectly suggests whether carotid stenosis or occlusion is present by demonstrating flow reversal in the homolateral supraorbital artery from blood flow via an ECA collateral [151–153]. Duplex scanning seems to have a high correlation with arteriography [154] but the results depend on many factors, including the performance and interpretation of the study [155]. Ultrasonography can also be used to document progression of ICA stenosis [156].

Thermography seems to have similar utility to ODM and OPG. As expected, thermography is abnormal when the carotid stenosis is greater than or equal to 75%, but plaques of the ICA are not revealed [157].

8.3.5 Angiography

Since intra-arterial angiography visualizes the entire length of the ICA, it will demonstrate whether there is more than one site of stenosis, such as in the cervical and siphon segments. Demonstration of the extent of vascular pathology is particularly important prior to surgical intervention. Though the carotid artery can be directly imaged using a digital intravenous angiogram, the poor resolution and technical difficulties have limited the utility of this test. Digital intra-arterial angiography uses a tiny catheter and a small volume of contrast material and reveals the extent of carotid disease with less risk than routine cerebral angiography [158]. Magnetic resonance (MR) angiography has been used to provide a three-dimensional image of the ICA without risk and will probably replace arteriography for investigating atherosclerosis of the ICA [159, 160, 160a] (Fig. 8.14).

8.3.6 Cardiac Evaluation

In addition to the evaluation of the carotid artery disease, patients with ocular or retinal emboli should undergo cardiac and systemic evaluation. This is essential because a cardiac source of cerebral emboli causes approximately 24% of the cerebral infarcts in young adults [30]. An electrocardiogram or Holtor monitoring is performed when an arrhythmia is suspected of causing embolization.

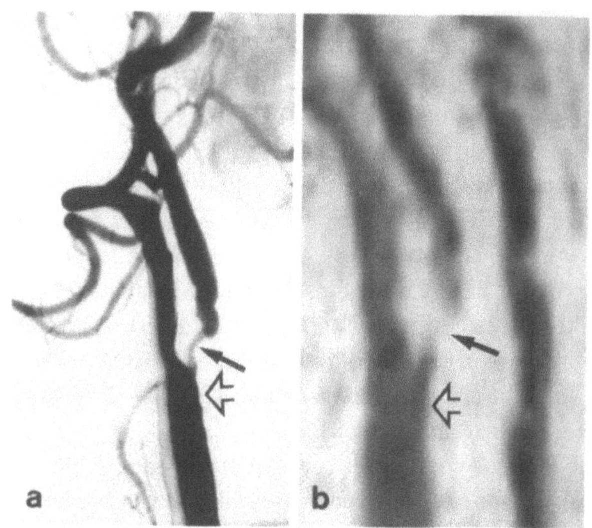

Fig. 8.14a,b. Both contrast (**a**) and magnetic resonance (**b**) angiograms in the same patient show similar alterations. There is severe carotid stenosis and some flattening of the carotid bulb (*open arrow*) proximal to the bifurcation with faint visualization of the region of stenosis (*closed arrow*). (From [160])

An echocardiogram will determine whether the emboli originate from a degenerated or calcified valvular lesion, an enlarged adynamic left atrium or ventricle, a patent foramen ovale, or an atrial myxoma [161, 162]. Although transesophogeal echocardiography may reveal a source of emboli not seen by other studies, the yield is still less than 3% in individuals without clinical evidence of heart disease [160b]. Additional tests such as a nuclear gated pool, cardiac MR imaging, stress test, and cardiac catheterization are performed for the evaluation of atherosclerotic heart disease when the cardiologist considers these tests to be required. Blood cultures are indicated when subacute bacterial endocarditis is suspected. Marantic emboli from collagen vascular disease or cancer are never the presenting problem in patients with these disorders, but emboli should be considered in patients with well-established disease.

It is important to distinguish a cardiac from a carotid origin of emboli, but this is not always possible. A cardiac source for emboli may be differentiated from a carotid arterial origin if CT or MR imaging of the brain or the neurologic signs and symptoms suggest ischemic events in multiple arterial circulations. However, this may not reliably dis-

Table 8.7. Nonatherosclerotic causes of cerebrovascular occlusions

---

Children
Migraine
Cardiac emboli
Sickle cell anemia (venous infarct)
Homocystinuria
Protein C or S deficiency
MELAS

Young adults (modified from [30])
Premature atherosclerosis
Arterial dissection (spontaneous and traumatic)
Migraine
Cardiac emboli
Coagulation disorders
Fibromuscular dysplasia
Cerebral aneurysm
Vasculitis
Syphilitic arteritis
Radiation vasculopathy
Sickle-cell disease

---

tinguish the source in cases where the posterior cerebral artery (PCA) has a direct ICA origin and there are carotid distribution and occipital episodes.

8.3.7 Differential Diagnosis of Cerebrovascular Occlusive Disease

Occlusive cerebral vascular disease can be caused by many disorders, but they are infrequent in comparison to the atherosclerosis and arteriosclerosis that typically occurs in the elderly. In contrast, most younger patients are affected by these other diseases (Table 8.7).

Though not strictly an occlusive disorder, amyloid angiopathy is a common cause of stroke-like symptoms. Recurrent intracerebral hemorrhages in the elderly can be caused by amyloid angiopathy. In this disorder, the walls of small and medium-sized cerebral arteries are prone to rupture because they are weakened by infiltration of β-amyloid protein [163, 164].

8.4 Aortic Arch Occlusive Disease

Although significant vascular lesions of the aortic arch are much less common than disease of the carotid and cerebral arteries and cardiac disorders, both visual and neurologic manifestations are numerous when there is severe occlusive disease of the arch. Atherosclerosis, syphilis, Takayasu's disease (see Sect. 9.10), and trauma can all cause a "pulseless" disease. It is so termed because of the absence of a radial or brachial pulse resulting from occlusion of the subclavian arteries. These patients often have bilateral visual complaints such as blurring or amaurosis, particularly in the face of orthostatic hypotension. Cerebral hemisphere dysfunction is recurrent and either transient or permanent. Blindness can develop from acute or chronic ocular ischemia [165, 166] or bilateral occipital infarcts.

8.5 Neuro-ophthalmologic Disorders Caused by Brain Infarcts

8.5.1 Sensory Visual Disturbances

8.5.1.1 Occipital Lobe

8.5.1.1.1 Visual Field Defects

When homonymous visual field defects are present without other neurologic deficits, the lesion is most likely located in the occipital lobe and the patient will often seek ophthalmologic assistance. Acute occlusion of the posterior cerebral artery (PCA) is sometimes associated with ipsilateral first-division trigeminal or ocular pain, which is referred, possibly from ischemia of the trigeminal nerve branches in the adventitia of the PCA or the adjacent tentorium [167]. Occipital infarcts from emboli originating in the ICA are rare and occur in patients where the origin of the PCA persists from the ICA [168, 169].

The field defects are typically homonymous and almost always congruous. Depending on the level of the arterial occlusion and the premorbid vascular anatomy, the defect can be a partial or complete quadranopia, a partial or complete hemianopia, or a homonymous scotoma [170–173] (see Sects.

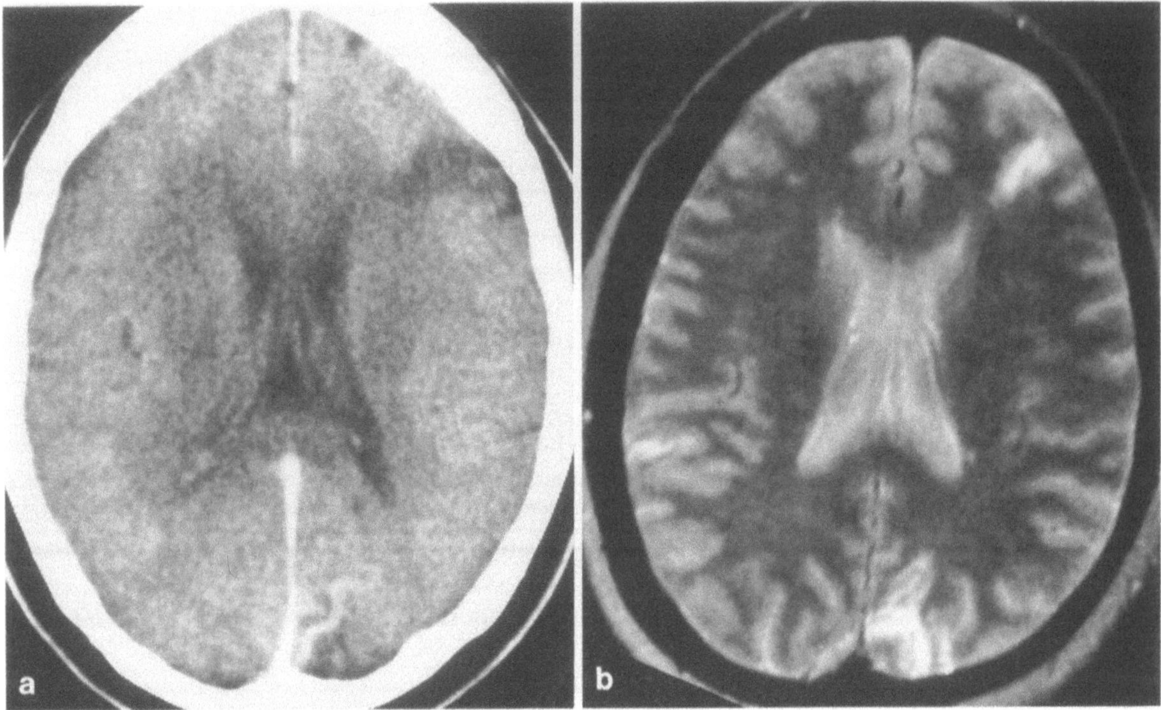

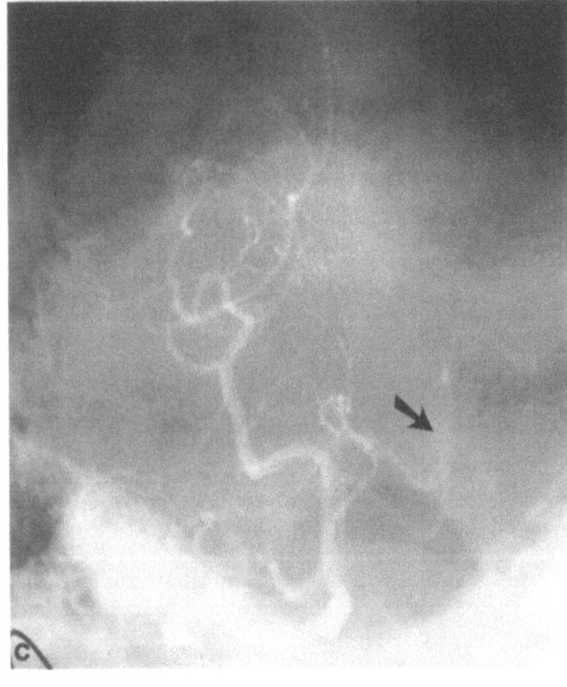

Fig. 8.15a–c. A 60-year-old hypertensive woman had a visual complaint that involved either the right eye or the right field. She was found to have 20/20 acuity but a dense complete right homonymous visual field defect. **a** The contrast-enhanced axial view computed tomogram demonstrates an area of lucency in the left occipital lobe with a ropelike enhancing structure without any mass effect. In retrospect, this was thought to suggest luxury perfusion in a gyrus adjacent to an infarct. There was also an additional area of lucency noted in the left frontal lobe. **b** The axial view T2-weighted magnetic resonance, performed 1 week later, demonstrates a bright signal in the corresponding frontal and occipital areas compatible with ischemic disease. **c** The frontal view right vertebral artery angiogram demonstrates occlusion (*arrow*) of the left posterior cerebral artery prior to the origin of the calcarine artery

1.5.2.7.1, 1.5.2.8). Thrombosis of the PCA or a distal branch can result from occlusive disease or emboli from occlusion of thrombosis of the basilar or vertebral arteries [174, 175]. A complete hemianopia occurs when the PCA is occluded within 2 cm of its origin and there is no middle cerebral artery collateral (Fig. 8.15). Occlusion of the calcarine artery or one or more or its branches can result in a homonymous paracentral or sector scotoma or a pie-shaped or quadrantic field defect [171, 176, 177]

The field defect is incongruous when the peripheral temporal field crescent is preserved. This occurs in cases where the infarct spares the anterior margin of the calcarine cortex [178]. It follows that an infarct of this same location will impair the peripheral temporal field [179].

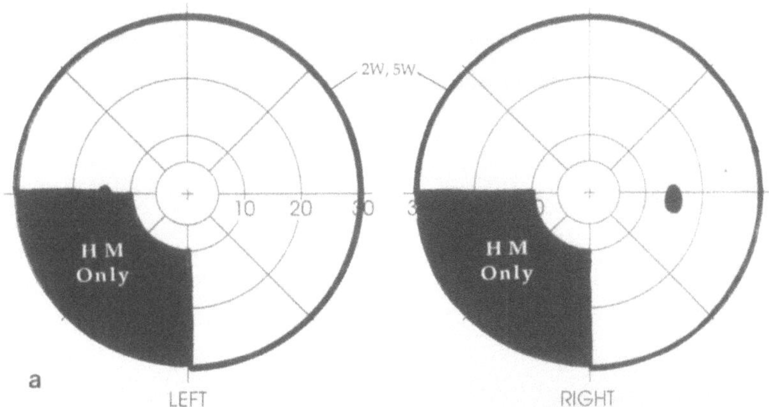

Fig. 8.16. a A 59-year-old hypertensive diabetic man complained of blurred vision in the left lower field. He was found to have a left inferior quadrant visual field loss to all isopters peripheral to the central 10° in this quadrant. **b** The T2-weighted axial view magnetic resonance scan demonstrates an increased signal that involved the anterior occipital cortex and the medial visual radiation but spared the more posterior locale of the macula fibers

8.5.1.1.2 Macula Sparing

The macula area on the side of the field defect is reported to be spared with most occipital infarcts. The frequency of macula sparing is undoubtedly overestimated by the artifact of a fixation shift in measuring the visual fields. Clearly, if the infarct spares the macula area (frequently the medial and posterior occipital lobe), preservation of macula function is expected (Fig. 8.16). However, there are definitive cases with diffuse occipital lobe disease, even following occipital lobectomy, where macula sparing does occur [180, 181].

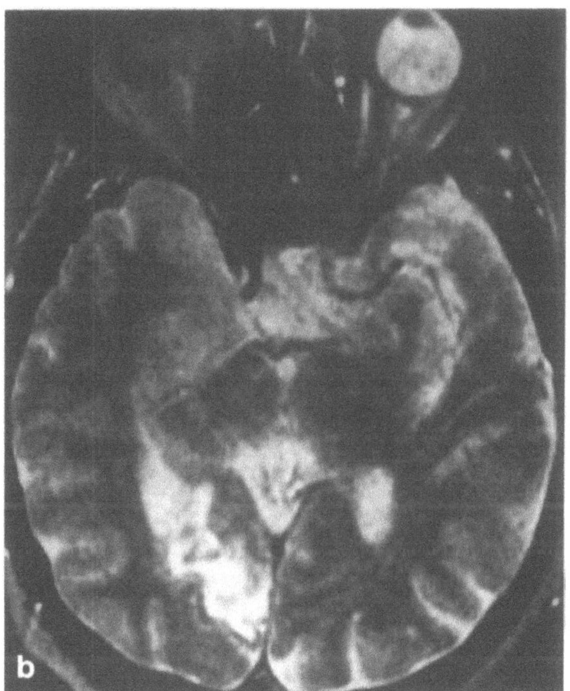

There is anatomical evidence that provides a rationale for why some patients have macula sparing. The vascular territory of the middle cerebral artery and PCA overlap in the occipital region (see Sects. 1.5.2.9, 1.5.2.10). Collateral blood flow to the cortical macula representation is provided via the middle cerebral artery [182] when there is an occlusion of the proximal homolateral PCA. Additionally, branches from the PCA often primarily supply or provide collaterals to the macula representation region.

It has also been theorized that some patients have bilateral representation of the macula in the occipital lobe [183]. Support for this hypotheses is found in horseradish peroxidase experiments in monkeys. In these studies, axons from some of the macula area ganglion cells, located within 3° of the fovea, cross to join the axonal projections from the opposite visual field [184].

8.5.1.1.3 Additional Occipital Lobe Dysfunction

A unilateral occipital lobe infarct may cause a hemianopia for color (hemidyschromatopsia) while the field to pattern vision or white test objects is preserved. In other cases, the field defect may also be present for stationary targets but absent for moving or flickering targets. An infarct of the anterior occipital lobe that includes the posterior corpus callosum in the dominant hemisphere can cause alexia without either agraphia or a hemianopia [185–188].

8.5.1.1.4 Bilateral Occipital Dysfunction

Approximately 25% of patients with a unilateral occipital infarct will experience an infarct in the contralateral occipital lobe within 3 years [189]. Bilateral lesions of the occipital lobes can cause diminished acuity, achromatopsia, bilateral homonymous field defects, or, in the extreme case, cortical blindness. Bilateral altitudinal field defects, though more characteristic of bilateral anterior ischemic optic neuropathy, can rarely occur with bilateral striate cortex infarcts [190–193]. Bilateral quadrantic visual field defects are infrequently seen. Bilateral infarcts can result from consecutive infarcts or a basilar artery occlusion that involves both PCAs [190, 194, 195]. Denial of the blindness (Anton's syndrome) is associated with extension of the infarcts into the parietal lobes [196]. Though most patients with Anton's syndrome have bilateral parietal occipital lesions, a denial of blindness has also been described in patients with a loss of vision from an anterior visual pathway disturbance who also had diffuse cortical dysfunction [197]. Bilateral lesions that extend into the parietal lobes can also cause the visual scene to appear fragmented beyond the areas affected by visual field loss (simultagnosia) [198].

On occasion, patients will present with the complaint of acute bilateral visual loss, because they do not recall or are unaware of a prior loss of one homonymous field from an earlier cerebral infarct. In most of these cases, CT or MR imaging will reveal the prior occipital damage. In general, patients with bilateral occipital TIAs have a poor prognosis. Patients with bilateral simultaneous blindness which lasts under 24 h have a much higher risk of having a cerebral infarct than patients with TIAs in other areas [199]. Bilateral transient dimness or gray-out of vision, generally lasting less than several minutes, may result from vertebral insufficiency [200, 201]. Flashes and other positive visual phenomena similar to migrainous visual disturbances

are rare [202], unless they occur in relation an evolving infarct of the occipital lobe with a field defect. Rarely, occlusion of the PCA will cause visual hallucinations (see Sects. 7.1.10.1.1, 7.1.10.1.2 for discussion).

Though occipital infarcts generally result from emboli or atherosclerosis in the vertebrobasilar arterial system, unilateral or bilateral posterior visual radiation and occipital cortex infarcts are also not uncommon in patients following coronary bypass surgery. The altered cortical microcirculation and hypotension that occurs during cardiopulmonary bypass may cause ischemia when poor collaterals and atherosclerosis have already subclinically compromised the circulation to the parieto-occipital lobes [203, 204]. In other cases, postoperative infarcts result from emboli caused by arrhythmia and intracardiac clot.

Following acute cortical blindness from occipital infarcts, recovery is frequent, but the normalization of visual function is usually incomplete. The ability to distinguish motion or the peripheral field tends to recover first [205, 206]. Residual dyschromatopsia and bilateral field defects are common. Actually, the color vision loss is often not noted until the patient recovers some vision [207–210]. The dyschromatopsia almost always persists, even when there is recovery of most of the visual field and acuity. This suggests persistent damage in the occipitotemporal lobe junction [211–213].

8.5.1.2 Parietal Lobe

A cerebral infarct of the dominant hemisphere angular gyrus in the parietal lobe causes the Gerstmann's syndrome, which includes finger agnosia, left–right disorientation, dyscalculia, and dysgraphia [214]. Dyslexia, which is usually accompanied by dysgraphia and receptive dysphasia, develops even when the visual fields are normal. A visual field defect is only present if the infarct extends into the visual radiations in the subcortical white matter [215]. The field defect is typically a homonymous quadranopia which is more congruous when the lesion is located more posteriorly. Rarely, a horizontal wedge-shaped defect develops that is similar to the field defect caused by a lesion of the lateral geniculate nucleus [216].

Infarcts of the nondominant parietal lobe that cause disorders of neglect in the space contralateral to the affected hemisphere, constructional apraxia, spatial agnosia, and neglect of the contralateral extremities are often associated with many visual disturbances. These patients may have little or no spontaneous gaze into the visual field contralateral to the lesion [217]. Similar visual dysfunction probably occurs with lesions of the dominant parietal lobe but the patients' communication disorder makes it difficult to reveal these subtler defects. Most of the cases with visual disturbances have lesions that extend into the ipsilateral occipital lobe [218].

The nature of the visual dysfunction also depends on whether the lesions are unilateral or bilateral. When patients have visual agnosias, which include an inability to recognize objects, pictures, symbols, colors, or faces (prosopagnosia), multiple lesions are typically present [219, 220]. The examiner must be careful to demonstrate that the visual disturbance is not merely secondary to a loss of primary vision in the homonymous field. The assessment must also consider that the patient may give inaccurate responses because of inattention, confusion, denial of a deficit, or suggestion by the examiner. Metamorphopsia, micropsia, macropsia, teleopsia (reduction of objects at a distance), fragmentation of linear components of an object, and tilted vision are all suggestive of bilateral disease. Other visual abnormalities include a loss of stereopsis, faulty recognition of object position in space (or projection into the wrong space – allesthesia), perseveration of objects and images in place and time (palinposia) (see Sects. 7.1.10.1.1, 7.1.10.1.2 for discussion), and visual inattention or extinction of the target in the field opposite the lesioned parietal lobe. Extinction can be demonstrated when stimuli are simultaneously presented in both fields. The stimulus in the field contralateral to a lesion located in the posterior parietal lobe rapidly fades out or disappears (visual images extinguish).

8.5.1.3 Anterior Visual Radiation

An infarct of the posterior internal capsule region will often damage the anterior visual radiation (possibly from obstruction of the anterior choroidal artery or a lateral lenticulostriate artery; see Sect. 8.5.1.4). Patients with this lesion manifest a partial or complete homonymous quadrantanopia or hemianopia in addition to a hemiparesis, hemisensory loss, and hemispatial neglect [221]. Ischemia of the medial posterior temporal lobe from occlusion of the posterior temporal artery can cause a quadranopia (Fig. 8.17).

Infarcts of the anterior portion of the visual radiation (Meyer's loop passes into the anterior temporal lobe) are rare. When a field defect occurs, it is usually a homonymous superior quandrantic incongruous field defect. Typically, the ipsilateral eye has a nasal field defect that is larger than the superotemporal defect in the contralateral eye.

8.5.1.4 Lateral Geniculate Nucleus and Anterior Choroidal Artery

The first recorded case of a visual field defect caused by an infarct in the territory of the anterior choroidal artery was a patient with syphilitic endarteritis of this vessel described by Abbie. The patient had a superior quadranopia caused by infarction of the lateral aspect of the lateral geniculate nucleus [222]. In other cases, occlusion of the anterior choroidal artery infarcts the lateral and spares the medial aspect of the lateral geniculate nucleus, resulting in an incongruous homonymous hemianopic field defect that spares the macula and the horizontal sector (see Fig. 1.28 in Chap. 1) [223]. The hilum or intermediate area of the lateral geniculate nucleus (which contains the macula projection) has the most extensive arterial input so it is less susceptible to ischemia.

Depending on the premorbid arterial supply and whether the closure is proximal or distal in the nutrient vessels, occlusion of the anterior or the lateral posterior choroidal artery can result in an infarct of the hilum of the lateral geniculate nucleus (see Sect. 1.5.2.8). A lesion of this region can cause a characteristic visual field defect, described as sectoropia (loss of horizontal sector) [224, 225]. Occlusion of the lateral posterior choroidal artery can also cause a homonymous superior quadranopia when it is the dominant artery to this region (Fig. 8.18) [226]. Following an infarct of the lateral geniculate nucleus, optic disc pallor, with bow-tie atrophy in the eye opposite to the infarction, and

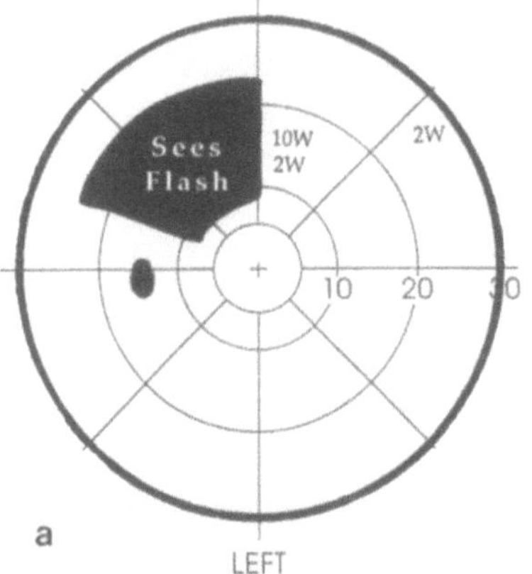

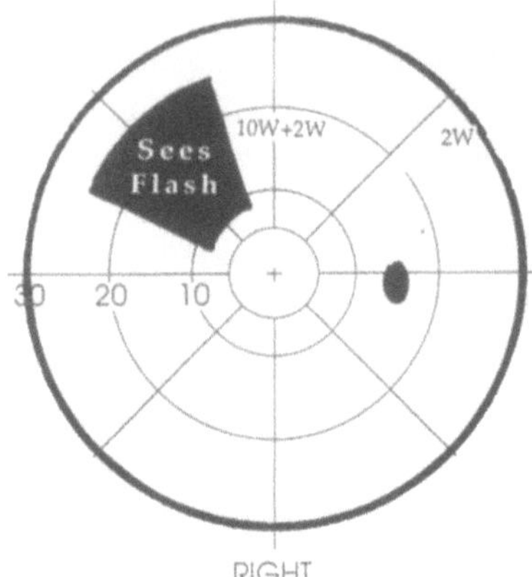

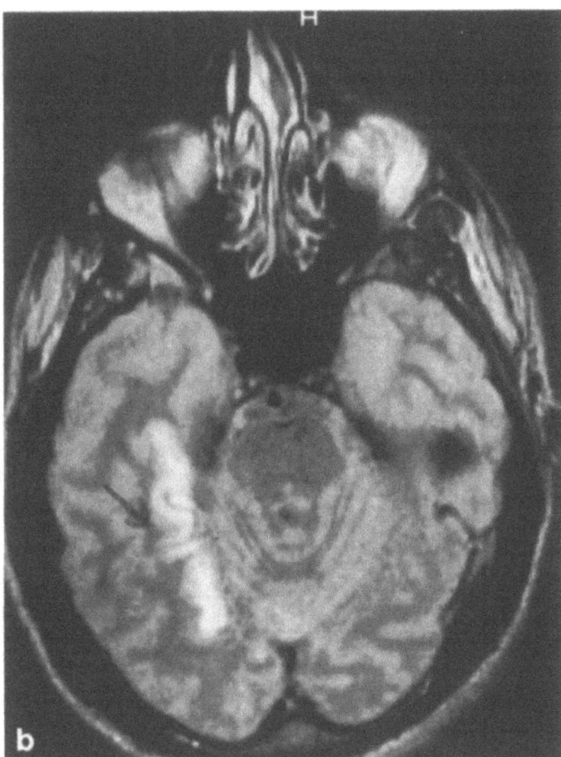

Fig. 8.17. a A 45-year-old hypertensive man had the sudden onset of blurred vision, dizziness, and numbness of the left arm and leg. The neurologic examination was normal except for an absolute left superior quadrant scotoma in the field peripheral to 8°. He subjectively had some spontaneous flashes in this scotoma. **b** The axial view T2-weighted magnetic resonance image reveals an infarct (presumably from occlusion of the posterior temporal artery) of the medial cortex and visual radiations in the right temporal lobe

Fig. 8.18a–f. The extent of a visual field defect resulting ▶ from occlusion of the posterior lateral choroidal artery depends on whether this vessel is the dominant blood supply to the lateral geniculate nucleus.

a The lateral view vertebral artery subtraction angiogram demonstrates an AVM supplied by the right posterior lateral choroidal artery (*arrow*).

b Superselective arteriography, in the frontal plane, demonstrates the microcatheter in the right posterior cerebral artery (*open arrow*) passing into the right posterior lateral choroidal artery (*arrows*) to visualize the AVM.

c Following embolization of the right posterior lateral choroidal artery, occlusion (*arrow*) of this vessel is demonstrated and the AVM is not visualized. The *open arrow* demonstrates patency of the more proximal posterior cerebral artery similar to the segment demonstrated by the *open arrow* in **b**.

d Intravascular cyanoacrylate and thrombus deposition is seen in the region of the right lateral geniculate nucleus possibly occluding small arterioles in the area. (Contd. p. 378)

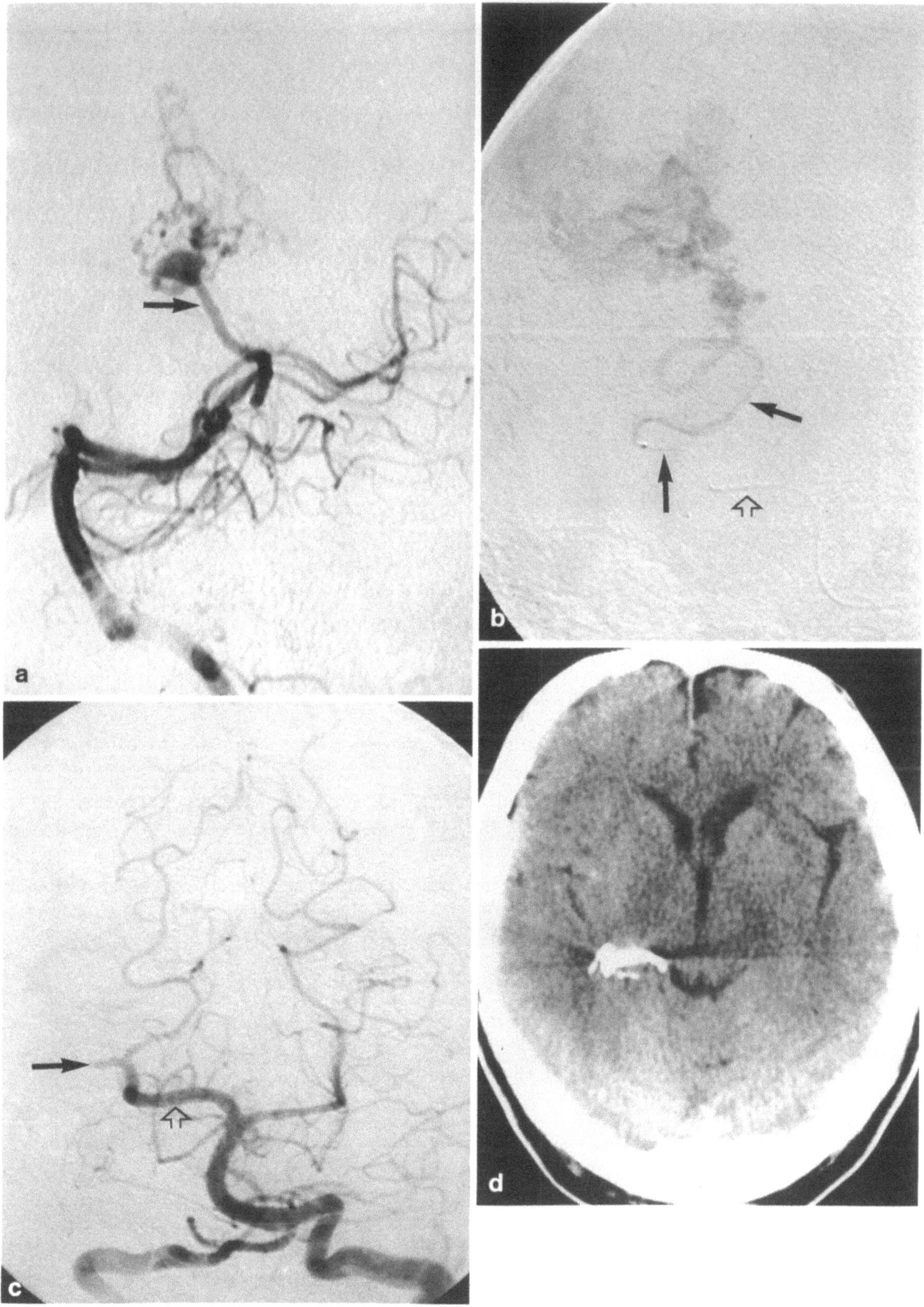

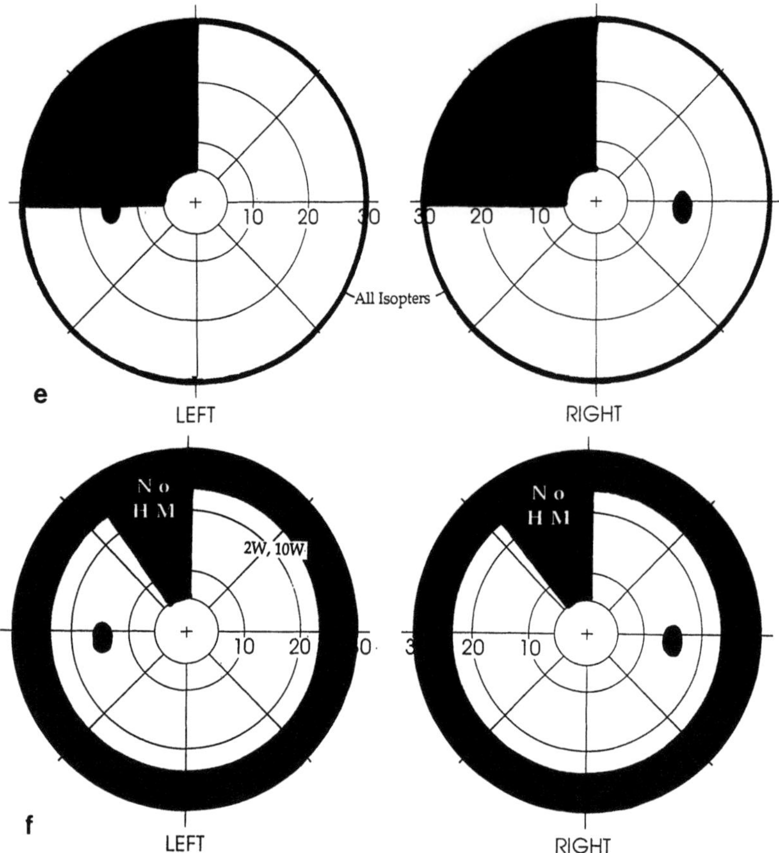

Fig. 3.18. e A left superior quadrantanopia which spared the central 5° was noted at the time.
f The visual field improved within 1 month, but he was left with an absolute left superior homonymous pie-shaped scotoma

retinal nerve fiber layer defects develop because of retrograde degeneration of the optic tract, which has been damaged as it enters the lateral geniculate nucleus.

Although ischemia in the territory supplied by the anterior choroidal artery has been hypothesized to cause hemisensory defects, hemipareses, and homonymous hemianopsia [227], pathological confirmation is often lacking. Most cases actually manifest only a few of these neurologic deficits [228]. In the few cases of occlusion of the anterior choroidal artery studied, lesions were most frequent in the posterior internal capsule and globus pallidus [229, 230]. CT has also demonstrated infarcts of the globus pallidus and the posterior limb of the internal capsule in patients with presumed occlusion of this artery [230]. However, ligation of the anterior choroidal artery (which was previously performed to suppress the tremor of Parkinson's disease) rarely caused hemianopia or a permanent hemiparesis. These latter cases suggest the existence of an additional functional arterial blood supply to this region from collateral circulation arising either from the posterior communicating artery [231, 232] or through arteries that originate from the P2 segment of the PCA (such as the posterior lateral choroidal artery; see Sect. 1.5.2.8). Whether an occlusion of the anterior choroidal artery causes significant impairment clearly depends on whether this vessel provides the dominant arterial blood supply to the area.

Rarely, occlusion of the anterior choroidal artery will result in an eye movement disorder typical of midbrain ischemia (see Sect. 1.5.3.3.3, 8.5.2.1.4).

8.5.2 Efferent Pathway Disturbances

8.5.2.1 Brainstem and Cerebellum

Infarcts and TIAs in the brainstem (see Fig. 1.42 in Chap. 1 for anatomic artery distribution) are commonly caused by atherosclerosis and thrombosis of the branches of the vertebral and basilar arteries. The symptoms and findings depend upon the location of the ischemia. Embolic disease to the vertebrobasilar system from extracranial lesions (except to the top of the basilar artery) is far less common than in the ICA territory. The symptoms of TIAs and vertebral insufficiency in an individual are often different for each attack. Vertigo, dysarthria, numbness and paresthesias on one or both sides of the face, weakness in an arm or leg, loss of coordination, diplopia, and blurring or gray-out of vision in both eyes are all common in varying combinations [233, 234]. Rarely, these symptoms result from a subclavian "steal" syndrome [235, 236]. Lacunar infarcts can cause discrete lesions with specific neurologic syndromes that are of localizing value (see below), while others such as a pure motor hemiparesis are less diagnostic [237, 238].

8.5.2.1.1 Medulla

Infarcts of the medulla can cause nystagmus as a result of the disruption of the integration of the vestibular and cerebellar systems (see Fig. 1.42d in Chap. 1 and Sect. 1.5.3.3.1). The symptoms and signs may be indistinguishable from the manifestations of a cerebellar infarct, particularly in the distribution of the distal posterior inferior cerebellar artery [239]. Infarcts of the medial medulla from anterior spinal artery occlusion cause major motor deficits but are associated with few ocular motor abnormalities except upbeat nystagmus [240].

An infarction of the lateral medulla, called Wallenberg's syndrome, causes characteristic findings (Table 8.8). Additional eyelid and eye movement defects are often present. Palpebral nystagmus may occur synchronously with the nystagmus [241]. The eyes tend to drift towards the side of the lesion (lateropulsion) and the saccades are hypermetric in the same direction [242]. Monocular downbeat nystagmus can appear in the homolateral eye when the gaze is directed towards the lesion. Rare cases develop an irregular multidirectional nystagmoid movement [243]. A skew devi-

Table 8.8. Clinical signs of Wallenberg's syndrome

Deficit ipsilateral face pain and temperature sensation
Ataxia ipsilateral extremity
Horner's syndrome, ipsilateral face
Dysarthria
Conjugate nystagmus with rapid phase away from lesion

ation, with the lower eye ipsilateral to the lesion, can also occur [244].

In addition to the typical symptoms of vertigo, horizontal oscillopsia, and visual blurring, the patients may have the curious complaint of tilting of the environment where the visual scene is tilted to the side by 90° or even upside down [245, 246a, 246b]. The tilting of the environment often abates within 24 h. The oscillopsia usually persists until the nystagmus remits, which is usually within weeks. Originally, Wallenberg's syndrome was considered pathognomonic for occlusion of the posterior inferior cerebellar artery, but further investigation showed this lesion to be more indicative of generalized ischemic disease in the vertebral artery [247].

8.5.2.1.2 Cerebellum

Infarcts and hemorrhages of the cerebellum are associated with a variety of ocular motor disorders. Since the symptoms and findings of ataxia, tremor, dysmetria, and vestibular dysfunction are also found with infarcts in the brainstem, accurate localization may be difficult. Nystagmus, saccadic overshoot, and various macrosaccadic oscillations are common, but a gaze palsy towards the lesion, skew deviation, and a sixth nerve pareses can also occur (probably because of secondary brainstem compression) [247a, b]. Recognizing signs of early brainstem compression such as a gaze paresis prior to the development of lethargy or corticospinal tract dysfunction is imperative. In this situation, a surgical decompression of the swollen cerebellum pressing on the brainstem can be life-saving. The nystagmus of an acute cerebellar infarct is typically conjugate, horizontal, and bidirectional but coarser or of larger amplitude towards the lesion [247c, d]. Conjugate upbeat nystagmus occurs when the anterior superior vermis region is affected [247e].

8.5.2.1.3 Pons

Most of the ocular motor deficits that arise from ischemia of the pons result from lesions in the paramedian territory (see Fig. 1.42c in Chap. 1 and Sect. 1.5.3.3.2). Paresis of the sixth nerve associated with a crossed hemiparesis localizes the lesion to the basis pontis (Millard–Gubler syndrome) because of damage to the nerve fascicle and the descending cortical spinal tract. In addition to the ipsilateral gaze paresis, a lesion of the sixth nerve nucleus concomitantly disrupts the genu of the ipsilateral seventh nerve, causing an ipsilateral facial paresis. An isolated abducens palsy that results from an infarct of the nerve fascicle in the pons has been considered rare [248], but MR imaging may reveal more brainstem lesions than previously diagnosed [249]. When an infarct includes the parapontine reticular formation (which sends projections to the sixth nerve nucleus, where interneurons project to the medial longitudinal fasciculus), a paresis of conjugate gaze to the side of the lesion that can be permanent results (in contrast to the gaze paresis away from a frontal lobe infarct which generally lasts only for hours). A partial lesion of the parapontine reticular formation may only cause slow conjugate saccades towards the side of the lesion without an obvious gaze paresis. It may be difficult to determine immediately the localization of the lesion in cases with a severe gaze paresis. Oculocephalic or caloric maneuvers can drive the eyes away from a frontal lobe lesion or towards the lesion in patients with a parapontine reticular formation defect (either both eyes or the adducting eye), but not with an abducens nucleus lesion [250].

Medial Longitudinal Fasciculus Lesions. An infarct can selectively injure the medial longitudinal fasciculus (MLF). This causes a paresis of the homolateral medial rectus and nystagmus of the yoke abducting eye on attempted gaze away from the lesion (an internuclear ophthalmoplegia) [251–253]. Internuclear ophthalmoplegia is probably the most common ocular motor defect in patients with pontine infarcts. Despite the ocular misalignment, diplopia is not a usual complaint in patients with an MLF lesion. The patients typically complain of blurred vision, and infrequently oscillopsia, when gazing in the direction of the paretic medial rectus. On occasion an obvious paresis of the medial rectus is not apparent, but the affected medial rectus does have a slow saccadic movement [254]. When the rostral MLF is damaged, the adducting paresis occurs but the abducting nystagmus is absent. An MLF lesion can be isolated or can occur in combination with other signs of pontine infarction such as motor weakness, ataxia,

small pupils, facial numbness, skew deviation, and nystagmus. A depressed level of consciousness and ocular bobbing are present when the damage to the pons is extensive. In the elderly an infarct is the most common etiology of a MLF lesion, while in younger patients multiple sclerosis and glioma are the most frequent causes. Bilateral MLF lesions are not uncommon in patients with cerebrovascular disease [255].

One-and-a-Half Syndrome. When the MLF and parapontine reticular formation on the same side are disrupted, a one-and-a-half syndrome results. The only horizontal eye movement preserved is abduction of the contralateral eye because the contralateral sixth nerve nucleus is spared [131, 256, 257]. Other signs of pontine level dysfunction are common in these patients. When there is extensive bilateral damage of the pontine paramedian reticular formation, patients with a one-and-a-half syndrome may also have a "locked-in" syndrome [258, 259].

While an infarct of the lateral caudal pons causes few ocular motor signs, a hemorrhage in this area often extends to cause a one-and-a-half syndrome [260]. An ipsilateral Horner's lid and pupil, a contralateral hemiparesis, and an ipsilateral limb ataxia can result from an infarct or hemorrhage in this locale.

Skew Deviation. Vertical diplopia results from a skew deviation (see Sect. 8.8.4 for comparison to fourth nerve palsy). Skew deviation occurs, with or without other neurologic deficits, due to lesions in the medulla, pons, or midbrain. The lesion is commonly ipsilateral to the lower eye if there is no associated MLF lesion [261, 262]. The hypertropia can be concomitant, vary with eye position, or even alternate. An infarct is the etiology in more than 50% of cases with a skew [263], though hemorrhage, multiple sclerosis, and neoplasms are not infrequent causes. A lesion of the labyrinth is an infrequent cause of a skew deviation which results because of an imbalance in the ascending otolith inputs in the MLF to the interstitial nucleus of Cajal [264].

Vertical Eye Movement Disorders. Disorders of voluntary vertical eye movements are not typically caused by pontine lesions. However, slowing of upward saccades may result from bilateral lesions of the parapontine reticular formation [265]. Ocular bobbing, the spontaneous downward conjugate or dysconjugate jerks of the eyes followed by a slow drift back to midposition, is predominantly found in comatose patients who have a large pontine infarct or hemorrhage [260, 266–269]. Ocular bobbing has also been observed in patients with a cerebellar or extra-axial posterior fossa hemorrhage, toxic encephalopathy, or a pontine tumor [270].

8.5.2.1.4 Midbrain

Ocular motor dysfunction is frequently associated with the alteration of consciousness, motor pareses, tremor, and ataxia when an infarct or a hemorrhage affects the midbrain. The ocular motor disturbance depends on the extent of the infarct in the midbrain and whether the lesion extends into the thalamus (see Fig. 1.42a, b, Sect. 1.5.3.3.3). The size of the infarct is related to the underlying anatomy of the arterial segment between the upper portion of the basilar artery and the ostium of the posterior communicating artery [271, 272]. Occlusion of a P1 segment of the PCA will not cause a paramedian territory infarct if the contralateral P1 segment is the origin of the paramedian vessels or the dominant artery. However, if the embolus or thrombus closes the dominant vessel, the findings can be bilateral. When occlusion of a paramedian artery occurs in a patient with an anatomical variant, such as hypoplasia or absence of the ipsilateral posterior communicating artery, the ischemic damage may involve additional regions not ordinarily supplied by the paramedian artery. In this situation, thrombus in the P1 segment or basilar artery can propagate further, occluding the segment of artery which does not receive posterior communicating artery blood [273]. Infarcts in the paramedian territory of the thalamus or midbrain are usually widespread, involving structures which normally receive blood from other vessels.

A third nerve paresis that is associated with a crossed hemiparesis localizes the lesion to the paramedian territory supplied by the unpaired mesencephalic artery, also termed the posterior thalamosubthalamic paramedian artery [274]. Ischemic damage to the descending corticospinal tract in the crus cerebri and the third nerve fascicles (which pass through the medial portion of the crus) accounts for the clinical syndrome (Weber's syndrome). A third nerve paresis with ataxia and tremor of the ipsilateral extremities (Nothnagel's syndrome) or the contralateral extremities (Benedikt's syndrome) develops with an infarct located in the intermediolateral territory that disrupts the brachium conjunctivum or red nucleus, respectively.

There are rare cases with an isolated partial paresis or a complete isolated third nerve paresis with pupillary sparing [275] caused by a discrete hemorrhage or infarct in the midbrain. These lesions disrupts the third nerve fascicles that project from the ventral portion of the oculomotor nuclei through the MLF and the red nucleus [276].

Infarcts of the dorsal aspect of the midbrain can cause complex extraocular movement disorders because of the proximity of both third nerve nuclei, the inputs from the ascending MLF and descending frontal eye fields pathways into the rostral interstitial nucleus of the MLF, the posterior commissure supranuclear center for conjugate upgaze, the prerubral area for conjugate down gaze, the interstitial nucleus of Cajal, and the nucleus of Darkschewitsch. In addition, there are numerous connections between these nuclei. For example, the interstitial nucleus of Cajal sends projections for elevation to the superior rectus and inferior oblique subnuclei of the third nerve nucleus and for depression to the inferior rectus subnucleus of the third nerve nucleus and the superior oblique nucleus. Not infrequently, patients with a midbrain infarct will have a concomitant nuclear third nerve palsy and a supranuclear gaze paresis [277].

Ischemic damage in the dorsal median territory [278] affects the ipsilateral third nerve subnuclei, causing a partial or complete nerve dysfunction. In these cases, upgaze paresis develops in the contralateral eye [279–281] because of disruption of the third nerve subnucleus innervation to the contralateral superior rectus [282]. Infarcts of the third nerve nuclei frequently cause ipsilateral pupillary dilation, bilateral paresis of the superior rectus, and bilateral ptosis. Lesions of the third nerve nucleus which cause an isolated superior branch (ptosis and reduced sursumduction) or an inferior branch (pupillary dilation, reduced adduction and depression) paresis that simulates a peripheral nerve injury are also rarely seen [283].

Prenuclear disturbances include internuclear ophthalmoplegia from a lesion in the rostral MLF, paresis of upgaze in one eye because of disruption of the projection of the contralateral subnucleus to the affected superior rectus, and a combination of up- and downgaze pareses [284–286]. Occlusion of the anterior choroidal artery has also been reported to cause a monocular elevation paresis and other midbrain eye movement dysfunction [287, 288].

Conjugate paresis of upgaze, a bilateral light–near pupillary dissociation, retraction nystagmus to downward movement of optokinetic stimuli, and convergence insufficiency are typical signs of dorsal midbrain compression, but similar deficits can result from an infarct [289–291] or a hemorrhage (usually includes the thalamus) [292] in this area. Bilateral infarcts in the midbrain tegmentum can also cause a conjugate upgaze paresis [293]. A paresis of conjugate upgaze has been reported with a unilateral lesion of the posterior commissure. This lesion disrupts the input, possibly from the rostral interstitial nucleus MLF, to both third nerve nuclei that pass in the decussations within the posterior commissure [294–297].

Selective downgaze paresis is an uncommon phenomenon. In almost all cases, conjugate downgaze paresis is associated with some degree of upgaze dysfunction, even if the disturbance is only slowing of the upward saccades [298]. Bilateral lesions of the rostral interstitial nucleus are typically found in these cases [293, 299] (Fig. 8.19). In fact, downgaze paresis is almost always a result of bilateral infarcts [300–304]. These bilateral lesions can result from occlusion of the dominant posterior thalamosubthalamic paramedian artery [272]. However, a paresis of conjugate downgaze has been described in a patient with a unilateral lesion of the midbrain that included the rostral interstitial nucleus of the MLF [298]. Bilateral midbrain tegmental infarcts can also cause a conjugate paresis of downgaze as well as a monocular elevation paresis with normal horizontal excursions (vertical one-and-a-half syndrome) [305].

A slow, aperiodic, intermittent, alternating hypertropia or skew deviation can develop in patients with a pretectal region vertical gaze disturbance [306]. An alternating skew is readily distin-

Fig. 8.19a–d. Two days after coronary bypass, this ▶ 72-year-old man developed atrial fibrillation and became stuporous. When he awoke, it was noted that he had a mild paresis on upgaze and a striking paresis in downgaze. Within several months, his paresis in upgaze improved but he was left with a left hypertropia in all fields of gaze, representing a skew deviation and a complete paresis of downgaze.
a Initially the computed tomogram was normal, but within several days bilateral medial diencephalic/midbrain lucencies (*arrows*) that corresponded to the region of the rostral interstitial nucleus of the medial longitudinal fasciculus were noted.
b,c Axial view T2-weighted magnetic resonance scan demonstrated the corresponding areas of bilateral ischemic damage (*arrows*) in the midbrain (**b**) and the midbrain/thalamic junction. This case was presumed to have had an embolus to a single trunk paramedian perforating artery.
d *Top,* Upgaze demonstrates slight underaction with the right eye; *middle,* in primary position a left hypertropia is seen; *bottom,* on attempted downgaze, there is bilateral severe limitation and the left eye does not depress passed the midline

guished from seesaw nystagmus, a rapid, rhythmic movement most often found in patients with suprasellar tumors and bitemporal visual field defects. Seesaw nystagmus has rarely been associated with a brainstem infarct [307]. An alternating skew deviation must be differentiated from a congenital dissociated vertical deviation, which is accompanied by an esotropia and has characteristic findings on the alternate cover test. With the latter congenital disorder but not with a skew, the occluded eye elevates and often intorts while the viewing eye maintains fixation; occluding the viewing eye and uncovering the elevated eye causes elevation and depression of each eye, respectively. An alternating skew can also be differentiated from overaction of the inferior oblique muscles associated with a congenital esotropia. The causes of an alternating skew in order of frequency are hydrocephalus, tumor, and infarct [309].

Because of disruption of the descending prefrontal and superior colliculus projections to the midbrain and pons, a unilateral infarct of the midbrain tegmentum can cause transient slowing of the saccades or a frank paresis of contralateral gaze [309]. Dysfunction in the oculomotor nucleus or rostral MLF may partially account for the adduction deficit. The abduction weakness is clearly su-

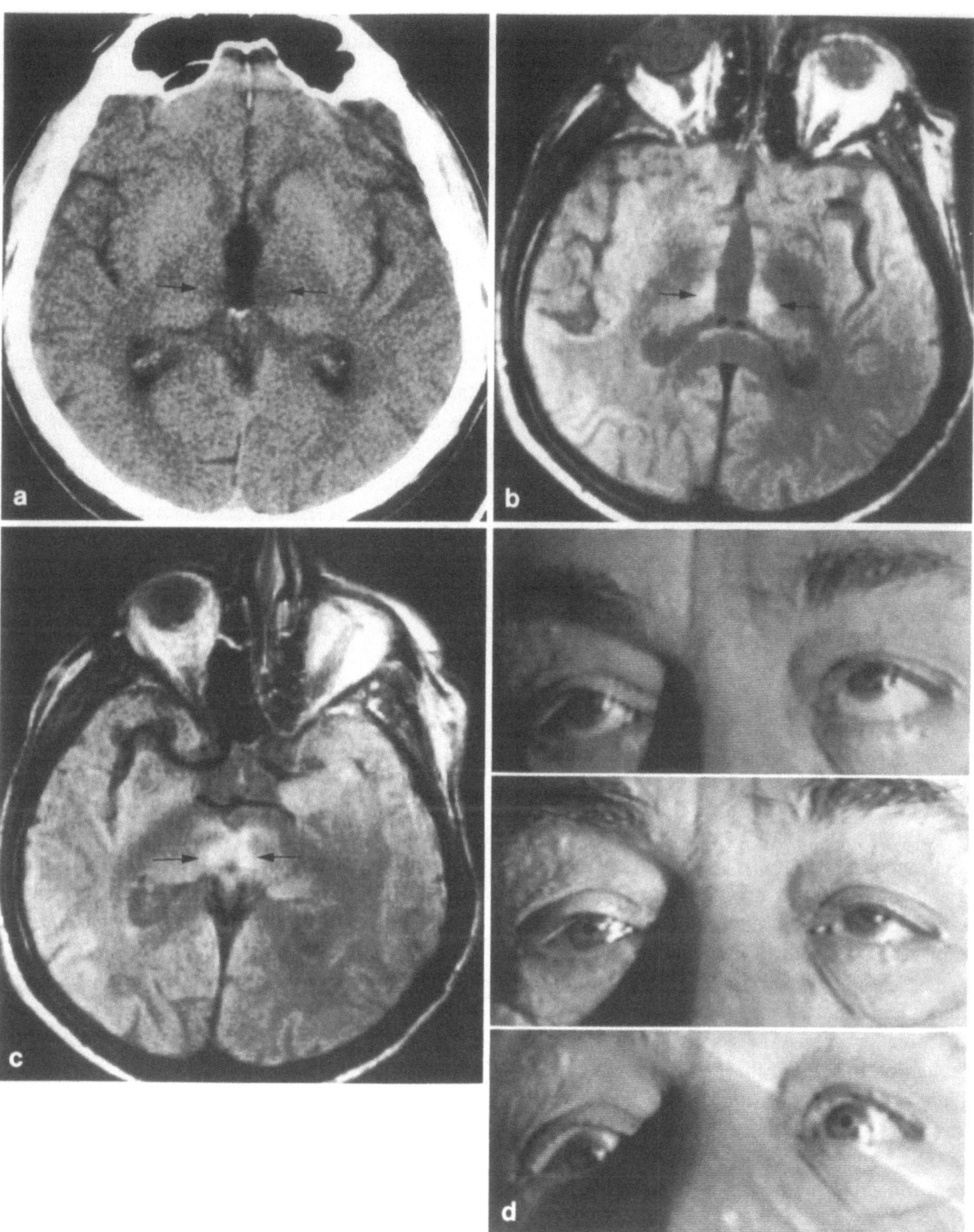

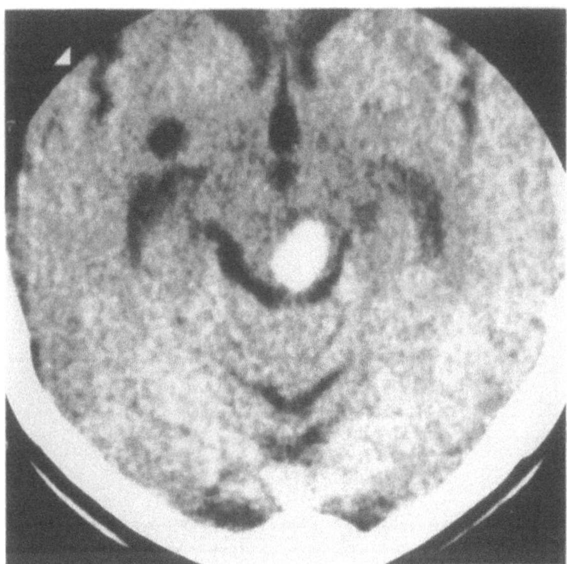

Fig. 8.20. This 50-year-old woman had the acute onset of double vision and was found to have a paresis of both up- and downgaze which was worse in downgaze. In addition, she had a right pseudo-sixth nerve paresis. The pupils were 3 mm, round, and reactive to light. The neurologic examination was otherwise normal, but the axial computed tomogram demonstrates a hemorrhage involving the tegmentum and periaqueductal region of the midbrain. Cerebral angiography was unrevealing

pranuclear because the oculocephalic maneuver overcomes the deficit [310–312].

Although an isolated convergence excess with small pupils is commonly a sign of nonorganic disease, it can be another manifestation of a midbrain or subthalamic region infarct. If the pupils are normal, the ocular motor dysfunction may be called a pseudo-sixth nerve paresis [131]. Disruption of bilateral descending frontopontine projections for horizontal gaze paresis has been suggested as a mechanism to account for the supranuclear bilateral lateral rectus underaction [313]. When this disorder occurs from ischemia it is typically transient. In contrast, convergence insufficiency is more common and often permanent with midbrain lesions (Fig. 8.20).

A lesion of the nucleus or fascicle of the fourth nerve is almost always accompanied by alteration of consciousness or cerebellar dysfunction or both because the causative infarcts are large. In most of these cases there is also a paresis of one or both third nerves [314]. Bilateral fourth nerve pareses from dorsal midbrain ischemia are rare [315].

Another rare visual disturbance that has been described with a midbrain infarct is peduncular hallucinosis [316, 317]. The hallucinations can be either formed or unformed, and are typically transient, lasting from minutes to hours. Compressive lesions, rather than infarcts of the midbrain, are the usual cause of this disorder [318].

8.5.2.1.5 Top of the Basilar Artery Syndrome

A variety of symptoms result from embolic or thrombotic ischemia to the rostral brainstem, thalamus, and medial temporal lobes, which are areas that normally receive blood from distal branches of the basilar artery [319]. The components of this "top of the basilar" syndrome include ocular motor and pupillary dysfunction, behavioral disorders, and visual sensory disturbances, with a relative preservation of motor function. The ocular movement abnormalities include paresis of up- or downgaze of one or both eyes, convergence insufficiency, an esotropia without a sixth nerve paresis, a skew deviation, as well as varying degrees of unilateral or bilateral third nerve pareses. Reflex vertical eye movements elicited by caloric or oculocephalic maneuvers and the Bell's phenomenon are absent. Nystagmus is not typically found in this syndrome. Unilateral or bilateral upper eyelid retraction or ptosis also occurs.

Small pupils (diencephalic) develop from interruption of both descending sympathetic pathways. However, one or both pupils may be dilated and unreactive to light because of unilateral or bilateral involvement of either the Edinger–Westphal nucleus or the third nerve. The pupils can also become eccentrically displaced in the iris (corectopia) by a midbrain lesion [320].

Though a homonymous hemianopia, palinopsia, visual agnosias, or cortical blindness are not classically considered part of the "top of the basilar syndrome," these sensory visual disturbances occur in approximately 50% of cases [321]. Cognitive abnormalities such as an inability to synthesize an entire visual scene so that only parts of the field are seen (Balint's syndrome) [322] occur when the ischemia extends into the territory supplied by the PCA. Additional disturbances of visual perception include image size distortion (metamorphopsia), poor hand–eye coordination, and problems in

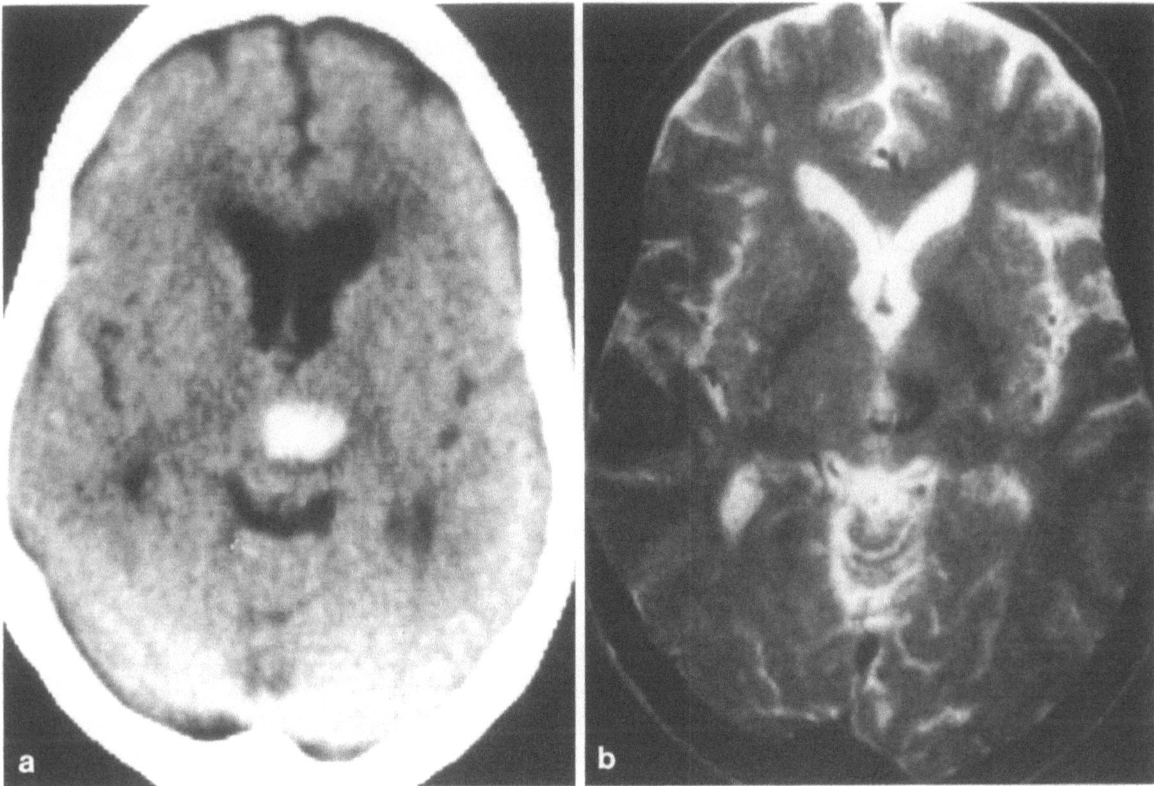

Fig. 8.21a, b. A 75-year-old woman had the onset of a sudden headache and confusion. There was mild limitation in upgaze at that time. **a** The axial view noncontrast computed tomogram demonstrated hemorrhage into the left thalamus extending to the midline. The eye movement limitation and confusion resolved within several days. **b** Fourteen months later, the axial view magnetic resonance scans for both proton density (not shown) and T2-weighting demonstrated the dark signal elicited by the residual hemosiderin

maintaining fixation on an image or scene. Other behavioral and cognitive abnormalities are common when the posterolimbic system and thalamus are involved.

8.5.2.1.6 Thalamus

Ocular motor disturbances are frequent with either infarcts or hemorrhages of the thalamus (Fig. 8.21). A hemorrhage (most often associated with a depressed level of consciousness) that disrupts the input from the frontal eye fields prior to their decussation results in the eyes conjugately deviating towards the side of the lesion [323]. If the hemorrhage does not involve the decussation or if the clot is located caudally, the eyes can be forced conjugately away from the lesion [131, 324]. Forced downward conjugate gaze with medial deviation of the eyes and miosis of the pupils or a paresis of upgaze can also develop with a thalamic hemor-

rhage. This vertical eye movement disorder often resolves after the local mass effect remits [292, 325, 326]. An acquired esotropia (pseudoabducens paresis) without miosis has also been reported with an infarct of the thalamus (see Sect. 8.5.2.1.4).

When the hemorrhage affects the posterior thalamus, a hemisensory and hemimotor defect are the predominant abnormalities. In these cases, the saccades are hypometric in the direction away from the lesion and pursuit is defective towards the lesion. A mild ptosis and miosis occurs ipsilateral to the lesion [327]. Numerous behavioral defects, including abnormalities of memory, visual spatial processing, and language, are reported with infarcts of the thalamus caused by occlusion of the tuberothalamic or deep interpeduncular arteries [328]. Hemisensory deficits, hemiataxia, hemi-involuntary movement disorders, and homonymous hemianopia develop when the infarct includes the lateral thalamus [329, 330]. However, the rich blood supply to these areas makes it unlikely that occlusion of single artery will cause a major syndrome. This is why the Dejerine–Roussy thalamic syndrome, which is associated with contralateral hyperesthesia and hemiparesis and, infrequently, a homonymous hemianopia, is rare [331].

8.5.2.2 Ocular Motor Dysfunction from Cerebral Hemisphere Infarcts and Hemorrhages

After a frontal cerebral infarct, the eyes can be conjugately deviated towards the lesion for several hours [332, 333]. Even when the gaze dysfunction is acute, the oculocephalic maneuver or caloric testing can drive the eyes towards the contralateral hemisphere. In contrast, an acute cerebral hemorrhage can irritate the frontal eye fields, driving the eyes towards the normal hemisphere [334]. After a frontal lobe injury the version eye movements recover, but the voluntary saccades are often slow and inaccurate and the latencies to initiate the saccades are prolonged. In one study, patients with acute cerebral infarction who demonstrated conjugate eye deviation had a higher rate of morbidity and mortality than stroke patients without deviation of the eyes [335].

Bilateral frontal cerebral hemisphere infarcts may cause periodic alternating horizontal gaze. Each horizontal excursion lasts approximately 2–5 s [336, 337]. However, this eye movement disorder is not diagnostic of lesions in this location because similar phenomena have been described in patients who have a hemorrhage in the mesencephalon or the cerebellar vermis [338, 339].

Pursuit abnormalities can occur with lesions in either the temporal or parietal lobes. Even when the visual field is unaffected, impairment of the slow pursuit of the optokinetic response can be seen when the stimulus is moved in the direction towards a parietal lesion [340]. The optokinetically induced rapid nystagmus in the direction contralateral to the parietal lesion may be lost if the optomotor area pathway to the lateral gaze center in the pons is disrupted.

The eye movements are not spared with occipital ischemic lesions. When an occipital infarct causes a homonymous hemianopsia the eyes may be deviated towards the lesion because the patient neglects looking into the defective field. Saccades into the hemianopic field are inaccurate. Patients alter their saccadic eye movements to compensate for the homonymous hemianopia [341]. Ocular pursuits toward the side of the occipital infarct are often, but not always, defective [342]. Even when no vision can be detected in the affected field, the optokinetic response is often maintained. This suggests that this reflex may be mediated by subcortical or extrastriate pathways [343].

8.6 Therapeutic Considerations

The treatment of patients with amaurosis fugax, TIAs, and cerebral infarcts is controversial and no single therapeutic approach has been shown to be overwhelmingly superior. Many of the therapeutic trials have a small number of patients, so statistical error clouds the conclusions. Since the incidence of stroke has diminished in the past 20 years [344], it may be difficult to judge the efficacy of the various treatment modalities in a long-term prospective study. A brief review of the current therapies will be helpful in contemplating the management of specific cases.

8.6.1 Cardiac Disease

Clearly, if there is a cardiac source for emboli, control of arrhythmias and management of valvular disturbances is required. Anticoagulation with heparin followed by warfarin is recommended for patients who have ocular or cerebral ischemic episodes with atrial fibrillation and valvular disease or a cardiac thrombus. In order to prevent a hemorrhagic stroke, patients with an acute cardiogenic stroke may receive anticoagulants after 48 h. Some, but not all, studies suggest that anticoagulation can prevent stroke in patients with atrial fibrillation and cerebral emboli, even if there is no obvious structural cardiac disease [345–349]. In some patients with atrial fibrillation, aspirin may also be effective in preventing stroke [349a]. Acute anticoagulation with heparin may also be preventive therapy in patients with a left ventricular thrombus that follows an acute myocardial infarction [31]. The treatment of other cardiac sources of emboli, such as infectious endocarditis and atrial myxoma are well established.

8.6.2 Carotid Atherosclerosis

In most investigations evaluating different therapies for symptomatic carotid atherosclerosis, patients with amaurosis fugax are typically combined with patients who have cerebral TIAs. Additionally, many of the studies on stroke prevention include numerous patients who had had remote rather than recent TIAs at entry. Despite these difficulties, two approaches, medication with platelet antiaggregation activity [350] and surgical endarterectomy of the carotid artery [351, 352], have demonstrated a benefit in reducing the stroke rate.

8.6.3 Antiplatelet Agents

Aspirin (acetylsalicylic acid) has been the principal agent used to block platelet aggregation. The antiaggregant affect of aspirin results from inhibition of the cyclo-oxygenase pathway for the production of thromboxane in platelets. However, aspirin has the negative effect of blocking the endothelial cell production of prostacyclin, which lyses thrombus. The dose of aspirin used in early studies of cerebrovascular disease ranged from 325 mg [353] to 1300 mg per day. Recently, two large-scale prospective stroke trials using aspirin in smaller doses (30 – 75 mg) in an attempt to prevent blockade of prostacyclin synthesis in the vessel walls [354] found a reduction in stroke comparable to that with higher doses but with few complications [354a, 354b].

One of the largest studies of antiplatelet therapy, and probably the most widely cited, the Canadian Cooperative Study, reported a combined stroke and death rate 31% lower in the aspirin-treated group compared to the controls [355]. In subgroup analysis, aspirin lowered the stroke rate in men but not in women [356]. In another study, if a stroke occurred in a patient taking aspirin, the severity of the neurologic deficit seemed to be reduced [357]. Following a cerebral infarct, aspirin may not prevent subsequent strokes [358]. Patients with an

asymptomatic Hollenhorst plaque who do not have a suggestion of high-grade carotid stenosis in the ipsilateral ICA should only receive antiplatelet therapy.

Aspirin does not seem to reduce the frequency of amaurosis fugax in young adults with normal carotid angiograms [359]. It is probable that these patients have etiologies other than carotid emboli to account for the visual episodes (see Sects. 8.2.3.3.1, 9.17, 10.4.2.1). Prophylaxis may be obtained with calcium channel blockade in those patients with a vasospastic etiology [359a].

Dipyramidole has been used as an antiaggregating agent because it decreases platelet adhesiveness by prolonging platelet survival, and older platelets are less "sticky." The American – Canadian study compared aspirin to a combination of aspirin and dypiramidole in patients with TIAs. Both groups had a reduced incidence of stroke and death, but there was no difference between the treatment groups [360]. Dipyramidole alone seems to have little if any benefit in preventing cerebral ischemic episodes or amaurosis fugax [361]. Another platelet antiaggregant, sulfinpyrazone, seems comparable to aspirin in the prevention of stroke [355].

A newer antiplatelet agent, ticlopidine, inhibits platelet aggregation induced by most laboratory stimuli by blocking the adenosine diphosphate pathway, but it does not alter prostacyclin production [362]. In a study conducted over a 2- to 6-year period, patients with amaurosis fugax, TIAs, or minor strokes treated with ticlopidine had a 21% risk reduction of stroke and a 12% reduction of nonfatal stroke or death compared to the group treated with aspirin 1300 mg/day [363]. After several weeks of ticlopidine, a few patients developed thrombotic thrombocytopenic purpura that regressed on withdrawal of the drug. Reinstitution of ticlopidine was not associated with a recurrence of the purpura [364]. However, gastrointestinal problems may occur more frequently, limiting the use of this agent in some patients.

8.6.4 Anticoagulation

Although numerous studies and case descriptions suggest that anticoagulation with warfarin can be beneficial, this therapy has not been proven to lower the stroke rate in patients with carotid or vertebral atherosclerosis [365–367]. The incidence of complications, which include systemic or intracerebral hemorrhages, are increased if the patient has systemic hypertension or when the prothrombin time exceeds two times the control [368, 369]. If warfarin is administered to prevent cardiac emboli, maintaining the prothrombin time at approximately 1.5 times the control is associated with a lower incidence of complications.

Although heparin is frequently administered to patients with progressive clinical deterioration during an acute stroke, this therapy remains controversial and unproven [370]. In these cases, heparin may precipitate a hemorrhagic infarct or a frank intraparenchymal hemorrhage at the site of the acute ischemic damage. However, heparin may be helpful in preventing a cerebral infarct in symptomatic patients with high-grade stenosis of the ICA who are awaiting a carotid endarterectomy.

Fibrinolytic or thrombolytic therapy has recently been reintroduced for consideration in treating acute cerebral ischemic episodes [371]. The complication of intracerebral hemorrhage has been well described with intravenous thrombolytic agents such as streptokinase, urokinase, and tissue plasminogen activator [372, 372a, 373]. Thrombolysis with intravenous [372a] or intra-arterial agents [373a, b] might be effective in patients with acute ischemic events who could be evaluated and treated before a brain infarct develops.

8.6.5 Surgery

It is overly simplistic to think of all cerebral infarcts as arising from embolic disease or thrombosis of major cerebral arteries. Clearly, the pattern of premorbid arterial collaterals, the opening of latent arterial collaterals, and the extent of atherosclerosis in the affected arterial system are all factors which result in a hemodynamic state that either protects or predisposes to infarction of the brain [374]. Following a TIA or infarct, impairment of local tissue autoregulation may further complicate the analysis and management of these cases. Vascular lesions, either single obstructions or multiple stenoses in one circulation, can result in a reduction of regional cerebral blood flow to below the level for which local autoregulation can compensate. Surgically restoring the patency of the proximal artery may not be effective if the distal lesions are hemodynamically significant.

Carotid endarterectomy (CEA) has been the principal surgical procedure performed in patients with cerebrovascular occlusive disease. CEA should be reserved for the appropriate patients. Clearly, it does not significantly reduce the morbidity and mortality caused by cerebrovascular disease in all patients with atherosclerosis of the ICA, particularly if the only symptom is a transient monocular visual disturbances [375, 376]. However, in one study of 80 patients with amaurosis fugax who had an endarterectomy, the stroke rate was favorably influenced. During the average 13-month follow-up, there were no nonfatal strokes, though amaurosis recurred in 10% and death occurred in 12% [377]. Unfortunately, CEA cannot be performed in many patients with CRA occlusions since these patients have a significant incidence of complete occlusion or inoperable lesions of the ICA [378]. Since the risk of stroke rate can be correlated with the degree and progression of carotid stenosis [100, 379, 380], the rationale for performing CEA seems sound. In fact, CEA has recently been shown to reduce the frequency and severity of infarcts in the ipsilateral cerebral hemisphere in patients with symptomatic high-grade stenosis (70%–99%) of the ICA [381]. In this study, the absolute risk of ipsilateral stroke and severe stroke or death were reduced by 17% and by more than 7%, respectively.

CEA complications include cerebral infarct and hemorrhage and myocardial infarct. Rarely, visual loss results from retinal ischemia [382, 383] or a carotid cavernous fistula (see Sect. 2.1.1). The risk of CEA must be less than the risk of stroke, which is approximately 12% in the first year, 8% in the second year, and 5% per year thereafter in patients with a cerebral TIA. Though the risk of stroke in asymptomatic patients is greater when both the carotid and vertebral arteries are affected by atherosclerosis [384, 385], the risk of stroke in these patients is still significantly lower than the incidence of complications in most endarterectomy series. CEA seems unwarranted in most patients with retinal emboli since the incidence of stroke is only 3% per year in these patients [386] unless high-

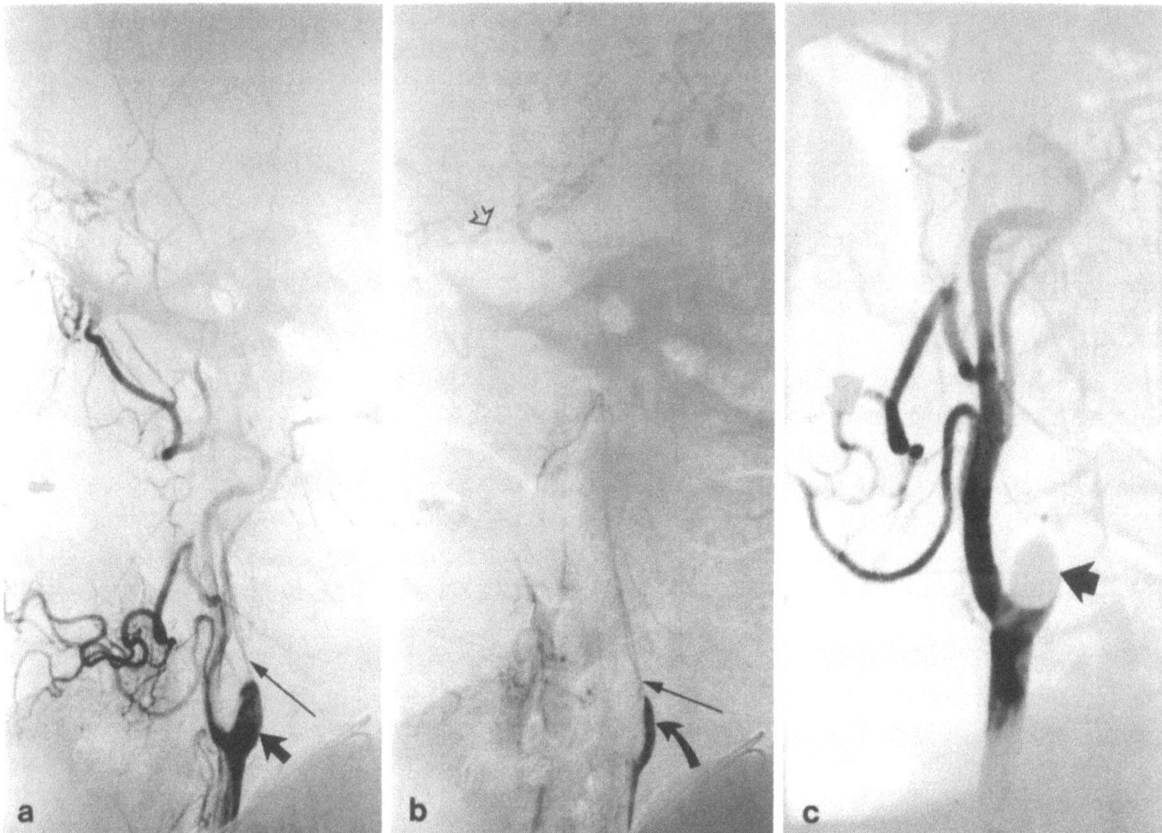

Fig. 8.22a–c. This 65-year-old man had recurrent amaurosis fugax despite aspirin therapy because of thrombosis at the origin of the internal carotid artery and stump emboli. **a** A lateral view common carotid artery subtraction angiogram demonstrates the stump of thrombus in internal carotid artery (*arrow*). Note the origin of the patient ascending pharyngeal artery (*long arrow*) from the internal carotid artery. **b** The late phase of the same contrast injection demonstrates stagnation of contrast material in the stump of the internal carotid artery (*curved arrow*). The ophthalmic artery (*open arrow*) fills via the external carotid artery. The stump in the internal carotid artery remained open because the thrombosed area still had an outflow via the ascending pharyngeal artery (*thin arrow*). **c** Following closure of the stump with a detachable balloon (*arrow*) the symptoms resolved

grade stenosis is present. Depending on many factors, including the experience of the surgical team, the morbidity and mortality of the surgical procedure can range from less than 3% to 21% and from 3% to 7%, respectively [387–390]. It has been suggested that the morbidity and mortality of angiography and endarterectomy in symptomatic patients without significant ICA stenosis must be less than 2.9% in order to exceed the benefit of medical therapy [391].

Other surgical vascular procedures have been performed with varying results. In patients with complete occlusion of the ICA who experience discrete ocular or hemispheric ischemic events, surgical excision or obliteration of the stump thrombus or an external CEA may be beneficial [392] (Fig. 8.22). In contrast, extracranial–intracranial (EC-IC) bypass is not useful for the treatment for atherosclerotic ICA disease. In the major randomized study on patients with retinal or hemispheric infarcts or TIAs who had a stenotic or occluded ICA, EC-IC bypass did not reduce the risk of subsequent brain or ocular ischemic infarcts [393, 394]. However, several anecdotal reports have described improvement in the signs of ischemic oculopathy following EC-IC bypass (see Sect. 8.6.8.2).

8.6.6 Antihypertensives

Control of systemic hypertension is essential because almost all studies report a higher incidence of strokes in hypertensive patients with TIAs, particularly when the diastolic pressure is 105 mmHg or higher [395–397]. The risk of stroke can be directly related to the severity of the hypertension [398]. Small vessel disease with lacunae infarcts is indicative of hypertension. The widespread treatment of hypertension has contributed to the decline in the incidence of stroke [399–402a, 402b]. Though cigarette use increases the risk of coronary artery disease, there is no definitive correlation with the development of stroke.

8.6.7 Posterior Circulation

The treatment of posterior circulation TIAs and infarcts is even less clear than the treatment of anterior circulation ischemia. Though anticoagulation has been reported to prevent progression of brainstem infarction in anecdotal cases [403], no controlled studies have proven the efficacy of this treatment. Surgery has successfully relieved atherosclerosis- related stenosis at the origin of the vertebral artery [404, 405], but the results of surgery in the prevention of stroke are even less clear than with CEA. In rare cases where turning the head causes symptomatic blockage of the intraosseus portion of the dominant vertebral artery, excision of the transverse process of the sixth cervical vertebrae may effectively prevent ischemia in the vertebrobasilar system.

8.6.8 Treatment of Ocular Ischemic Conditions

8.6.8.1 Central Retinal Artery Occlusion

The treatment of an acute nonarteritic CRA occlusion must increase the effective blood perfusion of the retina. The extent of the retinal infarct may be reduced if an embolus can be dislodged so that it migrates into one or more distal retinal arteries.

Emergency measures include ocular massage to mechanically lower the intraocular pressure (IOP), which may increase the ocular blood perfusion/IOP ratio. Inhalation techniques that increase the carbon dioxide concentration of the blood, such as by rebreathing into a bag or inhalation by mask of a 95% oxygen/5% carbon dioxide mixture, are used in order to dilate the retinal vessels. Systematic acetazolamide has been given in some cases to reduce the IOP, but the ocular blood perfusion/IOP ratio may actually be reduced because of an acute decrease in systemic blood pressure and blood flow into the eye. Drugs designed to directly vasodilate the retinal vessels have also been administered with limited success [406].

If the patient presents within the first 24 h of visual loss and the above measures are unsuccessful, paracentesis should be performed. Following topical anesthesia of the conjunctiva and cornea, a tuberculin syringe with a 27-gauge needle or a fine point scalpel is used to rapidly penetrate into the anterior chamber through the limbus in a direction parallel to the iris. The sudden reduction of the IOP may alter the blood in the CRA pushing the embolus of the main trunk.

The effectiveness of these treatments on visual function seems to depend on the level of visual impairment at the time of presentation. Patients with an initial visual acuity of hand motion at 1 m or worse are less likely to recover usable vision, than those with visual function of finger count at 1 m or better [76]. In patients where ophthalmoscopy suggests that a cilioretinal artery provides blood to the macula and the papillomacular bundle, the prognosis for recovery of visual acuity is improved [407, 408]. Spontaneous recovery without therapy, although rare, has also been described [74], even when the patient has no apparent cilioretinal artery [409]. Patients have also recovered even when the obstruction appeared to persist for longer than longer than 105 min, which is the duration of CRA occlusion associated with irreversible retinal damage in monkey experiments [410]. It is probable that in these patients the retinal ischemia was not absolute because a complete arrest of retinal arterial flow had not occurred.

In one patient with an embolic CRA occlusion, argon blue–green laser light was used to photocoagulate the embolus after ocular massage, acetazolamide, and rebreathing measures failed. The patient's vision did not recover despite the dislodging of the embolus [411].

8.6.8.2 Ischemic Oculopathy

The treatment of the various manifestations of the ischemic oculopathy syndrome are generally inadequate unless the arterial blood supply to the eye is significantly increased. Though the slow-flow hemorrhagic retinopathy can be either progressive or static, once there is neovascular glaucoma, painful blindness requiring enucleation usually results. Intensive medical, laser, and surgical therapies are often ineffective in cases with neovascular glaucoma. Surgical procedures, such as implanting a Molteno shunt, that allow drainage of the aqueous have been reported to prevent the ultimate destruction of the eye [412, 412a]. In patients with carotid occlusion, panretinal laser photocoagulation has been used with some success if done before neovascular glaucoma develops [413, 414]. However, the role of early retinal photocoagulation remains undecided because not all patients with venous stasis retinopathy due to carotid occlusive disease will develop neovascular retinopathy and glaucoma, particularly if capillary perfusion is maintained.

The light-induced visual loss and stasis retinopathy have been reported to resolve following surgical restoration of ICA lumen patency [415 – 419]. A superficial temporal artery to middle cerebral artery bypass has also been effective for these disturbances. An increase in the ODM pressure, a reduction of the fluorescein angiographic determined circulation time [420], and resolution of the iris neovascularization and retinopathy has been seen in some, but not all, patients after bypass [421, 422]. CEA has yielded similar beneficial results in most cases, but those with neovascular glaucoma were less likely to improve [423]. Unfortunately, no large controlled study on the effect of bypass surgery or CEA on ocular ischemia has been performed. Paradoxically, in one case, the iris neovascularization worsened following CEA [424]. This is difficult to explain unless more distal occlusive disease in the ophthalmic circulation prevented the postsurgically patent cervical ICA from increasing the ocular blood flow. It is possible that restoration of blood flow to the ciliary body resulted in reversal of the previous low aqueous production and caused an elevation of the IOP. Despite the lack of conclusive evidence, procedures that improve ophthalmic blood flow prior to severe visual loss seem to be beneficial for maintaining and restoring vision [425, 426].

8.7 Atypical Occlusive Disease of the Ophthalmic Artery

An unusual, and frequently overlooked, cause of ischemic visual loss results from prolonged external compression of the orbital contents. This complication can develop when a patient is positioned in the prone position, face down, while under general anesthesia. The mask for anesthetic gas delivery can press the globe into the orbit, causing ocular ischemia with visual loss, mild lid edema, and ocular motor dysfunction that are only noted when the patient awakens [427, 428]. This also causes patchy or generalized retinal ischemic changes seen on ophthalmoscopy. A similar problem has also been described in patients who remain immobile, lying face down on the eye and orbit following narcotic or ethanol ingestion. In these patients, cerebral angiography has failed to demonstrate evidence of ophthalmic artery occlusive disease, but the investigations were typically performed several days after the ictus [429]. Retinal ischemia seems to result from compression-induced hypoperfusion of the distal ophthalmic artery, but a complete occlusion of the CRA is typically absent. The visual recovery can be total or complete blindness can remain. Concomitant ischemia of the extraocular muscles accounts for the ocular motility disturbance. The muscle pareses are usually incomplete and frequently recover spontaneously.

8.8 Ischemic Cranial Nerve Syndromes

8.8.1 Introduction

When the patient presents with a complaint of diplopia or ptosis or both, specific historical details and the clinical examination will often clarify the localization and narrow the differential diagnosis. The first questions asked should ascertain whether the patient actually sees two distinct images that do not overlap and whether the diplopia is present binocularly and absent monocularly. Often the patient cannot accurately characterize the diplopia, but if

Table 8.9. Differential diagnosis of monocular diplopia

Astigmatism
Corneal opacity
Lens opacity
Macula lesion
Cerebral polyopia (rare)

the symptom is binocular it frequently arises from an extraocular muscle misalignment. Monocular diplopia (Table 8.9) is not a symptom of a cranial neuropathy and is rarely a symptom of neurologic disease.

The presence, character, duration, and localization of pain associated with diplopia can also narrow the differential diagnosis. For example, although myasthenia gravis, thyroid orbitopathy, and multiple sclerosis all cause diplopia, they are not typically associated with a painful ophthalmoplegia. The clinical evaluation should always determine whether the patient truly has an isolated nerve paresis or whether there is dysfunction of other cranial nerves or the central nervous system. When unilateral upper lid closure is present, the examiner must determine whether the ptosis results from levator palpebrae or Mueller's muscle weakness or whether the patient is involuntarily closing the lid to avoid the diplopia.

8.8.2 Third Cranial Nerve Vascular Neuropathy

A third nerve paresis may be partial or complete depending on the extent of the microvascular infarct. The levator weakness results in a ptosis of the upper lid that can cover a portion of the pupil or the entire globe. On attempted lid elevation, the patient uses the frontalis muscle to raise the lid. A true assessment of levator function is made by measuring the palpebral fissure opening in primary gaze and in upgaze while the frontalis is fixed to the brow by the examiner's finger. The extent of the involvement of the superior rectus, inferior oblique, medial rectus, and inferior oblique muscles may be obvious from the versions and ductions, but the prism alternate cover test may uncover abnormalities otherwise missed. Slowing of upward, downward, and adducting saccades of the involved eye

will also reveal subtle defects. Difficulty in evaluating individual muscle dysfunction can be encountered in cases with a longstanding paresis and a spread of comitance [430a] (see Sect. 6.3.2.1 for complete discussion of the pupil examination and the various combinations of third nerve innervated muscle dysfunction).

8.8.2.1 Microvascular Dysfunction

The probability that an isolated third nerve palsy is caused by microvascular ischemic disease varies with the age of the patient. Hypertension and diabetes mellitus are underlying systemic diseases commonly associated with a third nerve vasculopathic paresis, more often found, but not exclusively, in patients more than 50 years of age. Except in juvenile diabetics, an isolated third nerve paresis is usually not considered vascular related in patients less than 30 years of age. In one large study, approximately 30% of all third nerve pareses clearly had an ischemic microvascular etiology [431], but the incidence of ischemic cranial neuropathy is no doubt higher. The pupil is spared in 86% [432] to 91% [433] of diabetic third nerve palsies. The associated pain in and around the eye can be severe at the onset of the diplopia or ptosis, but it typically remits within several days and infrequently lasts more than 1 week. Ischemic third nerve pareses show some improvement within 4 weeks in 68% of cases, and by 12 weeks almost 100% of cases have improved [434]. However, complete normalization of extraocular movements does not occur in all patients. In diabetics, subsequent cranial nerve palsies are reported to occur in from less than 20% [432] to 27% [435] of patients.

Confirmation of the microvascular etiology in diabetes has been obtained from several pathological specimens. However, the exact site of arterial or arteriole occlusion has rarely been seen. Infarcts have been demonstrated in segments of the third nerve in the subarachnoid space [436, 437] and in the cavernous sinus [438, 439]. Dreyfus demonstrated ischemic necrosis of the larger axons and myelin located predominantly in the central portion of the nerve. Asbury et al. confirmed that the center region of the nerve is the typical site of myelin damage. However, the latter investigation also revealed signs of Wallerian degeneration in the third nerve in the cavernous sinus. This site of injury was far removed from the documented occlusion of a single epineural arteriole in the subarachnoid space (Dreyfus also found one occluded arteriole in his cases). The most prevalent vascular abnormalities of the intracavernous arterioles found were hyalinization of the vessel walls and endothelial proliferation that narrowed the arteriole lumens. However, there were no occlusions of the intraneural vessels in the cavernous sinus [438, 439]. The ischemic third nerve damage occurred in regions supplied by end vessels which seem to be watershed areas between the arterial supplies from the tentorial branch of the meningohypophyseal trunk of the cavernous carotid [440], the inferior lateral trunk superior branch, and branches from the posterior cerebral, posterior communicating, and basilar arteries (see Sect. 1.5.3.2.1).

One explanation for the phenomenon of pupillary sparing in the above cases may be related to the location of the pupillary fibers in the peripheral portion of the nerve. In this position, the pupillary efferents are not prone to ischemia since they are not supplied by end vessels. Additionally, the small parasympathetic unmyelinated fibers to the ciliary ganglion may be more resistant to ischemia than the larger myelinated central fibers of the nerve that innervate the eye muscles and lids.

Patients with iatrogenic vascular occlusive third nerve pareses have also provided clinical confirmation of the arterial supply to the nerve. Some of the reported cases that developed third nerve pareses during arteriography or embolization of tumors or vascular malformations had dysfunction of multiple cranial nerves [441] (see Sect. 8.8.5). However, we have observed a patient who developed an isolated third nerve pareses (including mydriasis) following cyanoacrylate embolization of a cerebral arteriovenous malformation through the inferior lateral trunk. Also, this patient did not develop fourth, fifth, or sixth nerve dysfunction, suggesting that these nerves received arterial blood from another source. In contrast, other cases did not develop a cranial neuropathy after inferior lateral trunk occlusion by embolization, indicating that in some cases the blood supply to the third nerve in the cavernous sinus can be provided by other arteries, such as the middle meningeal artery if this is the dominant vessel to this area.

Table 8.10. Etiologies of "isolated" third nerve paresis

Aneurysm (posterior communicating, internal carotid, basilar)
Microvascular infarct of the cranial nerve
Cavernous sinus neoplasm
Syphilis
Meningeal process (tuberculosis, fungal, carcinomatous, lymphomatous)
Migraine
Vasculitis (systemic lupus erythematosus)
Herpes zoster
Sarcoidosis
Giant cell arteritis
Postviral or postimmunization
Trauma or iatrogenic injury
Arteriovenous shunt to the cavernous sinus
Hemorrhage or infarct of the midbrain fascicle or nucleus

8.8.2.2 Differential Diagnosis

The differential diagnosis (Table 8.10) of an oculomotor nerve paresis is extensive but most patients with an isolated third nerve paresis have an aneurysm of the posterior communicating, cavernous carotid, or basilar arteries, a microinfarct of the peripheral nerve, or a neoplasm in the parasellar region. The third nerve is often the first cranial nerve affected by pituitary adenoma invasion into the cavernous sinus, so the visual fields should be evaluated in all patients with an "isolated" third nerve paresis.

The frequency of each etiology varies depending on the age of the patient [431, 442]. Further considerations in evaluating the patient with a third nerve paresis are discussed in Sect. 6.3.2.1.

The differential diagnosis of an upper lid ptosis depends on whether it is unilateral or bilateral and whether the ptosis is isolated or occurs with other ocular or orbital signs. In addition to third nerve or sympathetic dysfunction, myasthenia gravis, and dysthyroid orbitopathy, local lid disease can cause ptosis. Local lid disorders include local lid or tarsal- -conjunctival edema or inflammation (for example, contact lens induced), and partial or total levator dehiscence. Inflammation or neoplastic infiltration of the levator–superior rectus muscle sheath can also cause ptosis with subtle orbital signs.

8.8.2.3 Course and Treatment

Since most ischemic third nerve palsies recover over several weeks to months, long-term treatment is not needed in most patients. Initially, the ptosis temporarily serves as a natural patch, relieving the diplopia. When diplopia is present, occluding the affected eye (assuming usable vision is present in the unaffected eye) will help the patient function. While the patient is sitting or viewing television, covering the unaffected eye and using the affected eye will decrease the incidence of a secondary contracture. Once the deviation in primary gaze, downgaze, or both is reduced, fresnel (early) or ground (later) prisms may be useful for restoring binocular fusion in these fields. Botulinum toxin injections can also be beneficial but it should be administered only after some recovery has been documented (recall that signs of recovery are an important indicator of a benign process). Eye muscle surgery is undertaken only when the measurements are stable and not improving for a minimum of 4–6 months. The probability of restoring fusion in primary and downgaze is decreased when the dysfunction remains in multiple extraoxular muscles. As mentioned in Sects. 6.2.2 and 6.3.4.5, misdirection occurs most frequently with compression from a cavernous sinus mass or a posterior communicating aneurysm and it rarely develops following ischemia [443, 444].

8.8.3 Sixth Cranial Nerve Vascular Neuropathy

As with third nerve palsies, the etiology for an isolated sixth nerve palsy is age dependent. Microvascular disease associated with hypertension, diabetes mellitus, or atherosclerosis is the most common etiology in patients more than 50 years of age. The clinical examination of the ocular movements in a patient with a sixth nerve paresis reveals the underaction of the lateral rectus by versions and ductions. The lateral rectus saccade is slow and often hypometric. This is contrasted with the normal or minimally slowed saccade found in patients with a myopathy or mechanical restriction of the lateral rectus. An esotropia which is largest in the field of action of the lateral rectus is typically present. However, in longstanding pareses, a spread of com-

itance can alter this pattern, occasionally resulting in a similar amount of esotropia in all directions of gaze. Also, recall that because of Hering's law, if the patient prefers to fixate with the paretic eye, the deviation will be larger and a pseudoparesis of the opposite lateral rectus may be observed [430b].

8.8.3.1 Differential Diagnosis

A microvasculopathic etiology is found in less than 30% of isolated sixth nerve pareses in patients between the 15 and 50 years of age [431, 442, 445]. Neoplasms that invade or are located at the skull base (examples include petrous, clivus, or cavernous sinus meningioma, chordoma, epidermoid, metastatic neoplasm, nasopharyngeal carcinoma, and pituitary tumor) are less frequent causes of an isolated sixth nerve paresis in the elderly, but these disorders are more prevalent in young adults. Infiltration of the sixth nerve in the subarachnoid space or cavernous sinus by a lymphoma or carcinomatous meningitis accounts for less than 20% of all cases. The incidence of traumatic sixth nerve palsies is probably less than 10%, but this figure varies with the referral pattern of the reported series. Multiple sclerosis has been reported to be the etiology of an isolated paresis in approximately 15% of cases. However, this number is undoubtedly too high because concomitant subtle signs of gaze pareses and other features of multiple sclerosis are often overlooked. Suspected inflammatory disease in the cavernous sinus or superior orbital fissure is the cause in less than 10% of patients. A postlumbar puncture complication is reported as the etiology in less than 5% of cases but we have seen this only once in a patient given epidural anesthesia. Approximately 20% of sixth nerve pareses have been reported to have no known cause. The incidence of the etiologies in the reviewed series varies in part because of changing patterns of disease (e.g., syphilis, lymphoma, infectious meningitis), but also because of improved laboratory and neuroimaging capabilities.

Bilateral sixth nerve pareses are more often associated with tumors than microvascular disease [446]. Chronic unilateral paresis of the sixth nerve paresis, even when the patients are more than 50 years of age, have a higher probability of being caused by a neoplasm or a sphenoid sinus infection than the acute sixth nerve paresis that recovers over weeks or months [447, 448].

A forced duction test [430c] may be required to diagnose a restrictive myopathy, such as occurs with infiltration of the medial rectus in dysthyroid orbitopathy. Patients suspected of having myasthenia gravis and those who follow an atypical course should have an edrophonium test. Neuroimaging of the cavernous sinus or skull base with MR imaging or CT should be performed in young adults with an isolated sixth nerve paresis. Unless the dysfunction is atypical or chronic, neuroimaging is not essential in patients with an acute, truly isolated sixth nerve paresis who are older than 50 years or younger than 12 years (the latter typically have para- or postviral etiology). The cerebrospinal fluid should be examined following a scan in patients suspected of harboring a meningeal process or in the rare case of pseudotumor cerebri without papilledema.

8.8.3.2 Course and Treatment

Most patients with an ischemic sixth nerve paresis will demonstrate improvement in lateral rectus function within 3–12 weeks. Restoration of orthophoria in primary gaze occurs within 6 months in the majority of patients. Diplopia and esotropia in the field of gaze of the affected lateral rectus may be permanent. A spread of comitance infrequently causes a residual esotropia in all cardinal fields of gaze, despite recovery of lateral rectus function.

In acute cases, the treatment is designed to relieve the diplopia. A temporary patch of the affected eye can be used as long as there is no lateral rectus function. Once there is some recovery of the abduction, the patch can be alternated between the eyes. The unaffected eye is used when the patient is outside or mobile. The unaffected eye is covered for 1–2 h daily while the patient performs activities in familiar surroundings or in a sitting position. A fresnel prism can be used for correcting the diplopia in primary gaze when less than 15 diopters are required and the prism-induced blur is tolerated by the patient. Botulinum toxin is not indicated in acute cases where the diagnosis is not certain. Following the clinical course for possible recovery of

lateral rectus function is part of the diagnostic evaluation since spontaneous recovery will eliminate the need for neuroimaging in most cases. Also, an initial CT or MR image that is normal does not absolutely exclude the presence of a compressive or infiltrative lesion in the subarachnoid space or cavernous sinus or at the base of the skull.

Patients who have diplopia in primary gaze can be treated with strabismus surgery if they fail to demonstrate any improvement within 4–6 months or if recovery does not progress during a 3-month follow-up. Earlier surgery may be performed if a spread of comitance develops, but with alternating patch therapy this complication is unlikely to occur. A permanent prism can be used to alleviate the diplopia in the patient who does not desire surgical correction. In patients with double vision on lateral gaze that interferes with driving or crossing the street, a fresnel prism can be placed over the outer one third of a glass to relieve the diplopia that exists only in this direction.

8.8.4 Fourth Cranial Nerve Vascular Neuropathy

The true incidence of fourth nerve pareses is unknown because subtle defects are often misdiagnosed. Patients may complain of binocular blurring of vision rather than diplopia, which is worse on downgaze or when the chin elevated. The diplopia is typically obliquely vertical and variable with the direction of gaze and the position of the head. For example, tilting the head to the side of the paretic nerve displaces and tilts the images, while tilting the head towards the contralateral shoulder alleviates the diplopia. Because there is excyclic rotation of the globe, the patient may note a tilting of one of the images. This is most noticeable when viewing objects with a straight vertically oriented edge, such as a doorway frame.

8.8.4.1 Examination

In the acute case, the examination often reveals underaction of the affected superior oblique muscle in comparison to the excursion of the contralateral superior oblique muscle. However, following recovery of some function or the development of spread of comitance, definitive weakness may not be seen with version and duction testing.

The three-step method [430d] uses the prism alternate cover test to measure the hypertropia in primary (first step) and other fields of gaze. The hypertropia increases in the field of adduction and decreases in the field of abduction of the affected eye (second step). The third step in diagnosing a fourth nerve paresis takes advantage of the fact that the superior rectus and superior oblique muscles are intorters and the inferior rectus and oblique muscles are extorters. In order to maintain the proper visual orientation when the head is tilted towards the shoulder, there is intorsion (by the superior rectus and the superior oblique muscles) of the globe on the side of the tilt and extorsion of the opposite globe (by the inferior rectus and inferior oblique muscles). Normally, no vertical deviation develops with head tilting because the elevation induced by contraction of the superior rectus is counteracted by the depression induced by the superior oblique muscle. When a superior oblique paresis is present, tilting the head towards the side of the weak muscle causes the eye on the lesion side to elevate because the superior rectus pull is not balanced by the weak superior oblique muscle. Thus, the hypertropia worsens when the head is tilted towards the side of the paresis. The hypertropia diminishes or disappears upon tilting the head towards the opposite shoulder, because in the latter position the affected globe extorts and the superior rectus and superior oblique muscles on this eye do not contract. The three-step method is particularly useful in diagnosing a fourth nerve paresis in cases when underaction of the superior oblique is not obvious, when there is overaction of the antagonist inferior oblique muscle, or when bilateral fourth nerve pareses (usually traumatic) are present. Patients with other causes of a hypertropia, such as skew deviation or orbital myopathies, will not have a positive three-step examination.

Subjective documentation of the tilting of the image can be measured with double Maddox rods in trial frames.

8.8.4.2 Differential Diagnosis

A vasculopathic etiology is the most common cause of a fourth nerve paresis in patients over age 50, while the cryptogenic or traumatic types are common in younger patients. Fourth nerve palsies can also be caused by neoplasms such as ependymoma, medulloblastoma, or acoustic neuroma that invade the anterior vermis of the cerebellum or lesions, such as glioma or arteriovenous malformation, which extend directly into the quadrigeminal plate of the midbrain. An isolated fourth nerve paresis almost never occurs with a structural or mass lesion. Patients who have these tumors will have an abnormal neurologic examination within weeks of the onset of the fourth nerve paresis, if not at the initial clinical examination [449]. Isolated congenital fourth nerve pareses are not rare. However, in some cases the defect may not be noted until later in life when the patient is found to have an amblyopic eye or a secondary overaction of the antagonist inferior oblique muscle. A fourth nerve paresis (not isolated) can also be found as part of a superior orbital fissure or cavernous sinus syndrome [431, 442, 450]. We have also seen one patient with weakness of the superior oblique as the initial manifestation of myasthenia gravis.

8.8.4.3 Course and Treatment

Recovery of function occurs within 12 weeks in almost all cases with a microvascular etiology. Partial pareses often resolve entirely, but whether an initial complete palsy has a different probability of totally recovering is unknown. The underaction of the superior oblique may disappear, but secondary overaction of the antagonist inferior oblique muscle may contribute to a spread of comitance, causing a persistence of a significant hypertropia. We have seen patients who had this complication, but later became normal approximately 1 year after the neuropathy.

If the acute hypertropia in primary and downgaze is large, the patient is instructed to patch the affected eye while he or she is mobile. When the patient is sitting or in familiar surroundings, the unaffected eye is covered for a minimum of $1-2$ h in order to minimize secondary contracture of the antagonist muscles or the development of a spread of comitance. If the hypertropia is small enough, fresnel prisms can be used to correct the diplopia in primary gaze and downgaze for reading. A permanent prism can be used once the measurements are stabile. Surgical correction of the diplopia can be considered after a minimum of 4 months when the measurements show no recovery or later in cases where there is a stable residual deficit. The procedure should be designed to alleviate the diplopia in primary and downgaze. The type of surgery depends on the extent and pattern of the spread of comitance, the degree of weakness in the superior oblique muscle, and the amount of overaction of the inferior oblique muscle.

8.8.5 Ischemia of Multiple Cranial Nerves

Diabetic microvascular disease can cause consecutive cranial nerve pareses but the first paresis typically improves or resolves before the second nerve becomes involved [451–453]. Some combination of simultaneous third and sixth nerve palsies occurs in less than 7.6% of patients with an ocular motor dysfunction related to ischemic cranial neuropathy [431]. Rarely, concomitant infarction of the cranial nerves, resulting from diabetes mellitus [454] or giant cell arteritis, causes a cavernous sinus or superior orbital fissure syndrome. In these cases, a microvascular etiology is a diagnosis of exclusion that is made after eliminating the entities discussed in the differential diagnosis of a superior orbital fissure or cavernous sinus syndrome [441] (see Sect. 6.2.3).

8.9 Fibromuscular Dysplasia

Cerebral and ocular ischemic events are rarely caused by fibromuscular dysplasia involvement of the cerebral arteries. Fibromuscular dysplasia primarily affects the renal arteries (approximately 90% of cases) but other arteries including the ICA can be affected. Lesions of the ICA or other cerebral arteries have been reported in from less than 1% [455–457] to 10% of patients with fibromuscular dysplasia [458]. Occlusions, dissections, intradural

aneurysms, and arteriovenous fistulas can develop in the ICA or any cerebral artery [459, 460] (see Sects. 2.1.2, 8.12.2). Ischemic disease can affect the visual system at any level. However, most patients with fibromuscular dysplasia of the ICA are asymptomatic and few develop hypoperfusion or embolic disease to the eye or cerebral hemisphere [461–464].

Angiography of the ICA demonstrates lesions such as alternating segments of widening and narrowing, aneurysmal dilatations, or areas of concentric stenosis (Fig. 8.23) [459]. Pathologically, the narrowing is caused by hypertrophy of the fibroblasts in the media and proliferation of the smooth muscle cells or hypertrophy of fibroblasts in the intima, or both. Less commonly, the adventitia of the arteries contains increased amounts of fibrous tissue.

Various therapies have been used for the patients with ischemic complications of cerebral FMD. Surgical correction of the stenotic carotid disease in an asymptomatic patient is generally not recommended because the evolution of the disorder is unknown. Also, the abnormal vessel wall increases the risk of developing a surgical complication such as an iatrogenic carotid cavernous fistula [464]. CEA is performed only when the patient becomes symptomatic as a result of ICA stenosis. EC-IC bypass surgery can also reduce cerebral ischemic complications in patients with occlusion of the ICA [465]. Many patients with embolic amaurosis fugax, TIA, or cerebral infarct benefit from treatment only with medication such as antiplatelet agents [466] or anticoagulation [467]. The therapies for an intradural aneurysm or a carotid cavernous fistula are similar to those outlined in Sects. 6.1.10 and 2.4.

8.10 Moyamoya Disease

A rare progressive neurologic occlusive disorder, Moyamoya disease, develops in children and young adults, primarily of Oriental origin. The deficits result from the occlusion of the distal intracranial ICA, less commonly the vertebral arteries, and the subsequent vascular consequences [468, 469]. The etiology of the occlusions is unknown. The inci-

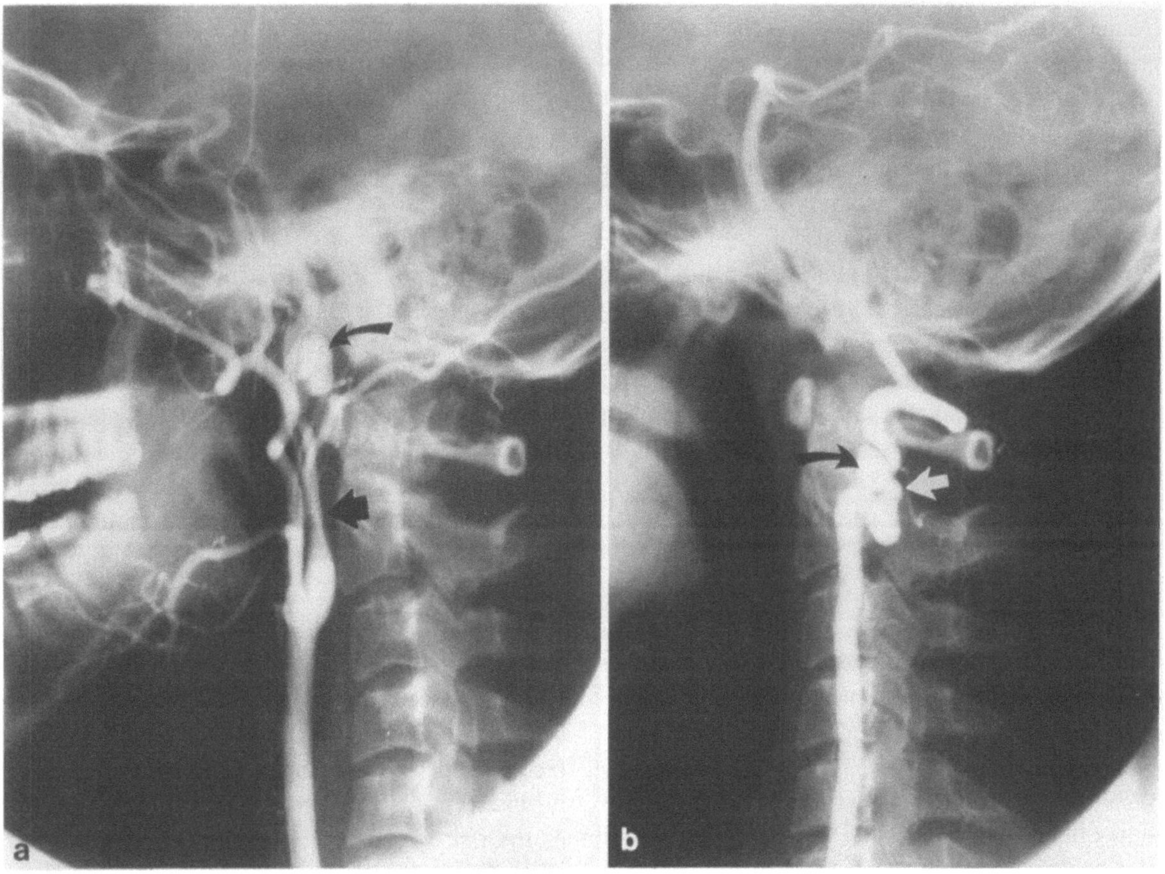

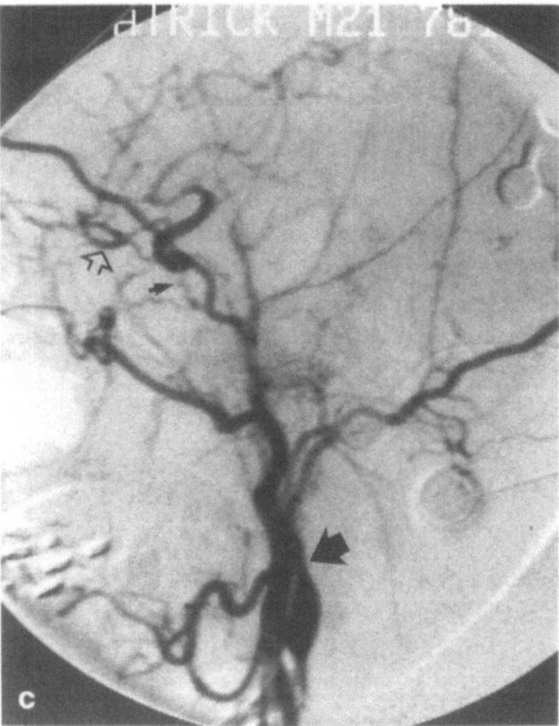

Fig. 8.23 a – c. A 35-year-old man had recurrent amaurosis to the right eye. a The right common carotid arteriogram demonstrates an area of narrowing (*arrow*) and aneurysmal dilatation (*curved arrow*) in the cervical portion of the internal carotid artery suggestive of fibromuscular dysplasia. However, similar findings can also be seen with a spontaneous dissection. b Investigation of another vessel such as the vertebral artery demonstrates the systemic nature of the vascular disorder. Areas of concentric narrowing (*arrow*) and dilatation (*curved arrow*) are characteristic of fibromuscular dysplasia. c The cervical portion of the internal carotid artery is totally thrombosed (*arrow*). The intracranial carotid artery is reconstituted via external carotid artery anastomosis with the inferior lateral trunk (*small arrow*) as well as an anastomosis with the ophthalmic artery (*open arrow*)

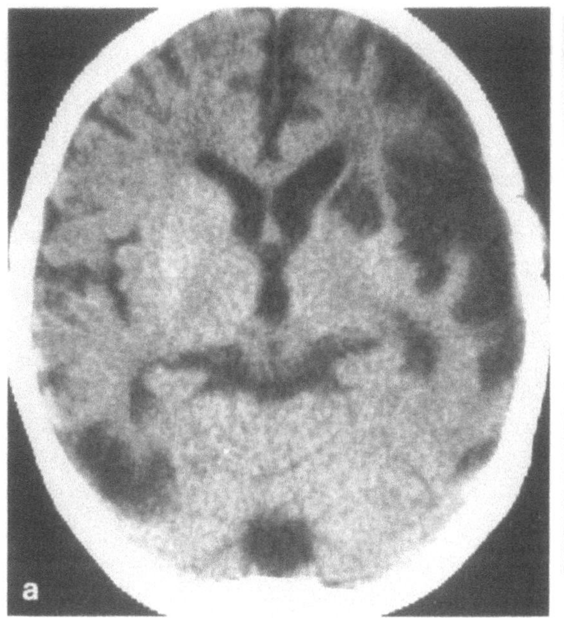

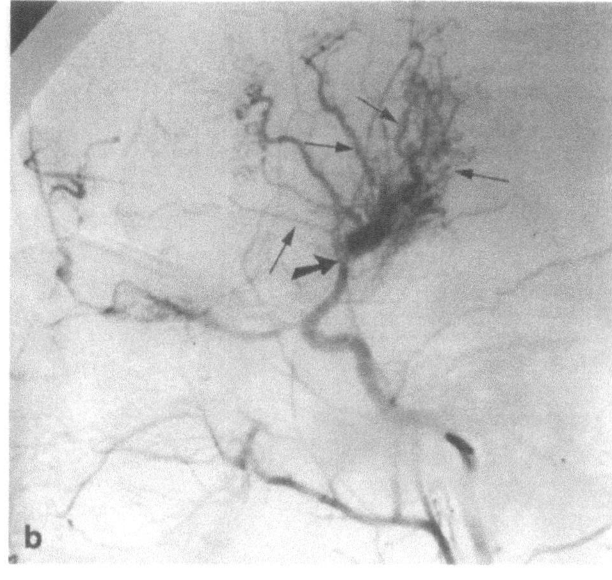

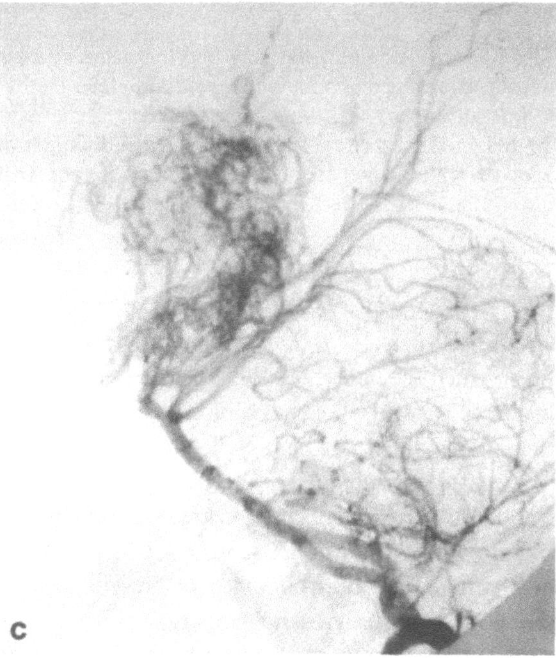

Fig. 8.24a–c. A 9-year-old Korean boy had recurrent ischemic episodes that had begun 2 years prior to evaluation. A significant right hemiparesis and a significant speech, learning, and memory disorder were present. **a** A noncontrast axial view computed tomogram demonstrates multiple cerebral infarcts. Cerebral angiography revealed findings diagnostic of Moyamoya disease. **b** The lateral view right common carotid subtraction angiogram demonstrates occlusion of the supraclinoid carotid (*arrow*) with massive dilatation of the medullary arteries (*small arrows*). **c** Similarly, the lateral view vertebral subtraction angiogram shows occlusive disease involving the posterior cerebral arteries with a large area of dilatation of the proximal perforating branches and choroidal arteries

dence of Moymoya disease is less than 1/100,000 persons. Females are slightly more frequently affected then males.

Compensatory collateral arteries massively enlarge in order to bypass the site of occlusion (Fig. 8.24). These collateral medullary vessels are numerous and entangled giving an appearance of a puff of smoke on angiography (hence the origin of the name in Japanese). These arteries and arterioles are inadequate to prevent distal ischemic events.

These vessels are also fragile so intracranial hemorrhage from their rupture is not rare [470]. Multiple cerebral infarcts are found in both cerebral hemispheres, with the watershed regions typically more frequently affected [469]. Motor and cognitive dysfunction are often severe. TIAs and seizures are also common, particularly in adults.

Visual disturbances typically result from the cerebral infarcts, but visual loss also occurs because of acute branch retinal artery or CRA occlusion

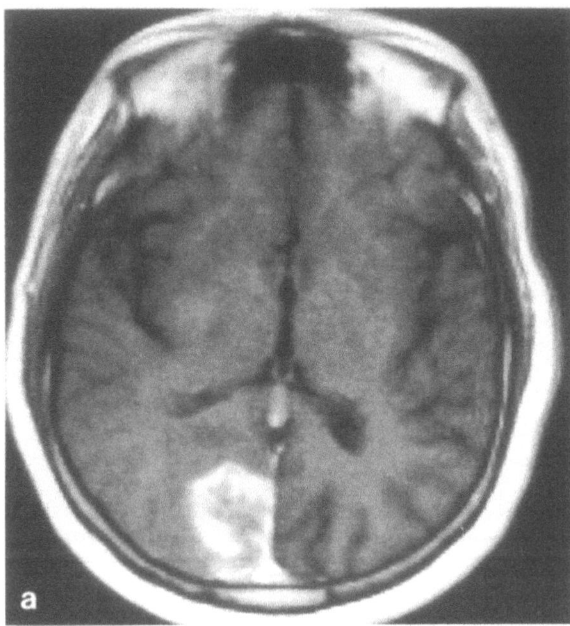
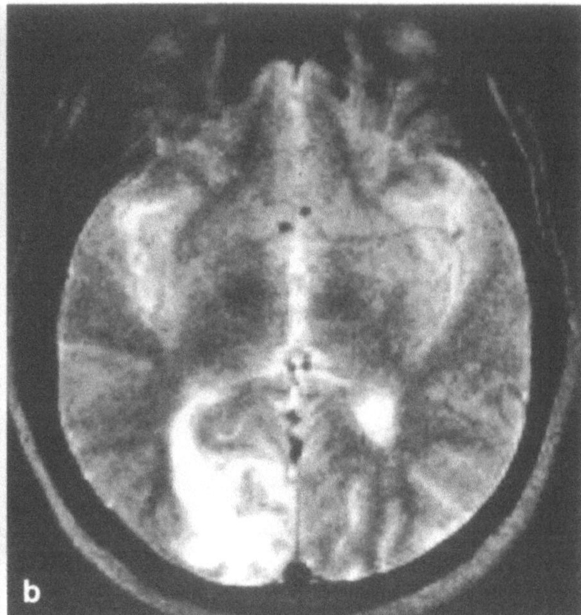

Fig. 8.25a, b. A 57-year-old man had the onset of a left homonymous hemianopia which was complete. He had a known history of lung cancer. **a** The axial view T1-weighted magnetic resonance image demonstrates an area of white signal involving the occipital lobe with questionable mass effect. **b** The T2-weighted image demonstrates possible white matter edema anterior to the right medial occipital mass. A biopsy failed to demonstrate any evidence of metastatic disease. The pathology was diagnostic of a cerebral infarct. The patient was found to have a hypercoagulative condition related to the lung neoplasia

[471, 472]. Patients also report episodes of amaurosis fugax [473]. Diplopia and nystagmus develop when the ischemia affects the posterior circulation [474, 475]. Cortical blindness is seen in patients with occlusion of both PCAs [476].

Treatment with various surgical procedures designed to bypass the ICA occlusion or provide blood supply from the ECA system to the cerebral hemisphere have been reported to prevent or slow the course of progressive cerebral injury [477–479]. The incidence of ICA territory TIAs and infarct is reduced most when both an STA-MCA anastomosis and a temporalis muscle flap is placed on the surface of the cerebral hemisphere [479a].

8.11 Hypercoagulation Disorders

8.11.1 Systemic Malignancy

Cerebral infarcts or hemorrhages can develop as a result of a hypercoagulation disorder associated with systemic cancer such as pancreatic carcinoma [480, 481]. As with an acute cerebral infarct of any etiology, CT can demonstrate a mass effect with an appearance similar to a metastatic tumor (Fig. 8.25). Treatment of the increased blood clotting factors with heparin and warfarin may prevent further systemic and cerebral thromboses.

8.11.2 Primary Coagulation System Deficiencies

Disorders that primarily increase blood clotting or platelet aggregation or block clot lysis are infrequent causes of cerebral and ocular ischemia. In contrast, defects in the normal hemostatic mechanisms or thrombocytopenia typically lead to hemorrhagic complications. Even when there is a severe derangement of the clotting mechanisms, such as with disseminated intravascular coagulation, neurologic symptoms are unusual [482, 483].

Recently, thrombotic disorders due to deficiencies of protein C or protein S have been recognized. Protein C is a normally occurring hepatic vitamin K-dependent protein with fibrinolytic and anticoagulant activity. The complete deficiency of protein C occurs as an autosomal recessive disorder. It is associated with massive systemic venous thrombosis and severe purpura that begins in the neonatal period [484]. Hemorrhagic cerebral infarcts (secondary to cerebral vein thrombosis) and recurrent vitreal hemorrhages account for the neurologic and visual disorders [485–487]. The heterozygote with the autosomal dominant form of this disorder can present with recurrent systemic or cerebral venous thromboses in early adulthood. The heterozygotic condition has also been associated with recurrent amaurosis fugax [488].

One recent study reported that 20% of young adults who had cerebral infarcts or amaurosis fugax without an obvious etiology had a deficiency in protein S, if the protein S level was measured close to the time of the vascular event. The protein S levels were normal when they were measured at a time remote to the episode [489]. Normally, the free, but not the bound, form of protein S reduces clotting by increasing the protein C-induced inactivation of blood factors Va and VIIIa [490, 491]. A protein S deficiency has also been associated with a CRA occlusion [492]. One patient with a protein C deficiency has been reported to have recurrent retinal artery occlusions [493].

8.11.3 Cerebral Venous Thrombosis

Cerebral vein thrombosis is an underdiagnosed cause of acute neurologic dysfunction. Venous occlusive disease is not often considered in the diagnosis except in cases where the patient is dehydrated, pregnant, or puerperal. Additional factors that predispose to cerebral venous thrombosis include blood dyscrasias (such as sickle cell anemia and polycythemia vera), congestive heart failure, sepsis, systemic malignancy, and the use of oral contraceptives. Patients with sickle cell disease are more likely to develop venous obstruction, but arterial occlusive disease in the central nervous system can also occur [494, 495]. Retinal vascular lesions are much more common in patients with the various sickle hemoglobinopathies (see Sect. 8.14.1.4).

Many patients with cerebral vein occlusion also have thrombosis of the adjacent dural venous sinus (see Sect. 3.3) or parenchymal hemorrhage or both.

The signs and symptoms are dependent on the location of the venous infarct. Multiple neurologic signs, including cortical visual loss, and seizures can be present. Recovery of neurologic and visual function depends on the whether adequate collateral venous channels to the affected region exist and whether recanalization of the occluded vein occurs.

Cortical blindness, without other neurologic dysfunction, has been described in only a few patients in the immediate postpartum period from venous thrombosis (which probably involves the posterior sagittal sinus) [496]. Fortunately, the vision spontaneously improves in these cases [497, 498] (see Fig. 3.19 in Chap. 3).

Unless there are extensive thromboses that include the venous sinuses, venous occlusive disease may not be apparent on CT or cerebral angiography. In some cases, CT and MR imaging may reveal a hemorrhagic infarct in the area of thrombosis (see Sect. 3.3). Thrombosis of a specific cortical vein may visualized on MR imaging. Treatment of the predisposing condition and anticoagulation in cases without hemorrhage may prevent further venous infarcts [499–503].

8.12 Dissection of the Carotid and Vertebral Arteries

8.12.1 Traumatic Dissection of the Cervical Carotid Artery

Though carotid occlusion most frequently results from trauma, a traumatic dissection of the ICA can develop which causes ocular and cerebral ischemic episodes. Intimal tears may lead to local thrombus formation, dissection, or a pseudoaneurysm (see Sect. 6.9.2). Subsequent emboli can cause a stroke [504]. In cases without a pseudoaneurysm, medical therapy with immediate heparinization followed by oral anticoagulation or antiplatelet therapy can prevent subsequent ischemic events [505]. When a pseudoaneurysm is present, distal embolization or expansion of a pseudoaneurysm requires surgical resection of the mass and repair of the vessel wall or percutaneous embolization closure of the ICA above and below the defect [506] (see Sect. 2.4.5).

8.12.2 Spontaneous Dissection of the Internal Carotid Artery

Spontaneous dissection of the cervical ICA can cause multiple cranial nerve pareses, monocular visual disturbances, and ischemic dysfunction of the cerebral hemisphere. This disorder was considered to be extremely rare before Fisher's series and review in 1978 [507]. The true incidence of this disorder is unknown but there are numerous reports documenting a variety of clinical signs and symptoms [508]. In one report of 4531 patients who underwent angiography for acute cerebrovascular episodes, 0.4% had dissection of the cervical ICA [509]. Other series describe approximately 2.5% of all patients with a cerebral infarct having a dissection in the ipsilateral ICA. Men and women are equally often affected. The patients range in age from early to late adulthood, with most cases between 40 and 70 years of age.

Patients present with the complaint of a constant nonthrobbing headache ipsilateral to the dissection. The pain can be periorbital, extending into the face, teeth, jaw, or deep in the orbit. Scalp tenderness can also be present. The cranial nerve dysfunction depends on the location and extent or length of the dissection. Because of the location of the twelfth and ninth cranial nerves in the adventitia of the cervical ICA, a unilateral tongue paralysis and dysphagia are often the first manifestations [510, 511]. The patients complain of a bad taste in the mouth which results from irritation of the chorda tympani or glossopharyngeal nerve as it passes forward between the internal jugular vein and the ICA. Disruption of the sympathetic chain in the carotid adventitia causes a third-order Horner's syndrome [512, 513]. Rostral extension of the dissection can result in facial muscle paralysis (seventh nerve) and facial numbness and paresthesias (fifth nerve) [514]. Rarely, an isolated sixth nerve paresis develops [515]. We have examined a patient with total ophthalmoplegia, associated with a traumatic ICA dissection, that spontaneously improved.

Emboli from thrombus formation in the dissection cause ocular and hemispheric dysfunction. In contrast to embolic carotid atherosclerosis, positive and negative monocular visual phenomena such as scintillations or complete amaurosis can develop concomitantly with cerebral TIA or infarct [516].

Transient visual disturbances occur in the ipsilateral eye in up to 30% of cases. Permanent monocular visual loss from infarction of the retina is rare [517]. Significant reduction of ocular perfusion pressure from narrowing of the ICA lumen or an embolus in the ophthalmic artery accounts for the lower ODM diastolic and systolic pressures in the ipsilateral eye. Monocular visual distortion with alterations of posture may develop in these cases [518]. Ocular hypoperfusion that caused ischemic oculopathy has been observed in only one elderly patient with occlusion of the ICA from a presumed dissection [519].

Cerebral TIAs occur in approximately 20% cases, and in a few reports as many as 50% of patients have been reported to suffer a cerebral infarct, although most studies have found this to be an infrequent complication. When cerebral ischemic events occur, the episodes typically develop within weeks of the onset of the dissection, but not after 1 month. Patients with recurrent TIAs over several weeks without infarction have been described. When a stroke occurs it is often associated with complete occlusion of the ICA [520]. A better prognosis for recovery is associated with early reopening of the occluded ICA. A poor prognosis is more likely in cases of persistent occlusion of the vessel.

The cause for the spontaneous dissection is unknown but the disorder seems to be unrelated to atherosclerosis. A prior undiagnosed external trauma or hyperextension of the neck that stretches the cervical carotid artery have been hypothesized as possible inciting events. Fibromuscular dysplasia has been reported in approximately 13% of cases [520, 521]. Cystic necrosis of the carotid wall media is found in some but not all cases [522]. Intimal tears are followed by hemorrhage into the media, which then progressively dissects in the rostral direction. Platelet aggregates and thrombi form in the subintimal space and vessel lumen. The cranial nerves (VIII and IX) may be mechanically compressed by a dilated artery. Disruption of the ICA contribution to the vasovasorum (for example, dorsal meningeal artery to VI) of the cranial nerves (XII, IX, VII, VI) can result in ischemia of the nerves.

8.12.2.1 Imaging

Cerebral angiography is frequently diagnostic. The typical angiogram demonstrates a long segment of stenosis in the ICA called the "string sign" (Fig. 8.26). A localized outpouching of the artery can be found in the area of narrowing or in the segment distal to it (Fig. 8.27). Either severe tapering of the lumen or abrupt closure of the vessel lumen is seen in patients with carotid occlusions. The incidence of occlusion of the ICA on angiography has been reported to be relatively infrequent [520, 521]. The above signs are not pathognomonic for a dissection since the string sign and tapering of the dye column have been seen in patients who did not have a dissection when examined at autopsy [523]. Nevertheless, the angiogram is helpful in that it distinguishes a dissection from fibromuscular dysplasia, vasospasm, and arteritis.

CT and MR imaging show typical findings when a cerebral infarct develops which are not specific for a dissection. Acute thrombus formation in the ICA at any level causes a hyperintense signal on a T1-weighted high-field study (Fig. 8.28). The normal dark flow void on the T1-weighted images is lost in the occluded vessel. MR imaging is useful for noninvasive documentation of the anatomical resolution of the dissection and reversal of the above-mentioned alterations in the vessel [524]. Thin-slice imaging in the coronal plane can demonstrate both cervical carotid arteries for comparison. MR angiography of the cervical carotid artery should be superior for demonstrating the anatomical alterations of the affected ICA.

Ultrasonography of the affected artery will give an intense systolic, low-frequency, high-amplitude Doppler signal that can be traced along the cervical ICA in approximately 76% of patients. Ultrasonography can also be used as a noninvasive method to follow the resolution or reduction in the carotid stenosis [525].

8.12.2.2 Course and Treatment

The dissection can progress over 3 weeks, but the clinical disease is self-limiting with most patients recovering cranial nerve dysfunction within several months. Spontaneous flattening of the dissection and widening of the ICA lumen can be demonstrated by ultrasonography, angiography, or MR imag-

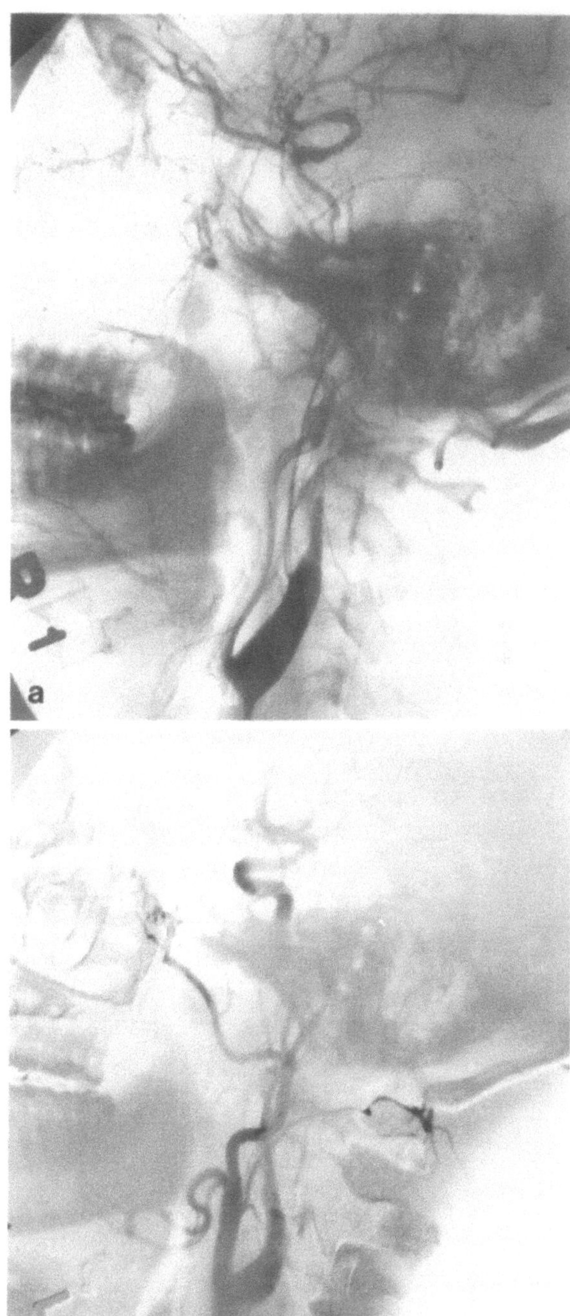

Fig. 8.26a, b. A 30-year-old right-handed man had an episode of expressive aphasia 1 week prior to angiography. Lateral view subtraction angiography of both common carotid arteries shows diffuse narrowing (typical of a "string" sign) of the upper cervical and petrous segments both the right (**a**) and left (**b**) internal carotid arteries. The cavernous segment of the right artery was also narrowed. Thus, a spontaneous bilateral internal carotid artery dissection had occurred

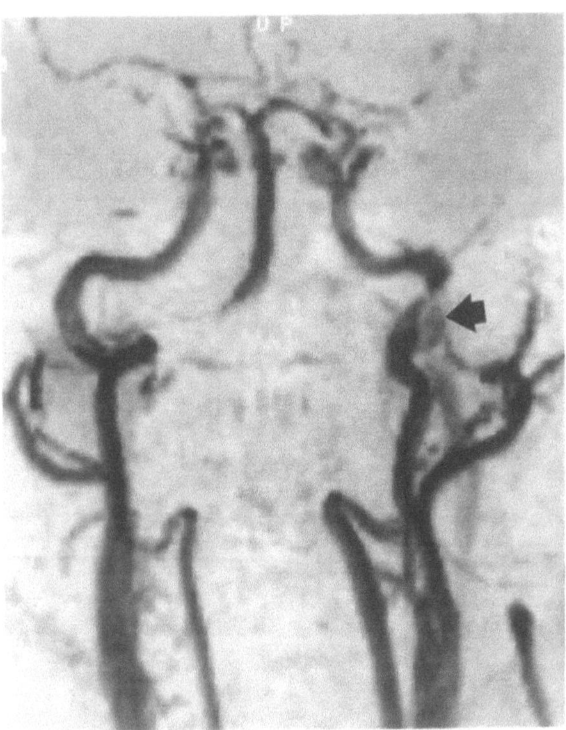

Fig. 8.27. Magnetic resonance angiography was performed on this 33-year-old woman who complained of the onset of a bad taste in her mouth followed by pain along the left forehead and development of a left third-order Horner's syndrome during pregnancy. Except for the Horner's syndrome, the patient was neurologically intact. Magnetic resonance angiography in the frontal view, performed 6 months after the onset of the symptoms, demonstrates an outpouching of the left petrous carotid artery (*arrow*) compatible with a residual defect from a spontaneous carotid dissection

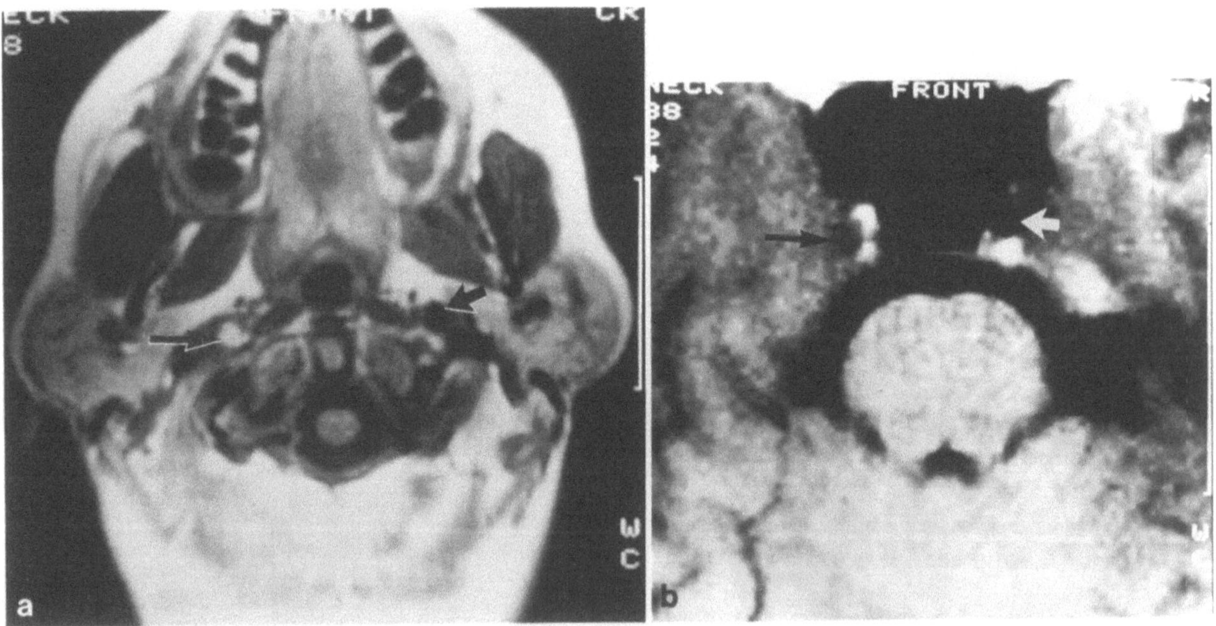

Fig. 8.28 a, b. An axial view magnetic resonance scan was performed in a patient with classic signs and symptoms of a spontaneous dissection of the internal carotid artery. **a** There is a suggestion of complete occlusion of the right petrous internal carotid artery (*thinner arrow*) compared to the flow void (*thicker arrow*) in the left internal carotid artery. **b** Narrowing of the right cavernous internal carotid artery is seen (*arrow*) compared to the left internal carotid artery flow void (*white arrow*). (Case provided by Michael Slavin)

ing. A cerebral infarct and complete occlusion of the ICA can be permanent.

Treatment is usually directed at preventing embolization or thrombus formation in the ICA territory. Heparin sulfate followed by warfarin is given for patients with recurrent TIAs or progression of the dissection or lumen narrowing. In patients with subarachnoid extension of the dissection, anticoagulant therapy may be contraindicated because of an increased risk of a subarachnoid hemorrhage. Antiplatelet drugs such as aspirin may be administered with minimal risk [526]. Endarterectomy is rarely required but has been performed in patients with acute occlusion of the cervical ICA [527]. Surgical occlusion, thrombectomy, and intimectomy, or balloon occlusion of the involved cervical ICA, including extracranial bypass to the middle cerebral artery when necessary, have been performed in patients with recurrent or progressive cerebral hemisphere dysfunction [528].

8.12.3 Dissection of the Vertebral Artery

Spontaneous dissection, with similar causes to that of ICA dissection, uncommonly develops in a vertebral artery. Occipital and suboccipital pain usually precede any neurologic deficits. Brainstem TIAs and infarcts and lower cranial nerve palsies are typical clinical manifestations [529]. Recurrent subarachnoid hemorrhages can also occur [530].

Distal embolization from traumatic occlusions and dissections of the cervical vertebral artery is a well-recognized complication of neck manipulation [531–533]. Brainstem signs such as a lateral medullary infarct can develop immediately or within several days of the injury [534, 535]. Diplopia and blurred vision are frequently noted. Homonymous visual field defects or cortical blindness from occipital infarcts result if emboli or thrombus from the damaged vertebral artery extend into the PCA [536, 537].

8.13 Metabolic Disorders Associated with Cerebral Vaso-occlusive Disease

8.13.1 Homocystinuria

Premature atherosclerosis and cerebral thromboembolic episodes occur in infants and children who are homozygous for a cystathionine synthase deficiency resulting in homocystinuria. The excess homocysteine accumulates in the vascular endothelial cells, damaging the vessels [538]. The heterozygous condition may also predispose to the development of premature atherosclerosis [539].

8.13.2 MELAS Syndrome

Another metabolic disorder is the syndrome of mitochondrial myopathy, encephalopathy, lactic acidosis, and stroke (MELAS syndrome). Multiple cerebral infarcts usually develop in children and young adults [540, 541]. Cortical blindness not infrequently results from bilateral parietal and occipital infarcts. Cerebral angiography is generally unremarkable in these patients.

8.13.3 Fabry's Disease

Recurrent cerebral infarcts occur in young adults with the rare X-linked sphingolipid storage disease caused by α-galactosidase A deficiency [542]. Ceramide trihexoside accumulates and damages the endothelium and smooth muscle of arterioles and peripheral nerve ganglion cells. The kidney, heart, and brain are commonly involved. Typical elevated red–brown skin lesions are found on the upper thighs, scrotum, and abdomen [543–545].

8.14 Specific Retinal Disorders

8.14.1 Arteriolar and Capillary Occlusive Disease

8.14.1.1 Hypertension

The retinal arteries are normally insulated from the effects of systemic hypertension for months. Eventually, patients with a moderate to severe elevation in blood pressure will develop diffuse and multifocal narrowing of the retinal arteries and arterioles (Fig. 8.29a) [546]. The relationship of vascular retinopathy to systemic hypertension first noted by Volhard and Fahr [547] has been the subject of numerous reports. Particularly in the elderly, hypertensive arteriosclerosis widens the arterial light reflex, increases arteriovenous crossing changes, and causes copper and silver wire alterations of the arterioles. Cotton wool spots and capillary alterations are found in areas of retinal ischemia. Microaneurysms, altered capillary permeability, remodelling, telangiectasis, as well as areas of capillary occlusion and loss are best seen on fluorescein angiography [548]. Focal intraretinal periarteriolar transudates may appear similar to the cotton wool spots, but fluorescein angiography reveals punctate areas of leakage from dilated precapillary retinal arterioles in the former abnormality [549]. Systemic hypertension is a risk factor in the development of branch retinal artery occlusion and anterior ischemic optic neuropathy (see Sect. 5.1.3). New retinal hemorrhages are often found in hypertensive patients who do not take their antihypertensive drugs.

When a severe elevation in blood pressure persists or a rapid malignant rise develops, vasospasm and smooth muscle necrosis develop in the arteries of the retina and optic nerve. The autoregulation in the retinal vessels is lost and the blood – retinal barrier, at the level of the endothelium of the arterioles and arteries, decompensates [549]. The leakage of plasma transudate (confirmed by fluorescein angiography) causes edema and exudates in the retina, which often result in a macula star (Fig. 8.29b) and a swollen optic disc [550]. The same process worsens the capillary occlusion, cotton wool spots, retinal hemorrhages, and papilledema [551]. Both eyes are usually involved and the patient complains of visual blurring or metamorphopsia or outright

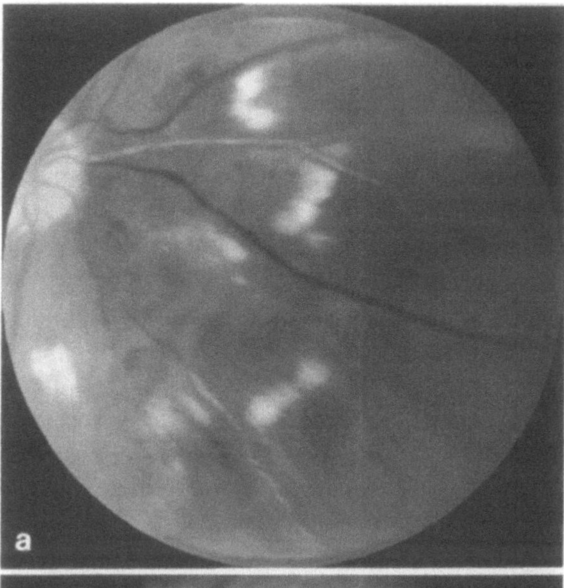
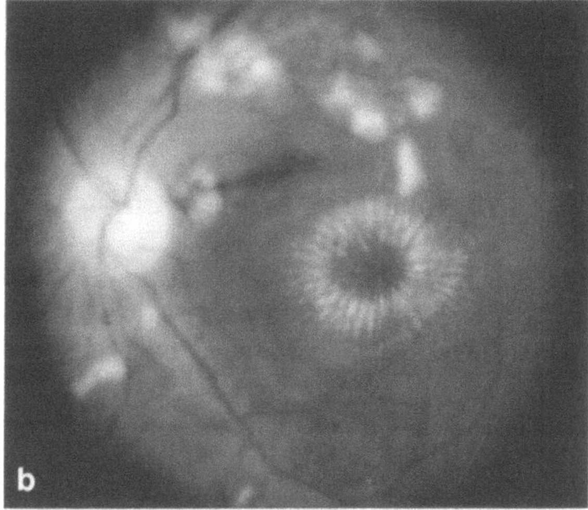

Fig. 8.29. a The arteries temporal to the disc in this markedly hypertensive patient are extremely narrowed and sheathed. There are also areas with multiple cotton-wool spots. **b** In another patient with high blood pressure and hypertensive encephalopathy, both eyes (left eye shown) had diffuse narrowing of the vessels, cotton wool spots, hemorrhages, and a circinate retinopathy because of the damage to the endothelium of the retinal arteries

visual loss depending on the extent of the macula photoreceptor distortion or ischemic damage to the optic nerve, respectively. Patients with malignant hypertension complain of headaches and develop encephalopathy and choroidal ischemia because of similar pathological changes in the vessels of the brain and choroid [552].

Table 8.11. Classification of hypertensive retinopathy (modified from [553])

	Group			
	1	2	3	4
Sclerosis	Mild	Mod.	Sev.	Sev.
General arteriolar narrowing	±	±	±	+
Focal arteriolar narrowing	±	±	±	+
Hemorrhage	−	±	±	±
Exudate	−	−	+	±
Papilledema	−	−	−	+

Mod., moderate; sev., severe; +, present; −, absent; ±, present or absent.

A uniform classification of hypertensive retinopathy has been difficult to establish, but the degree of retinopathy can useful for categorizing hypertensive patients into prognostic groups for the outcome of systemic disease [553]. For example, patients with severe retinopathy (group 4 in Table 8.11) have the highest incidence of cerebral, renal, and coronary occlusive disease.

Hypertensive disease also affects the choroid. The choroidopathy results from focal areas of occlusion of the choriocapillaris. Ischemic necrosis of the overlying retinal pigment epithelium causes focal leak which can be seen on fluorescein angiography. Eventually, these infarcted areas appear as small yellow dots on funduscopy (see Sect. 8.2.1.3) [554−556]. In young patients with malignant essential hypertension, toxemia of pregnancy, renal disease, and pheochromocytoma, more extensive multifocal areas of hypoperfusion of the choriocapillaris result from fibrinoid necrosis and occlusion of choroidal arteries and arterioles [557−559].

8.14.1.2 Diabetes Mellitus

The retinal disease of diabetes mellitus begins in the retinal capillaries and is related to the duration of the disease. After 10 years of diabetes, approximately 25% of juvenile onset and 50% of adult onset diabetics have some degree of retinopathy. The loss of the capillary pericytes leads to microaneurysm formation and areas of nonperfusion. Intraretinal microvascular abnormalities also include areas with dilatation of the capillaries

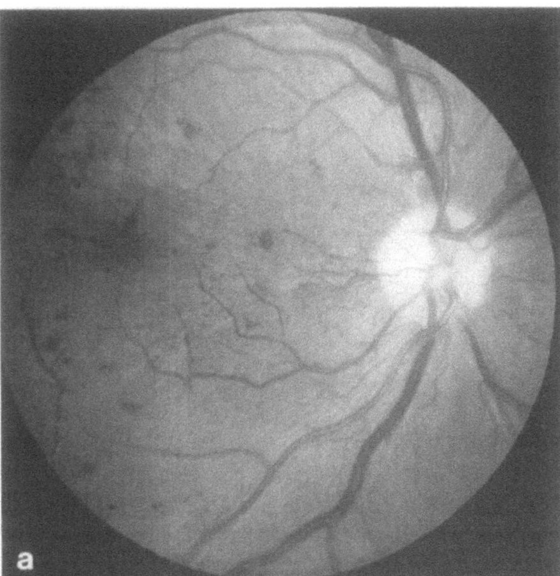

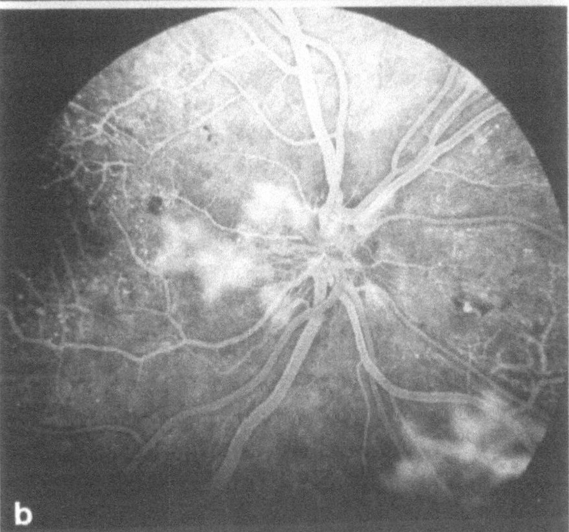

Fig. 8.30a, b. Diabetic retinopathy. **a** Ophthalmoscopy of the right eye reveals dot and blot hemorrhages and edema throughout the posterior pole in this diabetic patient. Dilated vessels are also seen on the disc. **b** Fluorescein angiography in the same patient shows multiple areas with intraretinal capillary loss, occlusions, and microaneurysms. The fluorescein study clearly shows leakage from the optic disc and from an area of the inferonasal retina diagnostic of neovascularization

(Fig. 8.30). The altered vessel permeability leads to edema, lipid exudates, and intraretinal hemorrhages, all of which perturb vision when the macula is affected. The vision can also be compromised by capillary occlusion, which causes an ischemic maculopathy. Areas of capillary nonperfusion and

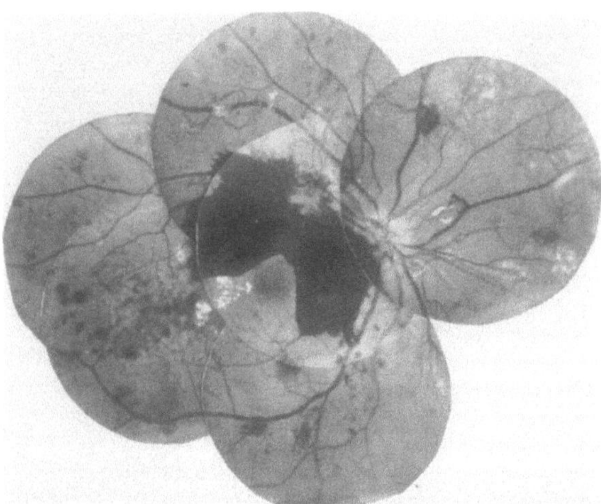

Fig. 8.31. Ophthalmoscopy in a patient shows severe diabetic neovascular retinopathy with a large area of massive retinal hemorrhage with areas of hemorrhage and exudates

retinal ischemia cause the release of angiogenesis factor, which stimulates the proliferation of endothelial cells.

The new endothelial cells lack pericytes and form true neovascularization. The neovascularization, which develops in approximately 10% of cases with retinopathy, is first apparent on the optic disc or along the major vascular arcades on the retinal surface (Fig. 8.31). Neovascularization can lead to devastating visual complications because of the associated vitreous hemorrhages, subhyaloid hemorrhages, and retinal traction detachments. If untreated, blindness develops in 80% of eyes with disc neovascularization and in 20% of eyes with retinal neovascularization that does not include the disc [560]. Both systemic hypertension and renal disease aggravate and accelerate the diabetic retinopathy. Successful treatment of some forms of diabetic retinopathy with laser retinal photocoagulation has been well documented.

8.14.1.3 Radiation Retinopathy

For details on radiation retinopathy, see Sect. 5.11.

8.14.1.4 Sickle Cell Retinopathy

Patients with a hemoglobinopathy caused by the sickle cell genetic defect can develop arteriolar and capillary occlusive disease in the peripheral retina [561] (see Sect. 8.11.3 Cerebral Venous Thrombosis). The structurally altered hemoglobin deforms the erythrocytes, increasing the blood viscosity and decreasing the rate of blood flow in the microcirculation. The slow blood flow worsens the deoxygenation in the erythrocytes, leading to further sickling and clogging of the arterioles, capillaries, and venules. Retinal lesions are most common in patients with hemoglobin C sickle cell disease. Patients with sickle cell and sickle thalassemia disease are less likely to develop severe retinopathy or a vitreous hemorrhage.

The vaso-occlusive disorder causes peripheral subretinal hemorrhages called "salmon patches." Deposits of residual hemosiderin and areas of hyperplasia in the overlying retinal pigment epithelium give an appearance of a sunburst pattern. Gliosis and retinal neovascularization, called "sea fans," develops at the junction of perfused and nonperfused areas of retina (Fig. 8.32). As in diabetes, these incompetent vessels can rupture to cause hemorrhage in the retina and vitreous [562]. A traction retinal detachment may follow a vitreous hemorrhage. Visual loss may also occur as a result of occlusion of the CRA [563, 564], central retinal vein, and choroidal arteries [565]. An optic neuropathy is extremely rare [566].

Laser photocoagulation and cryotherapy should be used to reduce or obliterate the neovascularization prior to a vitreous hemorrhage [567–569].

8.14.1.5 Retinopathy of Prematurity

Retinopathy of prematurity, also called retrolental fibroplasia, is the predominant cause of retinal arterial occlusive disease in infants. However, this disorder only occurs in approximately 500 new cases yearly [570]. Screening retinal examinations of patients at risk are best performed approximately 2 months post partum [571]. Though prematurity is a known risk factor, a greater risk is found in neonates with a birth weight less than 1100 g who require prolonged use of oxygen [572–574]. The

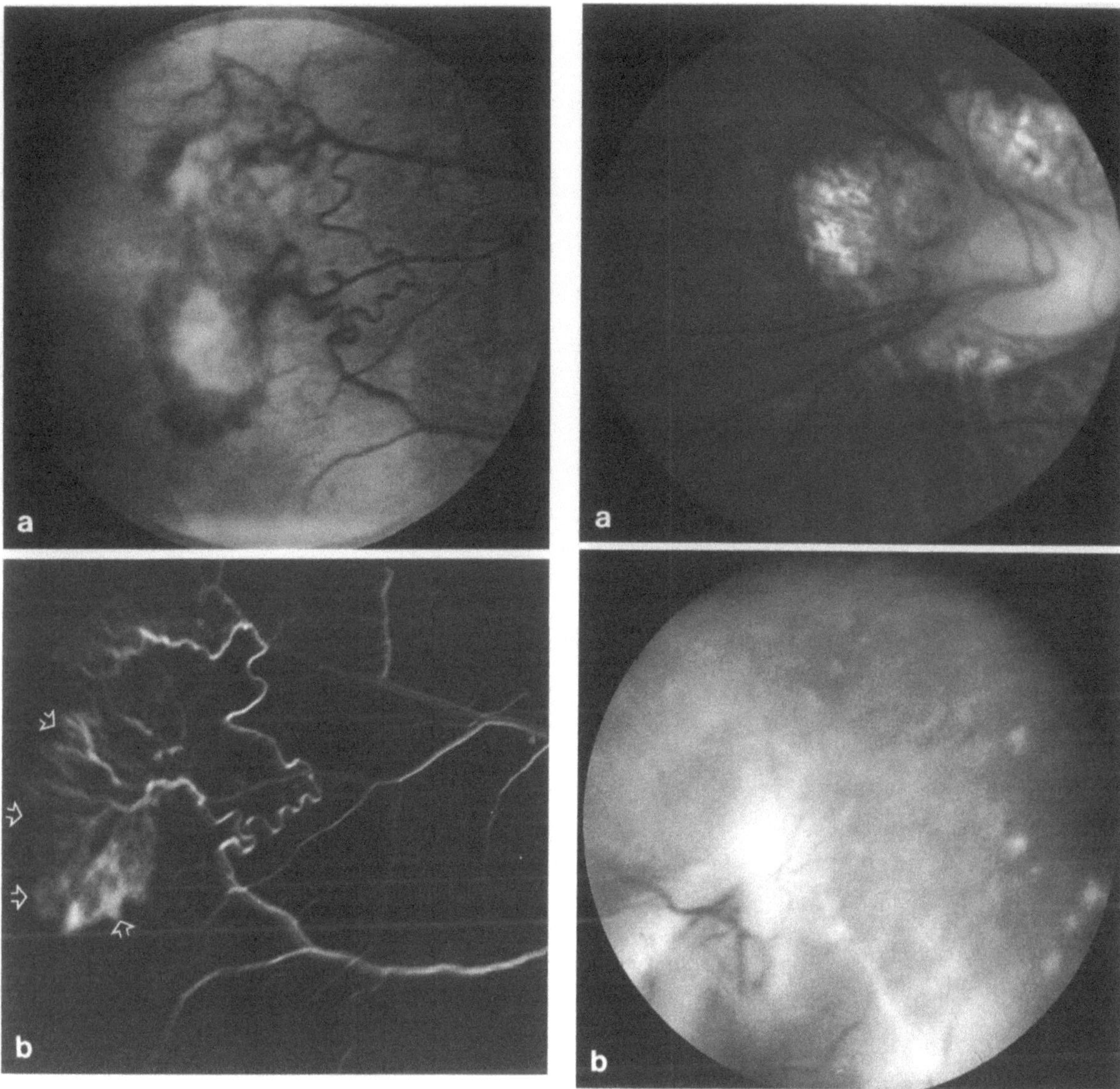

Fig. 8.32. a The posterior pole of this patient with sickle cell disease is normal (not shown) while the midperipheral retina shows a large gliotic neovascular membrane. b The fluorescein angiogram clearly demonstrates the seafans (*open white arrows*) typical of this disorder

Fig. 8.33 a, b. Retinopathy of prematurity. a The peripheral retina shows a line of demarcation between the normal retina and an areas with ischemic, possible neovascular, and gliotic changes. b In another patient, ophthalmoscopy demonstrates temporal dragging of the disc and the macula with straightened, stretched temporal retinal arteries (stage 3). There is a large area of peripapillary chorioretinal atrophy

above normal elevation of intra-arterial oxygen causes a failure of the normal growth of retinal vessels towards the periphery. The oxygen also induces vasoconstriction of immature retinal arterioles, resulting in retinal ischemia. However, not all infants with retinopathy of prematurity have prior exposure to oxygen [575].

The development of retinal disease is categorized by well-described stages [576]. An area of ischemic white demarcation in the retina is seen in stage 1. As this slightly raised tissue becomes revascularized, it appears pink (stage 2) (Fig. 8.33a). The retinal vessels migrate into the demarcation area and towards the peripheral retina. If the oxygen level remains high, retinal neovascularization, scarring, peripheral retinal breaks, and dragging of the macula temporally develop (stage 3) (Fig. 8.33b). Amblyopia and strabismus occur in cases with a displaced macula. Severe proliferation of the abnormal vessels with cicatricial retinal folds are seen in stage 4. Detachment of the retina due to traction, exudates, or retinal tears cause severe visual loss (stage 5). Fluorescein angiography confirms the vaso-occlusive and neovascular nature of this disease [577].

The best therapy is prevention. When oxygen is required, the arterial oxygen level is monitored frequently. The amount of oxygen administered is adjusted in order to maintain the pO_2 at approximately 60 mmHg. Once retinopathy of prematurity is present, recommended treatments have included vitamin E, photocoagulation, cryotherapy, or vitrectomy depending on the stage of the retinopathy. A recent multicenter study has demonstrated the benefit of cryoptherapy application to eyes with stage 3 in reducing the retinal folds and detachments [578, 578a].

8.14.2 Venous Disease

8.14.2.1 Retinal Vein Occlusion

Diseases that block or delay retinal venous blood flow cause findings commensurate with the degree and rapidity of occlusion and the availability of collateral venous channels. For example, a partial obstruction may cause venus dilatation and slight capillary endothelial cell damage that results in extravasation of transudate and red blood cells without severe retinal damage (Figs. 8.4–8.6). In contrast, complete occlusion of a vein causes hemorrhagic infarction in the retina adjacent to the affected vein. Once there is recanalization of the vessel or collateral veins permit adequate venous outflow, the retina may appear normal (Fig. 8.34). However, the edema may persist if the capillary endothelial cells are permanently damaged.

The majority of cases of retinal venous occlusive disease are unilateral and associated with arteriosclerosis. Branch vein thrombosis typically develops in patients 50 years of age or older, many of whom have systemic hypertension. The site of thrombosis is at the arteriovenous crossings where

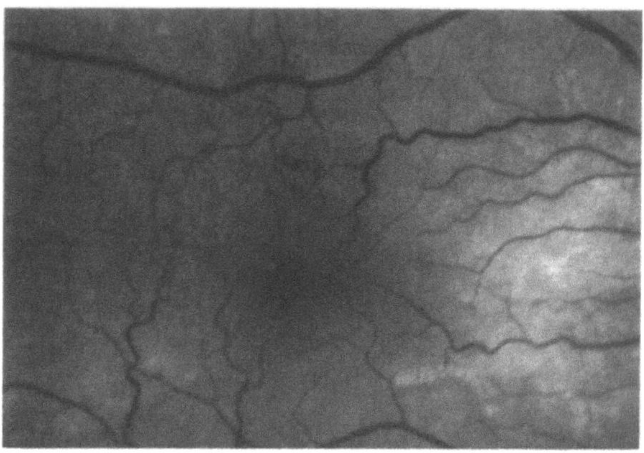

Fig. 8.34. Following a retinal branch vein occlusion, there is an appearance of a vertical oriented venous collateral between the superior and inferior temporal veins just temporal to the macula

both vessels can have a common wall [579]. Blood, edema, and cotton wool spots are seen in the retina along the thrombosed vein. The area of hemorrhage is wedge-shaped with the apex near the site of occlusion. Once the blood has cleared, fluorescein angiography often shows areas of capillary dilatation and loss and retinal edema. When the edema extends into the macula the vision is distorted or diminished. If significant retinal ischemia and areas of capillary nonperfusion persist, retinal neovascularization with the risk of severe ocular complications develop (Fig. 8.35) [580–582].

When the blood is seen in all four quadrants of the retina, the obstruction must be located in the central retinal vein. The extent of visual loss has been used to arbitrarily describe the severity of the central retinal vein occlusion. Patients with mild blurring and preserved acuity are often labeled as having a partial or impending occlusion (Fig. 8.5), while those with more extensive hemorrhages and significantly diminished acuity are said to have a central retinal vein occlusion. However, the presence of secondary retinal ischemia, which occurs in about 30% of cases, is probably more important because ischemia is associated with a poor prognosis for visual recovery [583, 584]. Almost two thirds of eyes with retinal ischemia will also develop focal or diffuse retinal neovascularization [585]. Retinal ischemia develops in an additional 5%–20% of cases that were initially diagnosed as having the nonischemic type of retinopathy [586]. Since multi-

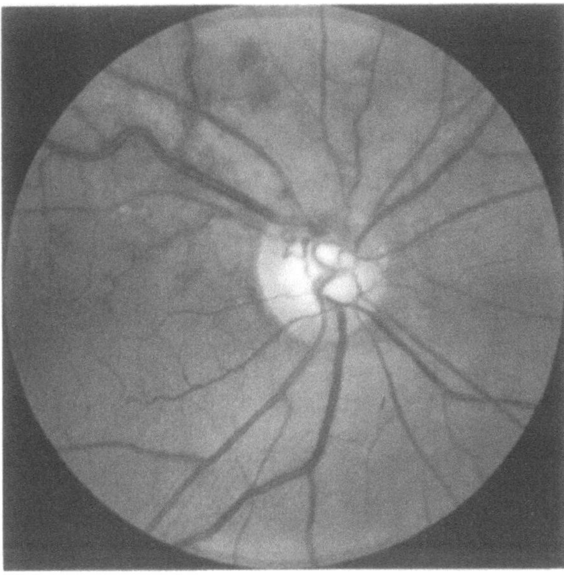

Fig. 8.35. Following a superior temporal retinal branch vein occlusion, neovascularization developed on the optic disc. The venous dilatation and residual hemorrhages were still seen along the superior temporal venous arcade

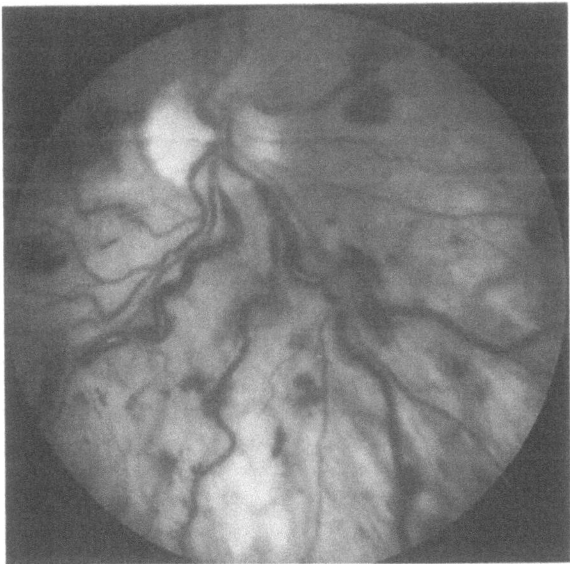

Fig. 8.36. The fundi of both eyes (left shown) reveals marked diffuse venous dilatation and hemorrhages secondary to macroglobulinemia

Table 8.12. Hyperviscosity disorders of the blood (Fig. 8.36) (modified from [590–593])

Polycythemia vera
Myeloproliferative disorders
Waldenstrom's macroglobulinemia
Multiple myeloma
Severe anemia
Thrombocytopenia

ple cotton wool spots are seen in cases without significant retinal ischemia, it can be difficult to determine an ischemic from a nonischemic case by ophthalmoscopy alone.

When performed soon after the onset of a vein occlusion, fluorescein angiography may not identify all of the patients at risk to develop neovascularization. However, the presence of a relative afferent pupillary defect in the involved eye seems to be helpful in differentiating an eye with the ischemic central retinal vein occlusion from the nonischemic type [587]. An abnormal electroretinogram can also be correlated with the presence of secondary retinal ischemia [588].

In one study, patients with an ischemic central retinal vein occlusion had a higher rate (50%) of atherosclerosis in both ICAs. However, these patients had no greater rate of symptomatic disease in the ICA ipsilateral to the vein occlusion than in patients with a nonischemic central retinal vein occlusion [589]. In contrast, other reports have found stenotic or nonobstructive atherosclerotic lesions in the ICA in only 8% of patients with a central retinal vein occlusion [583].

When areas of persistent capillary nonperfusion remain for several months, laser panretinal photocoagulation treatment is commonly employed to prevent neovascularization. Visual loss can persist in cases without ischemia if macula edema (cystoid macula edema) remains or cystic degeneration, retinal pigment atrophy, or a lamellar hole develop in the macula.

Other systemic diseases can also cause retinal venous disease, usually bilateral. Dilated veins and hemorrhages are found as a result of hyperviscosity of the blood (Table 8.12). On occasion a similar fundus picture can be seen in only one eye affected by retinal arterial insufficiency (see Sect. 8.2.1.2).

CHAPTER 9

Inflammatory Vascular Disorders of the Brain and Eye

9.1 Introduction

Systemic inflammatory disorders often cause a multifocal vasculitis that affects the retina, optic nerve, cranial nerves, or brain, so most patients with these illnesses have some neuro-ophthalmologic symptoms or findings during the course of their illness. Additionally, there are several inflammatory vascular diseases whose only clinical manifestations arise from the involvement of the eye or brain or both. All these inflammatory disorders are relatively uncommon to rare in the overall population. Despite the low frequency, the specific entities are discussed here to facilitate their early recognition. Once diagnosed, there are often treatments that can prevent permanent severe visual or neurologic impairment. In the patients with systemic vasculitis who present with neuro-ophthalmologic dysfunction, establishing the proper diagnosis may help in uncovering subclinical abnormalities affecting organ systems outside the nervous system. These findings can be monitored and treated before they become clinically obvious or debilitating.

Some of these disorders predominantly affect the large and medium-sized arteries (arteritis) while others principally involve the small arteries (visible on angiography). When the inflammation spares the arteries and affects the arterioles (approximately 100 µm in diameter and not visible on angiography) it is called an angiitis. Most of these disorders do not spare the capillaries, venules, or veins, so in fact the term vasculitis is more appropriate than arteritis for classifying most entities. Though cerebral angiography can demonstrate abnormalities in the arteries of patients with inflammatory disease, findings such as segmental blood flow or branch occlusion in these vessels are not pathognomonic of inflammation (Fig. 9.1). Similar angiographic findings are also seen with atherosclerosis and cerebral emboli. Even in the retina, where

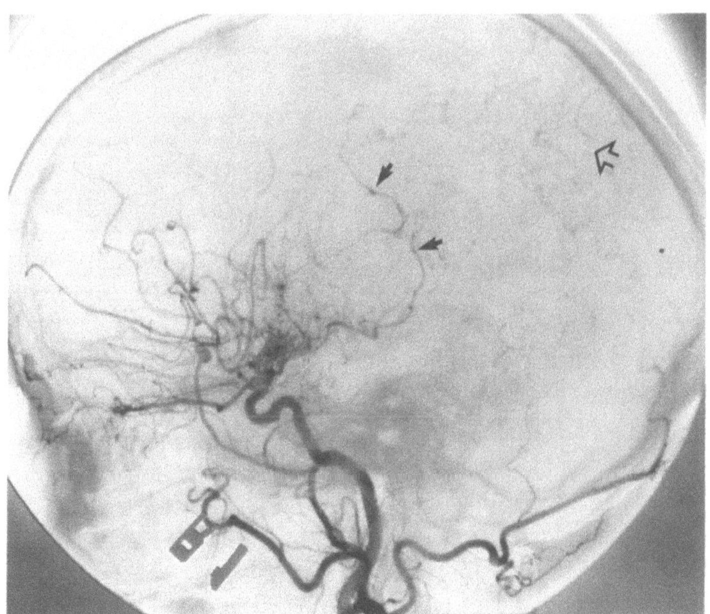

Fig. 9.1. This 21-year-old man experienced recurrent right cerebral hemorrhages. The lateral view right common carotid artery subtraction angiogram shows narrowing in many cerebral arteries with segments of widening and narrowing (*arrows*). Note the middle cerebral branches to the occipital region (*open arrow*). Irregular caliber cerebral arteries were also seen on the left internal carotid angiogram (not shown)

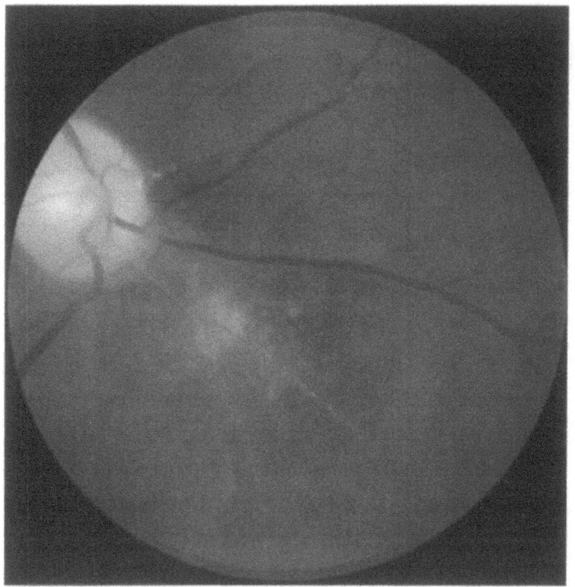

Fig. 9.2. Ophthalmoscopy in a patient with vasculitis shows diffuse narrowing of the retinal arteries, sheathing of these vessels, and large areas of secondary "gliosis" in some areas of the retina. There is also secondary (antegrade) atrophy of the optic disc

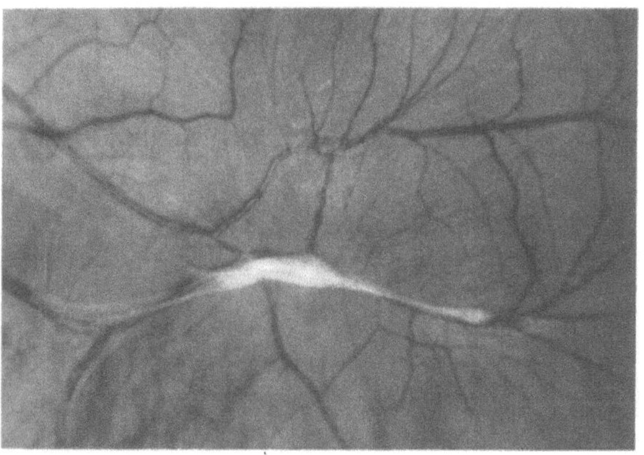

Fig. 9.3. Focal perivenous sheathing is highly suggestive of an inflammatory process in a patient with Behçet's syndrome

small arteries and arterioles can be viewed, the differentiation of vasculitis from noninflammatory occlusive disease is not always possible by ophthalmoscopy (Fig. 9.2). In fact, many patients with systemic vasculitis have retinal and brain arterial occlusive episodes that result from the secondary hypertension because of renal involvement. The presence of significant perivascular sheathing and

fluorescein dye leakage from the vessels on fluorescein angiography are strongly suggestive of an inflammatory process (Fig. 9.3).

Though infectious vasculitids are frequent in all immune-compromised patients and have been recognized in human immunodeficiency virus (HIV) positive patients [1], they are only briefly considered in this monograph. The hematogenous spread of bacteria, fungi, and viruses can involve the arteries, arterioles, and capillaries of the eye and brain (see Sects. 9.13.2, 6.9.3). Both tuberculosis and syphilis can cause ocular and neurologic symptoms associated with multiple vascular occlusions. A direct spread from suppurative processes in the meninges or paranasal sinuses can also cause a thrombophlebitis in the cerebral veins and dural venous channels (see Sect. 3.3).

9.2 Giant Cell Arteritis (Temporal Arteritis)

Giant cell arteritis predominantly affects the optic nerve: see Sect. 5.2 for further details.

9.3 Polyarteritis Nodosa

9.3.1 Incidence and Systemic Involvement

Polyarteritis nodosa (PAN) is an uncommon but not a rare disease. The onset is usually in the fifth decade of life, but it can affect both younger and older subjects. Men are affected 2.5 times more often than women. PAN is a systemic necrotizing vasculitis that can affect any organ system. Central nervous system involvement occurs in approximately 25% – 50% of cases [2]. This is about half as frequent as involvement of the kidney, which is the most frequently involved organ. Since the respiratory system is usually spared, PAN can be clinically differentiated from Wegener's granulomatosis.

9.3.2 Neurologic Manifestations

Most of the central nervous system dysfunction results from secondary hypertension-induced acceleration of cerebrovascular disease [3]. Approximately 10% of patients have neurologic symptoms or findings at their initial presentation. During the course of the illness, approximately 50% of patients develop a peripheral neuropathy, which is typically a mononeuritis multiplex that affects both sensory and motor nerves. However, a progressive distal polyneuropathy can also occur [4]. Cerebral infarcts, psychoses, and confusion are the most prevalent manifestations of central nervous system involvement. Seizures, visual disturbances and cranial neuropathy occur about half as frequently [5, 6]. Intracerebral hemorrhages are unusual but periarteritis of parenchymal penetrating arteries commonly causes lacunar infarcts. Polyarteritis rarely involves the major cerebral arteries.

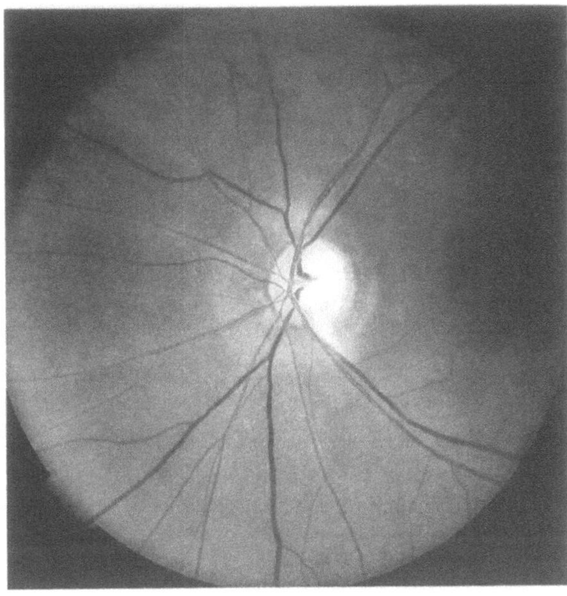

Fig. 9.4. Following an episode of uniocular visual loss which did not recover, optic atrophy and sheathed and narrow retinal arteries suggestive of vasculitis are seen in a patient with polyarteritis nodosa

9.3.3 Ophthalmologic and Neuro-ophthalmologic Manifestations

The ophthalmic manifestations of PAN are also frequently related to the complications of systemic hypertension (see Sects. 8.2, 8.14.1.1) [6–11]. Most of the findings result from retinal ischemia. Retinal edema or cotton wool spots are commonly seen on funduscopy, while retinal arterial sheathing or a true arteritis is infrequent (Fig. 9.4). An occlusion of the central retinal artery or a choroidal infarct is also uncommon.

Optic nerve dysfunction is seen in few cases. Rarely, an occlusion of a posterior ciliary artery can result in an anterior ischemic optic neuropathy. In these cases the optic disc is typically pale and swollen and the visual loss irreversible [10, 12]. Eventually, flat pale atrophy results. A posterior ischemic optic neuropathy is even rarer and has been described as the presenting sign of PAN in only one reported case [13]. An optic neuritis is also rare [13a]. At the onset of the visual disturbance, differentiating posterior ischemic optic neuropathy from a retrobulbar optic neuritis can be difficult since both may cause a central scotoma or an altitudinal field defect. However, with posterior ischemic optic neuropathy, the field defect may be

denser or absolute and the visual loss will not recover.

Other ophthalmic disorders such as an episcleritis, uveitis, or orbital inflammatory disease are also rare in PAN [14–16]. However, when a patient has recurrent or bilateral ocular or orbital inflammation, a systemic evaluation for PAN and Wegener's granulomatosis should be undertaken. In these cases, a biopsy of the orbital lesion is also appropriate in order to eliminate a neoplasm, such as lymphoma, as the cause.

9.3.4 Neuroimaging

There are no pathognomonic computed tomographic (CT) or magnetic resonance (MR) abnormalities for PAN. A cerebral infarct, no matter what the cause, has a typical appearance. Although angiography will visualize arterial aneurysms, segmental stenosis, and arterial occlusions, similar abnormalities are seen in cases as a result of systemic hypertension or emboli from an atrial myxoma [17]. Additionally, cerebral angiography is often

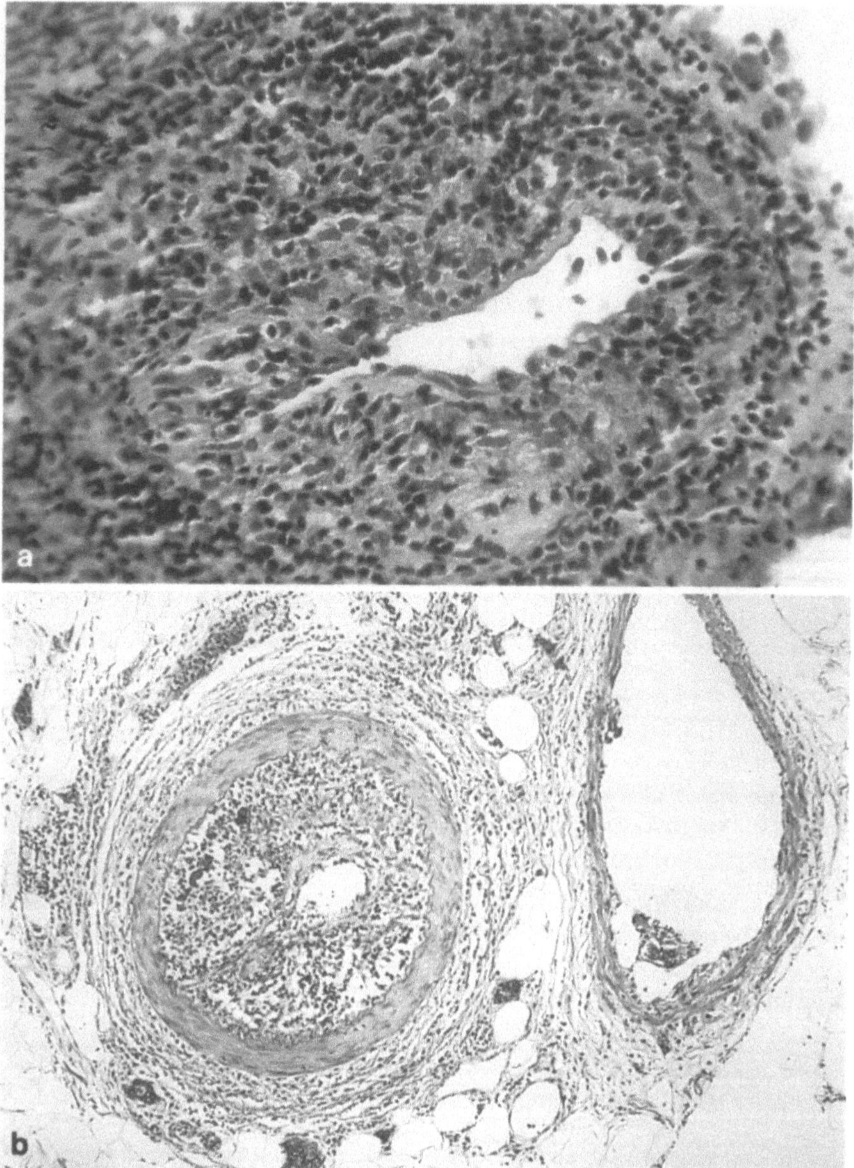

Fig. 9.5a, b. A 41-year-old woman developed a steroid responsive optic neuropathy after polyarteritis nodosa had caused cerebral dysfunction and ischemia of the bowel 3 and 15 years before the optic neuropathy, respectively. **a** Biopsy of a cerebral mass lesion revealed a "pseudotumor" of inflammation. The walls of the involved cerebral arteries were thickened by a diffuse monocytic inflammation, which included lymphocytes and plasma cells. **b** The bowel resection specimen also shows thrombus and inflammation in the arteries but not the veins. (From [13a])

normal in patients with PAN who have had cerebral ischemic episodes [18].

9.3.5 Pathology

Although the diagnosis of PAN is suggested by the presence of the laboratory abnormalities (see below), a biopsy which shows a vasculitis of the medium-sized and small arteries in muscle and skin is diagnostic (Fig. 9.5). The temporal artery is another vessel readily accessible for biopsy. On occasion, a renal biopsy is needed to secure the diagnosis. These biopsy specimens often show immune complexes deposited on endothelial cells and polymorphonuclear and mononuclear cell infiltrates in the arterial walls. Hepatitis B surface antigen, complement, and immunoglobulins can also be found in the arterial walls. The media and internal elastica can be focally or segmentally destroyed. The inflammatory process results in fibrinoid necrosis of the entire vessel wall. Nodular aneurysmal dilatations develop at arterial branchings that are affected by an acute or healed panmural necrosis. Thrombosis occurs in the lumen adjacent to the arteritic vessel wall. In active cases, arterial occlusions that cause infarcts are seen in most organs. The healed vessels may be free of inflammatory cells and may have an appearance that is quite similar to the changes caused by atherosclerosis.

9.3.6 Laboratory Evaluation

For details of the laboratory evaluation, see Table 9.1.

Eosinophilia is not part of the profile of PAN.

9.3.7 Treatment and Course

Treatment is essential since the prognosis for survival of untreated patients with PAN is only 13% after 5 years [19]. The survival rate improves to 48% if patients are treated with corticosteroid alone [19]. However, many cases, particularly those who have severe progressive PAN, require the addition of immunosuppressive agents such as cyclophosphamide or azathioprine. Immunosuppressive therapy increases the remission rate to 80% and permits the use of alternate day corticosteroids [20].

Table 9.1. Laboratory abnormalities in active polyasteritis nodosa

Abnormality	Percentage of patients
Elevated ESR	>90
Anemia or leukocytosis	70
Urine protein or casts	70
Reduced levels serum complement	50
Circulating immune complexes or positive rheumatoid factor	>60
Positive hepatitis B surface antigen	40
Saccular aneurysms renal angiogram	80

9.4 Systemic Lupus Erythematosus

9.4.1 Incidence and Systemic Involvement

Systemic lupus erythematosus (SLE) is the second most common collagen vascular disease, affecting women nine times more frequently than men. SLE begins mostly during early adulthood but patients can be afflicted at any age. Any organ can be damaged by the vasculitis and immune complex deposition. The clinical manifestations are listed in Table 9.2. In cases where the central nervous system or kidneys are affected, the survival rate is significantly reduced.

Table 9.2. Clinical manifestations of systemic lupus erythematosus

1. Fever
2. Skin rash
3. Arthritis and arthralgias
4. Renal dysfunction: glomerulonephritis, nephritis
5. Lymphadenopathy
6. Myalgias
7. Pleuritis, pericarditis
8. Hemolytic anemia
9. Neurological dysfunction
10. Hepatomegaly

9.4.2 Neurologic Manifestations

Neurologic and neuro-ophthalmologic manifestations are frequent in patients with well-established SLE. Any portion of the nervous system can be affected. Psychoses and episodic confusion [21] develop from what is clinically termed "lupus cerebritis," but most other neurologic dysfunction clearly results from the occlusive vasculopathy. Immune complex deposition and vasculitis are less frequently documented mechanisms for the neurologic symptoms [22]. Cerebral infarcts and hemorrhages are commonly associated with the presence of the lupus anticoagulant (see Sect. 9.17) and hypertension. Cerebral infarcts rarely result from emboli originating from noninfectious cardiac valvular thrombi, associated with Libman–Sacks lesions [23]. Rarely, thrombocytopenic purpura is associated with cerebral dysfunction and microangiopathic lesions in the brain [24].

Various cerebral hemispheric syndromes, movement disorders, seizures, aseptic meningitis, and increased intracranial pressure develop in 20%–65% of patients with SLE. A transverse myelopathy has been less frequently described [25], but it is probably more common than diagnosed. Neurologic dysfunction usually develops in patients with well-developed active SLE, but it can be the first manifestation of SLE [26]. Most of the patients with neurologic episodes also have evidence of systemic disease activity such as fever, hypocomplementemia, vasculitis of the skin or gastrointestinal tract, and a platelet count of less than 100,000/mm^3. Abnormal cerebrospinal fluid with elevated protein or pleocytosis occurs in 32% of patients with neurologic symptoms [27].

Cranial nerve palsies develop in approximately 8% of patients with SLE [27]. Ptosis, diplopia, pupillary abnormalities, and facial paralysis can all occur.

9.4.3 Ophthalmologic and Neuro-ophthalmologic Manifestations

9.4.3.1 Retinal and Choroidal Dysfunction

The most common ocular findings are nonspecific signs of a retinal vasculopathy, such as single or multiple retinal hemorrhages and cotton wool spots. Retinal exudates, retinal and disc edema, arteriolar narrowing, and retinal microaneurysms are also seen frequently [28, 29]. These retinal lesions can be either unilateral or bilateral and they suggest an active disease stage. Though retinal arterial and arteriolar occlusions are common, occlusion of the central retinal artery is uncommon in SLE [30, 31]. However, when significant levels of circulating immune complexes [32] or multiple arterial thromboses markedly reduce and slow the flow of blood through the retinal circulation [33], the retinal ischemic lesions can be extensive and the visual loss severe (Fig. 9.6). A drug-induced lupus syndrome that occurs with procainamide therapy rarely causes ophthalmic signs of vaso-occlusive disease of the retina [34].

Less frequently, a choroidopathy, manifested by serous retinal elevation and retinal pigment mottling, develops because of occlusions in the choriocapillaris and the resulting ischemic damage of the retina pigment epithelium. The etiology of the choroidal vasculopathy, either hypertension or a possible local vasculitis, cannot be differentiated clinically [35, 36]. However, in one pathologically studied case, thrombi and narrowing of the choroidal arteriolar lumens typical of hypertensive disease were found, and signs of vasculitis were not demonstrated [37].

9.4.3.2 Other Ophthalmologic Dysfunction

Other ophthalmologic abnormalities occur infrequently. Episcleritis, iritis, or vitritis rarely develops in patients without an accompanying retinal vasculopathy.

9.4.3.3 Optic Neuropathy

The incidence of optic neuropathy is low relative to the incidence of retinal or other central nervous system involvement in SLE. Both demyelinating optic neuritis and ischemic optic neuropathy have been recognized with increasing frequency [38, 39]. However, only in rare patients is an optic neuropathy the initial or the sole manifestation of SLE [40]. In some patients the optic nerve dysfunction is associated with a concomitant paresis of another cranial nerve [27].

When a unilateral or bilateral optic neuropathy

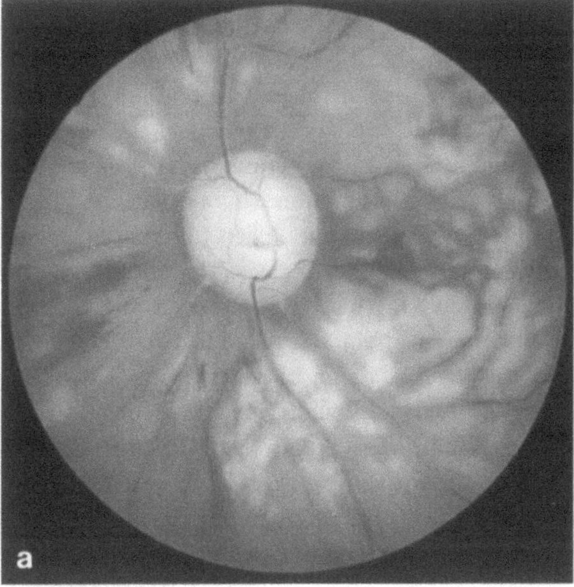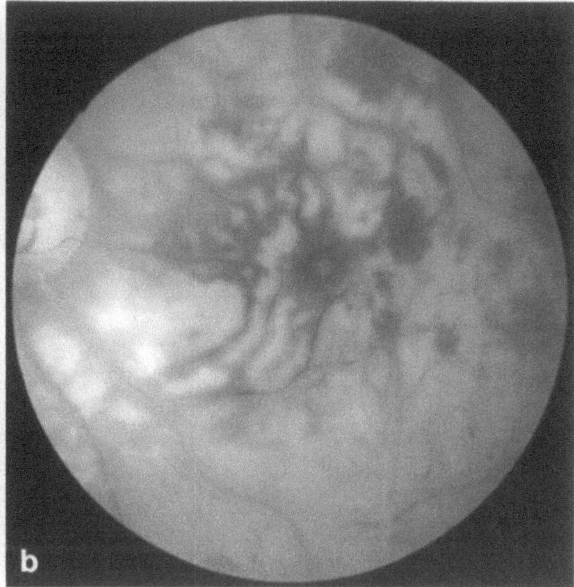

Fig. 9.6 a, b. Systemic lupus erythematosus typically causes retinal ischemia without significant visual loss unless there is confluence of the cotton wool spots. This patient with well-established SLE presented with bilateral poor vision with acuities of 15/200 that deteriorated to finger counting at 1 foot despite corticosteroid and plasmapheresis therapy. **a** One month after the onset of visual loss, ophthalmoscopy of both eyes (left eye shown) revealed pallor of the optic disc, severe narrowing of all the retinal arteries, and diffuse ischemia and hemorrhages of the inner layers of the retina. **b** Ophthalmoscopy of the macula region of the left eye reveals marked ischemic changes

is preceded or followed by a myelopathy, the patient with few signs of SLE may be diagnosed as having Devic's syndrome, a forme fruste of multiple sclerosis [41, 42]. Patients with Devic's syndrome, also called neuromyelitis optica, frequently have a virulent course. The progressive permanent visual loss and spinal cord dysfunction is atypical for early multiple sclerosis. Unquestionably, some of the patients receiving the clinical diagnosis of a demyelinating disease such as Devic's syndrome in fact have a collagen vascular disease such as SLE.

9.4.3.4 Retrochiasmal Disturbances

SLE can also cause retrochiasmal disorders of vision [43]. Visual hallucinations, migraine-type fortification imagery, visual seizures, or transient visual loss as severe as cortical blindness all occur [44]. Homonymous scotomas and visual field loss can be permanent [45]. As with other neurologic disturbances, these visual disturbances are rarely the presenting symptoms of a patient with SLE [46]. Also, these cerebral visual disorders are almost always associated with other signs and symptoms of central nervous system dysfunction.

9.4.3.5 Papilledema

Papilledema and increased intracranial pressure caused by SLE is an unusual cause of a corticosteroid responsive "pseudotumor cerebri" [47, 48]. An associated significant anemia may aggravate the papilledema and associated visual loss [49]. Occlusion of the dural venous sinus has been demonstrated in some of these patients, indicating that they do not have true idiopathic increased intracranial hypertension. Venous occlusive disease is the likely cause of the increased intracranial pressure in most patients, but thrombosis in the dural venous channels may be difficult to diagnose clinically or demonstrate by CT and MR imaging (see Sect. 3.3). The venous phase of a cerebral angiogram may be required to demonstrate an irregularity of the contrast column in the area of

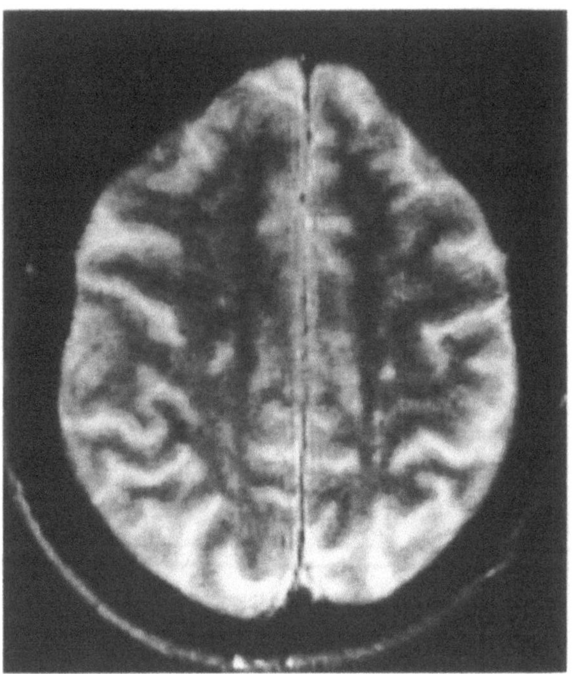

Fig. 9.7. A 35-year-old women with well-established systemic lupus erythematosus, headaches, and vague paresthesias had a magnetic resonance image which shows multiple bright spots in the cerebral white matter on the T2-weighted study

thrombus. In addition, if the occlusion is only transient, the lesion may not be revealed by any of the available imaging modalities.

9.4.3.6 Diplopia

Diplopia can develop from lesions affecting the efferent ocular motor pathways from the brainstem to the eye muscles. Painful ophthalmoplegia from inflammatory disease of the orbit or cavernous sinus is rarely seen with SLE [50–52]. An internuclear ophthalmoplegia [53, 54] and isolated cranial nerve palsies have also been rarely noted [55]. Cranial neuropathies are more frequent in patients who have additional neurologic deficits [27].

9.4.4 Neuroimaging

CT or MR imaging is frequently, but not always abnormal in patients with neurologic dysfunction. The most common finding on the CT is a focal or diffuse cerebral atrophy, which occurs most often when dementia or psychoses is present [60, 61]. The CT appearance of cerebral infarcts (common) and parenchymal hemorrhages (uncommon) support a vasculopathic etiology for much of the neurologic dysfunction. MR imaging can also demonstrate areas of cerebral infarction. More often, the MR image shows white matter defects that can be indistinguishable from the plaques of multiple sclerosis [62] (Fig. 9.7). However, in patients with SLE, the white matter bright areas tend to be located deeper in the cerebral hemisphere and less often in a periventricular distribution.

Cerebral arteriography is unnecessary in most cases. Abnormalities in the large cerebral arteries are infrequent, but angiography has demonstrated stenosis or complete occlusion of the supraclinoid internal carotid artery or middle cerebral artery in patients with major cerebral infarcts [63, 64].

9.4.5 Pathology

The classical report detailing the pathology of the nervous system in SLE by Johnson and Richardson showed microinfarcts and parenchymal hemorrhages resulting from an arteriolar and capillary vasculopathy to be the predominant lesions in the brain [55]. A true vasculitis was demonstrated histopathologically in only three of 24 cases, though in many cases, in this and other studies, perivascular inflammatory cells were seen [65]. Five of the 24 cases had hyaline degeneration or fibrinoid necrosis in the cerebral vessels. Active vasculitis of the cerebral, meningeal, choroidal, and retinal arterioles is more likely to be found in patients with acute fulminant systemic disease rather than in patients with a chronic indolent course [66]. Narrowing of the lumen and occlusion of the internal carotid artery or middle cerebral arteries are also unusual [64]. In patients with psychiatric manifestations but no neurologic deficits, the brain histopathology may not demonstrate significant vascular lesions [67].

The myelopathy of SLE can result from demyelination, a vasculopathy, or both [41, 68]. The histopathology of the affected optic nerve will show areas of demyelination or vasculitis or both, and cavitary necrosis may be present.

9.4.6 Laboratory Evaluation

In addition to the clinical manifestations (Table 9.2), the laboratory evaluation (Table 9.3) is essential in establishing a diagnosis of SLE. The cerebrospinal fluid is abnormal in approximately 50% of patients with neurologic dysfunction. Elevated protein levels or a lymphocytic pleocytosis are the most commonly found abnormalities. Other markers of an inflammatory disorder such as immune complexes, reduced levels of C4 complement, and oligoclonal bands are typically detected in the cerebrospinal fluid [56, 57, 58]. The presence of these abnormalities is not unexpected since the choroid plexus can accumulate immunoglobulins and complement factors that circulate in the blood [59].

Table 9.3. Laboratory abnormalities in systemic lupus erythomatosus

1. Anemia
2. Leukopenia
3. Thrombocytopenia
4. Lupus anticoagulant
5. Direct Coombs test
6. Positive ANA including single and double stranded DNA
7. Antibodies to cytoplasmic clements
8. LE cells
9. Rheumatoid factor
10. False positive test for syphilis (VDRL)
11. Reduced serum complement level
12. Elevated serum gamma globulin level

9.4.7 Treatment and Course

The neurologic and ophthalmologic manifestations of SLE may progress, remain static, or remit spontaneously. When treated with corticosteroids, patients usually improve from the initial as well as subsequent neurologic deficits. In some patients, corticosteroid treatment fails as a result of initiating treatment with inadequate doses. Often, these patients are then maintained on a moderate or low dose for a prolonged time period while side effects develop without suppression of the disease process. The starting dose of corticosteroid must be large enough (in some cases megadoses are required) to improve the clinical dysfunction. Thereafter, the corticosteroid dose is rapidly reduced to a level that maintains the benefit with the fewest complications. Alternate day therapy should not be given until significant improvement is produced. While large doses of corticosteroids may actually precipitate psychoses [69], in some patients this medication improves the lupus psychosis [27]. The use of short-term megadose corticosteroid therapy, when delivered properly, carries no increased risk of significant morbidity over conventional high-dose therapy.

Though corticosteroids often rapidly improve the neurologic or visual symptoms, severe or progressive disease can require the addition of immunosuppressive agents, particularly if prolonged therapy is required. Immunosuppressive agents affect a different mechanism of the disease and often allow the use of a lower dose of corticosteroids. A combination of corticosteroids and azathioprine or cyclophosphamide is often more effective than steroid therapy alone [69, 70]. Immunosuppressive therapy may not be beneficial if too small a dose is given. An effective dose of azathioprine (usually 150−250 mg daily) will cause an erythrocyte macrocytosis or reduce the lymphocyte count. Cyclophosphamide in a dosage of 1−2 mg/day (maintaining a leukocyte count less than 4000/mm^3) is usually required to have a beneficial effect.

Corticosteroids alone may only minimally improve an episode of fulminant retinal or neurologic dysfunction. In these cases plasmapheresis may reduce the levels of vessel-damaging circulating immune complexes and restore visual and neurologic function, if the damage is not severe. A combination therapy of corticosteroids and plasmapheresis followed by an immunosuppressive agent is occasionally warranted.

Various approaches have been used in treating the optic neuropathy of SLE. Though the vision can spontaneously improve, similar to the course of an idiopathic optic neuritis, in most cases moderate to high doses of corticosteroids are required [38]. The optic neuropathy may not always remit or may even worsen despite conventional dose of steroids. In these patients, the visual loss often responds to megadose corticosteroids followed by immunosuppressive therapy [40, 71]. When patients have a central or centrocecal scotoma pattern of visual field loss, they are more likely to have a neuritis and respond to steroids. In contrast, patients with a dense altitudinal field defect are more likely to have an infarct from vascular occlusion as the etiology [39, 72] and are less likely to respond to steroids.

Antiplatelet and anticoagulation therapies are considered in patients with recurrent arterial thrombotic episodes and a positive lupus anticoagulant (see Sect. 9.17).

9.5 Rheumatoid Arthritis

9.5.1 Incidence

Rheumatoid arthritis affects both adults and children. It is the most common connective tissue disorder [73], but neurologic and neuro-ophthalmologic complications are rare. Systemic vasculitis, rheumatoid granulomas, and circulating autoantibodies rarely affect the central nervous system.

9.5.2 Neurologic Manifestations

Seizures, dementia, or a noninfectious meningitis can develop. Congestion and occlusion of cortical and pial veins results from significant serum hyperviscosity. A cerebral vasculitis can be caused by immunoglobulin M, immunoglobulin G, and complement deposition in the arteriole walls [74–77]. Neurologic and visual dysfunction can also develop from an intracerebral hemorrhage that complicates anticoagulation treatment used for the systemic manifestations of occlusive vasculitis, such as ischemic bowel disease [78].

Neurologic dysfunction rarely develops because of an inflammatory or mass effect of intracranial rheumatoid nodules in the brain, meninges, or cerebrospinal fluid [77, 79, 80].

9.5.3 Ophthalmologic and Neuro-ophthalmologic Manifestations

A recurrent or chronic anterior uveitis is commonly seen in patients with juvenile rheumatoid arthritis. The iritis is silent in approximately 50% of cases with iritis and bilateral in almost 67% of patients [81]. Chronic uncontrolled intraocular inflammation often leads to glaucoma, corneal band keratopathy, and iris–lens synechiae. In contrast to adolescents, uveitis is uncommon in adult-onset rheumatoid arthritis. Episcleritis, scleritis, and progressive scleromalacia are less frequently found [82]. Retinal vascular occlusions are rare. Retinal lesions, particularly in the macula, arise as infrequent complications from prolonged use of chloroquine [83]. The chloroquine accumulates in the retinal pigment epithelial cells of the macula, which then degenerate.

Visual loss from central nervous system dysfunction is exceedingly rare. Only one patient has been reported with cortical blindness who also had diffuse cerebral dysfunction present [75]. Optic atrophy is a rare consequence of a chronic pachymeningitis [84]. An ischemic optic neuropathy secondary to vasculitis is also extremely rare [85].

9.5.4 Laboratory Evaluation

The laboratory evaluation is detailed in Table 9.4.

Table 9.4. Laboratory abnormalities in rheumatoid arthritis

Leukopenia or leukocytosis
Elevated ESR
Reduced serum albumin and elevated α_2-globulins
IgM antibody to Fc portion of IgG
  (rheumatoid factor)
HLA-DRw4
Anemia
Thrombocytopenia
Antinuclear antibodies

9.5.5 Treatment

Therapy for the arthritis, designed to suppress the inflammation and improve mobility in the joint, can be accomplished by nonsteroidal and steroidal anti-inflammatory agents. Gold salts, penicillamine, or hydroxycholoroquine can induce a remission. Immunosuppressive drugs can prevent progression or recurrences. The ocular lesions are usually controlled by ophthalmic topical corticosteroids but systemic therapy with corticosteroids or immunosuppressive agents may be required. Patients with a central nervous system vasculitis are treated with systemic corticosteroids or immunosuppressive agents or both.

9.6 Wegener's Granulomatosis

9.6.1 Incidence and Systemic Involvement

Wegener's granulomatosis is an uncommon disorder that affects men slightly more often than women and begins, in most cases, between the fourth and fifth decades of life. This necrotizing granulomatous vasculitis principally attacks the upper and lower respiratory tract and kidneys. Rarely, patients have a focal destructive granuloma of a kidney or lung or other organ as the sole mani-

Table 9.5. Differential diagnosis of systemic granulomatous disease

---

Sarcoid
Tuberculosis
Histoplasmosis, blastomycosis, coccidioidomycosis
Syphilis
Berylliosis
Midline granuloma[a]
Granulomatous angiitis[b]
Allergic vasculitis
Henoch-Schönlein purpura
Lymphatoid granulomatosis
Systemic lupus erythematosus
Sjögren's syndrome

---

[a] Differentiated from Wegener's granulomatosis by the absence of vasculitis.
[b] Limited to the central nervous system.

festation. Patients present with sinusitis in over 80% of cases, with otitis in 50% of cases, and with a productive cough in under 20% of cases. A fever, gum, nasal, or pharyngeal inflammation or hemorrhages, chest pain, joint pain, weight loss, malaise, or headache are noted in less than 10% of cases at the time of presentation. The necrotizing inflammation can destroy the nasal septum as well as the bones and cartilage of the nasal bridge (leading to a saddle nose deformity). These cases must be distinguished from those with other forms of systemic granulomatous disease (Table 9.5).

9.6.2 Neurologic Manifestations

Neurologic dysfunction is most often secondary to systemic hypertension, but a cerebral vasculitis or direct intracranial extension of the granulomatous inflammation from the adjacent sinuses are the causes in some cases [86]. A mononeuritis multiplex or polyneuropathy can also result from the systemic vasculitis. An intracerebral or subarachnoid hemorrhage or a myopathy occur less commonly [87]. Rarely, a massive cerebral infarct develops, and in these cases cerebral angiography reveals evidence suggestive of a cerebral arteritis [88]. Primary granulomatous inflammation of the cranial nerves is also unusual [89].

9.6.3 Ophthalmologic and Neuro-ophthalmologic Manifestations

Neuro-ophthalmologic symptoms are rarely the presenting problem but ocular or orbital involvement is not unusual in patients with a well-established illness. Ocular lesions have been described in patients with a limited form of Wegener's granulomatosis that spares the kidneys [90]. The inflammation can affect the cornea, conjunctiva, sclera, or orbit [91]. The associated perilimbal infiltration or marginal ulceration appear similar to a Mooren's ulcer, Terrien's degeneration, staphylococcal hypersensitivity, and corneoscleral lesions caused by other systemic occlusive vasculitids [92–94].

Orbital inflammatory disease is rarely the only clinical manifestation [95, 96], but Wegener's

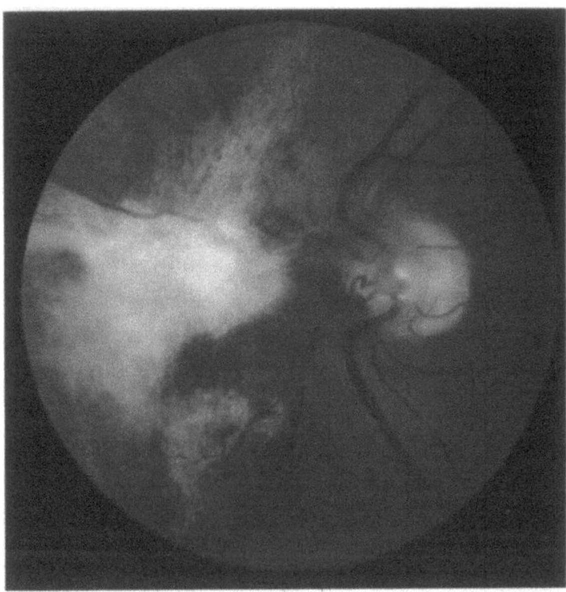

Fig. 9.8. Ophthalmoscopy in a patient with Wegener's granulomatosis shows a necrotizing retinal vasculitis

granulomatosis should be considered in cases of recurrent inflammation of the orbit or when the sinuses are involved. Lacrimal gland inflammation may occur alone or when there is involvement of the other orbital soft tissues. Secondary complaints arise if paranasal sinus granulomatous inflammation obstructs the nasal lacrimal duct and causes epiphora.

9.6.3.1 Retinal and Optic Nerve Lesions

Visual loss has been reported from both optic nerve and retinal lesions. An ischemic optic neuropathy can result from posterior ciliary artery vasculitis. Extension of granulomatous inflammation from adjacent sinuses into the orbit or optic canal can cause acute visual loss due to inflammation and compression of the retrobulbar or intracanalicular optic nerve [97, 98]. In one case, concomitant involvement of the cavernous sinus caused partial third and sixth nerve pareses and an anesthetic cornea [99]. Visual dysfunction infrequently results from uveitis, retinal phlebitis [100], occlusion of the retinal arteries or arterioles, including the central retinal artery [101], or branch artery retinal vasculitis [102] (Fig. 9.8).

9.6.4 Laboratory Evaluation and Pathology

The diagnosis is established by the clinical profile, the presence of serum antineutrophil cytoplasmic antibody (95% of cases), and biopsy of the lesion. There are no other laboratory abnormalities specific for Wegener's granulomatosis. Typical of many inflammatory disorders, an elevated erythrocyte sedimentation rate, normochromic normocytic anemia, thrombocytosis, and leukocytosis are commonly found.

The histopathological examination of the specimen is usually diagnostic. Fibrinoid necrosis develops in the affected small arteries and veins. In the acute lesions, the vessels are infiltrated by polymorphonuclear leukocytes. Chronic vascular lesions are dominated by monocytes and contain areas of fibrotic scarring. However, not all patients with clinical manifestations of central nervous system involvement have pathological evidence of vasculitis of the cerebral vessels [103]. Areas of necrosis and granulomas containing giant cells can be found in any organ of the body.

9.6.5 Treatment and Course

If Wegener's granulomatosis is untreated, the involved organs are progressively destroyed. Corticosteroids can be moderately successful therapy for some of the systemic manifestations. Cytotoxic immunosuppressive agents, particularly cyclophosphamide, are frequently required in patients with renal or pulmonary compromise [104, 105]. Other immunosuppressive agents, such as azathioprine [106] or chlorambucil [107], can also be effective. When using cyclophosphamide or chlorambucil, the total leukocyte count should not fall below 3000 cells/mm^3 and the polymorphonuclear cells should not be fewer than 1500 cells/mm^3. Recently, trimethoprim–sulfamethoxazole has been reported to be beneficial in selected patients [108].

9.7 Behçet's Syndrome

9.7.1 Incidence and Systemic Involvement

The classical triad of recurrent oral ulcers, genital ulcers, and intraocular inflammation involving both the anterior and posterior ocular chambers is found in patients with Behçet's syndrome [109]. The disease generally has a prevalence of less than 1 per 10,000, with a higher prevalence in middle eastern and far eastern countries. Men and women are equally often affected in the United States, but men of middle eastern origin have a greater risk. The first signs usually develop in the third decade, but the age of onset can range from 9 to 50 years. Ocular or oral lesions are the earliest problem for which medical attention is sought in approximately 50% of cases. Though symptoms of lesions in a single locale can occur, two of the three areas of the triad are needed for a certain diagnosis. Later in the course of the illness, most patients have lesions in multiple locations, although they are not necessarily present simultaneously. Because the triad of findings defines the illness, oral ulcers (aphthous stomatitis) are found in 98%, genital ulcers in 88%, and ocular inflammation in 76% of cases. Thrombophlebitis (35%), neurologic deficits (30%), and arthralgias or arthritis (30%) are common. Behçet's joint disease predominantly affects the knees, ankles, and wrists [110, 111]. Many of the clinical manifestations of Behçet's syndrome are caused by a multifocal vasculitis with an unknown etiology that principally affects the venules.

9.7.2 Neurologic Manifestations

Central nervous system dysfunction rarely precedes the onset of other systemic signs of the disease. The presence of a neurologic deficit is associated with a high rate of mortality. Brainstem and motor system defects occur more frequently than sensory system losses [112]. A myelopathy, with paraplegia or tetraplegia, and an encephalopathy are also common. The neurologic dysfunction can remit and relapse, with intervals relatively free of neurologic symptoms between relapses. Cerebral vein thromboses account for many cerebral deficits [113].

9.7.2.1 Neuropathology

The histopathology of the central nervous system lesions reveals focal softening in the brain or spinal cord along the distribution of the affected blood vessels. The inflamed walls of the veins, capillaries, and arterioles contain mononuclear and polymorphonuclear cell infiltration with deposition of immunoglobulins M, G, and A and complement. Perivascular lymphocytic infiltration and areas with minimal demyelination and perivascular cuffing are found in the tissue adjacent to the affected vessels [114].

9.7.3 Ophthalmologic and Neuro-ophthalmologic Manifestations

Because Behçet's syndrome is a multifocal disorder, neuro-ophthalmologic signs are frequently encountered. Unilateral and bilateral optic neuropathies are not uncommon. Although patients with optic nerve dysfunction typically improve, residual scotomas often follow a partial remission [112]. Blurred vision also occurs in patients with a brainstem lesion-associated nystagmus. The nystagmus can be either the "jellylike" or conjugate jerk type [115]. Partial or complete pareses of the sixth, fourth, third, or seventh cranial nerves develop from inflammation in the meninges or in the brainstem [116].

Papilledema and increased intracranial pressure result from the venous hypertension caused by dural venous sinus thrombophlebitis [117, 118]. Approximately 9% of patients with neurologic disease have been reported to have papilledema [119].

9.7.3.1 Intraocular Inflammation

The sterile hypopyon in the anterior chamber, which is classically described, is actually an uncommon complication of this syndrome. A granulomatous anterior uveitis with a viscous aqueous due to the elevated protein content is more common [116]. Panuveitis, with inflammation of both the retinal and choroidal vessels and a secondary vitritis that obscures the media, frequently develops and compromises vision. Cystoid macula edema also causes visual loss when a vitritis is present. In these cases, the fluorescein angiogram reveals focal or diffuse nonspecific leakage from small and large veins or the pooling of fluorescein dye in the macula. The retinal phlebitis (Fig. 9.3) causes retinal venous dilatation and thrombosis with retinal exudates and hemorrhages. Choroidal infarcts cause retinal and choroidal pigment loss and clumping. The ocular disease is bilateral in approximately 80% of patients with eye disease. Retinal lesions and bilateral ocular disease are more common in men with HLA type B5 [120]. Cataracts and secondary glaucoma are additional complications of the chronic uveitis, corticosteroid treatment, or both. Severe recurrent or continuous uveitis or keratitis often results in unilateral or bilateral blindness within 6 years of the onset of the ocular inflammation [121].

The inflammatory nature of the ocular involvement is confirmed by the histopathological demonstration of perivascular round cell infiltration in the optic nerve, central retinal artery, branch retinal arteries, and choroidal vessels [122].

9.7.4 Laboratory Evaluation and Neuroimaging

There are no pathognomonic laboratory abnormalities of Behçet's syndrome. During the phase of active disease, the erythrocyte sedimentation rate, the leukocyte count, and serum immunoglobulins are elevated. The cerebrospinal fluid contains an elevated concentration of monocytes in 80% of patients with central nervous system dysfunction and in less than 25% of those who are neurologically intact [123].

CT shows nondiagnostic changes with focal lucencies that do not enhance with contrast [124, 125]. MR imaging can demonstrate areas of atrophy and infarction. Hypertense white matter lesions are typically visualized on the T2-weighted images.

9.7.5 Treatment and Course

As mentioned previously, once neurologic dysfunction develops, the long-term prognosis for survival is poor. The prognosis for normal sight is also poor once a posterior uveitis or retinal phlebitis is pre-

sent. Corticosteroids, administered systemically or locally to the eye (via topical ophthalmic drops or ointment, or via a subconjunctival injection), are beneficial for the immediate treatment of the systemic or ocular disease, respectively. However, the long-term outcome of systemic and ocular inflammation is probably unaltered by corticosteroid treatment alone.

The addition of azathioprine increases the success rate of treatment of uveitis, but the results are better yet with the use of chlorambucil [126]. Both cyclophosphamide and chlorambucil have been reported to be effective treatments for the ocular, neurologic, and systemic dysfunction [127, 128].

In some patients a recurrence of the inflammation can be prevented by colchicine, which prevents polymorphonuclear neutrophil migration. Colchicine has the advantage of having less risk of severe complications than immunosuppressives [129]. Recently, cyclosporin A has been used to successfully treat the ocular lesions of Behçet's syndrome [130]. Thus, there are numerous therapeutic options in the treatment of Behçet's syndrome.

9.8 Granulomatous Angiitis Confined to the Central Nervous System

9.8.1 Neurologic Manifestations

Granulomatous angiitis is a rare disorder which is usually limited to the small arteries and arterioles of the central nervous system, particularly those located in the leptomeninges [131, 132]. Dementia, meningitis, depressed level of consciousness, and seizures are the most common manifestations. Focal cerebral deficits occur less frequently. One case causing visual loss [131] and another involving proptosis and ophthalmoplegia [133] have been described. Pareses of the cranial nerves, including the innervation of the extraocular and facial muscles and sensation, can develop late in the course of the disease [134].

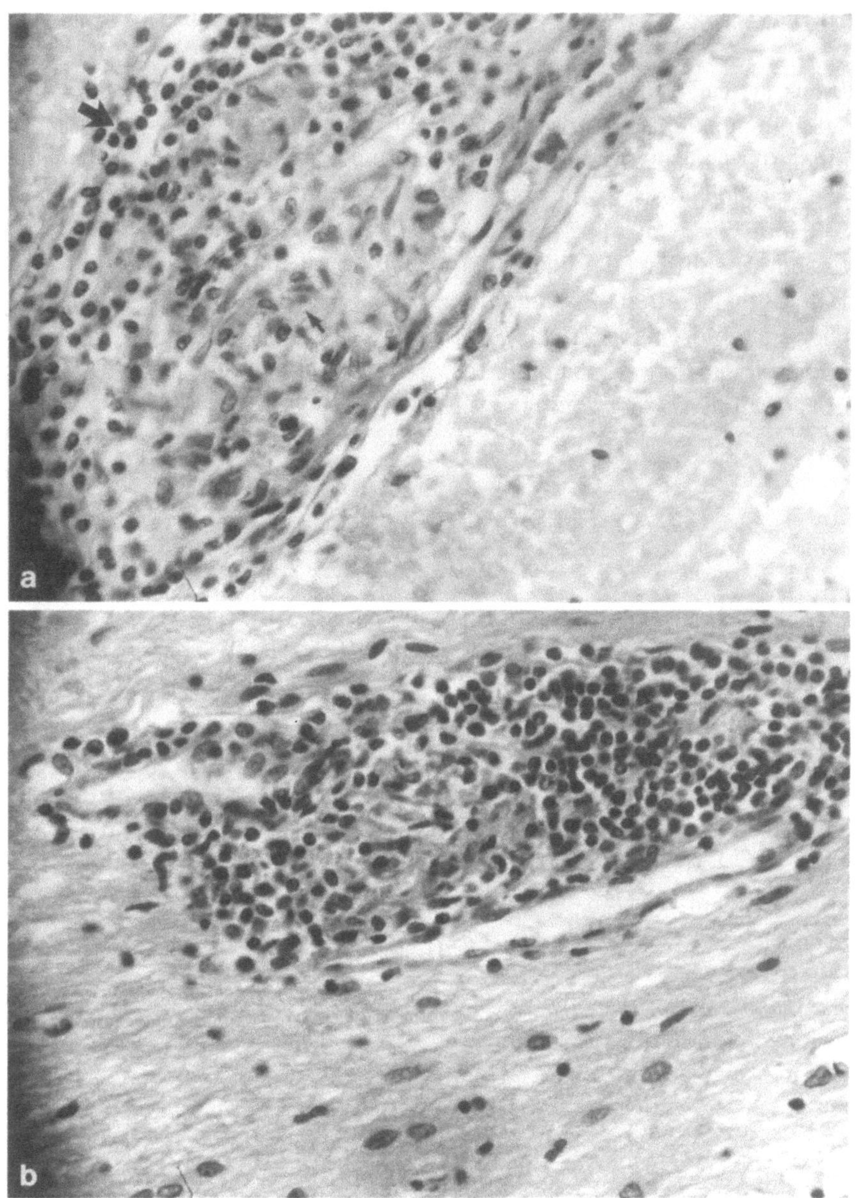

Fig. 9.9a, b. Histology of granulomatous angiitis. **a** The cerebral vessel wall in the brain is markedly expanded by granulomatous inflammation with epithelioid cells (area of *smaller arrow*) surrounded by monocytes (*larger arrow*). **b** A marked monocytic infiltration surrounds the epithelioid cells in another parenchymal vessel

9.8.2 Differential Diagnosis

The central nervous system dysfunction and the histopathology of granulomatous angiitis may appear similar to the arteritis caused by sarcoidosis [135, 136], allergic granulomatosis, or the granulomatous angiitis of Hodgkin's lymphoma [137, 138] (see Sect. 9.9 for complete differential diagnosis).

Rarely, patients with an atrial myxoma that causes an embolic cerebral occlusive vasculopathy will have an arteriogram that appears similar to the angiogram of patients with PAN or granulomatosis angiitis [17, 139].

9.8.3 Neuroimaging

Neither the CT or MR findings are specific for this disorder and in some patients the CT may appear normal. Multiple lesions with minimal ring enhancement that mimic abscesses can be seen on CT. MR lesions, which are often located in the periventricular white matter, appear as nonspecific hypointense lesions on T1-weighted images and hyperintense lesions on T2-weighted images [140]. On occasion, the lesions on MR images may appear as mass lesions [141].

Because the larger cerebral arteries are typically spared, cerebral angiography is usually normal [142]. However, on occasion narrowing and beading of the medium and small arteries are seen [143, 144, 145].

9.8.4 Pathology

The premorbid diagnosis is only established by a positive leptomeningeal or brain biopsy. Perivascular cuffing by monocytes and multinuclear giant cells and epithelioid cells in the arteries and the veins are typically seen (Fig. 9.9). Destruction of the internal elastic lamina and fibrinoid necrosis of the vessel walls occur in a multifocal distribution [131, 142, 146].

9.8.5 Laboratory Evaluation

The systemic laboratory evaluation is often normal. In contrast to many of the other vasculitides, the serum does not contain abnormal antibodies or immune complexes. The cerebrospinal fluid may have an elevated protein concentration or an increased number of monocytes or both.

9.8.6 Treatment and Course

The cerebral dysfunction, which is either episodic or progressive, initially responds to corticosteroids. Corticosteroids alone or in combination with azathioprine may successfully induce a remission, without a relapse [147]. Cyclophosphamide is the preferred treatment for recurrent disease [134, 145, 148]. If the angiitis arises in association with a lymphoma, the therapy is directed at the neoplastic process [149].

9.9 Central Nervous System Angiitis of Other Causes

9.9.1 Herpes Zoster

The incidence of cerebral arteritis in patients with herpes zoster is extremely low [150, 151]. However, systemic and dermatologic herpes zoster are common and numerous cases of central nervous system vasculitis have been reported [152]. Involvement of the first division of the trigeminal nerve which affects the eye (keratitis, uveitis, or both) often precedes, by days or weeks, the development of a clinical vasculitis in the ipsilateral cerebral hemisphere or cranial nerves [153]. Alhough we have found pathological evidence of a vasculitis associated with varicella zoster virus (Fig. 9.10), in one study of three patients with symptomatic cerebral vasculitis, only a minor inflammatory response was found in the affected vessels at autopsy. However, varicella zoster virus antigens were identified, using immunoperoxidase staining, within the media of the affected cerebral arteries in these patients [154]. Ocular motor palsies can develop either ipsilateral or contralateral to the ophthalmic zoster. The third

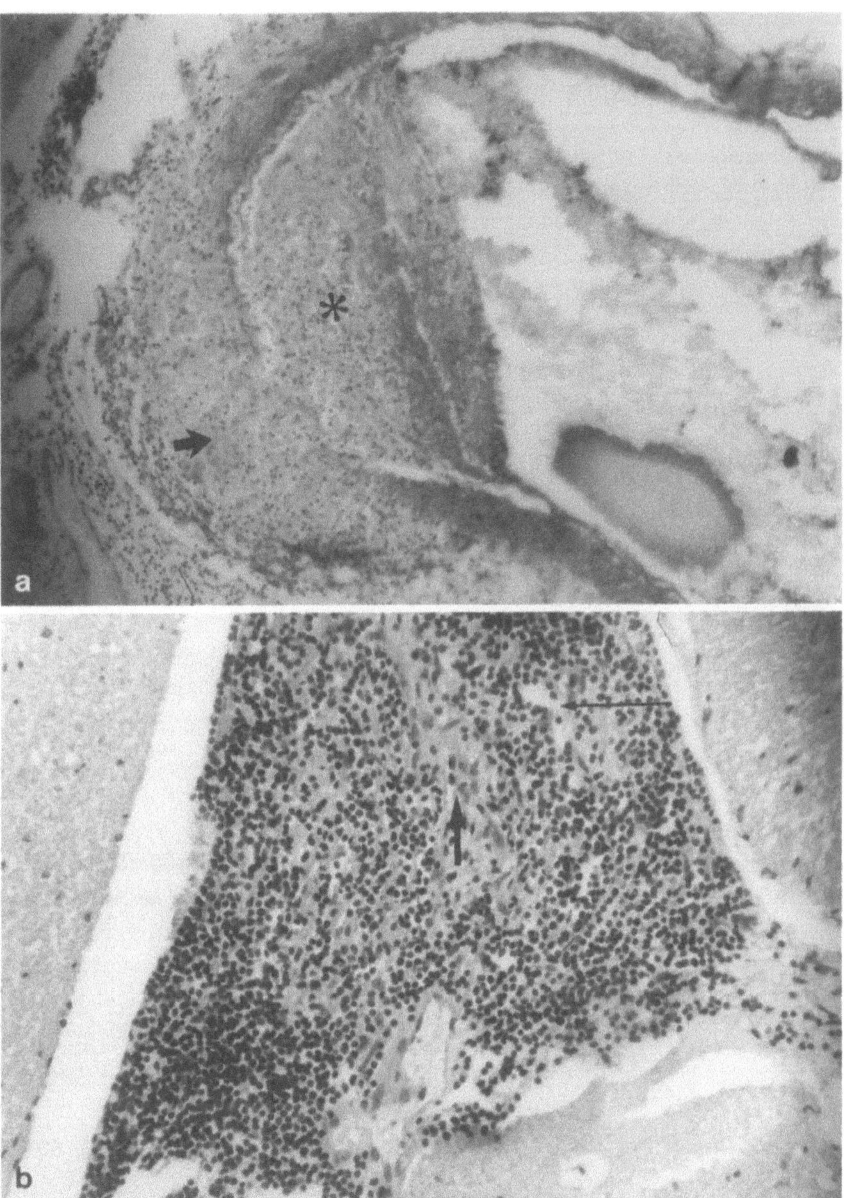

Fig. 9.10a, b. The vasculitis of the middle cerebral and basilar arteries associated with herpes zoster caused a diffuse encephalopathy. **a** The wall of a middle cerebral branch artery is markedly expanded by the fibrinoid and inflammatory process (*arrow*). Note the thrombus (*asterisk*) with inflammatory cells in the vessel lumen. **b** An intense monocytic inflammation involves the meninges, a pial artery (*arrow*), and pial veins (*narrow arrow*)

nerve is affected most commonly and the pupil is usually spared [155]. The ocular motor dysfunction can improve spontaneously or with corticosteroid therapy.

A vasculitis can affect the retinal arteries or veins in the eye in herpes zoster ophthalmicus [156, 157]. If the retinal vasculitis is severe, it can cause a fulminant acute retinal necrosis syndrome [158]. Occlusion of the central retinal artery with profound visual loss is a rare complication [159]. An optic neuropathy is also rare except in patients with diffuse orbital inflammation [160] or an orbital apex syndrome.

Treatment with systemic corticosteroids and acyclovir appear to speed the recovery and reduce the severity of the cerebral deficits associated with herpes zoster vasculitis.

9.9.2 Relapsing Polychondritis

Relapsing polychondritis, a disorder that causes inflammation and destruction of cartilage, is rarely associated with a cerebral vasculitis [161, 162]. One such case was noted to have visual loss associated with bilateral optic disc swelling without an increase of the intracranial pressure [163]. The cartilage in the head and neck are frequently affected and episcleritis is found in approximately 60% of cases. Corticosteroids can effectively suppress the disease.

9.9.3 Allergic Granulomatous Angiitis (Churg–Strauss Syndrome)

A rare disorder of systemic allergic granulomatous angiitis, associated with eosinophilia, fever, and bronchial asthma, can also affect the nervous system [164, 165]. Mononeuritis multiplex, ischemic optic neuropathy, pareses of the fourth or seventh nerves, and retinal vasculitis have all been reported in patients with Churg–Strauss syndrome [166].

9.9.4 Substance Abuse

A necrotizing angiitis of the cerebral arteries can arise as a complication of intravenous drug abuse,

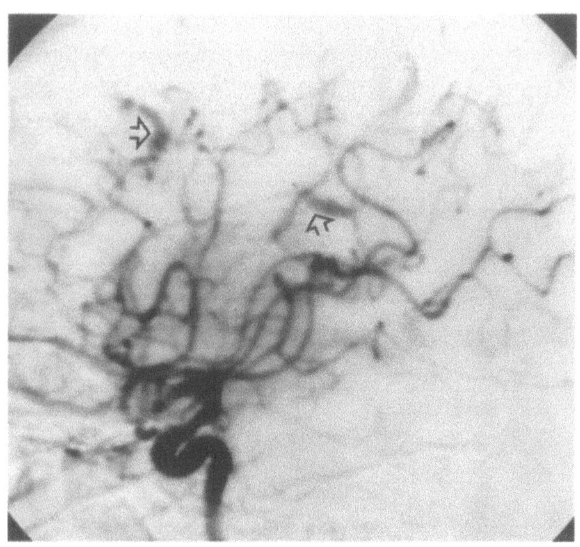

Fig. 9.11. This 25-year-old woman with a history of intravenous drug abuse had a right cerebral hemisphere hemorrhage. The lateral view right internal carotid angiogram shows middle cerebral artery branches of irregular caliber with areas of dilatation (*open arrows*)

particularly with the use of amphetamines [167–169]. In these cases cerebral angiography often reveals multiple areas of focal constriction, beading, and occlusion of the cerebral arteries [170] (Fig. 9.11). Rupture of the ectatic and thinned areas of the involved arterial wall results in an intracerebral hemorrhage [171]. Although the intravenous route is most frequently associated with this complication, a cerebral vasculitis can also follow oral or intranasal use of an amphetamine [172]. The arteritis may actually be caused by an inflammatory reaction to the additive materials and particulate matter typically found in illicit drugs. In fact, this particulate matter may become retinal and choroidal emboli in patients who abuse drugs (see Sect. 8.2.3.2). Transient amaurosis fugax associated with narrowing of all the retinal arterioles lasting 45–60 s and retinal phlebitis have also been associated with methamphetamine inhalation [173].

Other stimulant drugs are an unusual cause of a vasculopathy in the central nervous system. Chronic oral methylphenidate ingestion is an extremely rare cause of a cerebral vasculitis [174]. Stenosis and occlusion of the intracranial and cervical internal carotid arteries have rarely been reported in patients using lysergic acid [175, 176]. More recently, vasculitis-induced cerebral infarcts associated with cocaine use have been described [177–180].

9.10 Takayasu's Arteritis

9.10.1 Incidence and Systemic Involvement

Takayasu's arteritis is a rare disorder which affects women and men in a 9/1 female/male ratio and is most often seen in young women. Originally, Takayasu's arteritis was considered an illness limited to orientals, but in fact all races are affected. The symptoms result from an idiopathic inflammation that causes stenosis and occlusion of the aortic arch, subclavian arteries, and carotid arteries. The initial clinical manifestations include fever, malaise, myalgias and arthralgias, mild arthritis, iritis, and episcleritis. Pericarditis, pleuritis, and Raynaud's phenomenon are less common [181]. Painful arteries may also be present.

Later in the course, absent pulses and ischemia in the upper extremities develop from severe stenosis or thrombosis in the subclavian arteries. Bruits are often auscultated in the regions over the subclavian or carotid arteries. The patients complain of symptoms of claudication in the upper or lower extremities, or both. Systemic hypertension occurs in approximately 50% of cases. Right- and left-sided heart failure, coronary disease, and aortic insufficiency are potential devastating complications.

9.10.2 Neurologic and Neuro-ophthalmologic Manifestations

Syncope is the most common neurologic manifestation, occurring in at least half of the patients. Cerebral infarcts are less common, but patients can have multiple strokes in either the carotid or vertebral arterial territories. The neuro- ophthalmologic signs vary with the area of the brain affected. Retinal arterial alterations are the predominant ophthalmologic manifestation of Takayasu's disease and were noted in the first description of this ill-

ness. [182]. Most of the retinal arterial abnormalities are ischemic alterations that result from the associated systemic hypertension. Retinal arterial anastomoses have also been demonstrated. Signs of an ischemic oculopathy syndrome (see Sect. 8.2.5), including cataract, can also develop. The ocular and cerebral hypoperfusion induced by neck extension or lying down with the face up causes narrowing of both visual fields. This symptom is almost pathognomonic for aortic arch occlusive disease. The visual disturbance is relieved by lying the patient down in the prone position.

9.10.3 Imaging

Arteriography shows segmental and diffuse narrowing, occlusions, and fusiform aneurysms of the subclavian, aorta, renal, and carotid arteries (Fig. 9.12). Echocardiography demonstrates dilatation of the aortic root. Ophthalmodynamometry, Doppler flow studies, or duplex ultrasonography can be used to monitor progression of the carotid stenosis.

9.10.4 Pathology

A panarteritis of the large elastic arteries with all layers infiltrated by mononuclear and giant cells is seen in active cases. The inflammation causes secondary intimal damage and proliferation, elastica degeneration, fibrosis, small vessel proliferation into the arterial wall media, dissection and hemorrhage into the wall, and focal dilatations or aneurysms of the vessel walls [183, 184].

9.10.5 Laboratory Evaluation

A mild anemia and leukocytosis are seen in some cases but an elevated erythrocyte sedimentation rate is found in most patients. The serum immunoglobulins are typically abnormal but the antinuclear antibodies are negative. Patients with this disorder may have a genetic predisposition as demonstrated by an increased incidence of HLA-B5, Bw52, and A10 and the occurrence of disease in monozygotic twins.

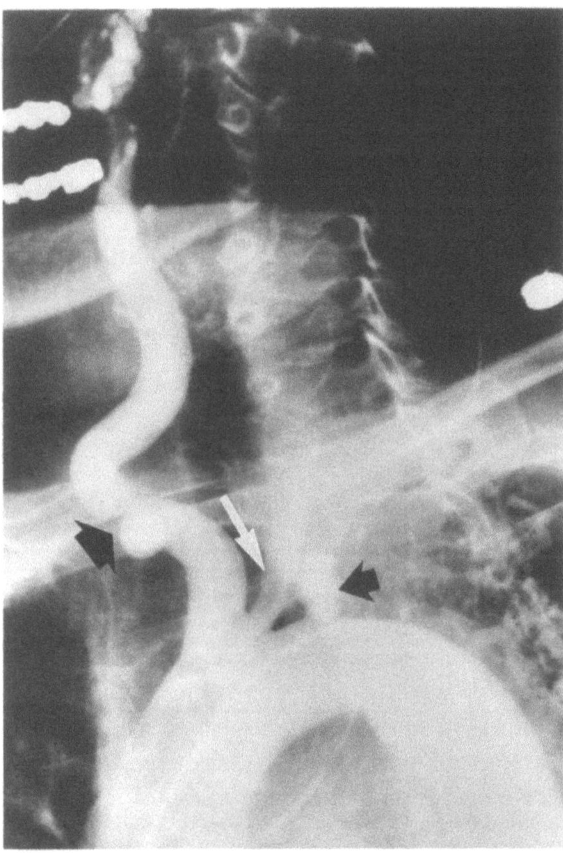

Fig. 9.12. An aortic arch angiogram in a patient with Takayasu's arteritis shows a high-grade stenosis (*broad arrow*) in the innominate artery, poor filling of the common carotid artery (*white arrow*), and occlusion of the subclavian artery with a residual stump (*arrow*). (Case provided by Larry Frohman)

9.10.6 Treatment

Corticosteroids are the most commonly used therapy for the prevention of long-term vascular occlusions [185]. Once the inflammation is suppressed, reconstructive vascular surgery for symptomatic stenotic lesions is performed [186]. Additionally, medical control of the systemic hypertension is essential.

9.11 Cogan's Syndrome

9.11.1 Incidence and Systemic Involvement

Cogan's syndrome is a rare disorder diagnosed by the presence of a nonsyphilitic interstitial keratitis and a vestibuloauditory deficit caused by a multifocal vasculitis [187]. Cogan's syndrome may actually be a form of PAN [188–190]. The age of onset ranges from the teenage years to the fifth decade. Aortitis and angiitis of the spleen, kidney, and gastrointestinal tract can also occur in this disorder [191, 192]. Fever, weight loss, a leukocytosis, and an elevated sedimentation rate are common.

9.11.2 Neurologic and Neuro-ophthalmologic Manifestations

The most common neurologic abnormality is, by definition, the loss of hearing, which is usually severe. Frequently, the process is bilateral and leads to deafness. Up to 50% of patients will have additional central nervous system dysfunction. The symptoms and signs include headache, seizures, cerebral ischemic episodes, and psychosis [193]. Peripheral neuropathy can also occur.

Though an interstitial keratitis with deep corneal neovascularization has been classically described, most patients have a bilateral peripheral subepithelial keratitis early in the course [194]. Recurrent keratitis is common but significant visual impairment does not usually develop. Scleritis, uveitis, and noninfectious conjunctivitis are infrequent.

9.11.3 Treatment

Treatment of the hearing loss requires prompt initiation of high-dose corticosteroids [195]. This and other forms of autoimmune sensorineural hearing loss without the ocular abnormalities may require a combination of corticosteroids and cyclophosphamide therapy in order to prevent deafness [196]. The early corneal lesions will resolve with topical ophthalmic corticosteroids. Once stromal keratitis and vascular ingrowth develops topical treatment is of little benefit. Systemic and neurological dysfunction may also require corticosteroids.

9.12 Lymphatoid Granulomatosis

9.12.1 Incidence and Systemic Involvement

Lymphatoid granulomatosis is a rare, relatively recently described entity characterized by pulmonary disease (cough, shortness of breath, chest pain). General malaise and weight loss occur in approximately 33% of cases. Men are affected slightly more often than women. The onset is usually during the fifth decade of life, but cases have been noted from the first to the ninth decade. Peripheral or central nervous system dysfunction occurs in more than 25% of patients, while arthralgias, myalgias, and gastrointestinal problems are found in fewer cases [197–199]. A poor survival rate has been associated with the onset of neurologic deficits. In 10%–50% of patients, the inflammatory lesion degenerates to become a non-Hodgkin's lymphoma. The worst prognosis occurs in those cases with lymphoma.

9.12.2 Neurologic and Neuro-ophthalmologic Manifestations

The nervous system can be attacked at any site, but the cerebral hemisphere and brainstem are more commonly affected than the cranial or peripheral nerves. Bilateral visual loss from a probable optic neuropathy has also been described in one case [200]. A unilateral or bilateral retinal vasculitis involving the arterioles and veins can occur with a posterior uveitis [201].

9.12.3 Laboratory Evaluation and Pathology

The laboratory evaluation includes a chest roentgenogram, with most patients demonstrating multiple nodular infiltrates in the lower portion of the lungs. Biopsy of these pulmonary lesions reveals areas of mononuclear cell infiltration without giant cells, necrosis of arteries and veins, and atypical lymphatoid and plasmacytoid cells surrounding the arteries. The cerebrospinal fluid may have an elevated protein level or a monocytosis, but it is normal in approximately 50% of the cases with neurologic dysfunction. In some patients, cerebral arteriography reveals narrowing of the cerebral arteries suggestive of an arteritis [199, 200]. The vessels can also be displaced because of a mass effect from the inflammatory tissue in the brain.

9.12.4 Treatment

The mortality rate for untreated patients is approximately 90% within 3 years. A combination of corticosteroids and cyclophosphamide improves the survival rate and induces a remission in approximately 54% of treated patients [202]. Radiotherapy has also been tried but it rarely induces a remission. If a lymphoma is present, the appropriate chemotherapy or radiotherapy, or both, should be given.

9.13 Microangiopathy of the Retina and Brain

9.13.1 Ophthalmologic and Neurologic Manifestations

Numerous authors have described patients with a microangiopathy that affects the retina and brain, and clearly more than one etiology is responsible for the clinical constellation. As originally described, a microangiopathy of the retina and brain is a rare subacute multifocal central nervous system disorder affecting women, usually between the ages of 20 and 50 [29, 203, 204]. Personality changes or frank psychosis often precede the other signs of neurologic or ocular dysfunction. The onset of dementia is followed, within 3–9 months, by the development of multiple retinal branch artery occlusions. Additional signs include hearing loss, seizures, and brainstem and bilateral cortical spinal tract dysfunction. A brain biopsy in one patient was reported to show sclerosis of the media and adventitia of the small pial and cortical arteries. Only a minor inflammatory response was found [203].

Retinal arteriole occlusions are found predominantly in the peripheral retina. White and silver threadlike arteries and cotton wool spots are typically seen on ophthalmoscopy. Multiple scotomas in the visual field are probably related to the retinal infarcts (Fig. 9.13). The fluorescein angiogram

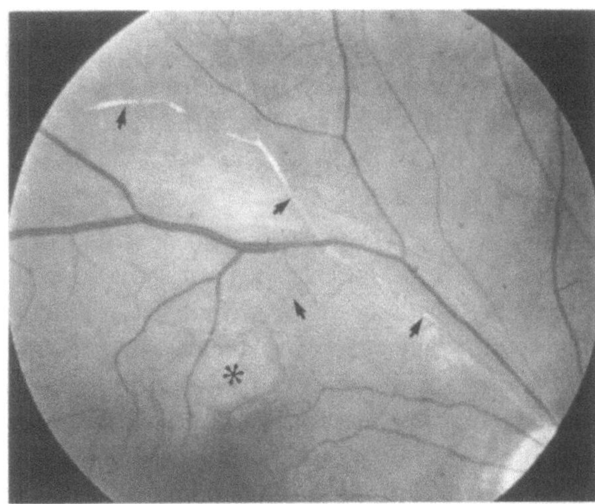

Fig. 9.13. Ophthalmoscopy in a patient with microangiopathy of the retina and brain shows an infarct in the retina (*asterisk*) associated with marked sheathing and narrowing of the retinal artery (*arrows*) in this area. (Case provided by J. Selhorst, from [203])

shows delayed filling or absent flow in some arterioles and irregularly narrowed lumens of the arterioles. Focal leakage is seen from some, but not all, of the involved arterioles. The sheathing and occlusions of branch arteries can be indistinguishable from an entity called idiopathic arteritis which causes recurrent retinal arterial occlusions and is not associated with neurologic or systemic disease [205] (Fig. 9.2). Another ischemic disorder of the retina, frosted branch angiitis, may be associated with vitreous cells overlying an inflammatory process, affecting both the retinal arteries and veins but not the brain. This process may be progressive, requiring corticosteroids, or self-limiting.

9.13.2 Differential Diagnosis

The differential diagnosis of microangiopathy of the retina and brain must include consideration of the following entities [206–211]:

1. Secondary meningovascular and retinal vascular syphilis
2. Acquired immunodeficiency syndrome-related retinal microvasculopathy
3. SLE
4. Behçet's syndrome
5. Herpes hominis or simplex or zoster retinitis

and encephalovasculitis (see below, this section)
6. Retinal emboli
7. Infectious vasculitis
8. Clotting disorders (see Eales' disease)
9. Crohn's disease (see below, this section)
10. Malignant angioendotheliomatosis (see below, this section)
11. Hereditary cerebroretinal vasculopathy (see below, this section)
12. Antiphospholipid antibody syndrome (see Sect. 9.17)

Herpes zoster or *herpes simplex* can cause a retinal vasculitis that produces a focal area of retinal ischemia or a diffuse acute retinal necrosis syndrome [213, 214]. In contrast to the microangiopathy of the brain and retina, an anterior chamber inflammation, vitritis, and papillitis are commonly seen in the patients with retinal necrosis syndrome [214–216] (see Sect. 9.9.1).

Crohn's disease (regional enteritis) is rarely associated with branch artery occlusions even when a systemic vasculitis is present [217]. However, a diffuse retinal arteritis and phlebitis can result in profound visual loss [218]. Central retinal vein occlusion [219] and ischemic optic neuropathy are additional rare causes of visual loss in patients with regional enteritis [220]. Episcleritis, iridocyclitis, and conjunctivitis are more common ocular complications in this disorder [221].

It may be difficult to distinguish an idiopathic microangiopathy of the eye and brain from the rare presentation of multiple branch retinal artery occlusions and a multifocal central nervous system disorder caused by *malignant angioendotheliomatosis* (Fig. 9.14). Malignant angioendotheliomatosis is actually a systemic lymphoma that affects multiple organs by occluding the lumens of arterioles [222]. A biopsy of the skin will usually make the diagnosis by demonstrating the intravascular anaplastic cells [223].

A *hereditary cerebroretinal vasculopathy* can be distinguished by the preponderance of retinal capillary abnormalities (such as telangiectases and occlusions) rather than arterial occlusions. Also, the CT scans of patients with this disorder show lesions that enhance with contrast and have a mass effect typical of a neoplasm. This is unlike the CT appearance in most other etiologies of microvascular

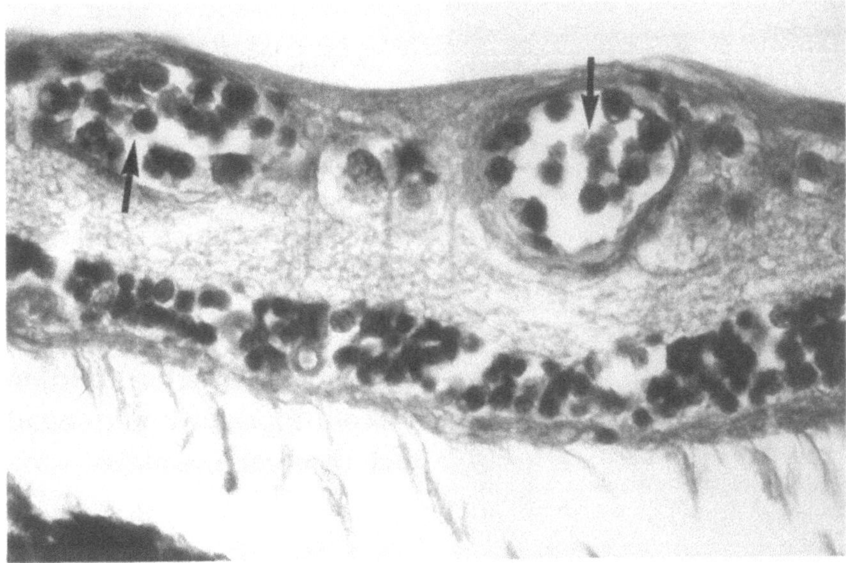

Fig. 9.14. A 65-year-old man had uniocular visual loss secondary to multiple retinal artery occlusions and a progressive myelopathy suggestive of a vasculitis, but the evaluation was nondiagnostic. Death resulted from pneumonia and respiratory failure. Histological section of the retina reveals the cells of malignant angioendothelioma (*arrows*) in the retinal arteries

occlusive disease. The histopathology of this hereditary disorder shows no sign of vasculitis in the vessels of the retina or brain [224].

9.13.3 Laboratory Evaluation

Except for biopsy of the skin, exhaustive evaluations for systemic vasculitis including serum complement, lupus erythematosus preparations, antinuclear antibodies, Australian antigen, and syphilis serology have failed to demonstrate laboratory abnormalities suggestive of a systemic disorder in most patients with microangiopathy of the retina and brain. In some cases, skin that appears clinically normal can demonstrate immunoglobulin M deposits in the walls of the blood vessel [225], which is suggestive of a collagen vascular disease. The cerebrospinal fluid may be normal or there may be a mild monocytic pleocytosis, with fewer than 20 lymphocytes/mm^3. On occasion the cerebrospinal fluid protein concentration is elevated to greater than 100 mg/100 ml. The neuroradiological studies including CT and cerebral angiography are typically normal.

9.13.4 Treatment

Treatment with daily oral corticosteroids (80 mg prednisone) or 100 IU parenteral adrenocorticotropic hormone has been effective in gradually improving the neurologic and retinal defects, usually within 2 months. Recurrences have not been reported following withdrawal of the corticosteroids.

9.14 Eales' Disease (Periphlebitis Retinae)

Eales' disease was originally described as a syndrome of recurrent retinal or vitreous hemorrhages and epistaxis. Men in the third or fourth decades who are otherwise healthy seem to be preferentially affected [226, 227]. The disease is considered idiopathic, but cases must be distinguished from other disorders associated with a retinal occlusive vasculopathy such as sickle cell disease, blood dyscrasias, sarcoidosis, and collagen vascular diseases. There are even patients who appear to have retinal vascular sheathing following calcific retinal emboli [228].

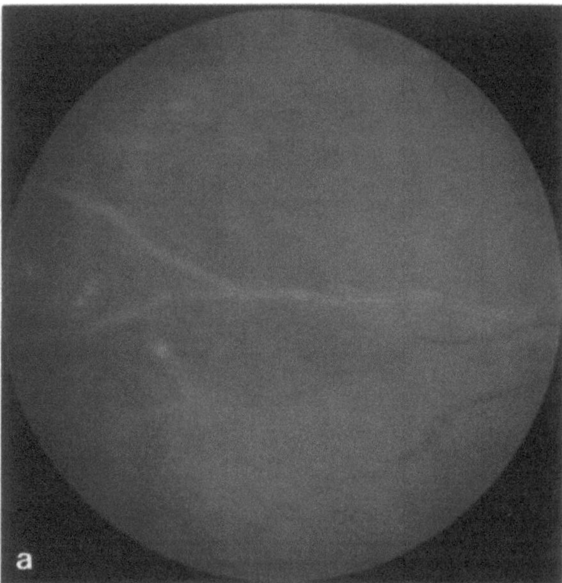

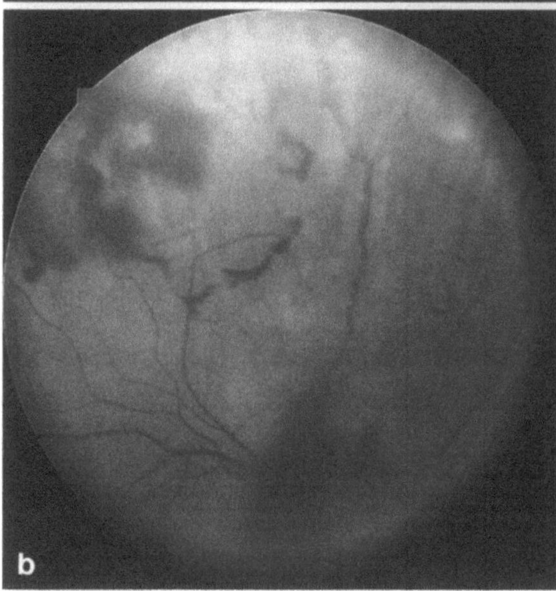

Fig. 9.15. a Sheathing of a peripheral vessel in a patient with Eale's disease. b The opposite eye has hemorrhages in the peripheral retina

Early in the course of Eale's disease, peripheral retinal capillary occlusions develop. The peripheral blood vessels may appear sheathed (Fig. 9.15). The secondary retinal ischemia leads to retinal neovascularization. The neovascular membrane can extend into the vitreous and causes retinal and vitreal hemorrhages [204] (Fig. 9.15). The clinical course is variable with some cases showing spontaneous resolution of neovascularization while others have recurrent bleeds that require laser photocoagulation.

Several, but not all, patients have shown fluorescein angiographic evidence (leakage) of a retinal vasculitis [228] and a vitritis, both suggestive of an inflammatory etiology. However, in many cases there are no additional signs of ocular inflammation [229]. There is no clear association of Eale's disease and tuberculosis, as was widely believed.

9.15 Acute Posterior Multifocal Placoid Pigment Epitheliopathy and Cerebral Vasculitis

Acute posterior multifocal placoid pigment epitheliopathy (AMPPE) (Fig. 9.16a) occurs in otherwise young healthy adults [230]. Despite the numerous case reports, this entity is uncommon in the population. The disorder typically consists of an acute bilateral (unusually unilateral) ocular process that appears as patchy cream-colored subretinal lesions, particularly in the posterior pole. These lesions are presumably caused by choroidal ischemia affecting the overlying retinal pigment (Fig. 9.16b). The acuity is diminished and multiple scotomas obscure the visual field. The disorder is usually self-limited.

AMPPE is infrequently associated with transient or permanent cerebral ischemic symptoms resulting from a vasculitis of the major cerebral arteries [231]. Rare deaths result from a diffuse cerebral vasculitis [232]. If the posterior cerebral artery is affected, a homonymous field defect can add to the visual deficit [233]. A retinal or optic disc vasculitis is another rare cause of visual loss in AMPPE [234–236]. The occasional patient with an associated small vessel nephropathy, dermal vasculitis, and thyroiditis suggests that a systemic vasculitis is the cause. A lymphocytosis is often present in the cerebrospinal fluid of patients with headaches even if no neurologic deficit is present.

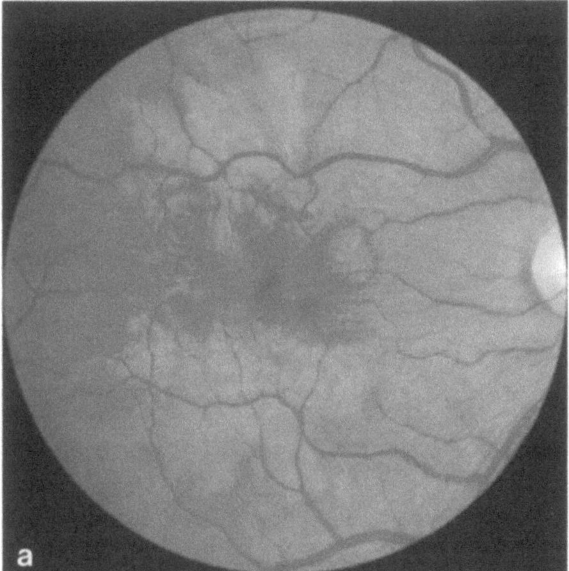

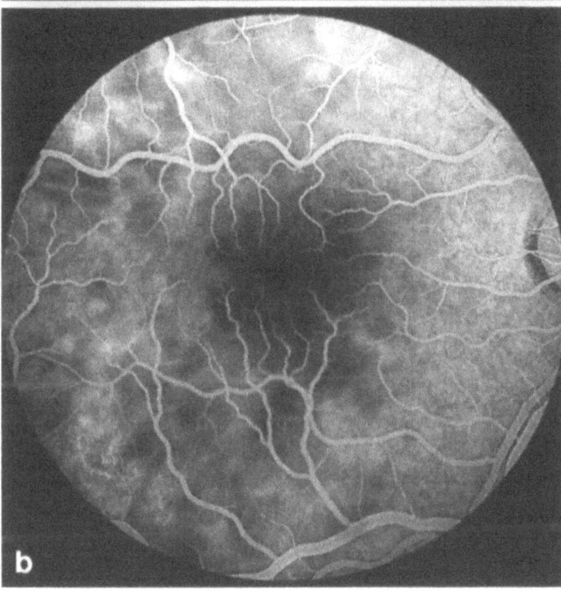

Fig. 9.16a,b. Acute multiplacoid pigment epitheliopathy was the cause of multiple scotomas despite 20/20 acuity. **a** Ophthalmoscopy reveals a confluence of swollen retinal pigment epithelium particularly around the macula. **b** In the same eye, fluorescein angiography shows multiple areas of either blockage or nonfilling of the choriocapillaris. (Case provided by Ken Noble)

9.15.1 Treatment and Course

The retinal and neurologic complications of AMP-PE are self- limiting and respond to oral prednisone in daily doses of 40 – 60 mg.

9.16 Sjögren's Syndrome

9.16.1 Incidence and Systemic Involvement

Sjögren's syndrome is a common connective tissue disorder that is frequently overlooked or poorly characterized [237]. The actual incidence of this disorder is unknown because patients with SLE, rheumatoid arthritis, and other inflammatory connective tissue illnesses have components of the sicca complex. Women are more frequently affected than men. Xerophthalmia and xerostomia form the sicca complex, which can occur primarily or as part of other connective tissue disorders [238]. Arthritis, Raynaud's phenomenon, and restrictive pulmonary disease are commonly seen. Approximately 25% of patients develop vasculitis, which usually involves the skin. Systemic vasculitis is less common [239].

9.16.2 Neurologic Manifestations

Recent reports have suggested that approximately 20% of patients develop neurologic deficits. Most dysfunction is secondary to demyelination, but some patients have a secondary vasculitis [240, 241]. There are some patients who develop neurologic symptoms prior to the confirmation of Sjögren's syndrome, while others have well-established disease [242].

Recurrent transient or permanent cerebral hemisphere dysfunction, aseptic meningitis, and transverse myelopathy can occur. Despite the relapsing and remitting course, patients with Sjögren's syndrome can be distinguished, by the associated systemic signs, from patients with another relapsing and remitting neurologic disorder, multiple sclerosis [243]. Vasculitis can also damage the peripheral nervous system, with a resulting polyneuropathy or mononeuritis multiplex [244]. However, some patients with peripheral nerve disease will have a ganglionitis, without a true vasculitis [245]. Dysfunction of the cranial nerves, with the sensory trigeminal nerve most often affected, is less common than central nervous system deficits [246].

9.16.3 Neuro-ophthalmologic Manifestations

The keratoconjunctivitis sicca, resulting from lymphocytic infiltration of the lacrimal gland, causes burning and blurred vision. Corneal and conjunctival erosions and filamentary keratitis are common. An optic neuropathy or transient monocular amaurosis can occur but they are not as common as paraparesis, hemiparesis, and ataxia [247]. Nystagmus, internuclear ophthalmoplegia, and other types of ocular motor dysfunction are also infrequent.

9.16.4 Laboratory Evaluation

The diagnosis of Sjögren's syndrome is based on the clinical examination and the history as well as on laboratory confirmation of inflammation of the lacrimal gland or salivary gland or the secondary effects (Table 9.6).

Serum abnormalities include the presence of serum anti-Ro (SSA) antibodies, hyperglobulinemia, antinuclear antibodies, a positive rheumatoid factor, and a mild to moderate elevation of the erythrocyte sedimentation rate.

Table 9.6. Sicca laboratory anomalities

1. Schirmer test, less than 10 mm tears in 5 min in unanesthetized eye
2. Rose bengal staining of conjunctiva, cornea [248]
3. Minor salivary gland mononuclear and plasma cell infiltration of vessel walls [249, 250]

9.16.5 Neuroimaging

MR imaging is more sensitive than CT in visualizing lesions in Sjögren's syndrome. MR reveals abnormalities in approximately 75% of patients with central nervous system dysfunction, but in only 9% of patients without clinical neurologic symptoms. The most prevalent lesions are small areas of increased signal on the T2-weighted images. These lesions affect both the gray and white matter at multiple sites in the brain [253]. If these lesions are periventricular or limited to the white matter, mul-tiple sclerosis may be erroneously diagnosed. Cerebral angiography, which is performed only in rare cases, rarely reveals a vasculitis.

9.16.6 Treatment

Tear replacement ophthalmic drops can relieve the symptoms of the dry eye. The neurologic deficits improve with moderate doses of corticosteroids, alone or in combination with plasmapheresis [254].

9.17 Occlusive Disease Associated with Antiphospholipid Antibody Syndromes: A Disorder in Evolution

A variety of neurologic and neuro-ophthalmologic disorders have been associated with the presence of serum antiphospholipid antibodies. The exact relationship of these antibodies and the clinical features of these diseases is not entirely understood, but normal patients should not have measurable antibody levels in their sera.

Transient and permanent cerebral and ocular ischemic episodes occur in patients with antibodies to various phospholipids, without carotid atherosclerosis or cardiac disease. These antibodies may be the cause of the false-positive serological test for syphilis (VDRL). One of these antibodies, the lupus anticoagulant, which increases platelet aggregation and prolongs the activated partial thromboplastin time [253], is associated with thrombotic events and cerebral infarcts [254–257]. The lupus anticoagulant has also been the most common antiphospholipid antibody associated with intrauterine death secondary to thrombosis in the placenta [258]. Anticardiolipin antibodies in patients with SLE may be the cause of retinal and cerebral infarcts [259–262]. A prolonged partial thromboplastin time is not present in all patients with antiphospholipid antibodies. Though amaurosis fugax in pregnant women has been considered to be caused by migraine, a recent report described the presence of anticardiolipin antibodies in four such patients [263]. In addition to retinal branch artery and arteriole occlusions, diffuse ischemic changes in the retina which can cause retinal neo-vascularization have recently been found [264]. An anterior ischemic optic neuropathy in young adults may also be associated with antiphospholipid antibodies [265].

9.17.1 Treatment

The vasculopathy has been treated with many agents. Success has been reported with aspirin, warfarin, corticosteroids, immunosuppressive agents, or some combination of these drugs.

CHAPTER 10

Migraine

10.1 Introduction and Epidemiology

Ophthalmologists and neurologists frequently see patients who complain of transient neurologic phenomena, most of which are visual, that are associated with a headache. The differentiation of migraine episodes from seizures and the vaso-occlusive disorders is not always immediately obvious, but an awareness of the types of migraines, migrainous phenomena, and migraine complications will help in formulating the appropriate clinical diagnosis. Migraine cannot be understood without knowledge of the normal ocular and cerebral blood flow and the factors that alter these circulations. A single approach to treatment of every case is not possible, but drugs or other therapies that modify or reverse these changes can be used to outline a rational plan of therapy.

Classic and common migraine can begin in children before 10 years of age, but the highest frequency of new cases is in the second and third decades of life. Migraine-equivalent episodes, usually visual phenomena without headache, may even develop during the sixth decade [1, 2].

The prevalence of migraine in the entire population varies between 5% and 20% [3, 4]. Women are affected twice as frequently as men [4]. In children younger than 16 years of age, the incidence of migraine is lower, approximately 6% [5].

10.2 Types of Migraine

The patterns of migraine attacks are numerous and most, but not all, episodes are classifiable using traditional criteria. An individual patient may have features of more than one type of migraine in a given episode or among attacks. Therefore, the categorization often merely reflects different manifestations of the same disorder.

10.2.1 Classic Migraine

Classic migraine is the best-recognized clinical type of migraine, yet only about 10% of migraineurs actually experience these episodes. The attack begins with a clearly defined prodrome which precedes a unilateral throbbing headache by approximately 20–60 min. The prodromes are frequently disturbances of visual, sensory, or motor function, usually but not exclusively contralateral to the headache. The most common symptoms are visual phenomena, most of which transiently disturb the visual scene. On occasion, the positive visual phenomena (see Sect. 10.2.3) are not prodromes, occurring concomitantly with or following the headache. Some attacks include only the prodromal symptoms and are termed "migraine equivalents."

In most patients, the headache is severe and is the worst aspect of an attack. The pain is often accompanied by combinations of nausea, vomiting, anorexia, photophobia, or sonophobia. The headache is sometimes relieved temporarily by compressing and massaging the temporal artery on the side of the headache [6]. Many patients also feel relief after emesis. The throbbing headache typically lasts approximately 1–2 h, but the generalized tension headache which follows can persist for hours. Ingestion of an alcoholic beverage can precipitate an attack. Psychological factors have a role in this disorder and patients with classic migraine are often high-achieving, perfectionistic, rigid individuals who are subject to emotional stress [7].

10.2.2 Common Migraine

Common migraine is the most frequent type of migraine. The prodrome is often vague, associated with visceral or autonomic complaints such as diarrhea, nausea, skin pallor, sweating, dizziness, bloating or generalized swelling in the extremities,

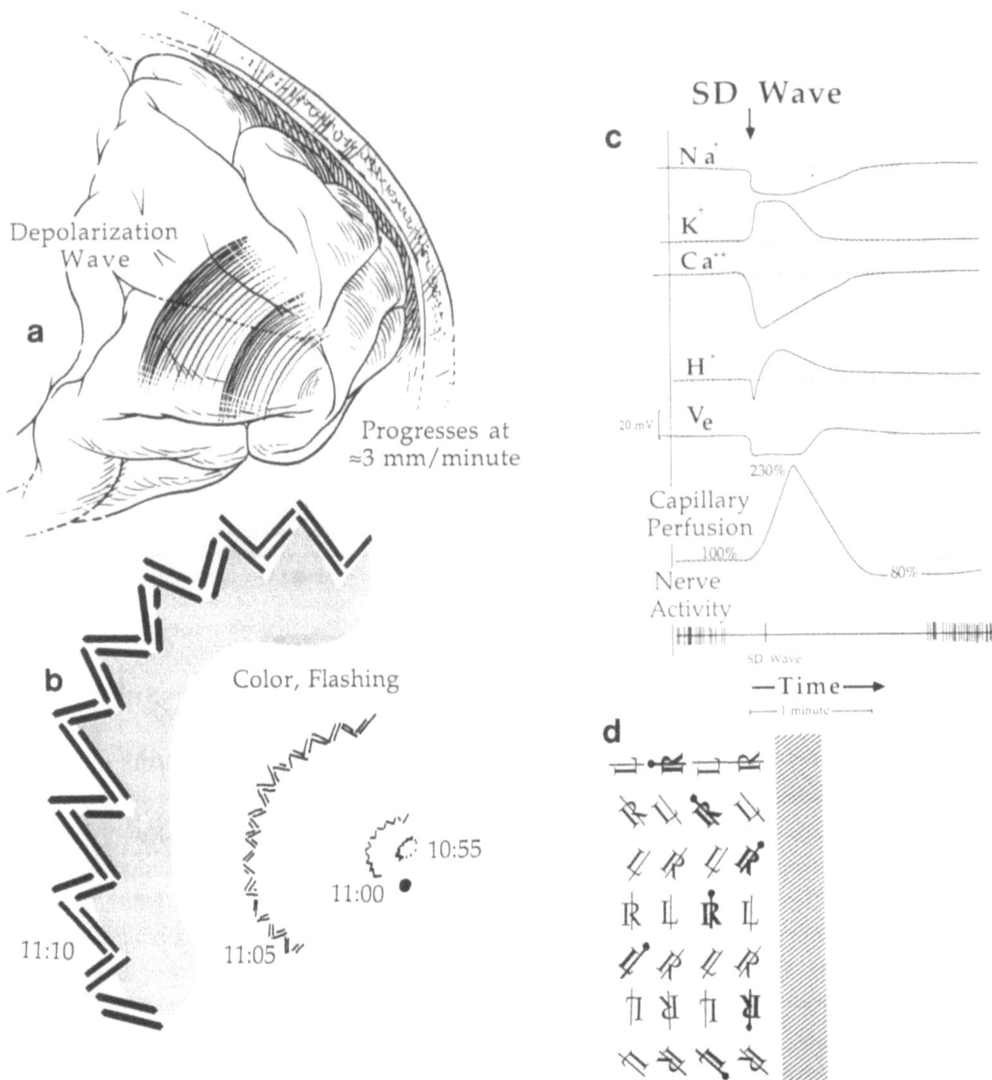

Fig. 10.1a–d. Diagrammatic representation of a migraine. **a** Waves of excitatory electrical activity followed by spreading depression travel from the occipital pole anteriorly at a rate of 3 mm/min, stimulating the neurons representing the central to peripheral visual field. **b** Flickering or wavering fortification images spread from a paracentral location peripherally. The images become larger in the peripheral field. **c** The spreading depression (*SD*) wave results in a period of neuronal silence following excitation of the nerve activity in the cortex. Interstitial levels of sodium (Na^+) and calcium (Ca^{++}) fall, while interstitial potassium (K^+) and hydrogen ($H+$) rise. The extracellular current amplitude (V_e) falls. Pial capillary perfusion briefly rises (for several minutes); this is followed by hypoperfusion for 20 min or longer. **d** The altered electrical activity of a migraine first stimulates the cortical functional columns with selective responses to right (*R*) or left (*L*) eye stimulation, orientation (*vert, *rslant, *lslant), or color (*bold letters*). The excitation is followed by a lack of neuronal cortical activity and a corresponding scotoma

and irritability that can last for hours to days before the onset of the headache. Certain clinical features are common to most types of migraine, including dilatation of the temporal artery, surrounded by local subcutaneous edema on one side of the head during the headache phase, or generalized water retention prior to the attack. Visual or neurologic phenomena similar to those that occur with classic migraine also regularly occur. The head pain can be unilateral, bilateral, or generalized. It can persist for hours to days, and frequently blends into a generalized tension headache. Not every headache has a definitive prodrome. The relationship of attacks to food or ethanol intake or specific personality types is unclear.

10.2.3 Positive Visual Disturbances in Classic and Common Migraine

Visual symptoms are the most frequent neurologic disturbance associated with common or classic migraine. Most are positive visual phenomena, the addition of imagery to the visual scene. Unformed visual imagery such as scintillating scotomas and jagged, flickering fortification may involve one or both eyes (Fig. 10.1b). The affected field can be less than a quadrant in size or the entire homonymous hemifield can be included. The images begin either in the central or peripheral field and then expand to encompass the entire field. The patient sees around and through areas of the visual scintillations that sparkle or dazzle, but areas of scotoma often border these positive images [8]. Since migraineurs still perceive the visual imagery with the eyes closed, it is difficult, if not impossible, to determine whether the disturbance is monocular or in both eyes. Less commonly, the visual distortion can appear as macropsia or micropsia. The positive visual phenomena usually last less than 30 min. The vision normalizes as the scotomas break up in an irregular pattern, usually beginning with the central field.

The visual imagery can appear as simple lines, arcs, or as complex geometric patterns, all with varying degrees of color. This reflects the stimulation or the alteration of metabolism of the cortical columns for orientation, spatial frequency content, and color vision in the occipital lobe (Fig. 10.1d). The development of migrainous visual phenomena

in patients with no eyes [9] further suggests a central nervous system origin of the visual symptoms.

When the visual disturbance affects one eye at a time rather than having a homonymous distribution, it implies that the visual system anterior to the chiasm is involved, most likely in the territory of the ophthalmic artery. Transient and permanent scotomas from dysfunction in the choroid, retina, and optic nerve in patients with complicated migraine support this hypothesis. As previously discussed, it may be difficult for the patient to distinguish whether the positive imagery affects one or both eyes, since it is present with either one or both eyes closed.

10.2.3.1 Differential Diagnosis of Positive Visual Disturbances

The differential diagnosis of a homonymous distribution of episodic positive visual imagery is brief. The positive visual phenomena of migraine usually last from several to less than 30 min. The positive visual images can be a manifestation of occipital seizures, but these symptoms usually last only a few minutes (see Sects. 7.1.5.2, 7.1.10.1.1). Infrequently, occlusion of a posterior cerebral artery causes homonymous scintillations prior to the loss of the same field (see Sect. 8.5.1.1.4). Positive symptoms also are occasionally seen following cerebral angiography and, although alarming, they generally resolve without a permanent visual defect or cerebral infarct. On occasion, distinguishing migraine from these other disorders can be difficult because complicated migraine is also associated with varying degrees of loss of vision.

Monocular flashes occur from etiologies other than migraine. The most common cause of recurrent unformed flashes in the temporal peripheral field is vitreous traction on the retina following posterior vitreous detachment [10]. The images, called Moore's lightning streaks, typically last several seconds or less and occur when the patient passes from a lighted to a dark room [11, 12]. The visual disturbance almost always ceases within 2 months. Flashes are also reported by patients with retinal tears or detachments or inflammation of the choroid and retina. Momentary flashes with eye movement, called phosphenes, are reported by patients with acute optic neuritis [13].

Repetitive episodes of amaurosis or a visual disturbance described as "falling water" over the visual scene, associated with an ipsilateral headache or eye pain, result from a recurrent anterior chamber microscopic hyphema following cataract extraction and either an anterior or posterior chamber intraocular lens implant [14]. Unless the patient is examined using biomicroscopy within several hours of the onset of the episode, the causative transient hemorrhage and elevation in the intraocular pressure will not be seen [14, 14a, b, c].

10.2.4 Complicated Migraine

Any type of migraine can become a complicated migraine when the episodes include negative phenomena such as visual loss or a neurologic deficit. Though these neurologic manifestations are usually prodromal, they can accompany or even follow the headache. In one study, approximately 10% of migraineurs experienced a neurologic deficit during at least one migraine episode [15]. Another report described neurologic complications in 25% of 3890 migraineurs [16]. Although major permanent visual or neurologic deficits are rare despite repeated attacks, because of the large number of patients with complicated migraine, residual deficits have been described in many cases.

10.2.4.1 Negative Visual Disturbances

Homonymous visual field disturbances, varying from less than a quadranopia to a complete hemianopia, are the most frequent negative symptoms [17]. Occasionally, the patient will complain of a tunneling of the vision with preservation of the central field. Transient cortical blindness (from bilateral hemianopia) is a rare manifestation of complicated migraine [18].

Partial or complete monocular amaurosis attacks can also occur. Recurrent transient monocular blackouts or whiteouts of vision are uncommon. Temporary retinal vein narrowing has been observed during an episode of migrainous transient visual loss [19, 20]. It is sometimes difficult to differentiate with certainty between migrainous episodes and amaurosis fugax caused by emboli. Migraine episodes usually last longer, from 3 to 30 min, and are associated with a significant headache around the eye. Nevertheless, patients with embolic amaurosis may have pain, albeit minor, around the involved eye or orbit. Also, about 10% of patients with embolic amaurosis experience scintillations [21].

Transient monocular obscurations are associated with a number of conditions that cause disc swelling and elevation, such as asymmetric papilledema of elevated intracranial pressure, optic disc drusen, optic nerve sheath meningioma, and giant cell arteritis-induced ischemia of the optic nerve prior to infarction. Transient retinal ischemia associated with emboli, collagen vascular disease, various hyperviscosity syndromes, severe elevation of the intraocular pressure, and systemic hypotension predominantly causes visual loss, but scintillations can be seen prior to the visual deficit. Reversible cataracts in diabetic patients are an extremely rare cause of monocular visual loss, which lasts for hours [22, 23] (see Sects. 5.3.2 and 8.2.3 for further discussion of transient visual disturbances).

10.2.4.2 Nonvisual Neurologic Disturbances

In contrast to the high incidence of visual dysfunction, sensory and motor disturbances are infrequent. Transient hemiplegia or hemianesthesia may precede, accompany, or follow the headache. These episodes, which usually last about 1 h, have a familial tendency [24], though spontaneous cases of hemiplegic migraine do occur [25]. The hemiparesis frequently recurs on the same side and can be part of a syndrome that results in episodes of diffuse or multifocal hemispheric dysfunction. Alteration of the state of consciousness can also occur. Complete recovery of the motor function usually occurs within several hours, but can take up to several days [26]. Permanent hemiparesis is rare in patients with complicated migraine, as are signs of meningeal irritation, including fever and neck stiffness. In these cases an elevation of the monocyte and protein concentrations can be found in the spinal fluid [27].

Basilar artery migraine is unusual and predominantly occurs in children. Episodic symptoms include vertigo, dysarthria, ataxia, and bilateral visual loss in addition to the headache [28, 29]. On occasion, an attack of either basilar artery or hemiplegic migraine can cause a transient loss of consciousness [30].

Acute confusional migraine is also found in children. The episodes consist of periods of agitated disorientation, which initially may not be associated with headache. Eventually, the confusional episodes abate and typical headaches develop [31, 32].

10.2.5 Ophthalmoplegic Migraine

Ophthalmoplegic migraine is a rare form of migraine, first described by Gubler [33]. It causes a specific and isolated visual disturbance. The episodic, throbbing headache is typically located behind the affected eye and is associated with paralysis of the muscles innervated by the ipsilateral third nerve, including the iris sphincter and the ciliary muscle [34]. Dysfunction limited to one branch of the third nerve is rare [35]. Although the third nerve is most often involved [36, 37], in rare cases isolated fourth or sixth nerves are affected. In the vast majority of initial attacks, the paresis resolves over several days or weeks, but with subsequent episodes a permanent oculomotor deficit and pupillary paresis are more likely. Ophthalmoplegic migraine typically begins prior to 10 years of age and may start as early as 8 months [38]. Rare patients begin experiencing the episodes during early adulthood [36]. The arteriogram is usually normal, but rare cases have shown narrowing of the cavernous ICA [39] or the basilar artery [34]. Neither of our two patients with this disorder showed an abnormality on computed tomography (CT) or cerebral angiography.

The differential diagnosis for this disorder includes an aneurysm in the vicinity of the third nerve, a congenital arteriovenous malformation located at the proximal posterior cerebral artery [40], Tolosa – Hunt syndrome, and a sphenoid mucocele. The incidence of diagnosed ophthalmoplegic migraine has fallen following the introduction of high-resolution CT. Any patient who has a chronic paresis of only one cranial nerve must be evaluated with high-resolution CT or magnetic resonance (MR) imaging and possibly a lumbar puncture. MR angiography or angiography should be considered if there is any doubt as to the diagnosis, because

subarachnoid aneurysms, although rare [41], can be symptomatic in children less than 10 years of age. In this age group these aneurysms usually present with a subarachnoid hemorrhage, however, and rarely cause a cranial neuropathy. It has been suggested that a recurrent painless cranial nerve paresis may represent another form of ophthalmoplegic migraine [42].

Cyclic pupillary dilation is another possible form of migraine that affects young adults [43]. It consists of attacks of transient unilateral pupillary dilation, with or without normal reactivity to light and without signs of external ophthalmoplegia. During the episode, the intraocular pressure is normal and the anterior chamber is clear. The pupillary dilation is often but not always followed by an ipsilateral retrobulbar headache [44]. Aneurysms, diabetes mellitus, or neoplasms are not in the differential diagnosis since they do not cause recurrent transient mydriasis alone. A fixed, dilated pupil without external oculomotor paresis has been described in only two patients with posterior communicating aneurysms [45, 46].

10.2.6 Acephalgic Migraine

Approximately 3% of migraineurs have acephalgic migraine prodromal symptoms, predominantly visual, without a headache [47, 48]. Many of the patients have a prior history of migraine headaches; if a headache history cannot be elicited, the diagnosis of acephalgic migraine should be made with caution. Any of the positive or negative neurologic symptoms or findings typical of a migrainous prodrome can occur as a migraine equivalent.

If these episodes are the first symptom of a migraine and are always on the same side or invariable, a structural lesion such as an arteriovenous malformation of the brain must be excluded (see Sect. 7.1.5.1). Attacks in patients more than 40 years of age that consist only of a neurologic deficit are similar to and highly suggestive of transient ischemic attacks caused by atherosclerosis or cardiac emboli [1, 2].

10.2.7 Other Headaches with Possible Vascular Dysfunction

10.2.7.1 Coital Cephalgia

One curious migraine syndrome is associated with sexual activity. An explosive, throbbing headache appears just before or during orgasm [49, 50]. Three times more men than women experience these headaches. The attack is sometimes misdiagnosed as subarachnoid hemorrhage from rupture of an intracranial arteriovenous malformation or aneurysm. The incidence of subarachnoid hemorrhage during intercourse, however, is low. It is only 3.8% for aneurysms and 4.1% for arteriovenous malformations [51], and patients with this form of migraine do not develop signs of meningeal irritation, increased intracranial pressure, or neurologic deficits.

10.2.7.2 Cluster Headache

Cluster headache, or Horton's histamine cephalgia [52], is an unusual type of vascular headache that probably should not be classified as true migraine. Men are affected five times more frequently than women [53], with a prevalence of 0.45% for men and 0.08% for women [54]. The age of onset is predominantly from late in the third decade to the sixth decade, giving rise to the notion that these headaches develop in middle-aged men. The pain is unilateral, severe, boring, nonthrobbing, and involves the orbit, face, temple, and upper teeth. The episodes last approximately 1 h and typically cluster in 6- to 8-week periods, separated by symptom-free intervals ranging from several months up to 2 years in duration. The attacks frequently wake the patient out of sleep, and are often accompanied by unilateral lacrimation, conjunctival injection, rhinorrhea, and nasal stuffiness. A partial third-order neuron Horner's syndrome can also occur. Pharmacological agents such as ethanol, histamine, and nitroglycerine which relax the smooth muscle of the vascular wall can precipitate cluster attacks.

10.2.7.3 Post-traumatic Headaches

Vascular headaches develop in more than 50% of head trauma patients, often as part of a postconcussion syndrome. The headaches are nonspecific, throbbing, unilateral or bilateral, and frequently localize around the eyes, the occiput or the site of the blow. Cervical pain may contribute to the occipital area discomfort and cause a secondary tension headache. The postconcussion syndrome also includes elements of depression, nonspecific dizziness or imbalance, anxiety, and a lack of motivation to return to employment. Numerous external and psychological factors contribute to an extremely variable clinical course.

10.3 Theories of the Causes of Migraine

The classical studies of Wolff and his collaborators suggested a vascular etiology for migraine and various other types of head pain. An inherent vasomotor instability is often blamed for an individual's susceptibility to migraine. Theoretically, vasoconstriction of the arteries of the internal carotid artery (ICA) system occurs during the prodrome and vasodilation of arteries of the external carotid artery (ECA) system during the headache phase. Blanching or flushing of the hands or face during the migraine also suggest a systemic vasomotor alteration. Painful dilatation of branches of the ECA in the scalp, dura, and meninges develops during the headache in certain patients. Histamine injection into the ICA induces painful dilatation of the large cerebral and pial arteries. Electrical stimulation of the internal carotid, anterior cerebral, middle cerebral, posterior cerebral, vertebral, and basilar arteries also causes pain. With stimulation of the intracranial internal carotid, proximal anterior, middle, and posterior cerebral arteries, the pain localizes in and around the ipsilateral eye. Electrical stimulation of the vertebral and basilar arteries induces pain in the occipital and suboccipital areas. Caffeine withdrawal, amyl nitrite or nitroglycerin, fever, and bacteremia all induce vasodilation of the ICA and ECA systems, resulting in a pulsatile headache [55]. Modern studies, however, suggest that the classical theory attrib-

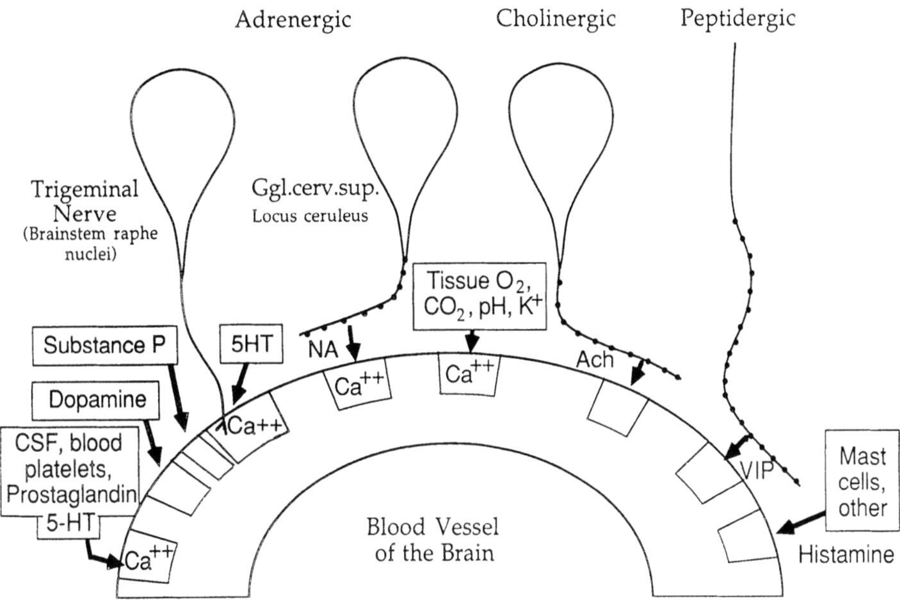

Fig. 10.2. Multiple factors influence the vasomotor state of the pial blood arteries in the brain. *Ggl. cerv. sup.,* superior cervical ganglion; *5-HT,* 5-hydroxytryptamine (serotonin); *NA,* noradrenaline (norepinephrine); *Ach,* acetylcholine; *VIP,* vasoactive intestinal polypeptide

uting migraine episodes solely to vasospasm of major cerebral arteries with cerebral hypoperfusion followed by dilatation of the vessels is too simplistic.

To understand those factors that can disturb the vasomotor tone in migraine, the normal complex control of the tone of the vessel wall and blood flow in the internal and external carotid arterial systems must be considered (Fig. 10.2). Various theories suggest that immediately before or at the onset of a clinical migraine, an alteration of neuronal activity and the release of vasoactive substances such as serotonin (5-hydroxytryptamine), circulating amines, and prostaglandins alter the pial arteriole vasomotor tone, overriding the normal autoregulation. Much of the experimental work has been performed on species other than man. Though many of the results are applicable to human vasomotor tone, some studies have reported conflicting data.

10.3.1 Neurotransmitter Maintenance and Altered Vasomotor Tone

The arteries of the circle of Willis, and those on the surface of the cortex, are innervated by noradrenergic nerve fibers originating from the ipsilateral superior cervical ganglion and the locus ceruleus [56, 57]. Experimentally applying norepinephrine to the pial vessels causes smooth muscle contraction and vasoconstriction of the arteries and arterioles [58], which is prevented by α_1-adrenergic receptor blockers [59–61] or calcium channel blockers [62].

Cholinergic fibers, some of which originate from the sphenopalatine ganglion, also innervate the cerebral capillaries, arterioles, and arteries [63, 64]. Experimental pharmacological stimulation of muscarinic receptors dilates cerebral arterioles. Blocking the receptor with atropine prevents acetylcholine-induced vasodilation [65, 66]. Despite these studies, however, it remains uncertain whether dilatation of the cerebral arteries in man is mediated through a cholinergic mechanism.

The fibers of the serotonin system (which originate from the midbrain brainstem raphe and pass with the trigeminal nerve fibers; see below) that innervate the large cerebral arteries and pial arterioles [67, 68] have long been thought to play a role in migraine. Serotonin induces a calcium-mediated vasoconstriction of human cerebral and pial arteries [69] which can be blocked by methysergide [70, 71] or β-adrenergic drugs [67, 72, 73]. Nitroglycerine reverses the serotonin-induced vasoconstriction of the basilar, middle, and cerebral arteries by directly relaxing the smooth muscle of the vascular wall [74]. Serotonin may actually dilate the smaller pial arterioles [75], and propranolol blocks this dilatation [76]. Serotonin is also present in platelets and in circulating blood. As long as the blood–brain barrier is intact, however, circulating serotonin should not alter cerebral blood flow [77].

Though β-adrenergic receptors are found in many mammalian species, it is uncertain whether there are β-adrenergic receptors on human intracranial arteries [78, 79]. The iontophoretic application of β-adrenergic agents such as isoproterenol has a minimal effect on arteries with normal resting tone [58], but when the muscle of the artery is already contracted, isoproterenol relaxes this spasm. Intracarotid injection of isoproterenol increases cerebral blood flow in dogs, and this increase is prevented by the β-adrenergic blocker propranolol [80]. The effect of β-adrenergic agonists and antagonists on the vasomotor tone clearly depends on the vascular tone when the agent is administered.

Histamine receptors (H2 receptors) are found on both the extracranial and intracranial arteries. Intracarotid injection of histamine vasodilates the ECA and ICA systems in monkeys [66]. Histamine, which is normally present in tissue macrophages and in the blood, can play a role in the generation of local tissue edema.

Numerous neuropeptides such as vasoactive intestinal peptide [81, 82] and bradykinin [66] have vasoactive capabilities that experimentally induce vasodilatation of the human cerebral arteries and may increase cerebral blood flow. Substance P is found in the cerebral blood vessels, the dura, and the trigeminal ganglion [83, 84]. Substance P seems to have an important role in migraine because it causes pain, relaxes serotonin-induced constriction of cerebral arteries, and dilates the pial arterioles and veins [85].

Despite the presence of complex neurotransmitter input into the control of vasomotor tone, questions and controversies remain as to the role of different neurotransmitters in altering vasomotor tone and initiating a migraine attack. It is also unclear whether therapeutic agents work by altering neurotransmission at the level of the blood vessels or on the neurons of the central nervous system, or both.

Various measures of systemic neurotransmitter metabolism reveal secondary alterations in migraineurs. Clearly, the changes in serum concentration of neurotransmitters such as serotonin, histamine [86] and norepinephrine [87] during a migraine are not causative. For example, serotonin is released from platelets during a migraine episode as a consequence and not as the cause of the attack [88]. The elevated plasma levels of norepinephrine found between attacks may suggest an underlying dysfunction of noradrenergic transmission as a predisposing factor [89]. Platelet monoamine oxidase B activity, which metabolizes circulating catecholamines, may also be reduced during an attack [90].

10.3.2 Neuronal Dysfunction

Recently, central nervous system dysfunction has been considered the primary initiating event of migraine. It is hypothesized that an alteration of the neuronal firing in the locus ceruleus, that alters the cerebral blood flow, has an important role in the pathophysiology of this disorder. Experimental evidence supports this hypothesis. Stimulation of the locus ceruleus via noradrenergic projections causes an increase in the ipsilateral cerebral arterial resistance and dilatation of the ECA system in monkeys [91]. Serotonin-containing neuronal pathways, originating in the raphe nuclei dorsalis and medianus in the midbrain, innervate the diencephalon [92] and project to the cerebral microvessels [93], possibly via the trigeminal nerve [94]. Stimulation of these brainstem raphe nuclei can increase the resistance of the cerebral arterioles [95].

The periodic nature of migraine seems to suggest a possible relation to the circadian rhythm of the hypothalamus. The relation of migraine attacks to stress may also arise from an alteration of limbic system input to the hypothalamus. Migraine alteration of neurotransmission or ischemia in the hypothalamus or thalamus may cause dysfunction in the autonomic nervous system, which would account

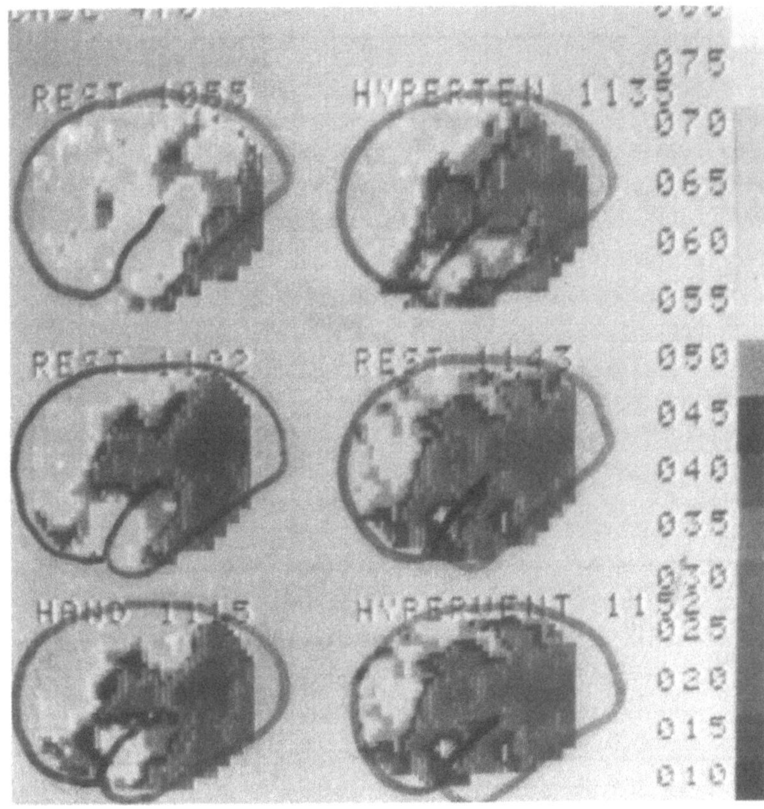

Fig. 10.3. Cerebral blood flow measured with the xenon 133 intracarotid injection technique during a classic migraine attack. The area of reduced cerebral blood flow (*darker gray to black*) began in the occipital region and propagated anteriorly over 60 min. Starting in the *upper left corner*, successive measurements (*top to bottom, then right column*) are shown. The time at which each individual measurement was performed is indicated. (Case 470 from [153])

for systemic visceral complaints such as nausea, anorexia, vomiting, and diarrhea; peripheral vasodilation and vasoconstriction; water retention; mood changes; and fever. The nausea, sonophobia, and hypersensitivity to smells may also indicate dysfunction in the various brainstem – hypothalamic catecholaminergic pathways. A dysfunction in the locus ceruleus or raphe nuclei may alter the neural control of the intracranial arteries, possibly through the trigeminal nerve efferents, which can also release substance P into the wall of pial arteries [96]. The trigeminal nerve also regulates pain sensation from the dura and large arteries.

Many drugs that successfully treat migraine seem to have direct effects on central nervous system neurotransmission via serotonin uptake mechanisms and receptors, β-adrenergic receptors, acetyl choline receptor blockade, and histamine receptor blockade. The migraine-induced reduction in cortical neuronal function (both neuronal firing and metabolism) may actually lead to a secondary reduction of local cerebral circulation.

10.3.3 Spreading Depression

Ischemia and spreading depression may be interrelated mechanisms which contributes to the pathophysiology of both the positive and negative phenomena of migraine. Spreading depression is a neurophysiological phenomenon, experimentally elicited by potassium perfusate, electrical stimulation, or trauma to an area of the cerebral cortex (Fig. 10.1c). Indirect elevation of the concentration of extracellular potassium, which follows hypoxia, inhibition of the cellular sodium – potassium pump, and hypoglycemia, also induce spreading depression [97, 98]. During experimental spreading depression, tissue levels of oxygen decrease, there is a shift towards anaerobic glycolysis, and the extracellular potassium concentration increases. The lactate concentration in the brain also increases despite the maintenance of a constant blood supply during the induction of the spreading depression [99].

The cerebral arteries change in caliber depending on the phase of the spreading depression. Blood flow reduction in the cortical arteriolar and capillary microcirculation follows the neuronal dysfunction of spreading depression [100] (Fig. 10.1c). Vasoconstriction precedes the potassium-related glial slow wave, whereas vasodilatation develops in the region approximately 1 min after the wave has passed. Hypoperfusion of the cerebral cortex and vasoconstriction of the cerebral arteries follow prolonged continuous spreading depression. During this latter stage, the arteries fail to dilate as they normally would in response to an increase in the blood CO_2 level or a lowered pH of the affected cortical tissue [101] (see Sect. 1.6).

Cerebral blood flow investigations in humans support the existence of both spreading depression and ischemia as mechanisms in classic migraine attacks. Regional cerebral blood flow (rCBF) studies in patients with classic migraine, but not common migraine, show a mild hyperemia followed by a reduction in the regional rCBF. The region of diminished flow begins posteriorly over the occipital lobes and travels anteriorly at a rate of 2–3 mm/min (this is similar to the 2–5 mm/min rate of spread for the depression wave) in a pattern unrelated to the distribution of the blood supply from the large cerebral arteries (Fig. 10.3) [102, 103]. Positron emission tomography may demonstrate an increase in cerebral oxygen extraction that reflects cerebral ischemia rather than metabolic depression [104]. Evidence for neuronal suppression during migraine has also been demonstrated with magnetoencephalography [104a]. Thus, depression of neuronal firing and metabolism, as well as capillary and arteriolar spasm-induced ischemia, seem to be important factors in migraine.

10.3.4 Platelet Hyperaggregability

Increased platelet aggregation may account for the elevation of the serum serotonin [105] and can aggravate the vasoconstriction in the pial arterial circulation that develops during the migraine prodrome. The rise in serum serotonin parallels the release of serotonin from platelets [106] during aggregation or as a result of the release of free fatty acids. Platelet-derived growth factor may also contribute to the focal vasoconstriction that occurs during aggregation [107]. The level of platelet hyperaggregability does not correlate with the type

or severity of migraine [108], and hyperaggregable platelets may even be found in the blood of migraineurs between attacks [109]. It has also been suggested that platelets from migraineurs contain more adenine nucleotides than those of controls [110].

Migrainelike headaches can develop in patients who have disorders that cause abnormal platelet activity and thrombocytopenia [111]. Patients who begin having migraines following the institution of estrogen-containing oral contraceptives often show an increase in serotonin-induced platelet aggregation. Platelet aggregation normalizes, and the attacks are less frequent, following withdrawal of the estrogens [112]. Even if the platelet hyperaggregability is a secondary phenomenon, platelet aggregates may worsen the ischemia in areas of reduced blood flow and thus may contribute to the development of permanent neurologic defects or infarcts.

10.3.5 Hormonal Influences

Estrogens alter vasomotor stability and clearly affect the course of migraine [113, 114]. Menstrual migraine attacks develop just before menstruation, as estradiol levels decrease. During pregnancy, when estrogen levels are elevated, the frequency of migraine attacks usually diminishes [115, 116]. A few women note worsening of their migraine during pregnancy, however, and others may experience their first vascular headache during the first trimester of pregnancy [117]. In contrast, blood progesterone level fluctuations do not alter the number or severity of migrainous episodes.

As noted above, oral contraceptives can increase the incidence and severity of migraine [118–122]. The onset or worsening of the vascular headaches usually develops within several months of starting an oral contraceptive, concomitant with an increase of clotting factors and platelet aggregability [123–126] and a decrease in fibrinolytic activity [127]. Other studies have found that oral contraceptives may not directly alter platelet aggregation [124]. The risk of vascular thrombosis may be increased with these agents under specific circumstances. A fall in plasma antithrombin III inhibitory activity on activated factor X, associated with the use of oral contraceptives, can promote peripheral venous thrombosis [125, 128]. Besides the complications associated with tobacco use, a history of migraine may be a risk factor for thrombotic stroke with the use of an oral contraceptive [129–133]. An estrogen-induced hypercoagulation tendency can persist for more than a month following cessation of oral contraceptives or during the post partum period. Because of the potential increase in clotting tendency, estrogen therapy is not recommended for control of menopausal or postmenopausal vascular headaches.

10.3.6 Psychological Factors

Emotional factors may be correlated with the development of migraine. Affected children are commonly sober, conscientious about school work, neat and clean dressers, diligent about organizing toys, and nearly always obedient to parental wishes. The adults have similar personalities and are meticulous, ambitious, tense, easily frustrated, rigid, and desirous of order in all facets of life. Failure or criticism (and even occasionally success) can precipitate a migraine attack. Migraines often occur at the end of the work week or on holidays. Situations that cause persistent tension, require excessive striving, or cause anxiety about work, family, and finances contribute to the development of migraine, possibly by triggering limbic system input to the hypothalamus [134].

10.3.7 Blood–Brain Barrier Alteration and Dietary Factors

Alteration in the blood–brain barrier has been proposed to have a role in the pathogenesis of migraine. One hypothesis suggests that a defect in the barrier permits passage of food allergens, vasoactive amines, or other substances that disturb neurotransmission into the central nervous system [135]. However, there is little evidence to support this theory. Dietary factors have been found to be significant in approximately 10% of migraineurs and in most patients with cluster headaches. An example is the extreme sensitivity to alcoholic beverages (a smooth muscle relaxer) [136]. In some patients foods such as chocolate [137], bananas, aged chees-

es, and red wine, which contain the vasoactive amines tyramine or β-phenylethylamine, can be associated with, or even acutely precipitate, a migraine. The theory suggests that the monoamine oxidase or phenolsulfotransferase in the gastrointestinal mucosa may not completely inactivate the amines or their phenolic metabolites in the foods and the increased plasma levels alter vasomotor tone [138]. Most of these foods contain only small amounts of amines, however, and attempts to induce migraines with the ingestion or injection of tyramine alone have been unsuccessful or produced marginal results [139–141].

Large amounts of the food additive monosodium glutamate may change vasomotor tone, leading to a headache in some patients. Hypoglycemia can alter hypothalamic activity, induce vasodilatation of cerebral vessels, and precipitate a vascular headache.

10.3.8 Allergy

In some studies, the onset of migraine has been correlated with the ingestion of certain foods in some migraineurs who have skin test reactions to possible allergens in these foods. Some patients report fewer and less severe migraine episodes when foods with these allergens are eliminated. When patients are fed these foods without knowledge of their content, however, no definite relationship can be established between these food allergens and migraine. Additionally, temporal artery biopsies from migraineurs who experienced temporal artery swelling showed no signs of inflammation, as would be expected if the swelling were secondary to an allergic reaction [55a].

Immune mechanisms may be secondarily activated during a migraine attack, however, as reflected by abnormalities of circulating immune and inflammatory factors in the blood. Serum immunoglobulins are variably elevated in migraineurs. Serum C4 and C5 complement components maybe reduced at the onset of the headache [142, 143]. Activation of the complement system may secondarily precipitate platelet aggregation and serotonin release and an increase in the serum and tissue histamine concentrations via degranulation of basophils and mast cells [55c].

Prostaglandins can be released by serotonin or histamine stimulation of the arterial wall [144]. Prostaglandin E causes experimental vasodilatation of cerebral and the ECA system arteries. This may contribute to the headache phase of a migraine [145, 146]. In addition, prostaglandin F_2-α (PGF_2-α) constricts cerebral arteries [147], which might worsen the visual or neurologic symptoms.

10.3.9 Trauma

No hypothesis concerning a specific mechanism for the role of trauma in developing migraine has gained favor. Post-traumatic complicated migraine seems to be more prevalent in juveniles [148]. Immediately following the trauma, transient neurologic deficit episodes, which can include episodes of cortical blindness, can occur. The episodes can be a result of epileptic discharge in the cortical neurons or a postictal Todd's phenomenon or both [149].

10.4 Theories of Visual and Neurologic Phenomena

10.4.1 Positive Phenomena

Prodromal symptoms of a migraine attack usually precede the headache, but similar symptoms can occur concomitantly with the headache phase or even follow the onset of the pain. The various visual and neurologic disturbances may develop as a result of vasospasm-induced cerebral ischemia or spreading depression following the ischemia, spreading depression in the cortex, or both. The interrelationship of ischemia and spreading depression may contribute to the pathophysiology of the positive and negative phenomena of migraine.

The mechanism of the migrainous homonymous scintillating scotoma can be understood in terms of a propagating cortical excitatory wave followed by a period of inhibition of cortical activity (Fig. 10.1) [150, 151]. Typically, the fortification images spread, migrating across the field with a propagation rate of approximately 3 mm/min [150, 152].

In about 60% of patients with classic migraine, rCBF is reduced by 50%, to below 23 ml/100 g per minute, which can account for the neurologic deficits during the prodrome. These deficits develop in 5–15 min following the focal rCBF reduction [153]. The normal increase in cerebral blood flow caused by CO_2 inhalation or movement of a contralateral limb is reduced during the attacks [102]. Normal rCBF regulation is found between episodes. Single photon emission computed tomography studies also confirm a mild reduction in the blood flow of the cerebral hemisphere appropriate to the neurologic dysfunction in patients with classic migraine [154].

Focal epilepsy also has been suggested to account for the positive neurologic disturbances. For example, migrainelike visual phenomena are controlled by anticonvulsants in patients with occipital arteriovenous malformations, suggesting excessive neuronal firing as a mechanism. Migraineurs, however, typically do not demonstrate seizure activity on electroencephalography and their episodic visual disturbances are not generally controlled by anticonvulsants.

10.4.2 Negative Phenomena

10.4.2.1 Vasospasm and Ischemia

Negative phenomena, the transient loss of visual, sensory, or motor function that may accompany migraine attacks, may result from temporary ischemia [15, 155]. Theoretically, vasospasm in the ICA or the ophthalmic artery may result in monocular amaurosis; vasospasm in the basilar artery or both posterior cerebral arteries may lead to visual loss in both eyes or both fields; vasospasm in one posterior cerebral artery may yield a homonymous hemianopic or quadranopic loss of vision; vasospasm in the middle or anterior cerebral artery may result in a hemiparesis [269]; and vasospasm in the middle cerebral artery may cause hemisensory deficits [24]. The cranial neuropathy in ophthalmoplegic migraine has been hypothesized to result from ischemia of the third nerve because of vasospasm and arterial wall edema, which narrow the lumen of nutrient arteries arising from the cavernous carotid (inferior lateral trunk) or basilar (interpeduncular branches) arteries [156] (see Sect. 1.5.3.2.1).

Indirect confirmation of the role of vasospasm is found in reports of transient reversal of migrainous field defects following the administration of agents that directly relax the smooth muscle of the arterial wall, such as amyl nitrite [157]. Negative visual field defects were similarly reversed by agents that indirectly cause vasodilation of the cerebral arteries, such as 10% CO_2 inhalation [155], that lower the pH in the extravascular and extracellular tissues [168], or that dilate vasoconstricted arteries by stimulating possible β-adrenergic vascular receptors. The vasoconstriction in the bulbar conjunctival vessels during the prodrome, and their dilatation when the headache begins, also reflect the vasomotor alterations in the ICA territory [159]. It is not clear why the retinal arteries rarely show a similar change in caliber during the prodrome or headache or with moderate doses of ergot drugs [160].

Cortical blood flow measurement reveals hypoperfusion of the cortex when negative symptoms such as a contralateral hemiparesis begin. Cerebral blood flow measurement by inhalation or intracarotid injection techniques shows a 20% – 50% reduction of perfusion during a classic migrainous prodrome [102, 161a, b]. When a neurologic deficit results, the focal cerebral blood flow falls below the level required to maintain cerebral function [153]. As expected, if a visual disturbance is the sole neurologic prodromal symptom, the vasomotor dysfunction is limited to the posterior cerebral circulation [162].

The decrease in blood flow is not always sufficient to cause the cortical dysfunction. Persistent hypoperfusion, despite recovery from the deficit, also suggests that ischemia does not account for all the neurologic dysfunction in migraine [163]. As mentioned earlier, the altered neuronal activity may secondarily cause the diminished regional blood flow. Despite experimental evidence that ergot drugs can cause vasoconstriction of cerebral arteries, normal doses of ergotamine do not alter the cerebral blood flow [164]. Ergotamine-induced vasoconstriction in the ECA supply to the scalp and dura can falsely suggest a decrease in cerebral flow when measured by the inhalation method, however.

Cerebral angiography has demonstrated only one case with definitive vasospasm followed by normalization of the cerebral arteries that could be correlated with the development of a neurologic deficit and subsequent normalization of function. During the angiogram, the patient developed bilateral visual scotomas and confusion, associated with the absence of filling of the anterior and middle cerebral arteries following injection of contrast material into the ipsilateral ICA. When a throbbing headache appeared and the neurologic dysfunction resolved, the cerebral arteries supplied by the ICA were visualized in the normal sequence. There were no signs of a catheter-induced embolus or dissection of the arterial wall [165].

Angiography has also rarely revealed an occlusion in a branch of the middle cerebral artery in patients with familial hemiplegic migraine [166]. In most cases of complicated migraine, cerebral angiography is normal [26].

10.4.2.2 Epilepsy

Clearly, an ischemic mechanism does not account for all the negative neurologic phenomena. In two patients with hemiplegic migraine, the electroencephalogram during the transient hemiparesis and dysphasia revealed slow sharp waves over the dominant hemisphere contralateral to the paresis [167], suggesting that the deficits were related to an epileptic discharge. As mentioned previously (see Sect. 10.3.9), the post-traumatic transient blindness, most frequently encountered in children, can be associated with migraine and accompanied by seizure activity on the electroencephalogram. In childhood migraine, particularly with basilar artery migraine, an abnormal electroencephalogram and occasional clinical response to anticonvulsants (see Sect. 10.7.3) suggest that seizure activity may cause some of the neurologic dysfunction [168]. On occasion, episodic slow spike and wave complexes recorded from over the occipital lobes can be associated with the visual scotomas in migraineurs [169]. Patients with basilar artery migraine who have an abnormal electroencephalogram seem to have a more benign prognosis than those who have a normal electroencephalogram [170].

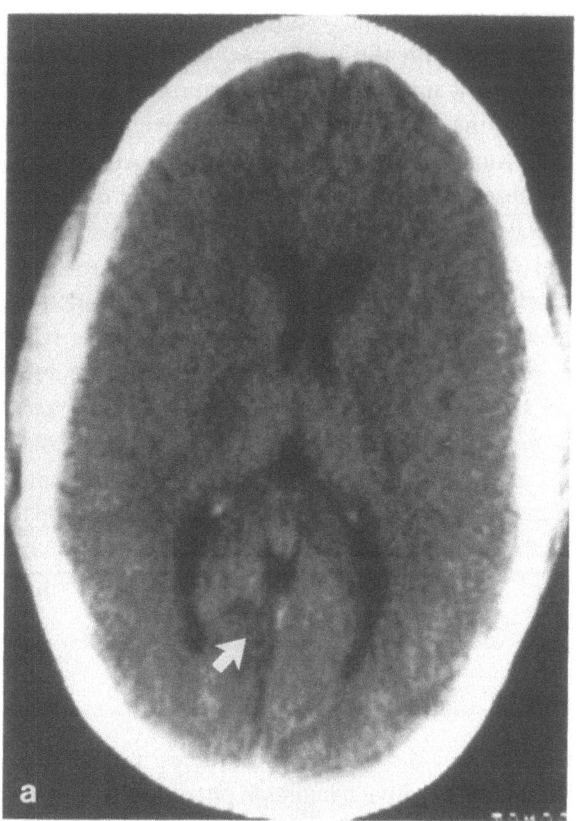

Fig. 10.4a, b. Cerebral infarct and visual field loss in a young patient with migraine attacks. **a** A noncontrast computed tomogram demonstrates a right anterior midline calcarine infarct (*arrow*), which spares the region representing macula function. **b** This patient had a dense partial left homonymous visual field defect to all isopters. Note the sparing of the macula field which corresponds to the preservation of the posterior occipital lobe

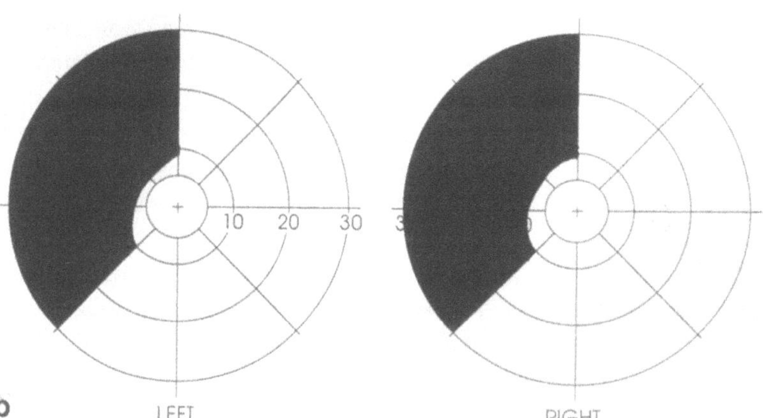

10.4.3 Permanent Neurologic or Visual Defects

Although uncommon, permanent neurologic defects, which most commonly involve the visual system, provide further evidence that ischemia induced by arterial stenosis or occlusion is responsible for many of the neural and visual deficits experienced by migraineurs [18]. Rarely, persistent visual dysfunction develops because of a retinal infarct from occlusion of a branch or central retinal artery [171, 172] or an anterior ischemic optic neuropathy [173–176]. An occipital lobe infarct with a homonymous field defect is the most common sequela of migraine [77, 177–180]. In one report of 46 patients with complicated migraine, 17 (most with homonymous visual field loss) had infarcts in the territory supplied by either the middle or posterior cerebral arteries [181] (Fig. 10.4). In patients with infarcts, angiography can reveal stenosis, beading, or occlusion of the posterior cerebral arteries or its

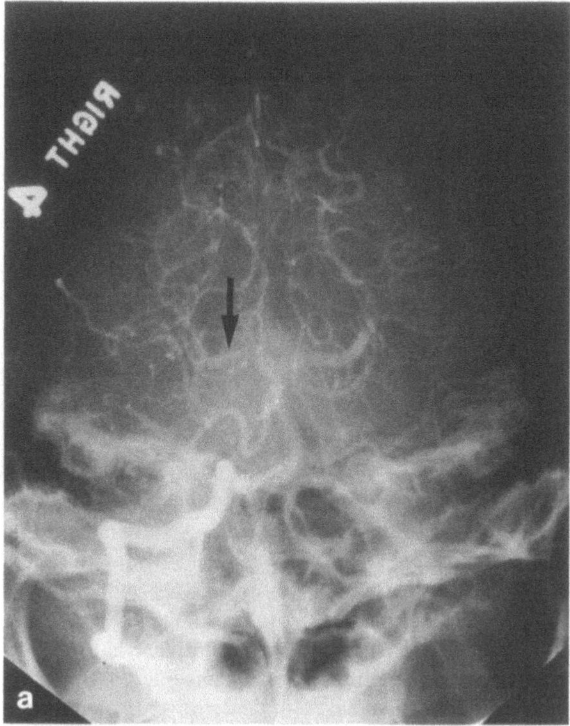

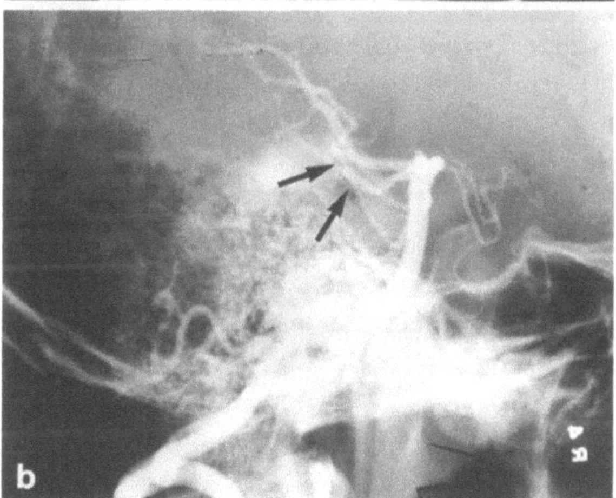

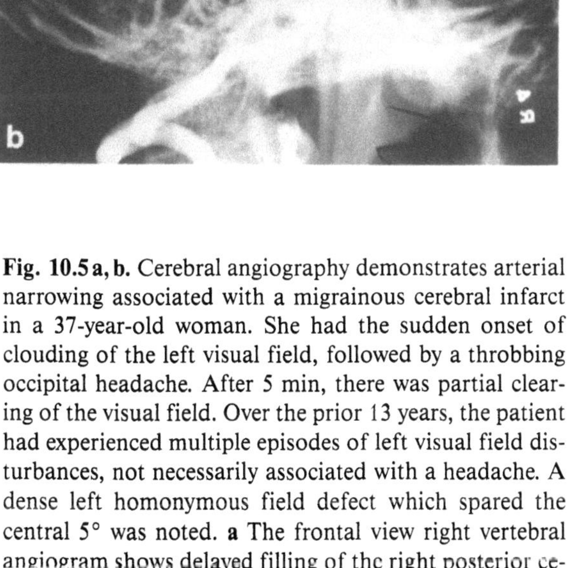

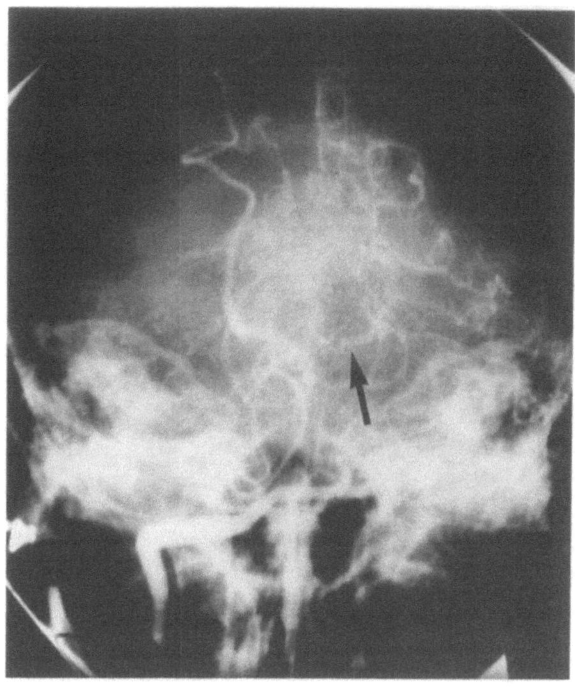

Fig. 10.6. A 26-year-old woman complained of a throbbing right-sided headache followed by poor vision in the right visual field of each eye and a mild weakness in the right arm and leg. During the previous 15 years she had experienced episodes of headaches with nausea, vomiting, photophobia, and sonophobia. The neurologic evaluation revealed mild weakness in the right arm and a dense complete left homonymous hemianopsia. The frontal view right vertebral angiogram shows marked narrowing (*arrow*) and reduced flow in the left posterior cerebral artery. The arm normalized, but the field defect was permanent

Fig. 10.5 a, b. Cerebral angiography demonstrates arterial narrowing associated with a migrainous cerebral infarct in a 37-year-old woman. She had the sudden onset of clouding of the left visual field, followed by a throbbing occipital headache. After 5 min, there was partial clearing of the visual field. Over the prior 13 years, the patient had experienced multiple episodes of left visual field disturbances, not necessarily associated with a headache. A dense left homonymous field defect which spared the central 5° was noted. **a** The frontal view right vertebral angiogram shows delayed filling of the right posterior cerebral artery (*arrow*). **b** The lateral view of the right vertebral angiogram shows an irregular filling of the right posterior cerebral artery (*arrows*). The visual field defect never resolved

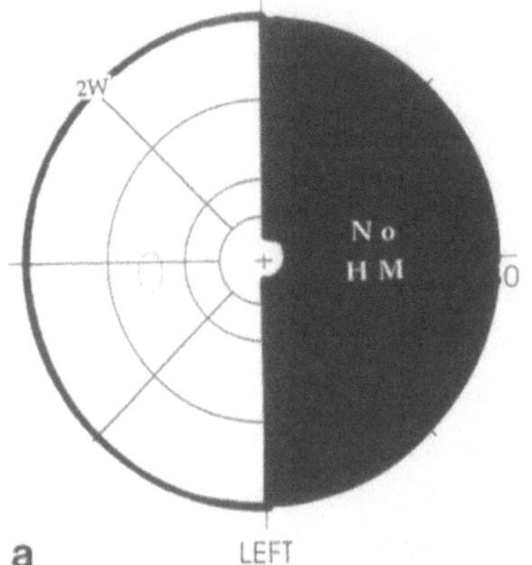
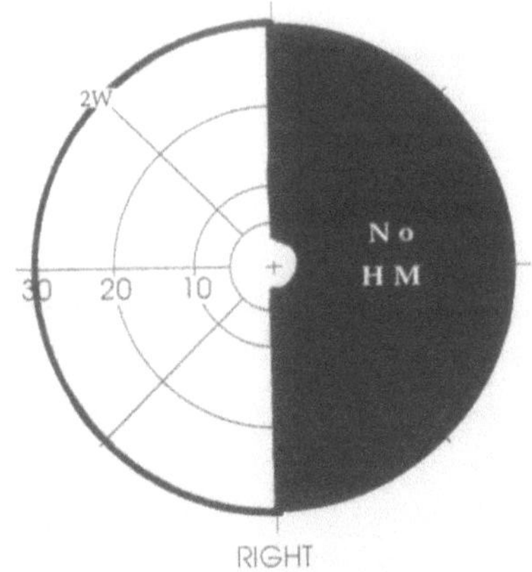

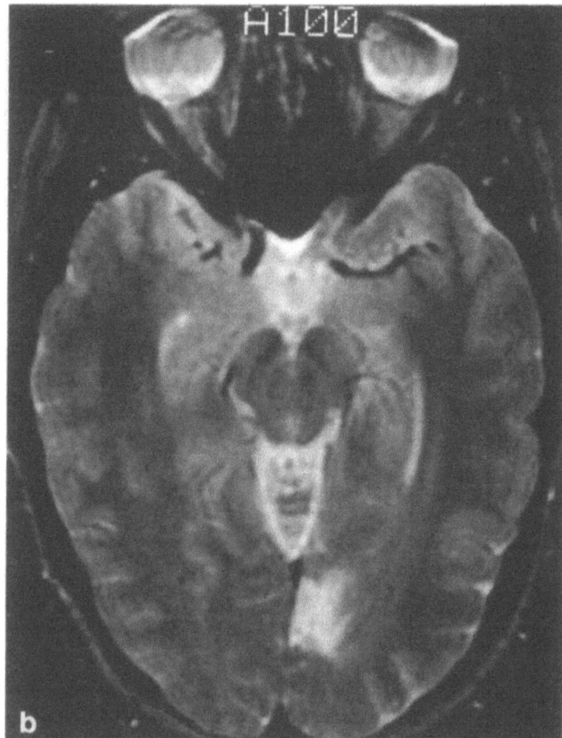

Fig. 10.7a, b. A 25-year-old woman with a long history of classic migraine experienced the sudden onset of painless loss of vision to the right side of the field while on oral contraceptives. No laboratory abnormalities were found. **a** The visual field demonstrates a dense right homonymous hemianopsia to all isopters, including hand motion, but there is sparing of the right central 3°. **b** The T2-weighted magnetic resonance image demonstrates an area of increased signal in the anterior half of the right occipital lobe. In this patient, the posterior half of the occipital lobe was not involved accounting for the remaining 3° of macula function

branches [179, 18], or it can be entirely normal [182] (Figs. 10.5, 10.6). Occlusion of a calcarine branch artery has been demonstrated in patients with an isolated homonymous quadranopia [183, 184]. CT or MR imaging confirms the common occurrence of cerebral infarcts, particularly in the occipital lobes [180, 185 – 187] (Figs. 10.4, 10.7). Prior to experiencing an infarct, there are no coagulation abnormalities, other systemic abnormalities, or specific migraine episodes that distinguished migraineurs who developed strokes from those who

did not experience permanent deficits [181]. Visual dysfunction may also occur in the rare cases with retinal hemorrhages [188, 189] or central retinal vein occlusion [190].

Migraine-associated cerebral infarcts have been reported in patients who have common, classic, or complicated migraine. An early CT study of 40 patients with long-term severe migraine showed definitive evidence of cerebral infarcts in three and cerebral atrophy in 11 cases. Cerebral infarcts were revealed in three cases and cerebral atrophy in an additional three cases in a series of 13 patients with hemiplegic migraine [185]. Later CT reports suggest that permanent neurologic deficits and cerebral ischemic damage generally follow the territory supplied by the major cerebral arteries [191, 192].

Fortunately, despite the recognition of cerebral infarction as a complication of migraine, there are only rare reports where the ischemic injury to the brain has lead to death. In one such patient the histopathology revealed severe and extensive ischemic damage involving the area of the brain supplied by the left anterior cerebral artery. Fewer ischemic changes were noted in the substantia nigra, the pontine nuclei reticularis, the nucleus of the trigeminal spinal tract, the occipital lobes, and Ammon's horns [193]. In another case, the histopathology demonstrated severe ischemia of multiple supratentorial regions areas supplied by the major arteries of the circle of Willis as a result of diffuse vasospasm [194].

When all the etiologies of cerebral infarct are considered, migraine remains a rare one. One study reported that the prevalence of clinical stroke was increased by migraine only in men who were either between 50 and 59 years old or more than 80 years old [195]. In the same investigation, the prevalence of hypertension and the risk of myocardial infarction were increased in migraineurs of all ages and both sexes.

Several different mechanisms may account for the central nervous system infarcts. Prolonged experimentally induced vasospasm seems to cause damage to the arterial endothelium with subsequent platelet adherence and aggregation at this site [196–198]. If a platelet–fibrin deposition adheres to a segment of a migraine-induced constricted artery, distal embolization can result, occluding the cerebral or ophthalmic circulation. Recall that migraineurs have increased platelet aggregation even if they do not have complicated attacks. Prolonged vasospasm may also primarily cause occlusion of the constricted vessel or irreversible ischemia in the affected tissue. It has been suggested that migraineurs who develop occlusive episodes have a different illness such as a focal arteritis which causes these events. Some of these patients who are younger than the age typically associated with atherosclerotic or hypertensive occlusive cerebrovascular disease have transient and permanent cerebral and ocular ischemic attacks caused by clotting abnormalities associated with a circulating lupus anticoagulant or other antiphospholipid antibodies [199, 200] (see Sect. 9.17).

10.5 Mechanisms of Headache

Classical theory suggests that distension of the large extracranial arteries in the scalp and face may cause the headache [201, 159a]. In fact, pressing on the temple temporarily decreases the pain. Another theory postulates that the intracranial dural venous sinuses distend because of increased blood flow from the shunting of cerebral blood. The cerebral blood flow is usually normal during the headache phase, but on occasion it can increase significantly above the preheadache rate [202, 161]. The cerebral or venous sinus hyperemia can persist beyond the duration of the headache, so vascular dilatation is unlikely to be the cause of the pain [203]. Additionally, the headache is not always on the side contralateral to the neurologic or visual deficit, as one would expect if the same arterial circulation was involved with the prodrome and the headache phases. Vasodilatation can occur in the ECA system while vasoconstriction persists or develops in the ICA system. The fact that some patients have evidence of systemic vascular constriction, which causes blanching of the skin that persists during the headache phase, also indicates that the different arterial systems do not necessarily constrict or dilate in phase.

The meninges and the meningeal arteries may be a source of pain in migraine [204]. Vasodilatation of meningeal arteries with release of various protein kinins can irritate the trigeminal nerve pain fibers innervating the dura. The pain of meningitis is mediated via the trigeminal nerve as well. The first division of this nerve provides sensory innervation to the large cerebral arteries and may transmit pain information from sudden or excessive vasodilatation [84]. This may explain why the pain is frequently referred to the eye or retrobulbar area and why photophobia is so common. Stimulation of the sensory trigeminal nerve also causes pain in other clinical entities, including trigeminal neuralgia [205] and "ice cream headaches" [206].

Though intravenous histamine causes painful dilatation of the extracranial arteries [207], the discomfort is unlike the pain of a classic or common migraine headache. However, histamine may play a role in cluster headaches [52, 86]. An increase in the number of perivascular mast cells and degranulation of these cells may be found in scalp biopsy specimens from patients with cluster attacks [208], suggesting that factors which causes local histamine release in the scalp can be responsible for the pain. The alteration of systemic histamine metabolism during attacks also suggests a role for histamine in this disorder [209, 210]. Injury of the facial nervus intermedius has also been hypothesized as a cause of pain in patients with cluster headaches [211].

tive of seizures or the disorder is basilar artery migraine, an electroencephalogram should obtained. A clinical cardiac evaluation and echocardiogram may uncover a potential source for ocular or cerebral emboli. However, in one study of 230 young women with a mitral valve prolapse, migraine was identified in 64 (28%) and complicated migraine in only six (2.6%) [212]. The numbers of migraineurs is clearly within the prevalence range (29%) of migraine for all women reported by Waters and O'Connor [4].

10.6 Laboratory Evaluation

Unless the headache is always located on the same side of the head, patients with typical common or classic migraine do not need a laboratory evaluation. MR imaging or contrast-enhanced CT of the head should be performed when the pain or neurologic phenomena remain in one location. Because of the rarity of coital cephalgia, most of these patients are also scanned. Patients with typical cluster headaches do not require neuroimaging studies unless the headache is always on the same side or a cranial nerve paresis develops. In the latter cases, MR imaging should image the cavernous sinus with thin coronal slices. In order to determine whether there is another etiology such as a possible vascular malformation (see Sect. 7.1 for distinguishing characteristics in the history), patients with complicated migraine should have a scan of the head, particularly if the recurrent neurologic deficits are stereotypical.

Other laboratory studies should also be considered in any patient with recurrent visual or neurologic deficits and headache. Rarely, patients with complicated migraine will have an elevation of monocytes in the cerebrospinal fluid [27]. The blood should be investigated for conditions that predispose to abnormal platelet aggregation, vasculitis, or spontaneous thromboses (see Sect. 8.11, Chaps. 8, 9). In view of the studies revealing a relationship between the various antiphospholipid antibodies and ocular and cerebral ischemic events (see Sect. 9.17), patients with severe complicated migraine or infarcts should have a complete evaluation for a systemic vasculitis or coagulopathy. If the clinical disorder has elements which are sugges-

10.7 Treatment

Evaluating therapies for the migraine headache is difficult because of the sporadic nature of the episodes and remissions as well as the notable placebo effect. In one study, 50% of 1644 migrainous patients improved with placebo administration [213]; while in other studies, 62% of patients improved initially and 75% benefitted after 4 weeks of placebo therapy [214]. The various antimigraine agents have the capability of directly or indirectly altering the mechanisms which influence vasomotor tone. Different categories of drugs may affect specific receptors, platelet function, inflammatory responses, and smooth muscle contractility. However, drugs which do not appear to be obviously similar may be effective because of their ability to modify similar or common mechanisms in the migraine process. For example, one recent study related the antimigraine effectiveness of methysergide, cyproheptadine, pizotifen, and propranolol to a uniform ability to bind to 5-hydroxytryptamine 1A receptors in the brain [215], a subset of serotonin receptors located in the hippocampus, in the dorsal raphe nuclei of the midbrain, and on intradural arteries [216, 217, 217a].

10.7.1 Acute Episodic Headache Management

10.7.1.1 Ergotamine Tartrate and Dihydroergotamine

Ergotamine tartrate and dihydroergotamine preparations have remained the agents most popular for the treatment of acute migraine attacks. Most studies describe a significant reduction in the number of episodes that progress to the headache phase when one of these drugs is taken at the onset of a prodrome. In contrast, some investigations report that low-dose ergot agents are no better than placebo in aborting a migraine. These drugs have been described as having a differential effect on the arteries of the ICA and ECA systems, because of differences in the type and distribution of neurotransmitter receptors and in the migraine-induced alterations of the existing vasomotor tone. Initially, the ergots were thought to abort a migraine attack by constricting the vasodilated and stretched arteries of the ECA and ICA systems [218]. However, the interaction of ergot agents with nonvascular brain receptors must also contribute to their antimigraine activity, because attacks can be aborted by low drug doses which do not cause significant alteration of the vasomotor tone [219]. Blocking 5-hydroxytryptamine receptors at the level of the brainstem dorsal raphe nuclei may be where these agents are effective in a migraine. It has also been proposed [220] that a therapeutic benefit of ergotamine and dihydroergotamine, independent of their vasoconstricting action, may also result via the prevention of the extravasation from the plasma of substances such as kinin which cause pain and inflammation in the dura [221–223].

The smooth muscle contracting capability of the different ergot preparations, which were first used to induce uterine contraction, still seems to be an important property responsible for some of the migraine-aborting effectiveness. At low doses, ergotamine increases the smooth muscle contraction induced by epinephrine, norepinephrine or serotonin of some experimental smooth muscle preparations [224]. Ergotamine vasoconstricts the ICA system, particularly if it is initially vasodilated [225]. This vasoconstriction seems to be mediated by α-adrenergic receptors because the effect is significantly antagonized by blocking the receptor with phenoxybenzamine or phentolamine [225]. The cerebral blood flow in normal humans is not altered by therapeutic doses of ergotamine. The observed decrease in blood flow following high doses in experimental animals probably has clinical relevance only in cases of ergot toxicity. Additionally, depending on the location of the tested vessel, low concentrations of ergotamine can have an opposite effect, antagonizing the vasoconstriction induced

by serotonin [226, 225]. In dogs and monkeys, ergotamine selectively vasoconstricts the ECA [227, 66].

Dihydroergotamine has actions which are qualitatively similar to those of ergotamine, but the hydrogenation decreases the vasoconstrictor activity [228] as well as the risk of the complications of ergotism (see Sect. 10.7.1.1.1). This agent blocks α-adrenergic receptors more effectively and causes hypotension (rather than the pressor effect of ergotamine) by altering the sympathetic outflow from the brainstem [228]. Dihydroergotamine also retains the capability of antagonizing serotonin-induced vasoconstriction.

The route of administration of these drugs influences the success and failure of treatment; therefore, patients and their doctors must work together to find an effective preparation. The medication is taken during the prodrome, prior to the migraine headache. Most patients are given ergot orally, in combination with caffeine, possibly to increase the intestinal absorption of the drug [229]. Oral ergotamine or dihydroergotamine, in doses from 1 to 2mg, may be repeated once in 30–60 min if the severe headache persists. However, when nausea and vomiting occur during the prodrome, the oral route is impractical since the drug can be expelled. Sublingual preparations can be effective, though the drug is absorbed inconsistently compared to those administered orally. The inhalation of ergotamine via an inhaler delivers approximately 0.3–0.75 mg per dose and provides a route for rapid absorption that avoids the difficulties caused by vomiting, abdominal migraine, and malabsorption. However, dose variability and pulmonary arterial constriction may be limiting. The rectal route, using a suppository preparation of 2 mg, gives excellent absorption. An initial dose of half the suppository may be effective, and if warranted, a 1- to 2-mg dose can be repeated in 1 h. Ergot agents should not be used more frequently than twice weekly. Intramuscular ergotamine or dihydroergotamine, usually in 0.5- or 1-mg doses, respectively, are generally given to patients with acute intractable headaches. Another agent which binds to 5-hydroxytryptamine receptors, sumatriptan, has also been reported to abort both acute migraine and cluster headache attacks when administered intramuscularly [230a, b].

10.7.1.1.1 Ergot Toxicity

Ergot toxicity is rare but as patients increase the frequency and dose of the various ergot preparations in order to alleviate their headaches, the risk of side effects increases. Since the ergots are metabolized by the hepatic system, if liver disease is present, regular use may lead to excessive serum drug levels. Most of the toxic complications are not visual or neurologic and result from the peripheral vasoconstriction induced by constant high-dose ergot ingestion. These drugs are relatively contraindicated in persons with stenotic peripheral vascular and coronary artery disease. Angiography has documented the induction by ergotamine tartrate of vasoconstriction of the limb arteries, which usually reverses following cessation of the drug [231, 232]. However, gangrene in the distal limbs can result from arterial occlusion and prolonged ischemia. The number and extent of arterial thromboses may increase with concomitant use of oral contraceptives [233]. Thrombosis in the peripheral arteries may also be worse if the patient is anemic. At times it may difficult to determine whether the patient's gastrointestinal complaints are abdominal migraine or secondary to drug-induced stenosis of the mesenteric arteries and ischemia of the bowel [234].

Clinical recovery usually follows cessation of the ergot alone. The institution of an α-adrenergic receptor blocker, phenoxybenzamine or phentolamine, antagonizes the ergot stimulation of these vascular receptors and may facilitate vasomotor relaxation. Vasodilation may also be induced by nitroprusside [235, 235a]. A percutaneously introduced Fogarty-tipped catheter has been used to mechanically dilate the segment of the artery in spasm [236]. However, none of the above therapies for ergot toxicity has been uniformly successful and amputations have been necessary. In patients with coronary atherosclerosis, ergots may cause coronary artery spasm, angina, and electrocardiogram changes, which may be reversed with nitroglycerine or isosorbide dinitrate.

Neurologic complications of ergotism are rare and have been reported in patients who have acutely ingested massive amounts of drug [237] or have taken chronic daily doses (1 – 6 mg rectally, or up to 8 mg orally) for as little as 3 weeks [238]. Neurologic dysfunction was reported more commonly in New England and medieval Europe following the ingestion of rye crop products that had been infected with *Claviceps pupurea*, which produces ergot. Reports described patients with delerium and behavioral problems, seizures, blindness, and focal neurologic deficits. Most of the recent literature consists of isolated cases with drug-induced neurologic deficits. For example, a 36-year-old women migraineur, who claimed to have taken half of a 1-mg ergotamine suppository daily for 3 weeks, developed a complete right hemispheric syndrome, with angiographic demonstration of severe stenosis of both ICAs. The patient became clinically normal within 10 days following cessation of the ergotamine and several days of treatment with corticosteroids and heparinization [239]. Hypalgesia of the left arm, a left facial paresis, nystagmus, and severe narrowing of the right ICA developed in another patient with ergotism [240]. Severe headaches and scalp tenderness mimicking temporal arteritis have also been described from prolonged vasoconstriction of the temporal artery.

The ophthalmologic complications of ergotism include visual loss from retinal artery occlusion [241]. Diffuse retinal dysfunction can also result from generalized narrowing of the retinal arteries [238]. In one case, bilateral optic neuropathy with optic disc swelling probably resulted from vascular occlusion, though the term "papillitis" was used to describe the findings [242]. Eventually, the optic discs appeared pale and the arteries narrowed. The associated visual field defects include generalized constriction [243] or ring scotomas [242]. If the optic nerve damage is not severe, vision can recover following cessation of the ergot.

Visual disturbances were also described in epidemic ergotism. Bilateral cataracts matured in as short a time interval as 6 weeks following the toxicity [244]. One study reported visual loss in approximately 25% of patients [245]. Because the various ergot drugs can cause vasoconstriction at high doses, they are probably not an advisable form of therapy in patients with complicated migraine (see Sect. 10.7.4).

10.7.1.2 Aspirin

Aspirin should be a potent agent for treating an acute attack because it inhibits platelet aggregation [246], prostaglandin synthesis in platelets and other tissues [247] and has analgesic properties. Though aspirin is generally not considered an effective medication for aborting a migraine once an attack begins, in some patients the headache phase can be prevented if it is taken during the prodrome.

We also consider aspirin prophylaxis of migraine here because some of the properties discussed may be applicable to both acute and prophylactic treatment. For example, the ingestion of aspirin in doses of 325–650 mg, 24–72 h before venipuncture, reverses the hyperaggregability of the platelets in migraineurs during headache-free periods [248]. Aspirin reduces the frequency of migraines by more than 50% in approximately 75% of patients [249]. The number of acephalgic or exercise-induced attacks will also be reduced with chronic aspirin therapy.

Aspirin also seems to prevent or reduce the incidence of transient or permanent visual or neurologic deficits, possibly by preventing platelet aggregation and adherence at the site of vasospasm-induced endothelial damage and subsequent distal embolization. The experimental evidence supports this concept since the rate of platelet aggregation is slowed in arteries with endothelial damage, experimentally induced by ultraviolet light [250] or by laser burns [251, 252], if the aspirin is administered prior to the vascular injury. Daily doses of 650 mg are given for migraine prophylaxis.

10.7.1.3 Isometheptene Mucate

Midrin is a brand name for a combination of three drugs that includes isometheptene mucate (65 mg), a sympathomimetic amine that vasoconstricts dilated cerebral and external carotid arteries [253]. Diclaorphenazone (325 mg), a mild sedative, and acetaminophen (325 mg) are also contained in this preparation. One or two capsules are taken orally at the onset of an attack. The dose can be repeated if necessary within 30–60 min. Generally, Midrin is not as effective as the ergot preparations in aborting migraine attacks.

10.7.1.4 Fiorinal

Fiorinal is a brand name for a combination of drugs, containing butalbital (50 mg), aspirin (325 mg), and caffeine (40 mg). Fiorinal is effective because of the sedative and anxiolytic properties from the barbiturate, as well as the effects of aspirin, and possible vasoconstriction from caffeine. Fiorinal can decrease the severity of an acute migraine but it does not generally abort the episode.

10.7.1.5 Corticosteroids

Corticosteroids are generally reserved for treating those patients with intractable migraine or cluster headaches. Cluster can cause edema and a secondary inflammation in the arteries of the external and internal carotid arteries. Corticosteroid inhibition of this inflammatory process can relieve the acute headache. Cluster headaches are particularly sensitive to corticosteroid therapy, if administered soon after an attack begins. Prednisone administered in daily doses of 60 mg for 3–4 days, followed by a rapid taper over 1 week, is often all that is required [254]. Long-term corticosteroid therapy, which has a significant number of potential systemic complications, is not prescribed to treat either migraine or cluster headaches.

10.7.1.6 Oxygen Inhalation

The inhalation of 100% oxygen via a mask (7 l/min for approximately 15 min) at the onset of the prodrome can prevent progression to the headache phase of a migraine [161]. Oxygen inhalation may be effective in aborting cluster headaches as well [255]. The therapeutic benefit may arise because the increase in the arterial oxygen concentration constricts the cerebral arteries and generally reduces cerebral blood flow [256–258] (Sect. 1.6).

10.7.1.7 Temporal Artery Compression

Compression of the dilated ECA system can alter the pain in some migraineurs [218]. Bilateral vigorous digital massage of one or both superficial temporal arteries, beginning at the onset of the visual prodrome and continued until the aura subsides, has been reported to prevent the headache phase in 81% of the attacks in one study [259]. This is contrasted with the very temporary improvement in the headache if compression of the superficial temporal artery is initiated after the headache starts [260]. It is possible that vigorous massage is effective because stimulation of the trigeminal nerve causes an inhibitory feedback to the central nervous system.

10.7.1.8 Narcotics

Narcotic agents nonspecifically alleviate pain. These agents, as well as the various anxiolytics and depressants, are not advocated for general use in migraine. However, the pain of some attacks may be severe enough to require narcotics.

10.7.2 Prophylaxis

Prophylaxis includes removing possible inciting or identifiable causes such as oral contraceptives, "colds" and nasal sprays that contain vasoconstricting agents, ethanol, diets containing potential vasoactive amines, and anxiety-provoking situations. Despite disclaimers of the lack of the activated amines in foods, those migraineurs who complain of attacks in relation to certain foods, experience a reduction in the number of episodes with abstinence from these food items. In a small number of patients, biofeedback, psychotherapy, meditation, hypnosis, and relaxation exercises reduce the frequency of attacks [55b, 261]. A regular meal schedule may be helpful in those cases where there is a potential relation to periods of presumed hypoglycemia.

10.7.2.1 β-Adrenergic Receptor Blockade

β-Adrenergic receptor blockers such as propanolol significantly reduce the attack rate as well as the duration and severity for each episode in 60% [262] to 96% of treated patients [263]. Yet another β-blocker, practolol, has no effect. The efficacy of propanolol is greater at 1 year compared to at 3 months of therapy [264]. Propanolol may inhibit the vasodilation of the external carotid system [265]. It is not useful if only taken to abort an acute attack since it has little effect on reversing acute alterations of vasomotor tone [266]. Propanolol also has antianxiety properties [267]. Propanolol may effect migraine prophylaxis via the alteration of brain function through the blockade of β-adrenergic receptors located in various regions of the brain as well as stabilization of vasomotor tone of both intra- and extracranial arteries [268].

Propanolol may also block the second phase of platelet aggregation [269]. Treatment with propanolol inhibits platelet synthesis of thromboxane and platelet aggregation in response to arachidonic acid or thrombin; but this alteration of platelet function is not accomplished through the blockade of the β-adrenergic receptor [270]. In contrast, at least one report suggests that propanolol may actually increase platelet aggregation by reducing platelet cyclic adenosine monophosphate levels [271].

Drugs that selectively block β_1-receptors (predominantly involved with cardiac and lipolysis function) are not as effective in preventing migraine as agents that block β_2-receptors (predominantly located on bronchioles and peripheral arteries). β-Adrenergic blocking agents such as metropolol, atenolol, and nadolol have also been reported to decrease the frequency and severity of migraine. However, propanolol remains the most commonly used β-adrenergic blocker in this disorder. The plasma half-life is 4 h so a long-acting preparation or at least a twice daily dosage is required. The starting dose is 20 mg twice daily to avoid unusual sensitivity that results in hypotension, bradycardia, or bronchospasm. A total daily dose from 120 mg to a maximum of 320 mg is often required.

The major side effects are cardiovascular, including bradycardia or an atrioventricular conduction block and systemic hypotension. Impotence, a poor sense of wellbeing, asthma or bronchospasm, depression and alteration of dreaming, and cold extremities are also common. Gradual reduction of propanolol over several days can prevent an adverse reaction following an abrupt withdrawal. Because of the potential inhibition of vasodilation in a vasoconstricted artery, propanolol may not be a first-line therapy in patients with complicated migraine (see Sect. 10.7.4).

10.7.2.2 Calcium Channel Blockers

The calcium channel blocking drugs may also prevent migraine. Depending on the agent used, vasoconstriction can be prevented and the vasomotor tone can be stabilized in the peripheral, coronary, or cerebral arteries [272]. For example, verapamil [273] is the most effective depressant of myocardial contractility, and nifedipine [274] causes the greatest vasodilatation of peripheral arteries. Blocking the contraction of smooth muscle by inhibiting calcium-dependent myofibrillar ATPase and preventing calcium influx lead to vasodilation and a reduction of the ischemia with coronary artery disease [275] and Raynaud's syndrome [276]. Nifedipine and nimodipine can reverse or block the cerebral artery vasospasm induced by experimental subarachnoid hemorrhage [277, 277a, 278], and possibly following a clinical subarachnoid hemorrhage [279]. Calcium channel blockers may also exert direct effects on neuronal function.

Various calcium channel blockers, including flunarizine, verapamil, and nifedipine, reduce the frequency and intensity of common and classic migraine attacks [280–282], providing further clinical evidence of a vascular etiology for migraine. Patients with severe or complicated migraine may be less responsive to verapamil, even when a doses from 80 mg to 160 mg three times daily are prescribed. Nifedipine is initially given in a dose of 10 mg three times a day, and the dose is gradually increased as needed to up to 90 mg three times a day. Dilitiazem is started with 30 mg twice daily and increased to up to 90 mg three times a day. The antimigraine effect of both dilitiazem and verapamil may also be related to alteration of platelet function since both drugs decrease collagen-induced but not adenosine diphosphate-induced platelet aggregation [283]. The calcium channel blockers do not seem as efficacious as propanolol in reducing the frequency or intensity of attacks. The vasomotor stability may also prevent episodes of vasospastic amaurosis in complicated migraine.

Cardiovascular side effects may limit the use of high doses of calcium channel blockers. Orthostatic hypotension, bradycardia, and reflex tachycardia are complications which are usually dose related.

10.7.2.3 Tricyclic Antidepressants

The various tricyclic antidepressants, such as amitriptyline, can reduce the frequency and intensity of tension, post-traumatic, and vascular headaches [284, 285]. These agents block the reuptake of catecholamines, including norepinephrine and serotonin, into the presynaptic nerve terminals in the brain. Another tricyclic compound, imipramine, may also prevent migraine. This drug binds to platelets and prevents the uptake of serotonin, which induces platelet aggregation [286, 287]. The tricyclic agents also have sedative and antianxiety effects which may be helpful if there is a relation between anxiety and the development of migraine [288]. Typically, a clinical benefit begins in approximately 4 weeks. The dose of amitriptyline is increased weekly, as tolerated, over several weeks to reach a dose of 25 mg three or four times daily.

The major side effects result from toxicity related to abuse and include cardiac arrythmia, sedation, and coma. Rarely, ophthalmoplegia develops as a toxic side effect. Dry mouth in most patients, urinary retention in patients with spinal cord or prostatic dysfunction, and angle closure glaucoma in patients with narrow angles have all been described, even with patients using the recommended doses. Patients with systemic complications rarely need treatment with physostigmine to reverse anticholinergic effects. However, the long (15 h) plasma half-life of a drug like amitriptyline may require the patient to receive supportive therapy for several days before recovery occurs.

10.7.2.4 Antihistamines

Antihistamines like cyproheptadine and pizotifen are beneficial in reducing the severity and frequency of migraines in some individuals. Both drugs block the effects of experimental exogenous serotonin and acetylcholine on vasomotor tone. Though, neither drug reverses the serotonin-induced vasoconstriction in the ECA system [227, 289, 290]. These agents can relax the tone of the arteries of the circle of Willis. Cyproheptadine also blocks the calcium-induced contraction of the dog basilar artery [266] and the experimental serotonin-induced contraction of the human ICA and basilar artery [291, 292].

Antihistamines may also alter a hypothalamic contribution to migraine. Stimulation of central nervous system H1 receptors (located in the limbic system) has a sedative effect. Pizotifen is both a depressant and an antidepressant, while cyproheptadine is a depressant. Both drugs may decrease the emotional factors that contribute to migraine.

Antihistamines may also be effective in preventing some of the systemic manifestations of a migraine attack. Peripheral arterial vasodilatation, which occurs because of stimulation of both H1 or H2 receptors can be blocked with antihistamines. Antihistamines also block the vascular mediated triple response. Typically, activation of both histamine receptors in postcapillary venules causes endothelial cell contraction with extravasation of fluid and macromolecules into the surrounding tissue. For example, the triple response, described by Lewis in 1927, is elicited by intradermal injection of histamine. A local red spot several millimeters in diameter immediately appears in the area around the injection from histamine induced vasodilatation. A brighter red flare slowly extends slowly approximately 1 cm beyond the red spot from the neuronal reflex-induced vasodilatation. A wheal, caused by edema in the tissue, replaces the red spot 1–2 min later [293]. Antihistamines may also decrease secondary local tissue inflammation by stabilizing lysosomes and reducing the effect of mast cell release of histamine.

The dose of cyproheptadine is 4–8 mg three times daily for adults. A dose of 4 mg three times daily may be very effective for childhood migraine [294].

10.7.2.5 Anticonvulsants

Recently, *sodium valproate*, an anticonvulsant that increases the concentration of brain γ-aminobutyric acid, has been used to prevent cluster and migraine attacks [295]. The success of this therapy may arise from an increase in inhibitory cortical interactions that reduce or prevent neuronal firing and the resulting spreading inhibition. Prior trials with other anticonvulsants such as phenytoin and phenobarbital (see Sect. 10.7.3) have not significantly prevented migraine episodes.

10.7.2.6 Methysergide Maleate

Methysergide maleate significantly reduces the incidence of migraine attacks in approximately 60% of patients and is effective in preventing cluster headache. Methysergide is an antiserotinergic agent, derived from lysergic acid, which blocks experimental serotonin-induced vasoconstriction, and reduces the number and severity of migraine attacks, but it will not terminate a headache in progress [296]. Methysergide probably also acts through preventing histamine release and blocking serotonin receptors in the brain. Methysergide also seems to have a mild direct vasoconstrictive action as suggested by studies on the ECA in dogs [227]. The antiinflammatory activity of methysergide can reduce the edema and injection induced by serotonin.

Methysergide is taken orally in divided doses daily. The amount is gradually increased until approximately 6 mg daily, in two or three divided doses, is administered.

Nausea is a common side effect which can be diminished by taking the medication with meals. Rarely, patients describe hallucinations or an alteration of dreams, attributed to the effects on central nervous system serotonin, which remit following drug withdrawal or dose reduction. Fibrosis of the retroperitoneal area, lungs, or endocardium is an infrequent, but potentially serious, side effect that can develop after approximately 6 months of use. This complication has caused this drug to fall out of favor [297]. Obstruction of the ureters and inferior vena cava, cardiac valvular dysfunction, and a restrictive pulmonary dysfunction can result. Methysergide is contraindicated in patients with other illnesses that can affect these organs. Fibrotic complications can be prevented if the medication is stopped for a period of 3–4 weeks every 3–4 months. Rarely, vasoconstriction of the arteries of the extremities can cause coldness and a reduced pulse in the limbs. Vasoconstriction of the mesenteric artery [298] can cause abdominal angina as another complication of methysergide [299].

10.7.2.7 Antiplatelet Agents

Regarding aspirin and other platelet aggregation blockers, see Sect. 10.7.1.2.

10.7.3 Formerly Used Agents

Clonidine is another prophylactic treatment which was shown to be considerably less efficacious than propanolol. The exact mechanism for its antimigraine property is unknown. Clonidine reduces the sympathetic outflow from the central nervous system by interacting with α_2-adrenergic receptors [300, 301].

In some, but not all patients, *anticonvulsants* such as phenobarbital can prevent basilar artery migraine episodes [302] which might be caused by an epileptic disorder [303, 304]. However, anticonvulsants do not reduce the incidence of other types of migraine despite the increased incidence of seizure activity in the electroencephalograms of all migraineurs [305–307]. In view of the cognitive impairment that can develop in children treated with phenobarbital, this agent is not recommended for migraine prophylaxis.

Antiprostaglandin drugs that block prostaglandin synthesis in platelets, arteries, and other tissues may also abort or prevent migraine. Recall that not all prostaglandins have the same effect on the vasomotor tone. One specific prostaglandin inhibitor, flufenamic acid, aborts but does not decrease the incidence of migraine episodes [308]. Indomethacin also blocks prostaglandin synthesis but it is no better than placebo in preventing attacks [309]. Tolfenamic acid may decrease the intensity and the duration of acute migraine attacks [310]. Daily nonsteroidal anti-inflammatory drugs may also reduce the recurrence rate of cluster headaches [311].

Though *lithium carbonate* has been used to prevent the recurrence of cluster headaches, the specific mechanism of action is unknown. Lithium's reduction of serotonin uptake in the platelets [312, 313], its central nervous system actions, or its blocking of the action of calcium may contribute to the antimigraine effect.

The side effects of lithium are generally related to an elevated lithium blood level, interference with the action of calcium, and hyponatremia. The lithium level needs to be closely monitored so the serum concentration does not exceed 1.2 mEq/l. Brainstem dysfunction can develop as a sign of toxicity, and on occasion the signs can progress even after withdrawal from lithium. Patients concomitantly taking haloperidol have a higher risk of developing neurologic dysfunction. Downbeat nystagmus has been described in several of these cases [314, 315].

Carbachol, a vasodilating drug, was also described as an agent that reduced the incidence of complicated migraine [316], but it is not in current use.

Haloperidol [317] and *flunarizine*, a calcium-channel blocker [318], have both been reported in uncontrolled studies to decrease the incidence of hemiplegic migraine in children.

10.7.4 Complicated Migraine Treatment

The treatment of the visual or neurologic deficits is rarely addressed. As in ordinary migraine, eliminating factors such as oral contraceptive that might worsen or precipitate the episodes is the first therapeutic measure. Once an infarct has occurred, treatment is designed to prevent subsequent attacks [181].

10.7.4.1 Aspirin

Aspirin (see Sect. 10.7.1.2) is an effective prophylactic for reducing the frequency of attacks and preventing permanent infarction. Exercise-induced complicated migraine is particularly responsive to aspirin when it is taken within 1 week of the exercise.

10.7.4.2 Vasodilating Agents

Drugs that can acutely vasodilate arteries which are vasoconstricted can reduce the duration of attacks or prevent the transient deficits. Calcium channel blocking agents may be prophylactic in patients with a variety of causes of vasospastic amaurosis (see Sects. 10.7.2.2, 8.6.3). Amyl nitrite, which has a nonspecific relaxing effect on the smooth muscle, can vasodilate cerebral arteries. It has been reported to temporarily reverse the homonymous quadranopia of migraine. The clinical use of amyl nitrite is limited by the induced hypotension [157]. Migrainous visual field loss has also been temporarily cleared by inhaling 10% carbon dioxide

[155]. The carbon dioxide vasodilatation may be related to its effect of lowering the pH in the extravascular tissues adjacent to the affected vessels.

Inhalation of isoproterenol, a β-adrenergic receptor agonist, at the onset of an episode, can reverse or abort a neurologic or visual deficit. Isoproterenol is effective in reversing all types of migraine-associated deficits, including those of basilar artery migraine [179]. Stimulation of the β-adrenergic receptors (see above) on vasoconstricted cerebral arteries during a migraine causes vasodilatation and reverses the ischemia of the affected region. Experimental evidence supports this hypothesis. β-Adrenergic receptor agonists vasodilate in vitro cerebral vessels that have been vasoconstricted by serotonin [72]. Also, the intra-arterial injection of isoproterenol increases the carotid blood flow in dogs [80].

Isoproterenol is delivered via an inhaler at the onset of the prodrome. The inhalation can be repeated within several minutes if the visual or neurologic defects persist. The patient sits down while taking the medication to guard against any untoward hypotension. Patients should have no signs of a cardiac arrhythmia and probably should have a normal Holter monitor before instituting therapy. The headache is not significantly altered by this agent.

10.7.4.3 Anticonvulsants

Anticonvulsants such as phenobarbital can decrease the number of basilar artery migrainous attacks in some but not all children, but it is not recommended (see above).

10.7.4.4 Ergots and β-Receptor Blockade

Controversy exists as to whether ergot preparations or prophylactic propanolol should be used in patients with complicated migraine. The complications from these agents have been reported in isolated cases. An increase in the incidence or severity of neurologic dysfunction has not been described in any series of patients on either treatment. Several anecdotal case reports describe young migraineurs who developed cerebral infarcts while taking propanolol. In one patient with complicated migraine, a left cerebral infarct with dysphasia, right hemiparesis, and right superior quadranopia occurred while the patient was treated with propanolol and aspirin [319]. Another patient with migraine began experiencing complicated migraine attacks and had a right cerebral infarct with a right cerebral hemispheric syndrome while being treated with propanolol [320].

Though the signs of ergot toxicity occur predominantly in the arteries of the extremities, and less commonly in the coronary and renal arteries, there are rare cases with documented occlusion of major cerebral arteries in patients who have abused one of the ergot drugs (see Sect. 10.7.1.1.1). There are no reports that definitively relate the use of the recommended dose of ergot agents to adversely affecting the number or severity of complicated migraine attacks or increasing the incidence of permanent visual or neurologic deficits. Nevertheless, ergotamine and dihydroergotamine should probably be avoided in patients with complicated migraine or a prior migraine-related infarct of the brain [26].

References

Chapter 1 Circulation

1. Graff-Radford NR, Damasio H, Yamada T, Eslinger PJ, Damasio AR. Nonhaemorrhagic thalamic infarction. Brain 108:485–516, 1985
2. Percheron G. The anatomy of the arterial supply of the human thalamus and its use for the interpretation of the thalamic vascular pathology. Z Neurol 205:1–13, 1973
3. Riggs HE, Rupp C. Variation in form of the circle of Willis. The relation of the variations to collateral circulation: anatomic analysis. Arch Neurol 8:24–30, 1963
4. Alpers BJ, Berry RG, Paddison RM. Anatomical studies of the circle of Willis in normal brain. Arch Neurol Psych 81:409–418, 1959
5. Krayenbuhl HA, Yasargil MG. Cerebral Angiography, 2nd edn. Philadelphia, JB Lippincott, 1968
6. Schlesinger B. The Upper Brainstem in the Human. Berlin, Springer-Verlag, 1976
7. Goto K, Tagawa K, Uemura K, Ishi K, Takahashi S. Posterior cerebral artery occlusion: clinical, computed tomographic, and angiographic correlation. Radiology 132:357–368, 1979
8. Page LK. The infratentorial-supracerebellar exposure of tumors in the pineal area. Neurosurgery 1:36–40, 1977
9. Yamamoto I, Kageyama N. Microsurgical anatomy of the pineal region. J Neurosurg 53:205–221, 1980
10. Johanson C. The central veins and deep dural sinuses of the brain. Acta Radiol Suppl 107:8–144, 1954
11. Wolf BS, Huang YP, Newman CM. Superficial sylvian venous drainage system. Am J Roentgen 89:398–410, 196
12. Oka K, Rhoton AL, Barry M, Rodriquez R. Microsurgical anatomy of the superficial veins of the cerebrum. Neurosurgery 17:711–748, 1985
13. Ono M, Rhoton AL, Peace D, Rodriquez RJ. Microsurgical anatomy of the deep venous system of the brain. Neurosurgery 15:621–657, 1984
14. Wolf BS, Huang YP. The subependymal veins of the lateral ventricles. Am J Roentgen 91:406–426, 1964
15. Hayreh SS. The ophthalmic artery. III. Branches. Br J Ophthalmol 46:212–247, 1962
16. Lasjaunias P, Berenstein A. Surgical Neuroangiography, Vol I, Springer-Verlag, Berlin, 1987, Table 2.5
16a. pp 33–54
16b. p 48
16c. p 43
16d. pp 129–143
16e. p 55
16f. p 223
16g. p 223
16h. p 225
16i. p 227
16j. pp 228–231
16k. pp 130–132
17. Padget DH. The development of the cranial arteries in the human embryo. Contr Embryol 32:205–262, 1948
18. Moret J, Lasjaunias P, Theron J, Merland JJ. The middle meningeal artery. Its contribution to the vascularization of the orbit. J Neuroradiol 4:225–248, 1977
19. Steele EJ, Blunt MJ. The blood supply of the optic nerve and chiasma in man. J Anat 90:486–493, 1956
20. Alterman M, Henkind P. Radial peripapillary capillaries of the retina. Br J Ophthalmol 52:26–31, 1968
21. Morrison JC, Van Buskirk M. Anterior collateral circulation in the primate eye. Ophthalmology 90:707–715, 1983
22. Henkind P. Radial peripapillary capillaries of the retina. I. Anatomy: human and comparative. BJO 51:115–123, 1967
23. Weiter JJ, Ernest JT. Anatomy of the choroidal vasculature. Am J Ophthalmol 78:583–590, 1974
24. Hayreh SS. Segmental nature of the choroidal vasculature. Br J Ophthalmol 59:631–648, 1975
25. Marsh RJ, Ford SM. Blood flow in the anterior segment of the eye. Trans Ophthalmol Soc UK 100:388–396, 1980
26. Strang R, Wilson TM, MacKenzie ET. Choroidal and cerebral blood flow in baboons measured by the external monitoring of radioactive inert gas. Invest Ophthalmol Vis Sci 16:571–576, 1977
27. Krey HF. Segmental vascular patterns of the choriocapillaries. Am J Ophthalmol 80:198–202, 1975
28. Ring HG, Fujino T. Observations on the anatomy and pathology. of the choroidal vasculature. Arch Ophthalmol 78:431–447, 1967
29. Hayreh SS. Physiological anatomy of the choridal vascular bed. Int Ophthalmol 6:85–93, 1983

30. Netleship E. Unusual distribution of retinal blood vessels. Br Med J 1:161–162, 1876
31. Randall B. Trans Amer Ophthal Soc 4:511–517, 1887
32. Duke-Elder S. Textbook of Ophthalmology, Vol 1. Klimpton, London, p 139, 1932
33. Hayreh SS. The cilio-retinal arteries. Br J Ophthalmol 47:71–89, 1963
34. Wybar KC. Anastomoses between the retinal and ciliary arterial circulation. Br J Ophthalmol 40:65–81, 1956
35. Wybar KC. Vacular anatomy of the choroid in relation to selective localization of ocular disease. BJO 38:513–527, 1954
36. Hayreh SS. Blood supply and vascular disorders of the optic. nerve. Ann Int Barraquer 4:7–109, 1963
37. Francois J, Neetens A. Central retinal artery and central optic nerve artery. Br J Ophthalmol 47:21–30, 1963
38. Kupersmith MJ, Shakib M. The blood ocular barrier. In: Neuwelt EA (ed) Implications of the Blood–Brain Barrier and its Manipulation. Plenum, New York, 1988, pp 369–390
39. Hayreh SS. Colour and fluorescence of the optic disc. Ophthalmologica 165:100–108, 1972
40. Henkind P, Levitsky M. Angioarchitecture of the optic nerve. I. The papilla. Am J Ophthamol 68:979–986, 1969
41. Henkind P, Levitsky M. Angioarchitecture of the optic nerve. II. Lamina cribrosa. Am J Ophthamol 68:986–996, 1969
42. Lieberman MF, Maumenee AE, Green WR. Histologic studies of the vasculature of the anterior optic nerve. Am J Ophthalmol 82:405–423, 1976
43. Zinn JG. Descriptio anatomica oculi humani. Wriesberg, Gottingen, 1755 quoted by Francois J, Neetens A. in # 44
44. Francois J, Neetens A. Vascularization of the optic pathway. I. lamina cribosa and optic nerve. Br J Ophthalmol 38:472–488, 1954
45. Haller A von. Arteriarum oculi historia et tabulae arteriaram oculi. Gottingen, 1754 cited # 44
46. Olver JM, Spalton DJ, McCartney ACE. Microvascular study of the retrolaminar optic nerve in man: the possible significance in anterior ischaemic optic neuropathy. Eye 4:7–24, 1990
47. Francois J, Neetens A, Collette JM. Vascular supply of the optic pathway. II. further studies by micro-arteriography of the optic nerve. Br J Ophthalmol 39:220–232, 1955
48. Hayreh SS. Blood supply of the optic nerve head and its role in optic atrophy, glaucoma and oedema of the optic disc. Br J Ophthalmol 53:721–748, 1969
49. Ernest T, Archer D. Fluorescein angiography of the optic disk. Am J Ophthalmol 75:973–978, 1973
50. Risco JM, Grimson BS, Johnson PT. Angioarchitecture of the ciliary artery of the posterior pole. Arch Ophthamol 99:864–868, 1981
51. Hayreh SS. Pathogenesis of visual field defects – role of the ciliary circulation. Br J Ophthalmol 54:289–311, 1970
52. David N, Nortorn E, Gass JD, Beauchamp J. Fluorescein angiography in central retinal artery. Arch Ophthal 77:619–629, 1967
53. O'Day D, Crock G, Galbraith JEK, Parel JM, Wigley A. Fluorescein angiography of normal and atrophic optic discs. Lancet 2:224–226, 1967
54. Shimizu K. Sectorial order of fluorescein filling in the disc area. In: Shimizu K (ed) Fluorescein Angiography. Tokyo International Symposium 1972. Igaku Shoin, 1974, p 349
55. Bill A. Intraocular pressure and blood flow through the uvea. Arch Ophthalmol 67:336–348, 1962
56. Bill A. The circulation in the eye. In: The Microcirculation, Part 2. Renkin EM, Michel CC (eds) Handbook of Physiology, Section 2. American Physiological Society, 1984, pp 1001–1035
57. Dollery CT, Hill DW, Hodge JV. The response of normal retinal blood vessels to angiotensin and noradrenaline. J Physiol 165:500–507, 1963
58. Robinson F, Riva CE, Grunwald JE, et al. Retinal blood flow autoregulation in response to an acute increase in blood pressure. Inv Ophth Vis Sci 27:722–726, 1986
59. Weinstein JM, Duckrow RB, Beard D, Brennan R. Regional optic nerve blood flow and its autoregulation. Inv Ophth 24:1559–1565, 1983
60. Fallon TJ, Maxwell D, Kohner EM. Retinal vascular autoregulation in conditions of hyperoxia and hypoxia using the blue field entoptic phenomenon. Ophthalmology 92:701–705, 1985
61. Hickam JB, Frayser R. Studies of the retinal circulation in man: observations on vessel diameter, arteriovenous oxygen difference, and mean circulation time. Circulation 33:302–316, 1966
62. Sossi N, Anderson DR. Effect of elevated intraocular pressure on blood flow. Occurrence in cat optic nerve head studies with iodoantipyrine I 125. Arch Ophthal 101:98–101, 1983
63. Ernest TJ. Optic disc blood flow. Trans Ophthalmol Soc UK 96:348–351, 1976
64. Geijer C, Bill A. Effect of raised intraocular pressure on retinal, prelaminar and retrolaminar optic nerve blood flow in monkeys. Invest Ophthalmol Vis Sci 18:1030–1042, 1979
65. Bill A. Some aspects of the ocular circulation. Inv Ophth Vis 26:410–424, 1985
66. Riva CE, Grunwald JE, Petrig BL. Autoregulation of human retinal blood flow. An investigation with laser doppler velocimetry. Inv Ophth Vis Sci 27:1706–1712, 1986
67. Riva CE, Grunwald JE, Sinclair SH. Laser doppler measurement of relative blood velocity in the human optic nerve head. Inv Oph Vis Sci 22:241–248, 1982
68. Hayreh SS. Pathogenesis of oedema of the optic discs. Doc Ophthal 24:289–411, 1968

69. Ernest JT, Potts AM. Pathophysiology of the distal portion of the optic nerve, II. vascular relationships. Am J Ophthalmol 66:380–387, 1969
70. Gitter KA, Blumenthal M, Galin MA, Best M. Origin of the peripapillary vascular network. Am J Ophthalmol 69:249–251, 1970
71. Francois J, Neetens A. Vascularization of the optic pathway. III. Study of intra-orbital and intracranial optic nerve by serial sections. Br J Ophthalmol 40:45–52, 1956
72. Hayreh SS, Dass R. The ophthalmic artery. II. Intra-orbital course. Br J Ophthal 46:165–185, 1962
73. Francois J, Neetens A, Collette JM. Vascularization of the primary optic pathways. Br J Ophthalmol 42:65–80, 1958
74. Magitot A. Sur l'origine intra-cranienne de l'atrophie optique glaucomateuse. Ann d'Oculist 180:321–341, 1947
75. Dawson BH. On the blood vessels of the human optic chiasma, hypophysis and hypothalamus. MD Thesis, Manchester, 1948
76. Marinkovic SV, Milisavljevic MM, Marinkovic ZD. Microanatomy and possible clinical significance of anastomoses among hypothalamic arteries. Stroke 20:1341–1352, 1989
77. Wolff E. Some aspects of the blood supply of the optic nerve. Trans Ophthalmol Soc UK 59:157–162, 1939
78. Dawson BH. The blood vessels of the human optic chiasma and their relation to those of hypophysis and hypothalamus. Brain 81:207–217, 1958
79. Gibo H, Lenkey C, Rhoton AL. Microsurgical anatomy of the supraclinoid portion of the internal carotid artery. J Neurosurg 55:560–574, 1981
80. Perlmutter D, Rhoton AL. Microsurgical anatomy of the anterior cerebral–anterior communicating–recurrent artery complex. J Neurosurg 45:259–272, 1976
81. Hughes B. Blood supply of the optic nerves and chiasma and its clinical significance. Br J Ophthalmol 42:106–125, 1958
82. Bergland R, Ray BS. The arterial supply of the human optic chiasm. J Neurosurg 31:327–334, 1969
83. Alicherif A, Raybaud CH, Riss M, et al. La vascularisation arterielle des voies optiques retrochiasmatiques chez l'homme: donnees recentes. Rev Neurol 133:339–352, 1977
84. Abbie AA. The blood supply of the lateral geniculate body, with a note on the morphology of the choroidal arteries. J Anatomy 67:491–527, 1933
85. Carpenter MB, Noback CR, Moss ML. The anterior choroidal artery. Its origins, course, distribution, and variations. Arch Neurol 54:714–722, 1954
86. Saeki N, Rhoton AL. Microsurgical anatomy of the upper basilar artery and the posterior circle of Willis. J Neurosurg 46:563–578, 1977
87. Abbie AA. The blood supply of the visual pathways. Med J Australia II:199–203, 1938
88. Percheron G. Les arteres du thalamus humain, les arteres choroidiennes. I. Etudes macroscopique des variations individuelles. II. Systematisation. Rev Neurol 133:533–545, 1977
89. Percheron G. Les arteres du thalamus humain, les arteres choroidiennes. III. Absence de territoire thalamique constitue de l'artere choroidienne anterieure. IV. Arteres et territoires thalamiques du systeme arteriel choroidien et thalamique posteromedian. V. Arteres et territoires thalamiques du systeme arteriel choroidien et thalamique posterolateral. Rev Neurol 133:547–558, 1977
90. Zeal A, Rhoton AL. Microsurgical anatomy of the posterior cerebral anatomy. J Neurosurg 48:534–559, 1978
91. Milisavljevic MM, Marinkovic SV, Gibo H, Puskas LF. The thalamogeniculate perforators of the posterior cerebral artery: the microsurgical anatomy. Neurosurgery 28:523–530, 1991
92. Abbie AA. The anatomy of capsular vascular disease. Med J Australia II:564–568, 1937
93. Salamon G. Atlas of the arteries of the human brain. Sandoz, Basel, 1971
94. Francois J, Neetens A, Collette JM. Vascularization of the optic radiation and the visual cortex. Br J Ophthalmol 43:394–407, 1959
95. Hoyt WF, Margolis MT. Arterial supply of the striate cortex: angiographic changes with occlusion of the posterior cerebral artery. In: Excerpta Medica Ophthalmology, Puig M (ed), pp 1323–1332, 1971
96. Hoyt WF, Newton TH, Margolis MT. The posterior cerebral artery. Section I, embryology and developmental anomalies. In: Newton TH, Potts DG (eds) Radiology of the skull and brain. Mosby, St Louis, pp 1540–1550, 1974
97. Smith CG, Richardson WFG. The course and distribution of the arteries supplying the visual (striate) cortex. Am J Ophthalmol 61:1391–1396, 1966
98. Margolis MT, Newton TH, Hoyt WF. Cortical branches of the posterior cerebral artery. anatomic–radiologic correlations. Neuroradiology 2:127–135, 1971
99. Lazorthes G, Gouaze A, Salamon G. Vascularisation et circulation de l'encephale. physiologie, exploration, angiographie. Masson, Paris, 1978
100. Duvernoy HM, Delon LS, Vannson JL. Cortical blood vessels of the human brain. Brain Res Bull 7:519–579, 1981
101. Fujino T, Yakabi R. The vascular architecture in the striate area. In: Excerpta Medica Ophthalmology, ed. M Puig Solanes. Amsterdam, pp 1316–1322, 1971
102. Holmes G, Lister WT. Disturbances of vision from cerebral lesions with special reference to the cortical representation of the macula. Brain 39:34–73, 1916
103. Holmes GM. The organization of the visual cortex in man. Proc R Soc Lond (Biol) 132:348–361, 1945
104. Hassler O. Deep cerebral venous system in man: A

microangiographic study on its area of drainage and its anastomoses with the superficial cerebral veins. Neurology 16:505–511, 1966

105. Serrano R, Babin E, Ben Amor M, Megret M. Radioanatomy of the internal occipital veins. Neuroradiology 3:153–154, 1972

106. Parkinson D. A surgical approach to the cavernous portion of the carotid artery. Anatomical studies and case report. J Neurosurg 23:474–483, 1965

107. MacConnel EM. The arterial blood supply of the human hypophysis cerebri. Anat Rec 115:175–203, 1953

108. Hosoya T, Kera M, Suzuki T, Yamaguchi K. Fat in the normal cavernous sinus. Neuroradiology 28: 264–266, 1986

109. Daroff case in Asbury AK, Fisher CM, Aldredge H, Hershberg R. Diabetic ophthalmolplegia: A clinico-pathologic investigation. Trans Am Neurol Ass 94: 64–68, 1969

110. Dreyfus PM, Hakim S, Adams RD. Diabetic ophthalmoplegia. Arch Neurol Psychiatr 77: 337–340, 1957

111. Asbury AK, Aldredge H, Hershberg R, Fisher CM. Oculomotor palsy in diabetes mellitus: A clinico-pathological study. Brain 93:555–566, 1970

112. Milisavljevic M, Marinkovic S, Lolic-Draganic V, Kovacevic M. Oculomotor, trochlear, and abducens nerves penetrated by cerebral vessels. Microanatomy and possible clinical significance. Arch Neurol 43:58–61, 1986

113. Bernasconi V, Cassinari V. Caratteristische angiografiche dei meningiomi del tenorio. Radiol Med 43:1015–1026, 1957

114. Harris FS, Rhoton AL. Anatomy of the cavernous sinus. a microsurgical study. J Neurosurg 45: 169–180, 1976

115. Calcaterra TC, Rand RW, Bentson JR. Ischemic paralysis of the facial nerve: a possible etiologic factor in Bell's palsy. Laryngoscope 86:92–97, 1976

116. LaPresle J, Lasjaunias P. Cranial nerve ischemic arterial syndromes. A review. Brain 109:207–215, 1986

117. Mitchell GAG. Anatomy of the Autonomic Nervous System. Edinburgh, E. and S. Livingstone, 1953

118. Lasjaunias P, Doyon D. The ascending pharyngeal artery and the blood supply of the lower cranial nerves. J Neuroradiology (Paris) 5:287–301, 1978

119. Bogousslavsky J, Meienberg O. Eye-movement disorders in brain-stem and cerebellar stroke. Arch Neurol 44:141–148, 1987

120. Gillilan LA. The correlation of the blood supply to the human brain stem lesions. J Neuropathol Exp Neurol 23:78–108, 1964

121. De Smet Y, Brucher JM, Gonsette RE. L'infarctus du territoire olivaire du bulbe. Rev Neurol 140: 559–566, 1984

122. Sieben G, de Reuck J, VanderEecken H. Thrombo-sis of the mesencephalic artery: A clinicopatholog-ical study of two cases and its correlation with the arterial vascularization. Acta Neurol Belg 77: 151–162, 1977

123. Segarra JM. Cerebral vascular disease and behavior. Arch Neurol 22:408–418, 1970

124a. Percheron G. Les arteres du thalamus humain. Arteres et territoires thalamique polaires de l'artere communicante posterieure. Rev Neurol 132: 297–307, 1976

124b. Percheron G. Les arteres du thalamus humain. Arteres et territoires thalamique paramedians de l'artere basilarie communicante. Rev Neurol 132:309–324, 1976

125. Duvernoy H. Human brainstem vessels. Springer, Berlin, 1978

126. Rhoton A, Kiyotaka F, Fradd B. Microsurgical anatomy of the anterior choroidal artery. Surg Neurol 12:171–187, 1979

127. Sieben G, de Reuck J, VanderEecken H. Thrombo-sis of the mesencephalic artery: A clinicopatholog-ical study of two cases and its correlation with the arterial vascularization. Acta Neurol Belg 77: 151–162, 1977

128. Duvernoy H. The superficial veins of the human brainstem. In: Salamon G (ed) Advances in cerebral angiography. Springer, Berlin, 1978

128a. Wall M, Slamovits TL, Weisberg LA, Trufant SA. Vertical gaze ophthalmoplegia from infarction in the area of the posterior thalamo-subthalamic paramedian artery. Stroke 17:546–555, 1986

129. Harper AM: Autoregulation of cerebral blood flow: influence of the arterial blood pressure on the blood flow through the cerebral cortex. J Neurol Neurosurg Psych 29:398–403, 1966

130. Lassen NA. Cerebral blood flow and oxygen consumption in man. Physiol Rev 39:183–238, 1959

131. Thomas DJ. Whole blood viscosity and cerebral blood flow. Stroke 13:285–287, 1982

132. Strandgaard S. Autoregulation of cerebral blood flow in hypertensive patients: the modifying influence of prolonged antihypertensive treatment on the tolerance to acute, drug-induced hypotension. Circulation 53:720–727, 1976

133. Thomas D, Crockard A. Cerebral metabolism and blood flow. In: Crockard A, Hayward R, Hoff JT (eds) Neurosurgery. The Scientific Basis of Clinical Practice. Blackwell Scientific, Oxford, 1985, pp 223–239

134. Dintenfass L. Blood Microrheology-Viscosity Factors in Blood Flow, Ischemia, and Thrombosis. Appleton-Century Crofts, New York, 1971, pp 5–129, 266–355

135. Fieschi C, Agnoli A, Battistini N, Bozzao L, Prencipe M. Derangement of regional cerebral blood flow and of its regulatory mechanisms in acute cerebrovascular lesions. Neurology 18:1166–1179, 1968

136. Symon L, Held K, Dorsch NWC. A study of regional autoregulation in the cerebral circulation to increased perfusion pressure in normocapnia and hypercapnia. Stroke 4:139–147, 1973

Chapter 2   Carotid Cavernous Fistulas

1. Hamby WB. Carotid Cavernous Fistulas. Charles C. Thomas, Springfield, 1966
2. Barker WF, Stern WE, Krayenbuhl H, et al. Carotid endarterectomy complicated by carotid cavernous fistula. Ann Surg 167:572, 1968
3. Eggers F, Lukin R, Chambers AA, et al. Iatrogenic carotid-cavernous fisutula following fogarty catheter thromboendarterectomy. case report. J Neuorsurg 51:543–545, 1979
4. Kakkasseril JS, Tomsick TA, Arbaugh JA, Cranley JJ. Carotid cavernous fistula following fogarty catheter thrombectomy. Arch Surg 119:1095–1096, 1984
5. Kupersmith MJ, Berenstein A, Flamm E, Ransohoff J. Neuroophthalmologic abnormalities and intravascular therapy of traumatic carotid cavernous fistulas. Ophthalmology 93:906–912, 1986
6. Takahashi M, Killeffer F, Wilson G. Iatrogenic carotid cavernous fistulas: case report. J Neurosurg 30:498–500, 1969
7. Kupersmith MJ, Berenstein A, Choi IS, Warren F, Flamm E. Management of nontraumatic vascular shunts involving the cavernous sinus. Ophthalmology 95:121–130, 1988
8. Lasjaunias P, Berenstein A. Surgical Neuroangiography, Vol II, Springer-Verlag, Berlin, 1987, pp 176–211
9. Graf CJ. Spontaneous carotid-cavernous fistula: Ehlers-Danlos syndrome and related conditions. Arch Neurol 13:662–672, 1965
10. Farley MK, Clark RD, Fallor MK, Geggel HS, Heckenlively JR. Spontaneous carotid-cavernous fistula and the Ehlers-Danlos syndromes. Ophthalmology 1983; 90:1337–1342
11. Koo AH, Newton TH. Pseudoxanthoma elasticum associated with carotid rate mirable. Am J Roentgenol Radium Ther Nucl Med 116:16–22, 1972
12. Kaufman HH, Lind TA, Mullan S. Spontaneous carotid-cavernous fistula with fibromuscular dysplasia. Acta Neurochir 40:123–129, 1978
13. Hollister DW. Heritable disorders of connective tissue: Ehler-Danlos syndrome. Pediatr. Clin. North Am. 25:575–591, 1978
14. McKusick VA. Heritable disorders of connective tissue, 4th edn. St Louis, Mosby, 1972
15. Henderson JW, Schneider RC. The ocular findings in carotid cavernous fistula in a series of 17 cases. AJO 48:585–597, 1959
16. deSchweinitz GA, Holloway TB. Pulsating exophthalmos. Philadelphia, W.B. Saunders, 1908, pp 11–120

17. Sanders MD, Hoyt WF. Hypoxic ocular sequelae of carotid- cavernous fistulae. Study of the causes and failure before and after neuro-surgical treatment in a series of 25 cases. Br J Ophthalmol 53:82–97, 1969
18. Harris MJ, Fine SL, Miller NR. Photocoagulation treatment of proliferative retinopathy secondary to carotid-cavernous fistula. AJO 90:515–518, 1980
19. LeWalt LT. Congenital absence of the superior orbital wall associated with pulsating exophthalmos. Am J Roentgenol 30:756–764, 1933
20. Rovit R, Sosman MC. Hemicranial aplasia with pulsating exophthalmos. An unusual manifestation of Von Recklinghausens's disease. J Neurosurg 17:104–121, 1960
21. Binet EF, Kiefer SA, Martin SH, Peterson HO. Orbital dysplasia in neurofibromatosis. Radiology 93:829–833, 1969
22. Zimmerman RA, Bilaniuk LT, Metzger RA, et al. Computed tomography of orbito-facial neurofibromatosis. Radiology 146:113–116, 1983
23. Moster MR, Kennerdell JS. B-scan ultrasonic evaluation of dilated superior ophthalmic vein in orbital and retro-orbital arteriovenous anomalies. J Clin Neuro-ophthalmol 3:105–198, 1983
24. Shammas HJ. Atlas of Ophthalmic Ultrasonography and Biometry. C. V. Mosby Company, St. Louis, 1984
25. De Keizer RJW. Carotid cavernous fistulas and doppler flow velocity measurements. Neuro-ophthalmology 8:205–211, 1988
25a. Huber P. Functional test in angiography of cerebrovascular disease. Neuroradiology 1:122–131, 1970
26. Guerry D, Harbison JW. Bilateral choroidal detachment and fluctuating proptosis secondary to bilateral dural arteriovenous fistulas treated with transcranial orbital decompression with resolution: Report of a case. Trans Am Ophthalmol Soc 73:64–73, 1975
27. Klein R, Meyers SM, Smith JL, et al. Abnormal choroidal circulation. Association with arteriovenous fistula in the cavernous sinus area. Arch Ophthalmol 96:1370–1373, 1978
28. Nordmann J, Lobstein A, Gerhard JP, Levy JP. A propos de 14 cas de glaucome par hypertension veineuse d'origine extraoculaire. Ophthalmologica 142 (suppl):501–505, 1981
29. Zeimer RC et al. A practical venomanometer. Arch Ophthalmol 101:1447–1449, 1983
30. Krakau CE, Widakowich J, Wilke K. Measurement of the episcleral venous pressure by means of an air-jet. Acta Ophthalmol 51:185–196, 1973
31. Podos SM, Minas TF, Macri JT. A new instrument to measure episcleral venous pressure: comparison of normal eyes and eyes with primary open angle glaucoma. Arch Ophthalmol 80:209–213, 1968

472   References

32. Phelps CD, Armaly MF. Measurement of episcleral venous pressure. Am J Ophthalmol 85:35–42, 1978
33. Friberg TR, Sandborn G, Weinreb RN. Episcleral venous pressure and intraocular pressure elevation during inverted posture. Am J Ophthalmol 103:523–526, 1987
34. Harris GJ, Rice PR. Angle closure in carotid cavernous fistula. Ophthalmology 86:1521–1529, 1979
35. Moses RA, Grodzki WJ. Mechanism of glaucoma secondary to increased venous pressure. Arch Ophthalmol 103:1701–1703, 1985
36. Spencer WH, Thompson SH, Hoyt WF. Ischemic ocular necrosis from carotid-cavernous fistula. Br J Ophthalmol 57:145–152, 1973
37. Crock G. Clinical syndromes of anterior segment ischaemia. Trans Ophthalmol Soc UK 87:513–533, 1967
38. Weiss DI, Shaffer RN. Neovascular glaucoma complicating carotid-cavernous fistula. Arch Ophthalmol 69:304–307, 1963
39. Madsen PH. Carotid cavernous fistulae: a study of 18 cases. Acta Ophthalmol 48:731–751, 1970
40. Leonard TJK, Mosley JF, Sanders MD. Ophthalmoplegia in carotid cavernous sinus fistula. Br J Ophthalmol 68:128–134, 1984
41. Mullan S. Carotid cavernous aneurysms and fistulae. In: Dolenc VV (ed), The Cavernous Sinus, pp 225–232, Springer Verlag, Wien, 1987
42. Turner DM, VanGilder JC, Mojthedi S, Dierson EW. Spontaneous intracerebral hematoma in carotid cavernous fistula. J Neurosurg 59:680–686, 1983
43. Hiramatsu K, Utsumi S, Kyoi K, et al. Intracerebral hemorrhage in carotid-cavernous fistula. Neuroradiology 33:67–69, 1990
44. Maurer JJ, Mills M, German WJ. Triad of unilateral blindness, orbital fracture and massive epistaxis after head injury. J Neurosurg 45:837–840, 1961
45. Sattler CH. Beitrag zur Kenntnis des pulsierenden Exophthalmus. Ztschr f. Augenh. 43:534–552, 1920
46. Palestine AG, Younge BR, Piepgras DG. Visual prognosis in carotid-cavernous fistulas. Arch Ophthalmol 99:1600–1603, 1981
47. Wettrell K, Wilke K, Pandolfi M. Effect of beta-adrenergic agonists and antagonists on repeated tonometry an episcleral venous pressure. Exp Eye Res 24:613–619, 1977
47a. Smith DA, Trope GE. Effect of a β-blocker on altered body position: induced ocular hypertension. Br J Ophthalmol 74:6–606, 1990
48. Kaskel D, Becker H, Rudolf H. Frühwirkungen von Clonidin, Adrenalin und Pilocarpin auf den Augeninnendruck und Episkleralvenendruck des gesunden menschlichen Auges. Albrecht von Graefes Arch Klin Exp Ophthlmol 213:251–259, 1980
49. Wilke K. Early effects of epinephrine and pilocarpine on the intraocular pressure and the episcleral venous pressure in the normal human eye. Acta Ophthalmol 52:231–241, 1974
50. Bain WES. Variations in the episcleral venous pressure in relation to glaucoma. Br J Ophthalmol 38:129–135, 1954
51. Bellows AR, Chylack LT, Epstein DL, Hutchinson BT. Choroidal effusion during glaucoma surgery in patients with prominent episcleral vessels. Arch Ophthalmol 97:403–497, 1979
52. Hieshima GB, Higashida RT, Halbach V. Advances in the diagnosis and treatment of carotid cavernous fistulae. Ophthalmology 93(supplement):69, 1986
53. Travers B. A case of aneurysm by anastomosis in the orbit, cured by the ligature of the common carotid artery. Med Chir Trans 2:1–16, 1811
54. Ismat F, Ferrer E, Twose J. Direct intracavernous obliteration of high-flow fistulas. J Neurosurg 65:770–775, 1986
55. Dandy WE. The treatment of carotid cavernous a.v. aneurysms. Ann Surg 102:916–210, 1935
56. Adson AW. Surgical treatment of vascular diseases altering the function of the eyes. J Am Acad Ophthalmol 46:95–111, 1942
57. Keltner JL, Satterfield D, Dublin AB, Lee BCP. Dural and carotid cavernous sinus fistulas. Ophthalmology 94:1585–1600, 1987
58. Kalina RE, Kelly WA. Proliferative retinopathy after treatment of carotid–cavernous fistulas. Arch Ophthalmol 96:2058–2060, 1978
59. Hamby WB, Gardner WJ. Treatment of pulsation exophthalmos with report of two cases. Arch Surg 27:676–685, 1933
60. Parkinson D. A surgical approach to the cavernous portion of the carotid artery. Anatomical studies and case report. J Neurosurg 23:474–483, 1965
61. Mullan S. Treatment of carotid-cavernous fistulas by cavernous sinus occlusion. J Neurosurg 50:131–144, 1979
62. Hosobuchi Y. Electrothrombosis of carotid-cavernous fistula. J Neurosurg 42:76–85, 1975
63. Brooks B. Discussion Noland L and Taylor AS. J S Surg 43:176–177, 1931
64. Serbinenko FA. Balloon catheterization and occlusion of major cerebral vessels. J Neurosurg 41:125–145, 1974
65. Debrun G, Lacour P, Vinuela F. Treatment of 54 traumatic carotid-cavernous fistulas. J Neurosurg 55:678–692, 1981
66. Debrun G, Lacour P, Caron JP, et al. Detachable balloon and calibrated leak balloon technique in the treatment of cerebral vascular lesions. J Neurosurg 49:635–649, 1975
67. Hieshima GB, Grinnell VS, Mehringer CM. A detachable balloon for therapeutic transcatheter occlusions. Radiology 138:227–228, 1981
68. Debrun GM, Vinuela FV, Fox AJ, Kan S. Two different calibrated leak balloons: experimental work and application in humans. AJNR 3:407, 1982

69. Taki W, Handa H, Miyake H, et al. New detachable balloon technique for traumatic carotid cavernous sinus fistulae. AJNR 6:961–964, 1985

70. Hieshima GB, Grinnell VS, Mehringer CM. A detachable balloon for therapeutic transcatheter occlusions. Radiology 138:227–228, 1981

71. Tsai FY, Hieshima GB, Mehringer CM, et al. Delayed effects in the treatment of carotid-cavernous fistulas. AJNR 4:357–361, 1983

72a. Berenstein A, Kricheff II. Catheter and material selection for transarterial embolization. Technical considerations I. Catheters. Radiology 1979, 132:619–630

72b. Berenstein A, Kricheff II. Catheter and material selection for transarterial embolization: Technical considerations II. Materials. Radiology 1979, 132:631–63

73. Manelfe G, Berenstein A. Treatment of carotid cavernous fistulas by venous approach. J Neuroradiol 7:13–21, 1980

74. Hanneken AM, Miller NR, Debrun GM, Nauta HJW. Treatment of carotid-cavernous sinus fisutlas using a detachable balloon catheter through the superior ophthalmic vein. Arch Ophthalmol 107:87–92, 1989

75. Uflacker R, Lima S, Ribas G, et al. Carotid-cavernous fistulas: embolization through the superior ophthalmic vein approach. Radiology 159:175–179, 1986

76. Walker AE, Allegre GE. Carotid-cavernous fistulas. Surgery 39:411–422, 1956

77. Stern WE, Brown WJ, Alksne JF. The surgical challenge of carotid-cavernous fistula: the critical role of intracranial circulatory dynamics. J Neurosurg 27:298–308, 1967

78. Morley RP. Appraisal of various forms of management in 41 cases of carotid-cavernous fistula. In: Morley TP (ed) Current Controversies in Neurosurgery. Philadelphia, WB Saunders, 1976, pp 223–236

79. Mullan S. Experience with surgical thrombosis of intracranial berry aneurysms and carotid cavernous fisutlas. J Neurosurg 41:657–670, 1974

80. Halbach VV, Higashida RT, Hieshima GB, et al. Transvenous embolization of direct carotid cavernous fistuals. AJNR 9:741–747, 1988

81. Barrow DL, Fleischer AS, Hoffman JC. Complications of detachable balloon catheter technique in the treatment of traumatic intracranial arteriovenous fistulas. J Neurosurg 56:396–403, 1982

82. Kendall B. Results of treatment of arteriovenous fistulas with the Debrun technique. AJNR 4:405–408, 1983

83. Scialfa G, Valsecchi F, Scotti G. Treatment of vascular lesions with balloon catheters. AJNR 4:395–398, 1983

84. Chalif DJ, Flamm ES, Berenstein A, Choi IS. Microsurgical removal of a balloon embolus to the internal carotid artery. J Neurosurg 58:112–116, 1983

85. Capo H, Kupersmith M, Berenstein A, Choi IS, Diamond G. The clinical importance of the inferolateral trunk of the internal carotid artery. Neurosurgery 28:733–738, 1991

86. Kerber CW, Bank WO, Cromwell LD. Calibrated leak balloon microcatheter: a new device for arterial exploration and occlusive therapy. AJR 132:202–207, 1979

Chapter 3 Dural Venous Sinus Disorders

1. Sachs E. The Diagnosis and Treatment of Brain Tumors. London, Klimpton, 1931, pp 168–171

2. Djindjian R, Merland JJ. Superselective arteriography of the external carotid artery. Springer-Verlag, New York, 1978

3. Newton TH, Cronqvist S. Involvement of dural arteries in intracranial arteriovenous malformations. Radiology 93:1071–1078, 1969

4. Ito J, Imamura H, Kobayashi K, et al. Dural arteriovenous malformations of the base of the anterior cranial fossa. Neuroradiology 24:149–154, 1983

5. Chaudhary MY, Sachdev VP, Cho SH, et al. Dural arteriovenous malformations of the major venous sinuses. An acquired lesion. AJNR 3:13–19, 1982

6. Houser OW, Baker HL, Rhoton AL, Okazaki H. Intracranial dural arteriovenous malformations. Radiology 105:55–64, 1972

7. Obrador S, Urquiza P. Antiome arterioveineux de la tente du cervelet. Folia Psychiatr Neurol et Neurochir Neerl 55:385–387, 1952

8. Epstein B, Platt N. Visualization of an intracranial arteriovenous fistula during angiocardiography in an infant with congestive heart failure. Radiology 79:625–627, 1962

9. Albright AL, Latchaw RE, Price RA. Posterior dural arteriovenous malformation in the infancy. Neurosurgery 13:129–135, 1983

10. Kosnik EJ, Hunt WE, Miller DA. Dural arteriovenous malformations. J Neurosurg 40:322–329, 1974

11. Fernandez Urdanibia J, Silvela J, Soto M. Occipital dural arteriovenous malformations. Neuroradiol 7:57–64, 1974

12. Obrador S, Soto M, Sileva J. Clinical syndromes of arteriovenous malformations of the transverse sigmoid sinus. JNNP 38:436–451, 1975

13. Grisoli F, Vincentelli F, Fuchs S, Baldinin M, Raybaud C, Leclerc TA, Vigouroux RP. Surgical treatment of tentorial arteriovenous malformations draining into the subarachnoid space. J Neurosurg 60:1059–1066, 1984

14. Choi IS, Berenstein A, Lee G, Kupersmith M, Jafar J, Benjamin V. Combined endovascular and surgical

treatment of dural AVMs with involvement of cortical veins. XIV Symposium Neuroradiology, London, June 17–23, 1990

15. Teradad T, Kikuchi H, Karasawa J, Nagata I. Intracerebral arteriovenous malformations fed by the anterior ethmoidal artery: case report. Neurosurgery 14:578–582, 1984

16. Gursoy G, Tolun R, Bahar S. Aneurysmal dilatation of torcular. Neuroradiology 18:285–288, 1979

17. Van de Werf AJ. Sur un cas d'anevrisme arterioveineux intradural bilateral de la fosse posterieure chez un enfant. Neurochirurgie 10:140–144, 1964

18. Bito S, Sasaki S. Spontaneous cure of dural arteriovenous malformation in the posterior fossa. Surg Neurol 12:111–114, 1979

19. Olutola PS, Eliam M, Molot M, et al. Spontaneous regression of a dural arteriovenous malformation. Neuorsurg 12:687–690, 1983

20. Magidson MA, Weinberg PE. Spontaneous closure of a dural arteriovenous malformation. Surg Neurol 6:107–110, 1976

21. Lasjaunias P, Berenstein A. Surgical Neuroangiography, Vol II. Embolization of craniofacial lesions. Springer-Verlag, Berlin, 1987

22. Newton TH, Weidner W, Grietz T. Dural arteriovenous malformation in the posterior fossa. Radiology 90:27–35, 1968

23. Gelwan MJ, Choi IS, Berenstein A, Pile-Spellamn JM, Kupersmith MJ. Dural arteriovenous malformations and papilledema. Neurosurgery 22:1079–1084, 1988

24. Beyer RA, McCarty GE. High jugular bulb and high carotid canal first observed as intracranial bruit. Arch Neurol 40:387–388, 1983

25. Adler JR, Ropper AH. Self-audible venous bruits and high jugular bulb. Arch Neurol 43:257–259, 1986

26. Rosenwasser H. Carotid body tumor of the middle ear and mastoid. Arch Otolaryngol 41:64–67, 1945

27. Kuhner A, Krastel A, Stolle W. Arteriovenous malformations of the transverse dural sinus. J Neurosurg 45:12–19, 1976

28. Sundt TM, Piepgras DG. The surgical approach to arteriovenous malformations of the lateral and sigmoid dural sinuses. J Neurosurg 59:32–39, 1983

29. Takahashi T, Okukubo K, Tagami Y, et al. A case of dural arteriovenous malformation with papilledema. Neuro-ophthalmology 3:199–204, 1983

30. Nazarian SM, Janati A, Angtuaco EJ, Jay WM. Oculomotor palsy and papilledema with pia-dural arteriovenous malformation. J Clin Neuro-ophthalmol 7:98–103, 1987

31. Laine E, Galibert P, Lopez C, et al. Anevrysmes arterio-veineux intraduraux (developpes dans l'espaisseur de la dure-mere) de la fosse posterieure. Neurochirurgie 9:147–158, 1963

32. Halbach VV, Higashida RT, Hieshima GB, et al. Dural arteriovenous fistulas supplied by ethmoidal arteries. Neurosurgery 26:816–823, 1990

33. Castaigne P, Bories J, Brunet P, et al. Les fistules arterioveineuses meninges pures a drainage veineux cortical. Rev Neurol 132:169–181, 1976

34. Lasjaunias P, Ming C, Terbrugge K, et al. Neurological manifestations of intracranial dural arteriovenous malformations. J Neurosurg 64:724–730, 1986

35. Vinuela F, Fox AJ, Pelz DM, Drake CG. Unusual clinical manifestations of dural arteriovenous malformations. J Neurosurg 64:554–558, 1986

36. Nakada T, Kwee IL, Ellis WG, et al. Subacute diencephalic necrosis and dural arteriovenous malformation. Neurosurg 17:653–656, 1985

37. Lamas E, Lobato RD, Esparza J, et al. Dural posterior fossa AVM producing raised sagittal sinus pressure. Case report. J Neurosurg 46:804–810, 1977

38. Obrador S, Urquiza P. Angioma arteriovenoso de la tienda del cerebelo. Rev Espanol Oto-neuro-oftamol Neurocirc 10:387–392, 1951

39. Laurent A, Guimarens L, Rufenacht D, Riche MC, Merland JJ. Five cases of unilateral exophthalmos associated with abnormalities in the lateral sinus area. J Neuroradiol 13:125–136, 1986

40. Grove AS. The dural shunt syndrome. Pathophysiology and clinical course. Ophthalmology 90:31–44, 1983

41. Sergott RC, Grossman RI, Savino PJ, Bosley TM, Schatz NJ. The syndrome of paradoxical worsening of dural-cavernous sinus arteriovenous malformations. Ophthalmology 94:205–212, 1987

42. Kataoka K, Taneda M. Angiographic disappearance of multiple dural arteriovenous malformations. J Neurosurg 60:1275–1278, 1984

43. Awad IA, Little JR, Akrawi WP, Ahl J. Intracranial dural arteriovenous malformations: factors predisposing to an aggressive neurological course. J Neurosurg 72:839–850, 1990

44. Turner DM, Vangilder JC, Mojtahedi S, et al. Spontaneous intracerebral hematoma in carotid-cavernous fistula. Report of three cases. J Neurosurg 59:680–686, 1983

45. Lasjaunias P, Berenstein A. Surgical Neuroangiography, Vol II. Embolization of craniofacial lesions. Springer-Verlag, Berlin, 1987, p 288

45a. Bogousslavsky J, Vinuela F, Barnett HJM, and Drake CG: Amaurosis fugax as the presenting manfestation of dural arteriovenous malformation. Stroke 10:891–893, 1985

46. Rowbotham GF, Little E. Circulations of the cerebral hemispheres. Brit J Surg 52:8–21, 1965

47. Kerber CW, Newton TH. The macro- and microvasculature of the dura mater. Neuroradiol 6:175–179, 1973

48. Vidyasagar C. Persistent embryonic veins in the arteriovenous malformations of the dura. Acta Neurochir (Wien) 48:199–216, 1979

49. Lie TA. Congenital Anomalies of the Carotid Arteries Including the Carotid-basilar and Carotid-vertebral Anastamoses; an Angiographic Study and Review of the Literature. Amsterdam, Excerpta Medica, 1968

50. Dahl RE, Kline DG. Intraparenchymal arteriovenous malformations with predominant external carotid artery occlusion. J Neurosurg 41:681–687, 1974

51. Balo J. The dural venous sinuses. Anat Rec 106:319–325, 1950

52. Verbiest H. L'anevrisme arterioveineux intraduraux. Rev Neurol 85:189–199, 1951

53. Dichigans J, Gottsehaldt M, Voigt K. Arteriovenöse Duraangiome am Sinus transversus. Klinische Symptome, charakteristische arterielle Versorgung und häufige venöse abflußstörungen. Zbl Neurochirurgia 33:1–18, 1972

54. Fontaine R, Dany A, Briot B. Fistules arterioveineuses posttraumatiques de la region parieto-occipitale droite du crane et de la dure mere sousjacente. Role probable des eanaux derivatifs du type sueguet. Neurochirurgie 3:30–39, 1957

55. McCormick WF, Boulter TR. Vascular malformations ("angiomas") of the dura mater. Report of two cases. J Neurosurg 25:309–311, 1966

56. Endo S, Koshu K, Kodama N, Okada H. Spontaneous regression of a posterior fossa dural arteriovenous malformation. Neurol Surg 7:1001–1004, 1979

57. Hook O, Johanson C. Intracranial arteriovenous aneurysms: a follow-up study with particular attention to their growth. Arch Neurol Psychiatr 80:39–54, 1958

58. Horton JA, Marano GD, Kerber CW, et al. Polyvinyl alcohol foam–gelfoam tor therapeutic embolization. A synergetic mixture. AJNR 4:143, 1983

59. Lasjaunias P, Halimi P, Lobez-Ibor L, et al. Traitement endovasculaire des malformations vasculaires durales (MVD) pures "spontanees." Revue de 23 cas explores et traites entre mai 1980 et octobre 1983. Neurochirurgie 30:207–223, 1984

60. Costin JA, Weinstein MA, Berlin AJ, Hardy RW, Gutman F. Dural arteriovenous malformations involving the cavernous sinus: A case report. Br J Ophthalmol 62:478–482, 1978

61. Halbach VV, Higashida RT, Hieshima GB, et al. Transvenous embolization of dural fistulas involving the transverse and sigmoid sinuses. AJNR 10:385–392, 1989

62. Mullan S. Treatment of carotid-cavernous fistulas by cavernous sinus occlusion. J Neurosurg 50:131–144, 1979

63. Grady MS, Pobereskin L. Arteriovenous malformation of the dura mater. Surg Neurol 28:135–140, 1987

64. Malik GM, Pearce JE, Ausman JL, Mehta B. Dural arteriovenous malformations and intracranial hemorrhage. Neurosurgery 15:332–339, 1984

65. Picard L, Bracard S, Moret J, Per A, Giacobbe HL, Roland J. Spontaneous dural arteriovenous fistula. Semin Intervent Radiol 4:219–241, 1987

66. Bito S, Hasegawa H, Fujiwara M, Nakao K. Irradiation of spontaneous carotid-cavernous fistulas. Surg Neurol 17:282–286, 1982

67. Kupersmith MJ, Berenstein A, Choi IS, Warren F, Flamm E. Management of nontraumatic vascular shunts involving the cavernous sinus. Ophthalmology 95:121–130, 1988

68. Kobayashi H, Hayashi M, Noguchi Y, et al. Dural arteriovenous malformations in the anterior cranial fossa. Surg Neurol 30:396–401, 1988

69. Phelps CD, Thompson HS, Ossoinig KC. The diagnosis and prognosis of atypical carotid-cavernous fistula (red-eyed shunt syndrome). Am J Ophthalmol 93:423–436, 1982

70. DeKeizer RJW. Spontaneous carotid-cavernous fistulas. The importance of the typical limbal vascular loops for the diagnosis, the recognition of glaucoma and the uses of conservative therapy in this condition. Doc Ophthalmologica 1979; 46:403–412

71. Newton TH, Hoyt WF. Dural arteriovenous shunts in the region of the cavernous sinus. Neuroradiology 1:71–81, 1970

72. Brismar G, Brismar J. Spontaneous carotid-cavernous fistulas, clinical symptomatology. Acta Ophthalmologica 1976; 54:542–552

72a. Fiore PM, Latina MA, Shingleton BJ, Rizzo JF, et al. The dural shunt syndrome. I. Management of glaucoma. Ophthalmol 97:56–62, 1990

73. Jorgensen JS and Gutthoff, RF. 24 cases of carotid cavernous fistulas: frequency, symptoms, diagnosis and treatment. Acta Ophthalmologica 63:67–71, 1985

74. Klein R, Meyers SM, Smith JL, Myers FL, Roth H, Becker B. Abnormal choroidal circulation. Association with arteriovenous fistula in the cavernous sinus area. Arch Ophthalmol 96:1370–1373

74a. Harbison JW, Guerry D, Weisinger H. Dural arteriovenous fistula and spontaneous choroidal detachment: New cause of an old disease. Br J Ophthalmol 62:483–490, 1978

75. Fourman S. Acute closed-angle glaucoma after arteriovenous fistulas. Am J Ophthalmol 107:156–159, 1989

76. Slusher MM, Lennington BR, Weaver RG, Davis CH. Ophthalmic findings in dural arteriovenous shunts. Ophthalmol AAO 86:720–731, 1979

77. Hawke SHB, Mullie MA, Hoyt WF, Hallinan JM, Halmagyi CM. Painful oculomotor nerve palsy due to dural-cavernous sinus shunt. Arch Neurol 46:1252–1255, 1989

78. Kapur A, Paukh NK, Sanghavz NG, Amen SK. Spontaneous carotid-cavernous fistula with ophthalmolplegia and facial palsy. Post Med J 58:773–775, 1983

79. Edwards MS, Connolly ES. Cavernous sinus syn-

drome produced by communication between the external carotid artery and cavernous sinus. J Neurosurg 46:92–96, 1977

80. Berenstein A, Kricheff I. Catheter and material selection for transarterial embolization: technical considerations I. Catheters. Radiology 132: 619–630, 1979

81. Berenstein A. Selective and superselective catheterization of the brachiocephalic vessels with a new catheter. Radiology 180:437–441, 1983

82. Kikuchi T, Strother CM, Boyar M. New catheter therapy for arteriovenous malformations. Radiology 165:870–871, 1987

83. Vinuela FV, Debrun GM, Fox AJ, Kan S. Detachable calibrated-leak balloon for superselective angiography and embolization of dural arteriovenous malformations. J Neurosug 58:817–823, 1983

84. Vinuela F, Fox AJ, Debrun GM, et al. Spontaneous carotid-cavernous fistulas: clinical, radiological, and therpeutic considerations. J Neurosurg 60: 976–984, 1984

85. Grossman RI, Sergott RC, Goldberg HI, Savino PJ, et al. Dural malformations with ophthalmic manifestations: Results of particulate embolization in seven patients. AJNR 1985; 6:809–813

86. Herrera M, Rysavy J, Kotula F, Rusnak B, Castaneda-Zuniga WR, Amplatz K. Ivalon shavings: technical considerations of a new embolic agent. Radiology 144:638–640, 1982

87. Higashida RT, Hieshima GB, Halbach VV, et al. Closure of carotid aneurysm sinus fistulae by external compression of carotid artery and jugular vein. Acta Radiol (Diag) (Suppl) 369:576–579, 1986

88. Beal MF, Park TS, Fisher CM. Cerebral atheromatous embolism following carotid sinus pressure. Arch Neurol 38:310–312, 1981

89. Barnett HJM, Hyland HH. Non-infective intracranial venous thrombosis. Brain 76:36–49, 1953

90. Towbin A. The syndrome of latent cerebral venous thrombosis: Its frequency and relation to age and congestive heart failure. Stroke 4:419–430, 1973

91. Brismar G, Brismar J. Aseptic thrombosis of orbital veins and cavernous sinus: clinical symptomatology. Acta Ophthalmol 55:9–22, 1977

92. Melamed E, Rachmilewitz EA, Reches A, et al. Aseptic cavernous sinus thrombosis after internal carotid arterial occlusion in polycythemia vera. J Neurol Neurosurg Pschiatr 39:320–324, 1976

93. Atkinson EA, Fairburn B, Heathfield KW. Intracranial venous thrombosis as complication of oral contraceptive. Lancet I:914–918, 1970

94. Buchanan DS, Brazinsky JH. Dural sinus and cerebral vein thrombosis: Incidence in young women receiving oral contraceptives. Arch Neurol 22:440–444, 1970

95. Schenk EA. Sickle cell trait and superior longitudinal sinus thrombosis. Ann Intern Med 60:465–470, 1964

96. Greitz T, Link H. Aseptic thrombosis of intracranial sinuses. Radiol Clin Biol 35:111–123, 1966

97. Dunsker SB, Torres-Reyes E, Peden JC. Pseudotumor cerebri associated with idiopathic cryofibrinogenemia: Report of a case. Arch Neurol 23: 120–127, 1970

98. Carroll JD, Leak D, Lee HA. Cerebral thrombophlebitis in pregnancy and the puerperium. Quart J Med 35:347–368, 1966

99. Dixon OJ. The pathologic examination in cavernous sinus thrombosis. JAMA 87:1088–1092, 1926

100. Eagleton WP. The carotid venous plexus as the path of infection in thrombophlebitis of the cavernous sinus: its relation to retropharyngeal and sphenoid suppuration. Arch Surg 12:275–287, 1926

101. Brown P. Septic cavernous sinus thrombosis. Bull Hopkins Hosp 109:68–75, 1961

102. Shaw E. Cavernous sinus thrombosis: a review. Br J Surg 40:40–48, 1952

103. Yarrington CT. Cavernous sinus thrombosis revisited. Proc R Soc Med 70:456–459, 1977

103a. DiNubile MJ. Septic thrombosis of the cavernous sinuses. Arch Neurol 45:567–572, 1988

104. Price CD, Hameroff SB, Richards RD. Cavernous sinus thrombosis and orbital cellulitis. South Med J 64:1243–1247, 1971

105. Geggel HS, Isenberg SJ. Cavernous sinus thrombosis as a cause of unilateral blindness. Ann Ophthalmol 14:569–574, 1982

106. Knapp H. Ueber Verstopfung der Blutgefässe des Auges. Albrecht Von Graefes Arch Klin Exp Ophthalmol 18:207–251, 1868

107. Kaplan HA, Browder J, Krieger AJ. Intercavernous connections of the cavernous sinuses. The superior and inferior circular sinuses. J Neurosurg 45:166–168, 1976

108. Friberg TR, Sogg RL. Ischemic optic neuropathy in cavernous sinus thrombosis. Arch Ophthalmol 96:453–456, 1978

109. Kasper LH, Bernat JL, Nordren RE, et al. Bilateral rhinocerebral phycomycosis. Ann Neurol 6: 131–133, 1976

110. Fox SL, West GB. Thrombosis of the cavernous sinus. JAMA 134:1452–1456, 1947

111. Qingli L, Orcutt JC, Seifter LS. Orbital mucormycosis with retinal and ciliary artery occlusions. Br J Ophthalmol 73:680–683, 1989

112. Mehra KS, Somani PN. Multiple emboli in central retinal artery following cavernous sinus thrombosis. J All India Ophthal Soc 15:71–72, 1967

112a. Gupta A, Jalali S, Bansal RK, Grewal SPS. Anterior ischemic optic neuropathy and branch retinal artery occlsuion in cavernous sinus thrombosis. J Clin Neuro-ophthalmol 10:193–196, 1990

113. Weisman AD. Cavernous sinus thrombophlebitis: Report of case with multiple cerebral infarcts and

necrosis of pituitary body. NEJM 231:118–122, 1944

114. Mathew NT, Abraham J, Taori GM, et al. Internal carotid artery occlusion in cavernous sinus thrombosis. Arch Neurol 24:11–16, 1971

115. MacNeal WJ, Frisbee FC, Blevins A. Thrombophlebitis of the cavernous sinus. Arch Ophthalmol 29:231–257, 1943

116. Kevin JI, Smith H. Hypopituitarism associated with cavernous sinus thrombosis. J Neurol Neurosurg Psychiatr 31:187–189, 1968

117. Oliven A, Harel D, Rosenfeld T, Spindel A, Gidron E. Neurology 30:897–899, 1980

118. Kilpatrick C, Ress B, King J. Computed tomography in rhinocerebral mucormycosis. Neuroradiology 26:71–73, 1984

119. Greenberg MR, Lippman SM, Grinnell VS. Computed tomographic findings in orbital mucor. West J Med 143:102–103, 1985

120. Curner JT, Creasy JL, Whaley RL, et al. Air in the cavernous sinus: A new sign of septic cavernous sinus thrombosis. Am J Neurol 8:176–177, 1987

121. Savino PJ, Grossman RI, Schatz NJ, Sergott RC, Bosley TM. High-field magnetic resonance imaging in the diagnosis of cavernous sinus thrombosis. Arch Neurol 43:1081–1082, 1986

122. Macchi PJ, Grossman RI, Gomori JM, et al. High-field MR imaging of cerebral venous thrombosis. J Comp Assist Tomogr 10:10–15, 1986

123. Valk PE, Hale JD, Crooks LE, et al. MRI of blood flow: correlation of image appearance with spin-echo phase shift and signal intensity. AJR 146:931–939, 1986

124. Sze G, Simmons B, Krol G, Walker R, Zimmerman RD, Deck MD. Dural sinus thrombosis: verification with spin-echo techniques. AJNR 9:679–686, 1988

125. Champion CK, Johnson TM. Rhino-orbital-cerebral phycomycosis. Mich Med 68:807–810, 1969

126. Friedman AH. Orbital phlebography in the study of cavernous sinus disease. AJO 76:712–715, 1973

127. Fiandaca MS, Spector RH, Hartman TM, et al. Unilateral septic cavernous sinus thrombosis: A case report with digital orbtial venographic documentation. J Clin Neuro Ophthalmol 6:35–38, 1986

128. Levine SR, Twyman RE, Gilman S. The role of anticoagulation in cavernous sinus thrombosis. Neurology 38:517–522, 1988

129. Pirkey WP. Thrombosis of the cavernous sinus. Arch Otolaryngol 45:917–924, 1949

130. Southwick FS, Richardson EP Jr, Swartz MN. Septic thrombosis of the dural venous sinuses. Medicine 65:82–106, 1986

131. Schwartz JN, Donnelly EH, Klintworth GK. Ocular and orbital phycomycosis. Surv Ophthalmol 22:3–28, 1977

132. Baum JL. Rhino-orbital mucormycosis. Am J Ophthalmol 63:335–339, 1976

133. Castelli JB, Pallin JL. Lethal rhinocerebral phycomycosis in a healthy adult: A case report and review of the literature. Ophthalmology 86:696–703, 1978

134. Reich H, Behr W, Barnert J. Rhinocerebral mucormycosis in a diabetic ketoacidotic patient. J Neurology 232:115–117, 1985

135. Abramson E, Wilson D, Arky RA. Rhinocerebral phycomycosis in association with diabetic ketoacidosis. Ann Intern Med 66:735–742, 1967

136. Reeves DL, Dickson DR, Benjamin EL. Phycomycosis (mucormycosis) of the central nervous system. J Neurosurg 23:82–84, 1965

137. Anaissie EJ, Shikhari AH. Rinocerebral mucormycosis with internal carotid occlusion. Report of two cases and review of the literature. Laryngoscope 85:1107–113, 1985

138. Lowe JT, Hudson WR. Rhinocerebral phycomycosis and internal carotid artery thrombosis. Arch Otolaryngol 101:100–103, 1975

139. Courey WR, New PFJ, Price DL. Angiographic manifestations of cranial phycomycosis. Radiology 103:329–334, 1972

140. Wasserman AJ, Shiels WS, Sporn IN. Cerebral mucormycosis. Southern Med J 54:403–410, 1961

141. Johnson EV, Kline LB, Bruce BA, Garcia JH. Bilateral cavernous sinus thrombosis due to mucormycosis. Arch Ophthalmol 106:1089–1092, 1988

142. Lehrer RI, Howard DH, Sypherd PS, et al. Mucormycosis. Ann Intern Med 93:93–108, 1980

143. Bullock JD, Jampol LM, Fezza AJ. Two cases of orbital phycomycosis with recovery. Am J Ophthalmol 78:811–813, 1974

144. Qingli L, Orcuttr JC, Seifter LS. Orbital mucormycosis with retinal and ciliary artery occlusions. Br J Ophthalmol 73:680–683, 1989

145. Ferry AP, Abedi S. Diagnosis and management of rhino-orbitocerebral mucormycosis. Ophthalmology 90:1096–1104, 1983

146. Blackwood W, Corsellis JAN. Greenfield's Neuropathology. London, Edward Arnold, 1976, pp 118–121

147. Bousser MG, Chiras J, Sauron B, et al. Cerebral venous thrombosis: A review of 38 cases. Stroke 16:199–213, 1985

148. Averback P. Primary cerbral venous thrombosis in young adults: The diverse manifestations of an underrecognized disease. Ann Neurol 3:81–86, 1978

149. Kalbag RM, Woolf AL. Cerebral Venous Thrombosis. London, Oxford University Press, 1967

150. Brant-Zawadzki M, Chang GY, McCarty GE. Computed tomography in dural sinus thrombosis. Arch Neurol 39:446–447, 1982

151. Goldberg AL, Rosenbaum AE, Wang H, et al. Computed tomography of dural sinus thrombosis. J Comput Assist Tomogr 10:16–20, 1986

152. Rao KCVG, Knipp HC, Wagner EJ. Computed tomographic findings in cerebral sinus and venous thrombosis. Radiology 140:391–398, 1981

153. Vines FS, Davis DO. Clinical-radiological correlation in cerebral venous occlusive disease. Radiology 98:9–22, 1971

154. Brismar J. Computer tomography in superior sagittal sinus thrombosis. Acta Radiol Diagn 21:321–326, 1980

155. Beal MF, Wechsler LR, Davis KR. Cerebral vein thrombosis and multiple intracranial hemorrhages by computed tomography. Arch Neurol 39:437–438, 1982

156. Bridgers SL, Strauss E, Smith EO, Reed D, Ezekowitz MD. Demonstration of superior sagittal sinus thrombosis by indium-111 platelet scintigraphy. Arch Neurol 43:1079–1081, 1986

157. Bradley WG Jr, Waluch V. Blood flow: Magnetic resonance imaging. Radiology 154:443–450, 1985

158. Erdman WA, Weinreb JC, Cohen JM, Maximilian Buja L, Chaney C, Peshock RM. Venous thrombosis: Clinical and experimental MR imaging. Radiology 161:233–238, 1986

159. White EM, Edelaman RR, Wedeen V, Brady TJ. Intravascular signal in MR imaging: Use of phase display for differentiation of blood-flow signal from intraluminal disease. Radiology 161:245–249, 1986

160. McMurdo SK, Brant-Zawacki M, Bradley WG Jr, Chang GY, Berg BO. Dural sinus thrombosis: Study using intermediate field strength MR imaging. Radiology 161:83–86, 1986

161. Askanasy HM, Kosary IZ, Braham J. Thrombosis of the longitudinal sinus: Diagnosis by carotid angiography. Neurology 12:288–292, 1962

162. Barnett HJM, Hyland HH. Non-infective intracranial venous thrombosis. Brain 76:36–49, 1953

162a. Einhaupl KM, Villringer A, Meister W, et al. Heparin treatment in sinus venous thrombosis. Lancet 338:597–600, 1991

163. Gettelfinger DM, Kokmen E. Superior sagittal sinus thrombosis. Arch Neurol 34:2–6, 1977

164. Alexander LF, Yamamoto Y, Ayoubi S, Al-Mefty O, Smith RR. Efficacy of tissue plasminogen activator in the lysis of thrombosis of the cerebral venous sinus. Neurosurgery 26:559–564, 1990

165. Di Rocco C, Iannelli A, Leone G, et al. Heparin–urokinase treatment in aseptic dural sinus thrombosis. Arch Neurol 38:431–435, 1981

166. Rousseaux P, Bernard MH, Scherpereel B, et al. Thrombose des sinus veineux intra-craniens. Neurochirurgie 24:1970203, 1978

167. Scott JA, Pascuzzi RM, Hall PV, Becker GJ. Treatment of dural sinus thrombosis with local urokinase infusion. J Neurosurg 68:284–287, 1988

168. Higashida RT, Helmer E, Halbach VV, Hieshima GB. Direct thrombolytic therapy for superior sinus thrombosis. AJNR 10 (suppl):4–6, 1989

169. Crowe SJ. An aid for the diagnosis of conditions associated with an obstruction to the outflow of blood from the brain: With special reference to sinus thrombosis of otitic origin. Johns Hopkins Med J 23:321–323, 1912

170. Foley J. Benign forms of intracranial hypertension: "Toxic" and "otitic" hydrocephalus. Brain 78:1–41, 1955

171. Kimmick H, Myers D. Lateral sinus thrombosis. Arch Otolaryng 68:156–159, 1958

172. Merei L, Donath T. Contributory data to the correlation of choked disk and otogenic thrombosis of the sigmoid sinus. Acta Otolaryng 47:336–345, 1957

173. Symonds C. Otitic hydrocephalus. Neurology 6:681–685, 1956

174. Greer M. Benign intracranial hypertension: I. Mastoiditis and lateral sinus obstruction. Neurology 12:472–476, 1962

175. Byers RK, Haas GM. Thrombosis of the dural venous sinuses in infancy and childhood. AJDC 45:1161–1183, 1933

176. Grauss F, Slatkin NE. Papilledema in the metastatic jugular foramen syndrome. Arch Neurol 40:816–818, 1983

177. Marr WG, Chambers JW. Occlusion of the cerebral dural sinuses. Am J Ophthalmol 61:45–49, 1966

178. Repka MX, Miller NR. Papilledema and dural sinus obstruction. J Clin Neuro-ophthalmol 4:247–250, 1984

179. Dung Truong D, Holgate RC, et al. Occlusion of the transverse sinus by meningioma simulating pseudotumor cerebri. Neuro-ophthalmology 7:113–117, 1987

180. Powers JM, Schnur JA, Baldree ME. Pseudotumor cerebri due to partial obstruction of the sigmoid sinus by a cholesteotoma. Arch Neurol 43:519–521, 1986

Chapter 4   Orbital Vascular Lesions

1. Cline RA, Rootman J. Enophthalmos: a clinical review. Ophthalmology 91:229–237, 1984

2. LeWalt LT. Congenital absence of the superior orbital wall associated with pulsating exophthalmos. Am J Roentgenol 30:756–764, 1933

3. Binet EF, Kiefer SA, Martin SH, Peterson HO. Orbital dysplasia in neurofibromatosis. Radiology 93:829–833, 1969

4. Rovit R, Sosman MC. Hemicranial aplasia with pulsating exophthalmos. An unusual manifestation of Von Recklinghausen's disease. J Neurosurg 17:104–121, 1960

5. Bienfang D, et al. Case 34–1984. Case records of

the Massachusetts General Hospital. NEJM 311: 520–527, 1984

6. Hook SR, Font RL, McCrary JA, Harper RL. Intraosseous capillary hemangioma of the frontal bone. Am J Ophthalmol 103:824–827, 1987
7. Wende S, Aulich A, Nover A, et al. Computerized tomography of orbital lesions. A cooperative study of 210 cases. Neuroradiolology 12:123–133, 1977
8. Wende S, Kazner E, Grumme T: The diagnostic value of computed tomography and orbital diseases: A cooperative study of 520 cases. Neurosurg Rev 3: 43–49, 1980
9. Zimmerman RA, Bilaniuk LT, Yanoff M, et al. Orbital magnetic resonance imaging. Am J Ophthalmol 100:312–317, 1985
10. Simon J, Szumowski J, Totterman S, et al. Fat-suppression MR imaging of the orbit. AJNR 9: 961–968, 1988
11. Hendrix LE, Kneeland JB, Haughton VM, et al. MR imaging of optic nerve lesions: value of gadopentetate dimeglumine and fat-suppression technique. AJNR 11:749–754, 1990
12. Ossoinig KC. The role of clinical periorbital and orbital lesions. In: Bleeker G (ed) Proceedings of the Third International Symposium on Orbial Disorders. Amsterdam, 1978, pp 496–537
13. Grizzard WS, Blackshear WM, Rush JA, Gordon SF. Noninvasive carotid evaluation using pulsed Doppler imaging for patients with ophthalmic disorders. Ophthalmology 89:1235–1240, 1982
14. Hanafee WN. Orbital venography. Radiol Clin NA 10:63–81, 1972
15. Krohel GB, Wright JE. Orbital hemorrhage. AJO 88:254–258, 1979
16. Bullock JD, Goldberg SH, Connelly PJ. Orbital varix thrombosis. Ophthalmology 97:251–256, 1990
17. Winter J, Centeno RS, Bentson JR. Maneuver to aid diagnosis of orbital varix by computed tomography. Am J Neuroradiol 3:30–40, 1982
18. Osborn RE, DeWitt JD, Lester PD, Yamanahsi WS. Magnetic resonance imaging of an orbital varix with CT and ultrasound correlation. Comput Radiol 10:155–159, 1986
19. Jinkins JR. Orbital varicose capillary angioma: radiologic description of a distinct vascular entity. AJNR 9:589–591, 1988
20. Handa H, Mori K. Large varix of the superior ophthalmic vein: demonstration by angular phlebography and removal by electrically induced thrombosis; case report. J Neurosurg 29:202–205, 1968
21. Vignaud J, Clay C, Bilaniuk LT. Venography of the orbit. An analytical report of 413 cases. Radiology 110:373–382, 1974
22. Du Boulay GH. Orbital phlebography in the diagnosis of intra-orbital hemangiomas. Trans Ophthalmol Soc UK 81:245–259, 1961
23. Fritz W, Klein HJ, Schmidt K. Arteriovenous malformation of the posterior ethmoidal artery as an unusual cause of amaurosis fugax. J Clin Neuro-ophthalmol 9:165–168, 1989
24. Dilenge D. Arteriography in angiomas of the orbit. Radiology 113:355–361, 1974
25. Howard GM, Jakobiec FA, Michelsen WJ. Orbital arteriovenous malformation with secondary capillary angiomatosis treated by embolization with Silastic liquid. Ophthalmology 90:1136–1139, 1983
26. Haik BG, Jakobiec FA, Ellsworth RM, et al. Capillary hemangioma of the lids and orbit. An analysis of the clinical features and therapeutic results in 101 cases. Ophthalmology 86:760–789, 1979
27. Robb RM. Refractive errors associated with hemangiomas of the eyelids and orbit in infancy. AJO 83:52–58, 1977
28. Stigmar G, Crawford JD, Ward EM et al. Ophthalmic sequelae of infantile hemangiomas of the eyelids and orbit. AJO 85:806–813, 1978
29. Kasaback HH, Merritt KK. Capillary hemangioma with extensive pupura. Am J Dis Child 59: 1063–1070, 1940
30. Henriksson P, Nilsson IM, Bergentz SE, et al: Giant hemangioma with a disorder of coagulation. Acta Paediatr Scand 60:227–234, 1971
31. Lang, PG, Dubin HV: Hemangioma-thrombocytopenia syndrome: a disseminated intravascular coagulopathy. Arch Dermatol 111:105–107, 1975
32. Iwamoto T, Jakobiec FA. Ultrastructural comparison of capillary and cavernous hemangioma of the orbit. Arch Ophthalmol 97:1144–1153, 1979
33. Steahly LP, Almquist HT: Steroid treatment of an orbital or periocular hemangioma. J Pediatr Ophthalmol 14:35–37, 1977
34. Kushner BJ. Local steroid therapy in adnexal hemangioma. Ann Ophthalmol 11:1005–1009, 1979
35. Kushner BJ: Intralesional corticosteroid injection for infantile adnexal hemangioma. Am J Ophthalmol 93:496–506, 1982.
36. Sutula FC, Glover AT. Eyelid necrosis following intralesional corticosteroid injection for capillary hemangioma. Ophthalm Surg 18:103–105, 1987
37. Pasyk KA, Dingman RO, Argenta LC, et al: The management of hemangiomas of the eyelid and orbit. Head Neck Surg 6:851–857, 1984.
38. Plesner-Rasmussen H, Marushak D, Goldschmidt E: Capillary hemangiomas of the eyelids and orbit. Acta Ophthalmol 61:645–654, 1983.
39. Neidhart JA, Roach R: Successful treatment of skeletal hemangioma and Kasabach-Merritt syndrome with aminocaproic acid. Am J Med 73:434–438, 1982.
40. Garden J, Geronemus. Dermatologic laser surgery. J Derm Surg Oncol 16:156–168, 1990
41. Kennedy RE. Arterial embolization of orbital hemangiomas. Trans Am Ophthalmol Soc 76: 266–277, 1978
42. Harris GJ, Jakobiec FA. Cavernous hemangioma of

480   References

the orbit: an analysis of 66 cases. J Neurosurg 51:219–228, 1979

43. Kopelow SM, Foos RY, Straatsma BR, et al: Cavernous hemangioma of the orbit. Int Ophthalmol Clin 11:113, 1971.

44. Unsold R, Hoyt WF. Blickinduzierte monokulare Obskurationen bei orbitalem Hämangiom. Klin Monatsbl Augenheilkd 174:715–721, 1979

45. Brown GG, Shields JA. Amaurosis fugax secondary to presumed cavernous hemangioma of the orbit. Ann Ophthalmol 13:1205–1209, 1981

46. Wright JE. Primary optic nerve meningiomas: clinical presentation and management. Trans Am Acad Ophthalmol Otolaryngol 83:OP 617–625, 1977

47. Hampton RG, Krohel GB. Gaze-evoked blindness. Ann Ophthalmol 15:73–76, 1983

48. Orcutt JC, Tucker WM, Mills RP, Smith CH. Gaze-evoked amaurosis. Ophthalmology 94:213–218, 1987

49. Davis KR, Hesselink JR, Dallow RL, Grove AS Jr. CT and ultrasound in the diagnosis of cavernous hemangioma and lymphangioma of the orbit. CT 4:98–104, 1980

50. Savoiardo M, Strada L, Passerini A. Cavernous hemangiomas of the orbit: value of CT, angiography, and phlebography. Am J Neurol Radiol 4:741–744, 1983

51. Kaplan PA, Williams SM. Mucocutaneous and peripheral soft-tissue hemangiomas: MR imaging. Radiology 163:163–166, 1987

52. Wright JE. Current concepts in orbital disease. Eye 2:1–11, 1988

53. Jones IS. Lymphangioma of the ocular adnexa: an analysis of 62 cases. Trans Am Ophthalmol Soc 57:602–665, 1959

54. Iliff WJ, Green WR. Orbital lymphangiomas. Trans Am Ophthalmol Soc 86:914–929, 1979

55. Haik BG, Saint Louis L, Smith ME, Ellsworth RM, Weiss R. Magnetic resonance imaging on orbital lymphangiomas. Am J Ophthalmology 103:724–725, 1987

56. Wright J. Orbital vascular anomalies. Trans Am Acad Ophthalmol Otolaryngol 78:606–616, 1974

57. Rootman J, Hay E, Graeb D, Miller R. Orbital-adnexal lymphangiomas. Ophthalmology 93:1558–1570, 1986

58. Wilson ME, Parker PL, Chavis RM. Conservative management of childhood orbital lymphangioma. Ophthalmology 96:484–490, 1989

59. Croxatto JO and Font RL. Hemangiopericytoma of the orbit. A clinicopathologic study of 30 cases. Hum Pathol 13:210–218, 1982

60. Brown DN, MacCarty CS, Soule EH: Orbital hemangiopericytoma. Review of the literature and report of four cases. J Neurosurg 22:354–361, 1965

61. Enzinger FM, Smith BH. Hemangiopericytoma: an analysis of 106 case. Hum Pathol 7:61–82, 1976

62. Jakobiec FA, Howard G, Jones IS, Wolff M. Hemangioperictyoma of the orbit. AJO 78:816–834, 1974

63. Angervall L, Kindnblom LG, Nielson JM, et al: Hemangiopericytoma: A clinicopathologic, angiographic and microangiographic study. Cancer 42:2412–2427, 1978

64. McMaster MJ, Soule EH, Ivins JC: Hemangiopericytoma: A clinico-pathologic study and long-term follow-up of 60 patients. Cancer 36:2232–2244, 1975

65. Sugar HS, Fishman GR, Kobernick S, et al: Orbital hemangiopericytoma or vascular meningioma? Am J Ophthalmol 70:103–109, 1970.

66. Henderson JW, Farrow GM. Primary orbital hemangiopericytoma: an aggressive and potentially malignant neoplasms. Arch Ophthalmol 96:666–673, 1978

67. Russman BA: Tumor of the orbit: A 33-year follow-up. Am J Ophthalmol 64:273–276, 1967.

68. Mira JG, Chu KCH, Fortner JG. The role of radiotherapy in the management of malignant hemangiopericytoma. Cancer 39:1254–1259, 1977

69. Stout AP: Hemangio-endothelioma: A tumor of blood vessels featuring vascular endothelial cells. Ann Surg 118:445–464, 1943.

70. Hufnagel T, Ma L, Kuo T. Orbital angiosarcoma with subconjunctival presentation. Ophthalmology 94:72–77, 1987

71. Carelli PV, Cangelosi JP: Angiosarcoma of the orbit. Am J Ophthalmol 31:453–456, 1948.

72. Nath K, Shukla BR: Orbital leiomyoma and its origin. Br J Ophthalmol 47:369–371, 1963.

73. Jakobiec FA, Howard G, Rosen M, Wolff M. Leiomayoma and leiomyosarcoma of the orbit. AJO 80:1028–1042, 1975

74. Duhig JJ, Ayer JP: Vascular leiomyoma: A study of 61 cases. Arch Pathol 68:424–430, 1959.

75. Sanborn GE, Valenzuela RE, Green RW: Leiomyoma of the orbit. Am J Ophthalmol 87:371–375, 1979.

76. Leeds N, Seeman WB. Fibrous dysplasia of the skull and its differential diagnosis, a clinical and roentgenographic study of 46 cases. Radiology 78:570–582, 1962

77. Retter RH. Unilateral proptosis due to monstotic fibrous dysplasia. Ann Ophthalmol 8:45–49, 1976

78. Liakos GM, Walker CB, Carruth JAS. Ocular complications in craniofacial fibrous dysplasia. Br J Ophthalmol 63:611–616, 1979

79. Weyand RD, Craig WM, Rucker CW. Unusual lesions involving the chiasm. Proc Mayo Clin 27:505–511, 1952

80. Calderon M, Brady HR. Fibrous dysplasia of bone with bilateral optic foramina involvement. Am J Ophthalmol 68:513–515, 1969

81. Donoso LA, Magargal LE, Eiferman RA. Fibrous dysplasia of the orbit with optic nerve decompression. Ann Ophthalmol 14:80–83, 1982

82. Sturge WA. A case of partial epilepsy, apparently due to a lesion of one of the vaso-motor centres of

the brain. Trans Clin Soc Lond 12:162–167, 1879

83. Jacobs AH, Walton RG. The incidence of birthmarks in the neonate. Pediatrics 58:218–222, 1976

84. Pratt AG. Birthmarks in infants. Arch Dermatol syphilol 67:302–305, 1953

85. Taly AB, Nagoraja D, Das S, Shankar SK, Pratibha NG. Sturge–Weber–Dimitri disease without facial nevus. Neurology 37:1063–1064, 1987

86. Gobbi G, Sorrenti G, Santucci M, Rossi G, et al. Epilepsy with bilateral occipital calcifications: a benign onset with progressive severity. Neurology 38:913–920, 1988

87. Hoyt WF, Rios-Montenegro EN, Behrens MM, Eckelhoff RJ. Homonymous hemioptic hypoplasia. Br J Ophthalmol 56:537–545, 1972

88. Pitta CG, Shingleton BJ, Harris PJ, et al. Solitary choroidal hemangioma. Am J Ophthalmol 88:694–698, 1979

89. Witschel H, Font RL. Hemangioma of the choroid. A clinicopathologic study of 71 cases and a review of the literature. Surv Ophthalmol 20:415–431, 1976

90. Barsky SH, Rosen S, Geer DE, Noe JM. The nature and evolution of port wine stains: a computer-assisted study. J Invest Dermatol 74:154–157, 1980

91. Wohlwill FJ, Yakovlev PI: Histopathology of meningo-facial angiomatosis (Sturge–Weber's disease). J Neuropathol Exp Neurol 16:341–364, 1957.

92. Norton EWD, Gutman F. Fluoresescein angiography and hemangiomas of the choroid. Arch Ophthalmol 78:121–125, 1967

93. Boltshauser E, Wilson J, Hoare RD. Sturge–Weber syndrome with bilateral intracranial calcifications. JNNP 39:429–435, 1976

94. Braffman BH, Bilaniuk LT, Zimmerman RA. The central nervous system manifestations of the phakomatoses on MR. Radiol Clin North Am 26:773–800, 1986

95. Chamberlain MC, Press GA, Hessenlink JR. MR imaging and CT in three cases of Sturge–Weber syndrome: prospective comparison. AJNR 10:491–496, 1989

96. Elster AD, Chen MYM. MR imaging of Sturge–Weber syndrome: role of gadopentate dimeglumine and gradient-echo techniques. AJNR 11:685–689, 1990

97. Lipski S, Aicardi J, Hirsch JF, Lallemand D. Gd-DOTA-enhanced MR imaging in two cases of Sturge–Weber syndrome. AJNR 11:690–692, 1990

98. Tan OT, Sherwood K, Gilchrest BA. Treatment of children wih port-wine stains using the flashlamp-pulsed tunable dye laser. NEJM 320:416–421, 1989

99. Reyes B, Geronemus R. Treatment of port wine stains with the flashlamp-pumped dye laser during childhood. J Am Acad Dermatol 23:1142–1148, 1990

100. Lindau A. Studien über Kleinhirncysten: Bau, Pathogenese und Beziehungen zur Angiomatosis retinae. Acta Pathol Microbiol Scand Suppl 1:1–128, 1926

101. Lindau A. Zur Frage der Angiomatosis retinae und ihrer Hirnkomplikationene. Acta Ophthalmol 4:193–226, 1926

102. Font RL, Ferry AP. The phakomatoses. Int Ophthalmol Clin 12:1–50, 1970

103. Hippel E von. Über eine sehr seltene Erkrankung der Netzhaut: Klinische Beobachtungen. Albrecht von Graefe's Arch Ophthalmol 59:83–106, 1904

104. Salazar FG, Lamiell JM. Early identification of retinal angiomas in a large kindred with von Hippel–Lindau disease. Am J Ophthalmol 89:540–545, 1980

105. Hardwig P, Robertson DM. von Hippel–Lindau disease: a familial, often lethal, multi-system phakomatosis. Ophthalmology 91:263–270, 1984

106. Imes RK, Monteiro ML, Hoyt WF. Incipient hemangioblastoma of the optic disc. Am J Ophthalmol 98:116, 1984

107. Nerad JA, Kersten RC, Anderson RL. Hemangioblastoma of the optic nerve. Report of a case and review of literature. Ophthalmology 95: 398–402, 1988

108. In S, Miyagi J, Kojho N, et al. Intraorbital optic nerve with von Hippel–Lindau disease. J Neurosurg 56:426–429, 1982

109. Cramer F, Kinsey WH. The cerebellar hemangioblastoma: a review of 53 cases. Arch Neurol Psychiatr 67:237–252, 1952

110. Ferrante L, Celli P, Fraioli B, Santoro A. Haemangioblastomas of the posterior cranial fossa. Acta Neurochir (Wien) 71:283–294, 1984

111. Guillain G, Bertrand I, Lereboullet J. Hemangioblastomes du systeme nerveux central a localisations multiples. Rev Neurol 47:432–441, 1932

112. Robinson RG. Aspects of the natural history of cerebellar haemangioblastomas. Acta Neurol Scand 41:372–380, 1965

113. Palmre JJ. Haemangioblastomas: a review of 81 cases. Acta Neurochir (Wien) 27:125–148, 1972

114. Kupersmith MJ, Berenstein A. Visual disturbances in von Hippel–Lindau disease. Ann Ophthalmol 13:195–197, 1981

115. Kinney TD, Fitzgerald PJ. Lindau–von Hippel disease with hemangioblastoma of the spinal cord and syringomyelia. Arch Path 43:439–455, 1947

116. Horton WA, Wong V, Eldridge R. Von Hippel–Lindau disease. Arch Intern Med 135:769–777, 1976

117. Waldmann TA, Levin EH, Baldwin M. The association of polycythemia with a cerebellar hemangioblastoma. The production of an erythropoiesis stimulating factor by the tumor. Am J Med 31:318–324, 1961

118. Nibbelink DW, Peters BH, McCormick WF. On the association of pheochromocytoma and cerebel-

lar hemangioblastoma. Neurology 19:455–460, 1969

119. Goldberg MF, Duke JR. Von Hippel–Lindau disease. Histopathological finding in a treated and an untreated eye. AJO 66:693–705, 1968

120. Nicholson DH, Green WR, Kenyon KR. Light and electron microscopic study of early lesions in angiomatosis retinae. AJO 82:193–204, 1976

121. Jesberg DO, Spencer WH, Hoyt WF. Incipient lesions of von Hippel–Lindau disease. Arch Ophthalmol 80:632–640, 1968

122. Welch RB. von Hippel–Lindau disease: the recognition and treatment of early angiomatosis retinae and the use of cryosurgery as an adjunct to therapy. Trans Am Ophthalmol Soc 68:367–420, 1970

123. Seeger JF, Burke DP, Knake JE, Gabrielsen TO. Computed tomographic and angiographic evaluation of hemangioblastomas. Radiology 138:65–70, 1981

124. Sato Y, Waziri M, Smith W, Frey E, Yuh WTC, Hanson J, Franken EA. Hippel–Lindau disease: MR imaging. Radiology 166:241–246, 1988

125. Annesley WJ Jr, et al. Fifteen year review of treated cases of retinal angiomatosis. Trans Am Acad Ophthalmol Otolaryngol 83:446–453, 1977

126. Frison L. The treatment of vascular retinal lesions. Trans Am Acad Ophthalmol Otololaryngol 77:25–26, 1973

127. Blodi CF, Russell SR, Pulido JS, Folk JC. Direct and feeder vessel photocoagulation of retinal angiomas with dye yellow laser. Ophthalmology 97:791–797, 1990

128. Cardosa RD, Brockhurst RJ. Perforating diathermy coagulation for retinal angiomas. Arch Ophthalmol 94:1702–1715, 1976

129. Peyman GA, Rednam KRV, Mottow–Lippa L, Flood T. Treatment of large von Hippel tumors by eye wall resection. Ophthalmology 90:840–847, 1983

130. Bonnet P, Dechaume J, Blanc E. L'anevrisme cirsoide de la retine. Le Journal de Medecine de Lyon 18:165–178, 1937

131. Lalonde G, Duquette P, Laflamme P, Vezina J: Bonnet–Dechaume–Blanc syndrome. Can J Ophthalmol 14:47–50, 1979

132. Wyburn-Mason RP. Arteriovenous aneurysm of mid-brain and retina, facial naevi, and mental changes. Brain 66:163–203, 1943

133. Langman J. Medical Embryology. Williams and Wilkins, Baltimore, 1975, p 368

134. Danis R, Appen RE. Optic atrophy and the Wyburn–Mason syndrome. J Clin Neuro-ophthalmol 4:91–95, 1984

135. Rathbun J, Hoyt WF, Beard C. Surgical management of orbitofrontal varix in Klippel–Trenaunay–Weber syndrome. AJO 70:109–112, 1970

136. Limaye SR, Doyle HA, Tang RA: Retinal varicosity in Klippel–Trenaunay syndrome. J Pediatr Ophthalmol Strabismus 16:371–373, 1979

137. Sutton H. Epistaxis as an indication of impaired nutrition and of degeneration of the vascular system. Med Mirror 1:769–781, 1864

138. Rendu M. Epistaxis repetees chez un sujet Pasteur de petits angiomes cutanes et muceux. Bull Mem Soc Med Hop 13:731–733, 1896

139. Osler W. On a family of recurring epistaxis associated with multiple telangiectases of the skin and mucous membranes. Bull Johns Hopkins Hosp 12:333–337, 1901

140. Weber F. Multiple hereditary developmental angiomata of the skin and the mucuous membranes associated with recurring haemorrhages. Lancet 2:160–162, 1907

141. Brant AM, Schachat AP, White RI. Ocular manifestations in hereditary hemorrhagic telangiectasia (Rendu–Osler–Weber disease). Am J Ophthalmol 107:642–646, 1989

142. Roman G, Fisher M, Perl DP, Poser CM. Neurological manifestations of hereditary hemorrhagic telangiectasis (Rendu–Osler–Weber disease). Ann Neurol 4:130–144, 1978

143. Boczko ML. Neurological implications of hereditary hemorrhagic telangiectasis. J Nerv Ment Dis 139:525–536, 1964

144. Adams HP, Subbiah B, Bosch EP. Neurologic aspects of hereditary hemorrhagic telangiectasia. report of two cases. Arch Neurol 34:101–104, 1977

145. Teplitz RL. Ataxia Telangiectasia. Arch Neurol 35:553–554, 1978

146. Peterson RD, Kelly WD, Good RA. Ataxia-telangiectasia: its association with a defective thymus, immunological deficiency disease, and malignancy. Lancet 1:1189–1193, 1964

147. Peterson RD, Cooper MD, Good RA. Lymphoid tissue abnormalities associated with ataxia-telangiectasia. Am J Med 41:342–259, 1966

148. Carbonari M, Cherchi M, Paganelli R, et al. Relative increase of T cells expressing the gamma/delta rather than the alpha/beta receptor in ataxia-telangiectasia. NEJM 322:73–76, 1990

149. Weskamp C, Cotlier I. Angioma del cerebro y de la retina con malformaciones capilares de la piel. Arch Oftalmol 15:1–10, 1940

150. Gass JDM. Cavernous hemangioma of the retina: a neuro-oculo-cutaneous syndrome. Am J Ophthalmol 71:799–814, 1971

151. Wallner EF jr, Moorman LT. Hemangioma of the optic disc. Arch Ophthalmol 53:115–117, 1955

152. Lewis RA, Cohen MH, Wise GN. Cavernous hemangioma of the retina and optic disc: a report of three cases and a review of the literature. Br J Ophthalmol 59:422–434, 1975

153. Schwartz AC, Weaver RG, Bloomfield R, Tyler ME. Cavernous hemangioma of the retina, cutaneous angiomas, and intracranial vascular lesion by com-

puted tomography and nuclear magnetic resonance imaging. Am J Ophthalmol 98:483–487, 1984

154. Bottoni F, Canevini MP, Canger R, Orzalesi N. Twin vessels in familial retinal cavernous hemangioma. Am J Ophthalmol 109:285–289, 1990

155. Takahashi T, Wada H, Tani E, Nakamura A, Hiramatsu K. Capillary hemangioma of the optic disc. J Clin Neuro-ophthamol 4:159–162, 1984

156. Gass JDM. A Stereocopic Atlas of Macular Diseases, Diagnosis and Treatment. C.V. Mosby Company, St Louis, 1987

Chapter 5   Vascular Optic Neuropathies

1. Cullen JF. Ischaemic optic neuropathy. Trans Ophthalmol Soc UK. 87:759–774, 1967

2. Kurz O. Vascular opticopathy. Doc Ophthalmol 26: 582–591, 1969

3. Guyer DR, Miller NR, Auer CL, Fine SL. The risk of cerebrovascular disease in patients with anterior ischemic optic neuropathy. Arch Ophthalmol 103: 1136–1142, 1985

4. Gowers WR. A Manual and Atlas of Medical Ophthalmology. London, J and A Churchill, 1882, pp 203–208

5. Uhthoff W. Zu den entzündlichen Sehnervenaffektionen bei Arteriosklerose (Atherosklerose). Ber Zusammenkunft Ophthalmol Dtsch Ges 44: 196–208, 1924

6. Hollenhorst RW, Wagener RW. Loss of vision after distant hemorrhage. Am J Med Sci 218:209–218, 1950

7. Kurz O. Über Papillitis arteriosclerotica. Ophthalmologica 116:281–285, 1948

8. Miller GR, Smith JL. Ischemic optic neuropathy. Am J Ophthalmol 62:103–115, 1966

9. Boghen DR, Glaser JS. Ischemic optic neuropathy. the clinical profile and natural history. Brain 98:689–708, 1975

10. Hayreh SS. Anterior ischemic optic neuropathy. Arch Ophthalmol 99:1030–1040, 1981

11. Francois J, Verriest G, Neetens A, et al. Pseudopapillites vasculaire. Bull Soc Ophthalmol Fr 69: 36–57, 1956

12. Hollenhorst RW, Brown JR, Wagener HP, Schick RM. Neurologic aspects of temporal arteritis. Neurology 10:490–498, 1960

13. Repka MX, Savino PJ, Schatz NJ, Sergott RC. Clinical profile and long-term implications of anterior ischemic optic neuropathy. AJO 96:478–483, 1983

14. Ellenberger C Jr, Keltner JL, Burde RM. Acute optic neuropathy in older patients. Arch Neurol 28: 182–185, 1973

15. Hayreh SS, Podhajsky P. Visual field defects in anterior ischemic optic neuropathy. Doc Ophthalmol 19:53–71, 1979

16. Burde RM. Ischemic optic neuropathies. In: Symposium on neuro-ophthalmology. Trans New Orleans Acad Ophthalmol, Burde RM, Glaser JS, Hollenhorst RW, et al (eds). CV Mosby, St Louis, 1976, pp 25–37

17. Stys PK, Ranson BR Waxman SG, Davis PK. Proc Natl Acad Sci 87:4212–4216, 1990

18. Quigley H, Anderson DR. Cupping of the optic in ischemic optic neuropathy. Trans Am Acad Ophthalmol Otolaryngol 83:755–762, 1977

19. Eagling FM, Sanders MD, Miller SJH. Ischaemic papillopathy. Clinical and fluorescein angiographic review of forty cases. Br J Ophthalmol 58: 990–1008, 1974

20. Miller S. The enigma of glaucoma simplex. Trans Ophthalmol Soc UK 92:563–581, 1972

21. Lavin PJM, Ellenberger C. Recurrent ischemic optic neuropathy. Neuro-ophthalmology 3:193–198, 1983

22. Kollarits CR, McCarthy RW, Corrie WS, Swann ER. Norepinephrine therapy of ischemic optic neuropathy. J Clin Neuro-ophthalmol I:283–288, 1981

23. Borchert M, Lessell S. Progressive and recurrent nonarteritic anterior ischemic optic neuropathy. Am J Ophthalmol 106:443–449, 1988

24. Smith JL, Goldhammer. Hypertensive optic neuropathy. Trans Am Acad Ophthalmol Otolaryngol 79:520–523, 1975

25. Beck RW, Savino PJ, Repka MX, Schatz NJ, Sergott RC. Optic disc structure in anterior ischemic optic neuropathy. Ophthalmology 91:1334–1337, 1984

26. Kline LB. Progression of visual defects in ischemic optic neuropathy. Am J Ophthalmol 106:199–203, 1988

27. Foulds WS. Visual disturbances in systemic disorders, optic neuropathy and systemic disease. Trans Ophthalmol Soc UK 89:125–146, 1969

28. Beck RW, Gamel JW, Willcourt RJ, Berman G. Acute ischemic optic neuropathy in severe preeclampsia. Am J Ophthalmol 90:342–346, 1980

29. Feit RH, Tomsak RL, Ellenberger C. Structure factors in the pathogenesis of ischemic optic neuropathy. Am J Ophthalmol 98:105–108, 1984

30. Doro S, Lessell S. Cup-disc ratio and ischemic optic neuropathy. Arch Ophthalmol 103:1143–1144, 1985

31. Reese AB, Carroll FD. Optic neuritis following cataract extraction. Am J Ophthalmol 45:659–662, 1958

32. Carroll FD. Optic nerve complications of cataract extraction. Trans Am Acad Ophthalmol Otolaryngol 77:623–629, 1973

33. Hayreh SS. Anterior ischemic optic neuropathy: IV. Occurrence after cataract extraction. Arch Ophthalmol 98:1410--1416, 1980

34. Gass JDM, Norton EWD. Cystoid macular edema and papilledema following cataract extraction: a fluorescein funduscopic and angiographic study. Arch Ophthalmol 76:646–661, 1966

35. Lieberman MF, Shahi A, Green WR. Embolic

484   References

ischemic optic neuropathy. Am J Ophthalmol 86:206–210, 1978

36. Burde RM, Smith ME, Black JT. Retinal artery occlusion in the absence of a cherry red spot. Surv Ophthalmol 27:181–186, 1982

37. Tomsak RL. Ischemic optic neuropathy associated with retinal embolism. AJO 99:590–592, 1985

38. Portnoy SL, Beer PM, Packer AJ, Van Dyk HJL. Embolic anterior ischemic optic neuropathy. J Clin Neuro-ophthalmol 9:21–25, 1989

39. Knox DL, Duke JR. Slowly progressive ischemic optic neuropathy. Trans Am Acad Ophthalmol Otolaryngol 75:1065–1068, 1971

40. Waybright EA, Selhorst JB, Combs J. Anterior ischemic optic neuropathy with internal carotid artery occlusion. Am J Ophthalmol 93:42–47, 1982

41. Brown GC. Anterior ischemic optic neuropathy occurring in association with carotid artery obstruction. J Clin Neuro-ophthalmol 6:39–42, 1986

42. Francosis J, Verriest G, Neetens A, et al. Pseudo-papillites vasculaire. Ann Oculist 195:83–885, 1962

43. Cogan DG. Neurology of the Visual System. Charles C. Thomas, Springfield, Illinois, 1966, p 186

44. Quigley HA, Miller NR, Green WR. The pattern of optic nerve fiber loss in anterior ischemic optic neuropathy. Am J Ophthalmol 100:769–776, 1985

45a. Hayreh SS. Anterior ischaemic optic neuropathy I. Fundus on ophthalmoscopy and fluorescein angiography. Br J Ophthal 58:955–963, 1974

45b. Hayreh SS. Anterior ischaemic optic neuropathy II. Fundus on ophthalmoscopy and fluorescein angiography. Br J Ophth 58:964–980, 1974

45c. Hayreh SS. Anterior ischaemic optic neuropathy III. Fundus on ophthalmoscopy and fluorescein angiography. Br J Ophth 58:981–989, 1974

46. Hayreh SS. Pathogenesis of visual field defects. Br J Ophthalmol 54:289–311, 1970

47. Hayreh SS. Inter-individual variation in blood supply of the optic nerve head. Its importance in various ischemic disorders of the optic nerve head, and glaucoma, low-tension glaucoma and allied disorders. Doc Ophthalmol 59:217–246, 1985

48. Fry CL, Carter JE, Kanter MC, Tegeler CH. Is there value in evaluating carotid artery patency in patients with anterior ischemic optic neuropathy. North American Neuro-ophthalmology Society, February 24–28, 1991, Utah

49. Ellenberger C. Ischemic optic neuropathy as a possible early complication of vascular hypertension. AJO 88:1045–1051, 1979

50. Keltner JL, Stamper RL, Becker B. Diphenylhydantoin in ischemic optic neuritis. Trans Am Ophthalmol Soc 70:113–130, 1973

51. Hayreh SS. Anterior ischemic optic neuropathy. Arch Neurol 38:675–678, 1981

52. Sergott RC, Cohen MS, Bosley TM, Savino PJ. Optic nerve decompression may improve the progressive form of nonarteritic ischemic optic neuropathy. Arch Ophthalmol 107:1743–1754, 1989

53. Hamilton CR Jr, Shelley WM, Tumulty PA. Giant cell arteritis: including temporal arteritis and polymyalgia rheumatica. Medicine 50:1–27, 1971

54. Huston KA, Hunder GG, Lie JT, Kennedy RH, Elveback LR. Temporal arteritis: a 25 year epidemiologic, clinical and pathologic study. Ann Int Med 188:162–167, 1978

55. Hauser WA, Ferguson RH, Holley KE, Kurland LT. Temporal arteritis in Rochester, Minnesota, 1951 to 1967. Mayo Clin Proc 46:597–692, 1971

56. Biller J, Asconape J, Weinblatt ME, Toole JF. Temporal arteritis associated with normal sedimentation rate. JAMA 247:486–487, 1982

57. Bethlenfalvay NC, Nusynowitz ML. Temporal arteritis. A rarity in the young adults. Arch Int Med 114:487–489, 1964

58. Stanford RG, Berney SN. Polymyalgia rheumatica and temporal arteritis in blacks: clinical features and HLA typing. J Rheum 4:435–442, 1977

59. Bielroy L, Ogunkoya A, Frohman LP. Temporal arteritis in blacks. Am J Med 86:707–708, 1989

60. Jennings GH. Arteritis of temporal arteries. Lancet 234:424–428, 1938

61. Ali Ibn Isa. Memorandum book of a 10th century oculist (translated by CA Woods). Chicago, Northwestern University Press, 1936

62. Hutchinson J. Diseases of the arteries. On a peculiar form of thrombotic arteritis of the aged which is sometimes productive of gangrene. Arch Surg 1:323–329, 1890

63. Grahm E, Holland A, Avery A, Russell RWR. Prognosis in giant cell arteritis. Br Med J 282:269–271, 1981

64. Jonasson F, Cullen JF, Elton RA. Temporal arteritis. A 14 year epidemiological, clinical and prognostic study. Scott Med J 24:111–118, 1979

65. Hollenhorst RW. Effects of posture on retinal ischemia from temporal arteritis. Arch Ophthalmol 78:569–577, 1967

66. Nykes WN, Adams GGW, Cullen JF. Temporal arteritis: visual loss associated with posture. Neuro-ophthalmology 4:107–109, 1984

67. Raymond LA, Sacks JG, Choromkos E, Khodad G. Short posterior ciliary artery insufficiency with hyperthermia (Uhtoff's symptom). AJO 90:619–623, 1980

68. Goldstein JE, Cogan DG. Exercise and the optic nerve of multiple sclerosis. Arch Ophthalmol 72:168–170, 1964

69. Lipton RB, Solomon S, Wertenbaker C. Gradual loss and recovery of vision in temporal arteritis. Arch Int Med 145:2252–2253, 1985

70. Symonds C, MacKenzie I. Bilateral loss of vision from cerebral infarction. Brain 80:415–455, 1957

71. Wilkinson IMS, Russell R, Ross W. Arteries of the head and neck in giant cell arteritis. A pathological

study to show the pattern of arterial involvement. Arch Neurol 27:378–391, 1972

72. Gibb WRG, Urry PA, Lees AJ. Giant cell arteritis with spinal cord infarction and basilar artery thrombosis. J Neurol Neurosurg Psych 48:945–948, 1985

73. Paulley JW, Hughes JP. Giant-cell arteritis, or arteritis of the aged. Br J Med 2:1562–1567, 1960

74. Hunder GG, Disney TF, Ward LE. Polymyalgia rheumatica. Mayo Clin Proc 44:849–879, 1969

75. Fauchald P, Rygvold O, Oystese B. Temporal arteritis and polymyalgia rheumatica. Ann Intern Med 77:845–852, 1972

76. Ettlinger RE, Hunder GG, Ward LE. Polymalgia rheumatica and giant cell arteritis. Annu Rev Med 29:15–22, 1978

77. Healey LA, Wilske KR. Polymyalgia rheumatica and giant cell arteritis. West J Med 141:64–67, 1984

78. Fessel WJ, Pearson CM. Polymyalgia rheumatic and blindness. N Engl J Med 276:1403–1405, 1967

79. Spiera H, Davison S. Vision loss in cranial arteritis: relation to corticosteroid therapy. Mt Sinai J Med 49:55–58, 1982

80. Speira H, Davison S. Long-term follow-up of polymyalgia rheumatica. Mt Sinai J Med 45:225–229, 1978

81. Meadows SP. Temporal or giant-cell arteritis. Proc Roy Soc Med 59:329–333, 1966

82. Strachman RW, How J, Bewsher PD. Masked giant-cell arteritis. Lancet 1:194–196, 1980

83. Simmons RG, Cogan DG. Occult temporal arteritis. Arch Ophthalmol 68:8–18, 1962

84. Kansu T, Corbett JJ, Savino P, Schatz NJ. Giant cell arteritis with normal sedimentation rate. Arch Neurol 34:624– 625, 1977

85. Caselli RJ. Giant-cell (temporal) arteritis: a treatable cause of multi-infarct dementia. Neurology 40:753–755, 1990

86. Monteiro MLR, Coppeto JR, Greco P. Giant cell arteritis of the posterior cerebral circulation presenting with ataxia and ophthalmolplegia. Arch Ophthalmol 102:407–409, 1984

87. Enzmann D, Scott WR. Intracranial involvement of giant- cell arteritis. Neurology 27:794–797

88. Paulley JW. Coronary ischaemia and occlusion in giant cell (temporal) arteritis. Acta Med Scand 208:257–263, 1980

89. Greene GM, Lain D, Sherwin RM, Wilson JE, McManus BM. Giant cell arteritis of the legs: clinical isolation of severe disease with gangrene and amputations. Am J Med 81:727–733, 1986

90. Ghose MK, Shensa S, Lerner PI. Arteritis of the aged (giant cell arteritis) and fever of unexplained origin. Am J Med 60:429–436, 1976

91. Bevan AT, Dunnill MS, Harrison MJK. Clinical and biopsy findings in temporal arteritis. Ann Rheum Dis 27:271–277, 1968

92. Stofferman RA. Lingual infarction in cranial arteritis. JAMA 243:2422–2423, 1980

93. Wagener HP, Hollenhorst RW. Ocular lesions of temporal arteritis. Am J Ophthalmol 45:617–630, 1958

94. Walsh FB, Hoyt WF. Clinical Neuro-Ophthalmology, Vol 2, 3rd edn. Baltimore, Williams & Wilkins 1969, pp 1878–1881

95. Sebag J, Thomas JV, Epstein DL, Grant WM. Optic disc cupping in arteritic anterior ischemic optic neuropathy resembles glaucomatious cupping. Ophthalmology 93:357–361, 1986

96. Henkind P, Charles NC, Pearson J. Histopathology of ischemic optic neuropathy. Am J Ophthalmol 69:78–90, 1970

97. Tang RA, Kaldis LC. Retinopathy in temporal arteritis. Ann Ophthalmol 14:652–654, 1982

98. Barricks ME, Traviesa DB, Glaser JS, Levy IS. Ophthalmolplegia in cranial arteritis. Brain 100:209–221, 1977

99. Sibony PA, Lessell S. Transient oculomotor synkinesis in temporal arteritis. Arch Neurol 41:87–88, 1984

100. Thomson GTD, Johnston JL, Sharpe JA, Inman RD. Internuclear ophthalmoplegia in giant cell arteritis. J Rheumatol 16:693–695, 1989

101. Dimant J, Grob D, Brunner NG. Ophthalmoplegia, ptosis, and miosis in temporal arteritis. Neurology 30:1054–1058, 1980

102. Currie J, Lessell S. Tonic pupils with giant cell arteritis. Br J Ophthalmol 68:135–138, 1984

103. Davis RH, Daroff RB, Hoyt WF. Tonic pupil after temporal arteritis. Lancet I:822, 1968

104. Wong RL, Korn JH. Temporal arteritis without an elevated erythrocyte sedimentation rate. Am J Med 80:959–963, 1986

105. Keltner JL. Giant-cell arteritis. Ophthalmology 89:1101–1110, 1982

106. Bull BS, Brecher G. An evaluation of the relative merits of the Wintrobe and Westergreen sedimentation methods, including hematocrit correction. AJCP 62:502–510, 1974

107. Jacobson DM, Slamovits TL. Erythrocyte sedimentation rate and it's relationship to hematocrit in giant cell arteritis. Arch Ophthalmol 105:965–967, 1987

108. Bottinger LE, Svedberg CA. Normal erythrocyte sedimentation rate and age. Br Med J 2:85–87, 1967

109. Bruk MI. Articular and vascular manifestations of polymyalgia rheumatica. Ann Rheum Dis 26:103–116, 1967

110. Hedges TR III, Gieger GL, Albert DM. The clinical value of negative temporal artery biopsy specimens. Arch Ophthalmol 101:1251–1254, 1983

111. Eschaghian J. Cranial arteritis: a review of 79 cases. Stroke 11:9, 1980

112. Malmvall BE, Bengtsson BA, Kaijser B, Nilsson LA, Alestig K. Serum levels of immunoglobulin and complement in giant-cell arteritis. JAMA 236:1876–1878, 1976

113. Chess J, Albert DM, Bhan AK, et al. Serologic and immunopathologic findings in temporal arteritis. AJO 96:283–289, 1983

114. Horton BT, Brown GE. Undescribed form of arteritis of the temporal vessels. Mayo Clin Proc 7:700–701, 1932

115. Elliot PD, Baker HL, Brown AL. The superficial temporal artery angiogram. Radiology 102: 635–638, 1972

116. Alpert DM, Searl SS, Craft JL. Histologic and ultrastructural characteristics of temporal arteritis. the value of the temporal artery biopsy. Ophthalmology 89:1111–1126, 1982

117. Kimmelstiel P, Gilmour MT, Hodges HH. Degeneration of elastic fibers in granulomatous giant cell arteritis (temporal arteritis). Arch Pathol 54:157–168, 1952

118. Albert DM, Ruchman MC, Keltner JL. Skip lesions in temporal arteritis. Arch Ophthalmol 94:2072–2077, 1976

119. Cohen DN, Smith TR. Skip areas in temporal arteritis: myth versus fact. Trans Am Acad Ophthalmol Otolaryngol 78:772–783, 1974

120. Wadman B, Werner I. Observations on temporal arteritis. Act Med Scand 192:377–383, 1972

121. Sorenson S, Lorenzen I. Giant-cell arteritis, temporal arteritis and polymyalgia rheumatica. A retrospective study of 63 patients. Acta Medica Scandinavica 201:207–213, 1977

122. Allsop CJ, Gallagher PJ. Temporal artery biopsy in giant cell arteritis: a reppraisal. Am J Surg Pathol 5:317–323, 1981

123. Egge K, Mitbo A, Westby R. Arteritis temporalis. Acta Ophthalmol 44:49–56, 1966

124. Klein RG, Campbell RJ, Hunder GG, Carney JA. Skip lesions in temporal arteritis. Mayo Clin Proc 51:504–510, 1976

125. Goodman BW. Temporal arteritis. Am J Med 67:839–852, 1979

126. McDonnell PJ, Moore GW, Miller NR, et al. Temporal arteritis: a clinicopathologic study. Ophthalmology 93:518–530, 1986

127. Allison MC, Gallagher PJ. Temporal artery biopsy and corticosteroid treatment. Ann Rheum Dis 43:416–417, 1984

128. Siemssen SJ. On the occurrence of necrotising lesions in arteritis temporalis: review of the literature with a note on the potential risk of a biopsy. Br J Plastic Surgery 40:73–82, 1987

129. Lie JT, Gordon LP, Titus JL. Juvenile temporal arteritis. biopsy study of four cases. JAMA 234:496–499, 1975

130. Parker JR, Jones HG, Harkiss GD, Hazleman BL. Circulating immune complexes in polymyalgia rheumatica and giant cell arteritis. Ann Rheum Dis 40:360–365, 1980

131. Wells KK, Folberg R, Goeken JA, Kemp JD. Temporal artery biopsies. correlation of light microscopy and immunofluorescence microscopy. Ophthalmol 96:1058–1064, 1989

132. Liang GC, Simkin PA, Mannik M. Immunoglobulins in temporal arteries: an immunofluorescent study. Ann Int Med 81:19–24, 1974

133. Waaler E, Tonder O, Milde EJ. Immunological and histological studies of temporal arteries from patients with temporal arteritis and/or polymlyalgia rheumatica. Acta Pathol Microbiol Scand (A) 84:55–63, 1976

134. Spencer WH, Hoyt WF. A fatal case of giant-cell arteritis (temporal or cranial arteritis) with ocular involvement. Arch Ophthalmol 64:862–867, 1960

135. Wilkse KR, Healey LA. Polymyalgia rheumatica. a manifestation of systemic giant-cell arteritis. Ann Intern Med 66:77–86, 1967

136. Rodenhauser JH. Über pathologisch-anatomische Augenveränderungen bei generalisierter Riesenzellarteriitis. Klin Mbl Augenheilk 145:414–429, 1964

137. MacFaul PA. Ciliary artery involvement in giant cell arteritis. Br J Ophthalmol 51:505–512, 1967

138. Blumberg S, Giansiracusa DF, Docken WP, Kantrowitz FG. Recurrence of temporal arteritis. Clinical recurrence nine years after initial illness. JAMA 244:1713–1714, 1980

139. Cullen JF. Temporal arteritis occurrences of ocular complications seven years after diagnosis. Br J Ophthalmol 56:584–588, 1972

140. Beevers DG, Harpur JE, Turk KAD. Giant cell arteritis: the need for prolonged treatment. J Chron J Dis 26:571–584, 1973

141. Mosher HA. The prognosis in temporal arteritis. Arch Ophthalmol 62:641–644, 1959

142. Bengtsson BA. Eye complications in giant cell arteritis. Acta Med Scand 658:38–43, 1982

143. Russell RWR. Giant-cell arteritis: a review of 35 cases. Q J Med 28:471–478, 1959

144. Birkhead NC, Wagener HP, Shick RM. Treatment of temporal arteritis with adrenal corticosteroids. Results in fifty-five case in which lesion was proved at biopsy. JAMA 163:821–827, 1957

145. Whitfield AGW, Cooke WT, Jameson-Evans P, Rudd C. Temporal arteritis and its treatment with cortisone and ACTH. Lancet I:408–410, 1953

146. Lockshin MD. Diplopia as early sign of temporal arteritis. Arthritis Rheum 13:419–421, 1970

147. Schneider HA, Weber AA, Ballen PH. The visual prognosis in temporal arteritis. Ann Ophthalmol 3:1215–1227, 1971

148. Grahm E. Survival in temporal arteritis. Trans Ophthalmol Soc UK 100:108–110, 1980

149. Rosenfeld SI, Kosmorsky GS, Klingele TG, Burder RM, Cohn EM. Treatment of temporal arteritis with ocular involvement. Am J Med 80:143–145, 1986

150. Model DG. Reversal of blindness in temporal arteritis with methylprednisolone. Lancet I:340, 1978

151. Slavin ML, Margolis AJ. Progressive anterior ischemic optic neuropathy due to giant cell arteritis despite high-dose intravenous corticosteroids. Arch Ophthalmol 106:1167, 1988

152. Hunder GG, Sheps SG, Allen GL, Joyce JW. Daily and alternate-day corticosteroid regimens in treatment of giant cell arteritis. Comparison in a prospective study. Ann Int Med 82:613–618, 1975

153. Doury P, Fabresse FX, Pattin S, Eulry F, Celton H, Larroque P. La place de la dapsone dans le traitement de la maladie de Horton et de la pseudopolyarthrite rhizomelique. Ann Med Interne 135:31–35, 1984

154. Reinitz E, Aversa A. Long-term treatment of temporal arteritis with dapsone. Am J Med 85:456–457, 1988

155. Graisely B, Audebert AA, Offret H, de Saint Maur PP. Evolutivite de la maladie de Horton sous corticotherapie. Nouv Presse Med 9:2736, 1980

156. Green GJ, Lessell S, Loewenstein JI. Ischemic optic neuropathy in chronic papilledema. Arch Ophthalmol 98:502–504, 1980

157. Orcutt JC, Page NGR, Sanders MD. Factors affecting visual loss in benign intracranial hypertension. Ophthalmology 91:1303–1312, 1984

157a. Wall M, Hart W, Burde R. Visual field defects in idiopathic intracranial hypertension (pseudotumor cerebri). Am J Ophthalmol 96:654–669, 1983

158. Jamison RR. Subretinal neovascularization and papilledema associated with pseudotumor cerebri. Am J Ophthalmol 85:78–81, 1978

159. Ernest JT, Potts AM. Pathophysiology of the distal portion of the optic nerve: I. tissue pressure relationships., Am J Ophthalmol 66:373–380, 1968

160. Ernest JT, Potts AM. Pathophysiology of the distal portion of the optic nerve: II vascular relationships. Am J Ophthalmol 66:380–387, 1968

161. Hayreh SS. Optic disc edema in raised intracranial pressure. V. Pathogenesis. Arch Ophthalmol 95:1553–1565, 1977

162. Hayreh SS. Optic disc edema in raised intracranial pressure. VI. Associated visual disturbances and their pathogenesis. Arch Ophthalmol 95:1566–1579, 1977

163. Hedges TR, Zaren HA. The relationship of optic nerve tissue pressure to intracranial and systemic arterial pressure. Am J Ophthalmol 75:90–98, 1973

164. Tso MOM, Hayreh SS. Optic disc edema in raised intracranial pressure. III. A pathologic study of experimental papilledema. Arch Ophthalmol 95:1448–1457, 1977

165. Tso MOM, Hayreh SS. Optic disc edema in raised intracranial pressure. IV. Axoplasmic transport in experimental papilledema. Arch Ophthalmol 95:1458–1462, 1977

166. Sadun AA, Currie JN, Lessell S. Transient visual obscurations with elevated optic discs. Ann Neurol 16:489–494, 1984

167. Li JC, Magargal LE, Carabasi A. Dynamic amaurosis fugax secondary to compression of vertebral artery. Ann Ophthalmol 20:219–224, 1988

168. Nikoskelainen EK, Savontaus ML, Wanne OP, Katila MJ, Nummelin KU. Leber's hereditary optic neuroretinopathy, a maternaly inherited disease. A genealogic study in four pedigrees. Arch Ophthalmol 105:665–671, 1987

169. Novotny EJ, Gurparkash S, Wallace DC, et al. Leber's disease and dystonia: a mitochondrial disease. Neurology 36:1053–1060, 1986

170. Parker WD, Oley CA, Parks JK. A defect in mitochondrial electron-transport activity (NADH-coenzyme Q oxidoreductase) in Leber's hereditary optic neuropathy. NEJM 320:1331–1333, 1989

171. Wallace DC, Singh G, Lott MT, et al. Mitochondrial DNA mutation associated with Leber's hereditary optic neuropathy. Science 242:1427–1430, 1988

172. Pages M, Pages AM. Leber's disease with spastic paraplegia and peripheral neuropathy: case report with nerve biopsy study. Eur Neurol 22:181–185, 1983;

173. Smith JL, Hoyt WF, Susac J. Ocular fundus in acute Leber optic neuropathy. Arch Ophthalmol 90:349–354, 1973

174. Nikoskelainen E, Hoyt WF, Nummelin K. Ophthalmoscopic findings in Leber's hereditary optic neuropathy. I. Fundus findings in asymptomatic family members. Arch Ophthalmol 100:1597–1602, 1982

175. Nikoskelainen E, Sogg RL, Rosenthal AR, et al. The early phase in Leber hereditary optic atrophy. Arch Ophthalmol 95:969–978, 1977

176. Carroll WM, Mastaglia FL. Leber's optic neuropathy: a clinical and visual evoked potential study of affected and asymptomatic members of a six generation family. Brain 102:559–580, 1979

177. Nikoskelainen E, Hoyt WF, Nummelin K, Schatz H. Fundus findings in Leber's hereditary optic neuropathy. III. Fluorescein angiographic studies. Arch Ophthamol 102:981–989, 1984

178. Lessell S, Gise MD, Krohel GB. Bilateral optic neuropathy with remission in young men. Variation on a theme by Leber? Arch Neurol 40:2–6, 1983

179. Brunette JR, Bernier RG. Diagnostic et prognostic de la maladie de Leber: incidence de la recuperation totale spontanee. Union Med Canada 99:643–652, 1970

180. Holt IJ, Miller DH, Harding AE. Genetic heterogeneity and mitochondrial DNA heteroplasmy in Leber's hereditary optic neuropathy. J Med Genetics 26:739–743, 1989

181. Singer KO. Ueber Sehstoerungen nach Blutverlust. Beitr Augenheilkd 6:1–30, 1904

182. Pears MA and Pickering GW: Changes in the fun-

dus oculi after hemorrhage. Q J Med 29:153–178, 1960.

183. Prescencia AC, Hernandez AM, and Guia ED: Amaurosis and blood loss. Ophthalmologica 191:119–121, 1985.

184. Chisholm IA. Optic neuropathy of recurrent blood loss. Br J Ophthalmol 53:289–295, 1969

185. Castaigne P, Brunet P, Derovesne C, et al. Atrophie optique post-hemorragique. Nouv Presse Med 5: 1631–1633, 1976

186. Levatin P. Atrophy of the optic nerve following hemorrhage. Arch Ophthalmol 37:18–24, 1947

187. Klewin KM, Appen RE, Kaufman PL. Amaurosis and blood loss. AJO 86:669–672, 1978

188. Drance SM, Morgan RW, Sweeney VP. Shock-induced optic neuropathy. A cause of nonprogressive glaucoma. NEJM 288:392–395, 1973

189. Sweeney PJ, Breuer AC, Selhorst JB, et al. Ischemic optic neuropathy: a complication of cardiopulmonary bypass surgery. Neurology 32:560–562, 1982

190. Hayreh SS. Posthemorrhagic amaurosis. Invest Ophthalmol Vis Sci 28:321, 1987

191. Jampol LM, Board RJ, Maumenee AE. Systemic hypotension and glaucomatous changes. Am J Ophthalmol 85:154–159, 1978

192. Rizzo JF, Lessell S. Posterior optic neuropathy during general surgery. AJO 103:808–811, 1987

193. Drance SM. The visual field of low tension glaucoma and shock-induced optic neuropathy. Arch Ophthalmol 95:1359–1361, 1977

194. Peck V, Lieberman A, Pinto R, Culliford A. Pituitary apoplexy following open heart surgery. N Y State J Med 641–643, 1980

195. Slavin ML, Budabin M. Pituitary apoplexy associated with cardiac surgery. Am J Ophthalmol 98:291–126, 1984

196. Drance SM, Sweeney VP, Morgan RW, Feldman F. Studies of factors involved in the production of low tension glaucoma. Arch Ophthalmol 89:457–465, 1973

197. Graefe A von. Über die Iridectomie bei Glaucoma und über den glaucomatosen Progress. Albrecht von Graefes Arch Ophthal 3:456–465, 1857

198. Schnabel I. Das glaucomatose Sehnervenleiden. Arch Augenheilk 24:273–292, 1892

199. Levene RZ. Low tension glaucoma: A critical review and new material. Surv Ophthalmol 24:621–664, 1980

200. Caprioli J, Spaeth GL. Comparison of visual field defects in the low-tension glaucomas with those in the high-tension glaucomas. Am J Ophthalmol 97:730–737, 1984

201. Drance SM, Douglas GR, Airaksinen, et al. Diffuse visual field loss in chronic open-angle and low-tension glaucoma. Am J Ophthalmol 104:577–580, 1987

202. Caprioli J, Spaeth GL. Comparison of optic nerve head in high- and low-tension glaucoma. Arch Ophthalmol 103:1145–1149, 1985

203. Spaeth GL. Fluorescein angiography: Its contributions toward understanding the mechanisms of visual loss in glaucoma. Trans Am Ophthalmol Soc 73:491–553, 1975

204. Fishbein SL, Schwartz B. Optic disc in glaucoma: topography and extent of fluorescein filling defects. Arch Ophthalmol 95:1975–1979

205. Nanba K, Schwartz B. Nerve fiber layer and optic disc fluorescein defects in glaucoma and ocular hypertension. Ophthalmology 95:1227–1233, 1988

206. Lewis RA, Hayreh SS, Phelps CD. Optic disk and visual field correlations in primary open-angle and low-tension glaucoma. Am J Ophthalmol 96: 148–152, 1983

207. Hayreh SS. Pathogenesis of visual field defects. Role of the ciliary circulation. Br J Ophthalmol 54:289–311, 1970

208. Salmon ML, Gay AJ. Ophthalmodynamometry. In Gay AJ, Burde RM (ed). Clinical concepts in neuro-ophthalmology. Int Ophthalmol Clin 7:745–764, 1967

209. Phelps CD, Corbett JJ. Migraine and low-tension glaucoma. A case-control study. Inves Ophthalmol Vis Sci 26:1105–1108, 1985

210. Aulhorn E, Tanzil M. Comparison of visual field defects in glaucoma and in acute anterior ischemic optic neuropathy. Docum Ophthal Proc Series 19:73–79, 1979

211. Hayreh SS. Pathogenesis of optic nerve head changes in glaucoma. Seminars Ophthal 1:1–13, 1986

212. Quigley HA, Addicks EM, Green WR, et al. Optic nerve damage in human glaucoma II. The site of injury and susceptibility to damage. Arch Ophthalmol 99:635–649, 1981

213. Quigley HA, Hohman RM, Sanchez R, et al. Optic nerve head blood flow in chronic experimental glaucoma. Arch Ophthal 103:956–962, 1985

214. Cristini G. Common pathological basis of the nervous ocular symptoms in chronic glaucoma. Br J Ophthalmol 35:11–20, 1951

215. Carter CJ, Brooks DE, Doyle DL, Drance SM. Investigations into a vascular etiology for low-tension glaucoma. Ophthalmol 97:49–55, 1990

216. Quigley HA, Addicks EM, Green WR. Optic nerve damage in human glaucoma III. Quantitative correlation of nerve fiber loss and visual field defect in glaucoma, ischemic neuropathy, papilledema, and toxic neuropathy. Arch Ophthalmol 100:135–146, 1982

217. Brownstein S, Font RL, Zimmerman LE, Murphy SB. Nonglautomaous cavernous degeneration of the optic nerve. report of two cases. Arch Ophthalmol 98:354–358, 1980

218. Giarelli L, Melato M, Campos E. Fourteen cases of

cavernous degeneration of the optic nerve. Ophthalmologica 174:316–321, 1977
219. Hilton GF, Hoyt WF. An arteriosclerotic chiasmal syndrome. Bitemporal hemianopia associated with fusiform dilatation of the anterior cerebral arteries. JAMA 196:1018–1020
220. Portney GL, Roth AM: Optic cupping caused by an intracranial aneurysm. Am J Ophthal 84:98–103, 1977
221. Kupersmith MJ, Krohn D. Cupping of the optic disc with compressive lesions of the anterior visual pathway. Ann Ophthalmol 16:948–952, 1984
222. Cartwright MJ, Anderson DR. Correlation of asymmetric damage with asymmetric intraocular pressure in normal-tension glaucoma (low-tension glaucoma). Arch Ophthalmol 106:898–900, 1988
223. Begg IS, Drance SM, Goldman H. Fluorescein angiography in the evaluation of focal circulatory ischaemia of the optic nerve head in relation to the arcuate scotoma in glaucoma. Cand J Ophthalmol 7:68–74, 1972
224. Chumbley LC, Brubaker RF. Low tension glaucoma. Am J Ophthalmol 81:761–767, 1976
225. Drance SM. Some factors in the production of low tension glaucoma. Br J Ophthalmol 56:229–242, 1972
226. O'Brien IAD, O'Hare JP, Lewin IG, Corrall RJM. Papilledema in diabetes: an ischaemic optic mononeuropathy. Lancet 1:267–269, 1984
227. Barjon P, Lestradet H, Labauge R. Atrophie optique primitive et surdite neurogene dans le diabete junvenile (a propos de trois observations). Presse Med 72:983–985, 1964
228. Marquardt JL, Loriaux L. Diabetes mellitus and optic atrophy: with associated findings of diabetes insipidus and neurosensory hearing loss in two siblings. Arch Intern Med 134:32–37, 1974
229. Lessell S, Rosman NP. Juvenile diabetes mellitus and optic atrophy. Arch Neurol 34:759–765, 1977
230. Lyle TK, Wybar K. Retinal vasculitis. BJO 45:778–788, 1961
231. Lonn LI, Hoyt WF. Papillophlebitis. A cause of protracted yet benign optic disc edema. Eye Ear Nose Throat Mon. 45:62–68, 1966
232. Hart CD, Sanders MD, Miller SJH. Benign retinal vasculitis. clinical and fluorscein angiographic study. Br J Ophthalmol 55:721–723, 1971
233. Ellenberger C, Messner KH. Papillophlebitis: Benign retinopathy resembling papilledema or papillitis. Ann Neurol 3:438–440, 1978
234. Miller N. The big blind spot syndrome: unilateral optic disc edema without visual loss or increased intracranial pressure. In: Smith JL (ed) Neuro-ophthalmology Update. New York, Masson, pp 163–169, 1977
235. Hayreh SS. Central retinal vein occlusion. In: The eye and systemic disease. CV Mosby, St Louis, 1980, pp 223–275
236. Kirkham TH, Sanders MD, Sapp GA. Unilateral papilledema in benign intracranial hypertension. Cand J Ophthal 8:533–538, 1973
237. Fletcher WA, Imes RK, Goodman D, Hoyt WF. Acute idiopathic blind spot enlargement. A big blind spot syndrome without optic disc edema. Arch Ophthalmol 106:44–49, 1988
238. Kimmel AS, Folk JC, Thompson HS, Strnad LS. The multiple evanescent white-dot syndrome with acute blind spot enlargement. Am J Ophthal 107:425–428, 1989
239. D'Amato RJ, Miller NR, Fine SL, Enger C, Quinlan P, Elman MJ. Papillophlebitis and central retinal vein occlusion: a comparison of prognosis. Invest Ophthalmol Vis Sci 28:43, 1987
240. Coats G. Further cases of thrombosis of the central vein. Ophthalmol Rec 16:516–561, 1906
241. Cogan DG. Retinal and papillary vasculitis. In: Cant JS (ed) The William Mackenzie Centenary Symposium on the Ocular Circulation in Health and Disease. St. Louis, C.V. Mosby, 1969, pp 249–270
242. Appen RE, de Venecia G, Ferwerda J. Optic disk vasculitis. Am. J Ophthalmol 90:352–359, 1980
243. Hayreh SS. Optic disc vasculitis. Br J Ophthalmol 56:652–670, 1972
244. Shimo-Oku M, Miyazaki S. Acute anterior and posterior ischemic optic neuropathy. Jap J Ophthalmol 28:159–170, 1984
245. Isayama Y, Takahashi T. Posterior ischemic optic neuropathy. II Histopathology of the idiopathic form. Ophthalmolgica 187:8–18, 1983
246. Isayama Y, Takahashi T, Inoue M, Jimura T. Posterior ischemic optic neuropathy. III Clinical diagnosis. Ophthalmologica 187:141–147, 1983
247. Doron Y, Behar A. Pathology of the optic nerves. I Histopathological changes in the intracranial portions, associated with arteriolsclerotic and hypertensive cardiovascular diseases and with liver parenchymal damage. Acta Neuropathol 10:273–278, 1968
248. Hayreh SS. Posterior ischemic optic neuropathy. Ophthalmologica 182:29–41, 1981
249. Bogousslavsky J, Regli F, Zografos L, and Uske A: Optico- cerebral syndrome: Simultaneous hemodynamic infarction of optic nerve and brain. Neurology 37:263–268, 1987
250. Kramer S. The hazards of therapeutic irradiation of the central nervous system. Clin Neurosurg 15:301–318, 1968
251. Chan RC, Shukovsky LJ. Effects of radiation on the eye. Radiology 120:673–675, 1976
252. Price RA, Jamieson PA. The central nervous system in childhood leukemia. II. Subacute leukoencephalopathy. Cancer 35:306–318, 1975
253. Painter MJ, Chutorian AM, Hilal SK. Cerebrovasculopathy following irradiation in childhood. Neurology 25:189–194, 1975

254. Similia S, Heikkinen E, Blanco G, Taskinen PJ, Kouvalainen K, Lanning M, Saukkonen AL. Brain damage in relation to irradiation and chemotherapy of central nervous system. Lancet 1:1000–1001, 1977

255. Kramer S, Lee KF. Complications of radiation therapy: the central nervous system. Semin Roentgenol 9:75–83, 1974

256. Merchut MP, Haberland C, Naheedy MH, Rubino FA. Long survival of primary cerebral lymphoma with progressive radiation necrosis. Neurology 35:552–556, 1985

257. Buys NS, Kearns TC. Irradiation damage to the chiasm. Am J Ophthalmol 44:483–486, 1957

258. Forrest APM. Pituitary radon implant for advanced cancer. Lancet i:399–401, 1956

259. Peck FC, McGovern ER. Radiation of the brain in acromegaly. J Neurosurg 15:536–542, 1966

260. Rider WD. Radiation damage to the brain – a new syndrome. J Can Assoc Radiol 14:67–69, 1963

261. Stallard HB. Radiant energy as (A) a pathogenic (B) a therapeutic agent in ophthalmic disorders. Br J Ophthalmol 70 (mono suppl VI):67–79, 133

262. Moore RF. Presidential address. Trans Ophthalmolo Soc UK 55:3–26, 1935

263. Chee PHY. Radiation retinopathy. Am J Ophthal 66:860–865, 1968

264. Shukorsky LJ, Fletcher LJ. Effect of irradiation on the eye. Radiology 120:673–675, 1976

265. Howard GM. Ocular effects of radiation and photocoagulation. A study of 100 globes with retinoblastoma, some of which were treated prior to enucleation. Arch Ophthalmol 76:7–10, 1966

266. Egbert PR, Fajardo LF, Donaldson SS, Moazed K. Posterior ocular abnormalities after irradiation for retinoblastoma: a histopathological study. Br J Ophthalmol 64:660–665, 1980

267. Char DH, Lonn LI, Margolis IW. Complications of cobalt plaque therapy of choroidal melanoma. AJO 84:536–547, 1977

268. Bedford MA, Bedotto C, MacFaul PA. Radiation retinopathy after the application of a cobalt plaque. Report of three cases. Br J Ophthalmol 54:505–509, 1970

269. Bagan SM, Hollenhorst RW. Radiation retinopathy after irradiation of intracranial lesions. AJO 88:694–697, 1979

270. Rush SC, Kupersmith MJ, Lerch I, et al. Neuro-ophthalmological assessment of vision before and after radiation therapy alone for pituitary macroadenomas. J Neurosurg 72:594–599, 1990

271. Hayreh SS. Postradiation retinopathy: a fluorescein fundus angiographic study. Br J Ophthalmol 54:765–774, 1970

272. Brown GC, Shields JA, Sanborn G, Augsberger JJ, Savino PJ, Schatz NJ. Radiation retinopathy. Ophthalmology 89:1494–1501, 1982

273. Noble K, Kupersmith MJ. Retinal vascular remodelling in radiation retinopathy. Br J Ophthalmol 68:475–478, 1984

274. Ross H, Rosenberg S, Friedman A. Delayed radiation necrosis of the optic nerve. Am J Ophthalmol 76:683–686, 1973

275. Singh J, Vashist S. Postirradiation optic neuropathy in antral carcinoma. J Clin Neuro-ophthalmol 4:103–104, 1984

276. Hashimoto N, Handa H, Yamashita J, et al. Long-term follow-up of large or invasive pituitary adenomas. Surg Neurol 25:49–54, 1986

277. Crompton MR, Layton DD. Delayed radionecrosis of the brain following therapeutic X-radiation of the pituitary. Brain 84:84–101, 1961

278. Schatz N, Lichtenstein S, Corbett JJ. Delayed radiation necrosis of the optic nerves and chiasm. In: Neuro-ophthalmology VIII, Glaser JS, Smith JL (eds). C. V. Mosby, St. Louis, 1975, pp 131–139

279. Kupersmith MJ, Warren FA, Newall J, Ransohoff J. Irradiation of meningiomas of the intracranial anterior visual pathway. Ann Neurol 21:131–137, 1987

280. Harris JR, Levene MB. Visual complications following irradiation for pituitary adenomas and craniopharyngiomas. Ther Radiol 120:167–171, 1976

281. McDonald LW, Hayes TL. The role of capillaries in the pathognesis of delayed radionecrosis of brain. Am J Pathology 50:745–764, 1967

282. Eyster EF, Nielsen SL, Sheline GE, et al. Cerebral radiation necrosis simulating a brain tumor. J Neurosurg 39:267–271, 1974

283. Wright TL, Bresnan MJ. Radiation-induced cerebrovascular disease in children. Neurology 26:540–543, 1976

284. Hilal SK, Solomon GE, Gold AP, et al. Primary cerebral arterial occlusive disease in children. Radiology 99:87–94, 1971

285. Lee KF, Hodges PJ. Intracranial ischaemic lesions. Rad Clin North Am 5:363–393, 1973

286. Beyer RA, Paden P, Sobel DF, Flynn FG. Moyamoya pattern of vascular occlusion after radiotherapy for glioma of the optic chiasm. Neurology 36:1173–1178, 1986

287. McCready RA, Hyde GL, Bivins BA, et al. Radiation-induced arterial injuries. Surgery 93:306–312, 1973

288. Benson EP. Radiation injury to large arteries: 3 further examples with prolonged asymptomatic intervals. Radiology 106:195–197, 1973

289. Silverberg GD, Britt RH, Goffinet DR. Radiation induced carotid artery disease. Cancer 41:130–137, 1978

290. Butler MJ, Lane RHS, Webster JHH. Irradiation injury to large arteries. Br J Surg 93:341–343, 1980

291. St Louis EL, McLoughlin MJ, Wortzman G. Chronic damage to medium and large arteries fol-

lowing irradiation. J Can Assoc Radiol 25:94–104, 1974

292. Conomy JP, Kellermeyer R. Delayed cerebrovascular consequences of therapeutic radiation: A clinicopathologic study of a stroke association with radiation-related carotid arteriography. Cancer 36: 1702–1708, 1975

293. Elerding SC, Fernandez RN, Grotta JC, Lindberg RD, Causay LC, McMurtrey MJ. Carotid artery disease following external cervical irradiation. Ann Surg 194:609–615, 1981

294. Murros KE, Toole JF. The effect of radiation on carotid arteries. A review article. Arch Neurol 46:449–455, 1989

295. Werner MH, Burger PC, Heinz ER, Friedman AH, Halperin EC, Schold SC. Intracranial atherosclerosis following radiotherapy. Neurology 38: 1158–1160, 1988

296. Waltz TA, Brownell B. Sarcoma: a possible late result of effective radiation therapy for pituitary adenoma: report of two case. J Neurosurg 24:901–907, 1966

297. Amine ARC, Sugar O. Suprasellar osteogenic sarcoma following radiation for pituitary adenoma. J Neurosurg 44:88–91, 1976

298. Iacono P, Apuzzo MLJ, Davis RL. Multiple meningiomas following radiation for medulloblastomas: case report. J Neurosurg 55:282–286, 1981

299. Ushio Y, Arita N, Yoshimine T, et al. Glioblastoma after radiotherapy for craniopharyngioma: case report. Neurosurgery 21:33–38, 1987

300. Marus G, Levin V, Rutherford S. Malignant glioma following radiotherapy for unrelated primary tumors. Cancer 58:886–894, 1986

301. Liwnciz BH, Berger TS, Liwnciz RG, et al. Radiation-associated gliomas: a report of four case and analysis of postradiation tumors of the central nervous system. Neurosurgery 17:436–445, 1985

302. Hufnagel TJ, Kim JH, Lesser R, et al. Malignant glioma of the optic chiasm eight years after radiotherapy for prolactinoma. Arch Ophthalmol 106:1701–1705, 1988

303. Nakissa N, Rubin P, Strohl R, Keys H. Ocular and orbital complications following radiation therapy of paranasal sinus malignancies. Cancer 51: 980–986, 1983

304. Stephens LC, Schultheiss TE, Peters LJ, Ang KK, Gray KN. Acute radiation injury of ocular adnexa. Arch Ophthalmol 106:389–391, 1988

305. Gass JDM. A fluorescein angiographic study of macular dysfunction secondary to retinal vascular disease. Arch Ophthalmol 80:606–617, 1968

306. Chaudhuri PR, Austin DJ, Rosenthal AR. Treatment of radiation retinopathy. Br J Ophthalmol 65:623–625, 1981

307. Lampert PW, Tom MI, Rider WD. Disseminated demyelination of the brain following Co60 (gamma) radiation. Arch Pathol 68:322–330, 1959

308. Guy J, Schatz NJ. Hyperbaric oxygen in the treatment of radiation-induced neuropathy. Ophthalmology 93:1083–1088, 1986

309. Roden D, Bosley TM, Fowble B, et al. Delayed radiation injury to the retrobulbar optic nerves and chiasm: clinical syndrome and treatment with hyperbaric oxygen and corticosteroids. Ophthalmology 97:346–351, 1990

310. Tharp B, Heyman A, Pfeiffer JB, et al. Cerebral ischaemia. Arch Neurol 12:142–146, 1968

311. Nardelli E, Fiaschi A, Ferrari G. Delayed cerebrovascular consequences of radiation to the neck. Arch Neurol 35:538–540, 1978

312. Loftus CM, Biller J, Hart MN, Cornell SH, Hiratzka LF. Management of radiation-induced accelerated carotid atherosclerosis. Arch Neurol 44: 711–714, 1987

Chapter 6   Aneurysms

1. Sundt TM, Whisant JP. Subarachnoid hemorrhage from intracranial aneurysms. NEJM 299:116–122, 1978

2. Winn HR, Almaani WS, Berga SL, et al: The long-term outcome in patients with multiple aneurysms. Incidence of late hemorrhage and implications for treatment of incidental aneurysms. J Neurosurg 59:642–651, 1983

3. Fisher M, Davidson R, Marcus E. Transient focal cerebral ischemia as a presenting manifestation of unruptured cerebral aneurysm. Ann Neurol 8:367–372, 1980

4. McCormick WF. Intracranial arterial aneurysm: a pathologist's view. Current Concepts Cerebrovascular Disease, Stroke VIII:15–19, 1973

5. McCormick WF. Problems and pathogenesis of intracranial aneurysms. In: Cerebral Vascular Disease, Toole JF, Moosy J, Janeway R (eds). Grune & Stratton, New York, 1971, pp 219–231

6. Skultety FM, Nishioka H. The results of intracranial surgery in the treatment of aneurysm. In: Sahs AL, Perret GE, Locksbury HB, Nishuka H (eds) Intracranial aneurysm and subarachnoid hemorrhage. a cooperative study. JB Lippincott, Philadelphia, 1969, pp 173–193

7. McCormick WF, Schmalstieg EJ. The relationship of arterial hypertension to intracranial aneurysms. Arch Neurol 34:285–287, 1977

8. Stehbens WE. Histopathology of cerebral aneurysms. Arch Neurol 8:272–285, 1963

9. Stehbens WE. Ultrastructure of aneurysms. Arch Neurol 32:798–807, 1975

10. Neil-Dwyer G, Bartlett JR, Nicholls AC, et al. Collagen deficiency and ruptured cerebral aneurysms. A clinical and biochemical study. J Neurosurg 59: 16–20, 1983

11. Ostergaard JR, Oxlund H. Collagen type III defi-

ciency in patients with rupture of intracranial saccular aneurysms. J Neurosurg 67:690–696, 1987

12. Belber CJ, Hoffman RB. The syndrome of intracranial aneurysms associated with fibromuscular hyperplasia of the renal arteries. J Neurosurg 28: 556–559, 1969

13. Pyeritz RE, McKusick VA. The Marfan syndrome: diagnosis and management. NEJM 300:772–777, 1979

14. Ter Berg JWM, Bijlsma JB, Veiga Pires JA, Ludwig JW, Van der Heiden C, Tulleken CAF, Willemse J. Familial association of intracranial aneurysms and multiple congenital anomalies. Arch Neurol 43: 30–33, 1986

15. Hollister DW. Heritable disorders of connective tissue: Ehlers-Danlos syndrome. Pediatr Clin North Am 25:575–591, 1978

16. Green LN. Progeria with carotid artery aneurysms. Report of a case. Arch Neurol 38:659–661, 1981

17. Hatfield PM, Pfister RC. Adult polycystic disease of the kidneys (Potter type 3). JAMA 222: 1527–1531, 1972

18. Levey AS, Pauker SG, Kassirer JP. Occult intracranial aneurysms in polycystic kidney disease. When is cerebral angiography indicated? New Engl J Med 308:986–994, 1983

19. Reifenstern GH, Levine SA, Gross RE. Coarctation of the aorta. Am Heart J 33:146–168, 1947

20. Graf CJ. Familial intracranial aneurysms. Report of four cases. J Neurosurg 25:304–308, 1966

21. Acosta-Rua G. Familial incidence of ruptured intracranial aneurysms. Arch Neurol 35:656–677, 1978

22. Hillal F, Mohr G, Toussi T, Martinez SN. Intracranial aneurysms: a report of a large pedigree. Am J Med Genet 15:89–95, 1983

23. Norrgard O, Anquist K, Fodstad H, Forsell A, Lindberg M. Intracranial aneurysms and hereditary. Neurosurgery 20:236–239, 1987

24. Wiebers DO, Whisnant JP, O'Fallon WM. The natural history of unruptured intracranial aneurysms. NEJM 304:696–698, 1981

25. Wiebers DO, Whisnant JP, Sundt TM, O'Fallon WM. The significance of unruptured intracranial saccular aneurysms. J Neurosurg 66:23–29, 1987

26. Morley TP, Barr HWK. Giant intracranial aneurysms: diagnosis, course, and management. Clin Neurosurg 16:73–93, 1968

27. Drake CG: Giant intracranial aneurysms. Experience with surgical treatment in 174 patients. Clin Neurosurg 26:12–95, 1979.

28. Onuma and Suzuki J. Surgical treatment of giant intracranial aneurysms. J Neurosurgery 51:33–36, 1979

29. Pakarinen S. Incidence, aetiology, and prognosis of primary subarachnoid haemorrhage: a study of 589 cases diagnosed in a defined urban population during a defined period. Acta Neurol Scand 43:Suppl 29:1–127, 1967

30. Phillips LH, Whisnant JP, O'Fallon WM, et al. The unchanging pattern of subarachnoid hemorrhage in a community. Neurology 30:1034–1040, 1980

31. Winn HR, Taylor J, Kaiser D. The prevalence of asymptomatic incidental aneurysms: review of 4,568 angiograms. Stroke 14:121, 1983

32. Birse SH, Tom MI. Incidence of cerebral infarction associated with ruptured intracranial aneurysms. Neurology 10:101–106, 1960

33. Millikan CH. Cerebral vasospasm and ruptured intracranial aneurysm. Arch Neurol 32:433–449, 1975

34. Hijdra A, Vermeulen M, van Gijn J, van Crevel H. Rerupture of intracranial aneurysms: A clinicoanatomic study. J Neurosurg 67:29–33, 1987

35. Winn HR, Richardson AE, Jane JA: The long-term prognosis in untreated cerebral aneurysms: I. The incidence of late hemorrhage in cerebral aneurysm: A 10 year evaluation of 364 patients. Annals of Neurology 1:358–370, 1977.

36a. 36b. Locksley HB. Report on the cooperative study of intracranial aneursysms and subarachnoid hemorrhage. Section 5: Natural history of subarachnoid hemorrhage, intracranial aneurysms and arteriovenous malformations. J Neurosurg 25:219–239 (part I), 321–368 (part II), 1966

37. Lougheed WM, Marshall BM. Management of aneurysms of the anterior circulation procedures. In: Youmans JR (ed) Neurosurgical Surgery. WB Saunders, Philadelphia, 1973, Vol 2, pp 742–750

38. Sundt TM, Kobayashi S, Fode NC, et al. Results and complications of surgical management of 809 intracranial aneurysms in 722 cases. Related and unrelated to grade of patient, type of aneurysm, and timing of surgery. J Neurourg 56:753–765, 1982

39. Adams HP, Kassell NF, Kongable GA, Torner JC. Intracranial operation within seven days of aneurysmal subarachnoid hemorrhage. Results in 150 patients. Arch Neurol 45:1065–1069, 1988

39a. Masucci EF, Kurtzke JF. Diabetic superior branch palsy of the oculomotor nerve. Ann Neurol 7:493, 1980

40. Schatz NJ, Savino PJ, Corbett JJ. Primary aberrant oculomotor regeneration. a sign of intracavernous meningioma. Arch Neurol 34:29–32, 1977

41. Trobe JD, Glaser JS, Post JD: Meningiomas and aneurysms of the cavernous sinus. Neuro-ophthalmologic features. Arch Ophthalmol 96:457–467, 1978

41a. Laguna JF, Smith MS. Aberrant regeneration in idiopathic oculomotor nerve palsy. J Neurosurg 52: 854–856, 1980

42. Thompson SH, Mensher JH. Adrenergic mydriasis in Horner's syndrome. Am J Ophthalmol 72: 472–480, 1972

43. Peiris JB, Ross Russell RW. Giant aneurysms of the carotid system presenting as visual field defect. J Neurol Neurosurg Pschiatr 43:1053–1064, 1980

44. Walsh FB. Visual field defects due to aneurysms at the circle of Willis. Arch Ophthalmol 71:15–26, 1964

45. Ballantyne AJ: The ocular manifestations of spontaneous subarachnoid hemorrhage. Br J Ophthalmol 27:383–414, 1943

46. Terson A. De l'hemorragie dans le corps vitre au cours d'une hemorragie cerebrale. Clin Ophthalmol 6:309–312, 1900

47. Terson A. Le syndrome de l'hematome du corps vitre et de l'hemorrhagie intracranienne spontanes. Ann Oculist 147:410–417, 1912

48. Jefferson G. Compression of the chiasm, optic nerves, and optic tracts by intracranial aneurysms. Brain 60:444–497, 1937

49. Richardson JC, Hyland HH. Intracranial aneurysms. A clinical and pathological study of subarachnoid and intracerebral hemorrhage caused by berry aneurysms. Medicine (Balt) 20:1–83, 1941

50. Manschot WA. The fundus oculi in subarachnoid hemorrhage. Acta Ophthalmol 22:281–299, 1944

51. Shaw HE, Landers MB, Sydnor CF. The significance of intraocular hemorrhage due to subarachnoid hemorrhage. Ann Ophthalmol 9:1403–1405, 1977

52. Weingest TA, Goldman EJ, Folk JC, Packer AJ, Ossoinig KC. Terson's syndrome. Clinicopathologic correlations. Ophthalmology 93:1435–1442, 1986

53. Morris DA, Henkind P. Relationship of intracranial optic-nerve sheath and retinal hemorrhage. AJO 64:853–859, 1967

54. Bisland T, and Topilow A: Subhyaloid hemorrhage and exophthalmos due to ruptured intraventricular aneurysm: A case occurring in toxemia of pregnancy. Arch Ophthalmol 47:470–476, 1952

55. Vanderlinden RG, Chisholm LD. Vitreous hemorrhages and sudden increased intracranial pressure. J Neurosurg 41:167–176, 1974

56. Walsh FB, Hoyt WF. Clinical Neuro-ophthalmology, Vol 2, Baltimore, Williams & Wilkins, 1969, pp 1782–1791, 2375

57. Fujino T, Curtin VT, Norton EWD. Experimental central retinal vein occlusion: a comparison of intraocular and extraocular occlusion. Arch Ophthalmol 31:395–406, 1969

58. Muller PJ, Deck JHN. Intraocular and optic nerve sheath hemorrhage in cases of sudden intracranial hypertension. J Neurosurg 41:160–166, 1974

59. Weaver RG, Davis CH. Subhyaloid hemorrhage. AJO 52:257–259, 1961

60. Hedges TR, Weinstein JD, Crystal CD. Orbital vascular response to acutely increased intrcranial pressure in the rhesus monkey. Arch Ophthalmol 71:226–237, 1964

61. Clarkson JG, Flynn HW Jr, Daily MJ. Vitrectomy in Terson's syndrome. AJO 90:549–552, 1980

62. Shinoda J, Iwamura M, Iwai T. et al. Intraocular hemorrhage in ruptured intracranial aneurysm. Clinical study of 172 cases and references to Terson's syndrome. Neurol Med Chir (Tokyo) 23:349–354, 1983

63. White JV, Ballantyne HI: Intrasellar aneurysms simulating hypophyseal tumors. J Neurosurg 18:34–50, 1961.

64. Mitchell SW: Aneurysm of an anomalous artery causing antero-posterior division of the chiasm of the optic nerves and producing bitemporal hemianopia. J Nervous and Mental Dis 14:14–62, 1889.

65. Pinto RS, Kricheff II, Butler AR, et al. Correlation of computed tomographic, angiographic, and neuropathological changes in giant cerebral aneurysms. Radiology 132:85–92, 1979

66. Dujovny M, Kossovsky N, Kossowsy R, Valdivia R, et al. Aneurysm clip motion during magnetic resonance imaging: in vivo experimental study with metallurgical factor analysis. Neurosurgery 17:543–548, 1985

67. Becker RL, Norfray JF, Teitelbaum GP, Bradley JB, et al. MR imaging in patients with intracranial clips. AJNR 9:885–889, 1988

68. Romner B, Olsson M, Ljunggren B, Holtas S et al. Magnetic resonance imaging and aneurysm clips: magnetic properties and image artifacts. J Neurosurg 70:426–431, 1989

69. Matas R. Testing the efficiency of the collateral circulation as a preliminary to the occlusion of the great surgical arteries. Ann Surg 53:1–43, 1911

70. Wada J, Rasmussen T. Intracarotid injection of sodium Amytal for the lateralization of cerebral dominance. J Neurosurg 17:266–282, 1960

71. Horton JA, Dawson RC. Retinal Wada test. AJNR 9:1167–1168, 1988

72. Matthews WF, Frommeyer WB jr. The in vitro behavior of erythrocytes in human cerebrospinal fluid. J Lab Clin Med 45:508–515, 1955

73. Resurreccion EC, Rosenblum JA. Common causes of spurious xanthochromia in cerebrospinal fluid. Angiology 23:105–110, 1972

74. Hammes EM. Reaction of the meninges to blood. Arch Neurol 52:505–514, 1944

75. Ransohoff J, Goodgold A, Benjamin MV. Preoperative management of patients with ruptured intracranial aneurysms. J Neurosurg 36:525–530, 1972

76. Corkill G. Earlier operation and antifibrinolytic therapy in the management of aneurysmal subarachnoid haemorrhage. Med J Australia 1:468–470, 1974

77. Adams HP. Current status of antifibrinolytic therapy for treatment of patients with aneurysmal subarachnoid hemorrhage. Stroke 13:256–259, 1982

78. Allen GS, Ahn HS, Preziosi TJ, et al. Cerebral arteri-

494   References

rial spasm – a controlled trial of nimodipine in patients with subarachnoid hemorrhage. NEJM 308:619–624, 1983

79. Pickard JD, Murray JD, Illingworth R, et al. Effect of oral nimodipine on cerebral infarction and outcome after subarachnoid haemorrhage. British Aneurysm Nimodipine Trial. Br Med J 298:636–642, 1989

80. Gilsbach JM, Reulen HJ, Ljunggren B, Brandt L, von Holts H, Mokry M, von Essen C, and Conzen MA: Early aneurysm surgery and preventive therapy with intravenously administered nimodipine: A multicenter, double-blind, dose-comparison study. Neurosurgery 26:458–464, 1990

81. Kassel NF, Peerless SJ, Durward Q, et al. Treatment of ischemic deficits from vasospasm with hypervolemia and induced arterial hypertension. Neurosurgery 11:337–343, 1982

82. Solomon RA, Fink ME, Lennihan L. Prophylactic volume expansion therapy for the prevention of delayed cerebral ischemia after early aneurysm surgery. Results of a preliminary trial. Arch Neurol 45:325–332, 1988

82a. Choi IS, Berenstein A, Jafar J, Kupersmith MJ. Cerebral vasospasm dilatation with a calibrated leak microballoon catheter. XIV Symposium Neuroradiology, London, June 17–23, 1990

83. Ljunggren B, Brandt L. Timing of aneurysm surgery. Clin Neurosurg 33:159–175, 1986

84. Ljunggren B, Brandt L, Kagstrom E, et al. Results of early operations for ruptured aneurysms. J Neurosurg 54:473–479, 1981

85. Gillingham FJ. The management of ruptured intracranial aneurysms. Hunterion Lecture. Ann R Coll Surg Engl 23:89–117, 1958

86. Sachs E Jr, Erbengi A, Margolis G, et al. Fatality from ruptured intracranial aneurysm after coating with methyl 2-cyanoacrylate (Eastman 910 monomer. M2 C-1). Case report. J Neurosurg 24:889–891, 1966

87. Carney PG, Oatley PE. Muslin wrapping of aneurysms and delayed visual failure. a report of three cases. J Clin Neuroophthalmol 3:91–96, 1983

88. Chambi I, Tasker RR, Gentili F, et al. Gauze-induced granuloma ("gauzoma"): an uncommon complication of gauze reinforcement of berry aneurysms. J Neurosurg 72:163–170, 1990

89. Tomsak RL. Muslin optic neuropathy. J Clin Neuroophthamol 5:71–72, 1985

90. Repka MX, Miller NR, Penix JO, et al. Optic neuropathy from the use of intracranial muslin. J Clin Neuroophthalmol 4:147–150, 1984

91. Marcus AO, Demakas JJ, Ross HA, et al. Optochiasmatic arachnoiditis with treatment by surgical lysis of adhesions, corticosteroids, and cyclophosphamide: report of a case. Neurosurgery 19:101–103, 1986

92. Haisa T, Matsumiya K, Yoshimasu N, Kuribayashi N. Foreign-body granuloma as a complication of wrapping and coating an intracranial aneurysm. J Neurosurg 72:292–294, 1990

93. Ecker A, Riemenschneider P. Deliberate thrombosis of intracranial arterial aneurysm by partial occlusion of the carotid artery with angiographic control. J Neurosurg Neurol Psychiatr 8:348–353, 1951

94. Jawad K, Miller JD, Wyper DJ, Rowan JO. Measurement of CBF and carotid artery pressure compoared with cerebral angiography in assessing collateral supply after carotid ligation. J Neurosurg 46:185–196, 1977

95. Tindall GT, Goree JA, Lee JF, Odom LG. Effects of common carotid ligation on size of internal carotid aneurysms and distal intracarotid and retinal artery pressure. J Neurosurg 25:503–511, 1966

96. Mount L, Taveras JM. The results of surgical treatment of intracranial aneurysms as demonstrated by progressive arteriography. J Neurosurg 13: 618–626, 1956

97. McKissock W, Richardson A, Walsh L. Posterior communicating aneurysm. A controlled trial of the conservative and surgical treatment of ruptured aneurysms of the internal carotid at or near the point of origin of the posterior communicating artery. Lancet 1:1203–1208, 1960

98. VanderArk GC, Kempe LG, Kobrine A. Classification of internal carotid aneurysms as a basis for surgical approach. Neurochirurgia 3:81–85, 1972

99. Jha AN, Butler P, Lye RH, Fawcitt RA. Carotid ligation. What happens in the long term? J Neurosurg Neurol Psychiatr 49:893–898, 1986

100. Sahs AL, Perrett GE, Lochsley HB, et al. Intracranial aneurysms and subarachnoid hemorrahge. A cooperative study. JB Lippincott, Philadelphia, 1969

101. Landolt AM, Millikan CM. Pathogenesis of cerebral infarction secondary to mechanical carotid artery occlusion. Stroke 1:52–62, 1970

102. Hopkins LN, Grand W. Extracranial–intracranial arterial bypass in the treatment of aneurysms of the carotid and middle cerebral arteries. Neurosurgery 5:21–31, 1979

103. Meining G, Gunter R, Ulrich P, Suss W, et al. Reduced risk of ICA ligation after balloon occlusion test. Neurosurg Rev 5:95–98, 1982

104. Roski RA, Spetzler RF, Nulsen FE. Late complications of carotid ligation in the treatment of intracranial aneurysms. J Neurosurg 54:583–587, 1981

105. Yasargil MG. Microsurgery Applied to Neurosurgery. New York, Academic Press, 1969, pp 49–52

106. Gelber BR, Sundt TM. Treatment of intracavernous and giant carotid aneurysms by combined internal carotid ligation and extra-to-intracranial bypass. J Neurosurg 52:1–10, 1980

107. Sekhar LN, Sen CN, Jho HD. Saphenous vein graft

bypass of the cavernous internal carotid artery. J Neurosurg 72:35–41, 1990

108. Dyste GW, Beck DW. De novo aneurysm formation following carotid ligation: case report and review of literature. Neurosurgery 24:88–92, 1989

109. Spetzler RF, Fukushima T, Martin N, Zabramski JM. Petrous carotid-to-intradural carotid saphenous vein graft for intracavernous giant aneurysm, tumor, and occlusive cerebrovascular. J Neurosurg 73:496–501, 1990

110. Serbinenko FA. Balloon catheterization and occlusion of major cerebral vessels. J Neurosurg 41:125–145, 1974

111. Romodanov AP, Shcheglov VI. Intravascular occlusion of saccular aneurysms of the cerebral arteries by means of a detachable balloon catheter. Adv Tech Stand Neurosurg 9:25–49, 1982

112. Serbinenko FA, Filatov JM, Spallone A, Tchurilov MV, Lazarev VA. Management of giant intracranial ICA aneurysms with combined extracranial-intracranial anastamosis and endovascular occlusion. J Neurosurg 73:57–63, 1990

113. Taki W, Handa H, Yamagata S, Matsuda I, Yonekawa Y, Ikada Y, and Iwata H: Balloon embolization of a giant aneurysm using a newly developed catheter. Surg Neurol 12:363–365, 1979

114. Higashida RT, Halbach VV, Cahan LD, Hieshima GB, and Konishi Y: Detachable balloon embolization therapy of posterior circulation intracranial aneurysms. J Neurosurg 71:512–519, 1989

115. Strother CM, Lunde S, Graves V, Toutant S, and Hieshima GB: Late paraophthalmic aneurysm rupture following endovascular treatment. Case report. J Neurosurg 71:777–780, 1989

116. Hodes JE, Fox AJ, Pelz DM, Peerless SJ. Rupture of aneurysms following balloon embolization. J Neurosurg 72:567–571, 1990

117. Krayenbuhl H. Klassifikation und klinische Symptomatologie der zerbralen Aneurysmen. Ophthalmologica 167:122–164, 1973

118. Henderson JW. Intracranial arterial aneurysms: A study of 119 cases, with special reference to the ocular findings. Trans Am Ophthalmol Soc 53:349–462, 1955

119. Jefferson G. Concerning injuries, aneurysms, and tumors involving the cavernous sinus. Trans Ophthalmol Soc UK 73:117–152, 1953

120. Thomas JE, Yoss RE. The parasellar syndrome: Problems in determining etiology. Mayo Clin Proc 45:617–623, 1970

121. Cogan D, Mount HT. Intracranial aneurysms causing ophthalmoplegia. Arch Ophthalmol 70:757–771, 1963

122. Barr HWK, Blackwood W, Meadows SP. Intracavernous carotid aneurysms. A clinical-pathological report. Brain 94:607–622, 1971

123. Rapport R, Murtagh FR. Ophthalmolplegia due to spontaneous thrombosis in a patient with bilateral cavernous carotid aneurysms. J Clin Ophthalmol 1:225–229, 1981

124. Markwalder TM, Meienberg O. Acute painful cavernous sinus syndrome in unruptured intracavernous aneurysms of the internal carotid artery. J Clin Neuro-ophthalmol 3:31–35, 1983

125. Rush JA and Younge BR: Paralysis of cranial nerves III, IV, and VI: Cause and prognosis in 1, 000 cases. Arch Ophthalmol 99:76–79, 1981

126. Maurice-Williams RS and Harvey PK: Isolated palsy of the fourth cranial nerve caused by an intracranial aneurysm. J Neurol Neurosurg Psychiatr 52:679, 1989

127. Boghen D, Chartrand JP, Laflamme P, et al. Primary aberrant third nerve regeneration. Ann Neurol 6:415–418, 1979

128. Alper MG. The anesthetic eye: an investigation of changes in the anterior ocular segment of the monkey caused by interrupting the trigeminal nerve at various levels along its course. Tr Am Ophth Soc LXXIII, pp 323–363, 1975

129. Meadows SP: Intracavernous aneurysms of the internal carotid artery. Their clinical features and natural history. Arch Ophthalmol 62:566–579, 1959.

130. Gauthier G, Rohr J, Wildi E, Megret M. L'hematome dissequant spontane de l'artere carotide interne: revue generale de 205 cas publies dont 10 personnels. Arch Suisse Neurol Psychiatr 136:53–74, 1986

131. Moorthy G, Behrens MM, Drachman DM, et al. Ocular pseudomyasthenia or ocular myasthenia 'plus': a warning to clinicians. Neurology 39:1150–1154, 1989

132. Nutik S. Carotid paraclinoid aneurysms with intradural origin and intracavernous location. J Neurosurg 48:526–533, 1978

133. Berenstein A, Ransohoff J, Kupersmith M, Flamm E, Graeb D, Kricheff I. Endovascular treatment of giant aneurysms of the carotid and vertebral arteries. Functional investigation and embolization. Surg Neurol 21:3–12, 1984

134. Kupersmith MJ, Berenstein A, Choi IS, et al. Percutaneous transvascular treatment of giant carotid aneursyms: neuro-ophthalmologic findings. Neurology 34:328–335, 1984

135. Fein JN, Flamm E. Planned intracranial revascularization before proximal ligation for traumatic aneurysms. Neurosurgery 5:54–57, 1979

136. Spetzler RF, Rhodes RS, Roski RA, et al. Subclavian to middle cerebral artery saphenous vein bypass graft. J Neurosurg 53:465–469, 1980

137. van Beusekom GT, Luyendijk W, Huizing EH. Severe epistaxis caused by rupture of non traumatic infraclinoid aneurysm of the internal carotid artery. Acta Neurochir 15:269–284, 1966

138. Linskey ME, Sekhar LN, Hirsch W, et al. Aneurysms of the intracavernous carotid artery: clinical

presentation, radiographic features, and pathogenesis. Neurosurgery 26:71–79, 1990

139. Ding MX. Traumatic aneurysm of the intracavernous part of the internal carotid artery presenting with epistaxis. Case report. Surg Neurol 30:65–67, 1988

139a. McCormick WF, Beals JD. Severe epistaxis caused by ruptured aneurysm of the internal carotid artery. J Neurosurg 21:678–686, 1964

140. Hodes JE, Fletcher WA, Goodman DF, Hoyt W. Rupture of cavernous carotid artery aneurysm causing subdural hematoma and death. J Neurosurg 69:617–619, 1988

141. Mullan S. Carotid cavernous aneurysms and fistulae. In: Dolenc VV (ed) The Cavernous Sinus. Springer-Verlag, Wien, 1987, pp 225–232

142. Pendl G, Vorkapic P, Richling B, Koos WT. Strategies in intracavernous saccular aneurysms. In: Dolenc VV (ed) The Cavernous Sinus. Springer-Verlag, Wien, 1987, pp 240–251

143. Nishioka T, Kondo A, Aoyama I, et al. Subarachnoid hemorrhage possibly caused by a saccular carotid artery aneurysm within the cavernous sinus. J Neurosurg 73:301–304, 1990

144. Kupersmith MJ, Hurst R, Choi IS, Berenstein A, Jafar J, Ransohoff J. The benign course of cavernous carotid aneurysms. J Neurosurg, in press, 1992

145. Higashida RT, Halbach VV, Dowd C, et al. Endovascular detachable balloon embolization therapy of cavernous cartoid artery aneurysms: results in 87 cases. J Neurosurg 72:857–863, 1990

146. Dolenc VV, Cerk M, Sustersic J, Pregelj R, Skrap M. Treatment of intracavernous aneurysms of the ICA and CCF's by direct approach. In: Dolenc VV (ed) The Cavernous Sinus. Springer-Verlag, Wien, 1987, pp 297–310

147. Diaz FG, Ohaegbulam S, Dujovny M, Ausman JI. Surgical alternatives in the treatment of cavernous sinus aneurysms. J Neurosurg 71:846–853, 1989

148. Capo H, Kupersmith MJ, Berenstein A, Choi IS, Diamond G. The clinical importance of the inferolateral trunk of the internal carotid artery. Neurosurgery 28:733–738, 1991

149. Strenger J. Neurological deficits following therapeutic collapse of intracavernous carotid aneurysms. J Neurosurg 25:215–218, 1966

150. Cuatico W, Cook AW, Tyshchenko V, Khatib R. Massive enlargement of intracranial aneurysms following carotid ligation. Arch Neurol 17:609–613, 1967

151. Alpers BJ, Berry RG. Circle of Willis in cerebrovascular disorders. Arch Neurol 8:398–402, 1963

152. Green WR, Hackett ER, Schlezinger NS. Neuroophthalmologic evaluation of oculomotor nerve paralysis. Arch Ophthalmol 72:154–167, 1964

153. Botterell EH, Lloyd LA, Hoffman HJ. Oculomotor palsy due to supraclinoid internal carotid artery berry aneurysm. A long-term study of the results of surgical treatments on recovery of third-nerve function. Am J Ophthalmol 54:609–616, 1962

154. Hyland HH, Barnett HJM. The pathogenesis of cranial nerve palsies associated with intracranial aneurysms. Proc. R. Soc. Med. 47:141–146, 1954

155. Okawara SH. Warnings signs prior to rupture of an intracranial aneurysm. J Neurosurg 38:575–580, 1973

156. Rucker CW. Paralysis of third, fourth, and sixth cranial nerves. Am J Ophthalmol 46:787–794, 1958

157. Rucker CW. Paralysis of the third, fourth, and sixth cranial nerves. Am J Ophthalmol 61:1293–1298, 1966

158. Jaeger R. Aneurysm of the intracranial carotid artery: syndrome of frontal headache with oculomotor nerve paralysis. JAMA 142:304–310, 1950

159. Goldstein JE, Cogan DG. Diabetic ophthalmoplegia with special reference to pupil. Arch. Ophthalmol. 64:592–600, 1960

160. Capo H, Warren F, Kupersmith MJ. Evolution of oculomotor nerve palsies. J Clin Neuro-ophthalmol 12:21–25, 1992

161. Kissel JT, Burde RM, Klingele TG, Zeiger HE. Pupil- sparing oculomotor palsies with internal carotid–posterior communicating artery aneurysms. Ann Neurol 13:149–154, 1983

162. Keane JR. Aneurysms and third nerve palsies. Ann Neurol 14:696–697, 1983

163. O'Connor PS, Tredici TJ, Green RP. Pupil-sparing third nerve palsies caused by aneurysm. Am J Ophthalmol 95:395–397, 1983

164. Bartleson JD, Trautmann JC, Sundt TM. Minimal oculomotor nerve paresis secondary to unruptured intracranial aneurysm. Arch Neurol 43:1015–1020, 1986

165. Good EF. Ptosis as the sole manifestation of compression of the oculomotor nerve by an aneurysm of the posterior communicating artery. J Clin Neuro-ophthalmol 10:59–61, 1990

166. Kasoff I, Kelly DL. Pupillary sparing in oculomotor palsy from internal carotid aneurysm. J Neurosurg 42:713–717, 1975

167. Kerr FWL, Hollowell OW. Location of pupillomotor and accomodation fibres in the oculomotor nerve: Experimental observations on paralytic mydriasis. J Neurol Neurosurg Psych 27:473–481, 1964

168. Sunderland S, Hughes ESR. The pupilloconstrictor pathway and the nerves ot the ocular muscles in man. Brain 69:301–309, 1946

169. Oono S, Shinohara Y, Fuji K. Pupillary sparing in oculomotor palsy from internal carotid aneurysm. In: Ishikawa S (ed) Highlights in Neuro-ophthalmology. Aeolus Press, Amsterdam, 1987, pp 203–208

170. Guy J, Savino PJ, Schatz NJ, et al. Superior division paresis of the oculomotor nerve. Ophthalmology 92:777–784, 1985

171. Trobe JD, Glaser JS, Quencer RC. Isolated oculomotor paralysis: the product of saccular and fusiform aneurysms of the basilar artery. Arch Ophthalmol 96:1236–1240, 1978

172. Lustbader JM, Miller NR. Painless, pupil-sparing but otherwise complete oculomotor nerve paresis caused by basilar artery aneurysm. Arch Ophthalmol 106:583–584, 1988

173. Guy JR, Day AL. Intracranial aneurysms with superior division paresis of the oculomotor nerve. Ophthalmology 96:1071–1076, 1989

174. Masucci EF, Kurtzke JF. Diabetic superior branch of the oculomotor nerve. Ann Neurol 7:493, 1980

175. Bregman DK, Harbour R. Diabetic superior branch division oculomotor nerve palsy. Arch Ophthalmol 106:1169–1170, 1988

176. Pollard ZF. Aneurysm causing third nerve palsy in a 15-year-old boy. Arch Ophthalmol 106:1648–1649, 1988

177. Matson DD. Intracranial arterial aneurysms in childhood. J Neurosurg 25:578–583, 1965

178. Gabianelli EB, Klingele TG, Burde RM. Acute oculomotor nerve palsy in childhood. Is angiography necessary? J Clin Neuro-ophthalmol 9:33–36, 1989

179. Payne JW, Adamkiewicz J. Unilateral internal ophthalmoplegia with intracranial aneurysm. Report of a case. Am J Ophthalmol 68:349–352, 1969

180. Cromptom JL, Moore CE. Painful third nerve palsy: how not to miss an intra-cranial aneurysm. Aust J Ophthalmol 9:113–115, 1981

181. Bartleson JD, Trautman JC, Sundt TM. Minimal oculomotor nerve paresis secondary unruptured intracranial aneurysm. Arch Neurol 43:1015–1020, 1986

182. Coppeto JR, Lessell S. Dorsal midbrain syndrome from giant aneurysm of the posterior fossa: Report of two cases. Neurology 33:732–736, 1983

183. Takahashi T, Kanatani I, Isayama Y, Tamaki N, and Matsumoto S: Visual disturbance due to internal carotid aneurysm. Ann Ophthalmol 13:1014–1024, 1983

184. Wilson WB, Barmatz HE. Acute angle-closure glaucoma secondary to an aneurysm of the posterior communicating artery. Am J Ophthalmol 89:868–870, 1980

185a. Hassler O, Saltzman GF. Histologic changes in infundibular widening of the posterior communicating artery: a preliminary report. Acta Pathol 46:305–312, 1959

185b. Hassler O, Saltzman GF. Angiographic and histologic changes in infundibular widening of the posterior communicating artery. Acta Radiol 1:321–327, 1963

186. Epstein F, Ransohoff J, Budzilovich GN: The clinical significance of junctional dilatation of the posterior communicating artery. J Neurosurg 33:529–531, 1970.

187. Graff-Radford NR, Damasio H, Yamada T, Eslinger PJ, Damasio AR. Nonhaemorrhagic thalamic infarction. Brain 108:485–516, 1985

188. Kudo T: Postoperative oculomotor palsy due to vasospasm in a patient with a ruptured internal carotid artery aneurysm. A case report. Neurosurgery 19:274–277, 1986

189. Morris L. Arteriographic studies in aneurysm of the internal carotid artery treated by carotid occlusion. Acta Radiol 1:367–372, 1963

190. Winn HR, Richardson AE, Jane JA. Late morbidity and mortality of common carotid ligation of posterior communicating aneurysms (a comparison to conservative therapy). J Neurosurg 47: 727–736, 1977

191. Hamer J: Prognosis of oculomotor palsy in patients with aneurysms of the posterior communicating artery. Acta Neurochir 66:173–185, 1982

192. Kerns JM, Smith DR, Jannotta FS, Alper MG. Oculomotor nerve regeneration after aneursym surgery. Am J Ophthalmol 87:225–233, 1979

193. Gowers W. The movements of the eyelids. Med Chir Trans 62:429–449, 1879

194. Grayson MC, Soni SR, Spooner VA. Analysis of recovery of third nerve function after direct surgical intervention for posterior communicating aneurysms. Br J Ophthalmol 58:118– 125, 1974

195. Cox TA, Wurster JB, Godfrey WA. Primary aberrant oculomotor regeneration due to intracranial aneurysm. Arch Neurol 36:570–571, 1979

196. Czarnecki JSC, Thompson HS. The iris sphincter in aberrant regeneration of the third nerve. Arch Ophthalmol 96:1606–1610, 1978

197. Ohno S, Mukuno K. Studies on synkinetic pupillary phenomena resulting from aberrant regeneration of the third. Jap J Clin Ophthalmol 27: 229–239, 1973

198. Levin PM. Intracranial aneurysms. Arch. Neurol. Psych.67:771–786, 1952

199. Bielchowsky A. Lectures on motor anomalies of the eyes:II. Paralysis of individual eye muscles. Arch. Ophthalmol.13:33–59, 1935

200. Wartenberg R. Associated movements in the oculomotor and facial muscles. Arch. Neurol. Psych.55:439–488, 1946

201. Sibony P, Lessell S, Gittinger JW. Acquired oculomotor synkinesis. Survey of Ophthalmology 28:382–390, 1984

202. Pool L, Potts G. Aneurysms and arteriovenous anomalies of the brain. Harper and Row, New York, 1965

203. Sengupta RP, Gryspeerdt GL, Hankinson J. Carotid ophthalmic aneurysms. J Neurol Neurosurg Psychiatr 39:837–853, 1976

204. Yasargil MG, Gasser JG, Hodosh RM, Rankin TV. Carotid- ophthalmic aneurysms: direct microsurgical approach. Surg Neurol 8:155–165, 1977

205. Guidetti B, La Torre E. Carotid-ophthalmic aneurysms. A series of 16 cases treated by direct approach. Acta Neurochir (Wien) 22:289–304, 1970

206. Drake CG, Vanderlinden RG, Amacher AL. Carotid-ophthalmic aneurysms. J Neurosurg 29:24–31, 1968

207. Ferguson G, Drake CG. Carotid-ophthalmic aneurysms: visual abnormalities in 32 patients and the results of treatment. Surg Neurol 16:1–8, 1981

208. Almeida GM, Shibata MK, Bianco E. Carotid-ophthalmic aneurysms. Surg Neurol 5:41–45, 1976

209. Hook O and Norlen G: Aneurysms of the internal carotid artery. Acta Neurol Scand 40:200–212, 1964

210. Hosobuchi Y. Direct surgical treatment of giant intracranial aneurysms. J Neurosurg 51:743–756, 1979

211. Jefferson G. Further concerning compression of the optic pathways by intracranial aneurysms. Clin Neurosurg 1:55–103, 1955

212. Offret G, Godde-Jolly D. Les anevrismes de l'artere ophtalmique. Arch D'Ophthalmologie 16: 388–396, 1956

213. Arseni M C, Laseo F. L'anevrisme de l'artere ophtalmique. Rev Otoneuroophtalmol 39:83–90, 1967

214. Cullen JF, Haining WM, Crombie AL. Cerebral aneurisms presenting with visual field defects. Br J Ophthalmol 50:251–256, 1966

215. White JC. Aneurysms mistaken for hypophyseal tumors. Clin Neurosurg 10:224–250, 1961

216. Sanford HS, Craig WM. An unusual chiasmal lesion and its operative treatment. Proc Mayo Clin 10:721–725, 1935

217. de Martel MM, Francois J, Guillaume J. Dilatation aneurismale de l'artere ophtalmique ayant determine un retecissssement pseudohemianopsique superieur du champ visuel. Soc Francaise D'Ophthalmol Extrait 1–10, 1936

218. Day A. Aneurysms of the ophthalmic segment. A clinical and anatomical analysis. J Neurosurg 72: 677–691, 1990

219. Cox TA, Corbett JJ, Thompson HS, Kassell NF: Unilateral nasal hemianopia as a sign of intracranial optic nerve compression. AJO 92:230–232, 1981

220. Portney GL, Roth AM: Optic cupping caused by an intracranial aneurysm. AJO 84:98–103, 1977.

221. Norwood EG, Kline LB, Chandra-Sekar B, Harsh GR. Aneurysmal compression of the anterior pathways. Neurology 36:1035–1041, 1986

222. Stern WH, Ernest JT. Intracranial ophthalmic artery aneurysm. AJO 80:203–206, 1975

223. Ringel RA, Brick JF. Carotid-ophthalmic artery aneurysms masquerading as optic neuritis. J Neurol Neurosurg Psychiatr 49:460, 1986

224. Kuzniecky R, Melmed C, and Schipper H: Carotid-ophthalmic aneurysm: An uncommon cause of acute monocular blindness. Can Med Assoc J 136:727–728, 1987

225. Sadun AA, Smythe BA, Schaechter JD. Optic neuritis or ophthalmic artery aneurysm. J Clin Neuro-ophthalmol 4:265–273, 1984

226. Jain KK: Saccular aneurysm of the ophthalmic artery. Am J Ophthalmol 69:997–998, 1970

227. Tomsak RL, Costin JA, and Hanson M: Carotid-ophthalmic artery aneurysm presenting as unilateral disc edema with choroidal folds. In: Neuro-ophthalmology Focus 1980, Smith JL (ed), pp 181–187, New York, Masson, 1979

228. Carter JE and Montgomery RJ: The impossible aneurysm: Canalicular and orbital apex aneurysms of the ophthalmic artery. Neuro-ophthalmology 9:103–110, 1989

229. Meyerson L and Lazar SJ: Intraorbital aneurysm of the ophthalmic artery. Br J Ophthalmol 55: 199–204, 1971

230. Berson EL, Freeman MI, Gay AJ. Visual field defects in giant suprasellar aneurysms of internal carotid. Arch Ophthalmol 76:52–58, 1966

231. Raymond LA, Tew J. Large suprasellar aneurysms imitating pituitary tumor. J Neurol Neurosurg Psychiatr 41:83–87, 1978

232. Hayreh SS, Dass R. The ophthalmic artery. I. Origin and intra-cranial and intra-canalicular course. Br J Ophthal 46:65–98, 1962

233. Renn WH, Rhoton AL. Microsurgical anatomy of the sellar region. J Neurosurg 43:288–298, 1975

234. Gibo H, Lenkey C, Rhoton AL. Microsurgical anatomy of the supraclinoid portion of the internal carotid artery. J Neurosurg 55:560–574, 1981

235. Kothandaram P, Dawson BH, Kruyt RC. Carotid-ophthalmic aneurysms. a study of 19 patients. J Neurosurg 34:544–548, 1971

236. Flamm ES. Suction decompression of aneurysms. Technical note. J Neurosurg 54:275–276, 1981

237. Fox JL: Microsurgical treatment of ventral (paraclinoid) internal carotid artery aneurysms. Neurosurgery 22:32–39, 1988.

238. Sundt TM, Piepgras D. Surgical approach to giant intracranial aneursysms. J Neurosurg 51:731–742, 1979

239. Sengupta RP. Management of large and giant aneurysms. Neurosurg Rev 5:173–178, 1982

240. Thurel C, Rey A, Thiebaut JB, Chai N, Houdart R. Anevrismes carotido-optalmiques. Neurochirurgie 20:184–190, 1974

241. Shantharam VV, Clift GV. Suprasellar aneurysm: an unusual cause of hypopituitarism. JAMA 229: 1473, 1974

242. Kokoris N, Rothman LM, Wolintz AH. Computed tomography and angiography in the diagnosis of suprasellar mass lesions. Am J Ophthalmol 89: 278–283, 1980

243. Sakurai Y, Endo S, Suzuki J. Intra- and/or suprasellar communicating artery aneurysms associated with visual field defects. In Suzuki J (ed) Cerebral aneurysms. Neuron Publishing, Tokyo, 1979, pp 115–122

244. Durston JHJ and Parsons-Smith BG: Blindness due to aneurysms of anterior communicating artery: With recovery after carotid ligation. Br J Ophthalmol 54:170–176, 1970

245. Hojer-Pedersen E and Haase J: Giant anterior communicating artery aneurysm with bitemporal hemianopsia: Case report. Neurosurgery 8:703–706, 1981

246. Versavel M, Witmer JP, and Matricali B: Giant aneurysm arising from the anterior cerebral artery and causing an isolated homonymous hemianopia. Neurosurgery 22:560–563, 1988

247. Stark RJ: Supranuclear ophthalmoplegia with basilar artery aneurysms. Surg Neurol 12:447–452, 1979

248. Steinberger A, Ganti SR, McMurtry JG, and Hilal SK: Transient neurological deficits secondary to saccular vertebrobasilar aneurysms: Report of two cases. J Neurosurg 60:410–413, 1984

249. Plum F, Posner J. Stupor and Coma. FA Davis Publishing, Philadelphia, 1972, pp 1–52

250. Tulleken CA: Giant aneurysms of the posterior fossa presenting as space occupying lesions. Clin Neurosurg 79:161–186, 1976

251. Rad M, Piscol K: Parinaudisches Syndrom und internucleare Ophthalmoplegie bei raumforderndem Basilarisaneurysma. Z Neurol 199:319–331, 1971

252. Uihlein A, Hughes RA. The surgical treatment of vestigial aneurysms. Surg Clin North Am 35:1071–1083, 1955

253. Sharr MM, Kelvin FM. Vertebrobasilar aneurysm. Experience with 27 cases. Eur Neurol 10:129–143, 1973

254. Arseni C, Ghitescu N, Cristescu A, and Mihaila Gh: The pseudotumoral form of aneurysms of the posterior cranial fossa. Neurochirurgia 12:123–127, 1969

255. Thron A and Bockenheimer S: Giant aneurysms of the posterior fossa suspected as neoplasms on computed tomography. Neuroradiology 18:93–97, 1979

256. Coppeto JR and Chan YS: Abducens nerve paresis caused by unruptured vertebral artery aneurysm. Surg Neurol 18:385–387, 1982

257. Pia HW, Fontana H. Aneurysms of the posterior cerebral artery: locations and clinical pictures. Acta Neurochir 38:13–35, 1977

258. Drake CG and Amacher AL: Aneurysms of the posterior cerebral artery. J Neurosurg 30:468–474, 1969

259. Hanafee W and Jannetta PJ: Aneurysm as a cause of stroke. AJR 98:647–652, 1966

260. Mochimatsu Hook O and Norlen G: Aneurysms of the middle cerebral artery: A report of 80 cases. Acta Chir Scand 235 (Suppl):5–39, 1959

261. Fujitsu K, Hayashi A, and Inada Y: Giant aneurysm of the distal posterior cerebral artery: Case report. Neurol Med Chir 27:214–217, 1987

262. Drake CG. The treatment of aneurysms of the posterior circulation. Clin Neurosurg 26:96–144, 1979

263. Hook O, Norlen G, Guzman J. Saccular aneurysms of the vertebral-basilar arterial system. a report of 28 cases. Acta Psychiatr Scand 39:271, 1963

264. Winn HR, Berga SL, Richardson AE, et al. The natural history of vertebral basilar aneurysms. Stroke 12:120, 1981

265. Yu YL, Moseley IF, Pullicino P, McDonald WI. The clinical picture of ectasia of the intracerebral arteries. J Neurol Neurosurg Psychiatr 45:29–36, 1982

266. Amacher AL, Drake CG. Cerebral artery aneurysms in infancy, childhood and adolescence. Childs Brain 1:71–80, 1975

267. Nishizaki T, Tamaki N, Takeda N, et al. Dolichoectatic basilar artery: A review of 23 cases. Stroke 17:1277–1281, 1986

268. Smoker WRK, Corbett JJ, Gentry LR, Keyes WD, Price MJ, and McKusker S: High resolution computed tomography of the basilar artery: 2. Vertebrobasilar dolichoectasia: Clinical- pathologic correlation and review. AJNR 7:61–72, 1986

269. Fisher CM, Caplan LR. Basilar artery branch occlusion: a cause of pontine infarction. Neurology 21:900–905, 1971

270. Pessin MS, Chimowitz MI, Levine SR, Kwan ES, et al. Stroke in patients with fusiform vertebrobasilar aneurysms. Neurology 39:16–21, 1989

271. Huang C, Chan YW, Wang R. Senile dementia and hydrocephalus due to carotid dolichoectasia. Clin Exp Neurol 19:171–176, 1983

272. Castleman B, and Towne VW: Case records of the Massachusetts General Hospital. Case 39451. N Engl J Med 249:776–779, 1953

273. Johnsen SD, Okamoto G, Kooiker J. Fusiform basilar artery aneurysm in a child. Neurology 27:334–336, 1977

274. Echiverri HC, Rubino FA, Gupta SR, and Gujrati M: Fusiform aneurysm of the vertebrobasilar arterial system. Stroke 20:1741–1747, 1989

275. Gibson WP, Wallace D. Basilar artery ectasia. An unusual cause of a cerebello-pontine lesion with hemifacial spasm. J Laryngol Otol 89:721–731, 1975

276. Boeri R, Passerini A. The megadolichobasilar anomaly. J Neurol Sci 1:475–484, 1964

277. Scuccimarra A, Russo A, and Cafarelli F: Paralysie du III par megadolichoartere basilaire. Neurochirurgie 34:137–138, 1988

278. Durand JR, Samples JR. Dolichoectasia and cranial nerve palsies. A case report. J Clin Neuro-ophthalmol 9:249–253, 1989

279. Slavin ML, LoPinto RJ. Isolated environmental tilt associated with lateral medullary compression by dolichoectasia of the vertebral artery. Is there a cause and effect relationship? J Clin Neuro-ophthalmol 7:29–33, 1987

280. Hayes WT, Bernhardt H, Young JM. Fusiform arteriolsclerotic aneurysm of the basilar artery. Five cases including two ruptures. Vasc Surg 1:171–178, 1967

281. Ley A. Compression of the optic nerve by a fusiform aneurysm of the carotid artery. J Neurol Neurosurg Psychiatr 13:75–86, 1950

282. Knapp A: Association of sclerosis of the cerebral basal vessels with optic atrophy and cupping. Arch Ophthalmol 8:637–648, 1932

283. Knapp A: Course in certain cases of atrophy of the optic nerve with cupping and low tension. Arch Ophthalmol 23:41–47, 1940

284. McLean JM and Ray BS: Soft glaucoma and calcification of the internal carotid arteries. Arch Ophthalmol 38:154–158, 1947

285. de los Reyes RA, Boehm FH, Spatola MA, Goodrich JT, Moser FG, Llena JF, Spiro AJ. Treatment of dolichoectasia of the middle cerebral artery with bypass and exclusion. NY State J Med 89:629–633, 1989

286. Matsuo K, Kobayashi S, and Sugita K: Bitemporal hemianopsia associated with sclerosis of the intracranial internal carotid arteries. J Neurosurg 53:566–569, 1980

287. von Wild K and Busse H: Neurovaskuläre Kompression des N. opticus als Ursache fortschreitender Sehfunktionsstörungen und ihre Dekompressionsbehandlung. Klin Monatsbl Augenheilkd 196:460–465, 1990

288. Hilton GF, Hoyt WF. An arteriosclerotic chiasmal syndrome. Bitemporal hemianopia associated with fusiform dilatation of the anterior cerebral arteries. JAMA 196:1018– 1020, 1966

289. Read D, Esiri MM. Fusiform basilar artery aneurysm in a child. Neurology 29:1045–1049, 1979

290. Sacks JG, Lindenburg L. Dolicho-ectatic intracranial arteries: symptomatology and pathogenesis of arterial elongation and distension. Hopkins Med J 125:95–106, 1969

291. Frank E, Brown BM, and Wilson DF: Asymptomatic fusiform aneurysm of the petrous carotid artery in a patient with von Recklinghausen's neurofibromatosis. Surg Neurol 32:75–78, 1989

292. Benson EP. Radiation injury to large arteries: 3 further examples with prolonged asymptomatic intervals. Radiology 106:195–197, 1973

293. McCready RA, Hyde GL, Bivins BA, et al. Radiation-induced arterial injuries. Surgery 93:306–312, 1973

294. Handa J and Handa H: Severe epistaxis caused by traumatic aneurysm of cavernous and carotid artery. Surg Neurol 5:241–243, 1976

295. Wang AN, Winfield JA, and Gucer G: Traumatic internal carotid artery aneurysm with rupture in the sphenoid sinus. Surg Neurol 25:77–81, 1986

296. Simpson RK Jr, Harper RL, and Bryan RN: Emergency balloon occlusion for massive epistaxis due to traumatic carotid-cavernous aneurysm: Case report. J Neurosurg 68:142–144, 1988

297. Hannesson B, Sachs E Jr. Mycotic aneurysms following purulent meningitis. Report of a case with recovery and review of the literature. Acta Neurochir 24:305–313, 1971

298. Roach MR, Drake CG. Ruptured cerebral aneurysms caused by micro-organisms. NEJM 273:240–244, 1965

299. Bohmfalk GL, Story JL, Wissinger JP, Brown WE. Bacterial intracranial aneurysm. J Neurosurg 48:369–382, 1978

300. Horten BC, Abbott GF, Porro RS. Fungal aneurysms of intracranial vessels. Vasc Surg 4:570–577, 1976

301. Maass U: Die Syphilis als häufigste Ursache der Aneurysmen an der Gehirnbasis. Beitr Pathol Anat 98:307–322, 1937

302. Mitchell N and Angrist A: Intracranial aneurysms – a report of thirty six cases. Ann Intern Med 19:909–921, 1943

303. Suwanwela C, Suwanwela N, Charuchinda S, Hongsaprabhas C. Intracranial mycotic aneurysms of extravascular origin. J Neurosurg 36:552–559, 1972

304. Morriss FH, Spock A. Intracranial aneurysm secondary to mycotic orbital and sinus infection. Am J Dis Child 119:357–362, 1970

305. Shibuya S, Igarashi S, Amo T, et al. Mycotic aneurysms of the internal carotid artery. Case report. J Neurosurg 44:105–108, 1976

306. Micheli F, Schteinschnaider A, Plaghos LL, Melero M, Mattar D, and Parera IC: Bacterial cavernous sinus aneurysm treated by detachable balloon technique. Stroke 20:1751–1754, 1989

307. Bingham WF. Treatment of mycotic intracranial aneurysms. J Neurosurg 46:428–437, 1977

308. Ziment I, Johnson BL. Angiography in the management of intracranial mycotic aneurysms. Arch Int Med 122:349–352, 1968

309. Mielke B, Weir B, Oldring D, Von Westrap C. Fungal aneurysm: case report and review of the literature. Neurosurgery 9:578–582, 1981

Chapter 7   Intracranial Vascular Malformations

1. Yasargil MG. Microneurosurgery IIIA. AVM of the brain, history, embryology, pathological considerations, hemodiagnostic studies, microsurgical anatomy. Thieme, New York, 1987

2. Mohr JP, Caplan LR, Melski JW, et al. The Har-

vard cooperative stroke registry: a prospective registry. Neurology 28:754–762, 1978

3. McCormick WF. Vascular disorders of nervous tissue: anomalies, malformations, and aneurysms. In: GH Bourne (ed) The structure and function of nervous tissue. Academic Press, New York, 1969

4. Courville CB. Pathology of the nervous system, 2nd edn. Pacific Press, Mountain View, 1945

5. Olivecrona H, Riives J. Arteriovenous aneurysms of the brain. Arch Neurol Psych 59:567–602, 1948

6. Perret G, Nishioka H. Arteriovenous malformations. an analysis of 545 cases of cranio-cerebral arteriovenous malformations and fistulae reported in the cooperative study. J Neurosurg 25:467–490, 1966

7. Drake CG, Friedman AH, Peerless SJ. J Neurosurg 64:1–10, 1966

8. Berenstein A, Lasjaunias P, Surgical Neuroangiography, Vol IV. Endovascular Treatment of Brain, Spinal Cord, and Spine Lesions. Springer-Verlag, Berlin, 1992

9. Houser OW, Baker HL, Svien HJ, Okazaki H. Arteriovenous malformations of the parenchyma of the brain. Radiology 109:83–90, 1973

10. Mackenzie ICK. The clinical presentation of the cerebral angioma. Brain 76:184–214, 1953

11. Murphy MJ. Long-term follow-up of seizures associated with cerebral arteriovenous malformations. Arch Neurol 42:477–479, 1985

12. Graf CJ, Perret GE, Torner JC. Bleeding from cerebral arteriovenous malformations as part of their natural history. J Neurosurg 58:331–337, 1983

13. Parker JC, Gaede JT. Occurrence of vascular anomalies in unilateral cerebral hypoplasia 'cerebral hemiatrophy.' Arch Pathol Lab Med 90:265–270, 1970

14. Miller NR, Newman SA. Transynaptic degeneration. Arch Ophthalmol 99:1654, 1981

15. Hoyt WF, Rios-Montenegro EN, Behrens MM, Eckelhoff RJ. Homonymous hemioptic hypoplasia. Br J Ophthalmol 56:537–545, 1972

16. Tychsen L, Hoyt WF. Relative afferent pupillary defect in congenital occipital hemianopia. Am J Ophthalmol 100:345–346, 1985

16a. Hoyt WF. Congenital occipital hemianopia. Neuro-ophthalmol 2:252–259, 1985

17. Willinsky R, Lasjaunias P, Terbrudge K, Burrows: Multiple cerebral arteriovenous malformations. Review of our experience from 203 patients with cerebral vascular lesions. Neuroradiology 32, 207–210, 1990

18. Crawford PM, West CR, Chadwick DW, Shaw MDM. Arteriovenous malformations of the brain: natural history in unoperated patients. J Neurol Neurosurg Psychiatr 49:1–10, 1986

19. Paterson JH, McKissock W. A clinical survey of intracranial angiomas with special reference to their mode of progression and surgical treatment. Brain 79:233–266, 1956

20. Miyasaka K, Wolpert SM, Prager RJ. The association of cerebral aneurysms, infundibula and intracranial arteriovenous malformations. Stroke 13:196–203, 1982

21a. Lasjaunias P, Piske R, Terbrugge K, Willinsky R : Cerebral arteriovenous malformations and associated arterial aneurysms. Acta Neurochir 91, 29–36, 1988

21. Okamoto S, Hand H, Hashimoto N. Location of intracranial aneurysms associated with cerebral arteriovenous malformations: statistical analysis. Surg Neurol 22:335–340, 1984

22. Brown RD, Wiebers DO, Forbes GS. Unruptured intracranial aneurysms and arteriovenous malformations: frequency of intracranial hemorrhage and relationship of lesions. J Neurosurg 73:859–863, 1990

23. Horman RW, Devous MD, Stokely EM, Bonte FJ. Quantification of intracerebral steal in patients with arteriovenous malformations. Arch Neurol 43:770–785, 1986

24. Newton TH, Cronqvist S. Involvement of dural arteries in intracranial arteriovenous malformations. Radiology 93: 1071–1078, 1969

25. Faria MA, Fleischer AS. Dual cerebral and meningeal supply to giant arteriovenous malformations of the posterior cerebral hemisphere. J Neurosurg 52:153–161, 1980

26. Mawad ME, Hilal SK, Michelson WJ et al. Occlusive disease associated with cerebral arteriovenous malformations. Radiology 153: 401–408, 1984

27. Pile-Spellman J, Baker KF, Lisczak TM, Sandrew BB, et al. High-flow angiopathy: cerebral blood vessel changes in experimental chronic arteriovenous fistula. AJNR 7:811, 19

28. Folkman J, Klagsbrun M. Angiogenic factors. Science 235:442–447, 1987

29. McCormick WF. The pathology of vascular ('arteriovenous') malformations. J Neurosurg 24:807–816, 1966

30. Takashima S, Becker LE. Neuropathology of cerebral arteriovenous malformations in children. J Neurol Neurosurg Pschiatr 43:380–385, 1980

31. Garcia JH, Williams JP, Tanaka J. Spontaneous thrombosis of deep cerebral veins: a complication of arteriovenous malformation. Stroke 6:164–171, 1975

32. Lees F. The migrainous symptoms of cerebral angiomata. J Neurol Neurosurg Psychiatr 25: 45–50, 1962

33. Troost BT, Newman TH. Occipital lobe arteriovenous malformations: clinical and radiologic features in 26 cases with comments on differentiation from migraine. Arch Ophthalomol 93:250–256, 1975

34. Logue V, Monckton G. Posterior fossa angiomas. A

clinical presentation of nine cases. Brain 77: 252–273, 1954

35. Troost BT, Mark LE, Maroon JC. Resolution of classic migraine after removal of an occipital lobe AVM. Ann Neurol 5:199–201, 1979

36. Kattah JC, Luessenhop AJ, Kolsky M, Ferraz F. Removal of occipital arteriovenous malformations with sparing of visual fields. Arch Neurol 38: 307–309, 1981

37. Martin NA, Wilson CB. Medial occipital arteriovenous malformations. Surgical treatment. J Neurosurg 56:798–802, 1982

38. Baruk H. Migraines d'apparence psychogenique suivies d'epilepsie jacksonienne dans un cas d'angiome cerebrale. Les troubles fonctionnels initiaux dans certaines atteintes organiques cerebrales en particulier dans les tumeurs cerebrales. Encephale 26:42–44, 1931

39. Hyland HH, Douglas RP. Cerebral angioma arteriale, a case in which migrainous headache was the earliest manifestation. Arch Neurol Psychiatr 40:1220–1232, 1938

40. Deutsch L, Friedman J. Über ophthalmische Migräne bei Gefäßmißbildung (arteriovenöse Aneurysmen). Dtsch Z Nervenheilk 146:199, 1938

40a. Fritz WL, Klein HJ, and Schmidt K: Arteriovenous malformation of the posterior (sic) ethmoidal artery as an unusual cause of amaurosis fugax. The ophthalmic steal syndrome. J Clin Neuroophthalmol 9:165–168, 1989

41. Kupersmith MJ, Warren FA, Hass WK. The nonbenign aspects of migraine. Neuro-ophthalmology 7:1–10, 1987

42. Walsh F, Hoyt WF. Clinical Neuro-ophthalmology. Williams & Wilkins, Baltimore, 1969, p 1708

43. Forster DMC, Steiner L, Hakanson S. Arteriovenous malformations of the brain. A long-term clinical study. J Neurosurg 37:562–570, 1972

44. Brown RD, Wiebers DO, Forbes G, et al. The natural history of unruptured intracranial arteriovenous malformations. J Neurosurg 68:352–357, 1988

45. Ondra SL, Troupp H, George ED, Schwab K. The natural history of symptomatic arteriovenous malformations of the brain: a 24-year follow-up assessment. J Neurosurg 73:387–391, 1990

46. Heros RC, Korosue K, Diebold PM. Surgical excision of cerebral arteriovenous malformations: late results. Neurosurgery 26:570– 578, 1990

47. Morello G, Broghi GP. Cerebral angiomas. a report of 154 personal cases and a comparison between the results of surgical excision and conservative management. Acta Neurochir 28:135–155, 1973

48. Luessenhop AJ, Presper JH. Surgical embolizaion of cerebral arteriovenous malformations through internal carotid and vertebral arteries. long term results. J Neurosurg 42:443–451, 1975

49. Olivecrona H, Ladenheim J. Congenital arteriovenous aneurysms of the carotid and vertebral arterial systems. Springer-Verlag, Berlin, 1957

50. Perret GE. Conservative management of inoperable arteriovenous malformations. In: Pia HW, Gleaave JRW, Grote E, Zierski J (eds) Cerebral angiomas: advances in diagnosis and therapy. New York, Springer-Verlag, 1975, pp 268–270

51. Pelletieri L, Carlsson GA, Grevsten S, et al. Surgical versus conservative treatment of intracranial arteriovenous malformations. A study in surgical decision making. Acta Neurochir (suppl) 29:1–36, 1980

52. Cannell DE, Botterell EH. Subarachnoid hemorrhage and pregnancy. Am J Ob Gyn 72:844–853, 1956

53. Cannell DE. Subarachnoid hemorrhage in pregnancy. Proc Roy Soc Med 52:950–952, 1959

54. Amias AG. Cerebral vascular disease in pregnancy. I. Haemorrhage. J Obstr Gynaecol 77:100–120, 1970

55a. Robinson JL, Hall CJ, Sedzimir CB. Subarachnoid hemorrhage in pregnancy. J Neurosurg 36:27–32, 1972

55b. Robinson JL, Hall CS, Sedzimir CB. Arteriovenous malformations, aneurysms, and pregnancy. J Neurosurg 41:63–70, 1974

56. Horton JC, Chambers WA, Lyons S, Adams RD, Kjelberg RN. Pregnancy and the risk of hemorrhage from arteriovenous malformations. Neurosurgery 27:867–872, 1990

57. Dias MS, Sekhar LN. Intracranial hemorrhage from aneurysms and arteriovenous malformations during pregnancy and puerperium. Neurosurgery 27:855–866, 1990

58. Batjer H, Samson D. Arteriovenous malformations of the posterior fossa. Clinical presentation, diagnostic evaluation, and surgical treatment. J Neurosurg 64:849–856, 1986

59. Svien HJ, McRae JA. Arteriovenous anomalies of the brain. Fate of patients not having definitive surgery. J Neurosurg 23:23–28, 1965

60. Guidetti B, Delitala A. Intracranial arteriovenous malformations. Conservative and surgical treatment. J Neurosurg 53:149–152, 1980

61. Kusske JA, Kelly WA. Embolization and reduction of the "steal" syndrome in cerebral arteriovenous malformations. J Neurosurg 40:313–321, 1974

62. Overbeeke JJ, Bosma NJ, Verdonck MM, Huffelen AC. Higher cortical disorders: an unusual presentation of an arteriovenous malformation. Neurosurgery 21:839–842, 1987

63. Silber MH, Sandok BA, Earnest F. Vascular malformations of the posterior fossa. clinical and radiologic features. Arch Neurol 44:965–969, 1987

64. Solomon RA, Stein BM. Management of arteriovenous malformations of the brain stem. J Neurosurg 64:857–864, 1986

65. McCormick WF, Hardner JM, Boulter TR. Vascu-

lar malformations ('angiomas') of the brain, with special reference to those occurring in the posterior fossa. J Neurosurg 28:241–251, 1968

66. Martin NA, Stein BM, Wilson CB. Arteriovenous malformations of the posterior fossa. In: Wilson CB, Stein BM (eds), Intracranial Arteriovenous Malformations. Williams & Wilkins, Baltimore, 1984

67. Hierons R. Brain-stem angiomas confirmed by arteriography: Relapsing symptoms and signs strongly suggestive of disseminated sclerosis. Proc Roy Soc Med 46:195–196, 1953

68. Stahl SM, Johnson KP, Malamud N. The clinical and pathological spectrum of brain-stem vascular malformations. Long-term course simulates multiples sclerosis. Arch Neurol 37:25–29, 1980

69. Lessell S, Ferris EJ, Feldman RG, Hoyt WF. Brain stem arteriovenous malformations. Arch Ophthal 86:255–259, 1971

70. Eisbrey AB, Hegarty WM. Trigeminal neuralgia and arteriovenous aneurysm of the cerebellopontine angle. J Neurosurg 13:647–649, 1956

71. Johnson MC, Salmon JH. Arteriovenous malformation presenting as trigeminal neuralgia. Case report. J Neurosurg 29:287–289, 1968

72. Janetta PJ. Vascular decompression in trigeminal neuralgia. In: Samii M, Janetta PJ (eds) The Cranial Nerves. New York, Springer-Verlag, 1981, pp 331–340

73. Scott BB, Seeger JF, Schneider RC. Successful evacuation of a pontine hematoma secondary to rupture of a pathologically diagnosed "cryptic" vascular malformation. JNS 39:104–108, 1973

73a. Teilmann K. Hemangiomas of the pons. Arch Neurol Psychiatr 69:208–223, 1953

74. Mangiardi JR, Epstein FJ. Brainstem haematomas: review of the literature and presentation of five new cases. J Neurol Neurosurg Pschiatr 51:966–976, 1988

75. Kushner MJ, Bressman SB. The clinical manifestations of pontine. Neurology 35:637–643, 1985

76. Arseni C, Stanciu M. Primary hematomas of the brainstem. Acta Neurochir (Wien) 28:323–330, 1973

77. Sand JJ, Biller J, Corbett JJ, Adams HP, Dunn V. Partial dorsal mesencephalic hemorrhages: report of three cases. Neurology 36:529–533, 1986

78. Wisoff JH, Berenstein A, Choi IS, Friedman D, Madrid M. Management of vein of Galen malformations. In: Marlin AE (ed) Concepts in pediatric neurosurgery, Vol 10. Karger, Basel, 1990, pp 137–155

79. Gold AP, Ransohoff J, Carter S. Vein of Galen malformations. Acta Neurol Scand 40:(suppl), 5–31, 1964

80. Amacher AL, Shillito J. The syndromes and surgical treatment of aneurysms of the great vein of Galen. J Neurosurg 39:89–98, 1973

81. Pool JL. Treatment of arteriovenous malformations of the cerebral hemispheres. J Neurosurg 19:136–141, 1962

82. Heros R, Yong-Kwang T. Is surgical therapy needed for unruptured arteriovenous malformations? Neurology 37:279–286, 1987

83. Axel L. Blood flow effects in magnetic resonance imaging. Am J Radiol 143:1157–1165, 1984

84. Gomori JM, Grossman RI, Goldberg HI, Zimmerman RA, Bilaniuk LT. Intracranial hematomas: Imaging by high-field MR. Radiology 157:87--93, 1985

85. Schlesinger S. MRI of intracranial hematomas and closed head trauma. In: Pomeranz S (ed) Cranial Magnetic Resonance Imaging. W.B. Saunders, Philadelphia, 1989, pp 399–435

86. Smith HJ, Strother CM, Kikuchi Y, et al. MR imaging in the management of supratentorial intracranial AVMs. AJNR 150:1143–1153, 1988

87. Viale GL, Pau A, Viale ES. Surgical treatment of arteriovenous malformations of the posterior fossa. Surg Neurol 12:379–384, 1979

88. McFerran DJ, Chir B, Marks PV, Garvan NJ. Angiographically occult arteriovenous malformations of the brainstem. Surg Neurol 28:221–224, 1987

89. Crawford JV, Russell DS. Cryptic arteriovenous and venous hamartomas of the brain. J Neurol Neurosurg Psychiatr 19:1–11, 1956

90a. Stein BM, Wolpert SM. Arteriovenous malformations of the brain. I. Current concepts and treatment. Arch Neurol 37:1–5, 1980

90b. Stein BM, Wolpert SM. Arteriovenous malformations of the brain. II. Current concepts and treatment. Arch Neurol 37:69–75, 1980

91. Sutton D. Discussion on the neuro-ophthalmological aspects of the cerebral angiomas. Proc Roy Soc Med 50:91–92, 1957

92. Enoksson P, Bynke H. Visual field defects in arteriovenous aneurysms of the brain. Acta Ophthalmol 36:586–599, 1958

93. Holtzman R, Goldensohn ES. Sensations of ocular movement in seizures originating in occipital lobe. Neurol 27:544–556, 1977

94. Michel EM, Troost BT. Palinopsia: cerebral localization with computed tomography. Neurology 30:887–889, 1980

95. Kinsbourne M, Warrington EK. A study of visual perseveration. J Neurol Neurosurg Psychiat 26:468–475, 1963

96. Bender MB, Feldman M, Sobin AJ. Palinopsia. Brain 91:321–338, 1968

97. Brust JC, Behrens MM. "Release hallucinations" as the major symptom of posterior cerebral artery occlusion: a report of 2 cases. Ann Neurol 2:432–436, 1977

98. Eretto PA, Schoen FS, Krohel GB, Pechette. Palinoptic visual allesthesia AJO 93:801–802, 1982

99. Critchely M. Types of visual perservation: "palinopsia" and "illusory visual spread." Brain 74:267–299, 1951

100. Lance JW. Simple formed hallucinations confined to the area of a specific visual defect. Brain 99:719–734, 1976

101. Weinberger LM, Grant FC. Visual hallucinations and their neuro-optical correlates. Arch Ophthalmol 23:166–199, 1941

102. Symonds C, Mackenzie. Bilateral loss of vison from cerebral infarction. Brain 80:415–455, 1957

103. Penfield W, Perot P. The brain's record of auditory and visual experience. Brain 86:595–696, 1963

104. Holtzman R, Goldensohn E. Sensations of ocular movement in seizures originating in occipital lobbe. Neurology 27:554–556, 1977

105. Jacobs L. Visual allesthesia. Neurology 30:1059–1063, 1980

106. Adie WJ. Permanent hemianopia in migraine and subarachnoid haemorrhage. Lancet 2:237–238, 1930

107. Hoyt WF. Vascular lesions of the visual cortex with brain herniation through the tentorial incisura: Neuro-ophthalmologic considerations. Arch Ophthalmol 64:44–57, 1960

108. Laurent A, Guimaraens L, Rufenacht D, Riche MC, Merland JJ. Five case of unilateral exophthalmos associated with abnormalities in the lateral sinus area. J Neuroradiol 13:125–136, 1986

109. Wyburn-Mason RP. Arteriovenous aneurysm of mid-brain and retina, facial naevi, and mental changes. Brain 66:163–203, 1943

110. Riishede J, Seedorff HH. Spontaneous hematoma of the optic chiasma. Acta Ophthalmologica 52:317–322, 1974

111. Sibony PA, Lessell S, Wray S. Chiasmal syndrome caused by arteriovenous malformations. Arch Ophthalmol 100:438–442, 1982

112. Maitland CG, Abiko S, Hoyt WF, Wilson CB, Okamura T. Chiasm apoplexy: report of four cases. J Neurosurg 56:118–122, 1982

113. Lavin PJM, McCrary JA, Roessmann U, Ellenberger C. Chiasmal apoplexy: hemorrhage from a cryptic vascular malformation in the optic chiasm. Neurol 34:1007–1011, 1984

114. Burnbaum MD, Harbison JW, Selhorst JB, Young HF. Blue-domed cyst with optic nerve compression. J Neurol Neurosurg Psychiatr 41:987–991, 1978

115. Roski RA, Gardner JH, Spetzler RF. Intrachiasmatic arteriovenous malformation. Case report. J Neurosurg 54:540–541, 1981

116. Fermaglich J, Kattah J, Manz H. Venous angioma of the optic chiasm. Ann Neurol 4:470–471, 1978

117. Dandy WE. Arteriovenous aneurysms of the brain. Arch Surg 17:190–243, 1928

118. Cushing HB, Bailey PB. Tumors arising from the blood vessels of the brain: Angiomatous malformations and hemangioblastomas. Springfield, Ill, Charles C Thomas Publisher, 1928

119. Spector RH, Davidoff RA, Schwartzman RJ. Phenytoin-induced ophthalmoplegia. Neurol 26:1031–1034, 1976

120. Askenasy H, Wijsenbeek H, Herzenberger E. Retraction nystagmus and retraction of eyelids due to arteriovenous aneurysm of midbrain. Arch Neurol Psychiatr 69:236–241, 1953

121. Walshe TM, Davis KR, Miller Fisher CM. Thalamic hemorrhage: a computed tomographic-clinical correlation. Neurol 27:217–222, 1977

122. Imes RK, Monteiro MLR, Hoyt WF. Ophthalmoplegic migraine with proximal posterior cerebral artery vascular anomaly. J Clin Neuro-ophthalmol 4:221–223, 1984

123. Bergstrand H, Olivecrona H, Tonnis W. Gefäßmißbildungen und Gefäßgeschwülste des Gehirns. Leipzig, Georg Thieme, 1936

124. Chimowitz MI, Little JR, Awad IA, et al. Intracranial hypertension associated with unruptured cerebral arteriovenous malformations. Ann Neurol 27:474–479, 1990

125. Kashii S, Solomon S, Moser FG, et al. Progressive visual field defects in patients with intracranial malformations. Am J Ophthalmol 109:556–562, 1990

126. Lussenhop AJ, Rosa L. Cerebral arteriovenous malformations. indications for and results of surgery, and the role of intravascular techniques. J Neurosurg 60:14–22, 1984

127. Heros RC, Tu YK. Unruptured arteriovenous malformations: a dilemma in surgical decision making. Clin Neurosurg 33:187–236, 1986

128. Fischer WS. Decision analysis: a tool of the future: an application to unruptured arteriovenous malformations. Neurosurgery 24:129–136, 1989

129. Berenstein A, Choi IS, Kupersmith M, Flamm E, Kricheff II, Madrid MM. Complications of endovascular embolization in 202 patients with cerebral AVMs. AJNR 10:876, 1989

130. Jafar J, Berenstein A, Kupersmith M, Miller D. The effects of presurgical embolization with NBCA on the operative excision of cerebral A-V malformations. American Association of Neurological Surgeons. April 20–25, 1991, New Orleans

131. Jack CR, Nichols DA, Sharbrough FW, Marsh WR, Petersen. Selective posterior cerebral artery amytal test for evaluating memory function before surgery for temporal lobe seizure. Radiology 168:787–793, 1988

132. Horton JA, Dawson RC. Retinal Wada test. AJNR 9:1167–1168, 1988

133. Debrun G, Vinuela F, Fox A, Drake CG. Embolization of cerebral arteriovenous malformations with bucrylate. J Neurosurg 56:615–627, 1982

134. Vinuela F, Fox AJ, Debrun G, Pelz D. Preembolization superselective angiography: role in the treat-

ment of brain arteriovenous malformations with isobutyl-2-cyanoacrylate. AJNR 5:765–769, 1984

135. Picard L, Moret J, Leporie J. Endovascular treatment of intracerebral arteriovenous angiomas. J Neuroradiol 11:9–28, 1984

136. Wolpert SM, Barnett FJ, Prager RJ. Benefits of embolization without surgery for cerebral arteriovenous malformations. AJNR 2:535–538, 1981

137. Kjelberg RN, Hanamura T, Davis KR, Lyons SL, Adams RD. Bragg-peak proton beam therapy for arteriovenous malformations of the brain. NEJM 309:269–274, 1983

138. Steinberg GK, Fabrikant JI, Marks MP, et al. Stereotactic heavy-charged-particle Bragg-peak radiation for intracranial arteriovenous malformations. NEJM 323:96–101, 1990

139. Steiner L, Leksell L, Forster DMC, Greitz T, Backlund EO. Stereotactic radiosurgery in intracranial arterio-venous malformations. Acta Neurochir (suppl)(Wien) 21:195–209, 1974

140. Leksell DG. Stereotactic radiosurgery: present status and future trends. Neurol Res 9:60–68, 1987

140a. Steiner L, Lindquist C, Adler JR et al: Clinical outcome of radiosurgery for cerebral arteriovenous malformations. J Neurosurg 77:1–8, 1992

141. Drake CG. Cerebral arteriovenous malformations: considerations for experience with surgical treatment in 166 cases. Clin Neurosurg 26:145–208, 1979

142. Nornes H, Lundar T, Wikeby P. Cerebral arteriovenous malformations: results of microsurgical management. Acta Neurochir 50:243–257, 1979

143. Aminoff MJ. Treatment of unruptured cerebral arteriovenous malformations. Neurology 37:815–819, 1987

144. Spetzler RF, Wilson CB, Weinstein P, et al. Normal perfusion pressure break-through therory. Clin Neurosurg 25:651–672, 1978

145. Mullan S, Brown FD, Patronas NJ. Hyperemic and ischemic problems of surgical treatment of arteriovenous malformations. J Neurosurg 51:757–764, 1979

146. Morgan MK, Johnston I, Besser M, Baines D. Cerebral arteriovenous malformations. J Neurosurg 66:563–567, 1987

147. Spetzler RF, Martin RNA, Carter LP, Flam RA, Raudzens PA, et al. Surgical management of large AVMs by staged embolization and operative excision. J Neurosurg 67:17–28, 1987

148. Simard JM, Garcia-Bengochea F, Ballinger WF, Mickle JP, Quisling RG. Cavernous angioma: a review of 126 collected and 12 new clinical cases. Neurosurgery 18:162–172, 1986

149. Voigt K, Yasargil MG. Cerebral cavernous haemangiomas or cavernomas: Incidence, pathology, localization, diagnosis, clinical features and treatment. Review of the literature and report of an un-usual case. Neurochirurgia (Stuttg) 19:59–68, 1976

150. Bicknell JM, Carlow TJ, Kornfeld M, et al. Familial cavernous angiomas. Arch Neurol 35:746–749, 1978

151. Clark JV. Familial occurrence of cavernous angiomata of the brain. J Neurol Neurosurg Psych 33:871–876, 1970

152. Dobyns WB, Michels VV, Groover RV, Mokri B, Trautmann JC, Forbes GS, Laws ER. Familial cavernous malformations of the central nervous system and retina. Ann Neurol 21:578–583, 1987

153. Andermann E, Staunton T, Andermann F, et al. Familial cavernous angioma an autosomal dominant disorder presenting with epilepsy. Am J Hum Genet 36:415, 1984

154. Hayman LA, Evans RA, Ferrell RE, et al. Familial cavernous angiomas. natural history and genetic study over a 5-year period. Am J Med Genet 11:147–160, 1982

155. Vaquero J, Leunda G, Martinez R, Bravo G. Cavernomas of the brain. Neurosurgery 12:208–210, 1983

156. Lance JW, Smee RI. Partial seizures with visual disturbance treated by radiotherapy of cavernous hemangioma. Ann Neurol 26:782–785, 1989

157. Giombi S, Morrello G. Cavernous angiomas of the brain: Account of fourteen personal cases and review of the literature. Acta Neurochir (Wien) 40:61–82, 1978

158. Roda JM, Alvarez F, Isla A, Blazquez MG. Thalamic cavernous malformation. Case report. J Neurosurg 72:647–649, 1990

159. Hassler W, Zentner J, Wilhelm H. Cavernous angiomas of the anterior visual pathways. J Clin Neuro-ophthalmol 9:160–164, 1989

160. Harper DG, Buck DR, Early CB. Visual loss from cavernous hemangiomas of the middle cranial fossa. Arch Neurol 39:252–254, 1982

161. Mizutani T, Goldberg HI, Kersonl LA, Murtagh F. Cavernous hemangioma in the diencephalon. Arch Neurol 38:379–382, 1981

162. Manz HJ, Klein LH, Fermaglich J, et al. Cavernous hemangioma of optic chiasm, optic nerves, and right optic tract. Virch Arch Path Anat 383:225–231, 1979

163. Maruoka N, Yamakawa Y, Shimauchi M. Cavernous hemangioma of the optic nerve. Case report. J Neurosurg 69:292–294, 1988

164. Matias-guiu X, Alejo M, Sole T, et al. Cavernous angiomas of the cranial nerves. Report of two cases. J Neurosurg 73:620–622, 1990

165. Mohr G, Hardy J, Gauvin P. Chiasmal apoplexy due to ruptured cavernous hemangioma of the optic chiasm. Surg Neurol 24:636–640, 1985

166. Honegger J, Fahlbusch R, Lieb W, et al. Cavernous hemangioma of the optic chiasm. Neuro-ophthalmology 10:81–88, 1990

167. Rigamonti D, Drayer BP, Johnson PC, Hadley MN,

Zabramski J, Spetzler RF. The MRI appearance of cavernous malformations (angiomas). J Neurosurg 67:518–524, 1987

168. New PFJ, Ojemann RG, Davis KR, et al. MR and CT of vascular malformations of the brain. AJNR 7:771–779, 1986

169. Farmer JP, Cosgrove GR, Villemure JG, Meagher-Villemure K, Tampieri D, Melanson D. Intracerebral cavernous angiomas. Neurology 38:1699–1704, 1988

170. Little JR, Awad IA, Jones SC, Ebrahim ZY. Vascular pressures and cortical blood in cavernous angioma of the brain. J Neurosurg 73:555–559, 1990

171. Vaquero J, Salazar J, Martinez R, Martinez P, Bravo G. Cavernomas of the central nervous system: clinical syndromes, CT scan diagnosis, and prognosis after surgical treatment in 25 cases. Acta Neurochirurgia 85:39–33, 1987

172. Yamaski T, Handa H, Yamashita J, et al. Intracranial and orbital cavernous angiomas. J Neurosurg 64:197–208, 1986

173. Lasjaunias P, Burrows P, Planet C: Developmental venous anomalies: the so-called venous angiomas. Neurosurg Rev 9, 233–244, 1986

174. Yasargil MG. Microsurgery IIIB. George Thieme Verlag, Stuttgart, 1988, pp 405–438

175. Sarwar M, McCormick WR. Intracerebral venous angiomas: case report and review. Arch Neurol 35:323–325, 1978

176. Rothfus WE, Albright AL, Casey KF, Latchaw RE, Roppolo HMN. Cerebellar venous angioma: "benign" entity? AJNR 5:61–66, 1984

177. Numaguchi Y, Kitamura K, Fukui M, Ikeda J, Kishikawa T, Okudera T, Uemura K, Matsuura K. Intracranial venous angiomas. Surg Neurol 18:193–202, 1982

178. Rigamonti D, Spetzler RF, Medina M, Rigamonti K, Geckle DS, Pappas C. Cerebral venous malformation. J Neurosurg 73:560–564, 1990

Chapter 8   Embolic and Noninflammatory Occlusive Diseases

1. Ashton N, Coomes EN, Garner A, Oliver DO: Retinopathy due to progressive systemic sclerosis. J Pathol Bacteriol 96:259–268, 1968

2. Ashton N, Dollery CT, Henkind P, et al: Focal retinal ischaemia. Br J Ophthalmol 50:283–389, 1966

3. Dollery CT: Microcirculatory changes and the cotton-wool spot. Proc R Soc Med 62:1267–1269, 1969

4. McLeod D, Marshall J, Kohner EM, Bird AC: The role of axoplasmic transport in the pathogenesis of retinal cotton-wool spots. Br J Ophthalmol 61:177–191, 1977

5. Gaudric A, Coscas G, DeMargerie J: Acute sectorial choroidal ischemia. Am J Ophthalmol 98:707–716, 1984

6. Sacks JG and Choromokos E: Ophthalmodynamometry reveals retinal embolus. Ann Ophthalmol 15:413–414, 1983

7. Walsh FB and Goldberg HK: Blindness due to air embolism: A complication of extrapleural pneumolysis. JAMA 114:654–655, 1940

8. Hollenhorst RW: Ocular manifestations of insufficiency or thrombosis of the internal carotid artery. Am J Ophthalmol 47:753–767, 1959

9. Hollenhorst RW: Significance of bright plaques in the retinal arterioles. JAMA 178:23–29, 1961

10. Arruga J, Sanders MD: Ophthalmologic findings in 70 patients with evidence of retinal embolism. Ophthalmology 89:1336–1347, 1982

11. Henkind P: Ballotini occlusion of retinal arteries. Collateral vessels. Br J Ophthalmol 50:482–495, 1966

12. David NJ, Klintworth GK, Friedberg SJ, et al: Fatal atheromatous cerebral embolism associated with bright plaques in the retinal arterioles: Report of a case. Neurology 13:708–713, 1963

13. Fisher CM: Observations of the fundus oculi in transient monocular blindness. Neurology 9:333–347, 1959

14. Cogan DG and Wray SH: Vascular occlusions in the eye from cardiac myxomas. Am J Ophthalmol 80:396–403, 1975

15. Goulet Y, Mackay CG: Atheromatous embolism: An entity with polymorphous symptomatology. Can Med Assoc J 88:1067–1070, 1963

16. Darsee JR: Cholesterol embolism: The great masquerader. South Med J 72:174–180, 1979

17. Coppeto JR, Lessell S, Lessell I, Greco TP, Eisenberg MS: Diffuse disseminated atheroembolism. Three cases with neuro-ophthalmic manifestations. Arch Ophthalmol 102:225–228, 1984

18. Jampol LM, Wong SW, Albert DM: Atrial myxoma and central retinal artery occlusion. Am J Ophthalmol 75:242–249, 1973

19. Sandok BA, Estorff IV, Giuliani ER: CNS embolism due to atrial myxoma. Clinical features and diagnosis. Arch Neurol 37:485–488, 1980

20. Sandok BA, Estorff IV, Giuliani ER: Subsequent neurological events in patients with atrial myxoma. Ann Neurol 8:305–307, 1980

21. Kooiker JC, MacLean JM, Sumi SM: Cerebral embolism, marantic endocarditis, and cancer. Arch Neurol 33:260–264, 1977

22. Reagan TJ: Cerebral ischemia in nonbacterial thrombotic endocarditis. Current Concepts of Cerebrovascular Disease. Stroke 10:13–17, 1975

23. Williams IM: Retinal vascular occlusions in open heart surgery. Br J Ophthalmol 59:81–91, 1975

24. Caltrider ND, Irvine AR, Kline HJ, Rosenblatt A: Retinal emboli in patients with mitral valve prolapse. Am J Ophthalmol 90:534–539, 1980

25. Rush JA, Kearns TP, Danielson GK: Cloth-particle retinal emboli from artificial cardiac valves. Am J Ophthalmol 89:845–850, 1980

26. Blauth C, Arnold J, Kohner EM, and Taylor KM: Retinal microembolism during cardiopulmonary bypass demonstrated by fluorescein angiography. Lancet 2:837–839, 1986

27. Blauth CI, Arnold JV, Schulenberg WE, McCartney AC, and Taylor KM: Cerebral microembolism during cardiopulmonary bypass. Retinal microvascular studies in vivo with fluorescein angiography. J Thorac Cardiovasc Surg 95:668–676, 1988

28. Williams IM: Intravascular changes in the retina during open heart surgery. Lancet 2:688–691, 1971

29. Williams IM: Retinal vascular occlusions in open heart surgery. Br J Ophthalmol 59:81–91, 1975

30. Adams HP, Butler MJ, Biller J, Toffol GJ: Nonhemorrhagic cerebral infarction in young adults. Arch Neurol 43:793–796, 1986

31. Cerebral Embolism Task Force. Cardiogenic brain embolism. The second report of the cerebral embolism task force. Arch Neurol 46:727–743, 1989

32. Appen RE, Wray SH, Cogan DG. Central retinal artery occlusion. AJO 79:374–381, 1975

33. Chuang EL, Miller FS, Kalina RE: Retinal lesions following long bone fractures. Ophthalmology 92:370–374, 1985

34. Inkeles DM, Walsh JB: Retinal fat emboli as a sequela to acute pancreatitis. AJO 80:935–938, 1975

35. Percival SPB: The eye and Moschcowitz's (thrombotic thrombocytopenic purpura): A review of 182 cases. Trans Ophthalmol Soc UK 90:375–378, 1970

36. Murphy SB, Jackson WB, Pare JAP: Talc retinopathy. Canad J Ophthalmol 13:152–156, 1978

37. Friberg TR, Gragoudas ES, Regan CDJ: Talc emboli and macular ischemia in intravenous drug abuse. Arch Ophthalmol 97:1089–1091, 1979

38. Schatz H, Drake M: Self-injected retinal emboli. Ophthalmology 86:468–483, 1980

39. Tse DT, Ober RR: Talc retinopathy. AJO 90:624–640, 1980

40. Siepser SB, Magargal LE, Augsburger JJ: Acute bilateral retinal microembolization in a heroin addict. Ann Ophthalmol 13:699–702, 1981

41. Ellis PP: Occlusion of the central retinal artery after retrobulbar corticosteroid injection. Am J Ophthalmol 85:353–356, 1978

42. Byers B: Blindness secondary to steroid injection into the nasal turbinates. Arch Ophthalmol 97:79–80, 1979

43. Whitesman DW, Rosen DA, Pinkerton RMH: Retinal and choroidal microvascular embolism after intranasal corticosteroid injection. AJO 89:851–853, 1980

44. Shorr N, Seiff SR: Central retinal artery occlusion association with periocular corticosteroid injection for juvenile hemangioma. Ophthalmic Surg 17:229–231, 1986

45. Dreizen NG and Framm L: Sudden unilateral visual loss after autologous fat injection into the glabellar area. Am J Ophthalmol 107:85–87, 1989

46. Evans DE, Zahorchak JA, Kennerdell JS: Visual loss as a result of primary optic neuropathy after intranasal corticosteroid injection. AJO 90:641–644, 1980

47. Mames RN, Snady-McCoy L, Guy J. Central retinal and posterior ciliary artery occlusion after particle embolization of the external carotid artery system. Ophthalmology 98:527–531, 1991

48. McLean EB: Inadvertent injection of corticosteroid into the choroidal vasculature. Am J Ophthalmol 80:835–837, 1975

49. Aasved H: Visual disturbances following intracranial angiography. Acta Ophthalmol 40:297–302, 1962

50. Stefansson E, Coin JT, Lewis WR, Belkin RN, Behar VS, Morris JJ, Anderson WB: Central retinal artery occlusion during cardiac catheterization. Am J Ophthalmol 99:586–589, 1985

51. Francois J, Goes F, Stockmans L: Embolie de l'artere centrale de la retine apres angiographie carotidienne. Ann Ocul 205:993–1004, 1972

52. Carlson MR, Pilger IS, Rosenbaum AL: Central retinal occlusion after carotid angiography. Am J Ophthalmol 81:103–106, 1976

53. Johnson LN, Krohel GB, Hong YK, Wood G: Central retinal artery occlusion following transfemoral cerebral angiography. Ann Ophthalmol 17:359–362, 1985

54. Goodwin JA, Gorelick PB, Helgason CM: Symptoms of amaurosis fugax in atherosclerotic carotid artery disease. Neurology 37:829–832, 1987

55. Fisher CM: Transient monocular blindness versus amaurosis fugax. Neurology 39:1622–1624, 1989

56. Burnbaum MD, Selhorst JB, Harbison JW, Brush JJ: Amaurosis fugax from disease of the external carotid artery. Arch Neurol 34:532–535, 1977

57. Barnett HJM, Peerless SJ, Wei M: The "stump" of the internal carotid artery: A source for further cerebral embolic ischemia. Stroke 8:14–15, 1977

58. Bogousslavsky J, Regli F, Hungerbuhler JP, Chrzanowski R: Transient ischemic attacks and external carotid artery. A retrospective study of 23 patients with an occlusion of the internal carotid artery. Stroke 12:627–630, 1981

59. Picard L, Vignaud J, Lombardi G, et al: Radiological anatomy of the origin of the ophthalmic artery. Mod Probl Ophthalmol 14:164–169, 1975

60. Weinberg PE, Patronas NJ, Kim KS, Melen O: Anomalous origin of the ophthalmic artery in a patient with amaurosis fugax. Arch Neurol 38:315–317, 1981

61. Cogan DG: Blackouts not obviously due to carotid occlusion. Arch Ophthalmol 66:180–187, 1961

62. Tippin J, Corbett JJ, Kerber RE, Schroeder E, Thompson HS: Amaurosis fugax and ocular infarc-

tion in adolescents and young adults. Ann Neurol 26:69–77, 1989

62a. Burger SK, Saul RF, Selhorst JB, Thurston SE. Transient monocular blindness caused by vasospasm. NEJM 325:870–873, 1991

63. Nicholson D: Occult iris erosion: A treatable cause of recurrent hyphema in iris-supported intraocular lenses. Ophthalmology 89:115–119, 1982

64. Magargal LE, Goldberg RE, Uram M, Gonder JR, and Brown GC: Recurrent microhyphema in the pseudophakic eye. Ophthalmology 90:1231–1234, 1983

65. Jacobson DM, Glaser P: Recurrent transient monocular visual impairment caused by hyphema following intra-ocular surgery. Neurology 38:163, 1988

66. Coppeto JR, Kawalick M: Ocular pseudomigraine after posterior chamber intra-ocular lens implantation. Am J Ophthalmol 102:393–394, 1986

67. Degenhart W, Brown GC, Augsburger JJ, Magargal L: Precapillary vascular loops. Ophthalmology 88:1126, 1981

68. Epstein ED: Reversible unilateral lens opacities in a diabetic patient. Arch Ophthalmol 94:461–463, 1976

69. Paylor RR, Selhorst JB, Weinberg RS: Reversible monocular cataract simulating amaurosis fugax. Ann Ophthalmol 17:423–425, 1985

70. Manschot WA: Embolism of the central retinal artery. Am J Ophthalmol 48:381–385, 1959

71. Zimmerman L: Embolism of the central retinal artery. Arch Ophthalmol 73:822–826, 1965

72. Dahrling BE: The histopathology of early central retinal artery occlusion. Arch Ophthalmol 73: 506–510, 1965

73. Perkins SA, Magargal LE, Maizel RD, Robb-Doyle E: Resolved incomplete central retinal-artery obstruction simulating ischemic optic neuropathy. Ann Ophthalmol 20:61–67, 1988

74. Karjalainen K: Occlusion of the central retinal artery and retinal branch arterioles. A clinical, tonographic, and fluorescein angiographic study of 175 patients. Acta Ophthalmologica Suppl 109:9–96, 1971

75. Von Graefe A: Ueber Embolie der arteria centralis retinae als Ursache plötzlicher Erblindung. Albrecht von Graefe's Arch Ophthalmol 5:136–140, 1859

76. Augsburger JJ, Magargal LE: Visual prognosis following treatment of acute central retinal artery obstruction. Br J Ophthalmol 64:913–917, 1980

77. Liversedge LA, Smith WH. Neuromedical and ophthalmic aspects of central retinal artery occlusion. Trans Ophthalmol Soc UK 82:571–588, 1962

78. David NJ, Gilbert DS, Gass JDM: Fluorescein angiography in retinal arterial branch obstructions. AJO 69:43–55, 1970

79. David NJ, Norton EWD, Gass JDM, Beauchamp J: Fluorescein angiography in central retinal artery occlusion. Arch Ophthalmol 77:619–629, 1967

80. Henkind P, Wise GN: Retinal neovascularization, collaterals, and vascular shunts. Br J Ophthalmol 58:413–422, 1974

81. Hayreh SS, Podhajsky P: Ocular neovascularization with retinal vascular occlusion. II. Occurrence in central and branch retinal artery occlusion. Arch Ophthalmol 100:1585–1596, 1982

82. Duker JS, Brown GC: Iris neovascularization associated with obstruction of the central retinal artery. Ophthalmology 95:1244–1250, 1988

83. Duker JS, Sivalinagam A, Brown GC, Reber R. A prospective study of acute central retinal artery obstruction. Arch Ophthalmol 109:339–342, 1991

84. Duker JS, Brown GC: Neovascularization of the optic disc associated with obstruction of the central retinal artery. Ophthalmol 96:87–91, 1989

85. Fisher CM: Occlusion of the interal carotid artery. Arch Neurol Psych 65:346–377, 1951

86. Fisher CM: Transient monocular blindness associated with hemiplegia. Arch Ophthalmol 47: 167–203, 1952

87. Millikan CH, Siekert TG: Studies in cerebrovascular disease 4. The syndrome of intermittent insufficiency of the carotid arterial system. Proc Staff Meeting Mayo Clin 30:186–191, 1955

88. Kollarits CR, Lubow M, Hissong SL: Retinal strokes. I. Incidence of carotid atheromata. JAMA 222:1273–1275, 1972

89. Chawluk JB, Kusher MJ, Bank WJ, Silver FL, Jamieson DG, Bosley TM, Conway DJ, Cohen D, Savino PJ: Neurology 38:858–863, 1988

90. Merchut MP, Gupta SR, Naheedy MH: The relation of retinal artery occlusions and carotid artery stenosis. Stroke 19:1239–1242, 1988

91. Slepyan DH, Rankin RM, Stahler C, Gibbons GE: Amaurosis fugax: A clinical comparison. Stroke 6:493–496, 1975

92. Harrison MJG, Marshall J: Arteriographic comparison of amaurosis fugax and hemispheric transient ischaemic attacks. Stroke 16:795–707, 1985

93. Adams HP, Putnam AF, Corbett JJ, et al: Amaurosis fugax: The results of arteriography in 59 patients. Stroke 14:742–744, 1983

94. Pessin MS, Ducan GW, Mohr JP, Poskanzer DC: Clinical and angiographic features of carotid transient ischemic attacks. NEJM 296:358–362, 1977

95. Sheng FC, Quinones-Baldrich W, Machleder H, Moore WS, Baker JD, and Busuttil RW: Relationship of extracranial carotid occlusive disease and central retinal artery occlusion. Am J Surg 152:175–178, 1986

96. Hurwitz BJ, Heyman A, Wilkinson WE, Haynes CS, Utley CM: Comparison of amaurosis fugax and transient cerebral ischemia: A prospective clinical and arteriographic study. Ann Neurol 18:698–704, 1985

97. Marshall J, Meadows S: The natural history of amaurosis fugax. Brain 91:419–434, 1968

98. Hosshmand H, Vines FS, Lee HM, et al: Amaurosis fugax: Diagnostic and therapeutic aspects. Stroke 5:643–647, 1974

99. Toole JF, Yuson CP, Janeway R, et al: Transient ischemic attacks: A prospective study of 225 patients. Neurology 28:746–753, 1978

100. Autret A, Saudeau D, Bertrand P, Pourcelot L, Marchal C, de Boisvilliers S: Stroke risk in patients with carotid stenosis. Lancet 1:888–890, 1987

101. Pfaffenbach DD, Hollenhorst RW: Morbidity and survivorship of patients with embolic cholesterol crystals in the ocular fundus. Am J Ophthalmol 75:66–72, 1973

102. Schwarcz TH, Eton D, Ellenby MI, et al: Hollenhorst plaques: Retinal manifestations and the role of carotid endarterectomy. J Vasc Surg 11:635–641, 1990

103. Bunt TJ: The clinical significance of the asymptomatic Hollenhorst plaque. J Vasc Surg 4:559–562, 1986

104. Jorgensen R, Towne JB, Kay M, Bandyk DF: Monocular ischemia – the influence of carotid atherosclerosis versus primary ocular disease on prognosis. Surgery 104:507–511, 1988

105. Lorentzen SE: Occlusion of the central retinal artery. A follow up. Acta Ophthalmol 47:690–703, 1969

106. Savino PJ, Glaser JS, Cassady J. Retinal stroke. Is the patient at risk? Arch Ophthalmol 95:1185–1189, 1977

107. Whisant JP: Does carotid endarterectomy decrease stroke and death in patients with transient ischemic attacks? Ann Neurol 22:72–76, 1987

108. Hollenhorst RW: Vascular status of patients who have cholesterol emboli in the retina. Am J Ophthalmol 61:753–767, 1966

109. Pekkanen J, Linn S, Heiss G, et al: Ten-year mortality from cardiovascular disease in relation to cholesterol level among men with and without pre-existing cardiovascular diseases. NEJM 322:1700–1707, 1990

110. Ross MA, Magargal LE, Hedges TR, Simeone FA: Ocular ischemic syndrome: Long-term ocular complications. Ann Ophthalmol 19:270–272, 1987

111. Bullock JD, Falter RT, Downing JE, and Snyder HE: Ischemic ophthalmia secondary to an ophthalmic artery occlusion. Am J Ophthalmol 74:486–493, 1972

112. Knox DL: Ischemic ocular inflammation. Am J Ophthalmol 60:995–1002, 1965

113. Young LHY, Appen RE: Ischemic oculopathy: A manifestation of carotid artery disease. Arch Neurol 38:358–361, 1981

114. Hedges TR: Ophthalmoscopic findings in internal carotid artery occlusion. Bull Johns Hopkins Hosp 111:89–96, 1962

115. Kearns TP, Hollenhorst RW: Venous stasis retinopathy of occlusive disease of the carotid artery. Proc Staff Meeting Mayo Clinic 38:304–312, 1963

116. Kearns TP: Ophthalmology and the carotid artery. Am J Ophthalmol 88:714–722, 1979

117. Brown GC, Magargal LE, Simeone FA, Goldberg RE, Federman JL, and Benson WE: Arterial obstruction and ocular neovascularization. Ophthalmology 89:139–146, 1982

118. Sturrock GD and Mueller HR: Chronic ocular ischaemia. Br J Ophthalmol 68:716–723, 1984

119. Gass JDM: A fluorescein angiographic study of macular dysfunction secondary to retinal vascular disease: VI. X-ray irradiation, carotid artery occlusion, collagen vascular disease with vitritis. Arch Ophthalmol 80:606–617, 1968

120. Huckman MS and Haas J: Reversed flow through the ophthalmic artery as a cause of rubeosis iridis. Am J Ophthalmol 74:1094–1099, 1972

121. Wiebers DO, Swanson JW, Cascino TL and Whisnant JP: Bilateral loss of vision in bright light. Stroke 20:554–558, 1989

122. Russell RWR, Ikeda H: Clinical and electrophysiological observations in patients with low pressure retinopathy. Br J Ophthalmol 70:651–656, 1986

123. Jacobs NA and Ridgway AEA: Syndrome of ischaemic ocular inflammation: Six cases and a review. Br J Ophthalmol 69:681–687, 1985

124. Hollenhorst RW: Effect of posture on retinal ischemia from temporal arteritis. Trans Am Ophthalmol Soc 65:94–104, 1967

125. Hollenhorst RW, Kuhlin JG, Millikan CH: Ophthalmodynamometry in the diagnosis of intracerebral orthostatic hypotension. Proc Staff Meeting Mayo Clinic 38:532–547, 1963

126. Smith JL, Susac JO: Unilateral arcus senilis: Signs of occlusive disease of the carotid artery. JAMA 226:676, 1973

127. Bagla SK, Golden RL: Unilateral arcus corneae senilis and carotid occlusive disease. JAMA 233:450, 1975

128. O'Doherty DS, Green JB. Diagnostic value of Horner's syndrome in thrombosis of the carotid artery. Neurology 11:842–845, 1958

129. Hamburger LP: Head murmurs. Am J Med Sci 181:756–768, 1931

130. Rennie L, Ejrup B, McDowell F: Arterial bruits in cerebro-vascular disease. J Neurol Neurosurg Psych 24:750–756, 1964

131. Fisher CM: Some neuro-ophthalmological observations. J Neurol Neurosurg Psych 30:383–392, 1967

132. Fisher CM: Augmentation bruit of the vertebral artery. J Neurol Neurosurg Psych 29:343–345, 1966

133. Bousser MG, Touboul PJ, Cabanis E, Guillaumt L, Castaigne P. The significance of ocular bruits in

ischaemic cerebrovascular disease. Neuro-ophthalmology 1:211–218, 1981

134. Cohen JH, Miller S: Eyeball bruits. NEJM 255:459–464, 1956

135. Fisher CM: Cranial bruit associated with occlusion of the internal carotid artery. Neurology 7:299–306, 1957

136. Karanjia PN: The external carotid bruit – whither does she blow? Neurology 39:122, 1989

137. Ingall TJ, Homer D, Whisnant JP, Baker HL, O'Fallon WM: Predictive value of carotid bruit for carotid atherosclerosis. Neurology 46:418–422, 1989

138. Ford CS, Frye JL, Toole JF, Lefkowitz D: Asymptomatic carotid bruit and stenosis. A prospective follow up study. Arch Neurol 43:219–222, 1986

139. Yatsu FM, Fields WS: Asymptomatic carotid bruit. Stenosis or ulceration, a conservative approach. Arch Neurol 42:383–385, 1985

140. Chambers BR, Norris JW: Outcome in patients with asymptomatic neck bruits. NEJM 315:860–865, 1986

141. Bailliart P: La pression arterielle dans les branches de l'artere centrale de la retine. Nouvelle technique pour la determiner. Ann Oculist 154:648–666, 1917

142. Hollenhorst RW: Ocular manifestations of insufficiency or thrombosis of the internal carotid artery. Trans Am Ophthalmol Soc 56:474–506, 1958

143. Hollenhorst RW: Ophthalmodynamometry and intracranial vascular disease. Med Clin North Am 42:951–958, 1958

144. Thomas MH, Petrohelos MA: Diagnostic significance of retinal artery pressure in internal carotid involvement. Findings in eight cases with carotid-artery lesions compared to 50 controls and 210 cases from the literature. Am J Ophthalmol 36:335–346, 1953

145. Kearns TP, Siekeert RG, Sundt TM: The ocular aspects of bypass surgery of the carotid artery. Mayo Clin Proc 54:3–11, 1979

146. Mullie MA, Kirkham TH: Ophthalmodynamometry revisited. Can J Ophthalmol 18:165–168, 1983

147. Gee W: Clinical application of ocular pneumoplethysmography. In: Bernstein EF (ed) Noninvasive Diagnostic Techniques in Vascular Disease. St. Louis, CV Mosby Co, 1978, pp 195–200

148. Wiebers DO, Folger WN, Forbes GS, Younge BR, and O'Fallon WM: Ophthalmodynamometry and ocular pneumoplethysmography for detection of carotid occlusive disease. Arch Neurol 39:690–691, 1982

149. Blackshear WM, Phillips DJ, Thiele BL, et al: Detection of carotid occlusive disease by ultrasonic imaging and pulsed Doppler spectrum analysis. Surgery 86:698–706, 1979

150. Farber R, Bromer M, Anderson D, Loewenson R, Yock D, Larson D: B-mode real-time ultrasonic carotid imaging: Impact on decision making and prediction of surgical findings. Neurology 34:541–544, 1984

151. Barnes RW, Garrett WV, Slaymaker EE, Reinertson: Doppler ultrasound and supraorbital photoplethysmography for non-invasive screening of carotid occlusive disease. Am J Surg 134:183–186, 1977

152. Wise G, Parker J, Burjholder J: Supraorbital doppler studies, carotid bruits, and arteriography in unilateral ocular or cerebral ischemic disorders. Neurology 29:34–37, 1979

153. Grizzard WS, Blackshear WM, Rush JA, Gordon SF: Ophthalmology 89:1235–1240, 1982

154. Steinke W, Kloetzch C, and Hennerici M: Carotid artery disease assessed by color Doppler flow imaging: Correlation with standard Doppler sonography and angiography. AJNR 11:259–266, 1990

155. Bridgers SL: Clinical correlates of Doppler/ultrasound errors in the detection of internal carotid artery occlusion. Stroke 20:612–615, 1989

156. Caplan LR: Carotid-artery disease. NEJM 315:886–888, 1986

157. Wiebers DO, Folger WN, Forbes GS, Younge BR, O'Fallon WM. Ophthalmodynamometry and ocular pneumoplethysmography for detection of carotid disease. Arch Neurol 39:690–691, 1982

158. Lipchik EO and Mewissen MW: A comparison of normal intraarterial digital subtraction angiography with standard angiography in patients with symptomatic cerebrovascular ischemia. AJNR 11:837–838, 1990

159. Wagle WA, Dumoulin CL, Souza SP, Cline HE: 3DFT MR angiography of carotid and basilar arteries. Am J Neuroradiol 10:911–919, 1989

160. Litt AW, Eidelman EM, Pinto RS, Riles TS, et al. Diagnosis of carotid artery stenosis: comparison of 2DFT time-of-flight MR angiography with contrast angiography in 50 patients. AJNR 12:149–154, 1991

160a. Kido DK, Panzer RJ, Szumowski J, et al. Clinical evaluation of stenosis of the carotid bifurcation with magnetic resonance angiographic techniques. Arch Neurol 48:484–489, 1991

160b. Cujek B, Polasek P, Voll C, Shuaib A: Transesophageal echocardiography in the detection of potential cardiac source of embolism in stroke patients. Stroke 22:727–733, 1991

161. Lechat Ph, Mas JL, Lasault G, et al. Prevalence of patent foramen ovale in patients with stroke. NEJM 318:1148–1152, 1988

162. DeBono DP, Warlow CP: Mitral-annulus calcification and cerebral or retinal ischaemia. Lancet 2:383–386, 1979

163. Gilles C, Brucher JM, Khoubessarian P, et al: Cerebral amyloid angiopathy as a cause of multiple intracerebral hemorrhages. Neurol 34:730–735, 1984

164. Lee SS, Stemmermann GN: Congophilic angiopathy and cerebral hemorrhage. Arch Pathol Lab Med 102:317–321, 1978

165. Kahn M, Knox DL, and Green WR: Clinicopathologic studies of a case of aortic arch syndrome. Retina 6:228–233, 1986

166. Font RL and Naumann G: Ocular histopathology in pulseless disease. Arch Ophthalmol 82:784–788, 1969

167. Knox DL and Cogan DG: Eye pain and homonymous hemianopia. Am J Ophthalmol 54:1091–1093, 1962

168. Pessin MS, Kwan ES, Scott M, and Hedges TR, III: Occipital infarction with hemianopsia from carotid occlusive disease. Stroke 20:409–411, 1989

169. Cohen SN: Occipital infarction with hemianopia from carotid occlusive disease. Stroke 20:1433–1434, 1989

170. Trobe JD, Lorber ML, Schlezinger NS: Isolated homonymous hemianopia: A review of 104 cases. Arch Ophthalmol 89:377–381, 1973

171. Kaul SN, DuBoulay GH, Kendall BE, Ross Russell RW: Relationship between visual field defects and arterial occlusion in the posterior cerebral circulation. J Neurol Neurosurg Psychiatr 36: 1022–1030, 1974

172. Spector RH, Glaser JS, David NJ, Vining DQ: Occipital lobe infarctions: Perimetry and computed tomography. Neurology 31:1098–1106, 1981

173. Marinkovic SV, Milisavljevic MM, Lolic-Draganic V, Lovacevic MS: Distribution of the occipital branches of the posterior cerebral artery. Correlation with occipital lobe infarcts. Stroke 18:728–732, 1987

174. Fisher CM: Occlusion of the vertebral arteries. Arch Neurol 22:13–19, 1970

175. Caplan LR: Occlusion of the vertebral or basilar artery. Stroke 10:277–282, 1979

176. Hoyt WF, Newton TH: Angiographic changes with occlusion of arteries that supply the visual cortex. New Zealand Med J 72:310–317, 1970

177. Grossman M, Galetta SL, Nichols CW, Grossman RI: Horizontal field defect after ischemic infarction of the occipital cortex. Am J Ophthalmol 109:234–236, 1990

178. Benton S, Levy I, Swah M: Vision in the temporal crescent in occipital infarction. Brain 103:83–97, 1980

179. Bender MB, Kanzer MG. Dynamics of homonymous hemianopias and preservation of central vision. Brain 62:404–442, 1939

180. Huber A: Homonymous hemianopia after occipital lobectomy. Am J Ophthalmol 54:623–629, 1962

181. Huber A: Homonymous hemianopia after removal of one occipital lobe. Exc Med Inter Cong Ser (Ophthalmology) 222:1333–1343, 1970

182. Walsh FB, Hoyt WF. Clinical Neuro-Ophthalmology. Baltimore, Williams & Wilkins, 1969, pp 77–81

183. Morax V: Discussion des hypotheses faites sur les connexions corticales des faisceaux maculaires. Ann Oculist 156:25–35, 1919

184. Buhnt AH: Foveal sparing. New anatomical evidence for bilateral representation of the central retina. Arch Ophthalmol 95:1445–1447, 1977

185. Dejerine J: Contribution a l'etude anatomopathologique et clinique des differentes varietes de cecite verbale. Mem Soc Biol (Paris) 44:61–90, 1892

186. Greenblatt SH: Subangular alexia without agraphia or hemianopsia: Anatomical analysis of an autopsied case. Brain 96:307–316, 1973

187. Damasio AR, Damasio H: The anatomic basis of pure alexia. Neurology 33:1573–1583, 1983

188. Henderson VW, Friedman RB, Teng EL, Weiner JM: Left hemisphere pathways in reading: Inferences from pure alexia without hemianopia. Neurology 35:962–968, 1985

189. Bogousslavsky J, Regli F, van Melle G: Unilateral occipital infarction: Evaluating the risks of developing bilateral loss of vision. J Neurol Neurosurg Psychiatr 46:78–80, 1983

190. Symonds C, Mackenzie I: Loss of vision from cerebral infarction. Brain 80:415–455, 1957

191. Heller-Bettinger I, Kepes JJ, Preskorn SH, Wurster JB: Bilateral altitudinal anopia caused by infarction of the calcarine cortex. Arch Neurol 26:1176–1179, 1976

192. Newman RP, Kinkel WR, Jacobs L: Altitudinal hemianopia caused by occipital infarctions. Arch Neurol 41:413–418, 1984

193. Bogousslavsky J, Miklossy J, Deruaz JP, Assal G and Regli F: Lingual and fusiform gyri in visual processing: A clinico-pathologic study of superior altitudinal hemianopia. J Neurol Neurosurg Psychiatr 50:607–714, 1987

194. Melamed E, Abraham FA, Lavy S: Cortical blindness as a manifestation of basilar occlusion. Europ Neurol 11:22, 1974

195. Olbert D: Sehstörungen bei Insuffizienz und Verschluß der Arteria basilaris. Klin Monatsbl Augenheilkd 186:36–38, 1985

196. Anton G: Ueber die Selbstwahrnehmungen der Herderkrankungen des Gehirns durch den Kranken bei Rindenblindheit und Rindentaubheit. Arch Psychiat Nervenkr 32:86–127, 1899

197. Gloning I, Gloning K, and Tschabitscher H: Die occipitale Blindheit auf vasculärer Basis: Untersuchungsergebnisse von 16 eigenen Fällen. Albrecht v. Graefes Arch Ophthalmol 165:138–177, 1962

198. Rizzo M and Hurtig R: Looking but not seeing: attention, perception and eye movements in simultagnosia. Neurology 37:1642–1648, 1987

199. Dennis MS, Sandercock PAG, Bamford JM, Warlow CP: Lone bilateral blindness: A transient ischemic attack. Lancet 1:185–188, 1989

200. Hoyt WF: Some neuro-ophthalmologic consider-

ations in cerebral vascular insufficiency. Arch Ophthalmol 62:260–272, 1959

201. Dennis MS, Bamford JM, Sandercock PAG, and Warlow CP: Lone bilateral blindness: A transient ischemic attack. Lancet 1:185–188, 1989

202. Hoyt WF: Transient bilateral blurring of vision. Arch Ophthalmol 70:746–751, 1963

203. Aguilar MJ, Gerbode F and Hill JD: Neuropathologic complications of cardiac surgery. J Thorac Cardiovasc Surg 61:676– 685, 1971

204. Smith JL and Cross SA: Occipital lobe infarction after open heart surgery. J Clin Neuro-ophthalmol 3:23–30, 1983

205. Holmes G, Lister WT: Disturbances of vision from cerebral lesions with special reference to the cortical representation of the macula. Brain 39:34–73, 1916

206. Riddoch G: Dissociation of visual perceptions due to occipital injuries with special reference to appreciation of movement. Brain 40:15–57, 1917

207. MacKay G, Dunlop JC: The cerebral lesions in a case of complete acquired colour blindness. Scot Med Surg J 5:503–512, 1899

208. Heidenhain A: Beitrag zur Kenntnis der Seelenblindheit. Monatsschr Psychiatr Neurol 65: 61–116, 1927

209. Green GJ, Lessell S: Acquired cerebral dyschromatopsia. Arch Ophthalmol 95:121–128, 1977

210. Pearlman AL, Birch J, Meadows JC: Cerebral color blindness: An acquired defect in hue discrimination. Ann Neurol 5:253–261, 1979

211. Damasio A, Yamada T, Damasio H, et al: Central achromatopsia: Behavioral, anatomic, and physiologic aspects. Neurology 30:1064–1071, 1980

212. Zeki SM. Colour coding in Rhesus monkey prestriate cortex. Br Res 53:422–427, 1973

213. Zeki SM: Colour coding in the superior temporal sulcus of rhesus monkey visual cortex. Proc R Soc Lond (Biol) 197:195–223, 1977

214. Gerstmann J: Syndrome of finger angosia; disorientation for right and left, agraphia, acalculia. Arch Neurol Psychiatr 44:398–408, 1940

215. Critchley M: The Parietal Lobe. New York, Hafner, 1953, pp 203–224

216. Carter JE, O'Connor P, Shacklett D, Rosenberg M: Lesions of the optic radiations mimicking lateral geniculate nucleus visual field defects. J Neurol Neurosurg Psychiatr 48:982–988, 1985

217. DeRenzi E, Colombo A, Faglioni P, and Gilbertoni M: Conjugate gaze paresis in stroke patients with unilateral damage: An unexpected instance of hemispheric asymmetry. Arch Neurol 39:482–486, 1982

218. Albert ML, Soffer D, Silverberg R, Reches A: The anatomic basis of visual agnosia. Neurology 29: 876–879, 1979

219. Damasio AR, Damasio H, Van Hoesen GW: Prosopagnosia: Anatomic basis and behavioral mechanisms. Neurology 32:331–341, 1982

220. Landis T, Regard M, Bliestle A, and Kleihues P: Prosopagnosia and agnosia for noncanonical views. An autopsied case. Brain 111:1287–1297, 1988

221. Ferro JM, Kertesz A: Posterior internal capsule infarction associated with neglect. Arch Neurol 41:422–424, 1984

222. Abbie A: The blood supply of the lateral geniculate body with a note on the morphology of the choroidal arteries. J Anat 67:491–527, 1933

223. Gunderson GH and Hoyt WF: Geniculate hemianopia: Incongruous homonymous field defects in two patients with partial lesions of the lateral geniculate nucleus. J Neurol Neurosurg Psychiatr 34:1–6, 1971

224. Frisen L: Quadruple sectoranopia and sectorial optic atrophy: A syndrome of the distal anterior choroidal artery. J Neurol Neurosurg Psych 42: 590–594, 1979

225. Frisen L, Holmegaard L, Rosencrantz M: Sectorial optic atrophy and homonymous, horizontal sectoranopia: A lateral choroidal artery. J Neurol Neurosurg Psych 41:374–380, 1978

226. Bogousslavsky J, Regli F, and Uske A: Thalamic infarcts: Clinical syndromes, etiology and prognosis. Neurology 38:837–848, 1988

227. Foix C, Chavany H, Hillemand P, et al: Obliteration de l'artere choroidienne anterieure: Ramollissement cerebral, hemiplegie, hemianesthesie et hemianopsie. Soc Ophthalmol 27:221–223, 1925

228. Decroix JP, Graveleau Ph, Masson M, and Cambier J: Infarction in the territory of the anterior choroidal artery. A clinical and computerized tomographic study of 16 cases. Brain 109:1071–1085, 1986

229. Poppi U: Sindrome talamo capsulare per ramollimento nel territorio dell'arteria coroidea anteriore. Riv Patol Nerv Ment 33:505–542, 1928

230. Takahashi S, Kawata Y, Uemura K: CT findings on anterior choroidal artery occlusion. Rinsho Hoshasen 25:575–581, 1980

231. Cooper IS. Surgical occlusion of the anterior choroidal artery in parkinsonism. Surg Gynecol Obstet 99:207–219, 1955

232. Cooper IS. Anterior choroidal artery ligation for involuntary movements. Science 118:193, 1953

233. Millikan CH and Siekert RG: Studies in cerebrovascular disease: I. The syndrome of intermittent insufficiency of the basilar arterial system. Mayo Clin Proc 30:61–68, 1955

234. Bradshaw P and McQuaid P: The syndrome of vertebrobasilar insufficiency. W J Med 32:279–296, 1963

235. North RR, Fields WS, DeBakey ME, and Crawford ES: Brachial-basilar insufficiency syndrome. Neurology 12:810–812, 1962

236. Hennerici M, Klemm C, and Rautenberg W: The subclavian steal phenomenon: A common vascular

disorder with rare neurologic deficits. Neurology 38:669–673, 1988

237. Fisher CM: Pure motor hemiplegia of vascular origin. Arch Neurol 13:30–44, 1965

238. Fisher CM: Lacunar infarct of the tegmentum of the lower lateral pons. Arch Neurol 46:566–567, 1989

239. Duncan GW, Parker SW, Fisher CM: Acute cerebellar infarction in the PICA territory. Arch Neurol 32:364–368, 1975

240. Ho KL, Meyrer KR: The medial medullary syndrome. Arch Neurology 38:385–387, 1981

241. Daroff RB, Hoyt WF, Sanders MD, et al: Gaze-evoked eyelid and ocular nystagmus inhibited by the near reflex: Unusual ocular motor phenomena in a lateral medullary syndrome. J Neurol Neurosurg Psychiatr 31:362–367, 1968

242. Kommerell G, Hoyt WF: Lateropulsion of saccadic eye movements: Electro-oculographic studies in a patient with a Wallenberg's syndrome. Arch Neurol 28:313–318, 1973

243. Currier RD, Giles CL, DeJong RN: Some comments on Wallenberg's lateral medullary syndrome. Neurology 1:778–791, 1961

244. Silverskold BP: Skew deviation in Wallenberg's syndrome. Acta Neurol Scand 41:381–386, 1965

245. Hagstrom L, Hornsten G, Silverskold BP: Oculostatic and visual phenomena occurring in association with Wallenberg's syndrome. Acta Neurol Scan 45:568–582, 1969

246a. Hornsten G: Wallenberg's syndrome. Part I. General symptomatology with special reference to visual disturbance and imbalance. Acta Neurol Scand 50:434–446, 1974

246b. Hornsten G: Wallenberg's syndrome. Part II. Oculomotor and oculostatic disturbances. Acta Neurol Scand 50:447–468, 1974

247. Fisher CM, Karnes WE, Kubik CS: Lateral medullary infarction: The pattern of vascular occlusion. J Neuropathol Exp Neurol 20:323–379, 1961

247a. Malamud N, Satya-Murti S. Cerebellar hemorrhage: a review and reappraisal of benign cases. Arch Neurol 32:425–428, 1984

247b. Ott KH, Kase CS, Ojemann RG, Mohr JP. Cerebellar hemorrhage: diagnosis and treatment. Arch Neurol 31:160–167, 1974

247c. Tomaszek DE, Rosner MJ. Cerebellar infarction: analysis of 21 cases. Surg Neurol 24:223–226, 1985

247d. Duncan GW, Parker SW, Fisher CM. Acute cerebellar infarction in the PICA territory. Arch Neurol 32:364–368, 1975

247e. Kase C, White JL, Joslyn JN, et al. Cerebellar infarction in the superior cerebellar artery distribution. Neurology 35:705–711, 1985

248. Donaldson D, Rosenberg NL: Infarction of abducens nerve fasicle as cause of isolated sixth nerve palsy related to hypertension. Neurology 38:1654, 1988

249. Bronstein AM, Morris J, DuBoulay G, Gresty MA, and Rudge P: Abnormalities of horizontal gaze. Clinical, oculographic and magnetic resonance imaging findings. I. Abducens palsy. J Neurol Neurosurg Psychiatr 53:194–199, 1990

250. Leigh RJ, Zee DS: The Neurology of Eye Movements. FA Davis, Philadelphia, 1983, pp 219–221

251. Smith JL, Cogan DG: Internuclear ophthalmoplegia. A review of fifty-eight cases. Arch Ophthalmol 61:687–694, 1959

252. Smith JL, David NJ: Internuclear ophthalmoplegia: Two new clinical signs. Neurology 14:307–309, 1964

253. Kupfer C, Cogan DG: Unilateral internuclear ophthalmoplegia: A clinicopathological case report. Arch Ophthalmol 75:484–489, 1966

254. Cogan DG: Internuclear ophthalmoplegia, typical and atypical. Arch Ophthalmol 84:583–589, 1970

255. Gonyea EF: Bilateral internuclear ophthalmoplegia. Arch Neurol 31:168–173, 1974

256. Pierrot-Deseilligny C, Chain F, Serdaru M, et al: The one-and-a-half syndrome. Brain 104:665–669, 1981

257. Wall M, Wray SH: The one-and-a-half syndrome – A unilateral disorder of the pontine tegmentum: A study of 20 cases and review of the literature. Neurology 33:971–980, 1983

258. Bogousslavsky J, Miklossy J, Regli F, et al: One-and-a-half syndrome in ischaemic locked-in state: A clinicopathological study. J Neurol Neurosurg Psychiat 47:927–935, 1984

259. Nordren RE, Markesbery WR, Fukuda K, et al: Seven cases of cerebromedullospinal disconnection: The 'locked-in' syndrome. Neurology 21:1140–1148, 1971

260. Kase CS, Maulsby GO, Mohr JP: Partial pontine hematomas. Neurology 30:652–655, 1980

261. Keane JR: Ocular skew deviation. Analysis of 100 cases. Arch Neurol 32:185–190, 1975

262. Smith JL, David NJ, Klintworth G: Skew deviation. Neurology 14:96–105, 1964

263. Goldstein JE, Cogan DG: Lateralizing value of ocular motor dysmetria and skew deviation. Arch Ophthalmol 66:687–694, 1961

264. Halmagyi GM, Gresty MA, Gibson WOR: Ocular tilt reaction with peripheral vestibular lesion. Ann Neurol 6:80–83, 1979

265. Hennerici M, Fromm C: Isolated complete gaze palsy: An unusual ocular movement deficit probably due to a bilateral reticular formation (PPRF) lesion. Neuro-ophthalmology 1:165–173, 1981

266. Daroff RB and Waldman AL: Ocular bobbing. J Neurol Neurosurg Psychiatr 28:375–377, 1965

267. Fisher CM: Ocular bobbing. Arch Neurol 11:543–546, 1964

268. Susac JO, Hoyt WF, Daroff RB, Lawrence W: Clin-

ical spectrum of ocular bobbing. J Neurol Neurosurg Psychiatr 33:771–775, 1970

269. Rosa A, Masmoudi K, Mizon JP: Typical and atypical ocular bobbing. Pathology through five case reports. Neuro-ophthalmology 7:285–290, 1987

270. Bosch EP, Kennedy SS, Aschenbrenner CA: Ocular bobbing: The myth of its localizing value. Neurology 25:940–953, 1975

271. Percheron G: Etude anatomique du thalamus de l'homme adulte et de sa vascularisation arterielle. Paris, These de Medecine, 1966

272. Percheron G: Les arteres du thalamus humain. Arteres et territoires thalamiques paramedianes de l'artere basilaire communicante. Rev Neurol (Paris) 132:309–324, 1976

273. Castaigne P, Lhermitte F, Buge A, Escourolle R, Hauw JJ, Lyon-Cain O: Paramedian thalamic and midbrain infarcts: Clinical and neuropathological study. Ann Neurol 10:127–148, 1981

274. Segarra JM: Cerebral vascular disease and behavior. I. The syndrome of the mesencephalic artery (basilar artery bifurcation). Arch Neurol 22: 408–418, 1970

275. Breen LA, Hopf HC, Farris BK, Gutman L: Pupil-sparing oculomotor nerve palsy due to midbrain infarction. Arch Neurol 48:105–106, 1990

276. Morel-Maroger A, Metzger J, Bories J, et al: Les hematomes benins du tronc cerebral chez les hypertendus arteriels. Rev Neurol 138:438–445, 1982

277. Reagan TJ, Trautmann JC: Combined nuclear and supranuclear defects in ocular motility. A clinicopathologic study. Arch Neurol 35:133–137, 1978

278. Sieben G, de Reuck J, Vander Eecken H: Thrombosis of the mesencephalic artery: A clinicopathological study of two cases and its correlation with the arterial vascularization. Acta Neurol Belg 77: 151–162, 1977

279. Pierrot-Deseilligny C, Schaison M, Bousser MG, et al: Syndrome nucleaire du nerf moteur oculaire commun: A propos de deux observations cliniques. Rev Neurol 137:217–222, 1981

280. Warren W, Burde RM, Klingele TG, Roper-Hall G: Atypical oculomotor paresis. J Clin Neuro-ophthalmol 2:13–18, 1982

281. Biller J, Sand JJ, Corbett JJ, Adams HP jr, Dunn V: Syndrome of the paramedian thalamic arteries: Clinical and neuroimaging correlation. J Clin Neuro-ophthalmol 5:217–223, 1985

282. Warwick R: Representation of the extraocular muscles in the oculomotor nuclei of the monkey. J Comp Neurol 98:449–503, 1953

283. Ksiazek SM, Repka MX, Maguire A, Harbour RC, Savino PJ, Miller NR, Sergott RC, Bosley TM: Divisional oculomotor nerve paresis caused by intrinsic brainstem disease. Ann Neurol 26:714–718, 1989

284. Bogousslavsky J, Regli F, Ghika J, et al: Internuclear ophthalmoplegia, prenuclear paresis of contralateral superior rectus, and bilateral ptosis. J Neurol 230:197–203, 1983

285. Jampel RS, Fells P: Monocular elevation paresis caused by a central nervous system lesion. Arch Ophthalmol 80:45–57, 1968

286. Bogousslavsky J, Regli F: Nuclear and prenuclear syndromes of the oculomotor nerve. Neuro-ophthalmology 3:211–216, 1983

287. Viader F, Masson M, Marion MH, et al: Infarctus cerebral dans le territoire de l'artere choroidienne anterieure avec trouble oculomoteur. Rev Neurol 140:668–670, 1984

288. Buge A, Escourolle R, Hauw J, Rancurel G, Gray F, and Tempier P: Syndrome pseudobulbaire aigu par infarctus bilateral limite du territoire des arteres choroidiennes anterieures. Rev Neurol 135: 313–318, 1979

289. Hatcher MA, Klintworth GK: The Sylvian aqueduct syndrome. Arch Neurol 15:215–222, 1966

290. Sullivan HG: Parinaud syndrome: Cerebrovascular disease as a common etiology. Surg Neurol 6: 301–305, 1976

291. Thames PB, Trobe JD, Ballinger WE: Upgaze paralysis caused by lesions of the periaqueductal gray matter. Arch Neurol 41:437–440, 1984

292. Fisher CM: The pathologic and clinical aspects of thalamic hemorrhage. Trans Am Neurol Assoc 84:56–59, 1959

293. Pierrot-Deseilligny C, Chain F, Gray F, et al: Parinaud's syndrome: Electro-oculographic and anatomical analyses of six vascular cases with deductions about vertical gaze organization in the premotor structures. Brain 105:667–696, 1982

294. Nashold BS, Seaber JH: Defects of ocular motility after stereotactic midbrain lesions in man. Arch Ophthalmol 88:245–248, 1972

295. Pasik P, Pasik T, Bender MB: The pretectal syndrome in monkeys. I. Disturbances of gaze and body posture. Brain 92:521–534, 1969

296. Bogousslavsky J, Miklossy J, Deruaz JP, Regli F, Assal G: Unilateral left paramedian infarction of thalamus and midbrain: A clinico-pathological study. J Neurol Neurosurg Psychiatr 49:686–694, 1986

297. Buttner-Ennever JA, Buttner U, Cohen B, and Baumgartner G: Vertical gaze palsy and the rostral interstitial nucleus of the medial longitudinal fasciculus. Brain 105:125–149, 1982

298. Ranalli PL, Sharpe JA, Fletcher WA: Palsy of upward and downward saccadic, pursuit, and vestibular movements with a unilateral midbrain lesion: Pathophysiologic correlations. Neurology 38: 114–122, 1988

299. Wall M, Slamovits TL, Weisberg LA, Trufant SA: Vertical gaze ophthalmoplegia from infarction in the area of the posterior thalamo-subthalamic paramedian artery. Stroke 17:546–555, 1986

300. Halmagyi GM, Evans WA, Hallinan JM: Failure of

downward gaze. Arch Neurol 35:22–26, 1978

301. Jacobs L, Anderson PJ, Bender MB: The lesions producing paralysis downward but not upward gaze. Arch Neurol 28:319–323, 1973

302. Trojanowski JQ, Wray SH: Vertical gaze ophthalmoplegia: Selective paralysis of downgaze. Neurology 30:605–610, 1980

303. Trojanowski JQ, Lafontaine MH: Neuroanatomical correlates of selective downgaze paralysis. J Neurol Sci 52:91–101, 1981

304. Jacobs L, Heffner RR Jr, Newman RP: Selective paralysis of downward gaze caused by bilateral lesions of the mesencephalic periaqueductal gray matter. Neurology 35:516–521, 1985

305. Deleu D, Buissert T, Ebinger G: Vertical one-and-a-half syndrome. Supranuclear downgaze paralysis with monocular elevation palsy. Arch Neurol 46:1361–1363, 1989

306. Corbett JJ, Schatz NJ, Shults TW, Behrens M, Berry RGL Slowly alternating skew deviation: Description of a pretectal syndrome in three patients. Ann Neurol 10:540–546, 11981

307. Mastaglia FL: See-saw nystagmus: An unusual sign of brainstem infarction. J Neurol Sci 22:439–444, 1974

308. Keane JR: Alternating skew deviation: 47 patients. Neurology 35:725–728, 1985

309. Zackon DH, Sharpe JA: Midbrain paresis of horizontal gaze. Ann Neurol 16:495–504, 1984

310. Claude H: Syndrome pedonculaire de la region du Noyau rouge. Rev Neurol (Paris) 23:311–313, 1912

311. Masdeu JC, Rosenberg M: Midbrain-diencephalic horizontal gaze paresis. J Clin Neuro-ophthalmol 7:227–234, 1987

312. Namer IJ, Oztekin MF, Kansu T, Zileli T: Pseudosixth nerve palsy with thalamo-mesencephalic junction lesion. Report of two cases. Neuro-ophthalmol 10:69–72, 1990

313. Masdeu J, Brannegan R, Rosenberg M, et al: Pseudoabducens palsy with midbrain lesions. Trans Am Neurol Assoc 105:184–185, 1980

314. Shutt HK, David MJ, Smith JL: Complete bilateral III and IV nerve palsies due to basilar artery disease. In: Smith JL (ed) Neuro-ophthalmology. Hallandale, Fla, Hiffman, 1971, chapter 5, pp 356–362

315. Iwakiri M and Yoshida H: Bilateral superior oblique paresis: Lesion of the trochlear nerve fascicles. Folia Ophthalmol Jpn 41:97–101, 1990

316. Alajouanine T: Lesion protruberantielle basse d'origine vasculaire et hallucinose. Rev Neurol (Paris) 76:90–91, 1944

317. Geller TJ, Bellur SN: Peduncular hallucinosis: Magnetic resonance imaging confirmation of mesencephalic infarction during life. Ann Neurol 21:602–604, 1987

318. Cogan DG: Visual hallucinations as release phenomena. Albrecht Von Graefes Arch Klin Exp Ophthalmol 188:139–150, 1973

319. Kubik CS, Adams RD: Occlusion of the basilar artery: A clinical and pathological study. Brain 69:6–121, 1946

320. Selhorst J, Hoyt WF, Feinsod M, et al: Midbrain corectopia. Arch Neurol 33:193–195, 1976

321. Mehler MF: The neuro-ophthalmologic spectrum of the rostral basilar artery syndrome. Arch Neurol 45:966–971, 1988

322. Balint R: Seelenlähmung des Schauens, optische Ataxie, räumliche Störung der Aufmerksamkeit. Monatsschr Psychiat Neurol 25:51–81, 1909

323. Walshe TM, Davis KR, Fisher CM: Thalamic hemorrhage: A computed tomographic–clinical correlation. Neurology 27:217–222, 1977

324. Keane JR: Contralateral gaze deviation with supratentorial hemorrhage. Arch Neurol 32:119–122, 1975

325. Gliner LI, Avin B: A reversible ocular manifestation of thalamic hemorrhage. Arch Neurol 34:715–716, 1977

326. Waga S, Okada M, Yamamoto Y: Reversibility of Parinaud syndrome in thalamic hemorrhage. Neurology 29:407–409, 1979

327. Hirose G, Kosoegawa H, Saeki M, Kitagawa Y, Oda R, Kanda S, Matsuhira T: The syndrome of posterior thalamic hemorrhage. Neurology 35:998–1002, 1985

328. Graff-Radford NR, Eslinger PJ, Damasio AR, Yamada T: Nonhemorrhagic infarction of the thalamus: Behavioral, anatomic, and physiologic correlates. Neurology 34:14–23, 1984

329. Caplan LR, DeWitt D, Pessin MS, Gorelick PB, Adelman LS: Lateral thalamic infarcts. Arch Neurol 45:959–964, 1988

330. Bogousslavsky J, Regli F, Uske A: Thalamic infarcts: Clinical syndromes, etiology, and prognosis. Neurology 38:837–848, 1988

331. Dejerine J, Roussy G: Le syndrome thalamique. Revue Neurologique 14:521–532, 1906

332. DeRenzi E, Colombo A, Faglioni P, Gilbertoni M: Conjugate gaze in paresis in stroke patients with unilateral damage. An unexpected instance of hemispheric asymmetry. Arch Neurol 39:482–486, 1982

333. Kelley RE and Kovacs AG: Horizontal gaze paresis in hemispheric stroke. Stroke 17:1030–1032, 1986

334. Pessin MS, Adelman LS, Prager RG, Lathi ES, Lange DJ: "Wrongway eyes" in supratentorial hemorrhage. Ann Neurol 9:79–81, 1981

335. Tijssen CC, Schultze BPM, Leyten ACM. Prognostic significance of conjugate eye deviation in stroke patients. Stroke 22:200–202, 1991

336. Stewart JD, Kirkham RH, Mathieson G: Periodic alternating gaze. Neurology 29:222–224, 1979

337. Mascucci EF, Fabara JA, Saini N, Kurtzke JF: Periodic alternating ping-pong gaze. Ann Ophthalmol 113:1123–1127, 1981

338. Senelick RC: "Ping-pong gaze": Periodic alternating gaze deviation. Neurology 26:532–535, 1976

516   References

339. Lapresle J, Said G: Deviation forcee des yeux vers le bas et en dedans et mouvements oculaires periodiques au cours d'une hemorragie anevrismale de la calotte mesencephalique. Rev Neurol (Paris) 133:497–503, 1977

340. Baloh RW, Yee RD, Honrubia V: Optokinetic nystagmus and parietal lobe lesions. Ann Neurol 7:269–276, 1980

341. Meienberg O, Zangemeister WH, Rosenberg M, Hoyt WF, Stark L: Saccadic eye movement strategies in patients with homonymous hemianopia. Ann Neurol 9:537–544, 1981

342. Daroff R, Hoyt WF: Supranuclear disorders of ocular control systems in man. In: Bach-y-Rita P, Collins C (eds) Control of Eye Movements. New York, Academic Press, 1971, pp 175–235

343. Ter Braak JWG, Schenck VWD, Van Vliet AGM: Visual reactions in a case of long-lasting cortical blindness. J Neurol Neurosurg Psychiatr 34: 140–147, 1971

344. Wolf PA, Kannel WM, McGeer PC: Epidemiology of strokes in North America. In: Barnett HJM, Stein BM, Mohr JP, Yatsu FM (eds) Stroke: Pathophysiology, diagnosis, and management, Vol I. New York, Churchill Livingstone, 1986, pp 19–29

345. Starkey I, Warlow C: The secondary prevention of stroke in patients with atrial fibrillation. Arch Neurol 43:66–68, 1986

346. Sherman DG, Hart RG, Easton JD: The secondary prevention of stroke in patients with atrial fibrillation. Arch Neurol 43:68–70, 1986

347. Cerebral Embolism Task Force. Cardiogenic brain embolism. Arch Neurol 43:71–84, 1986

348. Petersen P, Boysen G, Godtfredsen J, Andersen ED, Andersen B: Placebo-controlled, randomized trial of warfarin and aspirin for prevention of thromboembolic complications in chronic atrial fibrillation. The Copenhagen APASAK Study. Lancet 1:175–179, 1989

349. The Stroke Prevention in Atrial Fibrillation Investigators: Preliminary report of the Stroke Prevention in Atrial Fibrillation Study. N Engl J Med 322:863–865, 1990

349a. The Stroke Prevention in Atrial Fibrillation Investigators. The Stroke Prevention in Atrial Fibrillation Trial: final results. Circulation 84:527–539, 1991

350. Antiplatelet Trialists' Collaboration. Secondary prevention of vascular disease by prolonged antiplatelet treatment. Br Med J 296:320–331, 1988

351. Kistler JP, Ropper AH, Heros RC: Therapy of ischemic cerebral vascular disease due to atherothrombosis. NEJM 311:100–105, 1984

352. Grotta JC: Current medical and surgical therapy fopr cerebrovascular disease. NEJM 317: 1505–1516, 1987

353. The UK-TIA Study Group: Design and protocol of the UK-TIA aspirin study: In: Tognoni G, Garattini S (eds) Drug treatment and prevention in cerebrovascular disorders: Proceedings of the International Seminar on Drug Treatment and Prevention in Cerebrovascular Disorders held in Milan, Italy, 2–4 May, 1979. Amsterdam: Elsevier/North-Holland, 1979, pp 387–394

354. Moncada J, Higgs EA, Vane JR: Human arterial and venous tissue generates prostacyclin, a potent inhibitor of platelet aggregation. Lancet 1:18–21, 1977

354a. Dutch TIA Trial Study Group. A comparison of two doses of aspirin (30 mg vs. 283 mg a day) in patients after a transient ischemic attack or minor ischemic stroke. New Engl J Med 325:1261–1266, 1991

354b. Salt Collaborative Group. Swedish aspirin low-dose trial (SALT) of 7 mg aspirin as secondary prophylaxis after cerbrovascular ischaemic events. Lancet 338:1345–1349, 1991

355. Canadian Cooperative Study Group. A randomized trial of aspirin and sulfinpyrazone in threatened stroke. NEJM 299:53–59, 1978

356. Fields WS, Lemak NA, Frankowski RF, Hardy RJ: Controlled trial of aspirin in cerebral ischemia. Stroke 8:301–314, 1977

357. Grotta JC, Lemak NA, Gary H, Fields WS, Vital D: Does platelet antiaggregant therapy lessen the severity of the stroke? Neurology 35:632–636, 1985

358. A Swedish Cooperative Study. High-dose acetylsalicyclic acid after cerebral infarct. Stroke 18:325–334, 1987

359. Poole CJM, Russell RWR, Harrison P, Savidge GF: Amaurosis fugax under the age of 40 years. J Neurol Neurosurg Psychiatr 50:81–84, 1987

359a. Winterkorn JMS, Teman AJ. Recurrent attacks of amaurosis fugax treated with calcium channel blocker. Ann Neurol 30:423–425, 1991

360. American-Canadian Cooperative Study Group. Persantine aspirin trial in cerebral ischemia II. Endpoint results. Stroke 16:406–415, 1985

361. Fitzgerald GA: Dipyramidole. NEJM 316: 1247–1257, 1987

362. Panak E, Maffrand JP, Picard-Fraire C, Vallee E, Blanchard J, Roncucci R: Ticlopidine: A promise for the prevention and treatment of thrombosis and its complications. Hemostasis 13 (suppl 1):1–54, 1983

363. Hass WK, Easton JD, Adams HP, Pryse-Phillips W, et al: A randomized trial comparing ticlopidine hydrochloride with aspirin for the prevention of stroke in high-risk patients. NEJM 321:501–507, 1989

364. Page Y, Tardy B, Zeni F, et al. Thrombotic thrombocytopenic purpura related to ticlopidine. Lancet 337:774–776, 1991

365. Link H, et al: Prognosis in patients with infarction

and TIA in carotid territory during and after anticoagulant therapy. Stroke 10:529–532, 1979

366. Terent A, Anderson B: The outcome of patients with transient ischemic attacks and stroke treated with anticoagulants. Acta Med Scand 208:359–365, 1980

367. Dyken JI: Anticoagulant and platelet-aggregating therapy in stroke and threatened stroke. Neurol Clin 1:223–242, 1983

368. Levine M, Hirsh J: Hemorrhagic complications of long-term anticoagulant therapy for ischemic cerebral vascular disease. Stroke 17:111–116, 1986

369. Kase CS, Robinson RK, Stein RW, et al: Anticoagulant-related intracerebral hemorrhage. Neurology 35:943–948, 1985

370. Sage JI: Stroke. The use and overuse of heparin in therapeutic trials. Arch Neurol 42:315–318, 1985

371. Sloan MA: Thrombolysis and stroke. Past and future. Arch Neurol 44:748–768, 1987

372. Aldrich MS, Sherman SA, Greenberg HS: Cerebrovascular complications of streptokinase infusion. JAMA 253:1777–1779, 1985

372a. del Zoppo GJ, Poeck K, Pessin MS, et al: Recombinant tissue plasminogen activa for inacute thrombotic and embolic stroke. Ann Neurol 32:78–86, 1992

373. Carlson SE, Aldrich MS, Greenberg HS, Topol EJ: Intracerebral hemorrhage complicating intravenous tissue plasminogen activator treatment. Arch Neurol 45:1070–1073, 1988

373a. del Zoppo GJ, Ferbert A, Otis S, et al: Local intra-arterial fibrinolytic therapy in acute carotid territory stroke: a pilot study. Stroke 19:307–313, 1988

373b. Mori E, Tabuchi M, Yoshida T, Yamadori: Intracarotid urokinase with thromboembolic occulusion of the middle cerebral artery. Stroke 19:802–812, 1988

374. Powers WJ. Cerebral hemodynamics in ischemic cerebrovascular disease. Ann Neurol 29:231–240, 1991

375. Fields WS, Maslenikov V, Meyer JS, Hass WK, Remington RD, Macdonald M: Joint study of extracranial arterial occlusion. V. Progress report of prognosis following surgery or nonsurgical treatment for transient cerebral ischemic attacks and cervical carotid artery lesions. JAMA 211:1993–2003, 1970

376. Shaw DA, Venables GS, Cartlidge NE, Bates D, Dickinson PH: Carotid endarterectomy in patients with transient cerebral ischaemia. J Neurol Sci 64:45–53, 1984

377. Mungas JE, Baker WH: Amaurosis fugax. Stroke 8:232, 1977

378. Wilson LA, Warlow CP, Russell RWR: Cardiovascular disease in patients with retinal arterial occlusion. Lancet 1:292–294, 1979

379. Durward QJ, Ferguson CG, Barr HWK: The natural history of asymptomatic carotid bifurcation plaques. Stroke 13:459–464, 1982

380. Javid H, Ostermiller WE, Hengesh JW, et al: Natural history of carotid bifurcation atheroma. Surgery 67:80–86, 1970

381. North American Symptomatic Carotid Endarterectomy Trial Collaborators. Beneficial effect of carotid endarterectomy in symptomatic patients with high-grade stenosis. NEJM 325: 445–453, 1991)

382. Ross Russell RW: Atheromatous retinal embolism. Lancet 2:1354–1356, 1963

383. Treiman RL, Bloemendal LC, Foram RF, Levin PM, and Cohen JL: Ipsilateral blindness. A complication of carotid endarterectomy. Arch Surg 112:928–932, 1977

384. Hennerici M, Rautenberg W, Mohr S: Stroke risk from symptomless extracranial arterial disease. Lancet 2:1180–1183, 1982

385. Hennerici M, Hulsbomer HM, Herter S, et al: Risk of stroke in asymptomatic persons extracranial arterial disease. Results of a long-term prospective study. Brain 110:777–791, 1987

386. Trobe JD: Carotid endarterectomy. Who needs it? Ophthalmology 94:725–730, 1987

387. Ennix CL Jr, Lawrie GM, Morris CG, et al: Improved results of carotid endarterectomy in patients with asymptomatic coronary disease: An analysis of 1,546 consecutive carotid operations. Stroke 10:122–125, 1979

388. Allen GS, Preziosi TJ: Carotid endarterectomy: A prospective study of its efficacy and safety. Medicine 60:298–309, 1981

389. Dyken JL: Carotid endarterectomy studies: A glimmering of science. Stroke 17:355–358, 1986

390. Winslow CM, Solomon DH, Chassin MR, Kosecoff J, Merrick NJ, Brook RH: The appropriateness of carotid endarterectomy. NEJM 318:721–727, 1988

391. Hass WK: Caution falling rock zone: An analysis of the medical and surgical management of threatened stroke. Proc Inst Med Chicago 33:80–84, 1980

392. Gertler JP, Cambria RP: The role of external carotid endarterectomy in the treatment of ipsilateral internal carotid occlusion. Collective review. J Vasc Surg 6:158–167, 1987

393. The EC/IC Bypass Study Group. Failure of extracranial–intracranial arterial bypass to reduce the risk of ischemic stroke. Results of an international randomized trial. NEJM 313: 1191–1200, 1985

394. Barnett HJM, Sackett D, Taylor DW, Haynes B, Peerless SJ, Meissner I, Hachinski V, Fox A: Was the international randomized trial of extracranial–intracranial arterial bypass representative of the population at risk? NEJM 316:814–824, 1987

395. Whisnant JP, Cartlidge NEF, Elveback LR: Carotid

and vertebral-basilar transient ischemic attacks: Effects of anticoagulants, hypertension, and cardiac disorders on survival and stroke occurrence – a population study. Ann Neurol 3:107–115, 1978

396. Baker RN, Ramseyer JC, Schwartz WS: Prognosis in patients with transient cerebral ischemic attacks. Arch Neurology 18:1157–1166, 1968

397. Toole JF, Janeway R, Choi K: Transient ischemic attacks due to atherosclerosis. A prospective study of 160 patients. Arch Neurol 32:5–12, 1975

398. Kamel WB, Wolf PA, Verter J, McNamara PM: Epidemiologic assessment of the role of blood pressure in stroke: The Framingham study. JAMA 214:301–310, 1970

399. Whisnant JP: The decline of stroke. Stroke 15:160–168, 1984

400. Shekelle RB, Ostfeld AM, Klawans HL: Hypertension and the risk of stroke in the elderly populations. Stroke 5:71–75, 1974

401. American Heart Association Special Report: Risk factors in stroke. Stroke 15:1105–1111, 1984

402a. Veterans Administration Cooperative Study Group on Anti-hypertensive Agents. Effects of treatment on morbidity in hypertension. I. Results in patients with diastolic blood pressures averaging 115 through 129 mm Hg. JAMA 202:1028–1034, 1967

402b. Veterans Administration Cooperative Study Group on Anti-hypertensive Agents. Effects of treatment on morbidity in hypertension. II. Results in patients with diastolic blood pressures averaging 90 through 114 mm Hg. JAMA 213: 1143–1152, 1970

403. Millikan CG, Siekert RG, Shick RM: Studies in cerebrovascular disease. III. The use of anticoagulant drugs in the treatment of insufficiency or thrombosis with the basilar system. Mayo Clin Proc 30:116–126, 1955

404. Rainer WG, Quanizon EP, Ligett MA: Surgical considerations in the treatment of vertebrobasilar artery insufficiency. Am J Surg 120:594–507, 1970

405. Imparato AM, Riles TSL Surgery of the vertebral artery: Overview and results. In: Moore WS (ed) Surgery for Cerebrovascular Disease. Churchill Livingstone, New York, 1987, pp 749–768

406. Judge TG: Successful remission of central retinal artery occlusion following the use of isoxsuprine. Br J Clin Pract 22:223–234, 1968

407. Wise G, Dollery C, Henkind P: The Retinal Circulation. New York, Harper & Row, 1971, p 298

408. Brown GC, Shields JA: Cilioretinal arteries and retinal occlusion. Arch Ophthalmol 97:84–92, 1979

409. Perkins SA, Marargal LE, Augsburger JJ, Sanborn GE: The idling retina: Reversible visual loss in central retinal artery obstruction. Ann Ophthalmol 19:3–6, 1987

410. Hayreh SS, Kolder HE, Weingeist T: Central retinal artery occlusion and retinal tolerance time. Ophthalmology 87:75–78, 1980

411. Dutton GN and Craig G: Treatment of a retinal embolus by photocoagulation. Br J Ophthalmol 73:580–581, 1989

412. Ancker E, Molteno ACB: Molteno drainage implant for neovascular glaucoma. Trans Ophthal Soc U.K. 102:122–124, 1982

412a. Melamed S, Fiore PM: Molteno implant surgery in refractory glaucoma. Surv Ophthal 34:441–448, 1990

413. Eggleston CA, Bohling MD, Eggleston HC, et al: Photocoagulation for ocular ischemia associated with carotid occlusion. Ann Ophthalmol 12: 84–87, 1980

414. Carter JE: Panretinal photocoagulation for progressive ocular nonvascularization secondary to occlusion of the common carotid artery. Ann Ophthalmol 16:572–576, 1984

415. Furian AJ, Whisnant JP, Kearns TP: Unilateral visual loss in bright light – An unusual manifestation of carotid occlusive disease. Arch Neurol 36:675–676, 1979

416. Donnan GA, Sharbrough FW, Whisnant JP: Carotid occlusive disease. Effect of bright light on visual evoked response. Arch Neurol 39:687–689, 1982

417. Kahn M, Green WF, Knox DM, Miller NR: Ocular features of carotid occlusive disease. Retina 6:239–252, 1986

418. Neupert JR, Brubaker RF, Kearns TP, and Sundt TM: Rapid resolution of venous stasis retinopathy after carotid endarterectomy. Am J Ophthalmol 81:600–602, 1976

419. McCrary JA, III: Venous stasis retinopathy of stenotic or occlusive carotid origin. J Clin Neuroophthalmol 9:195–199, 1989

420. Stadefer MJ, Little JR, Tomsak RL, et al: Improvement in retinal circulation after superficial temporal artery to middle cerebral artery bypass. Neurosurgery 16:525–529, 1985

421. Kearns TP, Young BR, Piepra DG: Resolution of venous stasis retinopathy after carotid bypass surgery. Mayo Clin Proc 55:342–346, 1980

422. Kiser WD, Gonder J, Magargal LE, Sanborn GE, Simeone F: Recovery of vision following treatment of the ocular ischemic syndrome. Ann Ophthalmol 15:305–310, 1983

423. Rubin JR, McIntyre KM, Lukens MC, et al: Carotid endarterectomy for chronic retinal ischemia. Surg Gynecol Obstr 171:497–501, 1990

424. Coppeto JR, Wand M, Bear L, Sciarra R: Neovascular glaucoma and carotid artery obstructive disease. Am J Ophthalmol 99:567–570, 1985

425. Carter JE: Chronic ocular ischemia and carotid vascular disease. Stroke 16:721–728, 1985

426. Johnston ME, Gonder JR, and Canny CLB: Successful treatment of the ocular ischemic syndrome

with panretinal photocoagulation and cerebrovascular surgery. Can J Ophthalmol 23:114–119, 1988

427. Slocum HD, O'Neal KC, Allen CR: Neurovascular complications of malposition on the operating table. Surg Gynec Obst 86:729–734, 1948

428. Hollenhorst RW, Svien HJ, Benoit CF: Unilateral blindness occurring during anesthesia for neurosurgical operations. Arch Ophthalmol 52:819–830, 1954

429. Jayam AV, Hass WK, Carr RE, Kumar AJ: Saturday night retinopathy. J Neurolog Sci 22:413–418, 1974

430a. Burian HM, von Noorden GK: Binocular vision and ocular motility. CV Mosby Comp, St. Louis, 1974, pp 343–344

430b. Burian HM, von Noorden GK: Binocular vision and ocular motility. CV Mosby Comp, St. Louis, 1974, pp 66–69, p 343

430c. Burian HM, von Noorden GK: Binocular vision and ocular motility. CV Mosby Comp, St. Louis, 1974, pp 351–352

430d. Burian HM, von Noorden GK: Binocular vision and ocular motility. CV Mosby Comp, St. Louis, 1974, pp 359–360

431. Rucker CW: Paralysis of the third, fourth, and sixth cranial nerves. Am J Ophthalmol 46:180–185, 1958

432. Green WR, Hackett ER, Schlezinger NS: Neuro-ophthalmologic evaluation of oculomotor nerve paralysis. Arch Ophthalmol 72:154–166, 1964

433. Martin MM. Diabetic neuropathy. Brain 76:594–624, 1953

434. Capo H, Warren FA, Kupersmith MJ: Evolution of oculomotor nerve palsies. J Clin Neuro-ophthalmol 12:21–25, 1992

435. Goldstein JE, Cogan DG: Diabetic ophthalmoplegia with special reference to the pupil. Arch Ophthalmol 64:592–600, 1960

436. Daroff R:. Discussion of paper by Asbury AK, Fisher CM, Aldredge H, Hershberg R. Trans Am Neurol Ass 94:66–67, 1969

437. Weber RB, Daroff RB, MacKey EA: Pathology of oculomotor palsy in diabetics. Neurology 20:835–838, 1970

438. Dreyfus PM, Hakim S, Adams RD: Diabetic Ophthalmoplegia. Arch Neurol Psychiatr 77:337–349, 1957

439. Asbury AK, Aldredge H, Hershberg R, Fisher CM: Oculomotor palsy in diabetes mellitus: a clinico-pathological study. Brain 93:555–566, 1970

440. Parkinson D: Collateral circulation of cavernous carotid artery: Anatomy. Canad J Surg 7:251–268, 1964

441. Laspresle J, Lasjaunias P: Cranial nerve ischemic arterial syndromes. A review. Brain 109:207–215, 1986

442. Rucker WC: The cause of paralysis of the third, fourth, and sixth cranial nerves. Am J Ophthalmol 61:1293–1298, 1966

443. Masucci EF, Kurtzke JF: Diabetic superior branch of the oculomotor nerve. Ann Neurol 7:493, 1980

444. Bregman DK, Harbour R: Diabetic superior division oculomotor palsy. Arch Ophthalmol 106:1169–1170, 1988

445. Moster ML, Savino PJ, Sergott RC, Bosley TM, Schatz NJ: Isolated sixth-nerve palsies in younger adults. Arch Ophthalmol 102:1328–1330, 1984

446. Keane J: Bilateral sixth nerve palsy. Analysis of 125 cases. Arch Neurol 33:681–683, 1976

447. Savino PJ, Hilliker JK, Casell GH, et al: Chronic sixth nerve palsies. Arch Ophthalmol 100:1442–1444, 1982

448. Currie J, Lubin JH, Lessell S: Chronic isolated abducens paresis from tumors at the base of the brain. Arch Neurol 40:226–229, 1983

449. Burger LJ, Kalvin NH, Smith JL: Acquired lesions of the fourth cranial nerve. Brain 93:567–574, 1970

450. Coppeto JM, Lessell S: Cryptogenic unilateral paralysis of the superior oblique muscle. Arch Ophthalmol 96:275–277, 1978

451. Larson DL, Auchincloss HJ: Multiple symmetric bilateral cranial nerve palsies in patients with unregulated diabetes mellitus. Arch Intern Med 85:265–271, 1950

452. Symmonds CP: Recurrent multiple cranial nerve palsies. J Neurol Neurosurg Psychiatr 21:95–100, 1958

453. Ross AT: Recurrent cranial nerve palsies in diabetes mellitus. Neurology 12:180–185, 1962

454. Kosmorsky GS, Tomsak RL: Ischemic ("diabetic") cavernous sinus syndrome. J Clin Neuro-ophthalmol 6:96–99, 1986

455. Manelfe C, Clarisse C, Fredy D, et al: Fibromuscular dysplasia of the cervical-cephalic arteries: Report of 70 cases. J Neuroradiol 1:149–231, 1974

456. Harrington OB, Crosby VG, Nicholas L: Fibromuscular hyperplasia of the internal carotid artery. Ann Thorac Surg 9:516–524, 1970

457. So EL, Toole JF, Dalal P, et al: Cephalic fibromuscular dysplasia in 32 patients. Clinical findings and radiologic features. Arch Neurol 38:619–622, 1981

458. Palubinskas AJ, Ripley HR: Fibromuscular hyperplasia in extrarenal arteries. Radiology 82:451–454, 1964

459. Osborne AG, Anderson RE: Angiographic spectrum of cervical and intracranial fibromuscular dysplasia. Stroke 8:617–626, 1977

460. Sandok BA: Fibromuscular dysplasia of the internal carotid artery. Neurol Clin 1:17–26, 1983

461. Rainer WG, Cramer GG, Newby JP, et al: Fibromuscular hyperplasia of the carotid artery causing positional ischemia. Ann Surg 167:444–446, 1968

462. Ehrenfeld WK, Wylie EJ: Fibromuscular dysplasia

of the internal carotid artery: Surgical management. Arch Surg 109:676–681, 1974

463. Palubinskas AJ, Perloff D, Newton TH: Fibromuscular hyperplasia: An arterial dysplasia of increasing clincial importance. Am J Roentgenol 98: 907–913, 1966

464. Kupersmith MJ, Berenstein A, Flamm E, Ransohoff J: Neuroophthalmologic abnormalities and intravascular therapy of traumatic carotid cavernous fistulas. Ophthalmology 93:906–912, 1986

465. Sundt TM Jr, Siekert RG, Peipegras DG, et al: Bypass surgery for vascular disease of the carotid system. Mayo Clin Proc 51:677–692, 1976

466. Mettinger KL, Ericson K: Fibromuscular dysplasia and the brain: Observations on angiographic, clinical and genetic characteristics. Stroke 13:46–58, 1982

467. Polin SG: Carotid artery fibromuscular hyperplasia: Three cases and review of the literature. Am Surg 35:501–504, 1964

468. Taversas JM: Multiple progressive intracranial arterial occlusions. A syndrome of children and young adults. Am J Roentgenol 106:235–268, 1969

469. Bruno A, Yuh WTC, Biller J, Adams HP Jr, Cornell SH: Magnetic resonance imaging in young adults with cerebral infarction due to moyamoya. Arch Neurol 45:303–306, 1988

470. Aoki N, Mizutani H: Does Moyamoya disease cause subarachnoid hemorrhage? A review of 54 cases with intracranial hemorrhage confirmed by computerized tomography. J Neurosurg 60:348–353, 1984

471. Slamovits TL, Klingele TG, Burde RM, Gado MH: Moyamoya disease with central retinal artery occlusion. J Clin Neuro-ophthalmol 1:123–127, 1981

472. Chace R, Hedges TR III: Retinal artery occlusion due to Moyamoya disease. J Clin Neuro-Ophthalmol 4:31–34, 1984

473. Noda S, Hayasaka S, Setogawa T, Matsumoto S: Ocular symptoms of Moyamoya disease. Am J Ophthalmol 103:812–816, 1987

474. Karasawa J, Kikuchi H, Furuse S, Sakaki T, Yamagata S, Nagata I, Tanaka H, Kawamura J: Visual disturbance in moyamoya disease. Jpn Rev Clin Ophthalmol 72:540–546, 1978

475. Yoshida Y, Sakai T, Ikeda N, Abe H, Ichikawa A: Impairment of visual function in moyamoya disease. Neuro-ophthalm (Jpn) 3:163–168, 1986

476. Schrager GO, Cohen SJ, Vigman MP: Acute hemiplegia and cortical blindness due to Moya Moya disease. Report of a case in a child with Down's syndrome. Pediatrics 60:33–37, 1977

477. Karasawa J, Kikuchi H, Furuse S, et al: Treatment of moyamoya disease with STA-MCA anastamosis. J Neurosurg 49:679–688, 1978

478. Yonekawa Y, Yasargil MG: Arterial extracranial intracranial anastamosis. Technical and clinical aspects. Results. In: Krayenbuhl H (ed) Advances and technical standards of neurosurgery, Vol 3. Springer, Wien, 1976, pp 47–78

479. Suzuki R: Changes in cerebral hemodynamics following encephalo-duro-arterio-synangiosis (EDAS) in young patients with Moyamoya disease. Stroke 20:1604, 1989

479a. Karasawa J, Touho H. Ohnishi H, et al: Long-term follow-up study after extracranial-intracranial bypass surgery for anterior circulation ischemia in childhood moymoya disease. J Neurosurg 77: 84–89, 1992

480. Regan TJ, Okazaki H: The thrombotic syndrome associated with carcinoma: A clinical and neuropathological study. Arch Neurol 31:390–395, 1974

481. Collins RC, Al-Mondhiry H, Chernik NL, et al: Neurologic manifestations of intravascular coagulation in patients with cancer: A clinicopathological analysis of 12 cases. Neurology 25:795–806, 1975

482. Siegel T, et al: Clinical and laboratory aspects of disseminated intravascular coagulation. Thromb Haemost 39:122–134, 1978

483. Mant MJ, King EG: Severe acute disseminated intravascular coagulation. Am J Med 67:557–563, 1979

484. Marlar RA, Montgomery RR, Broekmans AW, Working Party: Diagnosis and treatment of homozygous protein C deficiency. J Pediatr 114: 528–534

485. Winten AR, Broekmans AW, Bertina RM, et al: Cerebral hemorrhagic infarction in young patients with hereditary protein C deficiency: Evidence for "spontaneous" cerebral venous thrombosis. Br Med J 290:350–352, 1985

486. Seligsohn U, Berger A, Abend M, et al: Homozygous protein C deficiency manifested by massive thrombosis in the newborn. NEJM 310:559–562, 1984

487. Pulido JS, Lingua RW, Byrne S: Protein C deficiency associated with vitreous hemorrhage in a neonate. Am J Ophthalmol 104:546–547, 1987

488. Smith DB, Ens GE: Protein deficiency: A cause of amaurosis fugax. J Neurol Neurosurg Psychiatr 59:361–362, 1987

489. Sacco RL, Owen J, Mohr JP, Tatemichi TK, Grossman BA: Free protein S deficiency: A possible association with cerebrovascular occlusion. Stroke 20:1657–1661, 1989

490. Esmon CT: The regulation of natural anticoagulant pathways. Science 235:1348–1352, 1987

491. Clouse LH, Comp PC: The regulation of hemostasis: The protein C system. NEJM 314: 1298–1304, 1986

492. Golub BM, Sibony PA, Coller BS: Protein S deficiency associated with central retinal artery occlusion. Arch Ophthalmol 108:918, 1990

493. Nelson ME, Talbot JP, and Preston PE: Recurrent multiple-branch retinal arteriolar occlusions in a patient with protein C deficiency. Graefe's Arch Clin Exp Ophthalmol 227:443–447, 1989

494. Portnoy BA, Herion JC: Neurological manifestations in sickle-cell disease. With a review of the literature and emphasis on the prevalence of hemiplegia. Ann Int Med 76:643–652, 1972

495. Boros L, Thomas C, and Weiner WJ: Large cerebral vessel disease in sickle cell anemia. J Neurol Neurosurg Psychiatr 39:1236–1239, 1976

496. Singh BM, Morris LJ, Strobos RJ: Cortical blindness in puerperium. JAMA 243:1134, 1980

497. Beal MF, Chapman PH: Cortical blindness and homonymous hemianopia in the postpartum period. JAMA 244:2085–2087, 1980

498. Monteiro ML, Hoyt WF, Imes RK: Puerperal cerebral blindness. Transient bilateral occipital involvement from presumed cerebral venous thrombosis. Arch Neurol 41:1300–1301, 1984

499. Beal MF, Wechsler LR, Davis KR: Cerebral vein thrombosis and multiple intracranial haemorrhages by computed tomography. Arch Neurol 39:437–438, 1982

500. Kalbag RM, Woolf AL: Cerebral Venous Thrombosis. London, Oxford University Press, 1967

501. Krayenbuhl H: Cerebral venous thrombosis, diagnostic value of cerebral angiography. Schweiz Arch Neurochir Psychioatr 74:261–287, 1954

502. Martin JP: Venous thrombosis in the nervous system. Proc Roy Soc Med 37:383–386, 1944

503. Symonds CP: Hydrocephalic and focal cerebral symptoms in relation to thrombophlebitis of dural sinuses and cerebral veins. Brain 60:531–550, 1937

504. Clarke PR, Dickson J, Smith BJ: Traumatic thrombosis of the internal carotid artery following a nonpenetrating injury and leading to infarction of the brain. Br J Surg 43:215–216, 1955

505. Watridge CB, Muhlbauer MS, Lowery RD: Traumatic carotid artery dissection: Diagnosis and treatment. J Neurosurg 71:854–857, 1989

506. Sundt TM, Pearson BW, Piepgras DG, et al: Surgical management of aneurysms of the distal extracranial internal carotid artery. J Neurosurg 64:169–182, 1986

507. Fisher CM, Ojemann RG, Roberson GH: Spontaneous dissection of the cervico-cerebral arteries. Can J Neurol Sci 5:9–19, 1978

508. Gauthier G, Rohr J, Wildi E, et al: L'hematome dissequant de l'artere carotide interne: Revue generale de 205 cas publies dont dix personnels. Schweiz Arch Neurol Neurochir Psychiatr 136:53–74, 1985

509. Biller J, Hingten WL, Adams HP, Smoker WRK, Godersky JC, Toffol GJ: Cervicocephalic arterial dissections. A ten-year experience. Arch Neurol 43:1234–1238, 1986

510. Bradac GB, Kaernback A, Bolk-Weischedel D, et al: Spontaneous dissecting aneurysm of cervical cerebral arteries: Report of six cases and review of the literature. Neuroradiology 21:149–154, 1981

511. Goodman JM, Zink WL, Cooper DF: Hemilingual paralysis caused by spontaneous carotid dissection. Arch Neurol 40:653–654, 1983

512. Mokri B, Sundt TM, Houser OW: Spontaneous internal carotid dissection, hemicrania, and Horner's syndrome. Arch Neurol 36:677–680, 1979

513. Kline LB, Vitek JJ, Raymon BC: Painful Horner's syndrome due to spontaneous carotid artery dissection. Ophthalmology 94:226–230, 1987

514. Francis KR, Williams DP, Troost BT: Facial numbness and dysesthesias: New features of carotid artery dissection. Arch Neurol 44:345–346, 1987

515. Maitland CG, Black JL, Smith WA: Abducens palsy due to spontaneous dissection of the internal carotid artery. Arch Neurol 40:448–449, 1983

516. Ehrenfeld WK, Wylie EJ: Spontaneous dissection of the internal carotid artery. Arch Surg 111:1294–1301, 1976

517. Newman NJ, Kline LB, Leifer D, Lessell S: Ocular stroke and carotid artery dissection. Neurology 39:1462–1464, 1989

518. McNeill DH, Dreisbach J, Marsden RJ: Spontaneous dissection of the internal carotid artery: Its conservative management with heparin sodium. Arch Neurol 37:54–55, 1980

519. Duker JS, Belmont JB: Ocular ischemic syndrome secondary to carotid artery dissection. Am J Ophthalmol 106:750–752, 1988

520. Bogousslavsky J, Despland PA, Regli F: Spontaneous carotid dissection with acute stroke. Arch Neurol 44:137–140, 1987

521. Mokri B, Sundt TM, Houser OW, Piepgras DG: Spontaneous dissection of the cervical internal carotid artery. Ann Neurol 19:126–138, 1986

522. Ojemann RG, Fisher CM, Rich JC: Spontaneous dissecting aneurysm of the internal carotid artery. Stroke 3:434–440, 1972

523. Miller VT. Carotid pseudo-dissection from distal occlusion. Neurology 38(suppl):297, 1988

524. Rothrock JF, Lim V, Press G, Gosink B: Serial magnetic resonance and carotid duplex examinations in the management of carotid dissection. Neurology 39:686–692, 1989

525. Hennerici M, Steinke W, Rautenberg W: High-resistance doppler flow pattern in extracranial carotid dissection. Arch Neurol 46:670–672, 1989

526. Pozzati E, Gaist G, Poppi M: Resolution of occlusion in spontaneously dissected carotid arteries. J Neurosurg 56:857–860, 1982

527. Roome NS, Aberfeld DC: Spontaneous dissecting aneurysm of the internal carotid artery. Arch Neurol 34:251–252, 1977

528. Miyamoto S. Kikuchi H, Karasawa J, Kuriyama Y: Surgical treatment for spontaneous carotid dissec-

522   References

tion with impending stroke. J Neurosurg 61: 382–386, 1984

529. Mokri B, Houser OW, Sandok VA, Piepgras DG: Spontaneous dissections of the vertebral arteries. Neurology 38:880–885, 1988

530. Caplan LR, Baquis GD, Pessin MS, D'Alton J, Adelman LS, DeWitt LD, Ho K, Izukawa D, Kwan ES: Dissection of the intracranial vertebral artery. Neurology 38:868–877, 1988

531. Mehalic T, Farhat SM. Vertebral artery injury from chiropractic manipulation. Surg Neurol 2:125–129, 1974

532. Hanus SH, Homer TD, Harter DH: Vertebral artery occlusion complicated yoga exercises. Arch Neurol 34:574–575, 1977

533. Schellas KP, Latchas RE, Wendling LR, Gold LH: Vertebrobasilar injuries following cervical manipulation. Arch Neurol 26:1450–1453, 1980

534. Levy RL, Dugan TM, Bernat JL, and Keating J: Lateral medullary syndrome after neck surgery. Neurology 30:788–790, 1980

535. Frumkin LR, Baloh RW: Wallenberg's syndrome following neck manipulation. Neurology 40: 611–615, 1990

536. Gittinger JW Jr: Occipital infarction following chiropractic cervical manipulation. J Clin Neuro-ophthalmol 6:11–13, 1986

537. Brain L: Some unsolved problems of cervical spondylosis. Br Med J 1:771–777, 1963

538. Mudd SH, Skovby F, Levy HL, et al: The natural history of homocystinuria due to cystathionine B-synthase deficiency. Am J Hum Genet 37:1–31, 1985

539. Boers GHJ, Smals AGH, Trijbels FJM, Fowler B, Bakkeren JAJM, Schooderwaldt HC, Kleijer WJ, Kloppneberg PWC: Heterozygosity for homocystinuria in premature peripheral and cerebral occlusive arterial disease. NEJM 313:709–715, 1985

540. Pavlakis SG, Phillips PC, DiMauro S, et al: Mitochondrial myopathy, encephalopathy, lactic acidosis, and strokelike episodes: A distinctive clinical syndrome. Ann Neurol 16:481–488, 1984

541. Kuriyama M, Umezaki H, Fukuda Y, et al: Mitochondrial encephalopathy with lactate-pyruvate elevation and brain infarctions. Neurology 34:72–77, 1984

542. Bird TD and Lagunoff D: Neurological manifestations of Fabry disease in female carriers. Ann Neurol 4:537–540, 1978

543. Tabira T, Goto I, Kuroiwa Y, Kikuchi M: Neuropathological and biochemical studies in Fabry's disease. Acta Neuropathol 30:345–350, 1974

544. Colucci WS, Lorell BH, Schoen FJ, et al: Hypertrophic obstructive cardiomyopathy due to Fabry's disease. NEJM 307:926–928, 1982

545. Lou HO, Reske-Nielsen E: The central nervous system in Fabry's disease. A clinical, pathological, and biochemical investigation. Arch Neurol 25: 351–359, 1971

546. Gunn M: On ophthalmoscopic evidence of general arterial disease. Trans Ophthalmol Soc UK 18: 356–381, 1898

547. Volhard F, Fahr KT. Die Brightsche Nierenkrankheit; Klinik Pathologie und Atlas. Berlin, Springer, 1914

548. Dollery CT, Hodge JV: Hypertensive retinopathy studied with fluorescein. Trans Ophthalmol Soc UK 83:115–126, 1963

549. Hayreh SS, Servais GE, Virdi PS: Fundus lesions in malignant hypertension. IV. Focal intraretinal periarteriolar transudates. Ophthalmology 93: 60–73, 1986

550. Meadows SP: The swollen optic disc. Trans Ophthalmol Soc UK 79:121–143, 1959

551. Kincaid-Smith P, McMichael J, Murphy EA: The clinical course and pathology with papilledema (malignant hypertension). Q J Med 27:117–153, 1958

552. Goldblatt H: Studies on experimental hypertension. VII. The production of the malignant phase of hypertension. J Exp Med 67:809–826, 1938

553. Walsh JB: Hypertensive retinopathy. Description, classification, and prognosis. Ophthalmology 89:1127–1131, 1982

554. Elschnig A: Die diagnostische und prognostische Bedeutung der Netzhauterkrankungen bei Nephritis. Wien Med Wochenschr 54:446–448, 1904

555. Leishman R: The eye in general vascular disease: Hypertension and arteriosclerosis. Br J Ophthalmol 41:641–701, 1957

556. Tso MO, Jampol LM: Pathophysiology of hypertensive retinopathy. Ophthalmology 89:1132–1145, 1982

557. Klien BA: Ischemic infarcts of the choroid (Elschnig spots). A cause of retinal separation in hypertensive disease with renal insufficiency. A clinical and histopathologic study. Am J Ophthalmol 66: 1069–1074, 1968

558. Gitter KA, Houser BP, Sarin LK, et al: Toxemia of pregnancy. Arch Ophthalmol 80:449–454, 1968

559. Fastenberg DM, Fetkenhour CL, Choromokos E, Shock DE: Choroidal vascular changes in toxemia of pregnancy. Am J Ophthalmol 89:362–368, 1980

560. Berkow JW, Shugarman RG, Maumenee AE, Patz A: A retrospective study of blind diabetic patients. JAMA 193:867–870, 1965

561. Goldberg MF: Natural history of untreated proliferative sickle retinopathy. Arch Ophthalmol 85:428–437, 1971

562. Goldberg MF: Classification and pathogenesis of proliferative sickle retinopathy. Am J Ophthalmol 71:649–665, 1971

563. Acacio I, Goldberg MF: Peripapillary and macular vessel occlusions in sickle anemia. Am J Ophthalmol 75:861–871, 1973

564. Radius RL, Finkelstein D: Central retinal artery occlusion (reversible) in sickle trait with glaucoma. Br J Ophthalmol 60:428- -453, 1976

565. Dizon RV, Jampol LM, Goldberg MF, Juarez C: Choroidal occlusive disease in sickle cell hemoglobinopathies. Surv Ophthalmol 23:297–306, 1979

566. Slavin ML, Barondes MJ: Ischemic optic neuropathy in sickle cell disease. Am J Ophthalmol 105:212–213, 1988

567. Hanscom TA: Indirect treatment of peripheral retinal neovascularization. Am J Ophthalmol 93:88–95, 1982

568. Rednam KRV, Jampol LM, Goldberg MF: Scatter retinal photocoagulation for proliferative sickle cell retinopathy. Am J Ophthalmol 93:594–599, 1982

569. Jampol LM, Condon P, Farber M, et al: A randomized clinical trial of feeder vessel photocoagulation of proliferative sickle cell retinopathy. I. Preliminary results. Ophthalmology 90:540–545, 1983

570. Patz A: Symposium on retrolental fibroplasia. Summary. Ophthalmology 86:1761–1763, 1979

571. Palmer EA: Optimal timing of examination for acute retrolental fibroplasia. Ophthalmology 88:662–668, 1981

572. Kinsey VE: Retrolental fibroplasia: Cooperative study of retrolental fibroplasia and the use of oxygen. Arch Ophthalmol 56:481–543, 1956

573. Kinsey VE, Arnold HJ, Kalina RE, et al: PaO2 levels and retrolental fibroplasia: A report of the cooperative study. Pediatrics 60:655–668, 1977

574. Brown DR, Milley JR, Ripepi U, Biglan AW: Retinopathy of prematurity – risk factors in a five-year cohort of critically ill premature neonates. Am J Dis Child 141:154–160, 1987

575. Adamkin DH, Shott RJ, Cook LN, Andrews BF: Nonhyperoxic retrolental fibroplasia. Pediatrics 60:828 830, 1977

576. Quinn GE, Schaffer DB, Johnson L: A revised classification of retinopathy or prematurity. Am J Ophthalmol 94:744–749, 1982

577. Flynn JT, Cassady J, Essner D, et al: Fluorescein angiography in retrolental fibroplasia: Experience from 1969–1977. Ophthalmology 86:1700–1723, 1979

578. Cryotherapy for Retinopathy of Prematurity Cooperative Group. Multicenter trial of cryotherapy for retinopathy of prematurity: preliminary results. Arch Ophthalmol 106:471–479, 1988

578a. Cryotherapy for retinopathy of prematurity cooperative group: Multicenter trial of cryotherapy for retinopathy of prematurity. One-year outcome – structure and function. Arch Ophthalmol 108:1408–1416, 1990

579. Frangieh GT, Green WR, Barraquer-Somers E, Finkelstein D: Histopathologic study of nine branch retinal vein occlusions. Arch Ophthalmol 100:1132–1140, 1982

580. Clemett RS, Kohner EM, Hamilton AM: The visual prognosis in retinal branch vein occlusion. Trans Ophthalmol Soc UK 93:523–535, 1973

581. Michels RG, Gass JDM: The natural course of retinal branch vein obstruction. Trans Am Acad Ophthalmol Otolaryngol 78:OP-166–177, 1974

582. Sanborn GE, Magargal LE: Characteristics of the hemispheric retinal vein occlusion. Ophthalmology 91:1616–1626, 1984

583. Hayreh SS: Classification of central retinal vein occlusion. Ophthalmology 90:458–474, 1983

584. Quinlan PM, Elman MJ, Bhatt AK, Mardesich P, Enger C. The natural course of central retinal vein occlusion. Am J Ophthalmol 110:118–123, 1990

585. Hayreh SS, Rojas P, Podhajsky P, et al: Ocular neovascularization with retinal vascularization occlusion–III. Incidence of ocular neovascularization with retinal vein occlusion. Ophthalmology 90:488–506, 1983

586. Minturn J, Brown GC: Progression of nonischemic central retinal vein obstruction to the ischemic variant. Ophthalmology 93:1158–1162, 1986

587. Servais GE, Thompson HS, Hayreh SS: Relative afferent pupillary defect in central retinal vein occlusion. Ophthalmology 93:301–303, 1986

588. Sabates R, Hirose T, McMeel JW: Electroretinography in the prognosis and classification of central retinal vein occlusion. Arch Ophthalmol 101:232–235, 1983

589. Brown GC, Shah HG, Magargal LE, Savino PJ: Central retinal vein obstruction and carotid artery disease. Ophthalmology 91:1627–1633, 1984

590. Sanders TE, Podos SM, Rosenbaum LJ: Intraocular manifestations of multiple myeloma. Arch Ophthalmol 77:789–794, 1967

591. Carr RE, Henkind P: Retinal findings associated with serum hyperviscosity. Am J Ophthalmol 56:23–31, 1963

592. Merin S, Freund M: Retinopathy in severe anemia. Am J Ophthalmol 66:1102–1106, 1968

593. Aisen ML, Bacon BR, Goodman AM, Chester EM: Retinal abnormalities associated with anemia. Arch Ophthalmol 101:1049–1052, 1983

Chapter 9   Inflammatory Vascular Disorders of the Brain and Eye

1. Yankner BA, Skolnik PR, Shoukimas GM, et al: Cerebral angiomatous angiitis associated with isolation of human T-lymphotropic virus type III from the central nervous system. Ann Neurol 20:362–364, 1986

2. Ford RG, Siekert RG: Central nervous system manifestations of periarteritis nodosa. Neurology 15:114–122, 1965

3. Griffith GC, Vural IL: Polyarteritis nodosa. A correlation of clinical and postmortem findings in seventeen cases. Circulation 3:481–490, 1951

4. Kernohan JW, Woltman HW: Polyarteritis nodosa: A clinico-pathologic study with special reference to

524   References

the nervous system. Arch Neurol Psychiat 39:655–686, 1938

5. Parker HL, Kernohan JW: The central nervous system in periarteritis nodosa. Mayo Clin Proc 24:43–48, 1949

6. Benedict WL, Wagener HP, Hollenhorst RW, Henderson JW: The ocular manifestations of the diffuse collagen diseases. Am J Med Sci 221:211–222, 1951

7. Sheehan B, Harriman DGF, Bradshaw JPP. Polyarteritis nodosa with ophthalmic and neurological complications. Arch Ophthalmol 60:537–547, 1958

8. Goldstein I, Wexler D: Bilateral atrophy of the optic nerve in periarteritis nodosa. Arch Ophthalmol 18:767–773, 1937

9. Goldsmith J: Periarteritis nodosa with involvement of the choroidal and retinal arteries. Am J Ophthal 29:435–446, 1946

10. Solomon SM, Solomon JH: Bilateral central artery occlusions in polyarteritis nodosa. Ann Ophthalmol 10:567–569, 1978

11. Herson RN, Sampson R: The ocular manifestations of polyarteritis nodosa. Q J Med 18:123–132, 1949

12. Kimbrell OC, Wheliss JA: Polyarteritis nodosa complicated by bilateral optic neuropathy. J Am Med Assoc 201:61–62, 1967

13. Hutchinson CH: Polyarteritis nodosa presenting as posterior ischemic optic neuropathy. J Roy Soc Med 77:1043–1046, 1977

13a. Diamond G, Warren F, Miller D, Marks C, Kupersmith MJ. Steroid responsive optic neuropathy. Frank Walsh Society, Vancouver, May 7–8, 1988

14. Hope-Robertson WJ: Pseudo-tumour of the orbit as a presenting sign in periarteritis nodosa. Trans Ophthalmol Soc NZ 8:56–66, 1956

15. Van Wien S, Merz EH: Exophthalmos secondary to periarteritis nodosa. Am J Ophthalmol 56:204–208, 1963

16. Walton EW: Pseudo-tumour of the orbit and polyarteritis nodosa. J Clin Pathol 12:419–426, 1959

17. Leonhardt ETG, Kullenberg KPG: Bilateral atrial myxomas with multiple arterial aneurysms – a syndrome mimicking polyarteritis nodosa. Am J Med 62:792–794, 1977

18. Ferris EJ: Arteritis. In: Newton TH, Potts DG (ed) Radiology of the skull and brain. CV Mosby, St. Louis, 1974

19. Frohnert PP, Sheps SG: Long-term follow-up study of periarteritis nodosa. Am J Med 43:8–14, 1967

20. Fauci AS, Katz P, Haynes BF, Wolff SM. Cyclophosphamide therapy of severe systemic necrotizing vasculitis. NEJM 301:235–238, 1979

21. O'Connor JF, Musher DM: Central nervous system involvement in systemic lupus erythematosus. Arch Neurol 14:157–164, 1966

22. Bresnihan B, Hohmeister R, Cutting J, et al: The neuropsychiatric disorder in systemic lupus erythematosus: Evidence for both vascular and immune mechanisms. Ann Rheum Dis 38:301–306, 1979

23. Gorelick PB, Rusinowitz MS, Tiku M, McDonald LW, Robbins L: Embolic stroke complicating systemic erythematosus. Arch Neurol 42:813–815, 1985

24. Devinsky O, Petito CK, Alonso DR: Clinical and neuropathological findings in systemic lupus erythematosus: The role of vasculitis, heart emboli, and thrombotic thrombocytopenic purpura. Ann Neurol 23:380–384, 1988

25. Andriankanakos AA, Duffy J, Suzuki M, Sharp JT: Transverse myelopathy in systemic lupus erythematosus: Report of three cases and review of the literature. Ann Intern Med 83:616–624, 1975

26. Siekert RG, Clark EC: Neurological signs and symptoms as early manifestations of systemic lupus erythematosus. Neurology 5:84–88, 1955

27. Feinglass EJ, Arnett FC, Dorsch CA, Zizic TM, Stevens MB: Neuropsychiatric manifestations of systemic lupus erythematosus. Diagnosis, clinical spectrum, and relationship to other features of the disease. Medicine 55:323–339, 1976

28. Gold DH, Morris DA, Henkind P: Ocular findings in systemic lupus erythematosus. Br J Ophthalmol 56:800–804, 1972

29. Pfaffenbach DD, Hollenhorst RW: Microangiopathy of the retinal arterioles. JAMA 225:480–483, 1973

30. Appen R, Wray S, Cogan D: Central retinal artery occlusion. Am J Ophthalmol 79:374–381, 1975

31. Dougal MA, Evans LS, McClellan KR, Robinson J: Central retinal artery occlusion in systemic lupus erythematosus. Ann Ophthalmo 15:38–40, 1983

32. Wong K, Ai E, Jones JV, Young D: Visual loss as the initial symptom of systemic lupus erythematosus. Am J Ophthalmol 92:238–244, 1981

33. Coppeto J, Lessell S: Retinopathy in systemic lupus erythematosus. Arch Ophthalmol 95:794–797, 1977

34. Nichols CJ, Mieler WF: Severe retinal vaso-occlusive disease secondary to procainamide-induced lupus. Ophthalmology 96:1535–1540, 1989

35. Lessell S: Some ophthalmological and neurological aspects of systemic lupus erythematosus. J Rheumatol 7:398–404, 1980

36. Jabs DA, Hanneken AM, Schachat AP, Fine SL: Choroidopathy in systemic lupus erythematosus. Arch Ophthalmol 106:230–234, 1988

37. Cordes FC, Aiken SD: Ocular changes in acute disseminated lupus erythematosus. Am J Ophthalmol 30:1541–1555, 1947

38. Jabs DA, Miller NR, Newman SA: Optic neuropathy in systemic lupus erythematosus. Arch Ophthalmol 104:564–568, 1986

39. Hackett ER, Martinez RD, Larson PF, Paddison RM: Optic neuritis in systemic lupus erythematosus. Arch Neurol 31:9–11, 1974

40. Kupersmith MJ, Burde RM, Warren FA, et al: Auto-

immune optic neuropathy: evaluation and treatment. J Neurol Neurosurg Psychiatr 51:1381–1386, 1988

41. April RS, Vansonnenberg EA: A case of neuromyelitis optica (Devic's syndrome) in systemic lupus erythematosus. Neurology 26:1066–1070, 1976

42. Vitale C, Kahn MF, de Seze M, de Seze S: Nevrite optique, myelite et maladie lupique. A propos de deux observations. Ann Med Interne 124:211–216, 1973

43. Brandt KD, Lessell S, Cohen AS: Cerebral disorders of vision in systemic lupus erythematosus. Ann Intern Med 83:163–169, 1975

44. Brandt KD, Lessell D. Migrainous phenomena in systemic lupus erythematosus. Arthritis Rheum 21:7–16, 1978

45. Traboulsi EI, Mansour AM, Aswad MI, Gharzuddin W, Frayha RA: Homonymous hemianopia and systemic erythematosus. J Clin Neuro-Ophthalmol 5:63–66, 1985

46. Honda Y: Scintillating scotoma as the first symptom of systemic lupus erythematosus. Am J Ophthalmol 99:607, 1985

47. Silberberg D, Laties A: Increased intracranial pressure in disseminated lupus erythematosus. Arch Neurol 29:88–90, 1973

48. Bettman J, Daroff, Sanders M, et al: Papilledema and asymptomatic intracranial hypertension in systemic lupus erythematosus. Arch Ophthalmol 80:189–193, 1968

49. Carlow TJ, Glaser JS: Pseudotumor cerebri syndrome in systemic lupus erythematosus. JAMA 228:197–200, 1974

50. Brenner EH, Shock JP: Proptosis secondary to systemic lupus erythematosus. Arch Ophthalmol 91:81–82, 1974

51. Wilkinson LS, Panush RS: Exophthalmos associated with systemic lupus erythematosus. Arthritis Rheum 18:188–189, 1975

52. Evans OB, Lexow SS: Painful ophthalmoplegia in systemic lupus erythematosus. Ann Neurol 4:584–585, 1978

53. Cogan DG, Jubik CS, Smith JL: Unilateral internuclear ophthalmoplegia. Arch Ophthalmol 44:783–796, 1950

54. Meyer MW, Wild JH: Unilateral internuclear ophthalmoplegia in systemic lupus erythematosus. Arch Neurol 32:487, 1975

55. Johnson RT, Richardson EP: The neurological manifestations of systemic lupus erythematosus: A clinical-pathological study of 24 cases and review of the literature. Medicine 47:337–369, 1968

56. Petz LD, Sharp GC, Cooper NR, Irvin WS: Serum and cerebral spinal fluid complement and serum autoantibodies in systemic lupus erythematosus. Medicine 50:259–275, 1971

57. Harbeck RJ: Anti-DNA complexes in cerebrospinal fluid in patients with SLE. Arthritis Rheum 16:552, 1973

58. Buyon JP, Tamarius J, Kolb W, et al: Determination of plasma complement split products (CSP) in lupus is the most sensitive predictor of disease activity. Clin Res 1988; abstr 531A

59. Atkins CJ, Kondon JJ, Quismoro FP, Friou GJ: The choroid plexus in systemic lupus erythematosus. Ann Intern Med 76:65–72, 1972

60. Bilaniuk LT, Patel S, Zimmerman RA: Computed tomography of systemic lupus erythematosus. Radiology 124:119–121, 1977

61. Gonzalez-Scarano F, Lisak RP, Bilaniuk LT, Zimmerman RA, Atkins PC, Zweiman B: Cranial computed tomography in the diagnosis of systemic lupus erythematosus. Ann Neurol 5:158–165, 1978

62. Miller DH, Ormerod IEC, Gibson A, du Boulay EPGH, Rudge P, McDonald WI. MR brain scanning in patients with vasculitis: differentiation from multiple sclerosis. Neuroradiology 29:226–231, 1987

63. Diederischen H. Neurologiske komplikationer ved lupus erythematosus disseminatus. Ugeskr Laeg 127:1220, 1965

64. Trevor RP, Sondheimer FK, Fessel WJ, Wolpert SM: Angiographic demonstration of major cerebral vessel occlusion in systemic lupus erythematosus. Neuroradiol 4:202–207, 1972

65. Ellis SG, Verity MA: Central nervous system involvement in systemic lupus erythematosus: A review of neuropathologic findings in 157 cases. 1955–1977. Semin Arthritis Rheum 8:212–221, 1979

66. Graham EM, Spalton DJ, Barnard RO, Garner A, Russell WR: Cerebral and retinal vascular changes in systemic lupus erythematosus. Ophthalmology 92:444–448, 1985

67. Malamud N, Saver G: Neuropathologic findings in disseminated lupus erythematosus. Arch Neurol 54:723–731, 1954

68. Kinney EL, Berdoff RL, Nagbhushan SR, Fox LM: Devic's syndrome and systemic lupus erythematosus. A case report with necroscopy. Arch Neurol 36:643–644, 1979

69. Sergent JS, Lockshin MD, Klempner MS, Lipsky BA: Central nervous system disease in systemic lupus erythematosus. Am J Med 58:644–654, 1975

70. Austin HA, Klippel JH, Balo JE, leRiche NGH, et al. Therapy of lupus nephritis. controlled trial of prednisone and cytotoxic drugs. NEJM 314:614–619, 1986

71. Dutton JJ, Burde RM, Klingele TC. Autoimmune retrobulbar optic neuritis. Am J Ophthalmol 94:11–17, 1982

72. Lessell S. The neuro-ophthalmology of systemic lupus erythematosus. Doc Ophthalmol 47:13–42, 1979

73. Hochberg MC: Adult and juvenile rheumatoid arthritis: Current epidemiologic concepts. Epidemiol Rev 2:27–44, 1981

74. Beck DO and Corbett J. Seizures due to central nervous system rheumatoid meningovasculitis. Neurology 33:1058–1061, 1983

526   References

75. Ramos M, Mandybur TL: Cerebral vasculitis in rheumatoid arthritis. Arch Neurol 32:271−275, 1975

76. Gupta VP, Ehrlich GE: Organic brain syndrome in rheumatoid arthritis following corticosteroid withdrawal. Arthritis Rheum 19:1333−1338, 1976

77. Watson P, Fekete J, Deck J: Central nervous system vasculitis in rheumatoid arthritis. Can J Neurol Sci 4:269−272, 1977

78. Edwards KR. Hemmorhagic complications of cerebral arteritis. Arch Neurol 34:549−555, 1977

79. Pirani CL, Bennett GA: Rheumatoid arthritis: A report of three cases progressing from childhood and emphasizing certain systemic manifestations. Bull Hosp Joint Dis 12:335−367, 1951

80. Kim RC: Rheumatoid disease with encephalopathy. Ann Neurol 7:86−91, 1980

81. Kanski JJ: Juvenile arthritis and uveitis. Surv Ophthalmol 34:253−267, 1990

82. Smith JL: Ocular complications of rheumatic fever and rheumatoid arthritis. Am J Ophthalmol 43:575−582, 1957

83. Scherbel A, Mackenzie AH, Nousek JA, Atdigian M: Ocular lesions in rheumatoid arthritis and related disorders with particular reference to retinopathy. A study of 741 patients with and without chloroquine drugs. NEJM 273:360−366, 1965

84. Bathon JM, Moreland LW, DiBartolomeo AG: Inflammatory central nervous system involvement in rheumatoid arthritis. Seminars in Arthritis and Rheumatism 18:258−266, 1989

85. Crompton JL, Iyer P, Begg MW. Vasculitis and ischaemic optic neuropathy associated with rheumatoid arthritis. Aust J Ophthalmol 8:219−230, 1980

86. Drachman DA: Neurological complications of Wegener's granulomatosis. Arch Neurol 8:145−155, 1963

87. Stern GM, Hoffbrand AV, Urich H: The peripheral nerves and skeletal muscles in Wegener's granulomatosis. Brain 88:151−164, 1965

88. Yamashita Y, Takahashi M, Bussaka H, et al: Cerebral vasculitis secondary to Wegener's granulomatosis: Computed tomography and angiographic findings. J Comp Tomogr 10:115−120, 1986

89. Kornblut AD, Wolff SM, DeFries HO, Fauci AS: Wegener's granulomatosis. Laryngoscope 90:1453−1465, 1980

90. Spalton DJ, Sanders MD: Ocular manifestations of Wegener's granulomatosis. Am J Ophthalmol 65:553−563, 1981

91. Straatsma BR: Ocular manifestations of Wegener's granulomatosis. Am J Ophthalmol 44:789−799, 1957

92. Cogan DG: Corneoscleral lesions in periarteritis nodosa and Wegener's granulomatosis. Trans Am Ophthalmol Soc 53:321−344, 1965

93. Weiter J, Farkas TG: Pseudotumor of the orbit as a presenting sign in Wegener's granulomatosis. Surv Ophthalmol 17:106−119, 1972

94. Austin P, Green WR, Sallyer DC, et al: Peripheral corneal degeneration and occlusive vasculitis in Wegener's granulomatosis. Am J Ophthalmol 85:311−317, 1978

95. Blodi FC, Gass JDM: Inflammatory pseudotumor of the orbit. Br J Ophthalmol 52:79−93, 1968

96. Faulds JS, Wear AR: Pseudotumor of the orbit and Wegener's granuloma. Lancet 2:955−957, 1960

97. Haynes BF, Fishman MC, Fauci AS, Wolff SM: The ocular manifestations of Wegener's granulomatosis. Fifteen years experience and review of the literature. Am J Med 63:131−141, 1977

98. Vermess M, Haynes BF, Fauci AS, Wolff SM: Computer assisted tomography of orbital lesions in Wegener's granulomatosis. J Comp Assist Tomogr 2:45−48, 1978

99. Goldberg AL, Taievsky AL, Jamshidi S. Wegener granulomatosis invading the cavernous sinus: a CT demonstration. J Comp Assist Tomograph 7:701−703, 1983

100. Couto RE, Klein M, Lessell S, Friedman E, Snider GL. Limited form of Wegener granulomatosis. JAMA 33:868−871, 1975

101. Greenberg MH: Central retinal artery closure in Wegener's granulomatosis. AJO 63:515−516, 1967

102. Nolan J, Cullen JF: Retinal vasculitis associated with anterior uveitis. Br J Ophthalmol 51:361−374, 1967

103. Budzilovich GN, Wilens SL: Fulminating Wegener's granulomatosis. Arch Pathol 70:653−660, 1960

104. Reza MJ, Dornfeld L, Goldberg LS, Bluestone R, Pearson CM: Wegener's granulomatosis. Long term follow up of patients treated with cyclophosphamide. Arthritis Rheum 18:501, 1975

105. Fauci AS, Wolff SM: Wegener's granulomatosis: Studies in eighteen patients and a review of the literature. Medicine 52:535−561, 1973

106. Norton WL, Suki W, Strunk S: Combined corticosteroid and azathioprine therapy in two patients with Wegener's granulomatosis. Arch Intern Med 121:554−560, 1968

107. Israel HL, Parchefsky AS: Wegener's granulomatosis of lung: Diagnosis and treatment. Experience with 12 cases. Arch Intern Med 74:881−891, 1971

108. Israel HL. Sulfamethoxazole−timethoprim therapy for Wegener's granulmatosis. Arch Intern Med 148:2293−2295, 1988

109. Behçet H: Ueber die rezidivierende Aphthose durch ein Virus verursachte Geschwüre am Mund, am Auge und an den Genitalien. Dermat Wochenschr 105:1152−1157, 1937

110. Chajek T, Fainaru M: Report of 41 cases and a review of the literature. Medicine 54:179−196, 1975

111. Shimizu T, Ehrlich GE, Inaba G, Hayashi K: Behcet's disease (Behcet syndrome). Semin Arthritis Rheum 8:223−260, 1979

112. Whitty CWM. Neurologic implications of Behcet's syndrome. Neurology 8:364–373, 1958

113. Bousser MG, Bietry O, Launay M, et al: Thromboses veineuses cerebrales au cours de la maladie de Behcet. Rev Neurol 136:753–762, 1980

114. McMenemy WH, Lawrence BJ: Encephalopathy in Behcet's disease. Lancet 2:353–359, 1957

115. Evans AD, Pallis CA, Spillane JD. Involvement of the nervous system in Behcet's syndrome: report of three cases and isolation of virus. Lancet 2: 349–353, 1957

116. Colvard DM, Robertson DM, O'Duffy JD. The ocular manifestations of Behcet's disease. Arch Ophthalmol 95:1813–1817, 1977

117. Kalbian V, Challis M: Behcet's disease: Report of 12 cases with three manifesting as papilledema. Am J Med 49:823–829, 1970

118. Pamir MN, Kansu T, Erbengi A, Zileli T: Papilledema in Behcet's syndrome. Arch Neurol 38:643–645, 1981

119. Kawakita H, Nishimura M, Satoh Y, et al: Neurological aspects of Behcet's disease. J Neurol Sci 5:417–439, 1967

120. Yazici H, Tuzun H, Pazarli S, et al: Influence of age of onset and patient's sex on the prevalence and severity of manifestations of Behcet's syndrome. Am Rheum Dis 43:783–789, 1984

121. Pallis CA, Fudgek BJ. The neurologcial complications of Behcet's syndrome. Arch Neurology and Psych 75:1–14, 1956

122. Lakhanpal S, Tani K, Lie JT, et al: Pathologic features of Behcet's syndrome: A review of Japanese autopsy registry data. Human Pathology 16: 790–795, 1985

123. Schotland DL, Wolf SM, White HH, Dubin HV: Neurologic aspects of Behcet's disease. Case report and review of the literature. Am J Med 34:544–553, 1963

124. Kozin F, Haughton V, Bernhard GC. Neuro-Behcet disease: two cases and neuroradiologic findings. Neurology 27:1148–1152, 1977

125. Serdaroglu P, Yazici H, Ozdemir C, et al: Neurologic involvement in Behcet's syndrome. A prospective study. Arch Neurol 46:265–269, 1989

126. Dinning WJ, Perkins ES: Immunosuppressives in uveitis. A preliminary report of experience with chlorambucil. Br J Ophthalmol 59:397–403, 1975

127. Buckley CE, Gillis JP: Cyclophosphamide therapy of Behcet's disease. J Allergy 43:273–283, 1969

128. Zelenski JD, Capraro JA, Holden D, Calabrese LH: Central nervous system vasculitis in Behcet's syndrome: Angiographic improvement after therapy with cytotoxic agents. Arthritis and Rheumatism 32:217–220, 1989

129. Miyachi Y, Taniguchi S, Ozaki M, Horio T: Colchicine in the treatment of the cutaneous manifestations of Behcet's disease. Br J Dermatol 104:67–69, 1981

130. BenEzra D, Cohen E, Chajek T, et al: Evaluation of conventional therapy versus cyclosporine A in Behcet's syndrome. Transplant Proc 20 (suppl 4):136–143, 1988

131. Cravioto H, Feigin I: Non-infectious granulomatous angiitis with a predilection for the nervous system. Neurology 9:599–609, 1959

132. Budzilovich GN, Feigin I, Siegel H: Granulomatous angiitis of the central nervous system. Arch Pathol 76:250–256, 1963

133. Kolodny EH, Rebeiz JJ, Caviness VS, Richardson EP: Granulomatous angiitis of the central nervous system. Arch Neurol 19:510–524, 1968

134. Moore PM: Diagnosis and management of isolated angiitis of the central nervous system. Neurology 39:167–173, 1989

135. Herring AB, Urich H: Sarcoidosis of the central nervous system. J Neurol Sci 9:405–422, 1969

136. Caplan L, Corbett J, Goodwin J, et al. Neuro-ophthalmologic signs in the angiitic form of neurosarcoidosis. Neurology 33:1130–1135, 1983

137. Rewcastle NB, Tom MI: Non-infectious granulomatous angiitis of the nervous system associated with Hodgkin's disease. J Neurol Neurosurg Psychiatr 25:51–58, 1962

138. Younger DS, Hays AP, Brust JC, Rowland LP: Granulomatous angiitis of the brain. An inflammatory reaction of diverse etiology. Arch Neurol 45:514–518, 1988

139. Price DL, Harris JL, New PFJ, Cantu RC: Cardiac myxoma. A clinicopathologic and angiographic study. Arch Neurol 23:558–567, 1970

140. Levine DN, Davis KR, Richardson EP: Case records of the Massachusetts General Hospital. NEJM 320:514–524, 1989

141. Johnson M, Maciunas R, Dutt P, et al: Granulomatous angiitis masquerading as a mass lesion. Magnetic resonance imaging and stereotactic biopsy findings in a patient with occult Hodgkin's disease. Surg Neurol 31:49–53, 1989

142. Nurick S, Blackwood W, Mair WGP: Giant cell granulomatous angiitis of the central nervous system. Brain 95:133–142, 1972

143. Vincent FM: Granulomatous angiitis. NEJM 296:452, 1977

144. Cupps TR, Moore PM, Fauci AS: Isolated angiitis of the central nervous system: Prospective diagnostic and therapeutic experience. Am J Med 74:97–105, 1983

145. Launes J, Iivanainen M, Erkinjuntti T, Vuorialho M: Isolated angiitis of the central nervous system. Acta Neurolog Scand 74:108–114, 1986

146. Burger PC, Burch JG, Vogel FS. Granulomatous angiitis, an unusual etiology of stroke. Stroke 8:29–35, 1977

147. Craven RS, French JK: Isolated angiitis of the central nervous system. Ann Neurol 18:263–265, 1985

148. Vanderzant C, Bromberg M, MacGuire A, McCune

WJ: Isolated small-vessel angiitis of the central nervous system. Arch Neurol 45:683–687, 1988

149. Rajjoub RK, Wood JH, Ommaya AK: Granulomatous angiitis of the brain: A successfully treated case. Neurology 27:588–591, 1977

150. Rosenblum WI, Hadfield MG: Granulomatous angiitis of the nervous system in cases of herpes zoster and lymphosarcoma. Neurology 22: 348–354, 1972

151. MacKenzie RA, Forbes GS, Karnes WE: Angiographic findings in herpes zoster arteritis. Ann Neurol 10:458–464, 1981

152. Linneman CC, Alvira MM: Pathogenesis of varicella-zoster angiitis in the CNS. Arch Neurol 37: 239–240, 1980

153. Acers TE: Herpes zoster ophthalmicus with contralateral hemiplegia. Arch Ophthalmol 71:371–376, 1964

154. Eidelberg D, Sorrel A, Horoupian DS, Neumann PE, et al. Thrombotic cerebral vasculopathy associated with herpes zoster. Ann Neurol 19:7–14, 1986

155. Marsh RJ, Dulley B, Kelly V: External ocular motor palsies in ophthalmic zoster: A review. Br J Ophthalmol 61:677–682, 1977

156. Brown RM, Mendis U: Retinal arteritis complicating herpes zoster ophthalmicus. Br J Ophthalmol 57: 344–346, 1973

157. Hesse RJ: Herpes zoster ophthalmicus associated with delayed retinal thrombophlebitis. Am J Ophthalmol 84:329–331, 1977

158. Browning DJ, Blumenkranz MS, Culbertson WW, Clarkson JD, Tardif Y, Gourdeau A, Minturn J: Association of varicella zoster dermatitis with acute retinal necrosis syndrome. Ophthalmology 94: 602–606, 1987

159. Hall S, Carlin L, Roach ES, McLean WT: Herpes zoster and central retinal artery occlusion. Ann Neurol 13:217–218, 1983

160. Harrison EQ: Complications of herpes zoster ophthalmicus. Am J Ophthalmol 60:1111–1114, 1965

161. McAdam LP, O'Hanlan MA, Bluestone R, Pearson CM: Relapsing polychondritis: Prospective study of 23 patients and a review of the literature. Medicine 55:193–215, 1976

162. Michet CJ, McKenna CH, Luthra HS, O'Fallon WM: Relapsing polychondritis: Survival and predictive role of early disease manifestations. Ann Intern Med 104:74–78, 1986

163. Stewart SS, Ashizawa T, Dudley AW, Goldberg JW, Lidsky MD: Cerebral vasculitis in relapsing polychondritis. Neurology 38:150–152, 1988

164. Churg J, Strauss L. Allergic granulomatosis, allergic angiitis, and periarteritis nodosa. Am J Pathol 27:277–301, 1951

165. Chumbley LC, Harrison EG, DeRemee RA. Allergic granulomatosis and angiitis (Churg–Strauss syndrome): report and analysis of 30 cases. Mayo Clin Proc 52:477–484, 1977

166. Weinstein JM, Chui H, Lane S, Corbett J, Towfighi J: Churg–Strauss syndrome (allergic granulomatous angiitis). Neuro-ophthalmic manifestations. Arch Neurol 101:1217–1220, 1983

167. Citron BP, Halpern M, McCarron M, et al: Necrotizing angiitis associated with drug abuse. NEJM 283:1003–1011, 1970

168. Lignelli GJ, Buchhect WA: Angiitis in drug abusers. NEJM 284:112–113, 1971

169. Margolis MT, Newton TH: Methamphetamine ("speed") arteritis. Neuroradiology 2:170–182, 1971

170. Rumbaugh CL, Bergeron RT, Fang HCH, McCormick R: Cerebral angiographic changes in the drug abuse patient. Radiol 101:335–344, 1971

171. Delaney P, Estes M: Intracranial hemorrhage with amphetamine abuse. Neurology 30:1125–1128, 1980

172. Harrington H, Heller HA, Dawson D, Caplan L, Rumbaugh C: Intracerebral hemorrhage and oral amphetamine. Arch Neurol 40:503–507, 1983

173. Shaw HE, Lawson JG, Stulting RD: Amaurosis fugax and retinal vasculitis associated with methamphetamine inhalation. J Clin Neuro-ophthalmol 5:169–176, 1985

174. Trugman JM: Cerebral arteritis and oral methylphenidate. Lancet 1:584–585, 1988

175. Sobel J, Espinas OE, Friedman SA: Carotid artery obstruction following LSD capsule ingestion. Arch Intern Med 11:657–666, 1971

176. Lieberman AN, Bloom W, Kishore PS, Lin JP: Carotid artery occlusion following ingestion of LSD. Stroke 5:213–215, 1974

177. Klonoff DC, Andrews BT, and Obana WG: Stroke associated with cocaine use. Arch Neurol 46:989–993, 1989

178. Krendel DA, Ditter SM, Frankel MR, and Ross WK: Biopsy-proven cerebral vasculitis associated with cocaine abuse. Neurology 40:1092–1094, 1990

179. Kaye BR and Fainstat M: Cerebral vasculitis associated with cocaine abuse. JAMA 258:2104–2106, 1987

180. Levine SR, Brust JCM, Furrell N, Ho KL, Blake D, Millikan CH, Brass LM, Fayad P, Schultz LR, Selwa JF, and Welch KMA: Cerebrovascular complications of the use of the "crack" form of alkaloidal cocaine. N Engl J Med 323:699–704, 1990

181. Lupi-Herrera E, Sanchez-Torres G, Marcushamer J, et al: Takayasu's arteritis. Clinical study of 107 cases. Am Hear J 93:94–103, 1977

182. Takayasu M: A case with peculiar changes of the central retinal vessels. Acta Soc Ophthalmol Jpn 12:554–557, 1908

183. Vinijchaikul K: Primary arteritis of the aorta and its main branches (Takayasu's arteriopathy). A clinicopathologic autopsy study of eight cases. Am J Med 43:15–27, 1967

184. Lande A, Berkmen YA: Aortitis. Pathologic, clini-

cal and arteriographic review. Radiol Clin North Am 14:219–240, 1976

185. Fraga A, Mintz G, Valle L, Flores-Izquierdo G: Takayasu's arteritis: Frequency of systemic manifestations (study of 22 patients) and favorable response to maintenance steroid therapy with adrenocorticosteroids (12 patients). Arthritis Rheum 15:617–624, 1972

186. Bloss RS, Duncan JM, Cooley DA, Leatherman LL, Schnee MJ: Takayasu's arteritis: Surgical considerations. Ann Thor Surg 27:574–579, 1979

187. Cheson BD, Bluming AZ, Alroy J: Cogan's syndrome: A systemic vasculitis. Am J Med 60: 549–555, 1976

188. Cogan DG: Syndrome of nonsyphilitic interstitial keratitis and vestibuloauditory symptoms. Arch Ophthalmol 33:144–149, 1945

189. Cogan DG, Dickersin GR: Nonsyphilitic interstitial keratitis with vestibuloauditory symptoms. Arch Ophthalmol 71:172–175, 1964

190. Haynes BF, Kaiser-Kupfer M, Mason P, Fauci AS: Cogan syndrome: Studies in thirteen patients, long-term follow up, and a review of the literature. Medicine 59:426–441, 1980

191. Crawford WJ: Cogan's syndrome associated with polyarteritis nodosa. A report of 3 cases. Penn Med J 60:835–838, 1957

192. Vollersten RS, McDonald TJ, Younge BR, et al: Cogan's syndrome: 18 cases and a review of the literature. Mayo Clin Proc 61:344–361, 1986

193. Bicknell JM, Holland JV: Neurological manifestations of Cogan's syndrome. Neurology 28: 278–281, 1978

194. Cobo LM, Haynes BF: Early corneal findings in Cogan's syndrome. Ophthalmology 91:903–907, 1984

195. Haynes BF, Pikus A, Kaiser-Kupfer M, Fauci AS: Successful treatment of sudden hearing loss in Cogan's syndrome with corticosteroids. Arthritis Rheum 24:501–503, 1981

196. McCabe BF: Autoimmune sensorineural hearing loss. Ann Otol 88:585–590, 1970

197. Liebow AS, Carrington CRB, Friedman PJ: Lymphatoid granulomatosis. Hum Pathol 3:457–458, 1972

198. Katzenstein AA, Carrington CB, Liebow AA: Lymphatoid granulomatosis. A clinicopathologic study of 152 cases. Cancer 43:360–373, 1979

199. Hogan PJ, Greenberg MK, McCarty GE: Neurologic complications of lymphatoid granulomatosis. Neurology 31:619–620, 1981

200. Kokmen E, Billman JK, Abell MR: Lymphatoid granulomatosis clinically confined to the CNS. A case report. Arch Neurol 34:782–784, 1977

201. Tse DT, Mandelbaum S, Chuck DA, et al: Lymphatoid granulomatosis with ocular involvement. Retina 5:94–97, 1985

202. Fauci AS, Haynes BF, Costa J, et al: Lymphatoid granulomatosis. Prospective clinical and therapeutic experience over 10 years. NEJM 306:68–74, 1982

203. Susac JO, Hardman JM, Selhorst JB: Microangiopathy of the brain and retina. Neurology 29: 313–316, 1979

204. Delaney WV, Torris PF: Occlusive retinal vascular disease and deafness. AJO 82:323–336, 1976

205. Gass JDM, Tiedeman J, Thomas MA: Idiopathic recurrent branch retinal arterial occlusion. Ophthalmology 93:1148–1157, 1986

206. Willerson D, Ashberg TM, Reeser FH. Necrotizing vaso-occlusive retinitis. Am J Ophthalmol 84: 209–219, 1977

207. Minckler DS, McLean EB, Shaw CM, Hendrikson A: Herpes virus hominis encephalitis and retinitis. Arch Ophthalmol 94:89–95, 1976

208. Sanders M: Retinal arteritis, retinal vasculitis and autoimmune retinal vasculitis. Eye 1:1441–1446, 1987

209. Savir H, Wender T, Creter D, Djaldetti M, Stein R: Bilateral retinal vasculitis associated with clotting disorders. Am J Ophthalmol 84:542–547, 1977

210. Crouch ER Jr, Goldberg MF: Retinal periarteritis secondary to syphilis. Arch Ophthalmol 93: 384–387, 1975

211. Freeman WR, Chen A, Henderly DE, Levine AM, Luttrull JK, et al: Prevalence and significance of acquired immunodeficiency syndrome-related retinal microvasculopathy. Am J Ophthalmol 107: 229–235, 1989

212. Lewis ML, Culbertson WW, Post JD, Miller D, Kokame GT, Dix RD: Herpes simplex virus type I. A cause of the acute retinal necrosis syndrome. Ophthalmology 96:875–878, 1989

213. Culbertson WW, Blumenkranz MS, Pepose JS, et al: Varicella zoster virus is a cause of the acute retinal necrosis syndrome. Ophthalmology 93: 559–569, 1986

214. Sergott RC, Belmont JB, Savino PJ, et al: Optic nerve involvement in the acute retinal necrosis syndrome. Arch Ophthalmol 103:1160–1162, 1985

215. Margolis T, Irvine AR, Hoyt WF, Hyman R: Acute retinal necrosis syndrome presenting with papillitis and arcuate neuroretinitis. Ophthalmology 95: 937–940, 1988

216. Sergott RC, Anand R, Belmont JB, Fischer DH, Bosley TM, Savino PJ: Acute retinal necrosis neuropathy. Clinical profile and surgical therapy. Arch Ophthalmol 107:692–696, 1989

217. Schneiderman JH, Sharpe JA, Sutton DM: Cerebral and retinal vascular complications of inflammatory bowel disease. Ann Neurol 5:331–337, 1979

218. Duker JS, Brown GC, Brooks L: Retinal vasculitis in Crohn's disease. Am J Ophthalmol 103: 664–668, 1987

219. Zegarra H, Guttman FA, Zakov N, Carim M: Par-

tial occlusion of the central retinal vein. Am J Ophthalmol 96:330–337, 1983

220. Heuer HK, Gager WE, Reeser FH: Ischemic optic neuropathy associated with Crohn's disease. J Clin Neuro-Ophthalmol 2:175–181, 1982

221. Petrelli EA, McKinley M, Troncale FJ: Ocular manifestations of inflammatory bowel disease. Ann Ophthalmol 14:356–360, 1982

222. Sheibani K, Battifora H, Weinberg CD, Burke JS, et al: Further evidence that "malignant angioendotheliomatosis" is an angiotropic large cell lymphoma. NEJM 314:943–948, 1986

223. Elner VM, Hidayat AA, Charles NC, Davitz MA, Smith ME, Burgess D, Dawson N: Neoplastic angioendotheliomatosis. A variant of malignant lymphoma immunohistochemical and ultra-structural observations of three cases. Ophthalmology 93:1237–1245, 1986

224. Grand MG, Kaine J, Fulling K, Atkinson J, Dowton SB, Farber M, Craver J, Rice K: Cerebroretinal vasculopathy. A new hereditary syndrome. Ophthalmology 95:649–659, 1988

225. Coppeto JR, Currie JN, Monteiro MLR, Lessell S. A syndrome of arterial-occlusive retinopathy and encephalopathy. Am J Ophthalmol 98:189–202, 1984

226. Eales H: Primary retinal haemorrhage in young men. Ophthalmic Rev 1:41–46, 1882

227. Spitznas M, Meyer-Schwickerath G, Stephan B: The clinical picture of Eales' disease. Albrecht von Graefes Arch Klin Exp Ophthalmol 194:73–85, 1975

228. Patrinely JR, Green WR, Randolph ME: Retinal phlebitis with chorioretinal emboli. Am J Ophthalmol 94:49–57, 1982

229. Jampol LM, Isenberg SJ, Goldberg MF: Occlusive retinal arteriolitis with neovascularization. Am J Ophthalmol 81:583–589, 1976

230. Gass JDM: Acute posterior multifocal placoid pigment epitheliopathy. Arch Ophthalmol 80:177–185, 1968

231. Smith CH, Savino PJ, Beck RW, Schatz NJ, Sergott RC. Acute posterior multifocal placoid pigment epitheliopathy and cerebral vasculitis. Arch Neurol 40:48–50, 1983

232. Wilson CA, Choromokos EA, Sheppard R. Acute posterior multifocal pigment epitheliopathy. Arch Ophthalmol 106:796–800, 1988

233. Sigelman J, Behrens M, Hilal S: Acute posterior multifocal placoid pigment epitheliopathy associated with cerebral vasculitis and homonymous hemianopia. Am J Ophthalmol 88:919–924, 1979

234. Kirkham T, Ffytche T, Sanders M: Placoid pigment epitheliopathy with retinal vasculitis and papillitis. Br J Ophthalmol 56:875–880, 1972

235. Savino PJ, Weinberg RJ, Yassin JG, Pilkerton AR: Diverse manifestations of acute posterior multi-

focal placoid pigment epitheliopathy. Am J Ophthalmol 77:659–662, 1974

236. Jenkins RB, Savino PJ, Pilkerton AR: Placoid pigment with swelling of the optic discs. Arch Neurol 29:204–205, 1973

237. Strickland RW, Tesar JT, Berne BH, et al: The prevalence of Sjogren's syndrome and associated rheumatoid diseases in an elderly population. Arthritis Rheum 26:S45, 1984

238. Whaley K, Williamson J, Chisholm DM, et al: Sjogren's syndrome: I. Sicca components. Q J Med 42:279–304, 1977

239. Tsokos M, et al. Vasculitis in primary Sjogren's syndrome. histologic classification and clinical presentation. Am J Clin Pathol 88:26–31, 1987

240. Alexander EL, Arnett FC, Provost RR, Stevens MB: Sjogren's syndrome: Association of anti-Ro (SS-A) antibodies with vasculitis, hematologic abnormalities, and serologic hyperreactivity. Ann Intern Med 98:155–159, 1983

241. Binder A, Snaith ML, Isenberg D: Sjogren's syndrome: A study of its neurological complications. Br J Rheumatol 27:275–280, 1988

242. Alexander GE, Provost TT, Stevens MB, Alexander EL: Sjogren's syndrome: Central nervous system manifestations. Neurology 31:1391–1396, 1981

243. Noseworthy JH, Bass BH, Vandervoort MK, Ebers GC, Rice GPA, Weinshenker BG, McLay CJL, Bell DA: The prevalence of primary Sjogren's syndrome in a multiple sclerosis population. Ann Neurol 25:95–98, 1989

244. Shearn M: Sjogren's Syndrome. Philadelphia, WB Saunders, 1971

245. Malinow K, Yannakakis GD, Glusman SM, Edlow DW, Griffin J, Pestronk A, Powell DL, Ramsey-Goldman R, Eidelman BH, Medsger TA, Alexander EL: Subacute sensory neuronopathy secondary to dorsal root ganglionitis in primary Sjogren's syndrome. Ann Neurol 20:535–537, 1986

246. Kaltreider HB, Talal N: The neuropathy of Sjogren's syndrome: Trigeminal nerve involvement. Ann Intern Med 70:751–762, 1969

247. Alexander EL, Malinow K, Lejewski JE, Jerdan MS, Provost TT, Alexander GE: Primary Sjogren's syndrome with central nervous system disease mimicking multiple sclerosis. Ann Intern Med 104:323–330, 1986

248. Holm S: Keratoconjunctivitis sicca and the sicca syndrome. Acta Ophthalmol (KBH) (Suppl) 33:1–230, 1949

249. Tarpley TM, Anderson LC, White CL: Minor salivary gland involvement in Sjogren's syndrome. Oral Surg 37:64–74, 1974

250. Greenspan JS, Daniel Talal N, Sylvester RA: The histo-pathology of Sjogren's syndrome in labial salivary gland biopsies. Oral Pathol 37:217–229, 1974

251. Alexander EL, Beall SS, Gordon B, Selnes OA, Yannakakis GD, Patronas N, Provost T, McFarland

HF: Magnetic resonance imaging of cerebral lesions in patients with the Sjogren syndrome. Ann Intern Med 108:815–8223, 1988

252. Konttinen YTm Kinnunen E, von Bonsdorff M, Lillqvist P, Immonen I, Bergroth V, Segerberg-Konttinen M, Friman C: Acute transverse myelopathy successfully treated with plasmapheresis and prednisone in a patient with primary Sjogren's syndrome. Arthritis Rheumatism 30: 339–344, 1987

253. Hughes GRV: Autoantibodies in lupus and its variants: Experience in 1000 patients. Br J Med (Clin Res) 289:339–342, 1984

254. Kelley RE, Gilman PB, Kovacs AG: Cerebral ischemia in the presence of lupus anticoagulant. Arch Neurol 41:521–523, 1984

255. Levine SR, Welch RMA: The spectrum of neurologic disease associated with antiphospholipid antibodies. Arch Neurol 44:876–883, 1987

256. Landi G, Calloni NV, Sabbedini MG, et al: Recurrent ischemic attacks in two young adults with lupus anticoagulant. Stroke 14:377–379, 1983

257. Briley DP, Coull BM, Goodnight SH: Neurological disease associated with antiphospholipid antibodies. Ann Neurol 25, 221–227, 1989

258. Branch DW, Scot JR, Kochenour NK, et al: Obstetric complications associated with the lupus anticoagulant. NEJM 313:1322–1326, 1985

259. Hall S, Buettner H, Luthra HS: Occlusive retinal vascular disease in systemuc lupus erythematosus. J Rheum 11:846–849, 1984

260. Jabs DA, Fine SL, Hochberg MC, Newman SA, Heiner GG, Steven MB: Severe retinal vaso-occlusive disease in systemic lupus erythematosus. Arch Ophthalmol 104:558–563, 1986

261. Asherson RA, Lubbe WF: Cerebral and valve lesions in SLE in association with antiphospholipid antibodies. J Rheumatol 15:539–543, 1988

262. Asherson RA, Merry P, Acheson JF, Harris EN, Hughes GRV: Antiphospholipid antibodies: A risk factor for occlusive ocular vascular disease in systemic lupus erythematosus and the 'primary' antiphospholipid syndrome. Ann Rheumat Dis 48:358–361, 1989

263. Digre KB, Durcan J, Branch DW, Jacobson DM, Varner MW, Baringer JR: Amaurosis fugax associated with antiphospholipid antibodies. Ann Neurol 25:228–232, 1989

264. Kleiner RC, Najarian LV, Schatten S, Jabs DA, Patz A, Kaplan HJ: Vaso-occlusive retinopathy associated with antiphospholipid antibodies (lupus anticoagulant retinopathy). Ophthalmology 96: 896–904, 1989

265. Frohman LP, Rescigno R, Biclroy L: Neuro-ophthalmic manifestations of the antiphospholipid antibody syndromes. Ophthalmology 96 (suppl):105, 1989

Chapter 10 Migraine

1. Fisher CM. Transient migrainous accompaniments (TMAs) of late onset. Stroke 10:96–97, 1979

2. Fisher CM. Late-life migraine accompaniments as a cause of unexplained transient ischemic attacks. Can J Neurol Sci 7:9–17, 1980

3. Friedman AP. In: Baker AB, Baker LH (eds) Clinical Neurology. Hueber-Harper, Maryland, vol 2, p 5, 1971

4. Waters WE, O'Connor PJ. Prevalence of migraine. J Neurol Neurosurg Psychiat 38:613–616, 1975

5. Bille BS. Migraine in school children. A study of incidence and short-term prognosis, and a clinical, psychological and electrencephalographic comparison between children with migraine and age matched controls. Acta Paediat Scand (Suppl 136) 51:1–151, 1962

6. Hare F. Mechanism of the pain in migraine. Med Press 79:583–586, 1905

7. Wolff HG. Personality features and reactions of subjects with migraine. Arch Neurol Psychiatr 37:895–921, 1937

8. Gowers WR. Subjective visual sensations. Trans Ophthalmol Soc UK 15:1–38, 1895

9. Peatfield RC, Rose FR. Migrainous visual symptoms in a woman without eyes. Arch Neurol 38:466, 1981

10. Moore RF. Subjective "lightning streaks." Br J Ophthalmol 19:545–547, 1935

11. Verhoeff FH. Moore's subjective "lightning streaks." Am J Ophthalmol 25:265–268, 1942

12. Verhoeff FH. Are Moore's lightning streaks of serious portent? Am J Ophthalmol 41:837–840, 1956

13. Davis FA, Bergen D, Schauf C, et al. Movement phosphenes in optic neuritis: a new clinical sign. Neurology 26:1100–1104, 1976

14. Coppeto JR, Kawalick M. Ocular pseudomigraine after posterior chamber intraocular lens implantation. Am J Ophthalmol 102:393–394, 1986

14a. Jacobson DM, Glaser P: Recurrent transient monocular visual impairment caused by hyphema following intra-ocular surgery. Neurology 38:163, 1988

14b. Nicholson D: Occult iris erosion: A treatable cause of recurrent hyphema in iris-supported intraocular lenses. Ophthalmology 89:115–119, 1982

14c. Magargal LE, Goldberg RE, Uram M, Gonder JR, and Brown GC: Recurrent microhyphema in the pseudophakic eye. Ophthalmology 90:1231–1234, 1983

15. Friedman AP. Current concepts in the diagnosis and treatment of chronic recurrent headache. Med Clin NA 56:1257–1271, 1972

16. Heck H. Die neurologischen Begleiterscheinungen der Migräne und des Problems des "angiospastischen Hirninsults". Nervenarzt 33:193–203, 1962

17. Hollenhorst RW. Ocular manifestations of migraine: report of four cases of hemianopsia. Proc

532 References

Staff Meetings Mayo Clin 28:686–693, 1953
18. Connor RCR. Complicated migraine: A study of permanent neurological and visual defects caused by migraine. Lancet 2:1072–1075, 1962
19. Wolter JR, Burchfield WJ. Ocular migraine in a young man resulting in unilateral blindness and retinal edema. J Pediatr Ophthalmol 8:173–176, 1971
20. Kline LB, Kelly CL. Ocular migraine in a patient with cluster headaches. Headache 20:253–257, 1980
21. Goodwin J, Gorelick PB, Helgason CM. Symptoms of amaurosis fugax in atherosclerotic carotid artery disease. Neurology 37:829–832, 1987
22. Epstein ED. Reversible unilateral lens opacities in a diabetic patient. Arch Ophthalmol 94:461–463, 1976
23. Paylor RR, Selhorst JB, Weinberg RS. Reversible monocular cataract simulating amaurosis fugax. Ann Ophthalmol 17:423–425, 1985
24. Whitty CWM. Familial hemiplegic migraine. J Neurol Neurosurg Psychiatr 16:172–177, 1953
25. Ross RT. Hemiplegic migraine. Canad Med Assoc J 78:10–16, 1958
26. Bradshaw P, Parsons M. Hemiplegic migraine, a clinical study. Quart J Med 34:65–85, 1965
27. Pearce JMS, Foster JB. An investigation of complicated migraine. Neurol 15:333–340, 1965
28. Bickerstaff ER. Basilar artery migraine. Lancet I:15–17, 1961
29. Lapkin ML, Golden GS. Basilar artery migraine. Am J Dis Child 132:278–281, 1978
30. Ferguson KS, Robinson SS. Life-threatening migraine. Arch Neurol 39:374–376, 1982
31. Gascon G, Barlow C. Juvenile migraine presenting as an acute confusional state. Pediatrics 45:628–635, 1970
32. Ehyai A, Fenichel GM. The natural history of acute confusional migraine. Arch Neurol 35:368–369, 1978
33. Gubler M. Des paralysies de la troisieme paire droit, recidivant pour la troisieme fois. Gaz Hop (Paris) 33:65, 1860
34. Friedman AP, Harter DH, Merritt HH. Ophthalmolplegic migraine. Arch Neurol 7:82–87, 1962
35. Katz B, Rimmer S. Ophthalmoplegic migraine with superior ramus oculomotor paresis. J Clin Neuro-ophthalmol 9:181–183, 1989
36. Cruciger MP, Mazow ML. An unusual case of ophthalmoplegic migraine. Am J Ophthalmol 86:414–417, 1978
37. Pearce J. The ophthalmological complications of migraine. J Neurol Sci 6:73–81, 1968
38. Van Pelt W. On the early onset of ophthalmoloplegic migraine. Am J Dis Child 107:628–631, 1964
39. Bickerstaff E. Ophthalmolplegic migraine. Rev Neurol 110:582–588, 1964
40. Imes R, Monteiro M, Hoyt W. Ophthalmoplegic migraine with proximal posterior cerebral artery vascular anomaly. J Clin Neuro-ophthalmol 4:221–223, 1984
41. Matson D. Intracranial arterial aneurysms in childhood. J Neurosurg 23:578–583, 1965
42. Durkan GP, Troost BT, Slamovits TL, Spoor TC, Kennerdell JS. Recurrent painless oculomotor palsy in children. A variant of ophthalmoplegic migraine? Headache 21:58–62, 1981
43. Woods D, O'Connor PS, Fleming R. Episodic unilateral mydriasis and migraine. Am J Ophthalmol 98:229–234, 1984
44. Edelson RN, Levy DE. Transient benign unilateral pupillary dilation in young adults. Arch Neurol 31:12–14, 1974
45. Payne J, Adamkiewicz J. Unilateral internal ophthalmoplegia with intracranial aneurysm. Am J Ophthalmol 68:349–352, 1969
46. Walsh FB, Hoyt WF. Clinical Neuro-ophthalmology, Vol 1. Williams & Wilkins, Baltimore, 1969, p 252
47. Alvarez WC. The migrainous scotoma as studied in 618 persons. Am J Ophthalmol 49:489–504, 1964.
48. O'Connor PJ. Acephalgic migraine. Ophthalmology 88:999–1003, 1981
49. Lance JW. Headaches related to sexual activity. J Neurol Neurosurg Psychiatr 39:1226–1230, 1976
50. Porter M, Jankowic J. Benign coital cephalgia. Differential diagnosis and treatment. Arch Neurol 38:710–712, 1981
51. Longsley HB. Natural history of subarachnoid hemorrhage, intracranial aneurysms, and arteriovenous malformations: based on 6,368 cases in the cooperative study. J Neurosurg 25:219–239, 1966
52. Horton BT. Histaminic cephalgia: differential diagnosis and treatment. Proc Mayo Clin 31:325–333, 1956
53. Kunkle EC, Pfeiffer JB, Wilhoit WM, Hamrick LW. Recurrent brief headache in "cluster" pattern. Trans Am Neurol Ass 77:240–243, 1952
54. Kudrow L. Cluster Headache: Mechanisms and Management. London, Oxford University Press, 1980, pp 10–20
55. Dalessio DJ. Wolff's Headache and Other Head Pain. Oxford University Press, New York, 1972, pp 225–347
55a. Dalessio, op cit, pp 307–308
55b. Dalessio DJ. Wolff's headache and other head pain. Oxford University Press, New York, 1980, pp 131–162
55c. Dalessio, op cit, p 70
56. Edvinsson I. Sympathetic control of cerebral circulation. Trends Neuorsci 5:425–429, 1982
57. Edvinsson L, Lindvall M, Nielsen KC, Owman C. Are brain vessels innervated also by central (nonsympathetic) adrenergic neurones? Brain Res 63:496–499, 1973
58. Wahl M, Kuschinsky W. Autonomic receptors as studied in pial vessels – microapplication in situ.

In: Owman C, Edvinsson L (eds) Neurogenic control of the brain circulation. Pergamon Press, New York, pp 185–195, 1977

59. Zervas NT, Lavyne MH, Negovo M. Neurotransmitters and the normal and ischemic cerebral circulation. NEJM 293:812–816, 1975

60. Edvinsson L, Woman C. Sympathetic innervation and adrenergic receptors in intraparenchymal cerebral arterioles of baboon. Acta Neurol Scand 56:304–305, 1977

61. Owman C, Edvinsson L, Nielsen KC. Autonomic neuroreceptors mechanisms in brain vessels. Blood Vessel 11:2–31, 1974

62. Skarby T, Hogestatt ED, Andersson KE. Influence of extracellular calcium and nifedipine on alpha1- and alpha2-adrenergic receptor-mediated contractile responses in isolated rat and cat cerebral and mesenteric arteries. Acta Physiol Scand 123:445–456, 1984

63. Duckles SP. Evidence for a functional cholinergic innervation of cerebral arteries. J Pharmacol Exp Ther 217:544–548, 1981

64. Florence VM, Bevan JA. Biochemical determinations of cholinergic innervation in cerebral arteries. Circ Res 45:212–218, 1979

65. Kuschinsky W, Wahl M, Neiss A. Evidence for cholinergic dilatory receptors in pial arteries of cats: A microapplication study. Pfluegers Arch 347:199–208, 1974

66. Spira PJ, Mylecharane EJ, Misbach J, Duckworth J, Lance JW. Internal external carotid vascular responses to vasoactive agents in monkey. Neurol 28:162–174, 1978

67. Edvinsson L, Hardebo JE, Owman C. Pharmacological analysis of 5-hydroxytryptamine receptors in isolated intracranial and extracranial vessels of cat and man. Circ Res 42:143–151, 1978

68. Edvinsson L, Birath E, Uddman R, Lee TJ-F, Duverger D, MacKenzie ET, Scatton B. Indoleaminergic mechanisms in brain vessels: localization, concentration, uptake and in vitro responses of 5-hydroxytryptamine. Acta Physiol Scand 121:291–299, 1984

69. Rusch NJ, Chyatte D, Sundt TM Jr, VanHoute PM. 5-hydroxytryptamine: source of activator calcium in human basilar arteries. Stroke 16:718–720, 1985

70. Muller-Schweinitzer E, Engel G. Evidence for mediation by 5-HT2 receptors of 5-hydroxytryptamine-induced contraction of canine basilar artery. Naunyn Schmiedebergs Arch Pharmacol 324:287–292, 1983

71. Edvinsson L, Hardebo JE. Characterization of serotonin-receptors in intracranial and extracranial vessels. Acta Physiol Canad 97:523–525, 1976

72. Edvinsson L, Owman C. Pharmacological characterization of adrenergic alpha and beta receptors mediating the vasomotor responses of cerebral arteries in vitro. Circ Res 35:835–849, 1975

73. Hardebo JE, Edvinsson L, Owman C, Svendaard N. Potentiation and antagonism of serotonin on intracranial and extracranial vessels. Neurology 28:64–70, 1978

74. Shibata S, Cheng JB, Murakami W. Reactivity of isolated human cerebral arteries to biogenic amines. Blood Vessels 14:356–365, 1977

75. Harper AM, MacKenzie ET. Effects of 5-hydroxytryptamine on pial arteriole calibre in anaesthetized cats. J Physiol 271:735–746, 1977

76. Edvinsson L, Hardebo JE, MacKenzie ET, Stewart M. Dual action of serotonin on pial arterioles in situ and the effect of propanolol on the response. Blood Vessels 14:366–371, 1977

77. Harper AM, MacKenzie ET. Cerebral circulatory and metabolic effects of 5-hydroxytryptamine in anaesthetized baboons. J Physiol 271:721–733, 1977

78. Herbst TJ, Raichle ME, Ferrendelli JA. B-adrenergic regulation of adenosine 3',5'-monophosphate concentration in brain microvessels. Science 204:330–332, 1979

79. Peroutka SJ, Moskowitz MA, Reinhard J, Snyder SH. Neurotransmitter receptor binding in bovine cerebral microvessels. Science 208:610–612, 1980

80. O'Neill JT, Tryastman RJ. Adrenergic receptors in the intra and extracranial vasculature. In: Owman C, Edvinsson L (eds) Neurogenic control of the brain circulation. Pergamon Press, New York, pp 245–265, 1977

81. McCulloch J, Edvinsson L. Cerebral circulatory and metabolic effects of vasoactive intestinal polypeptide. Am J Physiol 238:H449–H456, 1980

82. Edvinsson L, Ekman R. Distribution and dilatory effect of vasoactive intestinal polypeptide (VIP) in human cerebral arteries. Peptides 5:329–331, 1984

83. Edvinsson L, Rosendahl-Helgesen S, Uddman R. Substance P: localization, concentration and release in cerebral arteries, choroid plexus and dura mater. Cell Tissue Res 234:1–7, 1983

84. Mayberg M, Langer RS, Zervas NT, Moskowitz MA. Perivascular meningeal projections from cat trigeminal ganglia: possible for vascular headaches in man. Science 213:228–230, 1981

85. Edvinsson L, McCulloch J, Uddman R. Substance P: immunohistochemical localization and effect upon cat pial arteries in vitro and in situ. J Physiol 318:251–258, 1981

86. Anthony M, Lance JW. Histamine and serotonin in cluster headache. Arch Neurol 25:225–231, 1971

87. Fog-Moller F, Genefke IK, Bryndum B. Changes in concentration of catecholamines in blood during spontaneous migraine attacks and reserpine-induced attacks. In: Greene R (ed) Current concepts in migraine research. New York, Raven, pp 115–119, 1978

88. Sjaastad O. The significance of blood serotonin levels in migraine. a critical review. Acta Neurol Scand 51:200–210, 1975

89. Gotoh F, Komatsumoto S, Araki N, Gomi S.

Noradrenergic nervous activity in migraine. Arch Neurol 41:951−955, 1984

90. Glover V, Sandler M, Grant E, et al: Transitory decrease in platelet monoamine-oxidase activity during migraine attacks. Lancet 1:391−393, 1977

91. Lance JW, Lambert GA, Goadsby PJ, Duckworth JW. Brainstem influences on the cephalic circulation: experimental data from cat and monkey of relevance to migraine. Headache 23:258−265, 1983

92. Moore RY, Halaris AE, Jones BE. Serotonin neurons of the midbrain raphe: ascending progjections. J Comp Neurol 180:417−438, 1978

93. Reinhard JF, Liebmann JE, Schlosberg AJ, Moskowitz MA. Serotonin neurons project to small blood vessels in the brain. Science 206:85−87, 1979

94. Mayberg MR, Zervas NT, Moskowitz MA. Trigeminal projections to supratentorial pial and dural blood vessels in cats demonstrated by horseradish peroxidase histochemistry. J Comp Neurol 223:46−56, 1984

95. Lance JW. Mechanism and Management of Headache, 4th edn. London, Butterworth Scientific, 1982

96. Moskowitz MA. The neurobiology of vascular head pain. Ann Neurol 16:157−168, 1984

97. Leao AAP. Spreading depression of activity in cerebral cortex. J Neurophysiol 7:359−390, 1944.

98. Marshall WH. Spreading cortical depression of Leao. Physiol Rev 39:239−279, 1959

99. Hansen AJ, Lauritzen M. The role of spreading depression in acute brain disorders. Anias da Acad Brasil de Ciencias 56:457− 479, 1984

100. Hansen AJ, Quirstoff B, Gjedde A. Relationships between changes in cortical blood flow and extracellular K+ during spreading depression. Acta Physiol Scand 109:1−6, 1980

101. Lauritzen M, Balslev Jorgensen M, Diemer NH, Gjedde A, Hansen AJ. Persistent oligemia of rat cerebral cortex in the wake of spreading depression. Ann Neurol 12:469−474, 1982

102. Olesen J, Larsen B, Lauritzen M. Focal hyperemia followed by spreading oligemia and impaired activation of rCBF in classic migraine. Ann Neurol 9:344−352, 1981

103. Olesen J, Tfelt-Hansen P, Henriksen L, Larsen B. The common migraine may not be initiated by cerebral ischemia. Lancet 2:438−440, 1981

104. Herold S, Gibbs FM, Jones AKP, et al. Oxygen metabolism in migraine. J Cereb Blood Flow Metab 5(suppl 1):445−446, 1985

104a. Welch KMA, D'Andrea GD, Tepley N, et al. The concept of migraine as a state of central neuronal hyperexcitability. Neurologic clinics 8:817−828, 1990

105. Deshmukh SV, Meyer JS. Cyclic changes in platelet dynamics and the pathogenesis and prophylaxis of migraine. Headache 17:101−108, 1977

106. Dvilansky A, Rishpon S, Nathan I, et al. Release of platelet 5-hydroxytryptamine by plasma taken from platelets during and between migraine attacks. Pain 2:315−318, 1976

107. Berk BC, Alexander RW, Brock TA, Gimbrone MA, Webb RC. Vasoconstriction: A new activity for platelet-derived growth factor. Science 232:87−90, 1986

108. Couch JR, Hassanein RS. Platelet aggregability in migraine. Neurology 27:843−848, 1977

109. Jones RJ, Amess JAL, Wachowicz. Migraine: a platelet disorder. Lancet 2:720−723, 1981

110. Rydzewski W, Wachowicz B. Adenine nucleotides in platelets in and between attacks. In: Greene R (ed) Current Concepts in Migraine Research. New York, Raven Press, 1978, pp 153−158

111. Damasio H, Beck D. Migraine, thrombocytopenia, and serotonin metabolism. Lancet 1:240−242, 1978

112. Hannington E, Jones RJ, Ames JAL. Platelet aggregation in response to 5-HT in migraine patients taking oral contraceptives. Lancet 1:967−968, 1982

113. Critchley M, Ferguson FR. Migraine. Lancet 224:123−126, 1933

114. Friedman AP, Merritt HH. Headache: Diagnosis and Treatment. FA Davis, Philadelphia, 1959

115. Lance JW, Anthony M. Some clinical aspects of migraine. A prospective survey of 500 patients. Arch Neurol 15:356−361, 1966

116. Somerville BW. A study of migraine in pregnancy. Neurology 22:824−828, 1972

117. Callaghan N. The migraine syndrome in pregnancy. Neurology 18:197−199, 1968

118. Carey HM. Principles of oral contraception. 2. Effects of oral contraception. Med J Australia 2:1242−1250, 1971

119. Kudrau L. The relationship of headache frequency to hormone use in migraine. Headache 15:36−40, 1975

120. Phillips BM. Oral contraceptive drugs and migraine. Brit med J 2:99, 1968

121. Raskin NH, Appenzeller O. Headache, Vol XIX, major problems in internal medicine. W B Saunders & Co, Philadelphia, 1980

122. Whitty CWM, Hockaday JM, Whitty MM. The effect of oral contraceptives on migraine. Lancet I:856−869, 1966

123. Egeberg O, Owren PA. Oral contraception and blood coagulability. Br Med J 1:220−221, 1963

124. Carvalho ACA, Vaillancourt RA, Cabral RB, Lees RS, Colman RW. Coagulation abnormalities in women taking oral contraceptives. JAMA 237:875−878, 1977

125. Wessler S, Gitel SN, Wan LS, Pasternack BS. Estrogen-containing oral contraceptive agents. A basis for their thrombogenicity. JAMA 236:2179−2182, 1976

126. Mettler L, Meseck-Selchow B. Oral contraceptives

and platelet function. Thromb Diatheis Haemorrh 28:213–220, 1972

127. Phillips LI, Turksoy RN, Southam AL. Influence of ovarian function on the fibrinolytic enzyme system II. Influence of exogenous steroids. Am J Obstr Gynec 82:1216–1220, 1961

128. Sagar S, Thomas DP, Stamatakis JD, Kakkar VV. Oral contraceptives, antithrombin-III activity, and postoperative deep-vein thrombosis. Lancet 1: 509–510, 1976

129. Collaborative group for the study of stroke in young women. Oral contraceptives and stroke in young women; associated risk factors. J Amer Med Assoc 231:718–722, 1975

130. Kalendovsky Z, Austin J, Steele P. Increased platelet aggregability in young patients with stroke. Arch Neurol 32:13–20, 1975

131. Salmon ML, Winkelman JZ, Gay AJ. Neuro-ophthalmologic sequelae in users of oral contraceptives. JAMA 206:85–91, 1968

132. Shafey S, Scheinberg P. Neurological syndromes occurring in patients receiving synthetic steroids (oral contraceptives). Neurology 16:205–211, 1966

133. Speccavento LJ, Solomon GD, Mani S. An association between strokes and migraine in young adults. Headache 21:121, 1981

134. Pearce JMS. Migraine: a cerebral disorder. Lancet II:86–89, 1984

135. Harper AM, McCulloch J, MacKenzie ET, Picard JD. Migraine and the blood–brain barrier. Lancet i:1034–1036, 1977

136. Glover V, Littlewood J, Sandler M, Peatfield R, Petty R, Rose FC. Dietary migraine: looking beyond tyramine. In: Rose FC (ed) Progress in Migraine Research 2. London, Pitman, 1984, pp 113–119

137. Schweitzer JW, Friedhoff AJ, Schwartz R. Chocolate, B-phenylethylamine and migraine re-examined. Nature 257:256, 1975

138. Littlewood J, Glover V, Sandler M, Petty R, Penfield R, Rose FC. Platelet phenolsulphotransferase deficiency in dietary migraine. Lancet i:983–986, 1982

139. Hannington E. Preliminary report on tyramine headache. Br Med J ii:550–551, 1967

140. Moffett A, Swash M, Scott DF. Effect of tyramine in migraine: a double-blind study. J Neurol Neurosurg Psychiatr 35:496–499, 1972

141. Zeigler DY, Stewart R. Failure of tyramine to induce migraine. Neurology 27:725–726, 1977

142. Lord GDS, Duckworth JW. Immunoglobulin and complement studies in migraine. Headache 17: 163–168, 1977

143. Lord GDS. Headache 18:255–260, 1978

144. Edvinsson L, Owman Ch. A pharmacologic comparison of histamine receptors in isolated extracranial and intracranial arteries in vitro. Neurology 25:271–276, 1975

145. Carlson LA, Ekelund LG, Oro L. Clinical and metabolic effects of different doses of prostaglandin E1 in man. Acta Med Scand 183:423–430, 1968

146. Welch KMA, Spira PJ, Knowles L, et al. Effects of prostaglandins on the internal and external carotid blood flow in the monkey. Neurology 24:705–710, 1974

147. Pennink M, White RP, Crockarell JT, et al. Role of prostaglandin F2 alpha in the genesis of experimental vasospasm. J Neurosurg 37:398–406, 1972

148. Haas DC, Pineda GS, Lourie H. Juvenile head trauma syndromes and their relationship to migraine. Arch Neurol 32:727–730, 1975

149. Greenblatt SH. Posttraumatic transient cerebral blindness. JAMA 225:1073–1076, 1973

150. Lashley KS. Patterns of cerebral integration indicated by the scotomas of migraine. Arch Neurol Pscyh 46:331–339, 1941

151. Milner PM. Note on a possible correspondence between the scotomas of migraine and spreading depression of Leao. EEG Clin Neurophysiol 10:705, 1958

152. Airy H. On a distinct form of transient hemianopia. Philosoph Trans for 1870:247–264, London, 1871

153. Olsen TS, Friberg L, Lassen NA. Ischemia may be the primary cause of the neurologic deficits in classic migraine. Arch Neurol 44:156–161, 1987

154. Andersen AR, Friberg L, Olsen TS, Olesen J. Delayed hyperemia following hypoperfusion in classic migraine. Single photon emission computed tomographic demonstration. Arch Neurol 45: 154–159, 1988

155. Marcussen RM, Wolff HG. Studies on headache. Arch Neurol Psych 63:42–51, 1950

156. Vijayan N. Ophthalmoplegic migraine: ischemic or compressive neuropathy? Headache 20: 300–304, 1980

157. Schumacher G, Wolff HG. Experimental studies on headache. Arch Neurol Psychiatr 45:199–214, 1941

158. Kuschinsky W, Wahl M. Local chemical and neurogenic regulation of cerebral vascular reistance. Physiol Rev 58:656–689, 1978

159. Wolff HG. Headache and other Facial Pain, 2nd edn. Oxford University Press, New York, p 243, 1963

160. Joffe SN. Retinal blood vessel diameter during migraine. The Eye, Ear, Nose and Throat Monthly 52:338–342, 1973

161a. Sakai F, Meyer JS. Regional cerebral hemodynamics during migraine and cluster headaches measured by the ^{133}Xe inhalation method. Headache 18:122–132, 1978

161b. Sakai F, Meyer JS. Abnormal cerebrovascular reactivity in patients with migraine and cluster headache. Headache 19:257–266, 1979

162. Olesen J, Tfelt-Hansen P, Henriksen L, Larsen B. The common migraine attack may not be initiated by cerebral ischaemia. Lancet i:438–440, 1981

163. Lauritzen M, Skyhoj Olsen T, Lassen NA, Paulson OB. The changes of regional cerebral blood flow during the course of classical migraine attacks. Ann Neurol 13:633–641, 1983

164. Norris JW, Hachinski VC, Cooper PW. Changes in cerebral blood flow during a migraine attack. Br Med J 3:676–677, 1975

165. Dukes HT, Vieth RG. Cerebral arteriography during migraine prodrome and headache. Neurology 14:636–639, 1964

166. Waddington MM, Ring BA. Syndromes of occlusions of middle cerebral artery branches. Brain 91:686–696, 1968

167. Gastaut JL, Yermenos E, Bonnefoy M, Cross D. Familial hemiplegic migraine: EEG and CT scan of two case. Ann Neurol 10:392–395, 1981

168. Camfield PR, Metrakos K, Andermann F. Basilar migraine, seizures, and severe epileptiform EEG abnormalities. A relatively benign syndrome in adolescents. Neurology 28:584–588, 1978

169. Panayiotopoulos CP. Basilar migraine? Seizures, and severe epileptic EEG abnormalities. Neurology 30:1122–1125, 1980

170. Lee CH, Lance JW. Migraine stupor. Headache 17:32–38, 1977

171. Graveson GS. Retinal arterial occlusion in migraine. Br Med J 2:838–840, 1949

172. Katz B. Migrainous central retinal artery occlusion. J Clin Neuro-ophthalmol 6:69–71, 1986

173. Galezowski X. Ophthalmic megrim. Lancet i:176–177, 1882

174. Cowan CL, Knox DL. Migraine optic neuropathy. Ann Ophthal 14:164–166, 1982

175. Weinstein JM, Feman SS. Ischemic optic neuropathy in migraine. Arch Ophthalmol 100:1097–1100, 1982

176. McDonald WI, Sanders MD. Migraine complicated by ischaemic papillopathy, Lancet 1:521–523, 1971

177. Ormond AW. Two cases of permanent hemianopsia following severe attacks of migraine. Ophthalmol Rev 32:193–201, 1913

178. Hunt JR. A contribution to the paralytic and other persistent sequelae of migraine. Am J Med Sci 150:313–314, 1915

179. Kupersmith MJ, Hass WK, Chase NE. Isoproternol treatment of visual symptoms in migraine. Stroke 10:299–305, 1979

180. Brain R. Cerebral vascular disorders. Ann Int Med 41:175–181, 1954

181. Kupersmith MJ, Warren FA, Hass WK. The non-benign aspects of migraine. Neuro-ophthalmology 7:1–10, 1987

182. Castaigne P, Brunet P, Peirrot-Deseilligny C, Roullet E. Accidents deficitaires neurologiques centraux et migraine. Ann Med Interne 134:306–309, 1983

183. Kaul SN, Du Boulay GH, Kendall BE, Russel WR. Relationship between visual field defect and arterial occlusion in the posterior circulation. J Neurol Neurosurg Psychiatr 37:1022–1030, 1974

184. Rich WM. Permanent quadrantanopsia after migraine. Br Med J 116:592–594, 1948

185. Hungerford GD, Du Boulay GH, Zikha KJ. Computed axial tomography in patients with severe migraine: a preliminary report. J Neurol Neursurg Psychiatr 39:990–994, 1976

186. Sandeau D, Larmande P, Odier F, Autret A. Migraine complique, epilepsie et hypodensite tomodensitometrique. Rev Neurol (Paris) 138:787–790, 1982

187. Bogusslavsky J, Regli F, Melle V, Payot M, Uske A. Migraine stroke. Neuorol 38:223–227, 1988

188. Dorfman LJ, Marshall WH, Enzmann DR. Cerebral infarction and migraine: clinical and radiologic correlations. Neurology 29:317–322, 1979

189. Cohen RJ, Taylor JR. Persistent neurologic sequelae of migraine: a case report. Neurology 29:1175–1177, 1979

190. Guest IA, Woolf AL. Fatal infarction of brain in migraine. Br Med J 1:225–226, 1964

191. Buckle RM, du Boulay G, Smith B. Death due to cerebral vasospasm. J Neurol Neurosurg Psychiat 27:440–444, 1964

192. Leviton A, Malvea B, Graham JR. Vascular diseases, mortality, and migraine in the parents of migraine patients. Neurol 24:669–672, 1974

193. Vallery-Radot P, Blamoutier P, Mawas L, Hamburger J. Acces de migraine ophtalmique suivis d'une hemorragie retinienne. Ann de Mede 42:132–137, 1937

194. Victor DL, Welch RB. Bilateral retinal hemorrhages and disc edema in migraine. AJO 84:555–558, 1977

195. Friedman MW. Occlusion of central retinal vein in migraine. Arch Opthalmol 45:678–682, 1951

196. Chien TI, Tsai C. The mechanism of haemostasis in peripheral vessels. J Physiol (London) 107:280–288, 1948

197. Honour AJ, Mitchell JRA. Platelet clumping in injured vessels. Br J Exp Pathol 45:75–87, 1964

198. Nelson E, Gertz SD, Rennels M, et al. Ultrastructural changes in the arterial endothelium; a possible relationship to atherosclerosis. In: Scheinberg P (ed) Cerebrovascular Diseases, pp 97–107, Raven Press, NY, 1976

199. Levine SR, Welch KMA. Cerebrovascular ischemia associated with lupus anticoagulant. Stroke 18:257–263, 1987

200. Digre K, Durcan F, Branch D, Jacobson D, Varner M, Baringer M. Amaurosis fugax associated with antiphospholipid antibodies. Ann Neurol 25:228–232, 1989

201. Tunis MM, Wolff HG. Long term observations of

the reactivaty of the cranial arteries in subjects with vascular headache of the migraine type. Arch Neurol Psychiat 70:551–557, 1953

202. Skinhoj E. Hemodynamic studies within the brain during migraine. Arch Neurol 29:95–98 1973

203. Andersen AR, Friberg L, Olsen TS, Olesen J. Delayed hyperemia following hypoperfusion in classic migraine. Arch Neurol 45:154–159, 1988

204. Blau JN. Migraine: a vasomotor instability of the meningeal circulation. Lancet 2:1136–1139, 1978

205. Dandy WE. Concerning the cause of trigeminal neuralgia. Am J Surg 24:447–445, 1934

206. Mumford JM. Thermography and ice cream headache. Acta Thermogr 4:33–37, 1979

207. Krabbe AA, Olesen J. Headache provocation by continuous intravenous infusion of histamine. Clinical results and receptor mechanisms. Pain 8:253–259, 1980

208. Appenzeller O, Becker WJ, Ragaz A. Cluster headache: ultrastructural aspects and pathogenetic mechanisms. Arch Neurol 38:302–306, 1981

209. Sjaastad O, Sjaastad OV. Urinary histamine excretion in migraine and cluster headache. J Neurol 216:91–104, 1977

210. Sjaastad O, Sjaastad OV. Histamine metabolism in cluster headache and migraine. J Neurol 216:105–117, 1977

211. Solomon S, Apfelbaum RI. Surgical decompression of the facial nerve in the treatment of chronic cluster headache. Arch Neurol 43:479–482, 1986

212. Litman GI, Friedman HM. Migraine and the mitral valve prolapse syndrome. Am Heart J 96:610–614, 1978

213. Friedman AP, Merritt H. Treatment of headache JAMA 163:1111–1117, 1957

214. Couch JR. Placebo effect and clinical trials in migraine therapy. Neuroepidemiology 6:178–185, 1987

215. Hiner BC, Roth JL, Peroutka SJ. Antimigraine drug interactions with 5-hydroxytryptamine 1A receptors. Ann Neurol 19:511–513, 1986

216. Gozlan H, El Mestikawy S, Pichat L, et al. Indentification of presynaptic serotonin autoreceptors using a new ligand: 3H-PAT. Nature 305:14–143, 1983

217. Peroutka SJ, Kuhar MJ. Autoradiographic localization of 5-HT1 receptors to human and canine basilar arteries. Brain Res 310:193–196, 1984

217a. Goadsby PJ, Gundlach AL. Localization of 3H-dihydroergotamine-binding sites in the cat central nervous system: relevance to migraine. Ann Neurol 29:91–94, 1991

218. Graham JR, Wolff HG. Mechanism of migraine headache and action of ergotamine tartrate. Arch Neurol Psychiatr 39:737–763, 1938

219. Saxena PR. Selective vasoconstriction in the carotid vascular bed by methysergide: possible relevance to its anti-migraine effect. Eur J Pharmac 27:99–105, 1974

220. Saito K, Markowitz S, Moskowitz MA. Ergot alkaloids block neurogenic extravasation in dura mater: proposed action in vascular headaches. Ann Neurol 24:732–737, 1988

221. Chapman LF, Ramos AO, Goodell H, Wolff HG. Neurohumoral features of afferent fibers in man. Arch Neurol 4:617–650, 1961

222. Jancso N, Jancso-Gabor A, Szolcsanyi J. Direct evidence for neurogenic inflammation and its prevention by denervation and by pretreatment with capsaicin. Br J Pharm Chemother 31:138–151, 1967

223. Foreman JC, Jordan CC. Neurogenic inflammation. Trends in Pharmac Sci 5:116–119, 1984

224. Salzmann R, Pacha W, Taeschler M, Weidmann H. Naunyn-Schmiedebergs Arch exp Path Pharmak 261:360–378, 1968

225. Saxena PR, de Vlaam-Schluter GM. Role of some biogenic substances in migraine and relevant mechanism in antimigraine action of ergotamine – studies in an experimental model for migraine. Headache 14:142–163, 1974

226. Gaddum JH, Hameed KA. Drugs which antagonize 5-hydroxytryptamine. Br J Pharmacol Chemother 9:240–248, 1954

227. Saxena PR. The effects of antimigraine drugs on the vascular responses evoked by 5-hydroxytripamine and related biogenic substances on the external carotid bed of dogs: possible pharmacalogical implications to their antimigraine action. Headache 12:44–54, 1972

228. Chu D, Sturmer E, Berde B. Studies on the mechanism of action of dihydroergotamine (DHE) on the vascular bed of the cat skeletal muscle. Br J Pharmac 48:331p–332p, 1973

229. Berde B, Cerletti A, Dengler HJ, Zoglio MA. In: Cochrane ALA (ed) Background to Migraine 3. London, Heinemann, 1970, pp 80–102

230a. The Subcutaneous Sumatriptin International Study Group. Treatment of migraine attacks with sumatriptan. NEJM 325:316–321, 1991

230b. The Subcutaneous Sumatriptin International Study Group. Treatment of acute cluster headache with sumatriptan. NEJM 325:322–326, 1991

231. Hessov I, Kronamann-Andersen C, Madsen B. Peripheral arterial insufficiency during ergotamine treatment: course of disease, angiographical findings, and treatment by phenoxybenzamine. Dan Med Bull 19:236–244, 1972

232. Bagby RJ, Cooper RD. Angiography in ergotism: report of two cases and review of the literature AJR 116:179–186, 1972

233. Brohult J, Forsberg O, Hellstrom R. Multiple arterial thrombosis after oral contraceptives and ergotamine. Acta Med Scand 181:453–456, 1972

234. Greene GL, Ariyan S, Stancel HC. Mesenteric and

peripheral vascular ischemia secondary to ergotism. Surgery 81:176–179, 1977

235. Cranley JJ, Krause RJ, Strauss ES, Hafner CD. Impending gangrene of four extremities secondary to ergotism. NEJM 269:727–729, 1963

235a. Carliner NH, Denune DP, Finch CS, Goldberg LI. Sodium nitropusside treatment of ergotamine induced peripheral ischemia. JAMA 227:308–309, 1974

236. Shifrin E, Olschwang D, Perel A, Diamant Y, Cotev S. Reversal of ergotamine-induced arteriospasm by mechanical intra-arterial dilatation. Lancet 2:1278–1279, 1980

237. Hudgson P, Hart JA. Acute ergotism. Med J Austral 2:589–591, 1964

238. Mindel JS, Rubenstein AE, Franklin B. Ocular ergotamine tartrate toxicity during treatment of vacor-induced orthostatic hypotension. Am J Ophthalmol 92:492–496, 1981

239. Senter H, Lieberman A, Pinto R. Cerebral manifestations of ergotism. Stroke 7:88–92, 1976

240. Richter AM, Banker VP. Carotid ergotism. Radiology 106:339–340, 1973

241. Merhoff GC, Porter JM. Ergot intoxication: historical review and description. Ann Surg 180:773–779, 1974

242. Gupta DR, Strobos RJ. Bilateral papillitis associated with Cafergot therapy. Neurology 22:793–797, 1972

243. Kravitz D. Neuroretinitis associated with symptoms of ergot poisoning. Report of a case. Arch Ophthalmol 13:201–206, 1935

244. Meier L. Ueber die Entwickelung des grauen Stars in folge der Kriebelkrankheit. Arch Ophthalmol 8:120, 1862

245. von Bechterew W. Ueber neuro-psychische Störungen bei chronischem Ergotismus; nach den Beobachtungen von Dr. N. Reformatzki. Neurol Centralbl 11:769–775, 1892

246. Weiss HJ, Aledort LM. Impaired platelet connective tissue reaction in man after aspirin ingestion. Lancet 2:495–497, 1967

247. Smith JB, Willis AL. Aspirin selectively inhibits prostaglandin production in human platelets. Nature 231:235–237, 1971

248. Deshmukh SV, Meyer JS, Mouche RJ. Platelet dysfunction in migraine: effect of self-medication with aspirin. Thrombos Haemostas 36:319–323, 1976

249. O'Neill BP, Mann JD. Aspirin prophylaxis in migraine. Lancet 2:1179–1181, 1978

250. Rosenblum WI, El-Sabban F. Platelet aggregation in the cerebral microcirculation. Effect of aspirin and other agents. Circulation Res 40:320–328, 1977

251. Kovacs IB, Csalay L, Gorog P. Laser induced thrombosis in the microcirculation of the hamster cheek pouch and its inhibition by acetylsalicylic acid. Microvasc Res 6:194–201, 1973

252. Wiederman MP. Vascular reactions to laser in vitro. Microvasc Res 8:132–138, 1974

253. Ryan RE. A study of midrin in the symptomatic relief of migraine headache. Headache 14:33–42, 1974

254. Couch JR, Ziegler DK. Prednisone therapy for cluster headache. Headache 18:219–222, 1978

255. Fogan L. Treatment of cluster headache: a double blind comparison of oxygen vs air inhalation. Arch Neurol 42:362–263, 1985

256. Kety SS, Schmidt CF. The effects of altered arterial tensions of carbon dioxide and oxygen on cerebral blood flow and cerebral oxygen consumption of normal young men. J Clin Invest 27:484–492, 1947

257. Obrist WD, Thompson HK, Wang HE, et al. Regional cerebral blood flow estimated by 133Xe inhalation. Stroke 6:245–256, 1975

258. Nakajima S, Meyer JS, Amano T, et al. Cerebral vasomotor responsiveness during 100% oxygen inhalation in cerebral ischemia. Arch Neurol 40:271–276, 1983

259. Lipton SA. Prevention of classic migraine headache by digital massage of the superficial temporal arteries during visual aura. Ann Neurol 19:515–516, 1986

260. Drummond PD, Lance WJ. Extracranial vascular changes and the source of pain in migraine headache. Ann Neurol 13:32–37, 1983

261. Lance J. Mechanism and Management of Headache. Butterworth, London, 1978

262. Borgesen SE, Nielsen JL, Moller CE. Prophylactic treatment of migraine with propanolol. Acta Neurol Scand 50:651–661, 1974

263. Wideroe TE, Vigander T. Propanolol in the treatment of migraine Br Med J 2:699–701, 1974

264. Rosen JA. Observations on the efficacy of propanolol for the prophylaxis. Ann Neurol 13:92–93, 1983

265. Brick J, Glover W, Hutchinson J, Roddie JC. Effects of propanolol on peripheral vessels in man. Am J Cardiol 18:329–332, 1966

266. Peroutka SJ, Allen GS. The calcium antagonist properties of cyproheptadine: implications for antimigraine action. Neurology 34:304–309, 1984

267. Bonn JA, Turner P. D-propanolol and anxiety. Lancet ii:1355–1356, 1971

268. Palacios JM, Kuhar MJ. Beta-adrenergic-receptor localization by light microscopic autoradiography. Science 208:1378–1380, 1980

269. Weksler BB, Gillick M, Pink J. Effect of propanolol on platelet function. Blood 49:185–196, 1977

270. Campbell WB, Callanhan KS, Johnson AR, Graham RM. Anti-platelet activity of beta-adrenergic antagonists: inhibition of thromboxane synthesis and platelet aggregation in patients receiving long-term propanolol treatement. Lancet 2:1382–1384, 1981

271. Hansen KW, Klysner R, Geisler A, Knudsen JB,

Glazer S, Gormsen J. Platelet aggregation and beta-blockers. Lancet 1:224–225, 1982

272. Bevan JA. Selective action of diltiazem on cerebral vascular smooth muscle in the rabbit: antagonism of extrinsic but not intrinsic maintained tone. Am J Cardiol 49:519–524, 1982

273. Singh BN, Ellradt G, Peter CT. Verapamil: a review of its pharmacological properties and therapeutic use. Drugs 15:169–197, 1978

274. Williams NJ (ed): Therapeutic advances in medicine: nifedipine. Br J Clin Prac (suppl 8) 3–84, 1980

275. Braunwald E. Mechanism of action of calcium-channel-blocking agents. NEJM 307:1618–1627, 1982

276. Rodeheffer RJ, Rommer JA, Wigley F, Smith CR. Controlled double-blind tral of nifedipine in the treatment of Raynaud's phenomenon. NEJM 308:880–883, 1982

277. Allen GS, Banghart SB. Cerebral arterial spasm, part 9: in vitro effects of nifedipine on serotonin-, phenylephrine-, and potassium-induced contractions of canine basilar and femoral arteries. Neurosurgery 4:37–42, 1979

277a. Allen GS, Bahr AL. Cerebral arterial spasm, part 10: reversal of acute and chronic spasm in dogs with orally administered nifedipine. Neurosurgery 4:43–47, 1979

278. Edvinsson L, Brandt L, Andersson KE, Bengtsson B. Effect of a calcium antagonist on experimental constriction of human brain vessels. Surg Neurol 11:32–330, 1979

279. Pickard JD, Murray JD, Illingworth R, et al. Effect of oral nimodipine on cerebral infarction and outcome after subarachnoid haemorrhage. British Aneurysm Nimodipine Trial. Br Med J 298:636–642, 1989

280. Louis P. A double-blind placebo-controlled prophylactic study of flunarizine in migraine. Headache 21:235–239, 1981

281. Solomon GD, Steel JG, Spaccavento LJ. Verapamil prophylaxis of migraine: a double-blind, placebo-controlled study. JAMA 250:2500–2502, 1983

282. Kahan A, Weber S, Amor B, Guerin F, Degeorges M. Nifedipine in the treatment of migraine in Raynaud's phenomenon. NEJM 308:1102–1103, 1983

283. Gotta AW, Capuano C, Hartung J, Sullivan CA. Calcium entry blockers and human platelet aggregation. N Y State J Med 88:132–133, 1988

284. Couch JR, Ziegler DK, Hassanein R. Amitriptyline in the prophylaxis of migraine, effectiveness, and relationship of antimigraine and antidepressants. Neurology 26:121–127, 1976

285. Couch JR, Ziegler DK, Hassanein R. Amitriptyline in migraine prophylaxis. Arch Neurol 36:695–699, 1979

286. Paul SM, Rehavi M, Skolnick P, Goodwin FK. Demonstration of specific "high affinity" binding sites for (3H)-imipramine on human platelets. Life Sci 26:953–959, 1980

287. Langer SZ, Zarifian E, Briley M, Raisman R, Sechter D. High affinity binding of (3H)-imipramine in brain and platelets and its relevance to the biochemistry of affective disorders. Life Sci 29:211–220, 1981

288. Saper JR. Daily chronic headache. Neurologic clinics 8:891–901, 1990

289. Fozard JR. The animal pharmacology of drugs used in the treatment of migraine. J Pharm Pharmac 27:297–321, 1975

290. Fozard JR. Comparative effects of four migraine prophylactic drugs on an isolated extracranial artery. Eur J Pharmacol 36:127–139, 1976

291. Forster C, Whalley ET. Analysis of the 5-hydroxytryptamine induced contraction of the human basilar arterial strip compared with the rat aortic strip in vitro. Arch Pharmacol 319:12–17, 1982

292. Boisvert DPJ, Weir BKA, Overton TR, Reiffenstien RJ, Grace MGA. Cerebrovascular responses to subarachnoid blood and serotonin in the monkey. J Neurosurg 50:441–448, 1979

293. Lewis T. The blood vessels of the human skin and their responses. Shaw & Sons, London, 1927

294. Saper JR. Headache Disorders. Current concepts and treatment strategies. Littleton, Massachusetts, PSG, 1983

295. Hering R, Kuritzky A. Sodium valproate in the treatment of cluster headache an open clinical trial. Cephalgia 9:195–198, 1989

296. Graham JR. Methysergide for prevention of headache. experience in 500 patients over 3 years. NEJM 270:67–72, 1960

297. Graham JR, Suby HI, Lecompte PR, Sadowsky NL. Fibrotic disorders associated with methysergide therapy for headache. NEJM 274:359–368, 1966

298. Buenger RE, Hunter JA. Reversible mesenteric artery artery complicating methysergide maleate therapy. JAMA 198:558–560, 1966

299. Katz J, Vogel RM. Abdominal angina as a complication of methysergide maleate therapy. JAMA 199:124–126, 1967

300. Schmitt H, Schmitt H, Fenard S. Evidence for an alpha–sympathomimetic component in the effects of catapresan on vasomotor centres: antagonism by piperoxane. Eur J Pharmacol 14:98–100, 1971

301. Day MD, Roach AG. Central alpha- and beta-adrenoreceptors modifying arterial blood pressure and heart rate in conscious cats. Br J Pharmac 51:325–333, 1974

302. Swanson JW, Vick NA. Basilar artery migraine. Neurology 28:782–786, 1978

303. Bickerstaff ER. The basilar artery and the migraine-epilepsy syndrome. Proc R Soc Med 55:167–169, 1962
304. Basser LS. The relation of migraine and epilepsy. Brain 92:285–300, 1969
305. Dow DJ, Whitty CWM. Electroencephalographic changes in migraine. Lancet ii:52–54, 1947
306. Hockaday JM, Whitty CWM. Factors determining the electorencephalogram in migraine. A study of 560 patients, according to clinical type of migraine. Brain 92:769–788, 1969
307. Prensky AL, Sommer D. Diagnosis and treatment of migraine in children. Neurology 29:506–510, 1979
308. Vardi Y, Rabey IM, Streifler M, Schwartz A, Lindner HR, Zor U. Migraine attacks. Alleviation by an inhibitor of prostaglandin synthesis and action Neurology 26:447–450, 1976
309. Anthony M, Lance JW. Indomethacin in migraine. Med J Austr 1:56, 1968
310. Hakkarainen H, Vapaatalo H, Gothoni G, Parantainen J. Tolfenamic acid is as effective as ergotamine during migraine attacks. Lancet 2:326–328, 1979
311. Medina JL, Diamond S. Cluster headache variant: spectrum of a new headache syndrome. Arch Neurol 38:705–709, 1981
312. Bille PE, Jensen MK, Jensen JPK et al. Studies on the hematologic and cytogenetic effect of lithium. Acta Med Scand 198:281–286, 1975
313. Genefke IK. The active uptake of 5-hydroxytryptamine in rat and human blood platelets under the influence of lithium in vivo and in vitro. Acta Psychiatr Scand 48:394–399, 1972
314. Coppeto JR, Monteiro MLR, Lessell S, Bear L, Martinez-Maldonado M. Downbeat nystagmus. long-term therapy with moderate-dose lithium carbonate. Arch Neurol 40:754–755, 1983
315. Williams DP, Troost BT, Rogers J. Lithium-induced downbeat nystagmus. Arch Neurol 45:1022–1023, 1988
316. James AK. Prevention of migraine by oral administrattion of carbachol: analysis of 12 cases. Br Med J 1:663–664, 1945
317. Salamon MA, Wilson J. Drugs for alternating hemiplegic migraine. Lancet 2:980, 1984
318. Casaer P, Azou M. Flunarizine in alternating hemiplegia in childhood. Lancet 2:579, 1984
319. Prendes JL. Considerations on the use of propanolol in complicated migraine. Headache 20:93–95, 1980
320. Gilbert GJ. An occurrence of complicated migraine during propanolol therapy. Headache 22:81–83, 1982

Subject Index

Page numbers in *italics* refer to figures.

If you have any concerns about our products,
you can contact us on
ProductSafety@springernature.com

In case Publisher is established outside the EU,
the EU authorized representative is:
**Springer Nature Customer Service Center GmbH
Europaplatz 3, 69115 Heidelberg, Germany**

Printed by Libri Plureos GmbH
in Hamburg, Germany